T0268950

Introduction to Classical Mechanics

This textbook covers all the standard introductory topics in classical mechanics, including Newton's laws, oscillations, energy, momentum, angular momentum, planetary motion, and special relativity. It also explores more advanced topics, such as normal modes, the Lagrangian method, gyroscopic motion, fictitious forces, 4-vectors, and general relativity.

It contains more than 250 problems with detailed solutions so students can easily check their understanding of the topic. There are also over 350 unworked exercises, which are ideal for homework assignments. Password-protected solutions are available to instructors at www.cambridge.org/9780521876223. The vast number of problems alone makes it an ideal supplementary book for all levels of undergraduate physics courses in classical mechanics. The text also includes many additional remarks which discuss issues that are often glossed over in other textbooks, and it is thoroughly illustrated with more than 600 figures to help demonstrate key concepts.

David Morin is a Lecturer in the Physics Department at Harvard University. He received his Ph.D. in theoretical particle physics from Harvard in 1996. When not writing physics limericks or thinking of new problems whose answers involve *e* or the golden ratio, he can be found running along the Charles River or hiking in the White Mountains of New Hampshire.

Introduction to Classical Mechanics

With Problems and Solutions

David Morin

Harvard University

CAMBRIDGE
UNIVERSITY PRESS

CAMBRIDGE
UNIVERSITY PRESS

Shaftesbury Road, Cambridge CB2 8EA, United Kingdom

One Liberty Plaza, 20th Floor, New York, NY 10006, USA

477 Williamstown Road, Port Melbourne, VIC 3207, Australia

314–321, 3rd Floor, Plot 3, Splendor Forum, Jasola District Centre, New Delhi – 110025, India

103 Penang Road, #05–06/07, Visioncrest Commercial, Singapore 238467

Cambridge University Press is part of Cambridge University Press & Assessment, a department of the University of Cambridge.

We share the University's mission to contribute to society through the pursuit of education, learning and research at the highest international levels of excellence.

www.cambridge.org
Information on this title: www.cambridge.org/9780521876223

© D. Morin 2007

This publication is in copyright. Subject to statutory exception and to the provisions of relevant collective licensing agreements, no reproduction of any part may take place without the written permission of Cambridge University Press & Assessment.

First published 2008 (version 24, November 2024)

Printed in Great Britain by CPI Group (UK) Ltd, Croydon CR0 4YY, November 2024

A catalogue record for this publication is available from the British Library

ISBN 978-0-521-87622-3 Hardback

Cambridge University Press & Assessment has no responsibility for the persistence or accuracy of URLs for external or third-party internet websites referred to in this publication and does not guarantee that any content on such websites is, or will remain, accurate or appropriate.

To Allen Gerry and Neil Tame,
who took the time
to give a group of kids
some really cool problems

There once was a classical theory,
Of which quantum disciples were leery.
They said, "Why spend so long
On a theory that's wrong?"
Well, it works for your everyday query!

Contents

Preface

This book grew out of Harvard University's honors freshman mechanics course. It is essentially two books in one. Roughly half of each chapter follows the form of a normal textbook, consisting of text, along with exercises suitable for homework assignments. The other half takes the form of a "problem book," with all sorts of problems (and solutions) of varying degrees of difficulty. I've always thought that doing problems is the best way to learn, so if you've been searching for a supply to puzzle over, I think this will keep you busy for a while.

This book is somewhat of a quirky one, so let me say right at the start how I imagine it being used:

- As the primary text for honors freshman mechanics courses. My original motivation for writing it was the fact that there didn't exist a suitable book for Harvard's freshman course. So after nine years of using updated versions in the class, here is the finished product.
- As a supplementary text for standard freshman courses for physics majors. Although this book starts at the beginning of mechanics and is self contained, it doesn't spend as much time on the introductory material as other freshman books do. I therefore don't recommend using this as the only text for a standard freshman mechanics course. However, it will make an extremely useful supplement, both as a problem book for all students, and as a more advanced textbook for students who want to dive further into certain topics.
- As a supplementary text for upper-level mechanics courses, or as the primary text which is supplemented with another book for additional topics often covered in upper-level courses, such as Hamilton's equations, fluids, chaos, Fourier analysis, electricity and magnetism applications, etc. With all of the worked examples and in-depth discussions, you really can't go wrong in pairing up this book with another one.
- As a problem book for anyone who likes solving physics problems. This audience ranges from advanced high-school students, who I think will have a ball with it, to undergraduate and graduate students who want some amusing problems to ponder, to professors who are looking for a new supply of problems to use in their classes, and finally to anyone with a desire to learn about physics by doing problems. If you want, you can consider this to be a problem book that also happens to have comprehensive

introductions to each topic's set of problems. With about 250 problems (with included solutions) and 350 exercises (without included solutions), in addition to all the examples in the text, I think you'll get your money's worth! But just in case, I threw in 600 figures, 50 limericks, nine appearances of the golden ratio, and one cameo of $e^{-\pi}$.

The prerequisites for the book are solid high-school foundations in mechanics (no electricity and magnetism required) and single-variable calculus. There are two minor exceptions to this. First, a few sections rely on multivariable calculus, so I have given a review of this in Appendix B. The bulk of it comes in Section 5.3 (which involves the curl), but this section can easily be skipped on a first reading. Other than that, there are just some partial derivatives, dot products, and cross products (all of which are reviewed in Appendix B) sprinkled throughout the book. Second, a few sections (4.5, 9.2–9.3, and Appendices D and E) rely on matrices and other elementary topics from linear algebra. But a basic understanding of matrices should suffice here.

A brief outline of the book is as follows. Chapter 1 discusses various problem-solving strategies. This material is extremely important, so if you read only one chapter in the book, make it this one. You should keep these strategies on the tip of your brain as you march through the rest of the book. Chapter 2 covers statics. Most of this will likely be familiar, but you'll find some fun problems. In Chapter 3, we learn about forces and how to apply $F = ma$. There's a bit of math here needed for solving some simple differential equations. Chapter 4 deals with oscillations and coupled oscillators. Again, there's a fair bit of math needed for solving linear differential equations, but there's no way to avoid it. Chapter 5 deals with conservation of energy and momentum. You've probably seen much of this before, but it has lots of neat problems.

In Chapter 6, we introduce the Lagrangian method, which will most likely be new to you. It looks rather formidable at first, but it's really not all that rough. There are difficult concepts at the heart of the subject, but the nice thing is that the technique is easy to apply. The situation here is analogous to taking a derivative in calculus; there are substantive concepts on which the theory rests, but the act of taking a derivative is fairly straightforward.

Chapter 7 deals with central forces and planetary motion. Chapter 8 covers the easier type of angular momentum situations, where the direction of the angular momentum vector is fixed. Chapter 9 covers the more difficult type, where the direction changes. Spinning tops and other perplexing objects fall into this category. Chapter 10 deals with accelerating reference frames and fictitious forces.

Chapters 11 through 14 cover relativity. Chapter 11 deals with relativistic kinematics – abstract particles flying through space and time. Chapter 12 covers relativistic dynamics – energy, momentum, force, etc. Chapter 13 introduces the important concept of "4-vectors." The material in this chapter could alternatively be put in the previous two, but for various reasons I thought it best to create a separate chapter for it. Chapter 14 covers a few topics from General Relativity. It's impossible for one chapter to do this subject justice, of course, so we'll just look at some basic (but still very interesting) examples. Finally, the appendices cover various useful, but slightly tangential, topics.

Throughout the book, I have included many "Remarks." These are written in a slightly smaller font than the surrounding text. They begin with a small-capital "REMARK" and end with a shamrock (♣). The purpose of these remarks is to say something that needs to be said, without disrupting the overall flow of the argument. In some sense these are "extra" thoughts, although they are invariably useful in understanding what is going on. They are usually more informal than the rest of the text, and I reserve the right to use them to occasionally babble about things that I find interesting, but that you may find tangential. For the most part, however, the remarks address issues that arise naturally in the course of the discussion. I often make use of "Remarks" at the ends of the solutions to problems, where the obvious thing to do is to check limiting cases (this topic is discussed in Chapter 1). However, in this case, the remarks are *not* "extra" thoughts, because checking limiting cases of your answer is something you should *always* do.

For your reading pleasure (I hope!), I have included limericks throughout the text. I suppose that these might be viewed as educational, but they certainly don't represent any deep insight I have into the teaching of physics. I have written them for the sole purpose of lightening things up. Some are funny. Some are stupid. But at least they're all physically accurate (give or take).

As mentioned above, this book contains a huge number of problems. The ones with included solutions are called "Problems," and the ones without included solutions, which are intended to be used for homework assignments, are called "Exercises." There is no fundamental difference between these two types, except for the existence of written-up solutions. I have chosen to include the solutions to the problems for two reasons. First, students invariably want extra practice problems, with solutions, to work on. And second, I had a thoroughly enjoyable time writing them up. But a warning on these problems and exercises: Some are easy, but many are very difficult. I think you'll find them quite interesting, but don't get discouraged if you have trouble solving them. Some are designed to be brooded over for hours. Or days, or weeks, or months (as I can attest to!).

The problems (and exercises) are marked with a number of stars (actually asterisks). Harder problems earn more stars, on a scale from zero to four. Of course, you may disagree with my judgment of difficulty, but I think that an arbitrary weighting scheme is better than none at all. As a rough idea of what I mean by the number of stars, one-star problems are solid problems that require some thought, and four-star problems are really, really, *really* hard. Try a few and you'll see what I mean. Even if you understand the material in the text backwards and forwards, the four-star (and many of the three-star) problems will still be extremely challenging. But that's how it should be. My goal was to create an unreachable upper bound on the number (and difficulty) of problems, because it would be an unfortunate circumstance if you were left twiddling your thumbs, having run out of problems to solve. I hope I have succeeded.

For the problems you choose to work on, be careful not to look at the solution too soon. There's nothing wrong with putting a problem aside for a while and coming back to it later. Indeed, this is probably the best way to learn things. If you head to the solution at the first sign of not being able to solve a problem, then you have wasted the problem.

REMARK: This gives me an opportunity for my first remark (and first limerick, too). A fact that often gets overlooked is that you need to know more than the correct way(s) to do a problem; you also need to be familiar with many *incorrect* ways of doing it. Otherwise, when you come upon a new problem, there may be a number of decent-looking approaches to take, and you won't be able to immediately weed out the poor ones. Struggling a bit with a problem invariably leads you down some wrong paths, and this is an essential part of learning. To understand something, you not only have to know what's right about the right things; you also have to know what's wrong about the wrong things. Learning takes a serious amount of effort, many wrong turns, and a lot of sweat. Alas, there are no shortcuts to understanding physics.

> The ad said, For one little fee,
> You can skip all that course-work ennui.
> So send your tuition,
> For boundless fruition!
> Get your mail-order physics degree! ♣

Any book that takes ten years to write is bound to contain the (greatly appreciated) input of many people. I am particularly thankful for Howard Georgi's help over the years, with his numerous suggestions, ideas for many problems, and physics sanity checks. I would also like to thank Don Page for his entertaining and meticulous comments and suggestions, and an eye for catching errors in earlier versions. Other friends and colleagues who have helped make this book what it is (and who have made it all the more fun to write) are John Bechhoefer, Wes Campbell, Michelle Cyrier, Alex Dahlen, Gary Feldman, Lukasz Fidkowski, Jason Gallicchio, Doug Goodale, Bertrand Halperin, Matt Headrick, Jenny Hoffman, Paul Horowitz, Alex Johnson, Yevgeny Kats, Can Kilic, Ben Krefetz, Daniel Larson, Jaime Lush, Rakhi Mahbubani, Chris Montanaro, Theresa Morin, Megha Padi, Dave Patterson, Konstantin Penanen, Courtney Peterson, Mala Radhakrishnan, Esteban Real, Daniel Rosenberg, Wolfgang Rueckner, Aqil Sajjad, Alexia Schulz, Daniel Sherman, Oleg Shpyrko, David Simmons-Duffin, Steve Simon, Joe Swingle, Edwin Taylor, Sam Williams, Alex Wissner-Gross, and Eric Zaslow. I'm sure that I have forgotten others, especially from the earlier years where my memory fades, so please accept my apologies.

I am also grateful for the highly professional work done by the editorial and production group at Cambridge University Press in transforming this into an actual book. It has been a pleasure working with Lindsay Barnes, Simon Capelin, Margaret Patterson, and Dawn Preston.

Finally, and perhaps most importantly, I would like to thank all the students (both at Harvard and elsewhere) who provided input during the past decade.

The names here are literally too numerous to write down, so let me simply say a big thank you, and that I hope other students will enjoy what you helped create.

Despite the painstaking proofreading and all the eyes that have passed over earlier versions, there is at most an exponentially small probability that the book is error free. So if something looks amiss, please check the webpage (www.cambridge.org/9780521876223) for a list of typos, updates, etc. And please let me know if you discover something that isn't already posted. I'm sure that eventually I will post some new problems and supplementary material, so be sure to check the webpage for additions. Information for instructors will also be available on this site.

Happy problem solving – I hope you enjoy the book!

Chapter 1
Strategies for solving problems

Physics involves a great deal of problem solving. Whether you are doing cutting-edge research or reading a book on a well-known subject, you are going to need to solve some problems. In the latter case (the presently relevant one, given what is in your hand right now), it is fairly safe to say that the true test of understanding something is the ability to solve problems on it. Reading about a topic is often a necessary step in the learning process, but it is by no means a sufficient one. The more important step is spending as much time as possible solving problems (which is inevitably an active task) beyond the time you spend reading (which is generally a more passive task). I have therefore included a very large number of problems/exercises in this book.

However, if I'm going to throw all these problems at you, I should at least give you some general strategies for solving them. These strategies are the subject of the present chapter. They are things you should always keep in the back of your mind when tackling a problem. Of course, they are generally not sufficient by themselves; you won't get too far without understanding the physical concepts behind the subject at hand. But when you add these strategies to your physical understanding, they can make your life a lot easier.

1.1 General strategies

There are a number of general strategies you should invoke without hesitation when solving a problem. They are:

1. **Draw a diagram, if appropriate.**

   In the diagram, be sure to label clearly all the relevant quantities (forces, lengths, masses, etc.). Diagrams are absolutely critical in certain types of problems. For example, in problems involving "free-body" diagrams (discussed in Chapter 3) or relativistic kinematics (discussed in Chapter 11), drawing a diagram can change a hopelessly complicated problem into a near-trivial one. And even in cases where diagrams aren't this crucial, they're invariably very helpful. A picture is definitely worth a thousand words (and even a few more, if you label things!).

2. **Write down what you know, and what you are trying to find.**

   In a simple problem, you may just do this in your head without realizing it. But in more difficult problems, it is very useful to explicitly write things out. For example, if there are three unknowns that you're trying to find, but you've written down only two facts, then you know there must be another fact you're missing (assuming that the problem is in fact solvable), so you can go searching for it. It might be a conservation law, or an $F = ma$ equation, etc.

3. **Solve things symbolically.**

   If you are solving a problem where the given quantities are specified numerically, you should immediately change the numbers to letters and solve the problem in terms of the letters. After you obtain an answer in terms of the letters, you can plug in the actual numerical values to obtain a numerical answer. There are many advantages to using letters:

   - IT'S QUICKER. It's much easier to multiply a g by an ℓ by writing them down on a piece of paper next to each other, than it is to multiply them together on a calculator. And with the latter strategy, you'd undoubtedly have to pick up your calculator at least a few times during the course of a problem.
   - YOU'RE LESS LIKELY TO MAKE A MISTAKE. It's very easy to mistype an 8 for a 9 in a calculator, but you're probably not going to miswrite a q for a g on a piece of paper. But if you do, you'll quickly realize that it should be a g. You certainly won't just give up on the problem and deem it unsolvable because no one gave you the value of q!
   - YOU CAN DO THE PROBLEM ONCE AND FOR ALL. If someone comes along and says, oops, the value of ℓ is actually 2.4 m instead of 2.3 m, then you won't have to do the whole problem again. You can simply plug the new value of ℓ into your final symbolic answer.
   - YOU CAN SEE THE GENERAL DEPENDENCE OF YOUR ANSWER ON THE VARIOUS GIVEN QUANTITIES. For example, you can see that it grows with quantities a and b, decreases with c, and doesn't depend on d. There is much, much more information contained in a symbolic answer than in a numerical one. And besides, symbolic answers nearly always look nice and pretty.
   - YOU CAN CHECK UNITS AND SPECIAL CASES. These checks go hand-in-hand with the previous "general dependence" advantage. But since they're so important, we'll postpone their discussion and devote Sections 1.2 and 1.3 to them.

   Having said all this, it should be noted that there are occasionally times when things get a bit messy when working with letters. For example, solving a system of three equations in three unknowns might be rather cumbersome unless you plug in the actual numbers. But in the vast majority of problems, it is highly advantageous to work entirely with letters.

4. **Consider units/dimensions.**

   This is extremely important. See Section 1.2 for a detailed discussion.

5. **Check limiting/special cases.**

   This is also extremely important. See Section 1.3 for a detailed discussion.

6. **Check order of magnitude if you end up getting a numerical answer.**

   If you end up with an actual numerical answer to a problem, be sure to do a sanity check to see if the number is reasonable. If you've calculated the distance along the ground that a car skids before it comes to rest, and if you've gotten an answer of a kilometer or a millimeter, then you know you've probably done something wrong. Errors of this sort often come from forgetting some powers of 10 (say, when converting kilometers to meters) or from multiplying something instead of dividing (although you should be able to catch this by checking your units, too).

You will inevitably encounter problems, physics ones or otherwise, where you don't end up obtaining a rigorous answer, either because the calculation is intractable, or because you just don't feel like doing it. But in these cases it's usually still possible to make an educated guess, to the nearest power of 10. For example, if you walk past a building and happen to wonder how many bricks are in it, or what the labor cost was in constructing it, then you can probably give a reasonable answer without doing any severe computations. The physicist Enrico Fermi was known for his ability to estimate things quickly and produce order-of-magnitude guesses with only minimal calculation. Hence, a problem where the goal is to simply obtain the nearest power-of-10 estimate is known as a "Fermi problem." Of course, sometimes in life you need to know things to better accuracy than the nearest power of 10 . . .

> How Fermi could estimate things!
> Like the well-known Olympic ten rings,
> And the one hundred states,
> And weeks with ten dates,
> And birds that all fly with one . . . wings.

In the following two sections, we'll discuss the very important strategies of checking units and special cases. Then in Section 1.4 we'll discuss the technique of solving problems numerically, which is what you need to do when you end up with a set of equations you can't figure out how to solve. Section 1.4 isn't quite analogous to Sections 1.2 and 1.3, in that these first two are relevant to basically any problem you'll ever do, whereas solving equations numerically is something you'll do only for occasional problems. But it's nevertheless something that every physics student should know.

In all three of these sections, we'll invoke various results derived later in the book. For the present purposes, the derivations of these results are completely irrelevant, so don't worry at all about the physics behind them – there will be plenty of opportunity for that later on! The main point here is to learn what to do with the result of a problem once you've obtained it.

1.2 Units, dimensional analysis

The units, or dimensions, of a quantity are the powers of mass, length, and time associated with it. For example, the units of a speed are length per time. The consideration of units offers two main benefits. First, looking at units before you start a problem can tell you roughly what the answer has to look like, up to numerical factors. Second, checking units at the end of a calculation (which is something you should *always* do) can tell you if your answer has a chance at being correct. It won't tell you that your answer is definitely correct, but it might tell you that your answer is definitely incorrect. For example, if your goal in a problem is to find a length, and if you end up with a mass, then you know it's time to look back over your work.

> "Your units are wrong!" cried the teacher.
> "Your church weighs six joules – what a feature!
> And the people inside
> Are four hours wide,
> And eight gauss away from the preacher!"

In practice, the second of the above two benefits is what you will generally make use of. But let's do a few examples relating to the first benefit, because these can be a little more exciting. To solve the three examples below exactly, we would need to invoke results derived in later chapters. But let's just see how far we can get by using only dimensional analysis. We'll use the "[ ]" notation for units, and we'll let M stand for mass, L for length, and T for time. For example, we'll write a speed as $[v] = L/T$ and the gravitational constant as $[G] = L^3/(MT^2)$ (you can figure this out by noting that Gm_1m_2/r^2 has the dimensions of force, which in turn has dimensions ML/T^2, from $F = ma$). Alternatively, you can just use the mks units, kg, m, s, instead of M, L, T, respectively.[1]

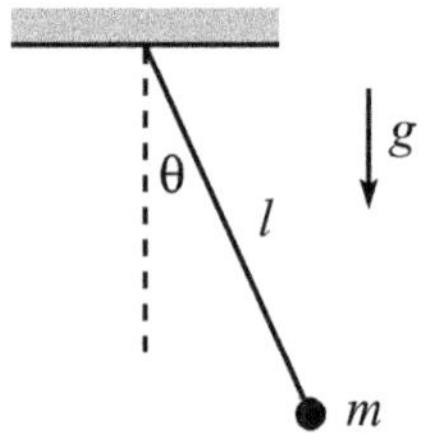

Fig. 1.1

Example (Pendulum): A mass m hangs from a massless string of length ℓ (see Fig. 1.1) and swings back and forth in the plane of the paper. The acceleration due to gravity is g. What can we say about the frequency of oscillations?

Solution: The only dimensionful quantities given in the problem are $[m] = M$, $[\ell] = L$, and $[g] = L/T^2$. But there is one more quantity, the maximum angle θ_0, which is dimensionless (and easy to forget). Our goal is to find the frequency, which

[1] When you check units at the end of a calculation, you will invariably be working with the kg,m,s notation. So that notation will inevitably get used more. But I'll use the M, L, T notation here, because I think it's a little more instructive. At any rate, just remember that the letter m (or M) stands for "meter" in one case, and "mass" in the other.

has units of $1/T$. The only combination of our given dimensionful quantities that has units of $1/T$ is $\sqrt{g/\ell}$. But we can't rule out any θ_0 dependence, so the most general possible form of the frequency is[2]

$$\omega = f(\theta_0)\sqrt{\frac{g}{\ell}}, \tag{1.1}$$

where f is a dimensionless function of the dimensionless variable θ_0.

REMARKS:

1. It just so happens that for small oscillations, $f(\theta_0)$ is essentially equal to 1, so the frequency is essentially equal to $\sqrt{g/\ell}$. But there is no way to show this by using only dimensional analysis; you actually have to solve the problem for real. For larger values of θ_0, the higher-order terms in the expansion of f become important. Exercise 4.23 deals with the leading correction, and the answer turns out to be $f(\theta_0) = 1 - \theta_0^2/16 + \cdots$.
2. Since there is only one mass in the problem, there is no way that the frequency (with units of $1/T$) can depend on $[m] = M$. If it did, there would be nothing to cancel the units of mass and produce a pure inverse-time.
3. We claimed above that the only combination of our given dimensionful quantities that has units of $1/T$ is $\sqrt{g/\ell}$. This is easy to see here, but in more complicated problems where the correct combination isn't so obvious, the following method will always work. Write down a general product of the given dimensionful quantities raised to arbitrary powers ($m^a \ell^b g^c$ in this problem), and then write out the units of this product in terms of a, b, and c. If we want to obtain units of $1/T$ here, then we need

$$M^a L^b \left(\frac{L}{T^2}\right)^c = \frac{1}{T}. \tag{1.2}$$

Matching up the powers of the three kinds of units on each side of this equation gives

$$M: \ a = 0, \qquad L: \ b + c = 0, \qquad T: \ -2c = -1. \tag{1.3}$$

The solution to this system of equations is $a = 0$, $b = -1/2$, and $c = 1/2$, so we have reproduced the $\sqrt{g/\ell}$ result. ♣

What can we say about the total energy of the pendulum (with the potential energy measured relative to the lowest point)? We'll talk about energy in Chapter 5, but the only thing we need to know here is that energy has units of ML^2/T^2. The only combination of the given dimensionful constants of this form is $mg\ell$. But again, we can't rule out any θ_0 dependence, so the energy must take the form $f(\theta_0)mg\ell$, where f is some function. That's as far as we can go with dimensional analysis. However, if we actually invoke a little physics, we can say that the total energy equals the potential energy at the highest point, which is $mg\ell(1 - \cos\theta_0)$. Using the Taylor expansion for $\cos\theta$ (see Appendix A for a discussion of Taylor series), we see that $f(\theta_0) = \theta_0^2/2 - \theta_0^4/24 + \cdots$. So in contrast with the frequency result above, the maximum angle θ_0 plays a critical role in the energy.

[2] We'll measure frequency here in radians per second, denoted by ω. So we're actually talking about the "angular frequency." Just divide by 2π (which doesn't affect the units) to obtain the "regular" frequency in cycles per second (hertz), usually denoted by ν. We'll talk at great length about oscillations in Chapter 4.

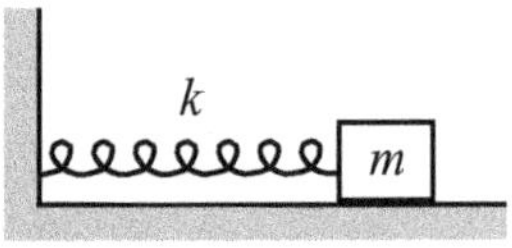

Fig. 1.2

Example (Spring): A spring with spring constant k has a mass m on its end (see Fig. 1.2). The spring force is $F(x) = -kx$, where x is the displacement from the equilibrium position. What can we say about the frequency of oscillations?

Solution: The only dimensionful quantities in this problem are $[m] = M$, $[k] = M/T^2$ (obtained by noting that kx has the dimensions of force), and the maximum displacement from the equilibrium, $[x_0] = L$. (There is also the equilibrium length, but the force doesn't depend on this, so there is no way it can come into the answer.) Our goal is to find the frequency, which has units of $1/T$. The only combination of our given dimensionful quantities with these units is

$$\omega = C\sqrt{\frac{k}{m}}, \tag{1.4}$$

where C is a dimensionless number. It just so happens that C is equal to 1 (assuming that we're measuring ω in radians per second), but there is no way to show this by using only dimensional analysis. Note that, in contrast with the pendulum above, the frequency cannot have any dependence on the maximum displacement.

What can we say about the total energy of the spring? Energy has units of ML^2/T^2, and the only combination of the given dimensionful constants of this form is Bkx_0^2, where B is a dimensionless number. It turns out that $B = 1/2$, so the total energy equals $kx_0^2/2$.

REMARK: A real spring doesn't have a perfectly parabolic potential (that is, a perfectly linear force), so the force actually looks something like $F(x) = -kx + bx^2 + \cdots$. If we truncate the series at the second term, then we have one more dimensionful quantity to work with, $[b] = M/LT^2$. To form a quantity with the dimensions of frequency, $1/T$, we need x_0 and b to appear in the combination $x_0 b$, because this is the only way to get rid of the L. You can then see (by using the strategy of writing out a general product of the variables, discussed in the third remark in the pendulum example above) that the frequency must be of the form $f(x_0 b/k)\sqrt{k/m}$, where f is some function. We can therefore have x_0 dependence in this case. This answer must reduce to $C\sqrt{k/m}$ for $b = 0$. Hence, f must be of the form $f(y) = C + c_1 y + c_2 y^2 + \cdots$. ♣

Example (Low-orbit satellite): A satellite of mass m travels in a circular orbit just above the earth's surface. What can we say about its speed?

Solution: The only dimensionful quantities in the problem are $[m] = M$, $[g] = L/T^2$, and the radius of the earth $[R] = L$.[3] Our goal is to find the speed, which has units of L/T. The only combination of our dimensionful quantities with these units is

$$v = C\sqrt{gR}. \tag{1.5}$$

It turns out that $C = 1$.

[3] You might argue that the mass of the earth, M_E, and Newton's gravitational constant, G, should be also included here, because Newton's gravitational force law for a particle on the surface of the earth is $F = GM_E m/R^2$. But since this force can be written as $m(GM_E/R^2) \equiv mg$, we can absorb the effects of M_E and G into g.

1.3 Approximations, limiting cases

As with units, the consideration of limiting cases (or perhaps we should say special cases) offers two main benefits. First, it can help you get started on a problem. If you're having trouble figuring out how a given system behaves, then you can imagine making, for example, a certain length become very large or very small, and then you can see what happens to the behavior. Having convinced yourself that the length actually affects the system in extreme cases (or perhaps you will discover that the length doesn't affect things at all), it will then be easier to understand how it affects the system in general, which will then make it easier to write down the relevant quantitative equations (conservation laws, $F = ma$ equations, etc.), which will allow you to fully solve the problem. In short, modifying the various parameters and observing the effects on the system can lead to an enormous amount of information.

Second, as with checking units, checking limiting cases (or special cases) is something you should *always* do at the end of a calculation. But as with checking units, it won't tell you that your answer is definitely correct, but it might tell you that your answer is definitely incorrect. It is generally true that your intuition about limiting cases is much better than your intuition about generic values of the parameters. You should use this fact to your advantage.

Let's do a few examples relating to the second benefit. The initial expressions given in each example below are taken from various examples throughout the book, so just accept them for now. For the most part, I'll repeat here what I'll say later on when we work through the problems for real. A tool that comes up often in checking limiting cases is the Taylor series approximations; the series for many functions are given in Appendix A.

Example (Dropped ball): A beach ball is dropped from rest at height h. Assume that the drag force from the air takes the form $F_{\text{d}} = -m\alpha v$. We'll find in Section 3.3 that the ball's velocity and position are given by

$$v(t) = -\frac{g}{\alpha}\left(1 - e^{-\alpha t}\right), \quad \text{and} \quad y(t) = h - \frac{g}{\alpha}\left(t - \frac{1}{\alpha}\left(1 - e^{-\alpha t}\right)\right). \tag{1.6}$$

These expressions are a bit complicated, so for all you know, I could have made a typo in writing them down. Or worse, I could have completely botched the solution. So let's look at some limiting cases. If these limiting cases yield expected results, then we can feel a little more confident that the answers are actually correct.

If t is very small (more precisely, if $\alpha t \ll 1$; see the discussion following this example), then we can use the Taylor series, $e^{-x} \approx 1 - x + x^2/2$, to make approximations to leading order in αt. The $v(t)$ in Eq. (1.6) becomes

$$v(t) = -\frac{g}{\alpha}\left(1 - \left(1 - \alpha t + \frac{(\alpha t)^2}{2} - \cdots\right)\right)$$
$$\approx -gt, \tag{1.7}$$

plus terms of higher order in αt. This answer is expected, because the drag force is negligible at the start, so we essentially have a freely falling body with acceleration g downward. For small t, Eq. (1.6) also gives

$$y(t) = h - \frac{g}{\alpha}\left[t - \frac{1}{\alpha}\left(1 - \left(1 - \alpha t + \frac{(\alpha t)^2}{2} - \cdots\right)\right)\right]$$
$$\approx h - \frac{gt^2}{2}, \tag{1.8}$$

plus terms of higher order in αt. Again, this answer is expected, because we essentially have a freely falling body at the start, so the distance fallen is the standard $gt^2/2$.

We can also look at large t (or rather, large αt). In this case, $e^{-\alpha t}$ is essentially zero, so the $v(t)$ in Eq. (1.6) becomes (there's no need for a Taylor series in this case)

$$v(t) \approx -\frac{g}{\alpha}. \tag{1.9}$$

This is the "terminal velocity." Its value makes sense, because it is the velocity for which the total force, $-mg - m\alpha v$, vanishes. For large t, Eq. (1.6) also gives

$$y(t) \approx h - \frac{gt}{\alpha} + \frac{g}{\alpha^2}. \tag{1.10}$$

Apparently for large t, g/α^2 is the distance (and this does indeed have units of length, because α has units of T^{-1}, because $m\alpha v$ has units of force) that our ball lags behind another ball that started out already at the terminal velocity, $-g/\alpha$.

Whenever you derive approximate answers as we just did, you gain something and you lose something. You lose some truth, of course, because your new answer is technically not correct. But you gain some aesthetics. Your new answer is invariably much cleaner (sometimes involving only one term), and this makes it a lot easier to see what's going on.

In the above example, it actually makes no sense to look at the limit where t is small or large, because t has dimensions. Is a year a large or small time? How about a hundredth of a second? There is no way to answer this without knowing what problem you're dealing with. A year is short on the time scale of galactic evolution, but a hundredth of a second is long on the time scale of a nuclear process. It makes sense only to look at the limit of a small (or large) *dimensionless* quantity. In the above example, this quantity is αt. The given constant α has units of T^{-1}, so $1/\alpha$ sets a typical time scale for the system. It therefore makes sense to look at the limit where $t \ll 1/\alpha$ (that is, $\alpha t \ll 1$), or where $t \gg 1/\alpha$ (that is, $\alpha t \gg 1$). In the limit of a small dimensionless quantity, a Taylor series can be used to expand an answer in powers of the small quantity, as we did above. We sometimes get sloppy and say things like, "In the limit of small t." But you know that we really mean, "In the limit of some small dimensionless quantity that has a t in the numerator," or, "In the limit where t is much smaller that a certain quantity that has the dimensions of time."

REMARK: As mentioned above, checking special cases tells you that either (1) your answer is consistent with your intuition, or (2) it's wrong. It never tells you that it's definitely correct. This is the same as what happens with the scientific method. In the real world, everything comes down to experiment. If you have a theory that you think is correct, then you need to check that its predictions are consistent with experiments. The specific experiments you do are the analog of the special cases you check after solving a problem; these two things represent what you know is true. If the results of the experiments are inconsistent with your theory, then you need to go back and fix your theory, just as you would need to go back and fix your answer. If, on the other hand, the results are consistent, then although this is good, the only thing it really tells you is that your theory *might* be correct. And considering the way things usually turn out, the odds are that it's not actually correct, but rather the limiting case of a more correct theory (just as Newtonian physics is a limiting case of relativistic physics, which is a limiting case of quantum field theory, etc.). That's how physics works. You can't prove anything, so you learn to settle for the things you can't disprove.

Consider, when seeking gestalts,
The theories that physics exalts.
It's not that they're known
To be written in stone.
It's just that we can't say they're false. ♣

When making approximations, how do you know how many terms in the Taylor series to keep? In the example above, we used $e^{-x} \approx 1 - x + x^2/2$. But why did we stop at the x^2 term? The honest (but slightly facetious) answer is, "Because I had already done this problem before writing it up, so I knew how many terms to keep." But the more informative (although perhaps no more helpful) answer is that before you do the calculation, there's really no way of knowing how many terms to keep. So you should just keep a few and see what happens. If everything ends up canceling out, then this tells you that you need to repeat the calculation with another term in the series. For example, in Eq. (1.8), if we had stopped the Taylor series at $e^{-x} \approx 1 - x$, then we would have obtained $y(t) = h - 0$, which isn't very useful, since the general goal is to get the leading-order behavior in the parameter we're looking at (which is t here). So in this case we'd know we'd have to go back and include the $x^2/2$ term in the series. If we were doing a problem in which there was still no t (or whatever variable) dependence at that order, then we'd have to go back and include the $-x^3/6$ term in the series. Of course, you could just play it safe and keep terms up to, say, fifth order. But that's invariably a poor strategy, because you'll probably never in your life have to go out that far in a series. So just start with one or two terms and see what it gives you. Note that in Eq. (1.7), we actually didn't need the second-order term, so we in fact could have gotten by with only $e^{-x} \approx 1 - x$. But having the extra term here didn't end up causing much heartache.

After you make an approximation, how do you know if it's a "good" one? Well, just as it makes no sense to ask if a dimensionful quantity is large or small without comparing it to another quantity, it makes no sense to ask if an approximation is "good" or "bad" without stating the accuracy you want. In the above example, if you're looking at a t value for which $\alpha t \approx 1/100$, then the term we ignored in Eq. (1.7) is smaller than gt by a factor $\alpha t/2 \approx 1/200$. So the error is on the order of 1%. If this is enough accuracy for whatever purpose you have in mind, then the approximation is a good one. If not, it's a bad one, and you should add more terms in the series until you get your desired accuracy.

The results of checking limits generally fall into two categories. Most of the time you know what the result should be, so this provides a double-check on your answer. But sometimes an interesting limit pops up that you might not expect. Such is the case in the following examples.

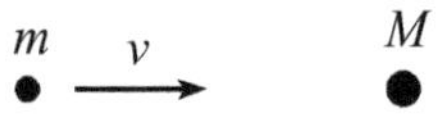

Fig. 1.3

Example (Two masses in 1-D): A mass m with speed v approaches a stationary mass M (see Fig. 1.3). The masses bounce off each other elastically. Assume that all motion takes place in one dimension. We'll find in Section 5.6.1 that the final velocities of the particles are

$$v_m = \frac{(m-M)v}{m+M}, \quad \text{and} \quad v_M = \frac{2mv}{m+M}. \tag{1.11}$$

There are three special cases that beg to be checked:

- If $m = M$, then Eq. (1.11) tells us that m stops, and M picks up a speed v. This is fairly believable (and even more so for pool players). And it becomes quite clear once you realize that these final speeds certainly satisfy conservation of energy and momentum with the initial conditions.
- If $M \gg m$, then m bounces backward with speed $\approx v$, and M hardly moves. This makes sense, because M is basically a brick wall.
- If $m \gg M$, then m keeps plowing along at speed $\approx v$, and M picks up a speed of $\approx 2v$. This $2v$ is an unexpected and interesting result (it's easier to see if you consider what's happening in the reference frame of the heavy mass m), and it leads to some neat effects, as in Problem 5.23.

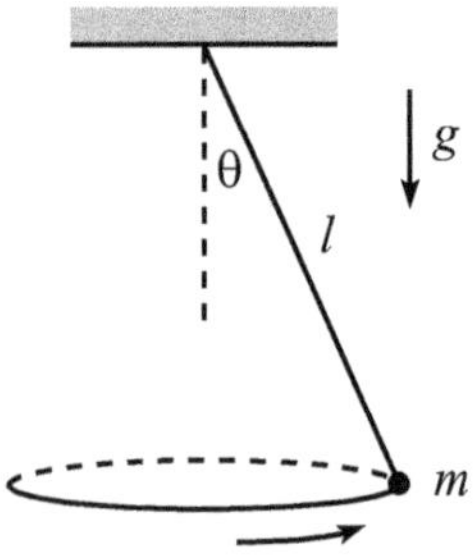

Fig. 1.4

Example (Circular pendulum): A mass hangs from a massless string of length ℓ. Conditions have been set up so that the mass swings around in a horizontal circle, with the string making a constant angle θ with the vertical (see Fig. 1.4). We'll find in Section 3.5 that the angular frequency, ω, of this motion is

$$\omega = \sqrt{\frac{g}{\ell \cos\theta}}. \tag{1.12}$$

As far as θ is concerned, there are two limits we should definitely check:

- If $\theta \to 90°$, then $\omega \to \infty$. This makes sense; the mass has to spin very quickly to avoid flopping down.
- If $\theta \to 0$, then $\omega \to \sqrt{g/\ell}$, which is the same as the frequency of a standard "plane" pendulum of length ℓ (for small oscillations). This is a cool result and not at all obvious. (But once we get to $F = ma$ in Chapter 3, you can convince yourself why this is true by looking at the projection of the force on a given horizontal line.)

In the above examples, we checked limiting and special cases of answers that were correct (I hope!). This whole process is more useful (and a bit more fun, actually) when you check the limits of an answer that is *incorrect*. In this case, you gain the unequivocal information that your answer is wrong. But rather than leading you into despair, this information is in fact something you should be quite happy about, considering that the alternative is to carry on in a state of blissful ignorance. Once you know that your answer is wrong, you can go back through your work and figure out where the error is (perhaps by checking limits at various stages to narrow down where the error could be). Personally, if there's any way I'd like to discover that my answer is garbage, this is it. At any rate, checking limiting cases can often save you a lot of trouble in the long run...

The lemmings get set for their race.
With one step and two steps they pace.
They take three and four,
And then head on for more,
Without checking the limiting case.

1.4 Solving differential equations numerically

Solving a physics problem often involves solving a differential equation. A differential equation is one that involves derivatives (usually with respect to time, in our physics problems) of the variable you're trying to solve for. The differential equation invariably comes about from using $F = ma$, and/or $\tau = I\alpha$, or the Lagrangian technique we'll discuss in Chapter 6. For example, consider a falling body. $F = ma$ gives $-mg = ma$, which can be written as $-g = \ddot{y}$, where a dot denotes a time derivative. This is a rather simple differential equation, and you can quickly guess that $y(t) = -gt^2/2$ is a solution. Or, more generally with the constants of integration thrown in, $y(t) = y_0 + v_0 t - gt^2/2$.

However, the differential equations produced in some problems can get rather complicated, so sooner or later you will encounter one that you can't solve exactly (either because it's in fact impossible to solve, or because you can't think of the appropriate clever trick). Having resigned yourself to not getting the exact answer, you should ponder how to obtain a decent approximation to it. Fortunately, it's easy to write a short program that will give you a very good numerical answer to your problem. Given enough computer time, you can obtain any desired accuracy (assuming that the system isn't chaotic, but we won't have to worry about this for the systems we'll be dealing with).

We'll demonstrate the procedure by considering a standard problem, one that we'll solve exactly and in great depth in Chapter 4. Consider the equation,

$$\ddot{x} = -\omega^2 x. \tag{1.13}$$

This is the equation for a mass on a spring, with $\omega = \sqrt{k/m}$. We'll find in Chapter 4 that the solution can be written, among other ways, as

$$x(t) = A\cos(\omega t + \phi). \tag{1.14}$$

But let's pretend we don't know this. If someone comes along and gives us the values of $x(0)$ and $\dot{x}(0)$, then it seems that somehow we should be able to find $x(t)$ and $\dot{x}(t)$ for any later t, just by using Eq. (1.13). Basically, if we're told how the system starts, and if we know how it evolves, via Eq. (1.13), then we should know everything about it. So here's how we find $x(t)$ and $\dot{x}(t)$.

The plan is to discretize time into intervals of some small unit (call it ϵ), and to then determine what happens at each successive point in time. If we know $x(t)$ and $\dot{x}(t)$, then we can easily find (approximately) the value of x at a slightly later time, by using the definition of $\dot{x}$. Similarly, if we know $\dot{x}(t)$ and $\ddot{x}(t)$, then we can easily find (approximately) the value of $\dot{x}$ at a slightly later time, by using the definition of $\ddot{x}$. Using the definitions of the derivatives, the relations are simply

$$\begin{aligned} x(t+\epsilon) &\approx x(t) + \epsilon\dot{x}(t), \\ \dot{x}(t+\epsilon) &\approx \dot{x}(t) + \epsilon\ddot{x}(t). \end{aligned} \tag{1.15}$$

These two equations, combined with (1.13), which gives us $\ddot{x}$ in terms of x, allow us to march along in time, obtaining successive values for x, $\dot{x}$, and $\ddot{x}$.[4]

Here's what a typical program might look like.[5] (This is a Maple program, but even if you aren't familiar with this, the general idea should be clear.) Let's say

[4] Of course, another expression for $\ddot{x}$ is the definitional one, analogous to Eqs. (1.15), involving the third derivative. But this would then require knowledge of the third derivative, and so on with higher derivatives, and we would end up with an infinite chain of relations. An *equation of motion* such as Eq. (1.13) (which in general could be an $F = ma$, $\tau = I\alpha$, or Euler–Lagrange equation) relates $\ddot{x}$ back to x (and possibly $\dot{x}$), thereby creating an intertwined relation among x, $\dot{x}$, and $\ddot{x}$, and eliminating the need for an infinite and useless chain.

[5] We've written the program in the most straightforward way, without any concern for efficiency, because computing time isn't an issue in this simple system. But in more complex systems that require programs for which computing time is an issue, a major part of the problem-solving process is developing a program that is as efficient as possible.

that the particle starts from rest at position $x = 2$, and let's pick $\omega^2 = 5$. We'll use the notation where x1 stands for $\dot{x}$, and x2 stands for $\ddot{x}$. And e stands for ϵ. Let's calculate x at, say, $t = 3$.

```
x:=2:              # initial position
x1:=0:             # initial velocity
e:=.01:            # small time interval
for i to 300 do    # do 300 steps (ie, up to 3 seconds)
x2:=-5*x:          # the given equation
x:=x+e*x1:         # how x changes, by definition of x1
x1:=x1+e*x2:       # how x1 changes, by definition of x2
end do:            # the Maple command to stop the do loop
x;                 # print the value of x
```

This procedure won't give the exact value for x, because x and $\dot{x}$ don't really change according to Eqs. (1.15). These equations are just first-order approximations to the full Taylor series with higher-order terms. Said differently, there is no way the above procedure can be exactly correct, because there are ambiguities in how the program can be written. Should line 5 come before or after line 7? That is, in determining $\dot{x}$ at time $t + \epsilon$, should we use the $\ddot{x}$ at time t or $t + \epsilon$? And should line 7 come before or after line 6? The point is that for very small ϵ, the order doesn't matter much. And in the limit $\epsilon \to 0$, the order doesn't matter at all.

If we want to obtain a better approximation, we can just shorten ϵ down to 0.001 and increase the number of steps to 3000. If the result looks basically the same as with $\epsilon = 0.01$, then we know we pretty much have the right answer. In the present example, $\epsilon = 0.01$ yields $x \approx 1.965$ after 3 seconds. If we set $\epsilon = 0.001$, then we obtain $x \approx 1.836$. And if we set $\epsilon = 0.0001$, then we get $x \approx 1.823$. The correct answer must therefore be somewhere around $x = 1.82$. And indeed, if we solve the problem exactly, we obtain $x(t) = 2\cos(\sqrt{5}\,t)$. Plugging in $t = 3$ gives $x \approx 1.822$.

This is a wonderful procedure, but it shouldn't be abused. It's nice to know that we can always obtain a decent numerical approximation if all else fails. But we should set our initial goal on obtaining the correct algebraic expression, because this allows us to see the overall behavior of the system. And besides, nothing beats the truth. People tend to rely a bit too much on computers and calculators nowadays, without pausing to think about what is actually going on in a problem.

> The skill to do math on a page
> Has declined to the point of outrage.
> Equations quadratica
> Are solved on Math'matica,
> And on birthdays we don't know our age.

1.5 Problems

Section 1.2: Units, dimensional analysis

1.1. **Escape velocity** ∗

As given below in Exercise 1.9, show that the escape velocity from the earth is $v = \sqrt{2GM_E/R}$, up to numerical factors. You can use the fact that the form of Newton's gravitation force law implies that the acceleration (and hence overall motion) of the particle doesn't depend on its mass.

1.2. **Mass in a tube** ∗

A tube of mass M and length ℓ is free to swing around a pivot at one end. A mass m is positioned inside the (frictionless) tube at this end. The tube is held horizontal and then released (see Fig. 1.5). Let η be the fraction of the tube that the mass has traversed by the time the tube becomes vertical. Does η depend on ℓ?

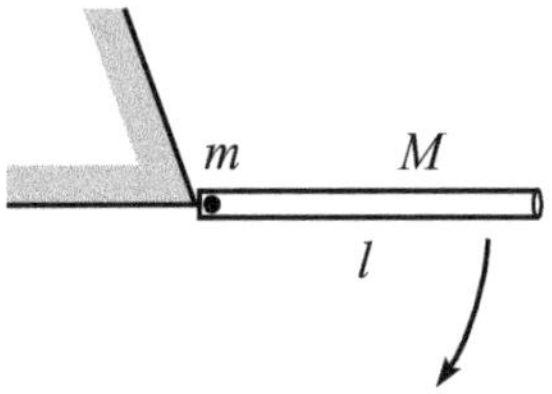

Fig. 1.5

1.3. **Waves in a fluid** ∗

How does the speed of waves in a fluid depend on its density, ρ, and "bulk modulus," B (which has units of pressure, which is force per area)?

1.4. **Vibrating star** ∗

Consider a vibrating star, whose frequency ν depends (at most) on its radius R, mass density ρ, and Newton's gravitational constant G. How does ν depend on R, ρ, and G?

1.5. **Damping** ∗∗

A particle with mass m and initial speed V is subject to a velocity-dependent damping force of the form bv^n.

(a) For $n = 0, 1, 2, \ldots$, determine how the stopping time depends on m, V, and b.
(b) For $n = 0, 1, 2, \ldots$, determine how the stopping distance depends on m, V, and b.

Be careful! See if your answers make sense. Dimensional analysis gives the answer only up to a numerical factor. This is a tricky problem, so don't let it discourage you from using dimensional analysis. Most applications of dimensional analysis are quite straightforward.

Section 1.3: Approximations, limiting cases

1.6. **Projectile distance** ∗

A person throws a ball (at an angle of her choosing, to achieve the maximum distance) with speed v from the edge of a cliff of height h. Assuming that one of the following quantities is the maximum horizontal distance the ball can travel, which one is it? (Don't solve the problem from scratch, just check special cases.)

$$\frac{gh^2}{v^2}, \quad \frac{v^2}{g}, \quad \sqrt{\frac{v^2h}{g}}, \quad \frac{v^2}{g}\sqrt{1+\frac{2gh}{v^2}}, \quad \frac{v^2}{g}\left(1+\frac{2gh}{v^2}\right), \quad \frac{v^2/g}{1-\frac{2gh}{v^2}}.$$

Section 1.4: Solving differential equations numerically

1.7. **Two masses, one swinging** **

Two equal masses are connected by a string that hangs over two pulleys (of negligible size), as shown in Fig. 1.6. The left mass moves in a vertical line, but the right mass is free to swing back and forth in the plane of the masses and pulleys. It can be shown (see Problem 6.4) that the equations of motion for r and θ (labeled in the figure) are

$$2\ddot{r} = r\dot{\theta}^2 - g(1-\cos\theta),$$
$$\ddot{\theta} = -\frac{2\dot{r}\dot{\theta}}{r} - \frac{g\sin\theta}{r}. \tag{1.16}$$

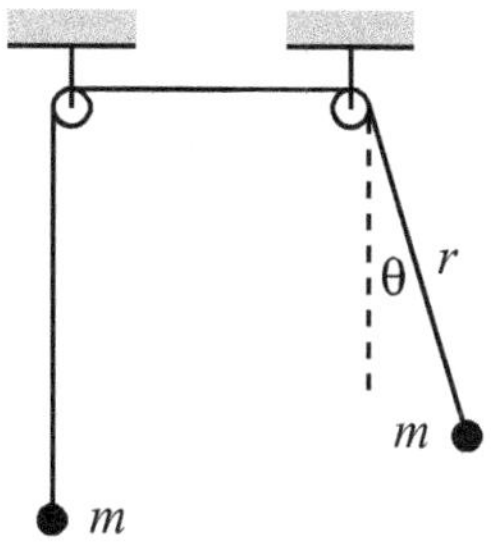

Fig. 1.6

Assume that both masses start out at rest, with the right mass making an initial angle of $10^\circ = \pi/18$ with the vertical. If the initial value of r is 1 m, how much time does it take for it to reach a length of 2 m? Write a program to solve this numerically. Use $g = 9.8\,\text{m/s}^2$.

1.6 Exercises

Section 1.2: Units, dimensional analysis

1.8. **Pendulum on the moon**

If a pendulum has a period of 3 s on the earth, what would its period be if it were placed on the moon? Use $g_M/g_E \approx 1/6$.

1.9. **Escape velocity** *

The *escape velocity* on the surface of a planet is given by

$$v = \sqrt{\frac{2GM}{R}}, \tag{1.17}$$

where M and R are the mass and radius of the planet, respectively, and G is Newton's gravitational constant. (The escape velocity is the velocity needed to refute the "What goes up must come down" maxim, neglecting air resistance.)

(a) Write v in terms of the average mass density ρ, instead of M.
(b) Assuming that the average density of the earth is four times that of Jupiter, and that the radius of Jupiter is 11 times that of the earth, what is v_J/v_E?

1.10. **Downhill projectile** *

A hill is sloped downward at an angle θ with respect to the horizontal. A projectile of mass m is fired with speed v_0 perpendicular to the hill. When it eventually lands on the hill, let its velocity make an angle β with respect to the horizontal. Which of the quantities θ, m, v_0, and g does the angle β depend on?

1.11. **Waves on a string** *

How does the speed of waves on a string depend on its mass M, length L, and tension (that is, force) T?

1.12. **Vibrating water drop** *

Consider a vibrating water drop, whose frequency ν depends on its radius R, mass density ρ, and surface tension S. The units of surface tension are (force)/(length). How does ν depend on R, ρ, and S?

Section 1.3: Approximations, limiting cases

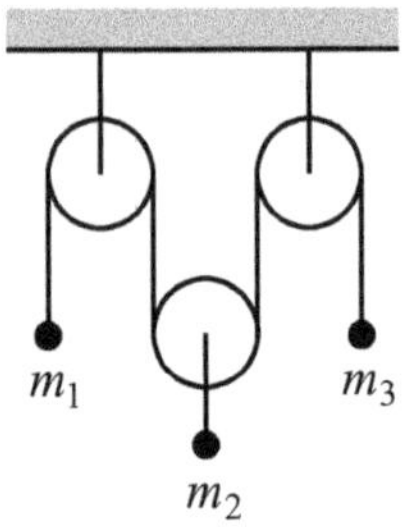

Fig. 1.7

1.13. **Atwood's machine** *

Consider the "Atwood's" machine shown in Fig. 1.7, consisting of three masses and three frictionless pulleys. It can be shown that the acceleration of m_1 is given by (just accept this):

$$a_1 = g\frac{3m_2m_3 - m_1(4m_3 + m_2)}{m_2m_3 + m_1(4m_3 + m_2)}, \tag{1.18}$$

with upward taken to be positive. Find a_1 in the following special cases:

(a) $m_2 = 2m_1 = 2m_3$.
(b) m_1 much larger than both m_2 and m_3.
(c) m_1 much smaller than both m_2 and m_3.
(d) $m_2 \gg m_1 = m_3$.
(e) $m_1 = m_2 = m_3$.

1.14. **Cone frustum** *

A cone frustum has base radius b, top radius a, and height h, as shown in Fig. 1.8. Assuming that one of the following quantities is the volume of the frustum, which one is it? (Don't solve the problem from scratch, just check special cases.)

$$\frac{\pi h}{3}(a^2+b^2), \quad \frac{\pi h}{2}(a^2+b^2), \quad \frac{\pi h}{3}(a^2+ab+b^2),$$
$$\frac{\pi h}{3}\cdot\frac{a^4+b^4}{a^2+b^2}, \quad \pi hab.$$

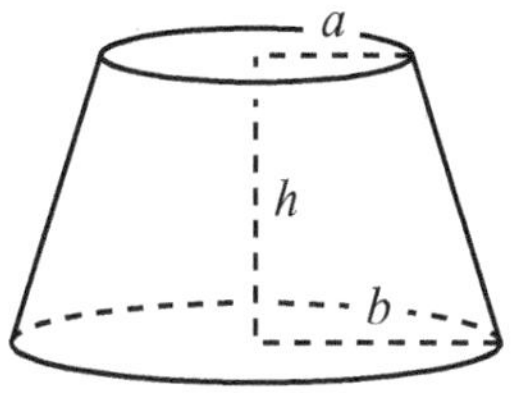

Fig. 1.8

1.15. **Landing at the corner** *

A ball is thrown at an angle θ up to the top of a cliff of height L, from a point a distance L from the base, as shown in Fig. 1.9. Assuming that one of the following quantities is the initial speed required to make the ball hit right at the edge of the cliff, which one is it? (Don't solve the problem from scratch, just check special cases.)

$$\sqrt{\frac{gL}{2(\tan\theta-1)}}, \quad \frac{1}{\cos\theta}\sqrt{\frac{gL}{2(\tan\theta-1)}}, \quad \frac{1}{\cos\theta}\sqrt{\frac{gL}{2(\tan\theta+1)}},$$
$$\sqrt{\frac{gL\tan\theta}{2(\tan\theta+1)}}.$$

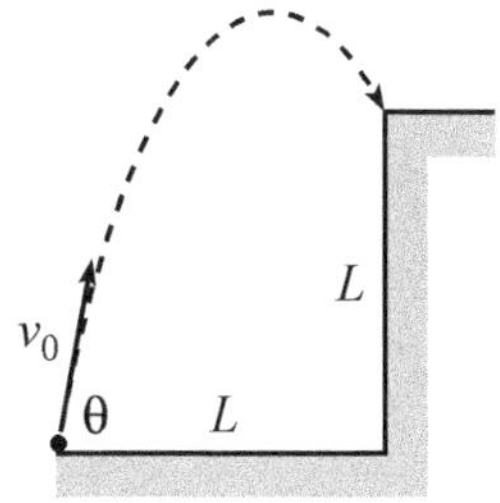

Fig. 1.9

1.16. **Projectile with drag** **

Consider a projectile subject to a drag force $\mathbf{F} = -m\alpha\mathbf{v}$. If it is fired with speed v_0 at an angle θ, it can be shown that the height as a function of time is given by (just accept this here; it's one of the tasks of Exercise 3.53)

$$y(t) = \frac{1}{\alpha}\left(v_0\sin\theta + \frac{g}{\alpha}\right)\left(1-e^{-\alpha t}\right) - \frac{gt}{\alpha}. \qquad (1.19)$$

Show that this reduces to the usual projectile expression, $y(t) = (v_0\sin\theta)t - gt^2/2$, in the limit of small α. What exactly is meant by "small α"?

Section 1.4: Solving differential equations numerically

1.17. **Pendulum** **

A pendulum of length ℓ is released from the horizontal position. It can be shown that the tangential $F = ma$ equation is (where θ is measured with respect to the vertical)

$$\ddot{\theta} = -\frac{g\sin\theta}{\ell}. \qquad (1.20)$$

If $\ell = 1\,\text{m}$, and $g = 9.8\,\text{m/s}^2$, write a program to show that the time it takes the pendulum to swing down through the vertical position is

$t \approx 0.592$ s. This happens to be about 1.18 times the $(\pi/2)\sqrt{\ell/g} \approx 0.502$ s it would take the pendulum to swing down if it were released from very close to the vertical (this is 1/4 of the standard period of $2\pi\sqrt{\ell/g}$ for a pendulum). It also happens to be about 1.31 times the $\sqrt{2\ell/g} \approx 0.452$ s it would take a mass to simply freefall a height ℓ.

1.18. **Distance with damping** **

A mass is subject to a damping force proportional to its velocity, which means that the equation of motion takes the form $\ddot{x} = -A\dot{x}$, where A is some constant. If the initial speed is 2 m/s, and if $A = 1\ \text{s}^{-1}$, how far has the mass traveled at 1 s? 10 s? 100 s? You should find that the distance approaches a limiting value.

Now assume that the mass is subject to a damping force proportional to the square of its velocity, which means that the equation of motion now takes the form $\ddot{x} = -A\dot{x}^2$, where A is some constant. If the initial speed is 2 m/s, and if $A = 1\ \text{m}^{-1}$, how far has the mass traveled at 1 s? 10 s? 100 s? How about some larger powers of 10? You should find that the distance keeps growing, but slowly like the log of t. (The results for these two forms of the damping are consistent with the results of Problem 1.5.)

1.7 Solutions

1.1. **Escape velocity**

It is tempting to use the same reasoning as in the low-orbit satellite example in Section 1.2. This reasoning gives the same result, $v = C\sqrt{gR} = C\sqrt{GM_E/R}$, where C is some number (it turns out that $C = \sqrt{2}$). Although this solution yields the correct answer, it isn't quite rigorous, in view of the footnote in the low-orbit satellite example. Because the particle isn't always at the same radius, the force changes, so it isn't obvious that we can absorb the M_E and G dependence into one quantity, g, as we did with the orbiting satellite. Let us therefore be more rigorous with the following reasoning.

The dimensionful quantities in the problem are $[m] = M$, the radius of the earth $[R] = L$, the mass of the earth $[M_E] = M$, and Newton's gravitational constant $[G] = L^3/MT^2$. These units for G follow from the gravitational force law, $F = Gm_1m_2/r^2$. If we use no information other than these given quantities, then there is no way to arrive at the speed of $C\sqrt{GM_E/R}$, because for all we know, there could be a factor of $(m/M_E)^7$ in the answer. This number is dimensionless, so it wouldn't mess up the units.

If we want to make any progress in this problem, we have to use the fact that the gravitational force takes the form of GM_Em/r^2. This then implies (as was stated in the problem) that the acceleration is independent of m. And since the path of the particle is determined by its acceleration, we see that the answer can't depend on m. We are therefore left with the quantities G, R, and M_E, and you can show that the only combination of these quantities that gives the units of speed is $v = C\sqrt{GM_E/R}$.

1.2. **Mass in a tube**

The dimensionful quantities are $[g] = L/T^2$, $[\ell] = L$, $[m] = M$, and $[M] = M$. We want to produce a dimensionless number η. Since g is the only constant involving time, η cannot depend on g. This then implies that η cannot depend on ℓ, which is the only length remaining. Therefore, η depends only on m and M (and furthermore only on the ratio m/M, since we want a dimensionless number). So the answer to the stated problem is, "No."

It turns out that you have to solve the problem numerically if you actually want to find η (see Problem 8.5). Some results are: If $m \ll M$, then $\eta \approx 0.349$. If $m = M$, then $\eta \approx 0.378$. And if $m = 2M$, then $\eta \approx 0.410$.

1.3. **Waves in a fluid**

We want to make a speed, $[v] = L/T$, out of the quantities $[\rho] = M/L^3$, and $[B] = [F/A] = (ML/T^2)/(L^2) = M/(LT^2)$. We can play around with these quantities to find the combination that has the correct units, but let's do it the no-fail way. If $v \propto \rho^a B^b$, then we have

$$\frac{L}{T} = \left(\frac{M}{L^3}\right)^a \left(\frac{M}{LT^2}\right)^b. \tag{1.21}$$

Matching up the powers of the three kinds of units on each side of this equation gives

$$M: \ 0 = a + b, \qquad L: \ 1 = -3a - b, \qquad T: \ -1 = -2b. \tag{1.22}$$

The solution to this system of equations is $a = -1/2$ and $b = 1/2$. Therefore, our answer is $v \propto \sqrt{B/\rho}$. Fortunately, there was a solution to this system of three equations in two unknowns.

1.4. **Vibrating star**

We want to make a frequency, $[\nu] = 1/T$, out of the quantities $[R] = L$, $[\rho] = M/L^3$, and $[G] = L^3/(MT^2)$. These units for G follow from the gravitational force law, $F = Gm_1m_2/r^2$. As in the previous problem, we can play around with these quantities to find the combination that has the correct units, but let's do it the no-fail way. If $\nu \propto R^a\rho^bG^c$, then we have

$$\frac{1}{T} = L^a \left(\frac{M}{L^3}\right)^b \left(\frac{L^3}{MT^2}\right)^c. \tag{1.23}$$

Matching up the powers of the three kinds of units on each side of this equation gives

$$M: \ 0 = b - c, \qquad L: \ 0 = a - 3b + 3c, \qquad T: \ -1 = -2c. \tag{1.24}$$

The solution to this system of equations is $a = 0$, and $b = c = 1/2$. Therefore, our answer is $\nu \propto \sqrt{\rho G}$. So it turns out that there is no R dependence.

REMARK: Note the difference in the given quantities in this problem (R, ρ, and G) and the ones in Exercise 1.12 (R, ρ, and S). In this problem with the star, the mass is large enough so that we can ignore the surface tension, S. And in Exercise 1.12 with the drop, the mass is small enough so that we can ignore the gravitational force, and hence G. ♣

1.5. **Damping**

(a) The constant b has units $[b] = [\text{Force}][v^{-n}] = (ML/T^2)(T^n/L^n)$. The other quantities are $[m] = M$ and $[V] = L/T$. There is also n, which is dimensionless. You can show that the only combination of these quantities that has units of T is

$$t = f(n)\frac{m}{bV^{n-1}}, \tag{1.25}$$

where $f(n)$ is a dimensionless function of n.

For $n = 0$, we have $t = f(0)\, mV/b$. This increases with m and V, and decreases with b, as it should.

For $n = 1$, we have $t = f(1)\, m/b$. So we *seem* to have $t \sim m/b$. This, however, cannot be correct, because t should definitely grow with V. A large initial speed V_1 requires some nonzero time to slow down to a smaller speed V_2, after which time we simply have the same scenario with initial speed V_2. So where did we go wrong? After all, dimensional analysis tells us that the answer *does* have to look like $t = f(1)\, m/b$, where $f(1)$ is a numerical factor. The resolution to this puzzle is that $f(1)$ is infinite. If we worked out the problem using $F = ma$, we would encounter an integral that diverges. So for any V, we would find an infinite t.[6]

Similarly, for $n \geq 2$, there is at least one power of V in the denominator of t. This certainly cannot be correct, because t should not decrease with V. So $f(n)$ must likewise be infinite for all of these cases.

The moral of this exercise is that sometimes you have to be careful when using dimensional analysis. The numerical factor in front of your answer nearly always turns out to be of order 1, but in some strange cases it turns out to be 0 or ∞.

REMARK: For $n \geq 1$, the expression in Eq. (1.25) still has relevance. For example, for $n = 2$, the $m/(Vb)$ expression is relevant if you want to know how long it takes to go from V to some final speed V_f. The answer involves $m/(V_f b)$, which diverges as $V_f \to 0$. ♣

(b) You can show that the only combination of the quantities that has units of L is

$$\ell = g(n)\frac{m}{bV^{n-2}}, \tag{1.26}$$

where $g(n)$ is a dimensionless function of n.

For $n = 0$, we have $\ell = g(0)\,mV^2/b$. This increases with V, as it should.

For $n = 1$, we have $\ell = g(1)\,mV/b$. This increases with V, as it should.

For $n = 2$ we have $\ell = g(2)\,m/b$. So we *seem* to have $\ell \sim m/b$. But as in part (a), this cannot be correct, because ℓ should definitely depend on V. A large initial speed V_1 requires some nonzero distance to slow down to a smaller speed V_2, after which point we simply have the same scenario with initial speed V_2. So, from the reasoning in part (a), the total distance is infinite for $n \geq 2$, because the function g is infinite.

REMARK: Note that for integral $n \neq 1$, t and ℓ are either both finite or both infinite. For $n = 1$, however, the total time is infinite, whereas the total distance is finite. This situation actually holds for $1 \leq n < 2$, if we want to consider fractional n. ♣

1.6. **Projectile distance**

All of the possible answers have the correct units, so we'll have to figure things out by looking at special cases. Let's look at each choice in turn:

$\dfrac{gh^2}{v^2}$: Incorrect, because the answer shouldn't be zero for $h = 0$. Also, it shouldn't grow with g. And even worse, it shouldn't be infinite for $v \to 0$.

$\dfrac{v^2}{g}$: Incorrect, because the answer should depend on h.

$\sqrt{\dfrac{v^2 h}{g}}$: Incorrect, because the answer shouldn't be zero for $h = 0$.

$\dfrac{v^2}{g}\sqrt{1 + \dfrac{2gh}{v^2}}$: Can't rule this out, and it happens to be the correct answer.

$\dfrac{v^2}{g}\left(1 + \dfrac{2gh}{v^2}\right)$: Incorrect, because the answer should be zero for $v \to 0$. But this expression goes to $2h$ for $v \to 0$.

$\dfrac{v^2/g}{1 - \frac{2gh}{v^2}}$: Incorrect, because the answer shouldn't be infinite for $v^2 = 2gh$.

[6] The total time t is actually undefined, because the particle never comes to rest. But t does grow with V, in the sense that if t is defined to be the time to slow down to some certain small speed, then t grows with V.

1.7. Two masses, one swinging

As in Section 1.4, we'll write a Maple program. We'll let q stand for θ, and we'll use the notation where q1 stands for $\dot{\theta}$, and q2 stands for $\ddot{\theta}$. Likewise for r. We'll run the program for as long as $r < 2$. As soon as r exceeds 2, the program will stop and print the value of the time.

```
r:=1:                             # initial r value
r1:=0:                            # initial r velocity
q:=3.14/18:                       # initial angle
q1:=0:                            # initial angular velocity
e:=.001:                          # small time interval
i:=0:                             # i counts the number of time steps
while r<2 do                      # run the program until r=2
i:=i+1:                           # increase the counter by 1
r2:=(r*q1^2-9.8*(1-cos(q)))/2:    # the first of the given eqs
r:=r+e*r1:                        # how r changes, by definition of r1
r1:=r1+e*r2:                      # how r1 changes, by definition of r2
q2:=-2*r1*q1/r-9.8*sin(q)/r:      # the second of the given eqs
q:=q+e*q1:                        # how q changes, by definition of q1
q1:=q1+e*q2:                      # how q1 changes, by definition of q2
end do:                           # the Maple command to stop the do loop
i*e;                              # print the value of the time
```

This yields a time of $t = 8.057$ s. If we instead use a time interval of 0.0001 s, we obtain $t = 8.1377$ s. And a time interval of 0.00001 s gives $t = 8.14591$ s. So the correct time must be somewhere around 8.15 s.

Chapter 2
Statics

The subject of statics often appears in later chapters in other books, after force and torque have been discussed. However, the way that force and torque are used in statics problems is fairly minimal, at least compared with what we'll be doing later in this book. Therefore, since we won't be needing much of the machinery that we'll be developing later on, I'll introduce here the bare minimum of force and torque concepts necessary for statics problems. This will open up a whole class of problems for us. But even though the underlying principles of statics are quick to state, statics problems can be unexpectedly tricky. So be sure to tackle a lot of them to make sure you understand things.

2.1 Balancing forces

A "static" setup is one where all the objects are motionless. If an object remains motionless, then Newton's second law, $F = ma$ (which we'll discuss in great detail in the next chapter), tells us that the total external force acting on the object must be zero. The converse is not true, of course. The total external force on an object is also zero if it moves with constant nonzero velocity. But we'll deal only with statics problems here. The whole goal in a statics problem is to find out what the various forces have to be so that there is zero net force acting on each object (and zero net torque, too, but that's the topic of Section 2.2). Because a force is a vector, this goal involves breaking the force up into its components. You can pick Cartesian coordinates, polar coordinates, or another set. It is usually clear from the problem which system will make your calculations easiest. Once you pick a system, you simply have to demand that the total external force in each direction is zero.

There are many different types of forces in the world, most of which are large-scale effects of complicated things going on at smaller scales. For example, the tension in a rope comes from the chemical bonds that hold the molecules in the rope together, and these chemical forces are electrical forces. In doing a mechanics problem involving a rope, there is certainly no need to analyze all the details of the forces taking place at the molecular scale. You just call the force in the rope a "tension" and get on with the problem. Four types of forces come up repeatedly:

Tension

Tension is the general name for a force that a rope, stick, etc., exerts when it is pulled on. Every piece of the rope feels a tension force in both directions, except the end points, which feel a tension on one side and a force on the other side from whatever object is attached to the end. In some cases, the tension may vary along the rope. The "Rope wrapped around a pole" example at the end of this section is a good illustration of this. In other cases, the tension must be the same everywhere. For example, in a hanging massless rope, or in a massless rope hanging over a frictionless pulley, the tension must be the same at all points, because otherwise there would be a net force on at least some part of the rope, and then $F = ma$ would yield an infinite acceleration for this (massless) piece.

Normal force

This is the force perpendicular to a surface that the surface applies to an object. The total force applied by a surface is usually a combination of the normal force and the friction force (see below). But for frictionless surfaces such as greasy ones or ice, only the normal force exists. The normal force comes about because the surface actually compresses a tiny bit and acts like a very rigid spring. The surface gets squashed until the restoring force equals the force necessary to keep the object from squashing in any more.

For the most part, the only difference between a "tension" and a "normal force" is the direction of the force. Both situations can be modeled by a spring. In the case of a tension, the spring (a rope, a stick, or whatever) is stretched, and the force on the given object is directed toward the spring. In the case of a normal force, the spring is compressed, and the force on the given object is directed away from the spring. Things like sticks can provide both normal forces and tensions. But a rope, for example, has a hard time providing a normal force. In practice, in the case of elongated objects such as sticks, a compressive force is usually called a "compressive tension," or a "negative tension," instead of a normal force. So by these definitions, a tension can point either way. At any rate, it's just semantics. If you use any of these descriptions for a compressed stick, people will know what you mean.

Friction

Friction is the force parallel to a surface that a surface applies to an object. Some surfaces, such as sandpaper, have a great deal of friction. Some, such as greasy ones, have essentially no friction. There are two types of friction, called "kinetic" friction and "static" friction. Kinetic friction (which we won't cover in this chapter) deals with two objects moving relative to each other. It is usually a good approximation to say that the kinetic friction between two objects is proportional to the normal force between them. The constant of proportionality is called μ_k (the "coefficient of kinetic friction"), where μ_k depends on the two surfaces involved. Thus, $F = \mu_k N$, where N is the normal force. The direction of the force is opposite to the motion.

Static friction deals with two objects at rest relative to each other. In the static case, we have $F \le \mu_s N$ (where μ_s is the "coefficient of static friction"). Note the inequality sign. All we can say prior to solving a problem is that the static friction force has a *maximum* value equal to $F_{\text{max}} = \mu_s N$. In a given problem, it is most likely less than this. For example, if a block of large mass M sits on a surface with coefficient of friction μ_s, and you give the block a tiny push to the right (tiny enough so that it doesn't move), then the friction force is of course not equal to $\mu_s N = \mu_s Mg$ to the left. Such a force would send the block sailing off to the left. The true friction force is simply equal and opposite to the tiny force you apply. What the coefficient μ_s tells us is that if you apply a force larger than $\mu_s Mg$ (the maximum friction force on a horizontal table), then the block will end up moving to the right.

Gravity

Consider two point objects, with masses M and m, separated by a distance R. Newton's gravitational force law says that the force between these objects is attractive and has magnitude $F = GMm/R^2$, where $G = 6.67 \cdot 10^{-11}\ \text{m}^3/(\text{kg s}^2)$. As we'll show in Chapter 5, the same law also applies to spheres of nonzero size. That is, a sphere may be treated like a point mass located at its center. Therefore, an object on the surface of the earth feels a gravitational force equal to

$$F = m\left(\frac{GM}{R^2}\right) \equiv mg, \tag{2.1}$$

where M is the mass of the earth, and R is its radius. This equation defines g. Plugging in the numerical values, we obtain $g \approx 9.8\ \text{m/s}^2$, as you can check. Every object on the surface of the earth feels a force of mg downward (g varies slightly over the surface of the earth, but let's ignore this). If the object is not accelerating, then there must be other forces present (normal forces, etc.) to make the total force be equal to zero.

Another common force is the Hooke's-law spring force, $F = -kx$. But we'll postpone the discussion of springs until Chapter 4, where we'll spend a whole chapter on them in depth.

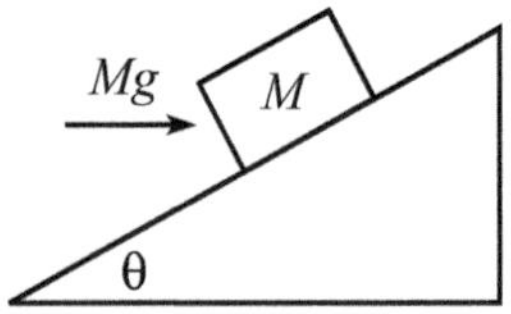

Fig. 2.1

Example (Block on a plane): A block of mass M rests on a fixed plane inclined at an angle θ. You apply a horizontal force of Mg on the block, as shown in Fig. 2.1. Assume that the friction force between the block and the plane is large enough to keep the block at rest. What are the normal and friction forces (call them N and F_f) that the plane exerts on the block? If the coefficient of static friction is μ, for what range of angles θ will the block in fact remain at rest?

Solution: Let's break the forces up into components parallel and perpendicular to the plane. (The horizontal and vertical components would also work, but the calculation would be a little longer.) The forces are N, F_f, the applied Mg, and the weight Mg, as shown in Fig. 2.2. Balancing the forces parallel and perpendicular to the plane gives, respectively (with upward along the plane taken to be positive),

$$\begin{aligned} F_f &= Mg\sin\theta - Mg\cos\theta, \\ N &= Mg\cos\theta + Mg\sin\theta. \end{aligned} \tag{2.2}$$

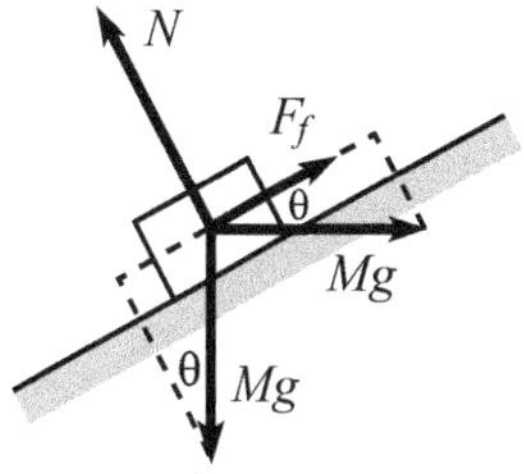

Fig. 2.2

INTERMEDIATE REMARKS:

1. If $\tan\theta > 1$, then F_f is positive (that is, it points up the plane). And if $\tan\theta < 1$, then F_f is negative (that is, it points down the plane). There is no need to worry about which way it points when drawing the diagram. Just pick a direction to be positive, and if F_f comes out to be negative (as it does in the figure above, because $\theta < 45°$), then it actually points in the other direction.
2. F_f ranges from $-Mg$ to Mg as θ ranges from 0 to $\pi/2$ (convince yourself that these limiting values make sense). As an exercise, you can show that N is maximum when $\tan\theta = 1$, in which case $N = \sqrt{2}Mg$ and $F_f = 0$.
3. The $\sin\theta$ and $\cos\theta$ factors in Eq. (2.2) follow from the angles θ drawn in Fig. 2.2. However, when solving problems like this one, it's easy to make a mistake in the geometry and then label an angle as θ when it really should be $90° - \theta$. So two pieces of advice: (1) Never draw an angle close to 45° in a figure, because if you do, you won't be able to tell the θ angles from the $90° - \theta$ ones. (2) Always check your results by letting θ go to 0 or 90° (in other words, does virtually all of a force, or virtually none of it, act in a certain direction when the plane is, say, horizontal). Once you do this a few times, you'll realize that you probably don't even need to work out the geometry in the first place. Since you know that any given component is going to involve either $\sin\theta$ or $\cos\theta$, you can just pick the one that works correctly in a certain limit. ♣

The coefficient μ tells us that $|F_f| \le \mu N$. Using Eq. (2.2), this inequality becomes

$$Mg|\sin\theta - \cos\theta| \le \mu Mg(\cos\theta + \sin\theta). \tag{2.3}$$

The absolute value here signifies that we must consider two cases:

- If $\tan\theta \ge 1$, then Eq. (2.3) becomes

$$\sin\theta - \cos\theta \le \mu(\cos\theta + \sin\theta) \quad \Longrightarrow \quad \tan\theta \le \frac{1+\mu}{1-\mu}. \tag{2.4}$$

  We divided by $1 - \mu$, so this inequality is valid only if $\mu < 1$. But if $\mu \ge 1$, we see from the first inequality here that any value of θ (subject to our assumption, $\tan\theta \ge 1$) works.
- If $\tan\theta \le 1$, then Eq. (2.3) becomes

$$-\sin\theta + \cos\theta \le \mu(\cos\theta + \sin\theta) \quad \Longrightarrow \quad \tan\theta \ge \frac{1-\mu}{1+\mu}. \tag{2.5}$$

Putting these two ranges for θ together, we have

$$\frac{1-\mu}{1+\mu} \le \tan\theta \le \frac{1+\mu}{1-\mu}. \tag{2.6}$$

REMARKS: For very small μ, these bounds both approach 1, which means that θ must be very close to 45°. This makes sense. If there is very little friction, then the components along the plane of the horizontal and vertical Mg forces must nearly cancel; hence, $\theta \approx 45°$. A special value for μ is 1, because from Eq. (2.6), we see that $\mu = 1$ is the cutoff value that allows θ to reach both 0 and $\pi/2$. If $\mu \ge 1$, then any tilt of the plane is allowed. We've been assuming throughout this example that $0 \le \theta \le \pi/2$. The task of Exercise 2.20 is to deal with the case where $\theta > \pi/2$, where the block is under an overhang. ♣

Let's now do an example involving a rope in which the tension varies with position. We'll need to consider differential pieces of the rope to solve this problem.

Example (Rope wrapped around a pole): A rope wraps an angle θ around a pole. You grab one end and pull with a tension T_0. The other end is attached to a large object, say, a boat. If the coefficient of static friction between the rope and the pole is μ, what is the largest force the rope can exert on the boat, if the rope is not to slip around the pole?

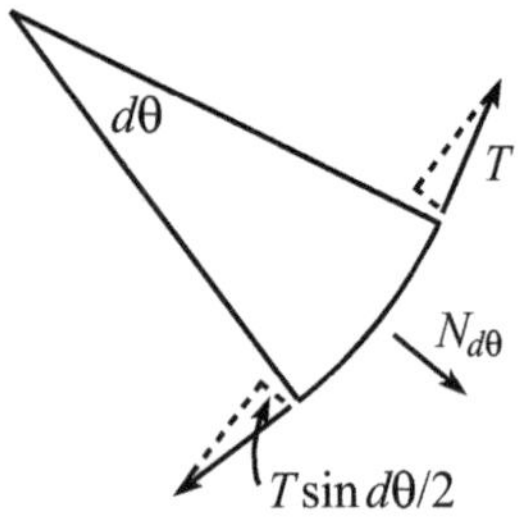

Fig. 2.3

Solution: Consider a small piece of the rope that subtends an angle $d\theta$. Let the tension in this piece be T (which varies slightly over the small length). As shown in Fig. 2.3, the pole exerts a small outward normal force, $N_{d\theta}$, on the piece. This normal force exists to balance the "inward" components of the tensions at the ends. These inward components have magnitude $T\sin(d\theta/2)$.[1] Therefore, $N_{d\theta} = 2T\sin(d\theta/2)$. The small-angle approximation, $\sin x \approx x$, allows us to write this as $N_{d\theta} = T\,d\theta$.

The friction force on the little piece of rope satisfies $F_{d\theta} \le \mu N_{d\theta} = \mu T\,d\theta$. This friction force is what gives rise to the difference in tension between the two ends of the piece. In other words, the tension, as a function of θ, satisfies

$$\begin{aligned}
T(\theta + d\theta) &\le T(\theta) + \mu T\,d\theta \\
\implies \quad dT &\le \mu T\,d\theta \\
\implies \quad \int \frac{dT}{T} &\le \int \mu\,d\theta \\
\implies \quad \ln T &\le \mu\theta + C \\
\implies \quad T &\le T_0 e^{\mu\theta},
\end{aligned} \tag{2.7}$$

[1] One of them actually has magnitude $(T + dT)\sin(d\theta/2)$, where dT is the increase in tension along the small piece. But the extra term this produces, $(dT)\sin(d\theta/2)$, is a second-order small quantity, so it can be ignored.

where we have used the fact that $T = T_0$ when $\theta = 0$. The exponential behavior here is quite strong (as exponential behaviors tend to be). If we let $\mu = 1$, then just a quarter turn around the pole produces a factor of $e^{\pi/2} \approx 5$. One full revolution yields a factor of $e^{2\pi} \approx 530$, and two full revolutions yield a factor of $e^{4\pi} \approx 300\,000$. Needless to say, the limiting factor in such a case is not your strength, but rather the structural integrity of the pole around which the rope winds.

2.2 Balancing torques

In addition to balancing forces in a statics problem, we must also balance torques. We'll have much more to say about torque in Chapters 8 and 9, but we'll need one important fact here. Consider the situation in Fig. 2.4, where three forces are applied perpendicular to a stick, which is assumed to remain motionless. F_1 and F_2 are the forces at the ends, and F_3 is the force in the interior. We have, of course, $F_3 = F_1 + F_2$, because the stick is at rest. But we also have the following relation:

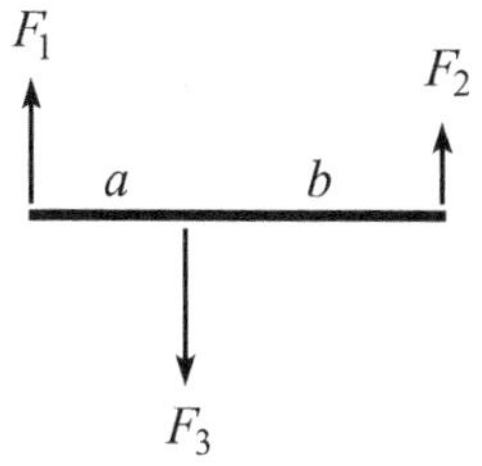

Fig. 2.4

Claim 2.1 *If the system is motionless, then $F_3a = F_2(a + b)$. In other words, the torques (force times distance) around the left end cancel.*[2] *And you can show that they cancel around any other point, too.*

We'll prove this claim in Chapter 8 by using angular momentum, but let's give a short proof here.

Proof: We'll make one reasonable assumption, namely, that the correct relationship between the forces and distances is of the form,

$$F_3 f(a) = F_2 f(a + b), \tag{2.8}$$

where $f(x)$ is a function to be determined.[3] Applying this assumption with the roles of "left" and "right" reversed in Fig. 2.4 gives

$$F_3 f(b) = F_1 f(a + b). \tag{2.9}$$

Adding Eqs. (2.8) and (2.9), and using $F_3 = F_1 + F_2$, yields

$$f(a) + f(b) = f(a + b). \tag{2.10}$$

This equation implies that $f(rx) = rf(x)$ for any x and for any rational number r, as you can show (see Exercise 2.28). Therefore, assuming $f(x)$ is continuous,

[2] Another proof of this claim is given in Problem 2.11.

[3] What we're doing here is simply assuming linearity in F. That is, two forces of F applied at a point should be the same as a force of $2F$ applied at that point. You can't really argue with that.

it must be a linear function, $f(x) = Ax$, as we wanted to show. The constant A is irrelevant, because it cancels in Eq. (2.8). ■

Note that dividing Eq. (2.8) by Eq. (2.9) gives $F_1 f(a) = F_2 f(b)$, and hence $F_1 a = F_2 b$, which says that the torques cancel around the point where F_3 is applied. You can show that the torques cancel around any arbitrary pivot point. When adding up all the torques in a given physical setup, it is of course required that you use the same pivot point when calculating each torque.

In the case where the forces aren't perpendicular to the stick, the above claim applies to the components of the forces perpendicular to the stick. This makes sense, because the components parallel to the stick have no effect on the rotation of the stick around the pivot point. Therefore, referring to Figs. 2.5 and 2.6, the equality of the torques can be written as

$$F_a a \sin\theta_a = F_b b \sin\theta_b. \tag{2.11}$$

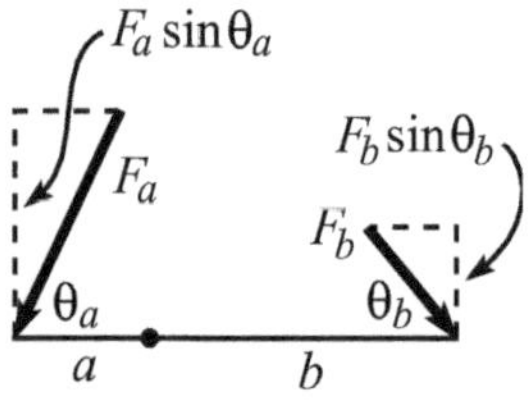

Fig. 2.5

This equation can be viewed in two ways:

- $(F_a \sin\theta_a)a = (F_b \sin\theta_b)b$. In other words, we effectively have smaller forces acting on the given "lever arms," as shown in Fig. 2.5.
- $F_a(a \sin\theta_a) = F_b(b \sin\theta_b)$. In other words, we effectively have the given forces acting on smaller "lever arms," as shown in Fig. 2.6.

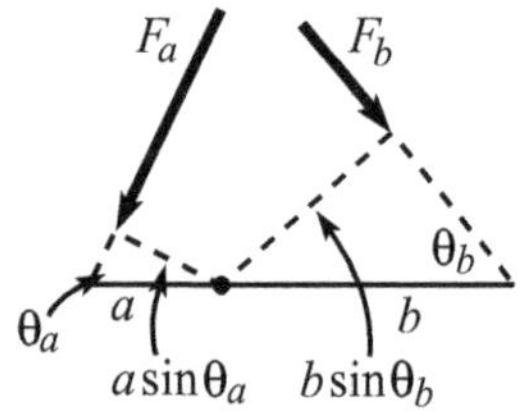

Fig. 2.6

Claim 2.1 shows that even if you apply only a tiny force, you can balance the torque due to a very large force, provided that you make your lever arm sufficiently long. This fact led a well-known mathematician of long ago to claim that he could move the earth if given a long enough lever arm.

> One morning while eating my Wheaties,
> I felt the earth move 'neath my feeties.
> The cause for alarm
> Was a long lever arm,
> At the end of which grinned Archimedes!

One handy fact that comes up often is that the gravitational torque on a stick of mass M is the same as the gravitational torque due to a point-mass M located at the center of the stick. The truth of this statement relies on the fact that torque is a linear function of the distance to the pivot point (see Exercise 2.27). More generally, the gravitational torque on an object of mass M may be treated simply as the gravitational torque due to a force Mg located at the center of mass.

We'll talk more about torque in Chapters 8 and 9, but for now we'll just use the fact that in a statics problem the torques around any given point must balance.

Example (Leaning ladder): A ladder leans against a frictionless wall. If the coefficient of friction with the ground is μ, what is the smallest angle the ladder can make with the ground and not slip?

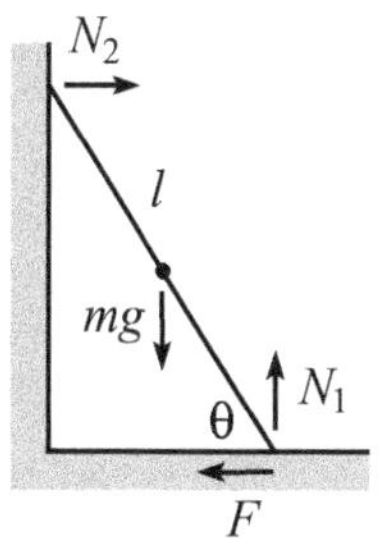

Fig. 2.7

Solution: Let the ladder have mass m and length ℓ. As shown in Fig. 2.7, we have three unknown forces: the friction force F, and the normal forces N_1 and N_2. And to solve for these three forces we fortunately have three equations: $\Sigma F_{\text{vert}} = 0$, $\Sigma F_{\text{horiz}} = 0$, and $\Sigma\tau = 0$ (τ is the standard symbol for torque). Looking at the vertical forces, we see that $N_1 = mg$. And then looking at the horizontal forces, we see that $N_2 = F$. So we have quickly reduced the unknowns from three to one.

We will now use $\Sigma\tau = 0$ to find N_2 (or F). But first we must pick the "pivot" point around which we will calculate the torques. Any stationary point will work fine, but certain choices make the calculation easier than others. The best choice for the pivot is generally the point at which the most forces act, because then the $\Sigma\tau = 0$ equation will have the smallest number of terms in it (because a force provides no torque around the point where it acts, since the lever arm is zero). In this problem, there are two forces acting at the bottom end of the ladder, so this is the point we'll choose for the pivot (but you should verify that other choices for the pivot, for example, the middle or top of the ladder, give the same result). Balancing the torques due to gravity and N_2, we have

$$N_2\ell\sin\theta = mg(\ell/2)\cos\theta \quad\Longrightarrow\quad N_2 = \frac{mg}{2\tan\theta}. \tag{2.12}$$

This is also the value of the friction force F. The condition $F \le \mu N_1 = \mu mg$ therefore becomes

$$\frac{mg}{2\tan\theta} \le \mu mg \quad\Longrightarrow\quad \tan\theta \ge \frac{1}{2\mu}. \tag{2.13}$$

Remarks: Note that the total force exerted on the ladder by the floor points up at an angle given by $\tan\beta = N_1/F = (mg)/(mg/2\tan\theta) = 2\tan\theta$. We see that this force does *not* point along the ladder. There is simply no reason why it should. But there *is* a nice reason why it should point upward with twice the slope of the ladder. This is the direction that causes the lines of the three forces on the ladder to be concurrent (that is, pass through a common point), as shown in Fig. 2.8. This concurrency is a neat little theorem for statics problems involving three forces. The proof is simple. If the three lines weren't concurrent, then one force would produce a nonzero torque around the intersection point of the other two lines of force.[4]

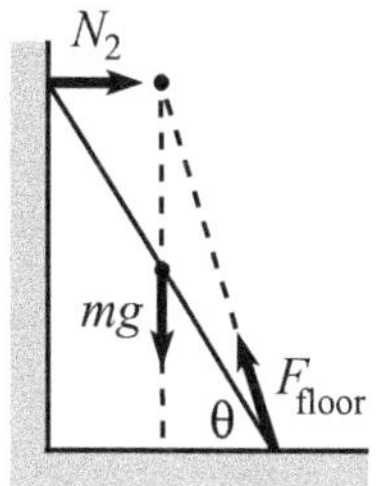

Fig. 2.8

This theorem provides a quick way to solve the ladder problem in the more general case where the center of mass is a fraction f of the way up. In this case, the concurrency theorem tells us that the slope of the total force from the floor is $(1/f)\tan\theta$, consistent with the $f = 1/2$ result from above. The vertical component is still mg, so the horizontal (friction) component is now $fmg/\tan\theta$. Demanding that this be less than or equal to μmg gives $\tan\theta \ge f/\mu$, consistent with the $f = 1/2$ result. Since this result depends

[4] The one exception to this reasoning is where no two of the lines intersect, that is, where all three lines are parallel. Equilibrium is certainly possible in such a scenario, as we saw in Claim 2.1. But you can hang on to the concurrency theorem in this case if you consider the parallel lines to meet at infinity.

only on the location of the center of mass, and not on the exact distribution of mass, a corollary is that if you climb up a ladder (resting on a frictionless wall), your presence makes the ladder more likely to slip if you are above the center of mass (because you have raised the center of mass of the entire system and thus increased f), and less likely if you are below. ♣

The examples we've done in this chapter have consisted of only one object. But many problems involve more than one object (as you'll find in the problems and exercises for this chapter), and there's one additional fact you'll often need to invoke for these, namely Newton's third law. This states that the force that object A exerts on object B is equal and opposite to the force that B exerts on A (we'll talk more about Newton's laws in Chapter 3). So if you want to find, say, the normal force between two objects, you might be able to figure it out by looking at forces and torques on either object, depending on how much you already know about the other forces acting on each. Once you've found the force by dealing with, say, object A, you can then use the equal and opposite force to help figure out things about B. Depending on the problem, one object is often more useful than the other to use first.

Note, however, that if you pick your subsystem (on which you're going to consider forces and torques) to include both A and B, then this won't tell you anything at all about the normal force (or friction) between them. This is true because the normal force is an *internal* force between the objects (when considered together as a system), whereas only *external* forces are relevant in calculating the total force and torque on the system (because all the internal forces cancel in pairs, by Newton's third law). The only way to determine a given force is to deal with it as an external force on some subsystem(s).

Statics problems often involve a number of decisions. If there are various parts to the system, then you must decide which subsystems you want to balance the external forces and torques on. And furthermore, you must decide which point to use as the origin for calculating the torques. There are invariably many choices that will give you the information you need, but some will make your calculations much cleaner than others (Exercise 2.35 is a good example of this). The only way to know how to choose wisely is to start solving problems, so you may as well tackle some . . .

2.3 Problems

Section 2.1: Balancing forces

2.1. **Hanging rope**

A rope with length L and mass density per unit length ρ is suspended vertically from one end. Find the tension as a function of height along the rope.

2.2. Block on a plane

A block sits on a plane that is inclined at an angle θ. Assume that the friction force is large enough to keep the block at rest. What are the horizontal components of the friction and normal forces acting on the block? For what θ are these horizontal components maximum?

2.3. Motionless chain $*$

A frictionless tube lies in the vertical plane and is in the shape of a function that has its endpoints at the same height but is otherwise arbitrary. A chain with uniform mass per unit length lies in the tube from end to end, as shown in Fig. 2.9. Show, by considering the net force of gravity along the curve, that the chain doesn't move.

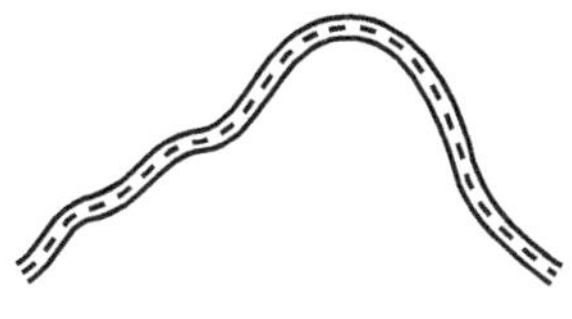

Fig. 2.9

2.4. Keeping a book up $*$

A book of mass M is positioned against a vertical wall. The coefficient of friction between the book and the wall is μ. You wish to keep the book from falling by pushing on it with a force F applied at an angle θ with respect to the horizontal ($-\pi/2 < \theta < \pi/2$), as shown in Fig. 2.10.

(a) For a given θ, what is the minimum F required?

(b) For what θ is this minimum F the smallest? What is the corresponding minimum F?

(c) What is the limiting value of θ, below which there does not exist an F that keeps the book up?

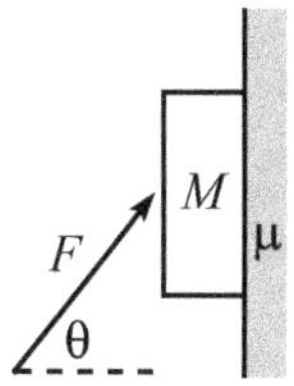

Fig. 2.10

2.5. Rope on a plane $*$

A rope with length L and mass density per unit length ρ lies on a plane inclined at an angle θ (see Fig. 2.11). The top end is nailed to the plane, and the coefficient of friction between the rope and the plane is μ. What are the possible values for the tension at the top of the rope?

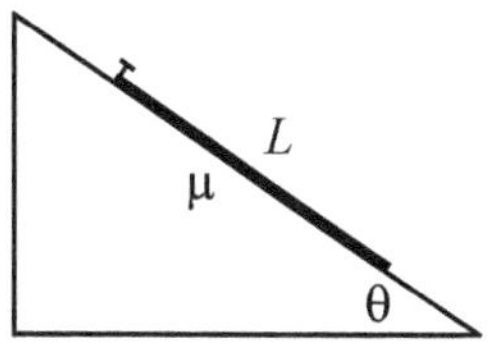

Fig. 2.11

2.6. Supporting a disk $**$

(a) A disk of mass M and radius R is held up by a massless string, as shown in Fig. 2.12. The surface of the disk is frictionless. What is the tension in the string? What is the normal force per unit length that the string applies to the disk?

(b) Let there now be friction between the disk and the string, with coefficient μ. What is the smallest possible tension in the string at its lowest point?

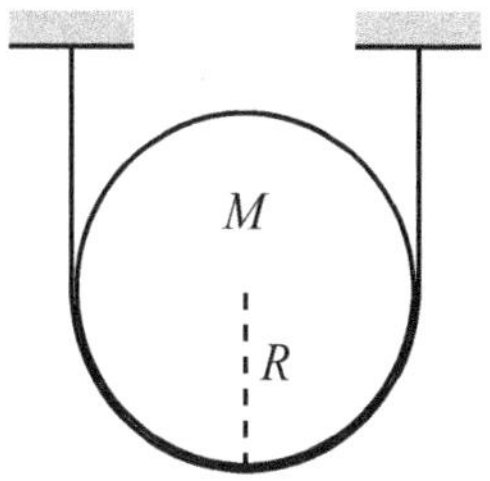

Fig. 2.12

2.7. Objects between circles $**$

Each of the following planar objects is placed, as shown in Fig. 2.13, between two frictionless circles of radius R. The mass density per unit

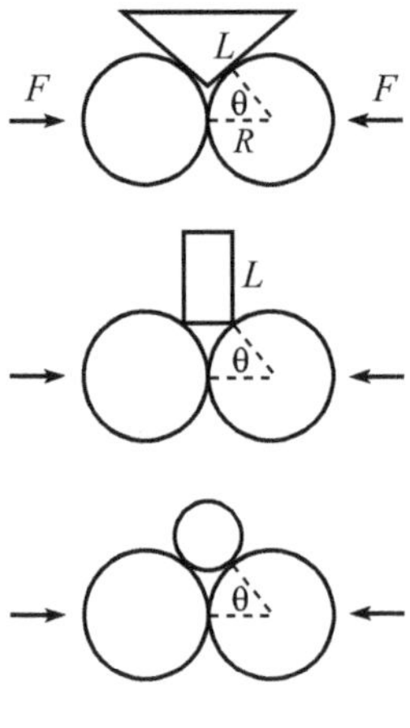

Fig. 2.13

area of each object is σ, and the radii to the points of contact make an angle θ with the horizontal. For each case, find the horizontal force that must be applied to the circles to keep them together. For what θ is this force maximum or minimum?

(a) An isosceles triangle with common side length L.
(b) A rectangle with height L.
(c) A circle.

2.8. **Hanging chain** ****

(a) A chain with uniform mass density per unit length hangs between two given points on two walls. Find the general shape of the chain. Aside from an arbitrary additive constant, the function describing the shape should contain one unknown constant. (The shape of a hanging chain is known as a *catenary*.)
(b) The unknown constant in your answer depends on the horizontal distance d between the walls, the vertical distance λ between the support points, and the length ℓ of the chain (see Fig. 2.14). Find an equation involving these given quantities that determines the unknown constant.

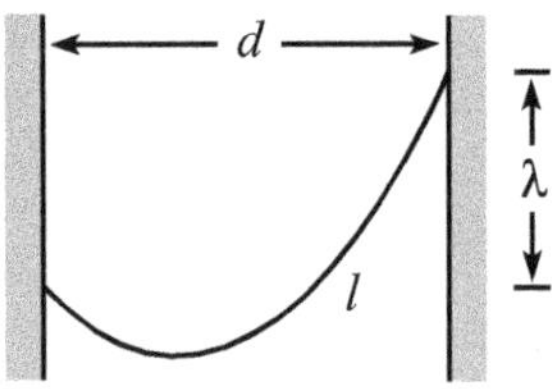

Fig. 2.14

2.9. **Hanging gently** **

A chain with uniform mass density per unit length hangs between two supports located at the same height, a distance $2d$ apart (see Fig. 2.15). What should the length of the chain be so that the magnitude of the force at the supports is minimized? You may use the fact that a hanging chain takes the form, $y(x) = (1/\alpha)\cosh(\alpha x)$. You will eventually need to solve an equation numerically.

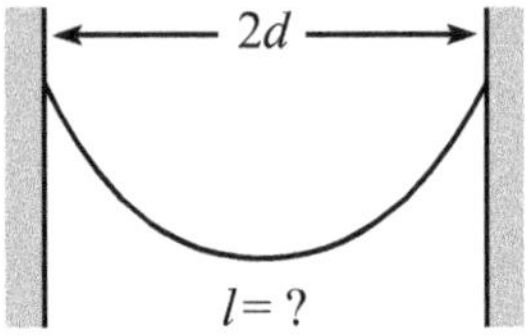

Fig. 2.15

2.10. **Mountain climber** ****

A mountain climber wishes to climb up a frictionless conical mountain. He wants to do this by throwing a lasso (a rope with a loop) over the top and climbing up along the rope. Assume that the climber is of negligible height, so that the rope lies along the mountain, as shown in Fig. 2.16. At the bottom of the mountain are two stores. One sells "cheap" lassos (made of a segment of rope tied to a loop of *fixed* length); see Fig. 2.17. The other sells "deluxe" lassos (made of one piece of rope with a loop of *variable* length; the loop's length may change without any friction of the rope with itself). When viewed from the side, the conical mountain has an angle α at its peak. For what angles α can the climber climb up along the mountain if he uses a "cheap" lasso? A "deluxe" lasso? (*Hint*: The answer in the "cheap" case isn't $\alpha < 90°$.)

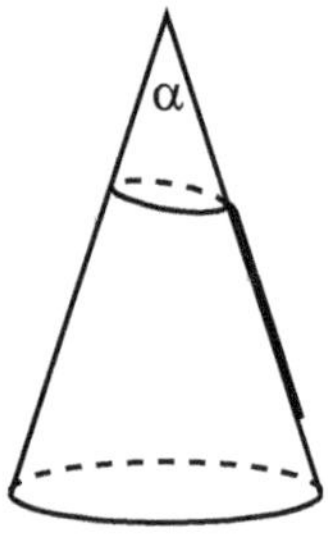

Fig. 2.16

Section 2.2: Balancing torques

2.11. **Equality of torques** **

This problem gives another way of demonstrating Claim 2.1, using an inductive argument. We'll get you started, and then you can do the general case.

Consider the situation where forces F are applied upward at the ends of a stick of length ℓ, and a force $2F$ is applied downward at the midpoint (see Fig. 2.18). The stick doesn't rotate (by symmetry), and it doesn't translate (because the net force is zero). If we wish, we may consider the stick to have a pivot at the left end. If we then erase the force F on the right end and replace it with a force $2F$ at the middle, then the two $2F$ forces in the middle cancel, so the stick remains at rest.[5] Therefore, we see that a force F applied at a distance ℓ from a pivot is equivalent to a force $2F$ applied at a distance $\ell/2$ from the pivot, in the sense that they both have the same effect in canceling out the rotational effect of the downwards $2F$ force.

Now consider the situation where forces F are applied upward at the ends, and forces F are applied downward at the $\ell/3$ and $2\ell/3$ marks (see Fig. 2.19). The stick doesn't rotate (by symmetry), and it doesn't translate (because the net force is zero). Consider the stick to have a pivot at the left end. From the above paragraph, the force F at $2\ell/3$ is equivalent to a force $2F$ at $\ell/3$. Making this replacement, we now have a total force of $3F$ at the $\ell/3$ mark. Therefore, we see that a force F applied at a distance ℓ is equivalent to a force $3F$ applied at a distance $\ell/3$.

Your task is to now use induction to show that a force F applied at a distance ℓ is equivalent to a force nF applied at a distance ℓ/n, and to then argue why this demonstrates Claim 2.1.

2.12. **Direction of the tension** *

Show that the tension in a completely flexible rope, massive or massless, points along the rope everywhere in the rope.

2.13. **Find the force** *

A stick of mass M is held up by supports at each end, with each support providing a force of $Mg/2$. Now put another support somewhere in the middle, say, at a distance a from one support and b from the other; see Fig. 2.20. What forces do the three supports now provide? Is this solvable?

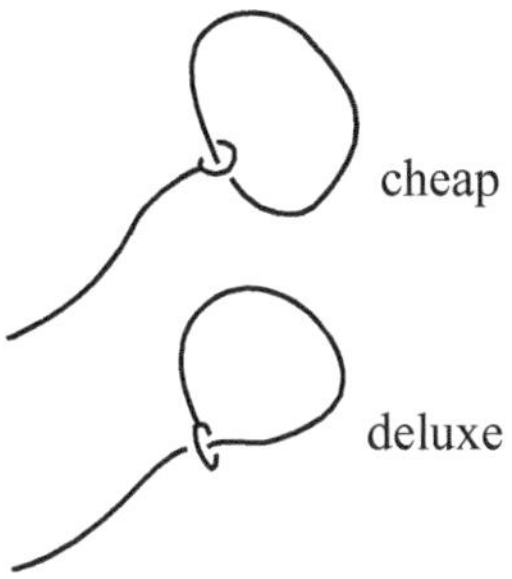

Fig. 2.17

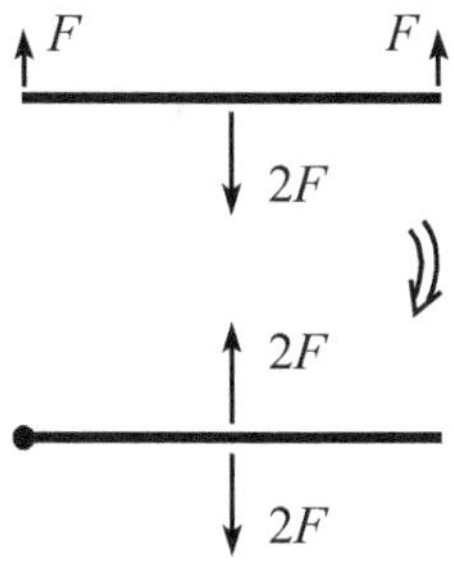

Fig. 2.18

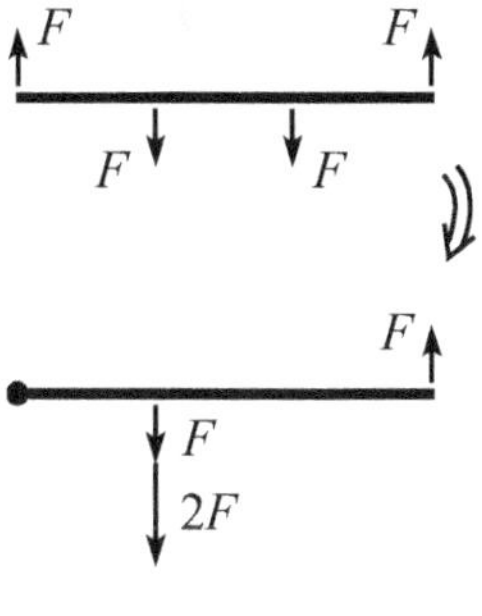

Fig. 2.19

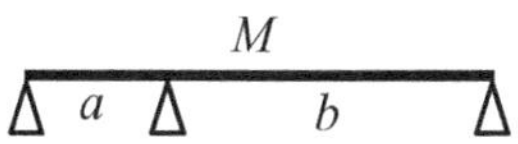

Fig. 2.20

[5] There is now a different force applied at the pivot, namely zero, but the purpose of the pivot is to simply apply whatever force is necessary to keep the left end motionless.

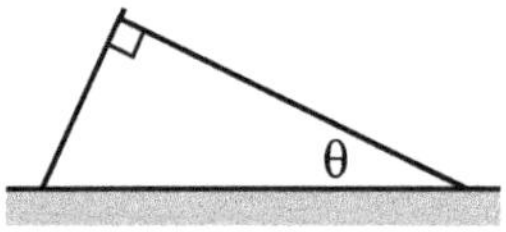

Fig. 2.21

2.14. Leaning sticks $*$

One stick leans on another as shown in Fig. 2.21. A right angle is formed where they meet, and the right stick makes an angle θ with the horizontal. The left stick extends infinitesimally beyond the end of the right stick. The coefficient of friction between the two sticks is μ. The sticks have the same mass density per unit length and are both hinged at the ground. What is the minimum angle θ for which the sticks don't fall?

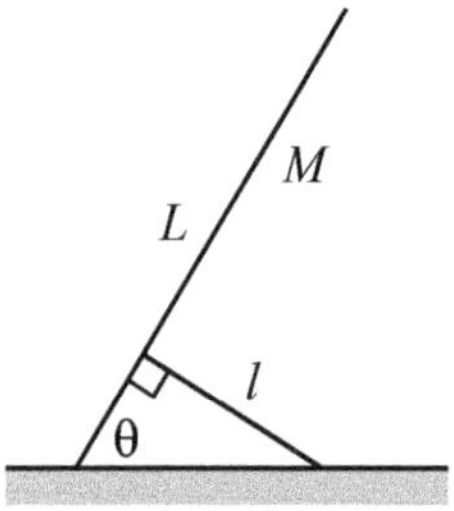

Fig. 2.22

2.15. Supporting a ladder $*$

A ladder of length L and mass M has its bottom end attached to the ground by a pivot. It makes an angle θ with the horizontal and is held up by a massless stick of length ℓ that is also attached to the ground by a pivot (see Fig. 2.22). The ladder and the stick are perpendicular to each other. Find the force that the stick exerts on the ladder.

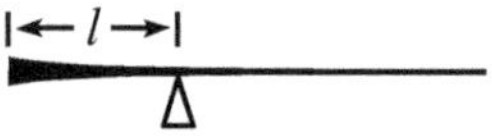

Fig. 2.23

2.16. Balancing the stick $**$

Given a semi-infinite stick (that is, one that goes off to infinity in one direction), determine how its density should depend on position so that it has the following property: If the stick is cut at an arbitrary location, the remaining semi-infinite piece will balance on a support that is located a distance ℓ from the end (see Fig. 2.23).

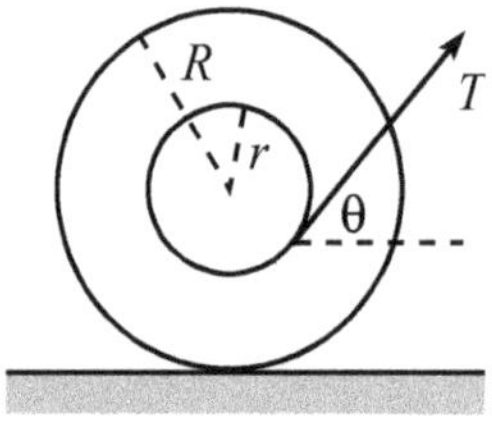

Fig. 2.24

2.17. The spool $**$

A spool consists of an axle of radius r and an outside circle of radius R which rolls on the ground. A thread is wrapped around the axle and is pulled with tension T at an angle θ with the horizontal (see Fig. 2.24).

(a) Given R and r, what should θ be so that the spool doesn't move? Assume that the friction between the spool and the ground is large enough so that the spool doesn't slip.

(b) Given R, r, and the coefficient of friction μ between the spool and the ground, what is the largest value of T for which the spool remains at rest?

(c) Given R and μ, what should r be so that you can make the spool slip from the static position with as small a T as possible? That is, what should r be so that the upper bound on T in part (b) is as small as possible? What is the resulting value of T?

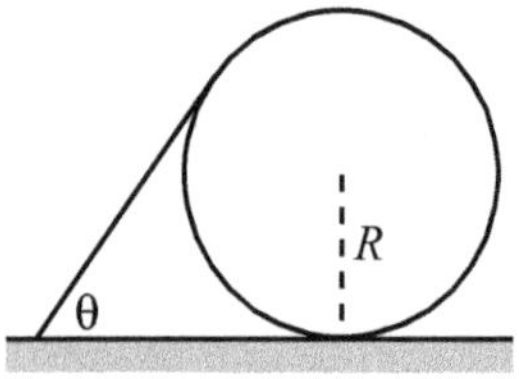

Fig. 2.25

2.18. Stick on a circle $**$

A stick of mass density per unit length ρ rests on a circle of radius R (see Fig. 2.25). The stick makes an angle θ with the horizontal and is tangent to the circle at its upper end. Friction exists at all points of contact, and assume that it is large enough to keep the system at rest. Find the friction force between the ground and the circle.

2.19. **Leaning sticks and circles** ***

A large number of sticks (with mass density per unit length ρ) and circles (with radius R) lean on each other, as shown in Fig. 2.26. Each stick makes an angle θ with the horizontal and is tangent to the next circle at its upper end. The sticks are hinged to the ground, and every other surface is *frictionless* (unlike in the previous problem). In the limit of a very large number of sticks and circles, what is the normal force between a stick and the circle it rests on, very far to the right? Assume that the last circle leans against a wall, to keep it from moving.

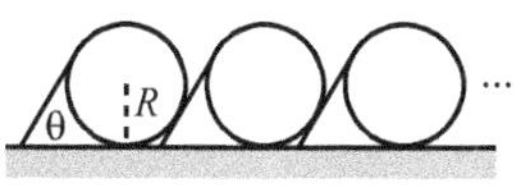

Fig. 2.26

2.4 Exercises

Section 2.1: Balancing forces

2.20. **Block under an overhang** *

A block of mass M is positioned underneath an overhang that makes an angle β with the horizontal. You apply a horizontal force of Mg on the block, as shown in Fig. 2.27. Assume that the friction force between the block and the overhang is large enough to keep the block at rest. What are the normal and friction forces (call them N and F_f) that the overhang exerts on the block? If the coefficient of static friction is μ, for what range of angles β does the block in fact remain at rest?

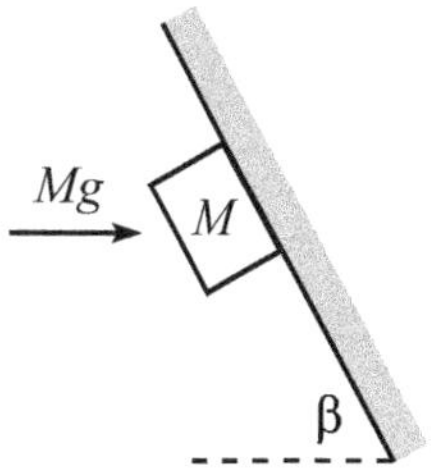

Fig. 2.27

2.21. **Pulling a block** *

A person pulls on a block with a force F, at an angle θ with respect to the horizontal. The coefficient of friction between the block and the ground is μ. For what θ is the F required to make the block slip a minimum? What is the corresponding F?

2.22. **Holding a cone** *

With two fingers, you hold an ice cream cone motionless upside down, as shown in Fig. 2.28. The mass of the cone is m, and the coefficient of static friction between your fingers and the cone is μ. When viewed from the side, the angle of the tip is 2θ. What is the minimum normal force you must apply with each finger in order to hold up the cone? In terms of θ, what is the minimum value of μ that allows you to hold up the cone? Assume that you can supply as large a normal force as needed.

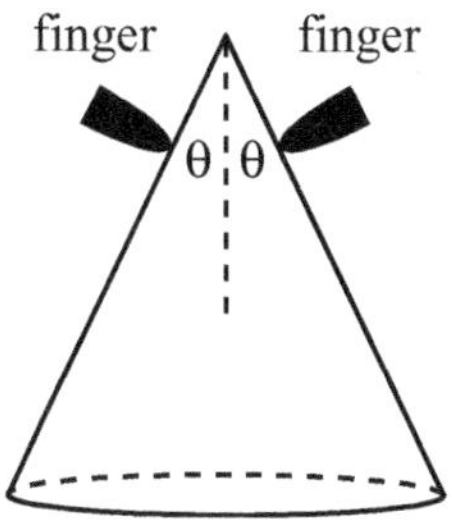

Fig. 2.28

2.23. **Keeping a book up** **

The task of Problem 2.4 is to find the minimum force required to keep a book up. What is the maximum allowable force, as a function of θ and μ? Is there a special angle that arises? Given μ, make a rough plot of the allowed values of F for $-\pi/2 < \theta < \pi/2$.

2.24. Bridges **

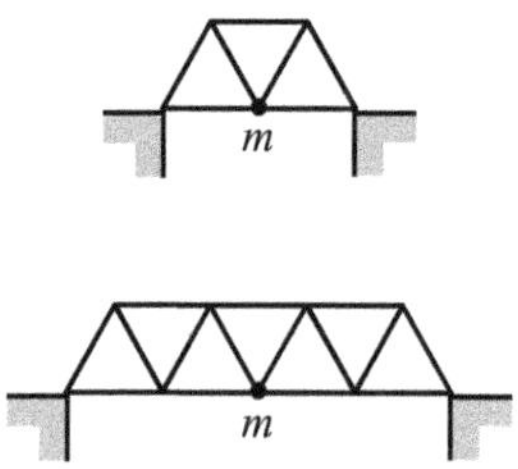

Fig. 2.29

(a) Consider the first bridge in Fig. 2.29, made of three equilateral triangles of beams. Assume that the seven beams are massless and that the connection between any two of them is a hinge. If a car of mass m is located at the middle of the bridge, find the forces (and specify tension or compression) in the beams. Assume that the supports provide no horizontal forces on the bridge.

(b) Same question, but now with the second bridge in Fig. 2.29, made of seven equilateral triangles.

(c) Same question, but now with the general case of $4n - 1$ equilateral triangles.

2.25. Rope between inclines **

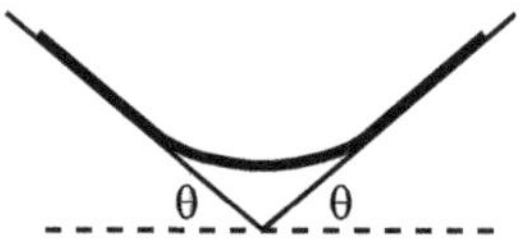

Fig. 2.30

A rope rests on two platforms that are both inclined at an angle θ (which you are free to pick), as shown in Fig. 2.30. The rope has uniform mass density, and the coefficient of friction between it and the platforms is 1. The system has left-right symmetry. What is the largest possible fraction of the rope that does not touch the platforms? What angle θ allows this maximum fraction?

2.26. Hanging chain **

A chain with mass M hangs between two walls, with its ends at the same height. The chain makes an angle θ with each wall, as shown in Fig. 2.31. Find the tension in the chain at the lowest point. Solve this in two different ways:

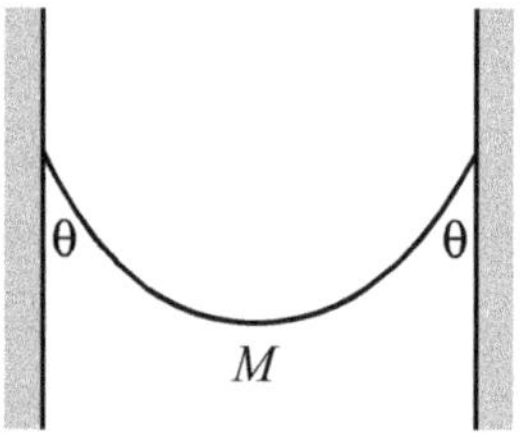

Fig. 2.31

(a) Consider the forces on half of the chain. (This is the quick way.)

(b) Use the fact (see Problem 2.8) that the height of a hanging chain is given by $y(x) = (1/\alpha)\cosh(\alpha x)$, and consider the vertical forces on an infinitesimal piece at the bottom. This will give you the tension in terms of α. Then find an expression for α in terms of the given angle θ. (This is the long way.)

Section 2.2: Balancing torques

2.27. Gravitational torque

A horizontal stick of mass M and length L is pivoted at one end. Integrate the gravitational torque along the stick (relative to the pivot), and show that the result is the same as the torque due to a mass M located at the center of the stick.

2.28. Linear function *

Show that if a function satisfies $f(a) + f(b) = f(a + b)$, then $f(rx) = rf(x)$ for any x and for any rational number r.

2.29. **Direction of the force** $*$

A stick is connected to other parts of a static system by hinges at its ends. Show that (1) if the stick is massless, then the forces it feels at the hinges are directed along the stick, but (2) if the stick is massive, then the forces need not point along the stick.

2.30. **Ball on a wall** $*$

A ball is held up by a string, as shown in Fig. 2.32, with the string tangent to the ball. If the angle between the string and the wall is θ, what is the minimum coefficient of static friction between the ball and the wall that keeps the ball from falling?

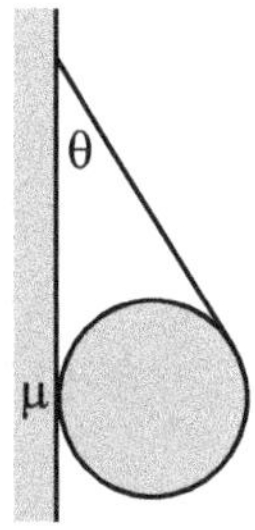

Fig. 2.32

2.31. **Cylinder and hanging mass** $*$

A uniform cylinder of mass M sits on a fixed plane inclined at an angle θ. A string is tied to the cylinder's rightmost point, and a mass m hangs from the string, as shown in Fig. 2.33. Assume that the coefficient of friction between the cylinder and the plane is sufficiently large to prevent slipping. What is m, in terms of M and θ, if the setup is static?

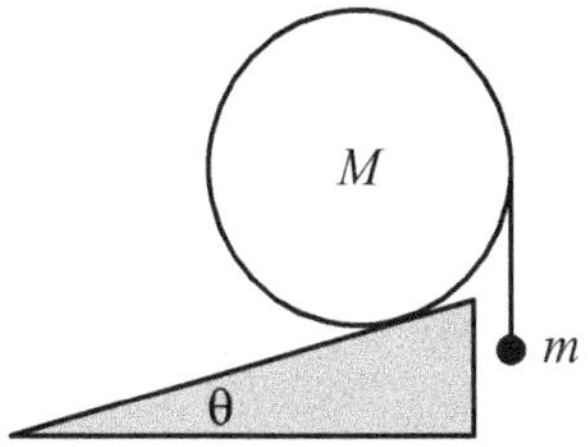

Fig. 2.33

2.32. **Ladder on a corner** $**$

A ladder of mass M and length L leans against a frictionless wall, with a quarter of its length hanging over a corner, as shown in Fig. 2.34. It makes an angle θ with the horizontal. What angle θ requires the smallest coefficient of friction at the corner to keep the ladder at rest? (Different values of θ require different ladder lengths, but assume that the mass is M for any length.)

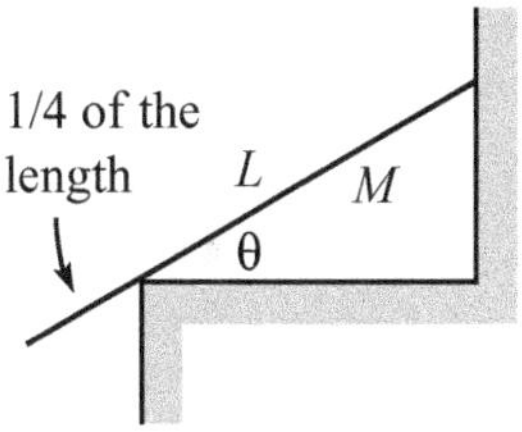

Fig. 2.34

2.33. **Stick on a corner** $**$

You support one end of a stick of mass M and length L with the tip of your finger. A quarter of the way up the stick, it rests on a frictionless corner of a table, as shown in Fig. 2.35. The stick makes an angle θ with the horizontal. What is the magnitude of the force your finger must apply to keep the stick in this position? For what angle θ does your force point horizontally?

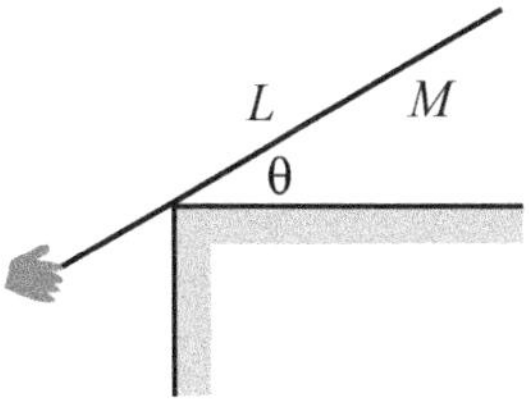

Fig. 2.35

2.34. **Stick and a cylinder** $**$

A horizontal stick of mass m has its left end attached to a pivot on a plane inclined at an angle θ, while its right end rests on the top of a cylinder also of mass m which in turn rests on the plane, as shown in Fig. 2.36. The coefficient of friction between the cylinder and both the stick and the plane is μ.

(a) Assuming that the system is at rest, what is the normal force from the plane on the cylinder?

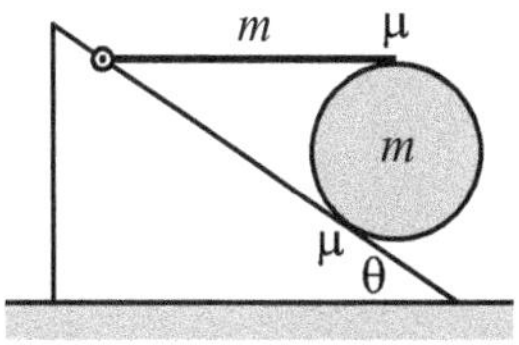

Fig. 2.36

(b) What is the smallest value of μ (in terms of θ) for which the system doesn't slip anywhere?

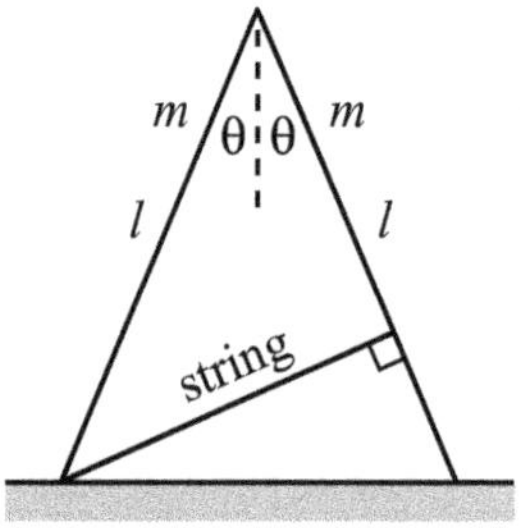

Fig. 2.37

2.35. Two sticks and a string **

Two sticks, each of mass m and length ℓ, are connected by a hinge at their top ends. They each make an angle θ with the vertical. A massless string connects the bottom of the left stick to the right stick, perpendicularly as shown in Fig. 2.37. The whole setup stands on a frictionless table.

(a) What is the tension in the string?

(b) What force does the left stick exert on the right stick at the hinge? *Hint*: No messy calculations required!

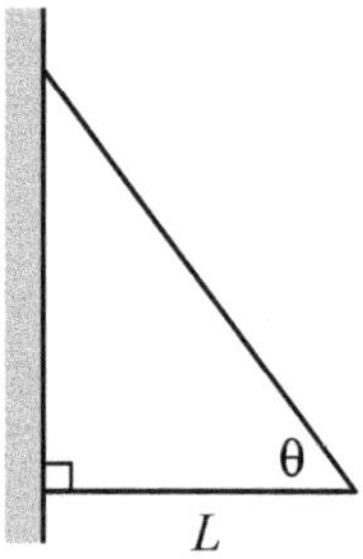

Fig. 2.38

2.36. Two sticks and a wall **

Two sticks are connected, with hinges, to each other and to a wall. The bottom stick is horizontal and has length L, and the sticks make an angle of θ with each other, as shown in Fig. 2.38. If both sticks have the same mass per unit length, ρ, find the horizontal and vertical components of the force that the wall exerts on the top hinge, and show that the magnitude goes to infinity for both $\theta \to 0$ and $\theta \to \pi/2$.[6]

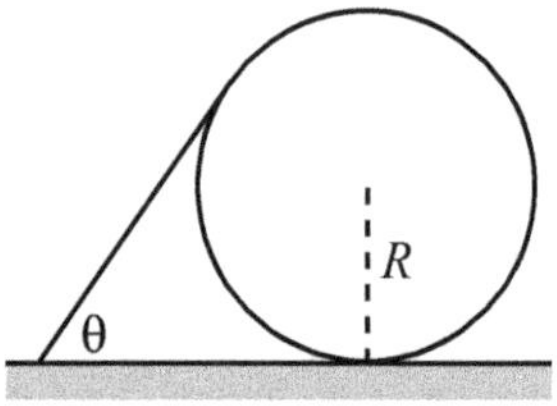

Fig. 2.39

2.37. Stick on a circle **

Using the results from Problem 2.18 for the setup shown in Fig. 2.39, show that if the system is to remain at rest, then the coefficient of friction:

(a) between the stick and the circle must satisfy

$$\mu \geq \frac{\sin\theta}{1+\cos\theta}. \tag{2.14}$$

(b) between the stick and the ground must satisfy[7]

$$\mu \geq \frac{\sin\theta\cos\theta}{(1+\cos\theta)(2-\cos\theta)}. \tag{2.15}$$

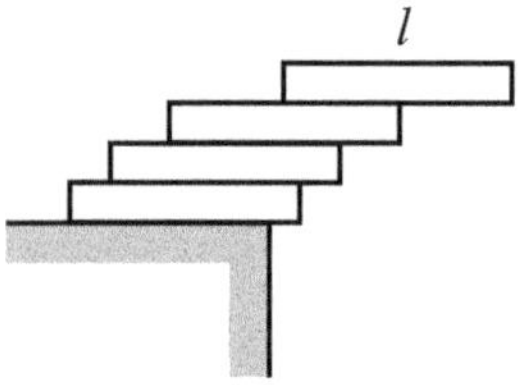

Fig. 2.40

2.38. Stacking blocks **

N blocks of length ℓ are stacked on top of each other at the edge of a table, as shown in Fig. 2.40 for $N = 4$. What is the largest horizontal

[6] The force must therefore achieve a minimum at some intermediate angle. If you want to go through the algebra, you can show that this minimum occurs when $\cos\theta = \sqrt{3} - 1$, which gives $\theta \approx 43°$.

[7] If you want to go through the algebra, you can show that the right-hand side achieves a maximum when $\cos\theta = \sqrt{3} - 1$, which gives $\theta \approx 43°$. (Yes, I did just cut and paste this from the previous footnote. But it's still correct!) This is the angle for which the stick is most likely to slip on the ground.

distance the rightmost point on the top block can hang out beyond the table? How does your answer behave for $N \to \infty$?[8]

2.5 Solutions

2.1. **Hanging rope**

Let $T(y)$ be the tension as a function of height. Consider a small piece of the rope between y and $y + dy$ ($0 \le y \le L$). The forces on this piece are $T(y + dy)$ upward, $T(y)$ downward, and the weight $\rho g\, dy$ downward. Since the rope is at rest, we have $T(y + dy) = T(y) + \rho g\, dy$. Expanding this to first order in dy gives $T'(y) = \rho g$. The tension in the bottom of the rope is zero, so integrating from $y = 0$ up to a position y gives

$$T(y) = \rho g y. \tag{2.16}$$

As a double-check, at the top end we have $T(L) = \rho g L$, which is the weight of the entire rope, as it should be.

Alternatively, you can simply write down the answer, $T(y) = \rho g y$, by noting that the tension at a given point in the rope is what supports the weight of all the rope below it.

2.2. **Block on a plane**

Balancing the forces shown in Fig. 2.41, parallel and perpendicular to the plane, we see that $F = mg \sin\theta$ and $N = mg\cos\theta$. The horizontal components of these are $F\cos\theta = mg \sin\theta\cos\theta$ (to the right), and $N \sin\theta = mg \cos\theta \sin\theta$ (to the left). These are equal, as they must be, because the net horizontal force on the block is zero. To maximize the value of $mg \sin\theta\cos\theta$, we can either take the derivative, or we can write it as $(mg/2)\sin 2\theta$, from which it is clear that the maximum occurs at $\theta = \pi/4$. The maximum value is $mg/2$.

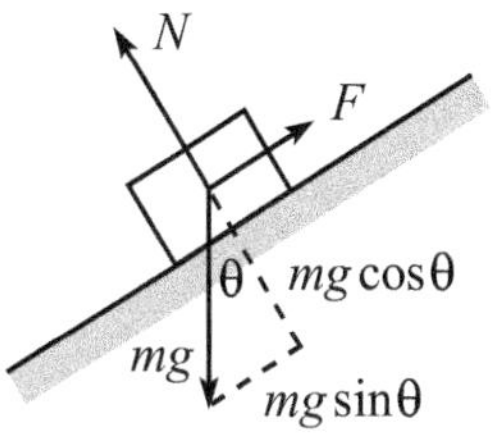

Fig. 2.41

2.3. **Motionless chain**

Let the curve be described by the function $f(x)$, and let it run from $x = a$ to $x = b$. Consider a little piece of the chain between x and $x + dx$ (see Fig. 2.42). The length of this piece is $\sqrt{1 + f'^2}\, dx$, so its mass is $\rho\sqrt{1 + f'^2}\, dx$, where ρ is the mass per unit length. The component of the gravitational acceleration along the curve is $-g \sin\theta = -gf'/\sqrt{1 + f'^2}$ (using $\tan\theta = f'$), with positive corresponding to moving along the curve from a to b. The total force along the curve is therefore

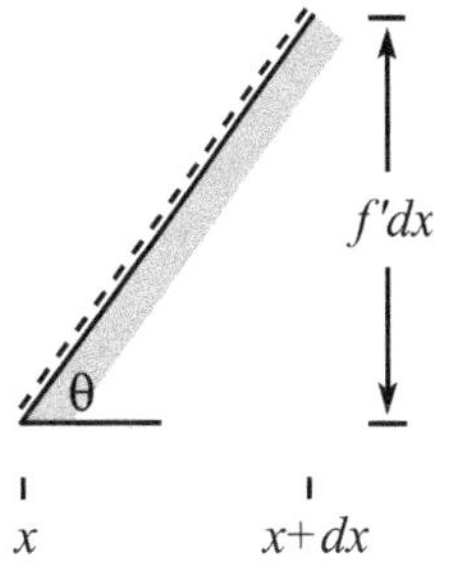

Fig. 2.42

$$\begin{aligned} F = \int_a^b (-g\sin\theta)\, dm &= \int_a^b \frac{-gf'}{\sqrt{1+f'^2}} \cdot \rho\sqrt{1+f'^2}\, dx \\ &= -g\rho \int_a^b f'\, dx \\ &= -g\rho\big(f(a) - f(b)\big) \\ &= 0. \end{aligned} \tag{2.17}$$

2.4. **Keeping a book up**

(a) The normal force from the wall is $F\cos\theta$, so the friction force F_f holding the book up is at most $\mu F\cos\theta$. The other vertical forces on the book are the

[8] It turns out that the method of stacking shown in Fig. 2.40 (with the blocks simply stacked on top of each other) doesn't yield the optimal overhang. See Hall (2005) for an interesting discussion of other methods.

gravitational force, which is $-Mg$, and the vertical component of F, which is $F \sin\theta$. If the book is to stay at rest, we must have $F \sin\theta + F_f - Mg = 0$. Combining this with the condition $F_f \le \mu F \cos\theta$ gives

$$F(\sin\theta + \mu\cos\theta) \ge Mg. \tag{2.18}$$

Therefore, F must satisfy

$$F \ge \frac{Mg}{\sin\theta + \mu\cos\theta}, \tag{2.19}$$

assuming that $\sin\theta + \mu\cos\theta$ is positive. If it is negative, then there is no solution for F.

(b) To minimize this lower bound, we must maximize the denominator. Taking the derivative gives $\cos\theta - \mu\sin\theta = 0$, so $\tan\theta = 1/\mu$. Plugging this value of θ back into Eq. (2.19) gives

$$F \ge \frac{mg}{\sqrt{1+\mu^2}} \quad (\text{with } \tan\theta = 1/\mu). \tag{2.20}$$

This is the smallest possible F that keeps the book up, and the angle must be $\theta = \tan^{-1}(1/\mu)$ for it to work. We see that if μ is very small, then to minimize your F, you should push essentially vertically with a force mg. But if μ is very large, you should push essentially horizontally with a force mg/μ.

(c) There is no possible F that satisfies the condition in Eq. (2.19) if the right-hand side is infinite (more precisely, there is no F that satisfies Eq. (2.18) if the coefficient of F is zero or negative). This occurs when

$$\tan\theta = -\mu. \tag{2.21}$$

If θ is more negative than this, then it is impossible to keep the book up, no matter how hard you push.

2.5. **Rope on a plane**

The component of the gravitational force along the plane is $(\rho L)g\sin\theta$, and the maximum value of the friction force is $\mu N = \mu(\rho L)g\cos\theta$. Therefore, you might think that the tension at the top of the rope is $\rho Lg\sin\theta - \mu\rho Lg\cos\theta$. However, this is not necessarily the case. The tension at the top depends on how the rope is placed on the plane. If, for example, the rope is placed on the plane without being stretched, then the friction force points upwards, and the tension at the top does indeed equal $\rho Lg\sin\theta - \mu\rho Lg\cos\theta$. Or it equals zero if $\mu\rho Lg\cos\theta > \rho Lg\sin\theta$, in which case the friction force need not achieve its maximum value.

If, on the other hand, the rope is placed on the plane after being stretched (or equivalently, it is dragged up along the plane and then nailed down at its top end), then the friction force points downwards, and the tension at the top equals $\rho Lg\sin\theta + \mu\rho Lg\cos\theta$.

Another special case occurs when the rope is placed on a frictionless plane, and then the coefficient of friction is "turned on" to μ. The friction force is still zero. Changing the plane from ice to sandpaper (somehow without moving the rope) doesn't suddenly cause there to be a friction force. Therefore, the tension at the top equals $\rho Lg\sin\theta$.

In general, depending on how the rope is placed on the plane, the tension at the top can take any value from a maximum of $\rho Lg\sin\theta + \mu\rho Lg\cos\theta$, down to a minimum of $\rho Lg\sin\theta - \mu\rho Lg\cos\theta$ (or zero, whichever is larger). If the rope is replaced by a stick (which can support a compressive force), then the tension can achieve negative values down to $\rho Lg\sin\theta - \mu\rho Lg\cos\theta$, if this happens to be negative.

2.6. **Supporting a disk**

(a) The gravitational force downward on the disk is Mg, and the force upward is $2T$. These forces must balance, so

$$T = \frac{Mg}{2}. \tag{2.22}$$

We can find the normal force per unit length that the string applies to the disk in two ways.

FIRST METHOD: Let $N\,d\theta$ be the normal force on an arc of the disk that subtends an angle $d\theta$. Such an arc has length $R\,d\theta$, so N/R is the desired normal force per unit arclength. The tension in the string is the same throughout it, because the string is massless. So all points are equivalent, and hence N is constant, independent of θ. The upward component of the normal force is $N\,d\theta\cos\theta$, where θ is measured from the vertical (that is, $-\pi/2 \le \theta \le \pi/2$ here). Since the total upward force is Mg, we must have

$$\int_{-\pi/2}^{\pi/2} N\cos\theta\,d\theta = Mg. \tag{2.23}$$

The integral equals $2N$, so we have $N = Mg/2$. The normal force per unit length, N/R, therefore equals $Mg/2R$.

SECOND METHOD: Consider the normal force, $N\,d\theta$, on a small arc of the disk that subtends an angle $d\theta$. The tension forces on each end of the corresponding small piece of string almost cancel, but they don't exactly, because they point in slightly different directions. Their nonzero sum is what produces the normal force on the disk. From Fig. 2.43, we see that the two forces have a sum of $2T\sin(d\theta/2)$, directed "inward". Since $d\theta$ is small, we can use $\sin x \approx x$ to approximate this as $T\,d\theta$. Therefore, $N\,d\theta = T\,d\theta$, and so $N = T$. The normal force per unit arclength, N/R, therefore equals T/R. Using $T = Mg/2$ from Eq. (2.22), we arrive at $N/R = Mg/2R$.

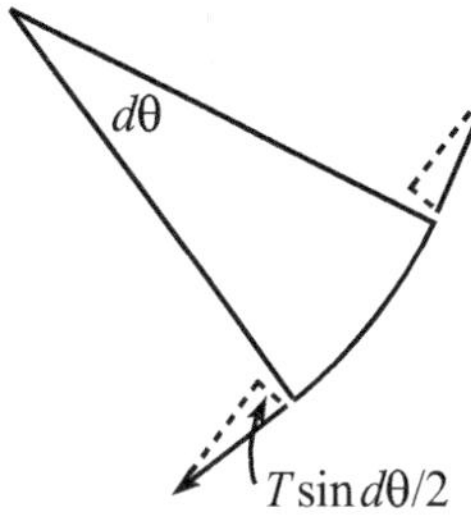

Fig. 2.43

(b) Let $T(\theta)$ be the tension, as a function of θ, for $-\pi/2 \le \theta \le \pi/2$. T now depends on θ, because there is a tangential friction force. Most of the work for this problem was already done in the "Rope wrapped around a pole" example in Section 2.1. We'll simply invoke Eq. (2.7), which in the present language says[9]

$$T(\theta) \le T(0)e^{\mu\theta}. \tag{2.24}$$

Letting $\theta = \pi/2$, and using $T(\pi/2) = Mg/2$, gives $Mg/2 \le T(0)e^{\mu\pi/2}$. We therefore see that the tension at the bottom point must satisfy

$$T(0) \ge \frac{Mg}{2}e^{-\mu\pi/2}. \tag{2.25}$$

REMARK: This minimum value of $T(0)$ goes to $Mg/2$ as $\mu \to 0$, as it should. And it goes to zero as $\mu \to \infty$, as it should (imagine a very rough surface, so that the friction force from the rope near $\theta = \pi/2$ accounts for essentially all the weight). But interestingly, the tension at the bottom doesn't exactly equal zero, no matter now large μ is. Basically, the smaller T is, the smaller N is. But the smaller N is, the smaller the change in T is (because N determines the friction force). So T doesn't decrease much when it's small, and this results in it never being able to reach zero. ♣

2.7. **Objects between circles**

(a) Let N be the normal force between the circles and the triangle. The goal in this problem is to find the horizontal component of N, that is, $N\cos\theta$. From Fig. 2.44, we see that the upward force on the triangle from the normal forces is $2N\sin\theta$. This must equal the weight of the triangle, which is $g\sigma$ times the area. Since the bottom angle of the isosceles triangle is 2θ, the top side has length

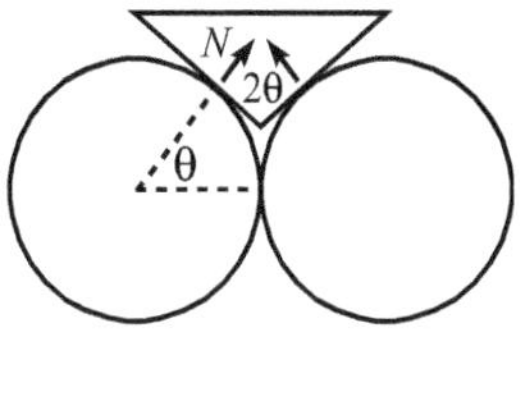

Fig. 2.44

[9] This holds for $\theta > 0$. There would be a minus sign on the right-hand side if $\theta < 0$. But since the tension is symmetric around $\theta = 0$ in the case we're concerned with, we'll just deal with $\theta > 0$.

$2L \sin\theta$, and the altitude to this side is $L\cos\theta$. So the area of the triangle is $L^2 \sin\theta \cos\theta$. The mass is therefore $\sigma L^2 \sin\theta \cos\theta$. Equating the weight with the upward component of the normal forces gives $N = (g\sigma L^2/2)\cos\theta$. The horizontal component of N is therefore

$$N\cos\theta = \frac{g\sigma L^2 \cos^2\theta}{2}. \tag{2.26}$$

This equals zero when $\theta = \pi/2$, and it increases as θ decreases, even though the triangle is getting smaller. It has the interesting property of approaching the finite value $g\sigma L^2/2$, as $\theta \to 0$.

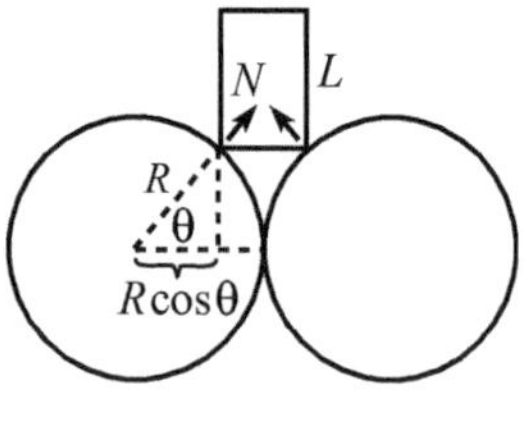

Fig. 2.45

(b) In Fig. 2.45, the base of the rectangle has length $2R(1-\cos\theta)$. Its mass is therefore $2\sigma RL(1-\cos\theta)$. Equating the weight with the upward component of the normal forces, $2N\sin\theta$, gives $N = \sigma gRL(1-\cos\theta)/\sin\theta$. The horizontal component of N is therefore

$$N\cos\theta = \frac{\sigma gRL(1-\cos\theta)\cos\theta}{\sin\theta}. \tag{2.27}$$

This equals zero for both $\theta = \pi/2$ and $\theta = 0$ (because $1-\cos\theta \approx \theta^2/2$ goes to zero faster than $\sin\theta \approx \theta$, for small θ). Taking the derivative to find where it reaches a maximum, we obtain (using $\sin^2\theta = 1-\cos^2\theta$),

$$\cos^3\theta - 2\cos\theta + 1 = 0. \tag{2.28}$$

Fortunately, there is an easy root of this cubic equation, namely $\cos\theta = 1$, which we know is not the maximum. Dividing through by the factor $(\cos\theta - 1)$ gives $\cos^2\theta + \cos\theta - 1 = 0$. The roots of this quadratic equation are

$$\cos\theta = \frac{-1\pm\sqrt{5}}{2}. \tag{2.29}$$

We must choose the plus sign, because we need $|\cos\theta| \le 1$. So our answer is $\cos\theta \approx 0.618$, which is the inverse of the golden ratio. The angle θ is $\approx 51.8°$.

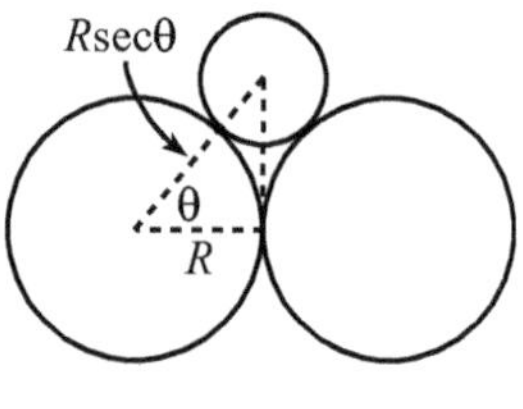

Fig. 2.46

(c) In Fig. 2.46, the length of the hypotenuse shown is $R\sec\theta$, so the radius of the top circle is $R(\sec\theta - 1)$. Its mass is therefore $\sigma\pi R^2(\sec\theta-1)^2$. Equating the weight with the upward component of the normal forces, $2N\sin\theta$, gives $N = \sigma g\pi R^2(\sec\theta-1)^2/(2\sin\theta)$. The horizontal component of N is therefore

$$N\cos\theta = \frac{\sigma g\pi R^2\cos\theta}{2\sin\theta}\left(\frac{1}{\cos\theta}-1\right)^2 = \frac{\sigma g\pi R^2(1-\cos\theta)^2}{2\sin\theta\cos\theta}. \tag{2.30}$$

This equals zero when $\theta = 0$ (using $\cos\theta \approx 1-\theta^2/2$ and $\sin\theta \approx \theta$, for small θ). For $\theta \to \pi/2$, it behaves like $1/\cos\theta$, which goes to infinity. In this limit, N points almost vertically, but its magnitude is so large that the horizontal component still approaches infinity.

2.8. **Hanging chain**

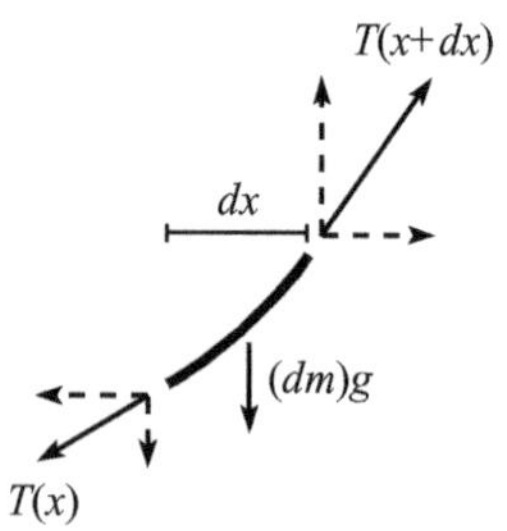

Fig. 2.47

(a) The key fact to note is that the horizontal component, T_x, of the tension is the same throughout the chain. This is true because the net horizontal force on any subpart of the chain must be zero. Label the constant value as $T_x \equiv C$.

Let the shape of the chain be described by the function $y(x)$. Since the tension points along the chain at all points (see Problem 2.12), its components satisfy $T_y/T_x = y'$, which gives $T_y = Cy'$. In other words, T_y is proportional to the slope of the chain.

Now consider a little piece of the chain, with endpoints at x and $x+dx$, as shown in Fig. 2.47. The difference in the T_y values at the endpoints is what balances the weight of the little piece, $(dm)g$. The length of the piece is $ds = dx\sqrt{1+y'^2}$, so if ρ is the density, we have

$$dT_y = (\rho\, ds)g = \rho g\, dx\sqrt{1+y'^2} \implies \frac{dT_y}{dx} = \rho g\sqrt{1+y'^2}. \tag{2.31}$$

Using the $T_y = Cy'$ result from above, this becomes $Cy'' = \rho g\sqrt{1+y'^2}$. Letting $z \equiv y'$, we can separate variables and integrate to obtain

$$\int \frac{dz}{\sqrt{1+z^2}} = \int \frac{\rho g\, dx}{C} \quad \Longrightarrow \quad \sinh^{-1} z = \frac{\rho g x}{C} + A, \tag{2.32}$$

where A is a constant of integration. We can make this look a little cleaner if we define constants α and a such that $\alpha \equiv \rho g/C$ and $a \equiv A/\alpha$. We then obtain

$$\sinh^{-1} z = \alpha(x+a) \quad \Longrightarrow \quad z = \sinh \alpha(x+a). \tag{2.33}$$

Recalling that $z \equiv dy/dx$, we can integrate again to obtain

$$y(x) = \frac{1}{\alpha}\cosh \alpha(x+a) + h. \tag{2.34}$$

The shape of the chain is therefore a hyperbolic cosine function. The constant h isn't too important, because it depends simply on where we pick the $y = 0$ height. Furthermore, we can eliminate the need for the constant a if we pick $x = 0$ to be where the lowest point of the chain is (or where it would be, in the case where the slope is always nonzero). In this case, using Eq. (2.34), we see that $y'(0) = 0$ implies $a = 0$, as desired. We then have (ignoring the constant h) the nice simple result,

$$y(x) = \frac{1}{\alpha}\cosh(\alpha x). \tag{2.35}$$

(b) The constant α can be determined from the locations of the endpoints and the length of the chain. As stated in the problem, the position of the chain may be described by giving (1) the horizontal distance d between the two endpoints, (2) the vertical distance λ between the two endpoints, and (3) the length ℓ of the chain, as shown in Fig. 2.48. Note that it isn't obvious what the horizontal distances between the ends and the minimum point (which we have chosen as the $x = 0$ point) are. If $\lambda = 0$, then these distances are $d/2$, by symmetry. But otherwise, they aren't so clear.

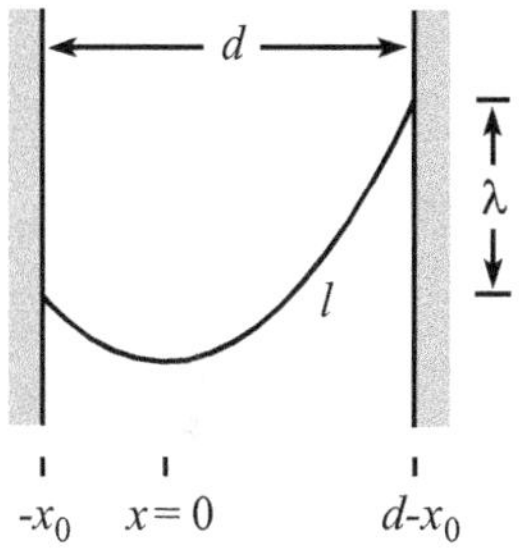

Fig. 2.48

If we let the left endpoint be located at $x = -x_0$, then the first of the above three facts says that the right endpoint is located at $x = d - x_0$. We now have two unknowns, x_0 and α. The second fact tells us that (we'll take the right end to be higher than the left end, without loss of generality)

$$y(d-x_0) - y(-x_0) = \lambda. \tag{2.36}$$

And the third fact gives, using Eq. (2.35),

$$\ell = \int_{-x_0}^{d-x_0} \sqrt{1+y'^2}\, dx = \frac{1}{\alpha}\sinh(\alpha x)\Big|_{-x_0}^{d-x_0}, \tag{2.37}$$

where we have used $(d/du)\cosh u = \sinh u$, and $1 + \sinh^2 u = \cosh^2 u$, and $\int \cosh u = \sinh u$. Writing out Eqs. (2.36) and (2.37) explicitly, we have

$$\begin{aligned} \cosh\big(\alpha(d-x_0)\big) - \cosh(-\alpha x_0) &= \alpha\lambda, \\ \sinh\big(\alpha(d-x_0)\big) - \sinh(-\alpha x_0) &= \alpha\ell. \end{aligned} \tag{2.38}$$

We can eliminate x_0 by taking the difference of the squares of these two equations. Using the hyperbolic identities $\cosh^2 u - \sinh^2 u = 1$ and $\cosh u \cosh v - \sinh u \sinh v = \cosh(u - v)$, we obtain

$$2\cosh(\alpha d) - 2 = \alpha^2(\ell^2 - \lambda^2). \tag{2.39}$$

This is the desired equation that determines α. Given d, λ, and ℓ, we can numerically solve for α. Using a "half-angle" formula, you can show that Eq. (2.39) may also be written as

$$2\sinh(\alpha d/2) = \alpha\sqrt{\ell^2 - \lambda^2}. \tag{2.40}$$

REMARK: Let's check a couple limits. If $\lambda = 0$ and $\ell = d$ (that is, the chain forms a horizontal straight line), then Eq. (2.40) becomes $2\sinh(\alpha d/2) = \alpha d$. The solution to this is $\alpha = 0$, which does indeed correspond to a horizontal straight line, because for small α, we can use $\cosh\epsilon \approx 1 + \epsilon^2/2$ to say that the $y(x)$ in Eq. (2.35) behaves like $\alpha x^2/2$ (up to an additive constant), which varies slowly with x for small α. Another limit is where ℓ is much larger than both d and λ. In this case, Eq. (2.40) becomes $2\sinh(\alpha d/2) \approx \alpha\ell$. The solution to this is a large α (or more precisely, $\alpha \gg 1/d$), which corresponds to a "droopy" chain, because the $y(x)$ in Eq. (2.35) varies rapidly with x for large α. ♣

2.9. **Hanging gently**

We must first find the mass of the chain by calculating its length. Then we must determine the slope of the chain at the supports, so we can find the components of the force there. Using the given information, $y(x) = (1/\alpha)\cosh(\alpha x)$, the slope of the chain as a function of x is

$$y' = \frac{d}{dx}\left(\frac{1}{\alpha}\cosh(\alpha x)\right) = \sinh(\alpha x). \tag{2.41}$$

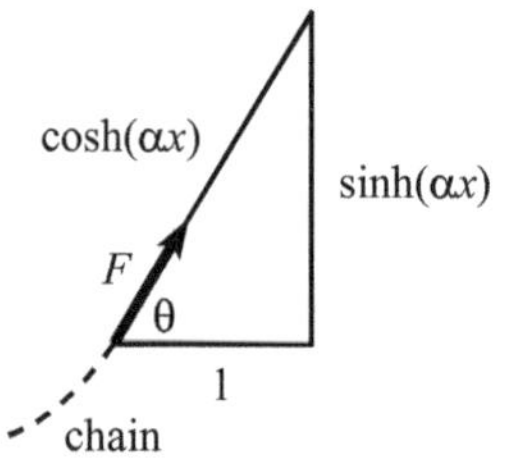

Fig. 2.49

The total length is therefore (using $1 + \sinh^2 z = \cosh^2 z$)

$$\ell = \int_{-d}^{d} \sqrt{1 + y'^2}\, dx = \int_{-d}^{d} \cosh(\alpha x) = \frac{2}{\alpha}\sinh(\alpha d). \tag{2.42}$$

The weight of the rope is $W = \rho\ell g$, where ρ is the mass per unit length. Each support applies a vertical force of $W/2$. So this equals $F\sin\theta$, where F is the magnitude of the force at each support, and θ is the angle it makes with the horizontal. Since $\tan\theta = y'(d) = \sinh(\alpha d)$, we see from Fig. 2.49 that $\sin\theta = \tanh(\alpha d)$. Therefore,

$$F = \frac{1}{\sin\theta}\cdot\frac{W}{2} = \frac{1}{\tanh(\alpha d)}\cdot\frac{\rho g \sinh(\alpha d)}{\alpha} = \frac{\rho g}{\alpha}\cosh(\alpha d). \tag{2.43}$$

Taking the derivative of this (as a function of α), and setting the result equal to zero to find the minimum, gives $\tanh(\alpha d) = 1/(\alpha d)$. This must be solved numerically. The result is

$$\alpha d \approx 1.1997 \equiv \eta. \tag{2.44}$$

So α is given by $\alpha = \eta/d$, and the shape of the chain that requires the minimum F is thus

$$y(x) \approx \frac{d}{\eta}\cosh\left(\frac{\eta x}{d}\right). \tag{2.45}$$

From Eqs. (2.42) and (2.44), the length of the chain is $\ell = (2d/\eta)\sinh(\eta) \approx (2.52)d$. To further get an idea of what the chain looks like, we can calculate the ratio of the height, h, to the width, $2d$.

$$\frac{h}{2d} = \frac{y(d) - y(0)}{2d} = \frac{\cosh(\eta) - 1}{2\eta} \approx 0.338. \tag{2.46}$$

We can also calculate the angle of the rope at the supports, using $\tan\theta = \sinh(\alpha d)$. This gives $\tan\theta = \sinh\eta$, and so $\theta \approx 56.5°$.

REMARK: We can also ask what shape the chain should take in order to minimize the horizontal or vertical component of F. The vertical component, F_y, is simply half the weight, so we want the shortest possible chain, namely a horizontal one (which requires an infinite F). This corresponds to $\alpha = 0$. The horizontal component, F_x, equals $F\cos\theta$. From Fig. 2.49, we see that $\cos\theta = 1/\cosh(\alpha d)$. Therefore, Eq. (2.43) gives $F_x = \rho g/\alpha$. This goes to zero as $\alpha \to \infty$, which corresponds to a chain with infinite length, that is, a very "droopy" chain. ♣

2.10. **Mountain climber**

CHEAP LASSO: We will take advantage of the fact that a cone is "flat," in the sense that we can make one out of a piece of paper, without crumpling the paper. Cut the cone along a straight line emanating from the peak and passing through the knot of the lasso, and roll the cone flat onto a plane. Call the resulting figure, which is a sector of a circle, S (see Fig. 2.50). If the cone is very sharp, then S looks like a thin "pie piece." If the cone is very wide, with a shallow slope, then S looks like a pie with a piece taken out of it. Points on the straight-line boundaries of the sector S are identified with each other. Let P be the location of the lasso's knot. Then P appears on each straight-line boundary, at equal distances from the tip of S. Let β be the angle of the sector S.

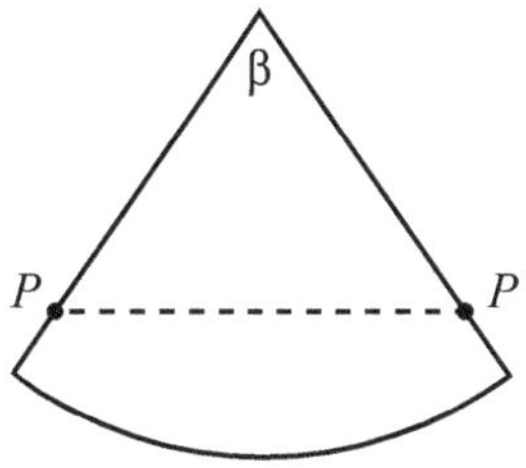

Fig. 2.50

The key to this problem is to realize that the path of the lasso's loop must be a straight line on S, as shown by the dotted line in Fig. 2.50. This is true because the rope takes the shortest distance between two points because there is no friction, and rolling the cone onto a plane doesn't change distances. But a straight line between the two identified points P is possible if and only if the sector S is smaller than a semicircle. The condition for a climbable mountain is therefore $\beta < 180°$.

What is this condition, in terms of the angle of the peak, α? Let C denote a cross-sectional circle of the mountain, a distance d (measured along the cone) from the top. (We are considering this circle for geometrical convenience. It is *not* the path of the lasso; see the remark below.) A semicircular S implies that the circumference of C equals πd. This then implies that the radius of C equals $d/2$. Therefore,

$$\sin(\alpha/2) < \frac{d/2}{d} = \frac{1}{2} \quad \Longrightarrow \quad \alpha < 60°. \tag{2.47}$$

This is the condition under which the mountain is climbable. In short, having $\alpha < 60°$ guarantees that there is a loop around the cone with shorter length than the distance straight to the peak and back.

REMARK: When viewed from the side, the rope will appear perpendicular to the side of the mountain at the point opposite the lasso's knot. A common mistake is to assume that this implies that the climbable condition is $\alpha < 90°$. This is not the case, because the loop does not lie in a plane. Lying in a plane, after all, would imply an elliptical loop. But the loop must certainly have a kink in it where the knot is, because there must exist a vertical component to the tension there to hold the climber up. If we had posed the problem with a planar, triangular mountain, then the condition would have been $\alpha < 90°$. ♣

DELUXE LASSO: If the mountain is very steep, the climber can slide down the mountain by means of the loop growing larger. If the mountain has a shallow slope, the climber can slide down by means of the loop growing smaller. The only situation in which the climber doesn't slide down is the one where the change in position of the knot along the mountain is exactly compensated by the change in length of the loop.

Roll the cone onto a plane as we did in the cheap-lasso case. In terms of the sector S in a plane, the above condition requires that if we move P a distance ℓ up (or down) along the mountain, the distance between the identified points P must decrease (or increase) by ℓ. In Fig. 2.50, we must therefore have an equilateral triangle, so $\beta = 60°$.

What peak-angle α does this correspond to? As in part the cheap-lasso case, let C be a cross-sectional circle of the mountain, a distance d (measured along the cone) from the top. Then $\beta = 60°$ implies that the circumference of C equals $(\pi/3)d$. This then implies that the radius of C equals $d/6$. Therefore,

$$\sin(\alpha/2) = \frac{d/6}{d} = \frac{1}{6} \quad \Longrightarrow \quad \alpha \approx 19°. \tag{2.48}$$

This is the condition under which the mountain is climbable. We see that there is exactly one angle for which the climber can climb up along the mountain. The cheap lasso is therefore much more useful than the fancy deluxe lasso, assuming, of course, that you want to use it for climbing mountains, and not, say, for rounding up cattle.

REMARK: Another way to see the $\beta = 60°$ result is to note that the three directions of rope emanating from the knot must all have the same tension, because the deluxe lasso is one continuous piece of rope. They must therefore have 120° angles between themselves (to provide zero net force on the massless knot). This implies that $\beta = 60°$ in Fig. 2.50. ♣

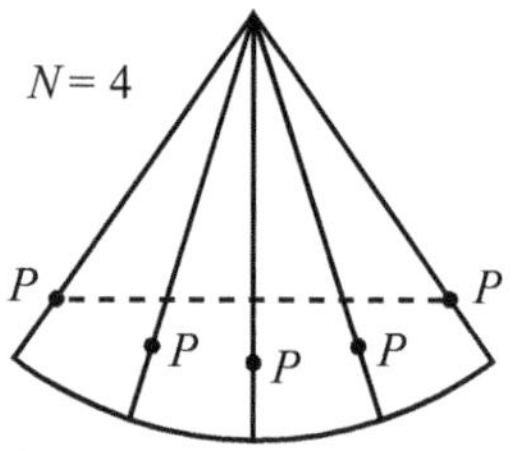

Fig. 2.51

FURTHER REMARKS: For each type of lasso, we can also ask the question: For what angles can the mountain be climbed if the lasso is looped N times around the top of the mountain? The solution here is similar to that above.

For the cheap lasso, roll the cone N times onto a plane, as shown in Fig. 2.51 for $N = 4$. The resulting figure, S_N, is a sector of a circle divided into N equal sectors, each representing a copy of the cone. As above, S_N must be smaller than a semicircle. The circumference of the circle C (defined above) must therefore be less than $\pi d/N$. Hence, the radius of C must be less than $d/2N$. Thus,

$$\sin(\alpha/2) < \frac{d/2N}{d} = \frac{1}{2N} \quad \Longrightarrow \quad \alpha < 2\sin^{-1}\left(\frac{1}{2N}\right). \tag{2.49}$$

For the deluxe lasso, again roll the cone N times onto a plane. From the original reasoning above, we must have $N\beta = 60°$. The circumference of C must therefore be $\pi d/3N$, and so its radius must be $d/6N$. Therefore,

$$\sin(\alpha/2) = \frac{d/6N}{d} = \frac{1}{6N} \quad \Longrightarrow \quad \alpha = 2\sin^{-1}\left(\frac{1}{6N}\right). \ ♣ \tag{2.50}$$

2.11. **Equality of torques**

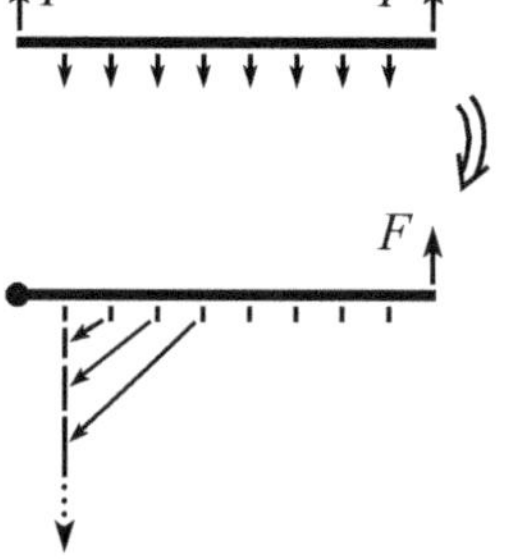

Fig. 2.52

The proof by induction is as follows. Assume that we have shown that a force F applied at a distance d is equivalent to a force kF applied at a distance d/k, for all integers k up to $n-1$. We now want to show that the statement holds for $k = n$.

Consider the situation in Fig. 2.52. Forces F are applied at the ends of a stick, and forces $2F/(n-1)$ are applied at the $j\ell/n$ marks (for $1 \le j \le n-1$). The stick doesn't rotate (by symmetry), and it doesn't translate (because the net force is zero). Consider the stick to have a pivot at the left end. Replacing the interior forces by their equivalent ones at the ℓ/n mark (see Fig. 2.52) gives a total force there equal to

$$\frac{2F}{n-1}\Big(1+2+3+\cdots+(n-1)\Big) = \frac{2F}{n-1}\left(\frac{n(n-1)}{2}\right) = nF. \tag{2.51}$$

We therefore see that a force F applied at a distance ℓ is equivalent to a force nF applied at a distance ℓ/n, as was to be shown.

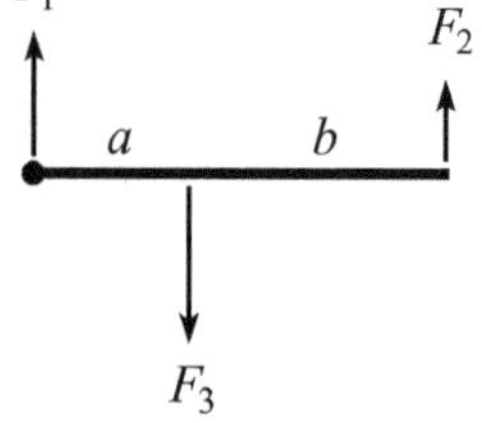

Fig. 2.53

We can now show that Claim 2.1 holds, for arbitrary distances a and b (see Fig. 2.53). Consider the stick to be pivoted at its left end, and let ϵ be a tiny distance (small compared with a). Then a force F_3 at a distance a is equivalent to a force $F_3(a/\epsilon)$ at a distance ϵ.[10] But a force $F_3(a/\epsilon)$ at a distance ϵ is equivalent to a force $F_3(a/\epsilon)(\epsilon/(a+b)) = F_3a/(a+b)$ at a distance $(a+b)$. This equivalent force at the distance $(a+b)$ must cancel the force F_2 there, because the stick is motionless. Therefore, we have $F_3a/(a+b) = F_2$, which proves the claim.

2.12. **Direction of the tension**

Consider an infinitesimal piece of the rope, and look at the torque around one end. Any forces acting at this end provide no torque around it. If the tension at the other

[10] Technically, we can use the reasoning in the previous paragraph to say this only if a/ϵ is an integer, but since a/ϵ is very large, we can simply pick the closest integer to it, and there will be negligible error.

end is directed at a finite angle away from the direction of the rope, then this produces a certain torque. But this torque can't be canceled by the much smaller torque from the tiny gravitational force, because this force is proportional to the length of the tiny piece. Therefore, the tension must point along the rope. It actually points along the direction of the rope at the end of the little piece it acts on, which isn't quite along the direction of the rope at the end we're considering torques around, because the rope bends (assuming it's not vertical). So the tension ends up producing a very small torque which cancels the very small torque from gravity.

This argument doesn't work for a rigid stick, because the stick can produce finite torques around the end of a piece via forces *at* that end, because the end is really a cross section of finite size. There is a shearing action in the stick, and the large shearing forces act with tiny lever arms (relative to, say, a point at the middle of the cross section) to produce finite torques.

2.13. **Find the force**

In Fig. 2.54, let the supports at the ends exert forces F_1 and F_2, and let the support in the interior exert a force F. Then

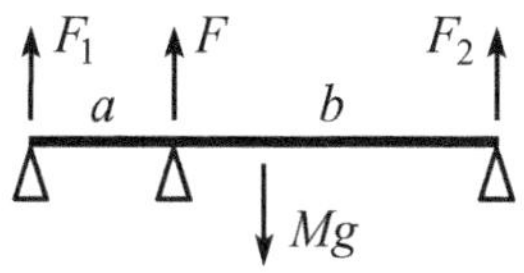

Fig. 2.54

$$F_1 + F_2 + F = Mg. \tag{2.52}$$

Balancing torques around the left and right ends gives, respectively,

$$\begin{aligned} Fa + F_2(a+b) &= Mg\frac{a+b}{2}, \\ Fb + F_1(a+b) &= Mg\frac{a+b}{2}, \end{aligned} \tag{2.53}$$

where we have used the fact that the stick can be treated like a point mass at its center. Note that the equation for balancing the torques around the center of mass is redundant; it is obtained by taking the difference of the two previous equations and then dividing by 2. And balancing torques around the middle pivot also takes the form of a linear combination of these equations, as you can show.

It appears as though we have three equations and three unknowns, but we really have only two equations, because the sum of Eqs. (2.53) gives Eq. (2.52). Therefore, since we have two equations and three unknowns, the system is underdetermined. Solving Eqs. (2.53) for F_1 and F_2 in terms of F, we see that any forces of the form

$$(F_1, F, F_2) = \left(\frac{Mg}{2} - \frac{Fb}{a+b},\ F,\ \frac{Mg}{2} - \frac{Fa}{a+b}\right) \tag{2.54}$$

are possible. In retrospect, it makes sense that the forces are not determined. By changing the height of the new support an infinitesimal distance, we can make F be anything from 0 up to $Mg(a+b)/2b$, which is when the stick comes off the left support (assuming $b \geq a$).

2.14. **Leaning sticks**

Let M_l be the mass of the left stick, and let M_r be the mass of the right stick. Then $M_l/M_r = \tan\theta$. Let N and F_f be the normal and friction forces between the sticks (see Fig. 2.55). F_f has a maximum value of μN. Balancing the torques on the left stick (around the contact point with the ground) gives $N = (M_l g/2)\sin\theta$. Balancing the torques on the right stick (around the contact point with the ground) gives $F_f = (M_r g/2)\cos\theta$. The condition $F_f \leq \mu N$ is therefore

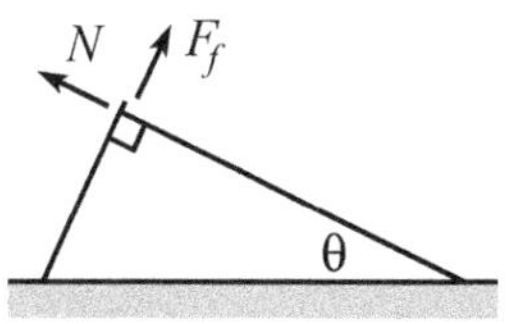

Fig. 2.55

$$M_r \cos\theta \leq \mu M_l \sin\theta \quad \Longrightarrow \quad \tan^2\theta \geq \frac{1}{\mu}, \tag{2.55}$$

where we have used $M_l/M_r = \tan\theta$. This answer checks in the two extremes: In the limit $\mu \to 0$, we see that θ must be very close to $\pi/2$, which makes sense. And in the limit $\mu \to \infty$ (that is, very sticky sticks), we see that θ can be very small, which also makes sense.

2.15. **Supporting a ladder**

Let F be the desired force. F must be directed along the stick, because otherwise there would be a net torque on the (massless) stick relative to the pivot at its right end, and this would contradict the fact that it is at rest. Look at torques on the ladder around the pivot at its bottom. The gravitational force provides a clockwise torque of $Mg(L/2)\cos\theta$, and the force F from the stick provides a counterclockwise torque of $F(\ell/\tan\theta)$. Equating these two torques gives

$$F = \frac{MgL}{2\ell}\sin\theta. \tag{2.56}$$

REMARKS: F goes to zero as $\theta \to 0$, as it should.[11] And F increases to $MgL/2\ell$ as $\theta \to \pi/2$, which isn't so obvious (the required torque from the stick is very small, but the lever arm is also very small). However, in the special case where the ladder is exactly vertical, no force is required. You can see that our calculations above are not valid in this case, because we divided by $\cos\theta$, which is zero when $\theta = \pi/2$.

The normal force at the pivot of the stick (which equals the vertical component of F, because the stick is massless) is equal to $MgL\sin\theta\cos\theta/2\ell$. This has a maximum value of $MgL/4\ell$ at $\theta = \pi/4$. ♣

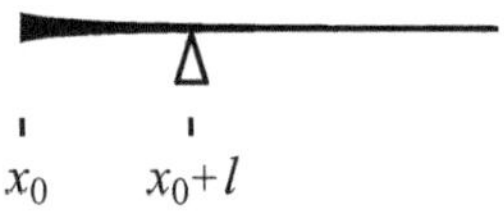

Fig. 2.56

2.16. **Balancing the stick**

Let the stick go off to infinity in the positive x direction, and let it be cut at $x = x_0$. Then the pivot point is located at $x = x_0 + \ell$ (see Fig. 2.56). Let the density be $\rho(x)$. The condition that the total gravitational torque relative to $x_0 + \ell$ be zero is

$$\tau = \int_{x_0}^{\infty} \rho(x)\big(x - (x_0 + \ell)\big)g\,dx = 0. \tag{2.57}$$

We want this to equal zero for all x_0, so the derivative of τ with respect to x_0 must be zero. τ depends on x_0 through both the limits of integration and the integrand. In taking the derivative, the former dependence requires finding the value of the integrand at the x_0 limit, while the latter dependence requires taking the derivative of the integrand with respect to x_0, and then integrating. (To derive these two contributions, just replace x_0 with $x_0 + dx_0$ and expand things to first order in dx_0.) We obtain

$$0 = \frac{d\tau}{dx_0} = g\ell\rho(x_0) - g\int_{x_0}^{\infty} \rho(x)\,dx. \tag{2.58}$$

Taking the derivative of this equation with respect to x_0 gives $\ell\rho'(x_0) = -\rho(x_0)$. The solution to this is (rewriting the arbitrary x_0 as x)

$$\rho(x) = Ae^{-x/\ell}. \tag{2.59}$$

We therefore see that the density decreases exponentially with x. The smaller ℓ is, the quicker it falls off. Note that the density at the pivot is $1/e$ times the density at the left end. And you can show that $1 - 1/e \approx 63\,\%$ of the mass is contained between the left end and the pivot.

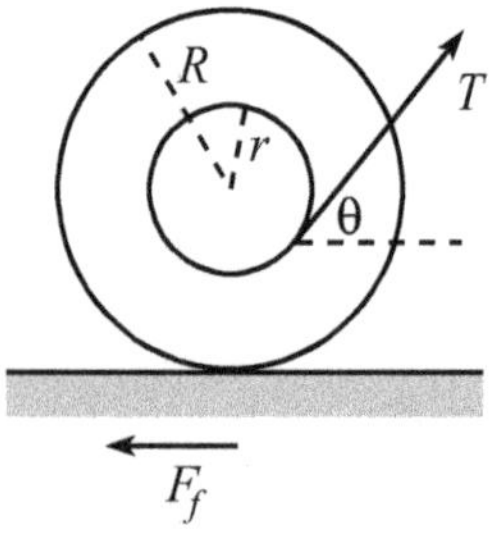

Fig. 2.57

2.17. **The spool**

(a) Let F_f be the friction force the ground provides. Balancing the horizontal forces on the spool gives (see Fig. 2.57)

$$T\cos\theta = F_f. \tag{2.60}$$

[11] For $\theta \to 0$, we would need to lengthen the ladder with a massless extension, because the stick would have to be very far to the right to remain perpendicular to the ladder.

Balancing torques around the center of the spool gives

$$Tr = F_f R. \tag{2.61}$$

These two equations imply

$$\cos\theta = \frac{r}{R}. \tag{2.62}$$

The niceness of this result suggests that there is a quicker way to obtain it. And indeed, we see from Fig. 2.58 that $\cos\theta = r/R$ is the angle that causes the line of the tension to pass through the contact point on the ground. Since gravity and friction provide no torque around this point, the total torque around it is therefore zero, and the spool remains at rest.

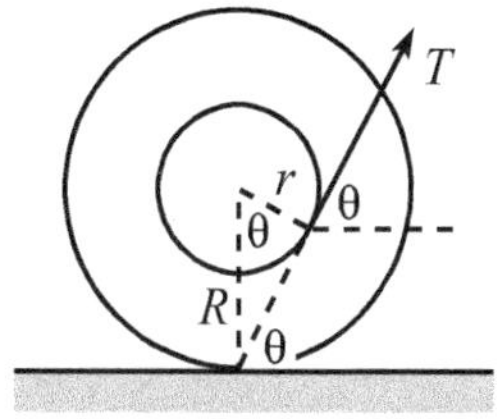

Fig. 2.58

(b) The normal force from the ground is

$$N = Mg - T\sin\theta. \tag{2.63}$$

Using Eq. (2.60), the statement $F_f \le \mu N$ becomes

$$T\cos\theta \le \mu(Mg - T\sin\theta) \quad \Longrightarrow \quad T \le \frac{\mu Mg}{\cos\theta + \mu\sin\theta}, \tag{2.64}$$

where θ is given in Eq. (2.62).

(c) The maximum value of T is given in (2.64). This depends on θ, which in turn depends on r. We want to find the r that minimizes this maximum T. Taking the derivative with respect to θ, we find that the θ that maximizes the denominator in Eq. (2.64) is given by $\tan\theta_0 = \mu$. You can then show that the value of T for this θ_0 is

$$T_0 = \frac{\mu Mg}{\sqrt{1+\mu^2}}. \tag{2.65}$$

To find the corresponding r, we can use Eq. (2.62) to write $\tan\theta = \sqrt{R^2 - r^2}/r$. The relation $\tan\theta_0 = \mu$ then yields

$$r_0 = \frac{R}{\sqrt{1+\mu^2}}. \tag{2.66}$$

This is the r that yields the smallest upper bound on T. In the limit $\mu = 0$, we have $\theta_0 = 0$, $T_0 = 0$, and $r_0 = R$. And in the limit $\mu = \infty$, we have $\theta_0 = \pi/2$, $T_0 = Mg$, and $r_0 = 0$.

2.18. **Stick on a circle**

Let N be the normal force between the stick and the circle, and let F_f be the friction force between the ground and the circle (see Fig. 2.59). Then we immediately see that the friction force between the stick and the circle is also F_f, because the torques from the two friction forces on the circle must cancel. We've drawn all forces as acting on the circle. By Newton's third law, N and F_f act in the opposite directions on the stick at its top end.

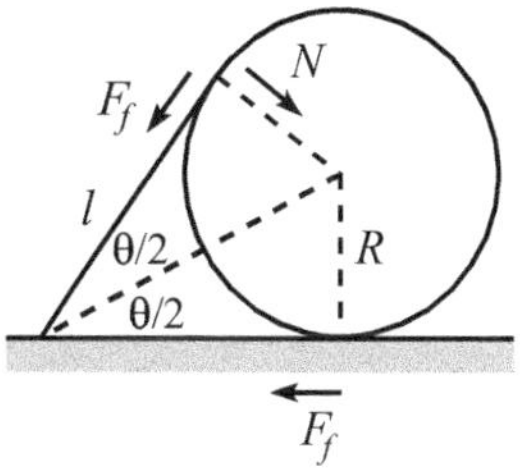

Fig. 2.59

Looking at torques on the stick around the point of contact with the ground, we have $Mg(\ell/2)\cos\theta = N\ell$, where $M = \rho\ell$ is the mass of the stick, and ℓ is its length. Therefore, $N = (\rho\ell g/2)\cos\theta$. Balancing the horizontal forces on the circle gives $N\sin\theta = F_f + F_f\cos\theta$, so we have

$$F_f = \frac{N\sin\theta}{1+\cos\theta} = \frac{\rho\ell g\sin\theta\cos\theta}{2(1+\cos\theta)}. \tag{2.67}$$

But from Fig. 2.59 we have $\ell = R/\tan(\theta/2)$. Using the identity $\tan(\theta/2) = \sin\theta/(1+\cos\theta)$, we finally obtain

$$F_f = \frac{1}{2}\rho g R\cos\theta. \tag{2.68}$$

In the limit $\theta \to \pi/2$, F_f approaches zero, which makes sense. In the limit $\theta \to 0$ (which corresponds to a very long stick), the friction force approaches $\rho g R/2$, which isn't so obvious.

2.19. **Leaning sticks and circles**

Let S_i be the ith stick, and let C_i be the ith circle. The normal forces that C_i feels from S_i and S_{i+1} are equal in magnitude, because these two forces provide the only horizontal forces on the frictionless circle, so they must cancel. Let N_i be this normal force.

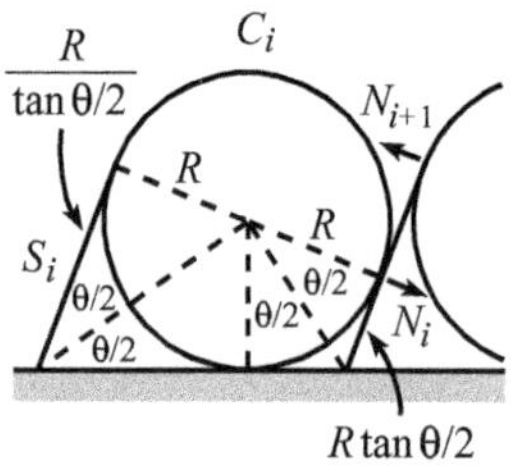

Fig. 2.60

Look at the torques on S_{i+1}, relative to the hinge on the ground. The torques come from N_i, N_{i+1}, and the weight of S_{i+1}. From Fig. 2.60, we see that N_i acts at a point which is a distance $R\tan(\theta/2)$ away from the hinge. Since the stick has a length $R/\tan(\theta/2)$, this point is a fraction $\tan^2(\theta/2)$ up along the stick. Therefore, balancing the torques on S_{i+1} gives

$$\frac{1}{2}Mg\cos\theta + N_i\tan^2\frac{\theta}{2} = N_{i+1}. \tag{2.69}$$

N_0 is by definition zero, so we have $N_1 = (Mg/2)\cos\theta$ (as in the previous problem). If we successively use Eq. (2.69), we see that N_2 equals $(Mg/2)\cos\theta\big(1+\tan^2(\theta/2)\big)$, and N_3 equals $(Mg/2)\cos\theta\big(1+\tan^2(\theta/2)+\tan^4(\theta/2)\big)$, and so on. In general,

$$N_i = \frac{Mg\cos\theta}{2}\left(1+\tan^2\frac{\theta}{2}+\tan^4\frac{\theta}{2}+\cdots+\tan^{2(i-1)}\frac{\theta}{2}\right). \tag{2.70}$$

In the limit $i \to \infty$, we may write this infinite geometric sum in closed form as

$$N_\infty \equiv \lim_{i\to\infty} N_i = \frac{Mg\cos\theta}{2}\left(\frac{1}{1-\tan^2(\theta/2)}\right). \tag{2.71}$$

Note that this is the solution to Eq. (2.69), with $N_i = N_{i+1}$. So if a limit exists, it must be this. Using $M = \rho R/\tan(\theta/2)$, we can rewrite N_∞ as

$$N_\infty = \frac{\rho R g\cos\theta}{2\tan(\theta/2)}\left(\frac{1}{1-\tan^2(\theta/2)}\right). \tag{2.72}$$

The identity $\cos\theta = \cos^2(\theta/2) - \sin^2(\theta/2)$ may then be used to write this as

$$N_\infty = \frac{\rho R g\cos^3(\theta/2)}{2\sin(\theta/2)}. \tag{2.73}$$

Remarks: N_∞ goes to infinity for $\theta \to 0$, which makes sense, because the sticks are very long. All of the N_i are essentially equal to half the weight of a stick (in order to cancel the torque from the weight relative to the pivot). For $\theta \to \pi/2$, we see from Eq. (2.73) that N_∞ approaches $\rho R g/4$, which is not at all obvious; the N_i start off at $N_1 = (Mg/2)\cos\theta \approx 0$, but gradually increase to $\rho R g/4$, which is a quarter of the weight of a stick. Note that the horizontal force that must be applied to the last circle far to the right is $N_\infty \sin\theta = \rho R g\cos^4(\theta/2)$. This ranges from $\rho R g$ for $\theta \to 0$, to $\rho R g/4$ for $\theta \to \pi/2$. ♣

Chapter 3

Using $F = ma$

The general goal of classical mechanics is to determine what happens to a given set of objects in a given physical situation. In order to figure this out, we need to know what makes the objects move the way they do. There are two main ways of going about this task. The first one, which you are undoubtedly familiar with, involves Newton's laws. This is the subject of the present chapter. The second one, which is more advanced, is the *Lagrangian* method. This is the subject of Chapter 6. It should be noted that each of these methods is perfectly sufficient for solving any problem, and they both produce the same information in the end. But they are based on vastly different principles. We'll talk more about this in Chapter 6.

3.1 Newton's laws

In 1687 Newton published his three laws in his *Principia Mathematica*. These laws are fairly intuitive, although I suppose it's questionable to attach the adjective "intuitive" to a set of statements that weren't written down until a mere 300 years ago. At any rate, the laws may be stated as follows.

- **First law:** A body moves with constant velocity (which may be zero) unless acted on by a force.
- **Second law:** The time rate of change of the momentum of a body equals the force acting on the body.
- **Third law:** For every force on one body, there is an equal and opposite force on another body.

We could discuss for days on end the degree to which these statements are physical laws, and the degree to which they are definitions. Sir Arthur Eddington once made the unflattering remark that the first law essentially says that "every particle continues in its state of rest or uniform motion in a straight line except insofar as it doesn't." However, although the three laws might seem somewhat light on content at first glance, there's actually more to them than Eddington's comment implies. Let's look at each in turn.[1]

[1] A disclaimer: This section represents my view on which parts of the laws are definitions and which parts have content. But you should take all of this with a grain of salt. For further reading, see Anderson (1990), Keller (1987), O'Sullivan (1980), and Eisenbud (1958).

First law

One thing this law does is give a definition of zero force. Another thing it does is give a definition of an *inertial frame*, which is defined simply as a frame of reference in which the first law holds; since the term "velocity" is used, we have to state what frame we're measuring the velocity with respect to. The first law does *not* hold in an arbitrary frame. For example, it fails in the frame of a rotating turntable.[2] Intuitively, an inertial frame is one that moves with constant velocity. But this is ambiguous, because we have to say what the frame is moving with constant velocity *with respect to*. But all this aside, an inertial frame is defined as the special type of frame in which the first law holds.

So, what we now have are two intertwined definitions of "force" and "inertial frame." Not much physical content here. But the important point is that the law holds for *all* particles. So if we have a frame in which one free particle moves with constant velocity, then *all* free particles move with constant velocity. This is a statement with content. We can't have a bunch of free particles moving with constant velocity while another one is doing a fancy jig.

Second law

Momentum is defined[3] to be $m\mathbf{v}$. If m is constant,[4] then the second law says that

$$\mathbf{F} = m\mathbf{a}, \tag{3.1}$$

where $\mathbf{a} \equiv d\mathbf{v}/dt$. This law holds only in an inertial frame, which is defined by the first law.

> For things moving free or at rest,
> Observe what the first law does best.
> It defines a key frame,
> "Inertial" by name,
> Where the second law then is expressed.

You might think that the second law merely gives a definition of force, but there is more to it than that. There is a tacit implication in the law that this "force" is something that has an existence that isn't completely dependent on the particle whose "m" appears in the law (more on this in the third law below). A spring force, for example, doesn't depend at all on the particle on which it acts. And the gravitational force, GMm/r^2, depends partly on the particle and partly on something else (another mass).

[2] It's possible to modify things so that Newton's laws hold in such a frame, provided that we introduce the so-called "fictitious" forces. But we'll save this discussion for Chapter 10.

[3] We're doing everything nonrelativistically here, of course. Chapter 12 gives the relativistic modification to the $m\mathbf{v}$ expression.

[4] We'll assume in this chapter that m is constant. But don't worry, we'll get plenty of practice with changing mass (in rockets and such) in Chapter 5.

If you feel like just making up definitions, then you can define a new quantity, $\mathbf{G} = m^2\mathbf{a}$. This is a perfectly legal thing to do; you can't really go wrong in making a definition (well, unless you've already defined the quantity to be something else). However, this definition is completely useless. You can define it for every particle in the world, and for any acceleration, but the point is that the definitions don't have anything to do with each other. There is simply no (uncontrived) quantity in this world that gives accelerations in the ratio of 4 to 1 when "acting" on masses m and $2m$. The quantity $\mathbf{G}$ has nothing to do with anything except the particle you defined it for. The main thing the second law says is that there does indeed exist a quantity $\mathbf{F}$ that gives the same $m\mathbf{a}$ when acting on different particles. The statement of the existence of such a thing is far more than a definition.

Along this same line, note that the second law says that $\mathbf{F} = m\mathbf{a}$, and not, for example, $\mathbf{F} = m\mathbf{v}$, or $\mathbf{F} = m\,d^3\mathbf{x}/dt^3$. In addition to being inconsistent with the real world, these expressions are inconsistent with the first law. $\mathbf{F} = m\mathbf{v}$ would say that a nonzero velocity requires a force, in contrast with the first law. And $\mathbf{F} = md^3\mathbf{x}/dt^3$ would say that a particle moves with constant acceleration (instead of constant velocity) unless acted on by a force, also in contrast with the first law.

As with the first law, it is important to realize that the second law holds for *all* particles. In other words, if the same force (for example, the same spring stretched by the same amount) acts on two particles with masses m_1 and m_2, then Eq. (3.1) says that their accelerations are related by

$$\frac{a_1}{a_2} = \frac{m_2}{m_1}. \tag{3.2}$$

This relation holds regardless of what the common force is. Therefore, once we've used one force to find the relative masses of two objects, then we know what the ratio of their a's will be when they are subjected to any other force. Of course, we haven't really defined *mass* yet. But Eq. (3.2) gives an experimental method for determining an object's mass in terms of a standard (say, 1 kg) mass. All we have to do is compare its acceleration with that of the standard mass, when acted on by the same force.

Note that $\mathbf{F} = m\mathbf{a}$ is a vector equation, so it is really three equations in one. In Cartesian coordinates, it says that $F_x = ma_x$, $F_y = ma_y$, and $F_z = ma_z$.

Third law

One thing this law says is that if we have two isolated particles interacting through some force, then their accelerations are opposite in direction and inversely proportional to their masses. Equivalently, the third law essentially postulates that the total momentum of an isolated system is conserved (that is, independent of time). To see this, consider two particles, each of which interacts only with the other particle and nothing else in the universe. Then we have

$$\begin{aligned}\frac{d\mathbf{p}_{\text{total}}}{dt} &= \frac{d\mathbf{p}_1}{dt} + \frac{d\mathbf{p}_2}{dt}\\ &= \mathbf{F}_1 + \mathbf{F}_2,\end{aligned} \tag{3.3}$$

where $\mathbf{F}_1$ and $\mathbf{F}_2$ are the forces acting on m_1 and m_2, respectively. This demonstrates that momentum conservation (that is, $d\mathbf{p}_{\text{total}}/dt = 0$) is equivalent to Newton's third law (that is, $\mathbf{F}_1 = -\mathbf{F}_2$). Similar reasoning holds with more than two particles, but we'll save this more general case, along with many other aspects of momentum, for Chapter 5.

There isn't much left to be defined via this law, so this is a law of pure content. It can't be a definition, anyway, because it's actually not always valid. It holds for forces of the "pushing" and "pulling" type, but it fails for the magnetic force, for example. In that case, momentum is carried off in the electromagnetic field (so the total momentum of the particles *and* the field is conserved). But we won't deal with fields here. Just particles. So the third law will always hold in any situation we'll be concerned with.

The third law contains an extremely important piece of information. It says that we will never find a particle accelerating unless there's some other particle accelerating somewhere else. The other particle might be far away, as with the earth–sun system, but it's always out there somewhere. Note that if we were given only the second law, then it would be perfectly possible for a given particle to spontaneously accelerate with nothing else happening in the universe, as long as a similar particle with twice the mass accelerated with half the acceleration when placed in the same spot, etc. This would all be fine, as far as the second law goes. We would say that a force with a certain value is acting at the point, and everything would be consistent. But the third law says that this is simply not the way the world (at least the one we live in) works. In a sense, a force without a counterpart seems somewhat like magic, whereas a force with an equal and opposite counterpart has a "cause and effect" nature, which seems (and apparently is) more physical.

In the end, however, we shouldn't attach too much significance to Newton's laws, because although they were a remarkable intellectual achievement and work spectacularly for everyday physics, they are the laws of a theory that is only approximate. Newtonian physics is a limiting case of the more correct theories of relativity and quantum mechanics, which are in turn limiting cases of yet more correct theories. The way in which particles (or waves, or strings, or whatever) interact on the most fundamental level surely doesn't bear any resemblance to what we call forces.

3.2 Free-body diagrams

The law that allows us to be quantitative is the second law. Given a force, we can apply $\mathbf{F} = m\mathbf{a}$ to find the acceleration. And knowing the acceleration, we can determine the behavior of a given object (that is, the position and velocity), provided that we are given the initial position and velocity. This process sometimes takes a bit of work, but there are two basic types of situations that commonly arise.

- In many problems, all you are given is a physical situation (for example, a block resting on a plane, strings connecting masses, etc.), and it is up to you to find all the forces acting on all the objects, using $\mathbf{F} = m\mathbf{a}$. The forces generally point in various directions, so it's easy to lose track of them. It therefore proves useful to isolate the objects and draw all the forces acting on each of them. This is the subject of the present section.
- In other problems, you are *given* the force explicitly as a function of time, position, or velocity, and the task immediately becomes the mathematical one of solving the $F = ma \equiv m\ddot{x}$ equation (we'll just deal with one dimension here). These *differential equations* can be difficult (or impossible) to solve exactly. They are the subject of Section 3.3.

Let's consider here the first of these two types of scenarios, where we are presented with a physical situation and we must determine all the forces involved. The term *free-body diagram* is used to denote a diagram with all the forces drawn on a given object. After drawing such a diagram for each object in the setup, we simply write down all the $F = ma$ equations they imply. The result will be a system of linear equations in various unknown forces and accelerations, for which we can then solve. This procedure is best understood through an example.

Example (A plane and masses): Mass M_1 is held on a plane with inclination angle θ, and mass M_2 hangs over the side. The two masses are connected by a massless string which runs over a massless pulley (see Fig. 3.1). The coefficient of kinetic friction between M_1 and the plane is μ. M_1 is released from rest. Assuming that M_2 is sufficiently large so that M_1 gets pulled up the plane, what is the acceleration of the masses? What is the tension in the string?

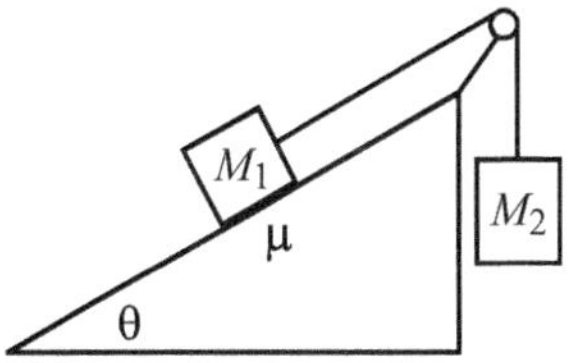

Fig. 3.1

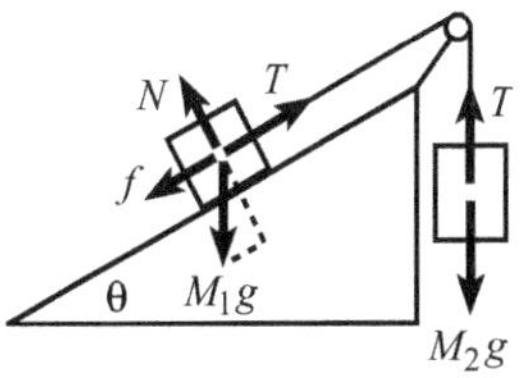

Fig. 3.2

Solution: The first thing to do is draw all the forces on the two masses. These are shown in Fig. 3.2. The forces on M_2 are gravity and the tension. The forces on M_1 are gravity, friction, the tension, and the normal force. Note that the friction force points down the plane, because we are assuming that M_1 moves up the plane.

Having drawn all the forces, we can now write down all the $F = ma$ equations. When dealing with M_1, we could break things up into horizontal and vertical components, but it is much cleaner to use the components parallel and perpendicular to the plane.[5] These two components of $\mathbf{F} = m\mathbf{a}$, along with the vertical $F = ma$ equation for M_2, give

$$\begin{aligned} T - f - M_1 g \sin\theta &= M_1 a, \\ N - M_1 g \cos\theta &= 0, \\ M_2 g - T &= M_2 a, \end{aligned} \tag{3.4}$$

where we have used the fact that the two masses accelerate at the same rate (and we have defined the positive direction for M_2 to be downward). We have also used the fact that the tension is the same at both ends of the string, because otherwise there would be a net force on some part of the string which would then undergo infinite acceleration, because it is massless.

There are four unknowns in Eq. (3.4) (namely T, a, N, and f), but only three equations. Fortunately, we have a fourth equation: $f = \mu N$, because we are assuming that M_1 is in fact moving, so we can use the expression for kinetic friction. Using this in the second equation above gives $f = \mu M_1 g \cos\theta$. The first equation then becomes $T - \mu M_1 g \cos\theta - M_1 g \sin\theta = M_1 a$. Adding this to the third equation leaves us with only a, so we find

$$a = \frac{g(M_2 - \mu M_1 \cos\theta - M_1 \sin\theta)}{M_1 + M_2} \quad \Longrightarrow \quad T = \frac{M_1 M_2 g(1 + \mu\cos\theta + \sin\theta)}{M_1 + M_2}. \tag{3.5}$$

Note that in order for M_1 to in fact accelerate upward (that is, $a > 0$), we must have $M_2 > M_1(\mu\cos\theta + \sin\theta)$. This is clear from looking at the forces along the plane.

REMARK: If we instead assume that M_1 is sufficiently large so that it slides down the plane, then the friction force points up the plane, and we find (as you can check),

$$a = \frac{g(M_2 + \mu M_1 \cos\theta - M_1 \sin\theta)}{M_1 + M_2}, \quad \text{and} \quad T = \frac{M_1 M_2 g(1 - \mu\cos\theta + \sin\theta)}{M_1 + M_2}. \tag{3.6}$$

In order for M_1 to in fact accelerate downward (that is, $a < 0$), we must have $M_2 < M_1(\sin\theta - \mu\cos\theta)$. Therefore, the range of M_2 for which the system doesn't accelerate (that is, it just sits there, assuming that it started at rest) is

$$M_1(\sin\theta - \mu\cos\theta) \le M_2 \le M_1(\sin\theta + \mu\cos\theta). \tag{3.7}$$

If μ is very small, then M_2 must essentially be equal to $M_1 \sin\theta$ if the system is to be static. Equation (3.7) also implies that if $\tan\theta \le \mu$, then M_1 won't slide down, even if $M_2 = 0$. ♣

In problems like the one above, it's clear which things you should pick as the objects you're going to draw forces on. But in other problems, where there are

[5] When dealing with inclined planes, it's usually the case that one of these two coordinate systems works much better than the other. Sometimes it isn't clear which one, but if things get messy with one system, you can always try the other.

various different subsystems you can choose, you must be careful to include all the relevant forces on a given subsystem. Which subsystems you want to pick depends on what quantities you're trying to find. Consider the following example.

Example (Platform and pulley): A person stands on a platform-and-pulley system, as shown in Fig. 3.3. The masses of the platform, person, and pulley[6] are M, m, and μ, respectively.[7] The rope is massless. Let the person pull up on the rope so that she has acceleration a upward. (Assume that the platform is somehow constrained to stay level, perhaps by having the ends run along some rails.) Find the tension in the rope, the normal force between the person and the platform, and the tension in the rod connecting the pulley to the platform.

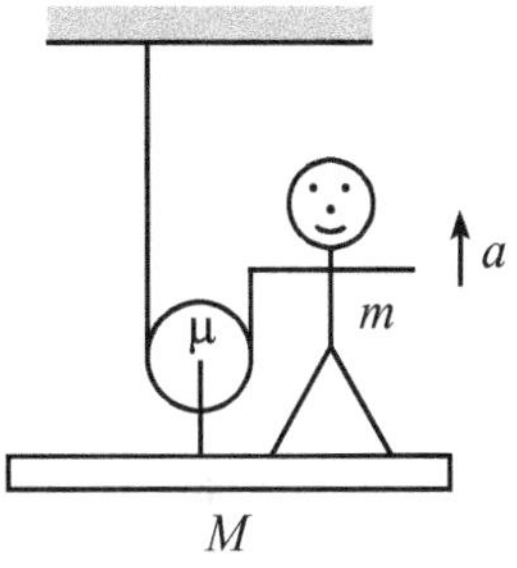

Fig. 3.3

Solution: To find the tension in the rope, we simply want to let our subsystem be the whole system (except the ceiling). If we imagine putting the system in a black box (to emphasize the fact that we don't care about any internal forces within the system), then the forces we see "protruding" from the box are the three weights (Mg, mg, and μg) downward, and the tension T upward. Applying $F = ma$ to the whole system gives

$$T - (M + m + \mu)g = (M + m + \mu)a \quad \Longrightarrow \quad T = (M + m + \mu)(g + a). \tag{3.8}$$

To find the normal force N between the person and the platform, and also the tension f in the rod connecting the pulley to the platform, it is not sufficient to consider the system as a whole. This is true because these forces are internal forces to this system, so they won't show up in any $F = ma$ equations (which involve only external forces to a system). So we must consider subsystems:

- Let's apply $F = ma$ to the person. The forces acting on the person are gravity, the normal force from the platform, and the tension from the rope (pulling downward on her hand). So we have

$$N - T - mg = ma. \tag{3.9}$$

- Now apply $F = ma$ to the platform. The forces acting on the platform are gravity, the normal force from the person, and the force upward from the rod. So we have

$$f - N - Mg = Ma. \tag{3.10}$$

- Now apply $F = ma$ to the pulley. The forces acting on the pulley are gravity, the force downward from the rod, and *twice* the tension in the rope (because it pulls

[6] Assume that the pulley's mass is concentrated at its center, so that we don't have to worry about any rotational dynamics (the subject of Chapter 8).

[7] My apologies for using μ as a mass here, since it usually denotes a coefficient of friction. Alas, there are only so many symbols for "m."

up on both sides). So we have

$$2T - f - \mu g = \mu a. \tag{3.11}$$

Note that if we add up the three previous equations, we obtain the $F = ma$ equation in Eq. (3.8), as should be the case, because the whole system is the sum of the three above subsystems. Equations (3.9)–(3.11) are three equations in the three unknowns, T, N, and f. Their sum yields the T in (3.8), and then Eqs. (3.9) and (3.11) give, respectively, as you can show,

$$N = (M + 2m + \mu)(g + a), \quad \text{and} \quad f = (2M + 2m + \mu)(g + a). \tag{3.12}$$

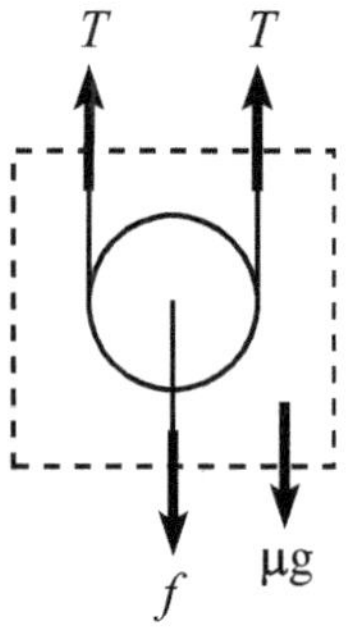

Fig. 3.4

Remarks: You can also obtain these results by considering subsystems different from the ones we chose above. For example, you can choose the pulley-plus-platform subsystem, etc. But no matter how you choose to break up the system, you will need to produce three independent $F = ma$ equations in order to solve for the three unknowns, T, N, and f.

In problems like this one, it's easy to forget to include certain forces, such as the second T in Eq. (3.11). The safest thing to do is to always isolate each subsystem, draw a box around it, and then draw all the forces that "protrude" from the box. In other words, draw the free-body diagram. Figure 3.4 shows the free-body diagram for the subsystem consisting of only the pulley. ♣

Another class of problems, similar to the above example, goes by the name of *Atwood's machines*. An Atwood's machine is the name used for any system consisting of a combination of masses, strings, and pulleys.[8] In general, the pulleys and strings can have mass, but we'll deal only with massless ones in this chapter. As we'll see in the following example, there are two basic steps in solving an Atwood's problem: (1) write down all the $F = ma$ equations, and (2) relate the accelerations of the various masses by noting that the length of the string(s) doesn't change, a fact that we call "conservation of string."

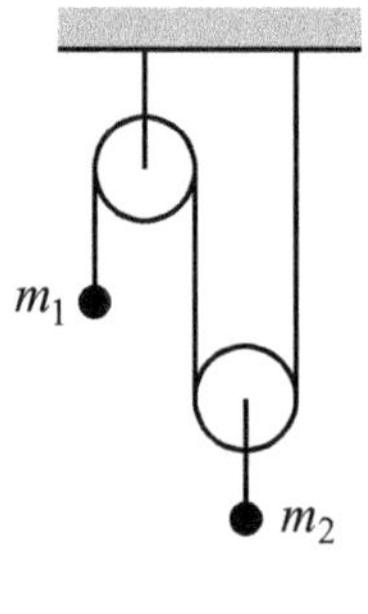

Fig. 3.5

Example (Atwood's machine): Consider the pulley system in Fig. 3.5, with masses m_1 and m_2. The strings and pulleys are massless. What are the accelerations of the masses? What is the tension in the string?

Solution: The first thing to note is that the tension T is the same everywhere throughout the massless string, because otherwise there would be an infinite acceleration of some part of the string. It then follows that the tension in the short string connected to m_2 is $2T$. This is true because there must be zero net force on the massless right pulley, because otherwise it would have infinite acceleration. The $F = ma$

[8] George Atwood (1746–1807) was a tutor at Cambridge University. He published the description of the first of his machines in Atwood (1784). For a history of Atwood's machines, see Greenslade (1985).

equations for the two masses are therefore (with upward taken to be positive)

$$\begin{aligned} T - m_1 g &= m_1 a_1, \\ 2T - m_2 g &= m_2 a_2. \end{aligned} \tag{3.13}$$

We now have two equations in the three unknowns, a_1, a_2, and T. So we need one more equation. This is the "conservation of string" fact, which relates a_1 and a_2. If we imagine moving m_2 and the right pulley up a distance d, then a length $2d$ of string has disappeared from the two parts of the string touching the right pulley. This string has to go somewhere, so it ends up in the part of the string touching m_1 (see Fig. 3.6). Therefore, m_1 goes down by a distance $2d$. In other words, $y_1 = -2y_2$, where y_1 and y_2 are measured relative to the initial locations of the masses. Taking two time derivatives of this statement gives our desired relation between a_1 and a_2,

$$a_1 = -2a_2. \tag{3.14}$$

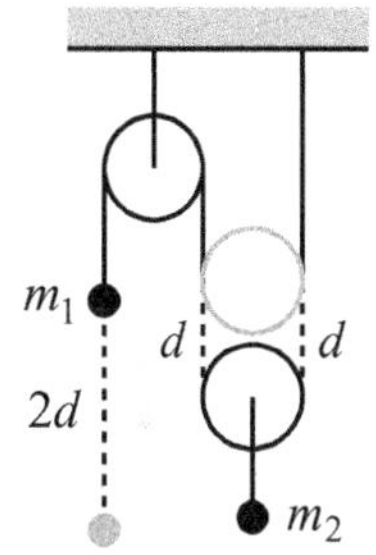

Fig. 3.6

Combining this with Eq. (3.13), we can now solve for a_1, a_2, and T. The result is

$$a_1 = g\frac{2m_2 - 4m_1}{4m_1 + m_2}, \quad a_2 = g\frac{2m_1 - m_2}{4m_1 + m_2}, \quad T = \frac{3m_1 m_2 g}{4m_1 + m_2}. \tag{3.15}$$

REMARKS: There are all sorts of limits and special cases that we can check here. A couple are: (1) If $m_2 = 2m_1$, then Eq. (3.15) gives $a_1 = a_2 = 0$, and $T = m_1 g$. Everything is at rest. (2) If $m_2 \gg m_1$, then Eq. (3.15) gives $a_1 = 2g$, $a_2 = -g$, and $T = 3m_1 g$. In this case, m_2 is essentially in free fall, while m_1 gets yanked up with acceleration $2g$. The value of T is exactly what is needed to make the net force on m_1 equal to $m_1(2g)$, because $T - m_1 g = 3m_1 g - m_1 g = m_1(2g)$. You can check the case where $m_1 \gg m_2$.

For the more general case where there are N masses instead of two, the "conservation of string" statement is a single equation that relates all N accelerations. It is most easily obtained by imagining moving $N - 1$ of the masses, each by an arbitrary amount, and then seeing what happens to the last mass. Note that these arbitrary motions undoubtedly do *not* correspond to the actual motions of the masses. This is fine; the single "conservation of string" equation has nothing to do with the N $F = ma$ equations. The combination of all $N + 1$ equations is needed to constrain the motions down to a unique set. ♣

In the problems and exercises for this chapter, you will encounter some strange Atwood's setups. But no matter how complicated they get, there are only two things you need to do to solve them, as mentioned above: write down the $F = ma$ equations for all the masses (which may involve relating the tensions in various strings), and then relate the accelerations of the masses, using "conservation of string."

It may seem, with the angst it can bring,
That an Atwood's machine's a cruel thing.
But you just need to say
That F is *ma*,
And use conservation of string!

3.3 Solving differential equations

Let's now consider the type of problem where we are *given* the force as a function of time, position, or velocity, and our task is to solve the $F = ma \equiv m\ddot{x}$ differential equation to find the position, $x(t)$, as a function of time.[9] In what follows, we will develop a few techniques for solving differential equations. The ability to apply these techniques dramatically increases the number of systems we can understand.

It's also possible for the force F to be a function of higher derivatives of x, in addition to the quantities t, x, and $v \equiv \dot{x}$. But these cases don't arise much, so we won't worry about them. The $F = ma$ differential equation we want to solve is therefore (we'll just work in one dimension here)

$$m\ddot{x} = F(t, x, v). \tag{3.16}$$

In general, this equation cannot be solved exactly for $x(t)$.[10] But for most of the problems we'll deal with, it can be solved. The problems we'll encounter will often fall into one of three special cases, namely, where F is a function of t only, or x only, or v only. In all of these cases, we must invoke the given initial conditions, $x_0 \equiv x(t_0)$ and $v_0 \equiv v(t_0)$, to obtain our final solutions. These initial conditions will appear in the limits of the integrals in the following discussion.[11]

Note: You may want to just skim the following page and a half, and then refer back as needed. Don't try to memorize all the different steps. We present them only for completeness. The whole point here can basically be summarized by saying that sometimes you want to write $\ddot{x}$ as dv/dt, and sometimes you want to write it as $v\,dv/dx$ (see Eq. (3.20)). Then you "simply" have to separate variables and integrate. We'll go through the three special cases, and then we'll do some examples.

- *F is a function of t only: $F = F(t)$.*
  Since $a = d^2x/dt^2$, we just need to integrate $F = ma$ twice to obtain $x(t)$. Let's do this in a very systematic way, to get used to the general procedure. First, write $F = ma$ as

$$m\frac{dv}{dt} = F(t). \tag{3.17}$$

[9] In some setups, such as in Problem 3.11, the force isn't given, so you have to figure out what it is. But the main part of the problem is still solving the resulting differential equation.

[10] You can always solve for $x(t)$ *numerically*, to any desired accuracy. This topic is discussed in Section 1.4.

[11] It is no coincidence that we need *two* initial conditions to completely specify the solution to our *second*-order (meaning the highest derivative of x that appears is the second one) $F = m\ddot{x}$ differential equation. It is a general result (which we'll just accept here) that the solution to an nth-order differential equation has n free parameters, which are determined by the initial conditions.

Then separate variables and integrate both sides to obtain[12]

$$m\int_{v_0}^{v(t)} dv' = \int_{t_0}^{t} F(t')\,dt'. \qquad (3.18)$$

We have put primes on the integration variables so that we don't confuse them with the limits of integration, but in practice we usually don't bother with them. The integral of dv' is just v', so Eq. (3.18) yields v as a function of t, that is, $v(t)$. We can then separate variables in $dx/dt \equiv v(t)$ and integrate to obtain

$$\int_{x_0}^{x(t)} dx' = \int_{t_0}^{t} v(t')\,dt'. \qquad (3.19)$$

This yields x as a function of t, that is, $x(t)$. This procedure might seem like a cumbersome way to simply integrate something twice. That's because it is. But the technique proves more useful in the following case.

- *F is a function of x only: $F = F(x)$.*
  We will use

$$a = \frac{dv}{dt} = \frac{dx}{dt}\frac{dv}{dx} = v\frac{dv}{dx} \qquad (3.20)$$

to write $F = ma$ as

$$mv\frac{dv}{dx} = F(x). \qquad (3.21)$$

Now separate variables and integrate both sides to obtain

$$m\int_{v_0}^{v(x)} v'\,dv' = \int_{x_0}^{x} F(x')\,dx'. \qquad (3.22)$$

The integral of v' is $v'^2/2$, so the left-hand side involves the square of $v(x)$. Taking the square root, this gives v as a function of x, that is, $v(x)$. Separating variables in $dx/dt \equiv v(x)$ then yields

$$\int_{x_0}^{x(t)} \frac{dx'}{v(x')} = \int_{t_0}^{t} dt'. \qquad (3.23)$$

Assuming that we can do the integral on the left-hand side, this equation gives t as a function of x. We can then (in principle) invert the result to obtain x as a function of t, that is, $x(t)$. The unfortunate thing about this case is that the integral in Eq. (3.23) might not be doable. And even if it is, it might not be possible to invert $t(x)$ to produce $x(t)$.

[12] If you haven't seen such a thing before, the act of multiplying both sides by the infinitesimal quantity dt might make you feel a bit uneasy. But it is in fact quite legal. If you wish, you can imagine working with the small (but not infinitesimal) quantities Δv and Δt, for which it is certainly legal to multiply both sides by Δt. Then you can take a discrete sum over many Δt intervals, and then finally take the limit $\Delta t \to 0$, which results in the integral in Eq. (3.18).

- *F is a function of v only: $F = F(v)$.*
  Write $F = ma$ as

$$m\frac{dv}{dt} = F(v). \tag{3.24}$$

Separate variables and integrate both sides to obtain

$$m\int_{v_0}^{v(t)} \frac{dv'}{F(v')} = \int_{t_0}^{t} dt'. \tag{3.25}$$

Assuming that we can do this integral, it yields t as a function of v, and hence (in principle) v as a function of t, that is, $v(t)$. We can then integrate $dx/dt \equiv v(t)$ to obtain $x(t)$ from

$$\int_{x_0}^{x(t)} dx' = \int_{t_0}^{t} v(t')\, dt'. \tag{3.26}$$

Note: In this $F = F(v)$ case, if we want to find v as a function of x, $v(x)$, then we should write a as $v(dv/dx)$ and integrate

$$m\int_{v_0}^{v(x)} \frac{v'\,dv'}{F(v')} = \int_{x_0}^{x} dx'. \tag{3.27}$$

We can then obtain $x(t)$ from Eq. (3.23), if desired.

When dealing with the initial conditions, we have chosen to put them in the limits of integration above. If you wish, you can perform the integrals without any limits, and just tack on a constant of integration to your result. The constant is then determined by the initial conditions.

Again, as mentioned above, you do *not* have to memorize the above three procedures, because there are variations, depending on what you're given and what you want to solve for. All you have to remember is that $\ddot{x}$ can be written as either dv/dt or $v\,dv/dx$. One of these will get the job done (namely, the one that makes only two of the three variables, t, x, and v, appear in your differential equation). And then be prepared to separate variables and integrate as many times as needed.[13]

> a is dv by dt.
> Is this useful? There's no guarantee.
> If it leads to "Oh, heck!" s,
> Take dv by dx,
> And then write down its product with v.

[13] We want only two of the variables to appear in the differential equation because the goal is to separate variables and integrate, and because equations have only two sides. If equations were triangles, it would be a different story.

Example (Gravitational force): A particle of mass m is subject to a constant force $F = -mg$. The particle starts at rest at height h. Because this constant force falls into all of the above three categories, we should be able to solve for $y(t)$ in two ways:

(a) Find $y(t)$ by writing a as dv/dt.
(b) Find $y(t)$ by writing a as $v\,dv/dy$.

Solution:

(a) $F = ma$ gives $dv/dt = -g$. Multiplying by dt and integrating yields $v = -gt + A$, where A is a constant of integration.[14] The initial condition $v(0) = 0$ gives $A = 0$. Therefore, $dy/dt = -gt$. Multiplying by dt and integrating yields $y = -gt^2/2 + B$. The initial condition $y(0) = h$ gives $B = h$. Therefore,

$$y = h - \frac{1}{2}gt^2. \tag{3.28}$$

(b) $F = ma$ gives $v\,dv/dy = -g$. Separating variables and integrating yields $v^2/2 = -gy + C$. The initial condition $v(h) = 0$ gives $v^2/2 = -gy + gh$. Therefore, $v \equiv dy/dt = -\sqrt{2g(h-y)}$. We have chosen the negative square root because the particle is falling. Separating variables gives

$$\int \frac{dy}{\sqrt{h-y}} = -\sqrt{2g}\int dt. \tag{3.29}$$

This yields $2\sqrt{h-y} = \sqrt{2g}\,t$, where we have used the initial condition $y(0) = h$. Hence, $y = h - gt^2/2$, as in part (a). In part (b) here, we essentially derived conservation of energy, as we'll see in Chapter 5.

Example (Dropped ball): A beach ball is dropped from rest at height h. Assume that the drag force[15] from the air takes the form of $F_{\rm d} = -\beta v$. Find the velocity and height as a function of time.

Solution: For simplicity in future formulas, let's write the drag force as $F_{\rm d} = -\beta v \equiv -m\alpha v$ (otherwise we'd have a bunch of $1/m$'s floating around). Taking upward to be the positive y direction, the force on the ball is

$$F = -mg - m\alpha v. \tag{3.30}$$

[14] We'll do this example by adding on constants of integration which are then determined by the initial conditions. We'll do the following example by putting the initial conditions in the limits of integration.

[15] The drag force is roughly proportional to v as long as the speed is fairly small (say, less than 10 m/s). For large speeds (say, greater than 100 m/s), the drag force is roughly proportional to v^2. But these approximate cutoffs depend on various things, and in any event there is a messy transition region between the two cases.

Note that v is negative here, because the ball is falling, so the drag force points upward, as it should. Writing $F = m\,dv/dt$ and separating variables gives

$$\int_0^{v(t)} \frac{dv'}{g+\alpha v'} = -\int_0^t dt'. \tag{3.31}$$

Integration yields $\ln(1+\alpha v/g) = -\alpha t$. Exponentiation then gives

$$v(t) = -\frac{g}{\alpha}\left(1 - e^{-\alpha t}\right). \tag{3.32}$$

Writing $dy/dt \equiv v(t)$, and then separating variables and integrating to obtain $y(t)$, yields

$$\int_h^{y(t)} dy' = -\frac{g}{\alpha}\int_0^t \left(1 - e^{-\alpha t'}\right) dt'. \tag{3.33}$$

Therefore,

$$y(t) = h - \frac{g}{\alpha}\left(t - \frac{1}{\alpha}\left(1 - e^{-\alpha t}\right)\right). \tag{3.34}$$

REMARKS:

1. Let's look at some limiting cases. If t is very small (more precisely, if $\alpha t \ll 1$), then we can use $e^{-x} \approx 1 - x + x^2/2$ to make approximations to leading order in t. You can show that Eq. (3.32) gives $v(t) \approx -gt$. This makes sense, because the drag force is negligible at the start, so the ball is essentially in freefall. And likewise you can show that Eq. (3.34) gives $y(t) \approx h - gt^2/2$, which is again the freefall result.

   We can also look at large t. In this case, $e^{-\alpha t}$ is essentially equal to zero, so Eq. (3.32) gives $v(t) \approx -g/\alpha$. (This is the "terminal velocity." Its value makes sense, because it is the velocity for which the total force, $-mg - m\alpha v$, vanishes.) And Eq. (3.34) gives $y(t) \approx h - (g/\alpha)t + g/\alpha^2$. Interestingly, we see that for large t, g/α^2 is the distance our ball lags behind another ball that started out already at the terminal velocity, $-g/\alpha$.
2. You might think that the velocity in Eq. (3.32) doesn't depend on m, since no m's appear. However, there is an m hidden in α. The quantity α (which we introduced just to make our formulas look a little nicer) was defined by $F_\text{d} = -\beta v \equiv -m\alpha v$. But the quantity $\beta \equiv m\alpha$ is roughly proportional to the cross-sectional area, A, of the ball. Therefore, $\alpha \propto A/m$. Two balls of the same size, one made of lead and one made of styrofoam, have the same A but different m's. So their α's are different, and they fall at different rates.

   If we have a solid ball with density ρ and radius r, then $\alpha \propto A/m \propto r^2/(\rho r^3) = 1/\rho r$. For large dense objects in a thin medium such as air, the quantity α is small, so the drag effects are not very noticeable over short times (because if we include the next term in the expansion for v, we obtain $v(t) \approx -gt + \alpha g t^2/2$). Large dense objects therefore all fall at roughly the same rate, with an acceleration essentially equal to g. But if the air were much thicker, then all the α's would be larger, and maybe it would have taken Galileo a bit longer to come to his conclusions.

   What would you have thought, Galileo,
   If instead you dropped cows and did say, "Oh!

To lessen the sound
Of the moos from the ground,
They should fall not through air, but through mayo!"[16] ♣

3.4 Projectile motion

Consider a ball thrown through the air, not necessarily vertically. We will neglect air resistance in the following discussion. Things get a bit more complicated when this is included, as Exercise 3.53 demonstrates.

Let x and y be the horizontal and vertical positions, respectively. The force in the x direction is $F_x = 0$, and the force in the y direction is $F_y = -mg$. So $\mathbf{F} = m\mathbf{a}$ gives

$$\ddot{x} = 0, \quad \text{and} \quad \ddot{y} = -g. \tag{3.35}$$

Note that these two equations are "decoupled." That is, there is no mention of y in the equation for $\ddot{x}$, and vice versa. The motions in the x and y directions are therefore completely independent. The classic demonstration of the independence of the x and y motions is the following. Fire a bullet horizontally (or, preferably, just imagine firing a bullet horizontally), and at the same time drop a bullet from the height of the gun. Which bullet will hit the ground first? (Neglect air resistance, the curvature of the earth, etc.) The answer is that they will hit the ground at the same time, because the effect of gravity on the two y motions is exactly the same, independent of what is going on in the x direction.

If the initial position and velocity are (X, Y) and (V_x, V_y), then we can easily integrate Eq. (3.35) to obtain

$$\dot{x}(t) = V_x, \quad \text{and} \quad \dot{y}(t) = V_y - gt. \tag{3.36}$$

Integrating again gives

$$x(t) = X + V_x t, \quad \text{and} \quad y(t) = Y + V_y t - \frac{1}{2}gt^2. \tag{3.37}$$

These equations for the speeds and positions are all you need to solve a projectile problem.

[16] It's actually much more likely that Galileo obtained his "all objects fall at the same rate in a vacuum" result by rolling balls down planes than by dropping balls from the Tower of Pisa; see Adler and Coulter (1978). So I suppose this limerick is relevant only in the approximation of the proverbial spherical cow.

Example (Throwing a ball):

(a) For a given initial speed, at what inclination angle should a ball be thrown so that it travels the maximum horizontal distance by the time it returns to the ground? Assume that the ground is horizontal, and that the ball is released from ground level.

(b) What is the optimal angle if the ground is sloped upward at an angle β (or downward, if β is negative)?

Solution:

(a) Let the inclination angle be θ, and let the initial speed be V. Then the horizontal speed is always $V_x = V\cos\theta$, and the initial vertical speed is $V_y = V\sin\theta$. The first thing we need to do is find the time t in the air. We know that the vertical speed is zero at time $t/2$, because the ball is moving horizontally at the highest point. So the second of Eqs. (3.36) gives $V_y = g(t/2)$. Therefore, $t = 2V_y/g$.[17] The first of Eqs. (3.37) tells us that the horizontal distance traveled is $d = V_x t$. Using $t = 2V_y/g$ in this gives

$$d = \frac{2V_x V_y}{g} = \frac{V^2(2\sin\theta\cos\theta)}{g} = \frac{V^2\sin 2\theta}{g}. \tag{3.38}$$

The $\sin 2\theta$ factor has a maximum at $\theta = \pi/4$. The maximum horizontal distance traveled is then $d_{\text{max}} = V^2/g$.

REMARK: For $\theta = \pi/4$, you can show that the maximum height achieved is $V^2/4g$. This is half the maximum height of $V^2/2g$ (as you can show) if the ball is thrown straight up. Note that any possible distance you might want to find in this problem must be proportional to V^2/g, by dimensional analysis. The only question is what the numerical factor is. ♣

(b) As in part (a), the first thing we need to do is find the time t in the air. If the ground is sloped at an angle β, then the equation for the line of the ground is $y = (\tan\beta)x$. The path of the ball is given in terms of t by

$$x = (V\cos\theta)t, \quad \text{and} \quad y = (V\sin\theta)t - \frac{1}{2}gt^2, \tag{3.39}$$

where θ is the angle of the throw, as measured with respect to the horizontal (not the ground). We must solve for the t that makes $y = (\tan\beta)x$, because this gives the time when the path of the ball intersects the line of the ground. Using Eq. (3.39), we find that $y = (\tan\beta)x$ when

$$t = \frac{2V}{g}(\sin\theta - \tan\beta\cos\theta). \tag{3.40}$$

[17] Alternatively, the time of flight can be found from the second of Eqs. (3.37), which says that the ball returns to the ground when $V_y t = gt^2/2$. We will have to use this second strategy in part (b), where the trajectory is not symmetric around the maximum.

(There is, of course, also the solution $t = 0$.) Plugging this into the expression for x in Eq. (3.39) gives

$$x = \frac{2V^2}{g}(\sin\theta\cos\theta - \tan\beta\cos^2\theta). \tag{3.41}$$

We must now maximize this value for x, which is equivalent to maximizing the distance along the slope. Setting the derivative with respect to θ equal to zero, and using the double-angle formulas, $\sin 2\theta = 2\sin\theta\cos\theta$ and $\cos 2\theta = \cos^2\theta - \sin^2\theta$, we find $\tan\beta = -\cot 2\theta$. This can be rewritten as $\tan\beta = -\tan(\pi/2 - 2\theta)$. Therefore, $\beta = -(\pi/2 - 2\theta)$, so we have

$$\theta = \frac{1}{2}\left(\beta + \frac{\pi}{2}\right). \tag{3.42}$$

In other words, the throwing angle should bisect the angle between the ground and the vertical.

REMARKS:

1. For $\beta \approx \pi/2$, we have $\theta \approx \pi/2$, as should be the case. For $\beta = 0$, we have $\theta = \pi/4$, as we found in part (a). And for $\beta \approx -\pi/2$, we have $\theta \approx 0$, which makes sense.
2. A quicker method of obtaining the time in Eq. (3.40) is the following. Consider the set of tilted axes parallel and perpendicular to the ground; let these be the x' and y' axes, respectively. The initial velocity in the y' direction is $V\sin(\theta-\beta)$, and the acceleration in this direction is $g\cos\beta$. The time in the air is twice the time it takes the ball to reach the maximum "height" above the ground (measured in the y' direction), which occurs when the velocity in the y' direction is instantaneously zero. The total time is therefore $2V\sin(\theta-\beta)/(g\cos\beta)$, which you can show is equivalent to the time in Eq. (3.40). Note that the $g\sin\beta$ acceleration in the x' direction is irrelevant in calculating this time. In the present example, using these tilted axes doesn't save a huge amount of time, but in some situations (see Exercise 3.50) the tilted axes can save you a lot of grief.
3. An interesting fact about the motion of the ball in the maximum-distance case is that the initial and final velocities are perpendicular to each other. The demonstration of this is the task of Problem 3.16.
4. Substituting the value of θ from Eq. (3.42) into Eq. (3.41), you can show (after a bit of algebra) that the maximum distance traveled along the tilted ground is

$$d = \frac{x}{\cos\beta} = \frac{V^2/g}{1+\sin\beta}. \tag{3.43}$$

Solving for V, we have $V^2 = g(d + d\sin\beta)$. This can be interpreted as saying that the minimum speed at which a ball must be thrown in order to pass over a wall of height h, at a distance L away on level ground, is given by $V^2 = g(\sqrt{L^2+h^2} + h)$. This checks in the limits of $h \to 0$ and $L \to 0$.
5. A compilation of many other projectile results can be found in Buckmaster (1985). ♣

Along with the bullet example mentioned above, another classic example of the independence of the x and y motions is the "hunter and monkey" problem. In it, a hunter aims an arrow (a toy one, of course) at a monkey hanging from a branch in a tree. The monkey, thinking he's being clever, tries to avoid the arrow by letting go of the branch right when he sees the arrow released. The unfortunate consequence of this action is that he in fact *will* get hit, because gravity acts on both him and the arrow in the same way; they both fall the same distance relative to where they would have been if there were no gravity. And the monkey *would* get hit in such a case, because the arrow is initially aimed at him. You can work this out in Exercise 3.44, in a more peaceful setting involving fruit.

If a monkey lets go of a tree,
The arrow will hit him, you see,
Because both heights are pared
By a half gt^2
From what they would be with no g.

3.5 Motion in a plane, polar coordinates

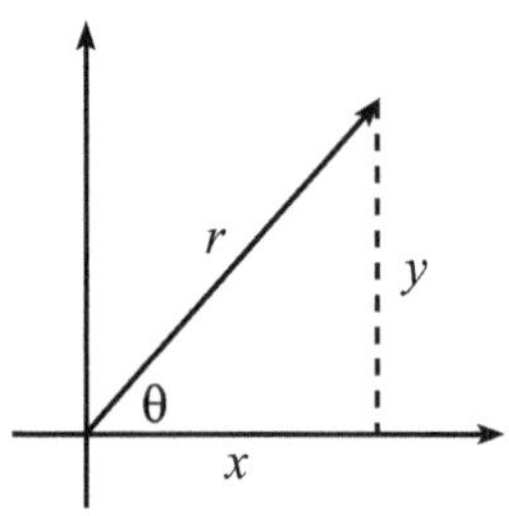

Fig. 3.7

When dealing with problems where the motion lies in a plane, it is often convenient to work with polar coordinates, r and θ. These are related to the Cartesian coordinates by (see Fig. 3.7)

$$x = r\cos\theta, \quad \text{and} \quad y = r\sin\theta. \tag{3.44}$$

Depending on the problem, either Cartesian or polar coordinates are easier to use. It is usually clear from the setup which is better. For example, if the problem involves circular motion, then polar coordinates are a good bet. But to use polar coordinates, we need to know what Newton's second law looks like when written in terms of them. Therefore, the goal of the present section is to determine what $\mathbf{F} = m\mathbf{a} \equiv m\ddot{\mathbf{r}}$ looks like when written in terms of polar coordinates.

At a given position $\mathbf{r}$ in the plane, the basis vectors in polar coordinates are $\hat{\mathbf{r}}$, which is a unit vector pointing in the radial direction; and $\hat{\boldsymbol{\theta}}$, which is a unit vector pointing in the counterclockwise tangential direction. In polar coordinates, a general vector may be written as

$$\mathbf{r} = r\hat{\mathbf{r}}. \tag{3.45}$$

Since the goal of this section is to find $\ddot{\mathbf{r}}$, we must, in view of Eq. (3.45), get a handle on the time derivative of $\hat{\mathbf{r}}$. And we'll eventually need the derivative of $\hat{\boldsymbol{\theta}}$, too. In contrast with the fixed Cartesian basis vectors ($\hat{\mathbf{x}}$ and $\hat{\mathbf{y}}$), the polar basis vectors ($\hat{\mathbf{r}}$ and $\hat{\boldsymbol{\theta}}$) change as a point moves around in the plane. We can find $\dot{\hat{\mathbf{r}}}$ and $\dot{\hat{\boldsymbol{\theta}}}$ in the following way. In terms of the Cartesian basis, Fig. 3.8 shows that

$$\begin{aligned}\hat{\mathbf{r}} &= \cos\theta\,\hat{\mathbf{x}} + \sin\theta\,\hat{\mathbf{y}},\\ \hat{\boldsymbol{\theta}} &= -\sin\theta\,\hat{\mathbf{x}} + \cos\theta\,\hat{\mathbf{y}}.\end{aligned} \tag{3.46}$$

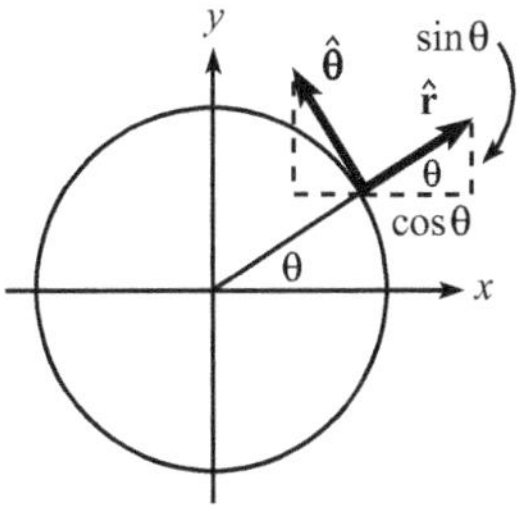

Fig. 3.8

Taking the time derivative of these equations gives

$$\begin{aligned}\dot{\hat{\mathbf{r}}} &= -\sin\theta\,\dot{\theta}\hat{\mathbf{x}} + \cos\theta\,\dot{\theta}\hat{\mathbf{y}},\\ \dot{\hat{\boldsymbol{\theta}}} &= -\cos\theta\,\dot{\theta}\hat{\mathbf{x}} - \sin\theta\,\dot{\theta}\hat{\mathbf{y}}.\end{aligned} \tag{3.47}$$

Using Eq. (3.46), we arrive at the nice clean expressions,

$$\dot{\hat{\mathbf{r}}} = \dot{\theta}\hat{\boldsymbol{\theta}}, \quad \text{and} \quad \dot{\hat{\boldsymbol{\theta}}} = -\dot{\theta}\hat{\mathbf{r}}. \tag{3.48}$$

These relations are fairly evident if you look at what happens to the basis vectors as **r** moves a tiny distance in the tangential direction. Note that the basis vectors do not change as **r** moves in the radial direction. We can now start differentiating Eq. (3.45). One derivative gives (yes, the product rule works fine here)

$$\dot{\mathbf{r}} = \dot{r}\hat{\mathbf{r}} + r\dot{\hat{\mathbf{r}}} = \dot{r}\hat{\mathbf{r}} + r\dot{\theta}\hat{\boldsymbol{\theta}}. \tag{3.49}$$

This makes sense, because $\dot{r}$ is the velocity in the radial direction, and $r\dot{\theta}$ is the velocity in the tangential direction, often written as $r\omega$ (where $\omega \equiv \dot{\theta}$ is the angular velocity, or "angular frequency").[18] Differentiating Eq. (3.49) then gives

$$\begin{aligned}\ddot{\mathbf{r}} &= \ddot{r}\hat{\mathbf{r}} + \dot{r}\dot{\hat{\mathbf{r}}} + \dot{r}\dot{\theta}\hat{\boldsymbol{\theta}} + r\ddot{\theta}\hat{\boldsymbol{\theta}} + r\dot{\theta}\dot{\hat{\boldsymbol{\theta}}}\\ &= \ddot{r}\hat{\mathbf{r}} + \dot{r}(\dot{\theta}\hat{\boldsymbol{\theta}}) + \dot{r}\dot{\theta}\hat{\boldsymbol{\theta}} + r\ddot{\theta}\hat{\boldsymbol{\theta}} + r\dot{\theta}(-\dot{\theta}\hat{\mathbf{r}})\\ &= (\ddot{r} - r\dot{\theta}^2)\hat{\mathbf{r}} + (r\ddot{\theta} + 2\dot{r}\dot{\theta})\hat{\boldsymbol{\theta}}.\end{aligned} \tag{3.50}$$

Finally, equating $m\ddot{\mathbf{r}}$ with $\mathbf{F} \equiv F_r\hat{\mathbf{r}} + F_\theta\hat{\boldsymbol{\theta}}$ gives the radial and tangential forces as

$$\begin{aligned}F_r &= m(\ddot{r} - r\dot{\theta}^2),\\ F_\theta &= m(r\ddot{\theta} + 2\dot{r}\dot{\theta}).\end{aligned} \tag{3.51}$$

(See Exercise 3.67 for a slightly different derivation of these equations.) Let's look at each of the four terms on the right-hand sides of Eqs. (3.51).

[18] For $r\dot{\theta}$ to be the tangential velocity, we must measure θ in radians and not degrees. Then $r\theta$ is by definition the position along the circumference, so $r\dot{\theta}$ is the velocity along the circumference.

- The $m\ddot{r}$ term is quite intuitive. For radial motion, it simply states that $F = ma$ along the radial direction.
- The $mr\ddot{\theta}$ term is also quite intuitive. For circular motion, it states that $F = ma$ along the tangential direction, because $r\ddot{\theta}$ is the second derivative of the distance $r\theta$ along the circumference.
- The $-mr\dot{\theta}^2$ term is also fairly clear. For circular motion, it says that the radial force is $-m(r\dot{\theta})^2/r = -mv^2/r$, which is the familiar force that causes the centripetal acceleration, v^2/r. See Problem 3.20 for an alternate (and quicker) derivation of this v^2/r result.
- The $2m\dot{r}\dot{\theta}$ term isn't so obvious. It is associated with the *Coriolis* force. There are various ways to look at this term. One is that it exists in order to keep angular momentum conserved. We'll have a great deal to say about the Coriolis force in Chapter 10.

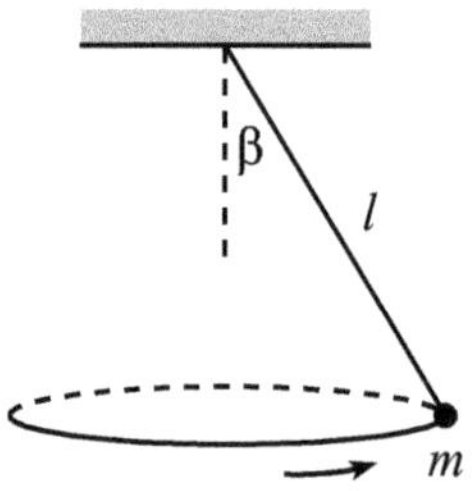

Fig. 3.9

Example (Circular pendulum): A mass hangs from a massless string of length ℓ. Conditions have been set up so that the mass swings around in a horizontal circle, with the string making a constant angle β with the vertical (see Fig. 3.9). What is the angular frequency, ω, of this motion?

Solution: The mass travels in a circle, so the horizontal radial force must be $F_r = mr\dot{\theta}^2 \equiv mr\omega^2$ (with $r = \ell \sin\beta$), directed radially inward. The forces on the mass are the tension in the string, T, and gravity, mg (see Fig. 3.10). There is no acceleration in the vertical direction, so $F = ma$ in the vertical and radial directions gives, respectively,

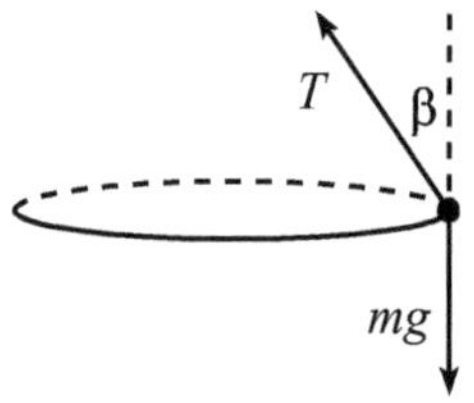

Fig. 3.10

$$\begin{aligned} T\cos\beta - mg &= 0, \\ T\sin\beta &= m(\ell\sin\beta)\omega^2. \end{aligned} \tag{3.52}$$

Solving for ω gives

$$\omega = \sqrt{\frac{g}{\ell\cos\beta}}. \tag{3.53}$$

If $\beta \approx 90^\circ$, then $\omega \to \infty$, which makes sense. And if $\beta \approx 0$, then $\omega \approx \sqrt{g/\ell}$, which happens to equal the frequency of a plane pendulum of length ℓ. The task of Exercise 3.60 is to explain why.

3.6 Problems

Section 3.2: Free-body diagrams

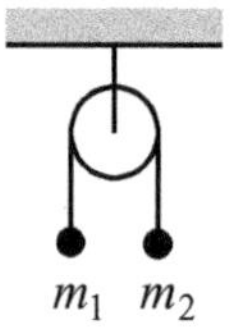

Fig. 3.11

3.1. **Atwood's machine** *

A massless pulley hangs from a fixed support. A massless string connecting two masses, m_1 and m_2, hangs over the pulley (see Fig. 3.11). Find the acceleration of the masses and the tension in the string.

3.2. Double Atwood's machine **

A double Atwood's machine is shown in Fig. 3.12, with masses m_1, m_2, and m_3. Find the accelerations of the masses.

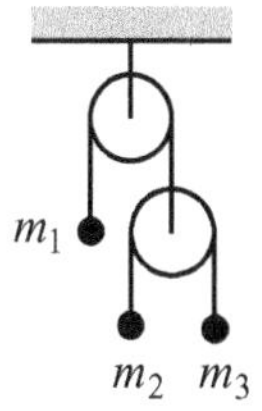

Fig. 3.12

3.3. Infinite Atwood's machine ***

Consider the infinite Atwood's machine shown in Fig. 3.13. A string passes over each pulley, with one end attached to a mass and the other end attached to another pulley. All the masses are equal to m, and all the pulleys and strings are massless. The masses are held fixed and then simultaneously released. What is the acceleration of the top mass? (You may define this infinite system as follows. Consider it to be made of N pulleys, with a nonzero mass replacing what would have been the $(N+1)$th pulley. Then take the limit as $N \to \infty$.)

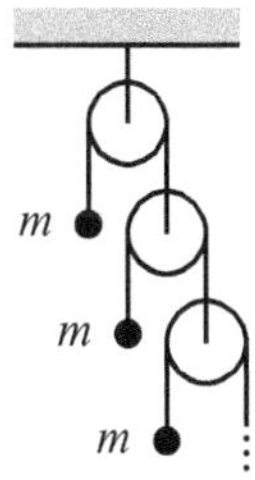

Fig. 3.13

3.4. Line of pulleys *

$N+2$ equal masses hang from a system of pulleys, as shown in Fig. 3.14. What are the accelerations of all the masses?

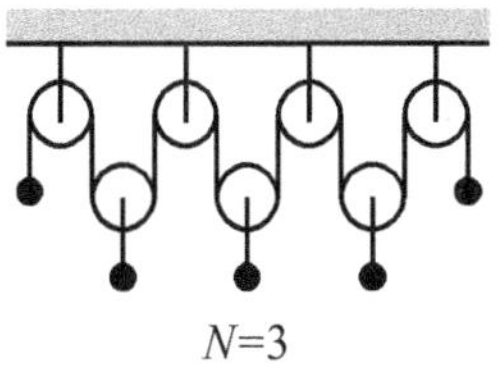

Fig. 3.14

3.5. Ring of pulleys **

Consider the system of pulleys shown in Fig. 3.15. The string (which is a loop with no ends) hangs over N fixed pulleys that circle around the underside of a ring. N masses, $m_1, m_2, \ldots, m_N$, are attached to N pulleys that hang on the string. What are the accelerations of all the masses?

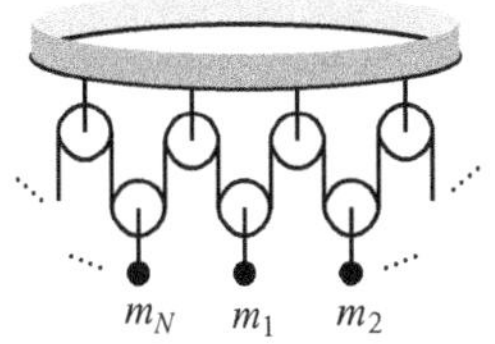

Fig. 3.15

3.6. Sliding down a plane **

(a) A block starts at rest and slides down a frictionless plane inclined at an angle θ. What should θ be so that the block travels a given horizontal distance in the minimum amount of time?

(b) Same question, but now let there be a coefficient of kinetic friction μ between the block and the plane.

3.7. Sliding sideways on a plane ***

A block is placed on a plane inclined at an angle θ. The coefficient of friction between the block and the plane is $\mu = \tan\theta$. The block is given a kick so that it initially moves with speed V horizontally along the plane (that is, in the direction perpendicular to the direction pointing straight down the plane). What is the speed of the block after a very long time?

3.8. Moving plane ***

A block of mass m is held motionless on a frictionless plane of mass M and angle of inclination θ (see Fig. 3.16). The plane rests on a frictionless horizontal surface. The block is released. What is the horizontal acceleration of the plane?

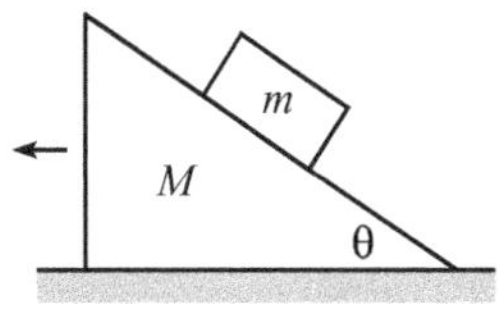

Fig. 3.16

Section 3.3: Solving differential equations

3.9. **Exponential force** $*$

A particle of mass m is subject to a force $F(t) = ma_0 e^{-bt}$. The initial position and speed are zero. Find $x(t)$.

3.10. **$-kx$ force** $**$

A particle of mass m is subject to a force $F(x) = -kx$, with $k > 0$. The initial position is x_0, and the initial speed is zero. Find $x(t)$.

3.11. **Falling chain** $**$

A chain with length ℓ is held stretched out on a frictionless horizontal table, with a length y_0 hanging down through a hole in the table. The chain is released. As a function of time, find the length that hangs down through the hole (don't bother with t after the chain loses contact with the table). Also, find the speed of the chain right when it loses contact with the table.[19]

3.12. **Throwing a beach ball** $***$

A beach ball is thrown upward with initial speed v_0. Assume that the drag force from the air is $F_{\text{d}} = -m\alpha v$. What is the speed of the ball, v_{f}, right before it hits the ground? (An implicit equation is sufficient.) Does the ball spend more time or less time in the air than it would if it were thrown in vacuum?

3.13. **Balancing a pencil** $***$

Consider a pencil that stands upright on its tip and then falls over. Let's idealize the pencil as a mass m sitting at the end of a massless rod of length ℓ.[20]

(a) Assume that the pencil makes an initial (small) angle θ_0 with the vertical, and that its initial angular speed is ω_0. The angle will eventually become large, but while it is small (so that $\sin\theta \approx \theta$), what is θ as a function of time?

(b) You might think that it should be possible (theoretically, at least) to make the pencil balance for an arbitrarily long time, by making the initial θ_0 and ω_0 sufficiently small. However, it turns out that due to Heisenberg's uncertainty principle (which puts a constraint on how well we can know the position and momentum of

[19] Assume that the hole is actually a short frictionless tube bent into a gradual right angle, so that the chain's horizontal momentum doesn't cause it to overshoot the hole. For a description of what happens in a similar problem when this constraint is removed, see Calkin (1989).

[20] It actually involves only a trivial modification to do the problem correctly using the moment of inertia and the torque. But the point-mass version is quite sufficient for the present purposes.

a particle), it is impossible to balance the pencil for more than a certain amount of time. The point is that you can't be sure that the pencil is initially both at the top *and* at rest. The goal of this problem is to be quantitative about this. The time limit is sure to surprise you.

Without getting into quantum mechanics, let's just say that the uncertainty principle says (up to factors of order 1) that $\Delta x \Delta p \geq \hbar$, where $\hbar = 1.05 \cdot 10^{-34}$ J s is Planck's constant. The implications of this are somewhat vague, but we'll just take it to mean that the initial conditions satisfy $(\ell\theta_0)(m\ell\omega_0) \geq \hbar$. With this constraint, your task is to find the maximum time it can take your $\theta(t)$ solution in part (a) to become of order 1. In other words, determine (roughly) the maximum time the pencil can balance. Assume $m = 0.01$ kg, and $\ell = 0.1$ m.

Section 3.4: Projectile motion

3.14. **Maximum trajectory area** *

A ball is thrown at speed v from zero height on level ground. At what angle should it be thrown so that the area under the trajectory is maximum?

3.15. **Bouncing ball** *

A ball is thrown straight upward so that it reaches a height h. It falls down and bounces repeatedly. After each bounce, it returns to a certain fraction f of its previous height. Find the total distance traveled, and also the total time, before it comes to rest. What is its average speed?

3.16. **Perpendicular velocities** **

In the maximum-distance case in part (b) of the example in Section 3.4, show that the initial and final velocities are perpendicular to each other.[21]

3.17. **Throwing a ball from a cliff** **

A ball is thrown with speed v from the edge of a cliff of height h. At what inclination angle should it be thrown so that it travels the maximum horizontal distance? What is this maximum distance? Assume that the ground below the cliff is horizontal.

[21] You can grind through this problem and explicitly find the final angle, but there's a quicker way. This quicker method makes use of the conservation-of-energy statement that the difference in the squares of the initial and final speeds depends only on the change in height (the relation happens to be $v_i^2 - v_f^2 = 2gh$, but you don't need this actual expression). *Hint*: Consider the reverse path.

3.18. Redirected motion $**$

A ball is dropped from rest at height h above level ground, and it bounces off a surface at height y (with no loss in speed). The surface is inclined so that the ball bounces off at an angle θ with respect to the horizontal. What should y and θ be so that the ball travels the maximum horizontal distance by the time it hits the ground?

3.19. Maximum trajectory length $***$

A ball is thrown at speed v from zero height on level ground. Let θ_0 be the angle at which the ball should be thrown so that the length of the trajectory is maximum. Show that θ_0 satisfies

$$\sin\theta_0 \ln\left(\frac{1+\sin\theta_0}{\cos\theta_0}\right) = 1. \tag{3.54}$$

You can show numerically that $\theta_0 \approx 56.5°$.

Section 3.5: Motion in a plane, polar coordinates

3.20. Centripetal acceleration $*$

Show that the magnitude of the acceleration of a particle moving in a circle at constant speed is v^2/r. Do this by drawing the position and velocity vectors at two nearby times, and then making use of some similar triangles.

3.21. Vertical acceleration $**$

A bead rests at the top of a fixed frictionless hoop of radius R that lies in a vertical plane. The bead is given a tiny push so that it slides down and around the hoop. At what points on the hoop is the bead's acceleration vertical?[22] What is this vertical acceleration? *Note*: We haven't studied conservation of energy yet, but use the fact that the bead's speed after it has fallen a height h is given by $v = \sqrt{2gh}$.

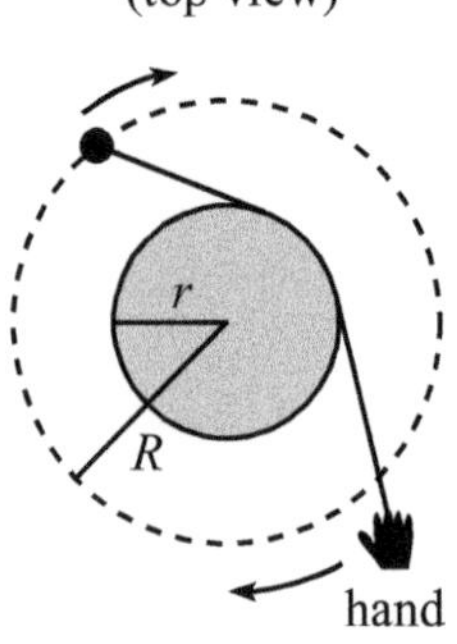

Fig. 3.17

3.22. Circling around a pole $**$

A mass, which is free to move on a horizontal frictionless surface, is attached to one end of a massless string that wraps partially around a frictionless vertical pole of radius r (see the top view in Fig. 3.17). You hold on to the other end of the string. At $t = 0$, the mass has speed v_0 in the tangential direction along the dotted circle of radius R shown. Your task is to pull on the string so that the mass keeps moving along the

[22] One such point is the bottom of the hoop. Another point is technically the top, where $a \approx 0$. Find the other two more interesting points (one on each side).

dotted circle. You are required to do this in such a way that the string remains in contact with the pole at all times. (You will have to move your hand around the pole, of course.) What is the speed of the mass as a function of time? There is a special value of the time; what is it and why is it special?

3.23. **A force $F_\theta = m\dot{r}\dot{\theta}$** **

Consider a particle that feels an angular force only, of the form $F_\theta = m\dot{r}\dot{\theta}$. Show that $\dot{r} = \sqrt{A \ln r + B}$, where A and B are constants of integration, determined by the initial conditions. (There's nothing all that physical about this force. It simply makes the $F = ma$ equations solvable.)

3.24. **Free particle** ***

Consider a free particle in a plane. With Cartesian coordinates, it is easy to use $F = ma$ to show that the particle moves in a straight line. The task of this problem is to demonstrate this result in a much more cumbersome way, using polar coordinates and Eq. (3.51). More precisely, show that $\cos\theta = r_0/r$ for a free particle, where r_0 is the radius at closest approach to the origin, and θ is measured with respect to this radius.

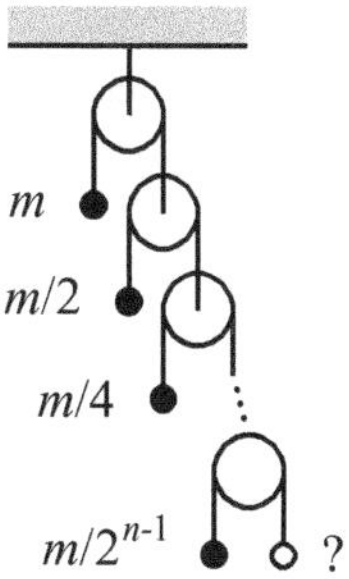

Fig. 3.18

3.7 Exercises

Section 3.2: Free-body diagrams

3.25. **A peculiar Atwood's machine**

(a) The Atwood's machine in Fig. 3.18 consists of n masses, m, $m/2$, $m/4, \ldots, m/2^{n-1}$. All the pulleys and strings are massless. Put a mass $m/2^{n-1}$ at the free end of the bottom string. What are the accelerations of all the masses?

(b) Remove the mass $m/2^{n-1}$ (which was arbitrarily small, for very large n) that was attached in part (a). What are the accelerations of all the masses, now that you've removed this infinitesimal piece?

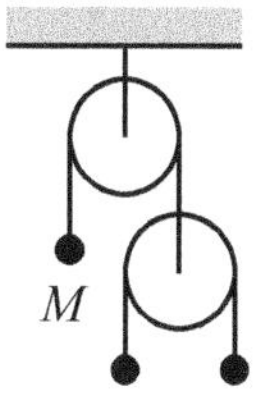

Fig. 3.19

3.26. **Keeping the mass still** *

In the Atwood's machine in Fig. 3.19, what should M be, in terms of m_1 and m_2, so that it doesn't move?

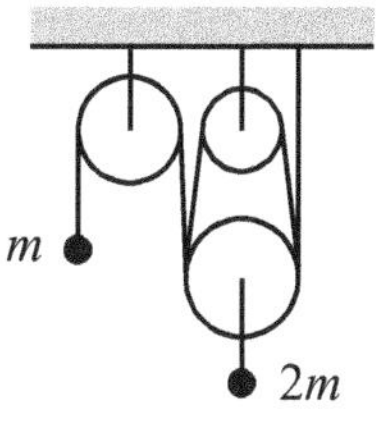

Fig. 3.20

3.27. **Atwood's 1** *

Consider the Atwood's machine in Fig. 3.20. It consists of three pulleys, a short piece of string connecting one mass to the bottom pulley, and a continuous long piece of string that wraps twice around the bottom side of the bottom pulley, and once around the top side of the top two pulleys. The two masses are m and $2m$. Assume that the parts of the string connecting the pulleys are essentially vertical. Find the accelerations of the masses.

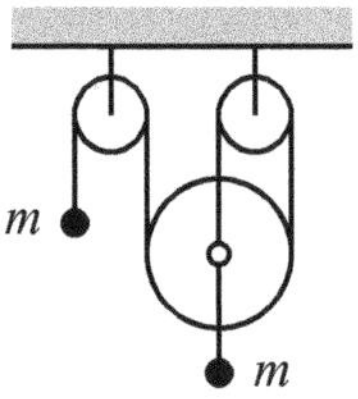

Fig. 3.21

3.28. Atwood's 2 *

Consider the Atwood's machine in Fig. 3.21, with two masses m. The axle of the bottom pulley has two string ends attached to it, as shown. Find the accelerations of the masses.

3.29. Atwood's 3 *

Consider the Atwood's machine in Fig. 3.22, with masses m, $2m$, and $3m$. Find the accelerations of the masses.

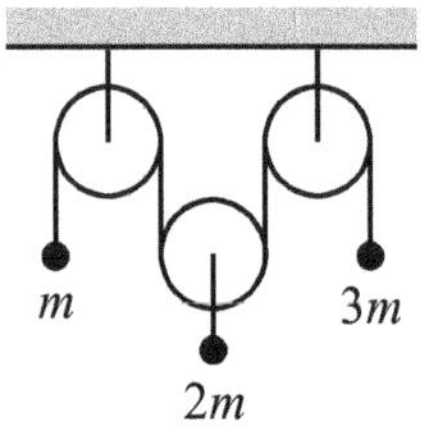

Fig. 3.22

3.30. Atwood's 4 **

Consider the Atwood's machine in Fig. 3.23 (and also on the front cover). If the number of pulleys that have string passing beneath them is N instead of the 3 shown, find the accelerations of the masses.

3.31. Atwood's 5 **

Consider the Atwood's machine in Fig. 3.24. The two shaded pulleys have mass m, and the string slides *frictionlessly* along all of the pulleys (so you don't have to worry about any rotational motion). Find the accelerations of the two shaded pulleys.

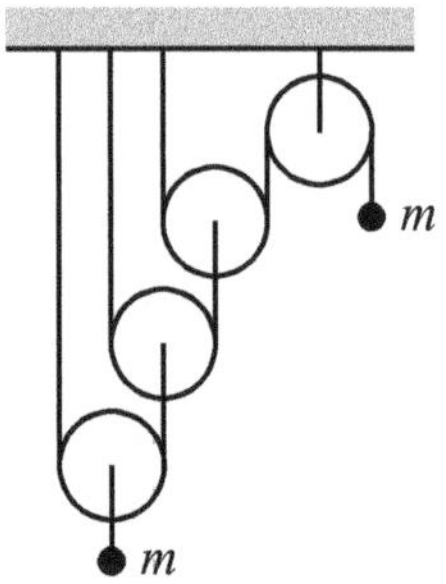

Fig. 3.23

3.32. Atwood's 6 **

Consider the Atwood's machine in Fig. 3.25. Find the accelerations of the masses. (This is a strange one.)

3.33. Accelerating plane **

A block of mass m rests on a plane inclined at an angle θ. The coefficient of static friction between the block and the plane is μ. The plane is accelerated to the right with acceleration a (which may be negative); see Fig. 3.26. For what range of a does the block remain at rest with respect to the plane? In terms of μ, there are two special values of θ; what are they, and why are they special?

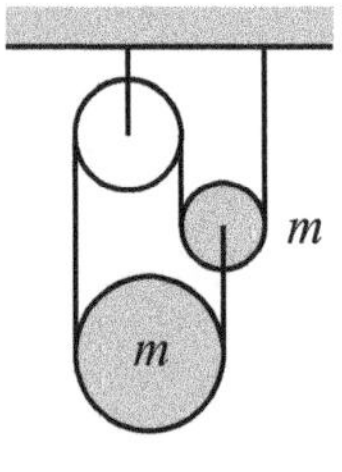

Fig. 3.24

3.34. Accelerating cylinders **

Three identical cylinders are arranged in a triangle as shown in Fig. 3.27, with the bottom two lying on the ground. The ground and the cylinders are frictionless. You apply a constant horizontal force (directed to the right) on the left cylinder. Let a be the acceleration you give to the system. For what range of a will all three cylinders remain in contact with each other?

3.35. **Leaving the sphere** **

A small mass rests on top of a fixed sphere of radius R. The coefficient of friction is μ. The mass is given a sideways kick that produces an initial angular speed ω_0. Let θ be the angle down from the top of the sphere. In terms of θ and its derivatives, what is the tangential $F = ma$ equation? Depending on the value of ω_0, the mass either comes to rest on the sphere or flies off it. If $g = 10\,\text{m/s}^2$, $R = 1\,\text{m}$, and $\mu = 1$, write a program to numerically determine the minimum ω_0 for which the mass leaves the sphere. For this cutoff case, give the angle at which the mass loses contact, and describe (roughly) what the plot of $\dot{\theta}$ versus θ looks like. See Prior and Mele (2007) for the exact solution for $\dot{\theta}$ in terms of θ.

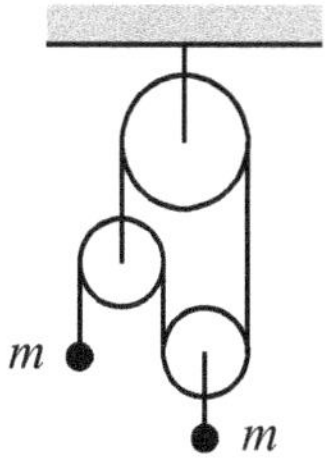

Fig. 3.25

3.36. **Comparing the times** ***

A block of mass m is projected up along the surface of a plane inclined at an angle θ. The initial speed is v_0, and the coefficients of static and kinetic friction are both equal to μ. The block reaches a highest point and then slides back down to the starting point.

(a) Show that for the block to in fact slide back down instead of remaining at rest at the highest point, $\tan\theta$ must be greater than μ.
(b) Assuming that $\tan\theta > \mu$, is the total up and down time longer or shorter than the total time it would take if the plane were frictionless? Or does the answer depend on what θ and μ are?
(c) Assuming that $\tan\theta > \mu$, show that for a given θ, the value of μ that yields the minimum total time is given by $\mu \approx (0.397)\tan\theta$. (You will need to solve something numerically.) This minimum time turns out to be about 90% of the time it would take if the plane were frictionless.

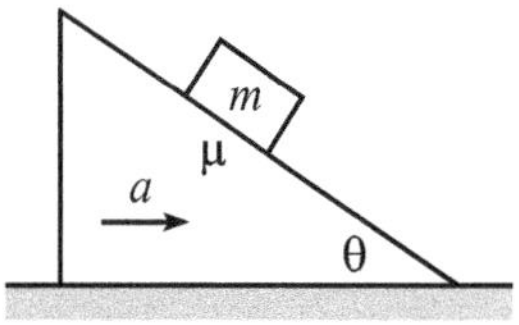

Fig. 3.26

Section 3.3: Solving differential equations

3.37. **$-bv^2$ force** *

A particle of mass m is subject to a force $F(v) = -bv^2$. The initial position is zero, and the initial speed is v_0. Find $x(t)$.

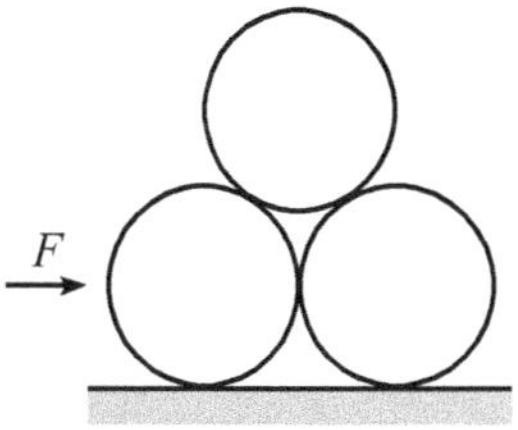

Fig. 3.27

3.38. **kx force** **

A particle of mass m is subject to a force $F(x) = kx$, with $k > 0$. The initial position is x_0, and the initial speed is zero. Find $x(t)$.

Section 3.4: Projectile motion

3.39. **Equal distances** *

At what angle should a ball be thrown so that its maximum height equals the horizontal distance traveled?

3.40. **Redirected motion** *

A ball is dropped from rest at height h. At height y, it bounces off a surface with no loss in speed. The surface is inclined at 45°, so the ball bounces off horizontally. What should y be so that the ball travels the maximum horizontal distance? What is this maximum distance?

3.41. **Throwing in the wind** *

A ball is thrown horizontally to the right, from the top of a vertical cliff of height h. A wind blows horizontally to the left, and assume (simplistically) that the effect of the wind is to provide a constant force to the left, equal in magnitude to the weight of the ball. How fast should the ball be thrown so that it lands at the foot of the cliff?

3.42. **Throwing in the wind again** *

A ball is thrown eastward across level ground. A wind blows horizontally to the east, and assume (simplistically) that the effect of the wind is to provide a constant force to the east, equal in magnitude to the weight of the ball. At what angle θ should the ball be thrown so that it travels the maximum horizontal distance?

3.43. **Increasing gravity** *

At $t = 0$ on the planet Gravitus Increasicus, a projectile is fired with speed v_0 at an angle θ above the horizontal. This planet is a strange one, in that the acceleration due to gravity increases linearly with time, starting with a value of zero when the projectile is fired. In other words, $g(t) = \beta t$, where β is a given constant. What horizontal distance does the projectile travel? What should θ be to maximize this distance?

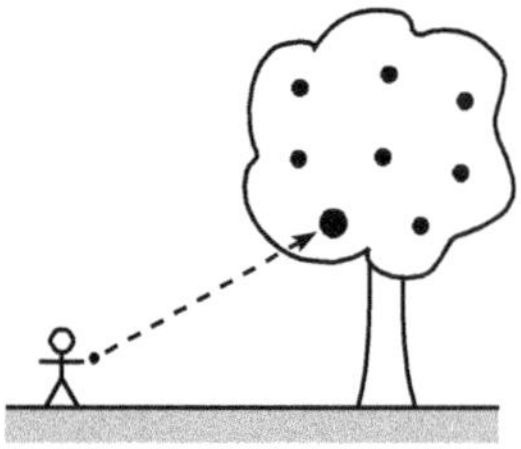

Fig. 3.28

3.44. **Newton's apple** *

Newton is tired of apples falling on his head, so he decides to throw a rock at one of the larger and more formidable-looking apples positioned directly above his favorite sitting spot. Forgetting all about his work on gravitation, he aims the rock directly at the apple (see Fig. 3.28). To his surprise, the apple falls from the tree just as he releases the rock. Show, by calculating the rock's height when it reaches the horizontal position of the apple, that the rock hits the apple.[23]

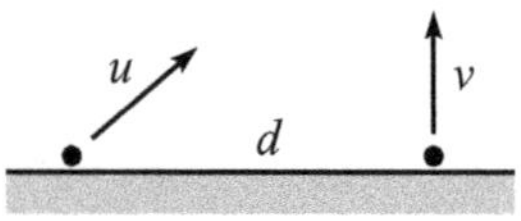

Fig. 3.29

3.45. **Colliding projectiles** *

Two balls are fired from ground level, a distance d apart. The right one is fired vertically with speed v (see Fig. 3.29). You wish to simultaneously

[23] This problem suggests a way in which William Tell and his son might survive their ordeal if they were plopped down, with no time to practice, on a planet with an unknown gravitational constant (provided that the son weren't too short or that g weren't too big).

fire the left one at the appropriate velocity **u** so that it collides with the right ball when they reach their highest point. What should **u** be (give the horizontal and vertical components)? Given d, what should v be so that the speed u is minimum?

3.46. **Equal tilts** *

A plane tilts down at an angle θ below the horizontal. On this plane, a projectile is fired with speed v at an angle θ above the horizontal, as shown in Fig. 3.30. What is the distance, d, along the plane that the projectile travels? What is d in the limit $\theta \to 90°$? What θ yields the maximum *horizontal* distance?

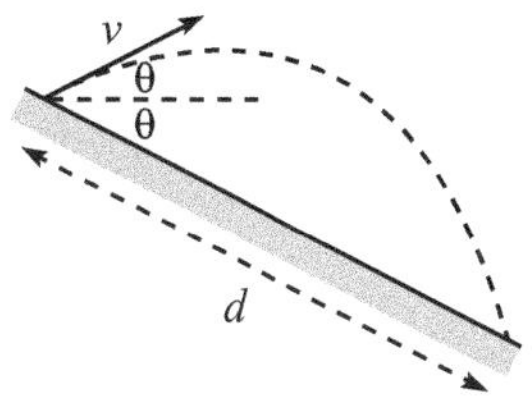

Fig. 3.30

3.47. **Throwing at a wall** *

You throw a ball with speed v_0 at a vertical wall, a distance ℓ away. At what angle should you throw the ball so that it hits the wall as high as possible? Assume $\ell < v_0^2/g$ (why?).

3.48. **Firing a cannon** **

A cannon, when aimed vertically, is observed to fire a ball to a maximum height of L. Another ball is then fired with this same speed, but with the cannon aimed up along a plane of length L, inclined at an angle θ, as shown in Fig. 3.31. What should θ be so that the ball travels the largest horizontal distance, d, by the time it returns to the height of the top of the plane?

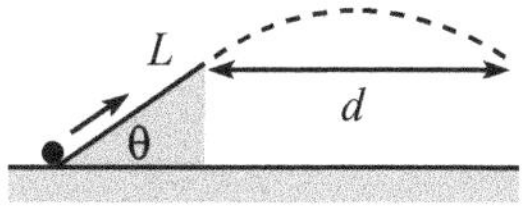

Fig. 3.31

3.49. **Perpendicular and horizontal** **

A plane is inclined at an angle θ below the horizontal. A person throws a ball with speed v_0 from the surface of the plane. How far down along the plane does the ball hit, if the person throws the ball (a) perpendicular to the plane? (b) horizontally?

3.50. **Cart, ball, and plane** **

A cart is held at rest on an inclined plane. A tube is positioned in the cart with its axis perpendicular to the plane. The cart is released, and at some later time a ball is fired from the tube. Will the ball eventually land back in the tube? *Hint*: Choose your coordinate system wisely.

3.51. **Perpendicular to plane** **

A hill is sloped downward at an angle β with respect to the horizontal. A projectile is fired with an initial velocity perpendicular to the hill. When it eventually lands on the hill, let its velocity make an angle θ with respect to the horizontal. What is θ? What β yields the minimum value of θ? What is this minimum θ?

3.52. **Increasing distance** **

(a) What is the maximum angle at which you can thrown a ball so that its distance from you never decreases during its flight?
(b) This maximum angle equals the minimum θ from Exercise 3.51. Explain why this is true. (It isn't necessary to have done that exercise.)

3.53. **Projectile with drag** ***

A ball is thrown with speed v_0 at an angle θ. Let the drag force from the air take the form $\mathbf{F}_\text{d} = -\beta\mathbf{v} \equiv -m\alpha\mathbf{v}$.

(a) Find $x(t)$ and $y(t)$.
(b) Assume that the drag coefficient takes the value that makes the magnitude of the initial drag force equal to the weight of the ball. If your goal is to have x be as large as possible when y achieves its maximum value (you don't care what this maximum value actually is), show that θ should satisfy $\sin\theta = (\sqrt{5}-1)/2$, which just happens to be the inverse of the golden ratio.

Section 3.5: Motion in a plane, polar coordinates

3.54. **Low-orbit satellite**

What is the speed of a satellite whose orbit is just above the earth's surface? Give the numerical value.

3.55. **Weight at the equator** *

A person stands on a scale at the equator. If the earth somehow stopped spinning but kept its same shape, would the reading on the scale increase or decrease? By what fraction?

3.56. **Banking an airplane** *

An airplane flies at speed v in a horizontal circle of radius R. At what angle should the plane be banked so that you don't feel like you are getting flung to the side in your seat? At this angle, what is your apparent weight (that is, what is the normal force from the seat)?

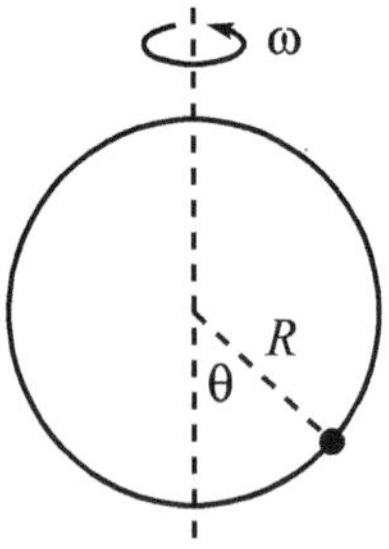

Fig. 3.32

3.57. **Rotating hoop** *

A bead lies on a frictionless hoop of radius R that rotates around a vertical diameter with constant angular frequency ω, as shown in Fig. 3.32. What should ω be so that the bead maintains the same position on the hoop, at an angle θ with respect to the vertical? There is a special value of ω; what is it, and why is it special?

3.58. **Swinging in circles** *

A large number of masses are attached by strings of various lengths to a point on the ceiling. All of the masses swing around in horizontal circles of various radii with the *same* frequency ω (one such circle is drawn in Fig. 3.33). If you take a picture (from the side) of the setup at an instant when all the masses lie in the plane of the paper (as shown for four masses), what does the "curve" formed by the masses look like?

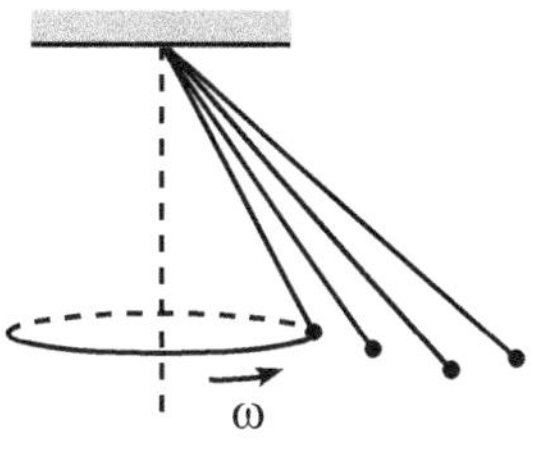

Fig. 3.33

3.59. **Swinging triangle** *

Two masses m are attached to two vertices of an equilateral triangle made of three massless rods of length ℓ. A pivot is located at the third vertex, and the triangle is free to swing back and forth in a vertical plane, as shown in Fig. 3.34. If it is initially released from rest when one of the rods is vertical (as shown), find the tensions in all three rods (and specify tension or compression), and also the accelerations of the masses, at the instant *right after* it is released.

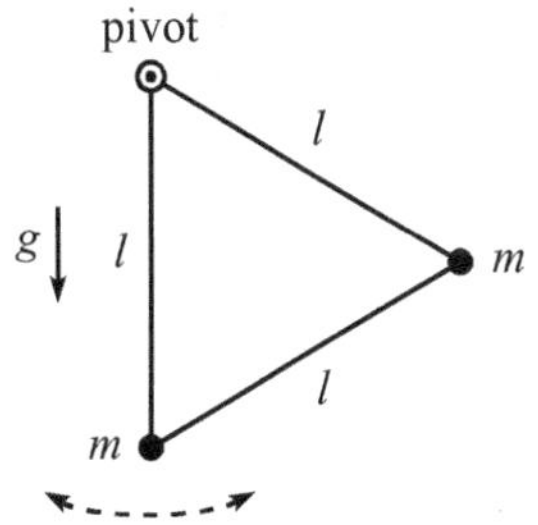

Fig. 3.34

3.60. **Circular and plane pendulums** *

Consider the circular pendulum in the example in Section 3.5. Let the x-y plane be the horizontal plane of the circle. For small β, what (approximately) is the F_x component of the force on the mass when it is at the position (x, y) on the circle?

Consider now a standard plane pendulum where the mass swings back and forth in the vertical plane containing the x axis. Let the maximum angle of the pendulum be the same small angle β.[24] What (approximately) is the F_x component of the force on the mass in terms of its x coordinate?

Your two results should be the same, which means that the x motions of the two systems are the same, because when they are each at their maximum x value ($\ell \sin\beta$), they have the same x speed (zero), and we just showed that they always have the same x acceleration, independent of any y motion. So the frequencies of the two systems must be equal. (We'll see in Section 4.2 that the frequency of a plane pendulum is $\sqrt{g/\ell}$, in agreement with this observation.)

3.61. **Rolling wheel** *

If you paint a dot on the rim of a rolling wheel, the coordinates of the dot may be written as[25]

$$(x, y) = (R\theta + R\sin\theta, R + R\cos\theta). \tag{3.55}$$

[24] It's actually not required that this be the same angle, as long as it's small. See Problem 3.10.

[25] This follows from writing (x, y) as $(R\theta, R) + (R\sin\theta, R\cos\theta)$. The first term here is the position of the center of the wheel, and the second term is the position of the dot relative to the center, where θ is measured clockwise from the top.

The path of the dot is called a *cycloid*. Assume that the wheel is rolling at constant speed, which implies $\theta = \omega t$.

(a) Find $\mathbf{v}(t)$ and $\mathbf{a}(t)$ of the dot.
(b) At the instant the dot is at the top of the wheel, what is the radius of curvature of its path? The radius of curvature is defined to be the radius of the circle that matches up with the path locally at a given point. *Hint*: You know v and a.

3.62. Radius of curvature ∗∗

A projectile is fired at speed v_0 and angle θ. What is the radius of curvature (defined in Exercise 3.61) of the parabolic motion

(a) at the top?
(b) at the beginning?
(c) At what angle should the projectile be fired so that the radius of curvature at the top equals half the maximum height, as shown in Fig. 3.35?

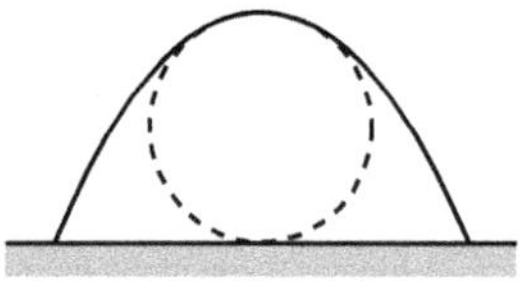

Fig. 3.35

3.63. Driving on tilted ground ∗∗

A driver encounters a large tilted parking lot, where the angle of the ground with respect to the horizontal is θ. The driver wishes to drive in a circle of radius R at constant speed. The coefficient of friction between the tires and the ground is μ.

(a) What is the largest speed the driver can have if he wants to avoid slipping?
(b) What is the largest speed the driver can have, assuming he is concerned only with whether or not he slips at one of the "side" points on the circle (that is, halfway between the top and bottom points; see Fig. 3.36)?

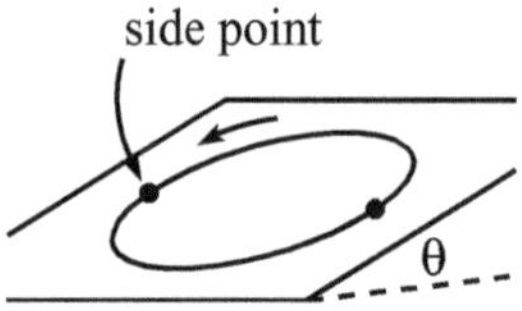

Fig. 3.36

3.64. Car on a banked track ∗∗

A car travels around a circular banked track of radius R. The angle of the bank is θ, and the coefficient of friction between the tires and the track is μ. For what range of speeds does the car not slip?

3.65. Horizontal acceleration ∗∗

A bead rests at the top of a fixed frictionless hoop of radius R that lies in a vertical plane. The bead is given a tiny push so that it slides down and around the hoop. At what points on the hoop is the bead's acceleration horizontal? *Note*: We haven't studied conservation of energy yet, but use the fact that the bead's speed after it has fallen a height h is given by $v = \sqrt{2gh}$.

3.66. **Maximum horizontal force** $**$

A bead rests at the top of a fixed frictionless hoop of radius R that lies in a vertical plane. The bead is given a tiny push so that it slides down and around the hoop. Consider the horizontal component of the force from the hoop on the bead. At what points on the hoop does this component achieve a local maximum or minimum? As in Exercise 3.65, use $v = \sqrt{2gh}$.

3.67. **Derivation of F_r and F_θ** $*$

In Cartesian coordinates, a general vector takes the form,

$$\mathbf{r} = x\hat{\mathbf{x}} + y\hat{\mathbf{y}} = r\cos\theta\,\hat{\mathbf{x}} + r\sin\theta\,\hat{\mathbf{y}}. \tag{3.56}$$

Derive Eq. (3.51) by taking two derivatives of this expression for $\mathbf{r}$, and then using Eq. (3.46) to show that the result can be written in the form of Eq. (3.50). Note that unlike $\hat{\mathbf{r}}$ and $\hat{\boldsymbol{\theta}}$, the vectors $\hat{\mathbf{x}}$ and $\hat{\mathbf{y}}$ do not change with time.

3.68. **A force $F_\theta = 3m\dot{r}\dot{\theta}$** $**$

Consider a particle that feels an angular force only, of the form $F_\theta = 3m\dot{r}\dot{\theta}$. Show that $\dot{r} = \pm\sqrt{Ar^4 + B}$, where A and B are constants of integration, determined by the initial conditions. Also, show that if the particle starts with $\dot{\theta} \neq 0$ and $\dot{r} > 0$, it reaches $r = \infty$ in a finite time. (As in Problem 3.23, there's nothing all that physical about this force. It simply makes the $F = ma$ equations solvable.)

3.69. **A force $F_\theta = 2m\dot{r}\dot{\theta}$** $**$

Consider a particle that feels an angular force only, of the form $F_\theta = 2m\dot{r}\dot{\theta}$. Show that $r = Ae^{\theta} + Be^{-\theta}$, where A and B are constants of integration, determined by the initial conditions. (This force is actually a physical one. If you put a bead on a stick and swing the stick around one end at a constant rate, then the normal force from the stick happens to be $2m\dot{r}\dot{\theta}$.[26])

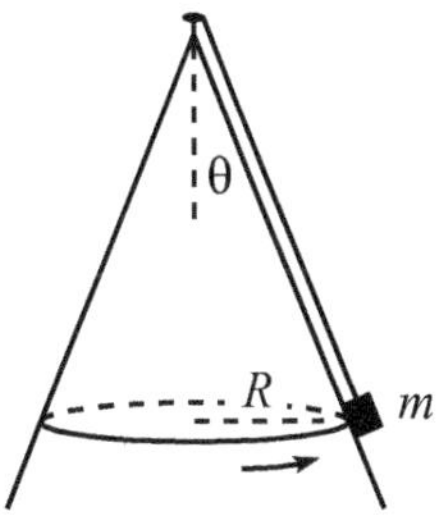

Fig. 3.37

3.70. **Stopping on a cone** $**$

When viewed from the side, the cone in Fig. 3.37 subtends an angle 2θ at its tip. A block of mass m is connected to the tip by a massless string and moves in a horizontal circle of radius R around the surface. If the

[26] Depending on what is meant by "physical," the forces in Exercise 3.68 and Problem 3.23 might also be considered to be physical. They correspond to putting a bead on a stick and swinging the stick around with angular speeds proportional to the bead's r or $1/r$, respectively (as is evident from the values of $\dot{\theta}$ in the solutions). This can also be deduced from $\tau = dL/dt$, but we won't get to torque until Chapter 8.

initial speed is v_0, and if the coefficient of kinetic friction between the block and the cone is μ, how much time does it take the block to stop? (The answer is a little messy, but there are some limits you can check that will make you feel better about it.)

3.71. **Motorcycle circle** ***

A motorcyclist wishes to travel in a circle of radius R on level ground. The coefficient of friction between the tires and the ground is μ. The motorcycle starts at rest. What is the minimum distance it must travel in order to achieve its maximum allowable speed, that is, the speed above which it will skid out of the circular path?[27] Solve this in two ways:

(a) Write down the radial and tangential $F = ma$ equations (you'll want to write a as $v\,dv/dx$), and then demand that the magnitude of the friction force equals μmg in the optimal case. Take it from there.

(b) Let the friction force make an angle $\beta(t)$ with respect to the tangential direction. Write down the radial and tangential $F = ma$ equations (you'll want to write a as dv/dt), and then take the derivative of the radial equation. Take it from there (this is the slick way).

3.8 Solutions

3.1. **Atwood's machine**

Let T be the tension in the string, and let a be the acceleration of m_1 (with upward taken to be positive). Then $-a$ is the acceleration of m_2. So the $F = ma$ equations are

$$T - m_1 g = m_1 a, \quad \text{and} \quad T - m_2 g = m_2(-a). \tag{3.57}$$

Solving these two equations for a and T gives

$$a = \frac{(m_2 - m_1)g}{m_2 + m_1}, \quad \text{and} \quad T = \frac{2m_1 m_2 g}{m_2 + m_1}. \tag{3.58}$$

Remarks: As a double-check, a has the correct limits when $m_2 \gg m_1$, $m_1 \gg m_2$, and $m_2 = m_1$ (namely $a \approx g$, $a \approx -g$, and $a = 0$, respectively). As far as T goes, if $m_1 = m_2 \equiv m$, then $T = mg$, as it should. And if $m_1 \ll m_2$, then $T \approx 2m_1 g$. This is correct, because it makes the net upward force on m_1 equal to $m_1 g$, which means that its acceleration is g upward, which is consistent with the fact that m_2 is essentially in free fall. ♣

3.2. **Double Atwood's machine**

Let the tension in the lower string be T. Then the tension in the upper string is $2T$ (by balancing the forces on the bottom pulley). The three $F = ma$ equations are therefore (with all the a's taken to be positive upward)

$$2T - m_1 g = m_1 a_1, \quad T - m_2 g = m_2 a_2, \quad T - m_3 g = m_3 a_3. \tag{3.59}$$

[27] This problem can be traced to an old edition of the Russian magazine *Kvant*.

And conservation of string says that the acceleration of m_1 is

$$a_1 = -\left(\frac{a_2 + a_3}{2}\right). \tag{3.60}$$

This follows from the fact that the average position of m_2 and m_3 moves the same distance as the bottom pulley, which in turn moves the same distance (but in the opposite direction) as m_1. We now have four equations in the four unknowns, a_1, a_2, a_3, and T. With a little work, we can solve for the accelerations,

$$\begin{aligned} a_1 &= g\frac{4m_2m_3 - m_1(m_2 + m_3)}{4m_2m_3 + m_1(m_2 + m_3)}, \\ a_2 &= -g\frac{4m_2m_3 + m_1(m_2 - 3m_3)}{4m_2m_3 + m_1(m_2 + m_3)}, \\ a_3 &= -g\frac{4m_2m_3 + m_1(m_3 - 3m_2)}{4m_2m_3 + m_1(m_2 + m_3)}. \end{aligned} \tag{3.61}$$

Remarks: There are many limits we can check here. A couple are: (1) If $m_2 = m_3 = m_1/2$, then all the a's are zero, which is correct. (2) If m_3 is much less than both m_1 and m_2, then $a_1 = -g$, $a_2 = -g$, and $a_3 = 3g$. To understand this $3g$, convince yourself that if m_1 and m_2 go down by d, then m_3 goes up by $3d$.

Note that a_1 can be written as

$$a_1 = g\left(\frac{4m_2m_3}{m_2 + m_3} - m_1\right)\Big/\left(\frac{4m_2m_3}{m_2 + m_3} + m_1\right). \tag{3.62}$$

In view of the result for a in Eq. (3.58) in Problem 3.1, we see that as far as m_1 is concerned here, the m_2, m_3 pulley system acts just like a mass of $4m_2m_3/(m_2 + m_3)$. This has the expected properties of equaling zero when either m_2 or m_3 is zero, and equaling $2m$ if $m_2 = m_3 \equiv m$. ♣

3.3. **Infinite Atwood's machine**

First solution: If the strength of gravity on the earth were multiplied by a factor η, then the tension in all of the strings in the Atwood's machine would likewise be multiplied by η. This is true because the only way to produce a quantity with the units of tension (that is, force) is to multiply a mass by g. Conversely, if we put the Atwood's machine on another planet and discover that all of the tensions are multiplied by η, then we know that the gravity there must be ηg.

Let the tension in the string above the first pulley be T. Then the tension in the string above the second pulley is $T/2$ (because the pulley is massless). Let the downward acceleration of the second pulley be a_2. Then the second pulley effectively lives in a world where gravity has strength $g - a_2$. Consider the subsystem of all the pulleys except the top one. This infinite subsystem is identical to the original infinite system of all the pulleys. Therefore, by the arguments in the above paragraph, we must have

$$\frac{T}{g} = \frac{T/2}{g - a_2}, \tag{3.63}$$

which gives $a_2 = g/2$. But a_2 is also the acceleration of the top mass, so our answer is $g/2$.

Remarks: You can show that the relative acceleration of the second and third pulleys is $g/4$, and that of the third and fourth is $g/8$, etc. The acceleration of a mass far down in the system therefore equals $g(1/2 + 1/4 + 1/8 + \cdots) = g$, which makes intuitive sense.

Note that $T = 0$ also makes Eq. (3.63) true. But this corresponds to putting a mass of zero at the end of a finite pulley system (see the following solution). ♣

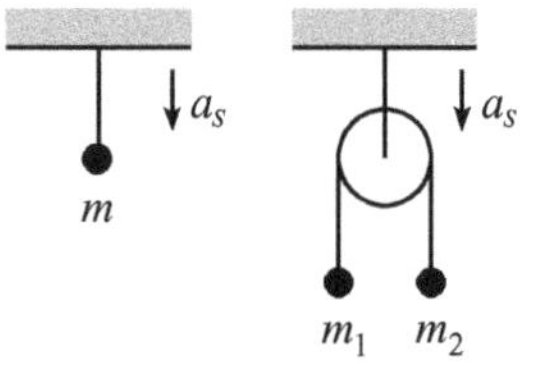

Fig. 3.38

SECOND SOLUTION: Consider the following auxiliary problem.

Problem: Two setups are shown in Fig. 3.38. The first contains a hanging mass m. The second contains two masses, m_1 and m_2, hanging over a pulley. Let both supports have acceleration a_s downward. What should m be, in terms of m_1 and m_2, so that the tension T in the top string is the same in both cases?

Answer: In the first case, we have

$$mg - T = ma_s. \tag{3.64}$$

In the second case, let a be the acceleration of m_2 relative to the support (with downward taken to be positive). Then we have

$$m_1 g - \frac{T}{2} = m_1(a_s - a), \quad \text{and} \quad m_2 g - \frac{T}{2} = m_2(a_s + a). \tag{3.65}$$

If we define $g' \equiv g - a_s$, then we may write the above three equations as

$$mg' = T, \quad m_1 g' - \frac{T}{2} = -m_1 a, \quad m_2 g' - \frac{T}{2} = m_2 a. \tag{3.66}$$

Eliminating a from the last two of these equations gives $T = 4m_1m_2g'/(m_1 + m_2)$. Using this value of T in the first equation then gives

$$m = \frac{4m_1m_2}{m_1 + m_2}. \tag{3.67}$$

Note that the value of a_s is irrelevant. We effectively have a fixed support in a world where the acceleration due to gravity is g' (see Eq. (3.66)), and the desired m can't depend on g', by dimensional analysis. This auxiliary problem shows that for any a_s the two-mass system in the second case can equivalently be treated like a mass m, given by Eq. (3.67), as far as the upper string is concerned. ■

Now let's look at our infinite Atwood's machine. Assume that the system has N pulleys, where $N \to \infty$. Let the bottom mass be x. Then the auxiliary problem shows that the bottom two masses, m and x, can be treated like an effective mass $f(x)$, where

$$f(x) = \frac{4mx}{m + x} = \frac{4x}{1 + (x/m)}. \tag{3.68}$$

We can then treat the combination of the mass $f(x)$ and the next m as an effective mass $f(f(x))$. These iterations can be repeated, until we finally have a mass m and a mass $f^{(N-1)}(x)$ hanging over the top pulley. So we must determine the behavior of $f^N(x)$, as $N \to \infty$. This behavior is clear if we look at the plot of $f(x)$ in Fig. 3.39. We see that $x = 3m$ is a fixed point of $f(x)$. That is, $f(3m) = 3m$. This plot shows that no matter what x we start with, the iterations approach $3m$ (unless we start at $x = 0$, in which case we remain there). These iterations are shown graphically by the directed lines in the plot. After reaching the value $f(x)$ on the curve, the line moves horizontally to the x value of $f(x)$, and then vertically to the value $f(f(x))$ on the curve, and so on. Therefore, since $f^N(x) \to 3m$ as $N \to \infty$, our infinite Atwood's machine is equivalent to (as far as the top mass is concerned) just two masses, m and $3m$. You can then quickly show that the acceleration of the top mass is $g/2$. Note that as far as the support is concerned, the whole apparatus is equivalent to a mass $3m$. So $3mg$ is the upward force exerted by the support.

3.4. **Line of pulleys**

Let m be the common mass, and let T be the tension in the string. Let a be the acceleration of the end masses, and let a' be the acceleration of the other N masses, with upward taken to be positive. These N accelerations are indeed all equal, because the same net force acts on all of the internal N masses, namely $2T$ upwards and mg downwards. The $F = ma$ equations for the end and internal masses are, respectively,

$$T - mg = ma, \quad \text{and} \quad 2T - mg = ma'. \tag{3.69}$$

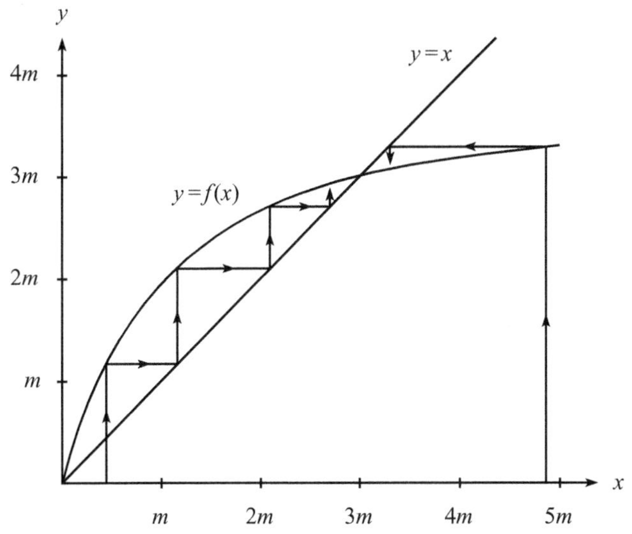

Fig. 3.39 Problem 3.3, second solution

But the string has fixed length. Therefore,

$$N(2a') + a + a = 0. \tag{3.70}$$

The "2" here comes from the fact that if one of the inside masses goes up by a distance d, then a length $2d$ of string has disappeared and must therefore appear somewhere else (namely, in the outer two segments). Eliminating T from Eq. (3.69) gives $a' = 2a + g$. Combining this with Eq. (3.70) then gives

$$a = -\frac{Ng}{2N+1}, \quad \text{and} \quad a' = \frac{g}{2N+1}. \tag{3.71}$$

REMARKS: For $N = 1$, we have $a = -g/3$ and $a' = g/3$. For larger N, a increases in magnitude and approaches $-g/2$ as $N \to \infty$, and a' decreases in magnitude and approaches zero as $N \to \infty$. The signs of a and a' in Eq. (3.71) may be surprising. You might think that if, say, $N = 100$, then these 100 masses will "win" out over the two end masses, so that the N masses will fall. But this is not correct, because there are many ($2N$, in fact) tensions acting up on the N masses. They do *not* behave like a mass Nm hanging below one pulley. In fact, two masses of $m/2$ on the ends will balance any number N of masses m in the interior (with the help of the upward forces from the top row of pulleys). ♣

3.5. **Ring of pulleys**

Let T be the tension in the string. Then $F = ma$ for m_i gives

$$2T - m_i g = m_i a_i, \tag{3.72}$$

with upward taken to be positive. The a_i's are related by the fact that the string has fixed length, which implies that the sum of the displacements of all the masses is zero. In other words,

$$a_1 + a_2 + \cdots + a_N = 0. \tag{3.73}$$

If we divide Eq. (3.72) by m_i, and then add the N such equations together and use Eq. (3.73), we find that T is given by

$$2T\left(\frac{1}{m_1}+\frac{1}{m_2}+\cdots+\frac{1}{m_N}\right)-Ng=0. \tag{3.74}$$

Therefore,

$$T=\frac{NMg}{2}, \quad \text{where} \quad \frac{1}{M}\equiv\frac{1}{m_1}+\frac{1}{m_2}+\cdots+\frac{1}{m_N} \tag{3.75}$$

is the so-called *reduced mass* of the system. Substituting this value for T into (3.72) gives

$$a_i=g\left(\frac{NM}{m_i}-1\right). \tag{3.76}$$

REMARK: A few special cases are: If all the masses are equal, then all the $a_i=0$. If $m_k=0$ (and all the others are not zero), then $a_k=(N-1)g$, and all the other $a_i=-g$. If $N-1$ of the masses are equal and much smaller than the remaining one, m_k, then $m_k\approx -g$, and all the other $a_i\approx g/(N-1)$. ♣

3.6. **Sliding down a plane**

(a) The component of gravity along the plane is $g\sin\theta$. The acceleration in the horizontal direction is therefore $a_x=(g\sin\theta)\cos\theta$. Our goal is to maximize a_x. By taking the derivative, or by noting that $\sin\theta\cos\theta=(\sin 2\theta)/2$, we obtain $\theta=\pi/4$. The maximum a_x is then $g/2$.

(b) The normal force from the plane is $mg\cos\theta$, so the kinetic friction force is $\mu mg\cos\theta$. The acceleration along the plane is therefore $g(\sin\theta-\mu\cos\theta)$, and so the acceleration in the horizontal direction is $a_x=g(\sin\theta-\mu\cos\theta)\cos\theta$. We want to maximize this. Setting the derivative equal to zero gives

$$(\cos^2\theta-\sin^2\theta)+2\mu\sin\theta\cos\theta=0 \implies \cos 2\theta+\mu\sin 2\theta=0$$
$$\implies \tan 2\theta=-\frac{1}{\mu}. \tag{3.77}$$

For $\mu\to 0$, this gives the $\pi/4$ result in part (a). For $\mu\to\infty$, we obtain $\theta\approx\pi/2$, which makes sense.

REMARKS: The time to travel a horizontal distance d is obtained from $a_x t^2/2=d$. In part (a), this gives a minimum time of $2\sqrt{d/g}$. In part (b), you can show that the maximum a_x is $(g/2)(\sqrt{1+\mu^2}-\mu)$, which then leads to a minimum time of $2\sqrt{d/g}\big(\sqrt{1+\mu^2}+\mu\big)^{1/2}$. This has the correct $\mu\to 0$ limit, and it behaves like $2\sqrt{2\mu d/g}$ for $\mu\to\infty$. ♣

3.7. **Sliding sideways on a plane**

The normal force from the plane is $N=mg\cos\theta$. Therefore, the friction force on the block is $\mu N=(\tan\theta)(mg\cos\theta)=mg\sin\theta$. This force acts in the direction opposite to the motion. The block also feels the gravitational force of $mg\sin\theta$ pointing down the plane.

Because the magnitudes of the friction force and the gravitational force along the plane are equal, the acceleration along the direction of motion equals the negative of the acceleration in the direction down the plane. Therefore, in a small increment of time, the speed that the block loses along its direction of motion exactly equals the speed that it gains in the direction down the plane. Letting v be the total speed of the block, and letting v_y be the component of the velocity in the direction down the plane, we therefore have

$$v + v_y = C, \tag{3.78}$$

where C is a constant. C is given by its initial value, which is $V + 0 = V$. The final value of C is $V_f + V_f = 2V_f$ (where V_f is the final speed of the block), because the block is essentially moving straight down the plane after a very long time. Therefore,

$$2V_f = V \quad \Longrightarrow \quad V_f = \frac{V}{2}. \tag{3.79}$$

3.8. **Moving plane**

Let N be the normal force between the block and the plane. Note that we *cannot* assume that $N = mg\cos\theta$, because the plane recoils. We can see that $N = mg\cos\theta$ is in fact incorrect, because in the limiting case where $M = 0$, we have no normal force at all.

The various $F = ma$ equations (vertical and horizontal for the block, and horizontal for the plane) are

$$\begin{aligned} mg - N\cos\theta &= ma_y, \\ N\sin\theta &= ma_x, \\ N\sin\theta &= MA_x, \end{aligned} \tag{3.80}$$

where we have chosen the positive directions for a_y, a_x, and A_x to be downward, rightward, and leftward, respectively. There are four unknowns here: a_x, a_y, A_x, and N, so we need one more equation. This fourth equation is the constraint that the block remains in contact with the plane. The horizontal distance between the block and its starting point on the plane is $(a_x + A_x)t^2/2$, and the vertical distance is $a_y t^2/2$. The ratio of these distances must equal $\tan\theta$ if the block is to remain on the plane (imagine looking at things in the frame of the plane). Therefore, we must have

$$\frac{a_y}{a_x + A_x} = \tan\theta. \tag{3.81}$$

Using Eq. (3.80) to solve for a_y, a_x, and A_x in terms of N, and then plugging the results into Eq. (3.81), gives

$$\frac{g - \frac{N}{m}\cos\theta}{\frac{N}{m}\sin\theta + \frac{N}{M}\sin\theta} = \tan\theta \quad \Longrightarrow \quad N = g\left(\sin\theta\tan\theta\left(\frac{1}{m} + \frac{1}{M}\right) + \frac{\cos\theta}{m}\right)^{-1}. \tag{3.82}$$

(In the limit $M \to \infty$, this reduces to $N = mg\cos\theta$, as it should.) Having found N, the third of Eqs. (3.80) gives A_x, which may be written as

$$A_x = \frac{N\sin\theta}{M} = \frac{mg\sin\theta\cos\theta}{M + m\sin^2\theta}. \tag{3.83}$$

REMARKS:

1. For given M and m, you can show that the angle θ_0 that maximizes A_x is $\tan\theta_0 = \sqrt{M/(M+m)}$. If $M \ll m$, then $\theta_0 \approx 0$; this makes sense, because the plane gets squeezed out very fast. If $M \gg m$, then $\theta_0 \approx \pi/4$; this is consistent with the $\pi/4$ result from Problem 3.6(a).
2. In the limit $M \ll m$, Eq. (3.83) gives $A_x \approx g/\tan\theta$. This makes sense, because m falls essentially straight down with acceleration g, and the plane gets squeezed out to the left.
3. In the limit $M \gg m$, Eq. (3.83) gives $A_x \approx g(m/M)\sin\theta\cos\theta$. The correctness of this is more transparent if we instead look at $a_x = (M/m)A_x \approx g\sin\theta\cos\theta$. Since the plane is essentially at rest in this limit, this value of a_x implies that the acceleration of m *along* the plane equals $a_x/\cos\theta \approx g\sin\theta$, as expected. ♣

3.9. **Exponential force**

$F = ma$ gives $\ddot{x} = a_0 e^{-bt}$. Integrating this with respect to time gives $v(t) = -a_0 e^{-bt}/b + A$. Integrating again gives $x(t) = a_0 e^{-bt}/b^2 + At + B$. The initial condition $v(0) = 0$ gives $-a_0/b + A = 0 \Longrightarrow A = a_0/b$. And the initial condition $x(0) = 0$ gives $a_0/b^2 + B = 0 \Longrightarrow B = -a_0/b^2$. Therefore,

$$x(t) = a_0 \left(\frac{e^{-bt}}{b^2} + \frac{t}{b} - \frac{1}{b^2} \right). \tag{3.84}$$

For $t \to \infty$ (more precisely, for $bt \to \infty$), v approaches a_0/b, and x approaches $a_0(t/b - 1/b^2)$. We see that the particle eventually lags a distance a_0/b^2 behind another particle that starts at the same position but moves with the constant speed $v = a_0/b$. For $t \approx 0$ (more precisely, for $bt \approx 0$), we can expand e^{-bt} in its Taylor series to obtain $x(t) \approx a_0 t^2/2$. This makes sense, because the exponential factor in the force is essentially equal to 1, so we essentially have a constant force with constant acceleration.

3.10. **$-kx$ force**

This is simply a Hooke's-law spring force, which we'll see much more of in Chapter 4. $F = ma$ gives $-kx = mv\,dv/dx$. Separating variables and integrating yields

$$-\int_{x_0}^{x} kx\,dx = \int_0^v mv\,dv \quad\Longrightarrow\quad \frac{1}{2}kx_0^2 - \frac{1}{2}kx^2 = \frac{1}{2}mv^2. \tag{3.85}$$

Solving for $v \equiv dx/dt$ and then separating variables and integrating again gives

$$\int_{x_0}^{x} \frac{dx}{\sqrt{x_0^2 - x^2}} = \pm \int_0^t \sqrt{\frac{k}{m}}\,dt. \tag{3.86}$$

You can look up this integral, or you can solve it with a trig substitution. Letting $x \equiv x_0 \cos\theta$ gives $dx = -x_0 \sin\theta\,d\theta$, and so we have

$$\int_0^{\theta} \frac{-x_0 \sin\theta\,d\theta}{x_0 \sin\theta} = \pm\sqrt{\frac{k}{m}}\,t \quad\Longrightarrow\quad \theta = \mp\sqrt{\frac{k}{m}}\,t. \tag{3.87}$$

From the definition of θ, the solution for $x(t)$ is therefore

$$x(t) = x_0 \cos\left(\sqrt{\frac{k}{m}}\,t\right). \tag{3.88}$$

We see that the particle oscillates back and forth sinusoidally. It completes a full oscillation when the argument of the cosine increases by 2π. So the period of the motion is $T = 2\pi\sqrt{m/k}$, which interestingly is independent of x_0. It increases with m and decreases with k, as expected.

3.11. **Falling chain**

Let the density of the chain be ρ, and let $y(t)$ be the length hanging down through the hole at time t. Then the total mass is $\rho\ell$, and the mass hanging below the hole is ρy. The net downward force on the chain is $(\rho y)g$, so $F = ma$ gives

$$\rho g y = (\rho\ell)\ddot{y} \quad\Longrightarrow\quad \ddot{y} = \frac{g}{\ell}y. \tag{3.89}$$

At this point, there are two ways we can proceed:

First method: Since we have a function whose second derivative is proportional to itself, a good bet for the solution is an exponential function. And indeed, a quick check shows that the solution is

$$y(t) = Ae^{\alpha t} + Be^{-\alpha t}, \quad \text{where } \alpha \equiv \sqrt{\frac{g}{\ell}}. \tag{3.90}$$

Taking the derivative of this to obtain $\dot{y}(t)$, and using the given information that $\dot{y}(0) = 0$, we find $A = B$. Using $y(0) = y_0$, we then find $A = B = y_0/2$. So the length that hangs below the hole is

$$y(t) = \frac{y_0}{2}\left(e^{\alpha t} + e^{-\alpha t}\right) \equiv y_0 \cosh(\alpha t). \tag{3.91}$$

And the speed is

$$\dot{y}(t) = \frac{\alpha y_0}{2}\left(e^{\alpha t} - e^{-\alpha t}\right) \equiv \alpha y_0 \sinh(\alpha t). \tag{3.92}$$

The time T that satisfies $y(T) = \ell$ is given by $\ell = y_0 \cosh(\alpha T)$. Using $\sinh x = \sqrt{\cosh^2 x - 1}$, we find that the speed of the chain right when it loses contact with the table is

$$\dot{y}(T) = \alpha y_0 \sinh(\alpha T) = \alpha\sqrt{\ell^2 - y_0^2} \equiv \sqrt{g\ell}\sqrt{1 - \eta_0^2}, \tag{3.93}$$

where $\eta_0 \equiv y_0/\ell$ is the initial fraction hanging below the hole. If $\eta_0 \approx 0$, then the speed at time T is $\sqrt{g\ell}$ (this quickly follows from conservation of energy, which is the subject of Chapter 5). Also, you can show that Eq. (3.91) implies that T goes to infinity logarithmically as $\eta_0 \to 0$.

SECOND METHOD: Write $\ddot{y}$ as $v\,dv/dy$ in Eq. (3.89), and then separate variables and integrate to obtain

$$\int_0^v v\,dv = \alpha^2 \int_{y_0}^y y\,dy \quad \Longrightarrow \quad v^2 = \alpha^2(y^2 - y_0^2), \tag{3.94}$$

where $\alpha \equiv \sqrt{g/\ell}$. Now write v as dy/dt and separate variables again to obtain

$$\int_{y_0}^y \frac{dy}{\sqrt{y^2 - y_0^2}} = \alpha \int_0^t dt. \tag{3.95}$$

The integral on the left-hand side is $\cosh^{-1}(y/y_0)$, so we arrive at $y(t) = y_0 \cosh(\alpha t)$, in agreement with Eq. (3.91). The solution then proceeds as above. However, an easier way to obtain the final speed with this method is to simply use the result for v in Eq. (3.94). This tells us that the speed of the chain when it leaves the table (that is, when $y = \ell$) is $v = \alpha\sqrt{\ell^2 - y_0^2}$, in agreement with Eq. (3.93).

3.12. **Throwing a beach ball**

On both the way up and the way down, the total force on the ball is

$$F = -mg - m\alpha v. \tag{3.96}$$

On the way up, v is positive, so the drag force points downward, as it should. And on the way down, v is negative, so the drag force points upward. Our strategy for finding v_f will be to produce two different expressions for the maximum height h, and then equate them. We'll find these two expressions by considering the upward and then the downward motion of the ball. In doing so, we will need to write the acceleration of the ball as $a = v\,dv/dy$. For the upward motion, $F = ma$ gives

$$-mg - m\alpha v = mv\frac{dv}{dy} \quad \Longrightarrow \quad \int_0^h dy = -\int_{v_0}^0 \frac{v\,dv}{g + \alpha v}. \tag{3.97}$$

where we have taken advantage of the fact that the speed of the ball at the top is zero. Writing $v/(g + \alpha v)$ as $[1 - g/(g + \alpha v)]/\alpha$, the integral yields

$$h = \frac{v_0}{\alpha} - \frac{g}{\alpha^2}\ln\left(1 + \frac{\alpha v_0}{g}\right). \tag{3.98}$$

Now consider the downward motion. Let v_f be the final speed, which is a positive quantity. The final velocity is then the negative quantity, $-v_f$. Using $F = ma$, we obtain

$$\int_h^0 dy = -\int_0^{-v_f} \frac{v\,dv}{g + \alpha v}. \tag{3.99}$$

Performing the integration (or just replacing the v_0 in Eq. (3.98) with $-v_f$) gives

$$h = -\frac{v_f}{\alpha} - \frac{g}{\alpha^2} \ln\left(1 - \frac{\alpha v_f}{g}\right). \tag{3.100}$$

Equating the expressions for h in Eqs. (3.98) and (3.100) gives an implicit equation for v_f in terms of v_0,

$$v_0 + v_f = \frac{g}{\alpha} \ln\left(\frac{g + \alpha v_0}{g - \alpha v_f}\right). \tag{3.101}$$

REMARKS: In the limit of small α (more precisely, in the limit $\alpha v_0/g \ll 1$), we can use $\ln(1 + x) = x - x^2/2 + \cdots$ to obtain approximate values for h in Eqs. (3.98) and (3.100). The results are, as expected,

$$h \approx \frac{v_0^2}{2g}, \quad \text{and} \quad h \approx \frac{v_f^2}{2g}. \tag{3.102}$$

We can also make approximations for large α (or large $\alpha v_0/g$). In this limit, the log term in Eq. (3.98) is negligible, so we obtain $h \approx v_0/\alpha$. And Eq. (3.100) gives $v_f \approx g/\alpha$, because the argument of the log must be very small in order to give a very large negative number, which is needed to produce a positive h on the left-hand side. There is no way to relate v_f and h in this limit, because the ball quickly reaches the terminal velocity of $-g/\alpha$ (which is the velocity that makes the net force equal to zero), independent of h. ♣

Let's now find the times it takes for the ball to go up and to go down. We'll present two methods for doing this.

FIRST METHOD: Let T_1 be the time for the upward path. If we write the acceleration of the ball as $a = dv/dt$, then $F = ma$ gives $-mg - m\alpha v = m\,dv/dt$. Separating variables and integrating yields

$$\int_0^{T_1} dt = -\int_{v_0}^0 \frac{dv}{g + \alpha v} \quad \Longrightarrow \quad T_1 = \frac{1}{\alpha} \ln\left(1 + \frac{\alpha v_0}{g}\right). \tag{3.103}$$

In a similar manner, we find that the time T_2 for the downward path is

$$T_2 = -\frac{1}{\alpha} \ln\left(1 - \frac{\alpha v_f}{g}\right). \tag{3.104}$$

Therefore,

$$T_1 + T_2 = \frac{1}{\alpha} \ln\left(\frac{g + \alpha v_0}{g - \alpha v_f}\right) = \frac{v_0 + v_f}{g}, \tag{3.105}$$

where we have used Eq. (3.101). This result is shorter than the time in vacuum (namely $2v_0/g$) because $v_f < v_0$.

SECOND METHOD: The very simple form of Eq. (3.105) suggests that there is a cleaner way of deriving it. And indeed, if we integrate $m\,dv/dt = -mg - m\alpha v$ with respect to time on the way up, we obtain $-v_0 = -gT_1 - \alpha h$ (because $\int v\,dt = h$). Likewise, if we integrate $m\,dv/dt = -mg - m\alpha v$ with respect to time on the way down, we obtain $-v_f = -gT_2 + \alpha h$ (because $\int v\,dt = -h$). Adding these two results gives Eq. (3.105). Note that this procedure works only because the drag force is proportional to v.

REMARK: The fact that the time here is shorter than the time in vacuum isn't obvious. On one hand, the ball doesn't travel as high in air as it would in vacuum, so you might think that $T_1 + T_2 < 2v_0/g$. But on the other hand, the ball moves slower in air on the way down, so you might think that $T_1 + T_2 > 2v_0/g$. It isn't obvious which effect wins, without doing a calculation.[28] For any α, you can use Eq. (3.103) to show that $T_1 < v_0/g$. But T_2 is harder to get a handle on, because it is given in terms of v_f. But in the limit of large α, the ball quickly reaches terminal velocity, so we have $T_2 \approx h/v_f$. Using the results from the previous remark, this becomes $T_2 \approx (v_0/\alpha)/(g/\alpha) = v_0/g$. Interestingly, this equals the downward (and upward) time for a ball thrown in vacuum. ♣

3.13. **Balancing a pencil**

(a) The component of gravity in the tangential direction is $mg\sin\theta \approx mg\theta$. Therefore, the tangential $F = ma$ equation is $mg\theta = m\ell\ddot{\theta}$, which may be written as $\ddot{\theta} = (g/\ell)\theta$. The general solution to this equation is[29]

$$\theta(t) = Ae^{t/\tau} + Be^{-t/\tau}, \quad \text{where} \quad \tau \equiv \sqrt{\ell/g}. \tag{3.106}$$

The constants A and B are found from the initial conditions,

$$\begin{aligned} \theta(0) = \theta_0 \quad &\Longrightarrow \quad A + B = \theta_0, \\ \dot{\theta}(0) = \omega_0 \quad &\Longrightarrow \quad (A - B)/\tau = \omega_0. \end{aligned} \tag{3.107}$$

Solving for A and B, and then plugging the results into Eq. (3.106) gives

$$\theta(t) = \frac{1}{2}(\theta_0 + \omega_0\tau)\, e^{t/\tau} + \frac{1}{2}(\theta_0 - \omega_0\tau)\, e^{-t/\tau}. \tag{3.108}$$

(b) The constants A and B will turn out to be small (they will each be of order $\sqrt{\hbar}$). Therefore, by the time the positive exponential has increased enough to make θ of order 1, the negative exponential will have become negligible. We will therefore ignore this term from here on. In other words,

$$\theta(t) \approx \frac{1}{2}(\theta_0 + \omega_0\tau)\, e^{t/\tau}. \tag{3.109}$$

The goal is to keep θ small for as long as possible. Hence, we want to minimize the coefficient of the exponential, subject to the uncertainty-principle constraint, $(\ell\theta_0)(m\ell\omega_0) \geq \hbar$. This constraint gives $\omega_0 \geq \hbar/(m\ell^2\theta_0)$. Therefore,

$$\theta(t) \geq \frac{1}{2}\left(\theta_0 + \frac{\hbar\tau}{m\ell^2\theta_0}\right) e^{t/\tau}. \tag{3.110}$$

Taking the derivative with respect to θ_0 to minimize the coefficient, we find that the minimum value occurs at $\theta_0 = \sqrt{\hbar\tau/m\ell^2}$. Substituting this back into Eq. (3.110) gives

$$\theta(t) \geq \sqrt{\frac{\hbar\tau}{m\ell^2}}\, e^{t/\tau}. \tag{3.111}$$

Setting $\theta \approx 1$, and then solving for t gives (using $\tau \equiv \sqrt{\ell/g}$)

$$t \leq \frac{1}{4}\sqrt{\frac{\ell}{g}}\ln\left(\frac{m^2\ell^3 g}{\hbar^2}\right). \tag{3.112}$$

[28] For a similar setup where it again isn't obvious (for good reason) which effect wins, see Exercise 3.36.

[29] If you want, you can derive this by separating variables and integrating. The solution is essentially the same as in the second method presented in the solution to Problem 3.11.

With the given values, $m = 0.01$ kg and $\ell = 0.1$ m, along with $g = 10\,\text{m/s}^2$ and $\hbar = 1.06 \cdot 10^{-34}$ J s, we obtain

$$t \le \frac{1}{4}(0.1\text{ s})\ln(9 \cdot 10^{61}) \approx 3.5\text{ s}. \tag{3.113}$$

No matter how clever you are, and no matter how much money you spend on the newest cutting-edge pencil balancing equipment, you can never get a pencil to balance for more than about four seconds.

REMARKS:

1. The smallness of this answer is quite amazing. It is remarkable that a quantum effect on a macroscopic object can produce an everyday value for a time scale. Basically, the point is that the fast exponential growth of θ (which gives rise to the log in the final result for t) wins out over the smallness of $\hbar$, and produces a result for t of order 1. When push comes to shove, exponential effects always win.
2. The above value for t depends strongly on ℓ and g, through the $\sqrt{\ell/g}$ term. But the dependence on m, ℓ, and g in the log term is very weak, because $\hbar$ is so small. If m is increased by a factor of 1000, for example, the result for t increases by only about 10%. This implies that any factors of order 1 that we neglected throughout this problem are completely irrelevant. They will appear in the argument of the log term, and will therefore have negligible effect.
3. Note that dimensional analysis, which is generally a very powerful tool, won't get you too far in this problem. The quantity $\sqrt{\ell/g}$ has dimensions of time, and the quantity $\eta \equiv m^2\ell^3 g/\hbar^2$ is dimensionless (it is the only such quantity), so the balancing time must take the form,

$$t \approx \sqrt{\frac{\ell}{g}}\, f(\eta), \tag{3.114}$$

where f is some function. If the leading term in f were a power (even, for example, a square root), then t would essentially be infinite ($t \approx 10^{30}$ s $\approx 10^{22}$ years for the square root). But f in fact turns out to be a log (which you can't know without solving the problem), which completely cancels out the smallness of $\hbar$, reducing an essentially infinite time down to a few seconds. ♣

3.14. **Maximum trajectory area**

Let θ be the angle at which the ball is thrown. Then the coordinates are given by $x = (v\cos\theta)t$ and $y = (v\sin\theta)t - gt^2/2$. The total time in the air is $2(v\sin\theta)/g$, so the area under the trajectory, $A = \int y\,dx$, is

$$\int_0^{x_{\text{max}}} y\,dx = \int_0^{2v\sin\theta/g}\left((v\sin\theta)t - \frac{gt^2}{2}\right)(v\cos\theta\,dt) = \frac{2v^4}{3g^2}\sin^3\theta\cos\theta. \tag{3.115}$$

Taking the derivative of this, we find that the maximum occurs when $\tan\theta = \sqrt{3}$, that is, when $\theta = 60°$. The maximum area is then $A_{\text{max}} = \sqrt{3}v^4/8g^2$. Note that by dimensional analysis we know that the area, which has dimensions of distance squared, must be proportional to v^4/g^2.

3.15. **Bouncing ball**

The ball travels $2h$ during the first up-and-down journey. It travels $2hf$ during the second, then $2hf^2$ during the third, and so on. Therefore, the total distance traveled is

$$D = 2h\big(1 + f + f^2 + f^3 + \cdots\big) = \frac{2h}{1-f}. \tag{3.116}$$

The time it takes to fall down during the first up-and-down is obtained from $h = gt^2/2$. Therefore, the time for the first up-and-down equals $2t = 2\sqrt{2h/g}$. Likewise, the time for the second up-and-down equals $2\sqrt{2(hf)/g}$. Each successive up-and-down time decreases by a factor of $\sqrt{f}$, so the total time is

$$T = 2\sqrt{\frac{2h}{g}}(1 + f^{1/2} + f^1 + f^{3/2} + \cdots) = 2\sqrt{\frac{2h}{g}} \cdot \frac{1}{1 - \sqrt{f}}. \tag{3.117}$$

The average speed is therefore

$$\frac{D}{T} = \frac{\sqrt{gh/2}}{1 + \sqrt{f}}. \tag{3.118}$$

Remark: The average speed for $f \approx 1$ is roughly half of the average speed for $f \approx 0$. This may seem counterintuitive, because in the $f \approx 0$ case the ball slows down far more quickly than in the $f \approx 1$ case. But the $f \approx 0$ case consists of essentially only one bounce, and the average speed for that one bounce is the largest of any bounce. Both D and T are smaller for $f \approx 0$ than for $f \approx 1$, but T is smaller by a larger factor. ♣

3.16. **Perpendicular velocities**

In the maximum-distance case, let v_i be the initial speed of the ball, and let v_f be the final speed right before it hits the plane (so $v_f = \sqrt{v_i^2 - 2gh}$, where h is the final height of the ball). Let the parabolic path be labeled P, and let the beginning and ending points be A and B, as shown in Fig. 3.40.

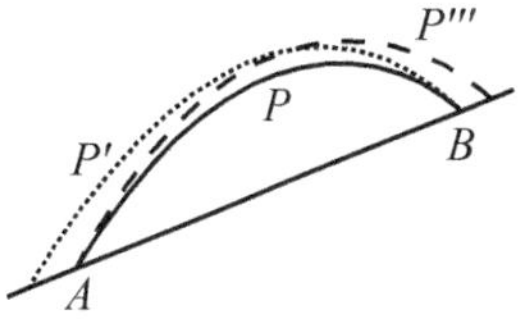

Fig. 3.40

Consider the question, "Given an initial speed v_f, at what inclination angle should a ball be thrown *down* the plane from point B so that it travels the maximum distance?" The answer is that it should be thrown along the same path P, tracing out the path backward. This is certainly a possible physical trajectory (reversing time still leads to a solution to $F = ma$ for projectile motion), and we claim that it does indeed yield the maximum distance. This can be seen in the following way.

Assume (in search of a contradiction) that with initial speed v_f, the maximum distance down the plane is obtained via some other path P' that lands farther down the plane, as shown in Fig. 3.40. Then if we decrease the initial speed v_f by an appropriate amount, we can have the ball land at point A via some path P'' (not shown, lest the figure get too cluttered). From the conservation-of-energy result, the final speed at A in this case is less than v_i. But if we simply reverse the motion along path P'', we see that we can get from A to B by using an initial speed that is less than v_i. So if we then increase the speed up to v_i, we can hit a point above B on the plane via some other path P''', contradicting the fact that B was the endpoint of the maximum-distance trajectory starting at A with initial speed v_i. This contradiction shows that the maximum distance down the plane, starting at B with speed v_f, must in fact be attained via the path P.

Now, from the example in Section 3.4, we know that in the maximum-distance case, the throwing angle bisects the angle between the ground and the vertical. We therefore have the situation shown in Fig. 3.41, because the same path gives the maximum distance for both the upward and downward throws. Since $2\alpha + 2\gamma = 180°$, we see that $\alpha + \gamma = 90°$, as we wanted to show.

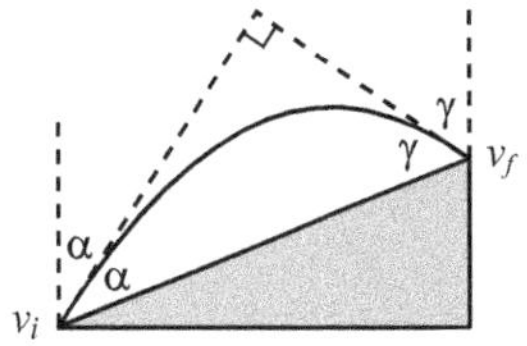

Fig. 3.41

3.17. **Throwing a ball from a cliff**

Let the inclination angle be θ. Then the horizontal speed is $v_x = v\cos\theta$, and the initial vertical speed is $v_y = v\sin\theta$. The time it takes for the ball to hit the ground is given by $h + (v\sin\theta)t - gt^2/2 = 0$. Therefore,

$$t = \frac{v}{g}\left(\sin\theta + \sqrt{\sin^2\theta + \beta}\right), \quad \text{where } \beta \equiv \frac{2gh}{v^2}. \tag{3.119}$$

(The "−" solution for t from the quadratic formula corresponds to the ball being thrown backward down through the cliff.) The horizontal distance traveled is $d = (v\cos\theta)t$, which gives

$$d = \frac{v^2}{g}\cos\theta\left(\sin\theta + \sqrt{\sin^2\theta + \beta}\right). \tag{3.120}$$

We want to maximize this function of θ. Taking the derivative, multiplying through by $\sqrt{\sin^2\theta + \beta}$, and setting the result equal to zero, gives

$$(\cos^2\theta - \sin^2\theta)\sqrt{\sin^2\theta + \beta} = \sin\theta\big(\beta - (\cos^2\theta - \sin^2\theta)\big). \tag{3.121}$$

Using $\cos^2\theta = 1 - \sin^2\theta$, and then squaring and simplifying this equation, gives an optimal angle of

$$\sin\theta_{\text{max}} = \frac{1}{\sqrt{2+\beta}} \equiv \frac{1}{\sqrt{2+2gh/v^2}}. \tag{3.122}$$

Plugging this into Eq. (3.120), and simplifying, gives a maximum distance of

$$d_{\text{max}} = \frac{v^2}{g}\sqrt{1+\beta} \equiv \frac{v^2}{g}\sqrt{1+\frac{2gh}{v^2}}. \tag{3.123}$$

If $h = 0$, then $\theta_{\text{max}} = \pi/4$ and $d_{\text{max}} = v^2/g$, in agreement with the example in Section 3.4. If $h \to \infty$ or $v \to 0$, then $\theta_{\text{max}} \approx 0$, which makes sense.

Remark: If we make use of conservation of energy (discussed in Chapter 5), it turns out that the final speed of the ball when it hits the ground is $v_{\text{f}} = \sqrt{v^2 + 2gh}$. The maximum distance in Eq. (3.123) can therefore be written as (with $v_{\text{i}} \equiv v$ being the initial speed)

$$d_{\text{max}} = \frac{v_{\text{i}} v_{\text{f}}}{g}. \tag{3.124}$$

Note that this is symmetric in v_{i} and v_{f}, as it must be, because we could imagine the trajectory running backward. Also, it equals zero if v_{i} is zero, and it reduces to v^2/g on level ground, as it should. We can also write the angle θ in Eq. (3.122) in terms of v_{f} (instead of h). You can show that the result is $\tan\theta = v_{\text{i}}/v_{\text{f}}$. This implies that the initial and final velocities are perpendicular to each other, because running the trajectory backward interchanges v_{i} and v_{f}, which means that the product of the tangents of the two angles equals 1. This is basically the same result as in Problem 3.16. ♣

3.18. **Redirected motion**

First solution: We will use the results of Problem 3.17, namely Eqs. (3.123) and (3.122), which say that an object projected from height y at speed v travels a maximum horizontal distance of

$$d_{\text{max}} = \frac{v^2}{g}\sqrt{1+\frac{2gy}{v^2}}, \tag{3.125}$$

and the optimal angle yielding this distance is

$$\sin\theta = \frac{1}{\sqrt{2+2gy/v^2}}. \tag{3.126}$$

In the problem at hand, the object is dropped from height h, so the kinematic relation $v_{\text{f}}^2 = v_{\text{i}}^2 + 2ad$ gives the speed at height y as $v = \sqrt{2g(h-y)}$. Plugging this into Eq. (3.125) tells us that the maximum horizontal distance, as a function of y, is

$$d_{\text{max}}(y) = 2\sqrt{h(h-y)}. \tag{3.127}$$

This is maximum when $y = 0$ (assuming we can't have negative y), in which case the distance is $d_{\max} = 2h$. Equation (3.126) gives the associated optimal angle as $\theta = 45°$.

SECOND SOLUTION: Assume that the greatest distance d_0 is obtained when the surface is at $y = y_0$. We will show that y_0 must be 0. We will do this by assuming $y_0 \neq 0$ and explicitly constructing a situation that yields a greater distance.

Let P be the point where the ball finally hits the ground after it bounces off the surface at height y_0. Consider a second scenario where a ball is dropped from height h directly above P. The speed of this ball at P will be the same as the speed of the original ball at P. This follows from conservation of energy. (Or, since we haven't covered energy yet, you can use the kinematic relation $v_\text{f}^2 = v_\text{i}^2 + 2ad$ for the y speeds of the two balls; apply it in two steps for the first ball.)

Now imagine putting a surface at P at the appropriate angle so that the second ball bounces off in the direction from which the first ball came. Then the second ball will travel backward along the parabolic trajectory of the first one. But this means that after the second ball gets to the location of the platform at y_0 (which we have now removed), it will cover more horizontal distance before it hits the ground. We have therefore constructed a setup in which the ball travels farther horizontally than in our proposed maximal case. So the optimal setup must have $y_0 = 0$, in which case the example in Section 3.4 says that the optimal angle is $\theta = 45°$. If we want the ball to go even farther, we can simply dig a (wide enough) hole in the ground and have the ball bounce from the bottom of the hole.

3.19. **Maximum trajectory length**

Let θ be the angle at which the ball is thrown. Then the coordinates are given by $x = (v\cos\theta)t$ and $y = (v\sin\theta)t - gt^2/2$. The ball reaches its maximum height at $t = v\sin\theta/g$, so the length of the trajectory is

$$\begin{aligned} L &= 2\int_0^{v\sin\theta/g} \sqrt{\left(\frac{dx}{dt}\right)^2 + \left(\frac{dy}{dt}\right)^2}\, dt \\ &= 2\int_0^{v\sin\theta/g} \sqrt{(v\cos\theta)^2 + (v\sin\theta - gt)^2}\, dt \\ &= 2v\cos\theta \int_0^{v\sin\theta/g} \sqrt{1 + \left(\tan\theta - \frac{gt}{v\cos\theta}\right)^2}\, dt. \end{aligned} \tag{3.128}$$

Letting $z \equiv \tan\theta - gt/v\cos\theta$, we obtain

$$L = -\frac{2v^2\cos^2\theta}{g}\int_{\tan\theta}^{0} \sqrt{1+z^2}\, dz. \tag{3.129}$$

We can look up this integral, or we can derive it by making a $z \equiv \sinh\alpha$ substitution. The result is

$$\begin{aligned} L &= \frac{2v^2\cos^2\theta}{g}\cdot\frac{1}{2}\left(z\sqrt{1+z^2} + \ln\left(z + \sqrt{1+z^2}\right)\right)\Bigg|_0^{\tan\theta} \\ &= \frac{v^2}{g}\left(\sin\theta + \cos^2\theta\,\ln\left(\frac{\sin\theta + 1}{\cos\theta}\right)\right). \end{aligned} \tag{3.130}$$

As an intermediate check, you can verify that $L = 0$ when $\theta = 0$, and $L = v^2/g$ when $\theta = 90°$. Taking the derivative of Eq. (3.130) to find the maximum, we obtain

$$\begin{aligned} 0 &= \cos\theta - 2\cos\theta\sin\theta\ln\left(\frac{1+\sin\theta}{\cos\theta}\right) \\ &\quad + \cos^2\theta\left(\frac{\cos\theta}{1+\sin\theta}\right)\frac{\cos^2\theta + (1+\sin\theta)\sin\theta}{\cos^2\theta}. \end{aligned} \tag{3.131}$$

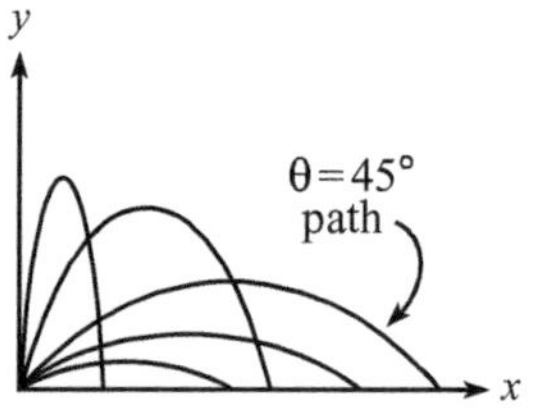

Fig. 3.42

This reduces to

$$1 = \sin\theta \ln\left(\frac{1+\sin\theta}{\cos\theta}\right). \tag{3.132}$$

You can show numerically that the solution for θ is $\theta_0 \approx 56.5°$.

REMARK: A few possible trajectories are shown in Fig. 3.42. Using the standard result that $\theta = 45°$ provides the maximum *horizontal* distance, it follows from the figure that the θ_0 that yields the maximum trajectory length must satisfy $\theta_0 \geq 45°$. The exact angle, however, requires the above detailed calculation. ♣

3.20. **Centripetal acceleration**

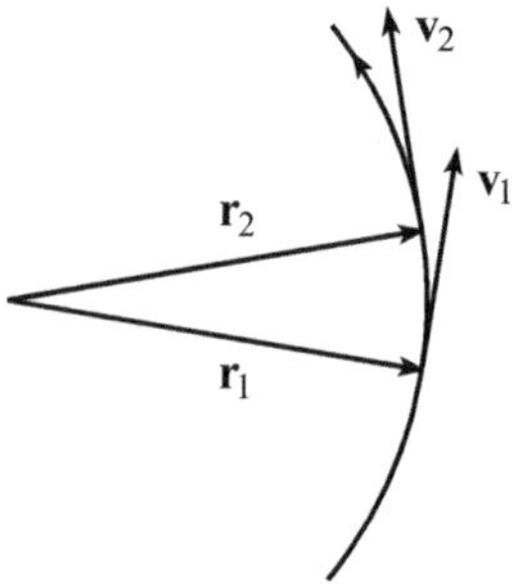

Fig. 3.43

The position and velocity vectors at two nearby times are shown in Fig. 3.43. Their differences, $\Delta\mathbf{r} \equiv \mathbf{r}_2 - \mathbf{r}_1$ and $\Delta\mathbf{v} \equiv \mathbf{v}_2 - \mathbf{v}_1$, are shown in Fig. 3.44. The angle between the **v**'s is the same as the angle between the **r**'s, because each **v** makes a right angle with the corresponding **r**. Therefore, the triangles in Fig. 3.44 are similar, so we have

$$\frac{|\Delta\mathbf{v}|}{v} = \frac{|\Delta\mathbf{r}|}{r}, \tag{3.133}$$

where $r \equiv |\mathbf{r}|$ and $v \equiv |\mathbf{v}|$. Dividing through by Δt gives

$$\frac{1}{v}\left|\frac{\Delta\mathbf{v}}{\Delta t}\right| = \frac{1}{r}\left|\frac{\Delta\mathbf{r}}{\Delta t}\right| \implies \frac{|\mathbf{a}|}{v} = \frac{|\mathbf{v}|}{r} \implies a = \frac{v^2}{r}. \tag{3.134}$$

We have assumed that Δt is infinitesimal here, which allows us to get rid of the Δ's in favor of instantaneous quantities.

3.21. **Vertical acceleration**

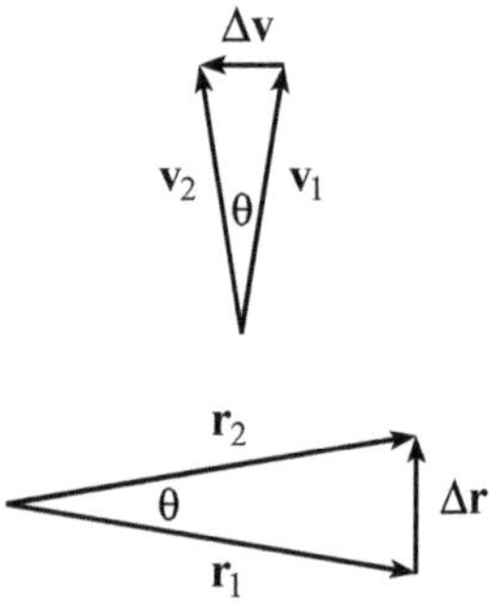

Fig. 3.44

Let θ be the angle down from the top of the hoop. The tangential acceleration is $a_t = g\sin\theta$, and the radial acceleration is $a_r = v^2/R = 2gh/R$. But the height fallen is $h = R - R\cos\theta$, so we have

$$a_r = \frac{2gR(1-\cos\theta)}{R} = 2g(1-\cos\theta). \tag{3.135}$$

We want the total acceleration to be vertical, which means that we want the horizontal components of $\mathbf{a}_t$ and $\mathbf{a}_r$ in Fig. 3.45 to cancel. That is, $a_t\cos\theta = a_r\sin\theta$. This gives

$$(g\sin\theta)\cos\theta = 2g(1-\cos\theta)\sin\theta \implies \sin\theta = 0 \;\text{ or }\; \cos\theta = 2/3. \tag{3.136}$$

The $\sin\theta = 0$ root corresponds to the top and bottom of the hoop ($\theta = 0$ and $\theta = \pi$). So we want the $\cos\theta = 2/3 \implies \theta \approx \pm 48.2°$ root. The vertical acceleration is the sum of the vertical components of $\mathbf{a}_t$ and $\mathbf{a}_r$, so

$$\begin{aligned} a_y = \; & a_t\sin\theta + a_r\cos\theta = (g\sin\theta)\sin\theta + 2g(1-\cos\theta)\cos\theta \\ & = g(\sin^2\theta + 2\cos\theta - 2\cos^2\theta). \end{aligned} \tag{3.137}$$

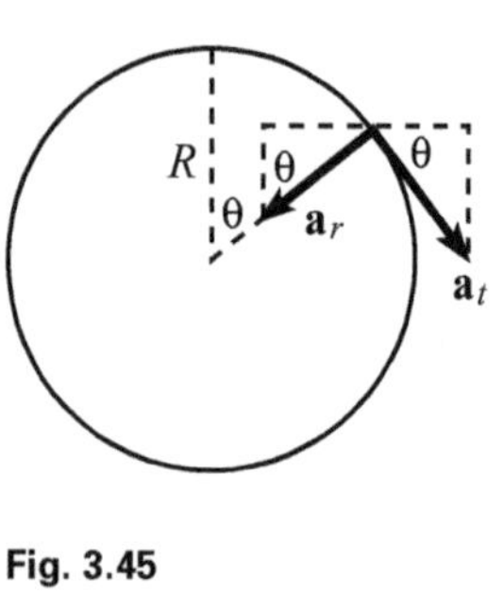

Fig. 3.45

Using $\cos\theta = 2/3$, and hence $\sin\theta = \sqrt{5}/3$, we have

$$a_y = g\left(\frac{5}{9} + 2\cdot\frac{2}{3} - 2\cdot\frac{4}{9}\right) = g. \tag{3.138}$$

REMARK: The reason for this nice answer is the following. If there is no horizontal acceleration, then the normal force from the hoop must have no horizontal component. In other words, $N\sin\theta = 0$. Therefore, either $\sin\theta = 0$ (which gives the top and bottom solutions of $\theta = 0$ and $\theta = \pi$), or $N = 0$, which means that there is no normal force, so the bead feels only gravity, so it's in freefall with $a_y = g$.

If we want, we can use this $N = 0$ requirement as the starting point for a second solution. Using the a_r from above, the radial $F = ma$ equation is $mg\cos\theta - N = 2mg(1 - \cos\theta)$, with positive N defined to point outward. Setting $N = 0$ gives the desired result, $\cos\theta = 2/3$. ♣

3.22. **Circling around a pole**

Let F be the tension in the string. At the mass, the angle between the string and the radius of the dotted circle is $\theta = \sin^{-1}(r/R)$. In terms of θ, the radial and tangential $F = ma$ equations are

$$F\cos\theta = \frac{mv^2}{R}, \quad \text{and} \quad F\sin\theta = m\dot{v}. \tag{3.139}$$

Dividing these two equations gives $\tan\theta = (R\dot{v})/v^2$. Separating variables and integrating gives

$$\int_{v_0}^{v} \frac{dv}{v^2} = \frac{\tan\theta}{R}\int_0^t dt \implies \frac{1}{v_0} - \frac{1}{v} = \frac{(\tan\theta)t}{R}$$
$$\implies v(t) = \left(\frac{1}{v_0} - \frac{(\tan\theta)t}{R}\right)^{-1}. \tag{3.140}$$

The speed v becomes infinite when

$$t = T \equiv \frac{R}{v_0\tan\theta}. \tag{3.141}$$

This means that you can keep the mass moving in the desired circle only up to time T. After that, it is impossible. (Of course, it will become impossible, for all practical purposes, long before v becomes infinite.) The total distance, $d = \int v\,dt$, is infinite, because this integral diverges (barely, like a log) as t approaches T.

3.23. **A force $F_\theta = m\dot{r}\dot{\theta}$**

With the given force, Eq. (3.51) becomes

$$0 = m(\ddot{r} - r\dot{\theta}^2), \quad \text{and} \quad m\dot{r}\dot{\theta} = m(r\ddot{\theta} + 2\dot{r}\dot{\theta}). \tag{3.142}$$

The second of these equations gives $-\dot{r}\dot{\theta} = r\ddot{\theta}$. Therefore,

$$\int \frac{\ddot{\theta}}{\dot{\theta}}\,dt = -\int \frac{\dot{r}}{r}\,dt \implies \ln\dot{\theta} = -\ln r + C \implies \dot{\theta} = \frac{D}{r}, \tag{3.143}$$

where $D = e^C$ is a constant of integration, determined by the initial conditions. Substituting this value of $\dot{\theta}$ into the first of Eqs. (3.142), and then multiplying through by $\dot{r}$ and integrating, gives

$$\ddot{r} = r\left(\frac{D}{r}\right)^2 \implies \int \ddot{r}\dot{r}\,dt = D^2\int \frac{\dot{r}}{r}\,dt \implies \frac{\dot{r}^2}{2} = D^2\ln r + E. \tag{3.144}$$

Therefore,

$$\dot{r} = \sqrt{A\ln r + B}, \tag{3.145}$$

where $A \equiv 2D^2$ and $B \equiv 2E$.

3.24. **Free particle**

For zero force, Eq. (3.51) gives

$$\ddot{r} = r\dot{\theta}^2, \quad \text{and} \quad r\ddot{\theta} = -2\dot{r}\dot{\theta}. \tag{3.146}$$

Separating variables in the second equation and integrating yields

$$\int \frac{\ddot{\theta}}{\dot{\theta}}\,dt = -\int \frac{2\dot{r}}{r}\,dt \implies \ln\dot{\theta} = -2\ln r + C \implies \dot{\theta} = \frac{D}{r^2}, \tag{3.147}$$

where $D = e^C$ is a constant of integration, determined by the initial conditions.[30] Substituting this value of $\dot{\theta}$ into the first of Eqs. (3.146), and then multiplying through by $\dot{r}$ and integrating, gives

$$\ddot{r} = r\left(\frac{D}{r^2}\right)^2 \quad \Longrightarrow \quad \int \ddot{r}\dot{r}\,dt = D^2 \int \frac{\dot{r}}{r^3}\,dt \quad \Longrightarrow \quad \frac{\dot{r}^2}{2} = -\frac{D^2}{2r^2} + E. \quad (3.148)$$

We want $\dot{r} = 0$ when $r = r_0$, which implies that $E = D^2/2r_0^2$. Therefore,

$$\dot{r} = V\sqrt{1 - \frac{r_0^2}{r^2}}, \quad (3.149)$$

where $V \equiv D/r_0$. Separating variables and integrating gives

$$\int \frac{r\dot{r}\,dt}{\sqrt{r^2 - r_0^2}} = \int V\,dt \quad \Longrightarrow \quad \sqrt{r^2 - r_0^2} = Vt \quad \Longrightarrow \quad r = \sqrt{r_0^2 + (Vt)^2}, \quad (3.150)$$

where the constant of integration is zero, because we have chosen $t = 0$ to correspond to $r = r_0$. Plugging this value for r into the $\dot{\theta} = D/r^2 \equiv Vr_0/r^2$ result in Eq. (3.147) gives

$$\int d\theta = \int \frac{Vr_0\,dt}{r_0^2 + (Vt)^2} \quad \Longrightarrow \quad \theta = \tan^{-1}\left(\frac{Vt}{r_0}\right) \quad \Longrightarrow \quad \cos\theta = \frac{r_0}{\sqrt{r_0^2 + (Vt)^2}}. \quad (3.151)$$

Finally, combining this with the result for r in Eq. (3.150) gives $\cos\theta = r_0/r$, as desired.

[30] The statement that $r^2\dot{\theta}$ is constant is simply the statement of conservation of angular momentum, because $r^2\dot{\theta} = r(r\dot{\theta}) = rv_\theta$, where v_θ is the tangential velocity. More on this in Chapters 7 and 8.

Chapter 4
Oscillations

In this chapter we will discuss oscillatory motion. The simplest examples of such motion are a swinging pendulum and a mass on a spring, but it is possible to make a system more complicated by introducing a damping force and/or an external driving force. We will study all of these cases.

We are interested in oscillatory motion for two reasons. First, we study it because we *can* study it. This is one of the few systems in physics where we can solve the motion exactly. There's nothing wrong with looking under the lamppost every now and then. Second, oscillatory motion is ubiquitous in nature, for reasons that will become clear in Section 5.2. If there was ever a type of physical system worthy of study, this is it. We'll start off by doing some necessary math in Section 4.1. And then in Section 4.2 we'll show how the math is applied to the physics.

4.1 Linear differential equations

A *linear differential equation* is one in which x and its time derivatives enter only through their first powers. An example is $3\ddot{x} + 7\dot{x} + x = 0$. An example of a nonlinear differential equation is $3\ddot{x} + 7\dot{x}^2 + x = 0$. If the right-hand side of the equation is zero, then we use the term *homogeneous* differential equation. If the right-hand side is some function of t, as in the case of $3\ddot{x} - 4\dot{x} = 9t^2 - 5$, then we use the term *inhomogeneous* differential equation. The goal of this chapter is to learn how to solve linear differential equations, both homogeneous and inhomogeneous. These come up again and again in physics, so we had better find a systematic method of solving them.

The techniques that we will use are best learned through examples, so let's solve a few differential equations, starting with some simple ones. Throughout this chapter, x is understood to be a function of t. Hence, a dot denotes time differentiation.

Example 1 ($\dot{x} = ax$): This is a very simple differential equation. There are two ways (at least) to solve it.

First method: Separate variables to obtain $dx/x = a\,dt$, and then integrate to obtain $\ln x = at + c$, where c is a constant of integration. Then exponentiate to obtain

$$x = Ae^{at}, \tag{4.1}$$

where $A \equiv e^c$. A is determined by the value of x at, say, $t = 0$.

Second method: Guess an exponential solution, that is, one of the form $x = Ae^{\alpha t}$. Substitution into $\dot{x} = ax$ immediately gives $\alpha = a$. Therefore, the solution is $x = Ae^{at}$. Note that we can't solve for A, due to the fact that our differential equation is homogeneous and linear in x (translation: A cancels out). A is determined by the initial condition.

This method may seem a bit silly, and somewhat cheap. But as we will see below, guessing these exponential functions (or sums of them) is actually the most general thing we can try, so the method is indeed quite general.

REMARK: Using this method, you might be concerned that although we have found one solution, we might have missed another one. But the general theory of differential equations says that a first-order linear equation has only one independent solution (we'll just accept this fact here). So if we find one solution, then we know that we've found the whole thing. ♣

Example 2 ($\ddot{x} = ax$): If a is negative, then we'll see that this equation describes the oscillatory motion of, say, a spring. If a is positive, then it describes exponentially growing or decaying motion. There are two ways (at least) to solve this equation.

First method: We can use the separation-of-variables method from Section 3.3 here, because our system is one in which the force depends only on the position x. But this method is rather cumbersome, as you found if you did Problem 3.10 or Exercise 3.38. It will certainly work, but in the case where our equation is *linear* in x, there is a much simpler method:

Second method: As in the first example above, we can guess a solution of the form $x(t) = Ae^{\alpha t}$ and then find out what α must be. Again, we can't solve for A, because it cancels out. Plugging $Ae^{\alpha t}$ into $\ddot{x} = ax$ gives $\alpha = \pm\sqrt{a}$. We have therefore found *two* solutions. The most general solution is an arbitrary linear combination of these,

$$x(t) = Ae^{\sqrt{a}\,t} + Be^{-\sqrt{a}\,t}, \tag{4.2}$$

which you can quickly check does indeed work. A and B are determined by the initial conditions. As in the first example above, you might be concerned that although we have found two solutions to the equation, we might have missed others. But the general theory of differential equations says that our second-order linear equation has only two independent solutions. Therefore, having found two independent solutions, we know that we've found them all.

VERY IMPORTANT REMARK: The fact that the sum of two different solutions is again a solution to our equation is a monumentally important property of *linear* differential equations.

This property does *not* hold for nonlinear differential equations, for example $\ddot{x}^2 = bx$, because the act of squaring after adding the two solutions produces a cross term which destroys the equality, as you should check (see Problem 4.1). This property is called the *principle of superposition*. That is, superposing two solutions yields another solution. In other words, *linearity* leads to superposition. This fact makes theories that are governed by linear equations *much* easier to deal with than those that are governed by nonlinear ones. General Relativity, for example, is based on nonlinear equations, and solutions to most General Relativity systems are extremely difficult to come by.

For equations with one main condition
(Those linear), you have permission
To take your solutions,
With firm resolutions,
And add them in superposition. ♣

Let's say a little more about the solution in Eq. (4.2). If a is negative, then it is helpful to define $a \equiv -\omega^2$, where ω is a real number. The solution then becomes $x(t) = Ae^{i\omega t} + Be^{-i\omega t}$. Using $e^{i\theta} = \cos\theta + i\sin\theta$, this can be written in terms of trig functions, if desired. Various ways of writing the solution are:

$$\begin{aligned} x(t) &= Ae^{i\omega t} + Be^{-i\omega t}, \\ x(t) &= C\cos\omega t + D\sin\omega t, \\ x(t) &= E\cos(\omega t + \phi_1), \\ x(t) &= F\sin(\omega t + \phi_2). \end{aligned} \tag{4.3}$$

Depending on the specifics of a given system, one of the above forms will work better than the others. The various constants in these expressions are related to each other. For example, $C = E\cos\phi_1$ and $D = -E\sin\phi_1$, which follow from the cosine sum formula. Note that there are two free parameters in each of the above expressions for $x(t)$. These parameters are determined by the initial conditions (say, the position and velocity at $t = 0$). In contrast with these free parameters, the quantity ω is determined by the particular physical system we're dealing with. For example, we'll see that for a spring, $\omega = \sqrt{k/m}$, where k is the spring constant. ω is independent of the initial conditions.

If a is positive, then let's define $a \equiv \alpha^2$, where α is a real number. The solution in Eq. (4.2) then becomes $x(t) = Ae^{\alpha t} + Be^{-\alpha t}$. Using $e^{\theta} = \cosh\theta + \sinh\theta$, this can be written in terms of hyperbolic trig functions, if desired. Various ways of writing the solution are:

$$\begin{aligned} x(t) &= Ae^{\alpha t} + Be^{-\alpha t}, \\ x(t) &= C\cosh\alpha t + D\sinh\alpha t, \\ x(t) &= E\cosh(\alpha t + \phi_1), \\ x(t) &= F\sinh(\alpha t + \phi_2). \end{aligned} \tag{4.4}$$

Again, the various constants are related to each other. If you are unfamiliar with the hyperbolic trig functions, a few facts are listed in Appendix A.

Although the solution in Eq. (4.2) is completely correct for both signs of a, it's generally more illuminating to write the negative-a solutions in either the trig forms or the $e^{\pm i\omega t}$ exponential form where the i's are explicit.

The usefulness of our method of guessing exponential solutions cannot be overemphasized. It may seem somewhat restrictive, but it works. The examples in the remainder of this chapter should convince you of this.

This is our method, essential,
For equations we solve, differential.
It gets the job done,
And it's even quite fun.
We just try a routine exponential.

Example 3 ($\ddot{x} + 2\gamma\dot{x} + ax = 0$): This will be our last mathematical example, and then we'll start doing some physics. As we'll see later, this example pertains to a damped harmonic oscillator. We've put a factor of 2 in the coefficient of $\dot{x}$ here to make some later formulas look nicer. The force in this example (if we switch from math to physics for a moment) is $-2\gamma\dot{x} - ax$ (times m), which depends on both v and x. Our methods of Section 3.3 therefore don't apply; we're not going to be able to use separation of variables here. This leaves us with only our method of guessing an exponential solution, $Ae^{\alpha t}$. So let's see what it gives us. Plugging $x(t) = Ae^{\alpha t}$ into the given equation, and canceling the nonzero factor of $Ae^{\alpha t}$, gives

$$\alpha^2 + 2\gamma\alpha + a = 0. \tag{4.5}$$

The solutions for α are $-\gamma \pm \sqrt{\gamma^2 - a}$. Call these α_1 and α_2. Then the general solution to our equation is

$$\begin{aligned} x(t) &= Ae^{\alpha_1 t} + Be^{\alpha_2 t} \\ &= e^{-\gamma t}\left(Ae^{t\sqrt{\gamma^2 - a}} + Be^{-t\sqrt{\gamma^2 - a}}\right). \end{aligned} \tag{4.6}$$

If $\gamma^2 - a < 0$, then we can write this in terms of sines and cosines, so we have oscillatory motion that decreases in time due to the $e^{-\gamma t}$ factor (or it increases, if $\gamma < 0$, but this is rarely physical). If $\gamma^2 - a > 0$, then we have exponential motion. We'll talk more about these different possibilities in Section 4.3.

In the first two examples above, the solutions were fairly clear. But in the present case, you're not apt to look at the above solution and say, "Oh, of course. It's obvious!" So our method of trying solutions of the form $Ae^{\alpha t}$ isn't looking so silly anymore.

In general, if we have an nth order homogeneous linear differential equation,

$$\frac{d^n x}{dt^n} + c_{n-1}\frac{d^{n-1}x}{dt^{n-1}} + \cdots + c_1\frac{dx}{dt} + c_0 x = 0, \tag{4.7}$$

then our strategy is to guess an exponential solution, $x(t) = Ae^{\alpha t}$, and to then (in theory) solve the resulting nth order equation, $\alpha^n + c_{n-1}\alpha^{n-1} + \cdots + c_1\alpha + c_0 = 0$, for α, to obtain the solutions $\alpha_1, \ldots, \alpha_n$. The general solution for $x(t)$ is then the superposition,

$$x(t) = A_1 e^{\alpha_1 t} + A_2 e^{\alpha_2 t} + \cdots + A_n e^{\alpha_n t}, \tag{4.8}$$

where the A_i are determined by the initial conditions. In practice, however, we will rarely encounter differential equations of degree higher than 2. (*Note*: if some of the α_i happen to be equal, then Eq. (4.8) is not valid, so a modification is needed. We will encounter such a situation in Section 4.3.)

4.2 Simple harmonic motion

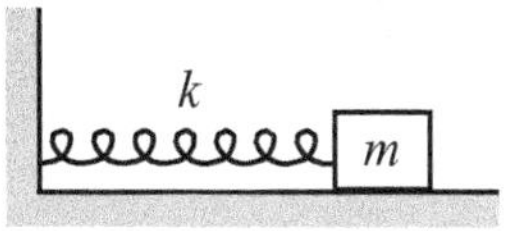

Fig. 4.1

Let's now do some real live physical problems. We'll start with simple harmonic motion. This is the motion undergone by a particle subject to a force $F(x) = -kx$. The classic system that undergoes simple harmonic motion is a mass attached to a massless spring, on a frictionless table (see Fig. 4.1). A typical spring has a force of the form $F(x) = -kx$, where x is the displacement from equilibrium (see Section 5.2 for the reason behind this). This is "Hooke's law," and it holds as long as the spring isn't stretched or compressed too far. Eventually this expression breaks down for any real spring. But if we assume a $-kx$ force, then $F = ma$ gives $-kx = m\ddot{x}$, or

$$\ddot{x} + \omega^2 x = 0, \quad \text{where } \omega \equiv \sqrt{\frac{k}{m}}. \tag{4.9}$$

This is simply the equation we studied in Example 2 in the previous section. From Eq. (4.3), the solution may be written as

$$x(t) = A\cos(\omega t + \phi). \tag{4.10}$$

This trig solution shows that the system oscillates back and forth forever in time. ω is the *angular frequency*. If t increases by $2\pi/\omega$, then the argument of the cosine increases by 2π, so the position and velocity are back to what they were. The *period* (the time for one complete cycle) is therefore $T = 2\pi/\omega = 2\pi\sqrt{m/k}$. The frequency in cycles per second (hertz) is $\nu = 1/T = \omega/2\pi$. The constant A (or rather the absolute value of A, if A is negative) is the *amplitude*, that is, the maximum distance the mass gets from the origin. Note that the velocity as a function of time is $v(t) \equiv \dot{x}(t) = -A\omega\sin(\omega t + \phi)$.

The constants A and ϕ are determined by the initial conditions. If, for example, $x(0) = 0$ and $\dot{x}(0) = v$, then we must have $A\cos\phi = 0$ and $-A\omega\sin\phi = v$. Hence, $\phi = \pi/2$, and so $A = -v/\omega$ (or $\phi = -\pi/2$ and $A = v/\omega$, but this leads to the same solution). Therefore, we have $x(t) = -(v/\omega)\cos(\omega t + \pi/2)$. This looks a little nicer if we write it as $x(t) = (v/\omega)\sin(\omega t)$. It turns out that if the facts you're given are the initial position and velocity, x_0 and v_0, then the $x(t) = C\cos\omega t + D\sin\omega t$ expression in Eq. (4.3) usually works best, because (as you can verify) it yields the nice clean results, $C = x_0$ and $D = v_0/\omega$. Problem 4.3 gives another setup that involves initial conditions.

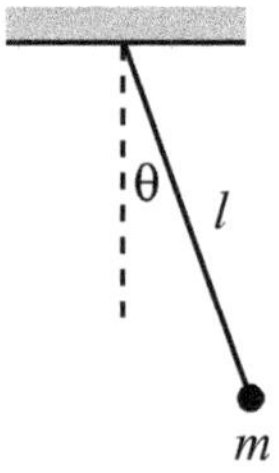

Fig. 4.2

Example (Simple pendulum): Another classic system that undergoes (approximately) simple harmonic motion is the simple pendulum, that is, a mass that hangs on a massless string and swings in a vertical plane. Let ℓ be the length of the string, and let $\theta(t)$ be the angle the string makes with the vertical (see Fig. 4.2). Then the gravitational force on the mass in the tangential direction is $-mg\sin\theta$. So $F = ma$ in the tangential direction gives

$$-mg\sin\theta = m(\ell\ddot{\theta}). \tag{4.11}$$

The tension in the string combines with the radial component of gravity to produce the radial acceleration, so the radial $F = ma$ equation serves only to tell us the tension, which we won't need here.

We will now enter the realm of approximations and assume that the amplitude of the oscillations is small. Without this approximation, the problem cannot be solved in closed form. Assuming θ is small, we can use $\sin\theta \approx \theta$ in Eq. (4.11) to obtain

$$\ddot{\theta} + \omega^2\theta = 0, \quad \text{where } \omega \equiv \sqrt{\frac{g}{\ell}}. \tag{4.12}$$

Therefore,

$$\theta(t) = A\cos(\omega t + \phi), \tag{4.13}$$

where A and ϕ are determined by the initial conditions. So the pendulum undergoes simple harmonic motion with a frequency of $\sqrt{g/\ell}$. The period is therefore $T = 2\pi/\omega = 2\pi\sqrt{\ell/g}$. The true motion is arbitrarily close to this, for sufficiently small amplitudes. Exercise 4.23 deals with the higher-order corrections to the motion in the case where the amplitude is not small.

There will be many occasions throughout your physics education where you will plow through a calculation and then end up with a simple equation of the form $\ddot{z} + \omega^2 z = 0$, where ω^2 is a positive quantity that depends on various parameters in the problem. When you encounter such an equation, you should jump for joy, because without any more effort you can simply write down the answer: the solution for z must be of the form $z(t) = A\cos(\omega t + \phi)$. No matter how complicated the system looks at first glance, if you end up with an equation that looks like $\ddot{z} + \omega^2 z = 0$, then you know that the system undergoes simple harmonic motion with a frequency equal to the square root of the coefficient of z, no matter what that coefficient is. If you end up with $\ddot{z} + (\text{zucchini})z = 0$, then the frequency is $\omega = \sqrt{\text{zucchini}}$ (well, as long as the zucchini is positive and has the dimensions of inverse time squared).

4.3 Damped harmonic motion

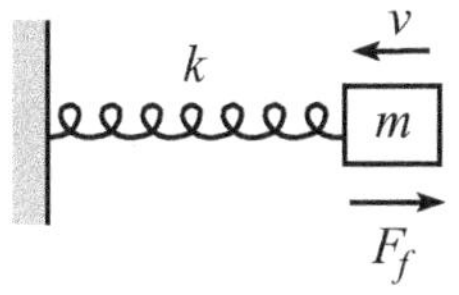

Fig. 4.3

Consider a mass m attached to the end of a spring with spring constant k. Let the mass be subject to a drag force proportional to its velocity, $F_\text{f} = -bv$ (the subscript f here stands for "friction"; we'll save the letter d for "driving" in the next section); see Fig. 4.3. Why do we study this $F_\text{f} = -bv$ damping force? Two reasons: First, it is linear in x, which will allow us to solve for the motion. And second, it is a perfectly realistic force; an object moving at a slow speed through a fluid generally experiences a drag force proportional to its velocity. Note that this $F_\text{f} = -bv$ force is *not* the force that a mass would feel if it were placed on a table with friction. In that case the drag force would be (roughly) constant.

Our goal in this section is to solve for the position as a function of time. The total force on the mass is $F = -b\dot{x} - kx$. So $F = m\ddot{x}$ gives

$$\ddot{x} + 2\gamma\dot{x} + \omega^2 x = 0, \tag{4.14}$$

where $2\gamma \equiv b/m$, and $\omega^2 \equiv k/m$. This is conveniently the same equation we already solved in Example 3 in Section 4.1 (with $a \to \omega^2$). Now, however, we have the physical restrictions that $\gamma > 0$ and $\omega^2 > 0$. Letting $\Omega^2 \equiv \gamma^2 - \omega^2$ for simplicity, we may write the solution in Eq. (4.6) as

$$x(t) = e^{-\gamma t}\left(Ae^{\Omega t} + Be^{-\Omega t}\right), \quad \text{where } \Omega \equiv \sqrt{\gamma^2 - \omega^2}. \tag{4.15}$$

There are three cases to consider.

Case 1: Underdamping ($\Omega^2 < 0$)

If $\Omega^2 < 0$, then $\gamma < \omega$. Since Ω is imaginary, let us define the real number $\tilde{\omega} \equiv \sqrt{\omega^2 - \gamma^2}$, so that $\Omega = i\tilde{\omega}$. Equation (4.15) then gives

$$\begin{aligned} x(t) &= e^{-\gamma t}\left(Ae^{i\tilde{\omega}t} + Be^{-i\tilde{\omega}t}\right) \\ &\equiv e^{-\gamma t} C\cos(\tilde{\omega}t + \phi). \end{aligned} \tag{4.16}$$

These two forms are equivalent. Using $e^{i\theta} = \cos\theta + i\sin\theta$, the constants in Eq. (4.16) are related by $A + B = C\cos\phi$ and $A - B = iC\sin\phi$. Note that in a physical problem, $x(t)$ is real, so we must have $A^* = B$, where the star denotes complex conjugation. The two constants A and B, or the two constants C and ϕ, are determined by the initial conditions.

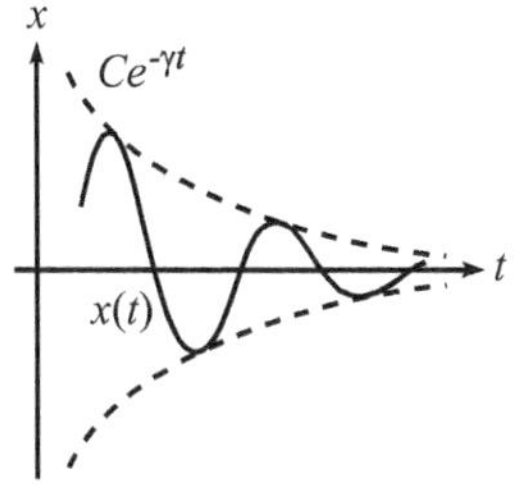

Fig. 4.4

Depending on the given problem, one of the expressions in Eq. (4.16) will inevitably work better than the other. Or perhaps one of the other forms in Eq. (4.3) (times $e^{-\gamma t}$) will be the most useful one. The cosine form makes it apparent that the motion is harmonic motion whose amplitude decreases in time, due to the $e^{-\gamma t}$ factor. A plot of such motion is shown in Fig. 4.4.[1] The frequency of the

[1] To be precise, the amplitude doesn't decrease exactly like $Ce^{-\gamma t}$, as Eq. (4.16) suggests, because $Ce^{-\gamma t}$ describes the envelope of the motion, and not the curve that passes through the extremes of

motion, $\tilde{\omega} = \sqrt{\omega^2 - \gamma^2}$, is smaller than the natural frequency ω of the undamped oscillator.

REMARKS: If γ is very small (more precisely, if $\gamma \ll \omega$), then $\tilde{\omega} \approx \omega$, which makes sense because we almost have an undamped oscillator. If γ is very close to ω, then $\tilde{\omega} \approx 0$. So the oscillations are very slow (more precisely, $\tilde{\omega} \ll \omega$). Of course, for very small $\tilde{\omega}$ it's hard to even tell that the oscillations exist, because they will damp out on a time scale of order $1/\gamma \approx 1/\omega$, which is short compared with the long time scale of the oscillations, $1/\tilde{\omega}$. ♣

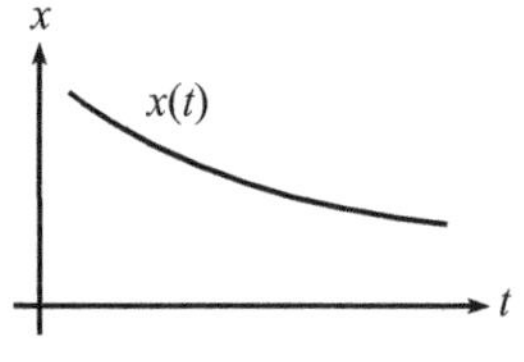

Fig. 4.5

Case 2: Overdamping ($\Omega^2 > 0$)

If $\Omega^2 > 0$, then $\gamma > \omega$. Ω is real (and taken to be positive), so Eq. (4.15) gives

$$x(t) = Ae^{-(\gamma-\Omega)t} + Be^{-(\gamma+\Omega)t}. \tag{4.17}$$

There is no oscillatory motion in this case; see Fig. 4.5. Since $\gamma > \Omega \equiv \sqrt{\gamma^2 - \omega^2}$, both of the exponents are negative, so the motion goes to zero for large t. This had better be the case, because a real spring is certainly not going to have the motion fly off to infinity. If we had obtained a positive exponent somehow, we'd know we had made a mistake.

REMARKS: If γ is just slightly larger than ω, then $\Omega \approx 0$, so the two terms in (4.17) are roughly equal, and we essentially have exponential decay, according to $e^{-\gamma t}$. If $\gamma \gg \omega$ (that is, strong damping), then $\Omega \approx \gamma$, so the first term in (4.17) dominates (it has the less negative exponent), and we essentially have exponential decay according to $e^{-(\gamma-\Omega)t}$. We can be somewhat quantitative about this by approximating Ω as

$$\Omega \equiv \sqrt{\gamma^2 - \omega^2} = \gamma\sqrt{1 - \omega^2/\gamma^2} \approx \gamma(1 - \omega^2/2\gamma^2). \tag{4.18}$$

Hence, the exponential behavior goes like $e^{-\omega^2 t/2\gamma}$. Because $\gamma \gg \omega$, this is slow decay (that is, slow compared with $t \sim 1/\omega$), which makes sense if the damping is very strong. The mass slowly creeps back to the origin, as in the case of a weak spring immersed in molasses. ♣

Case 3: Critical damping ($\Omega^2 = 0$)

If $\Omega^2 = 0$, then $\gamma = \omega$. Equation (4.14) therefore becomes $\ddot{x} + 2\gamma\dot{x} + \gamma^2 x = 0$. In this special case, we have to be careful in solving our differential equation. The solution in Eq. (4.15) is not valid, because in the procedure leading to Eq. (4.6), the roots α_1 and α_2 are equal (to $-\gamma$), so we have really found only one solution, $e^{-\gamma t}$. We'll just invoke here the result from the theory of differential equations that says that in this special case, the other solution is of the form $te^{-\gamma t}$.

REMARK: You should check explicitly that $te^{-\gamma t}$ solves the equation $\ddot{x} + 2\gamma\dot{x} + \gamma^2 x = 0$. Or if you want to, you can derive it in the spirit of Problem 4.2. In the more general case where there are n identical roots in the procedure leading to Eq. (4.8) (call them all α), the n independent solutions to the differential equation are $t^k e^{\alpha t}$, for $0 \le k \le (n-1)$. But more often than not, there are no repeated roots, so you don't have to worry about this. ♣

the motion. You can show that the amplitude in fact decreases like $Ce^{-\gamma t}\cos(\tan^{-1}(\gamma/\tilde{\omega}))$. This is the expression for the curve that passes through the extremes; see Castro (1986). But for small damping ($\gamma \ll \omega$), this is essentially equal to $Ce^{-\gamma t}$. And in any event, it is proportional to $e^{-\gamma t}$.

Our solution is therefore of the form,

$$x(t) = e^{-\gamma t}(A + Bt). \tag{4.19}$$

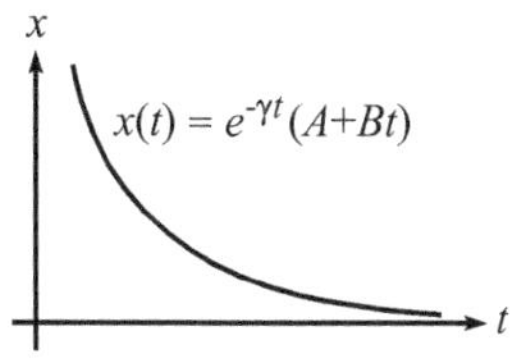

Fig. 4.6

The exponential factor eventually wins out over the Bt term, so the motion goes to zero for large t (see Fig. 4.6).

If we are given a spring with a fixed ω, and if we look at the system for different values of γ, then critical damping (when $\gamma = \omega$) is the case where the motion converges to zero in the quickest way (which is like $e^{-\omega t}$). This is true because in the underdamped case ($\gamma < \omega$), the envelope of the oscillatory motion goes like $e^{-\gamma t}$, which goes to zero slower than $e^{-\omega t}$, because $\gamma < \omega$. And in the overdamped case ($\gamma > \omega$), the dominant piece is the $e^{-(\gamma-\Omega)t}$ term. And as you can verify, if $\gamma > \omega$ then $\gamma - \Omega \equiv \gamma - \sqrt{\gamma^2 - \omega^2} < \omega$, so this motion also goes to zero slower than $e^{-\omega t}$. Critical damping is very important in many real systems, such as screen doors and shock absorbers, where the goal is to have the system head to zero (without overshooting and bouncing around) as fast as possible.

4.4 Driven (and damped) harmonic motion

Before we examine driven harmonic motion, we must learn how to solve a new type of differential equation. How can we solve something of the form

$$\ddot{x} + 2\gamma\dot{x} + ax = C_0 e^{i\omega_0 t}, \tag{4.20}$$

where γ, a, ω_0, and C_0 are given quantities? This is an inhomogeneous differential equation, due to the term on the right-hand side. It's not very physical, because the right-hand side is complex, but let's not worry about that for now. Equations of this sort come up again and again, and fortunately there's a straightforward (although sometimes messy) method for solving them. As usual, the method involves making a reasonable guess, plugging it in, and seeing what condition comes out. Since we have the $e^{i\omega_0 t}$ sitting on the right-hand side of Eq. (4.20), let's guess a solution of the form $x(t) = Ae^{i\omega_0 t}$. A will depend on ω_0, among other things, as we will see. Plugging this guess into Eq. (4.20) and canceling the nonzero factor of $e^{i\omega_0 t}$, we obtain

$$(-\omega_0^2)A + 2\gamma(i\omega_0)A + aA = C_0. \tag{4.21}$$

Solving for A, we find that our solution for x is

$$x(t) = \left(\frac{C_0}{-\omega_0^2 + 2i\gamma\omega_0 + a}\right)e^{i\omega_0 t}. \tag{4.22}$$

Note the differences between this technique and the one in Example 3 in Section 4.1. In that example, the goal was to determine the α in $x(t) = Ae^{\alpha t}$.

And there was no way to solve for A; the initial conditions determined A. But in the present technique, the ω_0 in $x(t) = Ae^{i\omega_0 t}$ is a given quantity, and the goal is to solve for A in terms of the given constants. Therefore, in the solution in Eq. (4.22), there are *no free constants* to be determined by the initial conditions. We've found one particular solution, and we're stuck with it. The term *particular solution* is used for Eq. (4.22).

With no freedom to adjust the solution in Eq. (4.22), how can we satisfy an arbitrary set of initial conditions? Fortunately, Eq. (4.22) does not represent the most general solution to Eq. (4.20). The most general solution is the sum of our particular solution in Eq. (4.22), *plus* the "homogeneous" solution we found in Eq. (4.6). This sum is certainly a solution, because the solution in Eq. (4.6) was explicitly constructed to yield zero when plugged into the left-hand side of Eq. (4.20). Therefore, tacking it on to our particular solution doesn't change the equality in Eq. (4.20), because the left side is linear. The principle of superposition has saved the day. The complete solution to Eq. (4.20) is therefore

$$x(t) = e^{-\gamma t}\left(Ae^{t\sqrt{\gamma^2-a}} + Be^{-t\sqrt{\gamma^2-a}}\right) + \left(\frac{C_0}{-\omega_0^2 + 2i\gamma\omega_0 + a}\right)e^{i\omega_0 t}, \tag{4.23}$$

where A and B are determined by the initial conditions.

With superposition in mind, it's clear what the strategy should be if we have a slightly more general equation to solve, for example,

$$\ddot{x} + 2\gamma\dot{x} + ax = C_1 e^{i\omega_1 t} + C_2 e^{i\omega_2 t}. \tag{4.24}$$

Simply solve the equation with only the first term on the right. Then solve the equation with only the second term on the right. Then add the two solutions. And then add on the homogeneous solution from Eq. (4.6). We are able to apply the principle of superposition because the left-hand side of Eq. (4.24) is linear.

Finally, let's look at the case where we have many such terms on the right-hand side, for example,

$$\ddot{x} + 2\gamma\dot{x} + ax = \sum_{n=1}^{N} C_n e^{i\omega_n t}. \tag{4.25}$$

We need to solve N different equations, each with only one of the N terms on the right-hand side. Then we add up all the solutions, and then we add on the homogeneous solution from Eq. (4.6). If N is infinite, that's fine; we just have to add up an infinite number of solutions. This is the principle of superposition at its best.

REMARK: The previous paragraph, combined with a basic result from Fourier analysis, allows us to solve (in principle) any equation of the form

$$\ddot{x} + 2\gamma\dot{x} + ax = f(t). \tag{4.26}$$

Fourier analysis says that any (nice enough) function $f(t)$ can be decomposed into its Fourier components,

$$f(t) = \int_{-\infty}^{\infty} g(\omega)e^{i\omega t}\,d\omega. \tag{4.27}$$

In this continuous sum, the functions $g(\omega)$ (times $d\omega$) take the place of the coefficients C_n in Eq. (4.25). So if $S_\omega(t)$ is the solution for $x(t)$ when there is only the term $e^{i\omega t}$ on the right-hand side of Eq. (4.26) (that is, $S_\omega(t)$ is the solution given in Eq. (4.22), without the C_0 factor), then the principle of superposition tells us that the complete particular solution to Eq. (4.26) is

$$x(t) = \int_{-\infty}^{\infty} g(\omega)S_\omega(t)\,d\omega. \tag{4.28}$$

Finding the coefficients $g(\omega)$ is the hard part (or rather, the messy part), but we won't get into that here. We won't do anything with Fourier analysis in this book, but it's nevertheless good to know that it *is* possible to solve (4.26) for any function $f(t)$. Most of the functions we'll consider will be nice functions like $\cos\omega_0 t$, which has a very simple Fourier decomposition, namely $\cos\omega_0 t = \frac{1}{2}(e^{i\omega_0 t} + e^{-i\omega_0 t})$. ♣

Let's now do a physical example.

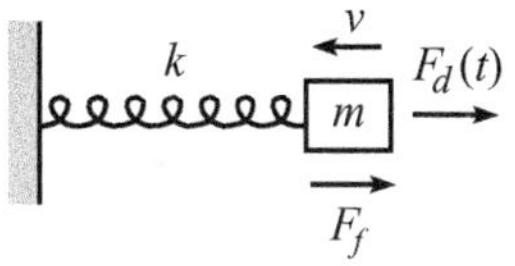

Fig. 4.7

Example (Damped and driven spring): Consider a spring with spring constant k. A mass m at the end of the spring is subject to a drag force proportional to its velocity, $F_\text{f} = -bv$. The mass is also subject to a driving force, $F_\text{d}(t) = F_\text{d}\cos\omega_\text{d} t$ (see Fig. 4.7). What is its position as a function of time?

Solution: The force on the mass is $F(x, \dot{x}, t) = -b\dot{x} - kx + F_\text{d}\cos\omega_\text{d} t$. So $F = ma$ gives

$$\begin{aligned}\ddot{x} + 2\gamma\dot{x} + \omega^2 x &= F\cos\omega_\text{d} t \\ &= \frac{F}{2}\left(e^{i\omega_\text{d} t} + e^{-i\omega_\text{d} t}\right).\end{aligned} \tag{4.29}$$

where $2\gamma \equiv b/m$, $\omega^2 \equiv k/m$, and $F \equiv F_\text{d}/m$. Note that there are two different frequencies here, ω and ω_d, which need not have anything to do with each other. Equation (4.22), along with the principle of superposition, tells us that our particular solution is

$$x_\text{p}(t) = \left(\frac{F/2}{-\omega_\text{d}^2 + 2i\gamma\omega_\text{d} + \omega^2}\right)e^{i\omega_\text{d} t} + \left(\frac{F/2}{-\omega_\text{d}^2 - 2i\gamma\omega_\text{d} + \omega^2}\right)e^{-i\omega_\text{d} t}. \tag{4.30}$$

The complete solution is the sum of this particular solution and the homogeneous solution from Eq. (4.15).

Let's now eliminate the i's in Eq. (4.30) (which we had better be able to do, because x must be real), and write x in terms of sines and cosines. Getting the i's out of the denominators (by multiplying both the numerator and the denominator by the complex conjugate of the denominator), and using $e^{i\theta} = \cos\theta + i\sin\theta$, we find, after a little work,

$$x_p(t) = \left(\frac{F\left(\omega^2 - \omega_d^2\right)}{\left(\omega^2 - \omega_d^2\right)^2 + 4\gamma^2\omega_d^2} \right) \cos\omega_d t + \left(\frac{2F\gamma\omega_d}{\left(\omega^2 - \omega_d^2\right)^2 + 4\gamma^2\omega_d^2} \right) \sin\omega_d t. \tag{4.31}$$

Remarks: If you want, you can solve Eq. (4.29) by taking the real part of the solution to Eq. (4.20) (with $C_0 \to F$), that is, the $x(t)$ in Eq. (4.22). This is true because if we take the real part of Eq. (4.20), we obtain

$$\begin{aligned} \frac{d^2}{dt^2}(\mathrm{Re}(x)) + 2\gamma\frac{d}{dt}(\mathrm{Re}(x)) + a(\mathrm{Re}(x)) &= \mathrm{Re}\big(C_0 e^{i\omega_0 t}\big) \\ &= C_0\cos(\omega_0 t). \end{aligned} \tag{4.32}$$

In other words, if x satisfies Eq. (4.20) with a $C_0 e^{i\omega_0 t}$ on the right-hand side, then $\mathrm{Re}(x)$ satisfies it with a $C_0\cos(\omega_0 t)$ on the right. At any rate, it's clear that the real part of the solution in Eq. (4.22) (with $C_0 \to F$) does indeed give the result in Eq. (4.31), because in Eq. (4.30) we simply took half of a quantity plus its complex conjugate, which is the real part.

If you don't like using complex numbers, another way of solving Eq. (4.29) is to keep it in the form with the $\cos\omega_d t$ on the right, and guess a solution of the form $A\cos\omega_d t + B\sin\omega_d t$, and then solve for A and B (this is the task of Problem 4.8). The result is Eq. (4.31). ♣

We can now write Eq. (4.31) in a very simple form. If we define

$$R \equiv \sqrt{\left(\omega^2 - \omega_d^2\right)^2 + (2\gamma\omega_d)^2}, \tag{4.33}$$

then we can rewrite Eq. (4.31) as

$$\begin{aligned} x_p(t) &= \frac{F}{R}\left(\frac{\omega^2 - \omega_d^2}{R}\cos\omega_d t + \frac{2\gamma\omega_d}{R}\sin\omega_d t \right) \\ &\equiv \frac{F}{R}\cos(\omega_d t - \phi), \end{aligned} \tag{4.34}$$

where ϕ (the *phase*) is defined by

$$\cos\phi = \frac{\omega^2 - \omega_d^2}{R}, \quad \sin\phi = \frac{2\gamma\omega_d}{R} \quad \Longrightarrow \quad \tan\phi = \frac{2\gamma\omega_d}{\omega^2 - \omega_d^2}. \tag{4.35}$$

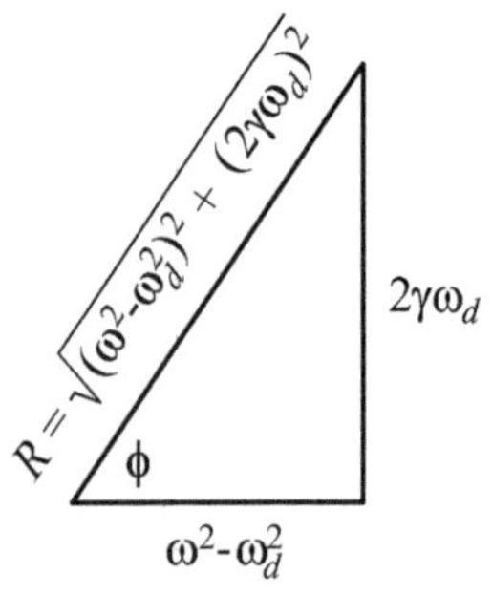

Fig. 4.8

The triangle describing the angle ϕ is shown in Fig. 4.8. Note that $0 \le \phi \le \pi$, because the $\sin\phi$ in Eq. (4.35) is greater than or equal to zero. See the end of this section for more discussion of ϕ.

Recalling the homogeneous solution in Eq. (4.15), we can write the complete solution to Eq. (4.29) as

$$x(t) = \frac{F}{R}\cos(\omega_d t - \phi) + e^{-\gamma t}\left(Ae^{\Omega t} + Be^{-\Omega t}\right). \tag{4.36}$$

The constants A and B are determined by the initial conditions. If there is any damping at all in the system (that is, if $\gamma > 0$), then the homogeneous part of the solution goes to zero for large t, and we are left with only the particular solution. In other words, the system approaches a definite $x(t)$, namely $x_p(t)$, independent of the initial conditions.

Resonance

The amplitude of the motion given in Eq. (4.34) is proportional to

$$\frac{1}{R} = \frac{1}{\sqrt{\left(\omega^2 - \omega_d^2\right)^2 + (2\gamma\omega_d)^2}}. \tag{4.37}$$

Given ω_d and γ, this is maximum when $\omega = \omega_d$. Given ω and γ, it is maximum when $\omega_d = \sqrt{\omega^2 - 2\gamma^2}$, as you can show in Exercise 4.29. But for weak damping (that is, $\gamma \ll \omega$, which is usually the case we are concerned with), this reduces to $\omega_d \approx \omega$ also. The term *resonance* is used to describe the situation where the amplitude of the oscillations is as large as possible. It is quite reasonable that this is achieved when the driving frequency equals the frequency of the spring. But what is the value of the phase ϕ at resonance? Using Eq. (4.35), we see that ϕ satisfies $\tan\phi \approx \pm\infty$ when $\omega_d \approx \omega$. Therefore, $\phi = \pi/2$ (it is indeed $\pi/2$, and not $-\pi/2$, because the $\sin\phi$ in Eq. (4.35) is positive), and the motion of the particle lags the driving force by a quarter of a cycle at resonance. For example, when the particle moves rightward past the origin (which means it has a quarter of a cycle to go before it hits the maximum value of x), the force is already at its maximum. And when the particle makes it out to the maximum value of x, the force is already back to zero.

The fact that the force is maximum when the particle is moving fastest makes sense from an energy point of view.[2] If you want the amplitude to become large, then you need to give the system as much energy as you can. That is, you must do as much work as possible on the system. And in order to do as much work as possible, you should have your force act over as large a distance as possible, which means that you should apply your force when the particle is moving fastest, that is, as it speeds past the origin. And similarly, you don't want to waste your force when the particle is barely moving near the endpoints of its motion. In short, v is the derivative of x and therefore a quarter cycle ahead of x (which is a general property of a sinusoidal function, as you can show). Since we want the force to be in phase with v at resonance (by the above energy argument), we see that the force is also a quarter cycle ahead of x.

Resonance has a large number of extremely important applications (both wanted and unwanted) in the real world. On the desirable side, resonance makes

[2] Energy is one of the topics of the next chapter, so you may want to come back and read this paragraph after reading that.

it possible to have a relaxing day at the beach at the Bay of Fundy, talking to a friend over your cell phone while pushing a child on a swing at low tide. On the undesirable side, your ride home on that newly discovered "washboard" dirt road will be annoyingly bumpy at a certain speed, and any attempt to take your mind off the discomfort by turning up the radio will result only in certain parts of your car rattling in perfect sync (well, actually 90° out of phase) with the bass line of your formerly favorite song.[3]

The phase ϕ

Equation (4.35) gives the phase of the motion as

$$\tan\phi = \frac{2\gamma\omega_\text{d}}{\omega^2 - \omega_\text{d}^2}, \tag{4.38}$$

where $0 \le \phi \le \pi$. Let's look at a few cases for ω_d (not necessarily at resonance) and see what the resulting phase ϕ is. Using Eq. (4.38), we have:

- If $\omega_\text{d} \approx 0$ (or more precisely, if $\gamma\omega_\text{d} \ll \omega^2 - \omega_\text{d}^2$), then $\phi \approx 0$. This means that the motion is in phase with the force. The mathematical reason for this is that if $\omega_\text{d} \approx 0$, then both $\ddot{x}$ and $\dot{x}$ are small, because they are proportional to ω_d^2 and ω_d, respectively. Therefore, the first two terms in Eq. (4.29) are negligible, so we end up with $x \propto \cos\omega_\text{d}t$. In other words, the phase is zero.

  The physical reason is that since there is essentially no acceleration, the net force is always essentially zero. This means that the driving force always essentially balances the spring force (that is, the two forces are 180° out of phase), because the damping force is negligible (since $\dot{x} \propto \omega_\text{d} \approx 0$). But the spring force is 180° out of phase with the motion (because of the minus sign in $F = -kx$). Therefore, the driving force is in phase with the motion.
- If $\omega_\text{d} \approx \omega$, then $\phi \approx \pi/2$. This is the case of resonance, discussed above.
- If $\omega_\text{d} \approx \infty$ (or more precisely, if $\gamma\omega_\text{d} \ll \omega_\text{d}^2 - \omega^2$), then $\phi \approx \pi$. The mathematical reason for this is that if $\omega_\text{d} \approx \infty$, then the $\ddot{x}$ term in Eq. (4.29) dominates, so we have $\ddot{x} \propto \cos\omega_\text{d}t$. Therefore, $\ddot{x}$ is in phase with the force. But x is 180° out of phase with $\ddot{x}$ (this is a general property of a sinusoidal function), so x is 180° out of phase with the force.

  The physical reason is that if $\omega_\text{d} \approx \infty$, then the mass hardly moves, because from Eq. (4.37) we see that the amplitude is proportional to $1/\omega_\text{d}^2$. This amplitude then implies that the velocity is proportional to $1/\omega_\text{d}$. Therefore, both x and v are always small. But if x and v are always small, then the spring and damping forces can be ignored. So we basically have a mass that feels only one force, the driving force. But we already understand very well a situation where a mass is subject to only one oscillating force: a mass on a spring. The mass in our setup can't tell if it's being driven by an oscillating

[3] Some other examples of resonance that are often cited are in fact not actually examples of resonance, but rather of "negative damping" (also known as positive feedback). Musical instruments fall into this category, as does the well-known Tacoma Narrows bridge failure. For a detailed discussion of this issue, see Billah and Scanlan (1991) and also Green and Unruh (2006).

driving force, or being pushed and pulled by an oscillating spring force. They both feel the same. Therefore, both phases must be the same. But in the spring case, the minus sign in $F = -kx$ tells us that the force is 180° out of phase with the motion. Hence, the same result holds in the $\omega_d \approx \infty$ case.

Another special case for the phase occurs when $\gamma = 0$ (no damping), for which we have $\tan\phi = \pm 0$, depending on the sign of $\omega^2 - \omega_d^2$. So ϕ is either 0 or π. The motion is therefore either exactly in phase or out of phase with the driving force, depending on which of ω or ω_d is larger.

4.5 Coupled oscillators

In the previous sections, we dealt with only one function of time, $x(t)$. What if we have two functions of time, say $x(t)$ and $y(t)$, that are related by a pair of "coupled" differential equations? For example, we might have

$$\begin{aligned} 2\ddot{x} + \omega^2(5x - 3y) &= 0, \\ 2\ddot{y} + \omega^2(5y - 3x) &= 0. \end{aligned} \tag{4.39}$$

For now, let's not worry about how these equations might arise. Let's just try to solve them (we'll do a physical example later in this section). We'll assume $\omega^2 > 0$ here, although this isn't necessary. We'll also assume there aren't any damping or driving forces, although a few of the problems and exercises for this chapter deal with these additions. We call the above equations "coupled" because there are x's and y's in both of them, and it isn't immediately obvious how to separate them to solve for x and y. There are two methods (at least) for solving these equations.

First method: Sometimes it is easy, as in this case, to find certain linear combinations of the given equations for which nice things happen. If we take the sum, we find

$$(\ddot{x} + \ddot{y}) + \omega^2(x + y) = 0. \tag{4.40}$$

This equation involves x and y only in the combination of their sum, $x + y$. With $z \equiv x + y$, Eq. (4.40) is just our old friend, $\ddot{z} + \omega^2 z = 0$. The solution is

$$x + y = A_1\cos(\omega t + \phi_1), \tag{4.41}$$

where A_1 and ϕ_1 are determined by the initial conditions. We may also take the difference of Eqs. (4.39), which results in

$$(\ddot{x} - \ddot{y}) + 4\omega^2(x - y) = 0. \tag{4.42}$$

This equation involves x and y only in the combination of their difference, $x - y$. The solution is

$$x - y = A_2\cos(2\omega t + \phi_2). \tag{4.43}$$

Taking the sum and difference of Eqs. (4.41) and (4.43), we find that $x(t)$ and $y(t)$ are given by

$$x(t) = B_1 \cos(\omega t + \phi_1) + B_2 \cos(2\omega t + \phi_2),$$
$$y(t) = B_1 \cos(\omega t + \phi_1) - B_2 \cos(2\omega t + \phi_2), \tag{4.44}$$

where the B_i's are half of the A_i's. The strategy of this solution was simply to fiddle around and try to form differential equations that involved only one combination of the variables. This allowed us to write down the familiar solution for these combinations, as we did in Eqs. (4.41) and (4.43).

We've managed to solve our equations for x and y. However, it turns out that the more interesting thing we've done is produce the equations (4.41) and (4.43). The combinations $(x + y)$ and $(x - y)$ are called the *normal coordinates* of the system. These are the combinations that oscillate with one pure frequency. The motion of x and y will in general look complicated, and it may be difficult to tell that the motion is really made up of just the two frequencies in Eq. (4.44). But if you plot the values of $(x + y)$ and $(x - y)$ as time goes by, for *any* motion of the system, then you will find nice sinusoidal graphs, even if x and y are each behaving in a rather unpleasant manner.

Second method: In the above method, it was fairly easy to guess which combinations of Eqs. (4.39) would produce equations involving only one combination of x and y. But surely there are problems in physics where the guessing isn't so easy. What do we do then? Fortunately, there is a fail-safe method for solving for x and y. It proceeds as follows.

In the spirit of Section 4.1, let's try a solution of the form $x = Ae^{i\alpha t}$ and $y = Be^{i\alpha t}$, which we will write, for convenience, as

$$\begin{pmatrix} x \\ y \end{pmatrix} = \begin{pmatrix} A \\ B \end{pmatrix} e^{i\alpha t}. \tag{4.45}$$

It isn't obvious that there should exist solutions for x and y that have the same t dependence, but let's try it and see what happens. We've explicitly put the i in the exponent, but there's no loss of generality here. If α happens to be imaginary, then the exponent is real. It's personal preference whether or not you put the i in. Plugging our guess into Eqs. (4.39), and dividing through by $e^{i\omega t}$, we find

$$2A(-\alpha^2) + 5A\omega^2 - 3B\omega^2 = 0,$$
$$2B(-\alpha^2) + 5B\omega^2 - 3A\omega^2 = 0, \tag{4.46}$$

or equivalently, in matrix form,

$$\begin{pmatrix} -2\alpha^2 + 5\omega^2 & -3\omega^2 \\ -3\omega^2 & -2\alpha^2 + 5\omega^2 \end{pmatrix} \begin{pmatrix} A \\ B \end{pmatrix} = \begin{pmatrix} 0 \\ 0 \end{pmatrix}. \tag{4.47}$$

This homogeneous equation for A and B has a nontrivial solution (that is, one where A and B aren't both 0) only if the matrix is *not* invertible. This is true because if it were invertible, then we could multiply through by the inverse to obtain $(A, B) = (0, 0)$. When is a matrix invertible? There is a straightforward (although tedious) method for finding the inverse. It involves taking cofactors, taking a transpose, and dividing by the determinant. The step that concerns us here is the division by the determinant, since this implies that the inverse exists if and only if the determinant is not zero. So we see that Eq. (4.47) has a nontrivial solution only if the determinant equals zero. Because we seek a nontrivial solution, we must therefore have

$$\begin{aligned} 0 &= \begin{vmatrix} -2\alpha^2 + 5\omega^2 & -3\omega^2 \\ -3\omega^2 & -2\alpha^2 + 5\omega^2 \end{vmatrix} \\ &= 4\alpha^4 - 20\alpha^2\omega^2 + 16\omega^4. \end{aligned} \tag{4.48}$$

This is a quadratic equation in α^2, and the roots are $\alpha = \pm\omega$ and $\alpha = \pm 2\omega$. We have therefore found four types of solutions. If $\alpha = \pm\omega$, then we can plug this back into Eq. (4.47) to obtain $A = B$. (Both equations give this same result. This was essentially the point of setting the determinant equal to zero.) And if $\alpha = \pm 2\omega$, then Eq. (4.47) gives $A = -B$. (Again, the equations are redundant.) Note that we cannot solve specifically for A and B, but only for their ratio. Adding up our four solutions according to the principle of superposition, we see that x and y take the general form (written in vector form for the sake of simplicity and bookkeeping),

$$\begin{aligned} \begin{pmatrix} x \\ y \end{pmatrix} &= A_1 \begin{pmatrix} 1 \\ 1 \end{pmatrix} e^{i\omega t} + A_2 \begin{pmatrix} 1 \\ 1 \end{pmatrix} e^{-i\omega t} \\ &\quad + A_3 \begin{pmatrix} 1 \\ -1 \end{pmatrix} e^{2i\omega t} + A_4 \begin{pmatrix} 1 \\ -1 \end{pmatrix} e^{-2i\omega t}. \end{aligned} \tag{4.49}$$

The four A_i are determined by the initial conditions. We can rewrite Eq. (4.49) in a somewhat cleaner form. If the coordinates x and y describe the positions of particles, they must be real. Therefore, A_1 and A_2 must be complex conjugates, and likewise for A_3 and A_4. If we then define some ϕ's and B's via $A_2^* = A_1 \equiv (B_1/2)e^{i\phi_1}$ and $A_4^* = A_3 \equiv (B_2/2)e^{i\phi_2}$, we may rewrite our solution in the form, as you can verify,

$$\begin{pmatrix} x \\ y \end{pmatrix} = B_1 \begin{pmatrix} 1 \\ 1 \end{pmatrix} \cos(\omega t + \phi_1) + B_2 \begin{pmatrix} 1 \\ -1 \end{pmatrix} \cos(2\omega t + \phi_2), \tag{4.50}$$

where the B_i and ϕ_i are real (and are determined by the initial conditions). We have therefore reproduced the result in Eq. (4.44).

It is clear from Eq. (4.50) that the combinations $x + y$ and $x - y$ (the normal coordinates) oscillate with the pure frequencies, ω and 2ω, respectively, because the combination $x + y$ makes the B_2 terms disappear, and the combination $x - y$ makes the B_1 terms disappear.

It is also clear that if $B_2 = 0$, then $x = y$ at all times, and they both oscillate with frequency ω. And if $B_1 = 0$, then $x = -y$ at all times, and they both oscillate with frequency 2ω. These two pure-frequency motions are called the *normal modes*. They are labeled by the vectors $(1, 1)$ and $(1, -1)$, respectively. In describing a normal mode, both the vector and the frequency should be stated. The significance of normal modes will become clear in the following example.

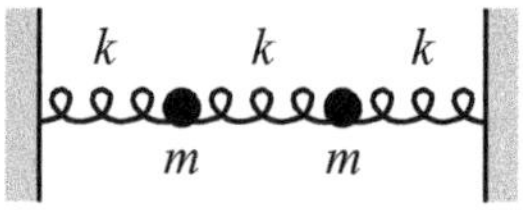

Fig. 4.9

Example (Two masses, three springs): Consider two masses m, connected to each other and to two walls by three springs, as shown in Fig. 4.9. The three springs have the same spring constant k. Find the most general solution for the positions of the masses as functions of time. What are the normal coordinates? What are the normal modes?

Solution: Let $x_1(t)$ and $x_2(t)$ be the positions of the left and right masses, respectively, relative to their equilibrium positions. Then the middle spring is stretched a distance $x_2 - x_1$ compared with the stretch at equilibrium. Therefore, the net force on the left mass is $-kx_1 + k(x_2 - x_1)$, and the net force on the right mass is $-kx_2 - k(x_2 - x_1)$. It's easy to make a mistake in the sign of the second term in these expressions, but you can check it by, say, looking at the force when x_2 is very big. At any rate, the second term must have the opposite sign in the two expressions, by Newton's third law. With these forces, $F = ma$ on each mass gives, with $\omega^2 = k/m$,

$$\begin{aligned} \ddot{x}_1 + 2\omega^2 x_1 - \omega^2 x_2 &= 0, \\ \ddot{x}_2 + 2\omega^2 x_2 - \omega^2 x_1 &= 0. \end{aligned} \tag{4.51}$$

These are rather friendly looking coupled equations, and we can see that the sum and difference are the useful combinations to take. The sum gives

$$(\ddot{x}_1 + \ddot{x}_2) + \omega^2(x_1 + x_2) = 0, \tag{4.52}$$

and the difference gives

$$(\ddot{x}_1 - \ddot{x}_2) + 3\omega^2(x_1 - x_2) = 0. \tag{4.53}$$

The solutions to these equations are the normal coordinates,

$$\begin{aligned} x_1 + x_2 &= A_+ \cos(\omega t + \phi_+), \\ x_1 - x_2 &= A_- \cos(\sqrt{3}\omega t + \phi_-). \end{aligned} \tag{4.54}$$

Taking the sum and difference of these normal coordinates, we have

$$\begin{aligned} x_1(t) &= B_+ \cos(\omega t + \phi_+) + B_- \cos(\sqrt{3}\omega t + \phi_-), \\ x_2(t) &= B_+ \cos(\omega t + \phi_+) - B_- \cos(\sqrt{3}\omega t + \phi_-), \end{aligned} \tag{4.55}$$

where the B's are half of the A's. Along with the ϕ's, they are determined by the initial conditions.

REMARK: For practice, let's also derive Eq. (4.55) by using the determinant method. Letting $x_1 = Ae^{i\alpha t}$ and $x_2 = Be^{i\alpha t}$ in Eq. (4.51), we see that for there to be a nontrivial solution for A and B, we must have

$$\begin{aligned} 0 &= \begin{vmatrix} -\alpha^2 + 2\omega^2 & -\omega^2 \\ -\omega^2 & -\alpha^2 + 2\omega^2 \end{vmatrix} \\ &= \alpha^4 - 4\alpha^2\omega^2 + 3\omega^4. \end{aligned} \tag{4.56}$$

This is a quadratic equation in α^2, and the roots are $\alpha = \pm\omega$ and $\alpha = \pm\sqrt{3}\omega$. If $\alpha = \pm\omega$, then Eq. (4.51) gives $A = B$. If $\alpha = \pm\sqrt{3}\omega$, then Eq. (4.51) gives $A = -B$. The solutions for x_1 and x_2 therefore take the general form

$$\begin{aligned} \begin{pmatrix} x_1 \\ x_2 \end{pmatrix} &= A_1 \begin{pmatrix} 1 \\ 1 \end{pmatrix} e^{i\omega t} + A_2 \begin{pmatrix} 1 \\ 1 \end{pmatrix} e^{-i\omega t} \\ &\quad + A_3 \begin{pmatrix} 1 \\ -1 \end{pmatrix} e^{\sqrt{3}i\omega t} + A_4 \begin{pmatrix} 1 \\ -1 \end{pmatrix} e^{-\sqrt{3}i\omega t} \\ &\equiv B_+ \begin{pmatrix} 1 \\ 1 \end{pmatrix} \cos(\omega t + \phi_+) + B_- \begin{pmatrix} 1 \\ -1 \end{pmatrix} \cos(\sqrt{3}\omega t + \phi_-), \end{aligned} \tag{4.57}$$

where the last line follows from the same substitutions that led to Eq. (4.50). This expression is equivalent to Eq. (4.55). ♣

The normal modes are obtained by setting either B_- or B_+ equal to zero in Eq. (4.55) or Eq. (4.57). Therefore, the normal modes are $(1, 1)$ and $(1, -1)$. How do we visualize these? The mode $(1, 1)$ oscillates with frequency ω. In this case (where $B_- = 0$), we have $x_1(t) = x_2(t) = B_+ \cos(\omega t + \phi_+)$ at all times. So the masses simply oscillate back and forth in the same manner, as shown in Fig. 4.10. It is clear that such motion has frequency ω, because as far as the masses are concerned, the middle spring is effectively not there, so each mass moves under the influence of only one spring, and therefore has frequency ω.

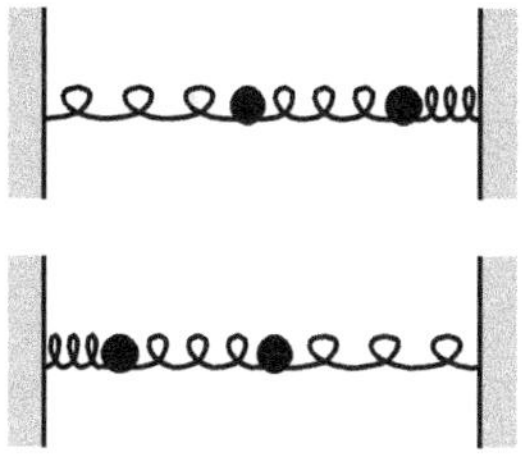

Fig. 4.10

The mode $(1, -1)$ oscillates with frequency $\sqrt{3}\omega$. In this case (where $B_+ = 0$), we have $x_1(t) = -x_2(t) = B_- \cos(\sqrt{3}\omega t + \phi_-)$ at all times. So the masses oscillate back and forth with opposite displacements, as shown in Fig. 4.11. It is clear that this mode should have a frequency larger than that for the other mode, because the middle spring is stretched (or compressed), so the masses feel a larger force. But it takes a little thought to show that the frequency is $\sqrt{3}\omega$.[4]

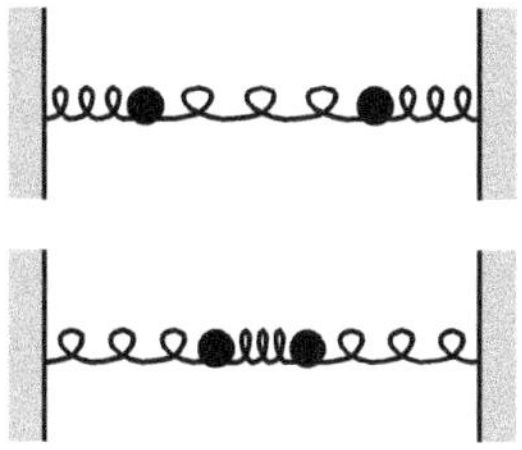

Fig. 4.11

The normal mode $(1, 1)$ above is associated with the normal coordinate $x_1 + x_2$. They both involve the frequency ω. However, this association is *not* due to the fact that the coefficients of both x_1 and x_2 in this normal coordinate are equal to 1.

[4] If you want to obtain this $\sqrt{3}\omega$ result without going through all of the above work, just note that the center of the middle spring doesn't move. Therefore, it acts like two "half springs," each with spring constant $2k$ (as you can verify). Hence, each mass is effectively attached to a "k" spring and a "$2k$" spring, yielding a total effective spring constant of $3k$. Thus the $\sqrt{3}$.

Rather, it is due to the fact that the *other* normal mode, namely $(x_1,x_2) \propto (1,-1)$, gives no contribution to the sum $x_1 + x_2$. There are a few too many 1's floating around in the above example, so it's hard to see which results are meaningful and which results are coincidence. But the following example should clear things up. Let's say we solved a problem using the determinant method, and we found the solution to be

$$\begin{pmatrix} x \\ y \end{pmatrix} = B_1 \begin{pmatrix} 3 \\ 2 \end{pmatrix} \cos(\omega_1 t + \phi_1) + B_2 \begin{pmatrix} 1 \\ -5 \end{pmatrix} \cos(\omega_2 t + \phi_2). \tag{4.58}$$

Then $5x + y$ is the normal coordinate associated with the normal mode $(3,2)$, which has frequency ω_1. (This is true because there is no $\cos(\omega_2 t + \phi_2)$ dependence in the combination $5x+y$.) And similarly, $2x-3y$ is the normal coordinate associated with the normal mode $(1,-5)$, which has frequency ω_2 (because there is no $\cos(\omega_1 t + \phi_1)$ dependence in the combination $2x - 3y$).

Note the difference between the types of differential equations we solved in the previous chapter in Section 3.3, and the types we solved throughout this chapter. The former dealt with forces that did not have to be linear in x or $\dot{x}$, but that had to depend on only x, or only $\dot{x}$, or only t. The latter dealt with forces that could depend on all three of these quantities, but that had to be linear in x and $\dot{x}$.

4.6 Problems

Section 4.1: Linear differential equations

4.1. **Superposition**

Let $x_1(t)$ and $x_2(t)$ be solutions to $\dot{x}^2 = bx$. Show that $x_1(t) + x_2(t)$ is *not* a solution to this equation.

4.2. **A limiting case** *

Consider the equation $\ddot{x} = ax$. If $a = 0$, then the solution to $\ddot{x} = 0$ is simply $x(t) = C + Dt$. Show that in the limit $a \to 0$, Eq. (4.2) reduces to this form. *Note*: $a \to 0$ is a sloppy way of saying what we mean. What is the proper way to write this limit?

Section 4.2: Simple harmonic motion

4.3. **Increasing the mass** **

A mass m oscillates on a spring with spring constant k. The amplitude is d. At the moment (let this be $t = 0$) when the mass is at position $x = d/2$ (and moving to the right), it collides and sticks to another mass m. The speed of the resulting mass $2m$ right after the collision is half the speed of the moving mass m right before the collision (from momentum conservation, discussed in Chapter 5). What is the resulting $x(t)$? What is the amplitude of the new oscillation?

4.4. **Average tension** $**$

Is the average (over time) tension in the string of a pendulum larger or smaller than mg? By how much? As usual, assume that the angular amplitude A is small.

4.5. **Walking east on a turntable** $**$

A person walks at constant speed v eastward with respect to a turntable that rotates counterclockwise at constant frequency ω. Find the general expression for the person's coordinates with respect to the ground (with the x direction taken to be eastward).

Section 4.3: Damped harmonic motion

4.6. **Maximum speed** $**$

A mass on the end of a spring (with natural frequency ω) is released from rest at position x_0. The experiment is repeated, but now with the system immersed in a fluid that causes the motion to be overdamped (with damping coefficient γ). Find the ratio of the maximum speed in the former case to that in the latter. What is the ratio in the limit of strong damping ($\gamma \gg \omega$)? In the limit of critical damping?

Section 4.4: Driven (and damped) harmonic motion

4.7. **Exponential force** $*$

A particle of mass m is subject to a force $F(t) = ma_0e^{-bt}$. The initial position and speed are both zero. Find $x(t)$. (This problem was already given as Problem 3.9, but solve it here by guessing an exponential function, in the spirit of Section 4.4.)

4.8. **Driven oscillator** $*$

Derive Eq. (4.31) by guessing a solution of the form $x(t) = A\cos\omega_d t + B\sin\omega_d t$ in Eq. (4.29).

Section 4.5: Coupled oscillators

4.9. **Unequal masses** $**$

Three identical springs and two masses, m and $2m$, lie between two walls as shown in Fig. 4.12. Find the normal modes.

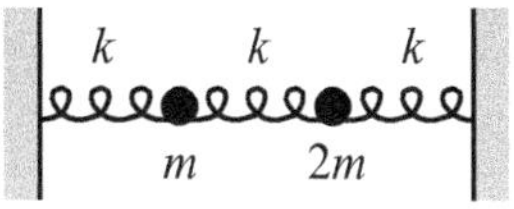

Fig. 4.12

4.10. **Weakly coupled** $**$

Three springs and two equal masses lie between two walls, as shown in Fig. 4.13. The spring constant, k, of the two outside springs is much

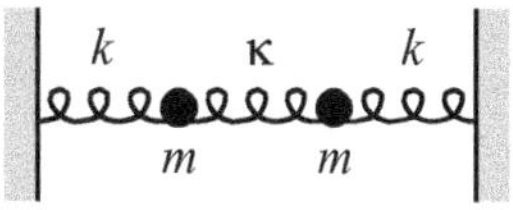

Fig. 4.13

larger than the spring constant, κ, of the middle spring. Let x_1 and x_2 be the positions of the left and right masses, respectively, relative to their equilibrium positions. If the initial positions are given by $x_1(0) = a$ and $x_2(0) = 0$, and if both masses are released from rest, show that x_1 and x_2 can be written as (assuming $\kappa \ll k$)

$$
\begin{aligned}
x_1(t) &\approx a\cos\big((\omega+\epsilon)t\big)\cos(\epsilon t), \\
x_2(t) &\approx a\sin\big((\omega+\epsilon)t\big)\sin(\epsilon t),
\end{aligned}
\tag{4.59}
$$

where $\omega \equiv \sqrt{k/m}$ and $\epsilon \equiv (\kappa/2k)\omega$. Explain qualitatively what the motion looks like.

Fig. 4.14

4.11. **Driven mass on a circle** **

Two identical masses m are constrained to move on a horizontal hoop. Two identical springs with spring constant k connect the masses and wrap around the hoop (see Fig. 4.14). One mass is subject to a driving force $F_d \cos\omega_d t$. Find the particular solution for the motion of the masses.

Fig. 4.15

4.12. **Springs on a circle** ****

(a) Two identical masses m are constrained to move on a horizontal hoop. Two identical springs with spring constant k connect the masses and wrap around the hoop (see Fig. 4.15). Find the normal modes.

(b) Three identical masses are constrained to move on a hoop. Three identical springs connect the masses and wrap around the hoop (see Fig. 4.16). Find the normal modes.

(c) Now do the general case with N identical masses and N identical springs.

Fig. 4.16

4.7 Exercises

Section 4.1: Linear differential equations

4.13. ***kx* force** *

A particle of mass m is subject to a force $F(x) = kx$, with $k > 0$. What is the most general form of $x(t)$? If the particle starts out at x_0, what is the one special value of the initial velocity for which the particle doesn't eventually get far away from the origin?

4.14. **Rope on a pulley** **

A rope with length L and mass density σ kg/m hangs over a massless pulley. Initially, the ends of the rope are a distance x_0 above and below their average position. The rope is given an initial speed. If you want the rope to not eventually fall off the pulley, what should this initial speed be? (Don't worry about the issue discussed in Calkin (1989).)

Section 4.2: Simple harmonic motion

4.15. **Amplitude** *

Find the amplitude of the motion given by $x(t) = C\cos\omega t + D\sin\omega t$.

4.16. **Angled rails** *

Two particles of mass m are constrained to move along two horizontal frictionless rails that make an angle 2θ with respect to each other. They are connected by a spring with spring constant k, whose relaxed length is at the position shown in Fig. 4.17. What is the frequency of oscillations for the motion where the spring remains parallel to the position shown?

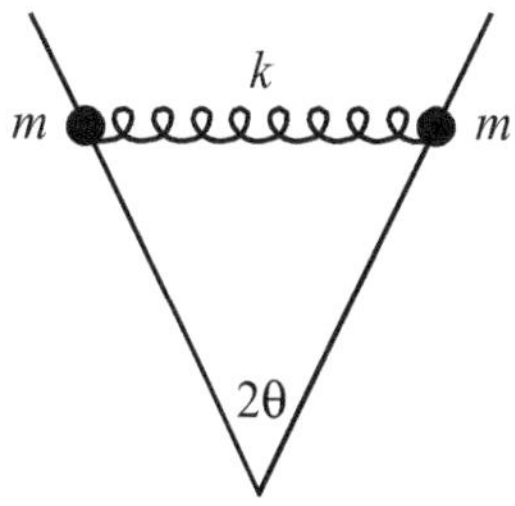

Fig. 4.17

4.17. **Effective spring constant** *

(a) Two springs with spring constants k_1 and k_2 are connected in parallel, as shown in Fig. 4.18. What is the effective spring constant, k_{eff}? In other words, if the mass is displaced by x, find the k_{eff} for which the force equals $F = -k_{\text{eff}}x$.

(b) Two springs with spring constants k_1 and k_2 are connected in series, as shown in Fig. 4.19. What is the effective spring constant, k_{eff}?

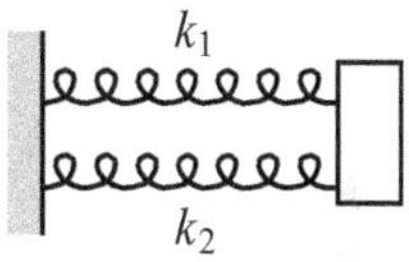

Fig. 4.18

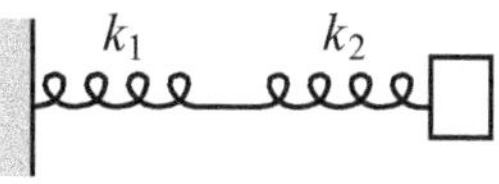

Fig. 4.19

4.18. **Changing k** **

Two springs each have spring constant k and equilibrium length ℓ. They are both stretched a distance ℓ and attached to a mass m and two walls, as shown in Fig. 4.20. At a given instant, the right spring constant is somehow magically changed to $3k$ (the relaxed length remains ℓ). What is the resulting $x(t)$? Take the initial position to be $x = 0$.

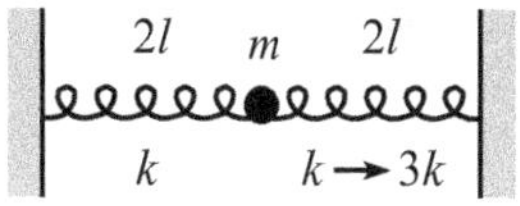

Fig. 4.20

4.19. **Removing a spring** **

The springs in Fig. 4.21 are at their equilibrium length. The mass oscillates along the line of the springs with amplitude d. At the moment (let this be $t = 0$) when the mass is at position $x = d/2$ (and moving to the right), the right spring is removed. What is the resulting $x(t)$? What is the amplitude of the new oscillation?

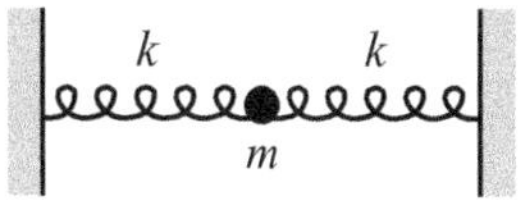

Fig. 4.21

4.20. **Springs all over** **

(a) A mass m is attached to two springs that have relaxed lengths of zero. The other ends of the springs are fixed at two points (see Fig. 4.22). The two spring constants are equal. The mass sits at its equilibrium position and is then given a kick in an arbitrary

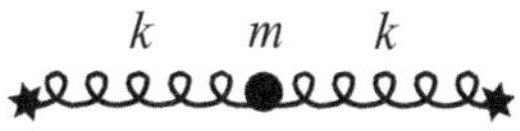

Fig. 4.22

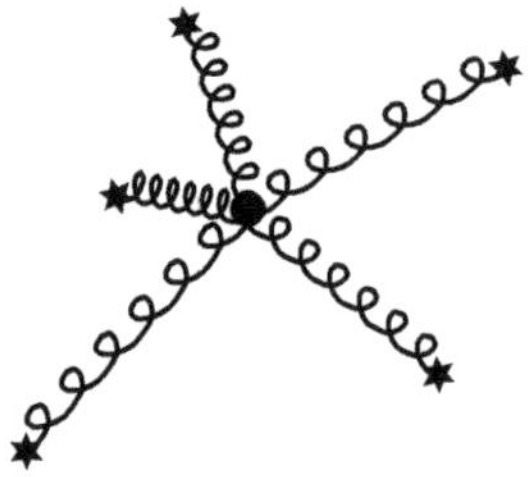

Fig. 4.23

direction. Describe the resulting motion. (Ignore gravity, although you actually don't need to.)

(b) A mass m is attached to n springs that have relaxed lengths of zero. The other ends of the springs are fixed at various points in space (see Fig. 4.23). The spring constants are $k_1, k_2, \ldots, k_n$. The mass sits at its equilibrium position and is then given a kick in an arbitrary direction. Describe the resulting motion. (Again, ignore gravity, although you actually don't need to.)

4.21. **Rising up** ***

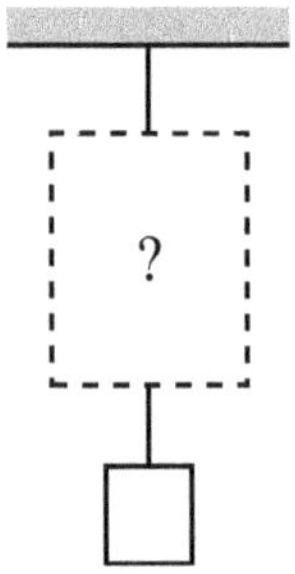

Fig. 4.24

In Fig. 4.24, a mass hangs from a ceiling. A piece of paper is held up to obscure three strings and two springs; all you see is two other strings protruding from behind the paper, as shown. How should the three strings and two springs be attached to each other and to the two visible strings (different items can be attached only at their endpoints) so that if you start with the system at its equilibrium position and then cut a certain one of the hidden strings, the mass will rise up?[5]

4.22. **Projectile on a spring** ***

A projectile of mass m is fired from the origin at speed v_0 and angle θ. It is attached to the origin by a spring with spring constant k and relaxed length zero.

(a) Find $x(t)$ and $y(t)$.

(b) Show that for small $\omega \equiv \sqrt{k/m}$, the trajectory reduces to normal projectile motion. And show that for large ω, the trajectory reduces to simple harmonic motion, that is, oscillatory motion along a line (at least before the projectile smashes back into the ground). What are the more meaningful statements that should replace "small ω" and "large ω"?

(c) What value should ω take so that the projectile hits the ground when it is moving straight downward?

4.23. **Corrections to the pendulum** ***

(a) For small oscillations, the period of a pendulum is approximately $T \approx 2\pi\sqrt{\ell/g}$, independent of the amplitude, θ_0. For finite oscillations, use $dt = dx/v$ to show that the exact expression for T is

$$T = \sqrt{\frac{8\ell}{g}} \int_0^{\theta_0} \frac{d\theta}{\sqrt{\cos\theta - \cos\theta_0}}. \tag{4.60}$$

[5] Thanks to Paul Horowitz for this extremely cool problem. For more applications of the idea behind it, see Cohen and Horowitz (1991).

(b) Find an approximation to this T, up to second order in θ_0^2, in the following way. Make use of the identity $\cos\phi = 1 - 2\sin^2(\phi/2)$ to write T in terms of sines (because it's more convenient to work with quantities that go to zero as $\theta \to 0$). Then make the change of variables, $\sin x \equiv \sin(\theta/2)/\sin(\theta_0/2)$ (you'll see why). Finally, expand your integrand in powers of θ_0, and perform the integrals to show that[6]

$$T \approx 2\pi\sqrt{\frac{\ell}{g}}\left(1 + \frac{\theta_0^2}{16} + \cdots\right). \tag{4.61}$$

Section 4.3: Damped harmonic motion

4.24. **Crossing the origin**

Show that an overdamped or critically damped oscillator can cross the origin at most once.

4.25. **Strong damping** *

In the strong damping ($\gamma \gg \omega$) case discussed in the remark in the overdamping subsection, we saw that $x(t) \propto e^{-\omega^2 t/2\gamma}$ for large t. Using the definitions of ω and γ, this can be written as $x(t) \propto e^{-kt/b}$, where b is the coefficient of the damping force. By looking at the forces on the mass, explain why this makes sense.

4.26. **Maximum speed** *

A critically damped oscillator with natural frequency ω starts out at position $x_0 > 0$. What is the maximum initial speed (directed toward the origin) it can have and not cross the origin?

4.27. **Another maximum speed** **

An overdamped oscillator with natural frequency ω and damping coefficient γ starts out at position $x_0 > 0$. What is the maximum initial speed (directed toward the origin) it can have and not cross the origin?

4.28. **Ratio of maxima** **

A mass on the end of a spring is released from rest at position x_0. The experiment is repeated, but now with the system immersed in a fluid that causes the motion to be critically damped. Show that the maximum speed of the mass in the first case is e times the maximum speed in the second case.[7]

[6] If you like this sort of thing, you can show that the next term in the parentheses is $(11/3072)\theta_0^4$. But be careful, this fourth-order correction comes from two terms.

[7] The fact that the maximum speeds differ by a fixed numerical factor follows from dimensional analysis, which tells us that the maximum speed in the first case must be proportional to ωx_0. And

Section 4.4: Driven (and damped) harmonic motion

4.29. **Resonance**

Given ω and γ, show that the R in Eq. (4.33) is minimum when $\omega_d = \sqrt{\omega^2 - 2\gamma^2}$ (unless this is imaginary, in which case the minimum occurs at $\omega_d = 0$).

4.30. **No damping force** *

A particle of mass m is subject to a spring force, $-kx$, and also a driving force, $F_d \cos \omega_d t$. But there is no damping force. Find the particular solution for $x(t)$ by guessing $x(t) = A \cos \omega_d t + B \sin \omega_d t$. If you write this in the form $C \cos(\omega_d t - \phi)$, where $C > 0$, what are C and ϕ? Be careful about the phase (there are two cases to consider).

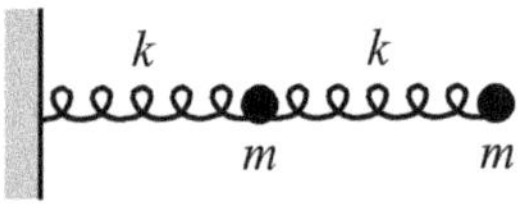

Fig. 4.25

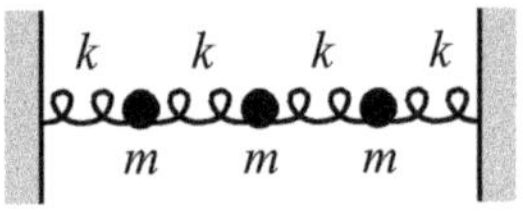

Fig. 4.26

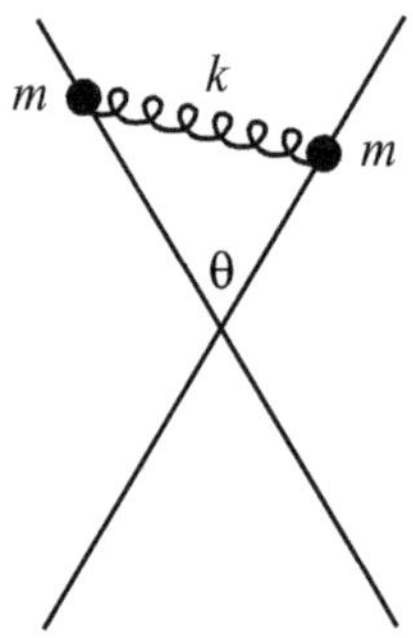

Fig. 4.27

Section 4.5: Coupled oscillators

4.31. **Springs and one wall** **

Two identical springs and two identical masses are attached to a wall as shown in Fig. 4.25. Find the normal modes, and show that the frequencies can be written as $\sqrt{k/m}(\sqrt{5} \pm 1)/2$. This numerical factor is the golden ratio (and its inverse).

4.32. **Springs between walls** **

Four identical springs and three identical masses lie between two walls (see Fig. 4.26). Find the normal modes.

4.33. **Beads on angled rails** **

Two horizontal frictionless rails make an angle θ with each other, as shown in Fig. 4.27. Each rail has a bead of mass m on it, and the beads are connected by a spring with spring constant k and relaxed length zero. Assume that one of the rails is positioned a tiny distance above the other, so that the beads can pass freely through the crossing. Find the normal modes.

4.34. **Coupled and damped** **

The system in the example in Section 4.5 is modified by immersing it in a fluid so that both masses feel a damping force, $F_f = -bv$. Solve for $x_1(t)$ and $x_2(t)$. Assume underdamping.

4.35. **Coupled and driven** **

The system in the example in Section 4.5 is modified by subjecting the left mass to a driving force $F_d \cos(2\omega t)$, and the right mass to a driving

since $\gamma = \omega$ in the critical-damping case, the damping doesn't introduce a new parameter, so the maximum speed has no choice but to again be proportional to ωx_0. But showing that the maximum speeds differ by the nice factor of e requires a calculation.

force $2F_d \cos(2\omega t)$, where $\omega = \sqrt{k/m}$. Find the particular solution for $x_1(t)$ and $x_2(t)$, and explain why your answer makes sense.

4.8 Solutions

4.1. **Superposition**

The sum $x_1 + x_2$ is a solution to $\ddot{x}^2 = bx$ if

$$\begin{aligned} \left(\frac{d^2(x_1+x_2)}{dt^2}\right)^2 &= b(x_1+x_2) \\ \Longleftrightarrow \quad (\ddot{x}_1+\ddot{x}_2)^2 &= b(x_1+x_2) \\ \Longleftrightarrow \quad \ddot{x}_1^2 + 2\ddot{x}_1\ddot{x}_2 + \ddot{x}_2^2 &= b(x_1+x_2). \end{aligned} \tag{4.62}$$

But $\ddot{x}_1^2 = bx_1$ and $\ddot{x}_2^2 = bx_2$, by assumption. So we are left with the $2\ddot{x}_1\ddot{x}_2$ term on the left-hand side, which destroys the equality. (Note that $2\ddot{x}_1\ddot{x}_2$ can't be zero, because if either $\ddot{x}_1$ or $\ddot{x}_2$ is identically zero, then either x_1 or x_2 is also, so we didn't really have a solution to begin with.)

4.2. **A limiting case**

The expression "$a \to 0$" is sloppy because a has units of inverse time squared, and the number 0 has no units. The proper statement is that Eq. (4.2) reduces to $x(t) = C + Dt$ when $\sqrt{a}\,t \ll 1$, or equivalently when $t \ll 1/\sqrt{a}$, which is now a comparison of quantities with the same units. The smaller a is, the larger t can be. Therefore, if "$a \to 0$," then t can basically be anything. Assuming $\sqrt{a}\,t \ll 1$, we can write $e^{\pm\sqrt{a}\,t} \approx 1 \pm \sqrt{a}\,t$, and Eq. (4.2) becomes

$$\begin{aligned} x(t) &\approx A(1+\sqrt{a}\,t) + B(1-\sqrt{a}\,t) \\ &= (A+B) + \sqrt{a}(A-B)t \\ &\equiv C + Dt. \end{aligned} \tag{4.63}$$

C is the initial position, and D is the speed of the particle. If these quantities are of order 1 in the units chosen, then if we solve for A and B, we see that they must be roughly negatives of each other, and both of order $1/\sqrt{a}$. So if the speed and initial position are of order 1, then A and B actually diverge in the "$a \to 0$" limit. If a is small but nonzero, then t will eventually become large enough so that $\sqrt{a}\,t \ll 1$ won't hold, in which case the linear form in Eq. (4.63) won't be valid.

4.3. **Increasing the mass**

The first thing we must do is find the velocity of the mass right before the collision. The motion before the collision looks like $x(t) = d\cos(\omega t + \phi)$, where $\omega = \sqrt{k/m}$. The collision happens at $t = 0$ (although it actually doesn't matter what time we plug in here), so we have $d/2 = x(0) = d\cos\phi$, which gives $\phi = \pm\pi/3$. The velocity right before the collision is therefore

$$v(0) \equiv \dot{x}(0) = -\omega d \sin\phi = -\omega d \sin(\pm\pi/3) = \mp(\sqrt{3}/2)\omega d. \tag{4.64}$$

We want the plus sign, because we are told that the mass is moving to the right. Finding the motion after the collision is now reduced to an initial conditions problem. We have a mass $2m$ on a spring with spring constant k, with initial position $d/2$ and initial velocity $(\sqrt{3}/4)\omega d$ (half of the result above). In situations where we know the initial position and velocity, it turns out that the best form to use for $x(t)$ from the expressions in Eq. (4.3) is

$$x(t) = C\cos\omega' t + D\sin\omega' t, \tag{4.65}$$

because the initial position at $t = 0$ is simply C, and the initial velocity at $t = 0$ is $\omega' D$. The initial conditions are therefore easy to apply. We have put a prime on the frequency in Eq. (4.65) to remind us that it is different from the initial frequency, because the mass is now $2m$. So we have $\omega' = \sqrt{k/2m} = \omega/\sqrt{2}$. The initial conditions therefore give

$$x(0) = d/2 \quad \Longrightarrow \quad C = d/2,$$

$$v(0) = (\sqrt{3}/4)\omega d \quad \Longrightarrow \quad \omega' D = (\sqrt{3}/4)\omega d \quad \Longrightarrow \quad D = (\sqrt{6}/4)d. \tag{4.66}$$

Our solution for $x(t)$ is therefore

$$x(t) = \frac{d}{2}\cos\omega' t + \frac{\sqrt{6}d}{4}\sin\omega' t, \quad \text{where } \omega' = \sqrt{\frac{k}{2m}}. \tag{4.67}$$

To find the amplitude, we must calculate the maximum value of $x(t)$. This is the task of Exercise 4.15, and the result is that the amplitude of the $x(t) = C\cos\omega' t + D\sin\omega' t$ motion is $A = \sqrt{C^2 + D^2}$. So we have

$$A = \sqrt{\frac{d^2}{4} + \frac{6d^2}{16}} = \sqrt{\frac{5}{8}}\,d. \tag{4.68}$$

This is smaller than the original amplitude d, because energy is lost to heat during the collision (but energy is one of the topics of the next chapter).

4.4. **Average tension**

Let the length of the pendulum be ℓ. We know that the angle θ depends on time according to

$$\theta(t) = A\cos(\omega t), \tag{4.69}$$

where $\omega = \sqrt{g/\ell}$. If T is the tension in the string, then the radial $F = ma$ equation is $T - mg\cos\theta = m\ell\dot{\theta}^2$. Using Eq. (4.69), this becomes

$$T = mg\cos\big(A\cos(\omega t)\big) + m\ell\big(-\omega A\sin(\omega t)\big)^2. \tag{4.70}$$

Since A is small, we can use the small-angle approximation $\cos\alpha \approx 1 - \alpha^2/2$, which gives

$$\begin{aligned} T &\approx mg\left(1 - \frac{1}{2}A^2\cos^2(\omega t)\right) + m\ell\omega^2 A^2\sin^2(\omega t) \\ &= mg + mgA^2\left(\sin^2(\omega t) - \frac{1}{2}\cos^2(\omega t)\right), \end{aligned} \tag{4.71}$$

where we have made use of $\omega^2 = g/\ell$. The average value of both $\sin^2\theta$ and $\cos^2\theta$ over one period is $1/2$ (you can show this by doing the integrals, or you can just note that the averages are equal and they add up to 1), so the average value of T is

$$T_{\text{avg}} = mg + \frac{mgA^2}{4}, \tag{4.72}$$

which is larger than mg, by $mgA^2/4$. It makes sense that $T_{\text{avg}} > mg$, because the average value of the vertical component of T equals mg (because the pendulum has no net rise or fall over a long period of time), and there is some nonzero contribution to the magnitude of T from the horizontal component.

4.5. **Walking east on a turntable**

The velocity of the person with respect to the ground is the sum of $v\hat{\mathbf{x}}$ and $\mathbf{u}$, where $\mathbf{u}$ is the velocity (at the person's position) of the turntable with respect to the ground.

In terms of the angle θ in Fig. 4.28, the velocity components with respect to the ground are

$$\dot{x} = v - u\sin\theta, \quad \text{and} \quad \dot{y} = u\cos\theta. \tag{4.73}$$

But $u = r\omega$. So we have, using $r\sin\theta = y$ and $r\cos\theta = x$,

$$\dot{x} = v - \omega y, \quad \text{and} \quad \dot{y} = \omega x. \tag{4.74}$$

Taking the derivative of the first equation, and then plugging in $\dot{y}$ from the second, gives $\ddot{x} = -\omega^2 x$. Therefore, $x(t) = A\cos(\omega t + \phi)$. The first equation then quickly gives $y(t)$, and the result is that the general expression for the person's position is

$$(x, y) = \big(A\cos(\omega t + \phi), \quad A\sin(\omega t + \phi) + v/\omega\big). \tag{4.75}$$

This describes a circle centered at the point $(0, v/\omega)$. The constants A and ϕ are determined by the initial x and y values. You can show that

$$A = \sqrt{x_0^2 + (y_0 - v/\omega)^2}, \quad \text{and} \quad \tan\phi = \frac{y_0 - v/\omega}{x_0}. \tag{4.76}$$

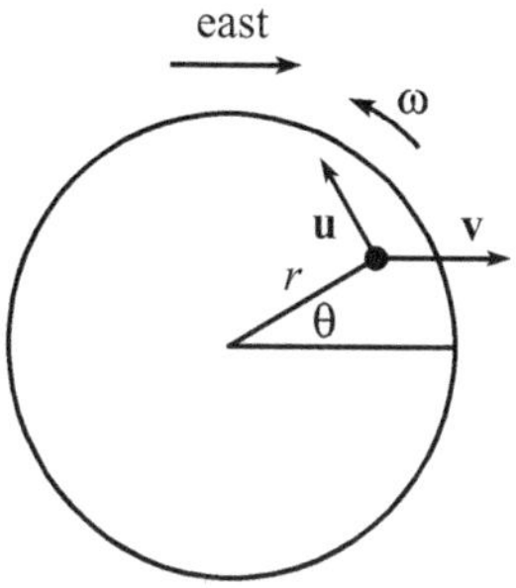

Fig. 4.28

REMARKS: It turns out that in the frame of the turntable, the person's path is also a circle. This can be seen in the following way. Imagine a distant object (say, a star) located in the eastward direction. In the frame of the turntable, this star rotates clockwise with frequency ω. And in the frame of the turntable, the person's velocity always points toward the star. Therefore, the person's velocity rotates clockwise with frequency ω. And since the magnitude of the velocity is constant, this means that the person travels clockwise in a circle in the frame of the turntable. From the usual expression $v = r\omega$, we see that this circle has a radius v/ω.

This result leads to another way of showing that the person's path is a circle as viewed in the ground frame. In short, when the person's clockwise circular motion at speed v with respect to the turntable is combined with the counterclockwise motion of the turntable with respect to the ground (with the same frequency ω, but in the opposite direction), the resulting motion of the person with respect to the ground is a circle with its center at the point $(0, v/\omega)$.

The situation is summarized in Fig. 4.29. From the above result (that the path is a circle in the turntable frame), we may characterize the person's motion in the turntable frame by imagining her riding on a merry-go-round that rotates clockwise with frequency ω with respect to the turntable. This merry-go-round is shown at five different times in the figure. The effect of the merry-go-round's clockwise rotational motion is to cancel the counterclockwise rotation of the turntable, so that the merry-go-round ends up not rotating at all with respect to the ground. Therefore, if the person (the dot in the figure) starts at the top of the merry-go-round (which she in fact does, because this corresponds to walking eastward with respect to the turntable), then she always remains at the top. She therefore travels in a circle that is simply shifted upward by the vertical radius of the merry-go-round (the dotted lines shown, which have length v/ω), relative to the center of the turntable. This agrees with the original result. You can also see from the figure how the values of A and ϕ in Eq. (4.76) arise. For example, A is the length of the solid lines shown. ♣

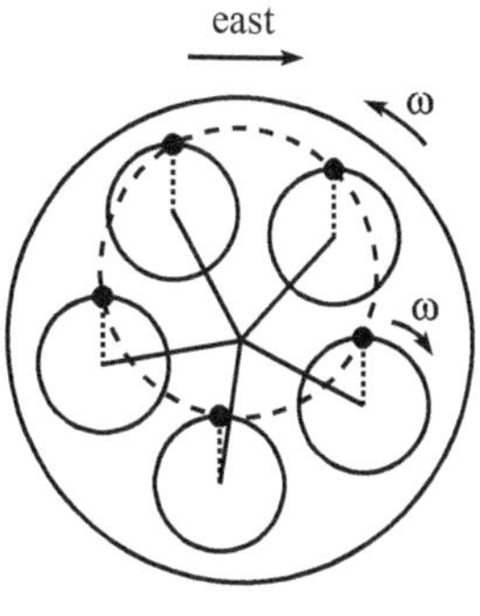

Fig. 4.29

4.6. **Maximum speed**

For the undamped case, the general form of x is $x(t) = C\cos(\omega t + \phi)$. The initial condition $v(0) = 0$ tells us that $\phi = 0$, and then the initial condition $x(0) = x_0$ tells us that $C = x_0$. Therefore, $x(t) = x_0\cos(\omega t)$, and so $v(t) = -\omega x_0\sin(\omega t)$. This has a maximum magnitude of ωx_0.

Now consider the overdamped case. Equation (4.17) gives the position as

$$x(t) = Ae^{-(\gamma-\Omega)t} + Be^{-(\gamma+\Omega)t}. \tag{4.77}$$

The initial conditions are

$$\begin{aligned} x(0) = x_0 \quad &\Longrightarrow \quad A + B = x_0, \\ v(0) = 0 \quad &\Longrightarrow \quad -(\gamma - \Omega)A - (\gamma + \Omega)B = 0. \end{aligned} \tag{4.78}$$

Solving these equations for A and B, and then plugging the results into Eq. (4.77), gives

$$x(t) = \frac{x_0}{2\Omega}\left((\gamma + \Omega)e^{-(\gamma-\Omega)t} - (\gamma - \Omega)e^{-(\gamma+\Omega)t}\right). \tag{4.79}$$

Taking the derivative to find $v(t)$, and using $\gamma^2 - \Omega^2 = \omega^2$, gives

$$v(t) = \frac{-\omega^2 x_0}{2\Omega}\left(e^{-(\gamma-\Omega)t} - e^{-(\gamma+\Omega)t}\right). \tag{4.80}$$

Taking the derivative again, we find that the maximum speed occurs at

$$t_{\text{max}} = \frac{1}{2\Omega}\ln\left(\frac{\gamma + \Omega}{\gamma - \Omega}\right). \tag{4.81}$$

Plugging this into Eq. (4.80), and taking advantage of the logs in the exponentials, gives

$$\begin{aligned} v(t_{\text{max}}) &= \frac{-\omega^2 x_0}{2\Omega}\exp\left(-\frac{\gamma}{2\Omega}\ln\left(\frac{\gamma+\Omega}{\gamma-\Omega}\right)\right)\left(\sqrt{\frac{\gamma+\Omega}{\gamma-\Omega}} - \sqrt{\frac{\gamma-\Omega}{\gamma+\Omega}}\right) \\ &= -\omega x_0\left(\frac{\gamma-\Omega}{\gamma+\Omega}\right)^{\gamma/2\Omega}. \end{aligned} \tag{4.82}$$

The desired ratio, R, of the maximum speeds in the two scenarios is therefore

$$R = \left(\frac{\gamma+\Omega}{\gamma-\Omega}\right)^{\gamma/2\Omega}. \tag{4.83}$$

In the limit of strong damping ($\gamma \gg \omega$), we have $\Omega \equiv \sqrt{\gamma^2 - \omega^2} \approx \gamma - \omega^2/2\gamma$. So the ratio becomes

$$R \approx \left(\frac{2\gamma}{\omega^2/2\gamma}\right)^{1/2} = \frac{2\gamma}{\omega}. \tag{4.84}$$

In the limit of critical damping ($\gamma \approx \omega$, $\Omega \approx 0$), we have, with $\Omega/\gamma \equiv \epsilon$,

$$R \approx \left(\frac{1+\epsilon}{1-\epsilon}\right)^{1/2\epsilon} \approx (1 + 2\epsilon)^{1/2\epsilon} \approx e, \tag{4.85}$$

in agreement with the result of Exercise 4.28 (the solution to which is much quicker than the one above, since you don't need to deal with all the Ω's). You can also show that in these two limits, t_{max} equals $\ln(2\gamma/\omega)/\gamma$ and $1/\gamma \approx 1/\omega$, respectively.

4.7. **Exponential force**

$F = ma$ gives $\ddot{x} = a_0 e^{-bt}$. Let's guess a particular solution of the form $x(t) = Ce^{-bt}$. Plugging this in gives $C = a_0/b^2$. And since the solution to the homogeneous equation $\ddot{x} = 0$ is $x(t) = At + B$, the complete solution for x is

$$x(t) = \frac{a_0 e^{-bt}}{b^2} + At + B. \tag{4.86}$$

The initial condition $x(0) = 0$ gives $B = -a_0/b^2$. And the initial condition $v(0) = 0$ applied to $v(t) = -a_0 e^{-bt}/b + A$ gives $A = a_0/b$. Therefore,

$$x(t) = a_0\left(\frac{e^{-bt}}{b^2} + \frac{t}{b} - \frac{1}{b^2}\right), \tag{4.87}$$

in agreement with Problem 3.9.

4.8. **Driven oscillator**

Plugging $x(t) = A\cos\omega_{\rm d}t + B\sin\omega_{\rm d}t$ into Eq. (4.29) gives

$$\begin{aligned} &- \omega_{\rm d}^2 A\cos\omega_{\rm d}t - \omega_{\rm d}^2 B\sin\omega_{\rm d}t \\ &- 2\gamma\omega_{\rm d}A\sin\omega_{\rm d}t + 2\gamma\omega_{\rm d}B\cos\omega_{\rm d}t \\ &+ \omega^2 A\cos\omega_{\rm d}t + \omega^2 B\sin\omega_{\rm d}t \;=\; F\cos\omega_{\rm d}t. \end{aligned} \tag{4.88}$$

If this is true for all t, the coefficients of $\cos\omega_{\rm d}t$ on both sides must be equal. And likewise for $\sin\omega_{\rm d}t$. Therefore,

$$\begin{aligned} -\omega_{\rm d}^2 A + 2\gamma\omega_{\rm d}B + \omega^2 A &= F, \\ -\omega_{\rm d}^2 B - 2\gamma\omega_{\rm d}A + \omega^2 B &= 0. \end{aligned} \tag{4.89}$$

Solving this system of equations for A and B gives

$$A = \frac{F\left(\omega^2 - \omega_{\rm d}^2\right)}{\left(\omega^2 - \omega_{\rm d}^2\right)^2 + 4\gamma^2\omega_{\rm d}^2}, \qquad B = \frac{2F\gamma\omega_{\rm d}}{\left(\omega^2 - \omega_{\rm d}^2\right)^2 + 4\gamma^2\omega_{\rm d}^2}, \tag{4.90}$$

in agreement with Eq. (4.31).

4.9. **Unequal masses**

Let x_1 and x_2 be the positions of the left and right masses, respectively, relative to their equilibrium positions. The forces on the two masses are $-kx_1 + k(x_2 - x_1)$ and $-kx_2 - k(x_2 - x_1)$, respectively, so the $F = ma$ equations are

$$\begin{aligned} \ddot{x}_1 + 2\omega^2 x_1 - \omega^2 x_2 &= 0, \\ 2\ddot{x}_2 + 2\omega^2 x_2 - \omega^2 x_1 &= 0. \end{aligned} \tag{4.91}$$

The appropriate linear combinations of these equations aren't obvious, so we'll use the determinant method. Letting $x_1 = A_1 e^{i\alpha t}$ and $x_2 = A_2 e^{i\alpha t}$, we see that for there to be a nontrivial solution for A and B, we must have

$$\begin{aligned} 0 &= \begin{vmatrix} -\alpha^2 + 2\omega^2 & -\omega^2 \\ -\omega^2 & -2\alpha^2 + 2\omega^2 \end{vmatrix} \\ &= 2\alpha^4 - 6\alpha^2\omega^2 + 3\omega^4. \end{aligned} \tag{4.92}$$

The roots of this quadratic equation in α^2 are

$$\alpha = \pm\omega\sqrt{\frac{3+\sqrt{3}}{2}} \equiv \pm\alpha_1, \quad \text{and} \quad \alpha = \pm\omega\sqrt{\frac{3-\sqrt{3}}{2}} \equiv \pm\alpha_2. \tag{4.93}$$

If $\alpha^2 = \alpha_1^2$, then the normal mode is proportional to $(\sqrt{3}+1, -1)$. And if $\alpha^2 = \alpha_2^2$, then the normal mode is proportional to $(\sqrt{3}-1, 1)$. So the normal modes are

$$\begin{aligned} \begin{pmatrix} x_1 \\ x_2 \end{pmatrix} &= \begin{pmatrix} \sqrt{3}+1 \\ -1 \end{pmatrix}\cos(\alpha_1 t + \phi_1), \quad \text{and} \\ \begin{pmatrix} x_1 \\ x_2 \end{pmatrix} &= \begin{pmatrix} \sqrt{3}-1 \\ 1 \end{pmatrix}\cos(\alpha_2 t + \phi_2). \end{aligned} \tag{4.94}$$

Note that these two vectors are not orthogonal (there is no need for them to be). The normal coordinates associated with these normal modes are $x_1 - (\sqrt{3}-1)x_2$ and $x_1 + (\sqrt{3}+1)x_2$, respectively, because these are the combinations that make the α_2 and α_1 frequencies disappear, respectively.

4.10. **Weakly coupled**

The magnitude of the force in the middle spring is $\kappa(x_2 - x_1)$, so the $F = ma$ equations are

$$\begin{aligned} m\ddot{x}_1 &= -kx_1 + \kappa(x_2 - x_1), \\ m\ddot{x}_2 &= -kx_2 - \kappa(x_2 - x_1). \end{aligned} \tag{4.95}$$

Adding and subtracting these equations gives

$$\begin{aligned} m(\ddot{x}_1 + \ddot{x}_2) = -k(x_1 + x_2) \quad &\Longrightarrow \quad x_1 + x_2 = A\cos(\omega t + \phi), \\ m(\ddot{x}_1 - \ddot{x}_2) = -(k + 2\kappa)(x_1 - x_2) \quad &\Longrightarrow \quad x_1 - x_2 = B\cos(\tilde{\omega} t + \tilde{\phi}), \end{aligned} \tag{4.96}$$

where

$$\omega \equiv \sqrt{\frac{k}{m}}, \quad \text{and} \quad \tilde{\omega} \equiv \sqrt{\frac{k+2\kappa}{m}}. \tag{4.97}$$

The initial conditions are $x_1(0) = a$, $\dot{x}_1(0) = 0$, $x_2(0) = 0$, and $\dot{x}_2(0) = 0$. The easiest way to apply these is to plug them into the normal coordinates in Eq. (4.96), before solving for $x_1(t)$ and $x_2(t)$. The velocity conditions quickly give $\phi = \tilde{\phi} = 0$, and then the position conditions give $A = B = a$. Solving for $x_1(t)$ and $x_2(t)$ yields

$$\begin{aligned} x_1(t) &= \frac{a}{2}\cos(\omega t) + \frac{a}{2}\cos(\tilde{\omega} t), \\ x_2(t) &= \frac{a}{2}\cos(\omega t) - \frac{a}{2}\cos(\tilde{\omega} t). \end{aligned} \tag{4.98}$$

Writing ω and $\tilde{\omega}$ as

$$\omega = \frac{\tilde{\omega}+\omega}{2} - \frac{\tilde{\omega}-\omega}{2}, \quad \text{and} \quad \tilde{\omega} = \frac{\tilde{\omega}+\omega}{2} + \frac{\tilde{\omega}-\omega}{2}, \tag{4.99}$$

and using the identity $\cos(\alpha+\beta) = \cos\alpha\cos\beta - \sin\alpha\sin\beta$, gives

$$\begin{aligned} x_1(t) &= a\cos\left(\frac{\tilde{\omega}+\omega}{2}\,t\right)\cos\left(\frac{\tilde{\omega}-\omega}{2}\,t\right), \\ x_2(t) &= a\sin\left(\frac{\tilde{\omega}+\omega}{2}\,t\right)\sin\left(\frac{\tilde{\omega}-\omega}{2}\,t\right). \end{aligned} \tag{4.100}$$

If we now approximate $\tilde{\omega}$ as

$$\tilde{\omega} \equiv \sqrt{\frac{k+2\kappa}{m}} = \sqrt{\frac{k}{m}}\sqrt{1+\frac{2\kappa}{k}} \approx \omega\left(1+\frac{\kappa}{k}\right) \equiv \omega + 2\epsilon, \tag{4.101}$$

where $\epsilon \equiv (\kappa/2k)\omega = (\kappa/2m)\sqrt{m/k}$, we can write x_1 and x_2 as

$$\begin{aligned} x_1(t) &\approx a\cos\big((\omega+\epsilon)t\big)\cos(\epsilon t), \\ x_2(t) &\approx a\sin\big((\omega+\epsilon)t\big)\sin(\epsilon t), \end{aligned} \tag{4.102}$$

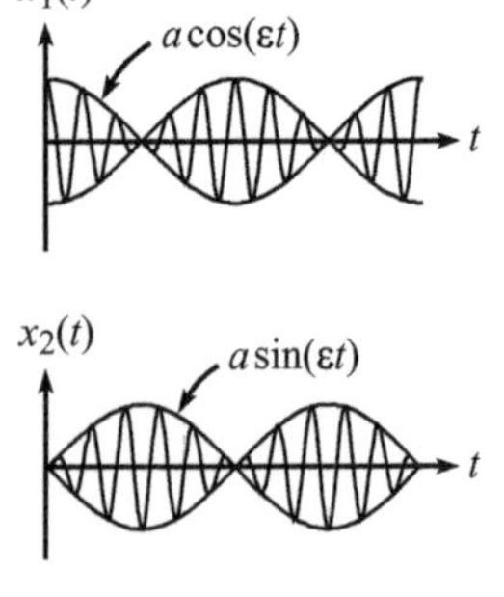

Fig. 4.30

as desired. To get an idea of what this motion looks like, let's examine x_1. Since $\epsilon \ll \omega$, the $\cos(\epsilon t)$ oscillation is much slower than the $\cos((\omega+\epsilon)t)$ oscillation. Therefore, $\cos(\epsilon t)$ is essentially constant on the time scale of the $\cos((\omega+\epsilon)t)$ oscillations. This means that for the time span of a few of these oscillations, x_1 essentially oscillates with frequency $\omega + \epsilon \approx \omega$ and amplitude $a\cos(\epsilon t)$. This $a\cos(\epsilon t)$ term is the "envelope" of the oscillation, as shown in Fig. 4.30, for $\epsilon/\omega = 1/10$. Initially, the amplitude of x_1 is a, but it decreases to zero when $\epsilon t = \pi/2$. By this time, the amplitude of the x_2

oscillation, which is $a \sin(\epsilon t)$, has grown to a. So at $t = \pi/2\epsilon$, the right mass has all the motion, and the left mass is at rest. This process keeps repeating. After each time period of $\pi/2\epsilon$, the motion of one mass gets transferred to the other. The weaker the coupling (that is, the spring constant κ) between the masses, the smaller the ϵ, and so the longer the time period.

REMARKS: The above reasoning also holds for two pendulums connected by a weak spring. All the above steps carry through, with the only change being that k is replaced by mg/ℓ, because the spring force, $-kx$, is replaced by the tangential gravitational force, $-mg \sin\theta \approx -mg(x/\ell)$. So after a time

$$t = \frac{\pi}{2\epsilon} = \frac{\pi}{2}\left(\frac{2m}{\kappa}\sqrt{\frac{k}{m}}\right) \longrightarrow \frac{\pi m}{\kappa}\sqrt{\frac{g}{\ell}}, \tag{4.103}$$

the pendulum that was initially oscillating is now momentarily at rest, and the other pendulum has all the motion. Since the time scale, T_s, of a single mass on the end of the weak spring is proportional to $\sqrt{m/\kappa}$, and since the time scale, T_p, of a simple pendulum is proportional to $\sqrt{\ell/g}$, we see that the above t is proportional to T_s^2/T_p.

The existence of the "beats" in Fig. 4.30 can be traced to the fact that the expressions in Eq. (4.98) are linear combinations of sinusoidal functions with two very close frequencies. The physics here is the same as the physics that produces the beats you hear when listening to two musical notes of nearly the same pitch, as when tuning a guitar.[8] The time between the zeros of, say, x_1 in Fig. 4.30 is π/ϵ, so the angular frequency of the beats is $2\pi/(\pi/\epsilon) = 2\epsilon$. ♣

4.11. **Driven mass on a circle**

Label two diametrically opposite points as the equilibrium positions. Let the positions of the masses relative to these points be x_1 and x_2, measured counterclockwise. If the driving force acts on mass "1," then the $F = ma$ equations are

$$\begin{aligned} m\ddot{x}_1 + 2k(x_1 - x_2) &= F_d \cos\omega_d t, \\ m\ddot{x}_2 + 2k(x_2 - x_1) &= 0. \end{aligned} \tag{4.104}$$

To solve these equations, we can treat the driving force as the real part of $F_d e^{i\omega_d t}$ and try solutions of the form $x_1(t) = A_1 e^{i\omega_d t}$ and $x_2(t) = A_2 e^{i\omega_d t}$, and then solve for A_1 and A_2. Or we can try some trig functions. If we take the latter route, we quickly find that the solutions can't involve any sine terms (this is due to the fact that there are no first derivatives of the x's in Eq. (4.104)). Therefore, the trig functions must look like $x_1(t) = A_1 \cos\omega_d t$ and $x_2(t) = A_2 \cos\omega_d t$. Using either of the two methods, Eq. (4.104) becomes

$$\begin{aligned} -\omega_d^2 A_1 + 2\omega^2(A_1 - A_2) &= F, \\ -\omega_d^2 A_2 + 2\omega^2(A_2 - A_1) &= 0, \end{aligned} \tag{4.105}$$

where $\omega \equiv \sqrt{k/m}$ and $F \equiv F_d/m$. Solving for A_1 and A_2, we find that the desired particular solution is

$$x_1(t) = \frac{-F\left(2\omega^2 - \omega_d^2\right)}{\omega_d^2\left(4\omega^2 - \omega_d^2\right)} \cos\omega_d t, \quad x_2(t) = \frac{-2F\omega^2}{\omega_d^2\left(4\omega^2 - \omega_d^2\right)} \cos\omega_d t. \tag{4.106}$$

[8] If the two frequencies involved aren't too close to each other, then you can actually hear a faint note with a frequency equal to the difference of the original frequencies (and possibly some other notes too, involving various combinations of the frequencies). But this is a different phenomenon from the above beats; it is due to the nonlinear way in which the ear works. See Hall (1981) for more details.

The most general solution is the sum of this particular solution and the homogeneous solution found in Eq. (4.111) in the solution to Problem 4.12 below.

REMARKS:

1. If $\omega_d = 2\omega$, the amplitudes of the motions go to infinity. This makes sense, considering that there is no damping, and that the natural frequency of the system (calculated in Problem 4.12) is 2ω.
2. If $\omega_d = \sqrt{2}\omega$, then the mass that is being driven doesn't move. The reason for this is that the driving force balances the force that the mass feels from the two springs due to the other mass's motion. And indeed, you can show that $\sqrt{2}\omega$ is the frequency that one mass moves at if the other mass is at rest (and thereby acts essentially like a brick wall). Note that $\omega_d = \sqrt{2}\omega$ is the cutoff between the masses moving in the same direction or in opposite directions.
3. If $\omega_d \to \infty$, then both motions go to zero. But x_2 is fourth-order small, whereas x_1 is only second-order small.
4. If $\omega_d \to 0$, then $A_1 \approx A_2 \approx -F/2\omega_d^2$, which is very large. The slowly changing driving force basically spins the masses around in one direction for a while, and then reverses and spins them around in the other direction. We essentially have the driving force acting on a mass $2m$, and two integrations of $F_d \cos\omega_d t = (2m)\ddot{x}$ show that the amplitude of the motion is $F/2\omega_d^2$, as above. Equivalently, you can calculate the $A_1 - A_2$ difference in the $\omega_d \to 0$ limit to show that the springs stretch just the right amount to cause there to be a net force of $(F_d/2)\cos\omega_d t$ on each mass. This leads to the same $F/2\omega_d^2$ amplitude. ♣

4.12. **Springs on a circle**

(a) Label two diametrically opposite points as the equilibrium positions. Let the positions of the masses relative to these points be x_1 and x_2, measured counterclockwise. Then the $F = ma$ equations are

$$\begin{aligned} m\ddot{x}_1 + 2k(x_1 - x_2) &= 0, \\ m\ddot{x}_2 + 2k(x_2 - x_1) &= 0. \end{aligned} \tag{4.107}$$

The determinant method works here, but let's just do it the easy way. Adding the equations gives

$$\ddot{x}_1 + \ddot{x}_2 = 0, \tag{4.108}$$

and subtracting them gives

$$(\ddot{x}_1 - \ddot{x}_2) + 4\omega^2(x_1 - x_2) = 0. \tag{4.109}$$

The normal coordinates are therefore

$$\begin{aligned} x_1 + x_2 &= At + B, \\ x_1 - x_2 &= C\cos(2\omega t + \phi). \end{aligned} \tag{4.110}$$

Solving these two equations for x_1 and x_2, and writing the results in vector form, gives

$$\begin{pmatrix} x_1 \\ x_2 \end{pmatrix} = \begin{pmatrix} 1 \\ 1 \end{pmatrix}(At + B) + C\begin{pmatrix} 1 \\ -1 \end{pmatrix}\cos(2\omega t + \phi), \tag{4.111}$$

where the constants A, B, and C are defined to be half of what they were in Eq. (4.110). The normal modes are therefore

$$\begin{aligned} &\begin{pmatrix} x_1 \\ x_2 \end{pmatrix} = \begin{pmatrix} 1 \\ 1 \end{pmatrix}(At + B), \quad \text{and} \\ &\begin{pmatrix} x_1 \\ x_2 \end{pmatrix} = C\begin{pmatrix} 1 \\ -1 \end{pmatrix}\cos(2\omega t + \phi). \end{aligned} \tag{4.112}$$

The first mode has frequency zero. It corresponds to the masses sliding around the circle, equally spaced, at constant speed. The second mode has both masses moving to the left, then both to the right, back and forth. Each mass feels a force of $4kx$ (because there are two springs, and each one stretches by $2x$), hence the $\sqrt{4} = 2$ in the frequency.

(b) Label three equally spaced points as the equilibrium positions. Let the positions of the masses relative to these points be x_1, x_2, and x_3, measured counterclockwise. Then the $F = ma$ equations are, as you can show,

$$\begin{aligned} m\ddot{x}_1 + k(x_1 - x_2) + k(x_1 - x_3) &= 0, \\ m\ddot{x}_2 + k(x_2 - x_3) + k(x_2 - x_1) &= 0, \\ m\ddot{x}_3 + k(x_3 - x_1) + k(x_3 - x_2) &= 0. \end{aligned} \tag{4.113}$$

The sum of all three of these equations definitely gives something nice. Also, differences between any two of the equations give something useful. But let's use the determinant method to get some practice. Trying solutions of the form $x_1 = A_1 e^{i\alpha t}$, $x_2 = A_2 e^{i\alpha t}$, and $x_3 = A_3 e^{i\alpha t}$, we obtain the matrix equation,

$$\begin{pmatrix} -\alpha^2 + 2\omega^2 & -\omega^2 & -\omega^2 \\ -\omega^2 & -\alpha^2 + 2\omega^2 & -\omega^2 \\ -\omega^2 & -\omega^2 & -\alpha^2 + 2\omega^2 \end{pmatrix} \begin{pmatrix} A_1 \\ A_2 \\ A_3 \end{pmatrix} = \begin{pmatrix} 0 \\ 0 \\ 0 \end{pmatrix}. \tag{4.114}$$

Setting the determinant equal to zero yields a cubic equation in α^2. But it's a nice cubic equation, with $\alpha^2 = 0$ as a solution. The other solution is the double root $\alpha^2 = 3\omega^2$.

The $\alpha = 0$ root corresponds to $A_1 = A_2 = A_3$. That is, it corresponds to the vector $(1, 1, 1)$. This $\alpha = 0$ case is the one case where our exponential solution isn't really an exponential. But α^2 equalling zero in Eq. (4.114) basically tells us that we're dealing with a function whose second derivative is zero, that is, a linear function $At + B$. Therefore, the normal mode is

$$\begin{pmatrix} x_1 \\ x_2 \\ x_3 \end{pmatrix} = \begin{pmatrix} 1 \\ 1 \\ 1 \end{pmatrix} (At + B). \tag{4.115}$$

This mode has frequency zero. It corresponds to the masses sliding around the circle, equally spaced, at constant speed.

The two $\alpha^2 = 3\omega^2$ roots correspond to a two-dimensional subspace of normal modes. You can show that any vector of the form (a, b, c) with $a + b + c = 0$ is a normal mode with frequency $\sqrt{3}\omega$. We will arbitrarily pick the vectors $(0, 1, -1)$ and $(1, 0, -1)$ as basis vectors for this space. We can then write the normal modes as linear combinations of the vectors

$$\begin{aligned} \begin{pmatrix} x_1 \\ x_2 \\ x_3 \end{pmatrix} &= C_1 \begin{pmatrix} 0 \\ 1 \\ -1 \end{pmatrix} \cos(\sqrt{3}\omega t + \phi_1), \\ \begin{pmatrix} x_1 \\ x_2 \\ x_3 \end{pmatrix} &= C_2 \begin{pmatrix} 1 \\ 0 \\ -1 \end{pmatrix} \cos(\sqrt{3}\omega t + \phi_2). \end{aligned} \tag{4.116}$$

Remarks: The $\alpha^2 = 3\omega^2$ case is very similar to the example in Section 4.5 with two masses and three springs oscillating between two walls. The way we've written the two modes in Eq. (4.116), the first one has the first mass stationary (so there could be a wall there, for all the other two masses know). Similarly for the second mode. Hence the $\sqrt{3}\omega$ result here, as in the example.

The normal coordinates in this problem are $x_1+x_2+x_3$ (obtained by adding the three equations in (4.113)), and also any combination of the form $ax_1+bx_2+cx_3$, where $a+b+c=0$ (obtained by taking a times the first equation in Eq. (4.113), plus b times the second, plus c times the third). The three normal coordinates that correspond to the mode in Eq. (4.115) and the two modes we chose in Eq. (4.116) are, respectively, $x_1+x_2+x_3$, $x_1-2x_2+x_3$, and $-2x_1+x_2+x_3$, because each of these combinations gets no contribution from the other two modes (demanding this is how you can derive the coefficients of the x_i's, up to an overall constant). ♣

(c) In part (b), when we set the determinant of the matrix in Eq. (4.114) equal to zero, we were essentially finding the eigenvectors and eigenvalues[9] of the matrix,

$$\begin{pmatrix} 2 & -1 & -1 \\ -1 & 2 & -1 \\ -1 & -1 & 2 \end{pmatrix} = 3I - \begin{pmatrix} 1 & 1 & 1 \\ 1 & 1 & 1 \\ 1 & 1 & 1 \end{pmatrix}, \tag{4.117}$$

where I is the identity matrix. We haven't bothered writing the common factor ω^2 here, because it doesn't affect the eigenvectors. As an exercise, you can show that for the general case of N springs and N masses on a circle, the above matrix becomes the $N \times N$ matrix,

$$3I - \begin{pmatrix} 1 & 1 & 0 & 0 & & 1 \\ 1 & 1 & 1 & 0 & \cdots & 0 \\ 0 & 1 & 1 & 1 & & 0 \\ 0 & 0 & 1 & 1 & & 0 \\ & \vdots & & & \ddots & \vdots \\ 1 & 0 & 0 & 0 & \cdots & 1 \end{pmatrix} \equiv 3I - M. \tag{4.118}$$

In the matrix M, the three consecutive 1's keep shifting to the right, and they wrap around cyclicly. We must now find the eigenvectors of M, which will require being a little clever.

We can guess the eigenvectors and eigenvalues of M if we take a hint from its cyclic nature. A particular set of things that are rather cyclic are the Nth roots of 1. If β is an Nth root of 1, you can verify that $(1, \beta, \beta^2, \dots, \beta^{N-1})$ is an eigenvector of M with eigenvalue $\beta^{-1}+1+\beta$. (This general method works for any matrix where the entries keep shifting to the right. The entries don't have to be equal.) The eigenvalues of the entire matrix in Eq. (4.118) are therefore $3-(\beta^{-1}+1+\beta) = 2-\beta^{-1}-\beta$. There are N different Nth roots of 1, namely $\beta_n = e^{2\pi i n/N}$, for $0 \le n \le N-1$. So the N eigenvalues are

$$\begin{aligned}\lambda_n = 2 - \left(e^{-2\pi i n/N} + e^{2\pi i n/N}\right) &= 2 - 2\cos(2\pi n/N) \\ &= 4\sin^2(\pi n/N). \end{aligned} \tag{4.119}$$

The corresponding eigenvectors are

$$V_n = \left(1, \beta_n, \beta_n^2, \dots, \beta_n^{N-1}\right). \tag{4.120}$$

Since the numbers n and $N-n$ yield the same value for λ_n in Eq. (4.119), the eigenvalues come in pairs (except for $n=0$, and $n=N/2$ if N is even). This is fortunate, because we can then form real linear combinations of the

[9] An eigenvector v of a matrix M is a vector that gets taken into a multiple of itself when acted upon by M. That is, $Mv = \lambda v$, where λ is some number (the eigenvalue). This can be rewritten as $(M-\lambda I)v = 0$, where I is the identity matrix. By our usual reasoning about invertible matrices, a nonzero vector v exists only if λ satisfies $\det|M-\lambda I| = 0$.

two corresponding complex eigenvectors given in Eq. (4.120). We see that the vectors

$$V_n^+ \equiv \frac{1}{2}(V_n + V_{N-n}) = \begin{pmatrix} 1 \\ \cos(2\pi n/N) \\ \cos(4\pi n/N) \\ \vdots \\ \cos\big(2(N-1)\pi n/N\big) \end{pmatrix} \tag{4.121}$$

and

$$V_n^- \equiv \frac{1}{2i}(V_n - V_{N-n}) = \begin{pmatrix} 0 \\ \sin(2\pi n/N) \\ \sin(4\pi n/N) \\ \vdots \\ \sin\big(2(N-1)\pi n/N\big) \end{pmatrix} \tag{4.122}$$

both have eigenvalue $\lambda_n = \lambda_{N-n}$ (as does any linear combination of these vectors). For the special case of $n = 0$, the eigenvector is $V_0 = (1, 1, 1, \dots, 1)$ with eigenvalue $\lambda_0 = 0$. And for the special case of $n = N/2$ if N is even, the eigenvector is $V_{N/2} = (1, -1, 1, \dots, -1)$ with eigenvalue $\lambda_{N/2} = 4$.

Referring back to the $N = 3$ case in Eq. (4.114), we see that we must take the square root of the eigenvalues and then multiply by ω to obtain the frequencies (because it was an α^2 that appeared in the matrix, and because we dropped the factor of ω^2). The frequency corresponding to the above two normal modes is therefore, using Eq. (4.119),

$$\omega_n = \omega\sqrt{\lambda_n} = 2\omega\sin(\pi n/N). \tag{4.123}$$

For even N, the largest value of the frequency is 2ω, with the masses moving in alternating equal positive and negative displacements. But for odd N, it is slightly less than 2ω.

To sum everything up, the N normal modes are the vectors in Eqs. (4.121) and (4.122), where n runs from 1 up to the greatest integer less than $N/2$. And then we have to add on the V_0 vector, and also the $V_{N/2}$ vector if N is even.[10] The frequencies are given in Eq. (4.123). Each frequency is associated with two modes, except the V_0 mode and the $V_{N/2}$ mode if N is even.

Remark: Let's check our results for $N = 2$ and $N = 3$. For $N = 2$: The values of n are the two "special" cases of $n = 0$ and $n = N/2 = 1$. If $n = 0$, we have $\omega_0 = 0$ and $V_0 = (1, 1)$. If $n = 1$, we have $\omega_1 = 2\omega$ and $V_1 = (1, -1)$. These results agree with the two modes in Eq. (4.112).

For $N = 3$: If $n = 0$, we have $\omega_0 = 0$ and $V_0 = (1, 1, 1)$, in agreement with Eq. (4.115). If $n = 1$, we have $\omega_1 = \sqrt{3}\omega$, and $V_1^+ = (1, -1/2, -1/2)$ and $V_1^- = (0, 1/2, -1/2)$. These two vectors span the same space we found in Eq. (4.116). And they have the same frequency as in Eq. (4.116). You can also find the vectors for $N = 4$. These are fairly intuitive, so try to write them down first without using the above results. ♣

[10] If you want, you can treat the $n = 0$ and $n = N/2$ cases the same as all the others. But in both of these cases, the V^- vector is the zero vector, so you can ignore it. So no matter what route you take, you will end up with exactly N nontrivial eigenvectors.

Chapter 5

Conservation of energy and momentum

Conservation laws are extremely important in physics. They are enormously helpful, both quantitatively and qualitatively, in figuring out what is going on in a physical system. When we say that something is "conserved," we mean that it is constant over time. If a certain quantity is conserved, for example, while a ball rolls around in a valley, or while a group of particles interact, then the possible final motions are greatly restricted. If we can write down enough conserved quantities (which we are generally able to do, at least for the systems we'll be concerned with), then we can restrict the final motions down to just one possibility, and so we have solved our problem. Conservation of energy and momentum are two of the main conservation laws in physics. A third, conservation of angular momentum, is discussed in Chapters 7–9.

It should be noted that it isn't *necessary* (in principle) to use conservation of energy and momentum when solving a problem. We'll derive these conservation laws from Newton's laws. Therefore, if you felt like it, you could always (in theory) simply start with first principles and use $F = ma$, etc. However, at best, you would soon grow weary of this approach. And at worst, you would throw in the towel after finding the problem completely intractable. For example, you would get nowhere trying to analyze the collision between two shopping carts (whose contents are free to shift around) by looking at the forces on all the various objects. But conservation of momentum can quickly give you a great deal of information. The point of conservation laws is that they make your calculations much easier, and they also provide a means for getting a good idea of the overall qualitative behavior of a system.

5.1 Conservation of energy in one dimension

Consider a force, in just one dimension for now, that depends only on position. That is, $F = F(x)$. If the force acts on a particle of mass m, and if we write a as $v\,dv/dx$, then $F = ma$ becomes

$$F(x) = mv\frac{dv}{dx}. \qquad (5.1)$$

We can separate variables here and integrate from a given point x_0 where the velocity is v_0 to an arbitrary point x where the velocity is v. The result is

$$\int_{x_0}^{x} F(x')\,dx' = \int_{v_0}^{v} mv'\,dv' \quad \Longrightarrow \quad \int_{x_0}^{x} F(x')\,dx' = \frac{1}{2}mv^2 - \frac{1}{2}mv_0^2$$
$$\Longrightarrow \quad E = \frac{1}{2}mv^2 - \int_{x_0}^{x} F(x')\,dx', \qquad (5.2)$$

where $E \equiv mv_0^2/2$. E depends on v_0, so it therefore also depends on the choice of x_0, because a different x_0 would yield (in general) a different v_0. What we've done here is simply follow the procedure in Section 3.3, for a function that depends only on x. If we now define the *potential energy*, $V(x)$, as

$$V(x) \equiv -\int_{x_0}^{x} F(x')\,dx', \qquad (5.3)$$

then Eq. (5.2) becomes

$$\frac{1}{2}mv^2 + V(x) = E. \qquad (5.4)$$

We define the first term here to be the *kinetic energy.* Since this equation is true at all points in the particle's motion, the sum of the kinetic energy and potential energy is constant. In other words, the total energy is conserved. If a particle loses (or gains) potential energy, then its speed increases (or decreases). Common examples of potential energy are $kx^2/2$ for a Hooke's-law spring force $(-kx)$, with x_0 chosen to be zero; and mgy for the gravitational force $(-mg)$, with y_0 chosen to be zero.

In Boston, lived Jack as did Jill,
Who gained *mgh* on a hill.
In their liquid pursuit,
Jill exclaimed with a hoot,
"I think we've just climbed a landfill!"

While noting, "Oh, this is just grand,"
Jack tripped on some trash in the sand.
He changed his potential
To kinetic, torrential,
But not before grabbing Jill's hand.

So that's what really happened on that hill. People don't just magically come "tumbling after" for no reason, of course.

For a particle undergoing a given motion, both E and $V(x)$ depend on the arbitrary choice of x_0 in Eq. (5.3). This implies that E and $V(x)$ have no real meaning by themselves. Only the *difference* between E and $V(x)$ is relevant (and it equals the kinetic energy); this difference is independent of the choice of x_0. But in order to be concrete in a given setup, we need to pick an arbitrary x_0, so we must remember to state which x_0 we've chosen. For example, it makes no sense to simply say that the gravitational potential energy of an object at height y above the ground is $-\int F\,dy = -\int(-mg)\,dy = mgy$. We have to say that the potential energy is mgy *with respect to the ground* (assuming that our y_0 is at ground level). If we wanted to, we could say that the potential energy is $mgy + mg(7\,\text{m})$ with respect to a point 7 meters below the ground. This is perfectly legitimate, although a bit unconventional.[1] But no matter what reference point we pick, the difference in the potential energies at the points, say, $y = 3$ m and $y = 5$ m equals $mg(2\,\text{m})$.

Note that although we introduced the x_0 in Eq. (5.2) as a point where the particle was at some time, the particle in fact need not ever be at the point x_0. For example, we can throw a ball up at 8 m/s from a height of 5 m while measuring the gravitational potential energy with respect to a point a kilometer high. The ball is certainly not going to reach a height of a kilometer, but that's fine. All that matters is the difference between E and $V(x)$ (both of which are very much negative here) throughout the motion, so a constant shift is irrelevant. We've simply added on the same negative quantity to both E and $V(x)$ in Eq. (5.4), compared with the values someone would measure if the ground were the reference point.

If we take the difference between Eq. (5.4) evaluated at two points, x_1 and x_2 (or if we just integrate Eq. (5.1) from x_1 to x_2), then we obtain

$$\begin{aligned}\frac{1}{2}mv^2(x_2) - \frac{1}{2}mv^2(x_1) &= V(x_1) - V(x_2)\\ &= \int_{x_1}^{x_2} F(x')\,dx' \equiv (\text{Work})_{x_1\to x_2}.\end{aligned} \qquad (5.5)$$

Here it is clear that only differences in potential energies matter. If we define the integral in this equation to be the *work* done on the particle as it moves from x_1 to x_2, then we have produced the *work–energy theorem*:[2]

Theorem 5.1 *The change in a particle's kinetic energy between points x_1 and x_2 equals the work done on the particle between x_1 and x_2.*

If the force points in the same direction as the motion (that is, if the $F(x)$ and the dx in Eq. (5.5) have the same sign), then the work is positive and the speed increases. If the force points in the direction opposite to the motion, then the work is negative and the speed decreases.

[1] It gets to be a pain to keep repeating "with respect to the ground." Therefore, whenever anyone talks about gravitational potential energy in a setup on the surface of the earth, it's generally understood that the ground is the reference point. If, on the other hand, the experiment reaches out to distances far from the earth, then $r = \infty$ is understood to be the reference point, for reasons of convenience we will see in the first example below.

[2] In the form stated here, this theorem holds only for a point particle with no internal structure. See the "Work vs. potential energy" subsection below for the general theorem.

Referring back to Eq. (5.4), and assuming we've chosen a reference point x_0 for the potential energy (and perhaps we've added on a constant to $V(x)$, just because we felt like it), let's draw in Fig. 5.1 the $V(x)$ curve and also the constant E line (which we can determine if we're given, say, the initial position and speed). Then the difference between E and $V(x)$ gives the kinetic energy. The places where $V(x) > E$ are the regions where the particle cannot go. The places where $V(x) = E$ are the "turning points" where the particle stops and changes direction. In the figure, the particle is trapped between x_1 and x_2, and oscillates back and forth. The potential $V(x)$ is extremely useful this way because it makes clear the general properties of the motion.

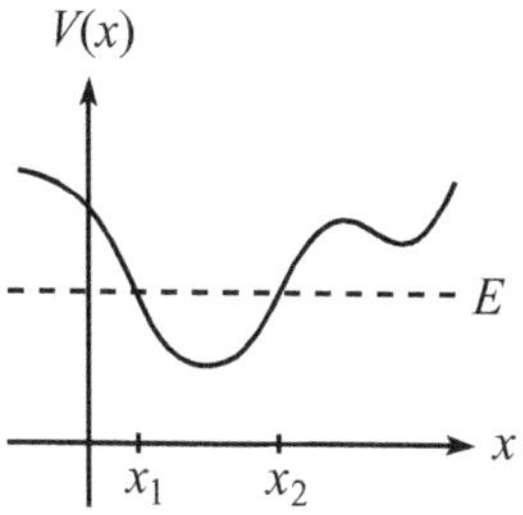

Fig. 5.1

REMARK: It may seem silly to introduce a specific x_0 as a reference point, considering that it's only the differences in the potential (which are independent of x_0) that have any meaning. It's sort of like taking the difference between 17 and 8 by first finding their sizes relative to 5, namely 12 and 3, and then subtracting 3 from 12 to obtain 9. However, since integrals are harder to do than simple subtractions, it's advantageous to do the integral once and for all and thereby label all positions with a definite number $V(x)$, and to then take differences between the V's when needed. ♣

The differential form of Eq. (5.3) is

$$F(x) = -\frac{dV(x)}{dx}. \tag{5.6}$$

Given $V(x)$, it is easy to take its derivative to obtain $F(x)$. But given $F(x)$, it may be difficult (or impossible) to perform the integration in Eq. (5.3) and write $V(x)$ in closed form. But this is not of much concern. The function $V(x)$ is well defined (assuming that the force is a function of x only), and if needed it can be computed numerically to any desired accuracy.

Example (Gravitational potential energy): Consider two point masses, M and m, separated by a distance r. Newton's law of gravitation says that the force between them is attractive and has magnitude GMm/r^2 (we'll talk more about gravity in Section 5.4.1). The potential energy of the system at separation r, measured relative to separation r_0, is therefore

$$V(r) - V(r_0) = -\int_{r_0}^{r} \frac{-GMm}{r'^2}\,dr' = \frac{-GMm}{r} + \frac{GMm}{r_0}, \tag{5.7}$$

where the minus sign in the integrand comes from the attractive nature of the force. A convenient choice for r_0 is ∞, because this makes the second term vanish. It will be understood from now on that this $r_0 = \infty$ reference point has been chosen. Therefore (see Fig. 5.2),

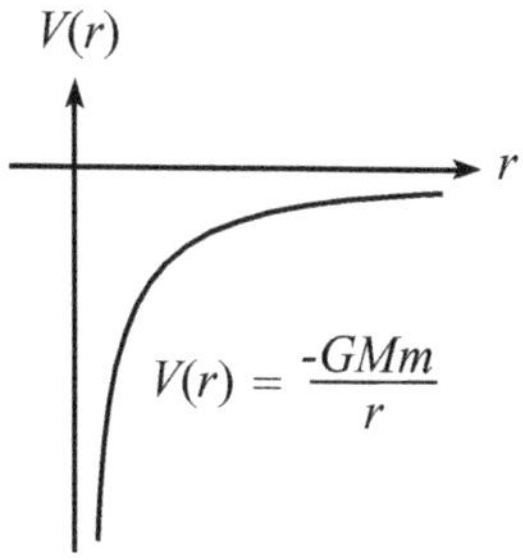

Fig. 5.2

$$V(r) = \frac{-GMm}{r}. \tag{5.8}$$

Example (Gravity near the earth): What is the gravitational potential energy of a mass m at height y, relative to the ground? We know, of course, that it is mgy, but let's do it the hard way. If M is the mass of the earth and R is its radius, then Eq. (5.8) gives (assuming $y \ll R$)

$$\begin{aligned} V(R+y) - V(R) &= \frac{-GMm}{R+y} - \frac{-GMm}{R} = \frac{-GMm}{R}\left(\frac{1}{1+y/R} - 1\right) \\ &\approx \frac{-GMm}{R}\Big((1-y/R) - 1\Big) \\ &= \frac{GMmy}{R^2}, \end{aligned} \tag{5.9}$$

where we have used the Taylor series approximation for $1/(1+\epsilon)$ to obtain the second line. We have also used the fact that a sphere can be treated like a point mass, as far as gravity is concerned. We'll prove this in Section 5.4.1.

Using $g \equiv GM/R^2$, we see that the potential energy difference in Eq. (5.9) equals mgy. We have, of course, simply gone around in circles here. We integrated in Eq. (5.7), and then we basically differentiated in Eq. (5.9) by taking the difference between the forces at nearby points. But it's good to check that everything works out.

A good way to visualize a potential $V(x)$ is to imagine a ball sliding around in a valley or on a hill. For example, the potential of a typical spring is $V(x) = kx^2/2$ (this produces the Hooke's-law force, $F(x) = -dV/dx = -kx$), and we can get a decent idea of what is going on if we imagine a valley with height given by $y = x^2/2$. The gravitational potential of the ball is then $mgy = mgx^2/2$. Choosing $mg = k$ gives the desired potential. If we then look at the projection of the ball's motion on the x axis, it seems like we have constructed a setup identical to the original spring.

However, although this analogy helps in visualizing the basic properties of the motion, the two setups are *not* the same. The details of this fact are left for Problem 5.7, but the following observation should convince you that they are indeed different. Let the ball be released from rest in both setups at a large value of x. Then the force kx due to the spring is very large. But the force in the x direction on the particle in the valley is only a fraction of mg, namely $(mg\sin\theta)\cos\theta$, where θ is the angle of the valley at that point. The setups are approximately the same, however, for small oscillations near the bottom of the valley. See Problem 5.7 for more details.

Conservative forces

Given any force (it can depend on x, v, t, and/or whatever), the work it does on a particle is defined by $W \equiv \int F\,dx$. If the particle starts at x_1 and ends up at x_2, then no matter how it gets there (it might speed up or slow down, or reverse direction a few times), we can calculate the total work done on it by all the forces in the setup, and then equate the result with the change in kinetic energy, via

$$W_{\text{total}} \equiv \int_{x_1}^{x_2} F_{\text{total}}\, dx = \int_{x_1}^{x_2} m\left(\frac{v\, dv}{dx}\right) dx = \frac{1}{2}mv_2^2 - \frac{1}{2}mv_1^2. \tag{5.10}$$

For some forces, the work done is independent of how the particle moves. A force that depends only on position (in one dimension) has this property, because the integral in Eq. (5.10) depends only on the endpoints. The $W = \int F\, dx$ integral is the (signed) area under the F vs. x graph, and this area is independent of how the particle goes from x_1 to x_2.

For other forces, the work done depends on how the particle moves. Such is the case for forces that depend on t or v, because it then matters *when* or *how quickly* the particle goes from x_1 to x_2. A common example of such a force is friction. If you slide a brick across a table from x_1 to x_2, then the work done by friction equals $-\mu mg|\Delta x|$. But if you slide the brick by wiggling it back and forth for an hour before you finally reach x_2, then the amount of negative work done by friction is very large. Since friction always opposes the motion, the contributions to the $W = \int F\, dx$ integral are always negative, so there is never any cancellation. The result is therefore a large negative number.

The issue with friction is that although the μmg force looks like a constant force (which is a subset of position-only dependent forces), it actually isn't. At a given location, the friction can point to the right or to the left, depending on which way the particle is moving. Friction is therefore a function of velocity. True, it's a function only of the *direction* of the velocity, but that's enough to ruin the position-only dependence.

We now define a *conservative force* as one for which the work done on a particle between two given points is independent of how the particle makes the journey. From the preceding discussion, we know that a one-dimensional force is conservative if and only if it depends only on x (or is constant).[3] The point we're leading up to here is that although we can calculate the work done by any force, it makes sense to talk about the potential energy associated with a force only if the force is conservative. This is true because we want to be able to label each value of x with a unique number, $V(x)$, given by $V(x) = -\int_{x_0}^{x} F\, dx$. If this integral were dependent on how the particle goes from x_0 to x, then it wouldn't be well defined, so we wouldn't know what number to assign to $V(x)$. We therefore talk about potential energies only if they are associated with conservative forces. In particular, it makes no sense to talk about the potential energy associated with a friction force.

A useful fact about the gravitational potential energy, mgz, is that it doesn't depend on the path the particle takes, even in two or three dimensions. This is true because even if the particle moves in a complicated direction, only the

[3] In two or three dimensions, however, we will see in Section 5.3 that a conservative force must satisfy another requirement, in addition to being dependent only on position.

vertical z component of the displacement is relevant in calculating the work done by gravity. If we break the path up into many little pieces, the total work done by gravity is obtained by adding up the many little $-mg(dz)$ terms. But the sum of all the dz's is always equal to the total z, independent of the path. Therefore, no matter what the particle is doing in the two horizontal directions, the change in gravitational potential energy is always just mgz. So the gravitational force is a conservative force in three dimensions. We'll see in Section 5.3 that this is a special case of a more general result.

Example (Unwinding string): A mass is connected to one end of a massless string, the other end of which is connected to a very thin frictionless vertical pole. The string is initially wound completely around the pole, in a very large number of tiny horizontal circles, so that the mass touches the pole. The mass is released, and the string gradually unwinds. What angle does the string make with the pole at the moment it becomes completely unwound?

Solution: Let ℓ be the length of the string, and let θ be the final angle it makes with the pole. Then the final height of the mass is $\ell\cos\theta$ below the starting point. So the mass loses a potential energy of $mg(\ell\cos\theta)$. Conservation of energy therefore gives (picking the initial height as $y = 0$, although this doesn't matter)

$$K_\text{i} + V_\text{i} = K_\text{f} + V_\text{f} \quad\Longrightarrow\quad 0 + 0 = \frac{1}{2}mv^2 - mg\ell\cos\theta \quad\Longrightarrow\quad v^2 = 2g\ell\cos\theta. \tag{5.11}$$

There are two unknowns here, v and θ, so we need one more equation. This will be the radial $F = ma$ equation for the (essentially) final circular motion. Because the pole is very thin, the motion can always be approximated by a horizontal circle, which very slowly lowers as time goes by. Because there is essentially no motion in the vertical direction, the total force in this direction is zero. Therefore, the vertical component of the tension is essentially mg. The horizontal component is then $mg\tan\theta$, so the $F = ma$ equation for the final circular motion (which has a radius of $\ell\sin\theta$) is

$$mg\tan\theta = \frac{mv^2}{\ell\sin\theta} \quad\Longrightarrow\quad v^2 = g\ell\sin\theta\tan\theta. \tag{5.12}$$

Equating our two expressions for v^2 gives $\tan\theta = \sqrt{2} \Longrightarrow \theta \approx 54.7°$. Interestingly, this angle is independent of ℓ and g.

Work vs. potential energy

When you drop a ball, does its speed increase because the gravitational force is doing work on it, or because its gravitational potential energy is decreasing? Well, both (or more precisely, either). Work and potential energy are two different ways of talking about the same thing (at least for conservative forces). Either method of reasoning gives the correct result. However, be careful not to use *both* reasonings and "double count" the effect of gravity on the ball. Which choice of terminology you use depends on what you call your "system." Just as with $F = ma$ and free-body diagrams, it is important to label your system when dealing with work and energy, as we'll see in the example below.

The work–energy theorem stated in Theorem 5.1 is relevant to one particle. What if we are dealing with the work done on a system that is composed of various parts? The general work–energy theorem states that the work done on a system by *external* forces equals the change in energy of the system. This energy may come in the form of (1) overall kinetic energy, (2) internal potential energy, or (3) internal kinetic energy (heat falls into this category, because it's simply the random motion of molecules). So we can write the general work–energy theorem as

$$W_{\text{external}} = \Delta K + \Delta V + \Delta K_{\text{internal}} \tag{5.13}$$

For a point particle, there is no internal structure, so we have only the first of the three terms on the right-hand side, in agreement with Theorem 5.1. But to see what happens when a system has internal structure, consider the following example.

Example (Raising a book): Assume that you lift a book up at constant speed, so there is no change in kinetic energy. Let's see what the general work–energy theorem says for various choices of the system.

- System = (book): Both you and gravity are external forces, and there is no change in the energy of the book as a system in itself. So the W–E theorem says

$$W_{\text{you}} + W_{\text{grav}} = 0 \quad \Longleftrightarrow \quad mgh + (-mgh) = 0. \tag{5.14}$$

- System = (book + earth): Now you are the only external force. The gravitational force between the earth and the book is an internal force which produces an internal potential energy. So the W–E theorem says

$$W_{\text{you}} = \Delta V_{\text{earth–book}} \quad \Longleftrightarrow \quad mgh = mgh. \tag{5.15}$$

- System = (book + earth + you): There is now no external force. The internal energy of the system changes because the earth–book gravitational potential energy increases, and also because *your* potential energy decreases. In order to lift the book, you have to burn some calories from the dinner you ate. So the W–E theorem says

$$0 = \Delta V_{\text{earth–book}} + \Delta V_{\text{you}} \quad \Longleftrightarrow \quad 0 = mgh + (-mgh). \tag{5.16}$$

Actually, a human body isn't 100% efficient, so what really happens here is that your potential energy decreases by more than *mgh*, but heat is produced. The sum of these two changes in energy equals $-mgh$. So, including an amount η of energy in the form of heat, we have

$$\begin{aligned} & 0 = \Delta V_{\text{earth–book}} + \Delta V_{\text{you}} + \Delta K_{\text{internal}} \\ \Longleftrightarrow \quad & 0 = mgh + (-mgh - \eta) + \eta. \end{aligned} \tag{5.17}$$

Another contribution to this η heat term can come from, for example, the heat from friction if you slide the book up a rough wall as you lift it.[4]

The moral of all this is that you can look at a setup in various ways, depending on what you pick as your system. Potential energy in one way might show up as work in another. In practice, it is usually more convenient to work in terms of potential energy. So for a dropped ball, people usually consider (consciously or not) gravity to be an internal force in the earth–ball system, as opposed to an external force on the ball system. In general, "conservation of energy," commonly used in setups involving gravity and/or springs, is a straightforward principle to apply (and you'll get plenty of practice with it in the problems and exercises for this chapter). So it turns out that you can usually ignore all these issues about work and about picking your system.

But let's look at one more example, just to make sure we're on the same page. Consider a car that is braking (but not skidding). The friction force from the ground on the tires is what causes the car to slow down. But this force does no work on the car, because the ground isn't moving; the force acts over zero distance. So the external work on the left side of Eq. (5.13) is zero. The right side is therefore also zero. That is, the total energy of the car doesn't change. This is indeed true, because although the overall kinetic energy of the car decreases, there is an equal increase in internal kinetic energy in the form of heat in the brake pads and discs. In other words, $\Delta K = -\Delta K_{\text{internal}}$, and the total energy remains constant. The unfortunate fact about this process is that the energy that goes into heat is lost and can't be converted back into overall kinetic energy of the car. It makes much more sense to convert the overall kinetic energy into some form of internal potential energy (that is, $\Delta K = -\Delta U$), which can then be converted back into overall kinetic energy. Such is the case with hybrid cars which convert overall kinetic energy into chemical potential energy in a battery.

Conversely, when a car accelerates, the friction force from the ground does no work (because the ground isn't moving), so the total energy of the car remains

[4] In this case we have included the wall as a fourth object in our system, because it might contain some of the heat. If you want to instead consider the wall as an external object providing a force, then things get tricky; see Problem 5.6.

the same. The internal potential energy of the gasoline (or battery) is converted into overall kinetic energy (along with some heat and sound). A similar thing happens when you stand at rest and then start walking. The friction force from the ground does no work on you, so your total energy remains the same. You're simply trading your breakfast for overall kinetic energy (plus some heat). For further discussion of work, see Mallinckrodt and Leff (1992).

5.2 Small oscillations

Consider an object in one dimension, subject to the potential $V(x)$. Let the object initially be at rest at a local minimum of $V(x)$, and then let it be given a small kick so that it moves back and forth around the equilibrium point. What can we say about this motion? Is it simple harmonic? Does the frequency depend on the amplitude?

It turns out that for small amplitudes, the motion is indeed simple harmonic, and the frequency can easily be found, given $V(x)$. To see this, expand $V(x)$ in a Taylor series around the equilibrium point, x_0,

$$\begin{aligned} V(x) = V(x_0) + V'(x_0)(x - x_0) + \frac{1}{2!}V''(x_0)(x - x_0)^2 \\ + \frac{1}{3!}V'''(x_0)(x - x_0)^3 + \cdots . \end{aligned} \tag{5.18}$$

This looks like a bit of a mess, but we can simplify it greatly. $V(x_0)$ is an irrelevant additive constant. We can ignore it because only differences in energy matter (or equivalently, because $F = -dV/dx$). And $V'(x_0) = 0$, by definition of the equilibrium point. So that leaves us with the $V''(x_0)$ and higher-order terms. But for sufficiently small displacements, these higher-order terms are negligible compared with the $V''(x_0)$ term, because they are suppressed by additional powers of $(x - x_0)$. So we are left with[5]

$$V(x) \approx \frac{1}{2}V''(x_0)(x - x_0)^2. \tag{5.19}$$

But this has exactly the same form as the Hooke's-law potential, $V(x) = (1/2)k(x - x_0)^2$, provided that we let $V''(x_0)$ be our "spring constant" k. Equivalently, the force is $F = -dV/dx = -V''(x_0)(x - x_0) \equiv -k(x - x_0)$. The frequency of small oscillations, $\omega = \sqrt{k/m}$, is therefore

$$\omega = \sqrt{\frac{V''(x_0)}{m}}. \tag{5.20}$$

[5] Even if $V'''(x_0)$ is much larger than $V''(x_0)$, we can always pick $(x - x_0)$ small enough so that the third-order term is negligible. The one case where this is not true is when $V''(x_0) = 0$. But the result in Eq. (5.20) is still valid in this case. The frequency ω just happens to be zero, in the limit of infinitesimal oscillations.

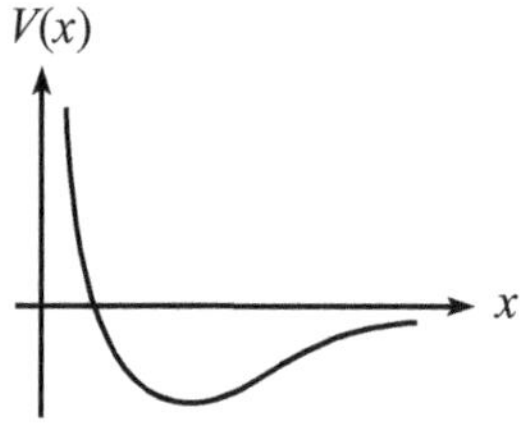

Fig. 5.3

Example: A particle moves under the influence of the potential $V(x) = A/x^2 - B/x$, where $A, B > 0$. Find the frequency of small oscillations around the equilibrium point. This potential is relevant to planetary motion, as we'll see in Chapter 7. The rough shape is shown in Fig. 5.3.

Solution: The first thing we need to do is calculate the equilibrium point x_0. The minimum occurs where

$$0 = V'(x) = -\frac{2A}{x^3} + \frac{B}{x^2} \quad \Longrightarrow \quad x = \frac{2A}{B} \equiv x_0. \tag{5.21}$$

The second derivative of $V(x)$ is

$$V''(x) = \frac{6A}{x^4} - \frac{2B}{x^3}. \tag{5.22}$$

Plugging in $x_0 = 2A/B$, we find

$$\omega = \sqrt{\frac{V''(x_0)}{m}} = \sqrt{\frac{B^4}{8mA^3}}. \tag{5.23}$$

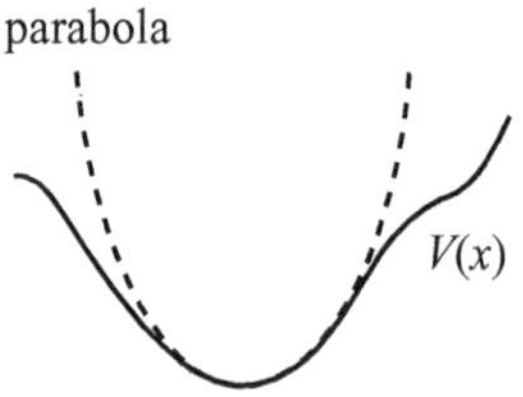

Fig. 5.4

Equation (5.20) is an important result, because *any* function $V(x)$ looks basically like a parabola (see Fig. 5.4) in a small enough region around a minimum (except in the special case where $V''(x_0) = 0$).

A potential may look quite erratic,
And its study may seem problematic.
But down near a min,
You can say with a grin,
"It behaves like a simple quadratic!"

5.3 Conservation of energy in three dimensions

The concepts of work and potential energy in three dimensions are somewhat more complicated than in one dimension, but the general ideas are the same.[6] As in the 1-D case, we'll start with Newton's second law, which now takes the vector form, $\mathbf{F} = m\mathbf{a}$. And as in the 1-D case, we'll deal only with forces that depend only on position, that is, $\mathbf{F} = \mathbf{F}(\mathbf{r})$, because these are the only ones that have any chance of being conservative. The $\mathbf{F} = m\mathbf{a}$ vector equation is shorthand for three equations analogous to Eq. (5.1), namely $mv_x(dv_x/dx) = F_x$, and likewise for y and z. The F_x here is a function of position, so we should really be writing

[6] We'll invoke a few results from vector calculus here. If you haven't seen such material before, a brief review is given in Appendix B.

$F_x(x, y, z)$, but we'll drop the arguments, lest our expressions get too cluttered. Multiplying through by dx, etc., in these three equations, and then adding them together gives

$$F_x\,dx + F_y\,dy + F_z\,dz = m(v_x\,dv_x + v_y\,dv_y + v_z\,dv_z). \tag{5.24}$$

The left-hand side is the work done on the particle. With $d\mathbf{r} \equiv (dx, dy, dz)$, this work can be written as $\mathbf{F} \cdot d\mathbf{r}$ (see Appendix B for the definition of the "dot product"). Using Eq. (B.2), we see that the work can also be written as $F|d\mathbf{r}|\cos\theta$, where θ is the angle between $\mathbf{F}$ and $d\mathbf{r}$. Grouping this as $(F\cos\theta)|d\mathbf{r}|$ shows that the work equals the distance moved times the component of the force along the displacement. Alternatively, grouping it as $F(|d\mathbf{r}|\cos\theta)$ shows that the work also equals the magnitude of the force times the component of the displacement in the direction of the force.

Integrating Eq. (5.24) from the point (x_0, y_0, z_0) to the point (x, y, z) yields[7]

$$E + \int_{x_0}^{x} F_x\,dx' + \int_{y_0}^{y} F_y\,dy' + \int_{z_0}^{z} F_z\,dz' = \frac{1}{2}m(v_x^2 + v_y^2 + v_z^2) = \frac{1}{2}mv^2, \tag{5.25}$$

where E is a constant of integration; it equals $mv_0^2/2$, where v_0 is the speed at (x_0, y_0, z_0). Note that the integrations on the left-hand side depend on the path in 3-D space that the particle takes in going from (x_0, y_0, z_0) to (x, y, z), because the components of $\mathbf{F}$ are functions of position. We'll address this issue below. In terms of the dot product, Eq. (5.25) can be written in the more compact form,

$$\frac{1}{2}mv^2 - \int_{\mathbf{r}_0}^{\mathbf{r}} \mathbf{F}(\mathbf{r}') \cdot d\mathbf{r}' = E. \tag{5.26}$$

Therefore, if we define the potential energy $V(\mathbf{r})$ as

$$V(\mathbf{r}) \equiv -\int_{\mathbf{r}_0}^{\mathbf{r}} \mathbf{F}(\mathbf{r}') \cdot d\mathbf{r}', \tag{5.27}$$

then we can write

$$\frac{1}{2}mv^2 + V(\mathbf{r}) = E. \tag{5.28}$$

In other words, the sum of the kinetic energy and potential energy is constant.

[7] We've put primes on the integration variables so that we don't confuse them with the limits of integration. And as mentioned above, F_x is really $F_x(x', y', z')$, etc.

Conservative forces in three dimensions

For a force that depends only on position (as we have been assuming), there is one complication that arises in 3-D that we didn't have to worry about in 1-D. In 1-D, there is only one route that goes from x_0 to x. The motion itself may involve speeding up or slowing down, or backtracking, but the path is always restricted to be along the line containing x_0 and x. But in 3-D, there is an infinite number of routes that go from $\mathbf{r}_0$ to $\mathbf{r}$. In order for the potential $V(\mathbf{r})$ to have any meaning and to be of any use, it must be well defined. That is, it must be path-independent. As in the 1-D case, we call the force associated with such a potential a *conservative force.* Let's now see what types of 3-D forces are conservative.

Theorem 5.2 *Given a force* $\mathbf{F}(\mathbf{r})$*, a necessary and sufficient condition for the potential,*

$$V(\mathbf{r}) \equiv -\int_{\mathbf{r}_0}^{\mathbf{r}} \mathbf{F}(\mathbf{r}') \cdot d\mathbf{r}', \tag{5.29}$$

to be well defined (that is, to be path-independent) is that the curl of $\mathbf{F}$ *is zero everywhere (that is,* $\nabla \times \mathbf{F} = \mathbf{0}$*; see Appendix B for the definition of the curl).*[8]

Proof: First, let us show that $\nabla \times \mathbf{F} = \mathbf{0}$ is a necessary condition for path-independence. In other words, "If $V(\mathbf{r})$ is path-independent, then $\nabla \times \mathbf{F} = \mathbf{0}$." This follows quickly from the discussion of the curl in Appendix B. We show in Eq. (B.24) that the integral $\int \mathbf{F} \cdot d\mathbf{r}$ (that is, the work done) around a little rectangle in the x-y plane equals the z component of the curl times the area. If the work done in going from corner A to corner B in Fig. 5.5 is the same for paths "1" and "2" (as we are assuming), then the round-trip integral around the rectangle is zero, because one of the paths is being traced out backwards, so it cancels the contribution from the other path. So path-independence implies that the round-trip integral $\int \mathbf{F} \cdot d\mathbf{r}$ is zero for any arbitrary rectangle in the x-y plane. Equation (B.24) therefore says that the z component of the curl must be zero everywhere. Likewise for the y and x components. We have therefore shown that $\nabla \times \mathbf{F} = \mathbf{0}$ is a necessary condition for path independence.

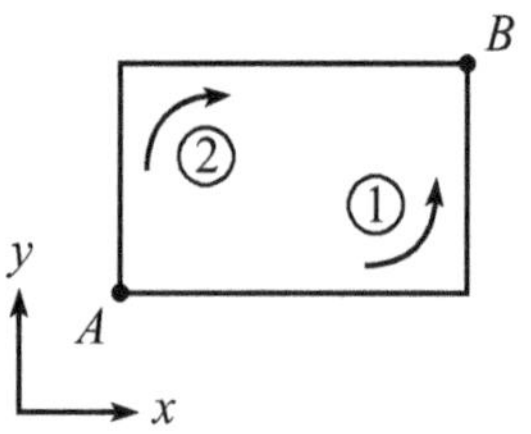

Fig. 5.5

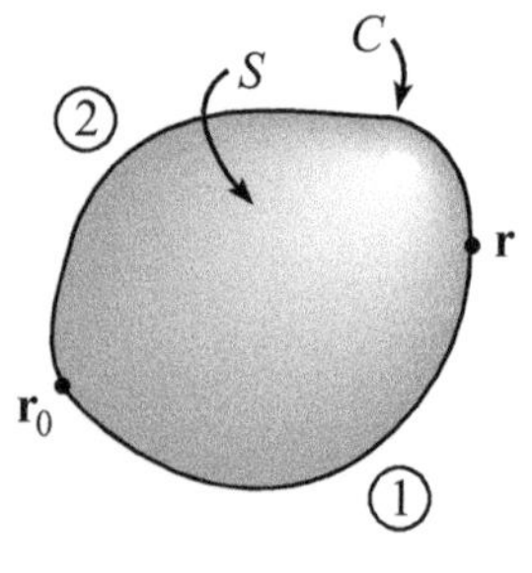

Fig. 5.6

Now let us show that it is sufficient. In other words, "If $\nabla \times \mathbf{F} = \mathbf{0}$, then $V(\mathbf{r})$ is path-independent." The proof of sufficiency follows immediately from Stokes' theorem, which is stated in Eq. (B.25). This theorem implies that if $\nabla \times \mathbf{F} = 0$ everywhere, then $\int_C \mathbf{F} \cdot d\mathbf{r} = 0$ for any closed curve. But Fig. 5.6 shows that traversing the loop C counterclockwise entails traversing path "1" in the "forward" direction, and then traversing path "2" in the "backward" direction. Therefore, from the same reasoning as in the previous paragraph, the integrals

[8] If the force is infinite at any point, then the proof of sufficiency below (which is based on Stokes' theorem) isn't valid, and it turns out that a second condition is required; see Feng (1969). But we won't worry about that here.

from $\mathbf{r}_0$ to $\mathbf{r}$ along paths "1" and "2" are equal. This holds for arbitrary points $\mathbf{r}_0$ and $\mathbf{r}$, and arbitrary curves C, so $V(\mathbf{r})$ is path-independent. ■

REMARK: Another way to show that $\nabla \times \mathbf{F} = 0$ is a necessary condition for path-independence (that is, "If $V(\mathbf{r})$ is path-independent, then $\nabla \times \mathbf{F} = 0$") is the following. If $V(\mathbf{r})$ is path-independent (and therefore well defined), then it is legal to write down the differential form of Eq. (5.27), namely

$$dV(\mathbf{r}) = -\mathbf{F}(\mathbf{r}) \cdot d\mathbf{r} \equiv -(F_x\, dx + F_y\, dy + F_z\, dz). \tag{5.30}$$

But another expression for dV is

$$dV(\mathbf{r}) = \frac{\partial V}{\partial x}\, dx + \frac{\partial V}{\partial y}\, dy + \frac{\partial V}{\partial z}\, dz. \tag{5.31}$$

These two expressions must be equivalent for arbitrary dx, dy, and dz. So we have

$$(F_x, F_y, F_z) = -\left(\frac{\partial V}{\partial x}, \frac{\partial V}{\partial y}, \frac{\partial V}{\partial z}\right) \quad \Longrightarrow \quad \mathbf{F}(\mathbf{r}) = -\nabla V(\mathbf{r}). \tag{5.32}$$

In other words, the force is the gradient of the potential. Therefore,

$$\nabla \times \mathbf{F} = -\nabla \times \nabla V(\mathbf{r}) = 0, \tag{5.33}$$

because the curl of a gradient is identically zero, as you can verify by using the definition of the curl in Eq. (B.20) and the fact that partial differentiation is commutative (that is, $\partial^2 V/\partial x\, \partial y = \partial^2 V/\partial y\, \partial x$). ♣

Example (Central force): A *central force* is defined to be a force that points radially and whose magnitude depends only on r. That is, $\mathbf{F}(\mathbf{r}) = F(r)\hat{\mathbf{r}}$. Show that a central force is conservative by explicitly showing that $\nabla \times \mathbf{F} = \mathbf{0}$.

Solution: The force $\mathbf{F}$ may be written as

$$\mathbf{F}(x,y,z) = F(r)\hat{\mathbf{r}} = F(r)\left(\frac{x}{r}, \frac{y}{r}, \frac{z}{r}\right). \tag{5.34}$$

Now, as you can verify,

$$\frac{\partial r}{\partial x} = \frac{\partial \sqrt{x^2 + y^2 + z^2}}{\partial x} = \frac{x}{r}, \tag{5.35}$$

and similarly for y and z. Therefore, the z component of $\nabla \times \mathbf{F}$ equals (writing F for $F(r)$, and F' for $dF(r)/dr$, and making use of the chain rule)

$$\begin{aligned}
\frac{\partial F_y}{\partial x} - \frac{\partial F_x}{\partial y} &= \frac{\partial (yF/r)}{\partial x} - \frac{\partial (xF/r)}{\partial y} \\
&= \left(\frac{y}{r}F'\frac{\partial r}{\partial x} - yF\frac{1}{r^2}\frac{\partial r}{\partial x}\right) - \left(\frac{x}{r}F'\frac{\partial r}{\partial y} - xF\frac{1}{r^2}\frac{\partial r}{\partial y}\right) \\
&= \left(\frac{yxF'}{r^2} - \frac{yxF}{r^3}\right) - \left(\frac{xyF'}{r^2} - \frac{xyF}{r^3}\right) = 0.
\end{aligned} \tag{5.36}$$

Likewise for the x and y components.

5.4 Gravity

5.4.1 Newton's universal law of gravitation

The gravitational force on a point mass m, located a distance r from a point mass M, is given by Newton's law of gravitation,

$$F(r) = \frac{-GMm}{r^2}, \tag{5.37}$$

where the minus sign indicates an attractive force. The numerical value of G is $6.67 \cdot 10^{-11}\ \text{m}^3/(\text{kg s}^2)$. We'll show how this value is obtained in Section 5.4.2 below.

What is the force if we replace the point mass M by a sphere of radius R and mass M? The answer (assuming that the sphere is spherically symmetric, that is, the density is a function only of r) is that it is still $-GMm/r^2$. A sphere acts just like a point mass at its center, for the purposes of gravity (as long as we're considering a mass m outside the sphere). This is an extremely pleasing result, to say the least. If it were not the case, then the universe would be a far more complicated place than it is. In particular, the motion of planets and such things would be much harder to describe.

To demonstrate that spheres behave like points, as far as gravity is concerned, it turns out to be much easier to calculate the potential energy due to a sphere, and to then take the derivative to obtain the force, rather than to calculate the force explicitly.[9] So this is the route we will take. It will suffice to demonstrate the result for a thin spherical shell, because a sphere is the sum of many such shells. Our strategy for calculating the potential energy at a point P, due to a spherical shell, will be to slice the shell into rings as shown in Fig. 5.7. Let the radius of the shell be R, let P be a distance r from the center of the shell, and let the ring make the angle θ shown. The distance, ℓ, from P to the ring is a function of R, r, and θ. It may be found as follows. In Fig. 5.8, segment AB has length $R\sin\theta$, and segment BP has length $r - R\cos\theta$. So the length ℓ in triangle ABP is

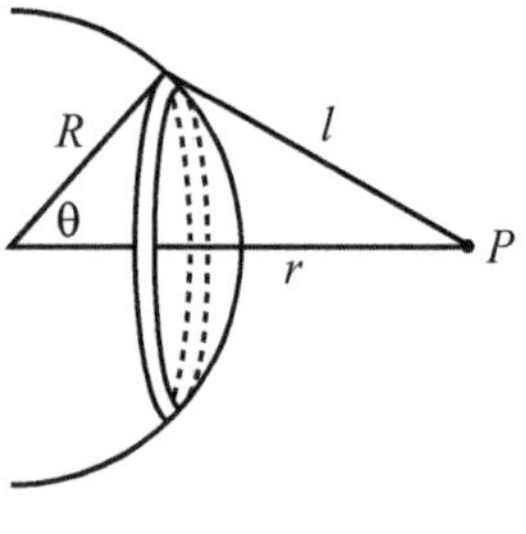

Fig. 5.7

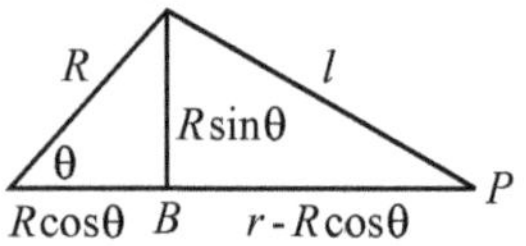

Fig. 5.8

$$\ell = \sqrt{(R\sin\theta)^2 + (r - R\cos\theta)^2} = \sqrt{R^2 + r^2 - 2rR\cos\theta}. \tag{5.38}$$

What we've done here is just prove the law of cosines.

The area of a ring between θ and $\theta + d\theta$ is its width (which is $R\,d\theta$) times its circumference (which is $2\pi R\sin\theta$). Letting $\sigma = M/(4\pi R^2)$ be the mass density per unit area of the shell, we see that the potential energy of a mass m at P due to a thin ring is $-Gm\sigma(R\,d\theta)(2\pi R\sin\theta)/\ell$. This is true because the gravitational potential energy,

$$V(\ell) = \frac{-Gm_1m_2}{\ell}, \tag{5.39}$$

[9] The reason for this is that the potential energy is a scalar quantity (just a number), whereas the force is a vector. If we tried to calculate the force, we would have to worry about forces pointing in all sorts of directions. With the potential energy, we simply have to add up a bunch of numbers.

is a scalar quantity, so the contributions from the little mass pieces simply add. Every piece of the ring is the same distance from P, and this distance is all that matters; the direction from P is irrelevant (unlike it would be with the force). The total potential energy at P is therefore

$$\begin{aligned} V(r) &= -\int_0^\pi \frac{2\pi\sigma GR^2 m \sin\theta\, d\theta}{\sqrt{R^2+r^2-2rR\cos\theta}} \\ &= -\frac{2\pi\sigma GRm}{r}\sqrt{R^2+r^2-2rR\cos\theta}\,\bigg|_0^\pi . \end{aligned} \tag{5.40}$$

The $\sin\theta$ in the numerator is what makes this integral nice and doable. We must now consider two cases. If $r > R$, then we have

$$V(r) = -\frac{2\pi\sigma GRm}{r}\Big((r+R)-(r-R)\Big) = -\frac{G(4\pi R^2\sigma)m}{r} = -\frac{GMm}{r}, \tag{5.41}$$

which is the potential due to a point mass M located at the center of the shell, as desired. If $r < R$, then we have

$$V(r) = -\frac{2\pi\sigma GRm}{r}\Big((r+R)-(R-r)\Big) = -\frac{G(4\pi R^2\sigma)m}{R} = -\frac{GMm}{R}, \tag{5.42}$$

which is independent of r. Having found $V(r)$, we can now find $F(r)$ by taking the negative of the gradient of V. The gradient is just $\hat{\mathbf{r}}(d/dr)$ here, because V is a function only of r. Therefore,

$$\begin{aligned} F(r) &= -\frac{GMm}{r^2}, \quad \text{if } r > R, \\ F(r) &= 0, \quad \text{if } r < R. \end{aligned} \tag{5.43}$$

These forces are directed radially, of course. A solid sphere is the sum of many spherical shells, so if P is outside a given sphere, then the force at P is $-GMm/r^2$, where M is the total mass of the sphere. This result holds even if the shells have different mass densities (but each one must have uniform density). Note that the gravitational force between two spheres is the same as if they were both replaced by point masses. This follows from two applications of our "point mass" result.

> Newton looked at the data, numerical,
> And then observations, empirical.
> He said, "But, of course,
> We get the same force
> From a point mass and something that's spherical!"

If P is inside a given sphere, then the only relevant material is the mass inside a concentric sphere through P, because all the shells outside this region give zero

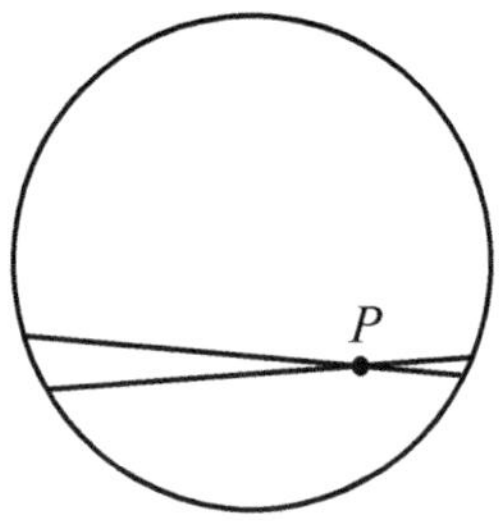

Fig. 5.9

force, from the second equation in Eq. (5.43). The material "outside" of P is, for the purposes of gravity, not there.

It isn't obvious that the force inside a spherical shell is zero. Consider the point P in Fig. 5.9. A piece of mass dm on the right side of the shell gives a larger force on P than a piece of mass dm on the left side, due to the $1/r^2$ dependence. But from the figure, there is more mass on the left side than the right side. These two effects happen to exactly cancel, as you can show in Problem 5.10.

Example (Supporting a tube): Imagine the following unrealistic undertaking. Drill a narrow tube, with cross-sectional area A, from the surface of the earth down to the center. Then line the cylindrical wall of the tube with a frictionless coating. Then fill the tube back up with the dirt (and magma, etc.) that you originally removed. What force is necessary at the bottom of the tube of dirt (that is, at the center of the earth) to hold it up? Let the earth's radius be R, and assume (incorrectly) a uniform mass density ρ.

Solution: The gravitational force on a mass dm at radius r is effectively due to the mass inside the radius r (call this M_r). The mass outside r is effectively not there. The gravitational force is therefore

$$F_{dm} = \frac{GM_r\, dm}{r^2} = \frac{G\big((4/3)\pi r^3 \rho\big)\, dm}{r^2} = \frac{4}{3}\pi G\rho r\, dm, \tag{5.44}$$

which we see increases linearly with r. The dirt in the tube between r and $r + dr$ has volume $A\,dr$, so its mass is $dm = \rho A\, dr$. The total gravitational force on the entire tube is therefore

$$\begin{aligned} F = \int F_{dm} &= \int_0^R \frac{4}{3}\pi G\rho r(\rho A\, dr) = \frac{4}{3}\pi G\rho^2 A \int_0^R r\, dr \\ &= \frac{4}{3}\pi G\rho^2 A \cdot \frac{R^2}{2} = \frac{2}{3}\pi G\rho^2 A R^2. \end{aligned} \tag{5.45}$$

The force at the bottom of the tube must be equal and opposite to this force. In terms of the mass of the earth, $M_{\mathrm{E}} = (4/3)\pi R^3 \rho$, and the total mass of the tube, $M_{\mathrm{t}} = \rho AR$, this result can be written as $F = GM_{\mathrm{E}}M_{\mathrm{t}}/2R^2$. So the required force is half of what a scale on the surface of the earth would read if all of the tube's dirt sat in a lump on top of it. The reason for this is basically that the force in Eq. (5.44) is linear in r.

An important subtopic of gravity is the *tidal force*, but because this is most easily discussed in the context of accelerating frames of reference and fictitious forces, we'll postpone it until Chapter 10.

5.4.2 The Cavendish experiment

How do we determine the numerical value of G in Eq. (5.37)? If we can produce a setup in which we know the values F, M, m, and r, then we can determine G.

The first strategy that comes to mind is to take advantage of the fact that the gravitational force on an object on the earth's surface is known to be $F = mg$. Combining this with Eq. (5.37) gives $g = GM_E/R^2$. We know the values of g and R,[10] so this tells us what the product GM_E is. However, this information unfortunately doesn't help us, because we don't know what the mass of the earth is (without knowing G first, which we're assuming we don't know yet; see the last paragraph in this section). For all we know, the mass of the earth might be 10 times larger than what we think it is, with G being 10 times smaller. The only way to find G is to use a setup with two known masses. But this then presents the problem that the resulting force is extremely small. So the task boils down to figuring out a way to measure the tiny force between two known masses. Henry Cavendish solved this problem in 1798 by performing an extremely delicate experiment (which was devised a few years earlier by John Michell, but he died before being able to perform it).[11] The basic idea behind the experiment is the following.

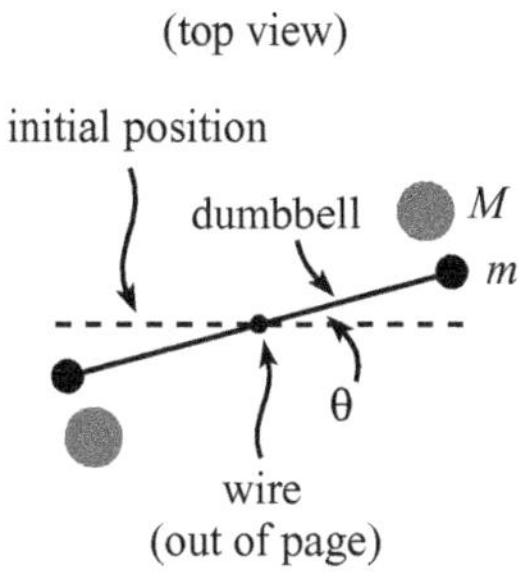

Fig. 5.10

Consider the setup in Fig. 5.10, which shows the top view. A dumbbell with two masses m on its ends hangs from a very thin wire. The dumbbell is free to twist, although if it twists, the wire will provide a tiny restoring torque.[12] The dumbbell starts with no twist in the wire, and then two other masses M are placed (fixed) at the positions shown. These masses produce attractive forces on the dumbbell masses and cause the dumbbell to twist counterclockwise. The dumbbell will oscillate back and forth before finally settling down at some tiny angle θ away from the initial position.

The torque on the dumbbell that arises from the twist in the wire takes the form of $\tau = -b\theta$ (with counterclockwise torque taken to be positive), where b is a constant that depends on the thickness and makeup of the wire. This linear relation between τ and θ holds for small θ for all the same reasons that the $F = -kx$ Hooke's-law result in Section 5.2 holds.

The gravitational force between each pair of masses is GMm/d^2, where d is the separation between the centers of the masses in each pair. So the torque on the dumbbell due to the two gravitational forces is $2(GMm/d^2)\ell$, where ℓ is half the length of the dumbbell. Demanding that the total torque on the dumbbell be zero gives

$$\frac{2GMm\ell}{d^2} - b\theta = 0 \quad \Longrightarrow \quad G = \frac{b\theta d^2}{2Mm\ell}. \tag{5.46}$$

[10] The radius of the earth has been known (at least roughly) since the time of Eratosthenes, about 250 BC. For an interesting way to measure it yourself, see Rawlins (1979).

[11] The purpose of the experiment, as intended by Michell and Cavendish, was actually to measure the density of the earth, and not G; see Clotfelter (1987). But as we'll see below, this is equivalent to measuring G.

[12] We won't talk about torque until Chapter 8, so you may want to come back and read this section after that. We'll invoke some results about rotational dynamics here, but the general setup should be clear even if you're not familiar with rotations.

We know all the parameters on the right-hand side except b, so if we can determine that, then we're done. It's very difficult to measure b directly with any reasonable accuracy, because the torque in the wire is so tiny. But fortunately there's a sneaky way to determine b that involves taking a page from our playbook on oscillations. The bread-and-butter equation for rotations is $\tau = I\ddot{\theta}$, where I is the moment of inertia (which we can calculate for the dumbbell); this is the rotational analog of Newton's second law, $F = m\ddot{x}$. Now, if the torque takes the form of $\tau = -b\theta$ for small θ, then $\tau = I\ddot{\theta}$ becomes $-b\theta = I\ddot{\theta}$. This is a good old simple-harmonic-oscillator equation, so we know that the frequency of the oscillations is $\omega = \sqrt{b/I}$. Therefore, all we need to do is measure the period, $T = 2\pi/\omega$, of the oscillations while the dumbbell is settling down, and we can determine b from $b = I\omega^2 = I(2\pi/T)^2$. (The time T is large, because b is small on the scale of things, because otherwise there wouldn't be any noticeable twist in the wire.) Plugging this value of b into Eq. (5.46) finally gives

$$G = \frac{4\pi^2 I\theta d^2}{2Mm\ell T^2}. \tag{5.47}$$

The Cavendish experiment is also known as the "weighing the earth" (or perhaps the "massing the earth") experiment, because now that we know G (and also g and R), we can use $g = GM_E/R^2$ to calculate the mass of the earth, M_E. The only possible way to determine M_E (without examining every cubic meter of the inside of the earth, which is obviously impossible) is to determine G first, as we have done.[13] Interestingly, the resulting value of M_E, which is roughly $6\cdot 10^{24}$ kg, leads to an average density of the earth of about $5.5\,\text{g/cm}^3$. This is larger than the density of the earth's crust and mantle, so we conclude that there must be something very dense deep down inside the earth. So the Cavendish experiment, which involves masses hanging from a wire, amazingly tells us something about the earth's core![14]

5.5 Momentum

5.5.1 Conservation of momentum

Newton's third law says that for every force there is an equal and opposite force. In other words, if $\mathbf{F}_{ab}$ is the force that particle a feels due to particle b, and if $\mathbf{F}_{ba}$ is the force that particle b feels due to particle a, then $\mathbf{F}_{ba} = -\mathbf{F}_{ab}$ at all times. This law has important implications concerning momentum, $\mathbf{p} \equiv m\mathbf{v}$. Consider two particles that interact over a period of time. Assume that they are isolated from outside forces. From Newton's second law, $\mathbf{F} = d\mathbf{p}/dt$, we see by integrating this that the total change in a particle's momentum equals the time

[13] If you want to determine M_E without using g or R (but still using G, of course, because M_E appears only through the combination GM_E), see Celnikier (1983).

[14] For a comprehensive discussion of the earth's core, see Brush (1980).

integral of the force acting on it. That is,

$$\mathbf{p}(t_2) - \mathbf{p}(t_1) = \int_{t_1}^{t_2} \mathbf{F}\,dt. \tag{5.48}$$

This integral is called the *impulse*. If we now invoke the third law, $\mathbf{F}_{ba} = -\mathbf{F}_{ab}$, we find

$$\mathbf{p}_a(t_2) - \mathbf{p}_a(t_1) = \int_{t_1}^{t_2} \mathbf{F}_{ab}\,dt = -\int_{t_1}^{t_2} \mathbf{F}_{ba}\,dt = -\big(\mathbf{p}_b(t_2) - \mathbf{p}_b(t_1)\big). \tag{5.49}$$

Therefore,

$$\mathbf{p}_a(t_2) + \mathbf{p}_b(t_2) = \mathbf{p}_a(t_1) + \mathbf{p}_b(t_1). \tag{5.50}$$

This is the statement that the total momentum of this isolated system of two particles is *conserved*; it does not depend on time. Note that Eq. (5.50) is a vector equation, so it is really three equations, namely conservation of p_x, p_y, and p_z.

REMARK: Newton's third law makes a statement about forces. But force is related to momentum via $F = dp/dt$. So the third law essentially *postulates* conservation of momentum. (The "proof" above in Eq. (5.49) is hardly a proof. It involves one simple integration.) So you might wonder if momentum conservation is something you can *prove*, or if it's something you have to *assume*, as we have basically done because we have simply accepted the third law.

The difference between a postulate and a theorem is rather nebulous. One person's postulate might be another person's theorem, and vice versa. You have to start *somewhere* in your assumptions. We chose to start with the third law. In the Lagrangian formalism in Chapter 6, the starting point is different, and momentum conservation is deduced as a consequence of translational invariance (as we will see). So it looks more like a theorem in that formalism.

But one thing is certain. Momentum conservation for two particles *cannot* be proved from scratch for arbitrary forces, because it is not necessarily true. For example, if two charged particles interact in a certain way through the magnetic fields they produce, then the total momentum of the two particles might *not* be conserved. Where is the missing momentum? It is carried off in the electromagnetic field. The total momentum of the system is indeed conserved, but the crucial point is that the system consists of the two particles *plus* the electromagnetic field. Said in another way, each particle actually interacts with the electromagnetic field, and not the other particle. Newton's third law does not necessarily hold for particles subject to such a force. ♣

Let's now look at momentum conservation for a system of many particles. As above, let $\mathbf{F}_{ij}$ be the force that particle i feels due to particle j. Then $\mathbf{F}_{ij} = -\mathbf{F}_{ji}$ at all times. Assume that the particles are isolated from outside forces. The change in momentum of the ith particle from t_1 to t_2 is (we won't bother writing all the t's in the expressions below)

$$\Delta\mathbf{p}_i = \int \Big(\sum_j \mathbf{F}_{ij}\Big)\,dt. \tag{5.51}$$

Therefore, the change in the total momentum of all the particles is (switching the order of the integration and the sum over i on the right-hand side here)

$$\Delta\mathbf{P} \equiv \sum_i \Delta\mathbf{p}_i = \int \Big(\sum_i \sum_j \mathbf{F}_{ij}\Big)\,dt. \tag{5.52}$$

But $\sum_i \sum_j \mathbf{F}_{ij} = 0$ at all times, because for every term $\mathbf{F}_{ab}$ there is a term $\mathbf{F}_{ba}$, and $\mathbf{F}_{ab} + \mathbf{F}_{ba} = 0$ (and also, $\mathbf{F}_{aa} = 0$). The forces all cancel in pairs. Therefore, the total momentum of an isolated system of particles is conserved.

Example (Snow on a sled): You are riding on a sled that is given an initial push and slides across frictionless ice. Snow is falling vertically (in the frame of the ice) on the sled. Assume that the sled travels in tracks that constrain it to move in a straight line. Which of the following three strategies causes the sled to move the fastest? The slowest?

A: You sweep the snow off the sled so that it leaves the sled in the direction perpendicular to the sled's tracks, as seen by you in the frame of the sled.

B: You sweep the snow off the sled so that it leaves the sled in the direction perpendicular to the sled's tracks, as seen by someone in the frame of the ice.

C: You do nothing.

First solution: The sideways motion of the snow after you sweep it off is irrelevant, because although Newton's third law says that the swept snow applies a sideways force on the sled, the normal force from the tracks keeps the sled from sliding off the tracks. Also, the vertical motion of the snow when it hits the sled is irrelevant, because the vertical normal force from the tracks keeps the sled from falling through the ground. We are therefore concerned only with the motion in the direction of the tracks. And since there are no external forces in this direction on the you/sled/snow system, the momentum in this direction is conserved.

In general, the speed of the sled can (possibly) change due to two kinds of events: (1) new snow hitting the sled (and eventually coming to rest with respect to the sled), and (2) snow being swept off the sled.

Let's first compare A with C. In strategy A, your sweeping action doesn't change the speed of the sled, because you are sweeping the snow directly sideways in your frame. Since the sideways motion of the snow is irrelevant as far as the forward momentum goes, you are essentially just reaching out and plopping a ball of snow on the ice. This snow then simply travels along next to the sled at the same speed; it might as well be connected to the sled (at least for a moment, until new snow hits the sled). In strategy C, your (non) action obviously don't change the speed of the sled. So in comparing A with C, the departure of snow doesn't differentiate them. We therefore need only consider what happens to the sled when new snow hits it. But this is easy to do: Since the sled is heavier in C than in A, a new snowflake slows it down less in C than in A. Therefore, C is faster than A.

Now let's compare B with C. B is faster than C because in B the snowflakes have zero forward momentum in the end, whereas in C they have nonzero forward momentum (because they are sitting on the moving sled). The momenta of the two systems must be the same (equal to the initial momentum of the sled plus you), so the sled must be moving faster in B.

Therefore, B is faster than C, which is faster than A. As a consistency check, it's easy to see that B is faster than A, because in B you must push the snow backward with respect to the sled. So by Newton's third law, the swept snow in B exerts a forward force on the sled.

Second solution: In the end, this solution is basically the same as the first one, but it's a slightly more systematic way of looking at things: In B, all the snow is moving *slower* than the sled; in fact, it is all at rest (at least in the forward direction) with respect to the ice. In C, all the snow is *on* the sled. And in A, all the snow is moving *faster* than the sled; it is moving along with various forward speeds, ranging from the initial speed of the sled down to the present speed, depending on when it was swept off.

By conservation of momentum, the total momentum of the sled (including you) plus the snow at any given time is the same in all three cases. The only consistent way to combine the facts in the previous paragraph with conservation of momentum is for the speed of the sled to satisfy $B > C > A$. This is true because if someone claimed that $C > B$, then the slowest object (namely everything, since all the snow is on the sled) in C would be going faster than the fastest object (the sled) in B; this would contradict the fact that the momenta of the two systems are equal. Therefore, we must have $B > C$. Likewise, if someone claimed that $A > C$, then the slowest object (the sled) in A would be going faster than the fastest object (everything) in C; this would again be a contradiction. Therefore, we must have $C > A$. Putting these together gives $B > C > A$. See Exercise 5.70 for a quantitative treatment of this setup.

5.5.2 Rocket motion

The quantitative application of momentum conservation can get a little tricky when the mass m is allowed to vary. Such is the case with rockets, because most of their mass consists of fuel which is eventually ejected.

Let mass be ejected backward with (constant) speed u relative to the rocket.[15] Since u is a speed here, it is defined to be positive. This means that the velocity of the ejected particles is obtained by subtracting u from the velocity of the rocket. Let the rocket have initial mass M, and let m be the (changing) mass at a later time. Then the rate of change of the rocket's mass is dm/dt, which is negative. So mass is ejected at a rate $|dm/dt| = -dm/dt$, which is positive. In other words, during a small time dt, a negative mass dm gets added to the rocket, and a positive mass $(-dm)$ gets shot out the back. (If you wanted, you could define dm to be positive, and then *subtract* it from the rocket's mass, and have dm get shot out the back. Either way is fine.) It may sound silly, but the hardest thing about rocket motion is picking a sign for these quantities and sticking with it.

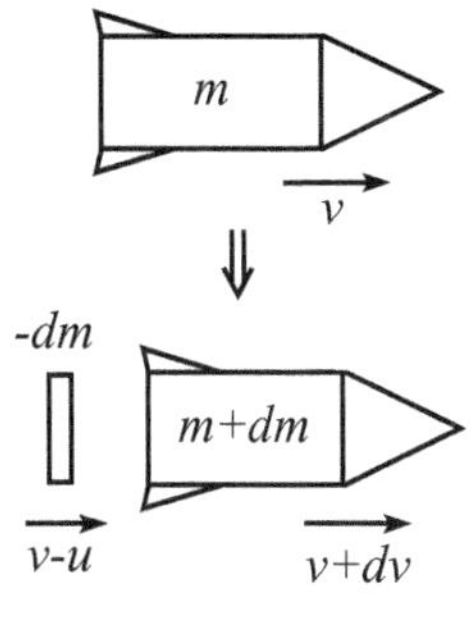

Fig. 5.11

Consider a moment when the rocket has mass m and speed v. Then at a time dt later (see Fig. 5.11), the rocket has mass $m + dm$ and speed $v + dv$, while the

[15] Just to emphasize, u is the speed with respect to the rocket. It wouldn't make sense to say "relative to the ground," because the rocket's engine shoots out the matter relative to itself, and the engine has no way of knowing how fast the rocket is moving with respect to the ground.

exhaust has mass $(-dm)$ and speed $v - u$ (which may be positive or negative, depending on the relative sizes of v and u). There are no external forces, so the total momentum at each of these times must be equal. Therefore,

$$mv = (m + dm)(v + dv) + (-dm)(v - u). \tag{5.53}$$

Ignoring the second-order term $dm\,dv$, this simplifies to $m\,dv = -u\,dm$. Dividing by m and integrating from t_1 to t_2 gives

$$\int_{v_1}^{v_2} dv = -\int_{m_1}^{m_2} u\frac{dm}{m} \quad \Longrightarrow \quad v_2 - v_1 = u \ln \frac{m_1}{m_2}. \tag{5.54}$$

For the case where the initial mass is M and the initial speed is 0, we have $v = u\ln(M/m)$. Note that we haven't assumed anything about dm/dt in this derivation. There is no need for it to be constant; it can change in any way it wants. The only thing that matters (assuming that M and u are given) is the final mass m. In the special case where dm/dt happens to be constant (call it $-\eta$, where η is positive), we have $v(t) = u\ln[M/(M - \eta t)]$.

The log in the result in Eq. (5.54) is not very encouraging. If the mass of the metal in the rocket is m, and if the mass of the fuel is $9m$, then the final speed is only $u \ln 10 \approx (2.3)u$. If the mass of the fuel is increased by a factor of 11 up to $99m$ (which is probably not even structurally possible, given the amount of metal required to hold it),[16] then the final speed only doubles to $u \ln 100 = 2(u \ln 10) \approx (4.6)u$. How do you make a rocket go significantly faster? Exercise 5.69 deals with this question.

REMARK: If you want, you can solve this rocket problem by using force instead of conservation of momentum. If a chunk of mass $(-dm)$ is ejected out the back, then its momentum changes by $u\,dm$ (which is negative). Therefore, because force equals the rate of change in momentum, the force on this chunk is $u\,dm/dt$. By Newton's third law, the remaining part of the rocket then feels a force of $-u\,dm/dt$ (which is positive). This force accelerates the remaining part of the rocket, so $F = ma$ gives $-u\,dm/dt = m\,dv/dt$,[17] which is equivalent to the $m\,dv = -u\,dm$ result above.

We see that this rocket problem can be solved by using either force or conservation of momentum. In the end, these two strategies are really the same, because the latter was derived from $F = dp/dt$. But the philosophies behind the approaches are somewhat different. The choice of strategy depends on personal preference. In an isolated system such as a rocket, conservation of momentum is usually simpler. But in a problem involving an external force, you have to use $F = dp/dt$. You'll get lots of practice with $F = dp/dt$ in the problems for this section and also in Section 5.8. Note that we used both $F = dp/dt$ and $F = ma$ in this second solution to the rocket problem. For further discussion of which expression to use in a given situation, see Appendix C. ♣

[16] The space shuttle's external fuel tank, just by itself, has a fuel-to-container mass ratio of only about 20.

[17] Whether we use m or $m + dm$ here for the mass of the rocket doesn't matter. Any differences are of second order.

5.6 The center of mass frame

5.6.1 Definition

When talking about momentum, it is understood that a certain frame of reference has been chosen. After all, the velocities of the particles have to be measured with respect to some coordinate system. Any inertial (that is, nonaccelerating) frame is legal to pick, but we will see that there is one particular reference frame that is often advantageous to use.

Consider a frame S and another frame S' that moves at constant velocity $\mathbf{u}$ with respect to S (see Fig. 5.12). Given a system of particles, the velocity of the ith particle in S is related to its velocity in S' by

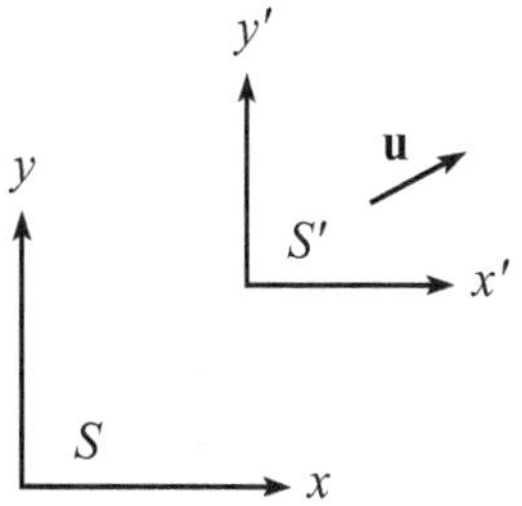

Fig. 5.12

$$\mathbf{v}_i = \mathbf{v}'_i + \mathbf{u}. \tag{5.55}$$

This relation implies that if momentum is conserved during a collision in frame S', then it is also conserved in frame S. This is true because both the initial and final momenta of the system in S are increased by the same amount, $(\sum m_i)\mathbf{u}$, compared with what they are in S'.[18]

Let us therefore consider the unique frame in which the total momentum of a system of particles is zero. This is called the *center of mass frame*, or CM frame. If the total momentum is $\mathbf{P} \equiv \sum m_i \mathbf{v}_i$ in frame S, then the CM frame is the frame S' that moves with velocity

$$\mathbf{u} = \frac{\mathbf{P}}{M} \equiv \frac{\sum m_i \mathbf{v}_i}{M} \tag{5.56}$$

with respect to S, where $M \equiv \sum m_i$ is the total mass. This follows from using Eq. (5.55) to write

$$\mathbf{P}' = \sum m_i \mathbf{v}'_i = \sum m_i \left(\mathbf{v}_i - \frac{\mathbf{P}}{M} \right) = \mathbf{P} - \mathbf{P} = \mathbf{0}. \tag{5.57}$$

The CM frame is extremely useful. Physical processes are generally much more symmetrical in this frame, and this makes the results more transparent. The CM frame is sometimes called the "zero momentum" frame. But the "center of mass" name is commonly used because the center of mass of the particles doesn't move in the CM frame, for the following reason. The position of the center of mass is defined by

$$\mathbf{R}_{\text{CM}} \equiv \frac{\sum m_i \mathbf{r}_i}{M}. \tag{5.58}$$

[18] Alternatively, nowhere in our earlier derivation of momentum conservation did we say what frame we were using. We assumed only that the frame wasn't accelerating. If it were accelerating, then $\mathbf{F}$ would *not* equal $m\mathbf{a}$. We will see in Chapter 10 how $\mathbf{F} = m\mathbf{a}$ is modified in a noninertial frame. But there's no need to worry about that here.

This is the location of the pivot upon which a rigid system would balance, as we'll see in Chapter 8. The fact that the CM doesn't move with respect to the CM frame follows from the fact that the derivative of $\mathbf{R}_{\text{CM}}$ is the velocity of the CM frame in Eq. (5.56). The center of mass may therefore be chosen as the origin of the CM frame.

If we take two derivatives of Eq. (5.58), we obtain

$$M\mathbf{a}_{\text{CM}} \equiv \sum m_i\mathbf{a}_i = \sum \mathbf{F}_i = \mathbf{F}_{\text{total}}. \tag{5.59}$$

So as far as the acceleration of the CM goes, we can treat the system of particles like a point mass at the CM, and then just apply $F = ma$ to this point mass. Since the internal forces cancel in pairs, we need only consider the external forces when calculating F_{total}.

Along with the CM frame, the other frame that people generally work with is the *lab frame*. There is nothing at all special about this frame. It is simply the frame (assumed to be inertial) in which the conditions of the problem are given. Any inertial frame can be called the "lab frame." Solving problems often involves switching back and forth between the lab and CM frames. For example, if the final answer is requested in the lab frame, then you may want to transform the given information from the lab frame to the CM frame where things are more obvious, and then transform back to the lab frame to give the answer.

Fig. 5.13

Example (Two masses in 1-D): A mass m with speed v approaches a stationary mass M (see Fig. 5.13). The masses bounce off each other without any loss in total energy. What are the final velocities of the particles? Assume that the motion takes place in 1-D.

Solution: Doing this problem in the lab frame would require a potentially messy use of conservation of energy (see the example in Section 5.7.1). But if we work in the CM frame, things are much easier. The total momentum in the lab frame is mv, so the CM frame moves to the right with speed $mv/(m+M) \equiv u$ with respect to the lab frame. Therefore, in the CM frame, the velocities of the two masses are

$$v_m = v - u = \frac{Mv}{m+M}, \quad \text{and} \quad v_M = 0 - u = -\frac{mv}{m+M}. \tag{5.60}$$

As a double-check, the difference in the velocities is v, and the ratio of the speeds is M/m, which makes the total momentum zero.

The important point to realize now is that in the CM frame, the two particles must simply reverse their velocities after the collision (assuming that they do indeed hit each other). This is true because the speeds must still be in the ratio M/m after the collision, in order for the total momentum to remain zero. Therefore, the speeds must either both increase or both decrease. But if they do either of these, then energy is not conserved.[19]

[19] So we *did* have to use conservation of energy in this CM-frame solution. But it was far less messy than it would have been in the lab frame.

If we now go back to the lab frame by adding the CM velocity of $mv/(m+M)$ to the two new velocities of $-Mv/(m+M)$ and $mv/(m+M)$, we obtain final lab velocities of

$$v_m = \frac{(m-M)v}{m+M}, \quad \text{and} \quad v_M = \frac{2mv}{m+M}. \tag{5.61}$$

REMARK: If $m = M$, then the left mass stops, and the right mass picks up a velocity of v (this should be familiar to pool players). If $M \gg m$, then the left mass bounces back with velocity $\approx -v$, and the right mass hardly moves (it's essentially a brick wall). If $m \gg M$, then the left mass keeps plowing along with velocity $\approx v$, and the right mass picks up a velocity of $\approx 2v$. This $2v$ is an interesting result (it is clearer if you consider things in the frame of the heavy mass m, which is essentially the CM frame), and it leads to some neat effects, such as in Problem 5.23. ♣

5.6.2 Kinetic energy

Given a system of particles, the relation between the total kinetic energy in two different frames is not very enlightening in general. But if one of the frames is the CM frame, then the relation turns out to be quite nice. Let S' be the CM frame, which moves at constant velocity $\mathbf{u}$ with respect to another frame S. Then the velocities of the particles in the two frames are related by $\mathbf{v}_i = \mathbf{v}'_i + \mathbf{u}$. The kinetic energy in the CM frame is

$$K_{\text{CM}} = \frac{1}{2}\sum m_i|\mathbf{v}'_i|^2. \tag{5.62}$$

And the kinetic energy in frame S is

$$\begin{aligned}
K_S &= \frac{1}{2}\sum m_i|\mathbf{v}'_i + \mathbf{u}|^2 \\
&= \frac{1}{2}\sum m_i(\mathbf{v}'_i \cdot \mathbf{v}'_i + 2\mathbf{v}'_i \cdot \mathbf{u} + \mathbf{u} \cdot \mathbf{u}) \\
&= \frac{1}{2}\sum m_i|\mathbf{v}'_i|^2 + \mathbf{u} \cdot \left(\sum m_i\mathbf{v}'_i\right) + \frac{1}{2}|\mathbf{u}|^2\sum m_i \\
&= K_{\text{CM}} + \frac{1}{2}Mu^2,
\end{aligned} \tag{5.63}$$

where M is the total mass of the system, and where we have used $\sum_i m_i\mathbf{v}'_i = 0$, by definition of the CM frame. Therefore, the K in any frame equals the K in the CM frame, plus the K of the whole system treated like a point mass M located at the CM, which moves with velocity $\mathbf{u}$. An immediate corollary of this fact is that if the K is conserved in a collision in one frame (which implies that K_{CM} is conserved, because conservation of momentum says that the CM speed u is the same before and after the collision), then it is conserved in any other frame (because again, the u relevant to that frame is the same before and after the collision).

5.7 Collisions

There are two basic types of collisions among particles, namely *elastic* ones (in which kinetic energy is conserved), and *inelastic* ones (in which kinetic energy is lost). In any collision, the *total* energy is conserved, but in inelastic collisions some of this energy goes into the form of heat (that is, relative motion of the molecules inside the particles) instead of showing up in the net translational motion of the particles.[20]

We'll deal mainly with elastic collisions here, although some situations are inherently inelastic, as we'll see in Section 5.8. For inelastic collisions where it is stated that a certain fraction, say 20%, of the kinetic energy is lost, only a trivial modification to the procedure is required. To solve any elastic collision problem, we just have to write down the conservation of energy and momentum equations, and then solve for whatever variables we want to find.

5.7.1 One-dimensional motion

Let's first look at one-dimensional motion. To see the general procedure, we'll solve the example from Section 5.6.1 again.

Fig. 5.14

Example (Two masses in 1-D, again): A mass m with speed v approaches a stationary mass M (see Fig. 5.14). The masses bounce off each other elastically. What are the final velocities of the particles? Assume that the motion takes place in 1-D.

Solution: Let v_f and V_f be the final velocities of the masses. Then conservation of momentum and energy give, respectively,

$$\begin{aligned} mv + 0 &= mv_f + MV_f, \\ \frac{1}{2}mv^2 + 0 &= \frac{1}{2}mv_f^2 + \frac{1}{2}MV_f^2. \end{aligned} \tag{5.64}$$

We must solve these two equations for the two unknowns v_f and V_f. Solving for V_f in the first equation and substituting into the second gives

$$\begin{aligned} mv^2 &= mv_f^2 + M\frac{m^2(v-v_f)^2}{M^2}, \\ \Longrightarrow \quad 0 &= (m+M)v_f^2 - 2mvv_f + (m-M)v^2, \\ \Longrightarrow \quad 0 &= \Big((m+M)v_f - (m-M)v\Big)(v_f - v). \end{aligned} \tag{5.65}$$

One solution is $v_f = v$, but this isn't the one we're concerned with. It is of course a solution, because the initial conditions certainly satisfy conservation of energy and

[20] We'll use the terminology where "kinetic energy" refers to the overall translational energy of a particle. That is, we'll exclude heat from this definition, even though heat is just the relative kinetic energy of molecules inside a particle.

momentum with the initial conditions (a fine tautology indeed). If you want, you can view $v_{\text{f}} = v$ as the solution where the particles miss each other. The fact that $v_{\text{f}} = v$ is always a root can often save you a lot of quadratic-formula trouble.

The $v_{\text{f}} = v(m - M)/(m + M)$ root is the one we want. Plugging this v_{f} back into the first of Eqs. (5.64) to obtain V_{f} gives

$$v_{\text{f}} = \frac{(m - M)v}{m + M}, \quad \text{and} \quad V_{\text{f}} = \frac{2mv}{m + M}, \tag{5.66}$$

in agreement with Eq. (5.61).

This solution was somewhat of a pain, because it involved a quadratic equation. The following theorem is extremely useful because it offers a way to avoid the hassle of quadratic equations when dealing with 1-D elastic collisions.

Theorem 5.3 *In a 1-D elastic collision, the relative velocity of the two particles after the collision is the negative of the relative velocity before the collision.*

Proof: Let the masses be m and M. Let v_{i} and V_{i} be the initial velocities, and let v_{f} and V_{f} be the final velocities. Conservation of momentum and energy give

$$\begin{aligned} mv_{\text{i}} + MV_{\text{i}} &= mv_{\text{f}} + MV_{\text{f}}, \\ \frac{1}{2}mv_{\text{i}}^2 + \frac{1}{2}MV_{\text{i}}^2 &= \frac{1}{2}mv_{\text{f}}^2 + \frac{1}{2}MV_{\text{f}}^2. \end{aligned} \tag{5.67}$$

Rearranging these yields

$$\begin{aligned} m(v_{\text{i}} - v_{\text{f}}) &= M(V_{\text{f}} - V_{\text{i}}), \\ m(v_{\text{i}}^2 - v_{\text{f}}^2) &= M(V_{\text{f}}^2 - V_{\text{i}}^2). \end{aligned} \tag{5.68}$$

Dividing the second equation by the first gives $v_{\text{i}} + v_{\text{f}} = V_{\text{i}} + V_{\text{f}}$. Therefore,

$$v_{\text{i}} - V_{\text{i}} = -(v_{\text{f}} - V_{\text{f}}), \tag{5.69}$$

as we wanted to show. In taking the quotient of these two equations, we have lost the $v_{\text{f}} = v_{\text{i}}$ and $V_{\text{f}} = V_{\text{i}}$ solution. But as stated in the above example, this is the trivial solution. ■

This is a splendid theorem. It has the quadratic energy-conservation statement built into it. Hence, using this theorem along with momentum conservation (both of which are linear equations and thus easy to deal with) gives the same information as the standard combination of Eqs. (5.67). Another quick proof is the following. It is fairly easy to see that the theorem is true in the CM frame (as we argued in the example in Section 5.6.1), so it is therefore true in any frame, because it involves only differences in velocities.

5.7.2 Two-dimensional motion

Let's now look at the more general case of two-dimensional motion. Three-dimensional motion is just more of the same, so we'll confine ourselves to 2-D. Everything is basically the same as in 1-D, except that there is one more momentum equation, and one more variable to solve for. This is best seen through an example.

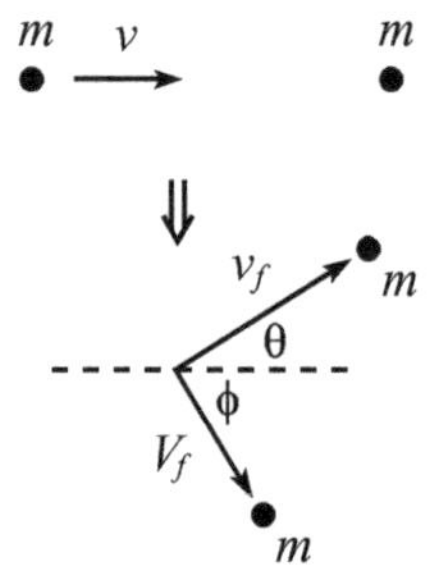

Fig. 5.15

Example (Billiards): A billiard ball with speed v approaches an identical stationary one. The balls bounce off each other elastically, in such a way that the incoming one gets deflected by an angle θ (see Fig. 5.15). What are the final speeds of the balls? What is the angle ϕ at which the stationary ball is deflected?

Solution: Let v_{f} and V_{f} be the final speeds of the balls. Then conservation of p_x, p_y, and E give, respectively,

$$\begin{aligned} mv &= mv_{\mathrm{f}}\cos\theta + mV_{\mathrm{f}}\cos\phi, \\ 0 &= mv_{\mathrm{f}}\sin\theta - mV_{\mathrm{f}}\sin\phi, \\ \frac{1}{2}mv^2 &= \frac{1}{2}mv_{\mathrm{f}}^2 + \frac{1}{2}mV_{\mathrm{f}}^2. \end{aligned} \tag{5.70}$$

We must solve these three equations for the three unknowns v_{f}, V_{f}, and ϕ. There are various ways to do this. Here's one. Eliminate ϕ by adding the squares of the first two equations (after putting the v_{f} terms on the left-hand side) to obtain

$$v^2 - 2vv_{\mathrm{f}}\cos\theta + v_{\mathrm{f}}^2 = V_{\mathrm{f}}^2. \tag{5.71}$$

Now eliminate V_{f} by combining this with the third equation to obtain[21]

$$v_{\mathrm{f}} = v\cos\theta. \tag{5.72}$$

The third equation then yields

$$V_{\mathrm{f}} = v\sin\theta. \tag{5.73}$$

The second equation then gives $m(v\cos\theta)\sin\theta = m(v\sin\theta)\sin\phi$, which implies $\cos\theta = \sin\phi$ (or $\theta = 0$, which corresponds to no collision). Therefore,

$$\phi = 90^\circ - \theta. \tag{5.74}$$

In other words, the balls bounce off at right angles with respect to each other. This fact is well known to pool players. Problem 5.19 gives another (cleaner) way to demonstrate this result. Note that we needed to specify one of the four quantities, v_{f}, V_{f}, θ, ϕ (we chose θ), because we have only three equations. Intuitively, we can't expect to solve for all four of these quantities, because we can imagine one ball hitting

[21] Another solution is $v_{\mathrm{f}} = 0$. In this case, ϕ must equal zero, and θ is not well defined. This is simply the 1-D motion in the example in Section 5.6.1.

the other at various distances away from directly head-on, which will cause the balls to be deflected at various angles.

As we saw in the 1-D example in Section 5.6.1, collisions are often much easier to deal with in the CM frame. Using the same reasoning (conservation of p and E) that we used in that example, we conclude that in 2-D (or 3-D) the final speeds of two elastically colliding particles must be the same as the initial speeds. The only degree of freedom in the CM frame is the angle of the line containing the final (oppositely directed) velocities. This simplicity in the CM frame invariably provides for a cleaner solution than the lab frame yields. A good example of this is Exercise 5.81, which gives yet another way to derive the above right-angle billiard result.

5.8 Inherently inelastic processes

There is a nice class of problems where the system has inherently inelastic properties, even if it doesn't appear so at first glance. In such a problem, no matter how you try to set it up, there will be inevitable kinetic energy loss that shows up in the form of heat. Total energy is conserved, of course, since heat is simply another form of energy. But the point is that if you try to write down a bunch of $(1/2)mv^2$'s and conserve their sum, then you're going to get the wrong answer. The following example is the classic illustration of this type of problem.

Example (Sand on conveyor belt): Sand drops vertically (from a negligible height) at a rate σ kg/s onto a moving conveyor belt.

(a) What force must you apply to the belt in order to keep it moving at a constant speed v?
(b) How much kinetic energy does the sand gain per unit time?
(c) How much work do you do per unit time?
(d) How much energy is lost to heat per unit time?

Solution:

(a) Your force equals the rate of change in momentum. If we let m be the combined mass of the conveyor belt plus the sand on the belt, then

$$F = \frac{dp}{dt} = \frac{d(mv)}{dt} = m\frac{dv}{dt} + \frac{dm}{dt}v = 0 + \sigma v, \tag{5.75}$$

where we have used the fact that v is constant.

(b) The kinetic energy gained per unit time is

$$\frac{d}{dt}\left(\frac{mv^2}{2}\right) = \frac{dm}{dt}\left(\frac{v^2}{2}\right) = \frac{\sigma v^2}{2}. \tag{5.76}$$

(c) The work done by your force per unit time is

$$\frac{d(\text{Work})}{dt} = \frac{F\,dx}{dt} = Fv = \sigma v^2, \tag{5.77}$$

where we have used Eq. (5.75).

(d) If work is done at a rate σv^2, and kinetic energy is gained at a rate $\sigma v^2/2$, then the "missing" energy must be lost to heat at a rate $\sigma v^2 - \sigma v^2/2 = \sigma v^2/2$.

In this example, it turned out that exactly the same amount of energy was lost to heat as was converted into kinetic energy of the sand. There is an interesting and simple way to see why this is true. In the following explanation, we'll just deal with one particle of mass M that falls onto the conveyor belt, for simplicity.

In the lab frame, the mass gains a kinetic energy of $Mv^2/2$ by the time it finally comes to rest with respect to the belt, because the belt moves at speed v. Now look at things in the conveyor belt's reference frame. In this frame, the mass comes flying in with an initial kinetic energy of $Mv^2/2$, and then it eventually slows down and comes to rest on the belt. Therefore, all of the $Mv^2/2$ energy is converted to heat. And since the heat is the same in both frames, this is the amount of heat in the lab frame, too.

We therefore see that in the lab frame, the equality of the heat loss and the gain in kinetic energy is a consequence of the obvious fact that the belt moves at the same rate with respect to the lab (namely v) as the lab moves with respect to the belt (also v).

In the solution to the above example, we did not assume anything about the nature of the friction force between the belt and the sand. The loss of energy to heat is an unavoidable result. You might think that if the sand comes to rest on the belt very "gently" (over a long period of time), then you can avoid the heat loss. This is not the case. In that scenario, the smallness of the friction force is compensated by the fact that the force must act over a very large distance. Likewise, if the sand comes to rest on the belt very abruptly, then the largeness of the friction force is compensated by the smallness of the distance over which it acts. No matter how you set things up, the work done by the friction force is the same nonzero quantity.

In other problems such as the following one, it is fairly clear that the process is inelastic. But the challenge is to correctly use $F = dp/dt$ instead of $F = ma$, because $F = ma$ will get you into trouble due to the changing mass.

Example (Chain on a scale): An "idealized" (see the comments following this example) chain with length L and mass density σ kg/m is held such that it hangs vertically just above a scale. It is then released. What is the reading on the scale, as a function of the height of the top of the chain?

First solution: Let y be the height of the top of the chain, and let F be the desired force applied by the scale. The net force on the entire chain is $F-(\sigma L)g$, with upward taken to be positive. The momentum of the entire chain (which just comes from the moving part) is $(\sigma y)\dot{y}$. Note that this is negative, because $\dot{y}$ is negative. Equating the net force on the entire chain with the rate of change in its momentum gives

$$\begin{aligned} F-\sigma L g &= \frac{d(\sigma y\dot{y})}{dt} \\ &= \sigma y\ddot{y}+\sigma\dot{y}^2. \end{aligned} \tag{5.78}$$

The part of the chain that is still above the scale is in free fall. Therefore, $\ddot{y}=-g$. And conservation of energy gives $\dot{y}=\sqrt{2g(L-y)}$, because the chain has fallen a distance $L-y$. Plugging these into Eq. (5.78) gives

$$\begin{aligned} F &= \sigma L g-\sigma y g+2\sigma(L-y)g \\ &= 3\sigma(L-y)g, \end{aligned} \tag{5.79}$$

which happens to be three times the weight of the chain already on the scale. This answer for F has the expected property of equaling zero when $y=L$, and also the interesting property of equaling $3(\sigma L)g$ right before the last bit touches the scale. Once the chain is completely on the scale, the reading suddenly drops down to the weight of the chain, namely $(\sigma L)g$.

If you used conservation of energy to do this problem and assumed that all of the lost potential energy goes into the kinetic energy of the moving part of the chain, then you would obtain a speed of infinity for the last infinitesimal part of the chain to hit the scale. This is certainly incorrect, and the reason is that there is inevitable heat loss that arises when the pieces of the chain inelastically smash into the scale.

Second solution: The normal force from the scale is responsible for doing two things. It holds up the part of the chain that already lies on the scale, and it also changes the momentum of the atoms that are suddenly brought to rest when they hit the scale. The first of these two parts of the force is simply the weight of the chain already on the scale, which is $F_{\text{weight}}=\sigma(L-y)g$.

To find the second part of the force, we need to find the change in momentum, dp, of the part of the chain that hits the scale during a given time dt. The amount of mass that hits the scale in a time dt is $dm=\sigma\,|dy|=\sigma\,|\dot{y}|\,dt=-\sigma\dot{y}\,dt$, since $\dot{y}$ is negative. This mass initially has velocity $\dot{y}$, and then it is abruptly brought to rest. Therefore, the change in its momentum is $dp=0-(dm)\dot{y}=\sigma\dot{y}^2\,dt$, which is positive. The force required to cause this change in momentum is

$$F_{dp/dt}=\frac{dp}{dt}=\sigma\dot{y}^2. \tag{5.80}$$

But as in the first solution, we have $\dot{y}=\sqrt{2g(L-y)}$. Therefore, the total force from the scale is

$$\begin{aligned} F=F_{\text{weight}}+F_{dp/dt} &= \sigma(L-y)g+2\sigma(L-y)g \\ &= 3\sigma(L-y)g. \end{aligned} \tag{5.81}$$

Note that $F_{dp/dt} = 2F_{\text{weight}}$ (until the chain is completely on the scale), independent of y.

In this example, we assumed that the chain was "ideal," in the sense that it was completely flexible, infinitesimally thin, and unstretchable. The simplest model that satisfies these criteria is a series of point masses connected by short massless strings. But in the above example, the strings actually don't even matter. You could instead start with many little unconnected point masses held in a vertical line, with the bottom one just above the scale. If you then dropped all of them simultaneously, they would successively smash into the scale in the same manner as if they were attached by little strings; the tension in the strings would all be zero. However, even though the strings aren't necessary in this chain and scale example, there are many setups involving idealized chains where they are in fact necessary, because a tension is required in them. This is evident in many of the problems and exercises for this chapter, as you will see.

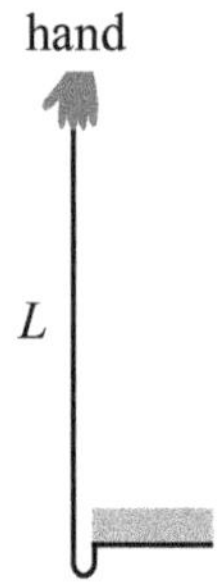

Fig. 5.16

An interesting fact is that even with the above definition of an ideal chain, there are some setups (in contrast with the one above) for which it is impossible to specify how the system behaves, without being given more information. This information involves the relative size of two specific length scales, as we'll see below. To illustrate this, consider the two following scenarios for the setup in Problem 5.28 (see Fig. 5.16), where a vertical ideal chain is dropped with its bottom end attached to the underneath of a support.

Fig. 5.17

- First scenario (energy nonconserving): Let the spacing between the point masses in our ideal chain be large compared with the horizontal span of the bend in the chain at its bottom; see Fig. 5.17. Then the system is for all practical purposes one dimensional. Each of the masses stops abruptly when it reaches the bend. This stoppage is a completely inelastic collision in the same way it was in the above example with the chain falling on the scale. Note that at any point in time, the bend consists of a massless piece of string folded back along itself (or perhaps it consists of one of the masses, if we happen to be looking at it right when a mass stops). There is no tension in this bottom piece of string (if there were, then the massless bend would have an infinite acceleration upward), so there is no tension pulling down the part of the chain on the left side of the bend. The left part of the chain is therefore in freefall.

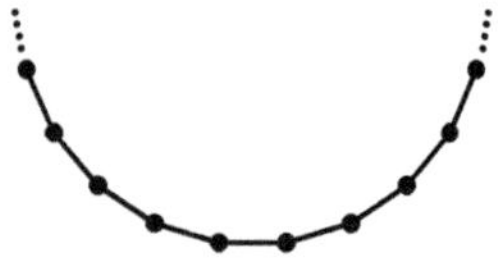

Fig. 5.18

- Second scenario (energy conserving): Let the spacing between the point masses in our ideal chain be small compared with the horizontal span of the bend in the chain at its bottom; see Fig. 5.18. The system is now inherently two dimensional, and the masses are essentially continuously distributed along the chain, as far as the bend is concerned. This has the effect of allowing each mass to gradually come to rest, so there is no abrupt inelastic stopping like there was in the first scenario. Each mass in the bend keeps the same distance from its two neighbors, whereas in the first scenario the mass that has just stopped soon sees the next mass fly directly past it before abruptly coming to rest. The process in this second scenario is elastic; no energy is lost to heat.

The basic difference between the two scenarios is whether or not there is slack in any of the strings in the bend. If there is, then the relative speed between a pair of masses changes abruptly at some point, which means that the relative kinetic energy of the masses goes into damped (perhaps very overdamped) vibrational motion in the connecting string, which then decays into the random motion of heat.[22]

If no energy is lost to heat in the second scenario, then you might think that the last infinitesimal piece of the chain will have an infinite speed. However, there isn't one *last* piece of the chain. When the left part of the chain has disappeared and we are left with only the bend and the right part, the small (but nonzero) bend is the last "piece," and it ends up swinging horizontally with a large speed. This then drags the whole chain to the side in a very visible motion (which can be traced to the horizontal force from the support), at which point we have a very noticeably two-dimensional system. The initial potential energy of the chain ends up as kinetic energy of the final wavy side-to-side motion.

A consequence of energy conservation in the second scenario is that for a given height fallen, the left part of the chain will be moving faster than the left part in the first scenario. In other words, the left part in the second scenario accelerates downward faster than the freefall g. But although this result follows quickly from energy considerations, it isn't so obvious in terms of a force argument. Apparently there exists a tension at the left end of the bend in the second scenario that drags down the left part of the chain to give it an acceleration greater than g. A qualitative way of seeing why a tension exists there is the following. A tiny piece of the chain that enters the bend from the left part slows down as it gradually joins the fixed right part of the chain. There must therefore be an upward force on this tiny piece. This upward force can't occur at the bottom of the piece, because any tension there pulls *down* on it. The force must therefore occur at the top of the piece. In other words, there is a tension at this point, and so by Newton's third law this tension pulls down on the left part of the chain, thereby causing it to accelerate faster than g. One of the tasks of Problem 5.29 is to find the tensions at the two ends of the bend.

There is a simple way to demonstrate the existence of a tension that pulls on the free part of the chain. The following setup is basically the falling-chain setup without gravity, but it still has all the essential parts. Place a rope on a

[22] If the strings were ideal springs with weak spring constants, then the energy would keep changing back and forth between potential energy of the springs and kinetic energy of the masses, causing the masses to bounce around and possibly run into each other. But we're assuming that the strings in our ideal chain are essentially very rigid overdamped springs.

(fairly smooth) table, in the shape of a very thin "U" so that it doubles back along itself. Then quickly yank on one of the ends, in the direction away from the bend. You will find that the other end moves backwards, in the direction *opposite* to the motion of your hand, toward the bend (at least until the bend reaches it and drags it forward). There must therefore exist a tension in the rope to drag the other end backwards. But there's no need to take my word for it – all you need is a piece of rope. This effect is essentially the same as the one (in a simplified version, since the rope here has constant density) that leads to the crack of a whip.

Note that a perfectly flexible thin *rope*, with its continuous mass distribution, does indeed behave elastically like the second scenario above. The continuous rope may be thought of as a series of point masses with infinitesimal separation, and so this separation is much smaller than the small (but finite) length of the bend. As long as the thickness of the rope is much smaller than the length of the bend, every piece of the rope slows to a stop gradually in the original falling-chain setup, or starts up from rest gradually in the preceding "U" setup. So there is no heat loss in either setup from abrupt changes in motion.

Returning to the falling-chain system with one of our ideal chains, you might think that if the bend is made *really* small, so that the system looks one-dimensional, then it should behave inelastically like the first scenario above. However, the only relevant fact is whether the bend is smaller than the spacing between the point masses in our ideal chain. The word "small" is meaningless, of course, because we are talking about the length of the bend, which is a dimensionful quantity. It makes sense only to use the word "smaller," that is, to compare one length with another. The other length here is the spacing between the masses. If the length of the bend is large compared with this spacing, then no matter what the actual length of the bend is, the system behaves elastically like the second scenario above.

So which of the two scenarios better describes a real chain? Details of an actual experiment involving a falling chain are given in Calkin and March (1989). The results show that a real chain behaves basically like the chain in the second scenario above, at least until the final part of the motion. In other words, it is energy conserving, and the left part accelerates faster than g.[23]

Having said all this, it turns out that the energy-conserving second scenario leads to complicated issues in problems (such as the numerical integration in Problem 5.29), so for all the problems and exercises in this chapter (with the exception of Problem 5.29), we'll assume that we're dealing with the inelastic first scenario.

[23] Spur-of-the-moment (but still plenty convincing) experiments were also performed by Wes Campbell in the physics laboratory of John Doyle at Harvard.

5.9 Problems

Section 5.1: Conservation of energy in one dimension

5.1. **Minimum length** *

The shortest configuration of string joining three given points is the one shown in the first setup in Fig. 5.19, where all three angles are 120°.[24] Explain how you could experimentally prove this fact by cutting three holes in a table and making use of three equal masses attached to the ends of strings, the other ends of which are connected as shown in the second setup in Fig. 5.19.

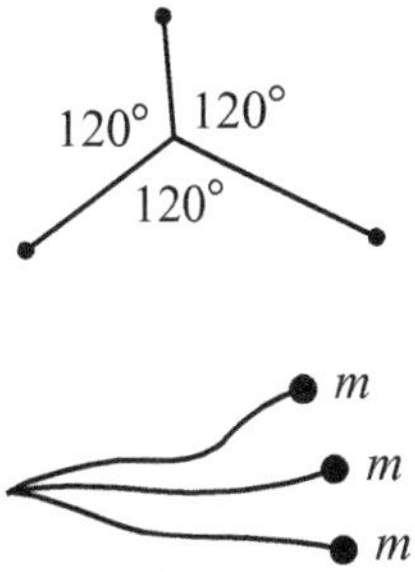

Fig. 5.19

5.2. **Heading to zero** *

A particle moves toward $x = 0$ under the influence of a potential $V(x) = -A|x|^n$, where $A > 0$ and $n > 0$. The particle has barely enough energy to reach $x = 0$. For what values of n will it reach $x = 0$ in a finite time?

5.3. **Leaving the sphere** *

A small mass rests on top of a fixed frictionless sphere. The mass is given a tiny kick and slides downward. At what point does it lose contact with the sphere?

5.4. **Pulling the pucks** **

(a) A massless string of length 2ℓ connects two hockey pucks that lie on frictionless ice. A constant horizontal force F is applied to the midpoint of the string, perpendicular to it (see Fig. 5.20). By calculating the work done in the transverse direction, find how much kinetic energy is lost when the pucks collide, assuming they stick together.

(b) The answer you obtained above should be very clean and nice. Find the slick solution that makes it transparent why the answer is so nice.

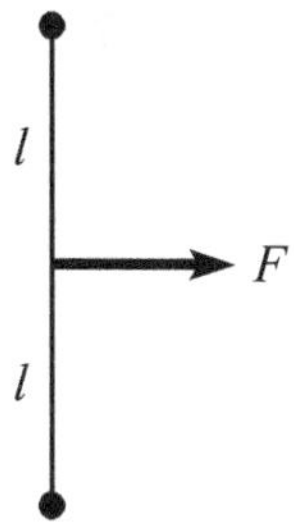

Fig. 5.20

5.5. **Constant $\dot{y}$** **

A bead, under the influence of gravity, slides down a frictionless wire whose height is given by the function $y(x)$. Assume that at position $(x, y) = (0, 0)$, the wire is vertical and the bead passes this point with a given speed v_0 downward. What should the shape of the wire be (that is, what is y as a function of x) so that the vertical speed remains v_0 at all times? Assume that the curve heads toward positive x.

[24] If the three points form a triangle that has an angle greater than 120°, then the string simply passes through the point where that angle is. We won't worry about this case.

5.6. **Dividing the heat** ***

A block rests on a table where the coefficient of kinetic friction is μ_k. You pull the block at a constant speed across the table by applying a force $\mu_k N$. Consider a period of time during which the block moves a distance d. How much work is done on the block? On the table? How much does each object heat up? Is it possible to answer these questions? *Hint*: You'll have to make some sort of model, however crude, for the way that friction works.

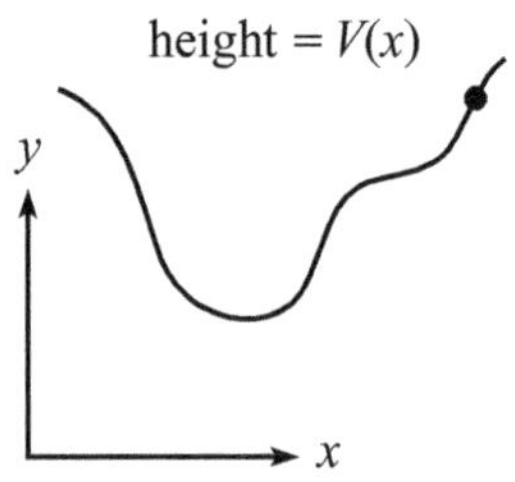

Fig. 5.21

5.7. **$V(x)$ vs. a hill** ***

A bead, under the influence of gravity, slides along a frictionless wire whose height is given by the function $V(x)$, as shown in Fig. 5.21. Find an expression for the bead's horizontal acceleration, $\ddot{x}$. (It can depend on whatever quantities you need it to depend on.) You should find that the result is *not* the same as the $\ddot{x}$ for a particle moving in one dimension in the potential $mgV(x)$, in which case $\ddot{x} = -gV'$. But if you grab hold of the wire, is there any way you can move it so that the bead's $\ddot{x}$ is equal to the $\ddot{x} = -gV'$ result for the one-dimensional potential $mgV(x)$?

Section 5.2: Small oscillations

5.8. **Hanging mass**

The potential for a mass hanging from a spring is $V(y) = ky^2/2 + mgy$, where $y = 0$ corresponds to the position of the spring when nothing is hanging from it. Find the frequency of small oscillations around the equilibrium point.

5.9. **Small oscillations** *

A particle moves under the influence of the potential $V(x) = -Cx^n e^{-ax}$. Find the frequency of small oscillations around the equilibrium point.

Section 5.4: Gravity

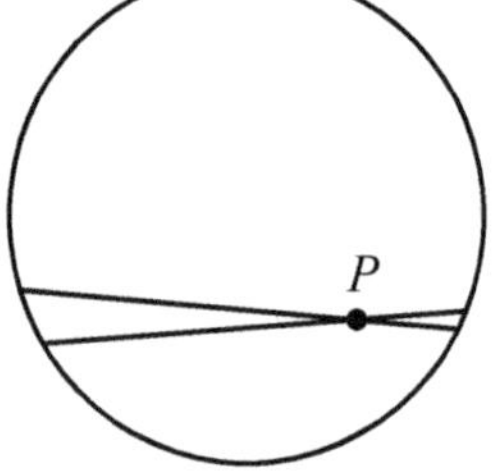

Fig. 5.22

5.10. **Zero force inside a sphere** *

Show that the gravitational force inside a spherical shell is zero by showing that the pieces of mass at the ends of the thin cones in Fig. 5.22 give canceling forces at point P.

5.11. **Escape velocity** *

(a) Find the escape velocity (that is, the velocity above which a particle escapes to $r = \infty$) for a particle on a spherical planet of radius R and mass M. What is the numerical value for the earth? The moon? The sun?

(b) Approximately how small must a spherical planet be in order for a human to be able to jump off? Assume a density roughly equal to the earth's.

5.12. **Ratio of potentials** $**$

Consider a cube of uniform mass density. Find the ratio of the gravitational potential energy of a mass at a corner to that of the same mass at the center. *Hint*: There's a slick way that doesn't involve any messy integrals.

5.13. **Through the hole** $**$

(a) A hole of radius R is cut out from an infinite flat sheet with mass density σ per unit area. Let L be the line that is perpendicular to the sheet and that passes through the center of the hole. What is the force on a mass m that is located on L, at a distance x from the center of the hole? *Hint*: Consider the plane to consist of many concentric rings.

(b) If a particle is released from rest on L, very close to the center of the hole, show that it undergoes oscillatory motion, and find the frequency of these oscillations.

(c) If a particle is released from rest on L, at a distance x from the sheet, what is its speed when it passes through the center of the hole? What is your answer in the limit $x \gg R$?

Section 5.5: Momentum

5.14. **Snowball** $*$

A snowball is thrown against a wall. Where does its momentum go? Where does its energy go?

5.15. **Propelling a car** $**$

For some odd reason, you decide to throw baseballs at a car of mass M that is free to move frictionlessly on the ground. You throw the balls at the back of the car at speed u, and they leave your hand at a mass rate of σ kg/s (assume the rate is continuous, for simplicity). If the car starts at rest, find its speed and position as a function of time, assuming that the balls bounce elastically directly backward off the back window.

5.16. **Propelling a car again** $**$

Do the previous problem, except now assume that the back window is open, so that the balls collect inside the car.

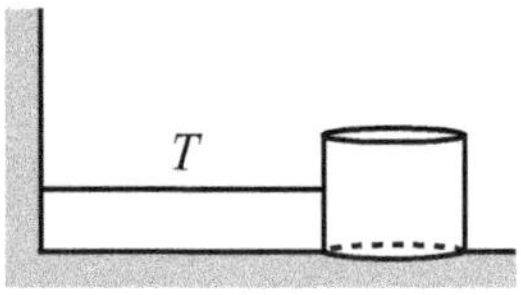

Fig. 5.23

5.17. **Leaky bucket** **

At $t = 0$, a massless bucket contains a mass M of sand. It is connected to a wall by a massless spring with constant tension T (that is, independent of length).[25] See Fig. 5.23. The ground is frictionless, and the initial distance to the wall is L. At later times, let x be the distance from the wall, and let m be the mass of sand in the bucket. The bucket is released, and on its way to the wall, it leaks sand at a rate $dm/dx = M/L$. In other words, the rate is constant with respect to distance, not time; and it ends up empty right when it reaches the wall. Note that dx is negative, so dm is also.

(a) What is the kinetic energy of the (sand in the) bucket, as a function of x? What is its maximum value?
(b) What is the magnitude of the momentum of the bucket, as a function of x? What is its maximum value?

5.18. **Another leaky bucket** ***

Consider the setup in Problem 5.17, but now let the sand leak at a rate proportional to the bucket's acceleration. That is, $dm/dt = b\ddot{x}$. Note that $\ddot{x}$ is negative, so dm is also.

(a) Find the mass as a function of time, $m(t)$.
(b) Find $v(t)$ and $x(t)$ during the time when the bucket contains a nonzero amount of sand. Also find $v(m)$ and $x(m)$. What is the speed right before all the sand leaves the bucket (assuming it hasn't hit the wall yet)?
(c) What is the maximum value of the bucket's kinetic energy, assuming it is achieved before it hits the wall?
(d) What is the maximum value of the magnitude of the bucket's momentum, assuming it is achieved before it hits the wall?
(e) For what value of b does the bucket become empty right when it hits the wall?

Section 5.7: Collisions

5.19. **Right angle in billiards** *

A billiard ball collides elastically with an identical stationary one. Use the fact that $mv^2/2$ may be written as $m(\mathbf{v} \cdot \mathbf{v})/2$ to show that the angle

[25] You can construct a constant-tension spring with a regular Hooke's-law spring in the following way. Pick the spring constant to be very small, and stretch the spring a very large distance. Have it pass through a hole in the wall, with its other end bolted down a large distance to the left of the wall. Any changes in the bucket's position will yield a negligible change in the spring force.

between the resulting trajectories is 90°. *Hint*: Take the dot product of the conservation of momentum equation with itself.

5.20. **Bouncing and recoiling** **

A ball of mass m and initial speed v_0 bounces back and forth between a fixed wall and a block of mass M, with $M \gg m$; see Fig. 5.24. The block is initially at rest. Assume that the ball bounces elastically and instantaneously. The coefficient of kinetic friction between the block and the ground is μ. There is no friction between the ball and the ground. What is the speed of the ball after the nth bounce off the block? How far does the block eventually move? How much total time does the block actually spend in motion? Work in the approximation where $M \gg m$, and assume that the distance to the wall is large enough so that the block comes to rest by the time the next bounce occurs.

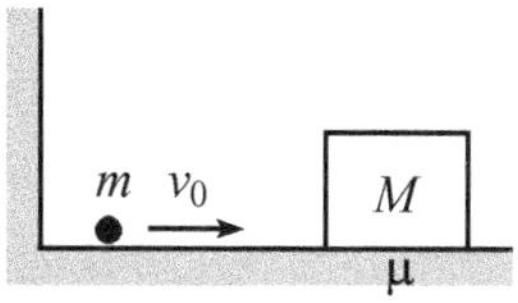

Fig. 5.24

5.21. **Drag force on a sheet** **

A sheet of mass M moves with speed V through a region of space that contains particles of mass m and speed v. There are n of these particles per unit volume. The sheet moves in the direction of its normal. Assume $m \ll M$, and assume that the particles do not interact with each other.

(a) If $v \ll V$, what is the drag force per unit area on the sheet?
(b) If $v \gg V$, what is the drag force per unit area on the sheet? Assume, for simplicity, that the component of every particle's velocity in the direction of the sheet's motion is exactly $\pm v/2$.[26]

5.22. **Drag force on a cylinder** **

A cylinder of mass M and radius R moves with speed V through a region of space that contains particles of mass m that are at rest. There are n of these particles per unit volume. The cylinder moves in a direction perpendicular to its axis. Assume $m \ll M$, and assume that the particles do not interact with each other. What is the drag force per unit length on the cylinder?

5.23. **Basketball and tennis ball** **

(a) A tennis ball with a small mass m_2 sits on top of a basketball with a large mass m_1 (see Fig. 5.25). The bottom of the basketball is a height h above the ground, and the bottom of the tennis ball is a height $h + d$ above the ground. The balls are dropped. To what height does the tennis ball bounce? *Note*: Work in the

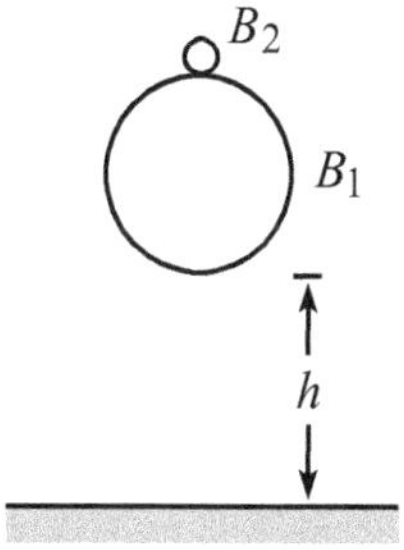

Fig. 5.25

[26] In reality, the velocities are randomly distributed, but this idealization actually gives the correct answer because the average speed in any direction is $\overline{|v_x|} = v/2$, as you can show.

approximation where m_1 is much larger than m_2, and assume that the balls bounce elastically. Also assume, for the sake of having a nice clean problem, that the balls are initially separated by a small distance, and that the balls bounce instantaneously.

(b) Now consider n balls, $B_1, \dots, B_n$, having masses $m_1, m_2, \dots, m_n$ (with $m_1 \gg m_2 \gg \cdots \gg m_n$), standing in a vertical stack (see Fig. 5.26). The bottom of B_1 is a height h above the ground, and the bottom of B_n is a height $h + \ell$ above the ground. The balls are dropped. In terms of n, to what height does the top ball bounce? *Note*: Make assumptions and approximations similar to the ones in part (a).

If $h = 1$ meter, what is the minimum number of balls needed for the top one to bounce to a height of at least 1 kilometer? To reach escape velocity? Assume that the balls still bounce elastically (which is a bit absurd here), and ignore wind resistance, etc., and assume that ℓ is negligible.

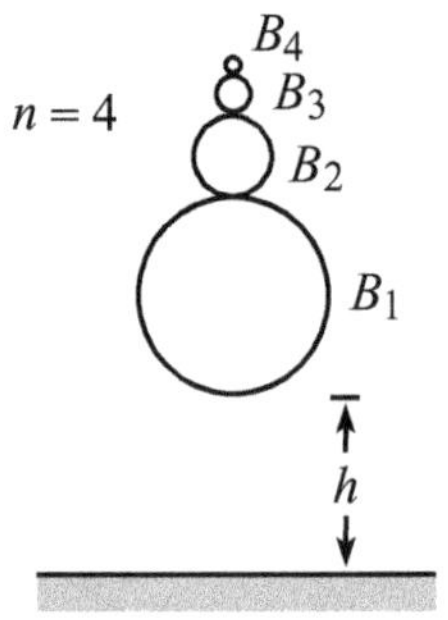

Fig. 5.26

5.24. Maximal deflection ***

A mass M collides with a stationary mass m. If $M < m$, then it is possible for M to bounce directly backward. However, if $M > m$, then there is a maximal angle of deflection of M. Show that this maximal angle equals $\sin^{-1}(m/M)$. *Hint*: It is possible to do this problem by working in the lab frame, but you can save yourself a lot of time by considering what happens in the CM frame, and then shifting back to the lab frame.

Section 5.8: Inherently inelastic processes

Note: In the problems involving chains in this section (with the exception of Problem 5.29), we'll assume that the chains are of the type described in the first scenario near the end of Section 5.8.

5.25. Colliding masses *

A mass M, initially moving at speed V, collides and sticks to a mass m, initially at rest. Assume $M \gg m$, and work in this approximation. What are the final energies of the two masses, and how much energy is lost to heat, in:

(a) The lab frame?
(b) The frame in which M is initially at rest?

5.26. Pulling a chain **

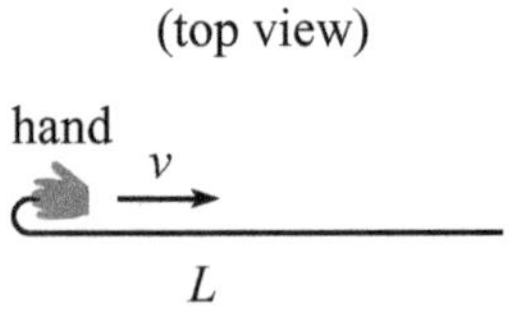

Fig. 5.27

A chain with length L and mass density σ kg/m lies straight on a frictionless horizontal surface. You grab one end and pull it back along itself, in a parallel manner (see Fig. 5.27). Assume that you pull it at constant speed v. What force must you apply? What is the total work that you do, by the time the chain is straightened out? How much energy is lost to heat, if any?

5.27. **Pulling a chain again** **

A chain with mass density σ kg/m lies in a heap on the floor. You grab an end and pull horizontally with constant force F. What is the position of the end of the chain, as a function of time, while it is unravelling? Assume that the chain is greased, so that it has no friction with itself.

5.28. **Falling chain** **

A chain with length L and mass density σ kg/m is held in the position shown in Fig. 5.28, with one end attached to a support. Assume that only a negligible length of the chain starts out below the support. The chain is released. Find the force that the support applies to the chain, as a function of time.

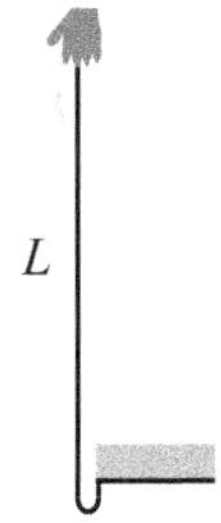

Fig. 5.28

5.29. **Falling chain (energy conserving)** ***

Consider the setup in the previous problem, but now let the chain be of the type in the second scenario described in Section 5.8. Show that the total time it takes the chain to straighten out is approximately 85% of the time it would take if the left part were in freefall (as it was in the previous problem); you will need to solve something numerically. Also, show that the tension at the left end of the infinitesimal bend equals the tension at the right end at all times.[27]

5.30. **Falling from a table** ***

(a) A chain with length L lies in a straight line on a frictionless table, except for a very small piece at one end which hangs down through a hole in the table. This piece is released, and the chain slides down through the hole. What is the speed of the chain at the instant it loses contact with the table? (See Footnote 3.19.)

(b) Answer the same question, but now let the chain lie in a heap on a table, except for a very small piece at one end which hangs down through the hole. Assume that the chain is greased, so that it has no friction with itself. Which of these two scenarios has the larger final speed?

[27] The "ends" of the bend actually aren't well defined, because the chain is at least a little bit curved everywhere. But since we're assuming that the horizontal span of the chain is very small, we can define the height of the bend to be, say, 100 times this horizontal span, and this height is still negligible compared with the total height of the chain.

5.31. **The raindrop** ****

Assume that a cloud consists of tiny water droplets suspended (uniformly distributed, and at rest) in air, and consider a raindrop falling through them. What is the acceleration of the raindrop? Assume that the raindrop is initially of negligible size and that when it hits a water droplet, the droplet's water gets added to it. Also, assume that the raindrop is spherical at all times.

5.10 Exercises

Section 5.1: Conservation of energy in one dimension

5.32. **Cart in a valley**

A cart containing sand starts at rest and then rolls, without any energy loss to friction, down into a valley and then up a hill on the other side. Let the initial height be h_1, and let the final height attained on the other side be h_2. If the cart leaks sand along the way, how does h_2 compare with h_1?

5.33. **Walking on an escalator**

An escalator moves downward at constant speed. You walk up the escalator at this same speed, so that you remain at rest with respect to the ground. In the ground frame, are you doing any work?

5.34. **Lots of work**

If you push on a wall with your hand, you don't do any work, because your hand doesn't move. But in the reference frame of a person moving past you (from front to back), you do in fact do work, because your hand moves. And since the person's speed can be made arbitrarily large, you can do an arbitrarily large amount of work in the person's frame. It therefore seems like you should quickly use up your dinner from the night before and become very hungry. But you don't. Why not?

5.35. **Spring energy**

Using the explicit form of the position of a mass on the end of a spring, $x(t) = A\cos(\omega t + \phi)$, verify that the total energy is conserved.

5.36. **Damping work** *

A damped oscillator (with $m\ddot{x} = -kx - b\dot{x}$) has initial position x_0 and speed v_0. After a long time, it will essentially be at rest at the origin. Therefore, by the work–energy theorem, the work done by the damping force must equal $-kx_0^2/2 - mv_0^2/2$. Verify that this is true. *Hint*: It's rather messy to explicitly find $\dot{x}$ in terms of the initial conditions and then calculate the desired integral. An easier way is to use the $F = ma$ equation to rewrite the $\dot{x}$ in your integral.

5.37. **Heading to infinity** *

A particle moves under the influence of a potential $V(x) = -A|x|^n$, where $A > 0$ and $n > 0$. It starts at a positive value of x with a velocity that points in the positive x direction. For what values of n will the particle reach infinity in a finite time? You may assume that $E > 0$, although this isn't necessary. (You can compare this exercise with Problem 5.2.)

5.38. **Work in different frames** *

An object, initially at rest, is subject to a force that gives it constant acceleration a for time t. Verify explicitly that $W = \Delta K$ in (a) the lab frame, and (b) a frame moving to the left at speed V.

5.39. **Roller coaster** *

A roller coaster car starts at rest and coasts down a frictionless track. It encounters a vertical loop of radius R. How much higher than the top of the loop must the car start if it is to remain in contact with the track at all times?

5.40. **Pendulum and peg** *

A pendulum of length L is held with its string horizontal, and then released. The string runs into a peg a distance d below the pivot, as shown in Fig. 5.29. What is the smallest value of d for which the string remains taut at all times?

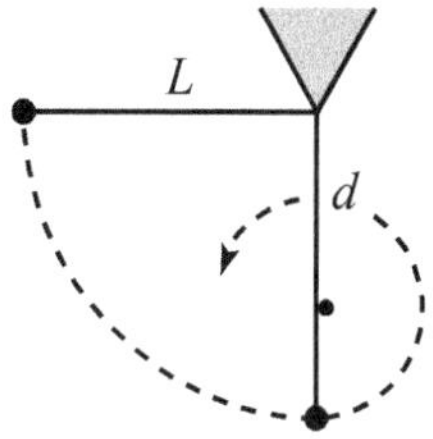

Fig. 5.29

5.41. **Circling around a cone** *

A fixed hollow frictionless cone is positioned with its tip pointing down. A particle is released from rest on the inside surface. After it has slid part way down to the tip, it bounces elastically off a platform. The platform is positioned at a 45° angle along the surface of the cone, so the particle ends up being deflected horizontally along the surface (in other words, into the page in Fig. 5.30). If the resulting motion of the particle is a horizontal circle around the cone, what is the ratio of the initial height of the particle to the height of the platform?

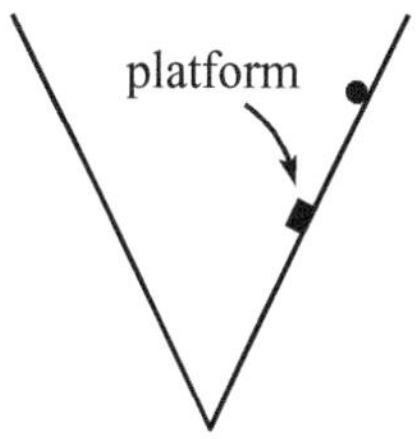

Fig. 5.30

5.42. **Hanging spring** *

A massless spring with spring constant k hangs vertically from a ceiling, initially at its relaxed length. A mass m is then attached to the bottom and is released.

(a) Calculate the potential energy V of the system, as a function of the height y (which is negative), relative to the initial position.

(b) Find y_0, the point at which the potential energy is minimum. Make a rough plot of $V(y)$.

(c) Rewrite the potential energy as a function of $z \equiv y - y_0$. Explain why your result shows that a hanging spring can be considered to be a spring in a world without gravity, provided that the new equilibrium point, y_0, is now called the "relaxed" length of the spring.

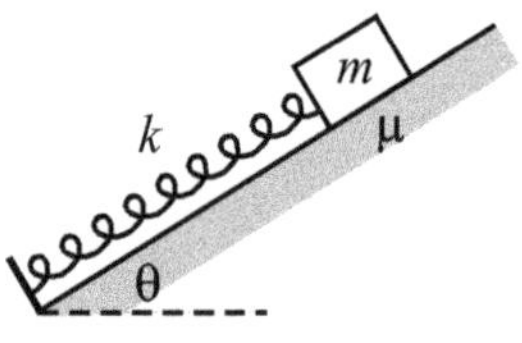

Fig. 5.31

5.43. Removing the friction *

A block of mass m is supported by a spring on an inclined plane, as shown in Fig. 5.31. The spring constant is k, the plane's angle of inclination is θ, and the coefficient of static friction between the block and the plane is μ.

(a) You move the block down the plane, compressing the spring. What is the maximum compression distance of the spring (relative to the relaxed length it has when nothing is attached to it) that allows the block to remain at rest when you let go of it?

(b) Assume that the block is at the maximum compression you found in part (a). At a given instant, you somehow cause the plane to become frictionless, and the block gets pushed up along the plane. What must the relation between θ and the original μ be, so that the block reaches its maximum height when the spring is at its relaxed length?

5.44. Spring and friction **

A spring with spring constant k stands vertically, and a mass m is placed on top of it. The mass is slowly lowered to its equilibrium position. With the spring held at this compression length, the system is rotated to a horizontal position. The left end of the spring is attached to a wall, and the mass is placed on a table with a coefficient of friction (both kinetic and static) of $\mu = 1/8$; see Fig. 5.32. The mass is released.

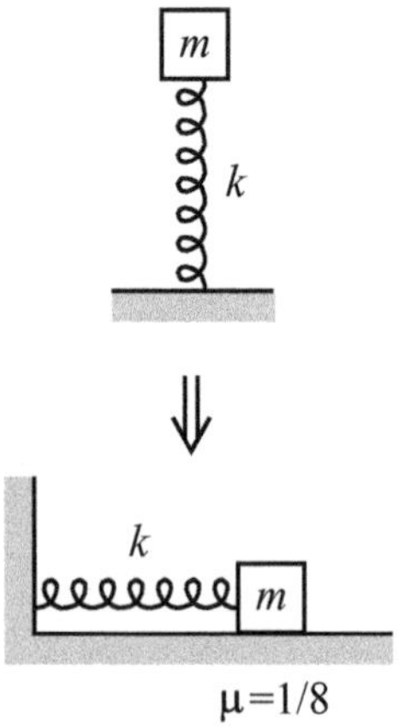

Fig. 5.32

(a) What is the initial compression of the spring?

(b) How much does the maximal stretch (or compression) of the spring decrease after each half-oscillation?

(c) How many times does the mass oscillate back and forth before coming to rest?

5.45. Keeping contact **

A frictionless circle of radius R is made out of a strip of metal and held fixed in a vertical plane. A massless spring with spring constant k has one end attached to the bottom point on the inside surface of the circle, and the other end attached to a mass m. The spring is compressed to zero length, with the mass touching the inside surface of the circle at the bottom. (Whatever negligible length of spring remains is essentially horizontal.) The spring is then released, and the mass gets pushed initially to the right and then up along the circle; the setup at a random later time is shown in Fig. 5.33. Let ℓ be the equilibrium length of the spring. What is the minimum value of ℓ for which the mass remains in contact with the circle at all times?

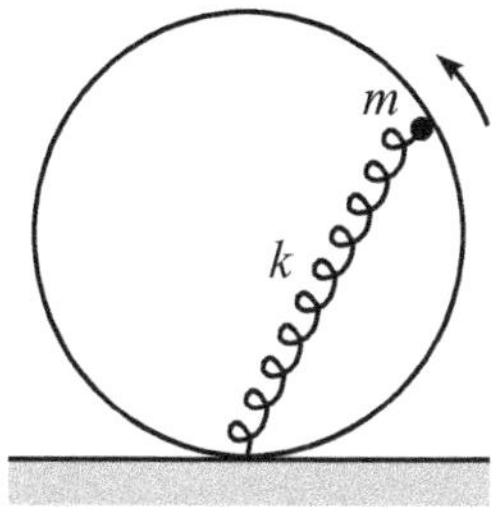

Fig. 5.33

5.46. **Spring and hoop** **

A fixed hoop of radius R stands vertically. A spring with spring constant k and relaxed length of zero is attached to the top of the hoop.

(a) A block of mass m is attached to the unstretched spring and dropped from the top of the hoop. If the resulting motion of the mass is a linear vertical oscillation between the top and bottom points on the hoop, what is k?

(b) The block is now removed from the spring, and the spring is stretched and connected to a bead, also of mass m, at the bottom of the hoop, as shown in Fig. 5.34. The bead is constrained to move along the hoop. It is given a rightward kick and acquires an initial speed v_0. Assuming that it moves frictionlessly, how does its speed depend on its position along the hoop?

Part (b):

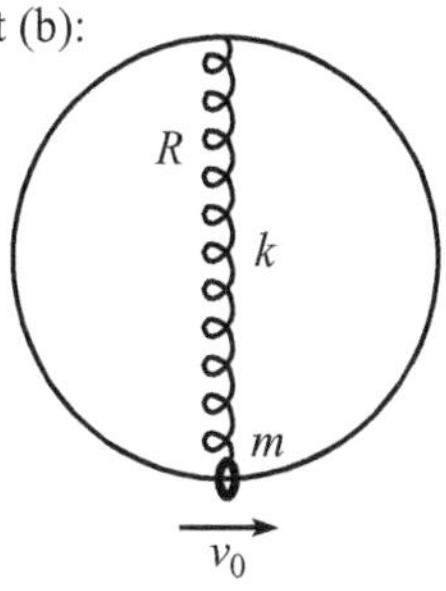

Fig. 5.34

5.47. **Constant $\dot{x}$** **

A bead, under the influence of gravity, slides along a frictionless wire whose height is given by the function $y(x)$. Assume that at position $(x, y) = (0, 0)$, the wire is horizontal and the bead passes this point with a given speed v_0 to the right. What should the shape of the wire be (that is, what is y as a function of x) so that the horizontal speed remains v_0 at all times? One solution is simply $y = 0$. Find the other.[28]

5.48. **Over the pipe** **

A frictionless cylindrical pipe with radius r is positioned with its axis parallel to the ground, at height h. What is the minimum speed at which a ball must be thrown (from ground level) in order to make it over the pipe? Consider two cases: (a) the ball is allowed to touch the pipe, and (b) the ball is not allowed to touch the pipe.

5.49. **Pendulum projectile** **

A pendulum is held with its string horizontal and is then released. The mass swings down, and then on its way back up, the string is cut when

[28] Solve this exercise in the spirit of Problem 5.5, that is, by solving a differential equation. Once you get the answer, you'll see that you could have just written it down without any calculations, based on your knowledge of a certain kind of physical motion.

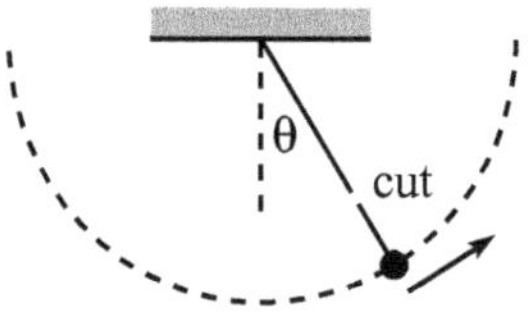

Fig. 5.35

it makes an angle θ with the vertical; see Fig. 5.35. What should θ be so that the mass travels the largest horizontal distance by the time it returns to the height it had when the string was cut?

5.50. **Centered projectile motion** ∗∗

A mass is attached to one end of a massless string, the other end of which is attached to a fixed support. The mass swings around in a vertical circle as shown in Fig. 5.36. Assuming that the mass has the minimum speed necessary at the top of the circle to keep the string from going slack, at what location should you cut the string so that the resulting projectile motion of the mass has its maximum height located directly above the center of the circle?

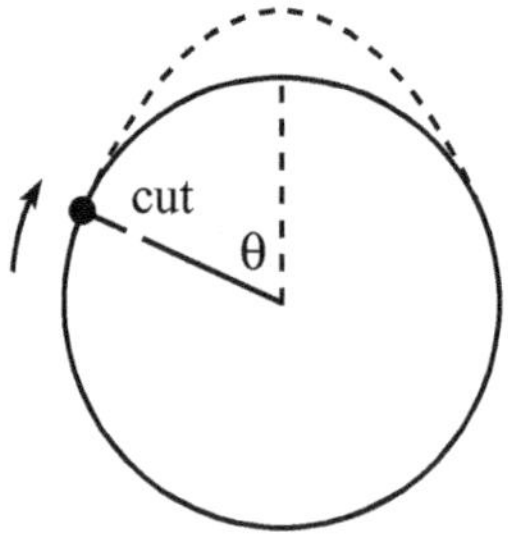

Fig. 5.36

5.51. **Beads on a hoop** ∗∗

Two beads of mass m are initially at rest at the top of a frictionless hoop of mass M and radius R, which stands vertically on the ground. The beads are given tiny kicks, and they slide down the hoop, one to the right and one to the left, as shown in Fig. 5.37. What is the largest value of m/M for which the hoop never rises up off the ground?

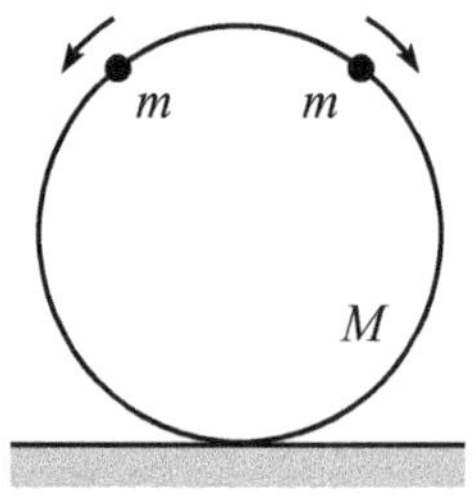

Fig. 5.37

5.52. **Stationary bowl** ∗∗∗

A hemispherical bowl of mass M rests on a table. The inside surface of the bowl is frictionless, while the coefficient of friction between the bottom of the bowl and the table is $\mu = 1$. A particle of mass m is released from rest at the top of the bowl and slides down into it, as shown in Fig. 5.38. What is the largest value of m/M for which the bowl never slides on the table? *Hint*: The angle you're concerned with is *not* 45°.

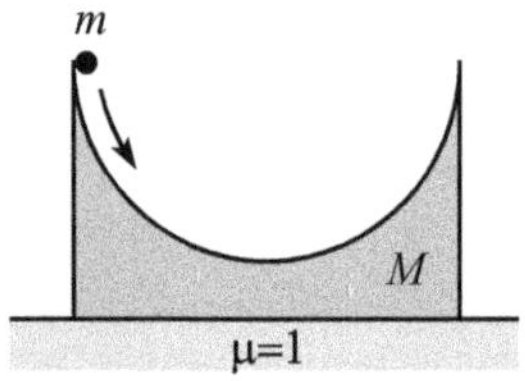

Fig. 5.38

5.53. **Leaving the hemisphere** ∗∗∗∗

A point particle of mass m sits at rest on top of a frictionless hemisphere of mass M, which rests on a frictionless table. The particle is given a tiny kick and slides down the (recoiling) hemisphere. At what angle θ (measured from the top of the hemisphere) does the particle lose contact with the hemisphere? In answering this question for $m \neq M$, it is sufficient for you to produce an equation that θ must satisfy (it's a cubic). However, for the special case of $m = M$, the equation can be solved without too much difficulty; find the angle in this case.

5.54. **Tetherball** ∗∗∗∗

A small ball is attached to a massless string of length L, the other end of which is attached to a very thin pole. The ball is thrown so that it initially travels in a horizontal circle, with the string making an angle θ_0 with the vertical. As time goes on, the string wraps itself around the pole. Assume that (1) the pole is thin enough so that the length of string in the air decreases very slowly, so that the ball's motion may always be approximated as a circle, and (2) the pole has enough friction so that the string does not slide on the pole, once it touches it. Show that the ratio of the ball's final speed (right before it hits the pole) to initial speed is $v_f/v_i = \sin\theta_0$.

Section 5.4: Gravity

5.55. **Projectile between planets** *

Two planets of mass M and radius R are at rest (somehow) with respect to each other, with their centers a distance $4R$ apart. You wish to fire a projectile from the surface of one planet to the other. What is the minimum firing speed for which this is possible?

5.56. **Spinning quickly** *

Consider a planet with uniform mass density ρ. If the planet rotates too fast, it will fly apart. Show that the minimum period of rotation is given by

$$T = \sqrt{\frac{3\pi}{G\rho}}.$$

What is the minimum T if $\rho = 5.5\,\mathrm{g/cm^3}$ (the average density of the earth)?

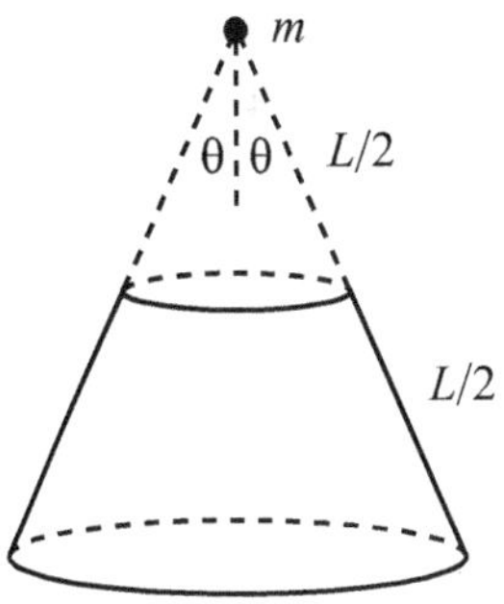

Fig. 5.39

5.57. **A cone** **

(a) A particle of mass m is located at the tip of a hollow cone (such as an ice cream cone without the ice cream) with surface mass density σ. The slant height of the cone is L, and the half-angle at the vertex is θ. What can you say about the gravitational force on the mass m due to the cone?

(b) If the top half of the cone is removed and thrown away (see Fig. 5.39), what is the gravitational force on the mass m due to the remaining part of the cone? For what angle θ is this force maximum?

5.58. **Sphere and cones** **

(a) Consider a thin hollow fixed spherical shell of radius R and surface mass density σ. A particle initially at rest falls in from infinity. What is its speed when it reaches the center of the shell? Assume that a tiny hole has been cut in the shell to let the particle through; see Fig. 5.40(a).

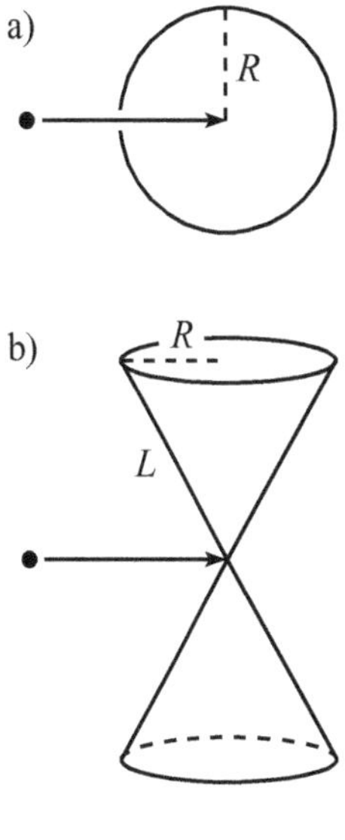

Fig. 5.40

(b) Consider two hollow fixed cones (such as ice cream cones without the ice cream), arranged as shown in Fig. 5.40(b). They have base radius R, slant height L, and surface mass density σ. A particle initially at rest falls in from infinity, along the perpendicular bisector line shown. What is its speed when it reaches the tip of the cones?

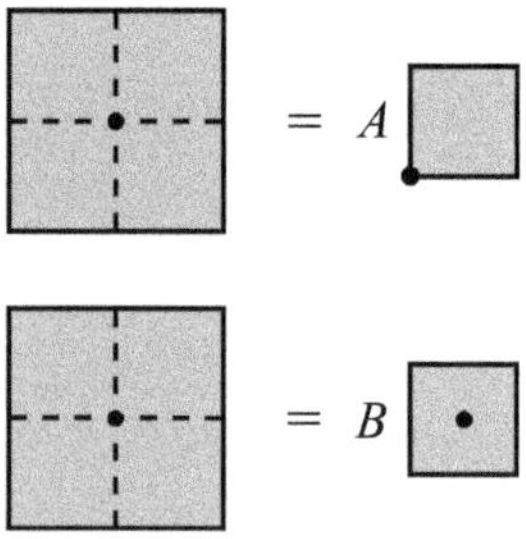

Fig. 5.41

5.59. Ratio of potentials **

Consider the following two systems: (1) a mass m is placed at a corner of a flat square sheet of mass M, and (2) a mass m is placed at the center of a flat square sheet of mass M. What is the ratio of the potential energies of m in the two systems? *Hint*: Find A and B in the suggestive relations in Fig. 5.41. You'll need to use a scaling argument to find B.

5.60. Solar escape velocity **

What is the minimum initial velocity (with respect to the earth) required for an object to escape from the solar system?[29] Take the orbital motion of the earth into account (but ignore the rotation of the earth, and ignore the other planets). You are free to choose (wisely) the firing direction. Make the (good) approximation that the process occurs in two separate steps: first the object escapes from the earth, and then it escapes from the sun (starting at the radius of the earth's orbit). Some useful quantities are given in the solution to Problem 5.11; also, the orbital speed of the earth is about 30 km/s. *Hint*: a common incorrect answer is 13.5 km/s.

5.61. Spherical shell **

(a) A spherical shell of mass M has inner radius R_1 and outer radius R_2. A particle of mass m is located a distance r from the center of the shell. Calculate (and make a rough plot of) the force on m, as a function of r, for $0 \le r \le \infty$.

(b) If the mass m is dropped from $r = \infty$ and falls down through the shell (assume that a tiny hole has been drilled in it), what will its speed be at the center of the shell? You can let $R_2 = 2R_1$ in this part of the problem, to keep things from getting too messy. Give your answer in terms of $R \equiv R_1$.

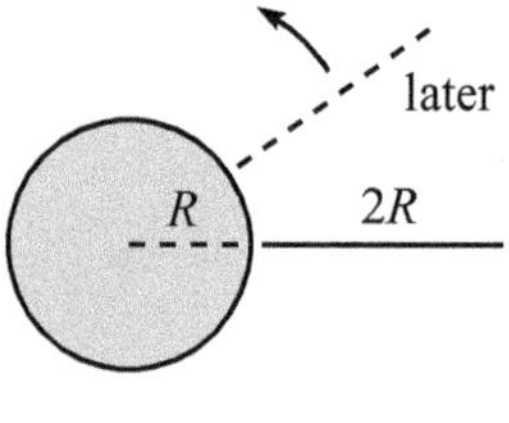

Fig. 5.42

5.62. Orbiting stick **

Consider a planet of mass M and radius R. A very long stick of length $2R$ extends from just above the surface of the planet out to a radius $3R$. If initial conditions have been set up so that the stick moves in a circular orbit while always pointing radially (see Fig. 5.42), what is the period

[29] This problem is discussed in Hendel and Longo (1988).

of this orbit? How does this period compare with the period of a satellite in a circular orbit of radius $2R$?

5.63. **Speedy travel** $**$

A straight tube is drilled between two points on the earth, as shown in Fig. 5.43. An object is dropped into the tube. What is the resulting motion? How long does it take to reach the other end? Ignore friction, and assume (incorrectly) that the density of the earth is constant ($\rho = 5.5\,\text{g/cm}^3$).

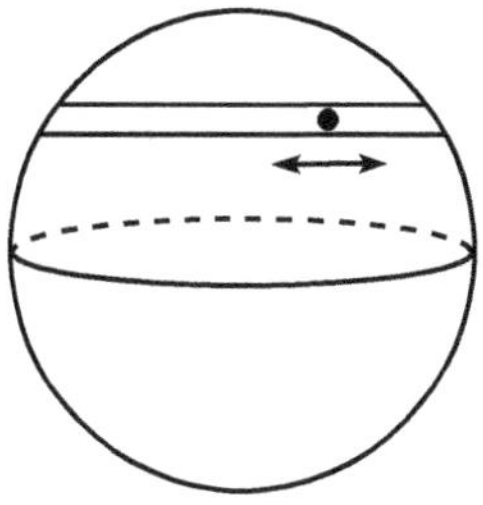

Fig. 5.43

5.64. **Mine shaft** $**$

(a) If the earth had constant density, the gravitational force would decrease linearly with radius as you descend in a mine shaft; see Eq. (5.44). However, the density of the earth is not constant, and in fact the gravitational force *increases* as you descend. Show that the general condition under which this is true is $\rho_c < (2/3)\rho_{avg}$, where ρ_{avg} is the average density of the earth, and ρ_c is the density of the crust at the surface. (The values for the earth are $\rho_c \approx 3\,\text{g/cm}^3$ and $\rho_{avg} \approx 5.5\,\text{g/cm}^3$.) See Zaidins (1972).

(b) A similar problem to the one in part (a), which actually turns out to be exactly the same, is the following. Consider a large flat horizontal sheet of material with density ρ and thickness x. Show that the gravitational force (from the earth plus the sheet) just below the sheet is larger than the force just above it if $\rho < (2/3)\rho_{avg}$, where ρ_{avg} is the average density of the earth. A sheet of wood (with a density roughly equal to that of water) satisfies this inequality, but a sheet of gold doesn't. A result from Problem 5.13 will be useful here.

(c) Assuming that the density of a planet is a function of radius only, what should $\rho(r)$ look like if you want the gravitational force to be independent of the depth in a mine shaft, all the way down to the center of the planet?

5.65. **Space elevator** $**$

(a) Let the earth's radius be R, its average density be ρ, and its angular frequency of rotation be ω. Show that if a satellite is to remain above the same point on the equator at all times, then it must travel in a circle of radius ηR, where

$$\eta^3 = \frac{4\pi G\rho}{3\omega^2}. \tag{5.82}$$

What is the numerical value of η?

(b) Instead of a satellite, consider a long rope with uniform mass density extending radially from just above the surface of the earth out to a radius $\eta' R$.[30] Show that if the rope is to remain above the same point on the equator at all times, then η' must be given by

$$\eta'^2 + \eta' = \frac{8\pi G\rho}{3\omega^2}. \tag{5.83}$$

What is the numerical value of η'? Where does the tension in the rope achieve its maximum value? *Hint*: no messy calculations required.

5.66. **Force from a straight wire** ***

A particle of mass m is placed a distance ℓ away from an infinitely long straight wire with mass density σ kg/m. Show that the force on the particle is $F = 2G\sigma m/\ell$. Do this in two ways:

(a) Integrate along the wire the contributions to the force.
(b) Integrate along the wire the contributions to the potential, and then differentiate to obtain the force. You will find that the potential due to the infinite wire is infinite,[31] but you can escape this difficulty by letting the wire have a large but finite length, then finding the potential and force, and then letting the length go to infinity.

5.67. **Maximal gravity** ***

Given a point P in space, and given a piece of malleable material of constant density, how should you shape and place the material in order to create the largest possible gravitational field at P?

Section 5.5: Momentum

5.68. **Maximum *P* and *E* of a rocket** *

A rocket that starts at rest with mass M ejects exhaust at a given speed u. What is the mass of the rocket (including unused fuel) when its momentum is maximum? What is the mass when its energy is maximum?

5.69. **Speedy rockets** *

Assume that it is impossible to build a structurally sound container that can hold fuel of more than, say, nine times its mass (the actual limit is higher than this, but let's use this number just to be concrete). It would then seem like the limit for the speed of a rocket is $u \ln 10$, from Eq. (5.54). How can you build a rocket that goes faster than this?

[30] Any proposed space elevator wouldn't have uniform mass density. But this simplified problem still gives a good idea of the general features. For more on the space elevator, see Aravind (2007).

[31] There's nothing bad about this. All that matters as far as the force is concerned is differences in the potential, and these differences are finite.

5.70. **Snow on a sled, quantitative** **

Consider the setup in the example in Section 5.5.1. At $t = 0$, let the mass of the sled (including you) be M, and let its speed be V_0. If the snow hits the sled at a rate of σ kg/s, find the speed as a function of time for the three cases.

5.71. **Leaky bucket** ***

Consider the setup in Problem 5.17, but now let the sand leak at a rate $dm/dt = -bM$. In other words, the rate is constant with respect to time, not distance. We've factored out an M here, just to make the calculations a little nicer.

(a) Find $v(t)$ and $x(t)$ during the time when the bucket contains a nonzero amount of sand.
(b) What is the maximum value of the bucket's kinetic energy, assuming it is achieved before it hits the wall?
(c) What is the maximum value of the magnitude of the bucket's momentum, assuming it is achieved before it hits the wall?
(d) For what value of b does the bucket become empty right when it hits the wall?

5.72. **Throwing a brick** ***

A brick is thrown from ground level, at an angle θ with respect to the (horizontal) ground. Assume that the long face of the brick remains parallel to the ground at all times, and that there is no deformation in the ground or the brick when the brick hits the ground. If the coefficient of friction between the brick and the ground is μ, what should θ be so that the brick travels the maximum total horizontal distance before finally coming to rest? Assume that the brick doesn't bounce. *Hint*: The brick slows down when it hits the ground. Think in terms of impulse.

Section 5.7: Collisions

5.73. **A one-dimensional collision** *

Consider the following one-dimensional collision. A mass $2m$ moves to the right, and a mass m moves to the left, both with speed v. They collide elastically. Find their final lab-frame velocities. Solve this by:

(a) Working in the lab frame.
(b) Working in the CM frame.

5.74. **Perpendicular vectors** *

A moving mass m collides elastically with a stationary mass $2m$. Let their resulting velocities be $\mathbf{v}_1$ and $\mathbf{v}_2$, respectively. Show that $\mathbf{v}_2$ must be perpendicular to $2\mathbf{v}_1 + \mathbf{v}_2$. *Hint*: See Problem 5.19.

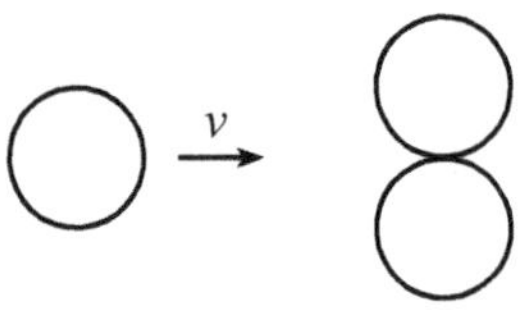

Fig. 5.44

5.75. **Three pool balls** $*$

A pool ball with initial speed v is aimed right between two other pool balls, as shown in Fig. 5.44. If the two right balls leave the (elastic) collision with equal speeds, find the final velocities of all three balls.

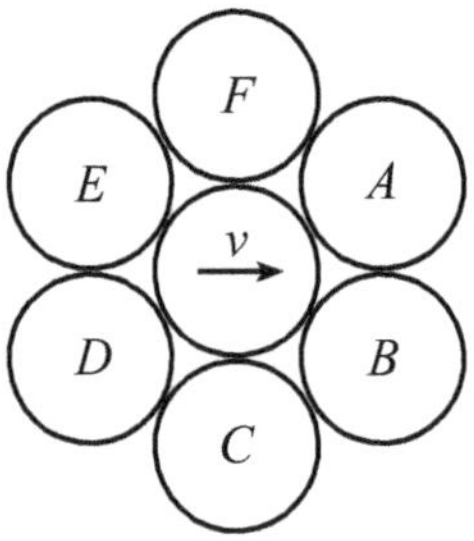

Fig. 5.45

5.76. **Seven pool balls** $**$

Seven pool balls are situated at rest as shown in Fig. 5.45. The middle ball suddenly somehow acquires a speed v to the right. Assume that starting with ball A, the balls spiral out an infinitesimal amount. So A is closer to the center ball than B is, and B is closer than C is, etc. This means that the center ball collides with A first, then it gets deflected into B, and then it gets deflected into C, and so on. But all the collisions happen in the blink of an eye. What will the center ball's velocity be after it collides (elastically) with all six balls? (You can use the results from the example in Section 5.7.2.)

5.77. **Midair collision** $**$

A ball is held and then released. At the instant it is released, an identical ball, moving horizontally with speed v, collides elastically with it and is deflected at an upward angle. What is the maximum horizontal distance the latter ball can travel by the time it returns to the height of the collision? (You can use the results from the example in Section 5.7.2.)

5.78. **Maximum number of collisions** $**$

N identical balls are constrained to move in one dimension. If you are allowed to pick their initial velocities, what is the maximum number of collisions you can arrange for the balls to have among themselves? Assume that the collisions are elastic.

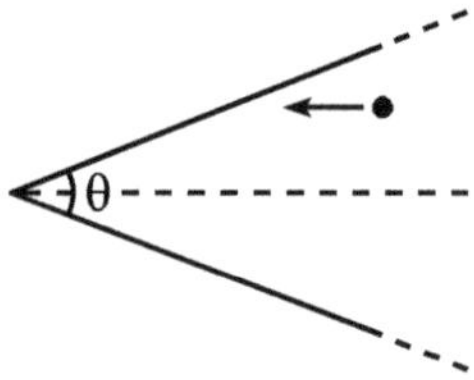

Fig. 5.46

5.79. **Triangular room** $**$

A ball is thrown against a wall of a very long triangular room which has vertex angle θ. The initial direction of the ball is parallel to the angle bisector (see Fig. 5.46). How many (elastic) bounces does the ball make? Assume that the walls are frictionless.

5.80. **Equal angles** $**$

(a) A mass $2m$ moving at speed v_0 collides elastically with a stationary mass m. If the two masses scatter at equal (nonzero) angles with respect to the incident direction, what is this angle?

(b) What is the largest number that the above "2" can be replaced with, if you want it to be possible for the masses to scatter at equal angles?

5.81. **Right angle in billiards** **

A billiard ball collides elastically with an identical stationary one. By looking at the collision in the CM frame, show that the angle between the resulting trajectories in the lab frame is 90°. (We proved this result by working in the lab frame in the example in Section 5.7.2.)

5.82. **Equal v_x's** **

A mass m moving with speed v in the x direction collides elastically with a stationary mass nm, where n is some number. After the collision, it is observed that both masses have equal x components of their velocities. What angle does the velocity of mass nm make with the x axis? (This can be solved by working in the lab frame or the CM frame, but the CM solution is slick.)

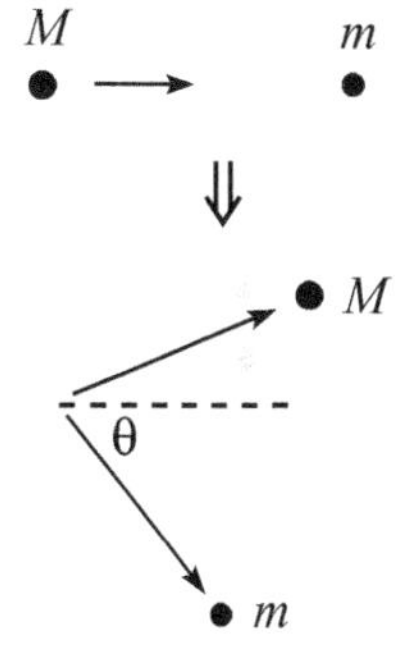

Fig. 5.47

5.83. **Maximum v_y** **

A mass M moving in the positive x direction collides elastically with a stationary mass m. The collision is not necessarily head-on, so the masses may come off at angles, as shown in Fig. 5.47. Let θ be the angle of m's resulting motion. What should θ be so that m has the largest possible speed in the y direction? *Hint*: Think about what the collision should look like in the CM frame.

5.84. **Bouncing between rings** **

Two fixed circular rings, in contact with each other, stand in a vertical plane. A ball bounces elastically back and forth between the rings (see Fig. 5.48). Assume that initial conditions have been set up so that the ball's motion forever lies in one parabola. Let this parabola hit the rings at an angle θ from the horizontal. Show that if you want the magnitude of the change in the horizontal component of the ball's momentum at each bounce to be maximum, then you should pick $\cos\theta = (\sqrt{5}-1)/2$, which just happens to be the inverse of the golden ratio.

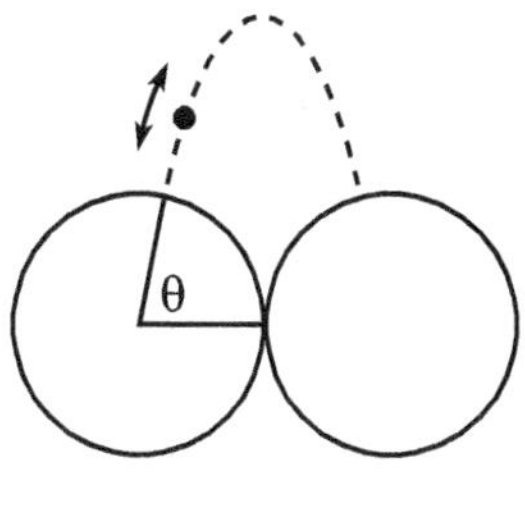

Fig. 5.48

5.85. **Bouncing between surfaces** **

Consider the following generalization of the previous exercise. A ball bounces back and forth between a surface defined by $f(x)$ and its reflection across the y axis (see Fig. 5.49). Assume that initial conditions have been set up so that the ball's motion forever lies in one parabola, with the contact points located at $\pm x_0$. For what function $f(x)$ is the magnitude of the change in the horizontal component of the ball's momentum at each bounce independent of x_0?

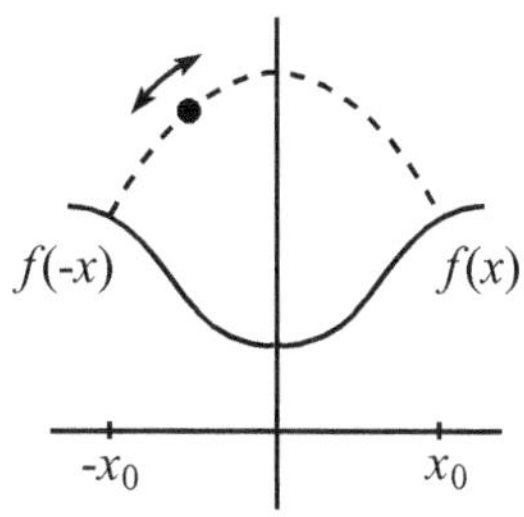

Fig. 5.49

5.86. **Drag force on a sphere** **

A sphere of mass M and radius R moves with speed V through a region of space that contains particles of mass m that are at rest. There are n of these particles per unit volume. Assume $m \ll M$, and assume that the particles do not interact with each other. What is the drag force on the sphere?

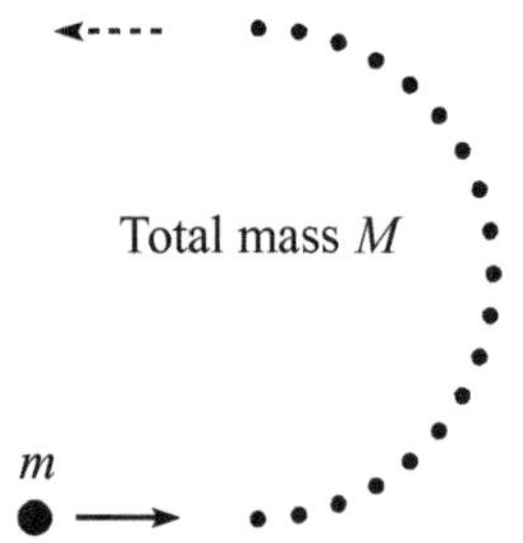

Fig. 5.50

5.87. **Balls in a semicircle** ***

N identical balls lie equally spaced in a semicircle on a frictionless horizontal table, as shown. The total mass of these balls is M. Another ball of mass m approaches the semicircle from the left, with the proper initial conditions so that it bounces (elastically) off all N balls and finally leaves the semicircle, heading directly to the left (see Fig. 5.50).

(a) In the limit $N \to \infty$ (so the mass of each ball in the semicircle, M/N, goes to zero), find the minimum value of M/m that allows the incoming ball to come out heading directly to the left. *Hint*: You'll need to do Problem 5.24 first.

(b) In the minimum M/m case found in part (a), show that the ratio of m's final speed to initial speed equals $e^{-\pi}$.

5.88. **Block and bouncing ball** ****

A block with large mass M slides with speed V_0 on a frictionless table toward a wall. It collides elastically with a ball with small mass m, which is initially at rest at a distance L from the wall. The ball slides toward the wall, bounces elastically, and then proceeds to bounce back and forth between the block and the wall.

(a) How close does the block come to the wall?

(b) How many times does the ball bounce off the block, by the time the block makes its closest approach to the wall?

Assume $M \gg m$, and give your answers to leading order in m/M.

Section 5.8: Inherently inelastic processes

Note: In the exercises involving chains in this section, we'll assume that the chains are of the type described in the first scenario near the end of Section 5.8.

5.89. **Slowing down, speeding up** *

A plate of mass M moves horizontally with initial speed v on a frictionless table. A mass m is dropped vertically onto it and soon comes to rest with respect to it. How much energy is required to bring the system back up to speed v? Explain intuitively your answer in the $M \gg m$ limit.

5.90. **Pulling a chain back** **

A chain with length L and mass density σ kg/m lies outstretched on a frictionless horizontal table. You grab one end and pull it back along itself, in a parallel manner, as shown in Fig. 5.51. If your hand starts from rest and has constant acceleration a, what is your force at the moment right before the chain is straightened out?

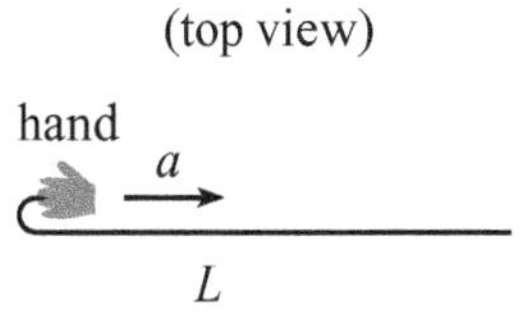

Fig. 5.51

5.91. **Falling chain** **

A chain with length L and mass density σ kg/m is held in a heap, and you grab an end that protrudes a tiny bit out of the top. The chain is then released. As a function of time, what is the force that your hand must apply to keep the top end of the chain motionless? Assume that the chain has no friction with itself, so that the remaining part of the heap is always in freefall. The setup at a random later time is shown in Fig. 5.52.

Fig. 5.52

5.92. **Pulling a chain down** **

A chain with mass density σ kg/m lies in a heap at the edge of a table. One end of the chain initially sticks out an infinitesimal distance from the heap. You grab this end and accelerate it downward with acceleration a. Assume that there is no friction of the chain with itself as it unravels. As a function of time, what force does your hand apply to the chain? Find the value of a that makes your force always equal to zero. (In other words, find the a with which the chain naturally falls.)

5.93. **Raising a chain** **

A chain with length L and mass density σ kg/m lies in a heap on the floor. You grab one end of the chain and pull upward with a force such that the chain moves at constant speed v. What is the total work you do, by the time the chain is completely off the floor? How much energy is lost to heat, if any? Assume that the chain is greased, so that it has no friction with itself.

5.94. **Downhill dustpan** **

A plane inclined at an angle θ is covered with dust. An essentially massless dustpan on wheels is released from rest and rolls down the plane, gathering up dust. The density of dust in the path of the dustpan is σ kg/m. What is the acceleration of the dustpan?

5.95. **Heap and block** **

A chain with mass density σ kg/m lies in a heap on the floor, with one end attached to a block of mass M. The block is given a sudden kick and instantly acquires a speed V_0. Let x be the distance traveled by the block. In terms of x, what is the tension in the chain, just to the right of the heap; that is, at the point P shown in Fig. 5.53? There is no friction in this problem; none with the floor, and none in the chain with itself.

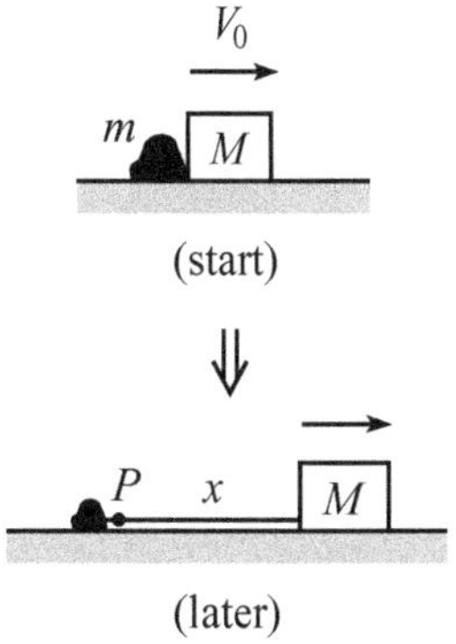

Fig. 5.53

5.96. **Touching the floor** ****

A chain with mass density σ kg/m hangs from a spring with spring constant k. In the equilibrium position, a length L is in the air, and the bottom part of the chain lies in a heap on the floor; see Fig. 5.54. The chain is raised by a very small distance, b, and then released. What is the amplitude of the oscillations, as a function of time?

Assume that (1) $L \gg b$, (2) the chain is very thin, so that the size of the heap on the floor is very small compared with b, (3) the length of the chain in the initial heap is larger than b, so that some of the chain always remains in contact with the floor, and (4) there is no friction of the chain with itself inside the heap.

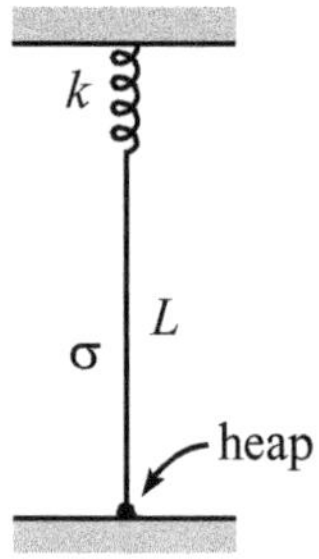

Fig. 5.54

5.11 Solutions

5.1. **Minimum length**

Drop the masses through the three holes, and let the system reach its equilibrium position. The equilibrium position is the one with the lowest potential energy of the masses, that is, the one with the most string hanging below the table. In other words, it is the one with the least string lying on the table. This is the desired minimum-length configuration.

What are the angles at the vertex of the string? The tensions in all three strings are equal to mg, because they are holding up the masses. The vertex of the string is in equilibrium, so the net force on it must be zero. This implies that each string must bisect the angle formed by the other two. Therefore, the angles between the strings must all be 120°.

5.2. **Heading to zero**

The energy of the particle is $E = mv^2/2 - A|x|^n$. The given information tells us that $v = 0$ when $x = 0$. Therefore, $E = 0$, which then implies that $v = -\sqrt{2Ax^n/m}$ (we'll assume $x > 0$; the $x < 0$ case works the same). We have chosen the minus sign because the particle is heading toward the origin. Writing v as dx/dt and separating variables gives

$$\int_{x_0}^{0} \frac{dx}{x^{n/2}} = -\sqrt{\frac{2A}{m}} \int_0^T dt = -T\sqrt{\frac{2A}{m}}, \qquad (5.84)$$

where x_0 is the initial position and T is the time to reach the origin. The integral on the left is finite if and only if $n/2 < 1$. Therefore, the condition that T is finite is $n < 2$.

REMARK: If $0 < n < 1$, then $V(x)$ has a cusp at $x = 0$ (infinite slope on either side), so it's clear that T is finite. If $n = 1$, then the slope is a finite constant, so it's also clear than T is finite. If $n > 1$, then the slope of $V(x)$ is zero at $x = 0$, so it's not obvious what happens with T. But the above calculation shows that $n = 2$ is the value where T becomes infinite.

The particle therefore takes a finite time to reach the top of a triangle or the curve $-Ax^{3/2}$. But it takes an infinite time to reach the top of a parabola, cubic, etc. A circle looks like a parabola at the top, so T is infinite in that case also. In fact, any nice polynomial function $V(x)$ requires an infinite T to reach a local maximum, because the Taylor series starts at order two (at least) around an extremum. ♣

5.3. **Leaving the sphere**

FIRST SOLUTION: Let R be the radius of the sphere, and let θ be the angle of the mass, measured from the top of the sphere. The radial $F = ma$ equation is

$$mg\cos\theta - N = \frac{mv^2}{R}, \tag{5.85}$$

where N is the normal force. The mass loses contact with the sphere when the normal force becomes zero (that is, when the normal component of gravity is barely large enough to account for the centripetal acceleration of the mass). Therefore, the mass loses contact when

$$\frac{mv^2}{R} = mg\cos\theta. \tag{5.86}$$

But conservation of energy gives $mv^2/2 = mgR(1 - \cos\theta)$. Hence, $v = \sqrt{2gR(1-\cos\theta)}$. Plugging this into Eq. (5.86) gives

$$\cos\theta = \frac{2}{3} \quad \Longrightarrow \quad \theta \approx 48.2°. \tag{5.87}$$

SECOND SOLUTION: Let's assume (incorrectly) that the mass always stays in contact with the sphere, and then find the point where the horizontal component of v starts to decrease, which it of course can't do, because the normal force doesn't have a "backward" component. From above, the horizontal component of v is

$$v_x = v\cos\theta = \sqrt{2gR(1-\cos\theta)}\,\cos\theta. \tag{5.88}$$

Taking the derivative of this, we find that the maximum occurs when $\cos\theta = 2/3$. So this is where v_x would start to decrease if the mass were constrained to remain on the sphere. But there is no such constraining force available, so the mass loses contact when $\cos\theta = 2/3$.

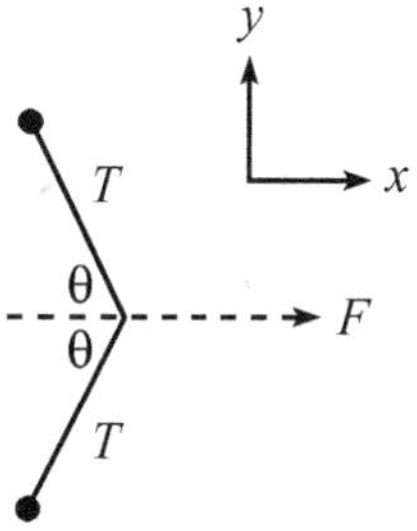

Fig. 5.55

5.4. **Pulling the pucks**

(a) Let θ be defined as in Fig. 5.55. Then the tension in the string is $T = F/(2\cos\theta)$, because the force on the massless kink in the string must be zero. Consider the "top" puck. The component of the tension in the y direction is $-T\sin\theta = -(F/2)\tan\theta$. The work done on the puck by this component is therefore

$$\begin{aligned} W_y &= \int_{\ell}^{0} \frac{-F\tan\theta}{2}\,dy = \int_{\pi/2}^{0} \frac{-F\tan\theta}{2}\,d(\ell\sin\theta) \\ &= \int_{\pi/2}^{0} \frac{-F\ell\sin\theta}{2}\,d\theta \\ &= \frac{F\ell\cos\theta}{2}\bigg|_{\pi/2}^{0} = \frac{F\ell}{2}. \end{aligned} \tag{5.89}$$

By the work–energy theorem (or equivalently, by separating variables and integrating $F_y = mv_y\,dv_y/dy$), this work equals the value of $mv_y^2/2$ right before the collision. There are two pucks, so the total kinetic energy lost when they stick together is twice this quantity (v_x doesn't change during the collision), which is $F\ell$.

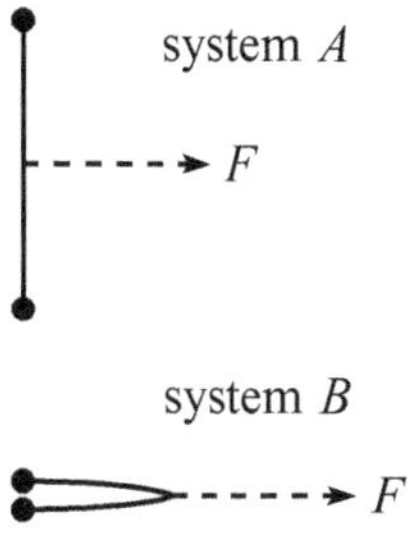

Fig. 5.56

(b) Consider two systems, A and B (see Fig. 5.56). A is the original setup, while B starts with θ already at zero. Let the pucks in both systems start simultaneously at $x = 0$. As the force F is applied, all four pucks will have the same $x(t)$, because the same force in the x direction, namely $F/2$, is applied to every puck at all times. After the collision, both systems will therefore look exactly the same.

Let the collision of the pucks occur at $x = d$. At this point, $F(d + \ell)$ work has been done on system A, because the center of the string (where the force is applied) ends up moving a distance ℓ more than the masses. However, only Fd work has been done on system B. Since both systems have the same kinetic energy after the collision, the extra $F\ell$ work done on system A must be what is lost in the collision.

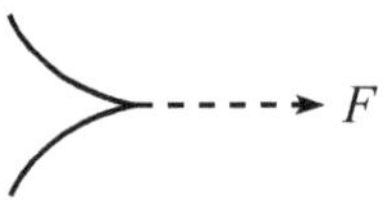

Fig. 5.57

REMARK: The reasoning in this second solution can be used to solve the problem in the case where we have a uniform massive rope (so the rope flops down, as in Fig. 5.57). The center of mass of the rope moves in exactly the same manner as the position of the two pucks in system B (assuming that the mass of each puck is chosen to be half the mass of the rope), because the same force F acts on both systems. You can show that this implies that the force acts over an extra distance of $\ell/2$ on the rope, compared with system B, by the time the rope has flopped into a straight line. From the reasoning above, $F\ell/2$ of work must therefore be lost to heat in the rope. ♣

5.5. **Constant $\dot{y}$**

By conservation of energy, the bead's speed at any time is given by (note that y is negative here)

$$\frac{1}{2}mv^2 + mgy = \frac{1}{2}mv_0^2 \implies v = \sqrt{v_0^2 - 2gy}\,. \tag{5.90}$$

The vertical component of the velocity is $\dot{y} = v\sin\theta$, where θ is the (negative) angle the wire makes with the horizontal. The slope of the wire is $\tan\theta = dy/dx \equiv y'$, which yields $\sin\theta = y'/\sqrt{1+y'^2}$. The requirement $\dot{y} = -v_0$, which is equivalent to $v\sin\theta = -v_0$, may therefore be written as

$$\sqrt{v_0^2 - 2gy}\cdot\frac{y'}{\sqrt{1+y'^2}} = -v_0. \tag{5.91}$$

Squaring both sides and solving for $y' \equiv dy/dx$ yields $dy/dx = -v_0/\sqrt{-2gy}$. Separating variables and integrating gives

$$\int\sqrt{-2gy}\,dy = -v_0\int dx \implies \frac{(-2gy)^{3/2}}{3g} = v_0x, \tag{5.92}$$

where the constant of integration has been set to zero, because $(x, y) = (0, 0)$ is a point on the curve. Therefore,

$$y = -\frac{(3gv_0x)^{2/3}}{2g}\,. \tag{5.93}$$

5.6. **Dividing the heat**

It turns out that it isn't possible to answer these questions without being given more information. The way that the work is divided up between the objects depends on what their surfaces look like. It's theoretically possible for one of the objects to gain all the heat, while the other doesn't heat up at all.

To understand this, we'll need to make a model of how friction works. The general way that friction works is that molecules from one surface rub against molecules from the other surface. The molecules stretch to the side and then bounce back and vibrate. This vibrational motion is the kinetic energy associated with heat. The model we'll use here will have a bunch of springs with masses on the ends, on both surfaces at the interface. When the surfaces rub against each other, the masses catch on each other for a short time (as shown in Fig. 5.58), and then they release, whereupon they vibrate back and forth on the springs. This is the kinetic energy of the heat we see. Though an oversimplification, this is basically the way friction works.

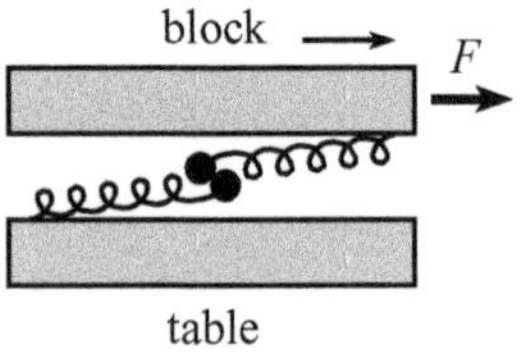

Fig. 5.58

Now, if everything is symmetrical between the two objects (that is, if the springs and masses on one object look like those on the other), then both objects will heat up by the same amount. But things need not be symmetrical. You can imagine the springs on one surface being much stiffer (that is, having a much larger k value) than the springs on the other surface. Or, you can even take the limit where one surface (say, the block) is made of completely rigid teeth, as shown (for one tooth) in Fig. 5.59. In this case, only the bottom surface (the table) will end up with heat from vibrational motion.

Is this asymmetrical result consistent with what we obtain from the work–energy theorem? Well, the net work done on the block is zero, because your pulling force does Fd positive work (where $F = \mu_k N$), while the friction force (the sum of all the forces from the masses on the little teeth) does Fd negative work. Therefore, since zero net work is done on the block, its total energy is constant. And since the kinetic energy due to its motion as a whole is constant, its internal thermal energy must also be constant. In other words, it doesn't heat up.

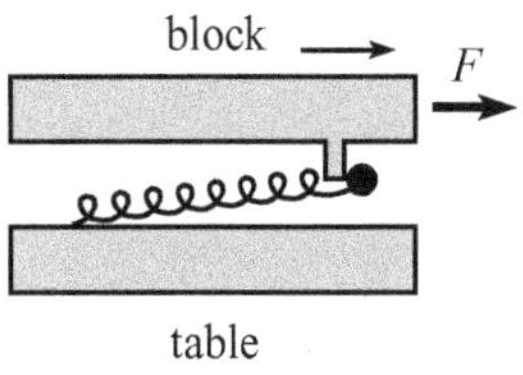

Fig. 5.59

The net work done on the table in this scenario comes from the force from the teeth on the little masses on the springs. These teeth do Fd positive work on all the little spring–mass systems, so Fd is the work done on the table. Therefore, its total energy increases by Fd. And since the kinetic energy due to its motion as a whole is constant (and zero, since the table is just sitting there), its internal thermal energy must increase by Fd. In other words, it heats up.

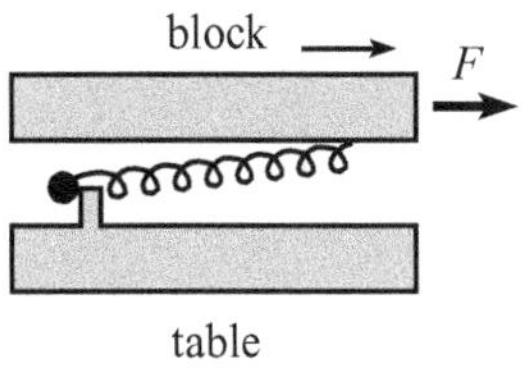

Fig. 5.60

Now consider the reverse situation, where the table has the rigid teeth and the block has the springs and masses, as shown in Fig. 5.60. The block is now the object that heats up, because it has the vibrational motion of the masses. And as above, we can show that this is consistent with the work–energy theorem, as follows. In the present case, the force from the table's teeth does no work on the block (because the teeth aren't moving), so the net work done on the block is simply the Fd from your pulling, so it heats up. Likewise, the little masses do no work on the teeth (because the teeth aren't moving), so the work done on the table is zero, so it doesn't heat up.

For the in between case where the spring constants on the two objects are equal, the net work done on each object is $Fd/2$, where d is the distance the block moves. This is true because the two masses in Fig. 5.58 each move half as far as the block moves. So the work done on the block is $Fd - Fd/2 = Fd/2$ (this is the positive work done by you, plus the negative work done by the table's little masses). And the work done on the table is $Fd/2$ (this is the positive work done by the block's little masses). So the objects heat up the same amount. For more discussion of the issues in this problem, see Sherwood (1984).

5.7. **$V(x)$ vs. a hill**

FIRST SOLUTION: Consider the normal force N acting on the bead at a given point. Let θ be the angle that the tangent to $V(x)$ makes with the horizontal, as shown in Fig. 5.61. The horizontal $F = ma$ equation is

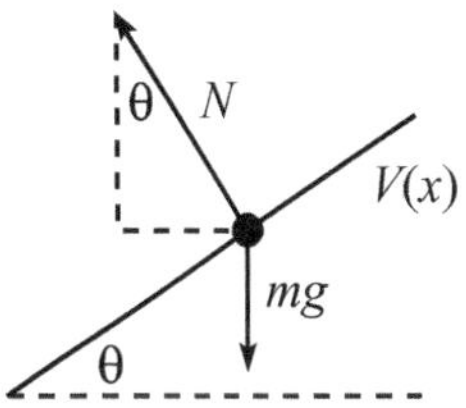

Fig. 5.61

$$-N \sin\theta = m\ddot{x}. \tag{5.94}$$

The vertical $F = ma$ equation is

$$N\cos\theta - mg = m\ddot{y} \implies N\cos\theta = mg + m\ddot{y}. \tag{5.95}$$

Dividing Eq. (5.94) by Eq. (5.95) gives

$$-\tan\theta = \frac{\ddot{x}}{g + \ddot{y}}. \tag{5.96}$$

But $\tan\theta = V'(x)$. Therefore,

$$\ddot{x} = -(g + \ddot{y})V'. \tag{5.97}$$

We see that this is not equal to $-gV'$. In fact, there is in general no way to construct a curve with height $z(x)$ that gives the same horizontal motion that a 1-D potential $mgV(x)$ gives, for all initial conditions. We would need $-(g + \ddot{z})z' = -gV'$, for all x. But at a given x, the quantities V' and z' are fixed, whereas $\ddot{z}$ depends on the initial conditions. For example, if there is a bend in the wire, then $\ddot{z}$ will be large if $\dot{z}$ is large. And $\dot{z}$ depends (in general) on how far the bead has fallen.

Equation (5.97) holds the key to constructing a situation that does give the $\ddot{x} = -gV'$ result. All we have to do is get rid of the $\ddot{y}$ term. So here's what we do. We grab our $y = V(x)$ wire and move it up and/or down in precisely the manner that makes the bead stay at the same height with respect to the ground. (Actually, constant vertical speed would be good enough.) This will make the $\ddot{y}$ term vanish, as desired. The vertical movement of the curve doesn't change the slope V' at a given value of x, so the θ in the above derivation is still the same θ.

Note that the quantity y here is the vertical position of the bead. It equals $V(x)$ if the curve is stationary, but not if the curve is being moved up and down.

REMARK: There is one case where $\ddot{x}$ is (approximately) equal to $-gV'$, even when the wire remains stationary. In the case of small oscillations of the bead near a minimum of $V(x)$, $\ddot{y}$ is small compared with g. Hence, Eq. (5.97) shows that $\ddot{x}$ is approximately equal to $-gV'$. Therefore, for small oscillations, it is reasonable to model a particle in a 1-D potential $mgV(x)$ as a particle sliding in a valley whose height is given by $y = V(x)$. ♣

SECOND SOLUTION: The component of gravity along the wire is what causes the change in velocity of the bead. That is,

$$-g\sin\theta = \frac{dv}{dt}, \tag{5.98}$$

where θ is given by

$$\tan\theta = V'(x) \quad \Longrightarrow \quad \sin\theta = \frac{V'}{\sqrt{1+V'^2}}, \quad \cos\theta = \frac{1}{\sqrt{1+V'^2}}. \tag{5.99}$$

We are, however, not concerned with the rate of change of v, but rather with the rate of change of $\dot{x}$. In view of this, let us write v in terms of $\dot{x}$. Since $\dot{x} = v\cos\theta$, we have $v = \dot{x}/\cos\theta = \dot{x}\sqrt{1+V'^2}$ (dots denote d/dt, primes denote d/dx). Therefore, Eq. (5.98) becomes

$$\begin{aligned} \frac{-gV'}{\sqrt{1+V'^2}} &= \frac{d}{dt}\left(\dot{x}\sqrt{1+V'^2}\right) \\ &= \ddot{x}\sqrt{1+V'^2} + \frac{\dot{x}V'(dV'/dt)}{\sqrt{1+V'^2}}. \end{aligned} \tag{5.100}$$

Hence, $\ddot{x}$ is given by

$$\ddot{x} = \frac{-gV'}{1+V'^2} - \frac{\dot{x}V'(dV'/dt)}{1+V'^2}. \tag{5.101}$$

We'll simplify this in a moment, but first a remark.

REMARK: A common incorrect solution to this problem is the following. The acceleration along the curve is $g\sin\theta = -g(V'/\sqrt{1+V'^2})$. Calculating the horizontal component of this acceleration brings in a factor of $\cos\theta = 1/\sqrt{1+V'^2}$. Therefore, we might think that

$$\ddot{x} = \frac{-gV'}{1+V'^2} \quad \text{(incorrect)}. \tag{5.102}$$

We have missed the second term in Eq. (5.101). Where is the mistake? The error is that we forgot to take into account the possible change in the curve's slope (Eq. (5.102) is true for straight lines). We addressed only the acceleration due to a change in *speed*. We forgot to consider the acceleration due to a change in the *direction* of motion (the term we missed is the one with dV'/dt). Intuitively, if we have a sharp enough bend in the wire, then $\dot{x}$ can change at an arbitrarily large rate, even if v is roughly constant. In view of this fact, Eq. (5.102) is definitely incorrect, because it is bounded (by $g/2$, in fact). ♣

To simplify Eq. (5.101), note that $V' \equiv dV/dx = (dV/dt)/(dx/dt) \equiv \dot{V}/\dot{x}$ ($\dot{V}$ is just the rate of change in the bead's height). Therefore, the numerator in the second term on the right-hand side of Eq. (5.101) is

$$\begin{aligned} \dot{x}V'\frac{dV'}{dt} &= \dot{x}V'\frac{d}{dt}\left(\frac{\dot{V}}{\dot{x}}\right) = \dot{x}V'\left(\frac{\dot{x}\ddot{V} - \dot{V}\ddot{x}}{\dot{x}^2}\right) \\ &= V'\ddot{V} - V'\ddot{x}\left(\frac{\dot{V}}{\dot{x}}\right) = V'\ddot{V} - V'^2\ddot{x}. \end{aligned} \tag{5.103}$$

Substituting this into Eq. (5.101) yields

$$\ddot{x} = -(g + \ddot{V})V', \tag{5.104}$$

in agreement with Eq. (5.97), because $y = V(x)$ if the wire is stationary. Equation (5.104) is valid only for a curve $V(x)$ that remains fixed. If we grab the wire and start moving it up and down, then the above solution is invalid, because the starting point, Eq. (5.98), rests on the assumption that gravity is the only force that does work on the bead. But if we move the wire, then the normal force also does work.

It turns out that for a moving wire, we simply need to replace the $\ddot{V}$ in Eq. (5.104) by $\ddot{y}$, which then gives Eq. (5.97). This can be seen by looking at things in the vertically accelerating frame in which the wire is at rest. We won't cover accelerating frames until Chapter 10, so we'll just invoke here the result that there is an extra fictitious "translational" force in this accelerating frame, and the consequence of this is that the bead thinks that it lives in a world where the acceleration from gravity is $g + \ddot{h}$ (if $\ddot{h}$ is positive, then the bead thinks gravity is larger), where h is the position of your hand that is accelerating the wire. In this new frame, the wire is at rest, so the above solution is valid. So with the g in Eq. (5.104) replaced by $g + \ddot{h}$, we have $\ddot{x} = -(g + \ddot{h} + \ddot{V})V'$. But $\ddot{V}$ (which is the vertical acceleration of the bead with respect to the new frame) plus $\ddot{h}$ (which is the vertical acceleration of the new frame with respect to the ground) equals $\ddot{y}$ (which, by our definition in the first solution, is the vertical acceleration of the bead with respect to the ground). We have therefore reproduced Eq. (5.97).

5.8. **Hanging mass**

We will calculate the equilibrium point y_0, and then use $\omega = \sqrt{V''(y_0)/m}$. The derivative of V is

$$V'(y) = ky + mg. \tag{5.105}$$

Therefore, $V'(y) = 0$ when $y = -mg/k \equiv y_0$. The second derivative of V is

$$V''(y) = k. \tag{5.106}$$

We therefore have

$$\omega = \sqrt{\frac{V''(y_0)}{m}} = \sqrt{\frac{k}{m}}. \tag{5.107}$$

Remark: This is independent of y_0, which makes sense because the only effect of gravity is to change the equilibrium position. More precisely, if y_r is the position relative to y_0 (so that $y \equiv y_0 + y_r$), then the total force as a function of y_r is

$$F(y_r) = -k(y_0 + y_r) - mg = -k\left(-\frac{mg}{k} + y_r\right) - mg = -ky_r, \tag{5.108}$$

so it still looks like a regular spring. (This works only because the spring force is linear.) Alternatively, you can think in terms of the potential energy; this is the task of Exercise 5.42. ♣

5.9. **Small oscillations**

We will calculate the equilibrium point x_0, and then use $\omega = \sqrt{V''(x_0)/m}$. The derivative of V is

$$V'(x) = -Ce^{-ax}x^{n-1}(n - ax). \tag{5.109}$$

Therefore, $V'(x) = 0$ when $x = n/a \equiv x_0$. The second derivative of V is

$$V''(x) = -Ce^{-ax}x^{n-2}\Big((n - 1 - ax)(n - ax) - ax\Big). \tag{5.110}$$

Plugging in $x_0 = n/a$ simplifies this a bit, and we find

$$\omega = \sqrt{\frac{V''(x_0)}{m}} = \sqrt{\frac{Ce^{-n}n^{n-1}}{ma^{n-2}}}. \tag{5.111}$$

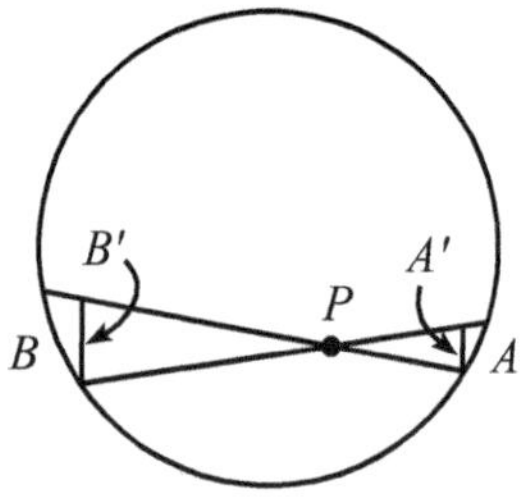

Fig. 5.62

5.10. **Zero force inside a sphere**

Let a be the distance from P to piece A, and let b be the distance from P to piece B (see Fig. 5.62). Draw the "perpendicular" bases of the cones, and call them A' and B'. The ratio of the areas of A' and B' is a^2/b^2. The key point here is that the angle between the planes of A and A' is the same as the angle between B and B'. This is true because the chord between A and B meets the circle at equal angles at its ends. So the ratio of the areas of A and B is also a^2/b^2. But the gravitational force decreases like $1/r^2$, and this effect exactly cancels the a^2/b^2 ratio of the areas. Therefore, the forces at P due to A and B (which can be treated like point masses, because the cones are assumed to be thin) are equal in magnitude; and opposite in direction, of course. If we draw enough cones to cover the whole shell, then the contributions to the force from little pieces over the whole shell cancel in pairs, so we are left with zero force at P. This holds for any point P inside the shell.

REMARK: Interestingly, the force inside an ellipsoidal shell of constant density (per volume) is also zero, assuming that the shell is defined to be the region between the surfaces described by $ax^2 + by^2 + cz^2 = k$, for two different values of k. In short, this is true because an ellipsoid is simply a stretched sphere. In detail: Let the above spherical shell have some thickness dr (which was irrelevant for the sphere, but it will be important in the ellipsoid case). From above, we know that the masses at the ends of the thin cones in the sphere have canceling forces. If we now stretch the sphere into an ellipsoid (uniformly in each direction, but the factors in the three directions can be different), then the distances from the end masses to P will still be in the ratio of a to b (as you can verify). And the end masses will still be in the ratio of a^2 to b^2, because both masses change by the same factor. This is true because every infinitesimal cube in the end masses has its volume changed by the same factor (namely $f_x f_y f_z$, where these f's are the stretching factors in each direction). So the masses are still in the ratio of a^2 to b^2, and the same argument about canceling forces holds as in the spherical case. Note that this zero-force result is *not* true for an ellipsoid with constant thickness, because such an object isn't the result of stretching a spherical shell (because a stretching results in the ellipsoidal shell being thicker near the ends where it is more "pointy"). ♣

5.11. **Escape velocity**

(a) The cutoff case is where the particle barely makes it to infinity, that is, where its speed is zero at infinity. Conservation of energy for this situation gives

$$\frac{1}{2}mv_{\text{esc}}^2 - \frac{GMm}{R} = 0 + 0. \tag{5.112}$$

In other words, the initial kinetic energy, $mv_{\text{esc}}^2/2$, must account for the gain in potential energy, GMm/R. Therefore,

$$v_{\text{esc}} = \sqrt{\frac{2GM}{R}}. \tag{5.113}$$

In terms of the acceleration, $g = GM/R^2$, at the surface of a planet, we can write this as $v_{\text{esc}} = \sqrt{2gR}$. Using $M = 4\pi\rho R^3/3$, we can also write it as $v_{\text{esc}} = \sqrt{8\pi GR^2\rho/3}$. So for a given density ρ, v_{esc} grows like R. Using the values of g and R given in Appendix J, we have:

For the earth, $v_{\text{esc}} = \sqrt{2gR} \approx \sqrt{2(9.8\,\text{m/s}^2)(6.4\cdot 10^6\,\text{m})} \approx 11.2\,\text{km/s}$.
For the moon, $v_{\text{esc}} = \sqrt{2gR} \approx \sqrt{2(1.6\,\text{m/s}^2)(1.7\cdot 10^6\,\text{m})} \approx 2.3\,\text{km/s}$.
For the sun, $v_{\text{esc}} = \sqrt{2gR} \approx \sqrt{2(270\,\text{m/s}^2)(7.0\cdot 10^8\,\text{m})} \approx 620\,\text{km/s}$.

REMARK: Another reasonable question to ask is: what is the escape velocity from the sun for an object located where the earth is (but imagine that the earth isn't there)? The answer is $\sqrt{2GM_S/R_{ES}}$, where R_{ES} is the earth–sun distance. Numerically, this is

$$\sqrt{2(6.67\cdot 10^{-11}\,\text{m}^3/\text{kg s}^2)(2\cdot 10^{30}\,\text{kg})/(1.5\cdot 10^{11}\,\text{m})} \approx 42\,\text{km/s}. \quad (5.114)$$

If you want to bring the earth back in (but let's assume it's at rest and not orbiting) and find the escape velocity (from both the sun and the earth) from a point on the earth's surface, you can't just add the 42 km/s and 11.2 km/s results. Instead, you have to take the square root of the sum of the squares. This follows from Eq. (5.112) and from the fact that potentials simply add. The result is about 43.5 km/s. The task of Exercise 5.60 is to find the escape velocity if the orbital motion of the earth is included. ♣

(b) To get a rough answer, we'll assume that the initial speed of a person's jump on the small planet is the same as it is on the earth. This probably isn't quite true, but it's close enough for the purposes here. A good jump on the earth is about a meter. For this jump, conservation of energy gives $mv^2/2 = mg(1\,\text{m})$. Therefore, $v = \sqrt{2g(1\,\text{m})} \approx \sqrt{20}\,\text{m/s}$. So we want $\sqrt{20}\,\text{m/s} = \sqrt{8\pi GR^2\rho/3}$. Using $\rho \approx 5500\,\text{kg/m}^3$, we find $R \approx 2.5\,\text{km}$. On such a planet, you should tread lightly.

5.12. **Ratio of potentials**

Let ρ be the mass density of the cube. Let V_ℓ^{cor} be the potential energy of a mass m at the corner of a cube of side ℓ, and let V_ℓ^{cen} be the potential energy of a mass m at the center of a cube of side ℓ. By dimensional analysis,

$$V_\ell^{\text{cor}} \propto \frac{G(\rho\ell^3)m}{\ell} \propto \ell^2. \quad (5.115)$$

Therefore,[32]

$$V_\ell^{\text{cor}} = 4V_{\ell/2}^{\text{cor}}. \quad (5.116)$$

But a cube of side ℓ can be built from eight cubes of side $\ell/2$. So by superposition, we have

$$V_\ell^{\text{cen}} = 8V_{\ell/2}^{\text{cor}}, \quad (5.117)$$

because the center of the larger cube lies at a corner of the eight smaller cubes (and because potentials just add). Therefore,

$$\frac{V_\ell^{\text{cor}}}{V_\ell^{\text{cen}}} = \frac{4V_{\ell/2}^{\text{cor}}}{8V_{\ell/2}^{\text{cor}}} = \frac{1}{2}. \quad (5.118)$$

[32] In other words, imagine expanding a cube of side $\ell/2$ to one of side ℓ. If we consider corresponding pieces of the two cubes, then the larger piece has $2^3 = 8$ times the mass of the smaller. But corresponding distances are twice as big in the large cube as in the small cube. Therefore, the larger piece contributes $8/2 = 4$ times as much to V_ℓ^{cor} as the smaller piece contributes to $V_{\ell/2}^{\text{cor}}$.

5.13. **Through the hole**

(a) By symmetry, only the component of the gravitational force perpendicular to the plane survives. A piece of mass dm at radius r on the plane provides a force equal to $Gm(dm)/(r^2+x^2)$. To obtain the component perpendicular to the plane, we must multiply this by $x/\sqrt{r^2+x^2}$. Slicing the plane up into rings with mass $dm = (2\pi r\, dr)\sigma$, we find that the total force is

$$F(x) = -\int_R^\infty \frac{Gm(2\pi r\sigma\, dr)x}{(r^2+x^2)^{3/2}} = 2\pi\sigma Gmx(r^2+x^2)^{-1/2}\Big|_{r=R}^{r=\infty}$$
$$= -\frac{2\pi\sigma Gmx}{\sqrt{R^2+x^2}}. \tag{5.119}$$

Note that if $R = 0$ (so that we have a uniform plane without a hole), then $F = -2\pi\sigma Gm$, which is independent of the distance from the plane.

(b) If $x \ll R$, then Eq. (5.119) gives $F(x) \approx -2\pi\sigma Gmx/R$, so $F = ma$ yields

$$\ddot{x} + \left(\frac{2\pi\sigma G}{R}\right)x = 0. \tag{5.120}$$

The frequency of small oscillations is therefore

$$\omega = \sqrt{\frac{2\pi\sigma G}{R}}, \tag{5.121}$$

which is independent of m.

Remark: For everyday values of R, this is a very small number because G is so small. Let's determine the rough size. If the sheet has thickness d, and if it is made of a material with density ρ (per unit volume), then $\sigma = \rho d$. So $\omega = \sqrt{2\pi\rho dG/R}$. In the above analysis, we assumed that the sheet was infinitely thin. In practice, we need d to be much smaller than the amplitude of the motion. But this amplitude must be much smaller than R in order for our approximation to hold. So we conclude that $d \ll R$. To get a rough upper bound on ω, let's pick $d/R = 1/10$. And let's make our sheet out of gold (with $\rho \approx 2 \cdot 10^4\,\text{kg/m}^3$). We then find $\omega \approx 1 \cdot 10^{-3}\,\text{s}^{-1}$, which corresponds to an oscillation about every 100 minutes. For the analogous system consisting of electrical charges, the frequency is much larger, because the electrical force is so much stronger than the gravitational force. ♣

(c) Integrating the force in Eq. (5.119) to obtain the potential energy (relative to the center of the hole) gives

$$V(x) = -\int_0^x F(x)\,dx = \int_0^x \frac{2\pi\sigma Gmx\,dx}{\sqrt{R^2+x^2}}$$
$$= 2\pi\sigma Gm\sqrt{R^2+x^2}\Big|_0^x = 2\pi\sigma Gm\left(\sqrt{R^2+x^2}-R\right). \tag{5.122}$$

By conservation of energy, the speed at the center of the hole is given by $mv^2/2 = V(x)$. Therefore,

$$v = 2\sqrt{\pi\sigma G\left(\sqrt{R^2+x^2}-R\right)}. \tag{5.123}$$

For $x \gg R$ this reduces to $v \approx 2\sqrt{\pi\sigma Gx}$.

Remark: You can also obtain this last result by noting that for large x, the force in Eq. (5.119) reduces to $F = -2\pi\sigma Gm$. This is constant, so it's basically just like a gravitational force $F = mg'$, where $g' \equiv 2\pi\sigma G$. But we know that in this familiar case, $v = \sqrt{2g'h} \to \sqrt{2(2\pi\sigma G)x}$, as above. ♣

5.14. **Snowball**

All of the snowball's momentum goes into the earth, which then translates (and rotates) a tiny bit faster (or slower, depending on which way the snowball was thrown).

What about the energy? Let m and v be the mass and initial speed of the snowball. Let M and V be the mass and final speed of the earth (with respect to the original rest frame of the earth). Since $m \ll M$, conservation of momentum gives $V \approx mv/M$. The kinetic energy of the earth is therefore

$$\frac{1}{2}M\left(\frac{mv}{M}\right)^2 = \frac{1}{2}mv^2\left(\frac{m}{M}\right) \ll \frac{1}{2}mv^2. \tag{5.124}$$

There is also a rotational kinetic-energy term of the same order of magnitude, but that doesn't matter. We see that essentially none of the snowball's energy goes into the earth. It therefore must all go into the form of heat, which melts some of the snow (and/or heats up the wall). This is a general result for a small object hitting a large object: The large object picks up essentially all of the momentum but essentially none of the energy (except possibly in the form of heat).

5.15. **Propelling a car**

Let the speed of the car be $v(t)$. Consider the collision of a ball of mass dm with the car. In the instantaneous rest frame of the car, the speed of the ball is $u - v$. In this frame, the ball reverses velocity when it bounces (because the car is so much more massive), so its change in momentum is $-2(u - v)\,dm$. This is also the change in momentum in the lab frame, because the two frames are related by a given speed at any instant. Therefore, in the lab frame the car gains a momentum of $2(u - v)\,dm$ from each ball that hits it. The rate of change in momentum of the car (that is, the force) is thus

$$\frac{dp}{dt} = 2\sigma'(u - v), \tag{5.125}$$

where $\sigma' \equiv dm/dt$ is the rate at which mass hits the car. σ' is related to the given σ by $\sigma' = \sigma(u - v)/u$, because although you throw the balls at speed u, the relative speed of the balls and the car is only $(u - v)$. We therefore have

$$\begin{aligned}
M\frac{dv}{dt} = \frac{2(u-v)^2\sigma}{u} \quad &\Longrightarrow \quad \int_0^v \frac{dv}{(u-v)^2} = \frac{2\sigma}{Mu}\int_0^t dt \\
&\Longrightarrow \quad \frac{1}{u-v} - \frac{1}{u} = \frac{2\sigma t}{Mu} \\
&\Longrightarrow \quad v(t) = \frac{\left(\frac{2\sigma t}{M}\right)u}{1 + \frac{2\sigma t}{M}}.
\end{aligned} \tag{5.126}$$

Note that $v \to u$ as $t \to \infty$, as it should. Writing this speed as $u\big(1 - 1/(1 + 2\sigma t/M)\big)$, we can integrate it to obtain the position,

$$x(t) = ut - \frac{Mu}{2\sigma}\ln\left(1 + \frac{2\sigma t}{M}\right), \tag{5.127}$$

where the constant of integration is zero because $x = 0$ at $t = 0$. We see that even though the speed approaches u, the car will eventually be an arbitrarily large distance behind an object that moves with constant speed u (for example, pretend that your first ball misses the car and continues forward at speed u).

5.16. **Propelling a car again**

We can carry over some of the results from the previous problem. The only change in the calculation of the force on the car is that since the balls don't bounce backward, we don't pick up the factor of 2 in Eq. (5.125). The force on the car is therefore

$$m\frac{dv}{dt} = \frac{(u-v)^2\sigma}{u}, \tag{5.128}$$

where $m(t)$ is the mass of the car-plus-contents, as a function of time. The main difference between this problem and the previous one is that this mass m changes because the balls are collecting inside the car. From the previous problem, the rate at which mass enters the car is $\sigma' = \sigma(u - v)/u$. Therefore,

$$\frac{dm}{dt} = \frac{(u - v)\sigma}{u}. \tag{5.129}$$

We must now solve the two preceding differential equations. Dividing Eq. (5.128) by Eq. (5.129), and separating variables, gives[33]

$$\int_0^v \frac{dv}{u - v} = \int_M^m \frac{dm}{m} \implies -\ln\left(\frac{u - v}{u}\right) = \ln\left(\frac{m}{M}\right) \implies m = \frac{Mu}{u - v}. \tag{5.130}$$

Note that $m \to \infty$ as $v \to u$, as it should. Substituting this value of m into either Eq. (5.128) or Eq. (5.129) gives

$$\int_0^v \frac{dv}{(u - v)^3} = \int_0^t \frac{\sigma\, dt}{Mu^2} \implies \frac{1}{2(u - v)^2} - \frac{1}{2u^2} = \frac{\sigma t}{Mu^2}$$
$$\implies v(t) = u - \frac{u}{\sqrt{1 + \frac{2\sigma t}{M}}}. \tag{5.131}$$

Note that $v \to u$ as $t \to \infty$, as it should. Integrating this speed to obtain the position gives

$$x(t) = ut - \frac{Mu}{\sigma}\sqrt{1 + \frac{2\sigma t}{M}} + \frac{Mu}{\sigma}, \tag{5.132}$$

where the constant of integration has been chosen so that $x = 0$ at $t = 0$. For a given t, the $v(t)$ in Eq. (5.131) is smaller than the $v(t)$ for the previous problem in Eq. (5.126), which is easy to see if the latter is written as $u\big(1 - 1/(1 + 2\sigma t/M)\big)$. This makes sense, because in the present problem the balls have less of an effect on $v(t)$, because (1) they don't bounce back, and (2) the mass of the car-plus-contents is larger.

5.17. **Leaky bucket**

(a) FIRST SOLUTION: The initial position is $x = L$. The given rate of leaking implies that the mass of the bucket at position x is $m = M(x/L)$. Therefore, $F = ma$ gives $-T = (Mx/L)\ddot{x}$. Writing the acceleration as $v\, dv/dx$, and separating variables and integrating, gives

$$-\frac{TL}{M}\int_L^x \frac{dx}{x} = \int_0^v v\, dv \implies -\frac{TL}{M}\ln\left(\frac{x}{L}\right) = \frac{v^2}{2}. \tag{5.133}$$

The kinetic energy at position x is therefore

$$E = \frac{mv^2}{2} = \left(\frac{Mx}{L}\right)\frac{v^2}{2} = -Tx\ln\left(\frac{x}{L}\right). \tag{5.134}$$

In terms of the fraction $z \equiv x/L$, we have $E = -TLz\ln z$. Setting $dE/dz = 0$ to find the maximum gives

$$z = \frac{1}{e} \implies E_{\text{max}} = \frac{TL}{e}. \tag{5.135}$$

[33] We can also quickly derive this equation by writing down conservation of momentum for the time interval when a mass dm enters the car: $dm\, u + mv = (m + dm)(v + dv)$. This yields the first equality in Eq. (5.130). But we will still need to use one of Eqs. (5.128) and (5.129) in what follows.

Note that the (fractional) location of E_{max} is independent of M, T, and L, but its value depends on T and L. These facts follow from dimensional analysis.

REMARK: We began this solution by writing down $F = ma$, where m is the mass of the bucket. You may be wondering why we didn't use $F = dp/dt$, where p is the momentum of the bucket. This would certainly give a different result, because $dp/dt = d(mv)/dt = ma + (dm/dt)v$. We used $F = ma$ because at any instant the mass m is what is being accelerated by the force F.

If you want, you can imagine the process occurring in discrete steps: The force pulls on the mass for a short period of time, then a little piece falls off. Then the force pulls again on the new mass, then another little piece falls off. And so on. In this scenario, it is clear that $F = ma$ is the appropriate formula, because it holds for each step in the process.

It is indeed true that $F = dp/dt$, if you let F be *total* force in the problem, and let p be the *total* momentum. The tension T is the only horizontal force in the problem, because we've assumed the ground to be frictionless. However, the total momentum consists of both the sand in the bucket *and* the sand that has leaked out and is sliding along on the ground. If we use $F = dp/dt$, where p is the total momentum, we obtain (remember that dm/dt is a negative quantity)

$$-T = \frac{dp_{\text{bucket}}}{dt} + \frac{dp_{\text{leaked}}}{dt} = \left(ma + \frac{dm}{dt}v\right) + \left(-\frac{dm}{dt}\right)v = ma, \quad (5.136)$$

as expected. See Appendix C for further discussion of the uses of $F = ma$ and $F = dp/dt$. ♣

SECOND SOLUTION: Consider a small time interval during which the bucket moves from x to $x + dx$ (where dx is negative). The bucket's kinetic energy changes by $(-T)\,dx$ (this is positive) due to the work done by the spring, and also changes by a fraction dx/x (this is negative) due to the leaking. Therefore, $dE = -T\,dx + E\,dx/x$, or

$$\frac{dE}{dx} = -T + \frac{E}{x}. \quad (5.137)$$

In solving this differential equation, it is convenient to introduce the variable $y \equiv E/x$. Then $E' = xy' + y$, where a prime denotes differentiation with respect to x. Equation (5.137) then becomes $xy' = -T$, which gives

$$\int_0^{E/x} dy = -T\int_L^x \frac{dx}{x} \implies E = -Tx\ln\left(\frac{x}{L}\right), \quad (5.138)$$

as in the first solution.

(b) From Eq. (5.133), the speed is $v = \sqrt{2TL/M}\,\sqrt{-\ln z}$, where $z \equiv x/L$. Therefore, the magnitude of the momentum is

$$p = mv = (Mz)v = \sqrt{2TLM}\,\sqrt{-z^2\ln z}. \quad (5.139)$$

Setting $dp/dz = 0$ to find the maximum gives

$$z = \frac{1}{\sqrt{e}} \implies p_{\max} = \sqrt{\frac{TLM}{e}}. \quad (5.140)$$

Note that the (fractional) location of p_{max} is independent of M, T, and L, but its value depends on all three. These facts follow from dimensional analysis.

REMARK: E_{max} occurs closer to the wall (that is, at a later time) than p_{max}. The reason for this is that v matters more in $E = mv^2/2$ than it does in $p = mv$. As far as E is concerned, it is beneficial for the bucket to lose a little more mass if it means being able to pick up a little more speed (up to a certain point). ♣

5.18. **Another leaky bucket**

(a) $F = ma$ gives $-T = m\ddot{x}$. Combining this with the given $dm/dt = b\ddot{x}$ equation yields $m\,dm = -bT\,dt$. Integration then gives $m^2/2 = C - bTt$. But $m = M$ when $t = 0$, so we have $C = M^2/2$. Therefore,

$$m(t) = \sqrt{M^2 - 2bTt}\,. \tag{5.141}$$

This holds for $t < M^2/2bT$, provided that the bucket hasn't hit the wall yet.

(b) The given equation $dm/dt = b\ddot{x} = b\,dv/dt$ integrates to $v = m/b + D$. But $v = 0$ when $m = M$, so we have $D = -M/b$. Therefore,

$$v(m) = \frac{m - M}{b} \quad \Longrightarrow \quad v(t) = \frac{\sqrt{M^2 - 2bTt}}{b} - \frac{M}{b}\,. \tag{5.142}$$

At the instant right before all the sand leaves the bucket, we have $m = 0$. Therefore, $v = -M/b$ at this point. Integrating $v(t)$ to obtain $x(t)$, we find

$$x(t) = \frac{-(M^2 - 2bTt)^{3/2}}{3b^2T} - \frac{M}{b}t + L + \frac{M^3}{3b^2T}\,, \tag{5.143}$$

where the constant of integration has been chosen so that $x = L$ when $t = 0$. Solving for t in terms of m from Eq. (5.141), substituting the result into Eq. (5.143), and simplifying, gives

$$x(m) = L - \frac{(M - m)^2(M + 2m)}{6b^2T}\,. \tag{5.144}$$

(c) Using Eq. (5.142), the kinetic energy is (it's easier to work in terms of m here)

$$E = \frac{1}{2}mv^2 = \frac{1}{2b^2}m(m - M)^2. \tag{5.145}$$

Taking the derivative dE/dm to find the maximum, we obtain

$$m = \frac{M}{3} \quad \Longrightarrow \quad E_{\text{max}} = \frac{2M^3}{27b^2}\,. \tag{5.146}$$

(d) Using Eq. (5.142), the momentum is

$$p = mv = \frac{1}{b}m(m - M). \tag{5.147}$$

Taking the derivative to find the maximum magnitude, we obtain

$$m = \frac{M}{2} \quad \Longrightarrow \quad |p|_{\text{max}} = \frac{M^2}{4b}\,. \tag{5.148}$$

(d) We want $x = 0$ when $m = 0$, so Eq. (5.144) gives

$$0 = L - \frac{M^3}{6b^2T} \quad \Longrightarrow \quad b = \sqrt{\frac{M^3}{6TL}}\,. \tag{5.149}$$

This is the only combination of M, T, and L that has the units of b, namely kg s/m. But we needed to do the calculation to find the numerical factor of $1/\sqrt{6}$.

5.19. **Right angle in billiards**

Let $\mathbf{v}$ be the initial velocity, and let $\mathbf{v}_1$ and $\mathbf{v}_2$ be the final velocities. Since the masses are equal, conservation of momentum gives $\mathbf{v} = \mathbf{v}_1 + \mathbf{v}_2$. Taking the dot product of this equation with itself gives

$$\mathbf{v} \cdot \mathbf{v} = \mathbf{v}_1 \cdot \mathbf{v}_1 + 2\mathbf{v}_1 \cdot \mathbf{v}_2 + \mathbf{v}_2 \cdot \mathbf{v}_2. \tag{5.150}$$

And conservation of energy gives (dropping the factors of $m/2$)

$$\mathbf{v} \cdot \mathbf{v} = \mathbf{v}_1 \cdot \mathbf{v}_1 + \mathbf{v}_2 \cdot \mathbf{v}_2. \tag{5.151}$$

Taking the difference of these two equations yields

$$\mathbf{v}_1 \cdot \mathbf{v}_2 = 0. \tag{5.152}$$

So $v_1 v_2 \cos\theta = 0$, which means that $\theta = 90°$. (Or $v_1 = 0$, which means the incoming mass stops because the collision is head-on. Or $v_2 = 0$, which means the masses miss each other.)

5.20. **Bouncing and recoiling**

Let v_i be the speed of the ball after the ith bounce, and let V_i be the speed of the block right after the ith bounce. Then conservation of momentum gives

$$mv_i = MV_{i+1} - mv_{i+1}. \tag{5.153}$$

But Theorem 5.3 says that $v_i = V_{i+1} + v_{i+1}$. Solving this system of two linear equations gives

$$v_{i+1} = \frac{(M-m)v_i}{M+m} \equiv \frac{(1-\epsilon)v_i}{1+\epsilon} \approx (1-2\epsilon)v_i, \quad \text{and} \quad V_{i+1} \approx 2\epsilon v_i, \tag{5.154}$$

where $\epsilon \equiv m/M \ll 1$. This expression for v_{i+1} implies that the speed of the ball after the nth bounce is

$$v_n = (1-2\epsilon)^n v_0, \quad \text{where} \quad \epsilon \equiv m/M. \tag{5.155}$$

The total distance traveled by the block can be obtained by looking at the work done by friction. Eventually, the ball has negligible energy, so all of its initial kinetic energy goes into heat from friction. Therefore, $mv_0^2/2 = F_f d = (\mu Mg)d$, which gives

$$d = \frac{mv_0^2}{2\mu Mg}. \tag{5.156}$$

To find the total time, we can add up the times, t_n, the block moves after each bounce. Since the product of force and time equals the change in momentum, we have $F_f t_n = MV_n$, and so $(\mu Mg)t_n = M(2\epsilon v_{n-1}) = 2M\epsilon(1-2\epsilon)^{n-1}v_0$. Therefore,

$$t = \sum_{n=1}^{\infty} t_n = \frac{2\epsilon v_0}{\mu g}\sum_{n=0}^{\infty}(1-2\epsilon)^n = \frac{2\epsilon v_0}{\mu g} \cdot \frac{1}{1-(1-2\epsilon)} = \frac{v_0}{\mu g}. \tag{5.157}$$

We've let the sum go to $n = \infty$ even though our $v_n = (1-2\epsilon)^n v_0$ approximation eventually breaks down for very large n (because we dropped terms of order ϵ^2 in deriving it). But by the time it breaks down, the terms are negligibly small anyway. The calculation of d above can also be done by adding up the geometric series of the distances moved after each bounce.

Remarks: This $t = v_0/\mu g$ result is much larger than the result obtained in the case where the ball sticks to the block on the first hit, in which case the answer is $t = mv_0/(\mu Mg)$. The total time is proportional to the total momentum that the block picks up, and the $t = v_0/\mu g$ answer is larger because the wall keeps transferring positive momentum to the ball, which then gets transferred to the block.

In contrast, d is the same as it would be in the case where the ball sticks to the block on the first hit. The total distance is proportional to the total energy that the block picks up, and in both cases the total energy given to the block is $mv_0^2/2$. The wall (which is attached to the very massive earth) transfers essentially no energy to the ball.

The $t = v_0/\mu g$ result is independent of the masses (as long as $M \gg m$), although it's not at all intuitively obvious that if we keep the same v_0, but decrease m by a factor of 100, that we'll end up with the same t. The distance d, on the other hand, would decrease by 100. ♣

5.21. **Drag force on a sheet**

(a) We will set $v = 0$ here. When the sheet hits a particle, the particle acquires a speed of essentially $2V$. This follows from Theorem 5.3, or by working in the frame of the heavy sheet. The momentum of the particle is then $2mV$. In time t, the sheet sweeps through a volume AVt, where A is the area of the sheet. Therefore, in time t, the sheet hits $AVtn$ particles. The sheet therefore loses momentum at a rate of $dP/dt = -(AVn)(2mV)$. But $F = dP/dt$, so the magnitude of the drag force per unit area is

$$\frac{F}{A} = 2nmV^2 \equiv 2\rho V^2, \tag{5.158}$$

where ρ is the mass density of the particles. We see that the force depends quadratically on V.

(b) If $v \gg V$, the particles now hit the sheet from various directions on both sides, but we need only consider the particles' motions along the line of the sheet's motion. As stated in the problem, we will assume that all velocities in this direction are equal to $\pm v/2$. Note that we won't be able to set V exactly equal to zero here, because we would obtain a force of zero and miss the lowest-order effect.

Consider a particle in front of the sheet, moving backward toward the sheet. The relative speed between the particle and the sheet is $v/2 + V$. This relative speed reverses direction during the collision, so the change in momentum of the particle is $2m(v/2 + V)$. We have used the fact that the speed of the heavy sheet is essentially unaffected by the collision. The rate at which particles collide with the sheet is $A(v/2 + V)(n/2)$, from the reasoning in part (a). The $n/2$ factor comes from the fact that half of the particles move toward the sheet, and half move away from it.

Now consider a particle behind the sheet, moving forward toward the sheet. The relative speed between the particle and the sheet is $v/2 - V$. This relative speed reverses direction during the collision, so the change in momentum of this particle is $-2m(v/2 - V)$. And the rate at which particles collide with the sheet is $A(v/2 - V)(n/2)$. Therefore, the magnitude of the drag force per unit area is

$$\begin{aligned}\frac{F}{A} &= \frac{1}{A} \cdot \left|\frac{dP}{dt}\right| \\ &= \left(\frac{n}{2}(v/2 + V)\right)\left(2m(v/2 + V)\right) + \left(\frac{n}{2}(v/2 - V)\right)\left(-2m(v/2 - V)\right) \\ &= 2nmvV \equiv 2\rho vV,\end{aligned} \tag{5.159}$$

where ρ is the mass density of the particles. We see that the force depends linearly on V. The fact that it agrees with the result in part (a) in the case of $v = V$ is coincidence. Neither result is valid when $v = V$.

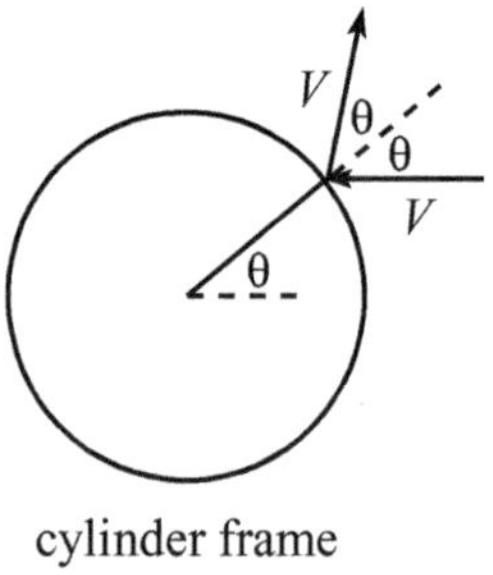

Fig. 5.63

5.22. **Drag force on a cylinder**

Consider a particle that makes contact with the cylinder at an angle θ with respect to the line of motion. In the frame of the heavy cylinder (see Fig. 5.63), the particle comes in with velocity $-V$ and then bounces off with a horizontal velocity component of $V \cos 2\theta$. So in this frame (and hence also in the lab frame), the particle increases its horizontal momentum by $mV(1 + \cos 2\theta)$. The cylinder must therefore lose this momentum.

The area on the cylinder that lies between θ and $\theta + d\theta$ sweeps out volume at a rate $(R\, d\theta \cos\theta)V\ell$, where ℓ is the length of the cylinder. The $\cos\theta$ factor here gives the projection orthogonal to the direction of motion. The force per unit length on the cylinder (that is, the rate of change in momentum per unit length) is therefore

$$\begin{aligned}\frac{F}{\ell} &= \int_{-\pi/2}^{\pi/2} \Big(n(R\,d\theta\,\cos\theta)V\Big)\Big(mV(1+\cos 2\theta)\Big)\\ &= 2nmRV^2 \int_{-\pi/2}^{\pi/2} \cos\theta(1-\sin^2\theta)\,d\theta\\ &= 2nmRV^2\left(\sin\theta - \frac{1}{3}\sin^3\theta\right)\Big|_{-\pi/2}^{\pi/2}\\ &= \frac{8}{3}nmRV^2 \equiv \frac{8}{3}\rho RV^2,\end{aligned} \tag{5.160}$$

where ρ is the mass density of the particles. Note that the average force per cross-sectional area, $F/(2R\ell)$, equals $(4/3)\rho V^2$. This is smaller than the result for the sheet in the previous problem, as it should be, because the particles bounce off somewhat sideways in the cylinder case.

5.23. **Basketball and tennis ball**

(a) Right before the basketball hits the ground, both balls move downward with speed (using $mv^2/2 = mgh$)

$$v = \sqrt{2gh}. \tag{5.161}$$

Right after the basketball bounces off the ground, it moves upward with speed v, while the tennis ball still moves downward with speed v. The relative speed is therefore $2v$. After the balls bounce off each other, the relative speed is still $2v$ (this follows from Theorem 5.3, or by working in the frame of the heavy basketball). Since the upward speed of the basketball essentially stays equal to v, the upward speed of the tennis ball is $2v+v = 3v$. By conservation of energy, it therefore rises to a height of $H = d + (3v)^2/(2g)$. But $v^2 = 2gh$, so we have

$$H = d + 9h. \tag{5.162}$$

(b) Right before B_1 hits the ground, all of the balls move downward with speed $v = \sqrt{2gh}$. We will inductively determine the speed of each ball after it bounces off the one below it. If B_i achieves a speed of v_i after bouncing off B_{i-1}, then what is the speed of B_{i+1} after it bounces off B_i? The relative speed of B_{i+1} and B_i (right before they bounce) is $v + v_i$. This is also the relative speed after they bounce. Therefore, since B_i is still moving upward at essentially speed v_i, we see that the final upward speed of B_{i+1} equals $(v + v_i) + v_i$. Thus,

$$v_{i+1} = 2v_i + v. \tag{5.163}$$

Since $v_1 = v$, we obtain $v_2 = 3v$ (in agreement with part (a)), and then $v_3 = 7v$, and then $v_4 = 15v$, etc. In general,

$$v_n = (2^n - 1)v, \tag{5.164}$$

which is easily seen to satisfy Eq. (5.163), with the initial value $v_1 = v$. From conservation of energy, B_n therefore rises to a height of

$$H = \ell + \frac{((2^n-1)v)^2}{2g} = \ell + (2^n-1)^2 h. \tag{5.165}$$

If h is 1 meter, and if we want this height to be 1000 meters, then (assuming ℓ is not very large) we need $2^n - 1 > \sqrt{1000}$. Five balls won't quite do the trick, but six will, in which case the height is almost 4 kilometers. Escape velocity from the earth (which is $v_{esc} = \sqrt{2gR} \approx 11\,200$ m/s) is reached when

$$v_n \geq v_{esc} \implies (2^n - 1)\sqrt{2gh} \geq \sqrt{2gR} \implies n \geq \ln_2\left(\sqrt{\frac{R}{h}} + 1\right). \tag{5.166}$$

With $R = 6.4 \cdot 10^6$ m and $h = 1$ m, we find $n \geq 12$. Of course, the elasticity assumption is absurd in this case, as is the notion that you can find 12 balls with the property that $m_1 \gg m_2 \gg \cdots \gg m_{12}$.

5.24. **Maximal deflection**

First solution: Let's figure out what the collision looks like in the CM frame. If M has initial speed V in the lab frame, then the CM moves with speed $V_{CM} = MV/(M+m)$. The speeds of M and m in the CM frame therefore equal, respectively,

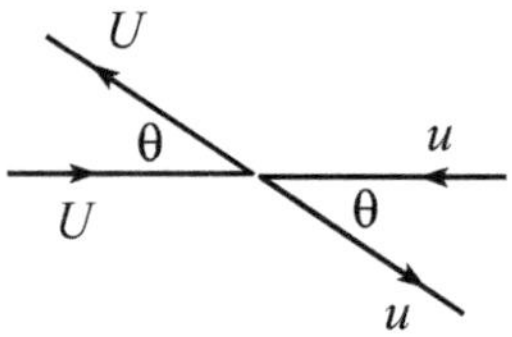

Fig. 5.64

$$U = V - V_{CM} = \frac{mV}{M+m}, \quad \text{and} \quad u = |-V_{CM}| = \frac{MV}{M+m}. \tag{5.167}$$

In the CM frame, the collision is simple. The particles keep the same speeds, but simply change their directions (while still moving in opposite directions), as shown in Fig. 5.64. The angle θ is free to have any value. This scenario clearly satisfies conservation of energy and momentum, so it must be what happens.

The important point to note is that since θ can have any value, the tip of the **U** velocity vector can be located anywhere on a circle of radius U. If we then shift back to the lab frame, we see that the final velocity of M with respect to the lab frame, $\mathbf{V}_{lab}$, is obtained by adding $\mathbf{V}_{CM}$ to the vector **U**, which can point anywhere on the dotted circle in Fig. 5.65. A few possibilities for $\mathbf{V}_{lab}$ are shown. The largest angle of deflection is obtained when $\mathbf{V}_{lab}$ is tangent to the dotted circle, in which case we have the situation shown in Fig. 5.66. The maximum angle of deflection, ϕ_{max}, is therefore given by

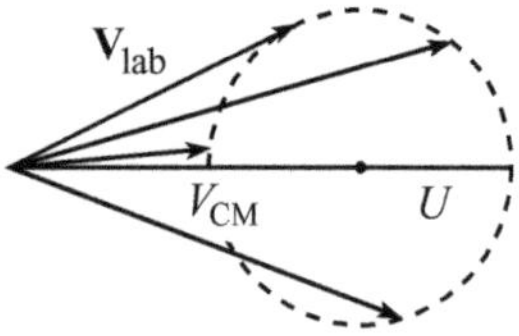

Fig. 5.65

$$\sin\phi_{max} = \frac{U}{V_{CM}} = \frac{mV/(M+m)}{MV/(M+m)} = \frac{m}{M}. \tag{5.168}$$

If $M < m$, then the dotted circle passes to the left of the left vertex of the triangle. This means that ϕ can take on any value. In particular, it is possible for M to bounce directly backward.

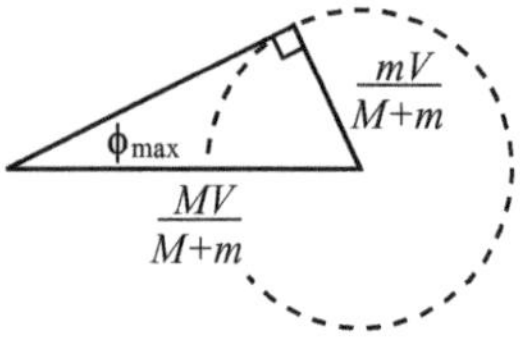

Fig. 5.66

Second solution: We'll work in the lab frame in this solution. Let V' and v' be the final speeds, and let ϕ and γ be the scattering angles of M and m, respectively, in the lab frame. Then conservation of p_x, p_y, and E give

$$MV = MV'\cos\phi + mv'\cos\gamma, \tag{5.169}$$

$$0 = MV'\sin\phi - mv'\sin\gamma, \tag{5.170}$$

$$\frac{1}{2}MV^2 = \frac{1}{2}MV'^2 + \frac{1}{2}mv'^2. \tag{5.171}$$

Putting the ϕ terms on the left-hand sides of Eqs. (5.169) and (5.170), and then squaring and adding these equations, gives

$$M^2(V^2 + V'^2 - 2VV'\cos\phi) = m^2v'^2. \tag{5.172}$$

Equating this expression for $m^2v'^2$ with the one obtained by multiplying Eq. (5.171) through by m gives

$$M(V^2 + V'^2 - 2VV'\cos\phi) = m(V^2 - V'^2)$$

$$\implies (M+m)V'^2 - (2MV\cos\phi)V' + (M-m)V^2 = 0. \tag{5.173}$$

A solution to this quadratic equation in V' exists if and only if the discriminant is non-negative. Therefore, we must have

$$
\begin{aligned}
(2MV\cos\phi)^2 - 4(M+m)(M-m)V^2 \geq 0 \\
\Longrightarrow \quad m^2 &\geq M^2(1-\cos^2\phi) \\
\Longrightarrow \quad m^2 &\geq M^2\sin^2\phi \\
\Longrightarrow \quad \frac{m}{M} &\geq \sin\phi. \qquad (5.174)
\end{aligned}
$$

5.25. **Colliding masses**

(a) By conservation of momentum, the final speed of the combined masses is $MV/(M+m) \approx (1-m/M)V$, plus higher-order corrections. The final energies are therefore

$$
\begin{aligned}
E_m &= \frac{1}{2}m\left(1-\frac{m}{M}\right)^2 V^2 \approx \frac{1}{2}mV^2, \\
E_M &= \frac{1}{2}M\left(1-\frac{m}{M}\right)^2 V^2 \approx \frac{1}{2}MV^2 - mV^2.
\end{aligned} \qquad (5.175)
$$

These energies add up to $MV^2/2 - mV^2/2$, which is $mV^2/2$ less than the initial energy of mass M, namely $MV^2/2$. Therefore, $mV^2/2$ is lost to heat.

(b) In this frame, mass m has initial speed V, so its initial energy is $E_{\text{i}} = mV^2/2$. By conservation of momentum, the final speed of the combined masses is $mV/(M+m) \approx (m/M)V$, plus higher-order corrections. The final energies are therefore

$$
\begin{aligned}
E_m &= \frac{1}{2}m\left(\frac{m}{M}\right)^2 V^2 = \left(\frac{m}{M}\right)^2 E_{\text{i}} \approx 0, \\
E_M &= \frac{1}{2}M\left(\frac{m}{M}\right)^2 V^2 = \left(\frac{m}{M}\right) E_{\text{i}} \approx 0.
\end{aligned} \qquad (5.176)
$$

This negligible final energy of zero is $mV^2/2$ less than E_{i}. Therefore, $mV^2/2$ is lost to heat, in agreement with part (a).

5.26. **Pulling a chain**

Let x be the distance your hand has moved. Then $x/2$ is the length of the moving part of the chain, because the chain gets "doubled up." The momentum of this moving part is therefore $p = (\sigma x/2)\dot{x}$. The force that your hand applies is found from $F = dp/dt$, which gives $F = (\sigma/2)(\dot{x}^2 + x\ddot{x})$. But since v is constant, the $\ddot{x}$ term vanishes. The change in momentum here is due simply to additional mass acquiring speed v, and not due to any increase in speed of the part already moving. Hence,

$$F = \frac{\sigma v^2}{2}, \qquad (5.177)$$

which is constant. Your hand applies this force over a total distance $2L$, so the total work you do is

$$F(2L) = \sigma L v^2. \qquad (5.178)$$

The mass of the chain is σL, so its final kinetic energy is $(\sigma L)v^2/2$. This is only half of the work you do. Therefore, an energy of $\sigma L v^2/2$ is lost to heat. Each atom in the chain goes abruptly from rest to speed v, and there is no way to avoid heat loss in such a process. This is clear when viewed in the reference frame of your hand. In this frame, the chain initially moves at speed v and then eventually comes to rest, piece by piece. So all of its initial kinetic energy, $(\sigma L)v^2/2$, goes into heat.

5.27. **Pulling a chain again**

Let x be the position of the end of the chain. The momentum of the chain is then $p = (\sigma x)\dot{x}$. $F = dp/dt$ gives (using the fact that F is constant) $Ft = p$, so we have $Ft = (\sigma x)\dot{x}$. Separating variables and integrating yields

$$\int_0^x \sigma x\, dx = \int_0^t Ft\, dt \implies \frac{\sigma x^2}{2} = \frac{Ft^2}{2} \implies x = t\sqrt{F/\sigma}. \tag{5.179}$$

The position therefore grows linearly with time. In other words, the speed is constant, and it equals $\sqrt{F/\sigma}$.

REMARK: Realistically, when you grab the chain, there is some small initial value of x (call it ϵ). The dx integral above now starts at ϵ instead of 0, so x takes the form, $x = \sqrt{Ft^2/\sigma + \epsilon^2}$. If ϵ is very small, the speed very quickly approaches $\sqrt{F/\sigma}$. Even if ϵ is not small, the position becomes arbitrarily close to $t\sqrt{F/\sigma}$ as t becomes large. The "head start" of ϵ therefore doesn't help you in the long run. ♣

5.28. **Falling chain**

FIRST SOLUTION: The left part of the chain is in freefall (because we're dealing with a chain described by our first scenario), so at time t it is moving at speed gt and has fallen a distance $gt^2/2$. The chain gets doubled up below the support, so a length of only $gt^2/4$ hangs at rest. Hence, a length of $L - gt^2/4$ is falling at speed gt. With upward taken to be positive, the momentum of the entire chain (which just comes from the moving part, of course) is therefore

$$p = \sigma(L - gt^2/4)(-gt) = -\sigma Lgt + \sigma g^2 t^3/4. \tag{5.180}$$

If F_s is the force from the support, then the net force on the entire chain is $F_s - \sigma Lg$. So $F = dp/dt$ for the entire chain gives

$$F_s - \sigma Lg = \frac{d}{dt}\left(-\sigma Lgt + \frac{\sigma g^2 t^3}{4}\right) \implies F_s = \frac{3\sigma g^2 t^2}{4}. \tag{5.181}$$

This result holds until the top of the chain has fallen a distance $2L$ (at $T = \sqrt{4L/g}$). Prior to time T, F_s equals three times the weight, $\sigma(gt^2/4)g$, of the part of the chain that hangs at rest. After time T, F_s simply equals the total weight of the chain, σLg. So at time T, F_s abruptly drops from $3\sigma Lg$ to σLg.

SECOND SOLUTION: F_s is responsible for two things: (1) It holds up the part of the chain that hangs at rest below the support, and (2) it changes the momentum of the atoms in the chain that are suddenly brought to rest at the kink in the chain. In other words, $F_s = F_{\text{weight}} + F_{dp/dt}$. From the first solution above, we have $F_{\text{weight}} = \sigma(gt^2/4)g$.

Now let's find $F_{dp/dt}$. At time t, the speed of the chain is gt, so in a small time interval dt, the top of the chain falls a distance $(gt)\,dt$. But due to the "doubling up" effect, only half of this length is brought to rest in the time dt. Therefore, a small piece of mass $\sigma(1/2)(gt)\,dt$ that was moving at speed gt is suddenly brought to rest. The momentum was $(\sigma/2)(gt)^2 dt$ downward, and then it becomes zero. Hence, $dp = +(\sigma/2)g^2t^2 dt$, and so $F_{dp/dt} = dp/dt = (\sigma/2)g^2t^2$. Therefore,

$$F_s = F_{\text{weight}} + F_{dp/dt} = \frac{\sigma g^2 t^2}{4} + \frac{\sigma g^2 t^2}{2} = \frac{3\sigma g^2 t^2}{4}. \tag{5.182}$$

5.29. **Falling chain (energy conserving)**

Let σ be the mass density of the chain, let L be its total length, and let x be the distance (defined to be positive) that the top of the chain has fallen. For a given x, a piece of chain with mass σx and CM position $L - x/2$ has effectively been replaced by a thin "U" below the support, with height $x/2$; so its CM position is $-x/4$. The loss in potential energy is therefore $(\sigma x)g(L - x/2) - (\sigma x)g(-x/4) = \sigma xg(L - x/4)$. Since we are assuming that energy is conserved in this setup, this loss in potential energy shows up as the gain in kinetic energy of the moving part of the chain. This part has length $L - x/2$ (because $x/2$ is the length hanging below the support), so conservation of energy gives

$$\frac{1}{2}\sigma(L - x/2)v^2 = \sigma xg(L - x/4) \quad \Longrightarrow \quad v = \sqrt{\frac{2gx(L - x/4)}{L - x/2}}. \tag{5.183}$$

This has the expected property of going to infinity for $x \to 2L$. However, the finite size of the bend becomes relevant when x approaches $2L$, so all of the energy never ends up being concentrated in an infinitely small piece. All speeds therefore remain finite, as they must.

Writing v as dx/dt in Eq. (5.183), and separating variables and integrating, gives

$$t = \frac{1}{\sqrt{g}}\int_0^{2L}\sqrt{\frac{L - x/2}{2x(L - x/4)}}\,dx \quad \Longrightarrow \quad t = \sqrt{\frac{L}{g}}\int_0^{2}\sqrt{\frac{1 - z/2}{2z(1 - z/4)}}\,dz, \tag{5.184}$$

where we have changed variables to $z \equiv x/L$. Numerically integrating this gives a total time of $t \approx (1.694)\sqrt{L/g}$. Since the freefall time in the previous problem is given by $gt^2/2 = 2L \Longrightarrow t = 2\sqrt{L/g}$, we see that the time in the present energy-conserving case is about 0.847 times the time in the freefall case.

To find the tension T at the left end of the bend, let's paint a little dot on the chain there. We'll find the acceleration of the falling part above the dot and then use $F = ma$. The acceleration of the falling part is $a = dv/dt$, where v is given in Eq. (5.183). Taking the derivative, you can show (don't forget the dx/dt from the chain rule) that a can be written in the form,

$$a = g\left(1 + \frac{(x/2)(L - x/4)}{(L - x/2)^2}\right). \tag{5.185}$$

The $F = ma$ equation for the part of the chain above the dot, with downward taken to be positive, gives $T + mg = ma \Longrightarrow T = m(a - g)$. Therefore, with $m = \sigma(L - x/2)$, we have

$$T = \sigma(L - x/2)g\left(\frac{(x/2)(L - x/4)}{(L - x/2)^2}\right) = \frac{\sigma gx(L - x/4)}{2(L - x/2)} = \frac{\sigma v^2}{4}, \tag{5.186}$$

where we have used Eq. (5.183). This has the expected properties of equaling zero when $x = 0$ and diverging when $x \to 2L$.

To find the tension at the right end of the bend, note that the total upward tension on the tiny bend equals the rate of change in momentum of the bend (the gravitational force is negligible). In a small time dt, a length $v\,dt/2$ of chain is brought to rest (the factor of 2 comes from the doubling-up effect). This length goes from moving downward at speed v to being at rest, so the change in momentum is $\sigma(v\,dt/2)v$ upward. Therefore, $dp/dt = \sigma v^2/2$. This must be the total upward force on the bend, and since we found above that the upward force at the left end is $\sigma v^2/4$, there must also be an upward force at the right end of $\sigma v^2/4$. The tensions at the two ends are therefore equal.

5.30. **Falling from a table**

(a) FIRST SOLUTION: Let σ be the mass density of the chain. From conservation of energy, we know that the chain's final kinetic energy, which is $(\sigma L)v^2/2$, equals the loss in potential energy. This loss equals $(\sigma L)(L/2)g$, because the center of mass falls a distance $L/2$. Therefore,

$$v = \sqrt{gL}. \tag{5.187}$$

This equals the speed obtained by an object that falls a distance $L/2$. Note that if the initial piece hanging down through the hole is arbitrarily short, then the chain will take an arbitrarily long time to fall down. But the final speed will still be (arbitrarily close to) $\sqrt{gL}$.

SECOND SOLUTION: Let x be the length that hangs down through the hole. The gravitational force on this length, which is $(\sigma x)g$, is responsible for changing the momentum of the entire chain, which is $(\sigma L)\dot{x}$. Therefore, $F = dp/dt$ gives $(\sigma x)g = (\sigma L)\ddot{x}$, which is simply the $F = ma$ equation. Hence, $\ddot{x} = (g/L)x$, and the general solution to this equation is[34]

$$x(t) = Ae^{t\sqrt{g/L}} + Be^{-t\sqrt{g/L}}. \tag{5.188}$$

Let T be the time for which $x(T) = L$. If ϵ is very small, then T will be very large. But for large t (more precisely, for $t \gg \sqrt{L/g}$), we may neglect the negative-exponent term in Eq. (5.188). We then have

$$x \approx Ae^{t\sqrt{g/L}} \quad \Longrightarrow \quad \dot{x} \approx \left(Ae^{t\sqrt{g/L}}\right)\sqrt{g/L} \approx x\sqrt{g/L} \quad \text{(for large } t\text{)}. \tag{5.189}$$

When $x = L$, we obtain

$$\dot{x}(T) = L\sqrt{g/L} = \sqrt{gL}, \tag{5.190}$$

in agreement with the first solution.

(b) Let σ be the mass density of the chain, and let x be the length that hangs down through the hole. The gravitational force on this length, which is $(\sigma x)g$, is responsible for changing the momentum of the chain. This momentum is $(\sigma x)\dot{x}$, because only the hanging part is moving. Therefore, $F = dp/dt$ gives

$$xg = x\ddot{x} + \dot{x}^2. \tag{5.191}$$

Note that $F = ma$ gives the wrong equation, because it neglects the fact that the amount of moving mass, σx, is changing. It therefore misses the second term on the right-hand side of Eq. (5.191). In short, the momentum of the chain increases because it is speeding up (which gives the $x\ddot{x}$ term) *and* because additional mass is continually being added to the moving part (which gives the $\dot{x}^2$ term, as you can show).

Let's now solve Eq. (5.191) for $x(t)$. Since g is the only parameter in the equation, the solution for $x(t)$ can involve only g's and t's.[35] By dimensional analysis, $x(t)$ must then be of the form $x(t) = bgt^2$, where b is a numerical constant to be determined. Plugging this expression for $x(t)$ into Eq. (5.191) and dividing by g^2t^2 gives $b = 2b^2 + 4b^2$. Therefore, $b = 1/6$, and our solution may be written as

$$x(t) = \frac{1}{2}\left(\frac{g}{3}\right)t^2. \tag{5.192}$$

This is the equation for something that accelerates downward with acceleration $g' = g/3$. The time the chain takes to fall a distance L is then given by $L = g't^2/2$, which yields $t = \sqrt{2L/g'}$. The final speed is thus

$$v = g't = \sqrt{2Lg'} = \sqrt{\frac{2gL}{3}}. \tag{5.193}$$

34 If ϵ is the initial value of x, then $A = B = \epsilon/2$ satisfies the initial conditions $x(0) = \epsilon$ and $\dot{x}(0) = 0$, in which case we can write $x(t) = \epsilon\cosh\left(t\sqrt{g/L}\right)$. But we won't need this information in what follows.

35 The other dimensionful quantities in the problem, L and σ, do not appear in Eq. (5.191), so they cannot appear in the solution. Also, the initial position and speed (which in general appear in the solution for $x(t)$, because Eq. (5.191) is a second-order differential equation) do not appear in this case, because they are equal to zero.

This is smaller than the $\sqrt{gL}$ result in part (a). We therefore see that although the total time for the scenario in part (a) is very large, the final speed in that case is in fact larger than that in the present scenario. You can show that the speed in part (a)'s scenario is smaller than the speed in part (b)'s scenario for x less than $2L/3$, but larger for x greater than $2L/3$.

REMARKS: Using Eq. (5.193), you can show that $1/3$ of the available potential energy is lost to heat. This inevitable loss occurs during the abrupt motions that suddenly bring the atoms from zero to nonzero speed when they join the moving part of the chain. The use of conservation of energy is therefore *not* a valid way to solve part (b). If you used conservation of energy, you would (as you can verify) obtain an incorrect acceleration of $g/2$. In view of the above solution based on $F = dp/dt$, this $g/2$ result cannot be correct, because there is simply not enough downward force in the setup to yield this acceleration. The only downward force comes from gravity, and we showed above that this leads to an acceleration of $g/3$.

If we want to try to get rid of the energy loss and somehow produce an acceleration of $g/2$, a plausible idea is to use a continuous piece of rope instead of the ideal kind of chain we've been using. As mentioned near the end of Section 5.8, a rope yields an energy-conserving system. However, from the reasoning on page 171, there is now a nonzero tension *everywhere* throughout the rope, even in the part inside the heap; this is the price we pay for having none of the points in the rope abruptly go from zero to nonzero speed. This tension then causes the rope (all of it) in the heap to move. The system is therefore completely different from the original one with our ideal chain, where it was understood that all the little strings connecting the ideal point masses in the heap are initially limp with zero tension; the tension in a given little string becomes nonzero only when it joins the moving part of the chain. The solution to this new energy-conserving setup therefore depends on exactly how the rope in the heap is initially situated; more information is therefore needed to solve the problem. But one thing is certain: this new setup is definitely not going to yield an acceleration of $g/2$, because the conservation-of-energy solution that leads to $g/2$ is based on the assumption that the entire loss in potential energy goes exclusively into the gain in kinetic energy of the part of the rope that has fallen below the table. The fact that the rope inside the heap is now also moving ruins this assumption. ♣

5.31. **The raindrop**

Let ρ be the mass density of the raindrop, and let λ be the average mass density in space of the water droplets. Let $r(t)$, $M(t)$, and $v(t)$ be the radius, mass, and velocity of the raindrop, respectively. We need three equations to solve for these three unknowns. The equations we will use are two different expressions for dM/dt, and the $F = dp/dt$ expression for the raindrop. The first expression for $\dot{M}$ is obtained by taking the derivative of $M = (4/3)\pi r^3\rho$, which gives

$$\dot{M} = 4\pi r^2 \dot{r}\rho \tag{5.194}$$

$$= 3M\frac{\dot{r}}{r}. \tag{5.195}$$

The second expression for $\dot{M}$ is obtained by noting that the change in M is due to the acquisition of water droplets. The raindrop sweeps out volume at a rate given by its cross-sectional area times its velocity. Therefore,

$$\dot{M} = \pi r^2 v\lambda. \tag{5.196}$$

The $F = dp/dt$ equation is found as follows. The gravitational force is Mg, and the momentum is Mv. Therefore, $F = dp/dt$ gives

$$Mg = \dot{M}v + M\dot{v}. \tag{5.197}$$

We now have three equations involving the three unknowns, r, M, and v.[36] Our goal is to find $\dot{v}$. We will do this by first finding $\ddot{r}$. Equating the expressions for $\dot{M}$ in Eqs. (5.194) and (5.196) gives

$$v = \frac{4\rho}{\lambda}\dot{r} \tag{5.198}$$

$$\Longrightarrow \quad \dot{v} = \frac{4\rho}{\lambda}\ddot{r}. \tag{5.199}$$

Plugging Eqs. (5.195), (5.198), and (5.199) into Eq. (5.197) gives

$$Mg = \left(3M\frac{\dot{r}}{r}\right)\left(\frac{4\rho}{\lambda}\dot{r}\right) + M\left(\frac{4\rho}{\lambda}\ddot{r}\right). \tag{5.200}$$

Therefore,

$$\tilde{g}r = 12\dot{r}^2 + 4r\ddot{r}, \tag{5.201}$$

where we have defined $\tilde{g} \equiv g\lambda/\rho$, for convenience. The only parameter in Eq. (5.201) is $\tilde{g}$. Therefore, $r(t)$ can depend only on $\tilde{g}$ and t.[37] Hence, by dimensional analysis, r must take the form

$$r(t) = A\tilde{g}t^2, \tag{5.202}$$

where A is a numerical constant, to be determined. Plugging this expression for r into Eq. (5.201) gives

$$\tilde{g}(A\tilde{g}t^2) = 12(2A\tilde{g}t)^2 + 4(A\tilde{g}t^2)(2A\tilde{g})$$

$$\Longrightarrow \quad A = 48A^2 + 8A^2. \tag{5.203}$$

Therefore, $A = 1/56$, and so $\ddot{r} = 2A\tilde{g} = \tilde{g}/28 = g\lambda/28\rho$. Eq. (5.199) then gives the acceleration of the raindrop as

$$\dot{v} = \frac{g}{7}, \tag{5.204}$$

independent of ρ and λ. For further discussion of the raindrop problem, see Krane (1981).

REMARK: A common invalid solution to this problem is the following, which (incorrectly) uses conservation of energy: The fact that v is proportional to $\dot{r}$ (shown in Eq. (5.198)) means that the volume swept out by the raindrop is a cone. The center of mass of a cone is $1/4$ of the way from the base to the apex (as you can show by integrating over horizontal circular slices). Therefore, if M is the mass of the raindrop after it has fallen a height h, then an (incorrect) application of conservation of energy gives

$$\frac{1}{2}Mv^2 = Mg\frac{h}{4} \quad \Longrightarrow \quad v^2 = \frac{gh}{2}. \tag{5.205}$$

Taking the derivative of this (or just using the general result, $v^2 = 2ah$) gives

$$\dot{v} = \frac{g}{4} \quad \text{(incorrect)}. \tag{5.206}$$

[36] Note that we *cannot* write down the naive conservation-of-energy equation (which would say that the decrease in the water's potential energy equals the increase in its kinetic energy), because mechanical energy is *not* conserved. The collisions between the raindrop and the droplets are completely inelastic. The raindrop will, in fact, heat up. See the remark at the end of the solution.

[37] The other dimensionful quantities in the problem, ρ and λ, do not appear in Eq. (5.201), except through $\tilde{g}$, so they cannot appear in the solution. Also, the initial values of r and $\dot{r}$ (which in general appear in the solution for $r(t)$, because Eq. (5.201) is a second-order differential equation) do not appear in this case, because they are equal to zero.

The reason why this solution is invalid is that the collisions between the raindrop and the droplets are completely inelastic. Heat is generated, and the overall kinetic energy of the raindrop is smaller than you would otherwise expect.

Let's calculate how much mechanical energy is lost (and therefore how much the raindrop heats up) as a function of the height fallen. The loss in mechanical energy is

$$E_{\text{lost}} = Mg\frac{h}{4} - \frac{1}{2}Mv^2. \tag{5.207}$$

Using $v^2 = 2(g/7)h$, this becomes

$$\Delta E_{\text{int}} = E_{\text{lost}} = \frac{3}{28}Mgh, \tag{5.208}$$

where ΔE_{int} is the gain in internal thermal energy. The energy required to heat 1 g of water by 1 °C is 1 calorie (= 4.18 joules). Therefore, the energy required to heat 1 kg of water by 1 °C is $\approx$4200 J. In other words,

$$\Delta E_{\text{int}} = 4200\, M\, \Delta T, \tag{5.209}$$

where M is measured in kilograms, and T is measured in degrees Celsius. Equations (5.208) and (5.209) give the increase in temperature as a function of h,

$$4200\, \Delta T = \frac{3}{28}gh. \tag{5.210}$$

How far must the raindrop fall before it starts to boil? If we assume that the water droplets' temperature is near freezing, then the height through which the raindrop must fall to have $\Delta T = 100\,°\text{C}$ is found from Eq. (5.210) to be

$$h \approx 400\,000\,\text{m} = 400\,\text{km}, \tag{5.211}$$

which is much larger than the height of the atmosphere. We have, of course, idealized the problem in a drastic manner. But needless to say, there is no need to worry about getting burned by the rain. A typical value for h is a few kilometers, which would raise the temperature by only about one degree. This effect is completely washed out by many other factors. ♣

Chapter 6
The Lagrangian method

In this chapter, we're going to learn about a whole new way of looking at things. Consider the system of a mass on the end of a spring. We can analyze this, of course, by using $F = ma$ to write down $m\ddot{x} = -kx$. The solutions to this equation are sinusoidal functions, as we well know. We can, however, figure things out by using another method which doesn't explicitly use $F = ma$. In many (in fact, probably most) physical situations, this new method is far superior to using $F = ma$. You will soon discover this for yourself when you tackle the problems and exercises for this chapter. We will present our new method by first stating its rules (without any justification) and showing that they somehow end up magically giving the correct answer. We will then give the method proper justification.

6.1 The Euler–Lagrange equations

Here is the procedure. Consider the following seemingly silly combination of the kinetic and potential energies (T and V, respectively),

$$L \equiv T - V. \tag{6.1}$$

This is called the *Lagrangian*. Yes, there is a minus sign in the definition (a plus sign would simply give the total energy). In the problem of a mass on the end of a spring, $T = m\dot{x}^2/2$ and $V = kx^2/2$, so we have

$$L = \frac{1}{2}m\dot{x}^2 - \frac{1}{2}kx^2. \tag{6.2}$$

Now write

$$\frac{d}{dt}\left(\frac{\partial L}{\partial \dot{x}}\right) = \frac{\partial L}{\partial x}. \tag{6.3}$$

Don't worry, we'll show you in Section 6.2 where this comes from. This equation is called the *Euler–Lagrange (E–L) equation*. For the problem at hand, we have $\partial L/\partial \dot{x} = m\dot{x}$ and $\partial L/\partial x = -kx$ (see Appendix B for the definition of a partial derivative), so Eq. (6.3) gives

$$m\ddot{x} = -kx, \tag{6.4}$$

which is exactly the result obtained by using $F = ma$. An equation such as Eq. (6.4), which is derived from the Euler–Lagrange equation, is called an *equation of motion*.[1] If the problem involves more than one coordinate, as most problems do, we just have to apply Eq. (6.3) to each coordinate. We will obtain as many equations as there are coordinates. Each equation may very well involve many of the coordinates (see the example below, where both equations involve both x and θ).

At this point, you may be thinking, "That was a nice little trick, but we just got lucky in the spring problem. The procedure won't work in a more general situation." Well, let's see. How about if we consider the more general problem of a particle moving in an arbitrary potential $V(x)$ (we'll stick to one dimension for now). The Lagrangian is then

$$L = \frac{1}{2}m\dot{x}^2 - V(x), \tag{6.5}$$

and the Euler–Lagrange equation, Eq. (6.3), gives

$$m\ddot{x} = -\frac{dV}{dx}. \tag{6.6}$$

But $-dV/dx$ is the force on the particle. So we see that Eqs. (6.1) and (6.3) together say exactly the same thing that $F = ma$ says, when using a Cartesian coordinate in one dimension (but this result is in fact quite general, as we'll see in Section 6.4). Note that shifting the potential by a given constant has no effect on the equation of motion, because Eq. (6.3) involves only the derivative of V. This is equivalent to saying that only differences in energy are relevant, and not the actual values, as we well know.

In a three-dimensional setup written in terms of Cartesian coordinates, the potential takes the form $V(x, y, z)$, so the Lagrangian is

$$L = \frac{1}{2}m(\dot{x}^2 + \dot{y}^2 + \dot{z}^2) - V(x, y, z). \tag{6.7}$$

It then immediately follows that the three Euler–Lagrange equations (obtained by applying Eq. (6.3) to x, y, and z) may be combined into the vector statement,

$$m\ddot{\mathbf{x}} = -\nabla V. \tag{6.8}$$

But $-\nabla V = \mathbf{F}$, so we again arrive at Newton's second law, $\mathbf{F} = m\mathbf{a}$, now in three dimensions.

Let's do one more example to convince you that there's really something nontrivial going on here.

[1] The term "equation of motion" is a little ambiguous. It is understood to refer to the second-order differential equation satisfied by x, and not the actual equation for x as a function of t, namely $x(t) = A\cos(\omega t + \phi)$ in this problem, which is obtained by integrating the equation of motion twice.

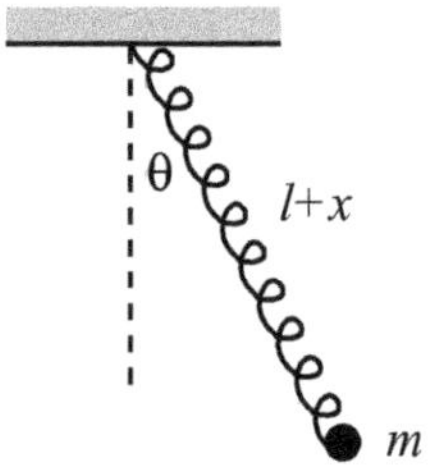

Fig. 6.1

Example (Spring pendulum): Consider a pendulum made of a spring with a mass m on the end (see Fig. 6.1). The spring is arranged to lie in a straight line (which we can arrange by, say, wrapping the spring around a rigid massless rod). The equilibrium length of the spring is ℓ. Let the spring have length $\ell + x(t)$, and let its angle with the vertical be $\theta(t)$. Assuming that the motion takes place in a vertical plane, find the equations of motion for x and θ.

Solution: The kinetic energy may be broken up into the radial and tangential parts, so we have

$$T = \frac{1}{2}m\left(\dot{x}^2 + (\ell + x)^2\dot{\theta}^2\right). \tag{6.9}$$

The potential energy comes from both gravity and the spring, so we have

$$V(x, \theta) = -mg(\ell + x)\cos\theta + \frac{1}{2}kx^2. \tag{6.10}$$

The Lagrangian is therefore

$$L \equiv T - V = \frac{1}{2}m\left(\dot{x}^2 + (\ell + x)^2\dot{\theta}^2\right) + mg(\ell + x)\cos\theta - \frac{1}{2}kx^2. \tag{6.11}$$

There are two variables here, x and θ. As mentioned above, the nice thing about the Lagrangian method is that we can just use Eq. (6.3) twice, once with x and once with θ. So the two Euler–Lagrange equations are

$$\frac{d}{dt}\left(\frac{\partial L}{\partial \dot{x}}\right) = \frac{\partial L}{\partial x} \quad \Longrightarrow \quad m\ddot{x} = m(\ell + x)\dot{\theta}^2 + mg\cos\theta - kx, \tag{6.12}$$

and

$$\begin{aligned}
\frac{d}{dt}\left(\frac{\partial L}{\partial \dot{\theta}}\right) = \frac{\partial L}{\partial \theta} \quad &\Longrightarrow \quad \frac{d}{dt}\left(m(\ell + x)^2\dot{\theta}\right) = -mg(\ell + x)\sin\theta \\
&\Longrightarrow \quad m(\ell + x)^2\ddot{\theta} + 2m(\ell + x)\dot{x}\dot{\theta} = -mg(\ell + x)\sin\theta \\
&\Longrightarrow \quad m(\ell + x)\ddot{\theta} + 2m\dot{x}\dot{\theta} = -mg\sin\theta. \qquad (6.13)
\end{aligned}$$

Equation (6.12) is simply the radial $F = ma$ equation, complete with the centripetal acceleration, $-(\ell + x)\dot{\theta}^2$. And the first line of Eq. (6.13) is the statement that the torque equals the rate of change of the angular momentum (this is one of the subjects of Chapter 8). Alternatively, if you want to work in a rotating reference frame, then Eq. (6.12) is the radial $F = ma$ equation, complete with the centrifugal force, $m(\ell + x)\dot{\theta}^2$. And the third line of Eq. (6.13) is the tangential $F = ma$ equation, complete with the Coriolis force, $-2m\dot{x}\dot{\theta}$. But never mind about this now. We'll deal with rotating frames in Chapter 10.[2]

[2] Throughout this chapter, I'll occasionally point out torques, angular momenta, centrifugal forces, and other such things when they pop up in equations of motion, even though we haven't covered

REMARK: After writing down the E–L equations, it is always best to double-check them by trying to identify them as $F = ma$ and/or $\tau = dL/dt$ equations (once we learn about that). Sometimes, however, this identification isn't obvious. And for the times when everything is clear (that is, when you look at the E–L equations and say, "Oh, of course!"), it is usually clear only *after* you've derived the equations. In general, the safest method for solving a problem is to use the Lagrangian method and then double-check things with $F = ma$ and/or $\tau = dL/dt$ if you can. ♣

At this point it seems to be personal preference, and all academic, whether you use the Lagrangian method or the $F = ma$ method. The two methods produce the same equations. However, in problems involving more than one variable, it usually turns out to be *much* easier to write down T and V, as opposed to writing down all the forces. This is because T and V are nice and simple scalars. The forces, on the other hand, are vectors, and it is easy to get confused if they point in various directions. The Lagrangian method has the advantage that once you've written down $L \equiv T - V$, you don't have to think anymore. All you have to do is blindly take some derivatives.[3]

When jumping from high in a tree,
Just write down del L by del z.
Take del L by z dot,
Then t-dot what you've got,
And equate the results (but quickly!)

But ease of computation aside, is there any fundamental difference between the two methods? Is there any deep reasoning behind Eq. (6.3)? Indeed, there is ...

6.2 The principle of stationary action

Consider the quantity,

$$S \equiv \int_{t_1}^{t_2} L(x, \dot{x}, t)dt. \tag{6.14}$$

S is called the *action*. It is a quantity with the dimensions of (Energy) × (Time). S depends on L, and L in turn depends on the function $x(t)$ via Eq. (6.1).[4] Given any

them yet. I figure it can't hurt to bring your attention to them. But rest assured, a familiarity with these topics is by no means necessary for an understanding of what we'll be doing in this chapter, so just ignore the references if you want. One of the great things about the Lagrangian method is that even if you've never heard of the terms "torque," "centrifugal," "Coriolis," or even "$F = ma$" itself, you can still get the correct equations by simply writing down the kinetic and potential energies, and then taking a few derivatives.

[3] Well, you eventually have to *solve* the resulting equations of motion, but you have to do that with the $F = ma$ method, too.

[4] In some situations, the kinetic and potential energies in $L \equiv T - V$ may explicitly depend on time, so we have included the "t" in Eq. (6.14).

function $x(t)$, we can produce the quantity S. We'll deal with only one coordinate, x, for now.

Integrals like the one in Eq. (6.14) are called *functionals*, and S is sometimes denoted by $S[x(t)]$. It depends on the entire function $x(t)$, and not on just one input number, as a regular function $f(t)$ does. S can be thought of as a function of an infinite number of values, namely all the $x(t)$ for t ranging from t_1 to t_2. If you don't like infinities, you can imagine breaking up the time interval into, say, a million pieces, and then replacing the integral by a discrete sum.

Let's now pose the following question: Consider a function $x(t)$, for $t_1 \le t \le t_2$, which has its endpoints fixed (that is, $x(t_1) = x_1$ and $x(t_2) = x_2$, where x_1 and x_2 are given), but is otherwise arbitrary. What function $x(t)$ yields a stationary value of S? A stationary value is a local minimum, maximum, or saddle point.[5]

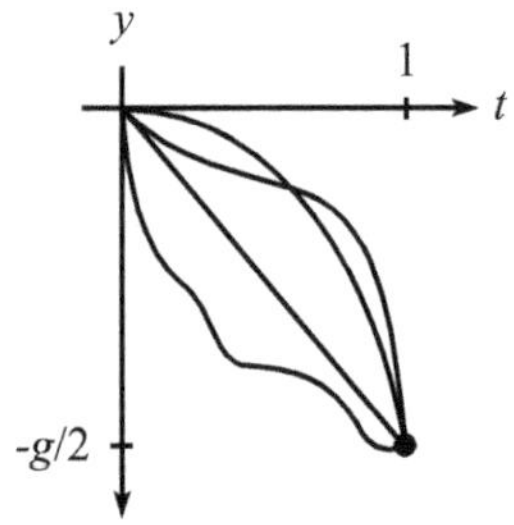

Fig. 6.2

For example, consider a ball dropped from rest, and consider the function $y(t)$ for $0 \le t \le 1$. Assume that we somehow know that $y(0) = 0$ and $y(1) = -g/2$.[6] A number of possibilities for $y(t)$ are shown in Fig. 6.2, and each of these can (in theory) be plugged into Eqs. (6.1) and (6.14) to generate S. Which one yields a stationary value of S? The following theorem gives us the answer.

Theorem 6.1 *If the function $x_0(t)$ yields a stationary value (that is, a local minimum, maximum, or saddle point) of S, then*

$$\frac{d}{dt}\left(\frac{\partial L}{\partial \dot{x}_0}\right) = \frac{\partial L}{\partial x_0}. \tag{6.15}$$

It is understood that we are considering the class of functions whose endpoints are fixed. That is, $x(t_1) = x_1$ and $x(t_2) = x_2$.

Proof: We will use the fact that if a certain function $x_0(t)$ yields a stationary value of S, then any other function very close to $x_0(t)$ (with the same endpoint values) yields essentially the same S, up to first order in any deviations. This is actually the definition of a stationary value. The analogy with regular functions is that if $f(b)$ is a stationary value of f, then $f(b+\epsilon)$ differs from $f(b)$ only at second order in the small quantity ϵ. This is true because $f'(b) = 0$, so there is no first-order term in the Taylor series expansion around b.

Assume that the function $x_0(t)$ yields a stationary value of S, and consider the function

$$x_a(t) \equiv x_0(t) + a\beta(t), \tag{6.16}$$

[5] A saddle point is a point where there are no first-order changes in S, and where some of the second-order changes are positive and some are negative (like the middle of a saddle, of course).

[6] This follows from $y = -gt^2/2$, but pretend that we don't know this formula.

where a is a number, and where $\beta(t)$ satisfies $\beta(t_1) = \beta(t_2) = 0$ (to keep the endpoints of the function fixed), but is otherwise arbitrary. When producing the action $S[x_a(t)]$ in (6.14), the t is integrated out, so S is just a number. It depends on a, in addition to t_1 and t_2. Our requirement is that there be no change in S at first order in a. How does S depend on a? Using the chain rule, we have

$$\begin{aligned}\frac{\partial}{\partial a}S[x_a(t)] &= \frac{\partial}{\partial a}\int_{t_1}^{t_2} L\,dt = \int_{t_1}^{t_2}\frac{\partial L}{\partial a}\,dt \\ &= \int_{t_1}^{t_2}\left(\frac{\partial L}{\partial x_a}\frac{\partial x_a}{\partial a} + \frac{\partial L}{\partial \dot{x}_a}\frac{\partial \dot{x}_a}{\partial a}\right)dt. \qquad (6.17)\end{aligned}$$

In other words, a influences S through its effect on x, and also through its effect on $\dot{x}$. From Eq. (6.16), we have

$$\frac{\partial x_a}{\partial a} = \beta, \quad \text{and} \quad \frac{\partial \dot{x}_a}{\partial a} = \dot{\beta}, \qquad (6.18)$$

so Eq. (6.17) becomes[7]

$$\frac{\partial}{\partial a}S[x_a(t)] = \int_{t_1}^{t_2}\left(\frac{\partial L}{\partial x_a}\beta + \frac{\partial L}{\partial \dot{x}_a}\dot{\beta}\right)dt. \qquad (6.19)$$

Now comes the one sneaky part of the proof. We will integrate the second term by parts (you will see this trick many times in your physics career). Using

$$\int \frac{\partial L}{\partial \dot{x}_a}\dot{\beta}\,dt = \frac{\partial L}{\partial \dot{x}_a}\beta - \int\left(\frac{d}{dt}\frac{\partial L}{\partial \dot{x}_a}\right)\beta\,dt, \qquad (6.20)$$

Eq. (6.19) becomes

$$\frac{\partial}{\partial a}S[x_a(t)] = \int_{t_1}^{t_2}\left(\frac{\partial L}{\partial x_a} - \frac{d}{dt}\frac{\partial L}{\partial \dot{x}_a}\right)\beta\,dt + \left.\frac{\partial L}{\partial \dot{x}_a}\beta\right|_{t_1}^{t_2}. \qquad (6.21)$$

But $\beta(t_1) = \beta(t_2) = 0$, so the last term (the "boundary term") vanishes. We now use the fact that $(\partial/\partial a)S[x_a(t)]$ must be zero for *any* function $\beta(t)$, because we are assuming that $x_0(t)$ yields a stationary value. The only way this can be true is if the quantity in parentheses above (evaluated at $a = 0$) is identically equal to zero, that is,

$$\frac{d}{dt}\left(\frac{\partial L}{\partial \dot{x}_0}\right) = \frac{\partial L}{\partial x_0}. \qquad \blacksquare \quad (6.22)$$

[7] Note that nowhere do we assume that x_a and $\dot{x}_a$ are independent variables. The partial derivatives in Eq. (6.18) are very much related, in that one is the derivative of the other. The use of the chain rule in Eq. (6.17) is still perfectly valid.

The E–L equation, Eq. (6.3), therefore doesn't just come out of the blue. It is a consequence of requiring that the action be at a stationary value. We may therefore replace $F = ma$ by the following principle.

- **The principle of stationary action:**
  The path of a particle is the one that yields a stationary value of the action.

This principle (also known as Hamilton's principle) is equivalent to $F = ma$ because the above theorem shows that if (and only if, as you can show by working backwards) we have a stationary value of S, then the E–L equations hold. And the E–L equations are equivalent to $F = ma$ (as we showed for Cartesian coordinates in Section 6.1, and as we'll prove for any coordinate system in Section 6.4). Therefore, "stationary action" is equivalent to $F = ma$.

If we have a multidimensional setup where the Lagrangian is a function of the variables $x_1(t), x_2(t), \ldots$, then the above principle of stationary action is still all we need. With more than one variable, we can now vary the path by varying each coordinate (or combinations thereof). The variation of each coordinate produces an E–L equation which, as we saw in the Cartesian case, is equivalent to an $F = ma$ equation.

Given a classical mechanics problem, we can solve it with $F = ma$, or we can solve it with the E–L equations, which are a consequence of the principle of stationary action (often called the principle of "least action" or "minimal action," but see the fourth remark below). Either method will get the job done. But as mentioned at the end of Section 6.1, it is often easier to use the latter, because it avoids the use of force which can get confusing if you have forces pointing in all sorts of complicated directions.

> It just stood there and did nothing, of course,
> A harmless and still wooden horse.
> But the minimal action
> Was just a distraction;
> The plan involved no use of force.

Let's now return to the example of a ball dropped from rest, mentioned above. The Lagrangian is $L = T - V = m\dot{y}^2/2 - mgy$, so Eq. (6.22) gives $\ddot{y} = -g$, which is simply the $F = ma$ equation (divided through by m), as expected. The solution is $y(t) = -gt^2/2 + v_0 t + y_0$, as we well know. But the initial conditions tell us that $v_0 = y_0 = 0$, so our solution is $y(t) = -gt^2/2$. You are encouraged to verify explicitly that this $y(t)$ yields an action that is stationary with respect to variations of the form, say, $y(t) = -gt^2/2 + \epsilon t(t-1)$, which also satisfies the endpoint conditions (this is the task of Exercise 6.30). There is, of course, an infinite number of other ways to vary $y(t)$, but this specific result should help convince you of the validity of Theorem 6.1.

Note that the stationarity implied by the Euler–Lagrange equation, Eq. (6.22), is a *local* statement. It gives information only about nearby paths. It says nothing about the *global* nature of how the action depends on all possible paths. If we find that a solution to Eq. (6.22) happens to produce a local minimum (as opposed to a maximum or a saddle), there is no reason to conclude that it is a global minimum, although in many cases it turns out to be (see Exercise 6.32, for the case of a thrown ball).

REMARKS:

1. Theorem 6.1 is based on the assumption that the ending time, t_2, of the motion is given. But how do we know this final time? Well, we don't. In the example of a ball thrown upward, the total time to rise and fall back to your hand can be anything, depending on the ball's initial speed. This initial speed will show up as an integration constant when solving the E–L equations. The motion must end sometime, and the principle of stationary action says that for whatever time this happens to be, the physical path has a stationary action.
2. Theorem 6.1 shows that we can explain the E–L equations by the principle of stationary action. This, however, simply shifts the burden of proof. We are now left with the task of justifying why we should want the action to have a stationary value. The good news is that there is a very solid reason for this. The bad news is that the reason involves quantum mechanics, so we won't be able to discuss it properly here. Suffice it to say that a particle actually takes all possible paths in going from one place to another, and each path is associated with the complex number $e^{iS/\hbar}$ (where $\hbar = 1.05 \cdot 10^{-34}$ J s is *Planck's constant*). These complex numbers have absolute value 1 and are called "phases." It turns out that the phases from all possible paths must be added up to give the "amplitude" of going from one point to another. The absolute value of the amplitude must then be squared to obtain the probability.[8]

   The basic point, then, is that at a nonstationary value of S, the phases from different paths differ (greatly, because $\hbar$ is so small compared with the typical size of the action for a macroscopic particle) from one another, which effectively leads to the addition of many random vectors in the complex plane. These end up canceling each other, yielding a sum of essentially zero. There is therefore no contribution to the overall amplitude from non-stationary values of S. Hence, we do not observe the paths associated with these S's. At a stationary value of S, however, all the phases take on essentially the same value, thereby adding constructively instead of destructively. There is therefore a nonzero probability for the particle to take a path that yields a stationary value of S. So this is the path we observe.
3. But again, the preceding remark simply shifts the burden of proof one step further. We must now justify why these phases $e^{iS/\hbar}$ should exist, and why the Lagrangian that appears in S should equal $T - V$. But here's where we're going to stop.
4. The principle of stationary action is sometimes referred to as the principle of "least" action, but this is misleading. True, it is often the case that the stationary value turns out to be a minimum value, but it need not be, as we can see in the following example. Consider a harmonic oscillator which has a Lagrangian equal to

$$L = \frac{1}{2}m\dot{x}^2 - \frac{1}{2}kx^2. \tag{6.23}$$

   Let $x_0(t)$ be a function that yields a stationary value of the action. Then we know that $x_0(t)$ satisfies the E–L equation, $m\ddot{x}_0 = -kx_0$. Consider a slight variation on this path,

[8] This is one of those remarks that is completely useless, because it is incomprehensible to those who haven't seen the topic before, and trivial to those who have. My apologies. But this and the following remarks are definitely not necessary for an understanding of the material in this chapter. If you're interested in reading more about these quantum mechanical issues, you should take a look at Richard Feynman's book (Feynman, 2006). Feynman was, after all, the one who thought of this idea.

$x_0(t) + \xi(t)$, where $\xi(t)$ satisfies $\xi(t_1) = \xi(t_2) = 0$. With this new function, the action becomes

$$S_\xi = \int_{t_1}^{t_2} \left(\frac{m}{2}\left(\dot{x}_0^2 + 2\dot{x}_0\dot{\xi} + \dot{\xi}^2\right) - \frac{k}{2}\left(x_0^2 + 2x_0\xi + \xi^2\right)\right) dt. \tag{6.24}$$

The two cross-terms add up to zero, because after integrating the $\dot{x}_0\dot{\xi}$ term by parts, their sum is

$$m\dot{x}_0\xi\Big|_{t_1}^{t_2} - \int_{t_1}^{t_2} (m\ddot{x}_0 + kx_0)\xi \, dt. \tag{6.25}$$

The first term is zero, due to the boundary conditions on $\xi(t)$. The second term is zero, due to the E–L equation. We've basically just reproduced the proof of Theorem 6.1 for the special case of the harmonic oscillator here.

The terms in Eq. (6.24) involving only x_0 give the stationary value of the action (call it S_0). To determine whether S_0 is a minimum, maximum, or saddle point, we must look at the difference,

$$\Delta S \equiv S_\xi - S_0 = \frac{1}{2}\int_{t_1}^{t_2} (m\dot{\xi}^2 - k\xi^2)\, dt. \tag{6.26}$$

It is always possible to find a function ξ that makes ΔS positive. Simply choose ξ to be small, but make it wiggle very fast, so that $\dot{\xi}$ is large. Therefore, it is *never* the case that S_0 is a maximum. Note that this reasoning works for any potential, not just a harmonic oscillator, as long as it is a function of position only (that is, it contains no derivatives, as we always assume).

You might be tempted to use the same line of reasoning to say that it is also always possible to find a function ξ that makes ΔS negative, by making ξ large and $\dot{\xi}$ small. If this were true, then we could put everything together and conclude that all stationary points are saddle points, for a harmonic oscillator. However, it is *not* always possible to make ξ large enough and $\dot{\xi}$ small enough so that ΔS is negative, due to the boundary conditions $\xi(t_1) = \xi(t_2) = 0$. If ξ changes from zero to a large value and then back to zero, then $\dot{\xi}$ may also have to be large, if the time interval is short enough. Problem 6.6 deals quantitatively with this issue. For now, let's just recognize that in some cases S_0 is a minimum, in some cases it is a saddle point, and it is never a maximum. "Least action" is therefore a misnomer.

5. It is sometimes said that nature has a "purpose," in that it seeks to take the path that produces the minimum action. In view of the second remark above, this is incorrect. In fact, nature does exactly the opposite. It takes *every* path, treating them all on equal footing. We end up seeing only the path with a stationary action, due to the way the quantum mechanical phases add. It would be a harsh requirement, indeed, to demand that nature make a "global" decision (that is, compare paths that are separated by large distances) and choose the one with the smallest action. Instead, we see that everything takes place on a "local" scale. Nearby phases simply add, and everything works out automatically.

   When an archer shoots an arrow through the air, the aim is made possible by all the other arrows taking all the other nearby paths, each with essentially the same action. Likewise, when you walk down the street with a certain destination in mind, you're not alone . . .

   When walking, I know that my aim
   Is caused by the ghosts with my name.
   And although I can't see
   Where they walk next to me,
   I know they're all there, just the same.

6. Consider a function, $f(x)$, of one variable (for ease of terminology). Let $f(b)$ be a local minimum of f. There are two basic properties of this minimum. The first is that $f(b)$ is smaller than all nearby values. The second is that the slope of f is zero at b. From the above remarks, we see that (as far as the action S is concerned) the first property is completely irrelevant, and the second one is the whole point. In other words, saddle points (and maxima, although we showed above that these never exist for S) are just as good as minima, as far as the constructive addition of the $e^{iS/\hbar}$ phases is concerned.
7. Given that classical mechanics is an approximate theory, while quantum mechanics is the (more) correct one, it is quite silly to justify the principle of stationary action by demonstrating its equivalence to $F = ma$, as we did above. We should be doing it the other way around. However, because our intuition is based on $F = ma$, it's easier to start with $F = ma$ as the given fact, rather than calling upon the latent quantum-mechanical intuition hidden deep within all of us. Maybe someday …

   At any rate, in more advanced theories dealing with fundamental issues concerning the tiny building blocks of matter (where actions are of the same order of magnitude as $\hbar$), the approximate $F = ma$ theory is invalid, and you *have* to use the Lagrangian method.
8. When dealing with a system in which a nonconservative force such as friction is present, the Lagrangian method loses much of its appeal. The reason for this is that nonconservative forces don't have a potential energy associated with them, so there isn't a specific $V(x)$ that you can write down in the Lagrangian. Although friction forces can in fact be incorporated in the Lagrangian method, you have to include them in the E–L equations essentially by hand. We won't deal with nonconservative forces in this chapter. ♣

6.3 Forces of constraint

A nice thing about the Lagrangian method is that we are free to impose any given constraints at the beginning of the problem, thereby immediately reducing the number of variables. This is always done (perhaps without thinking) whenever a particle is constrained to move on a wire or surface, etc. Often we are concerned not with the exact nature of the forces doing the constraining, but only with the resulting motion, given that the constraints hold. By imposing the constraints at the outset, we can find the motion, but we can't say anything about the constraining forces.

If we want to determine the constraining forces, we must take a different approach. The main idea of the strategy, as we will show below, is that we must not impose the constraints too soon. This leaves us with a larger number of variables to deal with, so the calculations are more cumbersome. But the benefit is that we are able to find the constraining forces.

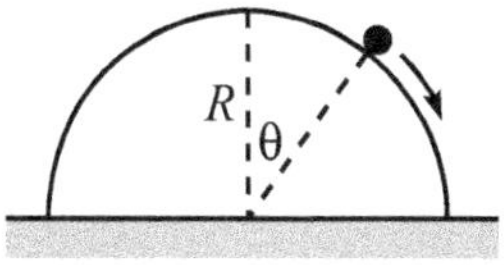

Fig. 6.3

Consider the setup of a particle sliding off a fixed frictionless hemisphere of radius R (see Fig. 6.3). Let's say that we are concerned only with finding the equation of motion for θ, and not the constraining force. Then we can write everything in terms of θ, because we know that the radial distance r is constrained to be R. The kinetic energy is $mR^2\dot{\theta}^2/2$, and the potential energy (relative to the bottom of the hemisphere) is $mgR\cos\theta$, so the Lagrangian is

$$L = \frac{1}{2}mR^2\dot{\theta}^2 - mgR\cos\theta, \tag{6.27}$$

and the equation of motion, via Eq. (6.3), is

$$\ddot{\theta} = (g/R)\sin\theta, \tag{6.28}$$

which is equivalent to the tangential $F = ma$ statement.

Now let's say that we want to find the constraining normal force that the hemisphere applies to the particle. To do this, let's solve the problem in a different way and write things in terms of both r and θ. Also (and here's the critical step), let's be really picky and say that r isn't *exactly* constrained to be R, because in the real world the particle actually sinks into the hemisphere a little bit. This may seem a bit silly, but it's really the whole point. The particle pushes and sinks inward a tiny distance until the hemisphere gets squashed enough to push back with the appropriate force to keep the particle from sinking in any more (just consider the hemisphere to be made of lots of little springs with very large spring constants). The particle is therefore subject to a (very steep) potential arising from the hemisphere's force. The constraining potential, $V(r)$, looks something like the plot in Fig. 6.4. The true Lagrangian for the system is thus

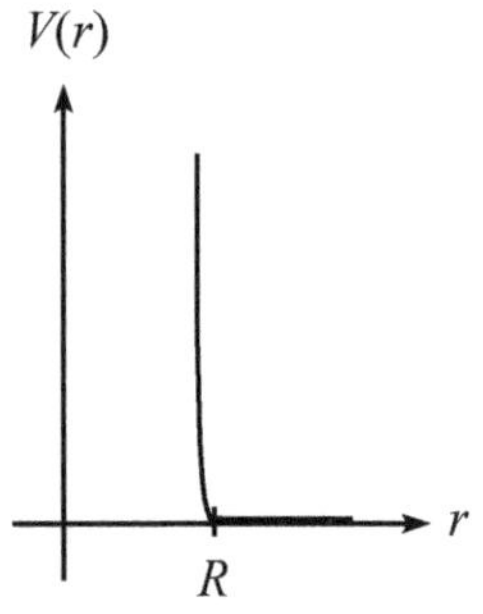

Fig. 6.4

$$L = \frac{1}{2}m(\dot{r}^2 + r^2\dot{\theta}^2) - mgr\cos\theta - V(r). \tag{6.29}$$

(The $\dot{r}^2$ term in the kinetic energy will turn out to be insignificant.) The equations of motion obtained from varying θ and r are therefore

$$\begin{aligned} mr^2\ddot{\theta} + 2mr\dot{r}\dot{\theta} &= mgr\sin\theta, \\ m\ddot{r} &= mr\dot{\theta}^2 - mg\cos\theta - V'(r). \end{aligned} \tag{6.30}$$

Having written down the equations of motion, we will *now* apply the constraint condition that $r = R$. This condition implies $\dot{r} = \ddot{r} = 0$. (Of course, r isn't *really* equal to R, but any differences are inconsequential from this point onward.) The first of Eqs. (6.30) then reproduces Eq. (6.28), while the second yields

$$-\left.\frac{dV}{dr}\right|_{r=R} = mg\cos\theta - mR\dot{\theta}^2. \tag{6.31}$$

But $F_N \equiv -dV/dr$ is the constraint force applied in the r direction, which is precisely the force we are looking for. The normal force of constraint is therefore

$$F_N(\theta, \dot{\theta}) = mg\cos\theta - mR\dot{\theta}^2. \tag{6.32}$$

This is equivalent to the radial $F = ma$ equation, $mg\cos\theta - F_N = mR\dot{\theta}^2$ (which is certainly a quicker way to find the normal force in the present problem). Note that this result is valid only if $F_N(\theta, \dot{\theta}) > 0$. If the normal force becomes zero, then this means that the particle has left the sphere, in which case r no longer equals R.

REMARKS:

1. What if we instead had (unwisely) chosen Cartesian coordinates, x and y, instead of polar coordinates, r and θ? Since the distance from the particle to the surface of the hemisphere is $\eta \equiv \sqrt{x^2+y^2} - R$, we obtain a true Lagrangian equal to

$$L = \frac{1}{2}m(\dot{x}^2+\dot{y}^2) - mgy - V(\eta). \tag{6.33}$$

The equations of motion are (using the chain rule)

$$m\ddot{x} = -\frac{dV}{d\eta}\frac{\partial \eta}{\partial x}, \quad \text{and} \quad m\ddot{y} = -mg - \frac{dV}{d\eta}\frac{\partial \eta}{\partial y}. \tag{6.34}$$

We now apply the constraint condition $\eta = 0$. Since $-dV/d\eta$ equals the constraint force F, you can show that the equations we end up with (namely, the two E–L equations and the constraint equation) are

$$m\ddot{x} = F\frac{x}{R}, \quad m\ddot{y} = -mg + F\frac{y}{R}, \quad \text{and} \quad \sqrt{x^2+y^2} - R = 0. \tag{6.35}$$

These three equations are sufficient to determine the three unknowns $\ddot{x}$, $\ddot{y}$, and F as functions of the quantities x, $\dot{x}$, y, and $\dot{y}$. See Exercise 6.37, which should convince you that polar coordinates are the way to go. In general, the strategy for solving for F is to take two time derivatives of the constraint equation and then eliminate the second derivatives of the coordinates by using the E–L equations (this process was trivial in the polar-coordinate case).
2. You can see from Eq. (6.35) that the E–L equations end up taking the form,

$$\frac{d}{dt}\left(\frac{\partial L}{\partial \dot{q}_i}\right) = \frac{\partial L}{\partial q_i} + F\frac{\partial \eta}{\partial q_i}, \tag{6.36}$$

for each coordinate q_i. The quantity η is what appears in the constraint equation, $\eta = 0$. In our hemisphere problem, we had $\eta = r - R$ in polar coordinates, and $\eta = \sqrt{x^2+y^2} - R$ in Cartesian coordinates. The E–L equations, combined with the $\eta = 0$ condition, give us exactly the number of equations ($N+1$ of them, where N is the number of coordinates) needed to determine all of the $N+1$ unknowns (all the $\ddot{q}_i$, and F) in terms of the q_i and $\dot{q}_i$.

   Writing down the equations in Eq. (6.36) is basically the method of Lagrange multipliers, where the Lagrange multiplier turns out to be the force. But if you're not familiar with this method, no need to worry; you can derive everything from scratch using the above technique involving the steep potential. If you do happen to be familiar with it, then there might in fact be a need to worry about how you apply it, as the following remark explains.
3. When trying to determine the forces of constraint, you can just start with Eq. (6.36), without bothering to write down $V(\eta)$. But you must be careful to make sure that η does indeed represent the distance the particle is from where it should be. In polar coordinates, if someone gives you the constraint condition for the hemisphere as $7(r-R) = 0$, and if you use the left-hand side of this as the η in Eq. (6.36), then you will get the wrong constraint force; it will be too small by a factor of 7. Likewise, in Cartesian coordinates, writing the constraint as $y - \sqrt{R^2-x^2} = 0$ will give you the wrong force. The best way to avoid this problem is, of course, to pick one of your variables as the distance the particle is from where it should be (up to an additive constant, as in the case of $r - R = 0$). ♣

6.4 Change of coordinates

When L is written in terms of Cartesian coordinates x, y, z, we showed in Section 6.1 that the Euler–Lagrange equations are equivalent to Newton's $\mathbf{F} = m\mathbf{a}$

equations; see Eq. (6.8). But what about the case where we use polar, spherical, or some other coordinates? The equivalence of the E–L equations and $\mathbf{F} = m\mathbf{a}$ isn't so obvious. As far as trusting the E–L equations for such coordinates goes, you can achieve peace of mind in two ways. You can accept the principle of stationary action as something so beautiful and profound that it simply has to work for any choice of coordinates. Or, you can take the more mundane road and show through a change of coordinates that if the E–L equations hold for one set of coordinates (and we know that they *do* hold for at least one set, namely Cartesian coordinates), then they also hold for any other coordinates (of a certain form, described below). In this section, we will demonstrate the validity of the E–L equations through the explicit change of coordinates.[9]

Consider the set of coordinates,

$$x_i : \quad (x_1, x_2, \ldots, x_N). \tag{6.37}$$

For example, if $N = 6$, then x_1, x_2, x_3 could be the Cartesian x, y, z coordinates of one particle, and x_4, x_5, x_6 could be the r, θ, ϕ polar coordinates of a second particle, and so on. Assume that the E–L equations hold for these variables, that is,

$$\frac{d}{dt}\left(\frac{\partial L}{\partial \dot{x}_i}\right) = \frac{\partial L}{\partial x_i} \quad (1 \le i \le N). \tag{6.38}$$

Consider a new set of variables that are functions of the x_i and t,

$$q_i = q_i(x_1, x_2, \ldots, x_N; t). \tag{6.39}$$

We will restrict ourselves to the case where the q_i do not depend on the $\dot{x}_i$. (This is quite reasonable. If the coordinates depended on the velocities, then we wouldn't be able to label points in space with definite coordinates. We'd have to worry about how the particles were behaving when they were at the points. These would be strange coordinates indeed.) We can, in theory, invert Eq. (6.39) and express the x_i as functions of the q_i and t,

$$x_i = x_i(q_1, q_2, \ldots, q_N; t). \tag{6.40}$$

Claim 6.2 *If Eq. (6.38) is true for the x_i coordinates, and if the x_i and q_i are related by Eq. (6.40), then Eq. (6.38) is also true for the q_i coordinates. That is,*

$$\frac{d}{dt}\left(\frac{\partial L}{\partial \dot{q}_m}\right) = \frac{\partial L}{\partial q_m} \quad (1 \le m \le N). \tag{6.41}$$

[9] This calculation is straightforward but a bit messy, so you may want to skip this section and just settle for the "beautiful and profound" reasoning.

Proof: We have

$$\frac{\partial L}{\partial \dot{q}_m} = \sum_{i=1}^{N} \frac{\partial L}{\partial \dot{x}_i} \frac{\partial \dot{x}_i}{\partial \dot{q}_m}. \tag{6.42}$$

(Note that if the x_i depended on the $\dot{q}_i$, then we would have to include the additional term, $\sum(\partial L/\partial x_i)(\partial x_i/\partial \dot{q}_m)$. But we have excluded such dependence.) Let's rewrite the $\partial \dot{x}_i/\partial \dot{q}_m$ term. From Eq. (6.40), we have

$$\dot{x}_i = \sum_{m=1}^{N} \frac{\partial x_i}{\partial q_m} \dot{q}_m + \frac{\partial x_i}{\partial t}. \tag{6.43}$$

Therefore,

$$\frac{\partial \dot{x}_i}{\partial \dot{q}_m} = \frac{\partial x_i}{\partial q_m}. \tag{6.44}$$

Substituting this into Eq. (6.42) and taking the time derivative of both sides gives

$$\frac{d}{dt}\left(\frac{\partial L}{\partial \dot{q}_m}\right) = \sum_{i=1}^{N} \frac{d}{dt}\left(\frac{\partial L}{\partial \dot{x}_i}\right) \frac{\partial x_i}{\partial q_m} + \sum_{i=1}^{N} \frac{\partial L}{\partial \dot{x}_i} \frac{d}{dt}\left(\frac{\partial x_i}{\partial q_m}\right). \tag{6.45}$$

In the second term here, it is legal to switch the order of the total derivative, d/dt, and the partial derivative, $\partial/\partial q_m$.

REMARK: In case you have your doubts, let's prove that this switching is legal.

$$\begin{aligned} \frac{d}{dt}\left(\frac{\partial x_i}{\partial q_m}\right) &= \sum_{k=1}^{N} \frac{\partial}{\partial q_k}\left(\frac{\partial x_i}{\partial q_m}\right) \dot{q}_k + \frac{\partial}{\partial t}\left(\frac{\partial x_i}{\partial q_m}\right) \\ &= \frac{\partial}{\partial q_m}\left(\sum_{k=1}^{N} \frac{\partial x_i}{\partial q_k} \dot{q}_k + \frac{\partial x_i}{\partial t}\right) \\ &= \frac{\partial \dot{x}_i}{\partial q_m}. \quad ♣ \end{aligned} \tag{6.46}$$

In the first term on the right-hand side of Eq. (6.45), we can use the given information in Eq. (6.38) and rewrite the $(d/dt)(\partial L/\partial \dot{x}_i)$ term. We then obtain

$$\begin{aligned} \frac{d}{dt}\left(\frac{\partial L}{\partial \dot{q}_m}\right) &= \sum_{i=1}^{N} \frac{\partial L}{\partial x_i} \frac{\partial x_i}{\partial q_m} + \sum_{i=1}^{N} \frac{\partial L}{\partial \dot{x}_i} \frac{\partial \dot{x}_i}{\partial q_m} \\ &= \frac{\partial L}{\partial q_m}, \end{aligned} \tag{6.47}$$

as we wanted to show. ■

We have therefore demonstrated that if the Euler–Lagrange equations are true for one set of coordinates, x_i (and they *are* true for Cartesian coordinates), then they are also true for any other set of coordinates, q_i, satisfying Eq. (6.39). If you're inclined to look at the principle of stationary action with distrust, thinking that it might be a coordinate-dependent statement, this proof should put you at ease. The Euler–Lagrange equations are valid in any coordinates.

Note that the above proof did not in any way use the precise form of the Lagrangian. If L were equal to $T + V$, or $8T + \pi V^2/T$, or any other arbitrary function, our result would still be true: If Eq. (6.38) is true for one set of coordinates, then it is also true for any other set of coordinates q_i satisfying Eq. (6.39). The point is that the only L for which the hypothesis is true at all (that is, for which Eq. (6.38) holds) is $L \equiv T - V$ (or any constant multiple of this).

REMARK: On one hand, it is quite amazing how little we assumed in proving the above claim. *Any* new coordinates of the very general form in Eq. (6.39) satisfy the E–L equations, as long as the original coordinates do. If the E–L equations had, say, a factor of 5 on the right-hand side of Eq. (6.38), then they would *not* hold in arbitrary coordinates. To see this, just follow the proof through with the factor of 5.

On the other hand, the claim is quite believable, if you make an analogy with a function instead of a functional. Consider the function $f(z) = z^2$. This has a minimum at $z = 0$, consistent with the fact that $df/dz = 0$ at $z = 0$. But let's now write f in terms of the variable y defined by, say, $z = y^4$. Then $f(y) = y^8$, and f has a minimum at $y = 0$, consistent with the fact that $df/dy = 0$ at $y = 0$. So $f' = 0$ holds in both coordinates at the corresponding points $y = z = 0$. This is the (simplified) analog of the E–L equations holding in both coordinates. In both cases, the derivative equation describes where the stationary value occurs.

This change-of-variables result may be stated in a more geometrical (and friendly) way. If you plot a function and then stretch the horizontal axis in an arbitrary manner (which is what happens when you change coordinates), then a stationary value (that is, one where the slope is zero) will still be a stationary value after the stretching.[10] A picture (or even just the thought of one) is worth a dozen equations, apparently.

As an example of an equation that does *not* hold for all coordinates, consider the preceding example, but with $f' = 1$ instead of $f' = 0$. In terms of z, $f' = 1$ when $z = 1/2$. And in terms of y, $f' = 1$ when $y = (1/8)^{1/7}$. But the points $z = 1/2$ and $y = (1/8)^{1/7}$ are not the same point. In other words, $f' = 1$ is not a coordinate-independent statement. Most equations are coordinate dependent. The special thing about $f' = 0$ is that a stationary point is a stationary point no matter how you look at it. ♣

6.5 Conservation laws

6.5.1 Cyclic coordinates

Consider the case where the Lagrangian does not depend on a certain coordinate q_k. Then

$$\frac{d}{dt}\left(\frac{\partial L}{\partial \dot{q}_k}\right) = \frac{\partial L}{\partial q_k} = 0 \quad \Longrightarrow \quad \frac{\partial L}{\partial \dot{q}_k} = C, \tag{6.48}$$

[10] There is, however, one exception. A stationary point in one coordinate system might be located at a kink in another coordinate system, so that f' is not defined there. For example, if we had instead defined y by $z = y^{1/4}$, then $f(y) = y^{1/2}$, which has an undefined slope at $y = 0$. Basically, we've stretched (or shrunk) the horizontal axis by a factor of infinity at the origin, and this is a process that can change a zero slope into an undefined one. But let's not worry about this.

where C is a constant, that is, independent of time. In this case, we say that q_k is a *cyclic* coordinate, and that $\partial L/\partial \dot{q}_k$ is a *conserved* quantity (meaning that it doesn't change with time). If Cartesian coordinates are used, then $\partial L/\partial \dot{x}_k$ is simply the momentum, $m\dot{x}_k$, because $\dot{x}_k$ appears only in the the kinetic energy's $m\dot{x}_k^2/2$ term (we exclude cases where the potential depends on $\dot{x}_k$). We therefore call $\partial L/\partial \dot{q}_k$ the *generalized momentum* corresponding to the coordinate q_k. And in cases where $\partial L/\partial \dot{q}_k$ does not change with time, we call it a *conserved momentum*. Note that a generalized momentum need not have the units of linear momentum, as the angular-momentum examples below show.

Example (Linear momentum): Consider a ball thrown through the air. In the full three dimensions, the Lagrangian is

$$L = \frac{1}{2}m(\dot{x}^2 + \dot{y}^2 + \dot{z}^2) - mgz. \tag{6.49}$$

There is no x or y dependence here, so both $\partial L/\partial \dot{x} = m\dot{x}$ and $\partial L/\partial \dot{y} = m\dot{y}$ are constant, as we well know. The fancy way of saying this is that conservation of $p_x \equiv m\dot{x}$ arises from spatial translation invariance in the x direction. The fact that the Lagrangian doesn't depend on x means that it doesn't matter if you throw the ball in one spot, or in another spot a mile down the road. The setup is independent of the x value. This independence leads to conservation of p_x. Likewise for p_y.

Example (Angular and linear momentum in cylindrical coordinates): Consider a potential that depends only on the distance to the z axis. In cylindrical coordinates, the Lagrangian is

$$L = \frac{1}{2}m(\dot{r}^2 + r^2\dot{\theta}^2 + \dot{z}^2) - V(r). \tag{6.50}$$

There is no z dependence here, so $\partial L/\partial \dot{z} = m\dot{z}$ is constant. Also, there is no θ dependence, so $\partial L/\partial \dot{\theta} = mr^2\dot{\theta}$ is constant. Since $r\dot{\theta}$ is the velocity in the tangential direction around the z axis, we see that our conserved quantity, $mr(r\dot{\theta})$, is the angular momentum (discussed in Chapters 7–9) around the z axis. In the same manner as in the preceding example, conservation of angular momentum around the z axis arises from rotation invariance around the z axis.

Example (Angular momentum in spherical coordinates): In spherical coordinates, consider a potential that depends only on r and θ. Our convention for spherical coordinates is that θ is the angle down from the north pole, and ϕ is the angle around the equator. The Lagrangian is

$$L = \frac{1}{2}m(\dot{r}^2 + r^2\dot{\theta}^2 + r^2\sin^2\theta\,\dot{\phi}^2) - V(r,\theta). \tag{6.51}$$

There is no ϕ dependence here, so $\partial L/\partial\dot{\phi} = mr^2 \sin^2\theta\,\dot{\phi}$ is constant. Since $r\sin\theta$ is the distance from the z axis, and since $r\sin\theta\,\dot{\phi}$ is the speed in the tangential direction around the z axis, we see that our conserved quantity, $m(r\sin\theta)(r\sin\theta\,\dot{\phi})$, is the angular momentum around the z axis.

6.5.2 Energy conservation

We will now derive another conservation law, namely conservation of energy. The conservation of momentum or angular momentum above arose when the Lagrangian was independent of x, y, z, θ, or ϕ. Conservation of energy arises when the Lagrangian is independent of time. This conservation law is different from those in the above momenta examples, because t is not a coordinate that the stationary-action principle can be applied to. You can imagine varying the coordinates x, θ, etc., which are functions of t. But it makes no sense to vary t. Therefore, we're going to have to prove this conservation law in a different way. Consider the quantity

$$E \equiv \left(\sum_{i=1}^{N} \frac{\partial L}{\partial \dot{q}_i}\dot{q}_i\right) - L. \tag{6.52}$$

E turns out (usually) to be the energy. We'll show this below. The motivation for this expression for E comes from the theory of Legendre transforms, but we won't get into that here. We'll just accept the definition in Eq. (6.52) and prove a very useful fact about it.

Claim 6.3 *If L has no explicit time dependence (that is, if $\partial L/\partial t = 0$), then E is conserved (that is, $dE/dt = 0$), assuming that the motion obeys the E–L equations (which it does).*

Note that there is one partial derivative and one total derivative in this statement.

Proof: L is a function of the q_i, the $\dot{q}_i$, and possibly t. Making copious use of the chain rule, we have

$$\begin{aligned}\frac{dE}{dt} &= \frac{d}{dt}\left(\sum_{i=1}^{N} \frac{\partial L}{\partial \dot{q}_i}\dot{q}_i\right) - \frac{dL}{dt}\\ &= \sum_{i=1}^{N}\left(\left(\frac{d}{dt}\frac{\partial L}{\partial \dot{q}_i}\right)\dot{q}_i + \frac{\partial L}{\partial \dot{q}_i}\ddot{q}_i\right) - \left(\sum_{i=1}^{N}\left(\frac{\partial L}{\partial q_i}\dot{q}_i + \frac{\partial L}{\partial \dot{q}_i}\ddot{q}_i\right) + \frac{\partial L}{\partial t}\right).\end{aligned} \tag{6.53}$$

There are five terms here. The second cancels with the fourth. And the first (after using the E–L equation, Eq. (6.3), to rewrite it) cancels with the third. We therefore arrive at the simple result,

$$\frac{dE}{dt} = -\frac{\partial L}{\partial t}. \tag{6.54}$$

In the event that $\partial L/\partial t = 0$ (that is, there are no t's sitting on the paper when you write down L), which is usually the case in the situations we consider (because we generally won't deal with potentials that depend on time), we have $dE/dt = 0$. ■

Not too many things are constant with respect to time, and the quantity E has units of energy, so it's a good bet that it's the energy. Let's show this in Cartesian coordinates (however, see the remark below). The Lagrangian is

$$L = \frac{1}{2}m(\dot{x}^2 + \dot{y}^2 + \dot{z}^2) - V(x, y, z), \tag{6.55}$$

so Eq. (6.52) gives

$$E = \frac{1}{2}m(\dot{x}^2 + \dot{y}^2 + \dot{z}^2) + V(x, y, z), \tag{6.56}$$

which is the total energy. The effect of the operations in Eq. (6.52) in most cases is just to switch the sign in front of the potential.

Of course, taking the kinetic energy T and subtracting the potential energy V to obtain L, and then using Eq. (6.52) to produce $E = T + V$, seems like a rather convoluted way of arriving at $T + V$. But the point of all this is that we used the E–L equations to *prove* that E is conserved. Although we know very well from the $F = ma$ methods in Chapter 5 that the sum $T + V$ is conserved, it's not fair to assume that it is conserved in our new Lagrangian formalism. We have to show that this *follows* from the E–L equations.

As with the translation and rotation invariance we observed in the examples in Section 6.5.1, we see that energy conservation arises from time translation invariance. If the Lagrangian has no explicit t dependence, then the setup looks the same today as it did yesterday. This fact leads to conservation of energy.

REMARK: The quantity E in Eq. (6.52) gives the energy of the system only if the entire system is represented by the Lagrangian. That is, the Lagrangian must represent a closed system with no external forces. If the system is not closed, then Claim 6.3 (or more generally, Eq. (6.54)) is still perfectly valid for the E defined in Eq. (6.52), but this E may simply not be the energy of the system. Problem 6.8 is a good example of such a situation.

Another example is the following. Imagine a long rod in the horizontal x-y plane. The rod points in the x direction, and a bead is free to slide frictionlessly along it. At $t = 0$, an external machine is arranged to accelerate the rod in the negative y direction (that is, transverse to itself) with acceleration $-g$. So $\dot{y} = -gt$. There is no internal potential energy in this system, so the

Lagrangian is just the kinetic energy, $L = m\dot{x}^2/2 + m(gt)^2/2$. Equation (6.52) therefore gives $E = m\dot{x}^2/2 - m(gt)^2/2$, which isn't the energy. But Eq. (6.54) is still true, because

$$\frac{dE}{dt} = -\frac{\partial L}{\partial t} \quad \Longleftrightarrow \quad m\dot{x}\ddot{x} - mg^2 t = -mg^2 t \quad \Longleftrightarrow \quad \ddot{x} = 0, \tag{6.57}$$

which is correct. However, this setup is exactly the same as projectile motion in the x-y plane, where y is now the vertical axis, provided that we eliminate the rod and consider gravity instead of the machine to be causing the acceleration in the y direction. But if we are thinking in terms of gravity, then the normal thing to do is to say that the particle moves under the influence of the potential $V(y) = mgy$. The Lagrangian for this closed system (bead plus earth) is $L = m(\dot{x}^2 + \dot{y}^2)/2 - mgy$, and so Eq. (6.52) gives $E = m(\dot{x}^2 + \dot{y}^2)/2 + mgy$, which is indeed the energy of the particle. But having said all this, most of the systems we'll deal with are closed, so you can usually ignore this remark and assume that the E in Eq. (6.52) gives the energy. ♣

6.6 Noether's theorem

We now present one of the most beautiful and useful theorems in physics. It deals with two fundamental concepts, namely *symmetries* and *conserved quantities*. The theorem (due to Emmy Noether) may be stated as follows.

Theorem 6.4 (Noether's theorem) *For each symmetry of the Lagrangian, there is a conserved quantity.*

By "symmetry," we mean that if the coördinates are changed by some small quantities, then the Lagrangian has no first-order change in these quantities. By "conserved quantity," we mean a quantity that does not change with time. The result in Section 6.5.1 for cyclic coordinates is a special case of this theorem.

Proof: Let the Lagrangian be invariant, to first order in the small number ϵ, under the change of coordinates,

$$q_i \longrightarrow q_i + \epsilon K_i(q). \tag{6.58}$$

Each $K_i(q)$ may be a function of all the q_i, which we collectively denote by the shorthand, q.

REMARK: As an example of what these K_i's might look like, consider the Lagrangian, $L = (m/2)(5\dot{x}^2 - 2\dot{x}\dot{y} + 2\dot{y}^2) + C(2x - y)$. We've just pulled this out of a hat, although it happens to be the type of L that arises in Atwood's machine problems; see Problem 6.9 and Exercise 6.40. This L is invariant under the transformation $x \to x + \epsilon$ and $y \to y + 2\epsilon$, because the derivative terms are unaffected, and the difference $2x - y$ is unchanged. (It's actually invariant to all orders in ϵ, and not just first order. But this isn't necessary for the theorem to hold.) Therefore, $K_x = 1$ and $K_y = 2$, which happen to be independent of the coordinates. In the problems we'll be doing, the K_i's can generally be determined by simply looking at the potential term.

Of course, someone else might come along with $K_x = 3$ and $K_y = 6$, which is also a symmetry. And indeed, any factor can be taken out of ϵ and put into the K_i's without changing the quantity $\epsilon K_i(q)$ in Eq. (6.58). Any such modification will just bring an overall constant factor (and hence not change the property of being conserved) into the conserved quantity in Eq. (6.61) below. It is therefore irrelevant. ♣

The fact that the Lagrangian does not change at first order in ϵ means that

$$0 = \frac{dL}{d\epsilon} = \sum_i \left(\frac{\partial L}{\partial q_i} \frac{\partial q_i}{\partial \epsilon} + \frac{\partial L}{\partial \dot{q}_i} \frac{\partial \dot{q}_i}{\partial \epsilon} \right)$$
$$= \sum_i \left(\frac{\partial L}{\partial q_i} K_i + \frac{\partial L}{\partial \dot{q}_i} \dot{K}_i \right). \tag{6.59}$$

Using the E–L equation, Eq. (6.3), we can rewrite this as

$$0 = \sum_i \left(\frac{d}{dt} \left(\frac{\partial L}{\partial \dot{q}_i} \right) K_i + \frac{\partial L}{\partial \dot{q}_i} \dot{K}_i \right)$$
$$= \frac{d}{dt} \left(\sum_i \frac{\partial L}{\partial \dot{q}_i} K_i \right). \tag{6.60}$$

Therefore, the quantity

$$P(q, \dot{q}) \equiv \sum_i \frac{\partial L}{\partial \dot{q}_i} K_i(q) \tag{6.61}$$

does not change with time. It is given the generic name of *conserved momentum*. But it need not have the units of linear momentum. ■

As Noether most keenly observed
(And for which much acclaim is deserved),
It's easy to see
That for each symmetry,
A quantity must be conserved.

Example 1: Consider the Lagrangian in the above remark, $L = (m/2)(5\dot{x}^2 - 2\dot{x}\dot{y} + 2\dot{y}^2) + C(2x - y)$. We saw that $K_x = 1$ and $K_y = 2$. The conserved momentum is therefore

$$P(x, y, \dot{x}, \dot{y}) = \frac{\partial L}{\partial \dot{x}} K_x + \frac{\partial L}{\partial \dot{y}} K_y = m(5\dot{x} - \dot{y})(1) + m(-\dot{x} + 2\dot{y})(2)$$
$$= m(3\dot{x} + 3\dot{y}). \tag{6.62}$$

The overall factor of $3m$ isn't important.

Example 2: Consider a thrown ball. We have $L = (m/2)(\dot{x}^2 + \dot{y}^2 + \dot{z}^2) - mgz$. This is invariant under translations in x, that is, $x \to x + \epsilon$; and also under translations in y, that is, $y \to y + \epsilon$. (Both x and y are cyclic coordinates.) We need invariance only to first order in ϵ for Noether's theorem to hold, but this L is invariant to all orders.

We therefore have two symmetries in our Lagrangian. The first has $K_x = 1$, $K_y = 0$, and $K_z = 0$. The second has $K_x = 0$, $K_y = 1$, and $K_z = 0$. Of course, the nonzero K_i's here can be chosen to be any constants, but we may as well pick them to be 1. The two conserved momenta are

$$\begin{aligned} P_1(x, y, z, \dot{x}, \dot{y}, \dot{z}) &= \frac{\partial L}{\partial \dot{x}} K_x + \frac{\partial L}{\partial \dot{y}} K_y + \frac{\partial L}{\partial \dot{z}} K_z = m\dot{x}, \\ P_2(x, y, z, \dot{x}, \dot{y}, \dot{z}) &= \frac{\partial L}{\partial \dot{x}} K_x + \frac{\partial L}{\partial \dot{y}} K_y + \frac{\partial L}{\partial \dot{z}} K_z = m\dot{y}. \end{aligned} \tag{6.63}$$

These are simply the x and y components of the linear momentum, as we saw in the first example in Section 6.5.1. Note that any combination of these momenta, say $3P_1 + 8P_2$, is also conserved. (In other words, $x \to x + 3\epsilon$, $y \to y + 8\epsilon$, $z \to z$ is a symmetry of the Lagrangian.) But the above P_1 and P_2 are the simplest momenta to choose as a "basis" for the infinite number of conserved momenta (which is how many you have, if there are two or more independent continuous symmetries).

Example 3: Consider a mass on a spring, with relaxed length zero, in the x-y plane. The Lagrangian, $L = (m/2)(\dot{x}^2 + \dot{y}^2) - (k/2)(x^2 + y^2)$, is invariant under the change of coordinates, $x \to x + \epsilon y$ and $y \to y - \epsilon x$, to first order in ϵ (as you can check). So we have $K_x = y$ and $K_y = -x$. The conserved momentum is therefore

$$P(x, y, \dot{x}, \dot{y}) = \frac{\partial L}{\partial \dot{x}} K_x + \frac{\partial L}{\partial \dot{y}} K_y = m(\dot{x}y - \dot{y}x). \tag{6.64}$$

This is the (negative of the) z component of the angular momentum. The angular momentum is conserved here because the potential, $V(x, y) \propto x^2 + y^2 = r^2$, depends only on the distance from the origin. We'll discuss such potentials in Chapter 7.

In contrast with the first two examples above, the $x \to x + \epsilon y$, $y \to y - \epsilon x$ transformation isn't so obvious here. How did we get this? Well, unfortunately there doesn't seem to be any fail-safe method of determining the K_i's in general, so sometimes you just have to guess around. But in many problems, the K_i's are simple constants which are easy to see.

REMARKS:

1. As we saw above, in some cases the K_i's are functions of the coordinates, and in some cases they are not.
2. The cyclic-coordinate result in Eq. (6.48) is a special case of Noether's theorem, for the following reason. If L doesn't depend on a certain coordinate q_k, then $q_k \longrightarrow q_k + \epsilon$ is certainly a symmetry. Hence $K_k = 1$ (with all the other K_i's equal to zero), and Eq. (6.60) reduces to Eq. (6.48).
3. We use the word "symmetry" to describe the situation where the transformation in Eq. (6.58) produces no first-order change in the Lagrangian. This is an appropriate choice of word, because the Lagrangian describes the system, and if the system essentially doesn't change when the coordinates are changed, then we say that the system is symmetric. For example, if we have a setup that doesn't depend on θ, then we say that the setup is symmetric under rotations. Rotate the system however you want, and it looks

the same. The two most common applications of Noether's theorem are the conservation of angular momentum, which arises from symmetry under rotations; and conservation of linear momentum, which arises from symmetry under translations.

4. In simple systems, as in Example 2 above, it is clear why the resulting P is conserved. But in more complicated systems, as in Example 1 above, the resulting P might not have an obvious interpretation. But at least you know that it is conserved, and this will invariably help in understanding a setup.
5. Although conserved quantities are extremely useful in studying a physical situation, it should be stressed that there is no more information contained in them than there is in the E–L equations. Conserved quantities are simply the result of integrating the E–L equations. For example, if you write down the E–L equations for Example 1 above, and then add the "x" equation (which is $5m\ddot{x} - m\ddot{y} = 2C$) to twice the "$y$" equation (which is $-m\ddot{x} + 2m\ddot{y} = -C$), then you arrive at $3m(\ddot{x} + \ddot{y}) = 0$. In other words, $3m(\dot{x} + \dot{y})$ is constant, as we found from Noether's theorem.

   Of course, you might have to do some guesswork to find the proper combination of the E–L equations that gives a zero on the right-hand side. But you'd have to do some guesswork anyway, to find the symmetry for Noether's theorem. At any rate, a conserved quantity is useful because it is an integrated form of the E–L equations. It puts you one step closer to solving the problem, compared with where you would be if you started with the second-order E–L equations.
6. Does every system have a conserved momentum? Certainly not. The one-dimensional problem of a falling ball ($m\ddot{z} = -mg$) doesn't have one. And if you write down an arbitrary potential in 3-D, odds are that there won't be one. In a sense, things have to contrive nicely for there to be a conserved momentum. In some problems, you can just look at the physical system and see what the symmetry is, but in others (for example, in the Atwood's-machine problems for this chapter), the symmetry is not at all obvious.
7. By "conserved quantity," we mean a quantity that depends on (at most) the coordinates and their first derivatives (that is, not on their second derivatives). If we don't make this restriction, then it is trivial to construct quantities that are independent of time. For example, in Example 1 above, the "x" E–L equation (which is $5m\ddot{x} - m\ddot{y} = 2C$) tells us that $5m\ddot{x} - m\ddot{y}$ has its time derivative equal to zero. Note that an equivalent way of excluding these trivial cases is to say that the value of a conserved quantity depends on the initial conditions (that is, the velocities and positions). The quantity $5m\ddot{x} - m\ddot{y}$ doesn't satisfy this criterion, because its value is always constrained to be $2C$. ♣

6.7 Small oscillations

In many physical systems, a particle undergoes small oscillations around an equilibrium point. In Section 5.2, we showed that the frequency of these small oscillations is

$$\omega = \sqrt{\frac{V''(x_0)}{m}}, \tag{6.65}$$

where $V(x)$ is the potential energy, and x_0 is the equilibrium point. However, this result holds only for *one-dimensional* motion (we'll see below why this is true). In more complicated systems, such as the one described below, it is necessary to use another procedure to obtain the frequency ω. This procedure is a fail-safe one, applicable in all situations. It is, however, a bit more involved than simply writing down Eq. (6.65). So in one-dimensional problems, Eq. (6.65) is still what you want to use. We'll demonstrate our fail-safe method through the following problem.

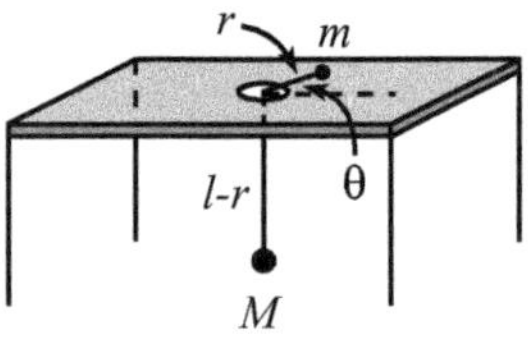

Fig. 6.5

Problem: A mass m is free to slide on a frictionless table and is connected, via a string that passes through a hole in the table, to a mass M that hangs below (see Fig. 6.5). Assume that M moves in a vertical line only, and assume that the string always remains taut.

(a) Find the equations of motion for the variables r and θ shown in the figure.
(b) Under what condition does m undergo circular motion?
(c) What is the frequency of small oscillations (in the variable r) about this circular motion?

Solution:

(a) Let the string have length ℓ (this length won't matter). Then the Lagrangian (we'll call it "$\mathcal{L}$" here, to save "L" for the angular momentum, which arises below) is

$$\mathcal{L} = \frac{1}{2}M\dot{r}^2 + \frac{1}{2}m(\dot{r}^2 + r^2\dot{\theta}^2) + Mg(\ell - r). \tag{6.66}$$

For the purposes of the potential energy, we've taken the table to be at height zero, but any other value could be chosen. The E–L equations of motion obtained from varying θ and r are

$$\begin{aligned} &\frac{d}{dt}(mr^2\dot{\theta}) = 0, \\ &(M+m)\ddot{r} = mr\dot{\theta}^2 - Mg. \end{aligned} \tag{6.67}$$

The first equation says that angular momentum is conserved. The second equation says that the Mg gravitational force accounts for the acceleration of the two masses along the direction of the string, plus the centripetal acceleration of m.

(b) The first of Eqs. (6.67) says that $mr^2\dot{\theta} = L$, where L is some constant (the angular momentum) which depends on the initial conditions. Plugging $\dot{\theta} = L/mr^2$ into the second of Eqs. (6.67) gives

$$(M+m)\ddot{r} = \frac{L^2}{mr^3} - Mg. \tag{6.68}$$

Circular motion occurs when $\dot{r} = \ddot{r} = 0$. Therefore, the radius of the circular orbit is given by

$$r_0^3 = \frac{L^2}{Mmg}. \tag{6.69}$$

Since $L = mr^2\dot{\theta}$, Eq. (6.69) is equivalent to

$$mr_0\dot{\theta}^2 = Mg, \tag{6.70}$$

which can also be obtained by letting $\ddot{r} = 0$ in the second of Eqs. (6.67). In other words, the gravitational force on M exactly accounts for the centripetal acceleration of m if the motion is circular. Given r_0, Eq. (6.70) determines what $\dot{\theta}$ must be in order to have circular motion, and vice versa.

(c) To find the frequency of small oscillations about the circular motion, we need to look at what happens to r if we perturb it slightly from its equilibrium value of r_0. Our fail-safe procedure is the following.

Let $r(t) \equiv r_0 + \delta(t)$, where $\delta(t)$ is very small (more precisely, $\delta(t) \ll r_0$), and expand Eq. (6.68) to first order in $\delta(t)$. Using

$$\frac{1}{r^3} \equiv \frac{1}{(r_0+\delta)^3} \approx \frac{1}{r_0^3 + 3r_0^2\delta} = \frac{1}{r_0^3(1+3\delta/r_0)} \approx \frac{1}{r_0^3}\left(1 - \frac{3\delta}{r_0}\right), \tag{6.71}$$

we obtain

$$(M+m)\ddot{\delta} \approx \frac{L^2}{mr_0^3}\left(1 - \frac{3\delta}{r_0}\right) - Mg. \tag{6.72}$$

The terms not involving δ on the right-hand side cancel, by the definition of r_0 given in Eq. (6.69). This cancellation always occurs in such a problem at this stage, due to the definition of the equilibrium point. We are therefore left with

$$\ddot{\delta} + \left(\frac{3L^2}{(M+m)mr_0^4}\right)\delta \approx 0. \tag{6.73}$$

This is a good old simple-harmonic-oscillator equation in the variable δ. Therefore, the frequency of small oscillations about a circle of radius r_0 is

$$\omega \approx \sqrt{\frac{3L^2}{(M+m)mr_0^4}} = \sqrt{\frac{3M}{M+m}}\sqrt{\frac{g}{r_0}}, \tag{6.74}$$

where we have used Eq. (6.69) to eliminate L in the second expression.

To sum up, the above frequency is the frequency of small oscillations in the variable r. In other words, if you have nearly circular motion, and if you plot r as a function of time (and ignore what θ is doing), then you will get a nice sinusoidal graph whose frequency is given by Eq. (6.74). Note that this frequency need not have anything to do with the other relevant frequency in this problem, namely the frequency of the circular motion, which is $\sqrt{M/m}\sqrt{g/r_0}$, from Eq. (6.70).

REMARKS: Let's look at some limits. For a given r_0, if $m \gg M$, then $\omega \approx \sqrt{3Mg/mr_0} \approx 0$. This makes sense, because everything is moving very slowly. This frequency equals $\sqrt{3}$ times the frequency of circular motion, namely $\sqrt{Mg/mr_0}$, which isn't at all obvious. For a given r_0, if $m \ll M$, then $\omega \approx \sqrt{3g/r_0}$, which isn't so obvious, either.

The frequency of small oscillations is equal to the frequency of circular motion if $M = 2m$, which, once again, isn't obvious. This condition is independent of r_0. ♣

The above procedure for finding the frequency of small oscillations can be summed up in three steps: (1) find the equations of motion; (2) find the equilibrium point; and (3) let $x(t) \equiv x_0 + \delta(t)$, where x_0 is the equilibrium point of the relevant variable, and expand one of the equations of motion (or a combination of them) to first order in δ, to obtain a simple-harmonic-oscillator equation for δ. If the equilibrium point happens to be at $x = 0$ (which is often the case), then everything is greatly simplified. There is no need to introduce δ, and the expansion in the third step above simply entails ignoring powers of x that are higher than first order.

REMARK: If you just use the potential energy for the above problem (which is Mgr, up to an additive constant) in Eq. (6.65), then you will obtain a frequency of zero, which is incorrect. You *can* use Eq. (6.65) to find the frequency, if you instead use the "effective potential" for this problem, namely $L^2/(2mr^2) + Mgr$, and if you use the total mass, $M + m$, as the mass in Eq. (6.65), as you can check. The reason why this works will become clear in Chapter 7 when we introduce the effective potential. In many problems, however, it isn't obvious what the appropriate modified potential is that should be used, or what mass goes in Eq. (6.65). So it's generally much safer to take a deep breath and go through an expansion similar to the one in part (c) of the example above. ♣

The one-dimensional result in Eq. (6.65) is, of course, a special case of our above expansion procedure. We can repeat the derivation of Section 5.2 in the present language. In one dimension, the E–L equation of motion is $m\ddot{x} = -V'(x)$. Let x_0 be the equilibrium point, so $V'(x_0) = 0$. And let $x(t) \equiv x_0 + \delta(t)$. Expanding $m\ddot{x} = -V'(x)$ to first order in δ, we have $m\ddot{\delta} = -V'(x_0) - V''(x_0)\delta$, plus higher-order terms. Since $V'(x_0) = 0$, we have $m\ddot{\delta} \approx -V''(x_0)\delta$, as desired.

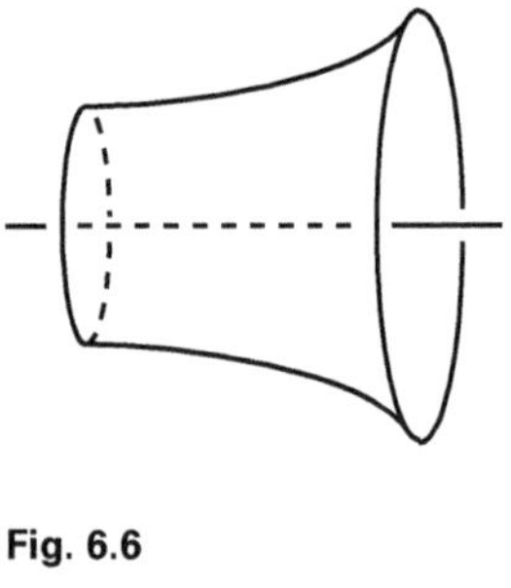

Fig. 6.6

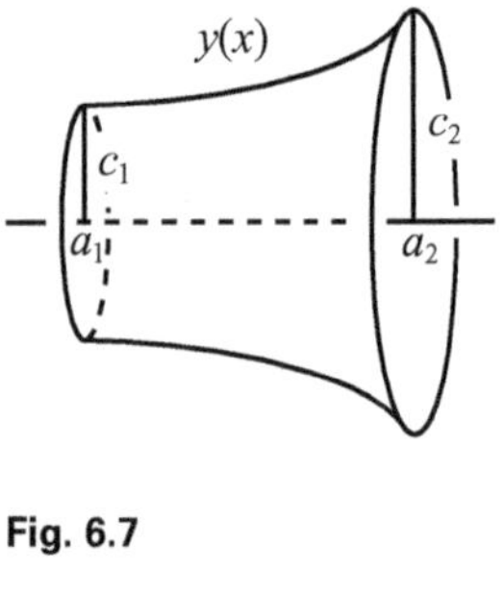

Fig. 6.7

6.8 Other applications

The formalism developed in Section 6.2 works for *any* function $L(x, \dot{x}, t)$. If our goal is to find the stationary points of $S \equiv \int L$, then Eq. (6.15) holds, no matter what L is. There is no need for L to be equal to $T - V$, or indeed, to have anything to do with physics. And t need not have anything to do with time. All that is required is that the quantity x depend on the parameter t, and that L depend only on x, $\dot{x}$, and t (and not, for example, on $\ddot{x}$; see Exercise 6.34). The formalism is very general and powerful, as the following example demonstrates.

Example (Minimal surface of revolution): A surface of revolution has two given parallel rings as its boundary; see Fig. 6.6. What should the shape of the surface be so that it has the minimum possible area? We will present three solutions. A fourth is left for Problem 6.22.

First solution: Let the surface be generated by rotating the curve $y = y(x)$ around the x axis. The boundary conditions are $y(a_1) = c_1$ and $y(a_2) = c_2$; see Fig. 6.7.

Slicing the surface up into vertical rings, we see that the area is given by

$$A = \int_{a_1}^{a_2} 2\pi y\sqrt{1 + y'^2}\, dx. \tag{6.75}$$

Our goal is to find the function $y(x)$ that minimizes this integral. We therefore have exactly the same situation as in Section 6.2, except that x is now the parameter (instead of t), and y is now the function (instead of x). Our "Lagrangian" is thus $L \propto y\sqrt{1 + y'^2}$. To minimize the integral A, we "simply" have to write down the E–L equation,

$$\frac{d}{dx}\left(\frac{\partial L}{\partial y'}\right) = \frac{\partial L}{\partial y}, \tag{6.76}$$

and calculate the derivatives. This calculation however, gets a bit tedious, so I've relegated it to Lemma 6.5 at the end of this section. For now we'll just use the result in Eq. (6.86) which gives (with $f(y) = y$ here)

$$1 + y'^2 = By^2. \tag{6.77}$$

At this point we can cleverly guess (motivated by the fact that $1 + \sinh^2 z = \cosh^2 z$) that the solution is

$$y(x) = \frac{1}{b}\cosh b(x + d), \tag{6.78}$$

where $b = \sqrt{B}$, and d is a constant of integration. Or we can separate variables to obtain (again with $b = \sqrt{B}$)

$$dx = \frac{dy}{\sqrt{(by)^2 - 1}}, \tag{6.79}$$

and then use the fact that the integral of $1/\sqrt{z^2 - 1}$ is $\cosh^{-1} z$, to obtain the same result. The answer to our problem, therefore, is that $y(x)$ takes the form of Eq. (6.78), with b and d determined by the boundary conditions,

$$c_1 = \frac{1}{b}\cosh b(a_1 + d), \quad \text{and} \quad c_2 = \frac{1}{b}\cosh b(a_2 + d). \tag{6.80}$$

In the symmetrical case where $c_1 = c_2$, we know that the minimum occurs in the middle, so we may choose $d = 0$ and $a_1 = -a_2$.

Solutions for b and d exist only for certain ranges of the a's and c's. Basically, if $a_2 - a_1$ is too large, then there is no solution. In this case, the minimal "surface" turns out to be the two given circles, attached by a line (which isn't a nice two-dimensional surface). If you perform an experiment with soap bubbles (which want to minimize their area), and if you pull the rings too far apart, then the surface will break and disappear as it tries to form the two circles. Problem 6.23 deals with this issue.

Second solution: Consider the curve that we rotate around the x axis to be described now by the function $x(y)$. That is, let x be a function of y. The area is then given by

$$A = \int_{c_1}^{c_2} 2\pi y\sqrt{1 + x'^2}\, dy, \tag{6.81}$$

where $x' \equiv dx/dy$. Note that the function $x(y)$ may be double-valued, so it may not really be a function. But it looks like a function locally, and all of our formalism deals with local variations.

Our "Lagrangian" is now $L \propto y\sqrt{1+x'^2}$, and the E–L equation is

$$\frac{d}{dy}\left(\frac{\partial L}{\partial x'}\right) = \frac{\partial L}{\partial x} \implies \frac{d}{dy}\left(\frac{yx'}{\sqrt{1+x'^2}}\right) = 0. \tag{6.82}$$

The nice thing about this solution is the "0" on the right-hand side, which arises from the fact that L doesn't depend on x (that is, x is a cyclic coordinate). Therefore, $yx'/\sqrt{1+x'^2}$ is constant. If we define this constant to be $1/b$, then we can solve for x' and then separate variables to obtain

$$dx = \frac{dy}{\sqrt{(by)^2 - 1}}, \tag{6.83}$$

in agreement with Eq. (6.79). The solution proceeds as above.

Third solution: The "Lagrangian" in the first solution above, $L \propto y\sqrt{1+y'^2}$, is independent of x. Therefore, in analogy with conservation of energy (which arises from a Lagrangian that is independent of t), the quantity

$$E \equiv y'\frac{\partial L}{\partial y'} - L = \frac{y'^2 y}{\sqrt{1+y'^2}} - y\sqrt{1+y'^2} = \frac{-y}{\sqrt{1+y'^2}} \tag{6.84}$$

is constant (that is, independent of x). This statement is equivalent to Eq. (6.77), and the solution proceeds as above. As demonstrated by the brevity of the second and third solutions here, it is highly advantageous to make use of conserved quantities whenever you can.

Let us now prove the following lemma, which we invoked in the first solution above. This lemma is very useful, because it is common to encounter problems where the quantity to be extremized depends on the arclength, $\sqrt{1+y'^2}$, and takes the form of $\int f(y)\sqrt{1+y'^2}\,dx$. We will give two proofs. The first proof uses the Euler–Lagrange equation. The calculation gets a bit messy, so it's a good idea to work through it once and for all and then just invoke the result whenever needed. This derivation isn't something you'd want to repeat too often. The second proof makes use of a conserved quantity. And in contrast with the first proof, this method is exceedingly clean and simple. It actually *is* something you'd want to repeat quite often. But we'll still do it once and for all.

Lemma 6.5 *Let $f(y)$ be a given function of y. Then the function $y(x)$ that extremizes the integral,*

$$\int_{x_1}^{x_2} f(y)\sqrt{1+y'^2}\,dx, \tag{6.85}$$

satisfies the differential equation,

$$1 + y'^2 = Bf(y)^2, \tag{6.86}$$

where B is a constant of integration.[11]

First proof: Our goal is to find the function $y(x)$ that extremizes the integral in Eq. (6.85). We therefore have exactly the same situation as in Section 6.2, except with x in place of t, and y in place of x. Our "Lagrangian" is thus $L = f(y)\sqrt{1+y'^2}$, and the Euler–Lagrange equation is

$$\frac{d}{dx}\left(\frac{\partial L}{\partial y'}\right) = \frac{\partial L}{\partial y} \implies \frac{d}{dx}\left(f \cdot y' \cdot \frac{1}{\sqrt{1+y'^2}}\right) = f'\sqrt{1+y'^2}, \tag{6.87}$$

where $f' \equiv df/dy$. We must now perform some straightforward (albeit tedious) differentiations. Using the product rule on the three factors on the left-hand side, and making copious use of the chain rule, we obtain

$$\frac{f'y'^2}{\sqrt{1+y'^2}} + \frac{fy''}{\sqrt{1+y'^2}} - \frac{fy'^2y''}{(1+y'^2)^{3/2}} = f'\sqrt{1+y'^2}. \tag{6.88}$$

Multiplying through by $(1 + y'^2)^{3/2}$ and simplifying gives

$$fy'' = f'(1 + y'^2). \tag{6.89}$$

We have completed the first step of the proof, namely producing the Euler–Lagrange differential equation. We must now integrate it. Equation (6.89) happens to be integrable for arbitrary functions $f(y)$. If we multiply through by y' and rearrange, we obtain

$$\frac{y'y''}{1+y'^2} = \frac{f'y'}{f}. \tag{6.90}$$

Taking the dx integral of both sides gives $(1/2)\ln(1 + y'^2) = \ln(f) + C$, where C is a constant of integration. Exponentiation then gives (with $B \equiv e^{2C}$)

$$1 + y'^2 = Bf(y)^2, \tag{6.91}$$

as we wanted to show. In an actual problem, we would solve this equation for y', and then separate variables and integrate. But we would need to be given a specific function $f(y)$ to be able to do this.

[11] The constant B and also one other constant of integration (arising when Eq. (6.86) is integrated to obtain y) are determined by the boundary conditions on $y(x)$; see, for example, Eq. (6.80). This situation, where the two constants are determined by the values of the function at two points, is slightly different from the situation in the physics problems we've done where the two constants are determined by the value (that is, the initial position) and the slope (that is, the speed) at just one point. But either way, two given facts must be used.

Second proof: Our "Lagrangian," $L = f(y)\sqrt{1+y'^2}$, is independent of x. Therefore, in analogy with the conserved energy given in Eq. (6.52), the quantity

$$E \equiv y'\frac{\partial L}{\partial y'} - L = \frac{-f(y)}{\sqrt{1+y'^2}} \tag{6.92}$$

is independent of x. Call it $1/\sqrt{B}$. Then we have easily reproduced Eq. (6.91). For practice, you can also prove this lemma by considering x to be a function of y, as we did in the second solution in the minimal-surface example above. ■

6.9 Problems

Section 6.1: The Euler–Lagrange equations

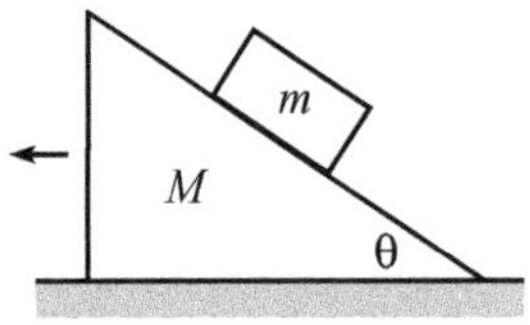

Fig. 6.8

6.1. **Moving plane** **

A block of mass m is held motionless on a frictionless plane of mass M and angle of inclination θ (see Fig. 6.8). The plane rests on a frictionless horizontal surface. The block is released. What is the horizontal acceleration of the plane? (This problem already showed up as Problem 3.8. If you haven't already done so, try solving it using $F = ma$. You will then have a greater appreciation for the Lagrangian method.)

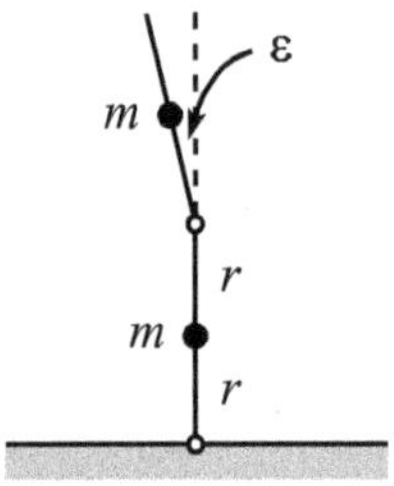

Fig. 6.9

6.2. **Two falling sticks** **

Two massless sticks of length $2r$, each with a mass m fixed at its middle, are hinged at an end. One stands on top of the other, as shown in Fig. 6.9. The bottom end of the lower stick is hinged at the ground. They are held such that the lower stick is vertical, and the upper one is tilted at a small angle ϵ with respect to the vertical. They are then released. At this instant, what are the angular accelerations of the two sticks? Work in the approximation where ϵ is very small.

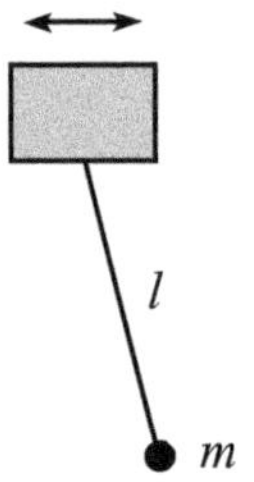

Fig. 6.10

6.3. **Pendulum with an oscillating support** **

A pendulum consists of a mass m and a massless stick of length ℓ. The pendulum support oscillates horizontally with a position given by $x(t) = A\cos(\omega t)$; see Fig. 6.10. What is the general solution for the angle of the pendulum as a function of time?

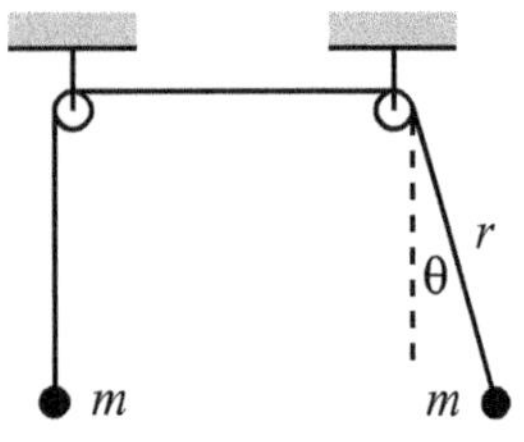

Fig. 6.11

6.4. **Two masses, one swinging** ***

Two equal masses m, connected by a massless string, hang over two pulleys (of negligible size), as shown in Fig. 6.11. The left one moves in a vertical line, but the right one is free to swing back and forth in the plane of the masses and pulleys. Find the equations of motion for r and θ, as shown.

Assume that the left mass starts at rest, and the right mass undergoes small oscillations with angular amplitude ϵ (with $\epsilon \ll 1$). What is the initial average acceleration (averaged over a few periods) of the left mass? In which direction does it move?

6.5. **Inverted pendulum** ∗∗∗∗

A pendulum consists of a mass m at the end of a massless stick of length ℓ. The other end of the stick is made to oscillate vertically with a position given by $y(t) = A\cos(\omega t)$, where $A \ll \ell$. See Fig. 6.12. It turns out that if ω is large enough, and if the pendulum is initially nearly upside-down, then surprisingly it will *not* fall over as time goes by. Instead, it will (sort of) oscillate back and forth around the vertical position. If you want to do the experiment yourself, see the 28th demonstration of the entertaining collection in Ehrlich (1994).

Find the equation of motion for the angle of the pendulum (measured relative to its upside-down position). Explain why the pendulum doesn't fall over, and find the frequency of the back and forth motion.

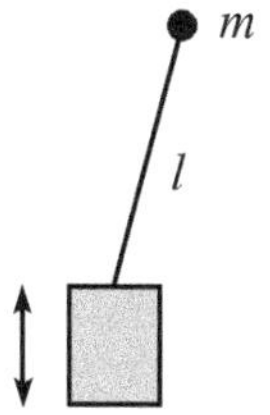

Fig. 6.12

Section 6.2: The principle of stationary action

6.6. **Minimum or saddle** ∗∗

(a) In Eq. (6.26), let $t_1 = 0$ and $t_2 = T$, for convenience. And let $\xi(t)$ be an easy-to-deal-with "triangular" function, of the form

$$\xi(t) = \begin{cases} \epsilon t/T, & 0 \le t \le T/2, \\ \epsilon(1 - t/T), & T/2 \le t \le T. \end{cases} \tag{6.93}$$

Under what condition is the harmonic-oscillator ΔS in Eq. (6.26) negative?

(b) Answer the same question, but now with $\xi(t) = \epsilon \sin(\pi t/T)$.

Section 6.3: Forces of constraint

6.7. **Normal force from a plane** ∗∗

A mass m slides down a frictionless plane that is inclined at an angle θ. Show, using the method in Section 6.3, that the normal force from the plane is the familiar $mg\cos\theta$.

Section 6.5: Conservation laws

6.8. **Bead on a stick** ∗

A stick is pivoted at the origin and is arranged to swing around in a horizontal plane at constant angular speed ω. A bead of mass m slides frictionlessly along the stick. Let r be the radial position of the bead. Find the conserved quantity E given in Eq. (6.52). Explain why this quantity is *not* the energy of the bead.

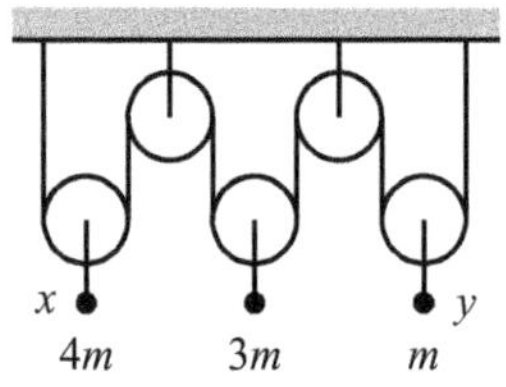

Fig. 6.13

Section 6.6: Noether's theorem

6.9. **Atwood's machine** **

Consider the Atwood's machine shown in Fig. 6.13. The masses are $4m$, $3m$, and m. Let x and y be the heights of the left and right masses, relative to their initial positions. Find the conserved momentum.

Section 6.7: Small oscillations

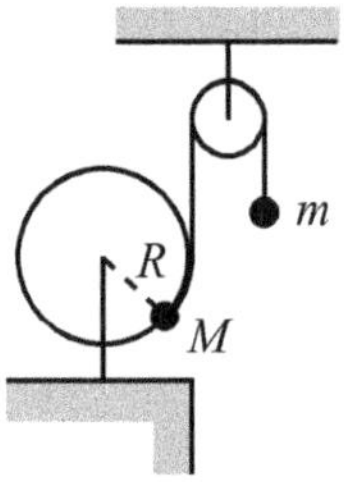

Fig. 6.14

6.10. **Hoop and pulley** **

A mass M is attached to a massless hoop of radius R that lies in a vertical plane. The hoop is free to rotate about its fixed center. M is tied to a string which winds part way around the hoop, then rises vertically up and over a massless pulley. A mass m hangs on the other end of the string (see Fig. 6.14). Find the equation of motion for the angle of rotation of the hoop. What is the frequency of small oscillations? Assume that m moves only vertically, and assume $M > m$.

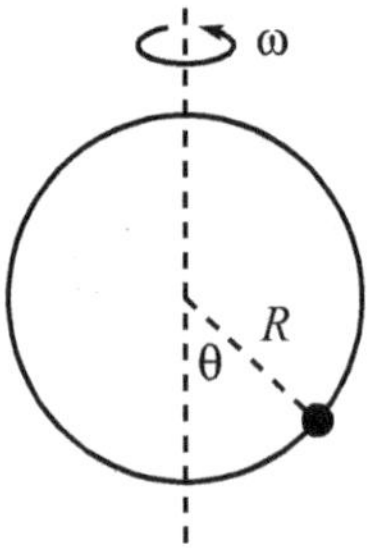

Fig. 6.15

6.11. **Bead on a rotating hoop** **

A bead is free to slide along a frictionless hoop of radius R. The hoop rotates with constant angular speed ω around a vertical diameter (see Fig. 6.15). Find the equation of motion for the angle θ shown. What are the equilibrium positions? What is the frequency of small oscillations about the stable equilibrium? There is one value of ω that is rather special; what is it, and why is it special?

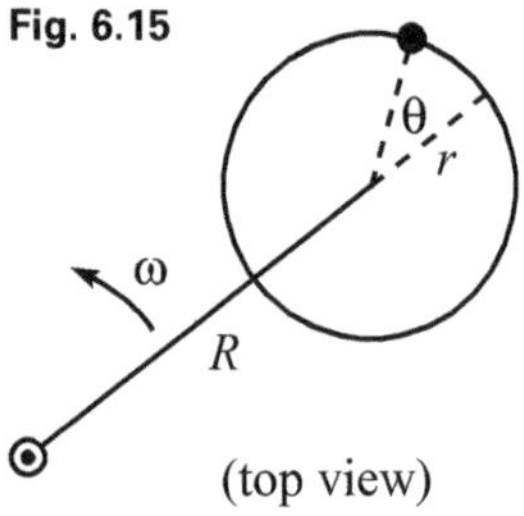

Fig. 6.16

6.12. **Another bead on a rotating hoop** **

A bead is free to slide along a frictionless hoop of radius r. The plane of the hoop is horizontal, and the center of the hoop travels in a horizontal circle of radius R, with constant angular speed ω, about a given point (see Fig. 6.16). Find the equation of motion for the angle θ shown. Also, find the frequency of small oscillations about the equilibrium point.

6.13. **Mass on a wheel** **

A mass m is fixed to a given point on the rim of a wheel of radius R that rolls without slipping on the ground. The wheel is massless, except for a mass M located at its center. Find the equation of motion for the angle through which the wheel rolls. For the case where the wheel undergoes small oscillations, find the frequency.

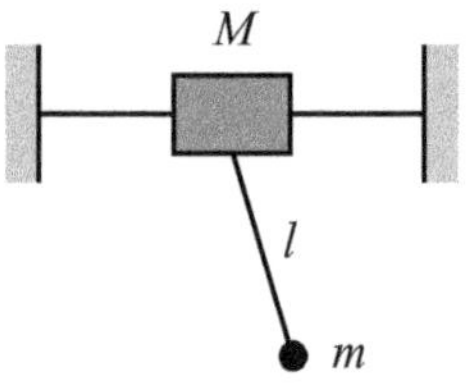

Fig. 6.17

6.14. **Pendulum with a free support** **

A mass M is free to slide along a frictionless rail. A pendulum of length ℓ and mass m hangs from M (see Fig. 6.17). Find the equations of motion. For small oscillations, find the normal modes and their frequencies.

6.15. Pendulum support on an inclined plane **

A mass M is free to slide down a frictionless plane inclined at an angle β. A pendulum of length ℓ and mass m hangs from M; see Fig. 6.18 (assume that M extends a short distance beyond the side of the plane, so the pendulum can hang down). Find the equations of motion. For small oscillations, find the normal modes and their frequencies.

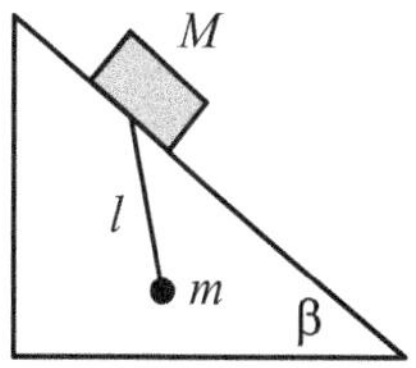

Fig. 6.18

6.16. Tilting plane ***

A mass M is fixed at the right-angled vertex where a massless rod of length ℓ is attached to a very long massless rod (see Fig. 6.19). A mass m is free to move frictionlessly along the long rod (assume that it can pass through M). The rod of length ℓ is hinged at a support, and the whole system is free to rotate, in the plane of the rods, about the hinge. Let θ be the angle of rotation of the system, and let x be the distance between m and M. Find the equations of motion. Find the normal modes when θ and x are both very small.

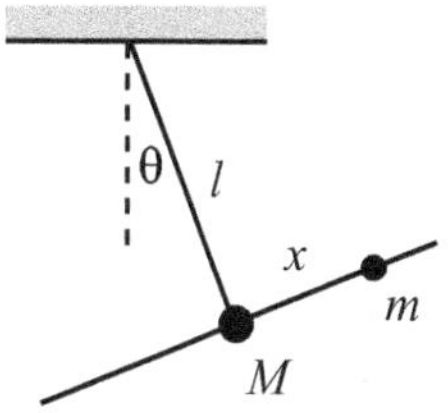

Fig. 6.19

6.17. Rotating curve ***

The curve $y(x) = b(x/a)^{\lambda}$ is rotated around the y axis with constant frequency ω (see Fig. 6.20). A bead moves frictionlessly along the curve. Find the frequency of small oscillations about the equilibrium point. Under what conditions do oscillations exist? (This problem gets a little messy.)

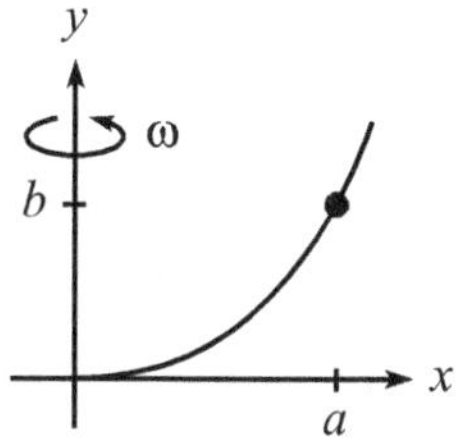

Fig. 6.20

6.18. Motion in a cone ***

A particle slides on the inside surface of a frictionless cone. The cone is fixed with its tip on the ground and its axis vertical. The half-angle at the tip is α (see Fig. 6.21). Let r be the distance from the particle to the axis, and let θ be the angle around the cone. Find the equations of motion.

If the particle moves in a circle of radius r_0, what is the frequency, ω, of this motion? If the particle is then perturbed slightly from this circular motion, what is the frequency, Ω, of the oscillations about the radius r_0? Under what conditions does $\Omega = \omega$?

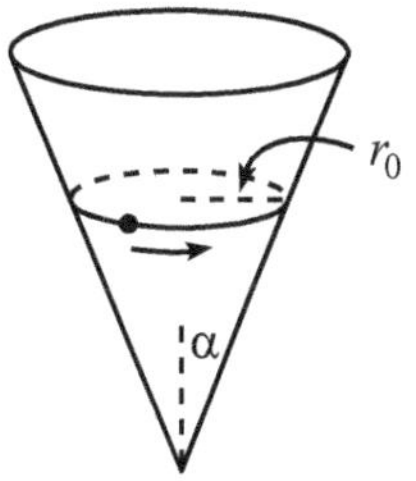

Fig. 6.21

6.19. Double pendulum ****

Consider a double pendulum made of two masses, m_1 and m_2, and two rods of lengths ℓ_1 and ℓ_2 (see Fig. 6.22). Find the equations of motion.

For small oscillations, find the normal modes and their frequencies for the special case $\ell_1 = \ell_2$ (and consider the cases $m_1 = m_2$, $m_1 \gg m_2$, and $m_1 \ll m_2$). Do the same for the special case $m_1 = m_2$ (and consider the cases $\ell_1 = \ell_2$, $\ell_1 \gg \ell_2$, and $\ell_1 \ll \ell_2$).

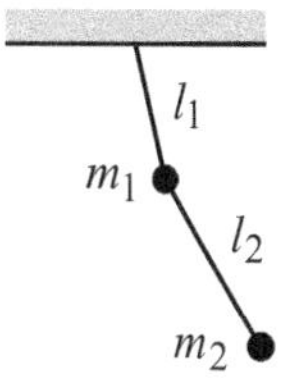

Fig. 6.22

Section 6.8: Other applications

6.20. Shortest distance in a plane *

In the spirit of Section 6.8, show that the shortest path between two points in a plane is a straight line.

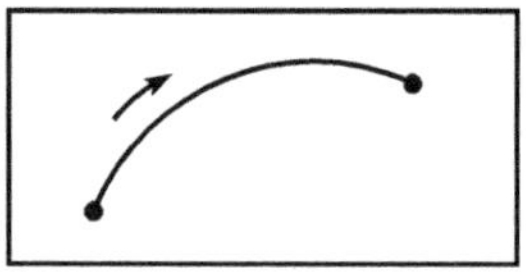

Fig. 6.23

6.21. Index of refraction **

Assume that the speed of light in a given slab of material is proportional to the height above the base of the slab.[12] Show that light moves in circular arcs in this material; see Fig. 6.23. You may assume that light takes the path of least time between two points (Fermat's principle of least time).

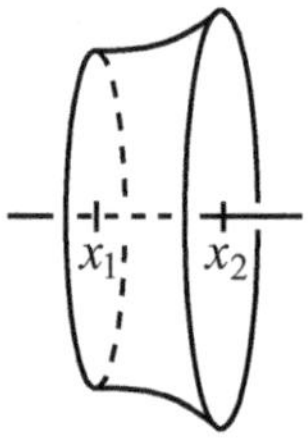

Fig. 6.24

6.22. Minimal surface **

Derive the shape of the minimal surface discussed in Section 6.8, by demanding that a cross-sectional "ring" (that is, the region between the planes $x = x_1$ and $x = x_2$) is in equilibrium; see Fig. 6.24. *Hint*: The tension must be constant throughout the surface (assuming that we're ignoring gravity, which we are).

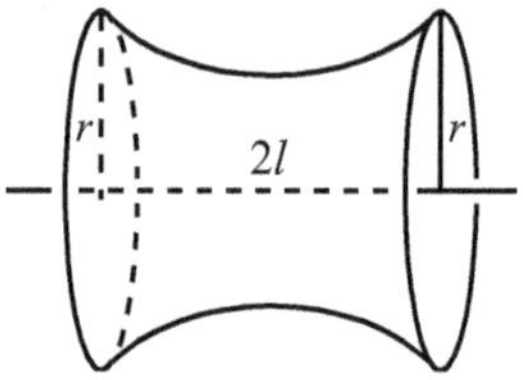

Fig. 6.25

6.23. Existence of a minimal surface **

Consider the minimal surface from Section 6.8, and look at the special case where the two rings have the same radius r (see Fig. 6.25). Let 2ℓ be the distance between the rings. What is the largest value of ℓ/r for which a minimal surface exists? You will need to solve something numerically here.

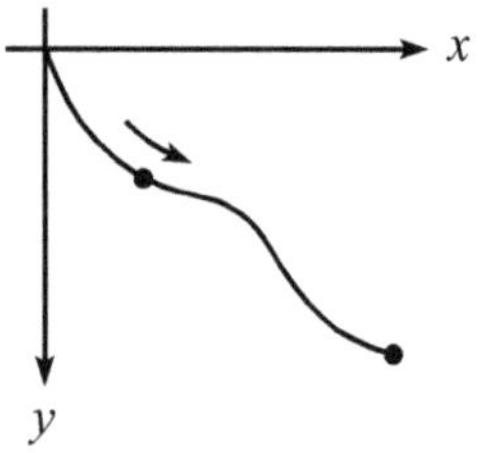

Fig. 6.26

6.24. The brachistochrone ***

A bead is released from rest at the origin and slides down a frictionless wire that connects the origin to a given point, as shown in Fig. 6.26. You wish to shape the wire so that the bead reaches the endpoint in the shortest possible time. Let the desired curve be described by the function $y(x)$, with downward taken to be positive. Show that $y(x)$ satisfies

$$1 + y'^2 = \frac{B}{y}, \tag{6.94}$$

where B is a constant. Then show that x and y may be written as

$$x = a(\theta - \sin\theta), \quad y = a(1 - \cos\theta). \tag{6.95}$$

[12] If you want to make the equivalent statement in terms of the material's "index of refraction," commonly denoted by n, then you can say: As a function of the height y, the index n is given by $n(y) = y_0/y$, where y_0 is some length that is larger than the height of the slab. This is equivalent to the original statement because the speed of light in a material equals c/n.

This is the parametrization of a *cycloid*, which is the path taken by a point on the rim of a rolling wheel.

6.10 Exercises

Section 6.1: The Euler–Lagrange equations

6.25. **Spring on a T** **

A rigid T consists of a long rod glued perpendicular to another rod of length ℓ that is pivoted at the origin. The T rotates around in a horizontal plane with constant frequency ω. A mass m is free to slide along the long rod and is connected to the intersection of the rods by a spring with spring constant k and relaxed length zero (see Fig. 6.27). Find $r(t)$, where r is the position of the mass along the long rod. There is a special value of ω; what is it, and why is it special?

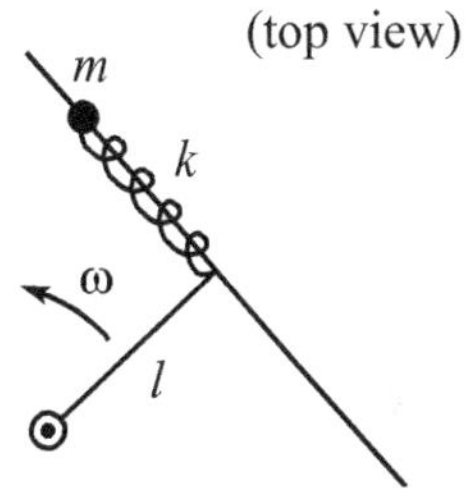

Fig. 6.27

6.26. **Spring on a T, with gravity** ***

Consider the setup in the previous exercise, but now let the T swing around in a vertical plane with constant frequency ω. Find $r(t)$. There is a special value of ω; what is it, and why is it special? (You may assume $\omega < \sqrt{k/m}$.)

6.27. **Coffee cup and mass** **

A coffee cup of mass M is connected to a mass m by a string. The coffee cup hangs over a frictionless pulley of negligible size, and the mass m is initially held with the string horizontal, as shown in Fig. 6.28. The mass m is then released. Find the equations of motion for r (the length of string between m and the pulley) and θ (the angle that the string to m makes with the horizontal). Assume that m somehow doesn't run into the string holding the cup up.

The coffee cup will initially fall, but it turns out that it will reach a lowest point and then rise back up. Write a program (see Section 1.4) that numerically determines the ratio of the r at this lowest point to the r at the start, for a given value of m/M. (To check your program, a value of $m/M = 1/10$ yields a ratio of about 0.208.)

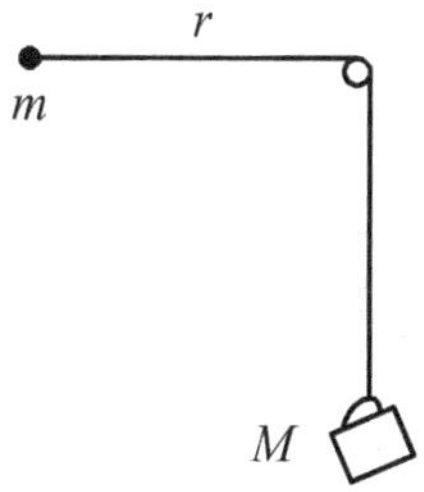

Fig. 6.28

6.28. **Three falling sticks** ***

Three massless sticks of length $2r$, each with a mass m fixed at its middle, are hinged at their ends, as shown in Fig. 6.29. The bottom end of the lower stick is hinged at the ground. They are held such that the lower two sticks are vertical, and the upper one is tilted at a small angle ϵ with respect to the vertical. They are then released. At this instant, what are the angular accelerations of the three sticks? Work

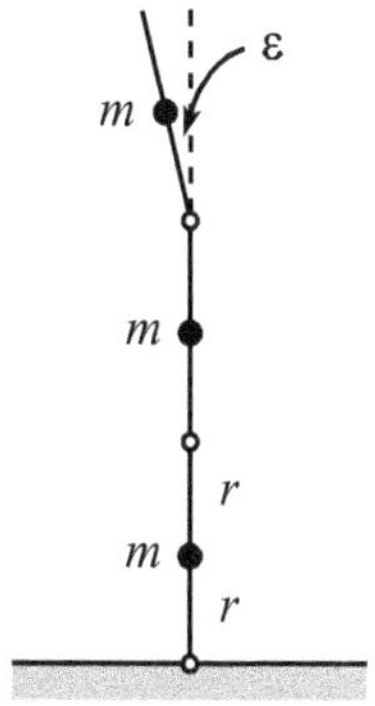

Fig. 6.29

in the approximation where ϵ is very small. (You may want to look at Problem 6.2 first.)

6.29. **Cycloidal pendulum** ****

The standard pendulum frequency of $\sqrt{g/\ell}$ holds only for small oscillations. The frequency becomes smaller as the amplitude grows. It turns out that if you want to build a pendulum whose frequency is independent of the amplitude, you should hang it from the cusp of a cycloid of a certain size, as shown in Fig. 6.30. As the string wraps partially around the cycloid, the effect is to decrease the length of string in the air, which in turn increases the frequency back up to a constant value. In more detail:

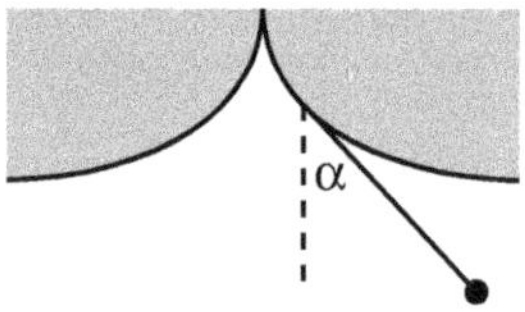

Fig. 6.30

A cycloid is the path taken by a point on the rim of a rolling wheel. The upside-down cycloid in Fig. 6.30 can be parametrized by $(x,y) = R(\theta - \sin\theta,\ -1+\cos\theta)$, where $\theta = 0$ corresponds to the cusp. Consider a pendulum of length $4R$ hanging from the cusp, and let α be the angle the string makes with the vertical, as shown.

(a) In terms of α, find the value of the parameter θ associated with the point where the string leaves the cycloid.
(b) In terms of α, find the length of string touching the cycloid.
(c) In terms of α, find the Lagrangian.
(d) Show that the quantity $\sin\alpha$ undergoes simple harmonic motion with frequency $\sqrt{g/4R}$, independent of the amplitude.
(e) In place of parts (c) and (d), solve the problem again by using $F = ma$. This actually gives a much quicker solution.

Section 6.2: The principle of stationary action

6.30. **Dropped ball** *

Consider the action, from $t = 0$ to $t = 1$, of a ball dropped from rest. From the E–L equation (or from $F = ma$), we know that $y(t) = -gt^2/2$ yields a stationary value of the action. Show explicitly that the particular function $y(t) = -gt^2/2 + \epsilon t(t-1)$ yields an action that has no first-order dependence on ϵ.

6.31. **Explicit minimization** *

For a ball thrown upward, guess a solution for y of the form $y(t) = a_2 t^2 + a_1 t + a_0$. Assuming that $y(0) = y(T) = 0$, this quickly becomes $y(t) = a_2(t^2 - Tt)$. Calculate the action between $t = 0$ and $t = T$, and show that it is minimized when $a_2 = -g/2$.

6.32. **Always a minimum** *

For a ball thrown up in the air, show that the stationary value of the action is always a global minimum.

6.33. **Second-order change** *

Let $x_a(t) \equiv x_0(t) + a\beta(t)$. Equation (6.19) gives the first derivative of the action with respect to a. Show that the second derivative is

$$\frac{d^2}{da^2}S[x_a(t)] = \int_{t_1}^{t_2}\left(\frac{\partial^2 L}{\partial x^2}\beta^2 + 2\frac{\partial^2 L}{\partial x\partial \dot{x}}\beta\dot{\beta} + \frac{\partial^2 L}{\partial \dot{x}^2}\dot{\beta}^2\right)dt. \qquad (6.96)$$

6.34. **$\ddot{x}$ dependence** *

Assume that there is $\ddot{x}$ dependence (in addition to $x, \dot{x}, t$ dependence) in the Lagrangian in Theorem 6.1. There will then be the additional term $(\partial L/\partial \ddot{x}_a)\ddot{\beta}$ in Eq. (6.19). It is tempting to integrate this term by parts twice, and then arrive at a modified form of Eq. (6.22):

$$\frac{\partial L}{\partial x_0} - \frac{d}{dt}\left(\frac{\partial L}{\partial \dot{x}_0}\right) + \frac{d^2}{dt^2}\left(\frac{\partial L}{\partial \ddot{x}_0}\right) = 0. \qquad (6.97)$$

Is this a valid result? If not, where is the error in the reasoning?

Section 6.3: Forces of constraint

6.35. **Constraint on a circle** *

A bead of mass m slides with speed v around a horizontal hoop of radius R. What force does the hoop apply to the bead? (Ignore gravity.)

6.36. **Atwood's machine** *

Consider the standard Atwood's machine in Fig. 6.31, with masses m_1 and m_2. Find the tension in the string.

Fig. 6.31

6.37. **Cartesian coordinates** **

In Eq. (6.35), take two time derivatives of the $\sqrt{x^2+y^2} - R = 0$ equation to obtain

$$R^2(x\ddot{x} + y\ddot{y}) + (x\dot{y} - y\dot{x})^2 = 0, \qquad (6.98)$$

and then combine this with the other two equations to solve for F in terms of $x, y, \dot{x}, \dot{y}$. Convert the result to polar coordinates (with θ measured from the vertical) and show that it agrees with Eq. (6.32).

6.38. **Constraint on a curve** ***

Let the horizontal plane be the x-y plane. A bead of mass m slides with speed v along a curve described by the function $y = f(x)$. What force does the curve apply to the bead? (Ignore gravity.)

Section 6.5: Conservation laws

6.39. **Bead on stick, using $F = ma$** *

After doing Problem 6.8, show again that the quantity E is conserved, but now use $F = ma$. Do this in two ways:

(a) Use the first of Eqs. (3.51). *Hint*: multiply through by $\dot{r}$.
(b) Use the second of Eqs. (3.51) to calculate the work done on the bead, and use the work–energy theorem.

Section 6.6: Noether's theorem

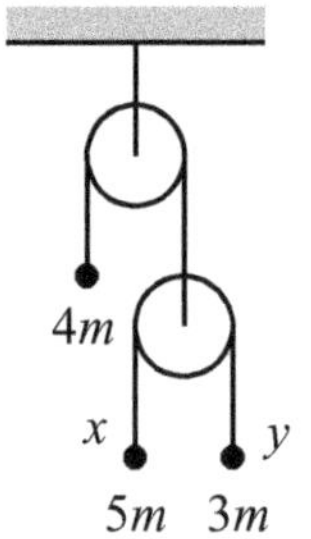

Fig. 6.32

6.40. **Atwood's machine** **

Consider the Atwood's machine shown in Fig. 6.32. The masses are $4m$, $5m$, and $3m$. Let x and y be the heights of the right two masses, relative to their initial positions. Use Noether's theorem to find the conserved momentum. (The solution to Problem 6.9 gives some other methods, too.)

Section 6.7: Small oscillations

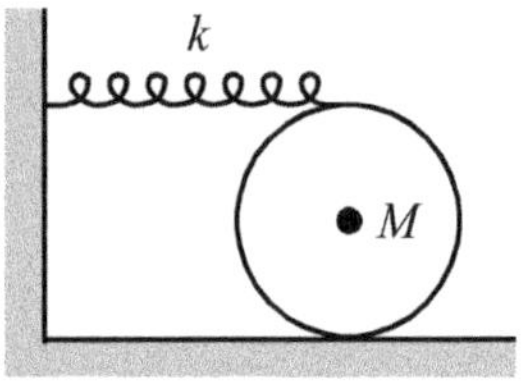

Fig. 6.33

6.41. **Spring and a wheel** *

The top of a wheel of mass M and radius R is connected to a spring (at its equilibrium length) with spring constant k, as shown in Fig. 6.33. Assume that all the mass of the wheel is at its center. If the wheel rolls without slipping, what is the frequency of (small) oscillations?

6.42. **Spring on a spoke** **

A spring with spring constant k and relaxed length zero lies along a spoke of a massless wheel of radius R. One end of the spring is attached to the center, and the other end is attached to a mass m that is free to slide along the spoke. When the system is in its equilibrium position with the spring hanging vertically, how far (in terms of R) should the mass hang down (you are free to adjust k) so that for small oscillations, the frequency of the spring oscillations equals the frequency of the rocking motion of the wheel? Assume that the wheel rolls without slipping.

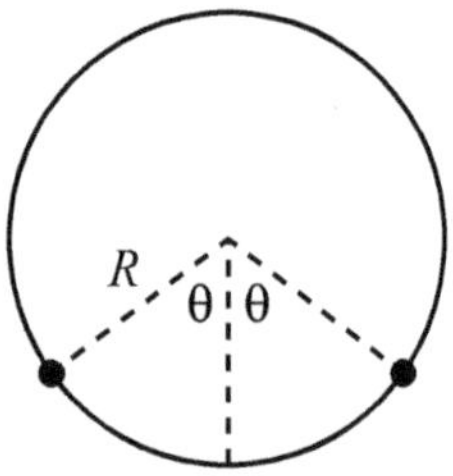

Fig. 6.34

6.43. **Oscillating hoop** **

Two equal masses are glued to a massless hoop of radius R that is free to rotate about its center in a vertical plane. The angle between the masses is 2θ, as shown in Fig. 6.34. Find the frequency of small oscillations.

6.44. **Oscillating hoop with a pendulum** ***

A massless hoop of radius R is free to rotate about its center in a vertical plane. A mass m is attached at one point, and a pendulum of length $\sqrt{2}R$

(and also of mass m) is attached at another point 90° away, as shown in Fig. 6.35. Let θ be the angle of the hoop relative to the position shown, and let α be the angle of the pendulum with respect to the vertical. Find the normal modes for small oscillations.

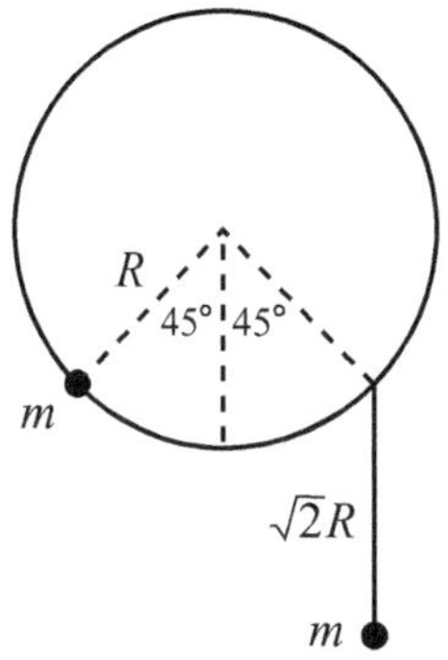

Fig. 6.35

6.45. **Mass sliding on a rim** **

A mass m is free to slide frictionlessly along the rim of a wheel of radius R that rolls without slipping on the ground. The wheel is massless, except for a mass M located at its center. Find the normal modes for small oscillations.

6.46. **Mass sliding on a rim, with a spring** ***

Consider the setup in the previous exercise, but now let the mass m be attached to a spring with spring constant k and relaxed length zero, the other end of which is attached to a point on the rim. Assume that the spring is constrained to run along the rim, and assume that the mass can pass freely over the point where the spring is attached to the rim. To keep things from getting too messy here, you can set $M = m$.

(a) Find the frequencies of the normal modes for small oscillations. Check the $g = 0$ limit, and (if you've done the previous exercise) the $k = 0$ limit.

(b) For the special case where $g/R = k/m$, show that the frequencies can be written as $\sqrt{k/m}(\sqrt{5} \pm 1)/2$. This numerical factor is the golden ratio (and its inverse). Describe what the normal modes look like.

6.47. **Vertically rotating hoop** ***

A bead is free to slide along a frictionless hoop of radius r. The plane of the hoop is vertical, and the center of the hoop travels in a vertical circle of radius R with constant angular speed ω about a given point (see Fig. 6.36). Find the equation of motion for the angle θ shown. For large ω (which implies small θ), find the amplitude of the "particular" solution with frequency ω. What happens if $r = R$?

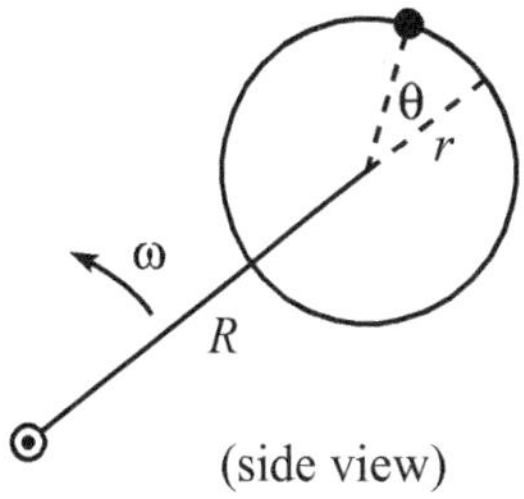

Fig. 6.36

6.11 Solutions

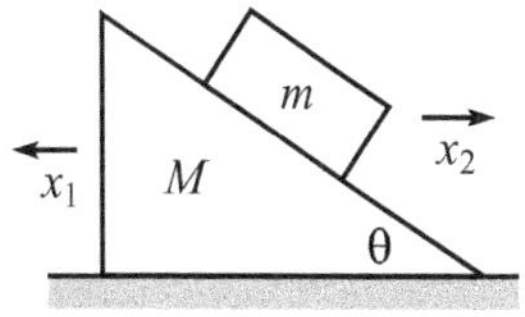

Fig. 6.37

6.1. **Moving plane**

Let x_1 be the horizontal coordinate of the plane (with positive x_1 to the left), and let x_2 be the horizontal coordinate of the block (with positive x_2 to the right); see Fig. 6.37. The relative horizontal distance between the plane and the block is $x_1 + x_2$, so the height fallen by the block is $(x_1 + x_2)\tan\theta$. The Lagrangian is therefore

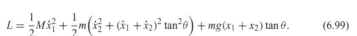

$$L = \frac{1}{2}M\dot{x}_1^2 + \frac{1}{2}m\left(\dot{x}_2^2 + (\dot{x}_1 + \dot{x}_2)^2\tan^2\theta\right) + mg(x_1 + x_2)\tan\theta. \quad (6.99)$$

The equations of motion obtained from varying x_1 and x_2 are

$$\begin{aligned} M\ddot{x}_1 + m(\ddot{x}_1 + \ddot{x}_2)\tan^2\theta &= mg\tan\theta, \\ m\ddot{x}_2 + m(\ddot{x}_1 + \ddot{x}_2)\tan^2\theta &= mg\tan\theta. \end{aligned} \tag{6.100}$$

Note that the difference of these two equations immediately yields conservation of momentum, $M\ddot{x}_1 - m\ddot{x}_2 = 0 \Longrightarrow (d/dt)(M\dot{x}_1 - m\dot{x}_2) = 0$. Equations (6.100) are two linear equations in the two unknowns, $\ddot{x}_1$ and $\ddot{x}_2$, so we can solve for $\ddot{x}_1$. After a little simplification, we arrive at

$$\ddot{x}_1 = \frac{mg\sin\theta\cos\theta}{M + m\sin^2\theta}. \tag{6.101}$$

For some limiting cases, see the remarks in the solution to Problem 3.8.

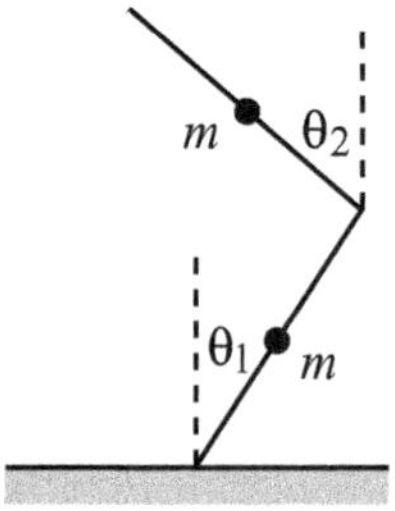

Fig. 6.38

6.2. **Two falling sticks**

Let $\theta_1(t)$ and $\theta_2(t)$ be defined as in Fig. 6.38. Then the position of the bottom mass in Cartesian coordinates is $(r\sin\theta_1, r\cos\theta_1)$, and the position of the top mass is $(2r\sin\theta_1 - r\sin\theta_2, 2r\cos\theta_1 + r\cos\theta_2)$. So the potential energy of the system is

$$V(\theta_1,\theta_2) = mgr(3\cos\theta_1 + \cos\theta_2). \tag{6.102}$$

The kinetic energy is somewhat more complicated. The kinetic energy of the bottom mass is simply $mr^2\dot{\theta}_1^2/2$. Taking the derivative of the top mass's position given above, we find that the kinetic energy of the top mass is

$$\frac{1}{2}mr^2\Big((2\cos\theta_1\dot{\theta}_1 - \cos\theta_2\dot{\theta}_2)^2 + (-2\sin\theta_1\dot{\theta}_1 - \sin\theta_2\dot{\theta}_2)^2\Big). \tag{6.103}$$

We can simplify this, using the small-angle approximations. The terms involving $\sin\theta$ are fourth order in the small θ's, so we can neglect them. Also, we can approximate $\cos\theta$ by 1, because this entails dropping only terms of at least fourth order. So the top mass's kinetic energy becomes $(1/2)mr^2(2\dot{\theta}_1 - \dot{\theta}_2)^2$. In retrospect, it would have been easier to obtain the kinetic energies of the masses by first applying the small-angle approximations to the positions, and then taking the derivatives to obtain the velocities. This strategy shows that both masses move essentially horizontally (initially). You will probably want to use this strategy when solving Exercise 6.28.

Using the small-angle approximation $\cos\theta \approx 1 - \theta^2/2$ to rewrite the potential energy in Eq. (6.102), we have

$$L \approx \frac{1}{2}mr^2\left(5\dot{\theta}_1^2 - 4\dot{\theta}_1\dot{\theta}_2 + \dot{\theta}_2^2\right) - mgr\left(4 - \frac{3}{2}\theta_1^2 - \frac{1}{2}\theta_2^2\right). \tag{6.104}$$

The equations of motion obtained from varying θ_1 and θ_2 are, respectively,

$$\begin{aligned} 5\ddot{\theta}_1 - 2\ddot{\theta}_2 &= \frac{3g}{r}\theta_1 \\ -2\ddot{\theta}_1 + \ddot{\theta}_2 &= \frac{g}{r}\theta_2. \end{aligned} \tag{6.105}$$

At the instant the sticks are released, we have $\theta_1 = 0$ and $\theta_2 = \epsilon$. Solving Eq. (6.105) for $\ddot{\theta}_1$ and $\ddot{\theta}_2$ gives

$$\ddot{\theta}_1 = \frac{2g\epsilon}{r}, \quad \text{and} \quad \ddot{\theta}_2 = \frac{5g\epsilon}{r}. \tag{6.106}$$

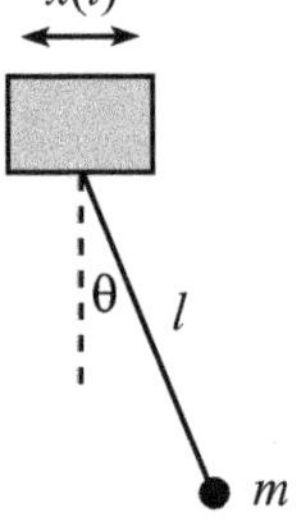

Fig. 6.39

6.3. **Pendulum with an oscillating support**

Let θ be defined as in Fig. 6.39. With $x(t) = A\cos(\omega t)$, the position of the mass m is given by

$$(X,Y)_m = (x + \ell\sin\theta, -\ell\cos\theta). \tag{6.107}$$

Taking the derivative to obtain the velocity, we find that the square of the speed is

$$V_m^2 = \dot{X}^2 + \dot{Y}^2 = \ell^2\dot{\theta}^2 + \dot{x}^2 + 2\ell\dot{x}\dot{\theta}\cos\theta, \tag{6.108}$$

which also follows from applying the law of cosines to the horizontal $\dot{x}$ and tangential $\ell\dot{\theta}$ parts of the velocity vector. The Lagrangian is therefore

$$L = \frac{1}{2}m(\ell^2\dot{\theta}^2 + \dot{x}^2 + 2\ell\dot{x}\dot{\theta}\cos\theta) + mg\ell\cos\theta. \tag{6.109}$$

The equation of motion for θ is

$$\begin{aligned} \frac{d}{dt}(m\ell^2\dot{\theta} + m\ell\dot{x}\cos\theta) &= -m\ell\dot{x}\dot{\theta}\sin\theta - mg\ell\sin\theta \\ \Longrightarrow \quad \ell\ddot{\theta} + \ddot{x}\cos\theta &= -g\sin\theta. \end{aligned} \tag{6.110}$$

Plugging in the explicit form of $x(t)$, we have

$$\ell\ddot{\theta} - A\omega^2\cos(\omega t)\cos\theta + g\sin\theta = 0. \tag{6.111}$$

In retrospect, this makes sense. Someone in the reference frame of the support, which has horizontal acceleration $\ddot{x} = -A\omega^2\cos(\omega t)$, may as well be living in a world where the acceleration from gravity has a component g downward and a component $A\omega^2\cos(\omega t)$ to the right. Equation (6.111) is just the $F = ma$ equation in the tangential direction in this accelerating world.

A small-angle approximation in Eq. (6.111) gives

$$\ddot{\theta} + \omega_0^2\theta = a\omega^2\cos(\omega t), \tag{6.112}$$

where $\omega_0 \equiv \sqrt{g/\ell}$ and $a \equiv A/\ell$. This equation is simply the equation for a driven oscillator, which we solved in Chapter 4. The solution is

$$\theta(t) = \frac{a\omega^2}{\omega_0^2 - \omega^2}\cos(\omega t) + C\cos(\omega_0 t + \phi), \tag{6.113}$$

where C and ϕ are determined by the initial conditions.

If ω happens to equal ω_0, then it appears that the amplitude goes to infinity. However, as soon as the amplitude becomes large, our small-angle approximation breaks down, and Eqs. (6.112) and (6.113) are no longer valid.

6.4. **Two masses, one swinging**

The Lagrangian is

$$L = \frac{1}{2}m\dot{r}^2 + \frac{1}{2}m(\dot{r}^2 + r^2\dot{\theta}^2) - mgr + mgr\cos\theta. \tag{6.114}$$

The last two terms are the (negatives of the) potentials of each mass, relative to where they would be if the right mass were located at the right pulley. The equations of motion obtained from varying r and θ are

$$\begin{aligned} 2\ddot{r} &= r\dot{\theta}^2 - g(1 - \cos\theta), \\ \frac{d}{dt}(r^2\dot{\theta}) &= -gr\sin\theta. \end{aligned} \tag{6.115}$$

The first equation deals with the forces and accelerations along the direction of the string. The second equation equates the torque from gravity with the change in angular momentum of the right mass. If we do a (coarse) small-angle approximation and keep only terms up to first order in θ, we find that at $t = 0$ (using the initial condition, $\dot{r} = 0$), Eqs. (6.115) become

$$\begin{aligned} \ddot{r} &= 0, \\ \ddot{\theta} + \frac{g}{r}\theta &= 0. \end{aligned} \tag{6.116}$$

These equations say that the left mass stays still, and the right mass behaves just like a pendulum.

If we want to find the leading term in the initial acceleration of the left mass (that is, the leading term in $\ddot{r}$), we need to be a little less coarse in our approximation. So let's keep terms in Eq. (6.115) up to second order in θ. We then have at $t = 0$ (using the initial condition, $\dot{r} = 0$)

$$2\ddot{r} = r\dot{\theta}^2 - \frac{1}{2}g\theta^2,$$
$$\ddot{\theta} + \frac{g}{r}\theta = 0. \tag{6.117}$$

The second equation still says that the right mass undergoes harmonic motion. We are told that the amplitude is ϵ, so we have

$$\theta(t) = \epsilon\cos(\omega t + \phi), \tag{6.118}$$

where $\omega = \sqrt{g/r}$. Plugging this into the first equation gives

$$2\ddot{r} = \epsilon^2 g\left(\sin^2(\omega t + \phi) - \frac{1}{2}\cos^2(\omega t + \phi)\right). \tag{6.119}$$

If we average this over a few periods, both $\sin^2\alpha$ and $\cos^2\alpha$ average to $1/2$, so we find

$$\ddot{r}_{\text{avg}} = \frac{\epsilon^2 g}{8}. \tag{6.120}$$

This is a small second-order effect. It is positive, so the left mass slowly begins to climb.

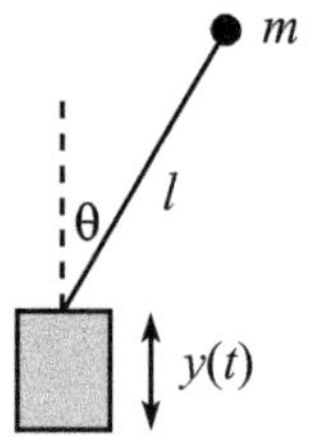

Fig. 6.40

6.5. **Inverted pendulum**

Let θ be defined as in Fig. 6.40. With $y(t) = A\cos(\omega t)$, the position of the mass m is given by

$$(X, Y) = (\ell\sin\theta,\ y + \ell\cos\theta). \tag{6.121}$$

Taking the derivative to obtain the velocity, we find that the square of the speed is

$$V^2 = \dot{X}^2 + \dot{Y}^2 = \ell^2\dot{\theta}^2 + \dot{y}^2 - 2\ell\dot{y}\dot{\theta}\sin\theta, \tag{6.122}$$

which also follows from applying the law of cosines to the vertical $\dot{y}$ and tangential $\ell\dot{\theta}$ parts of the velocity vector. The Lagrangian is therefore

$$L = \frac{1}{2}m(\ell^2\dot{\theta}^2 + \dot{y}^2 - 2\ell\dot{y}\dot{\theta}\sin\theta) - mg(y + \ell\cos\theta). \tag{6.123}$$

The equation of motion for θ is

$$\frac{d}{dt}\left(\frac{\partial L}{\partial\dot{\theta}}\right) = \frac{\partial L}{\partial\theta} \implies \ell\ddot{\theta} - \ddot{y}\sin\theta = g\sin\theta. \tag{6.124}$$

Plugging in the explicit form of $y(t)$, we have

$$\ell\ddot{\theta} + \sin\theta\left(A\omega^2\cos(\omega t) - g\right) = 0. \tag{6.125}$$

In retrospect, this makes sense. Someone in the reference frame of the support, which has vertical acceleration $\ddot{y} = -A\omega^2\cos(\omega t)$, may as well be living in a world where the acceleration from gravity is $g - A\omega^2\cos(\omega t)$ downward. Equation (6.125) is just the $F = ma$ equation in the tangential direction in this accelerating world.

Assuming θ is small, we may set $\sin\theta \approx \theta$, which gives

$$\ddot{\theta} + \theta\left(a\omega^2\cos(\omega t) - \omega_0^2\right) = 0, \tag{6.126}$$

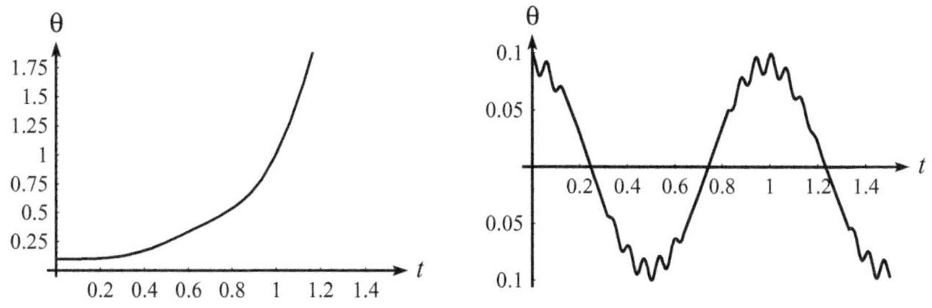

Fig. 6.41

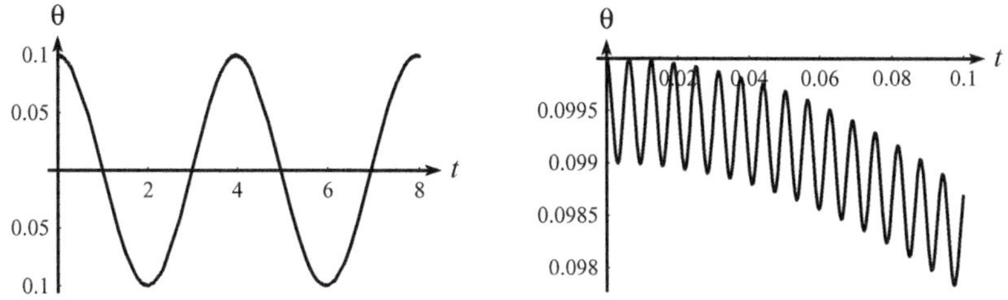

Fig. 6.42

where $\omega_0 \equiv \sqrt{g/\ell}$ and $a \equiv A/\ell$. Equation (6.126) cannot be solved exactly, but we can still get a good idea of how θ depends on time. We can do this both numerically and (approximately) analytically.

Figure 6.41 shows how θ depends on time for parameters with values $\ell = 1$ m, $A = 0.1$ m, and $g = 10\,\text{m/s}^2$. So $a = 0.1$, and $\omega_0^2 = 10\,\text{s}^{-2}$. We produced these plots numerically using Eq. (6.126), with the initial conditions of $\theta(0) = 0.1$ and $\dot{\theta}(0) = 0$. In the first plot, $\omega = 10\,\text{s}^{-1}$. And in the second plot, $\omega = 100\,\text{s}^{-1}$. The stick falls over in first case, but it undergoes oscillatory motion in the second case. Apparently, if ω is large enough the stick won't fall over.

Let's now explain this phenomenon analytically. At first glance, it's rather surprising that the stick stays up. It seems like the average (over a few periods of the ω oscillations) of the tangential acceleration in Eq. (6.126), namely $-\theta(a\omega^2\cos(\omega t) - \omega_0^2)$, equals the positive quantity $\theta\omega_0^2$, because the $\cos(\omega t)$ term averages to zero (or so it appears). So you might think that there is a net force making θ increase, causing the stick to fall over.

The error in this reasoning is that the average of the $-a\omega^2\theta\cos(\omega t)$ term is *not* zero, because θ undergoes tiny oscillations with frequency ω, as seen in the second plot in Fig. 6.42. Both of these plots have $a = 0.005$, $\omega_0^2 = 10\,\text{s}^{-2}$, and $\omega = 1000\,\text{s}^{-1}$ (we'll work with small a and large ω from now on; more on this below). The second plot is a zoomed-in version of the first one near $t = 0$. The important point here is that the tiny oscillations in θ shown in the second plot are correlated with $\cos(\omega t)$. It turns out that the θ value at the t where $\cos(\omega t) = 1$ is larger than the θ value at the t where $\cos(\omega t) = -1$. So there is a net negative contribution to the $-a\omega^2\theta\cos(\omega t)$ part of the acceleration. And it may indeed be large enough to keep the pendulum up, as we will now show.

To get a handle on the $-a\omega^2\theta\cos(\omega t)$ term, let's work in the approximation where ω is large and $a \equiv A/\ell$ is small. More precisely, we will assume $a \ll 1$ and $a\omega^2 \gg \omega_0^2$, for reasons we will explain below. Look at one of the little oscillations in the second plot in Fig. 6.42. These oscillations have frequency ω, because they are due to the support moving up and down. When the support moves up, θ increases; and when the support moves down, θ decreases. Since the average position of the pendulum doesn't change much over one of these small periods, we can look for an approximate solution to Eq. (6.126) of the form

$$\theta(t) \approx C + b\cos(\omega t), \tag{6.127}$$

where $b \ll C$. C will change over time, but on the scale of $1/\omega$ it is essentially constant, if $a \equiv A/\ell$ is small enough. Plugging this guess for θ into Eq. (6.126), and using $a \ll 1$ and $a\omega^2 \gg \omega_0^2$, we find $-b\omega^2\cos(\omega t) + Ca\omega^2\cos(\omega t) = 0$, to leading order.[13] So we must have $b = aC$. Our approximate solution for θ is therefore

$$\theta \approx C\big(1 + a\cos(\omega t)\big). \tag{6.128}$$

Let's now determine how C gradually changes with time. From Eq. (6.126), the average acceleration of θ, over a period $T = 2\pi/\omega$, is

$$\begin{aligned}
\overline{\ddot{\theta}} &= \overline{-\theta\Big(a\omega^2\cos(\omega t) - \omega_0^2\Big)} \\
&\approx \overline{-C\Big(1 + a\cos(\omega t)\Big)\Big(a\omega^2\cos(\omega t) - \omega_0^2\Big)} \\
&= -C\Big(a^2\omega^2\overline{\cos^2(\omega t)} - \omega_0^2\Big) \\
&= -C\left(\frac{a^2\omega^2}{2} - \omega_0^2\right) \\
&\equiv -C\Omega^2,
\end{aligned} \tag{6.129}$$

where

$$\Omega = \sqrt{\frac{a^2\omega^2}{2} - \frac{g}{\ell}}. \tag{6.130}$$

But if we take two derivatives of Eq. (6.127), we see that $\overline{\ddot{\theta}}$ simply equals $\ddot{C}$. Equating this value of $\overline{\ddot{\theta}}$ with the one in Eq. (6.129) gives

$$\ddot{C}(t) + \Omega^2 C(t) \approx 0. \tag{6.131}$$

This equation describes nice simple-harmonic motion. Therefore, C oscillates sinusoidally with the frequency Ω given in Eq. (6.130). This is the overall back and forth motion seen in the first plot in Fig. 6.42. Note that we must have $a\omega > \sqrt{2}\omega_0$ if this frequency is to be real so that the pendulum stays up. Since we have assumed $a \ll 1$, we see that $a^2\omega^2 > 2\omega_0^2$ implies $a\omega^2 \gg \omega_0^2$, which is consistent with our initial assumption above.

If $a\omega \gg \omega_0$, then Eq. (6.130) gives $\Omega \approx a\omega/\sqrt{2}$. This is the case if we change the setup and just have the pendulum lie flat on a horizontal table where the acceleration from gravity is zero. In this limit where g is irrelevant, dimensional analysis implies that the frequency of the C oscillations must be a multiple of ω, because ω is the only

[13] The reasons for the $a \ll 1$ and $a\omega^2 \gg \omega_0^2$ qualifications are the following. If $a\omega^2 \gg \omega_0^2$, then the $a\omega^2\cos(\omega t)$ term dominates the ω_0^2 term in Eq. (6.126). The one exception to this is when $\cos(\omega t) \approx 0$, but this occurs for a negligibly small amount of time if $a\omega^2 \gg \omega_0^2$. If $a \ll 1$, then we can legally ignore the $\ddot{C}$ term when Eq. (6.127) is substituted into Eq. (6.126). This is true because we will find below in Eq. (6.129) that our assumptions lead to $\ddot{C}$ being roughly proportional to $Ca^2\omega^2$. Since the other terms in Eq. (6.126) are proportional to $Ca\omega^2$, we need $a \ll 1$ in order for the $\ddot{C}$ term to be negligible. In short, $a \ll 1$ is the condition under which C varies slowly on the time scale of $1/\omega$.

quantity in the problem with units of frequency. It just so happens that the multiple is $a/\sqrt{2}$.

As a double check that we haven't messed up somewhere, the Ω value resulting from the parameters in Fig. 6.42 (namely $a = 0.005$, $\omega_0^2 = 10\,\text{s}^{-2}$, and $\omega = 1000\,\text{s}^{-1}$) is $\Omega = \sqrt{25/2 - 10} = 1.58\,\text{s}^{-1}$. This corresponds to a period of $2\pi/\Omega \approx 3.97$ s. And indeed, from the first plot in the figure, the period looks to be about 4 s (or maybe a hair less). For more on the inverted pendulum, see Butikov (2001).

6.6. **Minimum or saddle**

(a) For the given $\xi(t)$, the integrand in Eq. (6.26) is symmetric around the midpoint, so we obtain

$$\Delta S = \int_0^{T/2} \left(m\left(\frac{\epsilon}{T}\right)^2 - k\left(\frac{\epsilon t}{T}\right)^2 \right) dt = \frac{m\epsilon^2}{2T} - \frac{k\epsilon^2 T}{24}. \tag{6.132}$$

This is negative if $T > \sqrt{12m/k} \equiv 2\sqrt{3}/\omega$. Since the period of the oscillation is $\tau \equiv 2\pi/\omega$, we see that T must be greater than $(\sqrt{3}/\pi)\tau$ in order for ΔS to be negative, assuming that we are using the given triangular function for ξ.

(b) With $\xi(t) = \epsilon \sin(\pi t/T)$, the integrand in Eq. (6.26) becomes

$$\begin{aligned}\Delta S &= \frac{1}{2}\int_0^T \left(m\left(\frac{\epsilon\pi}{T}\cos(\pi t/T)\right)^2 - k\Big(\epsilon \sin(\pi t/T)\Big)^2 \right) dt.\\ &= \frac{m\epsilon^2\pi^2}{4T} - \frac{k\epsilon^2 T}{4},\end{aligned} \tag{6.133}$$

where we have used the fact that the average value of $\sin^2\theta$ and $\cos^2\theta$ over half of a period is $1/2$ (or you can just do the integrals). This result for ΔS is negative if $T > \pi\sqrt{m/k} \equiv \pi/\omega = \tau/2$, where τ is the period.

REMARK: It turns out that the $\xi(t) \propto \sin(\pi t/T)$ function gives the best chance of making ΔS negative. You can show this by invoking a theorem from Fourier analysis that says that any function satisfying $\xi(0) = \xi(T) = 0$ can be written as the sum $\xi(t) = \sum_1^\infty c_n \sin(n\pi t/T)$, where the c_n are numerical coefficients. When this sum is plugged into Eq. (6.26), you can show that all the cross terms (terms involving two different values of n) integrate to zero. Using the fact that the average value of $\sin^2\theta$ and $\cos^2\theta$ is $1/2$, the rest of the integral yields

$$\Delta S = \frac{1}{4}\sum_1^\infty c_n^2 \left(\frac{m\pi^2 n^2}{T} - kT \right). \tag{6.134}$$

In order to obtain the smallest value of T that can make this sum negative, we want only the $n = 1$ term to exist. We then have $\xi(t) = c_1 \sin(\pi t/T)$, and Eq. (6.134) reduces to Eq. (6.133), as it should.

As mentioned in Remark 4 in Section 6.2, it is always possible to make ΔS positive by picking a $\xi(t)$ function that is small but wiggles very fast. Therefore, we see that for a harmonic oscillator, if $T > \tau/2$, then the stationary value of S is a saddle point (some ξ's make ΔS positive, and some make it negative), but if $T < \tau/2$, then the stationary value of S is a minimum (all ξ's make ΔS positive). In the latter case, the point is that T is small enough so that there is no way for ξ to get large, without making $\dot{\xi}$ large also. ♣

6.7. **Normal force from a plane**

FIRST SOLUTION: The most convenient coordinates in this problem are w and z, where w is the distance upward along the plane, and z is the distance perpendicularly away from it. The Lagrangian is then

$$\frac{1}{2}m(\dot{w}^2 + \dot{z}^2) - mg(w\sin\theta + z\cos\theta) - V(z), \tag{6.135}$$

where $V(z)$ is the (very steep) constraining potential. The two equations of motion are

$$m\ddot{w} = -mg\sin\theta,$$
$$m\ddot{z} = -mg\cos\theta - \frac{dV}{dz}. \quad (6.136)$$

At this point we invoke the constraint $z = 0$. So $\ddot{z} = 0$, and the second equation gives

$$F_c \equiv -V'(0) = mg\cos\theta, \quad (6.137)$$

as desired. We also obtain the usual result, $\ddot{w} = -g\sin\theta$.

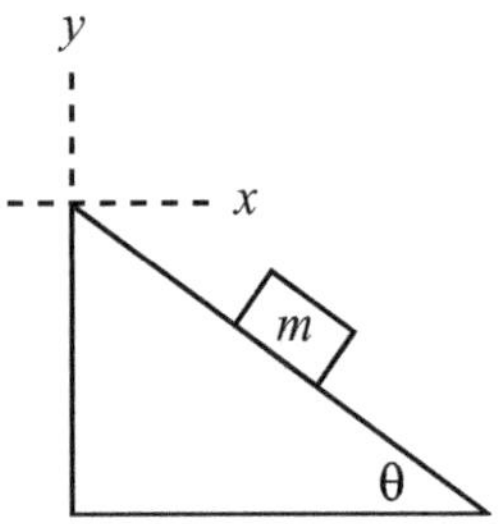

Fig. 6.43

SECOND SOLUTION: We can also solve this problem by using the horizontal and vertical coordinates, x and y. We'll choose $(x, y) = (0, 0)$ to be at the top of the plane; see Fig. 6.43. The (very steep) constraining potential is $V(z)$, where $z \equiv x\sin\theta + y\cos\theta$ is the distance from the mass to the plane (as you can verify). The Lagrangian is then

$$L = \frac{1}{2}m(\dot{x}^2 + \dot{y}^2) - mgy - V(z). \quad (6.138)$$

Keeping in mind that $z \equiv x\sin\theta + y\cos\theta$, the two equations of motion are (using the chain rule)

$$m\ddot{x} = -\frac{dV}{dz}\frac{\partial z}{\partial x} = -V'(z)\sin\theta,$$
$$m\ddot{y} = -mg - \frac{dV}{dz}\frac{\partial z}{\partial y} = -mg - V'(z)\cos\theta. \quad (6.139)$$

At this point we invoke the constraint condition $z = 0 \Longrightarrow x = -y\cot\theta$. This condition, along with the two E–L equations, allows us to solve for the three unknowns, $\ddot{x}$, $\ddot{y}$, and $V'(0)$. Using $\ddot{x} = -\ddot{y}\cot\theta$ in Eq. (6.139), we find

$$\ddot{x} = g\cos\theta\sin\theta, \quad \ddot{y} = -g\sin^2\theta, \quad F_c \equiv -V'(0) = mg\cos\theta. \quad (6.140)$$

The first two results here are simply the horizontal and vertical components of the acceleration along the plane, which is $g\sin\theta$.

6.8. **Bead on a stick**

There is no potential energy here, so the Lagrangian consists of just the kinetic energy, T, which comes from the radial and tangential motions:

$$L = T = \frac{1}{2}m\dot{r}^2 + \frac{1}{2}mr^2\omega^2. \quad (6.141)$$

Equation (6.52) therefore gives

$$E = \frac{1}{2}m\dot{r}^2 - \frac{1}{2}mr^2\omega^2. \quad (6.142)$$

Claim 6.3 says that this quantity is conserved, because $\partial L/\partial t = 0$. But it is *not* the energy of the bead, due to the minus sign in the second term.

The point here is that in order to keep the stick rotating at a constant angular speed, there must be an external force acting on it. This force in turn causes work to be done on the bead, thereby increasing its kinetic energy. The kinetic energy T is therefore *not* conserved. From Eqs. (6.141) and (6.142), we see that $E = T - mr^2\omega^2$ is the quantity that is constant in time. See Exercise 6.39 for some $F = ma$ ways to show that the quantity E in Eq. (6.142) is conserved.

6.9. **Atwood's machine**

FIRST SOLUTION: If the left mass goes up by x and the right mass goes up by y, then conservation of string says that the middle mass must go down by $x + y$. Therefore, the Lagrangian of the system is

$$L = \frac{1}{2}(4m)\dot{x}^2 + \frac{1}{2}(3m)(-\dot{x} - \dot{y})^2 + \frac{1}{2}m\dot{y}^2 - \Big((4m)gx + (3m)g(-x - y) + mgy\Big)$$
$$= \frac{7}{2}m\dot{x}^2 + 3m\dot{x}\dot{y} + 2m\dot{y}^2 - mg(x - 2y). \quad (6.143)$$

This is invariant under the transformation $x \to x + 2\epsilon$ and $y \to y + \epsilon$. Hence, we can use Noether's theorem, with $K_x = 2$ and $K_y = 1$. The conserved momentum is then

$$P = \frac{\partial L}{\partial \dot{x}} K_x + \frac{\partial L}{\partial \dot{y}} K_y = m(7\dot{x} + 3\dot{y})(2) + m(3\dot{x} + 4\dot{y})(1) = m(17\dot{x} + 10\dot{y}). \quad (6.144)$$

This P is constant. In particular, if the system starts at rest, then $\dot{x}$ always equals $-(10/17)\dot{y}$.

SECOND SOLUTION: The Euler–Lagrange equations are, from Eq. (6.143),

$$\begin{aligned} 7m\ddot{x} + 3m\ddot{y} &= -mg, \\ 3m\ddot{x} + 4m\ddot{y} &= 2mg. \end{aligned} \quad (6.145)$$

Adding the second equation to twice the first gives

$$17m\ddot{x} + 10m\ddot{y} = 0 \quad \Longrightarrow \quad \frac{d}{dt}\Big(17m\dot{x} + 10m\dot{y}\Big) = 0. \quad (6.146)$$

THIRD SOLUTION: We can also solve this problem using $F = ma$. Since the tension T is the same throughout the rope, we see that the three $F = dP/dt$ equations are

$$2T - 4mg = \frac{dP_{4m}}{dt}, \quad 2T - 3mg = \frac{dP_{3m}}{dt}, \quad 2T - mg = \frac{dP_m}{dt}. \quad (6.147)$$

The three forces depend on only two quantities (T and mg), so there must be some combination of them that adds up to zero. If we set $a(2T - 4mg) + b(2T - 3mg) + c(2T - mg) = 0$, then we have $a + b + c = 0$ and $4a + 3b + c = 0$, which is satisfied by $a = 2$, $b = -3$, and $c = 1$. Therefore,

$$\begin{aligned} 0 &= \frac{d}{dt}(2P_{4m} - 3P_{3m} + P_m) \\ &= \frac{d}{dt}\Big(2(4m)\dot{x} - 3(3m)(-\dot{x} - \dot{y}) + m\dot{y}\Big) \\ &= \frac{d}{dt}(17m\dot{x} + 10m\dot{y}). \end{aligned} \quad (6.148)$$

6.10. **Hoop and pulley**

Let the radius to M make an angle θ with the vertical (see Fig. 6.44). Then the coordinates of M relative to the center of the hoop are $R(\sin\theta, -\cos\theta)$. The height of m, relative to its position when M is at the bottom of the hoop, is $y = -R\theta$. The Lagrangian is therefore (and yes, we've chosen a different $y = 0$ reference point for each mass, but such a definition changes the potential only by a constant amount, which is irrelevant)

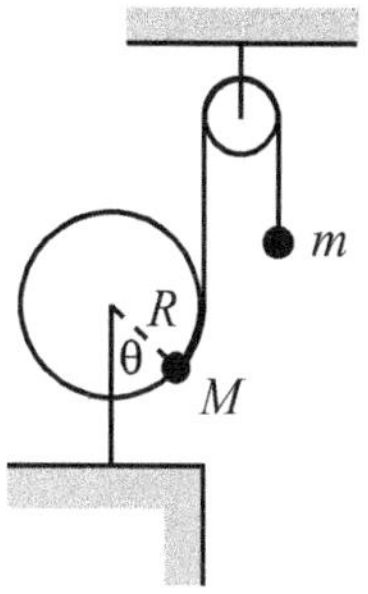

Fig. 6.44

$$L = \frac{1}{2}(M + m)R^2\dot{\theta}^2 + MgR\cos\theta + mgR\theta. \quad (6.149)$$

The equation of motion is then

$$(M + m)R\ddot{\theta} = g(m - M\sin\theta). \quad (6.150)$$

This is just $F = ma$ along the direction of the string (because $Mg\sin\theta$ is the tangential component of the gravitational force on M).

Equilibrium occurs when $\dot{\theta} = \ddot{\theta} = 0$. From Eq. (6.150), we see that this happens at $\sin\theta_0 = m/M$. Letting $\theta \equiv \theta_0 + \delta$, and expanding Eq. (6.150) to first order in δ, gives

$$\ddot{\delta} + \left(\frac{Mg\cos\theta_0}{(M + m)R}\right)\delta = 0. \quad (6.151)$$

The frequency of small oscillations is therefore

$$\omega = \sqrt{\frac{M\cos\theta_0}{M+m}}\sqrt{\frac{g}{R}} = \left(\frac{M-m}{M+m}\right)^{1/4}\sqrt{\frac{g}{R}}, \tag{6.152}$$

where we have used $\cos\theta_0 = \sqrt{1-\sin\theta_0^2}$.

Remarks: If $M \gg m$, then $\theta_0 \approx 0$, and $\omega \approx \sqrt{g/R}$. This makes sense, because m can be ignored, so M essentially oscillates around the bottom of the hoop, just like a pendulum of length R.

If M is only slightly greater than m, then $\theta_0 \approx \pi/2$, and $\omega \approx 0$. This also makes sense, because if $\theta \approx \pi/2$, then the restoring force $g(m - M\sin\theta)$ doesn't change much as θ changes (the derivative of $\sin\theta$ is zero at $\theta = \pi/2$), so it's as if we have a pendulum in a very weak gravitational field.

We can actually derive the frequency in Eq. (6.152) without doing any calculations. Look at M at the equilibrium position. The tangential forces on it cancel, and the radially inward force from the hoop must be $Mg\cos\theta_0$ to balance the radial outward component of the gravitational force. Therefore, for all the mass M knows, it is sitting at the bottom of a hoop of radius R in a world where gravity has strength $g' = g\cos\theta_0$. The general formula for the frequency of a pendulum (as you can quickly show) is $\omega = \sqrt{F'/M'R}$, where F' is the gravitational force (which is Mg' here), and M' is the total mass being accelerated (which is $M+m$ here). This gives the ω in Eq. (6.152). (This reasoning is a little subtle; food for thought.) ♣

6.11. **Bead on a rotating hoop**

Breaking the velocity up into the component along the hoop plus the component perpendicular to the hoop, we find

$$L = \frac{1}{2}m(\omega^2R^2\sin^2\theta + R^2\dot{\theta}^2) + mgR\cos\theta. \tag{6.153}$$

The equation of motion is then

$$R\ddot{\theta} = \sin\theta(\omega^2R\cos\theta - g). \tag{6.154}$$

The $F = ma$ interpretation of this is that the component of gravity pulling downward along the hoop accounts for the acceleration along the hoop plus the component of the centripetal acceleration along the hoop.

Equilibrium occurs when $\dot{\theta} = \ddot{\theta} = 0$. The right-hand side of Eq. (6.154) equals zero when either $\sin\theta = 0$ (that is, $\theta = 0$ or $\theta = \pi$) or $\cos\theta = g/(\omega^2R)$. Since $\cos\theta$ must be less than or equal to 1, this second condition is possible only if $\omega^2 \geq g/R$. So we have two cases:

- If $\omega^2 < g/R$, then $\theta = 0$ and $\theta = \pi$ are the only equilibrium points.

  The $\theta = \pi$ case is unstable. This is fairly intuitive, but it can also be seen mathematically by letting $\theta \equiv \pi + \delta$, where δ is small. Equation (6.154) then becomes

$$\ddot{\delta} - \delta(\omega^2 + g/R) = 0. \tag{6.155}$$

  The coefficient of δ is negative, so δ undergoes exponential instead of oscillatory motion.

  The $\theta = 0$ case turns out to be stable. For small θ, Eq. (6.154) becomes

$$\ddot{\theta} + \theta(g/R - \omega^2) = 0. \tag{6.156}$$

  The coefficient of θ is positive, so we have sinusoidal solutions. The frequency of small oscillations is $\sqrt{g/R - \omega^2}$. This goes to zero as $\omega \to \sqrt{g/R}$.
- If $\omega^2 \geq g/R$, then $\theta = 0$, $\theta = \pi$, and $\cos\theta_0 \equiv g/(\omega^2R)$ are all equilibrium points. The $\theta = \pi$ case is again unstable, by looking at Eq. (6.155). And the

$\theta = 0$ case is also unstable, because the coefficient of θ in Eq. (6.156) is now negative (or zero, if $\omega^2 = g/R$).

Therefore, $\cos\theta_0 \equiv g/(\omega^2 R)$ is the only stable equilibrium. To find the frequency of small oscillations, let $\theta \equiv \theta_0 + \delta$ in Eq. (6.154), and expand to first order in δ. Using $\cos\theta_0 \equiv g/(\omega^2 R)$, we find

$$\ddot{\delta} + \left(\omega^2 \sin^2\theta_0\right)\delta = 0. \tag{6.157}$$

The frequency of small oscillations is therefore $\omega \sin\theta_0 = \sqrt{\omega^2 - g^2/\omega^2 R^2}$.

The frequency $\omega = \sqrt{g/R}$ is the critical frequency above which there is a stable equilibrium at $\theta \neq 0$, that is, above which the mass wants to move away from the bottom of the hoop.

REMARK: This frequency of small oscillations goes to zero as $\omega \to \sqrt{g/R}$. And it approximately equals ω as $\omega \to \infty$. This second limit can be viewed in the following way. For very large ω, gravity isn't important, and the bead feels a centripetal force (the normal force from the hoop) essentially equal to $m\omega^2 R$ as it moves near $\theta = \pi/2$. So for all the bead knows, it is a pendulum of length R in a world where "gravity" pulls sideways with a force $m\omega^2 R \equiv mg'$ (outward, so that it is approximately canceled by the inward-pointing normal force, just as the downward gravitational force is approximately canceled by the upward tension in a regular pendulum). The frequency of such a pendulum is $\sqrt{g'/R} = \sqrt{\omega^2 R/R} = \omega$. ♣

6.12. **Another bead on a rotating hoop**

With the angles ωt and θ defined as in Fig. 6.45, the Cartesian coordinates for the bead are

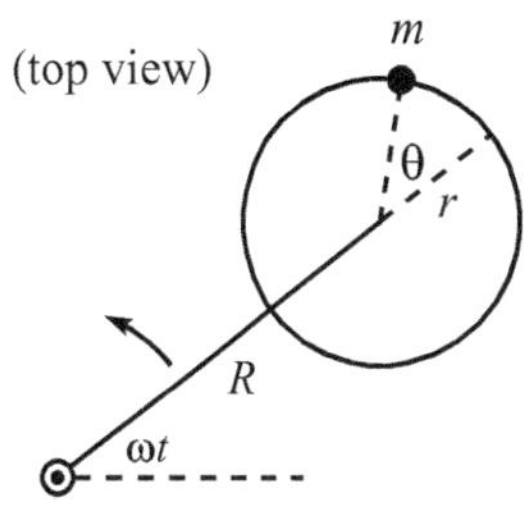

Fig. 6.45

$$(x, y) = \big(R\cos\omega t + r\cos(\omega t + \theta),\ R\sin\omega t + r\sin(\omega t + \theta)\big). \tag{6.158}$$

The velocity is then

$$(\dot{x},\dot{y}) = \big(-\omega R\sin\omega t - r(\omega + \dot{\theta})\sin(\omega t + \theta),$$
$$\omega R\cos\omega t + r(\omega + \dot{\theta})\cos(\omega t + \theta)\big). \tag{6.159}$$

The square of the speed is therefore

$$\begin{aligned} v^2 &= R^2\omega^2 + r^2(\omega + \dot{\theta})^2 \\ &\quad + 2Rr\omega(\omega + \dot{\theta})\big(\sin\omega t\sin(\omega t + \theta) + \cos\omega t\cos(\omega t + \theta)\big) \\ &= R^2\omega^2 + r^2(\omega + \dot{\theta})^2 + 2Rr\omega(\omega + \dot{\theta})\cos\theta. \end{aligned} \tag{6.160}$$

This speed can also be obtained by using the law of cosines to add the velocity of the center of the hoop to the velocity of the bead with respect to the center (as you can show).

There is no potential energy, so the Lagrangian is simply $L = mv^2/2$. The equation of motion is then, as you can show,

$$r\ddot{\theta} + R\omega^2\sin\theta = 0. \tag{6.161}$$

Equilibrium occurs when $\dot{\theta} = \ddot{\theta} = 0$, so Eq. (6.161) tells us that the equilibrium is located at $\theta = 0$, which makes intuitive sense. (Another solution is $\theta = \pi$, but that's an unstable equilibrium.) A small-angle approximation in Eq. (6.161) gives $\ddot{\theta} + (R/r)\omega^2\theta = 0$, so the frequency of small oscillations is $\Omega = \omega\sqrt{R/r}$.

REMARKS: If $R \ll r$, then $\Omega \approx 0$. This makes sense, because the frictionless hoop is essentially not moving. If $R = r$, then $\Omega = \omega$. If $R \gg r$, then Ω is very large. In this case, we can double-check the $\Omega = \omega\sqrt{R/r}$ result in the following way. In the accelerating frame of the hoop, the bead feels a centrifugal force (discussed in Chapter 10) of $m(R+r)\omega^2$. For all the bead knows, it is in a gravitational field with strength $g' \equiv (R+r)\omega^2$. So the bead (which acts like a pendulum of length r), oscillates with a frequency equal to

$$\sqrt{\frac{g'}{r}} = \sqrt{\frac{(R+r)\omega^2}{r}} \approx \omega\sqrt{\frac{R}{r}} \qquad (\text{for } R \gg r). \tag{6.162}$$

Note that if we try to use this "effective gravity" argument as a double check for smaller values of R, we get the wrong answer. For example, if $R = r$, we obtain an oscillation frequency of $\omega\sqrt{2R/r}$, instead of the correct value $\omega\sqrt{R/r}$. This is because in reality the centrifugal force fans out near the equilibrium point, while our "effective gravity" argument assumes that the field lines are parallel (and so it gives a frequency that is too large). ♣

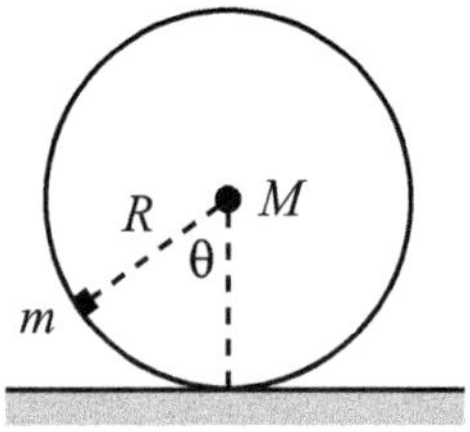

Fig. 6.46

6.13. **Mass on a wheel**

Let the angle θ be defined as in Fig. 6.46, with the convention that θ is positive if M is to the right of m. Then the position of m in Cartesian coordinates, relative to the point where m would be in contact with the ground, is

$$(x, y)_m = R(\theta - \sin\theta, 1 - \cos\theta). \tag{6.163}$$

We have used the nonslipping condition to say that the present contact point is a distance $R\theta$ to the right of where m would be in contact with the ground. Differentiating Eq. (6.163), we find that the square of m's speed is $v_m^2 = 2R^2\dot{\theta}^2(1-\cos\theta)$.

The position of M is $(x, y)_M = R(\theta, 1)$, so the square of its speed is $v_M^2 = R^2\dot{\theta}^2$. The Lagrangian is therefore

$$L = \frac{1}{2}MR^2\dot{\theta}^2 + mR^2\dot{\theta}^2(1-\cos\theta) + mgR\cos\theta, \tag{6.164}$$

where we have measured both potential energies relative to the height of M. The equation of motion is

$$MR\ddot{\theta} + 2mR\ddot{\theta}(1-\cos\theta) + mR\dot{\theta}^2\sin\theta + mg\sin\theta = 0. \tag{6.165}$$

In the case of small oscillations, we can use $\cos\theta \approx 1-\theta^2/2$ and $\sin\theta \approx \theta$. The second and third terms in Eq. (6.165) are then third order in θ and can be neglected (basically, the middle term in Eq. (6.164), which is the kinetic energy of m, is negligible), so we find

$$\ddot{\theta} + \left(\frac{mg}{MR}\right)\theta = 0. \tag{6.166}$$

The frequency of small oscillations is therefore

$$\omega = \sqrt{\frac{m}{M}}\sqrt{\frac{g}{R}}. \tag{6.167}$$

Remarks: If $M \gg m$, then $\omega \to 0$. This makes sense. If $m \gg M$, then $\omega \to \infty$. This also makes sense, because the huge mg force makes the situation similar to one where the wheel is bolted to the ground, in which case the wheel vibrates with a high frequency.

Equation (6.167) can actually be derived in a much quicker way, using torque. For small oscillations, the gravitational force on m produces a torque of $-mgR\theta$ around the contact point on the ground. For small θ, m has essentially no moment of inertia around the contact point, so the total moment of inertia is simply MR^2. Therefore, $\tau = I\alpha$ gives $-mgR\theta = MR^2\ddot{\theta}$, from which the result follows. ♣

6.14. **Pendulum with a free support**

Let x be the coordinate of M, and let θ be the angle of the pendulum (see Fig. 6.47). Then the position of the mass m in Cartesian coordinates is $(x + \ell \sin\theta, -\ell\cos\theta)$. Taking the derivative to find the velocity, and then squaring to find the speed, gives $v_m^2 = \dot{x}^2 + \ell^2\dot{\theta}^2 + 2\ell\dot{x}\dot{\theta}\cos\theta$. The Lagrangian is therefore

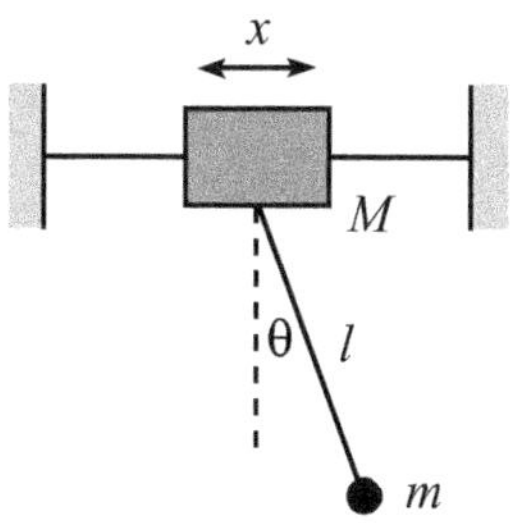

Fig. 6.47

$$L = \frac{1}{2}M\dot{x}^2 + \frac{1}{2}m(\dot{x}^2 + \ell^2\dot{\theta}^2 + 2\ell\dot{x}\dot{\theta}\cos\theta) + mg\ell\cos\theta. \tag{6.168}$$

The equations of motion obtained from varying x and θ are

$$\begin{aligned}(M+m)\ddot{x} + m\ell\ddot{\theta}\cos\theta - m\ell\dot{\theta}^2\sin\theta &= 0,\\ \ell\ddot{\theta} + \ddot{x}\cos\theta + g\sin\theta &= 0.\end{aligned} \tag{6.169}$$

If θ is small, we can use the small-angle approximations, $\cos\theta \approx 1 - \theta^2/2$ and $\sin\theta \approx \theta$. Keeping only the terms that are first-order in θ, we obtain

$$\begin{aligned}(M+m)\ddot{x} + m\ell\ddot{\theta} &= 0,\\ \ddot{x} + \ell\ddot{\theta} + g\theta &= 0.\end{aligned} \tag{6.170}$$

The first equation expresses momentum conservation. Integrating it twice gives

$$x = -\left(\frac{m\ell}{M+m}\right)\theta + At + B. \tag{6.171}$$

The second equation is $F = ma$ in the tangential direction. Eliminating $\ddot{x}$ from Eq. (6.170) gives

$$\ddot{\theta} + \left(\frac{M+m}{M}\right)\frac{g}{\ell}\theta = 0. \tag{6.172}$$

Therefore, $\theta(t) = C\cos(\omega t + \phi)$, where

$$\omega = \sqrt{1 + \frac{m}{M}}\sqrt{\frac{g}{\ell}}. \tag{6.173}$$

The general solutions for θ and x are therefore

$$\theta(t) = C\cos(\omega t + \phi), \quad x(t) = -\frac{Cm\ell}{M+m}\cos(\omega t + \phi) + At + B. \tag{6.174}$$

The constant B is irrelevant, so we'll ignore it. The two normal modes are:

- $A = 0$: In this case, $x = -\theta m\ell/(M+m)$. Both masses oscillate with the frequency ω given in Eq. (6.173), always moving in opposite directions. The center of mass does not move (as you can verify).
- $C = 0$: In this case, $\theta = 0$ and $x = At$. The pendulum hangs vertically, with both masses moving horizontally at the same speed. The frequency of oscillations is zero in this mode.

REMARKS: If $M \gg m$, then $\omega = \sqrt{g/\ell}$, as expected, because the support essentially stays still.

If $m \gg M$, then $\omega \to \sqrt{m/M}\sqrt{g/\ell} \to \infty$. This makes sense, because the tension in the rod is very large. We can actually be quantitative about this limit. For small oscillations and for $m \gg M$, the tension of mg in the rod produces a sideways force of $mg\theta$ on M. So the horizontal $F = Ma$ equation for M is $mg\theta = M\ddot{x}$. But $x \approx -\ell\theta$ in this limit, so we have $mg\theta = -M\ell\ddot{\theta}$, which gives the desired frequency. ♣

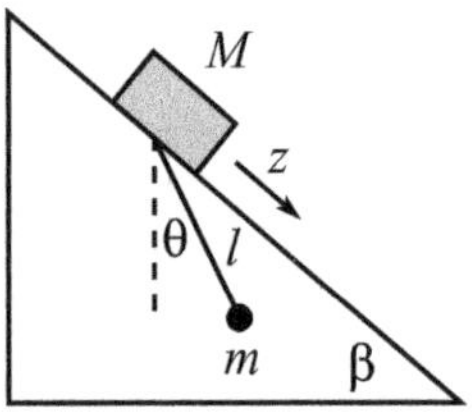

Fig. 6.48

6.15. **Pendulum support on an inclined plane**

Let z be the coordinate of M along the plane, and let θ be the angle of the pendulum (see Fig. 6.48). In Cartesian coordinates, the positions of M and m are

$$\begin{aligned}(x,y)_M &= (z\cos\beta, -z\sin\beta),\\ (x,y)_m &= (z\cos\beta + \ell\sin\theta, -z\sin\beta - \ell\cos\theta).\end{aligned} \tag{6.175}$$

Differentiating these positions, we find that the squares of the speeds are

$$\begin{aligned}v_M^2 &= \dot{z}^2,\\ v_m^2 &= \dot{z}^2 + \ell^2\dot{\theta}^2 + 2\ell\dot{z}\dot{\theta}(\cos\beta\cos\theta - \sin\beta\sin\theta).\end{aligned} \tag{6.176}$$

The Lagrangian is therefore

$$\frac{1}{2}M\dot{z}^2 + \frac{1}{2}m\Big(\dot{z}^2 + \ell^2\dot{\theta}^2 + 2\ell\dot{z}\dot{\theta}\cos(\theta+\beta)\Big) + Mgz\sin\beta + mg(z\sin\beta + \ell\cos\theta). \tag{6.177}$$

The equations of motion obtained from varying z and θ are

$$\begin{aligned}(M+m)\ddot{z} + m\ell\Big(\ddot{\theta}\cos(\theta+\beta) - \dot{\theta}^2\sin(\theta+\beta)\Big) &= (M+m)g\sin\beta,\\ \ell\ddot{\theta} + \ddot{z}\cos(\theta+\beta) &= -g\sin\theta.\end{aligned} \tag{6.178}$$

Let us now consider small oscillations about the equilibrium point (where $\ddot{\theta} = \dot{\theta} = 0$). We must first determine where this point is. The first equation above gives $\ddot{z} = g\sin\beta$. The second equation then gives $g\sin\beta\cos(\theta+\beta) = -g\sin\theta$. By expanding the cosine term, we find $\tan\theta = -\tan\beta$, so $\theta = -\beta$. ($\theta = \pi - \beta$ is also a solution, but this is an unstable equilibrium.) The equilibrium position of the pendulum is therefore where the string is perpendicular to the plane.[14]

To find the normal modes and frequencies for small oscillations, let $\theta \equiv -\beta + \delta$, and expand Eq. (6.178) to first order in δ. Letting $\ddot{\eta} \equiv \ddot{z} - g\sin\beta$ for convenience, we obtain

$$\begin{aligned}(M+m)\ddot{\eta} + m\ell\ddot{\delta} &= 0,\\ \ddot{\eta} + \ell\ddot{\delta} + (g\cos\beta)\delta &= 0.\end{aligned} \tag{6.179}$$

Using the determinant method (or using the method in Problem 6.14; either way works), we find the frequencies of the normal modes to be

$$\omega_1 = 0, \quad \text{and} \quad \omega_2 = \sqrt{1+\frac{m}{M}}\sqrt{\frac{g\cos\beta}{\ell}}. \tag{6.180}$$

These are the same as the frequencies in the previous problem (where M moves horizontally), but with $g\cos\beta$ in place of g; compare Eq. (6.179) with Eq. (6.170).[15] Looking at Eq. (6.174), and recalling the definition of η, we see that the general solutions for θ and z are

$$\theta(t) = -\beta + C\cos(\omega t + \phi), \quad z(t) = -\frac{Cm\ell}{M+m}\cos(\omega t+\phi) + \frac{g\sin\beta}{2}t^2 + At + B. \tag{6.181}$$

[14] This makes sense. The tension in the string is perpendicular to the plane, so for all the pendulum bob knows, it may as well be sliding down a plane parallel to the given one, a distance ℓ away. Given the same initial speed, the two masses slide down their two "planes" with equal speeds at all times.

[15] This makes sense, because in a frame that accelerates down the plane at $g\sin\beta$, the only external force on the masses is a gravitational force of $g\cos\beta$ perpendicular to the plane. As far as M and m are concerned, they live in a world where gravity pulls "downward" (perpendicular to the plane) with strength $g' = g\cos\beta$.

The constant B is irrelevant, so we'll ignore it. The basic difference between these normal modes and the ones in the previous problem is the acceleration down the plane. If you go to a frame that accelerates down the plane at $g \sin\beta$, and if you tilt your head at an angle β and accept the fact that $g' = g\cos\beta$ in your world, then the setup is identical to the one in the previous problem.

6.16. **Tilting plane**

Relative to the support, the positions of the masses are

$$\begin{aligned}(x, y)_M &= (\ell\sin\theta, -\ell\cos\theta),\\ (x, y)_m &= (\ell\sin\theta + x\cos\theta, -\ell\cos\theta + x\sin\theta).\end{aligned} \tag{6.182}$$

Differentiating these positions, we find that the squares of the speeds are

$$v_M^2 = \ell^2\dot\theta^2, \quad v_m^2 = (\ell\dot\theta + \dot x)^2 + x^2\dot\theta^2. \tag{6.183}$$

You can also obtain v_m^2 by noting that $(\ell\dot\theta + \dot x)$ is the speed along the long rod, and $x\dot\theta$ is the speed perpendicular to it. The Lagrangian is

$$L = \frac{1}{2}M\ell^2\dot\theta^2 + \frac{1}{2}m\Big((\ell\dot\theta + \dot x)^2 + x^2\dot\theta^2\Big) + Mg\ell\cos\theta + mg(\ell\cos\theta - x\sin\theta). \tag{6.184}$$

The equations of motion obtained from varying x and θ are

$$\begin{aligned}\ell\ddot\theta + \ddot x &= x\dot\theta^2 - g\sin\theta,\\ M\ell^2\ddot\theta + m\ell(\ell\ddot\theta + \ddot x) + mx^2\ddot\theta + 2mx\dot x\dot\theta &= -(M+m)g\ell\sin\theta - mgx\cos\theta.\end{aligned} \tag{6.185}$$

Let us now consider the case where both x and θ are small (or more precisely, $\theta \ll 1$ and $x/\ell \ll 1$). Expanding Eq. (6.185) to first order in θ and x/ℓ gives

$$\begin{aligned}(\ell\ddot\theta + \ddot x) + g\theta &= 0,\\ M\ell(\ell\ddot\theta + g\theta) + m\ell(\ell\ddot\theta + \ddot x) + mg\ell\theta + mgx &= 0.\end{aligned} \tag{6.186}$$

We can simplify these a bit. Using the first equation to substitute $-g\theta$ for $(\ell\ddot\theta + \ddot x)$, and also $-\ddot x$ for $(\ell\ddot\theta + g\theta)$, in the second equation gives

$$\begin{aligned}\ell\ddot\theta + \ddot x + g\theta &= 0,\\ -M\ell\ddot x + mgx &= 0.\end{aligned} \tag{6.187}$$

The normal modes can be found using the determinant method, or we can find them just by inspection. The second equation says that either $x(t) \equiv 0$, or $x(t) = A\cosh(\alpha t + \beta)$, where $\alpha = \sqrt{mg/M\ell}$. So we have two cases:

- If $x(t) = 0$, then the first equation in (6.187) says that the normal mode is

$$\begin{pmatrix}\theta\\ x\end{pmatrix} = B\begin{pmatrix}1\\ 0\end{pmatrix}\cos(\omega t + \phi), \tag{6.188}$$

  where $\omega \equiv \sqrt{g/\ell}$. This mode is fairly clear. With the proper initial conditions, m will stay right where M is. The normal force from the long rod will be exactly what is needed in order for m to undergo the same oscillatory motion as M. The two masses may as well be two pendulums of length ℓ swinging side by side.
- If $x(t) = A\cosh(\alpha t + \beta)$, then the first equation in (6.187) can be solved (by guessing a particular solution for θ of the same form) to give the normal mode,

$$\begin{pmatrix}\theta\\ x\end{pmatrix} = C\begin{pmatrix}-m\\ \ell(M+m)\end{pmatrix}\cosh(\alpha t + \beta), \tag{6.189}$$

  where $\alpha = \sqrt{mg/M\ell}$. This mode is not as clear. And indeed, its range of validity is rather limited. The exponential behavior will quickly make x and θ large, and thus outside the validity of our small-variable approximations. You can show that in this mode the center of mass remains directly below the pivot. This can occur, for example, by having m move down to the right as the rods rotate and swing M up to the left. There is no oscillation in this mode; the positions keep growing. The CM falls, to provide for the increasing kinetic energy.

6.17. **Rotating curve**

The speed along the curve is $\dot{x}\sqrt{1+y'^2}$, and the speed perpendicular to the curve is ωx. So the Lagrangian is

$$L = \frac{1}{2}m\left(\omega^2 x^2 + \dot{x}^2(1+y'^2)\right) - mgy, \tag{6.190}$$

where $y(x) = b(x/a)^\lambda$. The equation of motion is then

$$\frac{d}{dt}\left(\frac{\partial L}{\partial \dot{x}}\right) = \frac{\partial L}{\partial x} \quad \Longrightarrow \quad \ddot{x}(1+y'^2) + \dot{x}^2 y' y'' = \omega^2 x - gy'. \tag{6.191}$$

Equilibrium occurs when $\dot{x} = \ddot{x} = 0$, so Eq. (6.191) says that the equilibrium value of x satisfies

$$x_0 = \frac{gy'(x_0)}{\omega^2}. \tag{6.192}$$

The $F = ma$ explanation for this (writing $y'(x_0)$ as $\tan\theta$, where θ is the angle of the curve, and then multiplying through by $\omega^2\cos\theta$) is that the component of gravity along the curve accounts for the component of the centripetal acceleration along the curve. Using $y(x) = b(x/a)^\lambda$, Eq. (6.192) yields

$$x_0 = a\left(\frac{a^2\omega^2}{\lambda gb}\right)^{1/(\lambda-2)}. \tag{6.193}$$

As $\lambda \to \infty$, we see that x_0 goes to a. This makes sense, because the curve essentially equals zero up to a, and then it rises very steeply. You can check numerous other limits. Letting $x \equiv x_0 + \delta$ in Eq. (6.191), and expanding to first order in δ, gives

$$\ddot{\delta}\left(1 + y'(x_0)^2\right) = \delta\left(\omega^2 - gy''(x_0)\right). \tag{6.194}$$

The frequency of small oscillations is therefore given by

$$\Omega^2 = \frac{gy''(x_0) - \omega^2}{1 + y'(x_0)^2}. \tag{6.195}$$

Using the explicit form of y, along with Eq. (6.193), we find

$$\Omega^2 = \frac{(\lambda-2)\omega^2}{1 + \frac{a^2\omega^4}{g^2}\left(\frac{a^2\omega^2}{\lambda gb}\right)^{2/(\lambda-2)}}. \tag{6.196}$$

We see that λ must be greater than 2 in order for there to be oscillatory motion around the equilibrium point. For $\lambda < 2$, the equilibrium point is unstable, that is, to the left the force is inward, and to the right the force is outward.

For the case $\lambda = 2$, we have $y(x) = b(x/a)^2$, so the equilibrium condition, Eq. (6.192), gives $x_0 = (2gb/a^2\omega^2)x_0$. For this to be true for some x_0, we must have $\omega^2 = 2gb/a^2$. But if this holds, then Eq. (6.192) is true for all x. So in the special case of $\lambda = 2$, the bead happily sits anywhere on the curve if $\omega^2 = 2gb/a^2$. (In the rotating frame of the curve, the tangential components of the centrifugal and gravitational forces exactly cancel at all points.) If $\lambda = 2$ and $\omega^2 \neq 2gb/a^2$, then the particle feels a force either always inward or always outward.

REMARKS: For $\omega \to 0$, Eqs. (6.193) and (6.196) give $x_0 \to 0$ and $\Omega \to 0$. And for $\omega \to \infty$, they give $x_0 \to \infty$ and $\Omega \to 0$. In both cases $\Omega \to 0$, because in both cases the equilibrium position is at a place where the curve is very flat (horizontally or vertically, respectively), and the restoring force ends up being small.

For $\lambda \to \infty$, we have $x_0 \to a$ and $\Omega \to \infty$. The frequency is large here because the equilibrium position at a is where the curve has a sharp corner, so the restoring force changes quickly with position. Or, you can think of it as a pendulum with a very small length, if you approximate the "corner" by a tiny circle. ♣

6.18. **Motion in a cone**

If the particle's distance from the axis is r, then its height is $r/\tan\alpha$, and its distance up along the cone is $r/\sin\alpha$. Breaking the velocity into components up along the cone and around the cone, we see that the square of the speed is $v^2 = \dot{r}^2/\sin^2\alpha + r^2\dot{\theta}^2$. The Lagrangian is therefore

$$L = \frac{1}{2}m\left(\frac{\dot{r}^2}{\sin^2\alpha} + r^2\dot{\theta}^2\right) - \frac{mgr}{\tan\alpha}. \tag{6.197}$$

The equations of motion obtained from varying θ and r are

$$\begin{aligned} &\frac{d}{dt}(mr^2\dot{\theta}) = 0 \\ &\ddot{r} = r\dot{\theta}^2 \sin^2\alpha - g\cos\alpha\sin\alpha. \end{aligned} \tag{6.198}$$

The first of these equations expresses conservation of angular momentum. The second equation is more transparent if we divide through by $\sin\alpha$. With $x \equiv r/\sin\alpha$ being the distance up along the cone, we have $\ddot{x} = (r\dot{\theta}^2)\sin\alpha - g\cos\alpha$. This is the $F = ma$ statement for the diagonal x direction.

Letting $mr^2\dot{\theta} \equiv L$, we can eliminate $\dot{\theta}$ from the second equation to obtain

$$\ddot{r} = \frac{L^2\sin^2\alpha}{m^2r^3} - g\cos\alpha\sin\alpha. \tag{6.199}$$

We will now calculate the two desired frequencies.

- Frequency of circular oscillations, ω: For circular motion with $r = r_0$, we have $\dot{r} = \ddot{r} = 0$, so the second of Eqs. (6.198) gives

$$\omega \equiv \dot{\theta} = \sqrt{\frac{g}{r_0\tan\alpha}}. \tag{6.200}$$

- Frequency of oscillations about a circle, Ω: If the orbit were actually the circle $r = r_0$, then Eq. (6.199) would give (with $\ddot{r} = 0$)

$$\frac{L^2\sin^2\alpha}{m^2r_0^3} = g\cos\alpha\sin\alpha. \tag{6.201}$$

This is equivalent to Eq. (6.200), which can be seen by writing L as $mr_0^2\dot{\theta}$.

We will now use our standard procedure of letting $r(t) = r_0 + \delta(t)$, where $\delta(t)$ is very small, and then plugging this into Eq. (6.199) and expanding to first order in δ. Using

$$\frac{1}{(r_0+\delta)^3} \approx \frac{1}{r_0^3 + 3r_0^2\delta} = \frac{1}{r_0^3(1+3\delta/r_0)} \approx \frac{1}{r_0^3}\left(1 - \frac{3\delta}{r_0}\right), \tag{6.202}$$

we have

$$\ddot{\delta} = \frac{L^2\sin^2\alpha}{m^2r_0^3}\left(1 - \frac{3\delta}{r_0}\right) - g\cos\alpha\sin\alpha. \tag{6.203}$$

Recalling Eq. (6.201), the terms not involving δ cancel, and we are left with

$$\ddot{\delta} = -\left(\frac{3L^2\sin^2\alpha}{m^2r_0^4}\right)\delta. \tag{6.204}$$

Using Eq. (6.201) again to eliminate L we have

$$\ddot{\delta} + \left(\frac{3g}{r_0}\sin\alpha\cos\alpha\right)\delta = 0. \tag{6.205}$$

Therefore,

$$\Omega = \sqrt{\frac{3g}{r_0}\sin\alpha\cos\alpha}. \tag{6.206}$$

Having found the two desired frequencies in Eqs. (6.200) and (6.206), we see that their ratio is

$$\frac{\Omega}{\omega} = \sqrt{3}\sin\alpha. \tag{6.207}$$

This ratio Ω/ω is independent of r_0. The two frequencies are equal if $\sin\alpha = 1/\sqrt{3}$, that is, if $\alpha \approx 35.3^\circ \equiv \tilde{\alpha}$. If $\alpha = \tilde{\alpha}$, then after one revolution around the cone, r returns to the value it had at the beginning of the revolution. So the particle undergoes periodic motion.

REMARKS: In the limit $\alpha \to 0$ (that is, the cone is very thin), Eq. (6.207) says that $\Omega/\omega \to 0$. In fact, Eqs. (6.200) and (6.206) say that $\omega \to \infty$ and $\Omega \to 0$. So the particle spirals around many times during one complete r cycle. This seems intuitive.

In the limit $\alpha \to \pi/2$ (that is, the cone is almost a flat plane), both ω and Ω go to zero, and Eq. (6.207) says that $\Omega/\omega \to \sqrt{3}$. This result is not at all obvious.

If $\Omega/\omega = \sqrt{3}\sin\alpha$ is a rational number, then the particle undergoes periodic motion. For example, if $\alpha = 60^\circ$, then $\Omega/\omega = 3/2$, so it takes two complete circles for r to go through three cycles. Or, if $\alpha = \arcsin(1/2\sqrt{3}) \approx 16.8^\circ$, then $\Omega/\omega = 1/2$, so it takes two complete circles for r to go through one cycle. ♣

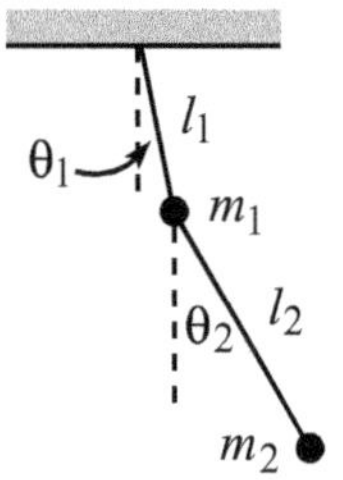

Fig. 6.49

6.19. **Double pendulum**

Relative to the pivot point, the Cartesian coordinates of m_1 and m_2 are, respectively (see Fig. 6.49),

$$\begin{aligned}(x,y)_1 &= (\ell_1\sin\theta_1, -\ell_1\cos\theta_1),\\ (x,y)_2 &= (\ell_1\sin\theta_1 + \ell_2\sin\theta_2, -\ell_1\cos\theta_1 - \ell_2\cos\theta_2).\end{aligned} \tag{6.208}$$

Taking the derivative to find the velocities, and then squaring, gives

$$\begin{aligned}v_1^2 &= \ell_1^2\dot{\theta}_1^2,\\ v_2^2 &= \ell_1^2\dot{\theta}_1^2 + \ell_2^2\dot{\theta}_2^2 + 2\ell_1\ell_2\dot{\theta}_1\dot{\theta}_2(\cos\theta_1\cos\theta_2 + \sin\theta_1\sin\theta_2).\end{aligned} \tag{6.209}$$

The Lagrangian is therefore

$$\begin{aligned}L = {}&\frac{1}{2}m_1\ell_1^2\dot{\theta}_1^2 + \frac{1}{2}m_2\Big(\ell_1^2\dot{\theta}_1^2 + \ell_2^2\dot{\theta}_2^2 + 2\ell_1\ell_2\dot{\theta}_1\dot{\theta}_2\cos(\theta_1-\theta_2)\Big)\\ &+ m_1 g\ell_1\cos\theta_1 + m_2 g(\ell_1\cos\theta_1 + \ell_2\cos\theta_2).\end{aligned} \tag{6.210}$$

The equations of motion obtained from varying θ_1 and θ_2 are

$$\begin{aligned}0 = {}&(m_1+m_2)\ell_1^2\ddot{\theta}_1 + m_2\ell_1\ell_2\ddot{\theta}_2\cos(\theta_1-\theta_2) + m_2\ell_1\ell_2\dot{\theta}_2^2\sin(\theta_1-\theta_2)\\ &+ (m_1+m_2)g\ell_1\sin\theta_1,\\ 0 = {}&m_2\ell_2^2\ddot{\theta}_2 + m_2\ell_1\ell_2\ddot{\theta}_1\cos(\theta_1-\theta_2) - m_2\ell_1\ell_2\dot{\theta}_1^2\sin(\theta_1-\theta_2)\\ &+ m_2 g\ell_2\sin\theta_2.\end{aligned} \tag{6.211}$$

This is a bit of a mess, but it simplifies greatly if we consider small oscillations. Using the small-angle approximations and keeping only the leading-order terms, we obtain

$$\begin{aligned} 0 &= (m_1 + m_2)\ell_1\ddot{\theta}_1 + m_2\ell_2\ddot{\theta}_2 + (m_1 + m_2)g\theta_1, \\ 0 &= \ell_2\ddot{\theta}_2 + \ell_1\ddot{\theta}_1 + g\theta_2. \end{aligned} \tag{6.212}$$

Consider now the special case, $\ell_1 = \ell_2 \equiv \ell$. We can find the frequencies of the normal modes by using the determinant method, discussed in Section 4.5. You can show that the result is

$$\omega_\pm = \sqrt{\frac{m_1 + m_2 \pm \sqrt{m_1 m_2 + m_2^2}}{m_1}}\sqrt{\frac{g}{\ell}}. \tag{6.213}$$

The normal modes turn out to be, after some simplification,

$$\begin{pmatrix} \theta_1(t) \\ \theta_2(t) \end{pmatrix}_\pm \propto \begin{pmatrix} \mp\sqrt{m_2} \\ \sqrt{m_1 + m_2} \end{pmatrix} \cos(\omega_\pm t + \phi_\pm). \tag{6.214}$$

Some special cases are:

- $m_1 = m_2$: The frequencies are

$$\omega_\pm = \sqrt{2 \pm \sqrt{2}}\sqrt{\frac{g}{\ell}}. \tag{6.215}$$

The normal modes are

$$\begin{pmatrix} \theta_1(t) \\ \theta_2(t) \end{pmatrix}_\pm \propto \begin{pmatrix} \mp 1 \\ \sqrt{2} \end{pmatrix} \cos(\omega_\pm t + \phi_\pm). \tag{6.216}$$

- $m_1 \gg m_2$: With $m_2/m_1 \equiv \epsilon$, the frequencies are (to leading nontrivial order in ϵ)

$$\omega_\pm = (1 \pm \sqrt{\epsilon}/2)\sqrt{\frac{g}{\ell}}. \tag{6.217}$$

The normal modes are

$$\begin{pmatrix} \theta_1(t) \\ \theta_2(t) \end{pmatrix}_\pm \propto \begin{pmatrix} \mp\sqrt{\epsilon} \\ 1 \end{pmatrix} \cos(\omega_\pm t + \phi_\pm). \tag{6.218}$$

In both modes, the upper (heavy) mass essentially stands still, and the lower (light) mass oscillates like a pendulum of length ℓ.

- $m_1 \ll m_2$: With $m_1/m_2 \equiv \epsilon$, the frequencies are (to leading order in ϵ)

$$\omega_+ = \sqrt{\frac{2g}{\epsilon\ell}}, \quad \omega_- = \sqrt{\frac{g}{2\ell}}. \tag{6.219}$$

The normal modes are

$$\begin{pmatrix} \theta_1(t) \\ \theta_2(t) \end{pmatrix}_\pm \propto \begin{pmatrix} \mp 1 \\ 1 \end{pmatrix} \cos(\omega_\pm t + \phi_\pm). \tag{6.220}$$

In the first mode, the lower (heavy) mass essentially stands still (from the x_2 in Eq. (6.208)), and the upper (light) mass vibrates back and forth at a high frequency (because there is a very large tension in the rods). In the second mode, the rods form a straight line, and the system is essentially a pendulum of length 2ℓ.

Consider now the special case, $m_1 = m_2$. Using the determinant method, you can show that the frequencies of the normal modes are

$$\omega_{\pm} = \sqrt{g}\sqrt{\frac{\ell_1 + \ell_2 \pm \sqrt{\ell_1^2 + \ell_2^2}}{\ell_1 \ell_2}}. \tag{6.221}$$

The normal modes turn out to be, after some simplification,

$$\begin{pmatrix} \theta_1(t) \\ \theta_2(t) \end{pmatrix}_{\pm} \propto \begin{pmatrix} \ell_2 \\ \ell_2 - \ell_1 \mp \sqrt{\ell_1^2 + \ell_2^2} \end{pmatrix} \cos(\omega_{\pm} t + \phi_{\pm}). \tag{6.222}$$

Some special cases are:

- $\ell_1 = \ell_2$: We already considered this case above. You can show that Eqs. (6.221) and (6.222) agree with Eqs. (6.215) and (6.216), respectively.
- $\ell_1 \gg \ell_2$: With $\ell_2/\ell_1 \equiv \epsilon$, the frequencies are (to leading order in ϵ)

$$\omega_+ = \sqrt{\frac{2g}{\ell_2}}, \quad \omega_- = \sqrt{\frac{g}{\ell_1}}. \tag{6.223}$$

The normal modes are

$$\begin{aligned} \begin{pmatrix} \theta_1(t) \\ \theta_2(t) \end{pmatrix}_{+} &\propto \begin{pmatrix} -\epsilon \\ 2 \end{pmatrix} \cos(\omega_+ t + \phi_+), \\ \begin{pmatrix} \theta_1(t) \\ \theta_2(t) \end{pmatrix}_{-} &\propto \begin{pmatrix} 1 \\ 1 \end{pmatrix} \cos(\omega_- t + \phi_-). \end{aligned} \tag{6.224}$$

In the first mode, the masses essentially move equal distances in opposite directions, at a high frequency (assuming ℓ_2 is small). The factor of 2 in the frequency arises because the angle of ℓ_2 is twice what it would be if m_1 were bolted in place; so m_2 feels double the tangential force. In the second mode, the rods form a straight line, and the masses move just like a mass of $2m$. The system is essentially a pendulum of length ℓ_1.

- $\ell_1 \ll \ell_2$: With $\ell_1/\ell_2 \equiv \epsilon$, the frequencies are (to leading order in ϵ)

$$\omega_+ = \sqrt{\frac{2g}{\ell_1}}, \quad \omega_- = \sqrt{\frac{g}{\ell_2}}. \tag{6.225}$$

The normal modes are

$$\begin{aligned} \begin{pmatrix} \theta_1(t) \\ \theta_2(t) \end{pmatrix}_{+} &\propto \begin{pmatrix} 1 \\ -\epsilon \end{pmatrix} \cos(\omega_+ t + \phi_+), \\ \begin{pmatrix} \theta_1(t) \\ \theta_2(t) \end{pmatrix}_{-} &\propto \begin{pmatrix} 1 \\ 2 \end{pmatrix} \cos(\omega_- t + \phi_-). \end{aligned} \tag{6.226}$$

In the first mode, the bottom mass essentially stands still, and the top mass oscillates at a high frequency (assuming ℓ_1 is small). The factor of 2 in the frequency arises because the top mass essentially lives in a world where the acceleration from gravity is $g' = 2g$ (because of the extra mg force downward from the lower mass). In the second mode, the system is essentially a pendulum of length ℓ_2. The factor of 2 in the angles is what is needed to make the tangential force on the top mass roughly equal to zero (because otherwise it would oscillate at a high frequency, since ℓ_1 is small).

6.20. **Shortest distance in a plane**

Let the two given points be (x_1, y_1) and (x_2, y_2), and let the path be described by the function $y(x)$. (Yes, we'll assume it can be written as a function. Locally, we don't have to worry about any double-valued issues.) Then the length of the path is

$$\ell = \int_{x_1}^{x_2} \sqrt{1 + y'^2}\, dx. \tag{6.227}$$

The "Lagrangian" is $L = \sqrt{1 + y'^2}$, so the Euler–Lagrange equation is

$$\frac{d}{dx}\left(\frac{\partial L}{\partial y'}\right) = \frac{\partial L}{\partial y} \quad \Longrightarrow \quad \frac{d}{dx}\left(\frac{y'}{\sqrt{1 + y'^2}}\right) = 0. \tag{6.228}$$

We see that $y'/\sqrt{1 + y'^2}$ is constant. Therefore, y' is also constant, so we have a straight line, $y(x) = Ax + B$, where A and B are determined by the endpoint conditions.

6.21. **Index of refraction**

Let the path be described by $y(x)$. The speed at height y is $v \propto y$. Therefore, the time to go from (x_1, y_1) to (x_2, y_2) is

$$T = \int_{x_1}^{x_2} \frac{ds}{v} \propto \int_{x_1}^{x_2} \frac{\sqrt{1 + y'^2}}{y}\, dx. \tag{6.229}$$

The "Lagrangian" is therefore

$$L \propto \frac{\sqrt{1 + y'^2}}{y}. \tag{6.230}$$

At this point, we could apply the E–L equation to this L, but let's just use Lemma 6.5, with $f(y) = 1/y$. Equation (6.86) gives

$$1 + y'^2 = Bf(y)^2 \quad \Longrightarrow \quad 1 + y'^2 = \frac{B}{y^2}. \tag{6.231}$$

We must now integrate this. Solving for y', and then separating variables and integrating, gives

$$\int dx = \pm \int \frac{y\, dy}{\sqrt{B - y^2}} \quad \Longrightarrow \quad x + A = \mp\sqrt{B - y^2}. \tag{6.232}$$

Therefore, $(x + A)^2 + y^2 = B$, which is the equation for a circle. Note that the circle is centered at a point with $y = 0$, that is, at a point on the bottom of the slab. This is the point where the perpendicular bisector of the line joining the two given points intersects the bottom of the slab.

6.22. **Minimal surface**

By "tension" in a surface, we mean the force per unit length in the surface. The tension throughout the surface must be constant, because it is in equilibrium. If the tension at one point were larger than at another, then some patch of the surface between these points would move.

The ratio of the circumferences of the circular boundaries of the ring is y_2/y_1. Therefore, the condition that the horizontal forces on the ring cancel is

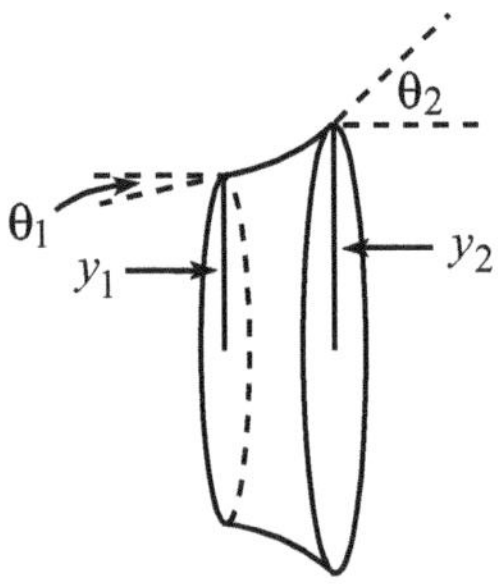

Fig. 6.50

$y_1 \cos\theta_1 = y_2 \cos\theta_2$, where the θ's are the angles of the surface, as shown in Fig. 6.50. In other words, $y \cos\theta$ is constant throughout the surface. But $\cos\theta = 1/\sqrt{1+y'^2}$, so we have

$$\frac{y}{\sqrt{1+y'^2}} = C. \tag{6.233}$$

This is equivalent to Eq. (6.77), and the solution proceeds as in Section 6.8.

6.23. **Existence of a minimal surface**

The general solution for $y(x)$ is given in Eq. (6.78) as $y(x) = (1/b)\cosh b(x+d)$. If we choose the origin to be midway between the rings, then $d = 0$. Both boundary condition are thus

$$r = \frac{1}{b}\cosh b\ell. \tag{6.234}$$

We will now determine the maximum value of ℓ/r for which the minimal surface exists. If ℓ/r is too large, then we will see that there is no solution for b in Eq. (6.234). If you perform an experiment with soap bubbles (which want to minimize their area), and if you pull the rings too far apart, then the surface will break and disappear as it tries to form the two boundary circles.

Define the dimensionless quantities,

$$\eta \equiv \frac{\ell}{r}, \quad \text{and} \quad z \equiv br. \tag{6.235}$$

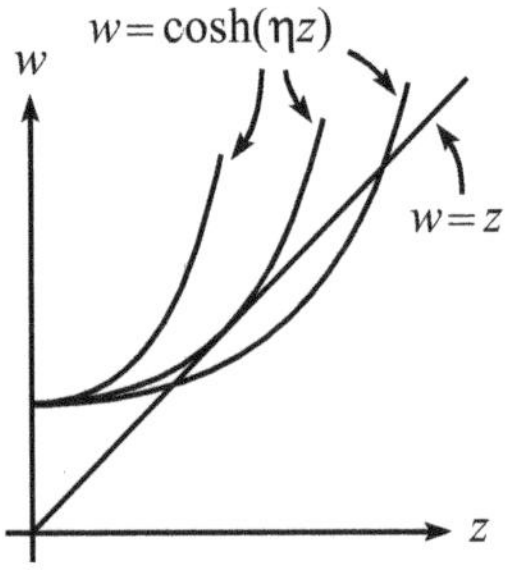

Fig. 6.51

Then Eq. (6.234) becomes

$$z = \cosh \eta z. \tag{6.236}$$

If we make a rough plot of the graphs of $w = z$ and $w = \cosh\eta z$ for a few values of η (see Fig. 6.51), we see that there is no solution to Eq. (6.236) if η is too large. The limiting value of η for which there exists a solution occurs when the curves $w = z$ and $w = \cosh\eta z$ are tangent; that is, when the slopes are equal in addition to the functions being equal. Let η_0 be the limiting value of η, and let z_0 be the place where the tangency occurs. Then equality of the values and the slopes gives

$$z_0 = \cosh(\eta_0 z_0), \quad \text{and} \quad 1 = \eta_0 \sinh(\eta_0 z_0). \tag{6.237}$$

Dividing the second of these equations by the first gives

$$1 = (\eta_0 z_0)\tanh(\eta_0 z_0). \tag{6.238}$$

This must be solved numerically. The solution is

$$\eta_0 z_0 \approx 1.200. \tag{6.239}$$

Plugging this into the second of Eqs. (6.237) gives

$$\left(\frac{\ell}{r}\right)_{\max} \equiv \eta_0 \approx 0.663. \tag{6.240}$$

(Note also that $z_0 = 1.200/\eta_0 = 1.810$.) We see that if ℓ/r is larger than 0.663, then there is no solution for $y(x)$ that is consistent with the boundary

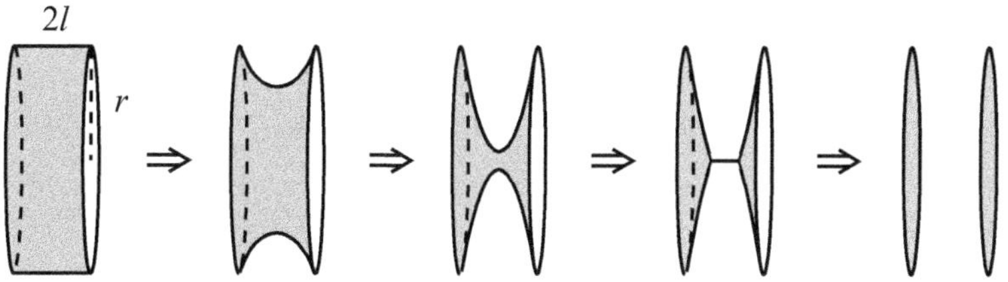

Fig. 6.52

conditions. Above this value of ℓ/r, the soap bubble minimizes its area by heading toward the shape of just two disks, but it will pop well before it reaches that configuration.

To get a sense of the rough shape of the limiting minimal surface, note that the ratio of the radius of the "middle" circle to the radius of the boundary rings is

$$\frac{y(0)}{y(\ell)} = \frac{\cosh(0)}{\cosh(b\ell)} = \frac{1}{\cosh(\eta_0 z_0)} = \frac{1}{z_0} \approx 0.55. \tag{6.241}$$

REMARKS:

1. We glossed over one issue above, namely that there may be more than one solution for the constant b in Eq. (6.234). In fact, Fig. 6.51 shows that for any $\eta < 0.663$, there are two solutions for z in Eq. (6.236), and hence two solutions for b in Eq. (6.234). This means that there are two possible surfaces that might solve our problem. Which one do we want? It turns out that the surface corresponding to the smaller value of b is the one that minimizes the area, while the surface corresponding to the larger value of b is the one that (in some sense) maximizes the area.

   We say "in some sense" because the surface with the larger b is actually a saddle point for the area. It can't be a maximum, after all, because we can always make the area larger by adding little wiggles to it. It's a saddle point because there does exist a class of variations for which it has the maximum area, namely ones where the "dip" in the curve is continuously made larger (just imagine lowering the midpoint in a smooth manner). Such a set of variations is shown in Fig. 6.52. If we start with a cylinder for a surface and then gradually pinch in the center, the area decreases at first (the decrease in the cross-sectional area is the dominant effect at the start). But then as the dip becomes very deep, the area increases because the surface starts to look like the two disks, and these two disks have a larger area than the original narrow cylinder. The surface eventually resembles two nearly flat cones connected by a line. As these cones finally flatten out to the two disks, the area decreases. Therefore, the area must have achieved a local maximum (at least with respect to this class of variations) somewhere in between. This local maximum (or rather, saddle point) arises because the Euler–Lagrange technique simply sets the "derivative" equal to zero and doesn't differentiate between maxima, minima, and saddle points.

   If $\eta \equiv \ell/r > 0.663$ (so that the initial cylinder in now wide instead of narrow), there exists at least one class of variations for which the area decreases monotonically from the area of the cylinder down to the area of the two disks. If you draw a series of pictures (for a wide cylinder) analogous to those in Fig. 6.52, it is quite believable that this is the case.

2. How does the area of the limiting surface (with $\eta_0 = 0.663$) compare with the area of the two disks? The area of the two disks is $A_d = 2\pi r^2$. And the area of

the limiting surface is

$$A_s = \int_{-\ell}^{\ell} 2\pi y\sqrt{1+y'^2}\,dx. \tag{6.242}$$

Using Eq. (6.234), this becomes

$$\begin{aligned} A_s &= \int_{-\ell}^{\ell} \frac{2\pi}{b}\cosh^2 bx\,dx = \int_{-\ell}^{\ell} \frac{\pi}{b}(1+\cosh 2bx)\,dx \\ &= \frac{2\pi\ell}{b} + \frac{\pi\sinh 2b\ell}{b^2}. \end{aligned} \tag{6.243}$$

But from the definitions of η and z, we have $\ell = \eta_0 r$ and $b = z_0/r$ for the limiting surface. Therefore, A_s can be written as

$$A_s = \pi r^2\left(\frac{2\eta_0}{z_0} + \frac{\sinh 2\eta_0 z_0}{z_0^2}\right). \tag{6.244}$$

Plugging in the numerical values ($\eta_0 \approx 0.663$ and $z_0 \approx 1.810$) gives

$$A_d \approx (6.28)r^2, \quad \text{and} \quad A_s \approx (7.54)r^2. \tag{6.245}$$

The ratio of A_s to A_d is approximately 1.2 (it's actually $\eta_0 z_0$, as you can show). The limiting surface therefore has a larger area. This is expected, because for $\ell/r > \eta_0$ the surface tries to run off to one with a smaller area, and there are no other stable configurations besides the cosh solution we found. ♣

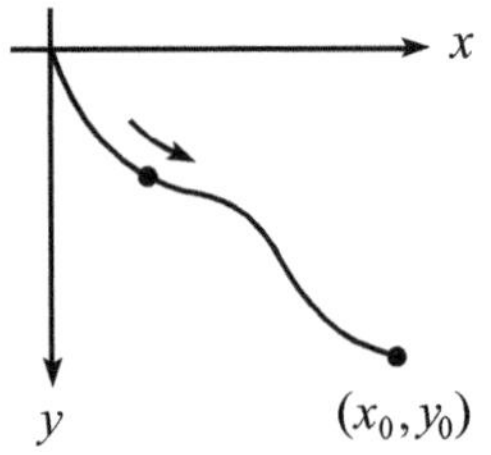

Fig. 6.53

6.24. **The brachistochrone**

FIRST SOLUTION: In Fig. 6.53, the boundary conditions are $y(0) = 0$ and $y(x_0) = y_0$, with downward taken to be the positive y direction. From conservation of energy, the speed as a function of y is $v = \sqrt{2gy}$. The total time is therefore

$$T = \int_0^{x_0} \frac{ds}{v} = \int_0^{x_0} \frac{\sqrt{1+y'^2}}{\sqrt{2gy}}\,dx. \tag{6.246}$$

Our goal is to find the function $y(x)$ that minimizes this integral, subject to the boundary conditions above. We can therefore apply the results of the variational technique, with a "Lagrangian" equal to

$$L \propto \frac{\sqrt{1+y'^2}}{\sqrt{y}}. \tag{6.247}$$

At this point, we could apply the E–L equation to this L, but let's just use Lemma 6.5, with $f(y) = 1/\sqrt{y}$. Equation (6.86) gives

$$1+y'^2 = Bf(y)^2 \quad \Longrightarrow \quad 1+y'^2 = \frac{B}{y}, \tag{6.248}$$

as desired. We must now integrate this. Solving for y' and separating variables gives

$$\frac{\sqrt{y}\,dy}{\sqrt{B-y}} = \pm\,dx. \tag{6.249}$$

A helpful change of variables to get rid of the square root in the denominator is $y \equiv B\sin^2\phi$. Then $dy = 2B\sin\phi\cos\phi\; d\phi$, and Eq. (6.249) simplifies to

$$2B\sin^2\phi\; d\phi = \pm\,dx. \tag{6.250}$$

We can now use $\sin^2\phi = (1-\cos 2\phi)/2$ to integrate this. After multiplying through by 2, the result is $B(2\phi - \sin 2\phi) = \pm 2x - C$, where C is a constant of integration. Now note that we can rewrite our definition of ϕ (which was $y \equiv B\sin^2\phi$) as $2y = B(1-\cos 2\phi)$. If we then define $\theta \equiv 2\phi$, we have

$$x = \pm\,a(\theta - \sin\theta) \pm d, \quad y = a(1-\cos\theta). \tag{6.251}$$

where $a \equiv B/2$, and $d \equiv C/2$. The particle starts at $(x, y) = (0, 0)$. Therefore, θ starts at $\theta = 0$, since this corresponds to $y = 0$. The starting condition $x = 0$ then implies that $d = 0$. Also, we are assuming that the wire heads down to the right, so we choose the positive sign in the expression for x. Therefore, we finally have

$$x = a(\theta - \sin\theta), \quad y = a(1-\cos\theta), \tag{6.252}$$

as desired. This is the parametrization of a *cycloid*, which is the path taken by a point on the rim of a rolling wheel. The initial slope of the $y(x)$ curve is infinite, as you can check.

REMARK: The above method derived the parametric form in (6.252) from scratch. But since Eq. (6.252) was given in the statement of the problem, another route is to simply verify that this parametrization satisfies Eq. (6.248). To this end, assume that $x = a(\theta - \sin\theta)$ and $y = a(1-\cos\theta)$, which gives

$$y' \equiv \frac{dy}{dx} = \frac{dy/d\theta}{dx/d\theta} = \frac{\sin\theta}{1-\cos\theta}. \tag{6.253}$$

Therefore,

$$1 + y'^2 = 1 + \frac{\sin^2\theta}{(1-\cos\theta)^2} = \frac{2}{1-\cos\theta} = \frac{2a}{y}, \tag{6.254}$$

which agrees with Eq. (6.248), with $B \equiv 2a$. ♣

SECOND SOLUTION: Let's use a variational argument again, but now with y as the independent variable. That is, let the chain be described by the function $x(y)$. The arclength is now given by $ds = \sqrt{1+x'^2}\,dy$. Therefore, instead of the Lagrangian in Eq. (6.247), we have

$$L \propto \frac{\sqrt{1+x'^2}}{\sqrt{y}}. \tag{6.255}$$

The Euler–Lagrange equation is

$$\frac{d}{dy}\left(\frac{\partial L}{\partial x'}\right) = \frac{\partial L}{\partial x} \quad\Longrightarrow\quad \frac{d}{dy}\left(\frac{1}{\sqrt{y}}\frac{x'}{\sqrt{1+x'^2}}\right) = 0. \tag{6.256}$$

The zero on the right-hand side makes things nice and easy because it means that the quantity in parentheses is a constant. If we define this constant to be $1/\sqrt{B}$, then we can solve for x' and then separate variables to obtain

$$\frac{\sqrt{y}\,dy}{\sqrt{B-y}} = \pm\, dx. \tag{6.257}$$

in agreement with Eq. (6.249). The solution proceeds as above.

THIRD SOLUTION: The "Lagrangian" in the first solution above, which is given in Eq. (6.247) as

$$L \propto \frac{\sqrt{1+y'^2}}{\sqrt{y}}, \tag{6.258}$$

is independent of x. Therefore, in analogy with conservation of energy (which arises from a Lagrangian that is independent of t), the quantity

$$E \equiv y'\frac{\partial L}{\partial y'} - L = \frac{y'^2}{\sqrt{y}\sqrt{1+y'^2}} - \frac{\sqrt{1+y'^2}}{\sqrt{y}} = \frac{-1}{\sqrt{y}\sqrt{1+y'^2}} \tag{6.259}$$

is constant (that is, independent of x). This statement is equivalent to Eq. (6.248), and the solution proceeds as above.

Chapter 7
Central forces

A *central force* is by definition a force that points radially and whose magnitude depends only on the distance from the source (that is, not on the angle around the source).[1] Equivalently, we may say that a central force is one whose potential depends only on the distance from the source. That is, if the source is located at the origin, then the potential energy is of the form $V(\mathbf{r}) = V(r)$. Such a potential does indeed yield a central force, because

$$\mathbf{F}(\mathbf{r}) = -\nabla V(r) = -\frac{dV}{dr}\hat{\mathbf{r}}, \tag{7.1}$$

which points radially and depends only on r. Gravitational and electrostatic forces are central forces, with $V(r) \propto 1/r$. The spring force is also central, with $V(r) \propto (r - \ell)^2$, where ℓ is the equilibrium length.

There are two important facts concerning central forces: (1) they are ubiquitous in nature, so we had better learn how to deal with them, and (2) dealing with them is much easier than you might think, because crucial simplifications occur in the equations of motion when V is a function of r only. These simplifications will become evident in the following two sections.

7.1 Conservation of angular momentum

Angular momentum plays a key role in dealing with central forces because, as we will show, it is constant over time. For a point mass, we define the *angular momentum* $\mathbf{L}$ by

$$\mathbf{L} = \mathbf{r} \times \mathbf{p}, \tag{7.2}$$

where the "cross product" is defined in Appendix B. $\mathbf{L}$ depends on $\mathbf{r}$, so it therefore depends on where you pick the origin of your coordinate system. Note that $\mathbf{L}$ is a vector, and that it is orthogonal to both $\mathbf{r}$ and $\mathbf{p}$, by nature of the cross product. You might wonder why we care enough about $\mathbf{r} \times \mathbf{p}$ to give it a name. Why not look at $r^3p^5\mathbf{r} \times (\mathbf{r} \times \mathbf{p})$, or something else? The answer is that $\mathbf{L}$ has some very nice properties, one of which is the following.

[1] Taken literally, the term "central force" would imply only the radial nature of the force. But a physicist's definition also includes the dependence solely on the distance from the source.

Theorem 7.1 *If a particle is subject to a central force only, then its angular momentum is conserved.*[2] *That is,*

$$\text{If } V(\mathbf{r}) = V(r), \quad \text{then } \frac{d\mathbf{L}}{dt} = 0. \tag{7.3}$$

Proof: We have

$$\begin{aligned} \frac{d\mathbf{L}}{dt} &= \frac{d}{dt}(\mathbf{r} \times \mathbf{p}) \\ &= \frac{d\mathbf{r}}{dt} \times \mathbf{p} + \mathbf{r} \times \frac{d\mathbf{p}}{dt} \\ &= \mathbf{v} \times (m\mathbf{v}) + \mathbf{r} \times \mathbf{F} \\ &= 0, \end{aligned} \tag{7.4}$$

because $\mathbf{F} \propto \mathbf{r}$, and the cross product of two parallel vectors is zero. ∎

We'll prove this theorem again in the next section, using the Lagrangian method. Let's now prove another theorem which is probably obvious, but good to show anyway.

Theorem 7.2 *If a particle is subject to a central force only, then its motion takes place in a plane.*

Proof: At a given instant t_0, consider the plane P containing the position vector $\mathbf{r}_0$ (with the source of the potential taken to be the origin) and the velocity vector $\mathbf{v}_0$. We claim that $\mathbf{r}$ lies in P at all times.[3] This is true because P is defined as the plane orthogonal to the vector $\mathbf{n}_0 \equiv \mathbf{r}_0 \times \mathbf{v}_0$. But in the proof of Theorem 7.1, we showed that the vector $\mathbf{r} \times \mathbf{v} \equiv (\mathbf{r} \times \mathbf{p})/m$ does not change with time. Therefore, $\mathbf{r} \times \mathbf{v} = \mathbf{n}_0$ for all t. Since $\mathbf{r}$ is certainly orthogonal to $\mathbf{r} \times \mathbf{v}$, we see that $\mathbf{r}$ is orthogonal to $\mathbf{n}_0$ for all t. Hence, $\mathbf{r}$ must always lie in P. ∎

An intuitive look at this theorem is the following. Since the position, velocity, and acceleration (which is proportional to $\mathbf{F}$, which in turn is proportional to the position vector $\mathbf{r}$) vectors initially all lie in P, there is a symmetry between the two sides of P. Therefore, there is no reason for the particle to head out of P on one side rather than the other. The particle therefore remains in P. We can then use this same reasoning again a short time later, and so on.

This theorem shows that we need only two coordinates, instead of the usual three, to describe the motion. But since we're on a roll, why stop there? We'll show below that we really need only *one* coordinate. Not bad, three coordinates reduced down to one.

[2] This is a special case of the fact that torque equals the rate of change of angular momentum. We'll talk about this in great detail in Chapter 8.

[3] The plane P is not well defined if $\mathbf{v}_0 = \mathbf{0}$, or if $\mathbf{r}_0 = \mathbf{0}$, or if $\mathbf{v}_0$ is parallel to $\mathbf{r}_0$. But in these cases, you can quickly show that the motion is always radial, which is even more restrictive than planar.

7.2 The effective potential

The *effective potential* provides a sneaky and useful method for simplifying a three-dimensional central-force problem down to a one-dimensional problem. Here's how it works. Consider a particle of mass m subject to a central force only, described by the potential $V(r)$. Let r and θ be the polar coordinates in the plane of the motion. In these polar coordinates, the Lagrangian (which we'll label as "$\mathcal{L}$", to save "L" for the angular momentum) is

$$\mathcal{L} = \frac{1}{2}m(\dot{r}^2 + r^2\dot{\theta}^2) - V(r). \tag{7.5}$$

The equations of motion obtained from varying r and θ are

$$\begin{aligned} m\ddot{r} &= mr\dot{\theta}^2 - V'(r), \\ \frac{d}{dt}(mr^2\dot{\theta}) &= 0. \end{aligned} \tag{7.6}$$

Since $-V'(r)$ equals the force $F(r)$, the first of these equations is the radial $F = ma$ equation, complete with the centripetal acceleration, in agreement with the first of Eqs. (3.51). The second equation is the statement of conservation of angular momentum, because $mr^2\dot{\theta} = r(mr\dot{\theta}) = rp_\theta$ (where p_θ is the momentum in the angular direction) is the magnitude of $\mathbf{L} = \mathbf{r} \times \mathbf{p}$, from Eq. (B.9). We therefore see that the magnitude of $\mathbf{L}$ is constant. And since the direction of $\mathbf{L}$ is always perpendicular to the fixed plane of the motion, the vector $\mathbf{L}$ is constant in time. We have therefore just given a second proof of Theorem 7.1. In the present Lagrangian language, the conservation of $\mathbf{L}$ follows from the fact that θ is a cyclic coordinate, as we saw in Example 2 in Section 6.5.1. Since $mr^2\dot{\theta}$ does not change with time, let us denote its constant value by

$$L \equiv mr^2\dot{\theta}. \tag{7.7}$$

L is determined by the initial conditions. It can be specified, for example, by giving the initial values of r and $\dot{\theta}$. Using $\dot{\theta} = L/(mr^2)$, we can eliminate $\dot{\theta}$ from the first of Eqs. (7.6). The result is

$$m\ddot{r} = \frac{L^2}{mr^3} - V'(r). \tag{7.8}$$

Multiplying by $\dot{r}$ and integrating with respect to time yields

$$\frac{1}{2}m\dot{r}^2 + \left(\frac{L^2}{2mr^2} + V(r)\right) = E, \tag{7.9}$$

where E is a constant of integration. E is simply the energy, which can be seen by noting that this equation could also have been obtained by using Eq. (7.7) to eliminate $\dot{\theta}$ in the energy equation, $(m/2)(\dot{r}^2 + r^2\dot{\theta}^2) + V(r) = E$.

Equation (7.9) is rather interesting. It involves only the variable r. And it looks a lot like the equation for a particle moving in one dimension (labeled by the coordinate r) under the influence of the potential,

$$V_{\text{eff}}(r) \equiv \frac{L^2}{2mr^2} + V(r). \tag{7.10}$$

The subscript "eff" here stands for "effective." $V_{\text{eff}}(r)$ is called the *effective potential*. The "effective force" is easily read off from Eq. (7.8) to be

$$F_{\text{eff}}(r) = \frac{L^2}{mr^3} - V'(r), \tag{7.11}$$

which agrees with $F_{\text{eff}} = -V'_{\text{eff}}(r)$, as it should. This "effective" potential concept is a marvelous result and should be duly appreciated. It says that if we want to solve a two-dimensional problem (which could have come from a three-dimensional problem) involving a central force, then we can recast the problem into a simple one-dimensional problem with a slightly modified potential. We can forget that we ever had the variable θ, and we can solve this one-dimensional problem (as we'll demonstrate below) to obtain $r(t)$. Having found $r(t)$, we can use $\dot{\theta}(t) = L/mr^2$ to solve for $\theta(t)$ (in theory, at least). This whole procedure works only because there is a quantity involving r and θ (or rather, $\dot{\theta}$) that is independent of time. The variables r and θ are therefore *not* independent, so the problem is really one-dimensional instead of two-dimensional.

To get a general idea of how r behaves with time, all we have to do is draw the graph of $V_{\text{eff}}(r)$. Consider the example where $V(r) = Ar^2$. This is the potential for a spring with relaxed length zero. Then

$$V_{\text{eff}}(r) = \frac{L^2}{2mr^2} + Ar^2. \tag{7.12}$$

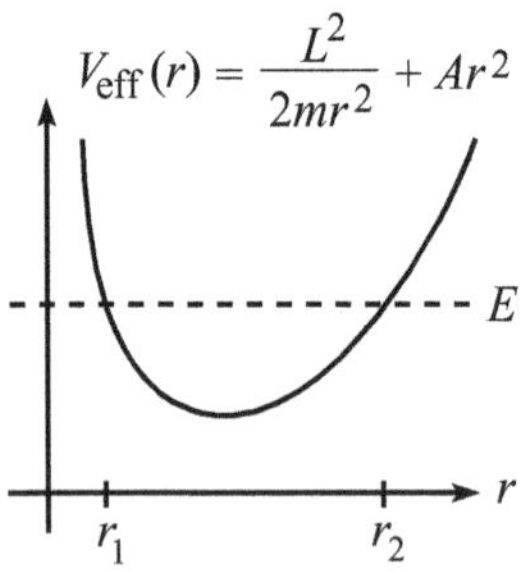

Fig. 7.1

To plot $V_{\text{eff}}(r)$, we must be given L (determined by the initial conditions), along with A and m (determined by the system we're dealing with). But the general shape looks like the curve in Fig. 7.1. The energy E (determined by the initial conditions), which must be given too, is also drawn. The coordinate r bounces back and forth between the turning points, r_1 and r_2, which satisfy $V_{\text{eff}}(r_{1,2}) = E$.[4] This is true because it is impossible for the particle to be located at an r for which $E < V_{\text{eff}}$, since Eq. (7.9) would then imply an imaginary value for $\dot{r}$. If E equals the minimum of $V_{\text{eff}}(r)$, then $r_1 = r_2$, so r is stuck at this one value, which means that the motion is a circle.

REMARK: The $L^2/2mr^2$ term in the effective potential is sometimes called the *angular momentum barrier*. It has the effect of keeping the particle from getting too close to the origin. Basically, the point is that $L \equiv mr^2\dot{\theta}$ is constant, so as r gets smaller, $\dot{\theta}$ gets bigger. But $\dot{\theta}$ increases at a

[4] It turns out that for our Ar^2 spring potential, the motion in space is an ellipse, with semi-axis lengths r_1 and r_2 (see Problem 7.5). But for a general potential, the motion isn't so nice.

greater rate than r decreases, due to the square of the r in $L = mr^2\dot{\theta}$. So eventually we end up with a tangential kinetic energy, $mr^2\dot{\theta}^2/2$, that is greater than what is allowed by conservation of energy.[5]

As he walked past the beautiful belle,
The attraction was easy to tell.
But despite his persistence,
He was kept at a distance
By that darn conservation of L! ♣

Note that it is by no means necessary to introduce the concept of the effective potential. You can simply solve the equations of motion in Eq. (7.6) as they are. But introducing V_{eff} makes it much easier to see what's going on in a central-force problem.

When using potentials, effective,
Remember the one main objective:
The goal is to shun
All dimensions but one,
And then view things with 1-D perspective.

7.3 Solving the equations of motion

If we want to be quantitative, we must solve the equations of motion in Eq. (7.6). Equivalently, we must solve their integrated forms, Eqs. (7.7) and (7.9), which are the conservation of L and E statements,

$$\begin{aligned} mr^2\dot{\theta} &= L, \\ \frac{1}{2}m\dot{r}^2 + \frac{L^2}{2mr^2} + V(r) &= E. \end{aligned} \tag{7.13}$$

The word "solve" is a little ambiguous here, because we should specify what quantities we want to solve for in terms of what other quantities. There are essentially two things we can do. We can solve for r and θ in terms of t. Or we can solve for r in terms of θ. The former has the advantage of immediately yielding velocities and, of course, the information of where the particle is at time t. The latter has the advantage of explicitly showing what the trajectory looks like in space, even though we don't know how quickly it is being traversed. We'll deal mainly with this latter case, particularly when we discuss the gravitational force and Kepler's laws below. But let's look at both cases now.

[5] This argument doesn't hold if $V(r)$ goes to $-\infty$ faster than $-1/r^2$. You can see this by drawing the graph of $V_{\text{eff}}(r)$, which heads to $-\infty$ instead of $+\infty$ as $r \to 0$. $V(r)$ decreases fast enough to allow for the increase in kinetic energy. But such potentials don't come up that often.

7.3.1 Finding $r(t)$ and $\theta(t)$

The value of $\dot{r}$ at any point is found from Eq. (7.13) to be

$$\frac{dr}{dt} = \pm\sqrt{\frac{2}{m}}\sqrt{E - \frac{L^2}{2mr^2} - V(r)}\,. \tag{7.14}$$

To get an actual $r(t)$ out of this, we must be supplied with E and L (determined by the initial values of r, $\dot{r}$, and $\dot{\theta}$), and also the function $V(r)$. To solve this differential equation, we "simply" have to separate variables and then (in theory) integrate:

$$\int \frac{dr}{\sqrt{E - \frac{L^2}{2mr^2} - V(r)}} = \pm \int \sqrt{\frac{2}{m}}\,dt = \pm\sqrt{\frac{2}{m}}\,(t - t_0). \tag{7.15}$$

We need to evaluate this (rather unpleasant) integral on the left-hand side, to obtain t as a function of r. Having found $t(r)$, we can then (in theory) invert the result to obtain $r(t)$. Finally, substituting this $r(t)$ into the relation $\dot{\theta} = L/mr^2$ from Eq. (7.13) gives $\dot{\theta}$ as a function of t, which we can (in theory) integrate to obtain $\theta(t)$.

As you might have guessed, this procedure has the potential to yield some stress. For most $V(r)$'s, the integral in Eq. (7.15) is not calculable in closed form. There are only a few "nice" potentials $V(r)$ for which we can evaluate it. And even in those cases, the remaining tasks are still a pain.[6] But the good news is that these "nice" potentials are precisely the ones we are most interested in. In particular, the gravitational potential, which goes like $1/r$ and which we will concentrate on during the remainder of this chapter, leads to a calculable integral (the spring potential $\sim r^2$ does also). However, having said all this, we're not going to apply this procedure to gravity. It's nice to know that it exists, but we won't be doing anything else with it. Instead, we'll use the following strategy to solve for r as a function of θ.

7.3.2 Finding $r(\theta)$

We can eliminate the dt from Eqs. (7.13) by getting the $\dot{r}^2$ term alone on the left side of the second equation, and then dividing by the square of the first equation. The dt^2 factors cancel, and we obtain

$$\left(\frac{1}{r^2}\frac{dr}{d\theta}\right)^2 = \frac{2mE}{L^2} - \frac{1}{r^2} - \frac{2mV(r)}{L^2}\,. \tag{7.16}$$

We can now (in theory) take a square root, separate variables, and integrate to obtain θ as a function of r. We can then (in theory) invert to obtain r as a function

[6] Of course, if you run out of patience or hit a brick wall, you always have the option of doing things numerically. See Section 1.4 for a discussion of this.

of θ. To do this, we must be given the function $V(r)$. So let's now finally give ourselves a $V(r)$ and do a problem all the way through. We'll study the most important potential of all, or perhaps the second most important one, gravity.[7]

7.4 Gravity, Kepler's laws

7.4.1 Calculation of $r(\theta)$

Our goal in this subsection is to obtain r as a function of θ, for a gravitational potential. Let's assume that we're dealing with the earth and the sun, with masses $M_\odot$ and m, respectively. The gravitational potential energy of the earth–sun system is

$$V(r) = -\frac{\alpha}{r}, \quad \text{where } \alpha \equiv GM_\odot m. \tag{7.17}$$

In the present treatment, we'll consider the sun to be bolted down at the origin of our coordinate system. Since $M_\odot \gg m$, this is approximately true for the earth–sun system. (If we want to do the problem exactly, we must use the *reduced mass*, which is the topic of Section 7.4.5.) Equation (7.16) becomes

$$\left(\frac{1}{r^2}\frac{dr}{d\theta}\right)^2 = \frac{2mE}{L^2} - \frac{1}{r^2} + \frac{2m\alpha}{rL^2}. \tag{7.18}$$

As stated above, we could take a square root, separate variables, integrate to find $\theta(r)$, and then invert to find $r(\theta)$. This method, although straightforward, is rather messy. So let's solve for $r(\theta)$ in a slick way.

With all the $1/r$ terms floating around, it might be easier to solve for $1/r$ instead of r. Using $d(1/r)/d\theta = -(dr/d\theta)/r^2$, and letting $y \equiv 1/r$ for convenience, Eq. (7.18) becomes

$$\left(\frac{dy}{d\theta}\right)^2 = -y^2 + \frac{2m\alpha}{L^2}y + \frac{2mE}{L^2}. \tag{7.19}$$

At this point, we could also use the separation-of-variables technique, but let's continue to be slick. Completing the square on the right-hand side gives

$$\left(\frac{dy}{d\theta}\right)^2 = -\left(y - \frac{m\alpha}{L^2}\right)^2 + \frac{2mE}{L^2} + \left(\frac{m\alpha}{L^2}\right)^2. \tag{7.20}$$

Defining $z \equiv y - m\alpha/L^2$ for convenience yields

$$\left(\frac{dz}{d\theta}\right)^2 = -z^2 + \left(\frac{m\alpha}{L^2}\right)^2\left(1 + \frac{2EL^2}{m\alpha^2}\right) \equiv -z^2 + B^2, \tag{7.21}$$

[7] The two most important potentials in physics are certainly the gravitational and harmonic-oscillator ones. They both lead to doable integrals, and interestingly both lead to elliptical orbits.

where

$$B \equiv \left(\frac{m\alpha}{L^2}\right)\sqrt{1+\frac{2EL^2}{m\alpha^2}}\,. \tag{7.22}$$

At this point, in the spirit of being slick, we can just look at Eq. (7.21) and observe that

$$z = B\cos(\theta - \theta_0) \tag{7.23}$$

is the solution, because $\cos^2 x + \sin^2 x = 1$. But lest we feel guilty about not doing separation-of-variables at least once in this problem, let's solve Eq. (7.21) that way, too. The integral is nice and doable, and we have

$$\int \frac{dz}{\sqrt{B^2 - z^2}} = \int d\theta \quad \Longrightarrow \quad \cos^{-1}\left(\frac{z}{B}\right) = \theta - \theta_0, \tag{7.24}$$

which gives $z = B\cos(\theta - \theta_0)$. It is customary to pick the axes so that $\theta_0 = 0$, so we'll drop the θ_0 from here on. Recalling our definition $z \equiv 1/r - m\alpha/L^2$ and also the definition of B from Eq. (7.22), Eq. (7.23) becomes

$$\frac{1}{r} = \frac{m\alpha}{L^2}(1 + \epsilon\cos\theta), \tag{7.25}$$

where

$$\epsilon \equiv \sqrt{1+\frac{2EL^2}{m\alpha^2}} \tag{7.26}$$

is the *eccentricity* of the particle's motion. We will see shortly exactly what ϵ signifies.

This completes the derivation of $r(\theta)$ for the gravitational potential, $V(r) \propto 1/r$. It was a little messy, but not unbearably painful. At any rate, we just discovered the basic motion of objects under the influence of gravity, which takes care of virtually all of the gazillion tons of stuff in the universe. Not bad for one page of work.

> Newton said as he gazed off afar,
> "From here to the most distant star,
> These wond'rous ellipses
> And solar eclipses
> All come from a 1 over r."

What are the limits on r in Eq. (7.25)? The minimum value of r is obtained when the right-hand side reaches its maximum value, which is $(m\alpha/L^2)(1+\epsilon)$. Therefore,

$$r_{\text{min}} = \frac{L^2}{m\alpha(1+\epsilon)}\,. \tag{7.27}$$

What is the maximum value of r? The answer depends on whether ϵ is greater than or less than 1. If $\epsilon < 1$ (which corresponds to circular or elliptical orbits, as we'll see below), then the minimum value of the right-hand side of Eq. (7.25) is $(m\alpha/L^2)(1-\epsilon)$. Therefore,

$$r_{\max} = \frac{L^2}{m\alpha(1-\epsilon)} \quad (\text{if } \epsilon < 1). \tag{7.28}$$

If $\epsilon \geq 1$ (which corresponds to parabolic or hyperbolic orbits, as we'll see below), then the right-hand side of Eq. (7.25) can become zero (when $\cos\theta = -1/\epsilon$). Therefore,

$$r_{\max} = \infty \quad (\text{if } \epsilon \geq 1). \tag{7.29}$$

7.4.2 The orbits

Let's examine in detail the various cases for ϵ.

- **Circle** ($\epsilon = 0$)

  If $\epsilon = 0$, then Eq. (7.26) says that $E = -m\alpha^2/2L^2$. The negative E means that the potential energy is more negative than the kinetic energy is positive, so the particle is trapped in the potential well. Equations (7.27) and (7.28) give $r_{\min} = r_{\max} = L^2/m\alpha$. Therefore, the particle moves in a circular orbit with radius $L^2/m\alpha$. Equivalently, Eq. (7.25) says that r is independent of θ.

  Note that it isn't necessary to do all the work of Section 7.4.1 if we just want to look at circular motion. For a given L, the energy $-m\alpha^2/2L^2$ is the minimum value that the E given by Eq. (7.13) can take. This is true because to achieve the minimum, we certainly want $\dot{r} = 0$. And you can show that minimizing the effective potential, $L^2/2mr^2 - \alpha/r$, yields this value for E. If we plot $V_{\text{eff}}(r)$, we have the situation shown in Fig. 7.2. The particle is trapped at the bottom of the potential well, so it has no motion in the r direction.

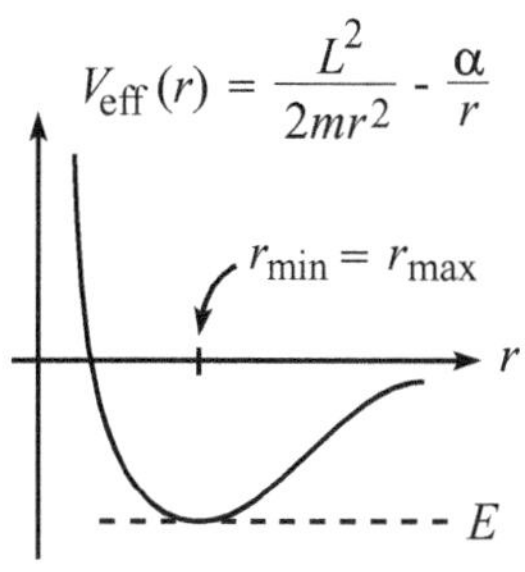

Fig. 7.2

- **Ellipse** ($0 < \epsilon < 1$)

  If $0 < \epsilon < 1$, then Eq. (7.26) says that $-m\alpha^2/2L^2 < E < 0$. Equations (7.27) and (7.28) give $r_{\min}$ and $r_{\max}$. It isn't obvious that the resulting motion is an ellipse. We'll demonstrate this below.

  If we plot $V_{\text{eff}}(r)$, we have the situation shown in Fig. 7.3. The particle oscillates between $r_{\min}$ and $r_{\max}$. The energy is negative, so the particle is trapped in the potential well.

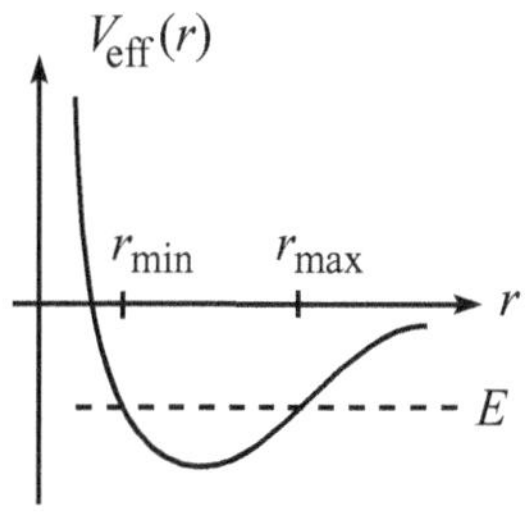

Fig. 7.3

- **Parabola** ($\epsilon = 1$)

  If $\epsilon = 1$, then Eq. (7.26) says that $E = 0$. This value of E implies that the particle barely makes it out to infinity (its speed approaches zero as $r \to \infty$). Equation (7.27) gives $r_{\min} = L^2/2m\alpha$, and Eq. (7.29) gives $r_{\max} = \infty$. Again, it isn't obvious that the resulting motion is a parabola. We'll demonstrate this below.

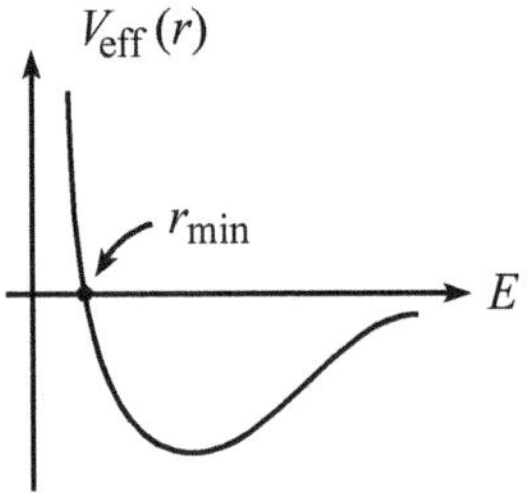

Fig. 7.4

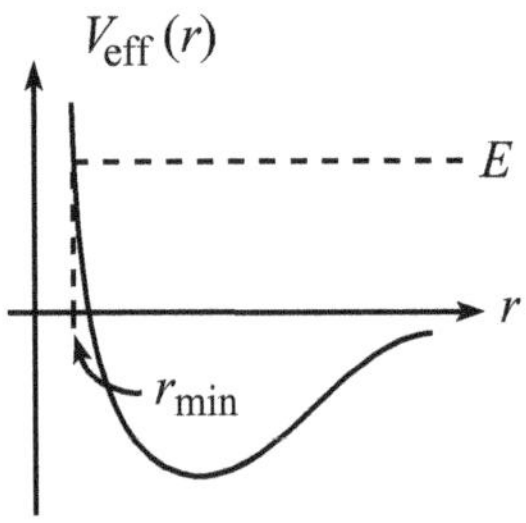

Fig. 7.5

If we plot $V_{eff}(r)$, we have the situation shown in Fig. 7.4. The particle does not oscillate back and forth in the r direction. It moves inward (or possibly not, if it was initially moving outward), turns around at $r_{min} = L^2/2m\alpha$, and then heads out to infinity forever.

- **Hyperbola** ($\epsilon > 1$)

  If $\epsilon > 1$, then Eq. (7.26) says that $E > 0$. This value of E implies that the particle makes it out to infinity with energy to spare. The potential goes to zero as $r \to \infty$, so the particle's speed approaches the nonzero value $\sqrt{2E/m}$ as $r \to \infty$. Equation (7.27) gives r_{min}, and Eq. (7.29) gives $r_{max} = \infty$. Again, it isn't obvious that the resulting motion is a hyperbola. We'll demonstrate this below.

  If we plot $V_{eff}(r)$, we have the situation shown in Fig. 7.5. As in the parabola case, the particle does not oscillate back and forth in the r direction. It moves inward (or possibly not, if it was initially moving outward), turns around at r_{min}, and then heads out to infinity forever.

7.4.3 Proof of conic orbits

Let's now prove that Eq. (7.25) does indeed describe the conic sections stated above. We'll also show that the origin (the source of the potential) is a focus of the conic section. These proofs are straightforward, although the ellipse and hyperbola cases get a bit messy. In what follows, we'll find it easier to work with Cartesian coordinates. For convenience, let

$$k \equiv \frac{L^2}{m\alpha}. \tag{7.30}$$

Multiplying Eq. (7.25) through by kr, and using $\cos\theta = x/r$, gives

$$k = r + \epsilon x. \tag{7.31}$$

Solving for r and squaring yields

$$x^2 + y^2 = k^2 - 2k\epsilon x + \epsilon^2 x^2. \tag{7.32}$$

Let's look at the various cases for ϵ. We will invoke without proof various facts about conic sections (focal lengths, etc.).

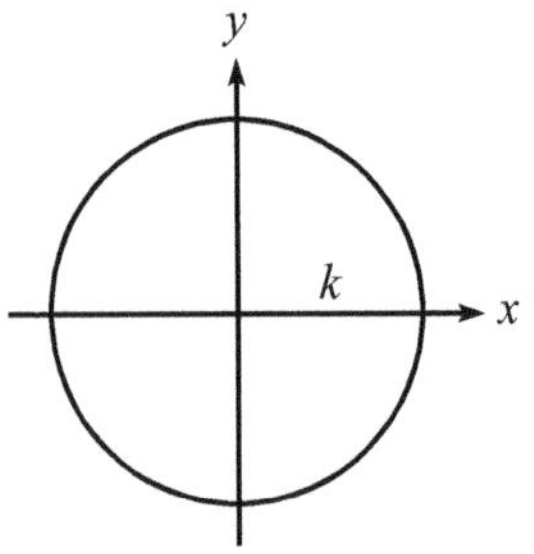

Fig. 7.6

- **Circle** ($\epsilon = 0$)

  In this case, Eq. (7.32) becomes $x^2 + y^2 = k^2$. So we have a circle with radius $k = L^2/m\alpha$, with its center at the origin (see Fig. 7.6).

- **Ellipse** ($0 < \epsilon < 1$)

  In this case, Eq. (7.32) can be written (after completing the square for the x terms, and expending some effort) as

$$\frac{\left(x + \frac{k\epsilon}{1-\epsilon^2}\right)^2}{a^2} + \frac{y^2}{b^2} = 1, \quad \text{where } a = \frac{k}{1-\epsilon^2}, \quad \text{and } b = \frac{k}{\sqrt{1-\epsilon^2}}. \tag{7.33}$$

This is the equation for an ellipse with its center located at $(-k\epsilon/(1-\epsilon^2), 0)$. The semi-major and semi-minor axes are a and b, respectively. And the focal length is $c = \sqrt{a^2 - b^2} = k\epsilon/(1-\epsilon^2)$. Therefore, one focus is located at the origin (see Fig. 7.7). Note that c/a equals the eccentricity, ϵ.

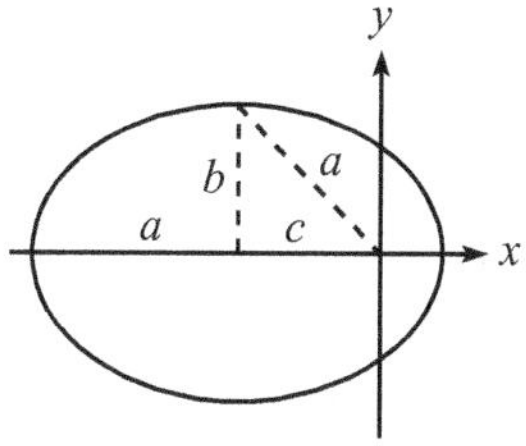

Fig. 7.7

- **Parabola** ($\epsilon = 1$)

  In this case, Eq. (7.32) becomes $y^2 = k^2 - 2kx$, which can be written as $y^2 = -2k(x - \frac{k}{2})$. This is the equation for a parabola with vertex at $(k/2, 0)$ and focal length $k/2$. (The focal length of a parabola written in the form $y^2 = 4ax$ is a.) So we have a parabola with its focus located at the origin (see Fig. 7.8).

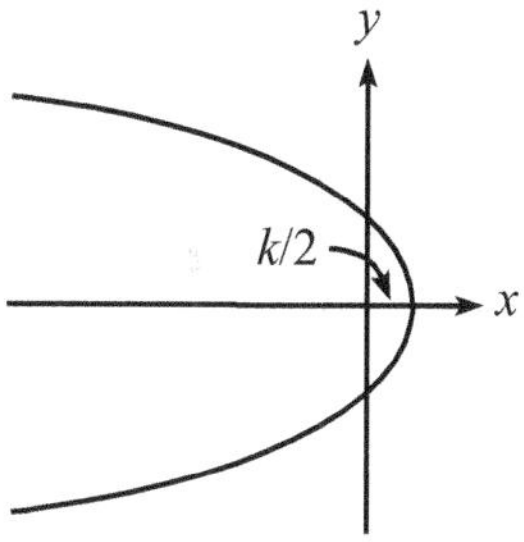

Fig. 7.8

- **Hyperbola** ($\epsilon > 1$)

  In this case, Eq. (7.32) can be written (after completing the square for the x terms)

$$\frac{\left(x - \frac{k\epsilon}{\epsilon^2 - 1}\right)^2}{a^2} - \frac{y^2}{b^2} = 1, \quad \text{where } a = \frac{k}{\epsilon^2 - 1}, \quad \text{and } b = \frac{k}{\sqrt{\epsilon^2 - 1}}. \tag{7.34}$$

  This is the equation for a hyperbola with its center (defined to be the intersection of the asymptotes) located at $(k\epsilon/(\epsilon^2 - 1), 0)$. The focal length is $c = \sqrt{a^2 + b^2} = k\epsilon/(\epsilon^2 - 1)$. Therefore, the focus is located at the origin (see Fig. 7.9). Note that c/a equals the eccentricity, ϵ.

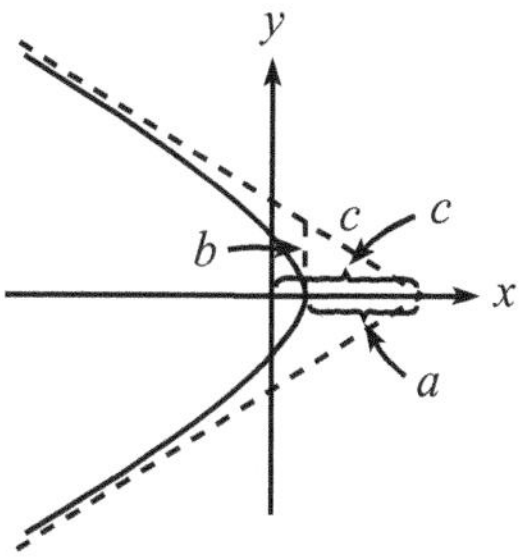

Fig. 7.9

The *impact parameter* (usually denoted by the letter b) of a trajectory is defined to be the closest distance to the origin the particle would achieve if it moved in the straight line determined by its initial velocity far from the origin (that is, along the dotted line in the Fig. 7.9). You might think that choosing the letter b here would cause a problem, because we already defined b in Eq. (7.34). However, it turns out that these two definitions are identical (see Exercise 7.14), so all is well.

Equation (7.34) actually describes an entire hyperbola, that is, it also describes a branch that opens up to the right with its focus located at $(2k\epsilon/(\epsilon^2 - 1), 0)$. However, this right branch was introduced in the squaring operation that produced Eq. (7.32). It isn't a solution to the original equation we wanted to solve, Eq. (7.31). It turns out that the right-opening branch (or its reflection across the y axis, depending on your sign convention for B and ϵ) is relevant for a repulsive, instead of attractive, $1/r$ potential; see Exercise 7.21.

7.4.4 Kepler's laws

We can now, with minimal extra work, write down Kepler's laws. Kepler (1571–1630) lived before Newton (1642–1727), so he didn't have Newton's laws at his disposal. Kepler arrived at his laws via observational data, which was a rather impressive feat. It was known since the time of Copernicus (1473–1543) that the planets move around the sun, but it was Kepler who first gave a quantitative description of the orbits. Kepler's laws assume that the sun is massive enough so that its position is essentially fixed in space. This is a very good approximation, but the following subsection on the *reduced mass* shows how to modify the laws and solve things exactly.

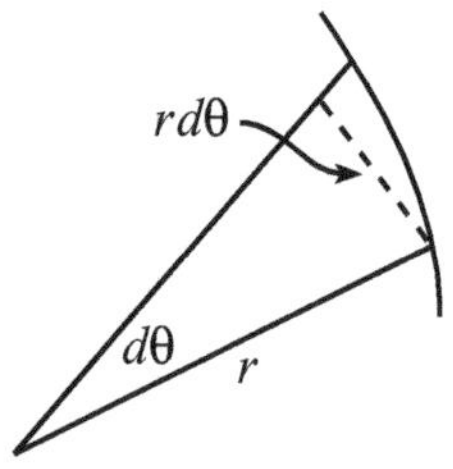

Fig. 7.10

- **First law:** *The planets move in elliptical orbits with the sun at one focus.*
  We proved this in Eq. (7.33).[8] There are undoubtedly also objects flying past the sun in hyperbolic orbits, but we don't call these things planets, because we never see the same one twice.
- **Second law:** *The radius vector to a planet sweeps out area at a rate that is independent of its position in the orbit.*
  This law is nothing other than a fancy way of stating conservation of angular momentum. The area swept out by the radius vector during a short period of time is $dA = r(r\,d\theta)/2$, because $r\,d\theta$ is the base of the thin triangle in Fig. 7.10. Therefore, we have (using $L = mr^2\dot{\theta}$)
$$\frac{dA}{dt} = \frac{r^2\dot{\theta}}{2} = \frac{L}{2m}, \tag{7.35}$$
  which is constant, because L is constant for a central force. This quick proof is independent of all the work we did in Sections 7.4.1–7.4.3.
- **Third law:** *The square of the period of an orbit, T, is proportional to the cube of the semi-major-axis length, a. More precisely,*
$$T^2 = \frac{4\pi^2 a^3}{GM_\odot}, \tag{7.36}$$
  where $M_\odot$ is the mass of the sun. Note that the planet's mass m doesn't appear in this equation.
  Proof: Integrating Eq. (7.35) over the time of a whole orbit gives
$$A = \frac{LT}{2m}. \tag{7.37}$$
  But the area of an ellipse is $A = \pi ab$, where a and b are the semi-major and semi-minor axes, respectively. Squaring Eq. (7.37) and using Eq. (7.33) to write $b = a\sqrt{1-\epsilon^2}$ gives
$$\pi^2 a^4 = \left(\frac{L^2}{m(1-\epsilon^2)}\right)\frac{T^2}{4m}. \tag{7.38}$$
  We have grouped the right-hand side in this way because we can now use the $L^2 \equiv m\alpha k$ relation from Eq. (7.30) to transform the term in parentheses into $\alpha k/(1-\epsilon^2) \equiv \alpha a$, where a is given in Eq. (7.33). But $\alpha a \equiv (GM_\odot m)a$, so we obtain
$$\pi^2 a^4 = \frac{(GM_\odot ma)T^2}{4m}, \tag{7.39}$$
  which gives Eq. (7.36), as desired. ■

[8] For an alternate geometrical proof due to Feynman (and also Maxwell), see Goodstein and Goodstein (1996).

These three laws describe the motion of all the planets (and asteroids, comets, and such) in the solar system. But our solar system is only the tip of the iceberg. There's a lot of other stuff out there, and it's all governed by gravity (although Newton's inverse square law must be supplanted by Einstein's General Relativity theory of gravitation). There's a whole universe around us, and as time passes on we see and understand more and more of it, both experimentally and theoretically. In recent years, we've even begun to look for friends we might have out there. Why? Because we can. There's nothing wrong with looking under the lamppost now and then. It just happens to be a very big one in this case.

As we grow up, we open an ear,
Exploring the cosmic frontier.
In this coming of age,
We turn in our cage,
All alone on a tiny blue sphere.

7.4.5 Reduced mass

We assumed in Section 7.4.1 that the sun is large enough so that it is only negligibly affected by the presence of the planets. That is, it is essentially fixed at the origin. But how do we solve a problem in which the masses of the two interacting bodies are comparable in size? Equivalently, how do we solve the earth–sun problem exactly? It turns out that the only modification required is a replacement of the earth's mass with the *reduced mass*, defined below. The following discussion actually holds for any central force, not just gravity.

The Lagrangian of a general central-force system consisting of the interacting masses m_1 and m_2 is

$$\mathcal{L} = \frac{1}{2}m_1\dot{\mathbf{r}}_1^2 + \frac{1}{2}m_2\dot{\mathbf{r}}_2^2 - V(|\mathbf{r}_1 - \mathbf{r}_2|). \tag{7.40}$$

We have written the potential in this form, dependent only on the distance $|\mathbf{r}_1 - \mathbf{r}_2|$, because we are assuming a central force. Let us define

$$\mathbf{R} \equiv \frac{m_1\mathbf{r}_1 + m_2\mathbf{r}_2}{m_1 + m_2}, \quad \text{and} \quad \mathbf{r} \equiv \mathbf{r}_1 - \mathbf{r}_2. \tag{7.41}$$

$\mathbf{R}$ and $\mathbf{r}$ are the position of the center of mass and the vector between the masses, respectively. Invert these equations to obtain

$$\mathbf{r}_1 = \mathbf{R} + \frac{m_2}{M}\mathbf{r}, \quad \text{and} \quad \mathbf{r}_2 = \mathbf{R} - \frac{m_1}{M}\mathbf{r}, \tag{7.42}$$

where $M \equiv m_1 + m_2$ is the total mass of the system. In terms of $\mathbf{R}$ and $\mathbf{r}$, the Lagrangian becomes

$$\begin{aligned}\mathcal{L} &= \frac{1}{2}m_1\left(\dot{\mathbf{R}} + \frac{m_2}{M}\dot{\mathbf{r}}\right)^2 + \frac{1}{2}m_2\left(\dot{\mathbf{R}} - \frac{m_1}{M}\dot{\mathbf{r}}\right)^2 - V(|\mathbf{r}|) \\ &= \frac{1}{2}M\dot{\mathbf{R}}^2 + \frac{1}{2}\left(\frac{m_1 m_2}{m_1 + m_2}\right)\dot{\mathbf{r}}^2 - V(r) \\ &= \frac{1}{2}M\dot{\mathbf{R}}^2 + \frac{1}{2}\mu\dot{\mathbf{r}}^2 - V(r),\end{aligned} \tag{7.43}$$

where the *reduced mass*, μ, is defined by

$$\frac{1}{\mu} \equiv \frac{1}{m_1} + \frac{1}{m_2}. \tag{7.44}$$

We now note that the Lagrangian in Eq. (7.43) depends on $\dot{\mathbf{R}}$, but not on $\mathbf{R}$. Therefore, the Euler–Lagrange equations say that $\dot{\mathbf{R}}$ is constant. That is, the CM moves at constant velocity (this is just the statement that there are no external forces). The CM motion is therefore trivial, so let's ignore it. Our Lagrangian therefore becomes

$$\mathcal{L} \to \frac{1}{2}\mu\dot{\mathbf{r}}^2 - V(r). \tag{7.45}$$

But this is simply the Lagrangian for a particle of mass μ that moves around a fixed origin under the influence of the potential $V(r)$. For gravity, we have

$$\mathcal{L} = \frac{1}{2}\mu\dot{\mathbf{r}}^2 + \frac{\alpha}{r} \quad (\text{where } \alpha \equiv GM_\odot m). \tag{7.46}$$

To solve the earth–sun system exactly, we therefore just need to replace (in the calculation in Section 7.4.1) the earth's mass, m, with the reduced mass, μ, given by

$$\frac{1}{\mu} \equiv \frac{1}{m} + \frac{1}{M_\odot}. \tag{7.47}$$

The resulting value of r in Eq. (7.25) is the distance between the earth and sun. The earth and sun are therefore distances of $(M_\odot/M)r$ and $(m/M)r$, respectively, away from the CM, from Eq. (7.42). These distances are just scaled-down versions of the distance r, which represents an ellipse, so we see that the earth and sun move in elliptical orbits (whose sizes are in the ratio $M_\odot/m$) with the CM as a focus. Note that the m's that are buried in L and ϵ in Eq. (7.25) must be changed to μ's. But α is still defined to be $GM_\odot m$, so the m in this definition does *not* get replaced with μ.

For the earth–sun system, the μ in Eq. (7.47) is essentially equal to m, because $M_\odot$ is so large. Using $m = 5.97 \cdot 10^{24}$ kg, and $M_\odot = 1.99 \cdot 10^{30}$ kg, we find that

μ is smaller than m by only one part in $3 \cdot 10^5$. Our fixed-sun approximation is therefore a very good one. You can show that the CM is about $5 \cdot 10^5$ m from the center of the sun, which is well inside the sun (about a thousandth of the radius).

How are Kepler's laws modified when we solve for the orbits exactly using the reduced mass?

- **First law:** The elliptical statement in the first law is still true, but with the CM (not the sun) located at a focus. The sun also travels in an ellipse with the CM at a focus.[9] Whatever is true for the earth must also be true for the sun, because they come into Eq. (7.43) symmetrically. The only difference is in the size of various quantities.
- **Second law:** In the second law, we need to consider the position vector from the CM (not the sun) to the earth. This vector sweeps out equal areas in equal times, because the angular momentum of the earth (and the sun, too) relative to the CM is constant. This is true because the gravitational force always points through the CM, so the force is a central force with the CM chosen as the origin.
- **Third law:** The period of the earth's orbit (and the sun's, too) is the same as the period of the orbit of our hypothetical particle of mass μ orbiting around a fixed origin under the influence of the potential $-\alpha/r \equiv -GM_{\odot}m/r$. This is true because the radius vectors in all three of these systems are always in the same ratio. To find the period of the particle's orbit, we can repeat the derivation leading up to Eq. (7.39). But in that equation, the m on the bottom is replaced by μ, while the m on the top remains m, because this is the m that appears in α. Therefore, we obtain[10]

$$T^2 = \frac{4\pi^2 a_\mu^3 \mu}{GM_{\odot}m} = \frac{4\pi^2 a_\mu^3}{GM}, \tag{7.48}$$

where we have used $\mu \equiv M_{\odot}m/(M_{\odot} + m) \equiv M_{\odot}m/M$. This result is symmetric in $M_{\odot}$ and m, as it must be because if we interchange the labels of $M_{\odot}$ and m, we still have the same system. And it also correctly reduces to Eq. (7.36) when $M_{\odot} \gg m$.

If you want to write Eq. (7.48) in terms of the semi-major axis of the earth's elliptical orbit, which is $a_{\rm E} = (M_{\odot}/M)a_\mu$, then just plug in $a_\mu = (M/M_{\odot})a_{\rm E}$ to obtain

$$T^2 = \frac{4\pi^2 (M/M_{\odot})^3 a_{\rm E}^3}{GM} = \left(\frac{M^2}{M_{\odot}^2}\right)\frac{4\pi^2 a_{\rm E}^3}{GM_{\odot}}. \tag{7.49}$$

Let's perform a check on this formula by considering the special case of equal masses m orbiting around the same circular path of radius r (at diametrically opposite points)

[9] Well, this statement is true only if there is just one planet. With many planets, the tiny motion of the sun is very complicated. This is perhaps the best reason to work in the approximation where it is essentially bolted down.

[10] We have put the subscript μ on the length a to remind us that it is the semi-major axis of our hypothetical particle's orbit, and not the semi-major axis of the earth's orbit.

with their CM at the center of the circle. For this simple system, we can solve for the period from scratch using $F = ma$:

$$\frac{mv^2}{r} = \frac{Gm^2}{(2r)^2} \implies \frac{m(2\pi r/T)^2}{r} = \frac{Gm^2}{(2r)^2} \implies T^2 = \frac{16\pi^2 r^3}{Gm}, \tag{7.50}$$

which agrees with Eq. (7.49), in different notation, when $M = 2M_\odot$.

REMARK: There's actually a fairly quick way to see where the $M^2/M_\odot^2$ factor in Eq. (7.49) comes from. Imagine a new system where the earth's orbit has the same dimensions but where the (slightly) moving sun is replaced by a stationary mass bolted down at the location of the earth–sun CM. This new mass is now a fraction $M_\odot/M$ as far away from the earth as the sun was. Therefore, since the gravitational force is proportional to $1/r^2$, if we make our new mass equal to $(M_\odot/M)^2 M_\odot$, then it will exert the same force on the earth that the sun exerted. So if mother earth has her eyes closed, she'll never know the difference. The periods of these two systems must therefore be the same. But from Eq. (7.36), the period of the second system, which has a bolted-down mass, is

$$T^2 = \frac{4\pi^2 a_E^3}{G \cdot (M_\odot/M)^2 M_\odot}, \tag{7.51}$$

in agreement with Eq. (7.49). ♣

7.5 Problems

Section 7.2: The effective potential

7.1. **Exponential spiral** *

Given L, find the $V(r)$ that leads to a spiral path of the form $r = r_0 e^{a\theta}$. Choose E to be zero. *Hint*: Obtain an expression for $\dot{r}$ that contains no θ's, and then use Eq. (7.9).

7.2. **Cross section** **

A particle moves in a potential, $V(r) = -C/(3r^3)$.

(a) Given L, find the maximum value of the effective potential.

(b) Let the particle come in from infinity with speed v_0 and impact parameter b. In terms of C, m, and v_0, what is the largest value of b (call it b_{max}) for which the particle is captured by the potential? In other words, what is the "cross section" for capture, πb_{max}^2, for this potential?

7.3. **Maximum L** ***

A particle moves in a potential, $V(r) = -V_0 e^{-\lambda^2 r^2}$.

(a) Given L, find the radius of the stable circular orbit. An implicit equation is fine.

(b) It turns out that if L is too large, then no circular orbit exists. What is the largest value of L for which a circular orbit does in fact exist?

If r_0 is the radius of the circle in this cutoff case, what is the value of $V_{\text{eff}}(r_0)$?

Section 7.4: Gravity, Kepler's laws

7.4. r^k potential ***

A particle of mass m moves in a potential given by $V(r) = \beta r^k$. Let the angular momentum be L.

(a) Find the radius, r_0, of the circular orbit.
(b) If the particle is given a tiny kick so that the radius oscillates around r_0, find the frequency, ω_r, of these small oscillations in r.
(c) What is the ratio of the frequency ω_r to the frequency of the (nearly) circular motion, $\omega_\theta \equiv \dot{\theta}$? Give a few values of k for which the ratio is rational, that is, for which the path of the nearly circular motion closes back on itself.

7.5. Spring ellipse ***

A particle moves in a $V(r) = \beta r^2$ potential. Following the general strategy in Sections 7.4.1 and 7.4.3, show that the particle's path is an ellipse.

7.6. β/r^2 potential ***

A particle is subject to a $V(r) = \beta/r^2$ potential. Following the general strategy in Section 7.4.1, find the shape of the particle's path. You will need to consider various cases for β.

7.7. Rutherford scattering ***

A particle of mass m travels in a hyperbolic orbit past a mass M, whose position is assumed to be fixed. The speed at infinity is v_0, and the impact parameter is b (see Exercise 7.14).

(a) Show that the angle through which the particle is deflected is

$$\phi = \pi - 2\tan^{-1}(\gamma b) \quad \Longrightarrow \quad b = \frac{1}{\gamma}\cot\left(\frac{\phi}{2}\right), \tag{7.52}$$

where $\gamma \equiv v_0^2/GM$.

(b) Let $d\sigma$ be the cross-sectional area (measured when the particle is initially at infinity) that gets deflected into a solid angle of size $d\Omega$ at angle ϕ.[11] Show that

$$\frac{d\sigma}{d\Omega} = \frac{1}{4\gamma^2 \sin^4(\phi/2)}. \tag{7.53}$$

[11] The *solid angle* of a patch on a sphere is the area of the patch divided by the square of the sphere's radius. So a whole sphere subtends a solid angle of 4π *steradians* (the name for one unit of solid angle).

This quantity is called the *differential cross section*. *Note*: the label of this problem, *Rutherford scattering*, actually refers to the scattering of charged particles. But since the electrostatic and gravitational forces are both inverse-square laws, the scattering formulas look the same, except for a few constants.

7.6 Exercises

Section 7.1: Conservation of angular momentum

7.8. **Wrapping around a pole** *

A puck of mass m sliding on frictionless ice is attached by a horizontal string of length ℓ to a thin vertical pole of radius R. The puck initially travels in (essentially) a circle around the pole at speed v_0. The string wraps around the pole, and the puck gets drawn in and eventually hits the pole. What quantity is conserved during this motion? What is the puck's speed right before it hits the pole?

7.9. **String through a hole** *

A block of mass m sliding on a frictionless table is attached to a horizontal string that passes through a tiny hole in the table. The block initially travels in a circle of radius ℓ around the hole at speed v_0. If you slowly pull the string down through the hole, what quantity is conserved during this motion? What is the block's speed when it is a distance r from the hole?

Section 7.2: The effective potential

7.10. **Power-law spiral** **

Given L, find the $V(r)$ that leads to a spiral path of the form $r = r_0\theta^k$. Choose E to be zero. *Hint*: Obtain an expression for $\dot{r}$ that contains no θ's, and then use Eq. (7.9).

Section 7.4: Gravity, Kepler's laws

7.11. **Circular orbit** *

For a circular orbit, derive Kepler's third law from scratch, using $F = ma$.

7.12. **Falling into the sun** *

Imagine that the earth is suddenly (and tragically) stopped in its orbit, and then allowed to fall radially into the sun. How long will this take? Use data from Appendix J, and assume that the initial orbit is essentially circular. *Hint*: Consider the radial path to be part of a very thin ellipse.

7.13. **Intersecting orbits** $**$

Two masses, m and $2m$, orbit around their CM. If the orbits are circular, they don't intersect. But if they are very elliptical, they do. What is the smallest value of the eccentricity for which they intersect?

7.14. **Impact parameter** $**$

Show that the distance b defined in Eq. (7.34) and Fig. 7.9 is equal to the impact parameter. Do this:

(a) Geometrically, by showing that b is the distance from the origin to the dotted line in Fig. 7.9.

(b) Analytically, by letting the particle come in from infinity at speed v_0 and impact parameter b', and then showing that the b in Eq. (7.34) equals b'.

7.15. **Closest approach** $**$

A particle with speed v_0 and impact parameter b starts far away from a planet of mass M.

(a) Starting from scratch (that is, without using any of the results from Section 7.4), find the distance of closest approach to the planet.

(b) Use the results of the hyperbola discussion in Section 7.4.3 to show that the distance of closest approach to the planet is $k/(\epsilon + 1)$, and then show that this agrees with your answer to part (a).

7.16. **Skimming a planet** $**$

A particle travels in a parabolic orbit in a planet's gravitational field and skims the surface at its closest approach. The planet has mass density ρ. Relative to the center of the planet, what is the angular velocity of the particle as it skims the surface?

7.17. **Parabola L** $**$

A mass m orbits around a planet of mass M in a parabolic orbit of the form $y = x^2/(4\ell)$, which has focal length ℓ. Find the angular momentum in three different ways:

(a) Find the speed at closest approach.

(b) Use Eq. (7.30).

(c) Consider the point $(x, x^2/4\ell)$, where x is very large. Find approximate expressions for the speed and impact parameter at this point.

7.18. **Circle to parabola** $**$

A spaceship travels in a circular orbit around a planet. It applies a sudden thrust and increases its speed by a factor f. If the goal is to change the orbit from a circle to a parabola, what should f be if the thrust points in the tangential direction? Is your answer any different if the thrust points in some other direction? What is the distance of closest approach if the thrust points in the radial direction?

7.19. **Zero potential** **

A particle is subject to a constant potential, which we will take to be zero (equivalently, consider the $\alpha \equiv GMm = 0$ limit). Following the general strategy in Section 7.4, show that the particle's path is a straight line.

7.20. **Ellipse axes** **

Taking it as given that Eq. (7.25) describes an ellipse for $0 < \epsilon < 1$, calculate the lengths of the semi-major and semi-minor axes, and show that the results agree with Eq. (7.33).

7.21. **Repulsive potential** **

Consider an "anti-gravitational" potential (or more mundanely, the electrostatic potential between two like charges), $V(r) = \alpha/r$, where $\alpha > 0$. What is the basic change in the analysis of Section 7.4? Show that circular, elliptical, and parabolic orbits do not exist. Draw the figure analogous to Fig. 7.9 for the hyperbolic orbit.

7.7 Solutions

7.1. **Exponential spiral**

The given information $r = r_0 e^{a\theta}$ yields (using $\dot{\theta} = L/mr^2$)

$$\dot{r} = a(r_0 e^{a\theta})\dot{\theta} = ar\left(\frac{L}{mr^2}\right) = \frac{aL}{mr}. \tag{7.54}$$

Plugging this into Eq. (7.9) gives

$$\frac{m}{2}\left(\frac{aL}{mr}\right)^2 + \frac{L^2}{2mr^2} + V(r) = E = 0. \tag{7.55}$$

Therefore,

$$V(r) = -\frac{(1+a^2)L^2}{2mr^2}. \tag{7.56}$$

7.2. **Cross section**

(a) The effective potential is

$$V_{\text{eff}}(r) = \frac{L^2}{2mr^2} - \frac{C}{3r^3}. \tag{7.57}$$

Setting the derivative equal to zero gives $r = mC/L^2$. Plugging this into $V_{\text{eff}}(r)$ gives

$$V_{\text{eff}}^{\max} = \frac{L^6}{6m^3C^2}. \tag{7.58}$$

(b) If the energy E of the particle is less than $V_{\text{eff}}^{\max}$, then the particle will reach a minimum value of r, and then head back out to infinity (see Fig. 7.11). If E is

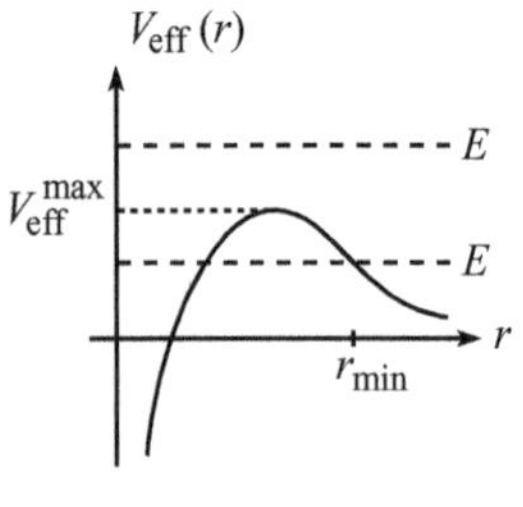

Fig. 7.11

greater than $V_{\text{eff}}^{\text{max}}$, then the particle will head all the way in to $r = 0$, never to return. The condition for capture is therefore $V_{\text{eff}}^{\text{max}} < E$. Using $L = mv_0 b$ and $E = E_\infty = mv_0^2/2$, this condition becomes

$$\frac{(mv_0 b)^6}{6m^3C^2} < \frac{mv_0^2}{2} \implies b < \left(\frac{3C^2}{m^2v_0^4}\right)^{1/6} \equiv b_{\text{max}}. \tag{7.59}$$

The cross section for capture is therefore

$$\sigma = \pi b_{\text{max}}^2 = \pi \left(\frac{3C^2}{m^2v_0^4}\right)^{1/3}. \tag{7.60}$$

It makes sense that this should increase with C and decrease with m and v_0.

7.3. **Maximum *L***

(a) The effective potential is

$$V_{\text{eff}}(r) = \frac{L^2}{2mr^2} - V_0 e^{-\lambda^2 r^2}. \tag{7.61}$$

A circular orbit exists at the value(s) of r for which $V'_{\text{eff}}(r) = 0$. Setting the derivative equal to zero and solving for L^2 gives

$$L^2 = (2mV_0\lambda^2)r^4 e^{-\lambda^2 r^2}. \tag{7.62}$$

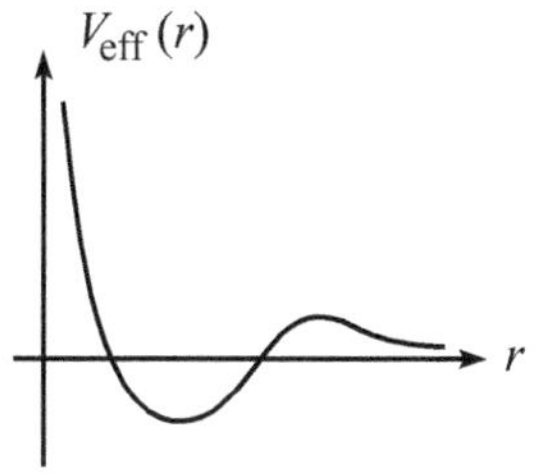

Fig. 7.12

This implicitly determines r. As long as L isn't too large, $V_{\text{eff}}(r)$ looks something like the graph in Fig. 7.12, although it doesn't necessarily dip down to negative values; see the remark below. You can arrive at this picture by noting that for any L, $V_{\text{eff}}(r)$ behaves like $1/r^2$ for both $r \to 0$ and $r \to \infty$; and for sufficiently small L, $V_{\text{eff}}(r)$ reaches negative values somewhere in between, due to the $-V_0$ term. The curve must therefore look like the one shown, which has two locations where $V'_{\text{eff}}(r) = 0$. The smaller solution is the one with the stable orbit. However, if L is too large, then there are no solutions to $V'_{\text{eff}}(r) = 0$, because $V_{\text{eff}}(r)$ decreases monotonically to zero (because $L^2/2mr^2$ does so). We'll be quantitative about this in part (b).

(b) The function $r^4 e^{-\lambda^2 r^2}$ on the right-hand side of Eq. (7.62) has a maximum value, because it goes to zero for both $r \to 0$ and $r \to \infty$. Therefore, there is a maximum value of L for which a solution for r exists. The maximum of $r^4 e^{-\lambda^2 r^2}$ occurs where

$$0 = \frac{d(r^4 e^{-\lambda^2 r^2})}{dr} = e^{-\lambda^2 r^2}\left(4r^3 + r^4(-2\lambda^2 r)\right) \implies r^2 = \frac{2}{\lambda^2} \equiv r_0^2. \tag{7.63}$$

Plugging r_0 into Eq. (7.62) gives

$$L_{\text{max}}^2 = \frac{8mV_0}{\lambda^2 e^2}. \tag{7.64}$$

Plugging r_0 and L_{max}^2 into Eq. (7.61) gives

$$V_{\text{eff}}(r_0) = \frac{V_0}{e^2} \quad (\text{for } L = L_{\text{max}}). \tag{7.65}$$

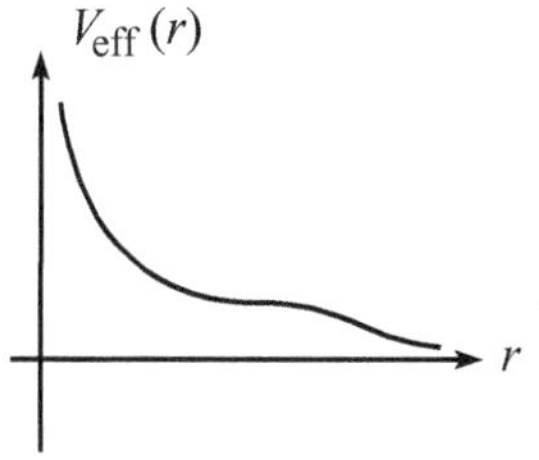

Fig. 7.13

Note that this is greater than zero. For the $L = L_{\text{max}}$ case, the graph of V_{eff} is shown in Fig. 7.13. This is the cutoff case between having a dip in the graph, and decreasing monotonically to zero.

REMARK: A common error in this problem is to say that the condition for a circular orbit to exist is that $V_{\text{eff}}(r) < 0$ at the point where $V_{\text{eff}}(r)$ is minimum.

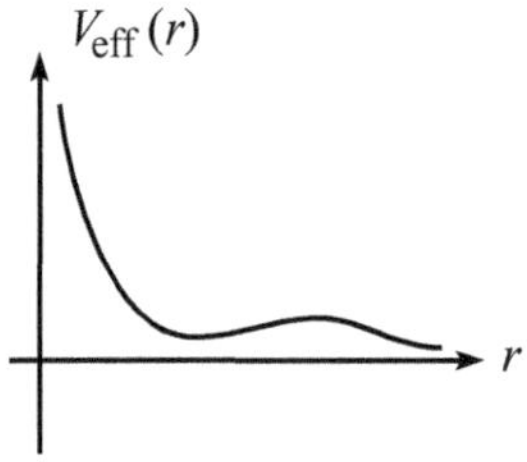

Fig. 7.14

The logic here is that since the goal is to have a well in which the particle can be trapped, it seems like we just need V_{eff} to achieve a value less than the value at $r = \infty$, namely 0. However, this gives the wrong answer ($L^2_{\text{max}} = 2mV_0/\lambda^2 e$, as you can show), because $V_{\text{eff}}(r)$ can look like the graph in Fig. 7.14. This has a local minimum with $V_{\text{eff}}(r) > 0$. ♣

7.4. **r^k potential**

(a) A circular orbit exists at the value of r for which the derivative of the effective potential (which is the negative of the effective force) is zero. This is simply the statement that the right-hand side of Eq. (7.8) equals zero, so that $\ddot{r} = 0$. Since $V'(r) = \beta k r^{k-1}$, Eq. (7.8) gives

$$\frac{L^2}{mr^3} - \beta k r^{k-1} = 0 \quad \Longrightarrow \quad r_0 = \left(\frac{L^2}{m\beta k}\right)^{1/(k+2)}. \tag{7.66}$$

If k is negative, then β must also be negative if there is to be a real solution for r_0.

(b) The long method of finding the frequency is to set $r(t) \equiv r_0 + \epsilon(t)$, where ϵ represents the small deviation from the circular orbit, and to then plug this expression for r into Eq. (7.8). The result (after making some approximations) is a harmonic-oscillator equation of the form $\ddot{\epsilon} = -\omega_r^2 \epsilon$. This general procedure, which was described in detail in Section 6.7, will work fine here (as you are encouraged to show), but let's use an easier method.

By introducing the effective potential, we have reduced the problem to a one-dimensional problem in the variable r. Therefore, we can make use of the result in Section 5.2, where we found in Eq. (5.20) that to find the frequency of small oscillations, we just need to calculate the second derivative of the potential. For the problem at hand, we must use the effective potential, because that is what determines the motion of the variable r. We therefore have

$$\omega_r = \sqrt{\frac{V''_{\text{eff}}(r_0)}{m}}. \tag{7.67}$$

If you work through the $r \equiv r_0 + \epsilon$ method described above, you will find that you are basically calculating the second derivative of V_{eff}, but in a rather cumbersome way.

Using the form of the effective potential, we have

$$V''_{\text{eff}}(r_0) = \frac{3L^2}{mr_0^4} + \beta k(k-1)r_0^{k-2} = \frac{1}{r_0^4}\left(\frac{3L^2}{m} + \beta k(k-1)r_0^{k+2}\right). \tag{7.68}$$

Using the r_0 from Eq. (7.66), this simplifies to

$$V''_{\text{eff}}(r_0) = \frac{L^2(k+2)}{mr_0^4} \quad \Longrightarrow \quad \omega_r = \sqrt{\frac{V''_{\text{eff}}(r_0)}{m}} = \frac{L\sqrt{k+2}}{mr_0^2}. \tag{7.69}$$

We could get rid of the r_0 here by using Eq. (7.66), but this form of ω_r will be more useful in part (c).

Note that we must have $k > -2$ for ω_r to be real. If $k < -2$, then $V''_{\text{eff}}(r_0) < 0$, which means that we have a local maximum of V_{eff}, instead of a local minimum. In other words, the circular orbit is unstable. Small perturbations grow, instead of oscillating around zero.

(c) Since $L = mr_0^2\dot{\theta}$ for the circular orbit, we have $\omega_\theta \equiv \dot{\theta} = L/(mr_0^2)$. Combining this with Eq. (7.69), we find

$$\frac{\omega_r}{\omega_\theta} = \sqrt{k+2}. \tag{7.70}$$

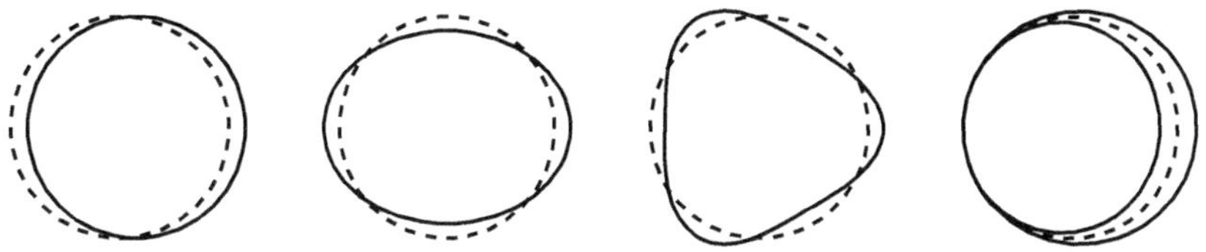

Fig. 7.15

A few values of k that yield rational values for this ratio are (the plots of the orbits are shown in Fig. 7.15):

- $k = -1 \Longrightarrow \omega_r/\omega_\theta = 1$: This is the gravitational potential. The variable r makes one oscillation for each complete revolution of the (nearly) circular orbit.
- $k = 2 \Longrightarrow \omega_r/\omega_\theta = 2$: This is the spring potential. The variable r makes two oscillations for each complete revolution.
- $k = 7 \Longrightarrow \omega_r/\omega_\theta = 3$: The variable r makes three oscillations for each complete revolution.
- $k = -7/4 \Longrightarrow \omega_r/\omega_\theta = 1/2$: The variable r makes half of an oscillation for each complete revolution. So we need to have two revolutions to get back to the same value of r.

There is an infinite number of k values that yield closed orbits. But note that this statement applies only to orbits that are nearly circular. Also, the "closed" nature of the orbits is only approximate, because it is based on Eq. (7.67) which is an approximate result based on small oscillations. The only k values that lead to exactly closed orbits for any initial conditions are $k = -1$ (gravity) and $k = 2$ (spring), and in both cases the orbits are ellipses. This result is known as Bertrand's theorem; see Brown (1978).

7.5. **Spring ellipse**

With $V(r) = \beta r^2$, Eq. (7.16) becomes

$$\left(\frac{1}{r^2}\frac{dr}{d\theta}\right)^2 = \frac{2mE}{L^2} - \frac{1}{r^2} - \frac{2m\beta r^2}{L^2}. \tag{7.71}$$

As stated in Section 7.4.1, we could take a square root, separate variables, integrate to find $\theta(r)$, and then invert to find $r(\theta)$. But let's solve for $r(\theta)$ in a slick way, as we did for the gravitational case, where we made the change of variables, $y \equiv 1/r$. Since there are lots of r^2 terms floating around in Eq. (7.71), it's reasonable to try the change of variables, $y \equiv r^2$ or $y \equiv 1/r^2$. The latter turns out to be the better choice. So, using $y \equiv 1/r^2$ and $dy/d\theta = -2(dr/d\theta)/r^3$, and multiplying Eq. (7.71) through by $1/r^2$, we obtain

$$\begin{aligned}\left(\frac{1}{2}\frac{dy}{d\theta}\right)^2 &= \frac{2mEy}{L^2} - y^2 - \frac{2m\beta}{L^2}.\\ &= -\left(y - \frac{mE}{L^2}\right)^2 - \frac{2m\beta}{L^2} + \left(\frac{mE}{L^2}\right)^2.\end{aligned} \tag{7.72}$$

Defining $z \equiv y - mE/L^2$ for convenience, we have

$$\begin{aligned}\left(\frac{dz}{d\theta}\right)^2 &= -4z^2 + 4\left(\frac{mE}{L^2}\right)^2\left(1 - \frac{2\beta L^2}{mE^2}\right)\\ &\equiv -4z^2 + 4B^2.\end{aligned} \tag{7.73}$$

As in Section 7.4.1, we can just look at this equation and observe that

$$z = B\cos 2(\theta - \theta_0) \tag{7.74}$$

is the solution. We can rotate the axes so that $\theta_0 = 0$, so we'll drop the θ_0 from here on. Recalling our definition $z \equiv 1/r^2 - mE/L^2$ and also the definition of B from Eq. (7.73), Eq. (7.74) becomes

$$\frac{1}{r^2} = \frac{mE}{L^2}(1 + \epsilon\cos 2\theta), \tag{7.75}$$

where

$$\epsilon \equiv \sqrt{1 - \frac{2\beta L^2}{mE^2}}. \tag{7.76}$$

It turns out, as we'll see below, that ϵ is *not* the eccentricity of the ellipse, as it was in the gravitational case.

We will now use the procedure in Section 7.4.3 to show that Eq. (7.76) represents an ellipse. For convenience, let

$$k \equiv \frac{L^2}{mE}. \tag{7.77}$$

Multiplying Eq. (7.75) through by kr^2, and using

$$\cos 2\theta = \cos^2\theta - \sin^2\theta = \frac{x^2}{r^2} - \frac{y^2}{r^2}, \tag{7.78}$$

and also $r^2 = x^2 + y^2$, we obtain $k = (x^2 + y^2) + \epsilon(x^2 - y^2)$. This can be written as

$$\frac{x^2}{a^2} + \frac{y^2}{b^2} = 1, \quad \text{where} \quad a = \sqrt{\frac{k}{1+\epsilon}}, \quad \text{and } b = \sqrt{\frac{k}{1-\epsilon}}. \tag{7.79}$$

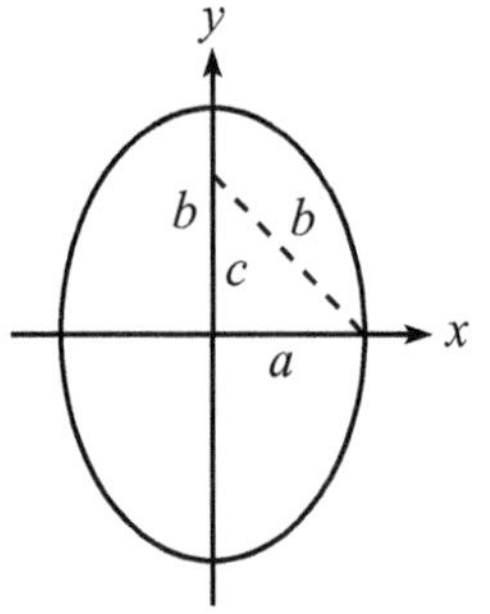

Fig. 7.16

This is the equation for an ellipse with its center located at the origin (as opposed to a focus located at the origin, as in the gravitational case). In Fig. 7.16, the semi-major and semi-minor axes are b and a, respectively, and the focal length is $c = \sqrt{b^2 - a^2} = \sqrt{2k\epsilon/(1-\epsilon^2)}$. The eccentricity is $c/b = \sqrt{2\epsilon/(1+\epsilon)}$.

REMARK: If $\epsilon = 0$, then $a = b$, which means that the ellipse is actually a circle. Let's see if this makes sense. Looking at Eq. (7.76), we see that we want to show that circular motion implies $2\beta L^2 = mE^2$. For circular motion, the radial $F = ma$ equation is $mv^2/r = 2\beta r \Longrightarrow v^2 = 2\beta r^2/m$. The energy is therefore $E = mv^2/2 + \beta r^2 = 2\beta r^2$. Also, the square of the angular momentum is $L^2 = m^2v^2r^2 = 2m\beta r^4$. Therefore, $2\beta L^2 = 2\beta(2m\beta r^4) = m(2\beta r^2)^2 = mE^2$, as we wanted to show. ♣

7.6. **β/r^2 potential**

With $V(r) = \beta/r^2$, Eq. (7.16) becomes

$$\begin{aligned}\left(\frac{1}{r^2}\frac{dr}{d\theta}\right)^2 &= \frac{2mE}{L^2} - \frac{1}{r^2} - \frac{2m\beta}{r^2L^2}\\ &= \frac{2mE}{L^2} - \frac{1}{r^2}\left(1 + \frac{2m\beta}{L^2}\right).\end{aligned} \tag{7.80}$$

Letting $y \equiv 1/r$, and using $dy/d\theta = -(1/r^2)(dr/d\theta)$, this becomes

$$\left(\frac{dy}{d\theta}\right)^2 + a^2y^2 = \frac{2mE}{L^2}, \quad \text{where } a^2 \equiv 1 + \frac{2m\beta}{L^2}. \tag{7.81}$$

We must now consider various possibilities for a^2. These possibilities depend on how β compares to L^2, which depends on the initial conditions of the motion. In what follows, note that the effective potential equals

$$V_{\text{eff}}(r) = \frac{L^2}{2mr^2} + \frac{\beta}{r^2} = \frac{a^2L^2}{2mr^2}. \tag{7.82}$$

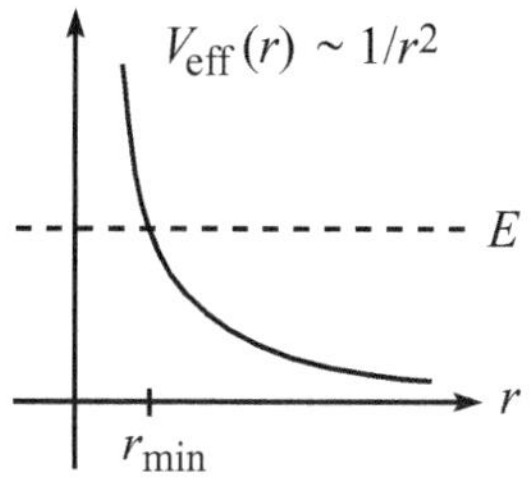

Fig. 7.17

CASE 1: $a^2 > 0$, or equivalently $\beta > -L^2/2m$. In this case, the effective potential looks like the graph in Fig. 7.17. The solution for y in Eq. (7.81) is a trig function, which we will take to be a "sin" by appropriately rotating the axes. Using $y \equiv 1/r$, we obtain

$$\frac{1}{r} = \frac{1}{a}\sqrt{\frac{2mE}{L^2}}\sin a\theta. \tag{7.83}$$

$\theta = 0$ and $\theta = \pi/a$ make the right-hand side equal to zero, so they correspond to $r = \infty$. And $\theta = \pi/2a$ makes the right-hand side maximum, so it corresponds to the minimum value of r, which is $r_{\text{min}} = a\sqrt{L^2/2mE}$. This minimum r can also be obtained in a much quicker manner by finding where $V_{\text{eff}}(r) = E$.

If the particle comes in from infinity at $\theta = 0$, it eventually heads back out to infinity at $\theta = \pi/a$. The angle that the outgoing path makes with the incoming path is therefore π/a. So if a is large (that is, if β is large and positive, or if L is small), then the particle bounces nearly straight backward. If a is small (that is, if β is negative and if L^2 is only slightly larger than $-2m\beta$), then the particle spirals around many times before it pops back out to infinity.

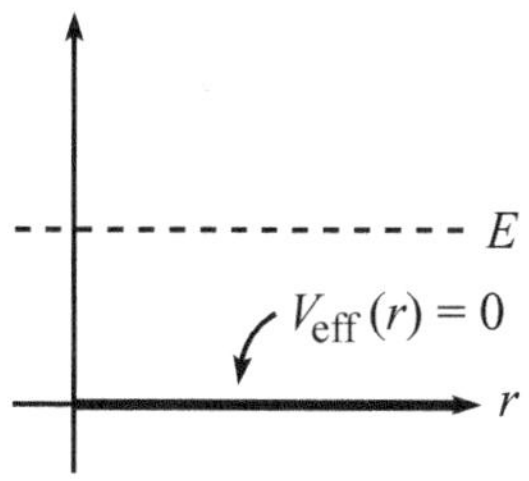

Fig. 7.18

A few special cases are: (1) $\beta = 0 \Longrightarrow a = 1$, which means that the total angle is π, that is, there is no net deflection. In fact, the particle's path is a straight line, because the potential is zero; see Exercise 7.19. (2) $L^2 = -8m\beta/3 \Longrightarrow a = 1/2$, which means that the total angle is 2π, that is, the final line of the particle's motion is (anti)parallel to the initial line. These lines are shifted sideways from one another, with the separation depending on the initial conditions.

CASE 2: $a = 0$, or equivalently $\beta = -L^2/2m$. In this case, the effective potential is identically zero, as shown in Fig. 7.18. Equation (7.81) becomes

$$\left(\frac{dy}{d\theta}\right)^2 = \frac{2mE}{L^2}. \tag{7.84}$$

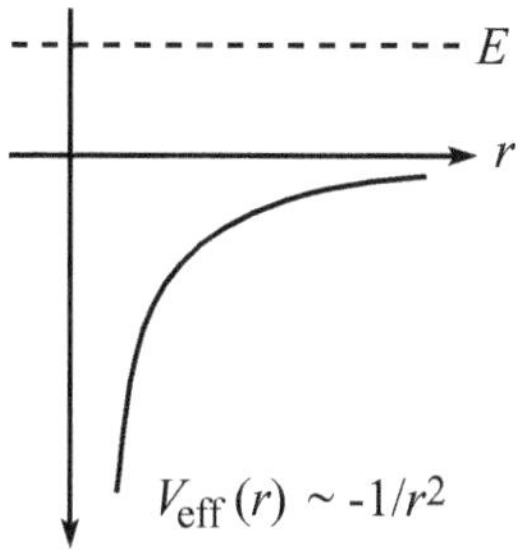

Fig. 7.19

The solution to this is $y = \theta\sqrt{2mE/L^2} + C$, which gives

$$r = \frac{1}{\theta}\sqrt{\frac{L^2}{2mE}}, \tag{7.85}$$

where we have set the integration constant C equal to zero by choosing $\theta = 0$ to be the angle that corresponds to $r = \infty$. Note that we can use $\beta = -L^2/2m$ to write r as $r = (1/\theta)\sqrt{-\beta/E}$.

Since the effective potential is flat, the rate of change of r is constant. Therefore, if the particle has $\dot{r} < 0$, it will reach the origin in finite time, even though Eq. (7.85) says that it will spiral around the origin an infinite number of times (because $\theta \to \infty$ as $r \to 0$).

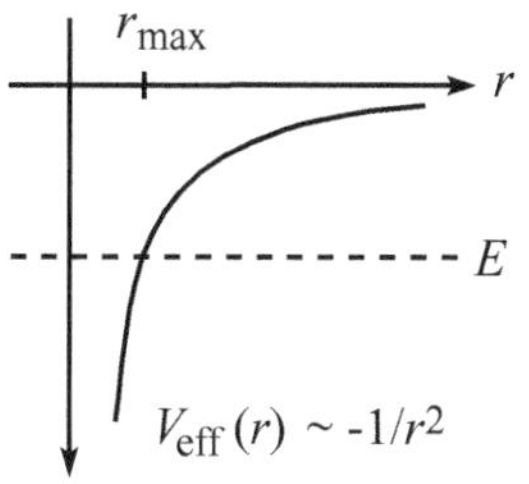

Fig. 7.20

CASE 3: $a^2 < 0$, or equivalently $\beta < -L^2/2m$. In this case, we have the situation shown in either Fig. 7.19 or Fig. 7.20, depending on the sign of E. For convenience, let b be the positive real number such that $b^2 = -a^2$. Then Eq. (7.81) becomes

$$\left(\frac{dy}{d\theta}\right)^2 - b^2y^2 = \frac{2mE}{L^2}. \tag{7.86}$$

The solution to this equation is a hyperbolic trig function. But we must consider two cases:

- $E > 0$: Using the identity $\cosh^2 z - \sinh^2 z = 1$, and recalling $y \equiv 1/r$, we see that the solution to Eq. (7.86) is[12]

$$\frac{1}{r} = \frac{1}{b}\sqrt{\frac{2mE}{L^2}}\sinh b\theta. \tag{7.87}$$

Unlike the $a^2 > 0$ case above, the sinh function has no maximum value. Therefore, the right-hand side can head to infinity, which means that r can head to zero, if the initial $\dot{r}$ is negative. It does so in a finite time, because $\dot{r}$ only becomes more negative as time goes by, from Fig. 7.19. For large z, we have $\sinh z \approx e^z/2$, so r heads to zero like $e^{-b\theta}$. The particle will therefore spiral around the origin an infinite number of times (because $\theta \to \infty$ as $r \to 0$).

- $E < 0$: In this case, Eq. (7.86) can be rewritten as

$$b^2y^2 - \left(\frac{dy}{d\theta}\right)^2 = \frac{2m|E|}{L^2}. \tag{7.88}$$

The solution to this equation is[13]

$$\frac{1}{r} = \frac{1}{b}\sqrt{\frac{2m|E|}{L^2}}\cosh b\theta. \tag{7.89}$$

As in the sinh case, the cosh function has no maximum value. Therefore, the right-hand side can head to infinity, which means that r can head to zero. But in the present cosh case, the right-hand side does achieve a nonzero minimum value, when $\theta = 0$. So r achieves a maximum value (if the initial $\dot{r}$ is positive) equal to $r_{\max} = b\sqrt{L^2/2m|E|}$. This is clear from Fig. 7.20. This maximum r can also be obtained by simply finding where $V_{\text{eff}}(r) = E$. After reaching $r_{\max}$, the particle heads back down to the origin with behavior similar (for large θ) to the sinh case.

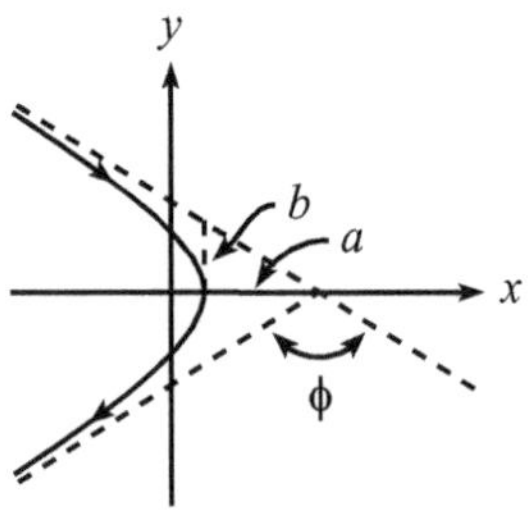

Fig. 7.21

7.7. **Rutherford scattering**

(a) From Exercise 7.14, we know that the impact parameter b equals the distance b shown in Fig. 7.9. Therefore, Fig. 7.21 tells us that the angle of deflection (the angle between the initial and final velocity vectors) is

$$\phi = \pi - 2\tan^{-1}\left(\frac{b}{a}\right). \tag{7.90}$$

But from Eqs. (7.34) and (7.26), we have

$$\frac{b}{a} = \sqrt{\epsilon^2 - 1} = \sqrt{\frac{2EL^2}{m\alpha^2}} = \sqrt{\frac{2(mv_0^2/2)(mv_0b)^2}{m(GMm)^2}} = \frac{v_0^2 b}{GM}. \tag{7.91}$$

Substituting this into Eq. (7.90), with $\gamma \equiv v_0^2/(GM)$, gives the first expression in Eq. (7.52). Dividing by 2 and taking the cotangent of both sides then gives the second expression,

$$b = \frac{1}{\gamma}\cot\left(\frac{\phi}{2}\right). \tag{7.92}$$

It actually isn't necessary to go through all the work of Section 7.4.3 to obtain this result via a and b. We can just use Eq. (7.25), which says that $r \to \infty$ when

[12] More generally, we should write $\sinh(\theta - \theta_0)$ here. But we can eliminate the need for θ_0 by picking $\theta = 0$ to be the angle that corresponds to $r = \infty$.

[13] Again, we should write $\cosh(\theta - \theta_0)$ here. But we can eliminate the need for θ_0 by picking $\theta = 0$ to be the angle that corresponds to the maximum value of r.

$\cos\theta \to -1/\epsilon$. This then implies that the dotted lines in Fig. 7.21 have slope $\tan\theta = \sqrt{\sec^2\theta - 1} = \sqrt{\epsilon^2 - 1}$, which reproduces Eq. (7.91).

(b) Imagine a wide beam of particles moving in the positive x direction toward the mass M. Consider a thin cross-sectional ring in this beam, with radius b and thickness db. Now consider a very large sphere centered at M. Any particle that passes through the cross-sectional ring of radius b will hit this sphere in a ring located at an angle ϕ relative to the x axis, with an angular spread of $d\phi$. The relation between db and $d\phi$ is found from Eq. (7.92). Using $d(\cot\beta)/d\beta = -1/\sin^2\beta$, we have

$$\left|\frac{db}{d\phi}\right| = \frac{1}{2\gamma\sin^2(\phi/2)}. \tag{7.93}$$

The area of the incident cross-sectional ring is $d\sigma = 2\pi b\,|db|$. What is the solid angle subtended by a ring at angle ϕ with thickness $d\phi$? Taking the radius of the large sphere to be R (which will cancel out), the radius of the ring is $R\sin\phi$, and the width is $R\,|d\phi|$. The area of the ring is therefore $2\pi(R\sin\phi)(R\,|d\phi|)$, and so the solid angle subtended by the ring is $d\Omega = 2\pi\sin\phi\,|d\phi|$ steradians. Therefore, the differential cross section is

$$\begin{aligned}\frac{d\sigma}{d\Omega} = \frac{2\pi b\,|db|}{2\pi\sin\phi\,|d\phi|} &= \left(\frac{b}{\sin\phi}\right)\left|\frac{db}{d\phi}\right| \\ &= \left(\frac{(1/\gamma)\cot(\phi/2)}{2\sin(\phi/2)\cos(\phi/2)}\right)\left(\frac{1}{2\gamma\sin^2(\phi/2)}\right) \\ &= \frac{1}{4\gamma^2\sin^4(\phi/2)}.\end{aligned} \tag{7.94}$$

Remarks: What does this "differential cross section" result tell us? It tells us that if we want to find out how much cross-sectional area gets mapped into the solid angle $d\Omega$ at the angle ϕ, then we can use Eq. (7.94) to say (recalling $\gamma \equiv v_0^2/GM$),

$$d\sigma = \frac{G^2M^2}{4v_0^4\sin^4(\phi/2)}\,d\Omega \quad\Longrightarrow\quad d\sigma = \frac{G^2M^2m^2}{16E^2\sin^4(\phi/2)}\,d\Omega, \tag{7.95}$$

where we have used $E = mv_0^2/2$ to obtain the second expression. Let's look at some special cases. If $\phi \approx 180°$ (that is, backward scattering), then the amount of area that gets scattered into a nearly backward solid angle of $d\Omega$ equals $d\sigma = (G^2M^2/4v_0^4)\,d\Omega$. If v_0 is small, then we see that $d\sigma$ is large, that is, a large area gets deflected nearly straight backward. This makes sense, because with $v_0 \approx 0$, the orbit is essentially parabolic, which means that the initial and final velocities at infinity are (anti)parallel. (If you release a particle from rest far away from a gravitational source, it will come back to you. Assuming it doesn't bump into the source, of course.) If v_0 is large, then we see that $d\sigma$ is small, that is, only a small area gets deflected backward. This makes sense, because the particle is more likely to fly past M without much deflection if it is moving fast, because the force has hardly any time to act. The particle needs to start with a very small b value (which corresponds to a very small area) in order to get close enough to M to allow there to be a large enough force to swing it around.

Another special case is $\phi \approx 0$, that is, negligible deflection. In this case, Eq. (7.95) tells us that the amount of area that gets scattered into a nearly forward solid angle of $d\Omega$ is $d\sigma \approx \infty$. This makes sense, because if the impact parameter b is large (and there is an infinite cross-sectional area for which this is true), then the particle will hardly feel the mass M, so it will continue to move essentially in a straight line.[14]

What if we consider the electrostatic force, instead of the gravitational force? What is the differential cross section in this case? To answer this, note that we can rewrite γ as

$$\gamma = \frac{v_0^2}{GM} = \frac{2(mv_0^2/2)}{GMm} \equiv \frac{2E}{\alpha}. \tag{7.96}$$

In the case of electrostatics, the force takes the form, $F_e = kq_1q_2/r^2$. This looks like the gravitational force, $F_g = Gm_1m_2/r^2$, except that the constant α is now kq_1q_2, instead of Gm_1m_2. Therefore, the γ in Eq. (7.96) becomes $\gamma_e = 2E/(kq_1q_2)$. Substituting this into Eq. (7.94), or equivalently replacing GMm by kq_1q_2 in Eq. (7.95), gives the differential cross section for electrostatic scattering,

$$\frac{d\sigma}{d\Omega} = \frac{k^2q_1^2q_2^2}{16E^2\sin^4(\phi/2)}. \tag{7.97}$$

This is the Rutherford-scattering differential-cross-section formula. Around 1910, Rutherford and his students bombarded metal foils with alpha particles. Their results for the distribution of scattering angles were consistent with the above formula. In particular, they observed backward scattering of the alpha particles. Since the above formula is based on the assumption of a point source for the potential, this led Rutherford to his theory that atoms contained a dense positively charged nucleus, as opposed to being made of a spread-out "plum pudding" distribution of charge, which (as a special case of not yielding the correct distribution of scattering angles in general) doesn't yield backward scattering. ♣

[14] Remember, all we care about is the angle here. So when you're picturing the large sphere of radius R centered at M, don't say, "If b is large (for example, $R/2$), then a straight-line trajectory will hit the sphere at a large angle up above the x axis (for example, 30°)." This is incorrect. If you want to think in terms of a physical sphere of radius R, it is understood that R is infinitely large. Or more precisely, $R \gg b$, for any impact parameter b you might choose. So even if b is "large," it is still small compared with R, so a straight-line trajectory will hit the sphere of radius R at an angle that is essentially zero. Alternatively, you can just think in terms of angles, and not visualize an actual large sphere centered at M.

Chapter 8
Angular momentum, Part I (Constant $\hat{\mathbf{L}}$)

The angular momentum of a point mass, relative to a given origin, is defined by

$$\mathbf{L} = \mathbf{r} \times \mathbf{p}. \tag{8.1}$$

For a collection of particles, the total $\mathbf{L}$ is simply the sum of the $\mathbf{L}$'s of all the particles. The vector $\mathbf{r} \times \mathbf{p}$ is a useful thing to study because it has many nice properties. One of these is the conservation law presented in Theorem 7.1, which allowed us to introduce the "effective potential" in Section 7.2. And later in this chapter we will introduce the concept of *torque*, $\boldsymbol{\tau}$, which appears in the bread-and-butter statement, $\boldsymbol{\tau} = d\mathbf{L}/dt$ (analogous to Newton's $\mathbf{F} = d\mathbf{p}/dt$ law).

There are two basic types of angular momentum problems in the world. Since the solution to any rotational problem invariably comes down to using $\boldsymbol{\tau} = d\mathbf{L}/dt$, as we will see, we must determine how $\mathbf{L}$ changes in time. And since $\mathbf{L}$ is a vector, it can change because (1) its length changes, or (2) its direction changes (or through some combination of these effects). In other words, if we write $\mathbf{L} = L\hat{\mathbf{L}}$, where $\hat{\mathbf{L}}$ is the unit vector in the $\mathbf{L}$ direction, then $\mathbf{L}$ can change because L changes, or because $\hat{\mathbf{L}}$ changes, or both.

The first of these cases, that of constant $\hat{\mathbf{L}}$, is the easily understood one. Consider a spinning record, with the center chosen as the origin. The vector $\mathbf{L} = \sum \mathbf{r} \times \mathbf{p}$ is perpendicular to the record, because every term in the sum has this property. If we give the record a tangential force in the proper direction, then it will speed up (in a precise way which we will soon determine). There is nothing mysterious going on here. If we push on the record, it goes faster. $\mathbf{L}$ points in the same direction as before, but now simply with a larger magnitude. In fact, in this type of problem, we can completely forget that $\mathbf{L}$ is a vector. We can deal just with its magnitude L, and everything will be fine. This first case is the subject of the present chapter.

The second case, however, where $\mathbf{L}$ changes direction, can get rather confusing. This is the subject of the following chapter, where we will talk about twirling tops and other kinds of spinning objects that have a tendency to make one's head spin too. In these situations, the entire point is that $\mathbf{L}$ is actually a vector. And unlike in the constant-$\hat{\mathbf{L}}$ case, we really have to visualize things in three dimensions to see what's going on.[1]

The angular momentum of a point mass is given by the simple expression in Eq. (8.1). But in order to deal with setups in the real world, which invariably consist of many particles, we must learn how to calculate the angular momentum of an extended object. This is the task of the Section 8.1. In this chapter, we'll deal only with rotations around the z axis, or an axis parallel to the z axis. We'll save the general 3-D motion for Chapter 9.

8.1 Pancake object in x-y plane

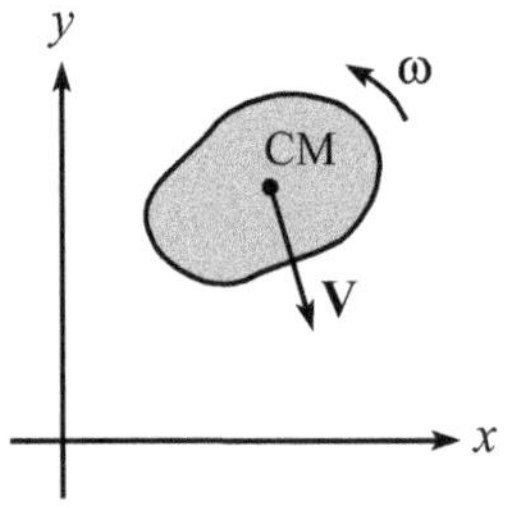

Fig. 8.1

Consider a flat rigid body undergoing arbitrary motion (both translating and spinning) in the x-y plane; see Fig. 8.1. What is the angular momentum of this body, relative to the origin of the coordinate system?[2] If we imagine the body to consist of particles of mass m_i, then the angular momentum of the entire body is the sum of the angular momenta of each m_i, which is $\mathbf{L}_i = \mathbf{r}_i \times \mathbf{p}_i$. So the total angular momentum is

$$\mathbf{L} = \sum_i \mathbf{r}_i \times \mathbf{p}_i. \tag{8.2}$$

For a continuous distribution of mass, we would have an integral instead of a sum. $\mathbf{L}$ depends on the locations and momenta of the masses. The momenta in turn depend on how fast the body is translating and spinning. Our goal here is to find the dependence of $\mathbf{L}$ on the distribution and motion of the constituent masses. The result will involve the geometry of the body in a specific way, as we will show.

In this section, we will deal only with pancake-like objects that move in the x-y plane. We will calculate $\mathbf{L}$ relative to the origin, and we will also derive an expression for the kinetic energy. We will deal with non-pancake objects in Section 8.2. Note that since both the $\mathbf{r}$ and $\mathbf{p}$ of all the masses in our pancake-like objects always lie in the x-y plane, the vector $\mathbf{L} = \sum \mathbf{r} \times \mathbf{p}$ always points in the $\hat{\mathbf{z}}$ direction. As mentioned above, this fact is what makes these pancake cases easy to deal with. $\mathbf{L}$ changes only because its length changes, not its direction. So when we eventually get to the $\boldsymbol{\tau} = d\mathbf{L}/dt$ equation, it will take on a simple form. Let's first look at a special case, and then we'll look at general motion in the x-y plane.

[1] The difference between these two cases is essentially the same as the difference between the two basic $\mathbf{F} = d\mathbf{p}/dt$ cases. The vector $\mathbf{p}$ can change because its magnitude changes, in which case we have $F = ma$ (assuming that m is constant). Or, $\mathbf{p}$ can change because its direction changes, in which case we have the centripetal-acceleration statement, $F = mv^2/r$. (Or there could be a combination of these effects.) The former case seems a bit more intuitive than the latter.

[2] Remember, $\mathbf{L}$ is defined relative to a chosen origin, because it has the vector $\mathbf{r}$ in it. So it makes no sense to ask what $\mathbf{L}$ is, without specifying what origin we've chosen.

8.1.1 Rotation about the z axis

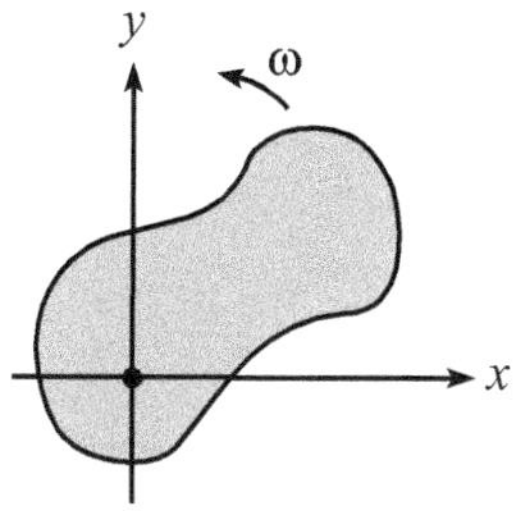

Fig. 8.2

The pancake in Fig. 8.2 is pivoted at the origin and rotates with angular speed ω around the z axis, in the counterclockwise direction (as viewed from above). Consider a little piece of the body, with mass dm and position (x, y). This little piece travels in a circle around the origin with speed $v = \omega r$, where $r = \sqrt{x^2 + y^2}$. Therefore, the angular momentum of this piece (relative to the origin) equals $\mathbf{L} = \mathbf{r} \times \mathbf{p} = r(v\,dm)\hat{\mathbf{z}} = dm\,r^2\omega\hat{\mathbf{z}}$. The $\hat{\mathbf{z}}$ direction arises from the cross product of the (orthogonal) vectors $\mathbf{r}$ and $\mathbf{p}$. The angular momentum of the entire body is therefore

$$\mathbf{L} = \int r^2\omega\hat{\mathbf{z}}\,dm = \int (x^2 + y^2)\omega\hat{\mathbf{z}}\,dm, \tag{8.3}$$

where the integration runs over the area of the body. If the density of the object is constant, as is usually the case, then we have $dm = \rho\,dx\,dy$. If we define the *moment of inertia* around the z axis to be

$$I_z \equiv \int r^2\,dm = \int (x^2 + y^2)\,dm, \tag{8.4}$$

then the z component of $\mathbf{L}$ is

$$L_z = I_z\omega, \tag{8.5}$$

and both L_x and L_y are zero. In the case where the rigid body is made up of a collection of point masses m_i in the x-y plane, the moment of inertia in Eq. (8.4) takes the discretized form,

$$I_z \equiv \sum_i m_i r_i^2. \tag{8.6}$$

Given any rigid body in the x-y plane, we can calculate I_z. And given ω, we can then multiply it by I_z to find L_z. In Section 8.3.1, we'll get some practice calculating various moments of inertia.

What is the kinetic energy of our object? We need to add up the energies of all the little pieces. A little piece has energy $dm\,v^2/2 = dm(r\omega)^2/2$. So the total kinetic energy is

$$T = \int \frac{r^2\omega^2}{2}\,dm. \tag{8.7}$$

With our definition of I_z in Eq. (8.4), this becomes

$$T = \frac{I_z\omega^2}{2}. \tag{8.8}$$

This is easy to remember, because it looks a lot like the kinetic energy of a point mass, $mv^2/2$.

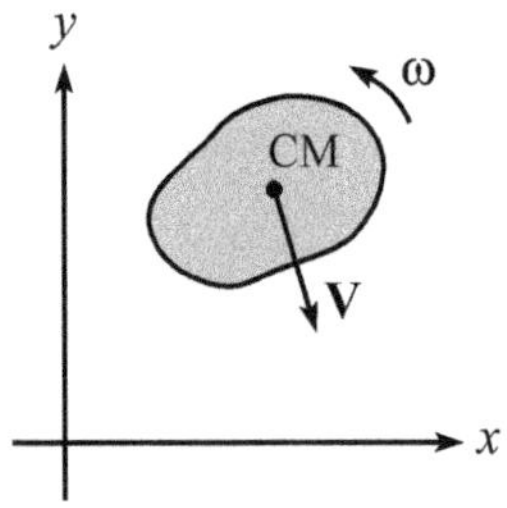

Fig. 8.3

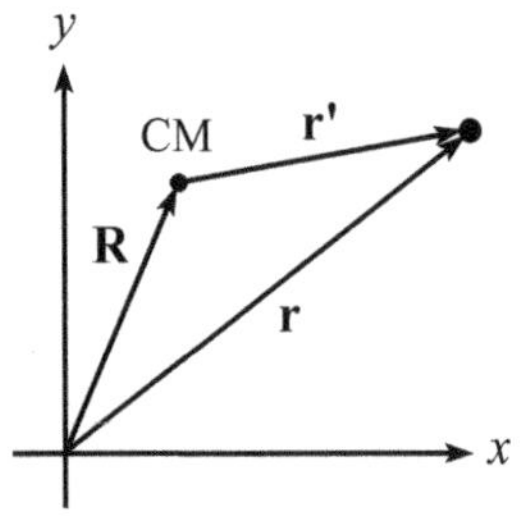

Fig. 8.4

8.1.2 General motion in x-y plane

How do we deal with general motion in the x-y plane? For the motion in Fig. 8.3, where the object is both translating and spinning, the various pieces of mass don't travel in circles around the origin, so we can't write $v = \omega r$ as we did above. It turns out to be highly advantageous to write the angular momentum, $\mathbf{L}$, and the kinetic energy, T, in terms of the center-of-mass (CM) coordinates and the coordinates relative to the CM. The expressions for $\mathbf{L}$ and T take on very nice forms when written this way, as we now show.

Let the position of the CM relative to a fixed origin be $\mathbf{R} = (X, Y)$. And let the position of a given point relative to the CM be $\mathbf{r}' = (x', y')$. Then the position of the given point relative to the fixed origin is $\mathbf{r} = \mathbf{R} + \mathbf{r}'$ (see Fig. 8.4). Let the velocity of the CM be $\mathbf{V}$, and let the velocity relative to the CM be $\mathbf{v}'$. Then $\mathbf{v} = \mathbf{V} + \mathbf{v}'$. Let the body rotate with angular speed ω' around the CM (around an instantaneous axis parallel to the z axis, so that the pancake remains in the x-y plane at all times).[3] Then $v' = \omega' r'$.

Let's look at $\mathbf{L}$ first. Let M be the total mass of the pancake. The angular momentum relative to the origin is

$$
\begin{aligned}
\mathbf{L} &= \int \mathbf{r} \times \mathbf{v}\, dm \\
&= \int (\mathbf{R} + \mathbf{r}') \times (\mathbf{V} + \mathbf{v}')\, dm \\
&= \int \mathbf{R} \times \mathbf{V}\, dm + \int \mathbf{r}' \times \mathbf{v}'\, dm \quad \text{(cross terms vanish; see below)} \\
&= M\mathbf{R} \times \mathbf{V} + \left(\int r'^2 \omega' dm \right) \hat{\mathbf{z}} \\
&\equiv \mathbf{R} \times \mathbf{P} + \left(I_z^{\text{CM}} \omega' \right) \hat{\mathbf{z}}.
\end{aligned} \tag{8.9}
$$

In going from the second to third line above, the cross terms, $\int \mathbf{r}' \times \mathbf{V}\, dm$ and $\int \mathbf{R} \times \mathbf{v}'\, dm$, vanish by definition of the CM, which says that $\int \mathbf{r}'\, dm = 0$. (See Eq. (5.58); basically, the position of the CM in the CM frame is zero.) This implies that $\int \mathbf{v}'\, dm = d(\int \mathbf{r}'\, dm)/dt$ also equals zero. And since we can pull the constant vectors $\mathbf{V}$ and $\mathbf{R}$ out of the above integrals, we are therefore left with zero. The quantity I_z^{CM} in the final result is the moment of inertia around an axis through the CM, parallel to the z axis. Equation (8.9) is a very nice result, and it is important enough to be called a theorem. In words, it says:

Theorem 8.1 *The angular momentum (relative to the origin) of a body can be found by treating the body like a point mass located at the CM and finding the*

[3] What we mean here is the following. Consider a coordinate system whose origin is the CM and whose axes are parallel to the fixed x and y axes. Then the pancake rotates with angular speed ω' with respect to this system.

angular momentum of this point mass relative to the origin, and by then adding on the angular momentum of the body relative to the CM.[4]

Note that if we have the special case where the CM travels around the origin in a circle with angular speed Ω (so that $V = \Omega R$), then Eq. (8.9) becomes $\mathbf{L} = \big(MR^2\Omega + I_z^{\text{CM}}\omega'\big)\hat{\mathbf{z}}$.

Now let's look at T. The kinetic energy is

$$\begin{aligned}
T &= \int \frac{1}{2}v^2\,dm \\
&= \int \frac{1}{2}|\mathbf{V}+\mathbf{v}'|^2\,dm \\
&= \frac{1}{2}\int V^2\,dm + \frac{1}{2}\int v'^2\,dm \quad \text{(cross term vanishes; see below)} \\
&= \frac{1}{2}MV^2 + \frac{1}{2}\int r'^2\omega'^2 dm \\
&\equiv \frac{1}{2}MV^2 + \frac{1}{2}I_z^{\text{CM}}\omega'^2. \qquad (8.10)
\end{aligned}$$

In going from the second to third line above, the cross term $\int \mathbf{V}\cdot\mathbf{v}'\,dm = \mathbf{V}\cdot\int \mathbf{v}'\,dm$ vanishes by definition of the CM, as in the above calculation of $\mathbf{L}$. Again, Eq. (8.10) is a very nice result. In words, it says:

Theorem 8.2 *The kinetic energy of a body can be found by treating the body like a point mass located at the CM, and by then adding on the kinetic energy of the body due to the motion relative to the CM.*[5]

To calculate E, my dear class,
Just add up two things, and you'll pass.
Take the CM point's E,
And then add on with glee,
The E 'round the center of mass.

Example (Cylinder on a ramp): A cylinder of mass m, radius r, and moment of inertia $I = (1/2)mr^2$ (this is the I for a solid cylinder around its center, as we'll see in Section 8.3.1) rolls without slipping down a plane inclined at an angle θ. What is the acceleration of the center of the cylinder?

[4] This theorem works only if we use the CM as the location of the imagined point mass. True, in the above analysis we could have chosen a point P other than the CM, and then written things in terms of the coordinates of P and the coordinates relative to P (which could also be described by a rotation). But then the cross terms in Eq. (8.9) wouldn't vanish, and we'd end up with an unenlightening mess.

[5] We already knew this from Section 5.6.2. It's just that now we know that the kinetic energy in the CM frame takes the form of $I_z^{\text{CM}}\omega'^2/2$.

Solution: We'll use conservation of energy to determine the speed v of the center of the cylinder after it has moved a distance d down the plane, and then we'll read off a from the standard constant-acceleration kinematic relation, $v = \sqrt{2ad}$.

The loss in potential energy of the cylinder is $mgd \sin\theta$. This shows up as kinetic energy, which equals $mv^2/2+I\omega^2/2$ from Theorem 8.2. But the nonslipping condition is $v = \omega r$. Therefore, $\omega = v/r$, and conservation of energy gives

$$\begin{aligned} mgd\sin\theta &= \frac{1}{2}mv^2 + \frac{1}{2}I\omega^2 \\ &= \frac{1}{2}mv^2 + \frac{1}{2}\left(\frac{1}{2}mr^2\right)\left(\frac{v}{r}\right)^2 \\ &= \frac{3}{4}mv^2. \end{aligned} \tag{8.11}$$

So the speed as a function of distance is $v = \sqrt{(4/3)gd\sin\theta}$. Hence, $v = \sqrt{2ad}$ gives $a = (2/3)g\sin\theta$, which is independent of r.

REMARKS: Our answer is 2/3 of the $g\sin\theta$ result for a block sliding down a frictionless plane. It is smaller because there is kinetic energy "wasted" in the rotational motion. Alternatively, it is smaller because there is a friction force pointing up the plane (to provide the torque necessary to get the cylinder rotating, but we'll talk about torque in Section 8.4), so this decreases the net force down the plane.

If we let I take the general form $I = \beta mr^2$, where β is a numerical factor, then you can show that the acceleration becomes $a = \sqrt{g\sin\theta/(1+\beta)}$. So if $\beta = 0$ (all the mass is at the center), then we simply have $a = g\sin\theta$, so the cylinder behaves just like a sliding block. If $\beta = 1$ (all the mass is on the rim), then $a = (1/2)g\sin\theta$. If $\beta \to \infty$,[6] then $a \to 0$. After reading Section 8.4, you can think about these special cases alternatively in terms of the forces and torques involved.

Although this problem can also be solved by using force and torque (the task of Exercise 8.37), the conservation-of-energy method is generally quicker in more complicated problems of this type, as you can see by doing, for example, Exercises 8.28 and 8.46. ♣

8.1.3 The parallel-axis theorem

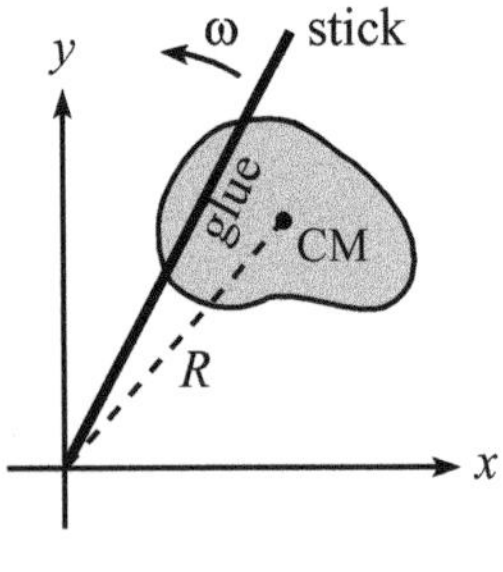

Fig. 8.5

Consider the special case where the CM rotates around the origin at the same rate as the body rotates around the CM. This can be achieved, for example, by gluing a stick across the pancake and pivoting one end of the stick at the origin; see Fig. 8.5. In this special case, we have the simplified situation where all points in the pancake travel in circles around the origin. Let their angular speed be ω. Then the speed of the CM is $V = \omega R$, so Eq. (8.9) gives the angular momentum around the origin as

$$L_z = (MR^2 + I_z^{\text{CM}})\omega. \tag{8.12}$$

[6] This can be obtained by adding on long spokes protruding from the cylinder and having them pass through a deep groove in the plane, or by having a spool with a very small inner radius roll down a thin plane with only its inner "axle" rolling on the plane.

In other words, the moment of inertia around the origin is

$$I_z = MR^2 + I_z^{\text{CM}}. \tag{8.13}$$

This is the *parallel-axis theorem*. It says that once you've calculated the moment of inertia of an object around the axis passing through the CM (namely I_z^{CM}), then if you want to calculate the moment of inertia around a parallel axis, you simply have to add on MR^2, where R is the distance between the two axes, and M is the mass of the object. Note that since the parallel-axis theorem is a special case of the result in Eq. (8.9), it is valid *only* with the CM, and not with any other point. The parallel-axis theorem actually holds for arbitrary nonplanar objects too, as we'll see in Section 8.2. And we'll also derive a more general form of the theorem in Chapter 9.

We can also look at the kinetic energy in this special case where the CM rotates around the origin at the same rate as the body rotates around the CM. Using $V = \omega R$ in Eq. (8.10), we find

$$T = \frac{1}{2}\left(MR^2 + I_z^{\text{CM}}\right)\omega^2 = \frac{1}{2}I_z\omega^2. \tag{8.14}$$

Example (A stick): Let's verify the parallel-axis theorem for a stick of mass m and length ℓ, in the case where we want to compare the moment of inertia around an axis through an end (perpendicular to the stick) with the moment of inertia around an axis through the CM (perpendicular to the stick).

For convenience, let $\rho = m/\ell$ be the density. The moment of inertia around an axis through an end is

$$I^{\text{end}} = \int_0^\ell x^2\, dm = \int_0^\ell x^2 \rho\, dx = \frac{1}{3}\rho\ell^3 = \frac{1}{3}(\rho\ell)\ell^2 = \frac{1}{3}m\ell^2. \tag{8.15}$$

The moment of inertia around an axis through the CM is

$$I^{\text{CM}} = \int_{-\ell/2}^{\ell/2} x^2\, dm = \int_{-\ell/2}^{\ell/2} x^2 \rho\, dx = \frac{1}{12}\rho\ell^3 = \frac{1}{12}(\rho\ell)\ell^2 = \frac{1}{12}m\ell^2. \tag{8.16}$$

This is consistent with the parallel-axis theorem, Eq. (8.13), because

$$I^{\text{end}} = m\left(\frac{\ell}{2}\right)^2 + I^{\text{CM}}. \tag{8.17}$$

Remember that this works only with the CM. If we instead want to compare I^{end} with the I around a point, say, $\ell/6$ from that end, then we cannot say that they differ by $m(\ell/6)^2$. But we *can* compare each of them to I^{CM} and say that they differ by $(\ell/2)^2 - (\ell/3)^2 = 5\ell^2/36$.

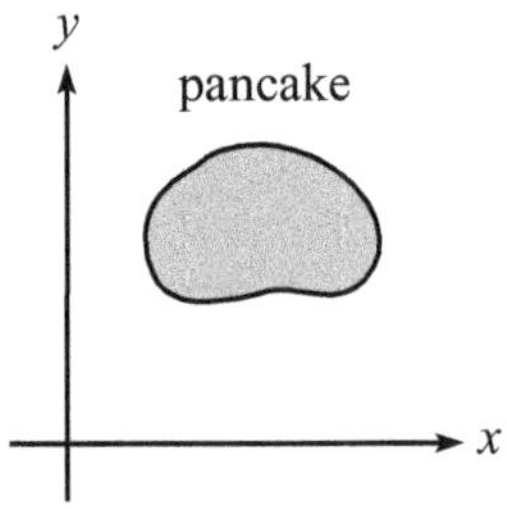

Fig. 8.6

8.1.4 The perpendicular-axis theorem

This theorem is valid *only* for pancake objects. Consider a pancake object in the x-y plane (see Fig. 8.6). Then the *perpendicular-axis theorem* says that

$$I_z = I_x + I_y, \tag{8.18}$$

where I_x and I_y are defined analogously to the I_z in Eq. (8.4). That is, to find I_x, imagine spinning the object around the x axis at angular speed ω, and then define $I_x \equiv L_x/\omega$. (Only the distance from the x axis matters in calculating the speed of a given point. So the fact that the object has extent along the x direction and is therefore not a pancake in the y-z plane is irrelevant. The following section has further discussion of this.) Likewise for I_y. In other words,

$$I_x \equiv \int (y^2 + z^2)\, dm, \quad I_y \equiv \int (z^2 + x^2)\, dm, \quad I_z \equiv \int (x^2 + y^2)\, dm. \tag{8.19}$$

To prove the perpendicular-axis theorem, we simply use the fact that $z = 0$ for our pancake object. Equation (8.19) then gives $I_z = I_x + I_y$. In the limited number of situations where this theorem is applicable, it can save you some trouble. A few examples are given in Section 8.3.1.

8.2 Nonplanar objects

In Section 8.1, we restricted the discussion to pancake objects in the x-y plane. However, nearly all the results we derived carry over to nonplanar objects, provided that the axis of rotation is parallel to the z axis, and provided that we are concerned only with L_z, and not L_x or L_y. So let's drop the pancake assumption and run through the results we obtained above.

First, consider an object rotating around the z axis. Let the object have extension in the z direction. If we imagine slicing the object into pancakes parallel to the x-y plane, then Eqs. (8.4) and (8.5) correctly give the L_z for each pancake. And since the L_z of the whole object is the sum of the L_z's of all the pancakes, we see that the I_z of the whole object is the sum of the I_z's of all the pancakes. The difference in the z values of the pancakes is irrelevant. Therefore, for *any* object rotating around the z axis, we have

$$I_z = \int (x^2 + y^2)\, dm, \quad \text{and} \quad L_z = I_z\omega, \tag{8.20}$$

where the integration runs over the entire volume of the body. We'll calculate the I_z for many nonplanar objects in Section 8.3.1. Note that even though Eq. (8.20) gives the L_z for an arbitrary object, the analysis in this chapter is still not completely general because (1) we are restricting the axis of rotation to be the (fixed) z axis, and (2) even with this restriction, an object outside the x-y plane might have nonzero x and y components of $\mathbf{L}$, but we found only the z component in Eq. (8.20). This second fact is strange but true. We'll deal with it in great detail in Section 9.2.

As far as the kinetic energy goes, the T for a nonplanar object rotating around the z axis is still given by Eq. (8.8), because we can obtain the total T by adding up the T's of all the pancake slices.

Also, Eqs. (8.9) and (8.10) continue to hold for a nonplanar object in the case where the CM is translating while the object is spinning around it (or more precisely, spinning around an axis parallel to the z axis and passing through the CM). The velocity $\mathbf{V}$ of the CM can actually point in any direction, and these two equations are still valid. But we'll assume in this chapter that all velocities are in the x-y plane.

Lastly, the parallel-axis theorem still holds for a nonplanar object (the derivation using Eq. (8.9) is the same). But as mentioned in Section 8.1.4, the perpendicular-axis theorem does *not* hold. This is the one instance where we need the planar assumption.

Finding the CM

The center of mass has come up repeatedly in this chapter. For example, when we used the parallel-axis theorem, we needed to know where the CM was. In some cases, such as with a stick or a disk, the location is obvious. But in other cases, it isn't so clear. So let's get a little practice calculating the location of the CM. Depending on whether the mass distribution is discrete or continuous, the position of the CM is defined by (see Eq. (5.58))

$$\mathbf{R}_{\text{CM}} = \frac{\sum \mathbf{r}_i m_i}{M}, \quad \text{or} \quad \mathbf{R}_{\text{CM}} = \frac{\int \mathbf{r}\, dm}{M}, \tag{8.21}$$

where M is the total mass. We'll do an example with a continuous mass distribution here. As is often the case with problems involving an integral, the main step in the solution is deciding how you want to slice up the object to do the integral.

Example (Hemispherical shell): Find the location of the CM of a hollow hemispherical shell, with uniform mass density and radius R.

Solution: By symmetry, the CM is located on the line above the center of the base. So our task reduces to finding the height, y_{CM}. Let the mass density be σ. We'll slice the hemisphere up into horizontal rings, described by the angle θ above the horizontal, as shown in Fig. 8.7. If the angular thickness of a ring is $d\theta$, then its mass is

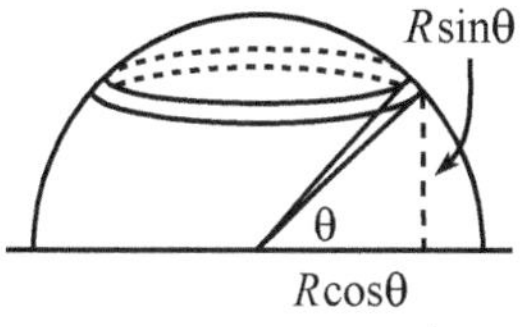

Fig. 8.7

$$dm = \sigma\, dA = \sigma(\text{length})(\text{width}) = \sigma(2\pi R\cos\theta)(R\, d\theta). \tag{8.22}$$

All points on the ring have a y value of $R\sin\theta$. Therefore,

$$\begin{aligned} y_{\text{CM}} = \frac{1}{M}\int y\,dm &= \frac{1}{(2\pi R^2)\sigma}\int_0^{\pi/2}(R\sin\theta)(2\pi R^2\sigma\cos\theta\,d\theta) \\ &= R\int_0^{\pi/2}\sin\theta\cos\theta\,d\theta \\ &= \frac{R\sin^2\theta}{2}\bigg|_0^{\pi/2} \\ &= \frac{R}{2}. \end{aligned} \tag{8.23}$$

The simple factor of $1/2$ here is nice, but it's not all that obvious. It comes from the fact that each value of y is represented equally. If you solved the problem by doing a dy integral instead of a $d\theta$ one, you would find that there is the same area (and hence the same mass) in each ring of vertical height dy. In short, as y increases, the larger tilt of the surface cancels out the smaller radius of the rings, yielding the same area. You are encouraged to work this out.

The calculation of a CM is very similar to the calculation of a moment of inertia. Both involve an integration over the mass of an object, but the former has one power of a length in the integrand, whereas the latter has two powers.

8.3 Calculating moments of inertia

8.3.1 Lots of examples

Let's now calculate the moments of inertia of various objects, around specified axes. We'll use ρ to denote the mass density (per unit length, area, or volume, as appropriate), and we'll assume that this density is uniform throughout the object. For the more complicated objects in the list below, it is generally a good idea to slice them up into pieces for which I is already known. The problem then reduces to integrating over these known I's. There is usually more than one way to do this slicing. For example, a sphere can be looked at as a series of concentric shells or a collection of disks stacked on top of each other. In the examples below, you may want to play around with slicings other than the ones given. Consider at least a few of these examples to be problems and try to work them out for yourself.

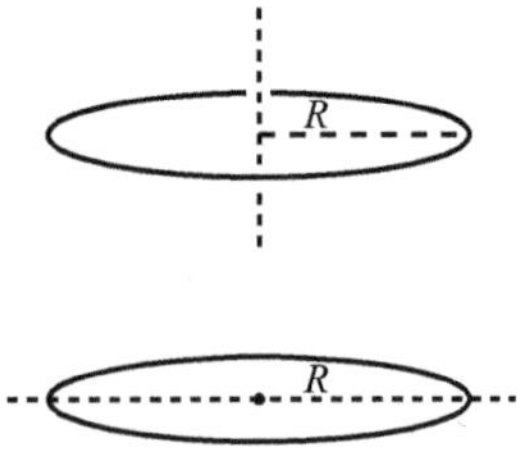

Fig. 8.8

1. A ring of mass M and radius R (axis through center, perpendicular to plane; Fig. 8.8):

$$I = \int r^2\,dm = \int_0^{2\pi} R^2\rho R\,d\theta = (2\pi R\rho)R^2 = \boxed{MR^2}, \tag{8.24}$$

as expected, because all of the mass is a distance R from the axis.

2. A ring of mass M and radius R (axis through center, in plane; Fig. 8.8): The distance from the axis is (the absolute value of) $R\sin\theta$. Therefore,

$$I = \int r^2\, dm = \int_0^{2\pi} (R\sin\theta)^2 \rho R\, d\theta = \frac{1}{2}(2\pi R\rho)R^2 = \boxed{\tfrac{1}{2}MR^2}, \qquad (8.25)$$

where we have used $\sin^2\theta = (1 - \cos 2\theta)/2$. You can also find I by using the perpendicular-axis theorem. In the notation of Section 8.1.4, we have $I_x = I_y$, by symmetry. Therefore, $I_z = 2I_x$. Using $I_z = MR^2$ from Example 1 then gives $I_x = MR^2/2$.

3. A disk of mass M and radius R (axis through center, perpendicular to plane; Fig. 8.9):

$$I = \int r^2\, dm = \int_0^{2\pi}\int_0^R r^2 \rho r\, dr\, d\theta = (R^4/4)2\pi\rho = \frac{1}{2}(\rho\pi R^2)R^2 = \boxed{\tfrac{1}{2}MR^2}. \qquad (8.26)$$

You can save the (trivial) step of integrating over θ by considering the disk to be made up of many concentric rings, and invoking Example 1. The mass of each ring is $\rho 2\pi r\, dr$. Integrating over the rings gives $I = \int_0^R (\rho 2\pi r\, dr) r^2 = \pi R^4 \rho/2 = MR^2/2$, as above. Slicing up the disk is fairly inconsequential in this example, but it will save you some trouble in others.

4. A disk of mass M and radius R (axis through center, in plane; Fig. 8.9): Slice the disk up into rings, and use Example 2.

$$I = \int_0^R (1/2)(\rho 2\pi r\, dr)r^2 = (R^2/4)\rho\pi = \frac{1}{4}(\rho\pi R^2)R^2 = \boxed{\tfrac{1}{4}MR^2}. \qquad (8.27)$$

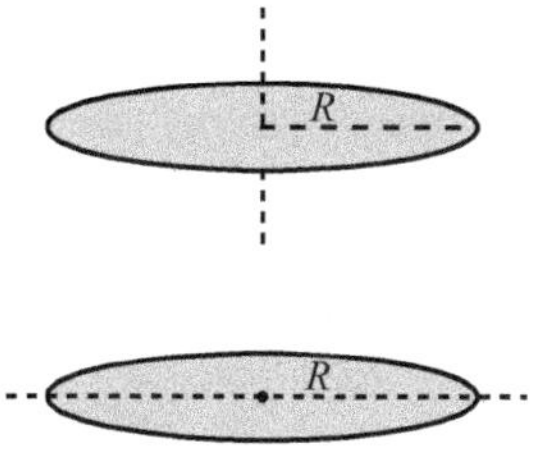

Fig. 8.9

Or just use Example 3 and the perpendicular-axis theorem.

5. A thin uniform rod of mass M and length L (axis through center, perpendicular to rod; Fig. 8.10): We already found this I and the next one in Section 8.1.3, but we'll include them here for completeness.

$$I = \int x^2 dm = \int_{-L/2}^{L/2} x^2 \rho\, dx = \frac{1}{12}(\rho L)L^2 = \boxed{\tfrac{1}{12}ML^2}. \qquad (8.28)$$

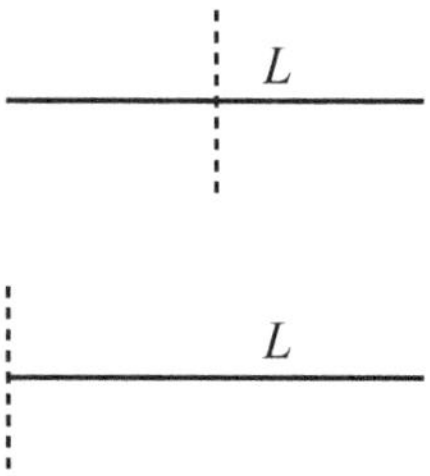

Fig. 8.10

6. A thin uniform rod of mass M and length L (axis through end, perpendicular to rod; Fig. 8.10):

$$I = \int x^2 dm = \int_0^L x^2 \rho\, dx = \frac{1}{3}(\rho L)L^2 = \boxed{\tfrac{1}{3}ML^2}. \qquad (8.29)$$

7. A spherical shell of mass M and radius R (any axis through center; Fig. 8.11): Let's slice the sphere into horizontal ring-like strips. In spherical coordinates, the radius of a ring is given by $r = R\sin\theta$, where θ is the angle down from the north pole.

The area of a strip is then $2\pi(R\sin\theta)R\,d\theta$. Using $\int\sin^3\theta = \int\sin\theta(1-\cos^2\theta) = -\cos\theta + \cos^3\theta/3$, we have

$$I = \int r^2\,dm = \int_0^\pi (R\sin\theta)^2\,2\pi\rho(R\sin\theta)R\,d\theta = 2\pi\rho R^4\int_0^\pi \sin^3\theta$$
$$= 2\pi\rho R^4(4/3) = \frac{2}{3}(4\pi R^2\rho)R^2 = \boxed{\tfrac{2}{3}MR^2}. \qquad (8.30)$$

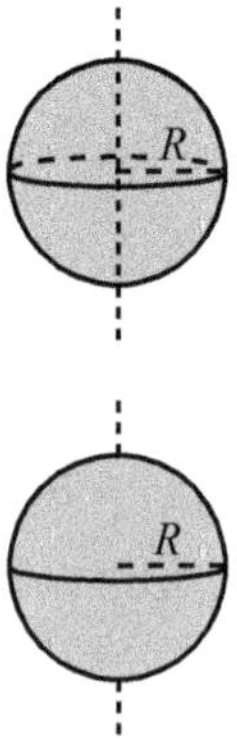

Fig. 8.11

An alternate and slick way of deriving this is to use the following result, similar in spirit to the perpendicular-axis theorem: Adding up the three moments of inertia in Eq. (8.19) gives

$$I_x + I_y + I_z = 2\int(x^2+y^2+z^2)\,dm = 2\int r^2\,dm. \qquad (8.31)$$

If we apply this to a spherical shell, where r always equals the radius R, then the right-hand side is $2MR^2$. And since the I's are all equal by symmetry, they must all be $2MR^2/3$.

8. A solid sphere of mass M and radius R (any axis through center; Fig. 8.11): A sphere is made up of concentric spherical shells. The volume of a shell is $4\pi r^2 dr$. Using Example 7, we have

$$I = \int_0^R (2/3)(4\pi\rho r^2 dr)r^2 = (R^5/5)(8\pi\rho/3) = \frac{2}{5}(4\pi R^3\rho/3)R^2 = \boxed{\tfrac{2}{5}MR^2}. \qquad (8.32)$$

The task of Exercise 8.33 is to derive this result by slicing the sphere into horizontal disks.

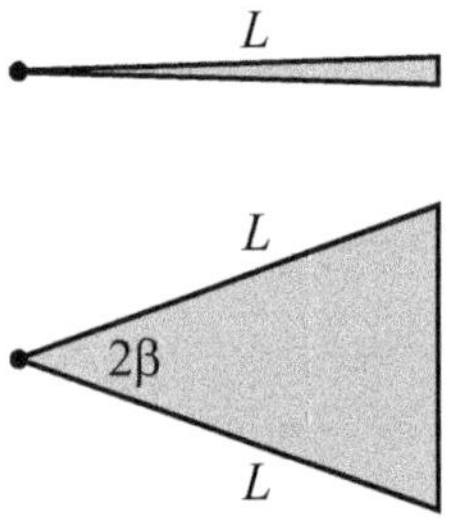

Fig. 8.12

9. An infinitesimally thin triangle of mass M and length L (axis through tip, perpendicular to plane; Fig. 8.12): Let the base have length a, where a is infinitesimally small. Then a tiny vertical slice a distance x from the tip has length $a(x/L)$. If the slice has thickness dx, then it is essentially a point mass of mass $dm = \rho ax\,dx/L$. Therefore,

$$I = \int x^2 dm = \int_0^L x^2\rho ax/L\,dx = \frac{1}{2}(\rho aL/2)L^2 = \boxed{\tfrac{1}{2}ML^2}, \qquad (8.33)$$

because $aL/2$ is the area of the triangle. This has the same form as the disk in Example 3, because a disk is made up of many of these triangles.

10. An isosceles triangle of mass M, vertex angle 2β, and common-side length L (axis through tip, perpendicular to plane; Fig. 8.12): Let h be the altitude of the triangle (so $h = L\cos\beta$). Slice the triangle into thin strips parallel to the base. Let x be the distance from a strip to the vertex. Then the length of a strip is $\ell = 2x\tan\beta$, and its mass is $dm = \rho(2x\tan\beta\,dx)$. Using Example 5 above, along with the parallel-axis theorem, we have

$$I = \int dm\left(\frac{\ell^2}{12}+x^2\right) = \int_0^h(\rho 2x\tan\beta\,dx)\left(\frac{(2x\tan\beta)^2}{12}+x^2\right)$$
$$= 2\rho\tan\beta\int_0^h\left(1+\frac{\tan^2\beta}{3}\right)x^3\,dx = 2\rho\tan\beta\left(1+\frac{\tan^2\beta}{3}\right)\frac{h^4}{4}. \qquad (8.34)$$

But the area of the entire triangle is $h^2 \tan\beta$, so we have $I = (Mh^2/2)\big(1 + (1/3)\tan^2\beta\big)$. In terms of $L = h/\cos\beta$, this is

$$I = (ML^2/2)(\cos^2\beta + \sin^2\beta/3) = \boxed{\tfrac{1}{2}ML^2\Big(1 - \tfrac{2}{3}\sin^2\beta\Big)}. \tag{8.35}$$

11. A regular N-gon of mass M and "radius" R (axis through center, perpendicular to plane; Fig. 8.13): The N-gon is made up of N isosceles triangles, so we can use Example 10, with $\beta = \pi/N$. The masses of the triangles simply add, so if M is the mass of the whole N-gon, we have

$$I = \boxed{\tfrac{1}{2}MR^2\Big(1 - \tfrac{2}{3}\sin^2\tfrac{\pi}{N}\Big)}. \tag{8.36}$$

We can list the values of I for a few N. With the shorthand notation $(N, I/MR^2)$, Eq. (8.36) gives $(3, \frac{1}{4})$, $(4, \frac{1}{3})$, $(6, \frac{5}{12})$, $(\infty, \frac{1}{2})$. These values of I form a nice arithmetic progression.

12. A rectangle of mass M and sides of length a and b (axis through center, perpendicular to plane; Fig. 8.13): Let the z axis be perpendicular to the plane. We know that $I_x = Mb^2/12$ and $I_y = Ma^2/12$ (because the extent of an object along an axis doesn't affect the moment around that axis, when written in terms of the total M). So the perpendicular-axis theorem gives

$$I_z = I_x + I_y = \boxed{\tfrac{1}{12}M(a^2 + b^2)}. \tag{8.37}$$

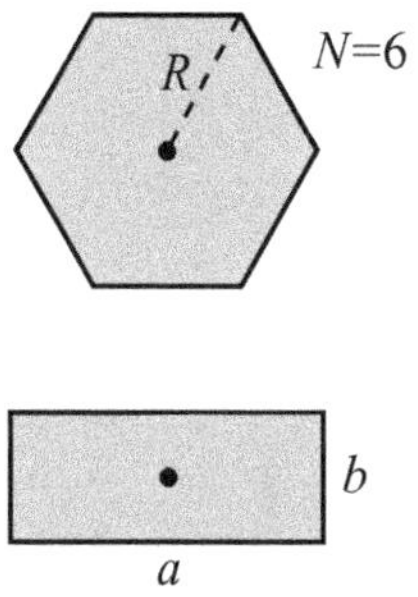

Fig. 8.13

8.3.2 A neat trick

For some objects with certain symmetries, it's possible to calculate the moment of inertia without doing any integrals. The only things we need are a scaling argument and the parallel-axis theorem. We'll illustrate this technique by finding the I for a stick around its center (Example 5 above). You'll find other applications in the problems for this chapter.

In the present example, the basic trick is to compare the I for a stick of length L with the I for a stick of length $2L$ (and same density ρ). A quick scaling argument shows that the latter is eight times the former. This is true because the integral $\int x^2\,dm = \int x^2\rho\,dx$ has three powers of x in it (yes, the dx counts). So a change of variables, $x = 2y$, brings in a factor of $2^3 = 8$. Equivalently, if we imagine expanding the smaller stick into the larger one, then a corresponding piece in the larger stick will be twice as far from the axis, and also twice as massive. The integral $\int x^2\,dm$ therefore increases by a factor of $2^2 \cdot 2 = 8$.

The technique is most easily illustrated with pictures. If we denote the moment of inertia of an object by a picture of the object, with a dot signifying the axis, then we have:

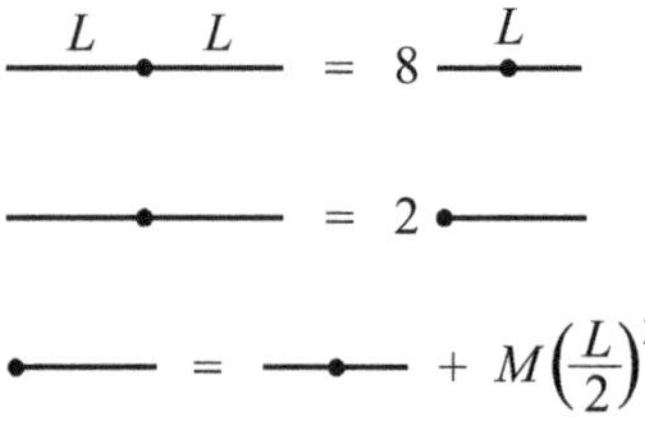

The first line comes from the scaling argument, the second line comes from the fact that moments of inertia simply add (the left-hand side is two copies of the right-hand side, attached at the pivot), and the third line comes from the parallel-axis theorem. Equating the right-hand sides of the first two equations gives

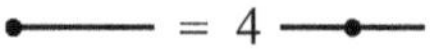

Plugging this expression for into the third equation gives the desired result,

 $= \frac{1}{12}ML^2$

Note that sooner or later we must use real live numbers, which enter here through the parallel-axis theorem. Using only scaling arguments isn't sufficient, because they provide only linear equations homogeneous in the I's, and therefore give no means of picking up the proper dimensions. (For an interesting account of Galileo's discovery of scaling laws, see Peterson (2002).)

Once you've mastered this trick and applied it to the fractal objects in Problem 8.8, you can impress your friends by saying that you know how to "use scaling arguments, along with the parallel-axis theorem, to calculate moments of inertia of objects with fractal dimension." And you never know when that might come in handy!

8.4 Torque

We will now show that under certain conditions (stated below), the rate of change of angular momentum is equal to a certain quantity, $\boldsymbol{\tau}$, which we call the *torque*. That is, $\boldsymbol{\tau} = d\mathbf{L}/dt$. This is the rotational analog of our old friend $\mathbf{F} = d\mathbf{p}/dt$ involving linear momentum. The basic idea here is straightforward, but there are two subtle issues. One deals with internal forces within a collection of particles. The other deals with the possible acceleration of the origin (the point relative to which the torque and angular momentum are calculated). To keep things straight, we'll prove the general result by dealing with three increasingly complicated situations.

Our derivation of $\boldsymbol{\tau} = d\mathbf{L}/dt$ here holds for completely general motion, so we can take the result and use it in the following chapter, too. If you wish, you can construct a more specific proof of $\boldsymbol{\tau} = d\mathbf{L}/dt$ for the special case where the axis of rotation is parallel to the z axis. But since the general proof is no more difficult, we'll present it here in this chapter and do it once and for all.

8.4.1 Point mass, fixed origin

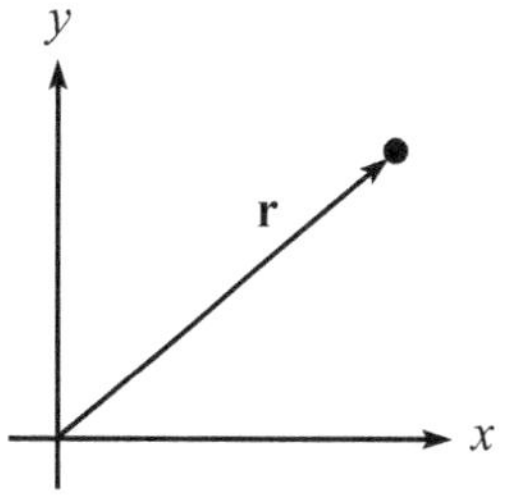

Fig. 8.14

Consider a point mass at position $\mathbf{r}$ relative to a fixed origin (see Fig. 8.14). The time derivative of the angular momentum, $\mathbf{L} = \mathbf{r} \times \mathbf{p}$, is

$$\begin{aligned} \frac{d\mathbf{L}}{dt} &= \frac{d}{dt}(\mathbf{r} \times \mathbf{p}) \\ &= \frac{d\mathbf{r}}{dt} \times \mathbf{p} + \mathbf{r} \times \frac{d\mathbf{p}}{dt} \\ &= \mathbf{v} \times (m\mathbf{v}) + \mathbf{r} \times \mathbf{F} \\ &= 0 + \mathbf{r} \times \mathbf{F}, \end{aligned} \tag{8.38}$$

where $\mathbf{F}$ is the force acting on the particle. This is the same calculation as in Theorem 7.1, except that here we are considering an arbitrary force instead of a central one. If we define the *torque* on the particle as

$$\boldsymbol{\tau} \equiv \mathbf{r} \times \mathbf{F}, \tag{8.39}$$

then Eq. (8.38) becomes

$$\boldsymbol{\tau} = \frac{d\mathbf{L}}{dt}. \tag{8.40}$$

It is understood that the $\mathbf{r}$ in the torque is measured with respect to the same origin as the $\mathbf{r}$ in the angular momentum.

8.4.2 Extended mass, fixed origin

In an extended object, there are internal forces acting on the various pieces of the object, in addition to whatever external forces exist. For example, the external force on a given molecule in a body might come from gravity, while the internal forces come from the adjacent molecules. How do we deal with these different types of forces?

In what follows, we will deal only with internal forces that are central forces, so that the force between two objects is directed along the line between them. This is a valid assumption for the pushing and pulling forces between molecules in a solid. (It isn't valid, for example, when dealing with magnetic forces. But we won't be interested in such things here.) We will invoke Newton's third law,

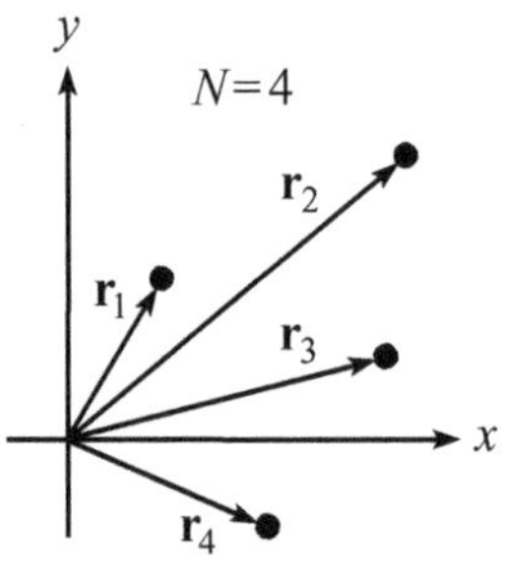

Fig. 8.15

which says that the force that particle 1 applies to particle 2 is equal and opposite to the force that particle 2 applies to particle 1.

For concreteness, let's assume that we have a collection of N discrete particles labeled by the index i (see Fig. 8.15). In the continuous case, we would need to replace the following sums with integrals. The total angular momentum of the system is

$$\mathbf{L} = \sum_{i=1}^{N} \mathbf{r}_i \times \mathbf{p}_i. \tag{8.41}$$

The force acting on each particle is $\mathbf{F}_i^{\text{ext}} + \mathbf{F}_i^{\text{int}} = d\mathbf{p}_i/dt$. Therefore,

$$\begin{aligned}\frac{d\mathbf{L}}{dt} &= \frac{d}{dt}\sum_i \mathbf{r}_i \times \mathbf{p}_i \\ &= \sum_i \frac{d\mathbf{r}_i}{dt} \times \mathbf{p}_i + \sum_i \mathbf{r}_i \times \frac{d\mathbf{p}_i}{dt} \\ &= \sum_i \mathbf{v}_i \times (m\mathbf{v}_i) + \sum_i \mathbf{r}_i \times (\mathbf{F}_i^{\text{ext}} + \mathbf{F}_i^{\text{int}}) \\ &= 0 + \sum_i \mathbf{r}_i \times \mathbf{F}_i^{\text{ext}} \\ &\equiv \sum_i \boldsymbol{\tau}_i^{\text{ext}}.\end{aligned} \tag{8.42}$$

The second-to-last line follows because $\mathbf{v}_i \times \mathbf{v}_i = 0$, and also because $\sum_i \mathbf{r}_i \times \mathbf{F}_i^{\text{int}} = 0$, as you can show in Problem 8.9. In other words, the internal forces provide no net torque. This is quite reasonable. It basically says that a rigid object with no external forces won't spontaneously start rotating.

Note that the right-hand side involves the *total* external torque acting on the body, which may come from forces acting at many different points. Note also that nowhere did we assume that the particles are rigidly connected to each other. Equation (8.42) still holds even if there is relative motion among the particles. But in that case, it's usually hard to get a handle on $\mathbf{L}$, because it doesn't take the nice $I\omega$ form.

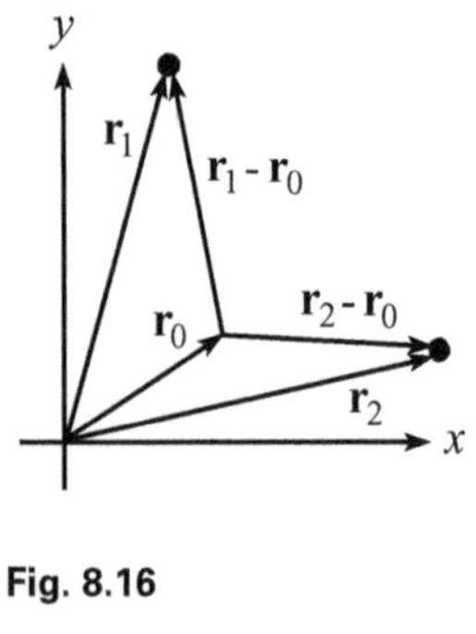

Fig. 8.16

8.4.3 Extended mass, nonfixed origin

Let the position of the origin be $\mathbf{r}_0$ (see Fig. 8.16), and let the positions of the particles be $\mathbf{r}_i$. The vectors $\mathbf{r}_0$ and $\mathbf{r}_i$ are measured with respect to a given fixed coordinate system. The total angular momentum of the system, relative to the (possibly accelerating) origin $\mathbf{r}_0$, is[7]

$$\mathbf{L} = \sum_i (\mathbf{r}_i - \mathbf{r}_0) \times m_i(\dot{\mathbf{r}}_i - \dot{\mathbf{r}}_0). \tag{8.43}$$

Therefore,

$$\begin{aligned}
\frac{d\mathbf{L}}{dt} &= \frac{d}{dt}\left(\sum_i (\mathbf{r}_i - \mathbf{r}_0) \times m_i(\dot{\mathbf{r}}_i - \dot{\mathbf{r}}_0)\right) \\
&= \sum_i (\dot{\mathbf{r}}_i - \dot{\mathbf{r}}_0) \times m_i(\dot{\mathbf{r}}_i - \dot{\mathbf{r}}_0) + \sum_i (\mathbf{r}_i - \mathbf{r}_0) \times m_i(\ddot{\mathbf{r}}_i - \ddot{\mathbf{r}}_0) \\
&= 0 + \sum_i (\mathbf{r}_i - \mathbf{r}_0) \times (\mathbf{F}_i^{\text{ext}} + \mathbf{F}_i^{\text{int}} - m_i\ddot{\mathbf{r}}_0),
\end{aligned} \tag{8.44}$$

because $m_i\ddot{\mathbf{r}}_i$ is the net force (namely $\mathbf{F}_i^{\text{ext}} + \mathbf{F}_i^{\text{int}}$) acting on the ith particle. But a quick corollary to Problem 8.9 is that the term involving $\mathbf{F}_i^{\text{int}}$ vanishes (as you should check). And since $\sum m_i\mathbf{r}_i = M\mathbf{R}$ (where $M = \sum m_i$ is the total mass, and $\mathbf{R}$ is the position of the center of mass), we have

$$\frac{d\mathbf{L}}{dt} = \left(\sum_i (\mathbf{r}_i - \mathbf{r}_0) \times \mathbf{F}_i^{\text{ext}}\right) - M(\mathbf{R} - \mathbf{r}_0) \times \ddot{\mathbf{r}}_0. \tag{8.45}$$

The first term here is the external torque, measured relative to the origin $\mathbf{r}_0$. The second term is something we wish would go away. And indeed, it usually does. It vanishes if any of the following three conditions is satisfied.

1. $\mathbf{R} = \mathbf{r}_0$, that is, the origin is the CM.
2. $\ddot{\mathbf{r}}_0 = \mathbf{0}$, that is, the origin is not accelerating.
3. $(\mathbf{R} - \mathbf{r}_0)$ is parallel to $\ddot{\mathbf{r}}_0$. This condition is rarely invoked.

If any of these conditions is satisfied, then we are free to write

$$\frac{d\mathbf{L}}{dt} = \sum_i (\mathbf{r}_i - \mathbf{r}_0) \times \mathbf{F}_i^{\text{ext}} \equiv \sum_i \boldsymbol{\tau}_i^{\text{ext}}. \tag{8.46}$$

In other words, we can equate the total external torque with the rate of change of the total angular momentum. An immediate corollary of this result is:

Corollary 8.3 *If the total external torque on a system is zero, then its angular momentum is conserved. In particular, the angular momentum of an isolated system (one that is subject to no external forces) is conserved.*

[7] More precisely, we are calculating the angular momentum relative to a coordinate system whose origin is $\mathbf{r}_0$ and whose axes remain parallel to the fixed axes. If we allowed for a rotation of the axes, then we would have to deal with all the fictitious-force issues that are the subject of Chapter 10. As it is, we will still end up dealing with one fictitious force (see the remark below).

Everything up to this point is valid for arbitrary motion. The particles can be moving relative to each other, and the various $\mathbf{L}_i$'s can point in different directions, etc. But let's now restrict the motion. In the present chapter, we are dealing only with cases where $\hat{\mathbf{L}}$ is constant (taken to point in the z direction). Therefore, $d\mathbf{L}/dt = d(L\hat{\mathbf{L}})/dt = (dL/dt)\hat{\mathbf{L}}$. If in addition we consider only rigid objects (where the relative distances among the particles are fixed) that undergo pure rotation around a given point, then $L = I\omega$, which gives $dL/dt = I\dot{\omega} \equiv I\alpha$. Taking the magnitude of both sides of Eq. (8.46) then gives

$$\tau_{\text{ext}} = I\alpha. \tag{8.47}$$

Invariably, we will calculate angular momentum and torque around either the CM or a fixed point (or a point that moves with constant velocity, but this doesn't come up often). These are the "safe" origins, in the sense that Eq. (8.46) holds. As long as you always use one of these safe origins, you can simply apply Eq. (8.46) and not worry much about its derivation.

REMARK: You'll probably never end up invoking the third condition above, but it's interesting to note that there's a simple way of understanding it in terms of accelerating reference frames. This is the topic of Chapter 10, so we're getting a little ahead of ourselves here, but the reasoning is as follows. Let $\mathbf{r}_0$ be the origin of a reference frame that is accelerating with acceleration $\ddot{\mathbf{r}}_0$. Then all objects in this accelerating frame feel a mysterious fictitious force of $-m\ddot{\mathbf{r}}_0$. For example, on a train accelerating to the right with acceleration a, you feel a strange force of ma pointing to the left. If you don't counter this with another force (by grabbing a handle, for example), then you will fall over. This fictitious force acts just like a gravitational force, because it is proportional to the mass. Therefore, it effectively acts at the CM, producing a torque of $(\mathbf{R} - \mathbf{r}_0) \times (-M\ddot{\mathbf{r}}_0)$. This is the second term in Eq. (8.45). This term vanishes if the CM is directly "above" or "below" (as far as the fictitious gravitational force is concerned) the origin, in other words, if $(\mathbf{R} - \mathbf{r}_0)$ is parallel to $\ddot{\mathbf{r}}_0$. See Problem 10.8 for further discussion of this in terms of fictitious forces.

There is one common situation where the third condition can be invoked. Consider a wheel rolling without slipping on the ground. Mark a dot on the rim. At the instant this dot is in contact with the ground, it is a valid choice for the origin. This is true because $(\mathbf{R} - \mathbf{r}_0)$ points vertically. And $\ddot{\mathbf{r}}_0$ also points vertically, because a dot on a rolling wheel traces out a cycloid. Right before the dot hits the ground, it is moving straight downward. And right after it hits the ground, it is moving straight upward. But having said this, there's usually nothing to be gained by picking such an origin. So the safe thing to do is to always pick your origin to be either the CM or a fixed point, even if the third condition holds.

For conditions that number but three,
We say, "Torque is dL by dt."
But though they're all true,
I'll stick to just two;
It's CM's and fixed points for me! ♣

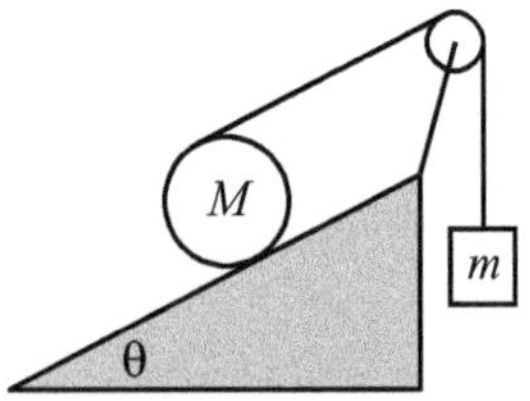

Fig. 8.17

Example: A string wraps around a uniform cylinder of mass M, which rests on a fixed plane. The string passes up over a massless pulley and is connected to a mass m, as shown in Fig. 8.17. Assume that the cylinder rolls without slipping on the plane, and that the string is parallel to the plane. What is the acceleration of the mass m?

What is the condition on the ratio M/m for which the cylinder accelerates down the plane?

First solution: The friction, tension, and gravitational forces are shown in Fig. 8.18. Define positive a_1, a_2, and α as shown. These three accelerations, along with T and F, are five unknowns. We therefore need to produce five equations. They are:

1. $F = ma$ on $m \implies T - mg = ma_2$.
2. $F = ma$ on $M \implies Mg\sin\theta - T - F = Ma_1$.
3. $\tau = I\alpha$ on M (around the CM) $\implies FR - TR = (MR^2/2)\alpha$.
4. Nonslipping condition $\implies \alpha = a_1/R$.
5. Conservation of string $\implies a_2 = 2a_1$.

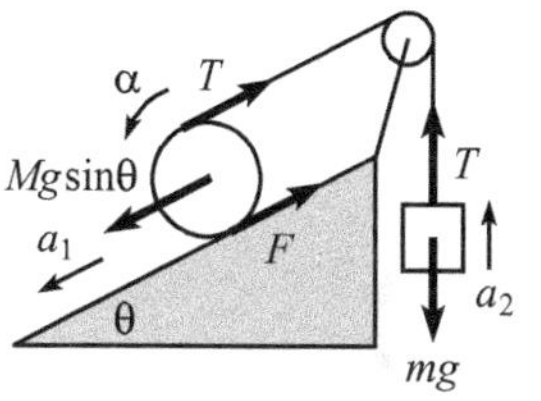

Fig. 8.18

A few comments on these equations: The normal force and the gravitational force perpendicular to the plane cancel, so we can ignore them. We have picked positive F to point up the plane, but if it happens to point down the plane and thereby turn out to be negative, that's fine (but it won't); we don't need to worry about which way it really points. In (3), we are using the CM of the cylinder as our origin, but we can also use a fixed point; see the second solution below. In (5), we have used the fact that the top of a rolling wheel moves twice as fast as the center. This is true because the top has the same speed relative to the center as the center has relative to the ground.

We can go about solving these five equations in various ways. Three of the equations involve only two variables, so it's not so bad. (3) and (4) give $F - T = Ma_1/2$. Adding this to (2) gives $Mg\sin\theta - 2T = 3Ma_1/2$. Using (1) to eliminate T, and using (5) to write a_1 in terms of a_2, then gives

$$Mg\sin\theta - 2(mg + ma_2) = \frac{3Ma_2}{4} \implies a_2 = \frac{(M\sin\theta - 2m)g}{(3/4)M + 2m}. \tag{8.48}$$

And $a_1 = a_2/2$. We see that a_1 is positive (that is, the cylinder rolls down the plane) if $M/m > 2/\sin\theta$. If $\theta \to 0$, then this gives $M/m \to \infty$, which makes sense. If $\theta \to \pi/2$, then $M/m \to 2$. The basic reason for this is that the friction force F, in addition to the tension T, is holding the cylinder up (the coefficient of friction must be very large in this case if no slipping is to occur).

Second solution: In using $\tau = dL/dt$, we can also pick a fixed point as our origin, instead of the CM. The most sensible point is one that is located somewhere along the plane. The $Mg\sin\theta$ force now provides a torque, but the friction does not. And the lever arm for the tension is now $2R$. The angular momentum of the cylinder with respect to a point on the plane is $L = I\omega + Mv_1R$, where the second term comes from the angular momentum due to the object being treated like a point mass at the CM. So $dL/dt = I\alpha + Ma_1R$, and $\tau = dL/dt$ gives

$$(Mg\sin\theta)R - T(2R) = (MR^2/2)\alpha + Ma_1R. \tag{8.49}$$

This turns out to be the sum of the third equation plus R times the second equation in the first solution. We therefore obtain the same result.

8.5 Collisions

In Section 5.7, we looked at collisions involving point particles or otherwise nonrotating objects. The fundamental ingredients there that enabled us to solve problems were conservation of linear momentum and conservation of energy (if the collision was elastic). With conservation of angular momentum now at our disposal, we can extend our study of collisions to ones with rotating objects. The additional fact of conservation of L is compensated for by the new degree of freedom of the rotation. Therefore, provided that the problem is set up properly, we will still have the same number of equations as unknowns.

Conservation of energy can be used in a collision only if it is elastic (by definition). But conservation of angular momentum is similar to conservation of linear momentum, in that it can *always* be used (see the remark below), assuming that the system is isolated. However, conservation of L is a little different from conservation of p, because we have to pick an origin before we can proceed. In view of the three conditions that are necessary for Corollary 8.3 to hold, we must pick our origin to be either a fixed point or the CM of the system (we'll ignore the third condition, since it's rarely used). If we unwisely choose an accelerating point, then $\boldsymbol{\tau} = d\mathbf{L}/dt$ does *not* hold, so we have no right to claim that $d\mathbf{L}/dt$ equals zero just because the torque is zero (as it is for an isolated system). In collision problems, it is easy to fall into the trap of picking an accelerating point as your origin. For example, you might choose the center of a stick as your origin. But if another object collides with the stick, then the center will accelerate, making it an invalid choice for the origin.

REMARK: As far as conservation goes, the way that E differs from $\mathbf{p}$ and $\mathbf{L}$ is that energy can be hidden in the microscopic motion of molecules in the body, in the form of heat. This motion consists of little vibrations with small amplitudes but high speeds. The energy of these vibrations can be large enough to be on the same order of magnitude as the overall energy of the system. But because the vibrations are too small to see, it appears that energy is lost. However, note that even though they're too small to see, you can still feel them with your hand, as heat.

Linear momentum, however, can't be hidden. If an object (not necessarily rigid) has nonzero momentum, then it has to be moving as a whole, and there's no way around that. In short, since $P = MV_{\text{CM}}$, we see that if P is nonzero, then V_{CM} is also. So the motion must be on a macroscopic scale; there's no way for it to be hidden on a microscopic scale.

With angular momentum, things are a little trickier. If the object is rigid, then it can't have hidden angular momentum, for reasons similar to those in the linear momentum case; since $L = I\omega$, we see that if L is nonzero, then ω is also. However, if the object isn't rigid (consider, say, a gas of particles), then it turns out that it *can* theoretically have hidden angular momentum in microscopic motion, although in practice it ends up being too small to notice. This hidden angular momentum can arise from little swirling regions throughout the system. In contrast with linear momentum, it is possible to have angular momentum without any overall motion. So in this respect, this microscopic swirling motion is similar to the microscopic vibrational motion that yields hidden energy. There are, however, three main differences.

First, if we're assuming that the swirling motion takes place on a microscopic scale, then the r in $L = mrv$ is very small, and this leads to a negligible L. This argument doesn't hold for E, because the energy of the vibrations doesn't involve r. Instead, it involves only v, in the form of

$mv^2/2$, so it can end up being large. Second, there's no easy way to start up the circular motion of many little swirls by means of a collision, in contrast with the easily started random linear motion that makes up the energy of heat; you just smash two things together. And third, in the case where the object is rigid, the molecules can easily vibrate, but they can't rotate indefinitely because this would involve ripping apart the bonds between adjacent molecules. The issue is that in vibrational motion all coordinates remain small, whereas in rotational motion they don't, because θ eventually becomes large.

There is, however, one very common phenomenon, namely magnetism, that (in a sense) is an exception to all three of the above points. Although magnetism isn't an angular momentum, it does come (roughly speaking) from the "circular" motion of electrons around the nuclei in atoms. (In general, it actually comes more from the "spin" of the electrons than their orbital motion around the nuclei, but let's not worry about that here. To get everything right, we'd have to think in terms of quantum mechanics, anyway. Let's just work in a rough classical approximation.) Electrons throughout a magnetic material move in tiny little correlated loops. We can escape the above three points of reasoning because first, the magnetic field involves the electric charge e, and this is large enough (on the scale of things) to cancel out the smallness of the r factor. (The actual angular momentum of the electrons in a magnetic material is negligible because their mass is so small. There isn't a large quantity like e to save the day.) Second, it is quite easy to get the electrons moving in correlated circles by means of magnetic forces; there's no need to smash things together. And third, the electrons are free to move around in little circles (in a classical sense) in atoms without ripping things apart. ♣

It is important to remember that you are free to choose your origin from the legal possibilities of fixed points or the CM. Since it is generally the case that one choice is better than others (in that it makes the calculations easier), you should take advantage of this freedom. Let's do two examples. First an elastic collision, and then an inelastic one.

Example (Elastic collision): A mass m travels perpendicular to a stick of mass m and length ℓ, which is initially at rest. At what location should the mass collide elastically with the stick, so that the mass and the center of the stick move with equal speeds after the collision?

Solution: Let the initial speed of the mass be v_0. We have three unknowns in the problem (see Fig. 8.19), namely the desired distance from the middle of the stick, h; the final (equal) speeds of the stick and the mass, v; and the final angular speed of the stick, ω. We can solve for these three unknowns by using our three available conservation laws:

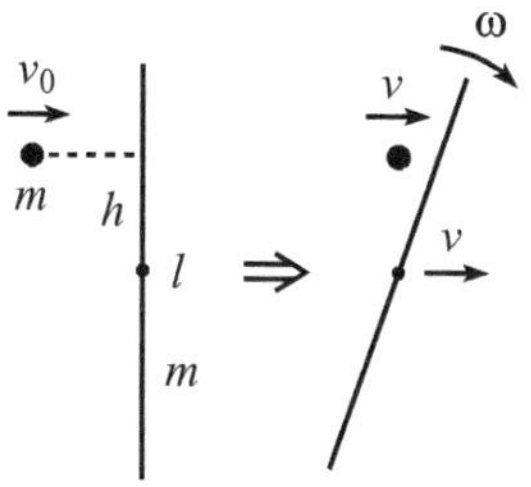

Fig. 8.19

- Conservation of p:

$$mv_0 = mv + mv \implies v = \frac{v_0}{2}. \tag{8.50}$$

- Conservation of E: Remembering that the energy of the stick equals the energy of the rotational motion around the center, plus the energy of the effective point mass at the center, we have

$$\frac{mv_0^2}{2} = \frac{m}{2}\left(\frac{v_0}{2}\right)^2 + \left[\frac{m}{2}\left(\frac{v_0}{2}\right)^2 + \frac{1}{2}\left(\frac{m\ell^2}{12}\right)\omega^2\right] \implies \omega = \frac{\sqrt{6}v_0}{\ell}. \tag{8.51}$$

- Conservation of L: Let's pick our origin to be the fixed point in space that coincides with the initial location of the center of the stick. Then conservation of L gives

$$mv_0 h = m\left(\frac{v_0}{2}\right)h + \left[\left(\frac{m\ell^2}{12}\right)\omega + 0\right]. \tag{8.52}$$

The zero here comes from the fact that the CM of the stick moves directly away from the origin, so there is no contribution to L from the first of the two parts in Theorem 8.1. Plugging the ω from Eq. (8.51) into Eq. (8.52) gives

$$\frac{1}{2}mv_0 h = \left(\frac{m\ell^2}{12}\right)\left(\frac{\sqrt{6}v_0}{\ell}\right) \implies h = \frac{\ell}{\sqrt{6}}. \tag{8.53}$$

You are encouraged to solve this problem again with a different choice of origin, for example, the fixed point that coincides with the spot where the mass hits the stick, or the CM of the entire system.

Remark: After the stick makes half of a revolution, it will hit the backside of m. The resulting motion will have the stick sitting at rest (both translationally and rotationally) and the mass moving with its initial speed v_0. You can show this by working though the second collision, using the quantities we found above. Or you can just use the fact that this scenario certainly satisfies conservation of p, E, and L with the initial conditions, so it must be what happens (because the quadratic conservation statements have only two solutions, and the other one corresponds to the intermediate motion). Note that the time it takes the stick to make half of a revolution is $\pi/\omega = \pi\ell/\sqrt{6}v_0$. So the stick travels a distance of $(v_0/2)(\pi\ell/\sqrt{6}v_0) = (\pi\ell/2\sqrt{6})$ before it ends up at rest. This distance is independent of v_0 (which follows from dimensional analysis). ♣

Let's now look at an inelastic collision, where one object sticks to another. We won't be able to use conservation of E now. But conservation of p and L will be sufficient since there is one fewer degree of freedom in the final motion, because the objects don't move independently.

Example (Inelastic collision): A mass m travels at speed v_0 perpendicular to a stick of mass m and length ℓ, which is initially at rest. The mass collides completely inelastically with the stick at one of its ends and sticks to it. What is the resulting angular velocity of the system?

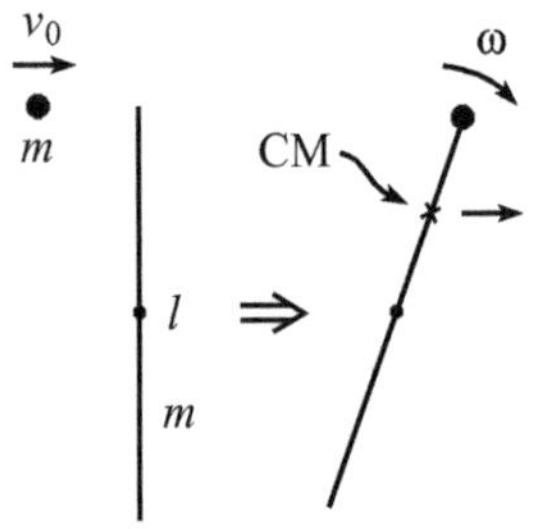

Fig. 8.20

Solution: The first thing to note is that the CM of the system is $\ell/4$ from the end, as shown in Fig. 8.20. After the collision, the system rotates about the CM as the CM moves in a straight line. Conservation of momentum quickly tells us that the speed of the CM is $v_0/2$. Also, using the parallel-axis theorem, the moment of inertia of the system around the CM is

$$I_{CM} = I_{CM}^{\text{stick}} + I_{CM}^{\text{mass}} = \left[\frac{m\ell^2}{12} + m\left(\frac{\ell}{4}\right)^2\right] + m\left(\frac{\ell}{4}\right)^2 = \frac{5}{24}m\ell^2. \tag{8.54}$$

There are now many ways to proceed, depending on what point we choose as our origin.

FIRST METHOD: Choose the origin to be the fixed point that coincides with the location of the CM right when the collision happens (that is, the point $\ell/4$ from the end of the stick). Conservation of L says that the initial L of the ball must equal the final L of the system. This gives

$$mv_0\left(\frac{\ell}{4}\right) = \left(\frac{5}{24}m\ell^2\right)\omega + 0 \quad \Longrightarrow \quad \omega = \frac{6v_0}{5\ell}. \tag{8.55}$$

The zero here comes from the fact that the CM of the stick moves directly away from the origin, so there is no contribution to L from the first of the two parts in Theorem 8.1. Note that we didn't need to use conservation of p in this method.

SECOND METHOD: Choose the origin to be the fixed point that coincides with the initial center of the stick. Then conservation of L gives

$$mv_0\left(\frac{\ell}{2}\right) = \left(\frac{5}{24}m\ell^2\right)\omega + (2m)\left(\frac{v_0}{2}\right)\left(\frac{\ell}{4}\right) \quad \Longrightarrow \quad \omega = \frac{6v_0}{5\ell}. \tag{8.56}$$

The right-hand side is the angular momentum of the system relative to the CM, plus the angular momentum (relative to the origin) of a point mass of mass $2m$ located at the CM.

THIRD METHOD: Choose the origin to be the CM of the system. This point moves to the right with speed $v_0/2$, along the line a distance $\ell/4$ below the top of the stick. Relative to the CM, the mass m moves to the right, and the stick moves to the left, both with speed $v_0/2$. Conservation of L gives

$$m\left(\frac{v_0}{2}\right)\left(\frac{\ell}{4}\right) + \left[0 + m\left(\frac{v_0}{2}\right)\left(\frac{\ell}{4}\right)\right] = \left(\frac{5}{24}m\ell^2\right)\omega \quad \Longrightarrow \quad \omega = \frac{6v_0}{5\ell}. \tag{8.57}$$

The zero here comes from the fact that the stick initially has no L around its center. A fourth reasonable choice for the origin is the fixed point that coincides with the initial location of the top of the stick. You can work this one out for practice.

8.6 Angular impulse

In Section 5.5.1, we defined the *impulse*, which we'll label as $\mathcal{I}$, to be the time integral of the force applied to an object. From Newton's second law, $\mathbf{F} = d\mathbf{p}/dt$, the impulse is therefore the net change in linear momentum. That is,

$$\mathcal{I} \equiv \int_{t_1}^{t_2} \mathbf{F}(t)\, dt = \Delta\mathbf{p}. \tag{8.58}$$

We now define the *angular impulse*, $\mathcal{I}_\theta$, to be the time integral of the torque applied to an object. From $\boldsymbol{\tau} = d\mathbf{L}/dt$, the angular impulse is therefore the net change in angular momentum. That is,

$$\mathcal{I}_\theta \equiv \int_{t_1}^{t_2} \boldsymbol{\tau}(t)\, dt = \Delta \mathbf{L}. \tag{8.59}$$

These are just definitions, devoid of any content. The place where the physics comes in is the following. Consider a situation where $\mathbf{F}(t)$ is always applied at the same position relative to the origin around which $\boldsymbol{\tau}(t)$ is calculated (this origin must be a legal one, of course). Let this position be $\mathbf{R}$. Then we have $\boldsymbol{\tau}(t) = \mathbf{R} \times \mathbf{F}(t)$. Plugging this into Eq. (8.59), and taking the constant $\mathbf{R}$ outside the integral, gives $\mathcal{I}_\theta = \mathbf{R} \times \mathcal{I}$. In other words,

$$\Delta \mathbf{L} = \mathbf{R} \times (\Delta \mathbf{p}) \quad \big(\text{for } \mathbf{F}(t) \text{ applied at one position}\big). \tag{8.60}$$

This is a very useful result. It gives the relation between the net changes in $\mathbf{L}$ and $\mathbf{p}$, as opposed to the individual values of each. Even if $\mathbf{F}(t)$ is changing in some arbitrary manner as time goes by, so that we have no idea what $\Delta\mathbf{L}$ and $\Delta\mathbf{p}$ themselves are, we still know that they are related by Eq. (8.60). In many cases, we don't have to worry about the cross product in Eq. (8.60), because the lever arm $\mathbf{R}$ is perpendicular to the change in momentum $\Delta\mathbf{p}$. In such cases, we have

$$|\Delta L| = R|\Delta p|. \tag{8.61}$$

Also, in many cases the object starts at rest, so we don't have to bother with the Δ's. The following example is a classic application of angular impulse and Eq. (8.61).

Example (Striking a stick): A stick of mass m and length ℓ, initially at rest, is struck with a hammer. The blow is made perpendicular to the stick at one end. Let the blow occur quickly, so that the stick doesn't have time to move much while the hammer is in contact. If the CM of the stick ends up moving at speed v, what are the velocities of the ends right after the blow?

Solution: Although we have no hope of knowing exactly what $F(t)$ looks like, or the length of time it is applied, we still know from Eq. (8.61) that $\Delta L = (\ell/2)\Delta p$, where we have chosen our origin to be the CM, which gives a lever arm of $\ell/2$. Therefore, $(m\ell^2/12)\omega = (\ell/2)mv$, so the final v and ω are related by $\omega = 6v/\ell$.

The velocities of the ends right after the blow are obtained by adding (or subtracting) the rotational motion to the CM's translational motion. The rotational velocities of the ends relative to the CM are $\pm\omega(\ell/2) = \pm(6v/\ell)(\ell/2) = \pm 3v$. Therefore, the end that is hit moves with velocity $v + 3v = 4v$, and the other end moves with velocity $v - 3v = -2v$ (that is, backward).

What L was, he just couldn't tell.
And p? He was clueless as well.
But despite his distress,
He wrote down the right guess
For their quotient: the lever arm's ℓ.

Impulse is also useful for "collisions" that occur over extended times. See, for example, Problem 8.24.

8.7 Problems

Section 8.1: Pancake object in x-y plane

8.1. **Massive pulley** *

Consider the Atwood's machine shown in Fig. 8.21. The masses are m and $2m$, and the pulley is a uniform disk of mass m and radius r. The string is massless and does not slip with respect to the pulley. Find the acceleration of the masses. Use conservation of energy.

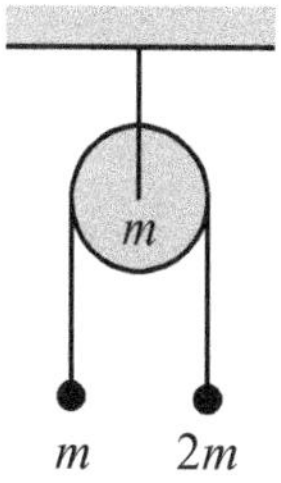

Fig. 8.21

8.2. **Leaving the sphere** **

A ball with moment of inertia βmr^2 rests on top of a fixed sphere. There is friction between the ball and the sphere. The ball is given an infinitesimal kick, and it rolls down without slipping. Assuming that r is much smaller than the radius of the sphere, at what point does the ball lose contact with the sphere? How does your answer change if the size of the ball is comparable to, or larger than, the size of the sphere? You may want to solve Problem 5.3 first, if you haven't already done so.

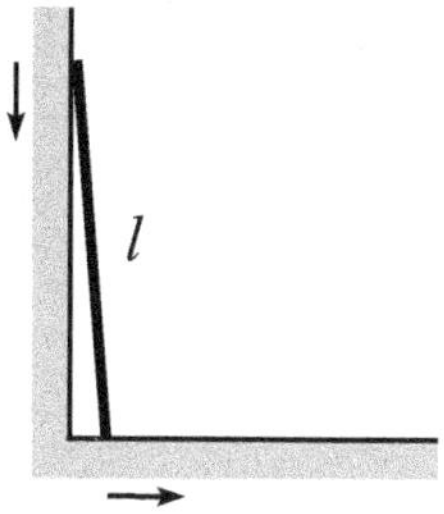

Fig. 8.22

8.3. **Sliding ladder** ***

A ladder of length ℓ and uniform mass density stands on a frictionless floor and leans against a frictionless wall. It is initially held motionless, with its bottom end an infinitesimal distance from the wall. It is then released, whereupon the bottom end slides away from the wall, and the top end slides down the wall (see Fig. 8.22). When it loses contact with the wall, what is the horizontal component of the velocity of the center of mass?

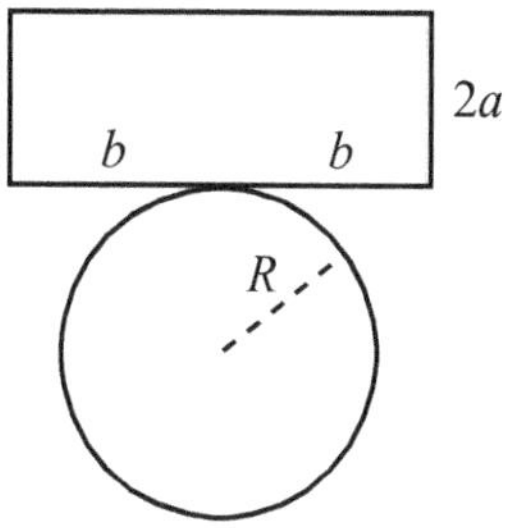

Fig. 8.23

8.4. **Leaning rectangle** ***

A rectangle of height $2a$ and width $2b$ rests on top of a fixed cylinder of radius R (see Fig. 8.23). The moment of inertia of the rectangle around its center is I. The rectangle is given an infinitesimal kick and then "rolls" on the cylinder without slipping. Find the equation of motion for the tilt angle of the rectangle. Under what conditions will the rectangle fall off the cylinder, and under what conditions will it oscillate back and forth? Find the frequency of these small oscillations.

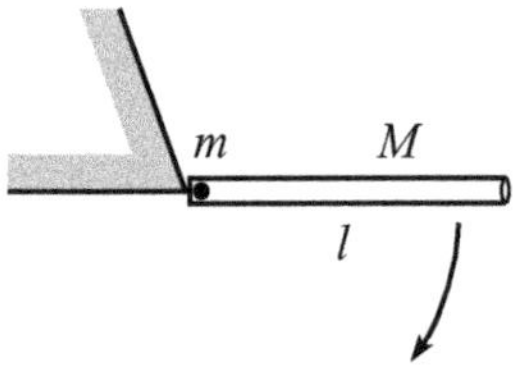

Fig. 8.24

8.5. **Mass in a tube** ***

A tube of mass M and length ℓ is free to swing around a pivot at one end. A mass m is positioned inside the (frictionless) tube at this end. The tube is held horizontal and then released (see Fig. 8.24). Let θ be the angle of the tube with respect to the horizontal, and let x be the distance the mass has traveled along the tube. Find the Euler–Lagrange equations for θ and x, and then write them in terms of θ and $\eta \equiv x/\ell$ (the fraction of the distance along the tube).

These equations can only be solved numerically, and you must pick a numerical value for the ratio $r \equiv m/M$ in order to do this. Write a program (see Section 1.4) that produces the value of η when the tube is vertical ($\theta = \pi/2$). Give this value of η for a few values of r.

Section 8.3: Calculating moments of inertia

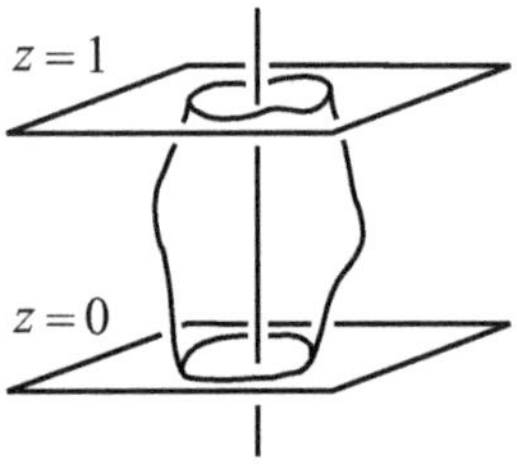

Fig. 8.25

8.6. **Minimum I** *

A moldable blob of matter of mass M is to be situated between the planes $z = 0$ and $z = 1$ (see Fig. 8.25) so that the moment of inertia around the z axis is as small as possible. What shape should the blob take?

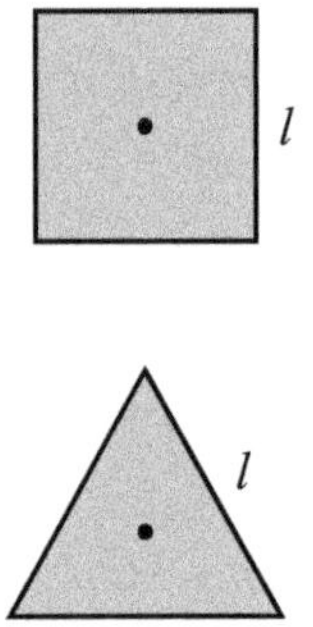

Fig. 8.26

8.7. **Slick calculations of I** **

In the spirit of Section 8.3.2, find the moments of inertia of the following objects (see Fig. 8.26).

(a) A uniform square of mass m and side ℓ (axis through center, perpendicular to plane).
(b) A uniform equilateral triangle of mass m and side ℓ (axis through center, perpendicular to plane).

8.8. **Slick calculations of I for fractal objects** ***

In the spirit of Section 8.3.2, find the moments of inertia of the following fractal objects. Be careful how the mass scales.

(a) Take a stick of length ℓ, and remove the middle third. Then remove the middle third from each of the remaining two pieces. Then remove the middle third from each of the remaining four pieces, and so on, forever. Let the final object have mass m, and let the axis be through the center, perpendicular to the stick; see Fig. 8.27.[8]

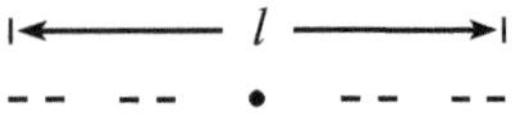

Fig. 8.27

[8] This object is the Cantor set, for those who like such things. It has no length, so the density of the remaining mass is infinite. If you suddenly develop an aversion to point masses with infinite density, simply imagine the above iteration being carried out only, say, a million times.

(b) Take an equilateral triangle of side ℓ, and remove the "middle" triangle (1/4 of the area). Then remove the "middle" triangle from each of the remaining three triangles, and so on, forever. Let the final object have mass m, and let the axis be through the center, perpendicular to the plane; Fig. 8.28.

(c) Take a square of side ℓ, and remove the "middle" square (1/9 of the area). Then remove the "middle" square from each of the remaining eight squares, and so on, forever. Let the final object have mass m, and let the axis be through the center, perpendicular to the plane; see Fig. 8.29.

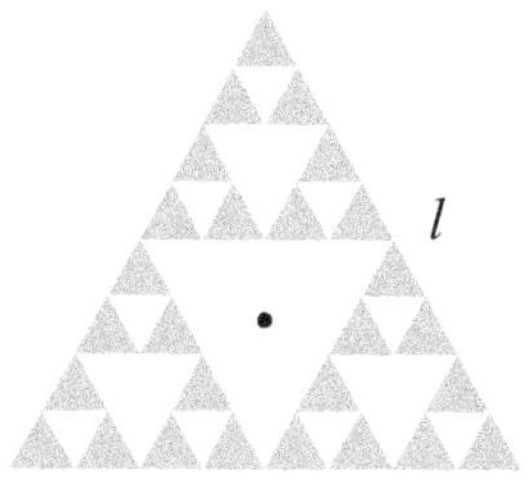

Fig. 8.28

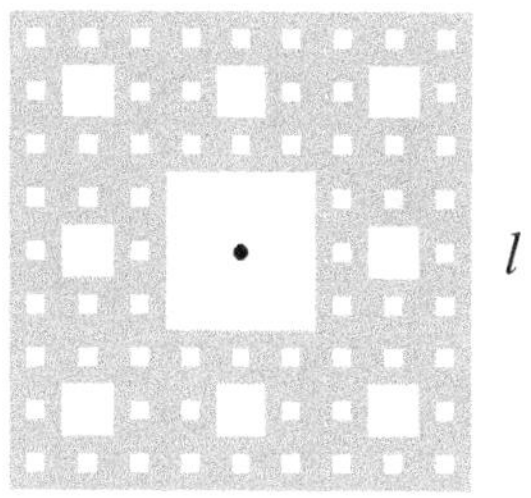

Fig. 8.29

Section 8.4: Torque

8.9. **Zero torque from internal forces** $**$

Given a collection of particles with positions $\mathbf{r}_i$, let the force on the ith particle due to all the others be $\mathbf{F}_i^{\text{int}}$. Assuming that the force between any two particles is directed along the line between them, use Newton's third law to show that $\sum_i \mathbf{r}_i \times \mathbf{F}_i^{\text{int}} = 0$.

8.10. **Removing a support** $*$

(a) A uniform rod of length ℓ and mass m rests on supports at its ends. The right support is quickly removed (see Fig. 8.30). What is the force from the left support immediately thereafter?

(b) A rod of length $2r$ and moment of inertia βmr^2 rests on top of two supports, each of which is a distance d away from the center. The right support is quickly removed (see Fig. 8.30). What is the force from the left support immediately thereafter?

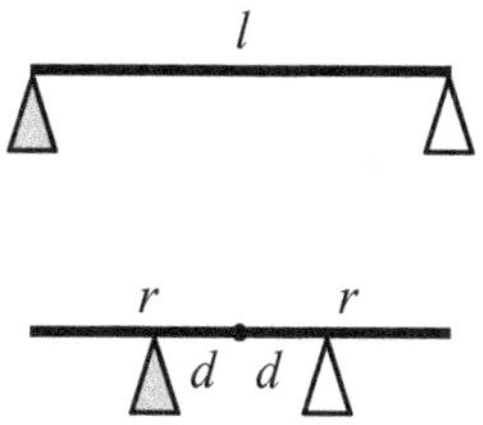

Fig. 8.30

8.11. **Falling stick** $*$

A massless stick of length b has one end pivoted on a support and the other end glued perpendicular to the middle of a stick of mass m and length ℓ.

(a) If the two sticks are held in a horizontal plane (see Fig. 8.31) and then released, what is the initial acceleration of the CM?

(b) If the two sticks are held in a vertical plane (see Fig. 8.31) and then released, what is the initial acceleration of the CM?

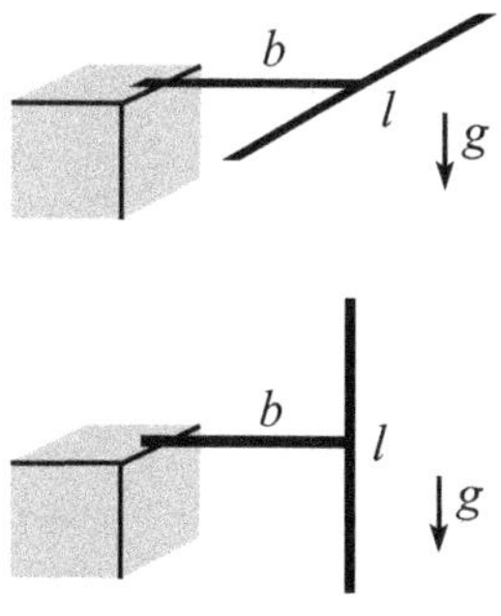

Fig. 8.31

8.12. **Pulling a cylinder** $**$

In Exercise 8.50 below, the cylinder moves directly to the right. The fact that it doesn't have any transverse motion follows from the fact that the two segments of the string pull only to the right and therefore cannot supply a transverse force. Demonstrate this result again by explicitly integrating the string's force on the cylinder over the semicircle of contact. (The $N = T\,d\theta$ result from the "Rope wrapped around a pole" example in Section 2.1 will come in handy.)

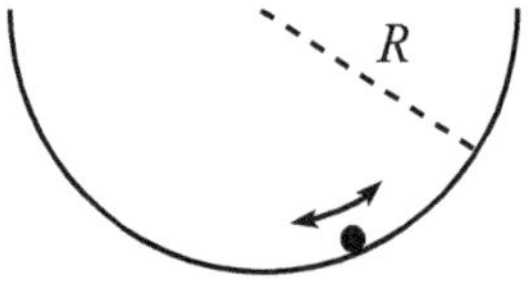

Fig. 8.32

8.13. Oscillating ball **

A small ball with radius r and uniform density rolls without slipping near the bottom of a fixed cylinder of radius R (see Fig. 8.32). What is the frequency of small oscillations? Assume $r \ll R$.

8.14. Oscillating cylinders **

A hollow cylinder of mass M_1 and radius R_1 rolls without slipping on the inside surface of another hollow cylinder of mass M_2 and radius R_2. Assume $R_1 \ll R_2$. Both axes are horizontal, and the larger cylinder is free to rotate about its axis. What is the frequency of small oscillations?

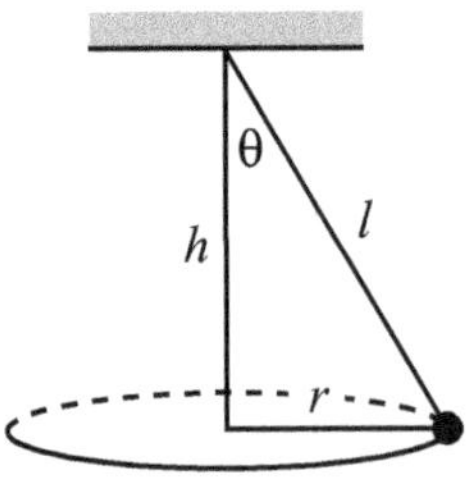

Fig. 8.33

8.15. Lengthening the string **

A mass hangs from a massless string and swings around in a horizontal circle, as shown in Fig. 8.33. The length of the string is then very slowly increased (or decreased). Let θ, ℓ, r, and h be defined as shown.

(a) Assuming that θ is very small, how does r depend on ℓ?
(b) Assuming that θ is very close to $\pi/2$, how does h depend on ℓ?

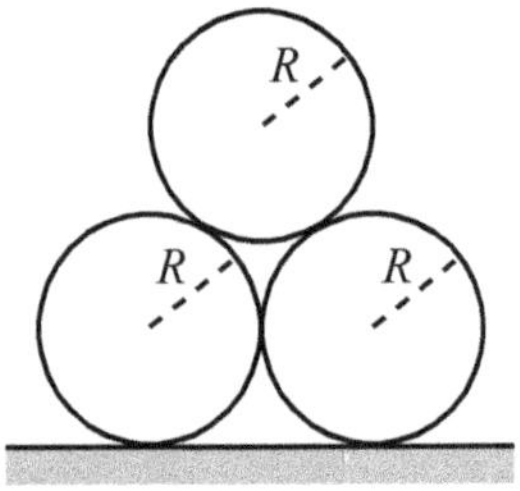

Fig. 8.34

8.16. A triangle of cylinders ***

Three identical cylinders with moments of inertia $I = \beta m R^2$ are situated in a triangle as shown in Fig. 8.34. Find the initial downward acceleration of the top cylinder for the following two cases. Which case has a larger acceleration?

(a) There is friction between the bottom two cylinders and the ground (so they roll without slipping), but there is no friction between any of the cylinders.
(b) There is no friction between the bottom two cylinders and the ground, but there is friction between the cylinders (so they don't slip with respect to each other).

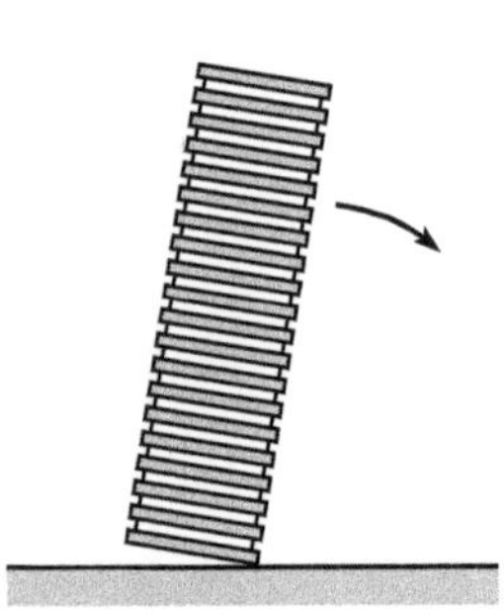

Fig. 8.35

8.17. Falling chimney ****

A chimney initially stands upright. It is given a tiny kick, and it topples over. At what point along its length is it most likely to break? In doing this problem, work with the following two-dimensional simplified model of a chimney. Assume that the chimney consists of boards stacked on top of each other, and that each board is attached to the two adjacent ones with tiny rods at each end (see Fig. 8.35). The goal is to determine which rod in the chimney achieves the maximum tension. Work in the approximation where the width of the chimney is very small compared with the height.

Section 8.5: Collisions

8.18. **Ball hitting stick** **

A ball of mass M collides with a stick with moment of inertia $I = \beta m\ell^2$ (relative to its center, which is its CM). The ball is initially traveling at speed V_0 perpendicular to the stick. The ball strikes the stick at a distance d from the center (see Fig. 8.36). The collision is elastic. Find the resulting translational and rotational speeds of the stick, and also the resulting speed of the ball.

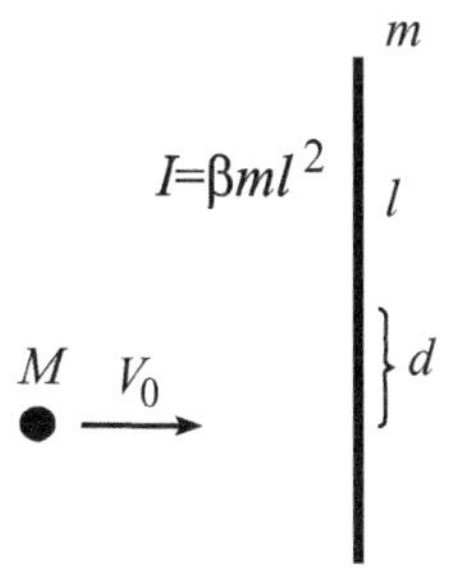

Fig. 8.36

8.19. **A ball and stick theorem** **

Consider the setup in Problem 8.18. Show that the relative speed of the ball and the point of contact on the stick is the same before and immediately after the collision. (This result is analogous to the "relative speed" result for a 1-D collision, Theorem 5.3 in Section 5.7.1.)

Section 8.6: Angular impulse

8.20. **The superball** **

A ball with radius R and $I = (2/5)mR^2$ is thrown through the air. It spins around the axis perpendicular to the (vertical) plane of the motion. Call this the x-y plane. The ball bounces off a floor without slipping during the time of contact. Assume that the collision is elastic, and that the magnitude of the vertical v_y is the same before and after the bounce. Show that v'_x and ω' after the bounce are related to v_x and ω before the bounce by

$$\begin{pmatrix} v'_x \\ R\omega' \end{pmatrix} = \frac{1}{7}\begin{pmatrix} 3 & -4 \\ -10 & -3 \end{pmatrix}\begin{pmatrix} v_x \\ R\omega \end{pmatrix}, \qquad (8.62)$$

where positive v_x is to the right, and positive ω is counterclockwise.

8.21. **Many bounces** *

Using the result of Problem 8.20, describe what happens over the course of many superball bounces.

8.22. **Rolling over a bump** **

A ball with radius R and moment of inertia $I = (2/5)mR^2$ rolls with speed V_0 without slipping on the ground. It encounters a step of height h and rolls up over it. Assume that the ball sticks to the corner of the step briefly (until the center of the ball is directly above the corner). Show that if the ball is to climb over the step, then V_0 must satisfy

$$V_0 \geq \sqrt{\frac{10gh}{7}} \left(1 - \frac{5h}{7R}\right)^{-1}. \qquad (8.63)$$

8.23. **Falling toast** **

After buttering your toast (assumed to be a uniform rigid square of side length ℓ, for the sake of having a doable problem) one morning and holding it in a horizontal position with buttered side up, you accidentally drop it from a height H above a counter, which itself is a height h above the ground. The toast is oriented "parallel" to the counter, and as it falls, one edge barely clips the counter (elastically), causing the toast to spin. What value of H, in terms of h and ℓ, yields the unfortunate scenario where the toast makes half of a revolution, landing buttered side down on the floor? There is a special value of ℓ in terms of h. What is it, and why is it special?

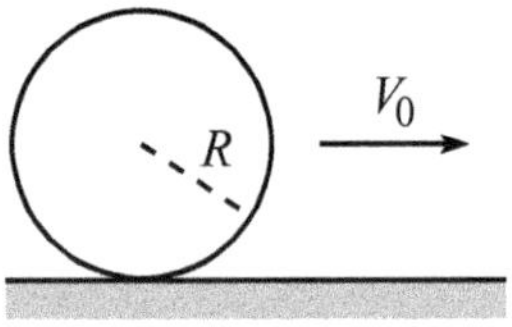

Fig. 8.37

8.24. **Sliding to rolling** **

A ball initially slides without rotating on a horizontal surface with friction (see Fig. 8.37). The initial speed is V_0, and the moment of inertia around the center is $I = \beta m R^2$.

(a) Without knowing anything about the nature of the friction force, find the speed of the ball when it begins to roll without slipping. Also, find the kinetic energy lost while sliding.

(b) Now consider the special case where the coefficient of kinetic friction is μ, independent of position. At what time, and at what distance, does the ball begin to roll without slipping? Verify that the work done by friction equals the energy loss calculated in part (a). (Be careful on this.)

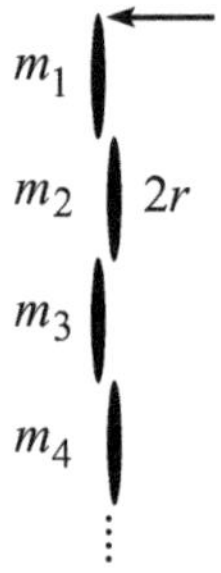

Fig. 8.38

8.25. **Lots of sticks** ***

Consider a collection of rigid sticks of length $2r$, masses m_i, and moments of inertia $\beta m_i r^2$, with $m_1 \gg m_2 \gg m_3 \gg \cdots$. The CM of each stick is located at its center. The sticks are placed on a horizontal frictionless surface, as shown in Fig. 8.38. The ends overlap a tiny distance in the y direction and are a tiny distance apart in the x direction. The first (heaviest) stick is given an instantaneous blow (as shown) which causes it to translate and rotate. The first stick will strike the second stick, which will then strike the third stick, and so on. Assume that all the collisions are elastic. Depending on the size of β, the speed of the nth stick will either (1) approach zero, (2) approach infinity, or

(3) be independent of n, as $n \to \infty$. Show that the special value of β corresponding to the third of these three scenarios is $\beta = 1/3$, which happens to correspond to a uniform stick.

8.8 Exercises

Section 8.1: Pancake object in x-y plane

8.26. **Swinging stick** **

A uniform stick of length L is pivoted at its bottom end and is initially held vertical. It is given an infinitesimal kick, and it swings down around the pivot. After three-quarters of a turn (in the horizontal position shown in Fig. 8.39), the pivot is somehow vaporized, and the stick flies freely up in the air. What is the maximum height of the center of the stick in the resulting motion? At what angle is the stick tilted when the center reaches this maximum height?

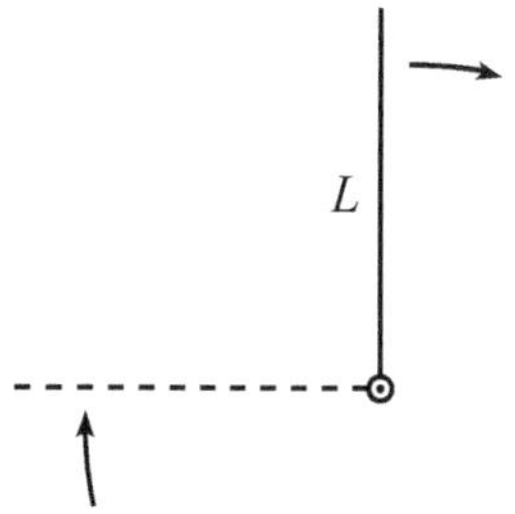

Fig. 8.39

8.27. **Atwood's with a cylinder** **

A massless string of negligible thickness is wrapped around a uniform cylinder of mass m and radius r. The string passes up over a massless pulley and is tied to a block of mass m at its other end, as shown in Fig. 8.40. The system is released from rest. What are the accelerations of the block and the cylinder? Assume that the string does not slip with respect to the cylinder. Use conservation of energy (after applying a quick $F = ma$ argument to show that the two objects move downward with the same acceleration).

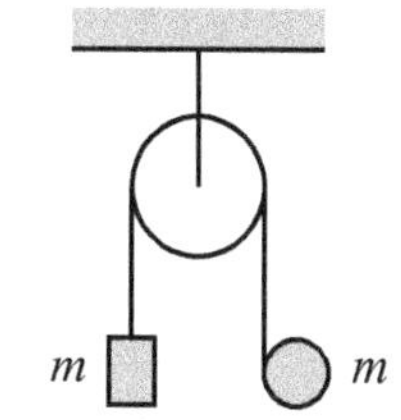

Fig. 8.40

8.28. **Board and cylinders** **

A board lies on top of two uniform cylinders that lie on a fixed plane inclined at an angle θ, as shown in Fig. 8.41. The board has mass m, and each of the cylinders has mass $m/2$. The system is released from rest. If there is no slipping between any of the surfaces, what is the acceleration of the board? Use conservation of energy.

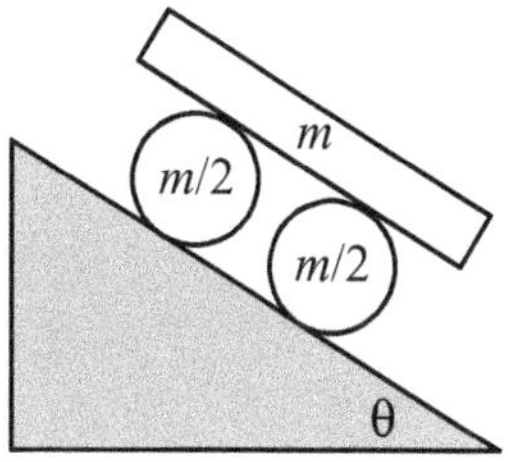

Fig. 8.41

8.29. **Moving plane** ***

A ball of mass m and moment of inertia $I = \beta mr^2$ is held motionless on a plane of mass M and angle of inclination θ (see Fig. 8.42). The plane rests on a frictionless horizontal surface. The ball is released. Assuming that it rolls without slipping on the plane, what is the horizontal acceleration of the plane? *Hint*: You might want to do Problem 3.8 first. But as with all the exercises in this section, use conservation of energy instead of force and torque; this problem gets extremely messy with the latter strategy.

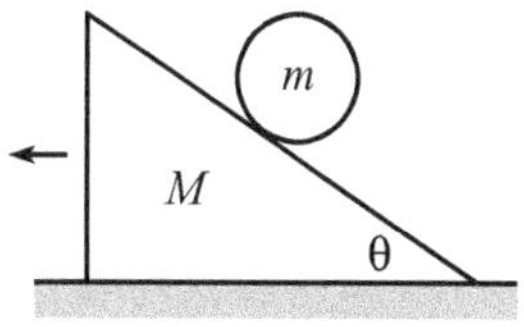

Fig. 8.42

Section 8.2: Nonplanar objects

8.30. **Semicircle CM** $*$

A wire is bent into a semicircle of radius R. Find the CM.

8.31. **Hemisphere CM** $*$

Find the CM of a solid hemisphere.

Section 8.3: Calculating moments of inertia

8.32. **A cone** $*$

Find the moment of inertia of a solid cone (mass M, base radius R) around its symmetry axis.

8.33. **A sphere** $*$

Find the moment of inertia of a solid sphere (mass M, radius R) around a diameter. Do this by slicing the sphere into disks.

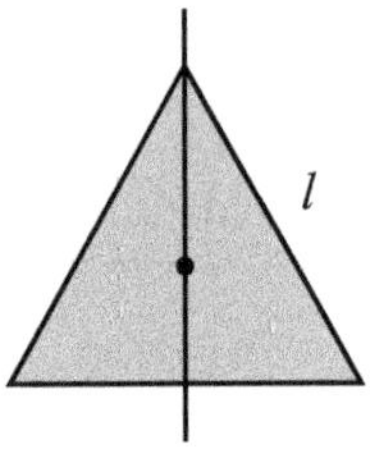

Fig. 8.43

8.34. **A triangle, the slick way** $**$

In the spirit of Section 8.3.2, find the moment of inertia of a uniform equilateral triangle of mass m and side ℓ, around a line joining a vertex and the opposite side; see Fig. 8.43.

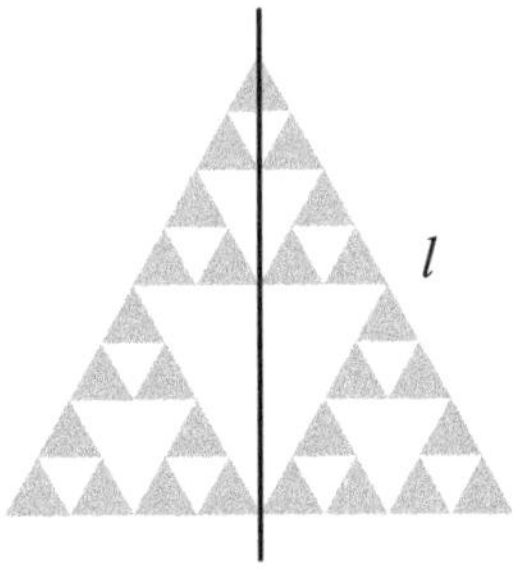

Fig. 8.44

8.35. **Fractal triangle** $**$

Take an equilateral triangle of side ℓ, and remove the "middle" triangle (1/4 of the area). Then remove the "middle" triangle from each of the remaining three triangles, and so on, forever. Let the final fractal object have mass m. In the spirit of Section 8.3.2, find the moment of inertia around a line joining a vertex and the opposite side; see Fig. 8.44.

Section 8.4: Torque

8.36. **Swinging your arms** $*$

You are standing on the edge of a step on some stairs, facing up the stairs. You feel yourself starting to fall backward, so you start swinging your arms around in vertical circles, like a windmill. This is what people tend to do in such a situation, but does it actually help you not to fall, or does it simply make you look silly? Explain your reasoning.

8.37. **Rolling down the plane** $*$

A ball with moment of inertia βmr^2 rolls without slipping down a plane inclined at an angle θ. What is its linear acceleration?

8.38. Coin on a plane *

A uniform coin rolls down a plane inclined at an angle θ. If the coefficient of static friction between the coin and the plane is μ, what is the largest angle θ for which the coin doesn't slip?

8.39. Accelerating plane *

A ball with $I = (2/5)MR^2$ is placed on a plane inclined at an angle θ. The plane is accelerated upwards (along its direction) with acceleration a; see Fig. 8.45. For what value of a does the CM of the ball not move? Assume that there is sufficient friction so that the ball doesn't slip with respect to the plane.

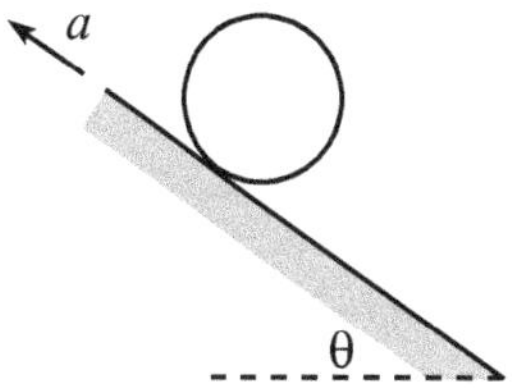

Fig. 8.45

8.40. Bowling ball on paper *

A bowling ball sits on a piece of paper on the floor. You grab the paper and pull it horizontally along the floor, with acceleration a_0. What is the acceleration of the center of the ball? Assume that the ball does not slip with respect to the paper.

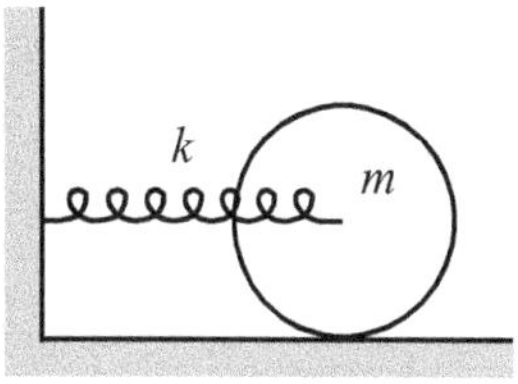

Fig. 8.46

8.41. Spring and cylinder *

The axle of a solid cylinder of mass m and radius r is connected to a spring with spring constant k, as shown in Fig. 8.46. If the cylinder rolls without slipping, what is the frequency of the oscillations?

8.42. Falling quickly *

A massless stick of length L is pivoted at one end and has a mass m attached to its other end. It is held in a horizontal position, as shown in Fig. 8.47. Where should a second mass m be attached to the stick, so that the stick falls as fast as possible when dropped?

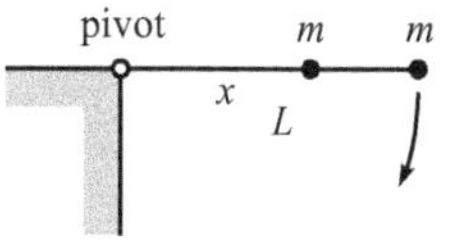

Fig. 8.47

8.43. Maximum frequency *

A pendulum is made of a uniform stick of length L. It is allowed to swing in a vertical plane. Where should the pivot be placed on the stick so that the frequency of (small) oscillations is maximum?

8.44. Massive pulley *

Solve Problem 8.1 again, but now use force and torque instead of conservation of energy.

8.45. Atwood's with a cylinder **

Solve Exercise 8.27 again, but now use force and torque instead of conservation of energy.

8.46. Board and cylinders **

Solve Exercise 8.28 again, but now use force and torque instead of conservation of energy.

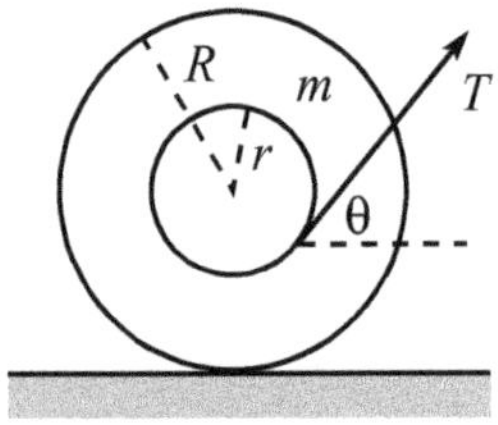

Fig. 8.48

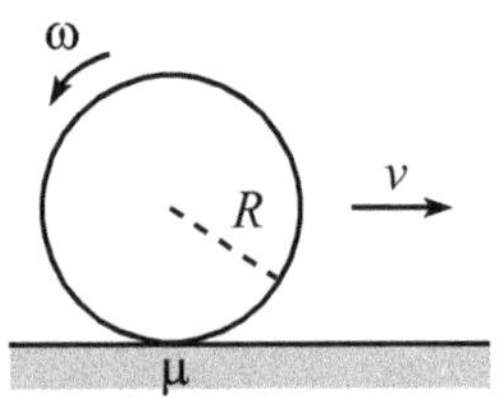

Fig. 8.49

8.47. **The spool** **

A spool of mass m and moment of inertia I is free to roll without slipping on a table. It has an inner radius r, and an outer radius R. If you pull on the string with tension T at an angle θ (see Fig. 8.48), what is the acceleration of the spool? Which way does it move?

8.48. **Stopping the coin** **

A coin stands vertically on a table. It is projected forward (in the plane of itself) with speed v and angular speed ω, as shown in Fig. 8.49. The coefficient of kinetic friction between the coin and the table is μ. What should v and ω be so that the coin comes to rest (both translationally and rotationally) a distance d from where it started?

8.49. **Measuring** ***g*** **

(a) Consider an extended pendulum whose CM is a distance ℓ from the pivot, and whose moment of inertia around the pivot is I. Show that the frequency of small oscillations is $\omega = \sqrt{mg\ell/I}$, which gives $T = 2\pi/\omega = 2\pi\sqrt{I/mg\ell}$, and hence $g = 4\pi^2 I/(m\ell T^2)$. Therefore, by measuring I, m, ℓ, and T, you can determine g. However, if the pendulum has an odd shape, it may be difficult to determine I. Consider, then, the following alternative method of measuring g.

(b) For simplicity, assume that the pendulum is planar. Pick an arbitrary point as the pivot and measure the period, T, of small oscillations. Then with the pendulum at rest, draw a vertical line through this pivot. By trial and error, find another pivot point on this line on the *same* side of the CM[9] (you may need to extend the line with a massless extension) that yields the same period T. Let L be the sum of the lengths from these two points to the CM.[10] Show that g is given by $g = 4\pi^2 L/T^2$, which is independent of m and I.

8.50. **Pulling a cylinder** **

A solid cylinder of mass m and radius r lies flat on a frictionless horizontal table, with a massless string running halfway around it, as shown in

[9] The CM can be found by hanging the pendulum from a point not on the drawn line, and then drawing a vertical line through this point. The intersection of the two lines is the CM.

[10] There are also two points on the other side of the CM that yield the same period (except in the special case where the two points coincide and produce the minimum period, as in Exercise 8.43, but don't worry about this). These two other points are the same distances from the CM as the original two points, as you can show. So if you happen to find all four points, you can obtain L by instead measuring the distance from the "inside" point on one side to the "outside" point on the other. This method doesn't require knowing the location of the CM.

Fig. 8.50. A mass also of mass m is attached to one end of the string, and you pull on the other end with a force T. The circumference of the cylinder is sufficiently rough so that the string does not slip with respect to it. What is the acceleration of the mass m attached to the end of the string?

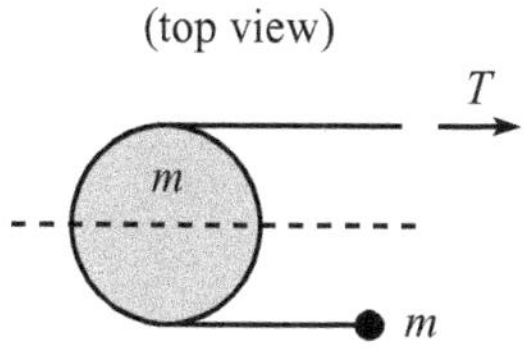

Fig. 8.50

8.51. **Coin and plank** $**$

A coin of mass M and radius R stands vertically on the right end of a horizontal plank of mass M and length L, as shown in Fig. 8.51. The system starts at rest. The plank is then pulled to the right with a constant force F. Assume that the coin does not slip with respect to the plank. What are the accelerations of the plank and coin? How far to the right does the coin move by the time the left end of the plank reaches it?

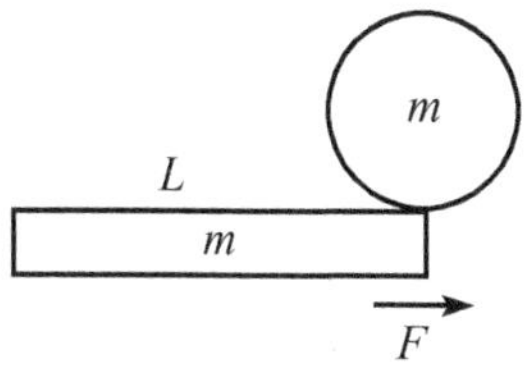

Fig. 8.51

8.52. **Cylinder, board, and spring** $**$

A board of mass m, which is free to slide on a frictionless floor, is connected by a spring (with spring constant k) to a wall. A cylinder, also of mass m (and $I = mR^2/2$), rests on top of the board, as shown in Fig. 8.52, and is free to roll without slipping on the board. If the board and the cylinder are pulled some distance to the left and then released from rest, what is the frequency of the resulting oscillatory motion?

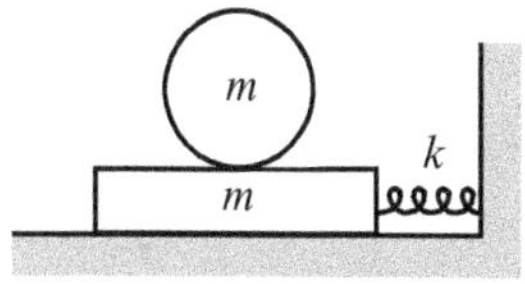

Fig. 8.52

8.53. **Swirling around a cone** $**$

A fixed hollow frictionless cone is positioned with its tip pointing down. A particle is released from rest on the inside surface. After it has slid halfway down to the tip, it bounces elastically off a platform. The platform is positioned at a 45° angle along the surface of the cone, so the particle ends up being deflected horizontally along the surface (in other words, into the page in Fig. 8.53). The particle then swirls up and around the cone before coming down. Measured from the tip of the cone, show that the ratio of the particle's maximum swirling height to the height of the platform is $(\sqrt{5}+1)/2$, which just happens to be the golden ratio.

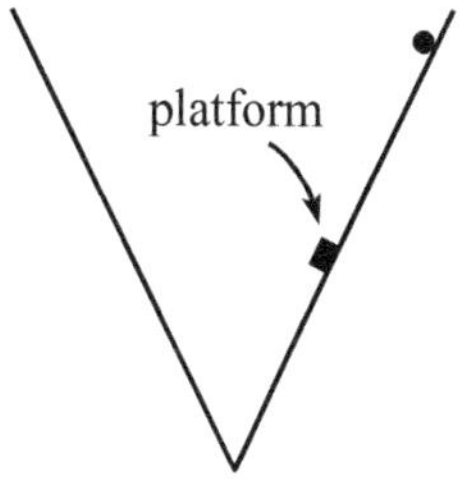

Fig. 8.53

8.54. **Raising a hoop** $**$

A bead of mass m is positioned at the top of a frictionless hoop of mass M and radius R, which stands vertically on the ground. A wall touches the hoop on its left, and a short wall of height R touches the hoop on its right, as shown in Fig. 8.54. All surfaces are frictionless. The bead is given a tiny kick, and it slides down the hoop, as shown. What is the largest value of m/M for which the hoop never rises up off the ground? *Note*: It's possible to solve this problem by using only force, but solve it here by using torque.

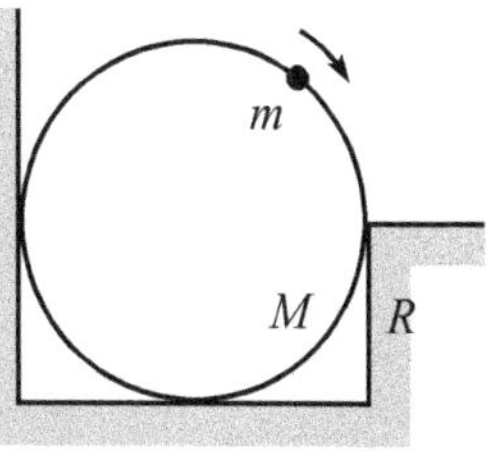

Fig. 8.54

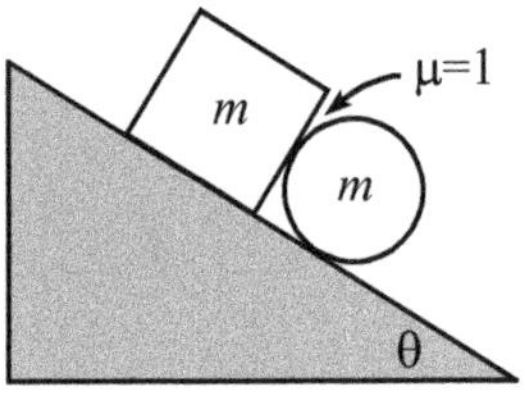

Fig. 8.55

8.55. **Block and cylinder** **

A block and a cylinder (with $I = \beta m R^2$), both of mass m, lie on a plane (inclined at an angle θ), touching each other as shown in Fig. 8.55. There is sufficient friction between the cylinder and the plane so that it rolls without slipping. But the bottom of the block is coated with grease, so there is no friction between it and the plane. However, there is a coefficient of kinetic friction $\mu = 1$ between the block's right face and the cylinder. What is the acceleration of the block? How does your answer compare with the result for a lone cylinder rolling down a plane? Assume that θ is small enough so that the block's bottom face remains in contact with the plane at all times; what is the condition on θ for which this is true?

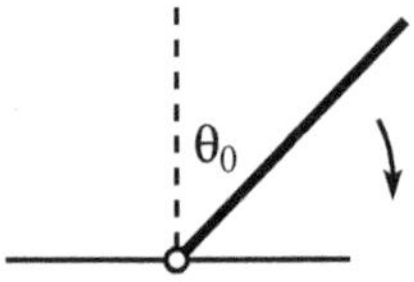

Fig. 8.56

8.56. **Falling and sliding stick** ***

One end of a uniform stick is attached to a pivot, and the pivot is free to slide along a frictionless horizontal rail. The stick is held at an initial angle θ_0 with respect to the (upward) vertical direction and then released; see Fig. 8.56. Assume that the stick can somehow swing down below the horizontal position without running into the rail (perhaps by having the pivot attached to the side of the rail, so that the stick is shifted horizontally a small distance from the rail).

(a) Show that when the stick is horizontal, the normal force N from the rail equals $mg/4$, independent of θ_0.
(b) If $\theta_0 = 0$ (a tiny kick is allowed), show that $N = 13mg$ when the stick is at the bottom of its motion (at $\theta = \pi$).
(c) If $\theta_0 = 0$, show that the minimum N occurs at $\theta \approx 61.5°$ and that the value is $N_{\text{min}} \approx (0.165)mg$. You will obtain a cubic; feel free to solve it numerically.

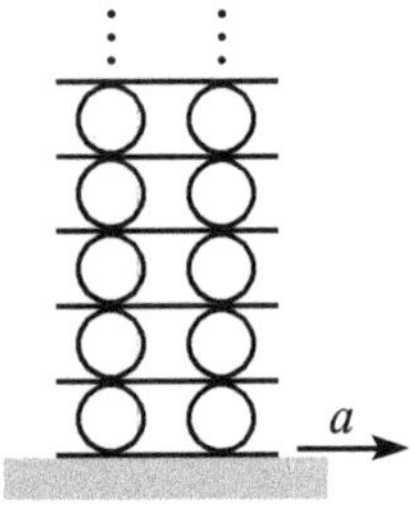

Fig. 8.57

8.57. **Tower of cylinders** ****

Consider the infinitely tall system of massless planks and identical massive cylinders shown in Fig. 8.57. The moment of inertia of the cylinders is $I = MR^2/2$. There are two cylinders at each level, and the number of levels is infinite. The cylinders do not slip with respect to the planks, but the bottom plank is free to slide on a table. If you pull on the bottom plank so that it accelerates horizontally with acceleration a, what is the horizontal acceleration of the bottom row of cylinders?

Section 8.5: Collisions

8.58. **Pendulum collision** *

A stick of mass m and length ℓ is pivoted at an end. It is held horizontal and then released. It swings down, and when it is vertical the free end elastically collides with a ball, as shown in Fig. 8.58. (Assume that the ball is initially held at rest and then released a split second before the stick strikes it.) If the stick loses half of its angular velocity during the collision, what is the mass of the ball? What is the speed of the ball right after the collision?

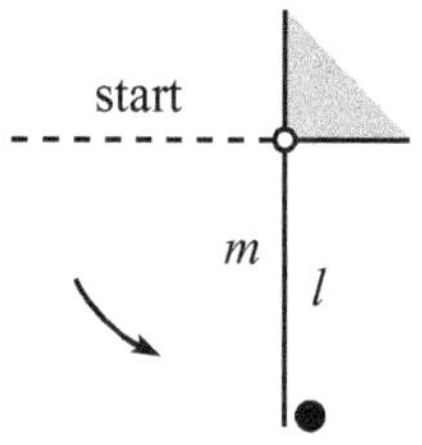

Fig. 8.58

8.59. **No final rotation** $*$

A stick of mass m and length ℓ spins around on a frictionless horizontal table, with its CM at rest (but not fixed by a pivot). A ball of mass M is placed on the table, and one end of the stick collides elastically with it, as shown in Fig. 8.59. What should M be so that after the collision the stick has translational motion, but no rotational motion?

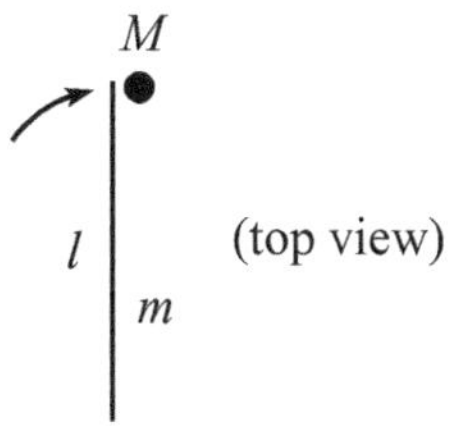

Fig. 8.59

8.60. **Same final speeds** $*$

A stick slides perpendicular to itself (without rotating) across a frictionless horizontal table and collides elastically at one of its ends with a stationary ball. Both the stick and the ball have mass m. The mass of the stick is distributed in such a way that the moment of inertia around the CM (which is at the center of the stick) is $I = Am\ell^2$, where A is some number. What should A be so that the ball moves at the same speed as the center of the stick after the collision?

8.61. **Perpendicular deflection** $**$

A mass M moves at speed V_0 perpendicular to a dumbbell at rest on a frictionless horizontal table, as shown in Fig. 8.60. The dumbbell consists of two masses m at the ends of a massless rod of length ℓ. The mass M collides elastically with one of the masses (not head-on), and afterwards it is observed that M moves perpendicular to its original direction, with speed u. What is u in terms of V_0, m, and M? What is the smallest value of m (in terms of M) for which this scenario is possible?

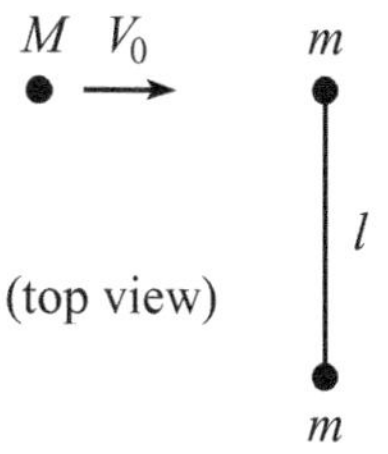

Fig. 8.60

8.62. **Glancing off a stick** $**$

A frictionless stick of mass m and length ℓ lies at rest on a frictionless horizontal table. A mass km (where k is some number) moves with speed v_0 at a 45° angle to the stick and collides elastically with it very close to an end; see Fig. 8.61. What should k be so that the mass ends up moving in the y-direction, as shown? *Hint*: Remember that the stick is frictionless.

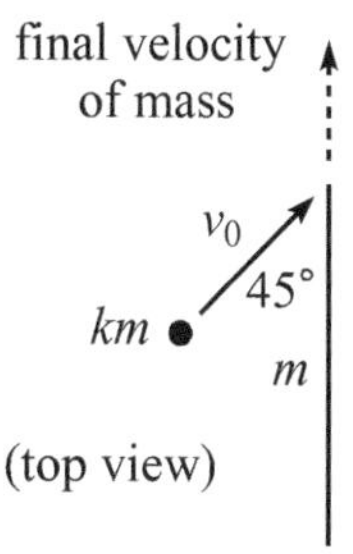

Fig. 8.61

8.63. **Sticking to a dumbbell** $*$

A mass m moves at speed v perpendicular to a dumbbell at rest on a frictionless horizontal table, as shown in Fig. 8.62. The dumbbell consists of two masses also of mass m at the ends of a massless rod of length ℓ. The moving mass collides and sticks to one of the masses.

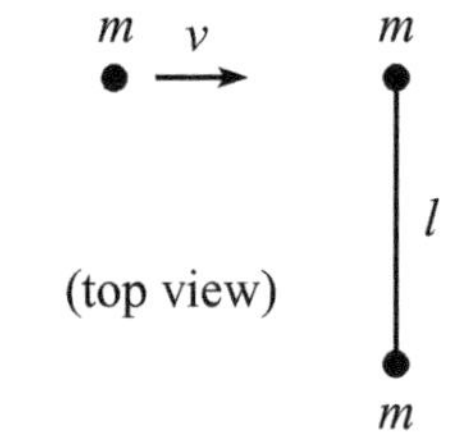

Fig. 8.62

What is the resulting ω of the system? What is the velocity of the end of the rod that has the two masses on it, after the rod has made one half of a revolution?

8.64. **Colliding sticks** **

On a frictionless horizontal table, a stick of mass m and length ℓ spins around a pivot at one of its ends with angular frequency ω. It collides and sticks to an identical stick, with an overlap length equal to x, as shown in Fig. 8.63. Immediately before the collision, the pivot is removed. What should x be so that after the collision the double-stick system has translational but no rotational motion?

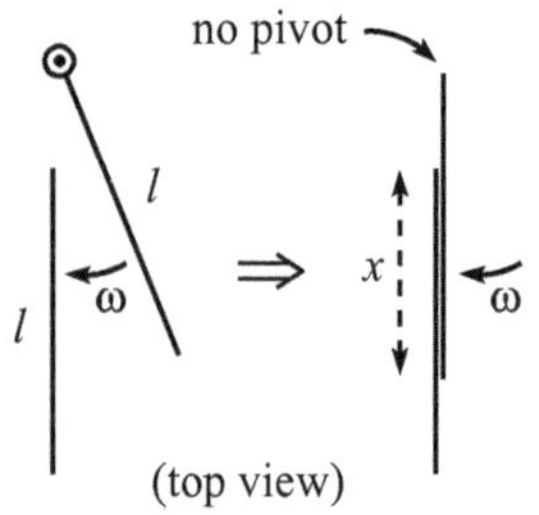

Fig. 8.63

8.65. **Lollipop** **

A hockey puck of mass m and radius R slides across frictionless ice, as shown in Fig. 8.64 (the view is from above). It has translational speed v to the right and rotational speed ω clockwise. It grazes the "top" end of a rod of mass m and length $2R$ which is initially at rest on the ice. It sticks to the rod, forming a rigid object that looks like a lollipop.

(a) In the special case of $v = R\omega$, what is the resulting angular speed of the lollipop?

(b) How much energy is lost during the collision? How do you explain the fact that energy is lost, given that the $v = R\omega$ condition implies that the contact point on the puck touches the rod with zero relative speed (and thus doesn't crash into it, as is commonly the case with inelastic collisions)?

(c) Given ω, show that v should equal $6R\omega/5$ if you want the minimum amount of energy to be lost.

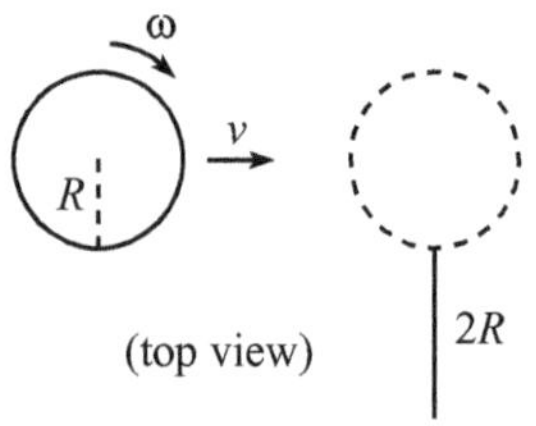

Fig. 8.64

8.66. **Pencil on a plane** ****

This exercise deals with the terminal velocity of a "pencil" rolling down a plane. To simplify things, we'll assume that the pencil has all its mass on its axis. And to avoid messy complications, we'll assume that the cross section of the pencil looks like a wheel with six equally spaced massless spokes and no rim (see Fig. 8.65).[11] Let the length of the spokes be r, and let the plane be inclined at an angle θ. Assume that there is sufficient friction to prevent the spokes from slipping on the plane, and assume that the pencil does not bounce when a spoke hits the plane.

(a) Explain qualitatively why the pencil reaches a terminal (average) velocity, assuming that it remains in contact with the plane at all times.

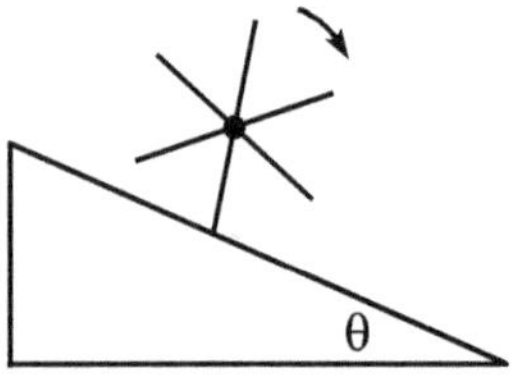

Fig. 8.65

[11] If the pencil instead looks like a hexagon with flat sides, then it is impossible to say how it behaves, because if the sides bow outward an infinitesimal amount then the system conserves energy, whereas if they bow inward an infinitesimal amount then it does not (for reasons you will figure out).

(b) Assume that conditions have been set up so that the pencil eventually reaches a nonzero terminal (average) velocity, while remaining in contact with the plane at all times. Describe this terminal velocity. You may do this by stating the maximum speed of the axis in the limiting steady state.

(c) What is the minimum value of θ for which a nonzero terminal velocity exists? An initial kick to the pencil is allowed.

(d) What is the maximum value of θ for which the pencil remains in contact with the plane at all times? As a check on your answer, the difference between the answers in parts (c) and (d) is about $5.09°$.

Section 8.6: Angular impulse

8.67. **Striking a pool ball** *

At what height should you horizontally strike a pool ball so that it immediately rolls without slipping?

8.68. **Center of percussion** *

You hold one end of a uniform stick of length L. The stick is struck with a hammer. Where should this blow occur so that the end you are holding doesn't move (immediately after the blow)? In other words, where should the blow occur so that you don't feel a "sting" in your hand? This point is called the *center of percussion*.

8.69. **Another center of percussion** *

You hold the top vertex of a solid equilateral triangle of side length L. The plane of the triangle is vertical. It is struck with a hammer, somewhere along the vertical axis. Where should this blow occur so that the point you are holding doesn't move (immediately after the blow)? Use the fact that the moment of inertia about any axis through the CM of an equilateral triangle is $mL^2/24$.

8.70. **Not hitting the pole** **

A (possibly non-uniform) stick of mass m and length ℓ lies on frictionless ice. Its midpoint (which is also its CM) touches a thin pole sticking out of the ice. One end of the stick is struck with a quick blow perpendicular to the stick, as shown in Fig. 8.66, so that the CM moves away from the pole. What is the minimum value of the stick's moment of inertia that allows the stick not to hit the pole?

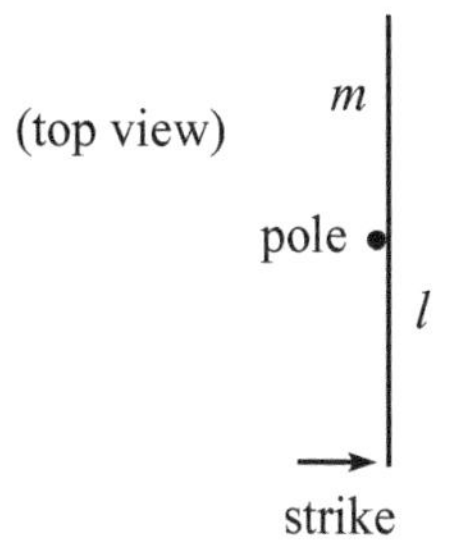

Fig. 8.66

8.71. **Pulling the paper** **

A ball sits at rest on a piece of paper on a table. You pull the paper in a straight line out from underneath the ball. You are free to pull the paper in an arbitrary (straight line) manner, frontward or backward. You may even give it abrupt, jerky motions, so that the ball slips with respect to it. After the ball comes off the paper, it will eventually roll without slipping on the table. Show (and you are encouraged to experimentally verify this) that the ball in fact ends up at rest. (The generalization of this fact is given in Problem 9.29.) Is it possible to pull the paper in such a way that the ball ends up exactly where it started?

8.72. **Up, down, and twisting** **

A uniform stick is held horizontally and then released. At the same instant, one end is struck with a quick upwards blow. If the stick ends up horizontal when it returns to its original height, what are the possible values for the maximum height to which the center rises?

8.73. **Doing work** **

(a) A pencil of mass m and length ℓ lies at rest on a frictionless table. You push on it at its midpoint (perpendicular to it), with a constant force F for a time t. Find the final speed and the distance traveled. Verify that the work you do equals the final kinetic energy.

(b) Assume that you apply the same force F for the same time t as above, but that you now apply it at one of the pencil's ends (perpendicular to the pencil). Assume that t is small, so that the pencil doesn't have much time to rotate (this means that you can assume that your force is always essentially perpendicular to the pencil, as far as the torque is concerned). Find the final CM speed, the final angular speed, and the distance your hand moves. Verify that the work you do equals the final kinetic energy.

8.74. **Bouncing between bricks** ***

A stick of length ℓ slides on frictionless ice. It bounces elastically between two parallel fixed bricks, a distance L apart, in such a way that the same end touches both bricks, and the stick hits the bricks at an angle θ each time. See Fig. 8.67. What is θ in terms of L and ℓ (an implicit equation is fine)? Draw a reasonably accurate picture of what the situation looks like in the limit $L \ll \ell$.

What should θ be in terms of L and ℓ (again, an implicit equation is fine), if the stick makes an additional n half revolutions between the bricks? What is the minimum value of L/ℓ for which n half revolutions are possible?

(top view)

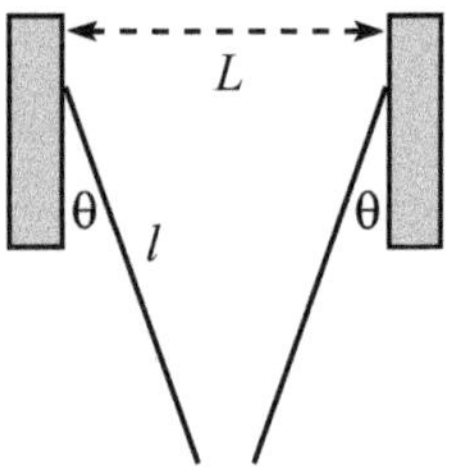

Fig. 8.67

8.75. **Repetitive bouncing** *

Using the result of Problem 8.20, what must the relation between v_x and $R\omega$ be so that a superball continually bounces back and forth between the same two points of contact on the ground?

8.76. **Bouncing under a table** **

You throw a superball so that it bounces off the floor, then off the underside of a table, then off the floor again. What must the initial relation between v_x and $R\omega$ be so that the ball returns to your hand, with the outward and return paths the same? Use the result of Problem 8.20, and modifications thereof.[12]

8.77. **Bouncing under a table again** ****

Consider the setup in the previous exercise, where we assumed that the outward and return paths were the same. Is this trajectory (where the path retraces itself) the only possible one for which the ball returns to your hand? Show[13] that the answer is "yes" unless $t_1 = 7t_2$, where t_1 is the time the ball spends between your hand and the floor (which is the same out and back, because the magnitude of v_y isn't changed by a bounce), and t_2 is the time the ball spends between the floor and the table (again, the same out and back). For the special case where $t_1 = 7t_2$, you will find that the ball returns to your hand for *any* initial relation between v_x and $R\omega$.[14]

For a ball with a general moment of inertia $I = \beta mr^2$, show that the answer is always "yes," without exception, if $\beta \leq 1/3$ (which corresponds to a wheel with massive spokes and a massless rim). Also, show that if $\beta = 1$ (a hoop), then the $t_1 = 7t_2$ condition becomes $t_1 = t_2$. In other words, if you throw a "super-hoop" from the same height as the table, then no matter how you throw it (as long as the throw is downward and in the plane of the hoop), it will return to your hand. This is a little more believable if you look at the remark in the solution to Problem 8.20.

8.9 Solutions

8.1. **Massive pulley**

The two masses have equal speeds at all times. Let v be their common speed after they have each moved a distance d. If $2m$ falls a distance d (so that m rises a distance d), then the change in potential energy is $-2mgd + 2mgd = -mgd$. The total kinetic energy is

$$K = \frac{1}{2}mv^2 + \frac{1}{2}(2m)v^2 + \frac{1}{2}I\omega^2$$

[12] You are strongly encouraged to bounce a ball in such a manner and have it magically come back to your hand. It turns out that the required value of ω is small, so a natural throw with $\omega \approx 0$ essentially gets the job done.

[13] You will want to use Mathematica or some other aid to keep track of the matrix multiplications, especially in the second part of this exercise, which would be completely intractable otherwise.

[14] I learned of this extremely bizarre fact from Howard Georgi.

$$
\begin{aligned}
&= \frac{1}{2}mv^2 + \frac{1}{2}(2m)v^2 + \frac{1}{2}\left(\frac{1}{2}mr^2\right)\left(\frac{v}{r}\right)^2 \\
&= \frac{7}{4}mv^2, \qquad (8.64)
\end{aligned}
$$

where we have used the nonslipping condition, $v = r\omega$. Conservation of energy therefore gives

$$
0 = \frac{7}{4}mv^2 - mgd \quad \Longrightarrow \quad v = \sqrt{\frac{4}{7}gd}\,. \qquad (8.65)
$$

The usual kinematic result $v = \sqrt{2ad}$ holds here, so we obtain $a = 2g/7$.

8.2. **Leaving the sphere**

In this setup, as in Problem 5.3, the ball leaves the sphere when the normal force becomes zero, that is, when

$$
\frac{mv^2}{R} = mg\cos\theta. \qquad (8.66)
$$

The only change from the solution to Problem 5.3 comes in the calculation of v. The ball now has rotational energy, so conservation of energy gives $mgR(1-\cos\theta) = mv^2/2 + I\omega^2/2 = mv^2/2 + \beta mr^2\omega^2/2$. But the nonslipping condition is $v = r\omega$, so we have

$$
\frac{1}{2}(1+\beta)mv^2 = mgR(1-\cos\theta) \quad \Longrightarrow \quad v = \sqrt{\frac{2gR(1-\cos\theta)}{1+\beta}}\,. \qquad (8.67)
$$

Plugging this into Eq. (8.66), we see that the ball leaves the sphere when

$$
\cos\theta = \frac{2}{3+\beta}\,. \qquad (8.68)
$$

Remarks: For $\beta = 0$, this equals $2/3$, as in Problem 5.3. For a uniform ball with $\beta = 2/5$, we have $\cos\theta = 10/17$, so $\theta \approx 54°$. For $\beta \to \infty$ (for example, a spool with a very thin axle rolling down the rim of a circle), we have $\cos\theta \to 0$, so $\theta \approx 90°$. This makes sense because v is always very small, since most of the energy takes the form of rotational energy. The coefficient of friction would have to be very large in this case, of course, to keep the spool from slipping near $\theta \approx 90°$. ♣

If the size of the ball is comparable to, or larger than, the size of the sphere, then we must take into account the fact that the CM of the ball does not move along a circle of radius R. Instead, it moves along a circle of radius $R + r$, so Eq. (8.66) becomes

$$
\frac{mv^2}{R+r} = mg\cos\theta. \qquad (8.69)
$$

Also, the conservation of energy equation takes the form, $mg(R+r)(1-\cos\theta) = mv^2/2 + \beta mr^2\omega^2/2$. But v still equals $r\omega$ (because the ball can be considered to be instantaneously rotating around the contact point with angular speed ω), so the kinetic energy still equals $(1+\beta)mv^2/2$. The conservation of energy statement is thus

$$
\frac{1}{2}(1+\beta)mv^2 = mg(R+r)(1-\cos\theta). \qquad (8.70)
$$

We therefore have the same equations as above, except that R is replaced everywhere by $R+r$. But R didn't appear in the result for θ in Eq. (8.68), so the answer is unchanged.

Remark: The method of the second solution to Problem 5.3 does *not* work in this problem, because there *is* a force available to make v_x decrease, namely the friction force. And indeed, v_x does decrease before the rolling ball leaves the sphere. At a given value of θ, the v in the present problem is simply $1/\sqrt{1+\beta}$ times the v in Problem 5.3, so the maximum v_x is achieved here at $\cos\theta = 2/3$, just as in Problem 5.3. But the angle in Eq. (8.68) is larger than this, so v_x decreases while the ball is between these two angles. (However, see the following problem for a setup involving rotations where the max v_x *is* relevant.) ♣

8.3. **Sliding ladder**

The important point to realize in this problem is that the ladder loses contact with the wall before it hits the ground. So we need to find where this loss of contact occurs. Let $r = \ell/2$, for convenience. While the ladder is in contact with the wall, its CM moves in a circle of radius r. This follows from the fact that the median to the hypotenuse of a right triangle has half the length of the hypotenuse. Let θ be the angle between the wall and the radius from the corner to the CM (see Fig. 8.68). This is also the angle between the ladder and the wall.

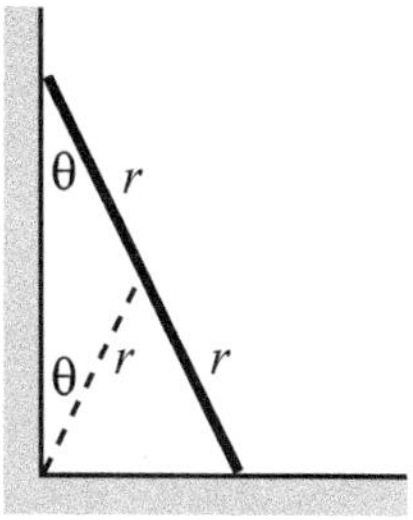

Fig. 8.68

We'll solve this problem by assuming that the CM always moves in a circle, and then determining the position at which the horizontal CM speed starts to decrease, that is, the point at which the normal force from the wall would have to become negative. Since the normal force of course can't be negative, this is the point where the ladder loses contact with the wall.

By conservation of energy, the kinetic energy of the ladder equals the loss in potential energy, which is $mgr(1-\cos\theta)$. This kinetic energy can be broken up into the CM translational energy plus the rotation energy. The CM translational energy is $mr^2\dot\theta^2/2$, because the CM travels in a circle of radius r. The rotational energy is $I\dot\theta^2/2$. The same $\dot\theta$ applies here as in the CM translational motion, because θ is the angle between the ladder and the vertical, and is thus the angle of rotation of the ladder. Letting $I \equiv \beta mr^2$ to be general ($\beta = 1/3$ for our ladder), the conservation of energy statement is $(1+\beta)mr^2\dot\theta^2/2 = mgr(1-\cos\theta)$. Therefore, the speed of the CM, which is $v = r\dot\theta$, equals

$$v = \sqrt{\frac{2gr(1-\cos\theta)}{1+\beta}}. \tag{8.71}$$

The horizontal component of this is

$$v_x = \sqrt{\frac{2gr}{1+\beta}}\sqrt{(1-\cos\theta)}\cos\theta. \tag{8.72}$$

Taking the derivative of $\sqrt{(1-\cos\theta)}\cos\theta$, we see that the horizontal speed is maximum when $\cos\theta = 2/3$. Therefore the ladder loses contact with the wall when

$$\cos\theta = \frac{2}{3} \implies \theta \approx 48.2^\circ, \tag{8.73}$$

which is independent of β. This means that, for example, a dumbbell (two masses at the ends of a massless rod, with $\beta = 1$) loses contact with the wall at the same angle. Plugging the value of θ from Eq. (8.73) into Eq. (8.72), and using $\beta = 1/3$, we obtain a final horizontal speed of

$$v_x = \frac{\sqrt{2gr}}{3} \equiv \frac{\sqrt{g\ell}}{3}. \tag{8.74}$$

Note that this is 1/3 of the $\sqrt{2gr}$ horizontal speed that the ladder would have if it were arranged (perhaps by having the top end slide down a curve) to eventually slide horizontally along the ground. You are encouraged to compare various aspects of this problem with those in Problem 8.2 and Problem 5.3.

REMARK: The normal force from the wall is zero at the start and zero at the finish, so it must reach a maximum at some intermediate value of θ. Let's find this θ. Taking the derivative of v_x in Eq. (8.72) to find the CM's horizontal acceleration a_x, and then using $\dot{\theta} \propto \sqrt{1-\cos\theta}$ from Eq. (8.71), we see that the force from the wall is proportional to

$$a_x \propto \frac{\sin\theta(3\cos\theta - 2)\dot{\theta}}{\sqrt{1-\cos\theta}} \propto \sin\theta(3\cos\theta-2). \tag{8.75}$$

Taking the derivative of this, we find that the force from the wall is maximum when $\cos\theta = (1+\sqrt{19})/6 \implies \theta \approx 26.7°$ ♣

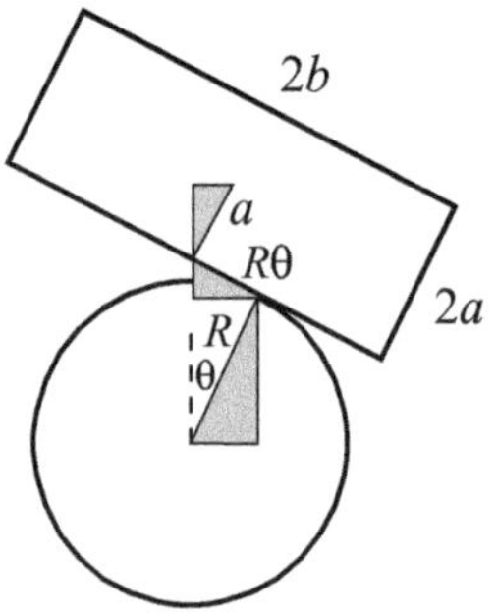

Fig. 8.69

8.4. **Leaning rectangle**

We must first find the position of the rectangle's CM when it has rotated through an angle θ. Using Fig. 8.69, we can obtain this position (relative to the center of the cylinder) by adding up the distances along the three shaded triangles. Because there is no slipping, the contact point has moved a distance $R\theta$ along the rectangle. We find that the position of the CM is

$$(x,y) = R(\sin\theta, \cos\theta) + R\theta(-\cos\theta, \sin\theta) + a(\sin\theta, \cos\theta). \tag{8.76}$$

We'll now use the Lagrangian method to find the equation of motion and the frequency of small oscillations. Using Eq. (8.76), you can show that the square of the speed of the CM is

$$v^2 = \dot{x}^2 + \dot{y}^2 = (a^2 + R^2\theta^2)\dot{\theta}^2. \tag{8.77}$$

The simplicity of this result suggests that there is a quicker way to obtain it. And indeed, the CM instantaneously rotates around the contact point with angular speed $\dot{\theta}$, and from Fig. 8.69, the distance to the contact point is $\sqrt{a^2+R^2\theta^2}$. Therefore, the speed of the CM is $\omega r = \dot{\theta}\sqrt{a^2+R^2\theta^2}$.

The Lagrangian is

$$\mathcal{L} = T - V = \frac{1}{2}m(a^2+R^2\theta^2)\dot{\theta}^2 + \frac{1}{2}I\dot{\theta}^2 - mg\Big((R+a)\cos\theta + R\theta\sin\theta\Big). \tag{8.78}$$

The equation of motion is, as you can check,

$$(ma^2 + mR^2\theta^2 + I)\ddot{\theta} + mR^2\theta\dot{\theta}^2 = mga\sin\theta - mgR\theta\cos\theta. \tag{8.79}$$

Let us now consider small oscillations. Using the small-angle approximations, $\sin\theta \approx \theta$ and $\cos\theta \approx 1 - \theta^2/2$, and keeping terms up to first order in θ, we obtain

$$(ma^2 + I)\ddot{\theta} + mg(R-a)\theta = 0. \tag{8.80}$$

The coefficient of θ is positive if $a < R$. Therefore, oscillatory motion occurs if $a < R$. Note that this condition is independent of b. The frequency of small oscillations is

$$\omega = \sqrt{\frac{mg(R-a)}{ma^2+I}}. \tag{8.81}$$

If $a \ge R$, then the rectangle falls off the cylinder.

Remarks: Let's look at some special cases. If $I = 0$ (that is, all of the rectangle's mass is located at the CM), we have $\omega = \sqrt{g(R-a)/a^2}$. If in addition $a \ll R$, then $\omega \approx \sqrt{gR/a^2}$. You can also derive these results by considering the CM to be a point mass sliding in a parabolic potential. If the rectangle is instead a uniform horizontal stick, so that $a \ll R$, $a \ll b$, and $I \approx mb^2/3$, then we have $\omega \approx \sqrt{3gR/b^2}$. If the rectangle is a vertical stick (satisfying $a < R$), so that $b \ll a$ and $I \approx ma^2/3$, then we have $\omega \approx \sqrt{3g(R-a)/4a^2}$. If in addition $a \ll R$, then $\omega \approx \sqrt{3gR/4a^2}$.

Without doing much work, there are two other ways we can determine the condition under which there is oscillatory motion. The first is to look at the height of the CM (although this is essentially what we ended up doing in the solution above). Using small-angle approximations in Eq. (8.76), the height of the CM is $y \approx (R+a) +$

$(R-a)\theta^2/2$. Therefore, if $a<R$, the potential energy increases with θ, so the rectangle wants to decrease its θ and fall back down to the middle. But if $a>R$, the potential energy decreases with θ, so the rectangle wants to increase its θ and fall off the cylinder.

The second way is to look at the horizontal positions of the CM and the contact point. Small-angle approximations in Eq. (8.76) show that the former equals $a\theta$ and the latter equals $R\theta$. Therefore, if $a<R$ then the CM is to the left of the contact point, so the torque from gravity (relative to the contact point) makes θ decrease, and the motion is stable. But if $a>R$ then the torque from gravity makes θ increase, and the motion is unstable. ♣

8.5. **Mass in a tube**

The Lagrangian is

$$\mathcal{L}=\frac{1}{2}\left(\frac{1}{3}M\ell^2\right)\dot{\theta}^2+\left(\frac{1}{2}mx^2\dot{\theta}^2+\frac{1}{2}m\dot{x}^2\right)+mgx\sin\theta+Mg\left(\frac{\ell}{2}\right)\sin\theta. \quad (8.82)$$

The Euler–Lagrange equations are then

$$\begin{aligned}\frac{d}{dt}\left(\frac{\partial\mathcal{L}}{\partial\dot{x}}\right)=\frac{\partial\mathcal{L}}{\partial x} &\implies m\ddot{x}=mx\dot{\theta}^2+mg\sin\theta,\\ \frac{d}{dt}\left(\frac{\partial\mathcal{L}}{\partial\dot{\theta}}\right)=\frac{\partial\mathcal{L}}{\partial\theta} &\implies \frac{d}{dt}\left(\frac{1}{3}M\ell^2\dot{\theta}+mx^2\dot{\theta}\right)=\left(mgx+\frac{Mg\ell}{2}\right)\cos\theta \quad (8.83)\\ &\implies \left(\frac{1}{3}M\ell^2+mx^2\right)\ddot{\theta}+2mx\dot{x}\dot{\theta}=\left(mgx+\frac{Mg\ell}{2}\right)\cos\theta.\end{aligned}$$

In terms of $\eta\equiv x/\ell$, these equations become

$$\begin{aligned}\ddot{\eta}&=\eta\dot{\theta}^2+\tilde{g}\sin\theta,\\ (1+3r\eta^2)\ddot{\theta}&=\left(3r\tilde{g}\eta+\frac{3\tilde{g}}{2}\right)\cos\theta-6r\eta\dot{\eta}\dot{\theta}, \quad (8.84)\end{aligned}$$

where $r\equiv m/M$ and $\tilde{g}\equiv g/\ell$. Below is a Maple program that numerically finds the value of η when θ equals $\pi/2$, in the case where $r=1$. As mentioned in Problem 1.2, this value of η does not depend on g or ℓ, and hence not $\tilde{g}$. In the program, we'll denote $\tilde{g}$ by `g`, which we'll give the arbitrary value of 10. We'll use `q` for θ, and `n` for η. Also, we'll denote $\dot{\theta}$ by `q1` and $\ddot{\theta}$ by `q2`, etc. Even if you don't know Maple, this program should still be understandable. See Section 1.4 for more discussion of solving differential equations numerically.

```
n:=0:                            # initial n value
n1:=0:                           # initial n speed
q:=0:                            # initial angle
q1:=0:                           # initial angular speed
e:=.0001:                        # small time interval
g:=10:                           # value of g/l
r:=1:                            # value of m/M
while q<1.57079 do               # do this process until
                                 # the angle is pi/2
n2:=n*q1^2+g*sin(q):             # the first E-L equation
q2:=((3*r*g*n+3*g/2)*cos(q)
    -6*r*n*n1*q1)/(1+3*r*n^2):   # the second E-L equation
n:=n+e*n1:                       # how n changes
n1:=n1+e*n2:                     # how n1 changes
q:=q+e*q1:                       # how q changes
q1:=q1+e*q2:                     # how q1 changes
end do:                          # stop the process
n;                               # print the value of n
```

The resulting value of η is 0.378. If you actually run this program on Maple with different values of g, you will find that the result for n doesn't depend on g, as stated above. A few results for η for various values of r are, in (r, η) notation: $(0, 0.349)$, $(1, 0.378)$, $(2, 0.410)$, $(10, 0.872)$, $(20, 3.290)$. It turns out that $r \approx 11.25$ yields $\eta \approx 1$. That is, the mass m gets to the end of the tube right when the tube becomes vertical.

For η values larger than 1, we could imagine attaching a massless tubular extension on the end of the given tube. It turns out that $\eta \to \infty$ as $r \to \infty$. In this case, the mass m drops nearly straight down, causing the tube to quickly swing down to a nearly vertical position. But m ends up slightly to one side and then takes a very long time to move over to become directly below the pivot.

8.6. **Minimum *I***

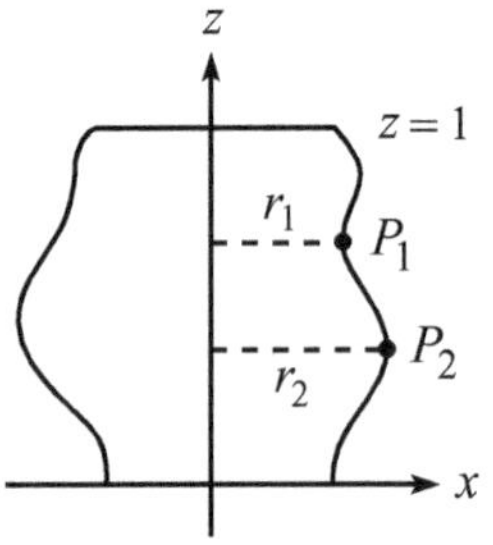

Fig. 8.70

The shape should be a cylinder with the z axis as its symmetry axis. A quick proof (by contradiction) is as follows. Assume that the optimal blob is not a cylinder, and consider the surface of the blob. If the blob is not a cylinder, then there exist two points on the surface, P_1 and P_2, that are located at different distances, r_1 and r_2, from the z axis. Assume $r_1 < r_2$ (see Fig. 8.70). If we move a small piece of the blob from P_2 to P_1, then we decrease the moment of inertia, $\int r^2\,dm$. Therefore, the proposed noncylindrical blob cannot be the one with the smallest I. In order to avoid this contradiction, all points on the surface must be equidistant from the z axis. The only blob with this property is a cylinder.

8.7. **Slick calculations of *I***

(a) We claim that the I for a square of side 2ℓ is 16 times the I for a square of side ℓ, assuming that the axes pass through any two corresponding points. This is true because dm is proportional to the area, which is proportional to length squared, so the corresponding dm's differ by a factor of 4. And then there are the two powers of r in the integrand. Therefore, when changing variables from one square to the other, there are four powers of 2 in the integral $\int r^2\,dm = \int r^2 \rho\,dx\,dy$.

As in Section 8.3.2, we can express the relevant relations in terms of pictures:

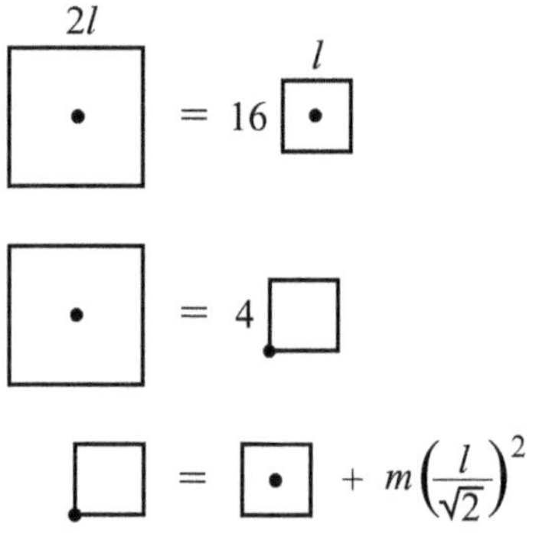

The first line comes from the scaling argument, the second comes from the fact that moments of inertia simply add, and the third comes from the parallel-axis theorem. Equating the right-hand sides of the first two, and then using the third to eliminate 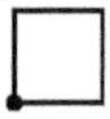 gives

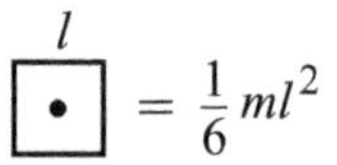

This agrees with the result of Example 12 in Section 8.3.1, with $a = b = \ell$.

(b) This is again a two-dimensional object, so the I for a triangle of side 2ℓ is 16 times the I for a triangle of side ℓ, assuming that the axes pass through any two corresponding points. With pictures, we have:

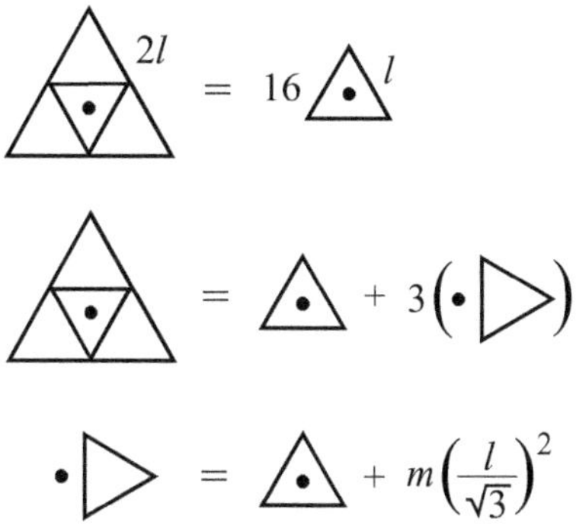

The first line comes from the scaling argument, the second comes from the fact that moments of inertia simply add, and the third comes from the parallel-axis theorem. Equating the right-hand sides of the first two, and then using the third to eliminate 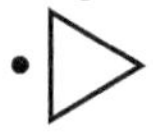 gives

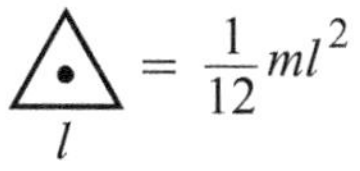

This agrees with the result of Example 11 in Section 8.3.1, with $N = 3$. The "radius" R used in that example equals $\ell/\sqrt{3}$ in the present notation.

8.8. **Slick calculations of I for fractal objects**

(a) The scaling argument here is a little trickier than the one in Section 8.3.2. Our object is self-similar to an object 3 times as big, so let's increase the length by a factor of 3 and see what happens to I. In the integral $\int x^2\,dm$, the x's pick up a factor of 3, so this gives a factor of 9. But what happens to the dm? Well, tripling the size of our object increases its mass by a factor of 2, because the new object is made up of two of the smaller ones, plus some empty space in the middle. So the dm picks up a factor of 2. Therefore, the I for an object of length 3ℓ is 18 times the I for an object of length ℓ, assuming that the axes pass through any two corresponding points. With pictures, we have (the following symbols denote our fractal object):

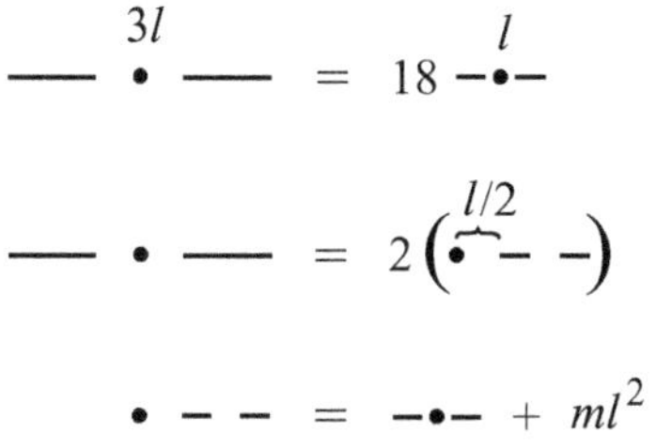

The first line comes from the scaling argument, the second comes from the fact that moments of inertia simply add, and the third comes from the parallel-axis theorem. Equating the right-hand sides of the first two, and then using the

third to eliminate • — — gives

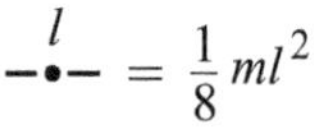

This is larger than the I for a uniform stick, namely $m\ell^2/12$, because the mass here is generally farther away from the center.

REMARK: When we increase the length of our object by a factor of 3, the factor of 2 in the dm is larger than the factor of 1 relevant to a zero-dimensional object, but smaller than the factor of 3 relevant to a one-dimensional object. So in some sense our object has a dimension between 0 and 1. It is reasonable to define the fractal dimension, d, of an object as the number for which r^d is the increase in "volume" when the dimensions are increased by a factor of r. In this problem, we have $3^d = 2$, so $d = \log_3 2 \approx 0.63$. ♣

(b) Again, the mass scales in a strange way. Let's increase the dimensions of our object by a factor of 2 and see what happens to I. In the integral $\int x^2\, dm$, the x's pick up a factor of 2, so this gives a factor of 4. But what happens to the dm? Doubling the size of our object increases its mass by a factor of 3, because the new object is made up of three of the smaller ones, plus an empty triangle in the middle. So the dm picks up a factor of 3. Therefore, the I for an object of side 2ℓ is 12 times the I for an object of side ℓ, assuming that the axes pass through any two corresponding points. With pictures, we have:

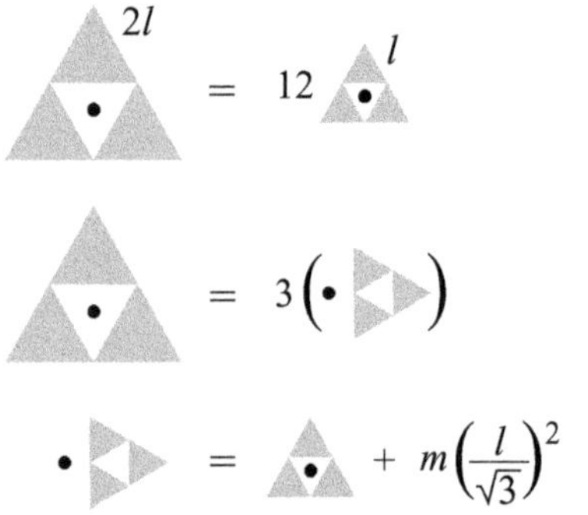

The first line comes from the scaling argument, the second comes from the fact that moments of inertia simply add, and the third comes from the parallel-axis theorem. Equating the right-hand sides of the first two, and then using the third

to eliminate  gives

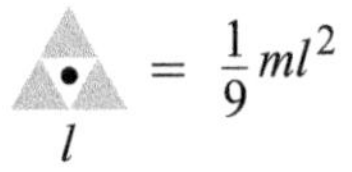

This is larger than the I for the uniform triangle in Problem 8.7, namely $m\ell^2/12$, because the mass here is generally farther away from the center. Increasing the size of our object by a factor of 2 increases the "volume" by a factor of 3, so the fractal dimension is given by $2^d = 3 \implies d = \log_2 3 \approx 1.58$.

(c) Again, the mass scales in a strange way. Let's increase the dimensions of our object by a factor of 3 and see what happens to I. In the integral $\int x^2\, dm$, the x's pick up a factor of 3, so this gives a factor of 9. But what happens to the dm?

Tripling the size of our object increases its mass by a factor of 8, because the new object is made up of eight of the smaller ones, plus an empty square in the middle. So the dm picks up a factor of 8. Therefore, the I for an object of side 3ℓ is 72 times the I for an object of side ℓ, assuming that the axes pass through any two corresponding points. With pictures, we have:

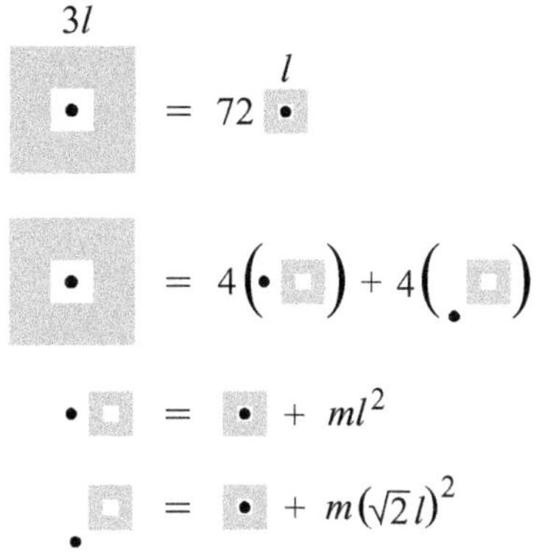

The first line comes from the scaling argument, the second comes from the fact that moments of inertia simply add, and the third and fourth come from the parallel-axis theorem. Equating the right-hand sides of the first two, and then using the third and fourth to eliminate 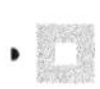 and  gives

$$= \frac{3}{16} ml^2$$

This is larger than the I for the uniform square in Problem 8.7, namely $m\ell^2/6$, because the mass here is generally farther away from the center. Increasing the size of our object by a factor of 3 increases the "volume" by a factor of 8, so the fractal dimension is given by $3^d = 8 \implies d = \log_3 8 \approx 1.89$.

8.9. **Zero torque from internal forces**

Let $\mathbf{F}_{ij}^{\text{int}}$ be the force that the ith particle feels due to the jth particle (see Fig. 8.71). Then

$$\mathbf{F}_i^{\text{int}} = \sum_j \mathbf{F}_{ij}^{\text{int}}. \tag{8.85}$$

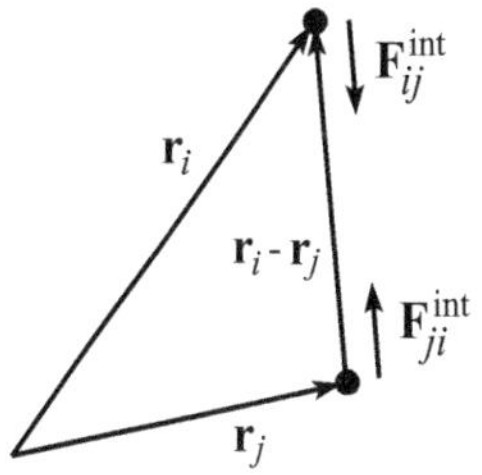

Fig. 8.71

The total internal torque on all the particles, relative to the chosen origin, is therefore

$$\boldsymbol{\tau}^{\text{int}} \equiv \sum_i \mathbf{r}_i \times \mathbf{F}_i^{\text{int}} = \sum_i \sum_j \mathbf{r}_i \times \mathbf{F}_{ij}^{\text{int}}. \tag{8.86}$$

But if we interchange the indices (which were labeled arbitrarily), we have

$$\boldsymbol{\tau}^{\text{int}} = \sum_j \sum_i \mathbf{r}_j \times \mathbf{F}_{ji}^{\text{int}} = -\sum_j \sum_i \mathbf{r}_j \times \mathbf{F}_{ij}^{\text{int}}, \tag{8.87}$$

where we have used Newton's third law, $\mathbf{F}_{ij}^{\text{int}} = -\mathbf{F}_{ji}^{\text{int}}$. Adding the two previous equations gives

$$2\boldsymbol{\tau}^{\text{int}} = \sum_i \sum_j (\mathbf{r}_i - \mathbf{r}_j) \times \mathbf{F}_{ij}^{\text{int}}. \tag{8.88}$$

But $\mathbf{F}_{ij}^{\text{int}}$ is parallel to $(\mathbf{r}_i - \mathbf{r}_j)$, by assumption. Therefore, each cross product in the sum equals zero.

The above sums might make this solution look a bit involved. But the idea is simply that the torques cancel in pairs. This is clear from Fig. 8.71, because the two forces shown are equal and opposite, and they have the same lever arm relative to the origin.

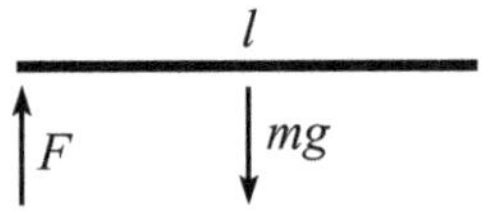

Fig. 8.72

8.10. **Removing a support**

(a) FIRST SOLUTION Let the desired force from the left support be F, and let the downward acceleration of the stick's CM be a. Then the $F = ma$ and $\tau = I\alpha$ (relative to the fixed support; see Fig. 8.72) equations, along with the circular-motion relation between a and α, are, respectively,

$$mg - F = ma, \quad mg\frac{\ell}{2} = \left(\frac{m\ell^2}{3}\right)\alpha, \quad a = \frac{\ell}{2}\alpha. \tag{8.89}$$

The second equation gives $\alpha = 3g/2\ell$. The third equation then gives $a = 3g/4$. And the first equation then gives $F = mg/4$. Note that the right end of the stick accelerates at $2a = 3g/2$, which is larger than g.

SECOND SOLUTION Looking at torques around the CM, and also torques around the fixed support, we have, respectively,

$$F\frac{\ell}{2} = \left(\frac{m\ell^2}{12}\right)\alpha, \quad \text{and} \quad mg\frac{\ell}{2} = \left(\frac{m\ell^2}{3}\right)\alpha. \tag{8.90}$$

Dividing the first of these equations by the second gives $F = mg/4$.

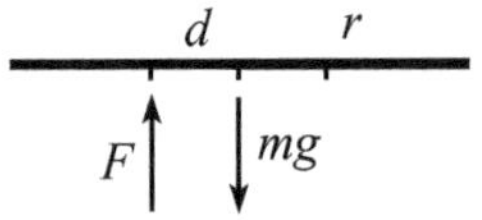

Fig. 8.73

(b) FIRST SOLUTION As in the first solution above, we have (using the parallel-axis theorem; see Fig. 8.73)

$$mg - F = ma, \quad mgd = (\beta mr^2 + md^2)\alpha, \quad a = \alpha d. \tag{8.91}$$

Solving for F gives $F = mg/(1 + d^2/\beta r^2)$. For $d = r$ and $\beta = 1/3$, we obtain the answer in part (a).

SECOND SOLUTION As in the second solution above, looking at torques around the CM, and also torques around the fixed support, we have, respectively,

$$Fd = (\beta mr^2)\alpha, \quad \text{and} \quad mgd = (\beta mr^2 + md^2)\alpha. \tag{8.92}$$

Dividing the first of these equations by the second gives $F = mg/(1+d^2/\beta r^2)$.

REMARKS: For the special case $d = r$, we have the following: If $\beta = 0$ (point mass in the middle) then $F = 0$; if $\beta = 1$ (dumbbell with masses at the ends) then $F = mg/2$; and if $\beta = \infty$ (masses at the ends of long massless extensions of the stick) then $F = mg$. These all make intuitive sense. In the limit $d = 0$, we have $F = mg$. And in the limit $d = \infty$ (using a massless extension), we have $F = 0$. Technically, we should be writing $d \ll \sqrt{\beta}\, r$ and $d \gg \sqrt{\beta}\, r$ here. ♣

8.11. **Falling stick**

(a) Let's calculate τ and L relative to the pivot point. The torque is due to gravity, which effectively acts on the CM and has magnitude mgb. The moment of inertia of the stick around the horizontal axis through the pivot (and perpendicular to the massless stick) is simply mb^2. So when the stick starts to fall, the $\tau = dL/dt$

equation is $mgb = (mb^2)\alpha$. Therefore, the initial acceleration of the CM, namely $b\alpha$, is

$$b\alpha = g, \tag{8.93}$$

which is independent of ℓ and b. This answer makes sense. The stick initially falls straight down, and the pivot provides no force because it doesn't know right away that the stick is moving.

(b) The only change from part (a) is the moment of inertia of the stick around the horizontal axis through the pivot (and perpendicular to the massless stick). From the parallel-axis theorem, this moment is $mb^2 + m\ell^2/12$. So when the stick starts to fall, the $\tau = dL/dt$ equation is $mgb = (mb^2 + m\ell^2/12)\alpha$. Therefore, the initial acceleration of the CM is

$$b\alpha = \frac{g}{1 + (\ell^2/12b^2)}\,. \tag{8.94}$$

For $\ell \ll b$, this goes to g, as it should. And for $\ell \gg b$, it goes to zero, as it should. In this case, a tiny movement of the CM corresponds to a very large movement of the points far out along the stick. Therefore, by conservation of energy, the CM must be moving very slowly.

8.12. **Pulling a cylinder**

Consider the net force in the y direction (the transverse direction) on an infinitesimal arclength of the cylinder. Measure θ clockwise from the bottom, as shown in Fig. 8.74. The tension increases as θ increases (assuming $T_1 > T_2$), and the friction force F_f on the cylinder equals the difference dT in the tension from one end of the small arc to the other (using Newton's third law, and noting that there can be no net force on the string because it is massless). The friction force on the cylinder therefore produces a force component $dT \sin\theta$ in the y direction. From the example in Section 2.1, the normal force on the cylinder is $N = T\,d\theta$, which yields a force component $T\,d\theta\cos\theta$ in the y direction. The net y-force on the cylinder at the little arc is therefore

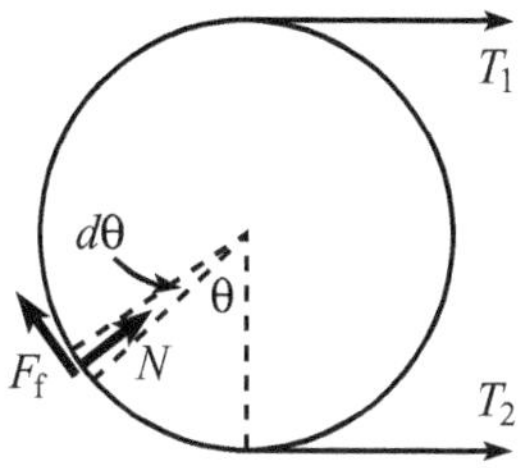

Fig. 8.74

$$dF_y = dT\sin\theta + T\,d\theta\cos\theta = d(T\sin\theta). \tag{8.95}$$

The total F_y on the cylinder is thus

$$F_y = \int dF_y = \int d(T\sin\theta) = \Delta(T\sin\theta). \tag{8.96}$$

But θ runs from 0 to π, which means that $\sin\theta$ starts and ends at zero. Therefore, the total change in $T\sin\theta$ equals zero, and so $F_y = 0$, as desired. Note that nowhere in this solution did we assume anything about T. The string might be slipping, and the cylinder might be rough in some places and smooth in others, and it doesn't matter. The total F_y is still zero (as we know it must be, from the simpler reasoning that T_1 and T_2 pull only in the x direction). Physically, what happens is that N is larger in the top half of the semicircle than in the bottom half, and the resulting net downward force cancels the upward force from the friction.

The above $F_y = \Delta(T\sin\theta)$ result holds in general, and not just when the string wraps around a semicircle. Because of our convention for θ, the signs work out so that $\Delta(T\sin\theta)$ is simply the sum of the y components of the tensions T_1 and T_2, which is the answer we expect.

8.13. **Oscillating ball**

Let the angle from the bottom of the cylinder to the ball be θ (see Fig. 8.75), and let F_f be the friction force. Then the tangential $F = ma$ equation is

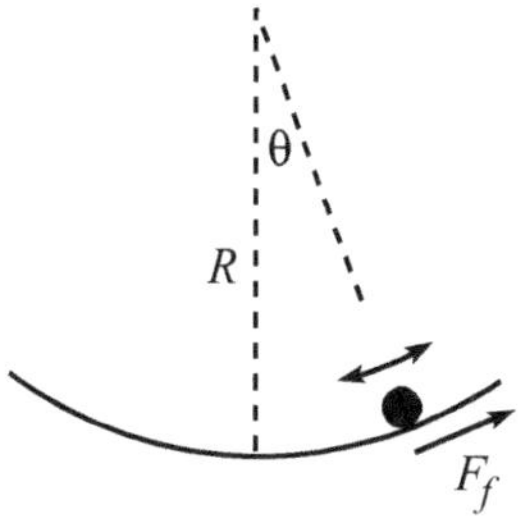

$$F_f - mg\sin\theta = ma, \tag{8.97}$$

where we have chosen rightward to be the positive direction for a and F_f. Also, the $\tau = I\alpha$ equation (relative to the CM) is

$$-rF_f = \frac{2}{5}mr^2\alpha, \tag{8.98}$$

Fig. 8.75

where we have chosen clockwise to be the positive direction for α. Using the non-slipping condition $r\alpha = a$, the torque equation becomes $F_f = -(2/5)ma$. Plugging this into Eq. (8.97), and using $\sin\theta \approx \theta$, we obtain $mg\theta + (7/5)ma = 0$. Under the assumption $r \ll R$, the center of the ball moves along a circle with radius essentially equal to R, so we have $a \approx R\ddot{\theta}$, We therefore arrive at

$$\ddot{\theta} + \left(\frac{5g}{7R}\right)\theta = 0. \tag{8.99}$$

This is the equation for simple harmonic motion with frequency

$$\omega = \sqrt{\frac{5g}{7R}}. \tag{8.100}$$

In general, if the ball has a moment of inertia equal to βmr^2, you can show that the frequency of small oscillations is $\sqrt{g/(1+\beta)R}$. Note that we needed to use two different expressions for a in this solution, namely $r\alpha$ and $R\ddot{\theta}$.

REMARKS: The answer in Eq. (8.100) is slightly smaller than the $\sqrt{g/R}$ answer for the case where the ball slides. In terms of forces, the reason for this is that the friction force causes there to be a smaller net tangential force. In terms of energy, the reason is that energy is "lost" in the rotational motion, so the ball ends up moving slower.

If we omit the $r \ll R$ assumption, then the $r\alpha = a$ relation still holds, because we can consider the ball to be instantaneously rotating around the contact point. But the $a = R\ddot{\theta}$ relation is replaced by $a = (R - r)\ddot{\theta}$, because the center of the ball moves along a circle of radius $R - r$. Therefore, the exact result for the frequency is $\omega = \sqrt{5g/7(R - r)}$. This goes to infinity as $r \to R$. ♣

8.14. **Oscillating cylinders**

The moments of inertia of the cylinders are $I_1 = M_1R_1^2$ and $I_2 = M_2R_2^2$. Let F be the force between the two cylinders, defined with rightward on the small cylinder being positive. Let θ_1 and θ_2 be the angles of rotation of the cylinders, with counterclockwise positive, relative to the position where the small cylinder is at the bottom of the big cylinder. Then the torque equations are

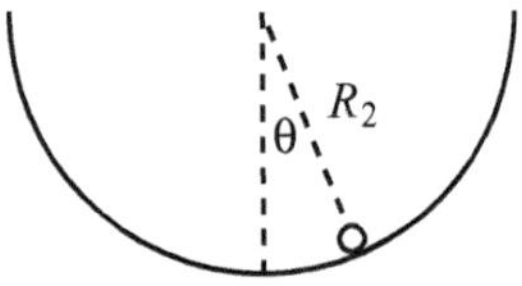

Fig. 8.76

$$FR_1 = M_1R_1^2\ddot{\theta}_1, \quad \text{and} \quad FR_2 = -M_2R_2^2\ddot{\theta}_2. \tag{8.101}$$

We are not so much concerned with θ_1 and θ_2 as we are with the angular position that M_1 makes with the vertical. Let this angle be θ (see Fig. 8.76). In the approximation $R_1 \ll R_2$, the nonslipping condition says that $R_2\theta \approx R_2\theta_2 - R_1\theta_1$, since both sides of this equation are expressions for the arclength away from the bottom of the big cylinder. Adding Eqs. (8.101) after dividing through by the masses then gives

$$F\left(\frac{1}{M_1} + \frac{1}{M_2}\right) = -R_2\ddot{\theta}. \tag{8.102}$$

The tangential force equation on M_1 is

$$F - M_1g\sin\theta = M_1(R_2\ddot{\theta}). \tag{8.103}$$

Substituting the F from (8.102) into this equation gives, with $\sin\theta \approx \theta$,

$$\left(M_1 + \frac{1}{\frac{1}{M_1} + \frac{1}{M_2}}\right)\ddot{\theta} + \left(\frac{M_1g}{R_2}\right)\theta = 0. \tag{8.104}$$

After simplifying, the frequency of small oscillations is

$$\omega = \sqrt{\frac{g}{R_2}}\sqrt{\frac{M_1 + M_2}{M_1 + 2M_2}}. \tag{8.105}$$

REMARKS: In the limit $M_2 \ll M_1$, we obtain $\omega \approx \sqrt{g/R_2}$. In this case, there is essentially no friction force between the cylinders, because otherwise the "massless" M_2 would have infinite angular acceleration. So there is only a normal force, and the small cylinder essentially acts like a pendulum of length R_2. In the limit $M_1 \ll M_2$, we obtain $\omega \approx \sqrt{g/2R_2}$. In this case, the big cylinder is essentially fixed, so we simply have the setup mentioned in the solution to Problem 8.13, with $\beta = 1$. ♣

8.15. **Lengthening the string**

Consider the angular momentum relative to the support point P. The forces on the mass are the tension in the string and gravity. The former provides no torque around P, and the latter provides no torque in the z direction. Therefore, L_z is constant. The motion is always approximately circular because the length of the string changes very slowly, so if we let ω_ℓ be the frequency of the circular motion when the string has length ℓ, then we can say that

$$L_z = mr^2\omega_\ell \tag{8.106}$$

is constant. The frequency ω_ℓ can be obtained by using $F = ma$ for the circular motion. The tension in the string is essentially $mg/\cos\theta$ (to make the forces in the y direction cancel), so the horizontal radial force is $mg\tan\theta$. Therefore,

$$mg\tan\theta = mr\omega_\ell^2 = m(\ell\sin\theta)\omega_\ell^2 \implies \omega_\ell = \sqrt{\frac{g}{\ell\cos\theta}} = \sqrt{\frac{g}{h}}. \tag{8.107}$$

Plugging this into Eq. (8.106), we see that the constant value of L_z is

$$L_z = mr^2\sqrt{\frac{g}{h}}. \tag{8.108}$$

This holds at all times, so the quantity $r^2/\sqrt{h}$ is constant. Let's look at the two cases.

(a) For $\theta \approx 0$, we have $h \approx \ell$, so Eq. (8.108) says that $r^2/\sqrt{\ell}$ is constant. Therefore,

$$r \propto \ell^{1/4}, \tag{8.109}$$

which means that r grows very slowly as you let the string out when $\theta \approx 0$.

(b) For $\theta \approx \pi/2$, we have $r \approx \ell$, so Eq. (8.108) says that $\ell^2/\sqrt{h}$ is constant. Therefore,

$$h \propto \ell^4, \tag{8.110}$$

which means that h grows very quickly as you let the string out when $\theta \approx \pi/2$.

Note that Eq. (8.108) says that $h \propto r^4$ for any value of θ. So no matter what θ is, if you slowly lengthen the string so that r doubles, then h increases by a factor of 16. Equivalently, if you draw the string in, the envelope of the motion of the mass is a surface of revolution generated by a curve of the form $y \propto -x^4$.

8.16. **A triangle of cylinders**

(a) Let N be the normal force between the cylinders, and let F be the friction force from the ground (see Fig. 8.77). Let a_x be the initial horizontal acceleration of the right bottom cylinder (so $\alpha = a_x/R$ is its angular acceleration), and let a_y be the initial vertical acceleration of the top cylinder, with downward taken to be positive.

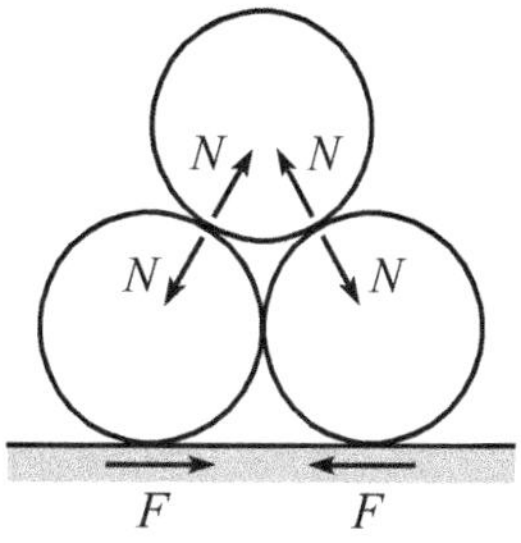

Fig. 8.77

If we consider the torque around the center of one of the bottom cylinders, then the only relevant force is F, because N, gravity, and the normal force from the ground all point through the center. The equations for $F_x = ma_x$ on the bottom right cylinder, $F_y = ma_y$ on the top cylinder, and $\tau = I\alpha$ on the bottom right cylinder are, respectively,

$$\begin{aligned} N\cos 60° - F &= ma_x, \\ mg - 2N\sin 60° &= ma_y, \\ FR &= (\beta mR^2)(a_x/R). \end{aligned} \tag{8.111}$$

We have four unknowns, N, F, a_x, and a_y, so we need one more equation. Fortunately, a_x and a_y are related. The contact surface between the top and bottom cylinders lies (initially) at an angle of 30° with the horizontal. Therefore, if the bottom cylinders move a distance d to the side, then the top cylinder moves a distance $d \tan 30°$ downward. Hence,

$$a_x = \sqrt{3} a_y. \tag{8.112}$$

We now have four equations and four unknowns. Solving for a_y by your method of choice gives

$$a_y = \frac{g}{7 + 6\beta}. \tag{8.113}$$

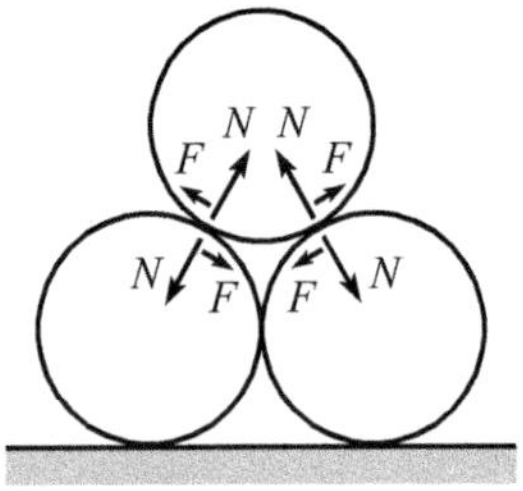

Fig. 8.78

(b) Let N be the normal force between the cylinders, and let F be the friction force between the cylinders, with positive directions shown in Fig. 8.78. Let a_x be the initial horizontal acceleration of the right bottom cylinder, and let a_y be the initial vertical acceleration of the top cylinder, with downward taken to be positive. Let α be the angular acceleration of the right bottom cylinder, with counterclockwise taken to be positive. Note that α is *not* equal to a_x/R, because the bottom cylinders slip on the ground.

If we consider the torque around the center of one of the bottom cylinders, then the only relevant force is F. And from the same reasoning as in part (a), we have $a_x = \sqrt{3}a_y$. Therefore, the four equations analogous to Eqs. (8.111) and (8.112) are

$$\begin{aligned} N\cos 60° - F\sin 60° &= ma_x, \\ mg - 2N\sin 60° - 2F\cos 60° &= ma_y, \\ FR &= (\beta mR^2)\alpha, \\ a_x &= \sqrt{3}a_y. \end{aligned} \tag{8.114}$$

We have five unknowns, N, F, a_x, a_y, and α, so we need one more equation. The tricky part is relating α to a_x. One way to do this is to ignore the y motion of the top cylinder and imagine the bottom right cylinder to be rotating up and around the top cylinder, which is held fixed. In this rotational motion, the center of the bottom cylinder moves at an angle of 30° with respect to the horizontal (at the start). So if it moves an infinitesimal distance d to the right, then its center moves a distance $d/\cos 30°$ up and to the right. So the angle through which the bottom cylinder rotates is $\theta = (d/\cos 30°)/R = (2/\sqrt{3})(d/R)$. Bringing back in the vertical motion of the top cylinder doesn't change this result. Therefore, taking two derivatives of this relation gives

$$\alpha = \frac{2}{\sqrt{3}}\frac{a_x}{R}. \tag{8.115}$$

We now have five equations and five unknowns. Solving for a_y by your method of choice gives

$$a_y = \frac{g}{7 + 8\beta}. \tag{8.116}$$

Remarks: If $\beta = 0$, that is, if all the mass is at the center of the cylinders, then the results in both parts (a) and (b) reduce to $g/7$. This $\beta = 0$ case is equivalent to the case of frictionless cylinders (of any mass distribution), because then nothing rotates. If $\beta \neq 0$, then the result in part (b) is smaller than that in part (a). This isn't so obvious, but the basic reason is that the bottom cylinders in part (b) take up more energy because they have to rotate slightly faster, because $\alpha = (2/\sqrt{3})(a_x/R)$ instead of $\alpha = a_x/R$. ♣

8.17. **Falling chimney**

Let θ be the angle through which the chimney has fallen. Before we start dealing with the forces in the rods, let's first determine $\ddot{\theta}$ as a function of θ. Let ℓ be the height of the chimney. Then the moment of inertia around the pivot point on the ground is $m\ell^2/3$ (if we ignore the width). And the torque (around the pivot point) due to gravity is $\tau = mg(\ell/2)\sin\theta$. Therefore, $\tau = dL/dt$ gives $mg(\ell/2)\sin\theta = (1/3)m\ell^2\ddot{\theta}$, and so

$$\ddot{\theta} = \frac{3g\sin\theta}{2\ell}. \tag{8.117}$$

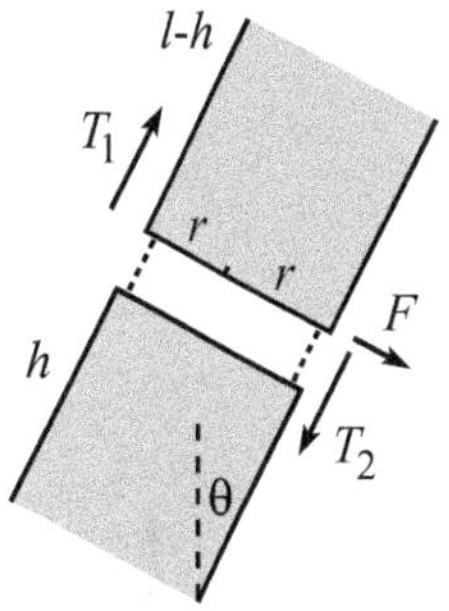

Fig. 8.79

Let's now determine the forces in the rods. Our strategy will be to imagine that the chimney consists of a chimney of height h, with another chimney of height $\ell - h$ placed on top of it. We'll find the forces in the rods connecting these two "sub-chimneys," and then we'll maximize one of these forces (T_2, defined below) as a function of h.

The forces on the top piece are gravity and also the forces from the two rods at each end of the bottom board. Let's break these latter forces up into transverse and longitudinal forces along the chimney. Let T_1 and T_2 be the two longitudinal components, and let F be the sum of the transverse components, as shown in Fig. 8.79. We have picked the positive directions for T_1 and T_2 so that positive T_1 corresponds to compression in the left rod, and positive T_2 corresponds to tension in the right rod (which is what the forces will turn out to be, as we'll see). It turns out that if the width (which we'll call $2r$) is much less than the height, then $T_2 \gg F$ (as we'll see below), so the tension in the right rod is essentially equal to T_2. We will therefore be concerned with maximizing T_2.

In writing down the force and torque equations for the top piece, we have three equations (the radial and tangential $F = ma$ equations, and $\tau = dL/dt$ around the CM), and three unknowns (F, T_1, and T_2). If we define the fraction $f \equiv h/\ell$, then the top piece has length $(1-f)\ell$ and mass $(1-f)m$, and its CM travels in a circle of radius $(1+f)\ell/2$, Therefore, our three force and torque equations are, respectively,

$$\begin{aligned}
T_2 - T_1 + (1-f)mg\cos\theta &= (1-f)m\left(\frac{(1+f)\ell}{2}\right)\dot{\theta}^2,\\
F + (1-f)mg\sin\theta &= (1-f)m\left(\frac{(1+f)\ell}{2}\right)\ddot{\theta},\\
(T_1+T_2)r - F\frac{(1-f)\ell}{2} &= (1-f)m\left(\frac{(1-f)^2\ell^2}{12}\right)\ddot{\theta}.
\end{aligned} \tag{8.118}$$

At this point, we could plow forward and solve this system of three equations in three unknowns. But things simplify greatly in the limit $r \ll \ell$. The third equation says that $T_1 + T_2$ is of order $1/r$, and the first equation says that $T_2 - T_1$ is of order 1. These imply that $T_1 \approx T_2$, to leading order in $1/r$. Therefore, we can set $T_1 + T_2 \approx 2T_2$ in the third equation. Using this approximation, along with the value of $\ddot{\theta}$ from Eq. (8.117), the second and third equations become

$$\begin{aligned}
F + (1-f)mg\sin\theta &= \frac{3}{4}(1-f^2)mg\sin\theta,\\
2rT_2 - F\frac{(1-f)\ell}{2} &= \frac{1}{8}(1-f)^3 mg\ell\sin\theta.
\end{aligned} \tag{8.119}$$

This first of these equations gives

$$F = \frac{mg\sin\theta}{4}(-1+4f-3f^2), \tag{8.120}$$

and then the second gives

$$T_2 \approx \frac{mg\ell\sin\theta}{8r}f(1-f)^2. \tag{8.121}$$

As stated above, this is much greater than F (because $\ell/r \gg 1$), so the tension in the right rod is essentially equal to T_2. Taking the derivative of T_2 with respect to f, we see that it is maximum at

$$f \equiv \frac{h}{\ell} = \frac{1}{3}. \tag{8.122}$$

Therefore, the chimney is most likely to break at a point one-third of the way up (assuming that the width is much less than the height). Interestingly, $f = 1/3$ makes the force F in Eq. (8.120) exactly equal to zero. For more on the falling chimney, see Madsen (1977) and Varieschi and Kamiya (2003).

8.18. **Ball hitting stick**

Let V, v, and ω be the speed of the ball, the speed of the stick's CM, and the angular speed of the stick, respectively, after the collision. Then conservation of momentum, angular momentum (around the fixed point that coincides with the initial center of the stick), and energy give

$$\begin{aligned} MV_0 &= MV + mv, \\ MV_0 d &= MVd + \beta m\ell^2\omega, \\ MV_0^2 &= MV^2 + mv^2 + \beta m\ell^2\omega^2. \end{aligned} \tag{8.123}$$

We must solve these three equations for V, v, and ω. The first two equations quickly give $vd = \beta\ell^2\omega$. Solving for V in the first equation and plugging the result into the third, and then eliminating ω through $vd = \beta\ell^2\omega$ gives

$$v = \frac{2V_0}{1 + \frac{m}{M} + \frac{d^2}{\beta\ell^2}} \quad \Longrightarrow \quad \omega = V_0 \frac{2\frac{d}{\beta\ell^2}V_0}{1 + \frac{m}{M} + \frac{d^2}{\beta\ell^2}}. \tag{8.124}$$

Having found v, the first equation above gives V as

$$V = V_0 \frac{1 - \frac{m}{M} + \frac{d^2}{\beta\ell^2}}{1 + \frac{m}{M} + \frac{d^2}{\beta\ell^2}}. \tag{8.125}$$

You are encouraged to check various limits of these answers. Another solution to Eq. (8.123) is of course $V = V_0$, $v = 0$, and $\omega = 0$. The initial conditions certainly satisfy conservation of p, L, and E with the initial conditions (a fine tautology, indeed). Nowhere in Eq. (8.123) does it say that the ball actually hits the stick.

8.19. **A ball and stick theorem**

As in the solution to Problem 8.18, we have

$$\begin{aligned} MV_0 &= MV + mv, \\ MV_0 d &= MVd + I\omega, \\ MV_0^2 &= MV^2 + mv^2 + I\omega^2. \end{aligned} \tag{8.126}$$

The speed of the contact point on the stick right after the collision equals the speed of the CM plus the rotational speed relative to the CM. In other words, it equals $v + \omega d$. The desired relative speed is therefore $(v + \omega d) - V$. We can determine the value of this relative speed by solving the above three equations for V, v, and ω. Equivalently, we can just use the results of Problem 8.18. There is, however, a much more appealing method, which is the following.

The first two equations quickly give $mvd = I\omega$. The last equation may then be written as, using $I\omega^2 = (I\omega)\omega = (mvd)\omega$,

$$M(V_0 - V)(V_0 + V) = mv(v + \omega d). \tag{8.127}$$

If we now write the first equation as

$$M(V_0 - V) = mv, \tag{8.128}$$

we can divide Eq. (8.127) by Eq. (8.128) to obtain $V_0 + V = v + \omega d$, or

$$V_0 = (v + \omega d) - V, \tag{8.129}$$

as we wanted to show. In terms of velocities, the correct statement is that the final relative velocity is the negative of the initial relative velocity. In other words, $V_0 - 0 = -\big(V - (v + \omega d)\big)$.

8.20. **The superball**

Since we are told that $|v_y|$ is unchanged by the bounce, we can ignore it when applying conservation of energy. And since the vertical impulse from the floor provides no torque around the ball's CM, we can completely ignore the y motion in this problem.

The horizontal impulse from the floor is responsible for changing both v_x and ω. With positive directions defined as in the statement of the problem, Eq. (8.61) gives

$$\begin{aligned} \Delta L &= R\Delta p \\ \Longrightarrow \quad I(\omega' - \omega) &= Rm(v_x' - v_x). \end{aligned} \tag{8.130}$$

And conservation of energy gives

$$\begin{aligned} \frac{1}{2}mv_x'^2 + \frac{1}{2}I\omega'^2 &= \frac{1}{2}mv_x^2 + \frac{1}{2}I\omega^2 \\ \Longrightarrow \quad I(\omega'^2 - \omega^2) &= m(v_x^2 - v_x'^2). \end{aligned} \tag{8.131}$$

Dividing this equation by Eq. (8.130) gives[15]

$$R(\omega' + \omega) = -(v_x' + v_x). \tag{8.132}$$

We can now combine this equation with Eq. (8.130) which can be rewritten as, using $I = (2/5)mR^2$,

$$\frac{2}{5}R(\omega' - \omega) = v_x' - v_x. \tag{8.133}$$

Given v_x and ω, the previous two equations are two linear equations in the two unknowns, v_x' and ω'. Solving for v_x' and ω', and writing the result in matrix notation, gives

$$\begin{pmatrix} v_x' \\ R\omega' \end{pmatrix} = \frac{1}{7}\begin{pmatrix} 3 & -4 \\ -10 & -3 \end{pmatrix}\begin{pmatrix} v_x \\ R\omega \end{pmatrix}, \tag{8.134}$$

as desired. Note that Eq. (8.132), when written in the form of $v_x + R\omega = -(v_x' + R\omega')$, says that the relative velocity of the ball's contact point and the ground simply changes sign during the bounce.

REMARK: For a ball with a general moment of inertia $I = \beta mR^2$, you can use the above procedure to show that the matrix in Eq. (8.134) takes the general form,

$$\frac{1}{1+\beta}\begin{pmatrix} 1-\beta & -2\beta \\ -2 & -(1-\beta) \end{pmatrix}. \tag{8.135}$$

[15] We have divided out the trivial $\omega' = \omega$ and $v_x' = v_x$ solution, which corresponds to slipping motion on a frictionless plane. The nontrivial solution we will find shortly is the nonslipping one. Basically, to conserve energy, there must be no work done by friction. And since work is force times distance, this means that either (1) the plane is frictionless, so that the force is zero, or (2) there is no relative motion between the ball's contact point and the plane, so that the distance is zero. The latter case is the one we are concerned with here.

For $\beta = 1$ (a hoop), this becomes

$$\begin{pmatrix} 0 & -1 \\ -1 & 0 \end{pmatrix}, \tag{8.136}$$

which means that the bounce simply interchanges and negates the values of v_x and $R\omega$. In particular, if you throw a "super-hoop" sideways with no spin (that is, $R\omega = 0$), then it will bounce straight up in the air (that is, $v'_x = 0$) while spinning. ♣

8.21. **Many bounces**

Equation (8.62) gives the result after one bounce, so the result after two bounces is

$$\begin{aligned}\begin{pmatrix} v''_x \\ R\omega'' \end{pmatrix} &= \begin{pmatrix} 3/7 & -4/7 \\ -10/7 & -3/7 \end{pmatrix}\begin{pmatrix} v'_x \\ R\omega' \end{pmatrix} \\ &= \begin{pmatrix} 3/7 & -4/7 \\ -10/7 & -3/7 \end{pmatrix}^2\begin{pmatrix} v_x \\ R\omega \end{pmatrix} \\ &= \begin{pmatrix} 1 & 0 \\ 0 & 1 \end{pmatrix}\begin{pmatrix} v_x \\ R\omega \end{pmatrix} \\ &= \begin{pmatrix} v_x \\ R\omega \end{pmatrix}. \end{aligned} \tag{8.137}$$

The square of the matrix turns out to be the identity. Therefore, after two bounces, both v_x and ω return to their original values. The ball then repeats the motion of the previous two bounces (and so on, after every two bounces). The only difference between successive pairs of bounces is that the ball may shift horizontally. You are strongly encouraged to experimentally verify this interesting periodic behavior.

8.22. **Rolling over a bump**

We will use the fact that the angular momentum of the ball with respect to the corner of the step (call this point P) is unchanged by the collision. This is true because any forces exerted at point P provide zero torque around P. (The torque from gravity will be relevant during the subsequent rising-up motion. But during the instantaneous collision, L does not change.) This fact will allow us to find the energy of the ball right after the collision, which we will then require to be greater than mgh.

Breaking the initial L into the contribution relative to the CM, plus the contribution from the ball treated like a point mass located at the CM, we see that the initial angular momentum is $L = (2/5)mR^2\omega_0 + mV_0(R - h)$, where ω_0 is the initial angular speed. But the nonslipping condition tells us that $\omega_0 = V_0/R$, so we can write L as

$$L = \frac{2}{5}mRV_0 + mV_0(R-h) = mV_0\left(\frac{7R}{5} - h\right). \tag{8.138}$$

Let ω' be the angular speed of the ball around point P immediately after the collision. The parallel-axis theorem says that the moment of inertia around P is equal to $(2/5)mR^2 + mR^2 = (7/5)mR^2$. Conservation of L around P during the collision then gives

$$mV_0\left(\frac{7R}{5} - h\right) = \frac{7}{5}mR^2\omega' \quad \Longrightarrow \quad \omega' = \frac{V_0}{R}\left(1 - \frac{5h}{7R}\right). \tag{8.139}$$

The energy of the ball right after the collision is therefore

$$E = \frac{1}{2}\left(\frac{7}{5}mR^2\right)\omega'^2 = \frac{7}{10}mV_0^2\left(1 - \frac{5h}{7R}\right)^2. \tag{8.140}$$

The ball will climb up over the step if $E \ge mgh$, which gives

$$V_0 \ge \sqrt{\frac{10gh}{7}}\left(1 - \frac{5h}{7R}\right)^{-1}. \tag{8.141}$$

REMARKS: It is possible for the ball to rise up over the step even if $h > R$, provided that the ball sticks to the corner, without slipping. (If $h > R$, the step would have to be "hollowed out" so that the ball doesn't collide with the side of the step.) But note that $V_0 \to \infty$ as $h \to 7R/5$. For $h \geq 7R/5$, it is impossible for the ball to make it up over the step, no matter how large V_0 is. The ball will get pushed down into the ground, instead of rising up, if $h > 7R/5$.

For an object with a general moment of inertia $I = \beta m R^2$ (so $\beta = 2/5$ in our problem), you can show that the minimum initial speed is

$$V_0 \geq \sqrt{\frac{2gh}{1+\beta}}\left(1 - \frac{h}{(1+\beta)R}\right)^{-1}. \tag{8.142}$$

This decreases as β increases. It is smallest when the "ball" is a wheel with all the mass on its rim (so that $\beta = 1$), in which case it is possible for the wheel to climb up over the step even if h is close to $2R$. ♣

8.23. **Falling toast**

Let v_0 and v be the speeds of the CM right before and right after the collision (so we know that $v_0 = \sqrt{2gH}$). The I for a square is the same as the I for a stick, $(1/12)m\ell^2$, so $\Delta L = (\ell/2)\Delta p$ gives[16]

$$\left(\frac{m\ell^2}{12}\right)\omega = -\frac{\ell}{2}(mv - mv_0) \quad \Longrightarrow \quad v_0 - v = \frac{\ell\omega}{6}. \tag{8.143}$$

Conservation of energy during the collision gives

$$\frac{1}{2}mv_0^2 = \frac{1}{2}mv^2 + \frac{1}{2}\left(\frac{m\ell^2}{12}\right)\omega^2 \quad \Longrightarrow \quad (v_0 + v)(v_0 - v) = \frac{\ell^2\omega^2}{12}. \tag{8.144}$$

Dividing this by Eq. (8.143) gives $v_0 + v = \ell\omega/2$. We now have two linear equations for v and ω. Solving gives $v = v_0/2$ and $\omega = 3v_0/\ell$.

After the collision, the time to hit the ground is given by $vt + gt^2/2 = h$. Solving for t, setting $\omega t = \pi$ for half a rotation, and using the v and ω we just found, yields

$$\frac{3v_0}{\ell} \cdot \frac{1}{g}\left(-\frac{v_0}{2} + \sqrt{\frac{v_0^2}{4} + 2gh}\right) = \pi. \tag{8.145}$$

Factoring out a v_0^2 and then using $v_0^2 = 2gH$ gives

$$\frac{3H}{\ell}\left(\sqrt{1 + \frac{4h}{H}} - 1\right) = \pi. \tag{8.146}$$

Isolating the square root, and then squaring and solving for H gives

$$H = \frac{\pi^2\ell^2}{6(6h - \pi\ell)}. \tag{8.147}$$

The special value of ℓ is $\ell = (6/\pi)h$ (which would be a huge piece of toast), in which case $H = \infty$. If ℓ is larger than $(6/\pi)h$, then there isn't enough time to make half a rotation before hitting the floor. The intuitive reason for this is that since $\omega = 6v/\ell$ (from above), there is no way to increase ω without also increasing v. The toast has the best chance of making half a rotation if v is very large, because then gravity doesn't have any time to increase v. In this limit, the toast makes half a rotation upon hitting the floor if $\pi/\omega = h/v$. Plugging in $\omega = 6v/\ell$ gives $6h = \pi\ell$, as desired.

[16] The minus sign on the right-hand side comes from the fact that we're defining the v's to be positive downward. Equivalently, the force from the counter increases the angular speed but decreases the linear speed.

For the reasonable values of $h = 1$ m and $\ell = 10$ cm, we obtain $H \approx 3$ mm, which is quite small. You should try this with a pencil to convince yourself that even this small distance can yield the desired (or rather, undesired) half rotation.

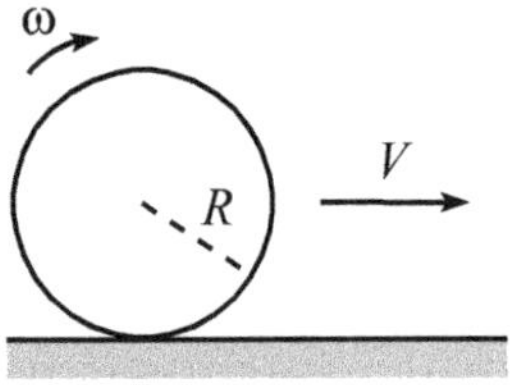

Fig. 8.80

8.24. **Sliding to rolling**

(a) Define all linear quantities to be positive to the right, and all angular quantities to be positive clockwise, as shown in Fig. 8.80. Then, for example, the friction force F_f is negative. The friction force slows down the translational motion and speeds up the rotational motion, according to

$$F_f = ma, \quad \text{and} \quad -F_f R = I\alpha. \tag{8.148}$$

Eliminating F_f, and using $I = \beta m R^2$, gives $a = -\beta R\alpha$. Integrating this over time, up to the time when the ball stops slipping, gives

$$\Delta V = -\beta R \Delta\omega. \tag{8.149}$$

Note that we could have obtained this by simply using the impulse equation, Eq. (8.61). Using $\Delta V = V_f - V_0$, and $\Delta\omega = \omega_f - \omega_0 = \omega_f$, and also $\omega_f = V_f/R$ (the nonslipping condition), Eq. (8.149) gives

$$V_f = \frac{V_0}{1+\beta}, \tag{8.150}$$

independent of the nature of F_f. F_f can depend on position, time, speed, or anything else. The relation $a = -\beta R\alpha$, and hence also Eq. (8.149), will still be true at all times.

Remark: We can also calculate τ and L relative to a dot painted on the ground that is the contact point at a given instant. There is zero torque relative to this point. To find L, we must add the L of the CM and the L relative to the CM. Therefore, $\tau = dL/dt$ gives $0 = (d/dt)(mvR + \beta m R^2\omega)$, and so $a = -\beta R\alpha$, as above. ♣

Using Eq. (8.150), and also the relation $\omega_f = V_f/R$, the loss in kinetic energy is

$$\begin{aligned}\Delta K &= \frac{1}{2}mV_0^2 - \left(\frac{1}{2}mV_f^2 + \frac{1}{2}I\omega_f^2\right)\\ &= \frac{1}{2}mV_0^2\left(1 - \frac{1}{(1+\beta)^2} - \frac{\beta}{(1+\beta)^2}\right)\\ &= \frac{1}{2}mV_0^2\left(\frac{\beta}{1+\beta}\right).\end{aligned} \tag{8.151}$$

For $\beta \to 0$, no energy is lost, which makes sense. And for $\beta \to \infty$ (a spool sliding on its axle), all the energy is lost, which also makes sense, because we essentially have a sliding block which can't rotate.

(b) Let's first find t. The friction force is $F_f = -\mu mg$, so $F = ma$ gives $-\mu g = a$. Therefore, $\Delta V = at = -\mu g t$. But Eq. (8.150) says that $\Delta V \equiv V_f - V_0 = -V_0\beta/(1+\beta)$. Therefore,

$$t = \frac{\beta}{(1+\beta)} \cdot \frac{V_0}{\mu g}. \tag{8.152}$$

For $\beta \to 0$, we have $t \to 0$, which makes sense. And for $\beta \to \infty$, we have $t \to V_0/(\mu g)$, which equals the time a sliding block would take to stop.

Let's now find d. We have $d = V_0 t + (1/2)at^2$. Using $a = -\mu g$, and plugging in the t from Eq. (8.152), we obtain

$$d = \frac{\beta(2+\beta)}{(1+\beta)^2} \cdot \frac{V_0^2}{2\mu g}. \tag{8.153}$$

The two extreme cases for β check here.

To calculate the work done by friction, we might be tempted to write down the product $F_f d$, with $F_f = -\mu mg$ and d given in Eq. (8.153). But the result doesn't equal the loss in kinetic energy calculated in Eq. (8.151). What's wrong with this reasoning? The error is that the friction force does not act over a distance d. To find the distance over which F_f acts, we must find how far the surface of the ball moves relative to the ground. The speed of a dot on the ball that is instantaneously the contact point is $V_{rel}(t) = V(t) - R\omega(t) = (V_0 + at) - R\alpha t$. Using $\alpha = -a/\beta R$ and $a = -\mu g$, this becomes

$$V_{rel}(t) = V_0 - \frac{1+\beta}{\beta}\mu g t. \tag{8.154}$$

Integrating this from $t = 0$ to the t given in Eq. (8.152) gives

$$d_{rel} = \int V_{rel}(t)\, dt = \frac{\beta}{1+\beta} \cdot \frac{V_0^2}{2\mu g}. \tag{8.155}$$

The work done by friction is $F_f d_{rel} = -\mu mg d_{rel}$, which does indeed give the loss in kinetic energy given in Eq. (8.151).

8.25. **Lots of sticks**

Consider the collision between two sticks. Let V be the speed of the contact point on the heavy one. Since this stick is essentially infinitely heavy, we may consider it to be an infinitely heavy ball, moving at speed V. The rotational degree of freedom of the heavy stick is irrelevant, as far as the light stick is concerned. We can therefore invoke the result of Problem 8.19 to say that the relative speed of the contact points is the same before and after the collision. This implies that the contact point on the light stick picks up a speed of $2V$, because the heavy stick is essentially unaffected by the collision and keeps moving at speed V.

Let us now find the speed of the other end of the light stick. This stick receives an impulse from the heavy stick, so we can apply Eq. (8.61) to the light stick to obtain

$$\Delta L = r\Delta p \quad \Longrightarrow \quad \beta m r^2 \omega = r(m v_{CM}) \quad \Longrightarrow \quad r\omega = \frac{v_{CM}}{\beta}. \tag{8.156}$$

The speed of the struck (top) end is $v_{top} = r\omega + v_{CM}$, because the CM speed adds to the rotational speed. The speed of the other (bottom) end is $v_{bot} = r\omega - v_{CM}$, because the CM speed subtracts from the rotational speed.[17] The ratio of these speeds is

$$\frac{v_{bot}}{v_{top}} = \frac{\frac{v_{CM}}{\beta} - v_{CM}}{\frac{v_{CM}}{\beta} + v_{CM}} = \frac{1-\beta}{1+\beta}. \tag{8.157}$$

This is a general result whenever you strike the end of a stick with any force. In the present problem, we have $v_{top} = 2V$. Therefore,

$$v_{bot} = V\left(\frac{2(1-\beta)}{1+\beta}\right). \tag{8.158}$$

[17] Since $\beta \le 1$ for any real stick, we have $r\omega = v_{CM}/\beta \ge v_{CM}$. Therefore, $r\omega - v_{CM}$ is greater than or equal to zero.

The same analysis holds for all the other collisions. Therefore, the bottom ends of the sticks move with speeds that form a geometric progression with ratio $2(1-\beta)/(1+\beta)$. If this ratio is less than 1 (that is, if $\beta > 1/3$), then the speeds go to zero as $n \to \infty$. If it is greater than 1 (that is, if $\beta < 1/3$), then the speeds go to infinity as $n \to \infty$. If it equals 1 (that is, if $\beta = 1/3$), then the speeds remain equal to V and are thus independent of n, as we wanted to show. A uniform stick has $\beta = 1/3$ relative to its center (which is usually written in the form $I = m\ell^2/12$, where $\ell = 2r$).

Chapter 9
Angular momentum, Part II (General $\hat{\mathbf{L}}$)

In Chapter 8, we discussed situations where the direction of the vector $\mathbf{L}$ remained constant, and only its magnitude changed. In this chapter, we will look at more general situations where the direction of $\mathbf{L}$ is allowed to change. The vector nature of $\mathbf{L}$ will prove to be vital here, and we will arrive at all sorts of strange results for spinning tops and such things. This chapter is rather long, alas, but the general outline is that the first three sections cover general theory, then Section 9.4 introduces some actual physical setups, and then Section 9.6 begins the discussion of tops.

9.1 Preliminaries concerning rotations

9.1.1 The form of general motion

Before getting started, we should make sure we're all on the same page concerning a few important things about rotations. Because rotations generally involve three dimensions, they can often be hard to visualize. A rough drawing on a piece of paper might not do the trick. For this reason, this chapter is one of the more difficult ones in this book. But to ease into it, the next few pages consist of some definitions and helpful theorems. This first theorem describes the general form of any motion. You might consider it obvious, but it's a little tricky to prove.

Theorem 9.1 *(Chasles' theorem) Consider a rigid body undergoing arbitrary motion. Pick any point P in the body. Then at any instant (see Fig. 9.1), the motion of the body can be written as the sum of the translational motion of P, plus a rotation around some axis (which may change with time) through P.*[1]

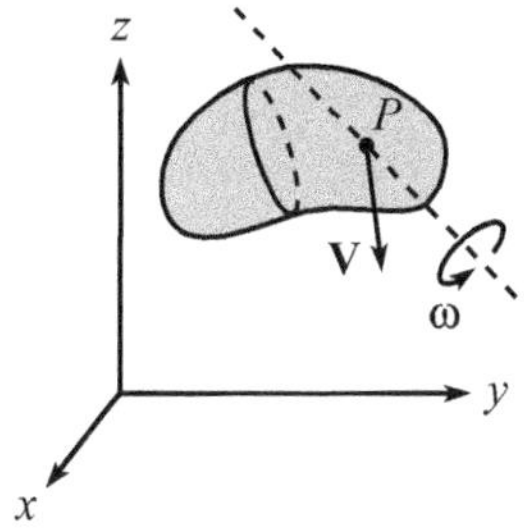

Fig. 9.1

Proof: The motion of the body can be written as the sum of the translational motion of P, plus some other motion relative to P (this is true because relative coordinates are additive quantities). We must show that this latter motion is a rotation. This seems quite plausible, and it holds because the body is rigid; that is,

[1] In other words, a person at rest with respect to a frame whose origin is P, and whose axes are parallel to the fixed-frame axes, sees the body undergoing a rotation around some axis through P.

all points keep the same distances relative to each other. If the body weren't rigid, then this theorem wouldn't be true.

To be rigorous, consider a spherical shell fixed in the body, centered at P. The motion of the body is completely determined by the motion of the points on this sphere, so we need only examine what happens to the sphere. Because distances are preserved in the rigid body, the points on the sphere must always remain the same radial distance from P. And because we are looking at motion relative to P, we have therefore reduced the problem to the following: In what manner can a rigid sphere transform into itself? We claim that any such transformation has the property that there exist two points that end up where they started.[2] These two points must then be diametrically opposite points (assuming that the whole sphere doesn't end up back where it started, in which case every point ends up where it started), because distances are preserved; given one point that ends up where it started, the diametrically opposite point must also end up where it started, to maintain the distance of a diameter.

If this claim is true, then we are done, because for an infinitesimal transformation, a given point moves in only one direction, because there is no time to do any turning. So a point that ends up where it started must have remained fixed for the whole (infinitesimal) time. Therefore, all the points on the diameter joining the two fixed points must also have remained fixed the whole time, because distances are preserved. So we are left with a rotation around this axis.

This "two points ending up where they started" claim is quite believable, but nevertheless tricky to prove. Claims with these properties are always fun to think about, so I've left this one as a problem (Problem 9.2). Try to solve it on your own. ■

We will invoke this theorem repeatedly in this chapter (often without bothering to say so). Note that we are assuming that P is a point in the body, because we used the fact that P keeps the same distances from other points in the body.

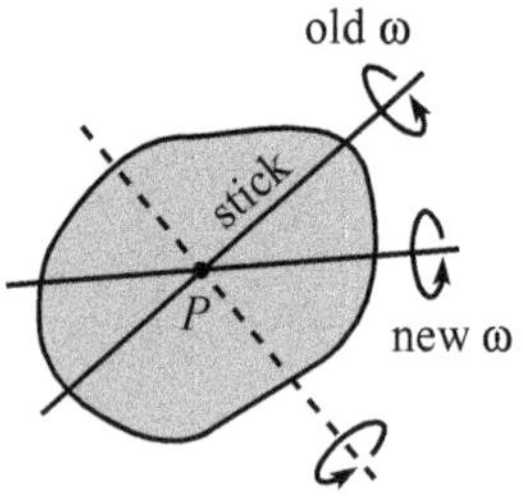

Fig. 9.2

REMARK: A situation where this theorem isn't so obvious is the following (this setup contains only rotation, with no translation of the point P). Consider an object rotating around a fixed axis, the stick shown in Fig. 9.2. But now imagine grabbing the stick and rotating it around some other axis (the dotted line shown). It isn't immediately obvious that the resulting motion is (instantaneously) a rotation around some new axis through the point P (which remains fixed). But indeed it is. We'll be quantitative about this in the "Rotating sphere" example later in this section. ♣

9.1.2 The angular velocity vector

It is extremely useful to introduce the angular velocity vector, $\boldsymbol{\omega}$, which is defined as the vector that points along the axis of rotation, and whose magnitude equals

[2] This claim is actually true for *any* transformation of a rigid sphere into itself, but for the present purposes we are concerned only with infinitesimal transformations, because we are looking only at what happens at a given instant in time.

the angular speed. The choice of the two possible directions along the axis is given by the right-hand rule: if you curl your right-hand fingers in the direction of the spin, then your thumb points in the direction of $\boldsymbol{\omega}$. For example, a spinning record has $\boldsymbol{\omega}$ perpendicular to the record, through the center (as shown in Fig. 9.3),[3] with its magnitude equal to the angular speed, ω. The points on the axis of rotation are the ones that (instantaneously) do not move. Of course, the direction of $\boldsymbol{\omega}$ may change over time, so the points that were formerly on the axis may now be moving.

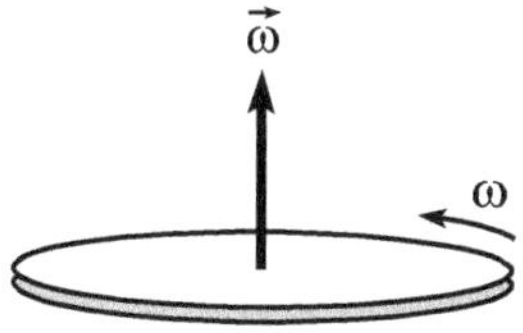

Fig. 9.3

REMARKS:

1. If you want, you can break the mold and use the left-hand rule to determine $\boldsymbol{\omega}$, as long as you use it consistently. The direction of $\boldsymbol{\omega}$ will be the opposite, but that doesn't matter, because $\boldsymbol{\omega}$ isn't really physical. Any physical result (for example, the velocity of a particle, given below in Theorem 9.2) will come out the same, independent of which hand you (consistently) use.

   When studying vectors in school,
   You'll use your right hand as a tool.
   But look in a mirror,
   And then you'll see clearer,
   It's just like the left-handed rule.

2. The fact that we can specify a rotation by specifying a vector $\boldsymbol{\omega}$ is a peculiarity to three dimensions. If we lived in one dimension, then there would be no such thing as a rotation. If we lived in two dimensions, then all rotations would take place in that plane, so we could label a rotation by simply giving its speed, ω. In three dimensions, rotations take place in $\binom{3}{2} = 3$ independent planes. And we choose to label these, for convenience, by the directions orthogonal to these planes, and by the angular speed in each plane. If we lived in four dimensions, then rotations could take place in $\binom{4}{2} = 6$ planes, so we would have to label a rotation by giving 6 planes and 6 angular speeds. Note that a vector, which has four components in four dimensions, would not do the trick. ♣

In addition to specifying the points that are instantaneously motionless, $\boldsymbol{\omega}$ also easily produces the velocity of any point in the rotating object. Consider the situation where the axis of rotation passes through the origin, which we'll generally assume to be the case in this chapter, unless otherwise stated. Then we have the following theorem.

Theorem 9.2 *Given an object rotating with angular velocity $\boldsymbol{\omega}$, the velocity of a point at position* $\mathbf{r}$ *is given by*

$$\mathbf{v} = \boldsymbol{\omega} \times \mathbf{r}. \tag{9.1}$$

Proof: Drop a perpendicular from the point in question (call it P) to the axis $\boldsymbol{\omega}$. Let Q be the foot of the perpendicular, and let $\mathbf{r}'$ be the vector from Q to P

[3] It's actually meaningless to say that $\boldsymbol{\omega}$ passes through the center of the record, because you can draw the vector anywhere, and it's still the same vector, as long as it has the correct magnitude and direction. Nevertheless, it's customary to draw $\boldsymbol{\omega}$ along the axis of rotation and to say things like, "An object rotates around $\boldsymbol{\omega}$..."

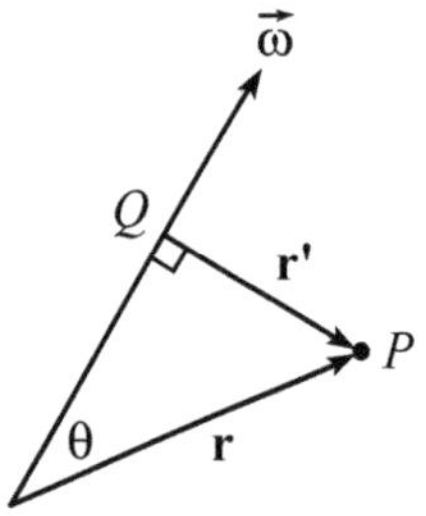

Fig. 9.4

(see Fig. 9.4). From the properties of the cross product (see Appendix B), $\mathbf{v} = \boldsymbol{\omega} \times \mathbf{r}$ is orthogonal to $\boldsymbol{\omega}$, $\mathbf{r}$, and also $\mathbf{r}'$ because $\mathbf{r}'$ is a linear combination of $\boldsymbol{\omega}$ and $\mathbf{r}$. Therefore, the direction of $\mathbf{v}$ is correct; it is always orthogonal to $\boldsymbol{\omega}$ and $\mathbf{r}'$, so it describes circular motion around the axis $\boldsymbol{\omega}$. Also, by the right-hand rule in the cross product (or the left-hand rule, if you had chosen to be different and defined $\boldsymbol{\omega}$ that way), $\mathbf{v}$ has the proper orientation around $\boldsymbol{\omega}$, namely into the page at the instant shown. And since

$$|\mathbf{v}| = |\boldsymbol{\omega}||\mathbf{r}|\sin\theta = \omega r', \tag{9.2}$$

we see that $\mathbf{v}$ has the correct magnitude, because $\omega r'$ is the speed of the circular motion around $\boldsymbol{\omega}$. So $\mathbf{v}$ is indeed the correct velocity vector. (If we have the special case where P lies along $\boldsymbol{\omega}$, then $\mathbf{r}$ is parallel to $\boldsymbol{\omega}$, so the cross product gives a zero result for $\mathbf{v}$, as it should.) ■

We'll make good use of Eq. (9.1) and apply it repeatedly throughout this chapter. Even if it's hard to visualize what's going on in a given rotation, all you have to do to find the speed of any point is calculate the cross product $\boldsymbol{\omega} \times \mathbf{r}$. Conversely, if the speed of every point in a body is given by $\mathbf{v} = \boldsymbol{\omega} \times \mathbf{r}$, then the body must be undergoing a rotation with angular velocity $\boldsymbol{\omega}$, because all points on the axis $\boldsymbol{\omega}$ are motionless, and all other points move with the proper speed for this rotation.

A very nice thing about angular velocities is that they simply add. Stated more precisely:

Theorem 9.3 *Let coordinate systems S_1, S_2, and S_3 have a common origin. Let S_1 rotate with angular velocity $\boldsymbol{\omega}_{1,2}$ with respect to S_2, and let S_2 rotate with angular velocity $\boldsymbol{\omega}_{2,3}$ with respect to S_3. Then S_1 rotates (instantaneously) with angular velocity*

$$\boldsymbol{\omega}_{1,3} = \boldsymbol{\omega}_{1,2} + \boldsymbol{\omega}_{2,3} \tag{9.3}$$

with respect to S_3.

Proof: If $\boldsymbol{\omega}_{1,2}$ and $\boldsymbol{\omega}_{2,3}$ point in the same direction, then the theorem is clear; the angular speeds just add. If, however, they don't point in the same direction, then things are a bit harder to visualize. But we can prove the theorem by making abundant use of the definition of $\boldsymbol{\omega}$.

Pick a point P_1 at rest in S_1. Let $\mathbf{r}$ be the vector from the origin to P_1. The velocity of P_1 (relative to a very close point P_2 at rest in S_2) due to the rotation of S_1 around $\boldsymbol{\omega}_{1,2}$ is $\mathbf{V}_{P_1P_2} = \boldsymbol{\omega}_{1,2} \times \mathbf{r}$. The velocity of P_2 (relative to a very close point P_3 at rest in S_3) due to the rotation of S_2 around $\boldsymbol{\omega}_{2,3}$ is $\mathbf{V}_{P_2P_3} = \boldsymbol{\omega}_{2,3} \times \mathbf{r}$, because P_2 is also located essentially at position $\mathbf{r}$. Therefore, the velocity of P_1 relative to P_3 is $\mathbf{V}_{P_1P_2} + \mathbf{V}_{P_2P_3} = (\boldsymbol{\omega}_{1,2} + \boldsymbol{\omega}_{2,3}) \times \mathbf{r}$. This holds for any point P_1 at rest in S_1, so the frame S_1 rotates with angular velocity $(\boldsymbol{\omega}_{1,2} + \boldsymbol{\omega}_{2,3})$ with respect to S_3. We see that the proof basically comes down to the facts that (1) the linear velocities just add, as usual, and (2) the angular velocities differ from the linear velocities by a cross product with $\mathbf{r}$. ■

If $\boldsymbol{\omega}_{1,2}$ is constant in S_2, then the vector $\boldsymbol{\omega}_{1,3} = \boldsymbol{\omega}_{1,2} + \boldsymbol{\omega}_{2,3}$ will change with respect to S_3 as time goes by, because $\boldsymbol{\omega}_{1,2}$, which is fixed in S_2, is changing with respect to S_3 (assuming that $\boldsymbol{\omega}_{1,2}$ and $\boldsymbol{\omega}_{2,3}$ aren't parallel). But at any instant, $\boldsymbol{\omega}_{1,3}$ may be obtained by adding the present values of $\boldsymbol{\omega}_{1,2}$ and $\boldsymbol{\omega}_{2,3}$. Consider the following example.

Example (Rotating sphere): A sphere rotates with angular speed ω_3 around a stick that initially points in the $\hat{\mathbf{z}}$ direction. You grab the stick and rotate it around the $\hat{\mathbf{y}}$ axis with angular speed ω_2. What is the angular velocity of the sphere, with respect to the lab frame, as time goes by?

Solution: In the language of Theorem 9.3, the sphere defines the S_1 frame; the stick and the $\hat{\mathbf{y}}$ axis define the S_2 frame; and the lab frame is the S_3 frame. The instant after you grab the stick, we are given that $\boldsymbol{\omega}_{1,2} = \omega_3\hat{\mathbf{z}}$, and $\boldsymbol{\omega}_{2,3} = \omega_2\hat{\mathbf{y}}$. Therefore, the angular velocity of the sphere with respect to the lab frame is $\boldsymbol{\omega}_{1,3} = \boldsymbol{\omega}_{1,2} + \boldsymbol{\omega}_{2,3} = \omega_3\hat{\mathbf{z}} + \omega_2\hat{\mathbf{y}}$, as shown in Fig. 9.5. Convince yourself that the combination of these two rotations yields zero motion for the points along the line of $\boldsymbol{\omega}_{1,3}$. As time goes by, the stick (and hence $\boldsymbol{\omega}_{1,2}$) rotates around the $\mathbf{y}$ axis, so $\boldsymbol{\omega}_{1,3} = \boldsymbol{\omega}_{1,2} + \boldsymbol{\omega}_{2,3}$ traces out a cone around the $\mathbf{y}$ axis, as shown.

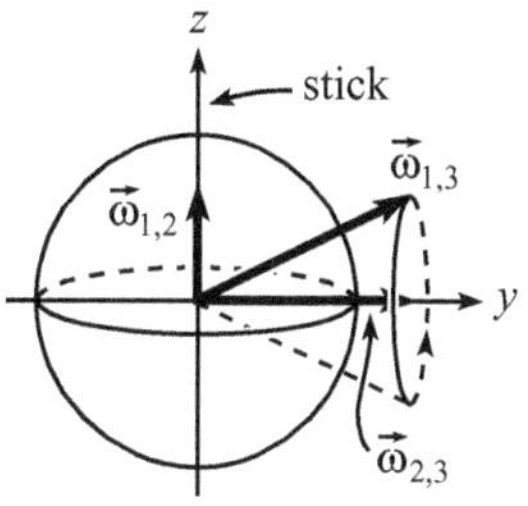

Fig. 9.5

REMARK: Note the different behavior of $\boldsymbol{\omega}_{1,3}$ for a slightly different statement of the problem: Let the sphere initially rotate with angular velocity $\omega_2\hat{\mathbf{y}}$ around a stick, and then grab the stick and rotate it with angular velocity $\omega_3\hat{\mathbf{z}}$. For this situation, $\boldsymbol{\omega}_{1,3}$ initially points in the same direction as in the original statement of the problem (it initially equals $\omega_2\hat{\mathbf{y}} + \omega_3\hat{\mathbf{z}}$). But as time goes by, it is now the horizontal component (defined by the stick) of $\boldsymbol{\omega}_{1,3}$ that changes, so $\boldsymbol{\omega}_{1,3} = \boldsymbol{\omega}_{1,2} + \boldsymbol{\omega}_{2,3}$ traces out a cone around the $\mathbf{z}$ axis, as shown in Fig. 9.6. ♣

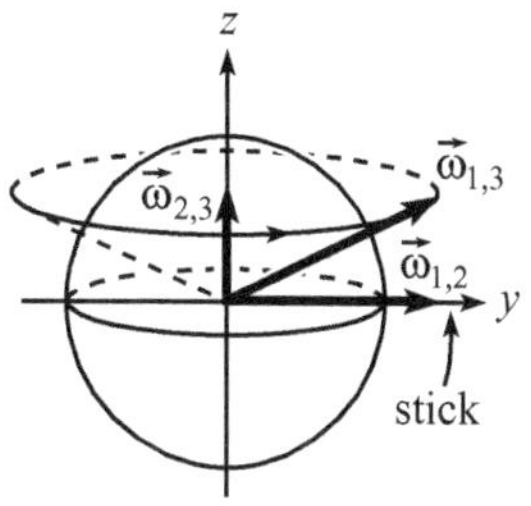

Fig. 9.6

An important point concerning rotations is that they are defined with respect to a *coordinate system*. It makes no sense to ask how fast an object is rotating with respect to a certain point, or even a certain axis. Consider, for example, an object rotating with angular velocity $\boldsymbol{\omega} = \omega_3\hat{\mathbf{z}}$ with respect to the lab frame. Saying only, "The object has angular velocity $\boldsymbol{\omega} = \omega_3\hat{\mathbf{z}}$," is not sufficient, because someone standing in the frame of the object would measure $\boldsymbol{\omega} = 0$, and would therefore be very confused by your statement. Throughout this chapter, we'll try to remember to state the coordinate system with respect to which $\boldsymbol{\omega}$ is measured. But if we forget, the default frame is the lab frame.

This section was definitely a bit abstract, so don't worry too much about it at the moment. The best strategy is perhaps to read on, and then come back for a second pass after digesting a few more sections. At any rate, we'll be discussing many other aspects (probably more than you'd ever want to know) of $\boldsymbol{\omega}$ in Section 9.7.2, so you're assured of getting a lot more practice with it. For now, if you want to strain some brain cells thinking about $\boldsymbol{\omega}$ vectors, you are encouraged to solve Problem 9.3, and also to look at the three given solutions.

9.2 The inertia tensor

Given an object undergoing general motion, the *inertia tensor* is what relates the angular momentum, $\mathbf{L}$, to the angular velocity, $\boldsymbol{\omega}$. This tensor (which is just a fancy name for "matrix" in this context) depends on the geometry of the object, as we'll see. In finding the $\mathbf{L}$ due to general motion, we'll follow the strategy of Section 8.1. We'll first look at the special case of rotation around an axis through the origin, then we'll look at the most general possible motion.

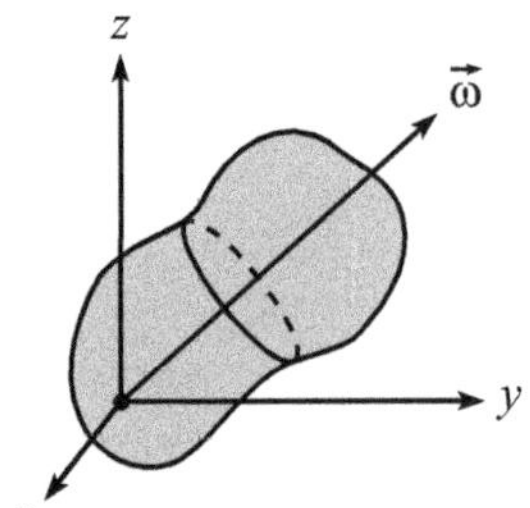

Fig. 9.7

9.2.1 Rotation around an axis through the origin

The three-dimensional object in Fig. 9.7 rotates with angular velocity $\boldsymbol{\omega}$. Consider a little piece of the body, with mass dm and position $\mathbf{r}$. The velocity of this piece is $\mathbf{v} = \boldsymbol{\omega}\times\mathbf{r}$, so its angular momentum (relative to the origin) is $\mathbf{r}\times\mathbf{p} = (dm)\mathbf{r}\times\mathbf{v} = (dm)\mathbf{r}\times(\boldsymbol{\omega}\times\mathbf{r})$. The angular momentum of the entire body is therefore

$$\mathbf{L} = \int \mathbf{r}\times(\boldsymbol{\omega}\times\mathbf{r})\,dm, \tag{9.4}$$

where the integration runs over the volume of the body. In the case where the rigid body is made up of a collection of point masses m_i, the angular momentum is

$$\mathbf{L} = \sum_i m_i\mathbf{r}_i\times(\boldsymbol{\omega}\times\mathbf{r}_i). \tag{9.5}$$

The double cross product in Eqs. (9.4) and (9.5) looks a bit intimidating, but it's actually not so bad. First, we have

$$\begin{aligned}\boldsymbol{\omega}\times\mathbf{r} &= \begin{vmatrix}\hat{\mathbf{x}} & \hat{\mathbf{y}} & \hat{\mathbf{z}}\\ \omega_1 & \omega_2 & \omega_3\\ x & y & z\end{vmatrix}\\ &= (\omega_2 z-\omega_3 y)\hat{\mathbf{x}} + (\omega_3 x-\omega_1 z)\hat{\mathbf{y}} + (\omega_1 y-\omega_2 x)\hat{\mathbf{z}}.\end{aligned} \tag{9.6}$$

We're using the notation ω_1 instead of ω_x, etc., because there are already enough x, y, z's floating around here. The double cross product is then

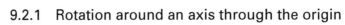

$$\begin{aligned}\mathbf{r}\times(\boldsymbol{\omega}\times\mathbf{r}) &= \begin{vmatrix}\hat{\mathbf{x}} & \hat{\mathbf{y}} & \hat{\mathbf{z}}\\ x & y & z\\ (\omega_2 z-\omega_3 y) & (\omega_3 x-\omega_1 z) & (\omega_1 y-\omega_2 x)\end{vmatrix}\\ &= \Big(\omega_1(y^2+z^2)-\omega_2 xy-\omega_3 zx\Big)\hat{\mathbf{x}}\end{aligned}$$

$$+\Big(\omega_2(z^2+x^2)-\omega_3 yz-\omega_1 xy\Big)\hat{\mathbf{y}}$$
$$+\Big(\omega_3(x^2+y^2)-\omega_1 zx-\omega_2 yz\Big)\hat{\mathbf{z}}. \tag{9.7}$$

The angular momentum in Eq. (9.4) may therefore be written in the concise matrix form,

$$\begin{pmatrix} L_1 \\ L_2 \\ L_3 \end{pmatrix} = \begin{pmatrix} \int(y^2+z^2) & -\int xy & -\int zx \\ -\int xy & \int(z^2+x^2) & -\int yz \\ -\int zx & -\int yz & \int(x^2+y^2) \end{pmatrix} \begin{pmatrix} \omega_1 \\ \omega_2 \\ \omega_3 \end{pmatrix}$$
$$\equiv \begin{pmatrix} I_{xx} & I_{xy} & I_{xz} \\ I_{yx} & I_{yy} & I_{yz} \\ I_{zx} & I_{zy} & I_{zz} \end{pmatrix} \begin{pmatrix} \omega_1 \\ \omega_2 \\ \omega_3 \end{pmatrix}$$
$$\equiv \mathbf{I}\boldsymbol{\omega}. \tag{9.8}$$

For the sake of clarity, we have not bothered to write the *dm* part of each integral (and we'll continue to drop it for most of the remainder of this section). The matrix **I** is called the *inertia tensor*. If the word "tensor" scares you, just ignore it. **I** is simply a matrix. It acts on a vector (the angular velocity) and produces another vector (the angular momentum).

Example (Cube with origin at corner): Calculate the inertia tensor for a solid cube of mass M and side length L, with the coordinate axes parallel to the edges of the cube, and the origin at a corner (see Fig. 9.8).

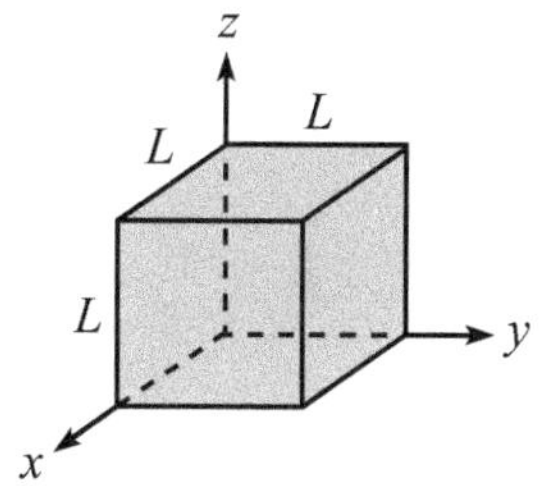

Fig. 9.8

Solution: Due to the symmetry of the cube, there are only two integrals we need to calculate in Eq. (9.8). The diagonal entries are all equal to $\int(y^2+z^2)\,dm$, and the off-diagonal entries are all equal to $-\int xy\,dm$. With $dm = \rho\,dx\,dy\,dz$, and $\rho = M/L^3$, these two integrals are

$$\int_0^L\int_0^L\int_0^L (y^2+z^2)\rho\,dx\,dy\,dz = \rho L^2\int_0^L y^2\,dy + \rho L^2\int_0^L z^2\,dz = \frac{2}{3}ML^2,$$
$$-\int_0^L\int_0^L\int_0^L xy\rho\,dx\,dy\,dz = -\rho L\int_0^L x\,dx\int_0^L y\,dy = -\frac{ML^2}{4}. \tag{9.9}$$

Therefore,

$$\mathbf{I} = ML^2\begin{pmatrix} 2/3 & -1/4 & -1/4 \\ -1/4 & 2/3 & -1/4 \\ -1/4 & -1/4 & 2/3 \end{pmatrix}. \tag{9.10}$$

Having found **I**, we can calculate the angular momentum associated with any given angular velocity. If, for example, the cube is rotating around the z axis with angular speed ω, then we can apply the matrix **I** to the vector $(0, 0, \omega)$ to find that the angular momentum is $\mathbf{L} = ML^2\omega(-1/4, -1/4, 2/3)$. Note the somewhat odd fact that L_x and L_y are nonzero, even though the rotation is only around the z axis. We'll discuss this issue after the following remarks.

REMARKS:

1. The inertia tensor in Eq. (9.8) is a rather formidable-looking object. You will therefore be very pleased to hear that you rarely have to use it. It's nice to know that it's there if you need it, but the concept of *principal axes* (discussed in Section 9.3) provides a way to avoid using the inertia tensor (or more precisely, to greatly simplify it) and is therefore much more useful in solving problems.
2. $\mathbf{I}$ is a symmetric matrix, which is a fact that will be important in Section 9.3. There are therefore only six independent entries, instead of nine.
3. In the case where the rigid body is made up of a collection of point masses m_i, the entries in the matrix are just sums. For example, the upper left entry is $\sum m_i(y_i^2 + z_i^2)$.
4. $\mathbf{I}$ depends only on the geometry of the object, and not on $\boldsymbol{\omega}$.
5. To construct an $\mathbf{I}$, you not only need to specify the origin, you also need to specify the x, y, z axes of your coordinate system. And the basis vectors must be orthogonal, because the cross product calculation above is valid only for an orthonormal basis. If someone else comes along and chooses a different orthonormal basis (but the same origin), then her $\mathbf{I}$ will have different *entries*, as will her $\boldsymbol{\omega}$, as will her $\mathbf{L}$. But her $\boldsymbol{\omega}$ and $\mathbf{L}$ will be exactly the same *vectors* as your $\boldsymbol{\omega}$ and $\mathbf{L}$. They will appear different only because they are written in a different coordinate system. A vector is what it is, independent of how you choose to look at it. If you each point your arm in the direction of what you calculate $\mathbf{L}$ to be, then you will both be pointing in the same direction.
6. For the case of a pancake object rotating in the x-y plane, we have $z = 0$ for all points in the object. And $\boldsymbol{\omega} = \omega_3\hat{\mathbf{z}}$, so $\omega_1 = \omega_2 = 0$. The only nonzero term in the $\mathbf{L}$ in Eq. (9.8) is therefore $L_3 = \int(x^2 + y^2)\,dm\,\omega_3$, which is simply the $L_z = I_z\omega$ result we found in Eq. (8.5). ♣

This is all perfectly fine. Given any rigid body, we can calculate $\mathbf{I}$ (relative to a given origin, using a given set of axes). And given $\boldsymbol{\omega}$, we can then apply $\mathbf{I}$ to it to find $\mathbf{L}$. But what do these entries in $\mathbf{I}$ really mean? How do we interpret them? Note, for example, that ω_3 appears not only in L_3 in Eq. (9.8), but also in L_1 and L_2. But ω_3 is relevant to rotations around the z axis, so what in the world is it doing in L_1 and L_2? Consider the following examples.

Example 1 (Point mass in the x-y plane): Consider a point mass m traveling in a circle of radius r (centered at the origin) in the x-y plane, with frequency ω_3, as shown in Fig. 9.9. Using $\boldsymbol{\omega} = (0, 0, \omega_3)$, $x^2 + y^2 = r^2$, and $z = 0$ in Eq. (9.8) (with a discrete sum of only one object, instead of the integrals), the angular momentum with respect to the origin is

$$\mathbf{L} = (0, 0, mr^2\omega_3). \tag{9.11}$$

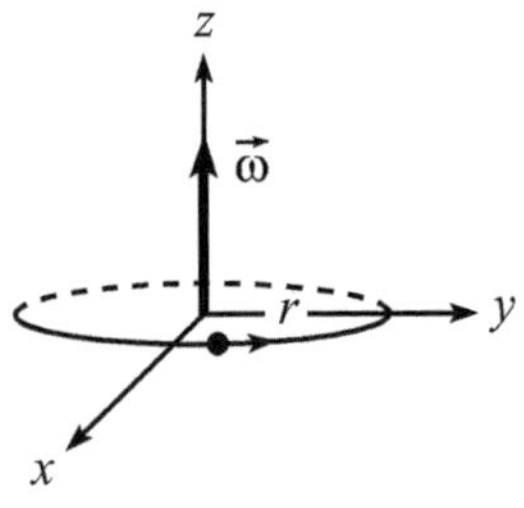

Fig. 9.9

The z component is $mr(r\omega_3) = mrv$, as it should be. And the x and y components are zero, as they should be. This case where $\omega_1 = \omega_2 = 0$ and $z = 0$ is simply the case we studied in Chapter 8, as mentioned in Remark 6 above.

Example 2 (Point mass in space): Consider a point mass m traveling in a circle of radius r, with frequency ω_3. But now let the circle be centered at the point $(0, 0, z_0)$, with the plane of the circle parallel to the x-y plane, as shown in Fig. 9.10. Using $\boldsymbol{\omega} = (0, 0, \omega_3)$, $x^2 + y^2 = r^2$, and $z = z_0$ in Eq. (9.8), the angular momentum with respect to the origin is

$$\mathbf{L} = m\omega_3(-xz_0,\ -yz_0,\ r^2). \tag{9.12}$$

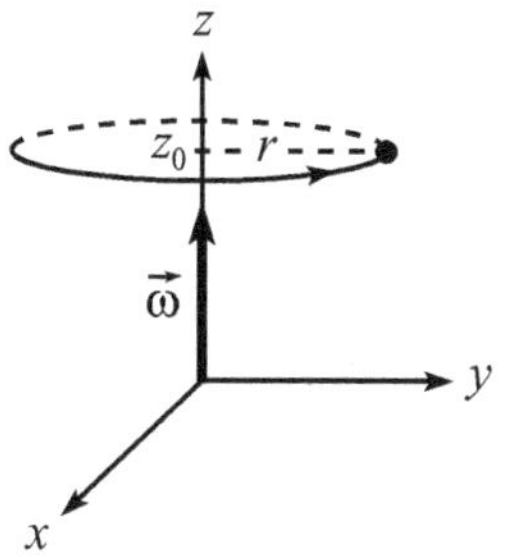

Fig. 9.10

The z component is mrv, as it should be. But surprisingly, we have nonzero L_1 and L_2, even though the mass is just rotating around the z axis. $\mathbf{L}$ does *not* point along $\boldsymbol{\omega}$ here. What's going on?

Consider an instant when the mass is in the y-z plane, as shown in Fig. 9.10. The velocity of the mass is then in the $-\hat{\mathbf{x}}$ direction. Therefore, the particle most certainly has angular momentum around the y axis, as well as the z axis. Someone looking at a split-second movie of the mass at this point can't tell whether it's rotating around the y axis, the z axis, or undergoing some complicated motion. But the past and future motion is irrelevant; at any instant in time, as far as the angular momentum goes, we are concerned only with what is happening at this instant.

At this instant, the angular momentum around the y axis is $L_2 = -mz_0v$, because z_0 is the distance from the y axis, and the minus sign comes from the right-hand rule. Using $v = \omega_3 r = \omega_3 y$, we have $L_2 = -mz_0\omega_3 y$, in agreement with Eq. (9.12). Also, at this instant, L_1 is zero, because the velocity is parallel to the x axis. This agrees with Eq. (9.12), since $x = 0$. As an exercise, you can check that Eq. (9.12) is also correct when the mass is at a general point (x, y, z_0).

We see that, for example, the $I_{yz} \equiv -\int yz$ entry in $\mathbf{I}$ tells us how much the ω_3 component of the angular velocity contributes to the L_2 component of the angular momentum. And due to the symmetry of $\mathbf{I}$, the $I_{yz} = I_{zy}$ entry in $\mathbf{I}$ also tells us how much the ω_2 component of the angular velocity contributes to the L_3 component of the angular momentum. In the former case, if we group the product of the various quantities as $-\int(\omega_3 y)z$, we see that this is simply the appropriate component of the velocity times the distance from the y axis. In the latter case with $-\int(\omega_2 z)y$, it is the opposite grouping. But in both cases there is one factor of y and one factor of z, hence the symmetry in $\mathbf{I}$.

REMARK: For a point mass, $\mathbf{L}$ is actually more easily obtained by just calculating $\mathbf{L} = \mathbf{r} \times \mathbf{p}$. The result for the instant shown in Fig. 9.10 is drawn in Fig. 9.11, where it is clear that $\mathbf{L}$ has both y and z components, and thus also clear that $\mathbf{L}$ doesn't point along $\boldsymbol{\omega}$. For a more complicated object, the tensor $\mathbf{I}$ is generally used, because it is necessary to perform the integral of the $\mathbf{L} = \mathbf{r} \times \mathbf{p}$ contributions over the entire object, and the tensor has this integral built into it. At any rate, whatever method you use, you will find that except in special circumstances (see Section 9.3), $\mathbf{L}$ doesn't point along $\boldsymbol{\omega}$.

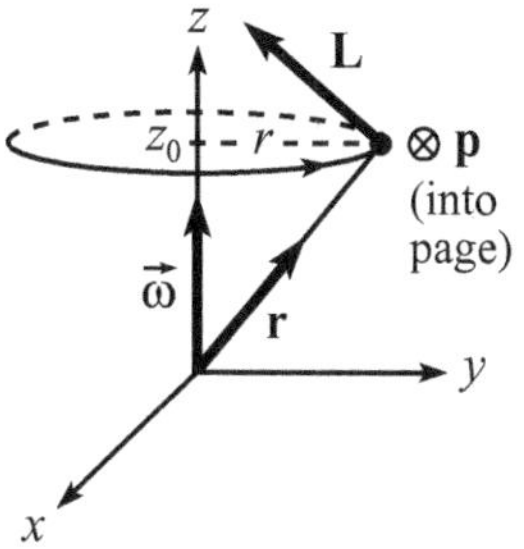

Fig. 9.11

Consider the vector of $\mathbf{L}$,
And that of $\boldsymbol{\omega}$ as well.
The erroneous claim
That they must aim the same
Is a view that you've got to dispel! ♣

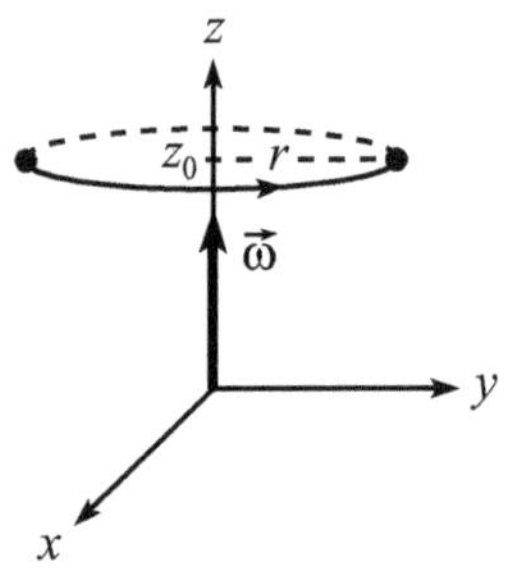

Fig. 9.12

Example 3 (Two point masses): Let's now add another point mass m to the previous example. Let it travel in the same circle, at the diametrically opposite point, as shown in Fig. 9.12. Using $\omega = (0, 0, \omega_3)$, $x^2 + y^2 = r^2$, and $z = z_0$ in Eq. (9.8), you can show that the angular momentum with respect to the origin is

$$\mathbf{L} = 2m\omega_3(0, 0, r^2). \tag{9.13}$$

Since $v = \omega_3 r$, the z component is $2mrv$, as it should be. And L_1 and L_2 are zero, unlike in the previous example, because these components of the $\mathbf{L}$'s of the two particles cancel. This occurs because of the symmetry of the masses around the z axis, which causes the I_{zx} and I_{zy} entries in the inertia tensor to vanish; they are each the sum of two terms, with opposite x values, or opposite y values. Alternatively, you can just note that adding on the mirror-image $\mathbf{L}$ vector in Fig. 9.10 produces canceling x and y components.

Let's now look at the kinetic energy of our object, which is rotating around an axis passing through the origin. To find this, we must add up the kinetic energies of all the little pieces. A little piece has energy $(dm)\,v^2/2 = dm\,|\boldsymbol{\omega} \times \mathbf{r}|^2/2$. Therefore, using Eq. (9.6), the total kinetic energy is

$$T = \frac{1}{2}\int \Big((\omega_2 z - \omega_3 y)^2 + (\omega_3 x - \omega_1 z)^2 + (\omega_1 y - \omega_2 x)^2\Big)\, dm. \tag{9.14}$$

Multiplying this out, we see (after a little work) that we can write T as

$$\begin{aligned} T &= \frac{1}{2}(\omega_1, \omega_2, \omega_3) \cdot \begin{pmatrix} \int(y^2+z^2) & -\int xy & -\int zx \\ -\int xy & \int(z^2+x^2) & -\int yz \\ -\int zx & -\int yz & \int(x^2+y^2) \end{pmatrix} \begin{pmatrix} \omega_1 \\ \omega_2 \\ \omega_3 \end{pmatrix} \\ &= \frac{1}{2}\boldsymbol{\omega} \cdot \mathbf{I}\boldsymbol{\omega} = \frac{1}{2}\boldsymbol{\omega} \cdot \mathbf{L}. \end{aligned} \tag{9.15}$$

If $\boldsymbol{\omega} = \omega_3\hat{\mathbf{z}}$, then this reduces to $T = I_{zz}\omega_3^2/2$, which agrees with the result in Eq. (8.8), with a slight change in notation.

9.2.2 General motion

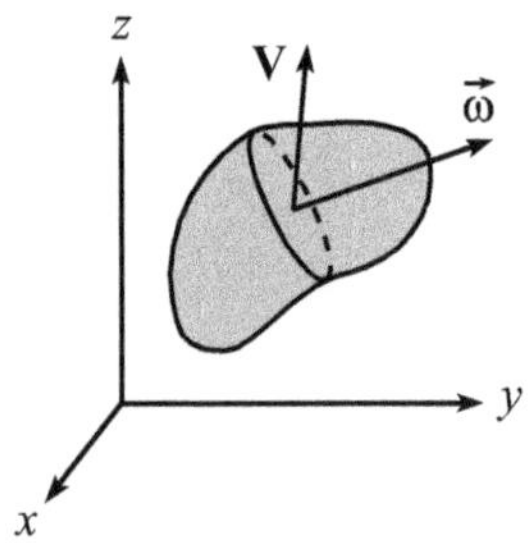

Fig. 9.13

How do we deal with general motion in space? That is, what if an object is both translating and rotating? For the motion in Fig. 9.13, the various pieces of mass aren't traveling in circles around the origin, so we can't write $\mathbf{v} = \boldsymbol{\omega} \times \mathbf{r}$, as we did prior to Eq. (9.4).

To determine $\mathbf{L}$ (relative to the origin), and also the kinetic energy T, we will use Theorem 9.1 to write the motion as the sum of a translation plus a rotation. In applying the theorem, we may choose any point in the body to be the point P in the theorem. However, only in the case where P is the object's CM can we extract anything useful, as we'll see. The theorem then says that the motion of the body is the sum of the motion of the CM plus a rotation around the CM. So let the CM move with velocity $\mathbf{V}$, and let the body instantaneously rotate with angular velocity $\boldsymbol{\omega}'$ around the CM (that is, with respect to the frame whose origin is the CM, and whose axes are parallel to the fixed-frame axes).[4]

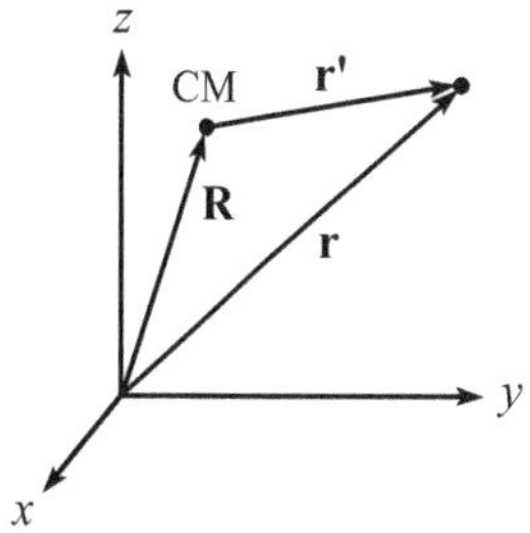

Fig. 9.14

Let the position of the CM relative to the origin be $\mathbf{R} = (X, Y, Z)$, and let the position of a given piece of mass relative to the CM be $\mathbf{r}' = (x', y', z')$. Then $\mathbf{r} = \mathbf{R} + \mathbf{r}'$ is the position of a piece of mass relative to the origin (see Fig. 9.14). Let the velocity of a piece of mass relative to the CM be $\mathbf{v}'$ (so $\mathbf{v}' = \boldsymbol{\omega}' \times \mathbf{r}'$). Then $\mathbf{v} = \mathbf{V} + \mathbf{v}'$ is the velocity relative to the origin.

Let's look at $\mathbf{L}$ first. The angular momentum is

$$\begin{aligned}\mathbf{L} &= \int \mathbf{r} \times \mathbf{v}\, dm = \int (\mathbf{R} + \mathbf{r}') \times \Big(\mathbf{V} + (\boldsymbol{\omega}' \times \mathbf{r}')\Big)\, dm \\ &= \int (\mathbf{R} \times \mathbf{V})\, dm + \int \mathbf{r}' \times (\boldsymbol{\omega}' \times \mathbf{r}')\, dm \\ &= M(\mathbf{R} \times \mathbf{V}) + \mathbf{L}_{\text{CM}},\end{aligned} \tag{9.16}$$

where the cross terms vanish because the integrands are linear in $\mathbf{r}'$. More precisely, the integrals involve $\int \mathbf{r}'\, dm$, which is zero by definition of the CM (because $\int \mathbf{r}'\, dm/M$ is the position of the CM relative to the CM, which is zero). $\mathbf{L}_{\text{CM}}$ is the angular momentum relative to the CM.[5]

We see that as in the pancake case in Section 8.1.2, the angular momentum (relative to the origin) of a body can be found by treating the body like a point mass located at the CM and finding the angular momentum of this point mass relative to the origin, and by then adding on the angular momentum of the body relative to the CM. Note that these two parts of the angular momentum need not point in the same direction, as they did in the case of the pancake moving in the x-y plane.

Now let's look at T. The kinetic energy is

$$\begin{aligned}T &= \int \frac{1}{2} v^2\, dm = \int \frac{1}{2} |\mathbf{V} + \mathbf{v}'|^2\, dm \\ &= \int \frac{1}{2} V^2\, dm + \int \frac{1}{2} v'^2\, dm \\ &= \frac{1}{2} M V^2 + \int \frac{1}{2} |\boldsymbol{\omega}' \times \mathbf{r}'|^2\, dm \\ &\equiv \frac{1}{2} M V^2 + \frac{1}{2} \boldsymbol{\omega}' \cdot \mathbf{L}_{\text{CM}},\end{aligned} \tag{9.17}$$

[4] It's not necessary to put the prime on the $\boldsymbol{\omega}$ here, because the angular velocity vector in the CM frame is the same as in the lab frame. But we'll use the prime just because we'll have primes on the other CM quantities below.

[5] By this, we mean the angular momentum as measured in the coordinate system whose origin is the CM, and whose axes are parallel to the fixed-frame axes.

where the last line follows from the steps leading to Eq. (9.15). The cross term $\int \mathbf{V} \cdot \mathbf{v}'\, dm = \int \mathbf{V} \cdot (\boldsymbol{\omega}' \times \mathbf{r}')\, dm$ vanishes because the integrand is linear in $\mathbf{r}'$ and thus yields a zero integral, by definition of the CM. As in the pancake case in Section 8.1.2, the kinetic energy of a body can be found by treating the body like a point mass located at the CM, and by then adding on the kinetic energy of the body due to the rotation around the CM.

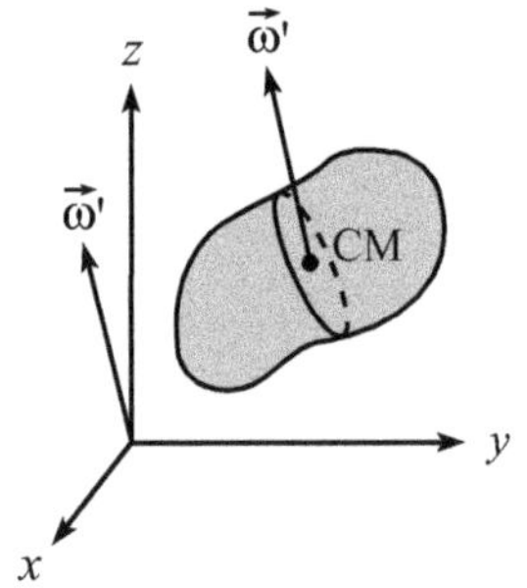

Fig. 9.15

9.2.3 The parallel-axis theorem

Consider the special case where the CM rotates around the origin with the same angular velocity at which the body rotates around the CM (see Fig. 9.15), that is, $\mathbf{V} = \boldsymbol{\omega}' \times \mathbf{R}$. This can be achieved, for example, by piercing the body with the base of a rigid "T" and then rotating the T and the body around the (fixed) line of the "upper" part of the T (the origin must pass through this line). We then have the nice situation where all points in the body travel in fixed circles around the axis of rotation. Mathematically, this follows from $\mathbf{v} = \mathbf{V} + \mathbf{v}' = \boldsymbol{\omega}' \times \mathbf{R} + \boldsymbol{\omega}' \times \mathbf{r}' = \boldsymbol{\omega}' \times \mathbf{r}$. Dropping the prime on $\boldsymbol{\omega}$, Eq. (9.16) becomes

$$\mathbf{L} = M\mathbf{R} \times (\boldsymbol{\omega} \times \mathbf{R}) + \int \mathbf{r}' \times (\boldsymbol{\omega} \times \mathbf{r}')\, dm \tag{9.18}$$

Expanding the double cross products as in the steps leading to Eq. (9.8), we can write this as

$$\begin{aligned}
\begin{pmatrix} L_1 \\ L_2 \\ L_3 \end{pmatrix} &= M \begin{pmatrix} Y^2 + Z^2 & -XY & -ZX \\ -XY & Z^2 + X^2 & -YZ \\ -ZX & -YZ & X^2 + Y^2 \end{pmatrix} \begin{pmatrix} \omega_1 \\ \omega_2 \\ \omega_3 \end{pmatrix} \\
&\quad + \begin{pmatrix} \int (y'^2 + z'^2) & -\int x'y' & -\int z'x' \\ -\int x'y' & \int (z'^2 + x'^2) & -\int y'z' \\ -\int z'x' & -\int y'z' & \int (x'^2 + y'^2) \end{pmatrix} \begin{pmatrix} \omega_1 \\ \omega_2 \\ \omega_3 \end{pmatrix} \\
&\equiv (\mathbf{I}_{\mathrm{R}} + \mathbf{I}_{\mathrm{CM}})\boldsymbol{\omega}.
\end{aligned} \tag{9.19}$$

This is the generalized parallel-axis theorem. It says that once you've calculated $\mathbf{I}_{\mathrm{CM}}$ relative to the CM, then if you want to calculate $\mathbf{I}$ relative to another point, you simply have to add on the $\mathbf{I}_{\mathrm{R}}$ matrix, obtained by treating the object like a point mass at the CM. So you have to compute six extra numbers (there are six, instead of nine, because the $\mathbf{I}_{\mathrm{R}}$ matrix is symmetric) instead of just the one MR^2 in the parallel-axis theorem in Chapter 8, given in Eq. (8.13). Problem 9.4 gives another derivation of the parallel-axis theorem, without mentioning the angular velocity.

REMARK: The name "parallel-axis" theorem is actually a misnomer here. The inertia tensor isn't associated with one particular axis, as the moment of inertia in Chapter 8 was. The moment of inertia is just one of the diagonal entries (associated with a given axis) in the inertia tensor. The inertia tensor depends on the entire coordinate system. So in that sense we should call this the "parallel-*axes*" theorem, because the coordinate axes in the CM frame are assumed to be parallel to the ones in the fixed frame. At any rate, the point is that the parallel-axis theorem in Chapter 8 dealt with shifting the axis, whereas the present theorem deals with shifting the origin (and hence all three axes in general). ♣

As far as the kinetic energy goes, if $\boldsymbol{\omega}$ and $\boldsymbol{\omega}'$ are equal, so that $\mathbf{V} = \boldsymbol{\omega}' \times \mathbf{R}$, then Eq. (9.17) gives (dropping the prime on $\boldsymbol{\omega}$)

$$T = \frac{1}{2}M|\boldsymbol{\omega} \times \mathbf{R}|^2 + \int \frac{1}{2}|\boldsymbol{\omega} \times \mathbf{r}'|^2\, dm. \tag{9.20}$$

Performing the steps leading to Eq. (9.15), this becomes

$$T = \frac{1}{2}\boldsymbol{\omega} \cdot (\mathbf{I}_\mathrm{R} + \mathbf{I}_\mathrm{CM})\boldsymbol{\omega} = \frac{1}{2}\boldsymbol{\omega} \cdot \mathbf{L}. \tag{9.21}$$

9.3 Principal axes

The cumbersome expressions in the previous section may seem a bit unsettling, but it turns out that usually we can get by without them. The strategy for avoiding all of the above mess is to use the *principal axes* of a body, which we will define below.

In general, the inertia tensor $\mathbf{I}$ in Eq. (9.8) has nine nonzero entries, of which six are independent due to the symmetry of $\mathbf{I}$. In addition to depending on the origin chosen, the inertia tensor depends on the set of orthonormal basis vectors chosen for the coordinate system; the x, y, z variables in the integrals in $\mathbf{I}$ depend, of course, on the coordinate system they're measured with respect to. Given a blob of material, and given an arbitrary origin,[6] any orthonormal set of basis vectors is usable, but there is one special set that makes all our calculations very nice. These special basis vectors are called the *principal axes*. They can be defined in various equivalent ways:

- The principal axes are the orthonormal basis vectors for which $\mathbf{I}$ is diagonal, that is, for which[7]

$$\mathbf{I} = \begin{pmatrix} I_1 & 0 & 0 \\ 0 & I_2 & 0 \\ 0 & 0 & I_3 \end{pmatrix}. \tag{9.22}$$

  I_1, I_2, and I_3 are called the *principal moments*. For many objects, it is quite obvious what the principal axes are. For example, consider a uniform rectangle in the x-y plane. Pick the origin to be the CM, and let the x and y axes be parallel to the sides. Then the

[6] The CM is often chosen to be the origin, but it need not be. There are principal axes associated with any origin.

[7] Technically, we should be writing I_{11} or I_{xx} instead of I_1, etc., in this matrix, because the one-index object I_1 looks like the component of a vector, not a matrix. But the two-index notation gets cumbersome, so we'll be sloppy and just use I_1, etc.

principal axes are clearly the x, y, and z axes, because all the off-diagonal elements in the inertia tensor in Eq. (9.8) vanish, by symmetry. For example, $I_{xy} \equiv -\int xy\,dm$ equals zero, because for every point (x, y) in the rectangle, there is a corresponding point $(-x, y)$, so the contributions to $\int xy\,dm$ cancel in pairs. Also, the integrals involving z are identically zero, because $z = 0$.

- A principal axis is an axis $\hat{\omega}$ for which $\mathbf{I}\hat{\omega} = I\hat{\omega}$. That is, a principal axis is a special direction with the property that if $\boldsymbol{\omega}$ points along it, then so does $\mathbf{L}$. The principal axes of an object are then the orthonormal set of three vectors $\hat{\omega}_1$, $\hat{\omega}_2$, $\hat{\omega}_3$ with the property that

$$\mathbf{I}\hat{\omega}_1 = I_1\hat{\omega}_1, \qquad \mathbf{I}\hat{\omega}_2 = I_2\hat{\omega}_2, \qquad \mathbf{I}\hat{\omega}_3 = I_3\hat{\omega}_3. \tag{9.23}$$

The three statements in Eq. (9.23) are equivalent to Eq. (9.22), because the vectors $\hat{\omega}_1$, $\hat{\omega}_2$, and $\hat{\omega}_3$ are simply $(1, 0, 0)$, $(0, 1, 0)$, and $(0, 0, 1)$ in the frame in which they are the basis vectors.

- Consider an object rotating around a fixed axis with constant angular speed. Then this axis is a principal axis if there is no need for any torque. So in some sense, the object is "happy" to spin around a principal axis. A set of three orthonormal axes, each of which has this property, is by definition what we call a set of principal axes.

This definition of a principal axis is equivalent to the previous definition for the following reason. Assume that the object rotates around a fixed axis $\hat{\omega}_1$ for which $\mathbf{L} = \mathbf{I}\hat{\omega}_1 = I_1\hat{\omega}_1$, as in Eq. (9.23). Then since $\hat{\omega}_1$ is assumed to be fixed, we see that $\mathbf{L}$ is also fixed. Therefore, $\boldsymbol{\tau} = d\mathbf{L}/dt = \mathbf{0}$.

Conversely, if the object is rotating around a fixed axis $\hat{\omega}_1$, and if $\boldsymbol{\tau} = d\mathbf{L}/dt = \mathbf{0}$, then we claim that $\mathbf{L}$ points along $\hat{\omega}_1$ (that is, $\mathbf{L} = I_1\hat{\omega}_1$). This is true because if $\mathbf{L}$ does *not* point along $\hat{\omega}_1$, then imagine painting a dot on the object somewhere along the line of $\mathbf{L}$. A little while later, the dot will have rotated around the fixed vector $\hat{\omega}_1$. But the line of $\mathbf{L}$ must always pass through the dot, because we could have rotated our axes around $\hat{\omega}_1$ and started the process at a slightly later time (this argument relies on $\hat{\omega}_1$ being fixed). Therefore, we see that $\mathbf{L}$ has changed, in contradiction to the assumption that $d\mathbf{L}/dt = \mathbf{0}$. Hence, $\mathbf{L}$ must in fact point along $\hat{\omega}_1$.

For a rotation around a principal axis $\hat{\omega}$, the lack of need for any torque means that if the object is pivoted at the origin, and if the origin is the only place where any force is applied (which implies that there is zero torque around it), then the object can undergo rotation with constant angular velocity $\boldsymbol{\omega}$. If you try to set up this scenario with a nonprincipal axis, it won't work.

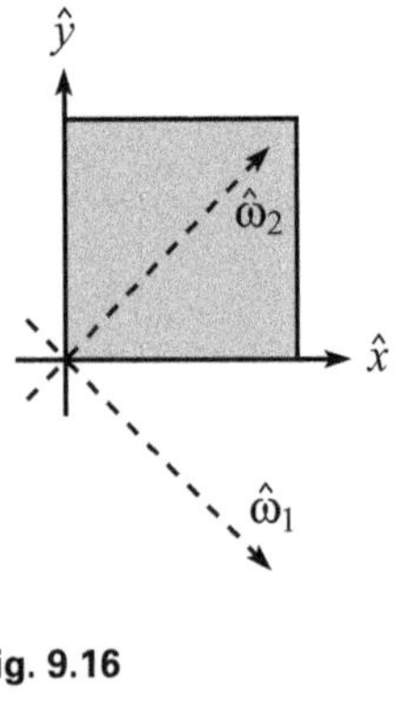

Fig. 9.16

Example (Square with origin at corner): Consider the uniform square in Fig. 9.16. In Appendix E, we show that the principal axes are the dotted lines drawn (and also the z axis perpendicular to the page). But there is no need to use the techniques in the appendix to see this, because in this new basis it is clear by symmetry that the integral $\int x_1x_2$ is zero; for every x_1 in the integral, there is a $-x_1$. And $x_3 \equiv z$

is identically zero, which makes all the other off-diagonal terms in $\mathbf{I}$ also equal to zero. Therefore, since $\mathbf{I}$ is diagonal in this new basis, these basis vectors are the principal axes.

Furthermore, it is intuitively clear that the square will be happy to rotate around any one of these axes indefinitely. During such a rotation, the pivot will certainly be applying a *force* (if the axis is $\hat{\boldsymbol{\omega}}_1$ or $\hat{\mathbf{z}}$, but not if it is $\hat{\boldsymbol{\omega}}_2$), to produce the centripetal acceleration of the CM in its circular motion. But it won't be applying a *torque* relative to the origin (because the $\mathbf{r}$ in $\mathbf{r} \times \mathbf{F}$ is $\mathbf{0}$). This is good, because for a rotation around one of these principal axes, $d\mathbf{L}/dt = \mathbf{0}$, so there is no need for any torque.

In contrast with the off-diagonal zeros in the new basis, the integral $\int xy$ in the old basis is *not* zero, because every point gives a positive contribution. So the inertia tensor is not diagonal in the old basis, which means that $\hat{x}$ and $\hat{y}$ are not principal axes. Consistent with this, it is reasonably clear that it is impossible to make the square rotate around, say, the x axis, assuming that its only contact with the outside world is through a pivot (for example, a ball and socket) at the origin. The square simply doesn't want to remain in this circular motion. Mathematically, $\mathbf{L}$ (relative to the origin) doesn't point along the x axis, so it therefore precesses around the x axis along with the square, tracing out the surface of a cone. This means that $\mathbf{L}$ is changing. But there is no torque available (relative to the origin) to provide for this change in $\mathbf{L}$. Hence, such a rotation cannot exist.

At the moment, it is not at all obvious that an orthonormal set of principal axes exists for an arbitrary object. But this is indeed the case, as stated in Theorem 9.4 below. Assuming for now that principal axes do exist, then in this basis the $\mathbf{L}$ and T in Eqs. (9.8) and (9.15) take on the particularly nice forms,

$$\begin{aligned} \mathbf{L} &= (I_1\omega_1, I_2\omega_2, I_3\omega_3), \\ T &= \frac{1}{2}\left(I_1\omega_1^2 + I_2\omega_2^2 + I_3\omega_3^2\right). \end{aligned} \tag{9.24}$$

The quantities ω_1, ω_2, and ω_3 here are the components of a general vector $\boldsymbol{\omega}$ written in the principal-axis basis; that is, $\boldsymbol{\omega} = \omega_1\hat{\boldsymbol{\omega}}_1 + \omega_2\hat{\boldsymbol{\omega}}_2 + \omega_3\hat{\boldsymbol{\omega}}_3$. Equation (9.24) is a vast simplification over the general formulas in Eqs. (9.8) and (9.15). We will therefore invariably work with principal axes in the remainder of this chapter.

Note that the directions of the principal axes (relative to the body) depend only on the geometry of the body. They may therefore be considered to be painted on (or in) it. Hence, they will generally move around in space as the body rotates. For example, if the object is rotating around a principal axis, then that axis stays fixed while the other two principal axes rotate around it. In relations like $\boldsymbol{\omega} = (\omega_1, \omega_2, \omega_3)$ and $\mathbf{L} = (I_1\omega_1, I_2\omega_2, I_3\omega_3)$, the components ω_i and $I_i\omega_i$ are measured along the *instantaneous* principal axes $\hat{\boldsymbol{\omega}}_i$. Since these axes change with time, it is quite possible that the components ω_i and $I_i\omega_i$ change with time, as we'll see in Section 9.5 (and onward).

Let's now state the theorem that implies that a set of principal axes does indeed exist for any body and any origin. The proof of this theorem involves a useful but rather slick technique, but it's slightly off the main line of thought, so we'll relegate it to Appendix D. Take a look at the proof if you wish, but if you want to just accept the fact that principal axes exist, that's fine.

Theorem 9.4 *Given a real symmetric* 3×3 *matrix,* $\mathbf{I}$*, there exist three orthonormal real vectors,* $\hat{\boldsymbol{\omega}}_k$*, and three real numbers,* I_k*, with the property that*

$$\mathbf{I}\hat{\boldsymbol{\omega}}_k = I_k \hat{\boldsymbol{\omega}}_k. \tag{9.25}$$

Proof: See Appendix D. ■

Since the inertia tensor in Eq. (9.8) is indeed symmetric for any body and any origin, this theorem says that we can always find three orthogonal basis vectors that satisfy Eq. (9.23). Or equivalently, we can always find three orthogonal basis vectors for which $\mathbf{I}$ is a diagonal matrix, as in Eq. (9.22). In other words, principal axes always exist. Problem 9.7 gives another way to demonstrate the existence of principal axes in the special case of a pancake object.

Invariably, it is best to work in a coordinate system that has principal axes as its basis, due to the simplicity of Eq. (9.24). And as mentioned in Footnote 6, the origin is generally chosen to be the CM, because from Section 8.4.3 the CM is one of the origins for which $\boldsymbol{\tau} = d\mathbf{L}/dt$ is a valid statement. But this choice is not necessary; there are principal axes associated with any origin.

For an object with a fair amount of symmetry, the principal axes are usually the obvious choices and can be written down by simply looking at the object (examples are given below). If, however, you are given an unsymmetrical body, then the only way to determine the principal axes is to pick an arbitrary basis, then find $\mathbf{I}$ in this basis, and then go through a diagonalization procedure. This diagonalization procedure basically consists of the steps at the beginning of the proof of Theorem 9.4 (given in Appendix D), with the addition of one more step to get the actual vectors, so we'll relegate it to Appendix E. There's no need to worry much about this method. Virtually every system you encounter will involve an object with sufficient symmetry to enable you to just write down the principal axes.

Let's now prove two very useful (and very similar) theorems.

Theorem 9.5 *If two principal moments are equal* ($I_1 = I_2 \equiv I$)*, then any axis (through the chosen origin) in the plane of the corresponding principal axes is a principal axis, and its moment is also* I*. Similarly, if all three principal moments are equal* ($I_1 = I_2 = I_3 \equiv I$)*, then any axis (through the chosen origin) in space is a principal axis, and its moment is also* I*.*

Proof: The first part was already proved at the end of the proof in Appendix D, but we'll do it again here. Since $I_1 = I_2 \equiv I$, we have $\mathbf{I}\mathbf{u}_1 = I\mathbf{u}_1$, and $\mathbf{I}\mathbf{u}_2 = I\mathbf{u}_2$, where the $\mathbf{u}$'s are the principal axes. Hence, $\mathbf{I}(a\mathbf{u}_1 + b\mathbf{u}_2) = I(a\mathbf{u}_1 + b\mathbf{u}_2)$, for any a and b. Therefore, any linear combination of $\mathbf{u}_1$ and $\mathbf{u}_2$ (that is, any vector in the plane spanned by $\mathbf{u}_1$ and $\mathbf{u}_2$) is a solution to $\mathbf{I}\mathbf{u} = I\mathbf{u}$ and is thus a principal axis, by definition.

The proof of the second part proceeds in a similar manner. Since $I_1 = I_2 = I_3 \equiv I$, we have $\mathbf{I}\mathbf{u}_1 = I\mathbf{u}_1$, $\mathbf{I}\mathbf{u}_2 = I\mathbf{u}_2$, and $\mathbf{I}\mathbf{u}_3 = I\mathbf{u}_3$. Hence, $\mathbf{I}(a\mathbf{u}_1 + b\mathbf{u}_2 + c\mathbf{u}_3) = I(a\mathbf{u}_1 + b\mathbf{u}_2 + c\mathbf{u}_3)$. Therefore, any linear combination of $\mathbf{u}_1$, $\mathbf{u}_2$, and $\mathbf{u}_3$ (that is, any vector in space) is a solution to $\mathbf{I}\mathbf{u} = I\mathbf{u}$ and is thus a principal axis, by definition.

In short, if $I_1 = I_2 \equiv I$, then $\mathbf{I}$ is the identity matrix (up to a multiple) in the space spanned by $\mathbf{u}_1$ and $\mathbf{u}_2$. And if $I_1 = I_2 = I_3 \equiv I$, then $\mathbf{I}$ is the identity matrix (up to a multiple) in the entire space. Note that it isn't required that the various $\mathbf{u}_i$ vectors be orthogonal. All we need is that they span the relevant space. ∎

If two or three moments are equal, so that there is freedom in choosing the principal axes, then it is possible to pick a nonorthogonal group of them. We will, however, always choose ones that are orthogonal. So when we say "a set of principal axes," we mean an orthonormal set.

Theorem 9.6 *If a pancake object is symmetric under a rotation through an angle $\theta \neq 180°$ in the x-y plane (such as a hexagon), then every axis in the x-y plane (with the origin chosen to be the center of the symmetry rotation) is a principal axis with the same moment.*

Proof: Let $\hat{\boldsymbol{\omega}}_0$ be a principal axis in the plane, and let $\hat{\boldsymbol{\omega}}_\theta$ be the axis obtained by rotating $\hat{\boldsymbol{\omega}}_0$ through the angle θ. Then $\hat{\boldsymbol{\omega}}_\theta$ is also a principal axis with the same principal moment, due to the symmetry of the object. Therefore, $\mathbf{I}\hat{\boldsymbol{\omega}}_0 = I\hat{\boldsymbol{\omega}}_0$, and $\mathbf{I}\hat{\boldsymbol{\omega}}_\theta = I\hat{\boldsymbol{\omega}}_\theta$.

Now, any vector $\boldsymbol{\omega}$ in the x-y plane can be written as a linear combination of $\hat{\boldsymbol{\omega}}_0$ and $\hat{\boldsymbol{\omega}}_\theta$, provided that $\theta \neq 180°$ (or zero, of course). That is, $\hat{\boldsymbol{\omega}}_0$ and $\hat{\boldsymbol{\omega}}_\theta$ span the x-y plane. Therefore, any vector $\boldsymbol{\omega}$ can be written as $\boldsymbol{\omega} = a\hat{\boldsymbol{\omega}}_0 + b\hat{\boldsymbol{\omega}}_\theta$, and so

$$\mathbf{I}\boldsymbol{\omega} = \mathbf{I}(a\hat{\boldsymbol{\omega}}_0 + b\hat{\boldsymbol{\omega}}_\theta) = aI\hat{\boldsymbol{\omega}}_0 + bI\hat{\boldsymbol{\omega}}_\theta = I\boldsymbol{\omega}. \tag{9.26}$$

Hence, $\boldsymbol{\omega}$ is also a principal axis. Problem 9.8 gives another proof of this theorem. ∎

The theorem actually holds even without the "pancake" restriction. That is, it holds for any object with a rotational symmetry around the z axis (excluding $\theta \neq 180°$). This can be seen as follows. The z axis is a principal axis, because if $\boldsymbol{\omega}$ points along $\hat{z}$, then $\mathbf{L}$ must also point along $\hat{z}$, by symmetry. There are therefore (at least) two principal axes in the x-y plane. Label one of these as $\hat{\boldsymbol{\omega}}_0$ and proceed as above.

Let's now do some quick examples. We'll state the principal axes for the objects listed below (relative to the origin). Your task is to show that they are correct. Usually, a quick symmetry argument shows that

$$\mathbf{I} \equiv \begin{pmatrix} \int(y^2+z^2) & -\int xy & -\int zx \\ -\int xy & \int(z^2+x^2) & -\int yz \\ -\int zx & -\int yz & \int(x^2+y^2) \end{pmatrix} \tag{9.27}$$

is diagonal. In all of these examples (see Fig. 9.17), the origin for the principal axes is understood to be the origin of the given coordinate system (which is not necessarily the CM). In describing the axes, they therefore all pass through the origin, in addition to having the other properties stated.

Example 1: Point mass at the origin.
principal axes: any axes.

Example 2: Point mass at the point (x_0, y_0, z_0).
principal axes: axis through point, any axes perpendicular to this.

Example 3: Rectangle centered at the origin, as shown.
principal axes: the x, y, and z axes.

Example 4: Cylinder with axis as z axis.
principal axes: z axis, any axes in x-y plane.

Example 5: Square with one corner at origin, as shown.
principal axes: z axis, axis through CM, axis perpendicular to this.

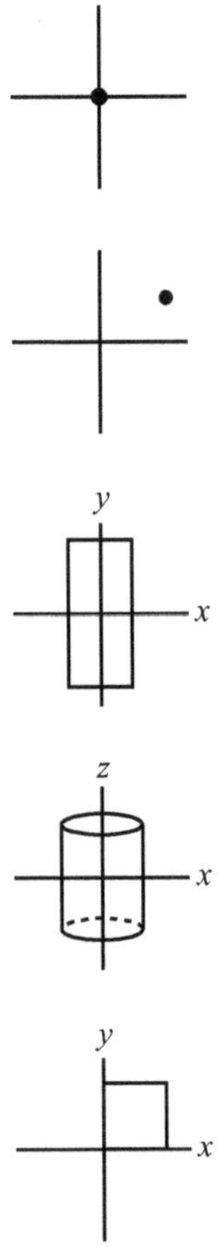

Fig. 9.17

9.4 Two basic types of problems

The previous three sections introduced a variety of abstract concepts. We will now finally look at some actual physical systems. The concept of principal axes gives us the ability to solve many kinds of problems. Two kinds, however, come up again and again. There are variations on these, of course, but they may be generally stated as follows.

- Strike a rigid object with an impulsive (that is, quick) blow. What is the motion of the object immediately after the blow?
- An object rotates around a fixed axis. A given torque is applied. What is the frequency of the rotation? Or conversely, given the frequency, what is the required torque?

We'll work through an example for each of these problems. In both cases, the solution involves a few standard steps, so we'll write them out explicitly.

9.4.1 Motion after an impulsive blow

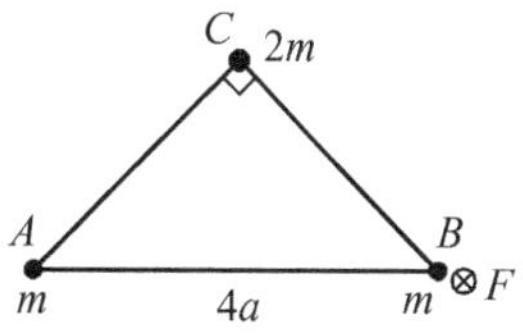

Fig. 9.18

Problem: Consider the rigid object in Fig. 9.18. Three masses are connected by three massless rods, in the shape of an isosceles right triangle with hypotenuse length $4a$. The mass at the right angle is $2m$, and the other two masses are m. Label them A, B, C, as shown. Assume that the object is floating freely in outer space. Mass B is struck with a quick blow, directed into the page. Let the imparted impulse have magnitude $\int F\,dt = P$. (See Section 8.6 for a discussion of impulse and angular impulse.) What are the velocities of the three masses immediately after the blow?

Solution: Our strategy will be to find the angular momentum of the system (relative to the CM) using the angular impulse, and then calculate the principal moments and find the angular velocity vector (which will give the velocities relative to the CM), and then finally add on the CM motion.

The altitude from the right angle to the hypotenuse has length $2a$, and the CM is easily seen to be located at its midpoint (see Fig. 9.19). Picking the CM as our origin, and letting the plane of the paper be the x-y plane, the positions of the three masses are $\mathbf{r}_A = (-2a, -a, 0)$, $\mathbf{r}_B = (2a, -a, 0)$, and $\mathbf{r}_C = (0, a, 0)$. There are now five standard steps that we must perform.

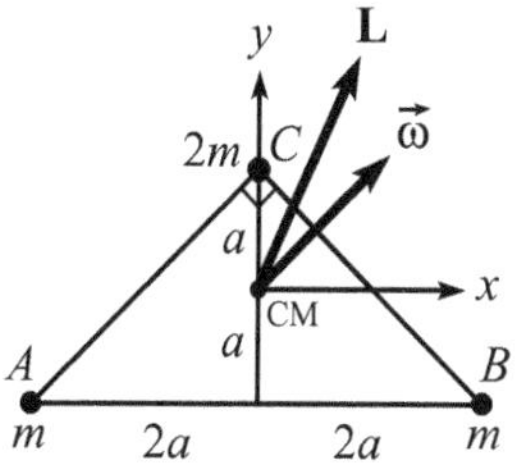

Fig. 9.19

- **Find L:** The positive z axis is directed out of the page, so the impulse vector is $\mathbf{P} \equiv \int \mathbf{F}\,dt = (0, 0, -P)$. Therefore, the angular momentum of the system (relative to the CM) is

$$\mathbf{L} = \int \boldsymbol{\tau}\,dt = \int (\mathbf{r}_B \times \mathbf{F})\,dt = \mathbf{r}_B \times \int \mathbf{F}\,dt$$
$$= (2a, -a, 0) \times (0, 0, -P) = aP(1, 2, 0), \qquad (9.28)$$

as shown in Fig. 9.19. We have used the fact that $\mathbf{r}_B$ is essentially constant during the blow (because the blow is assumed to happen very quickly) in taking $\mathbf{r}_B$ outside the integral.

- **Calculate the principal moments:** The principal axes are the x, y, and z axes, because the symmetry of the triangle makes $\mathbf{I}$ diagonal in this basis, as you can quickly check. The moments (relative to the CM) are

$$I_x = ma^2 + ma^2 + (2m)a^2 = 4ma^2,$$
$$I_y = m(2a)^2 + m(2a)^2 + (2m)0^2 = 8ma^2, \qquad (9.29)$$
$$I_z = I_x + I_y = 12ma^2.$$

We have used the perpendicular-axis theorem to obtain I_z, although it won't be needed to solve the problem.

- **Find ω:** We now have two expressions for the angular momentum of the system. One expression is in terms of the given impulse, Eq. (9.28). The other is in terms of the moments and the angular velocity components, Eq. (9.24). Equating these gives

$$\begin{aligned} (I_x\omega_x, I_y\omega_y, I_z\omega_z) &= aP(1,2,0) \\ \Longrightarrow \quad (4ma^2\omega_x, 8ma^2\omega_y, 12ma^2\omega_z) &= aP(1,2,0) \\ \Longrightarrow \quad (\omega_x, \omega_y, \omega_z) &= \frac{P}{4ma}(1,1,0), \end{aligned} \tag{9.30}$$

as shown in Fig. 9.19.

- **Calculate the velocities relative to the CM:** Right after the blow, the object rotates around the CM with the angular velocity found in Eq. (9.30). The velocities relative to the CM are then $\mathbf{u}_i = \boldsymbol{\omega} \times \mathbf{r}_i$. Thus,

$$\begin{aligned} \mathbf{u}_A &= \boldsymbol{\omega} \times \mathbf{r}_A = \frac{P}{4ma}(1,1,0) \times (-2a, -a, 0) = (0,0,P/4m), \\ \mathbf{u}_B &= \boldsymbol{\omega} \times \mathbf{r}_B = \frac{P}{4ma}(1,1,0) \times (2a, -a, 0) = (0,0,-3P/4m), \\ \mathbf{u}_C &= \boldsymbol{\omega} \times \mathbf{r}_C = \frac{P}{4ma}(1,1,0) \times (0, a, 0) = (0,0,P/4m). \end{aligned} \tag{9.31}$$

  As a check, it makes sense that $\mathbf{u}_B$ is three times as large as $\mathbf{u}_A$ and $\mathbf{u}_C$, because B is three times as far from the axis of rotation as A and C are, as you can verify by doing a little geometry in Fig. 9.19.

- **Add on the velocity of the CM:** The impulse (that is, the change in linear momentum) supplied to the whole system is $\mathbf{P} = (0,0,-P)$. The total mass of the system is $M = 4m$. Therefore, the velocity of the CM is

$$V_{\text{CM}} = \frac{\mathbf{P}}{M} = (0,0,-P/4m). \tag{9.32}$$

  The total velocities of the masses are therefore

$$\begin{aligned} \mathbf{v}_A &= \mathbf{u}_A + V_{\text{CM}} = (0,0,0), \\ \mathbf{v}_B &= \mathbf{u}_B + V_{\text{CM}} = (0,0,-P/m), \\ \mathbf{v}_C &= \mathbf{u}_C + V_{\text{CM}} = (0,0,0). \end{aligned} \tag{9.33}$$

REMARKS:

1. We see that masses A and C are instantaneously at rest immediately after the blow, and mass B acquires all of the imparted impulse. In retrospect, this is clear. Basically, it is possible for both A and C to remain at rest while B moves a tiny bit, so this is what happens. If B moves into the page by a small distance ϵ, then A and C won't know that B has moved, because their distances to B will change (assuming hypothetically that they don't move) by a distance of order only ϵ^2. If we changed the problem and added a mass D at, say, the midpoint of the hypotenuse, then it would *not* be possible for A, C, and D to remain at rest while B moved a tiny bit. So there would have to be some other motion in addition to B's. This setup is the topic of Exercise 9.38.
2. As time goes on, the system undergoes a rather complicated motion. What happens is that the CM moves with constant velocity while the masses rotate around it in

a messy manner. Since there are no torques acting on the system (after the initial blow), we know that **L** forever remains constant. It turns out that $\boldsymbol{\omega}$ moves around **L** while the masses rotate around this changing $\boldsymbol{\omega}$. These matters are the subject of Section 9.6, although in that discussion we restrict ourselves to symmetric tops, that is, ones with two equal moments. But these issues aside, it's good to know that we can, without too much difficulty, determine what's going on immediately after the blow.

3. The object in this problem was assumed to be floating freely in space. If we instead have an object that is pivoted at a given fixed point, then we should use this pivot as our origin. There is then no need to perform the last step of adding on the velocity of the origin (which was the CM, above), because this velocity is now zero. Equivalently, just consider the pivot to be an infinite mass, which is therefore the location of the (motionless) CM. ♣

9.4.2 Frequency of motion due to a torque

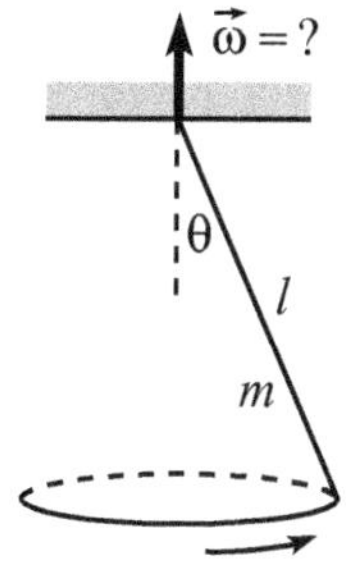

Fig. 9.20

Problem: Consider a stick of length ℓ, mass m, and uniform mass density. The stick is pivoted at its top end and swings around the vertical axis. Assume that conditions have been set up so that the stick always makes an angle θ with the vertical, as shown in Fig. 9.20. What is the frequency, ω, of this motion?

Solution: Our strategy will be to find the principal moments and then the angular momentum of the system (in terms of ω), and then find the rate of change of **L**, and then calculate the torque and equate it with $d\mathbf{L}/dt$. We will choose the pivot to be the origin.[8] Again, there are five standard steps that we must perform.

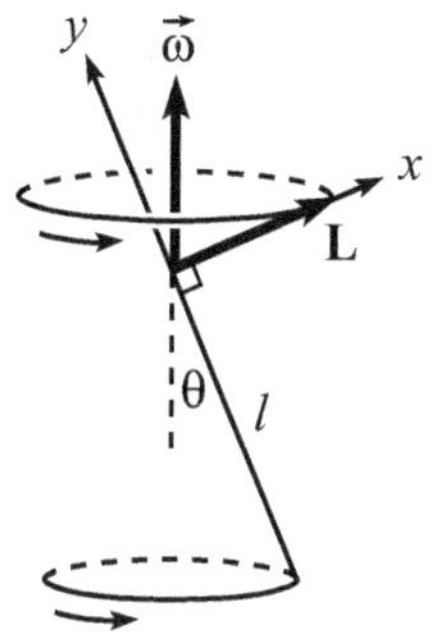

Fig. 9.21

- **Calculate the principal moments:** The principal axes are the axis along the stick, along with any two orthogonal axes perpendicular to the stick. So let the x and y axes be as shown in Fig. 9.21. The positive z axis then points out of the page. The moments (relative to the pivot) are $I_x = m\ell^2/3$, $I_y = 0$, and $I_z = m\ell^2/3$ (which won't be needed).
- **Find L:** The angular velocity vector points vertically (however, see the third remark following this solution), so in the basis of the principal axes, the angular velocity vector is $\boldsymbol{\omega} = (\omega \sin\theta, \omega\cos\theta, 0)$, where ω is yet to be determined. The angular momentum of the system (relative to the pivot) is therefore

$$\mathbf{L} = (I_x\omega_x,\ I_y\omega_y,\ I_z\omega_z) = \big((1/3)m\ell^2\omega\sin\theta, 0, 0\big). \tag{9.34}$$

- **Find $d\mathbf{L}/dt$:** The vector **L** in Eq. (9.34) points up to the right, along the x axis (at the instant shown in Fig. 9.21), with magnitude $L = (1/3)m\ell^2\omega\sin\theta$. As the stick rotates around the vertical axis, **L** traces out the surface of a cone. That is, the tip of **L** traces out a horizontal circle. The radius of this circle is the horizontal component of **L**, which is $L\cos\theta$. The speed of the tip (which is the magnitude of $d\mathbf{L}/dt$) is therefore

[8] This is a better choice than the CM because this way we won't have to worry about any messy forces acting at the pivot when computing the torque. The task of Exercise 9.41 is to work through the more complicated solution which has the CM as the origin.

$(L\cos\theta)\omega$, because $\mathbf{L}$ rotates around the vertical axis with the same frequency as the stick. So $d\mathbf{L}/dt$ has magnitude

$$\left|\frac{d\mathbf{L}}{dt}\right| = (L\cos\theta)\omega = \frac{1}{3}m\ell^2\omega^2\sin\theta\cos\theta, \tag{9.35}$$

and it points into the page.

REMARK: With more complicated objects where $I_y \neq 0$, $\mathbf{L}$ won't point nicely along a principal axis, so the length of its horizontal component (the radius of the circle that $\mathbf{L}$ traces out) won't immediately be obvious. In this case, you can either explicitly calculate the horizontal component (see the spinning-top example in Section 9.7.5), or you can just do things the formal way by finding the rate of change of $\mathbf{L}$ via the expression $d\mathbf{L}/dt = \boldsymbol{\omega}\times\mathbf{L}$, which holds for all the same reasons that $\mathbf{v} \equiv d\mathbf{r}/dt = \boldsymbol{\omega}\times\mathbf{r}$ holds. In the present problem, we obtain

$$\begin{aligned} d\mathbf{L}/dt &= (\omega\sin\theta, \omega\cos\theta, 0) \times \big((1/3)m\ell^2\omega\sin\theta, 0, 0\big) \\ &= \big(0, 0, -(1/3)m\ell^2\omega^2\sin\theta\cos\theta\big), \end{aligned} \tag{9.36}$$

in agreement with Eq. (9.35). And the direction is correct, because the negative z axis points into the page. Note that we calculated this cross product in the principal-axis basis. Although these axes are changing in time, they present a perfectly good set of basis vectors at any instant. ♣

- **Calculate the torque:** The torque (relative to the pivot) is due to gravity, which effectively acts on the CM of the stick. So $\boldsymbol{\tau} = \mathbf{r}\times\mathbf{F}$ has magnitude

$$\tau = rF\sin\theta = (\ell/2)(mg)\sin\theta, \tag{9.37}$$

and it points into the page.

- **Equate $\boldsymbol{\tau}$ with $d\mathbf{L}/dt$:** The vectors $d\mathbf{L}/dt$ and $\boldsymbol{\tau}$ both point into the page, which is good, because they had better point in the same direction. Equating their magnitudes gives

$$\frac{m\ell^2\omega^2\sin\theta\cos\theta}{3} = \frac{mg\ell\sin\theta}{2} \quad\Longrightarrow\quad \omega = \sqrt{\frac{3g}{2\ell\cos\theta}}. \tag{9.38}$$

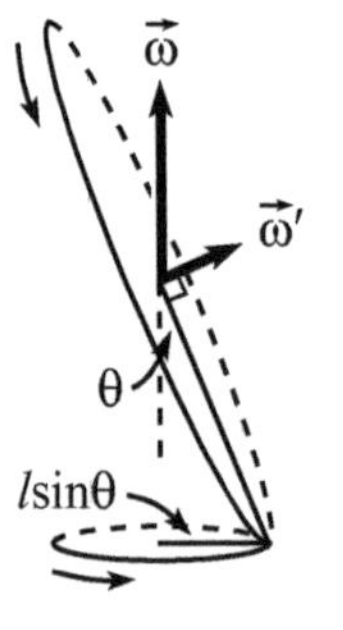

Fig. 9.22

REMARKS:

1. This frequency is slightly larger than the frequency that would arise if we instead had a mass on the end of a massless stick of length ℓ. From Problem 9.12, the frequency in that case is $\sqrt{g/\ell\cos\theta}$. So, in some sense, a uniform stick of length ℓ behaves like a mass on the end of a massless stick of length $2\ell/3$, as far as these rotations are concerned.
2. As $\theta \to \pi/2$, the frequency goes to ∞, which makes sense. And as $\theta \to 0$, it approaches $\sqrt{3g/2\ell}$, which isn't so obvious.
3. As explained in Problem 9.1, the instantaneous $\boldsymbol{\omega}$ is not uniquely defined in some situations. At the instant shown in Fig. 9.20, the stick is moving directly into the page. What if someone else wants to think of the stick as (instantaneously) rotating around the $\boldsymbol{\omega}'$ axis perpendicular to the stick (the x axis, in the above notation), instead of the vertical axis, as shown in Fig. 9.22. What is the angular speed ω'?

   Well, if ω is the angular speed of the stick around the vertical axis, then we may view the tip of the stick as instantaneously moving in a circle of radius $\ell\sin\theta$ around the vertical axis $\boldsymbol{\omega}$. So $\omega(\ell \sin\theta)$ is the speed of the tip of the stick. But we may also view the tip of the stick as instantaneously moving in a circle of radius ℓ around $\boldsymbol{\omega}'$, as shown. The speed of the tip is still $\omega(\ell \sin\theta)$, so the angular speed around this axis is given by $\omega'\ell = \omega(\ell \sin\theta)$. Hence $\omega' = \omega \sin\theta$, which is simply the x component of $\boldsymbol{\omega}$ that we found above, right before Eq. (9.34). The moment of inertia around $\boldsymbol{\omega}'$ is $m\ell^2/3$, so the angular momentum has magnitude $(m\ell^2/3)(\omega \sin\theta)$, in agreement with Eq. (9.34). And the direction is along the x axis, as it should be.

Note that although $\boldsymbol{\omega}$ is not uniquely defined at any instant, $\mathbf{L} \equiv \int (\mathbf{r} \times \mathbf{p})\, dm$ certainly is.[9] Choosing $\boldsymbol{\omega}$ to point vertically, as we did in the above solution, is in some sense the natural choice, because this $\boldsymbol{\omega}$ doesn't change with time. ♣

9.5 Euler's equations

Consider a rigid body instantaneously rotating around an axis $\boldsymbol{\omega}$. This $\boldsymbol{\omega}$ may change as time goes on, but all we care about for now is what it is at a given instant. The angular momentum is given by Eq. (9.8) as $\mathbf{L} = \mathbf{I}\boldsymbol{\omega}$, where $\mathbf{I}$ is the inertia tensor, calculated with respect to a given origin and a given set of axes (and $\boldsymbol{\omega}$ is written in the same basis, of course).

As usual, things are much nicer if we use the principal axes (relative to the chosen origin) as the basis vectors of our coordinate system. Since these axes are fixed with respect to the rotating object, they will rotate with respect to the fixed reference frame. In this basis, $\mathbf{L}$ takes the nice form,

$$\mathbf{L} = (I_1\omega_1, I_2\omega_2, I_3\omega_3), \tag{9.39}$$

where ω_1, ω_2, and ω_3 are the components of $\boldsymbol{\omega}$ along the principal axes. In other words, if you take the vector $\mathbf{L}$ in space and project it onto the instantaneous principal axes, then you get the components in Eq. (9.39).

On one hand, writing $\mathbf{L}$ in terms of the rotating principal axes allows us to write it in the nice form of Eq. (9.39). But on the other hand, writing $\mathbf{L}$ in this way makes it nontrivial to determine how it changes in time, because the principal axes themselves are changing. However, it turns out that the benefits outweigh the detriments, so we will invariably use the principal axes as our basis vectors.

The goal of this section is to find an expression for $d\mathbf{L}/dt$, and to then equate this with the torque. The result will be Euler's equations in Eq. (9.45).

Derivation of Euler's equations

If we write $\mathbf{L}$ in terms of the body frame, which we'll choose to be described by the principal axes painted on the body, then $\mathbf{L}$ can change (relative to the lab frame) due to two effects. It can change because its coordinates in the body frame change, and it can also change because of the rotation of the body frame. To be precise, let $\mathbf{L}_0$ be the vector $\mathbf{L}$ at a given instant. At this instant, imagine painting the vector $\mathbf{L}_0$ onto the body frame, so that $\mathbf{L}_0$ then rotates with the body. The rate

[9] The nonuniqueness of $\boldsymbol{\omega}$ arises from the fact that $I_y = 0$ here. If all the moments are nonzero, then $(L_x, L_y, L_z) = (I_x\omega_x, I_y\omega_y, I_z\omega_z)$ uniquely determines $\boldsymbol{\omega}$, given $\mathbf{L}$.

of change of $\mathbf{L}$ with respect to the lab frame may be written in the (identically true) way,

$$\frac{d\mathbf{L}}{dt} = \frac{d(\mathbf{L} - \mathbf{L}_0)}{dt} + \frac{d\mathbf{L}_0}{dt}. \tag{9.40}$$

The second term here is simply the rate of change of a body-fixed vector, which we know is $\boldsymbol{\omega} \times \mathbf{L}_0$, which equals $\boldsymbol{\omega} \times \mathbf{L}$ at this instant. The first term is the rate of change of $\mathbf{L}$ with respect to the body frame, which we'll denote by $\delta\mathbf{L}/\delta t$. This is what someone standing fixed on the body measures. So we end up with

$$\frac{d\mathbf{L}}{dt} = \frac{\delta\mathbf{L}}{\delta t} + \boldsymbol{\omega} \times \mathbf{L}. \tag{9.41}$$

This is actually a general statement, true for any vector in any rotating frame (we'll derive it in another more mathematical way in Chapter 10). There was nothing particular about $\mathbf{L}$ that we used in the above derivation. Also, there was no need to restrict ourselves to principal axes. In words, what we've shown is that the total change equals the change relative to the rotating frame, plus the change of the rotating frame relative to the fixed frame. This is just the usual way of adding velocities when one frame moves with respect to another.

Let us now make use of our choice of the principal axes as the body axes. This will put Eq. (9.41) in a usable form. Using Eq. (9.39), we can rewrite Eq. (9.41) as

$$\frac{d\mathbf{L}}{dt} = \frac{d}{dt}(I_1\omega_1, I_2\omega_2, I_3\omega_3) + (\omega_1, \omega_2, \omega_3) \times (I_1\omega_1, I_2\omega_2, I_3\omega_3). \tag{9.42}$$

The $\delta\mathbf{L}/\delta t$ term does indeed equal $(d/dt)(I_1\omega_1, I_2\omega_2, I_3\omega_3)$, because someone in the body frame measures the components of $\mathbf{L}$ with respect to the principal axes to be $(I_1\omega_1, I_2\omega_2, I_3\omega_3)$. And $\delta\mathbf{L}/\delta t$ is by definition the rate at which these components change.

Equation (9.42) equates two vectors. As is true for any vector, these (equal) vectors have an existence that is independent of the coordinate system we choose to describe them with (Eq. (9.41) makes no reference to a coordinate system). But since we've chosen an explicit frame on the right-hand side of Eq. (9.42), we should choose the same frame for the left-hand side. We can then equate the components on the left with the components on the right. Projecting $d\mathbf{L}/dt$ onto the instantaneous principal axes, we have

$$\left(\left(\frac{d\mathbf{L}}{dt}\right)_1, \left(\frac{d\mathbf{L}}{dt}\right)_2, \left(\frac{d\mathbf{L}}{dt}\right)_3\right) = \frac{d}{dt}(I_1\omega_1, I_2\omega_2, I_3\omega_3) + (\omega_1, \omega_2, \omega_3) \times (I_1\omega_1, I_2\omega_2, I_3\omega_3). \tag{9.43}$$

REMARK: The left-hand side looks nastier than it really is. The reason we've written it in this cumbersome way is the following (this is a remark that has to be read very slowly). We could have written the left-hand side as $(d/dt)(L_1, L_2, L_3)$, but this might cause confusion as to whether the L_i refer to the components with respect to the rotating axes, or the components with respect to the fixed set of axes that coincide with the rotating principal axes at this instant. That is, do we project **L** onto the principal axes to obtain components, and then take the derivative of these components? Or do we take the derivative of **L** and then project onto the principal axes to obtain components? The latter is what we mean in Eq. (9.43).[10] The way we've written the left-hand side of Eq. (9.43), it's clear that we're taking the derivative first. We are, after all, simply projecting Eq. (9.41) onto the principal axes. ♣

The time derivatives on the right-hand side of Eq. (9.43) are $d(I_1\omega_1)/dt = I_1\dot{\omega}_1$, etc., because the I's are constant. Performing the cross product and equating the corresponding components on each side yields the three equations,

$$\begin{aligned}
\left(\frac{d\mathbf{L}}{dt}\right)_1 &= I_1\dot{\omega}_1 + (I_3 - I_2)\omega_3\omega_2, \\
\left(\frac{d\mathbf{L}}{dt}\right)_2 &= I_2\dot{\omega}_2 + (I_1 - I_3)\omega_1\omega_3, \\
\left(\frac{d\mathbf{L}}{dt}\right)_3 &= I_3\dot{\omega}_3 + (I_2 - I_1)\omega_2\omega_1.
\end{aligned} \tag{9.44}$$

We will now invoke the results of Section 8.4.3 to say that if we have chosen the origin of our rotating frame to be either a fixed point or the CM (as we always do), then we can equate $d\mathbf{L}/dt$ with the torque, $\boldsymbol{\tau}$. We therefore have

$$\begin{aligned}
\tau_1 &= I_1\dot{\omega}_1 + (I_3 - I_2)\omega_3\omega_2, \\
\tau_2 &= I_2\dot{\omega}_2 + (I_1 - I_3)\omega_1\omega_3, \\
\tau_3 &= I_3\dot{\omega}_3 + (I_2 - I_1)\omega_2\omega_1.
\end{aligned} \tag{9.45}$$

These are *Euler's equations*. You need to remember only one of them, because the other two can be obtained by cyclic permutation of the indices.

REMARKS:

1. We repeat that the left- and right-hand sides of Eqs. (9.45) are components that are measured with respect to the instantaneous principal axes. Let's say we do a problem, for example, where τ_3 has a constant nonzero value, and τ_1 and τ_2 are always zero (as in the example in Section 9.4.2). This doesn't mean that $\boldsymbol{\tau}$ is a constant vector. On the contrary, $\boldsymbol{\tau}$ always points along the $\hat{\mathbf{x}}_3$ vector in the rotating frame, but this vector is changing in the fixed frame (unless $\hat{\mathbf{x}}_3$ points along $\boldsymbol{\omega}$).
2. The two types of terms on the right-hand sides of Eqs. (9.44) are the two types of changes that **L** can undergo. **L** can change because its components with respect to the rotating frame change, and **L** can also change because the body is rotating around $\boldsymbol{\omega}$.

[10] The former is $\delta\mathbf{L}/\delta t$, by definition. The two interpretations certainly give different results. For example, if instead of **L** we consider a vector fixed in the body (such as the $\mathbf{L}_0$ above), then the first interpretation gives a zero result, whereas the second interpretation gives a nonzero result. Considering what we mean by, say, the vector $(\omega_1, \omega_2, \omega_3)$, I think that the more logical interpretation of $(d/dt)(L_1, L_2, L_3)$ is the first one, so it should definitely be avoided.

3. Section 9.6.1 on the free symmetric top (viewed from the body frame) provides a good example of the use of Euler's equations. Another interesting application is the famed "Tennis racket theorem" (Problem 9.14).
4. It should be noted that you never *have* to use Euler's equations. You can simply start from scratch and use Eq. (9.41) each time you solve a problem. The point is that we've done the calculation of $d\mathbf{L}/dt$ once and for all, so you can just invoke the result in Eqs. (9.45). ♣

9.6 Free symmetric top

The free symmetric top is the classic example of an application of Euler's equations. Consider an object that has two of its principal moments equal, with the CM as the origin. Assume that the object is in outer space, far from any external forces. We will choose the object to have cylindrical symmetry around some axis (see Fig. 9.23), although this is not necessary; a square cross section, for example, would yield two equal moments. The principal axes are then the symmetry axis and any two orthogonal axes in the cross-section plane through the CM. Let the symmetry axis be chosen as the $\hat{\mathbf{x}}_3$ axis. Then the moments are $I_1 = I_2 \equiv I$, and I_3.

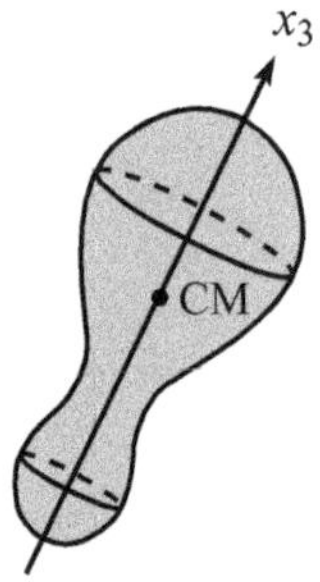

Fig. 9.23

We'll first look at things from the point of view of someone standing at rest on the body, and then we'll look at things from the point of view of someone standing at rest in an inertial frame. The mathematics involved here isn't so bad, but as with most types of top problems, it's hard to get an intuitive feel for what all the various vectors are doing. And the intuition is even more difficult in the following body-frame analysis because of the noninertial frame of reference. But let's see what we find.

9.6.1 View from the body frame

Plugging $I_1 = I_2 \equiv I$ into Euler's equations in Eq. (9.45), and using the fact that all the τ_i are zero (since there are no torques, because the top is "free"), we have

$$\begin{aligned} 0 &= I\dot{\omega}_1 + (I_3 - I)\omega_3\omega_2, \\ 0 &= I\dot{\omega}_2 + (I - I_3)\omega_1\omega_3, \\ 0 &= I_3\dot{\omega}_3. \end{aligned} \tag{9.46}$$

The last equation says that ω_3 is constant. If we then define

$$\Omega \equiv \left(\frac{I_3 - I}{I}\right)\omega_3, \tag{9.47}$$

the first two equations become

$$\dot{\omega}_1 + \Omega\omega_2 = 0, \quad \text{and} \quad \dot{\omega}_2 - \Omega\omega_1 = 0. \tag{9.48}$$

Taking the derivative of the first of these, and then using the second to eliminate $\dot{\omega}_2$, gives

$$\ddot{\omega}_1 + \Omega^2\omega_1 = 0, \tag{9.49}$$

and likewise for ω_2. This is a good old simple-harmonic-oscillator equation. We can therefore write $\omega_1(t)$ as, say, a cosine. And then Eq. (9.48) yields a sine for $\omega_2(t)$. So we have

$$\omega_1(t) = A\cos(\Omega t + \phi), \quad \omega_2(t) = A\sin(\Omega t + \phi). \tag{9.50}$$

We see that $\omega_1(t)$ and $\omega_2(t)$ are the components of a circle in the body frame. Therefore, the $\boldsymbol{\omega}$ vector traces out a cone around $\hat{\mathbf{x}}_3$ (see Fig. 9.24), with frequency Ω, as viewed by someone standing on the body. This frequency Ω in Eq. (9.47) depends on the value of ω_3 and on the geometry of the object (through I_3 and I). But the radius, A, of the $\boldsymbol{\omega}$ cone is determined by the initial values of ω_1 and ω_2.

The angular momentum is

$$\mathbf{L} = (I_1\omega_1, I_2\omega_2, I_3\omega_3) = \Big(IA\cos(\Omega t + \phi),\, IA\sin(\Omega t + \phi),\, I_3\omega_3\Big), \tag{9.51}$$

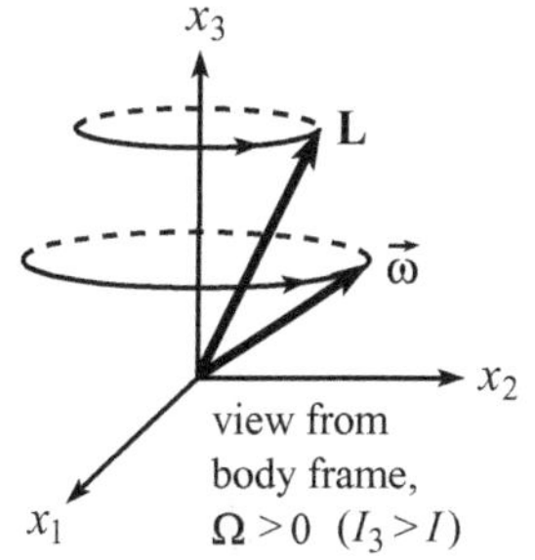

Fig. 9.24

so $\mathbf{L}$ also traces out a cone around $\hat{\mathbf{x}}_3$ with frequency Ω, as viewed by someone standing on the body. This is shown in Fig. 9.24 for the case $\Omega > 0$ (that is, $I_3 > I$). In this case, $I_3 > I$ implies $L_3/L_2 > \omega_3/\omega_2$, so the $\mathbf{L}$ vector lies above the $\boldsymbol{\omega}$ vector (that is, between $\boldsymbol{\omega}$ and $\hat{\mathbf{x}}_3$). An object with $I_3 > I$, such as a coin, is called an *oblate* top.

Figure 9.25 shows the case where $\Omega < 0$ (that is, $I_3 < I$). In this case, $I_3 < I$ implies $L_3/L_2 < \omega_3/\omega_2$, so the $\mathbf{L}$ vector lies below the $\boldsymbol{\omega}$ vector, as shown. And since Ω is negative, the $\boldsymbol{\omega}$ and $\mathbf{L}$ vectors precess around $\hat{\mathbf{x}}_3$ in the opposite direction (clockwise, as viewed from above). An object with $I_3 < I$, such as a carrot, is called a *prolate* top.

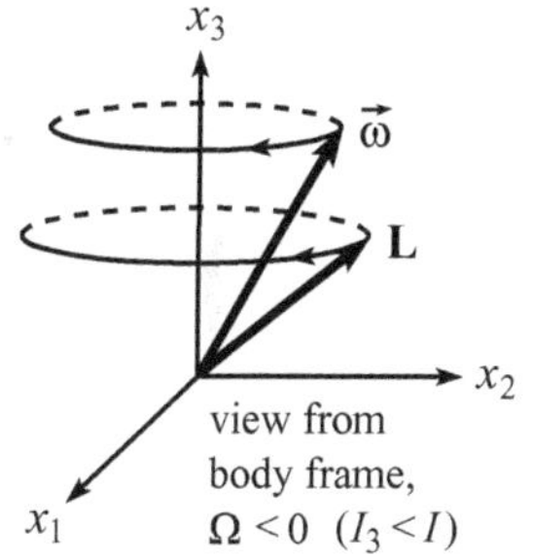

Fig. 9.25

Example (The earth): Let's consider the earth to be our top. Then $\omega_3 \approx 2\pi/(1 \text{ day})$.[11] The bulge at the equator (caused by the spinning of the earth) makes I_3 slightly larger than I, and it turns out that $(I_3 - I)/I \approx 1/320$. Therefore, Eq. (9.47) gives $\Omega \approx (1/320)\, 2\pi/(1 \text{ day})$. So the $\boldsymbol{\omega}$ vector should precess around in its cone once every 320 days, as viewed by someone on the earth. The true value is more like 430 days. The difference has to do with various things, including the nonrigidity of the earth, but at least we got an answer in the right ballpark. This precession of $\boldsymbol{\omega}$ is known as the "Chandler wobble."

In practice, how can we determine the direction of $\boldsymbol{\omega}$? Simply take an extended-time photograph exposure at night. The stars will form arcs of circles. At the center of all these circles is a point that doesn't move. This is the direction of $\boldsymbol{\omega}$. Fortunately, Ω is much smaller than ω, so the $\boldsymbol{\omega}$ vector doesn't change much during an exposure time of, say, an hour. So the center of the circles is essentially well defined.

How big is the $\boldsymbol{\omega}$ cone, for the earth? Equivalently, what is the value of A in Eq. (9.50)? Observation has shown that the $\boldsymbol{\omega}$ vector pierces the earth at a point on

[11] This isn't quite correct, since the earth rotates 366 times for every 365 days, due to the motion around the sun, but it's close enough for the purposes here.

the order of 10 m from the north pole, although this distance fluctuates over time.[12] Hence, $A/\omega_3 \approx (10\,\text{m})/R_E$. The half-angle of the $\boldsymbol{\omega}$ cone is therefore on the order of only 10^{-4} degrees. So if you use an extended-time photograph exposure one night to see which point in the sky stands still, and then if you do the same thing 200 nights later, you probably won't be able to tell that they're really two different points.

9.6.2 View from a fixed frame

Let's now see what our symmetric top looks like from a fixed frame. Euler's equations won't help much here, because they deal with the components of $\boldsymbol{\omega}$ in the body frame. But fortunately we can solve for the motion from scratch. In terms of the (changing) principal axes, $\hat{\mathbf{x}}_1, \hat{\mathbf{x}}_2, \hat{\mathbf{x}}_3$, we have

$$\begin{aligned} \boldsymbol{\omega} &= (\omega_1\hat{\mathbf{x}}_1 + \omega_2\hat{\mathbf{x}}_2) + \omega_3\hat{\mathbf{x}}_3, \\ \mathbf{L} &= I(\omega_1\hat{\mathbf{x}}_1 + \omega_2\hat{\mathbf{x}}_2) + I_3\omega_3\hat{\mathbf{x}}_3. \end{aligned} \tag{9.52}$$

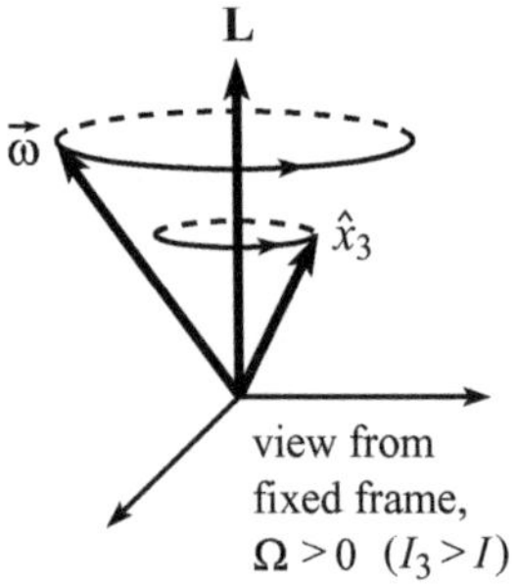

Fig. 9.26

Eliminating the $(\omega_1\hat{\mathbf{x}}_1 + \omega_2\hat{\mathbf{x}}_2)$ term from these equations gives (in terms of the Ω defined in Eq. (9.47))

$$\mathbf{L} = I(\boldsymbol{\omega} + \Omega\hat{\mathbf{x}}_3) \quad \Longrightarrow \quad \boldsymbol{\omega} = \frac{L}{I}\hat{\mathbf{L}} - \Omega\hat{\mathbf{x}}_3, \tag{9.53}$$

where $L = |\mathbf{L}|$, and $\hat{\mathbf{L}}$ is the unit vector in the $\mathbf{L}$ direction. The linear relationship among $\mathbf{L}$, $\boldsymbol{\omega}$, and $\hat{\mathbf{x}}_3$ implies that these three vectors lie in a plane. But $\mathbf{L}$ remains fixed, because there are no torques on the system. Therefore, $\boldsymbol{\omega}$ and $\hat{\mathbf{x}}_3$ precess (as we'll see below) around $\mathbf{L}$, with the three vectors always coplanar. See Fig. 9.26 for the case $\Omega > 0$, that is, $I_3 > I$ (an oblate top), and Fig. 9.27 for the case $\Omega < 0$, that is, $I_3 < I$ (a prolate top).

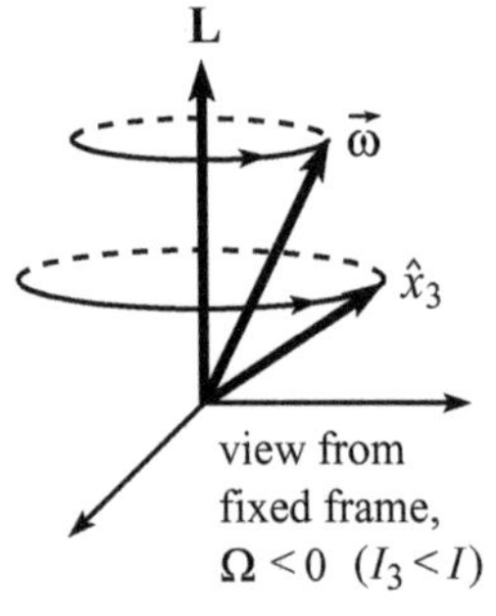

Fig. 9.27

What is the frequency of this precession, as viewed from the fixed frame? The rate of change of $\hat{\mathbf{x}}_3$ is $\boldsymbol{\omega} \times \hat{\mathbf{x}}_3$, because $\hat{\mathbf{x}}_3$ is fixed in the body frame, so its change comes only from rotation around $\boldsymbol{\omega}$. Therefore, Eq. (9.53) gives

$$\frac{d\hat{\mathbf{x}}_3}{dt} = \left(\frac{L}{I}\hat{\mathbf{L}} - \Omega\hat{\mathbf{x}}_3\right) \times \hat{\mathbf{x}}_3 = \left(\frac{L}{I}\hat{\mathbf{L}}\right) \times \hat{\mathbf{x}}_3. \tag{9.54}$$

But this is simply the expression for the rate of change of a vector rotating around the fixed vector $\tilde{\boldsymbol{\omega}} \equiv (L/I)\hat{\mathbf{L}}$. The frequency of this rotation is $|\tilde{\boldsymbol{\omega}}| = L/I$. Therefore, $\hat{\mathbf{x}}_3$ precesses around the fixed vector $\mathbf{L}$ with frequency

$$\tilde{\omega} = \frac{L}{I}, \tag{9.55}$$

[12] This distance could theoretically be much larger or much smaller than 10 m. It happens to be of this order due to the nature of the driving force. The present consensus for this force is pressure changes at the bottom of the ocean and in the atmosphere; see Gross (2000). Without a driving force, the amplitude would head to zero, due to the nonrigidity of the earth.

in the fixed frame. And therefore $\boldsymbol{\omega}$ does also, because it is coplanar with $\hat{\mathbf{x}}_3$ and $\mathbf{L}$.

REMARKS:

1. We just found that $\boldsymbol{\omega}$ precesses around $\mathbf{L}$ with frequency L/I. What, then, is wrong with the following reasoning: "Just as the rate of change of $\hat{\mathbf{x}}_3$ equals $\boldsymbol{\omega} \times \hat{\mathbf{x}}_3$, the rate of change of $\boldsymbol{\omega}$ should equal $\boldsymbol{\omega} \times \boldsymbol{\omega}$, which is zero. Therefore, $\boldsymbol{\omega}$ should remain constant." The error is that the vector $\boldsymbol{\omega}$ is not fixed in the body frame. A vector $\mathbf{A}$ must be fixed in the body frame in order for its rate of change to be given by $\boldsymbol{\omega} \times \mathbf{A}$.
2. We found in Eqs. (9.51) and (9.47) that a person standing on the rotating body sees $\mathbf{L}$ (and $\boldsymbol{\omega}$) precess with frequency $\Omega \equiv \omega_3(I_3 - I)/I$ around $\hat{\mathbf{x}}_3$. But we found in Eq. (9.55) that a person standing in the fixed frame sees $\hat{\mathbf{x}}_3$ (and $\boldsymbol{\omega}$) precess with frequency L/I around $\mathbf{L}$. Are these two facts compatible? Should we have obtained the same frequency from either point of view? (Answers: yes, no).

   These two frequencies are indeed consistent, as can be seen by the following reasoning. Consider the plane (call it S) containing the three vectors $\mathbf{L}$, $\boldsymbol{\omega}$, and $\hat{\mathbf{x}}_3$. We know from Eq. (9.51) that S rotates with frequency $\Omega\hat{\mathbf{x}}_3$ with respect to the body. Therefore, the body rotates with frequency $-\Omega\hat{\mathbf{x}}_3$ with respect to S. And from Eq. (9.55), S rotates with frequency $(L/I)\hat{\mathbf{L}}$ with respect to the fixed frame. Therefore, the total angular velocity of the body with respect to the fixed frame is (using the frame S as an intermediate step)

$$\boldsymbol{\omega}_{\text{total}} = \frac{L}{I}\hat{\mathbf{L}} - \Omega\hat{\mathbf{x}}_3. \tag{9.56}$$

   But from Eq. (9.53), this is simply $\boldsymbol{\omega}$, as it should be. So the two frequencies in Eqs. (9.47) and (9.55) are indeed consistent.

   For the earth, I_3 and I are nearly the same, so $\Omega \equiv \omega_3(I_3 - I)/I$ and L/I are quite different. L/I is roughly equal to L/I_3, which is essentially equal to ω_3. On the other hand, Ω is roughly equal to $(1/300)\omega_3$. Basically, an external observer sees $\boldsymbol{\omega}$ precess around its cone at roughly the rate at which the earth spins. But it's not exactly the same rate, and this difference is what causes the earth-based observer to see $\boldsymbol{\omega}$ precess with a nonzero Ω.
3. The fixed-frame precession of $\hat{\mathbf{x}}_3$ around $\mathbf{L}$ should not be confused with the "precession of the equinoxes" effect. See Problem 10.15 for a discussion of the latter. ♣

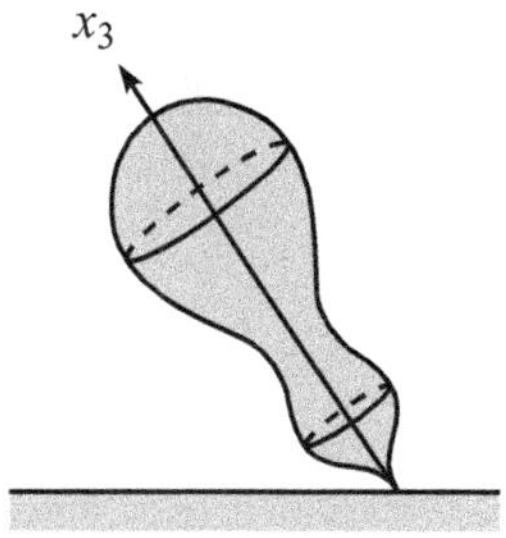

Fig. 9.28

9.7 Heavy symmetric top

Consider now a heavy symmetrical top, that is, one that spins on a table, under the influence of gravity (see Fig. 9.28). Assume that the tip of the top is fixed on the table by a pivot. We'll solve for the motion of the top in two different ways below in Sections 9.7.3 and 9.7.4. The first uses $\boldsymbol{\tau} = d\mathbf{L}/dt$, and the second uses the Lagrangian method.

9.7.1 Euler angles

For both of these methods, it is convenient to use the *Euler angles*, θ, ϕ, ψ, which are shown in Fig. 9.29 and defined as follows.

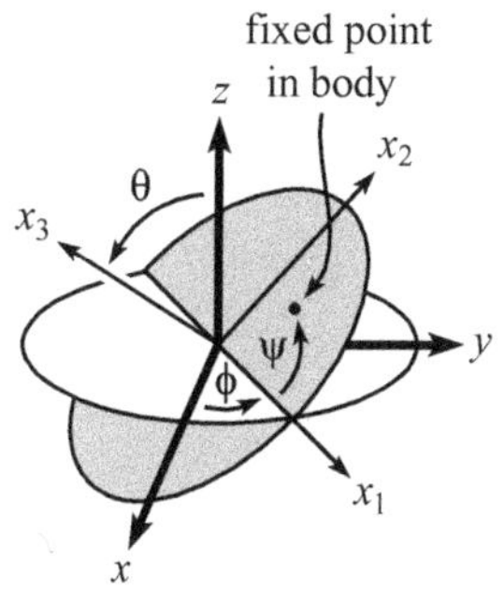

Fig. 9.29

- θ: Let $\hat{\mathbf{x}}_3$ be the symmetry axis of the top. Define θ to be the angle that $\hat{\mathbf{x}}_3$ makes with the vertical $\hat{\mathbf{z}}$ axis of the fixed frame.

- ϕ: Draw the plane orthogonal to $\hat{\mathbf{x}}_3$. Let $\hat{\mathbf{x}}_1$ be the intersection of this plane with the horizontal *x*-*y* plane. Define ϕ to be the angle that $\hat{\mathbf{x}}_1$ makes with the $\hat{\mathbf{x}}$ axis in the fixed frame. Note that $\hat{\mathbf{x}}_1$ is not necessarily fixed in the object.
- ψ: Let $\hat{\mathbf{x}}_2$ be orthogonal to $\hat{\mathbf{x}}_3$ and $\hat{\mathbf{x}}_1$, as shown. As with $\hat{\mathbf{x}}_1$, $\hat{\mathbf{x}}_2$ is not necessarily fixed in the object. Let frame S be the frame whose axes are $\hat{\mathbf{x}}_1$, $\hat{\mathbf{x}}_2$, and $\hat{\mathbf{x}}_3$. Define ψ to be the angle of rotation of the body around the $\hat{\mathbf{x}}_3$ axis in frame S. So $\dot{\psi}\hat{\mathbf{x}}_3$ is the angular velocity of the body with respect to S. And from the figure, we also see that the angular velocity of frame S with respect to the fixed frame is $\dot{\phi}\hat{\mathbf{z}} + \dot{\theta}\hat{\mathbf{x}}_1$.

The angular velocity of the body with respect to the fixed frame is equal to the angular velocity of the body with respect to frame S, plus the angular velocity of frame S with respect to the fixed frame. From above, we therefore have

$$\boldsymbol{\omega} = \dot{\psi}\hat{\mathbf{x}}_3 + (\dot{\phi}\hat{\mathbf{z}} + \dot{\theta}\hat{\mathbf{x}}_1). \tag{9.57}$$

It is often more convenient to rewrite $\boldsymbol{\omega}$ entirely in terms of the orthogonal $\hat{\mathbf{x}}_1$, $\hat{\mathbf{x}}_2$, $\hat{\mathbf{x}}_3$ basis vectors. Since $\hat{\mathbf{z}} = \cos\theta\hat{\mathbf{x}}_3 + \sin\theta\hat{\mathbf{x}}_2$, Eq. (9.57) gives

$$\boldsymbol{\omega} = (\dot{\psi} + \dot{\phi}\cos\theta)\hat{\mathbf{x}}_3 + \dot{\phi}\sin\theta\hat{\mathbf{x}}_2 + \dot{\theta}\hat{\mathbf{x}}_1. \tag{9.58}$$

This form of $\boldsymbol{\omega}$ is generally more useful, because $\hat{\mathbf{x}}_1$, $\hat{\mathbf{x}}_2$, $\hat{\mathbf{x}}_3$ are principal axes of the body. (We are assuming that we are working with a symmetrical top, with $I_1 = I_2 \equiv I$. This means that any axes in the $\hat{\mathbf{x}}_1$-$\hat{\mathbf{x}}_2$ plane are principal axes.) Although $\hat{\mathbf{x}}_1$ and $\hat{\mathbf{x}}_2$ are not fixed in the object, they are still good principal axes at any instant.

9.7.2 Digression on the components of $\boldsymbol{\omega}$

The above expressions for $\boldsymbol{\omega}$ might look a little scary, but there is a very helpful diagram we can draw (see Fig. 9.30) that makes it easier to see what's going on. Let's talk a bit about this before tackling the original problem of the spinning top. The diagram is rather dense (you might even say it looks scarier than the above $\boldsymbol{\omega}$), so we'll go through it slowly. In the following discussion, we'll simplify things by setting $\dot{\theta} = 0$. All the interesting features of $\boldsymbol{\omega}$ remain. The $\dot{\theta}\hat{\mathbf{x}}_1$ component of $\boldsymbol{\omega}$ in Eqs. (9.57) and (9.58) arises simply from the easily visualizable rising and falling of the top. We will therefore concentrate on the more complicated issues, namely the components of $\boldsymbol{\omega}$ in the plane of $\hat{\mathbf{x}}_3$, $\hat{\mathbf{z}}$, and $\hat{\mathbf{x}}_2$.

With $\dot{\theta} = 0$, Fig. 9.30 shows the vector $\boldsymbol{\omega}$ in the $\hat{\mathbf{x}}_3$-$\hat{\mathbf{z}}$-$\hat{\mathbf{x}}_2$ plane (the way we've drawn it, $\hat{\mathbf{x}}_1$ points into the page, in contrast with Fig. 9.29). We'll refer to this figure many times in the problems for this chapter. There are numerous comments to be made about it, so we'll just list them out. The following discussion deals with the *kinematics* of $\boldsymbol{\omega}$, that is, the meaning of the various components and how they relate to each other. The discussion of the *dynamics* of $\boldsymbol{\omega}$, that is, why the components take on the values they do, given a certain physical system, is the subject of Section 9.7.3 onward.

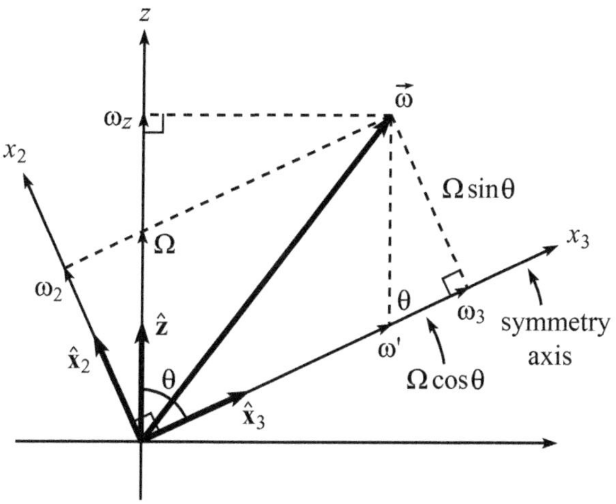

Fig. 9.30

1. If someone asks you to "decompose" $\boldsymbol{\omega}$ into pieces along $\hat{\mathbf{z}}$ and $\hat{\mathbf{x}}_3$, what would you do? Would you draw the lines perpendicular to these axes to obtain the lengths shown (which we'll label as ω_z and ω_3), or would you draw the lines parallel to these axes to obtain the lengths shown (which we'll label as Ω and ω')? There is no "correct" answer to this question. The four quantities, ω_z, ω_3, Ω, ω' simply represent different things. We will interpret each of these below, along with ω_2 (the projection of $\boldsymbol{\omega}$ along $\hat{\mathbf{x}}_2$). It turns out that Ω and ω' are the frequencies that your eye can see the easiest, while ω_2 and ω_3 are what you want to use when doing calculations involving the angular momentum. But as far as I can see, ω_z is not of much use.
2. Note that it is true that

$$\boldsymbol{\omega} = \omega' \hat{\mathbf{x}}_3 + \Omega \hat{\mathbf{z}}, \tag{9.59}$$

but it is *not* true that $\boldsymbol{\omega} = \omega_3 \hat{\mathbf{x}}_3 + \omega_z \hat{\mathbf{z}}$. Another true statement is

$$\boldsymbol{\omega} = \omega_3 \hat{\mathbf{x}}_3 + \omega_2 \hat{\mathbf{x}}_2. \tag{9.60}$$

3. In terms of the Euler angles, we see by comparing Eqs. (9.59) and (9.57), with $\dot{\theta} = 0$, that

$$\omega' = \dot{\psi}, \quad \text{and} \quad \Omega = \dot{\phi}. \tag{9.61}$$

And we also have, by comparing Eqs. (9.60) and (9.58), with $\dot{\theta} = 0$,

$$\begin{aligned} \omega_3 &= \dot{\psi} + \dot{\phi}\cos\theta \;=\; \omega' + \Omega\cos\theta, \\ \omega_2 &= \dot{\phi}\sin\theta \;=\; \Omega\sin\theta. \end{aligned} \tag{9.62}$$

These are also clear from Fig. 9.30. There is therefore technically no need to introduce the new ω_2, ω_3, Ω, ω' definitions in Fig. 9.30, because the Euler angles are quite sufficient. But we will be referring back to this figure many times, and it is a little easier to work with these omegas than the various combinations of the Euler angles.

4. Ω is the easiest of the frequencies to visualize. It is the frequency of precession of the top around the vertical $\hat{\mathbf{z}}$ axis.[13] In other words, the symmetry axis $\hat{\mathbf{x}}_3$ traces out a cone (assuming $\dot{\theta} = 0$) around the $\hat{\mathbf{z}}$ axis with frequency Ω. The reason for this is the following. The vector $\boldsymbol{\omega}$ is the vector that gives the speed of any point (at position $\mathbf{r}$) fixed in the top as $\boldsymbol{\omega} \times \mathbf{r}$. Therefore, since the vector $\hat{\mathbf{x}}_3$ is fixed in the top, we can write

$$\frac{d\hat{\mathbf{x}}_3}{dt} = \boldsymbol{\omega} \times \hat{\mathbf{x}}_3 = (\omega' \hat{\mathbf{x}}_3 + \Omega \hat{\mathbf{z}}) \times \hat{\mathbf{x}}_3 = (\Omega \hat{\mathbf{z}}) \times \hat{\mathbf{x}}_3. \qquad (9.63)$$

But this is precisely the expression for the rate of change of a vector rotating around the $\hat{\mathbf{z}}$ axis with frequency Ω. (This is exactly the same type of proof as the one leading to Eq. (9.54).) Note that the precession frequency around the $\hat{\mathbf{z}}$ axis is *not* ω_z. It certainly can't be ω_z, because we can imagine grabbing the symmetry axis and holding it in place, so that $\boldsymbol{\omega}$ points along $\hat{\mathbf{x}}_3$. This scenario has a nonzero ω_z, but no precession.

REMARK: In the derivation of Eq. (9.63), we basically just stripped off the part of $\boldsymbol{\omega}$ that points along the $\hat{\mathbf{x}}_3$ axis, because a rotation around $\hat{\mathbf{x}}_3$ contributes nothing to the motion of $\hat{\mathbf{x}}_3$. Note, however, that there is in fact an infinite number of ways to strip off a piece along $\hat{\mathbf{x}}_3$. For example, we can also break $\boldsymbol{\omega}$ up as, say, $\boldsymbol{\omega} = \omega_3 \hat{\mathbf{x}}_3 + \omega_2 \hat{\mathbf{x}}_2$. We then obtain $d\hat{\mathbf{x}}_3/dt = (\omega_2 \hat{\mathbf{x}}_2) \times \hat{\mathbf{x}}_3$, which means that $\hat{\mathbf{x}}_3$ is instantaneously rotating around $\hat{\mathbf{x}}_2$ with frequency ω_2. Although this is true, it isn't as useful as the result in Eq. (9.63), because the $\hat{\mathbf{x}}_2$ axis changes with time (it precesses around $\hat{\mathbf{z}}$). The point here is that the instantaneous angular velocity vector around which the symmetry axis rotates is not well defined (Problem 9.1 discusses this issue).[14] But the $\hat{\mathbf{z}}$ axis is the only one of these angular velocity vectors that is fixed. When we look at the top (or more precisely, the symmetry axis), we therefore see it precessing around the $\hat{\mathbf{z}}$ axis. ♣

5. ω' is also easy to visualize. Imagine that you are at rest in the frame that rotates around the $\hat{\mathbf{z}}$ axis with frequency Ω. Then you see the symmetry axis of the top remain perfectly still, and the only motion you see is the top spinning around this axis with frequency ω'. (This is true because $\boldsymbol{\omega} = \omega' \hat{\mathbf{x}}_3 + \Omega \hat{\mathbf{z}}$, and the rotation of your frame causes you not to see the $\Omega \hat{\mathbf{z}}$ part.) If you paint a dot somewhere on the top, then the dot traces out a fixed tilted circle, and the dot returns to, say, its maximum height at frequency ω'. A person in the lab frame sees this dot undergo a rather complicated motion but must observe the same frequency at which the dot returns to its highest point. So ω' is something quite physical in the lab frame also.

6. ω_3 is what you use to obtain the component of $\mathbf{L}$ along $\hat{\mathbf{x}}_3$, because $L_3 = I_3 \omega_3$. ω_3 is a little harder to visualize than Ω and ω', but it is the frequency with which the top

[13] Although we're using the same letter, this Ω doesn't have anything to do with the Ω defined in Eq. (9.47), except for the fact that they both represent the frequency of something precessing around an axis.

[14] The instantaneous angular velocity of the *whole body* is well defined, of course. There is a definite line of points in the body that are instantaneously at rest. But if you look at the symmetry axis by itself, then there is an ambiguity (see Footnote 9). In short, because only one point on the axis (the bottom end), instead of a whole line, is instantaneously at rest, the instantaneous angular velocity vector can point in any direction.

instantaneously rotates, as seen by someone at rest in the frame that rotates around the instantaneous $\hat{\mathbf{x}}_2$ axis with frequency ω_2. (This is true because $\boldsymbol{\omega} = \omega_2\hat{\mathbf{x}}_2 + \omega_3\hat{\mathbf{x}}_3$, and the rotation of the frame causes the person not to see the $\omega_2\hat{\mathbf{x}}_2$ part.) This rotation is harder to vizualize in the lab frame, because the $\hat{\mathbf{x}}_2$ axis changes with time.

There is one physical scenario in which ω_3 is the easily observed frequency. Imagine that the top is precessing around the $\hat{\mathbf{z}}$ axis at constant θ (we'll find in Section 9.7.5 that this is in fact a possible motion for the top), and imagine that the top has a frictionless rod protruding along its symmetry axis. If you grab the rod and stop the precession motion, so that the top is now spinning around its stationary symmetry axis, then this spinning has frequency ω_3. This is true because when you grab the rod, your torque has no component along the $\hat{\mathbf{x}}_3$ axis (because the rod lies along this axis, and because it is frictionless). Therefore, L_3 doesn't change, and so neither does ω_3.

7. ω_2 is similar to ω_3, of course. ω_2 is what you use to obtain the component of $\mathbf{L}$ along $\hat{\mathbf{x}}_2$, because $L_2 = I_2\omega_2$. It is the frequency with which the top instantaneously rotates, as seen by someone at rest in the frame that rotates around the instantaneous $\hat{\mathbf{x}}_3$ axis with frequency ω_3. (This is true because $\boldsymbol{\omega} = \omega_2\hat{\mathbf{x}}_2 + \omega_3\hat{\mathbf{x}}_3$, and the rotation of the frame causes the person not to see the $\omega_3\hat{\mathbf{x}}_3$ part.) Again, this rotation is harder to vizualize in the lab frame, because the $\hat{\mathbf{x}}_3$ axis changes with time. Note that by "instantaneous $\hat{\mathbf{x}}_3$ axis," we mean the fixed axis in space that coincides with the symmetry axis at a given instant. The symmetry axis will therefore move away from this fixed axis, consistent with the fact that the person in the above-mentioned rotating frame sees the top rotate around the $\hat{\mathbf{x}}_2$ axis.

   The physical scenario that produces ω_2, analogous to the scenario that produced ω_3 above, is the following. Imagine a frictionless rod glued to the top at its tip, perpendicular to the symmetry axis, so that they form a "T." As this rod is spinning around (ignore the fact that it has to keep passing through the table), grab it at the instant it points along $\hat{\mathbf{x}}_2$. The top will then rotate with frequency ω_2 around the fixed rod. This is true for reasons analogous to the ones in the ω_3 case above.

8. ω_z is not very useful, as far as I can see. The most important thing to note about ω_z is that it is *not* the frequency of precession around the $\hat{\mathbf{z}}$ axis, even though it is the projection of $\boldsymbol{\omega}$ onto $\hat{\mathbf{z}}$. The frequency of the precession is Ω, as we found above in Eq. (9.63). A true, but somewhat useless, fact about ω_z is that if someone is at rest in the frame that rotates around the $\hat{\mathbf{z}}$ axis with frequency ω_z, then she sees all points in the top instantaneously rotating around the horizontal $\hat{\mathbf{x}}$ axis with frequency ω_x, where ω_x is the projection of $\boldsymbol{\omega}$ onto the $\hat{\mathbf{x}}$ axis. (This is true because $\boldsymbol{\omega} = \omega_x\hat{\mathbf{x}} + \omega_z\hat{\mathbf{z}}$, and the rotation of the frame causes her not to see the $\omega_z\hat{\mathbf{z}}$ part.)

9.7.3 Torque method

Let's now finally solve for the motion of a heavy top. This first method involving torque is straightforward, although a bit tedious. We include it here to (1) show that the problem can be done without resorting to a Lagrangian, and to (2) get some practice using $\boldsymbol{\tau} = d\mathbf{L}/dt$. We'll make use of the form of $\boldsymbol{\omega}$ given in

Eq. (9.58), because there it is broken up into the principal-axis components. For convenience, define $\dot{\beta} = \dot{\psi} + \dot{\phi}\cos\theta$, so that

$$\boldsymbol{\omega} = \dot{\beta}\hat{\mathbf{x}}_3 + \dot{\phi}\sin\theta\hat{\mathbf{x}}_2 + \dot{\theta}\hat{\mathbf{x}}_1. \tag{9.64}$$

Note that we've returned to the most general motion, where $\dot{\theta}$ is not necessarily zero. For our origin, we'll choose the tip of the top, which is assumed to be fixed on the table.[15] Let the principal moments relative to this origin be $I_1 = I_2 \equiv I$, and I_3. The angular momentum of the top is then

$$\mathbf{L} = I_3\dot{\beta}\hat{\mathbf{x}}_3 + I\dot{\phi}\sin\theta\hat{\mathbf{x}}_2 + I\dot{\theta}\hat{\mathbf{x}}_1. \tag{9.65}$$

We must now calculate $d\mathbf{L}/dt$. What makes this nontrivial is the fact that the $\hat{\mathbf{x}}_1$, $\hat{\mathbf{x}}_2$, and $\hat{\mathbf{x}}_3$ unit vectors change with time (they change with θ and ϕ). But let's forge ahead and take the derivative of Eq. (9.65). Using the product rule (which works fine with the product of a scalar and a vector), we have

$$\begin{aligned}\frac{d\mathbf{L}}{dt} &= I_3\frac{d\dot{\beta}}{dt}\hat{\mathbf{x}}_3 + I\frac{d(\dot{\phi}\sin\theta)}{dt}\hat{\mathbf{x}}_2 + I\frac{d\dot{\theta}}{dt}\hat{\mathbf{x}}_1 \\ &\quad + I_3\dot{\beta}\frac{d\hat{\mathbf{x}}_3}{dt} + I\dot{\phi}\sin\theta\frac{d\hat{\mathbf{x}}_2}{dt} + I\dot{\theta}\frac{d\hat{\mathbf{x}}_1}{dt}.\end{aligned} \tag{9.66}$$

Using a little geometry, you can show that

$$\begin{aligned}\frac{d\hat{\mathbf{x}}_3}{dt} &= -\dot{\theta}\hat{\mathbf{x}}_2 + \dot{\phi}\sin\theta\hat{\mathbf{x}}_1, \\ \frac{d\hat{\mathbf{x}}_2}{dt} &= \dot{\theta}\hat{\mathbf{x}}_3 - \dot{\phi}\cos\theta\hat{\mathbf{x}}_1, \\ \frac{d\hat{\mathbf{x}}_1}{dt} &= -\dot{\phi}\sin\theta\hat{\mathbf{x}}_3 + \dot{\phi}\cos\theta\hat{\mathbf{x}}_2.\end{aligned} \tag{9.67}$$

As an exercise, you should verify these by making use of Fig. 9.29. In the first equation, for example, show that a change in θ causes $\hat{\mathbf{x}}_3$ to move a certain distance in the $\hat{\mathbf{x}}_2$ direction; and show that a change in ϕ causes $\hat{\mathbf{x}}_3$ to move a certain distance in the $\hat{\mathbf{x}}_1$ direction. Plugging the derivative expressions from Eq. (9.67) into Eq. (9.66) gives, after some algebra,

$$\begin{aligned}\frac{d\mathbf{L}}{dt} &= I_3\ddot{\beta}\hat{\mathbf{x}}_3 + \left(I\ddot{\phi}\sin\theta + 2I\dot{\theta}\dot{\phi}\cos\theta - I_3\dot{\beta}\dot{\theta}\right)\hat{\mathbf{x}}_2 \\ &\quad + \left(I\ddot{\theta} - I\dot{\phi}^2\sin\theta\cos\theta + I_3\dot{\beta}\dot{\phi}\sin\theta\right)\hat{\mathbf{x}}_1.\end{aligned} \tag{9.68}$$

Let's now look at the torque on the top. This arises from gravity pulling down on the CM. So from Fig. 9.29, $\boldsymbol{\tau}$ points in the $\hat{\mathbf{x}}_1$ direction and has magnitude

[15] We could use the CM as our origin, but then we would have to include the complicated forces acting at the pivot point, which is difficult. But see Problem 9.19 for the case where the tip is free to slide on a frictionless table.

$Mg\ell \sin\theta$, where ℓ is the distance from the pivot to the CM. Using Eq. (9.68), the third component of $\boldsymbol{\tau} = d\mathbf{L}/dt$ quickly gives

$$\ddot{\beta} = 0. \tag{9.69}$$

Therefore, $\dot{\beta}$ is a constant, which we'll call ω_3 (an obvious label, in view of Eq. (9.64)). The other two components of $\boldsymbol{\tau} = d\mathbf{L}/dt$ then give

$$\begin{aligned} I\ddot{\phi}\sin\theta + \dot{\theta}(2I\dot{\phi}\cos\theta - I_3\omega_3) &= 0, \\ (Mg\ell + I\dot{\phi}^2\cos\theta - I_3\omega_3\dot{\phi})\sin\theta &= I\ddot{\theta}. \end{aligned} \tag{9.70}$$

We'll wait to fiddle with these equations until we have derived them again using the Lagrangian method.

9.7.4 Lagrangian method

Equation (9.15) gives the kinetic energy of the top as $T = \boldsymbol{\omega} \cdot \mathbf{L}/2$. Using Eqs. (9.64) and (9.65), we have (writing $\dot{\psi} + \dot{\phi}\cos\theta$ instead of the shorthand $\dot{\beta}$)[16]

$$T = \frac{1}{2}\boldsymbol{\omega} \cdot \mathbf{L} = \frac{1}{2}I_3(\dot{\psi} + \dot{\phi}\cos\theta)^2 + \frac{1}{2}I(\dot{\phi}^2\sin^2\theta + \dot{\theta}^2). \tag{9.71}$$

The potential energy is

$$V = Mg\ell\cos\theta, \tag{9.72}$$

where ℓ is the distance from the pivot to the CM. The Lagrangian is $\mathcal{L} = T - V$ (we'll use "$\mathcal{L}$" here to avoid confusion with the angular momentum "L"), and so the equation of motion obtained from varying ψ is

$$\frac{d}{dt}\frac{\partial\mathcal{L}}{\partial\dot{\psi}} = \frac{\partial\mathcal{L}}{\partial\psi} \quad \Longrightarrow \quad \frac{d}{dt}(\dot{\psi} + \dot{\phi}\cos\theta) = 0. \tag{9.73}$$

Therefore, $\dot{\psi} + \dot{\phi}\cos\theta$ is a constant. Call it ω_3. The equations of motion obtained from varying ϕ and θ are then (making use of $\dot{\psi} + \dot{\phi}\cos\theta = \omega_3$)

$$\begin{aligned} \frac{d}{dt}\frac{\partial\mathcal{L}}{\partial\dot{\phi}} = \frac{\partial\mathcal{L}}{\partial\phi} \quad &\Longrightarrow \quad \frac{d}{dt}\Big(I_3\omega_3\cos\theta + I\dot{\phi}\sin^2\theta\Big) = 0, \\ \frac{d}{dt}\frac{\partial\mathcal{L}}{\partial\dot{\theta}} = \frac{\partial\mathcal{L}}{\partial\theta} \quad &\Longrightarrow \quad I\ddot{\theta} = (Mg\ell + I\dot{\phi}^2\cos\theta - I_3\omega_3\dot{\phi})\sin\theta. \end{aligned} \tag{9.74}$$

Taking the derivative in the first equation, we see that these equations are identical to those in Eq. (9.70).

[16] It was ok to use β in the previous subsection. We introduced it only because it was quicker to write. But we can't use it here, because it depends on the other coordinates, and the Lagrangian method requires the use of independent coordinates. The variational proof back in Chapter 6 assumed this independence.

Note that there are two conserved quantities, arising from the facts that $\partial\mathcal{L}/\partial\psi$ and $\partial\mathcal{L}/\partial\phi$ equal zero. The conserved quantities are the angular momenta in the $\hat{\mathbf{x}}_3$ and $\hat{\mathbf{z}}$ directions, respectively. This is true because from Eq. (9.65), the former is $L_3 = I_3\omega_3$ and the latter is $L_z = L_3\cos\theta + L_2\sin\theta = (I_3\omega_3)\cos\theta + (I\dot{\phi}\sin\theta)\sin\theta$. These angular momenta are conserved because the torque points in the $\hat{\mathbf{x}}_1$ direction, so there is no torque in the plane spanned by $\hat{\mathbf{x}}_3$ and $\hat{\mathbf{z}}$.

9.7.5 Spinning top with $\dot{\theta} = 0$

A special case of Eqs. (9.70) occurs when $\dot{\theta} = 0$. In this case, the first of Eqs. (9.70) says that $\dot{\phi}$ is constant. The CM of the top therefore undergoes uniform circular motion in a horizontal plane. Let $\Omega \equiv \dot{\phi}$ be the frequency of this motion (this is the same notation as in Eq. (9.61)). Then the second of Eqs. (9.70) becomes

$$I\Omega^2\cos\theta - I_3\omega_3\Omega + Mg\ell = 0. \tag{9.75}$$

This quadratic equation can be solved to yield two possible precessional frequencies, Ω, for the top. And yes, there are indeed two of them, provided that ω_3 is greater than a certain minimum value.

The previous pages in this "Heavy symmetric top" section have been a bit abstract, so let's now take a breather and rederive Eq. (9.75) from scratch. That is, we'll assume $\dot{\theta} = 0$ from the start of the solution, and then solve things by finding $\mathbf{L}$ and using $\boldsymbol{\tau} = d\mathbf{L}/dt$, in the spirit of Section 9.4.2. In practice, this strategy of starting from scratch is invariably the best route to take, as you'll see in the problems and exercises for this chapter. The procedures in Sections 9.7.3 and 9.7.4 are good to know, but the technique in the following example provides a much more intuitive way of looking at things. This example is the classic "top" problem. We'll warm up by solving it in an approximate way. Then we'll do it for real.

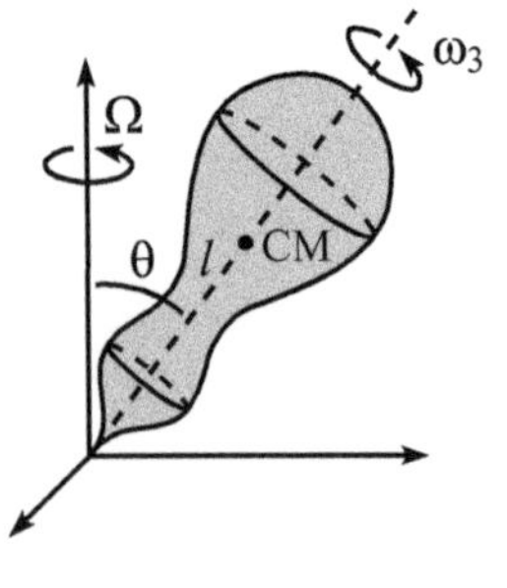

Fig. 9.31

Example (The top): A symmetric top of mass M has its CM a distance ℓ from its pivot. The moments of inertia relative to the pivot are $I_1 = I_2 \equiv I$, and I_3. The top spins around its symmetry axis with frequency ω_3 (in the language of Section 9.7.2), and initial conditions have been set up so that the CM precesses in a circle around the vertical axis. The symmetry axis makes a constant angle θ with the vertical (see Fig. 9.31).

(a) Assuming that the angular momentum due to ω_3 is much larger than any other angular momentum in the problem, find an approximate expression for the frequency, Ω, of precession.

(b) Now solve the problem exactly. That is, find Ω by considering all of the angular momentum.

Solution:

(a) The angular momentum (relative to the pivot) due to the spinning of the top has magnitude $L_3 = I_3\omega_3$, and it is directed along $\hat{\mathbf{x}}_3$. Let's label this angular momentum vector as $\mathbf{L}_3 \equiv L_3\hat{\mathbf{x}}_3$. As the top precesses, $\mathbf{L}_3$ traces out a cone around the vertical axis. So the tip of $\mathbf{L}_3$ moves in a circle of radius $L_3 \sin\theta$. The frequency of this circular motion is the frequency of precession, Ω. So $d\mathbf{L}_3/dt$, which is the velocity of the tip, has magnitude

$$\Omega(L_3 \sin\theta) = \Omega I_3\omega_3 \sin\theta, \tag{9.76}$$

and it is directed into the page.

The torque (relative to the pivot) is due to gravity acting on the CM, so it has magnitude $Mg\ell \sin\theta$, and it is directed into the page. Therefore, $\boldsymbol{\tau} = d\mathbf{L}/dt$ gives

$$\Omega = \frac{Mg\ell}{I_3\omega_3}. \tag{9.77}$$

This is independent of θ, and it is inversely proportional to ω_3.

(b) The error in the above analysis is that we omitted the angular momentum arising from the $\hat{\mathbf{x}}_2$ (defined in Section 9.7.1) component of the angular velocity due to the precession of the top around the $\hat{\mathbf{z}}$ axis. This component has magnitude $\Omega \sin\theta$.[17] Therefore, the angular momentum due to the angular velocity component in the $\hat{\mathbf{x}}_2$ direction has magnitude

$$L_2 = I\Omega \sin\theta. \tag{9.78}$$

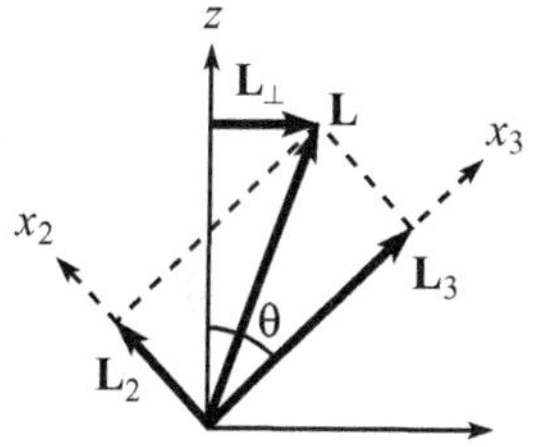

Fig. 9.32

Let's label this part of the angular momentum as $\mathbf{L}_2 \equiv L_2\hat{\mathbf{x}}_2$. The total $\mathbf{L} = \mathbf{L}_2 + \mathbf{L}_3$ is shown in Fig. 9.32. $\mathbf{L}$ precesses around in a cone, so only its horizontal component (call it $\mathbf{L}_\perp$) changes. From the figure, the length $L_\perp$ is the difference in lengths of the horizontal components of $\mathbf{L}_3$ and $\mathbf{L}_2$. Therefore,

$$L_\perp = L_3 \sin\theta - L_2 \cos\theta = I_3\omega_3 \sin\theta - I\Omega \sin\theta \cos\theta. \tag{9.79}$$

The magnitude of the rate of change of $\mathbf{L}$ is[18]

$$\left|\frac{d\mathbf{L}}{dt}\right| = \Omega L_\perp = \Omega(I_3\omega_3 \sin\theta - I\Omega \sin\theta \cos\theta). \tag{9.80}$$

Both $\boldsymbol{\tau}$ (which has magnitude $Mg\ell \sin\theta$) and $d\mathbf{L}/dt$ point into the page, so equating their magnitudes gives

$$I\Omega^2 \cos\theta - I_3\omega_3\Omega + Mg\ell = 0, \tag{9.81}$$

[17] The angular velocity due to the precession is $\Omega\hat{\mathbf{z}}$. We can break this up into components along the orthogonal directions $\hat{\mathbf{x}}_2$ and $\hat{\mathbf{x}}_3$. The $\Omega \cos\theta$ component along $\hat{\mathbf{x}}_3$ was already absorbed into the definition of ω_3 (see Fig. 9.30).

[18] This result can also be obtained in a more formal way. Since $\mathbf{L}$ precesses with angular velocity $\Omega\hat{\mathbf{z}}$, the rate of change of $\mathbf{L}$ is $d\mathbf{L}/dt = \Omega\hat{\mathbf{z}} \times \mathbf{L}$. If you compute this cross product in the x_1, x_2, x_3 basis, you will obtain the result in Eq. (9.80).

in agreement with Eq. (9.75), as we wanted to show. The quadratic formula quickly gives the two solutions for Ω, which may be written as

$$\Omega_{\pm} = \frac{I_3\omega_3}{2I\cos\theta}\left(1 \pm \sqrt{1 - \frac{4MIg\ell\cos\theta}{I_3^2\omega_3^2}}\right). \tag{9.82}$$

Note that if $\theta = \pi/2$, then Eq. (9.81) is actually a linear equation, so there is only one solution for Ω, which is the one in Eq. (9.77). The reason for this is that $\mathbf{L}_2$ points vertically, so it doesn't change. Only $\mathbf{L}_3$ contributes to $d\mathbf{L}/dt$, so the approximate solution in part (a) is in fact an exact solution. Because of this simplification, a top is much easier to deal with when its symmetry axis is horizontal.

The two solutions in Eq. (9.82) are known as the *fast-precession* and *slow-precession* frequencies. For large ω_3, you can show that the slow-precession frequency is

$$\Omega_{-} \approx \frac{Mg\ell}{I_3\omega_3}, \tag{9.83}$$

in agreement with the solution found in Eq. (9.77).[19] This task, along with many other interesting features of this problem (including the interpretation of the fast-precession frequency, Ω_+), is the subject of Problem 9.17, which you are encouraged to do.

9.7.6 An "explanation" of precession

The fact that a top can precess slowly around in a circle without simply falling down (as a simple pendulum would do) is rather bizarre. We showed above that this precession can be deduced perfectly well from $\boldsymbol{\tau} = d\mathbf{L}/dt$, but it would be nice if there was a more intuitive way to explain it, based at least somewhat on $\mathbf{F} = m\mathbf{a}$. Although a completely satisfactory intuitive explanation eludes me, I think the following discussion will clear some things up. This discussion will be qualitative (so no equations or numbers), but it should still suffice in explaining for the most part what's going on with precession.

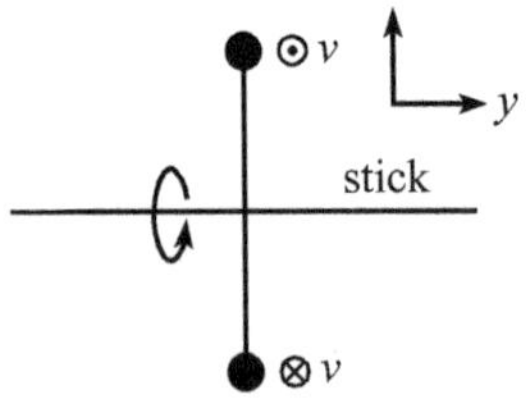

Fig. 9.33

Impulse applied to a dumbbell

Let's first look at a simple system consisting of a dumbbell with a massless stick glued perpendicular to it at its center. The dumbbell rotates around the stick which is held fixed; see Fig. 9.33. Let the y and z axes be defined as shown, with the x axis pointing out of the page. The total angular momentum points to the right, in the positive y direction. We'll ignore gravity for now. Equivalently, we can have the dumbbell pivoted at its center.

[19] This is fairly clear. If ω_3 is large enough compared with Ω, then we can ignore the first term in Eq. (9.81). That is, we can ignore the effects of $\mathbf{L}_2$, which is exactly what we did in the approximate solution in part (a).

At the instant the masses are in the plane of the paper, as shown (with the top one coming out of the page and the bottom one going in), we'll apply equal and opposite small impulses to the stick, upward on the right end and downward on the left; see Fig. 9.34.[20] Assume that the forces are applied for infinitesimal periods of time, but that the forces are large enough so that the impulse is nonzero. What happens to the masses? In particular, what happens to the plane of their rotation?

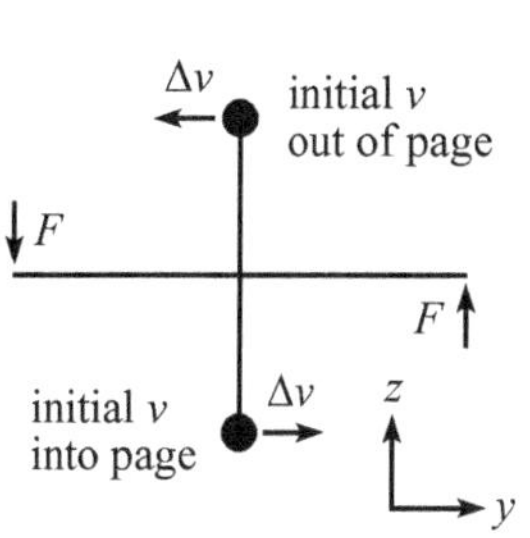

Fig. 9.34

Because the structure is rigid, the impulsive forces on the stick cause the two masses to pick up small velocity components in the $\pm y$ directions, as shown in Fig. 9.34. If the forces are applied for an infinitesimal time t, then these velocity components are of the form $v = at$ (with t very small and a very large). Additionally, the masses move a distance $d = at^2/2$ to the side. But because of the two powers of the infinitesimal time t here, this distance is negligible. In other words, the masses pick up a nonzero v, but essentially no d. This is a general result when an object is struck with a hammer: right after the blow, it has a nonzero speed, but essentially zero displacement.

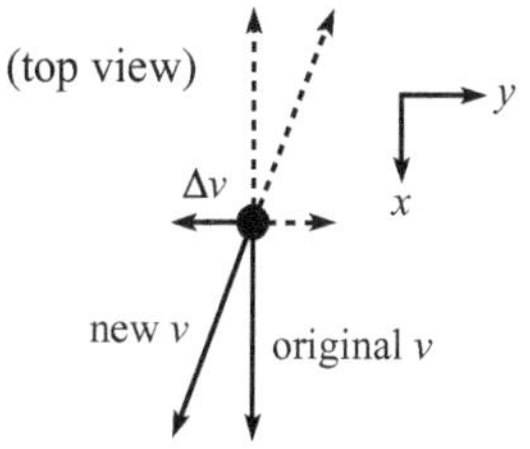

Fig. 9.35

A top view (with the z axis pointing out of the page) of the velocities of the two masses right after the blow is shown in Fig. 9.35. The dotted lines represent the velocity of the bottom mass which is behind (that is, below) the top mass. Now, if someone gives you these two velocities and doesn't tell you what was going on beforehand, then you will simply say that the dumbbell is rotating around in a circle that lies in the vertical plane defined by the line of the new velocity vectors in Fig 9.35. In other words, the plane of the circular motion has been rotated around the vertical z axis. A top view of the situation is shown in Fig. 9.36. From this time onward, the masses rotate around in the new vertical plane. This means that the angular momentum of the dumbbell has picked up a component in the positive x direction (out of the page in Fig. 9.34 and downward in Fig. 9.36), consistent with the fact that the torque from the two applied forces points in the positive x direction.

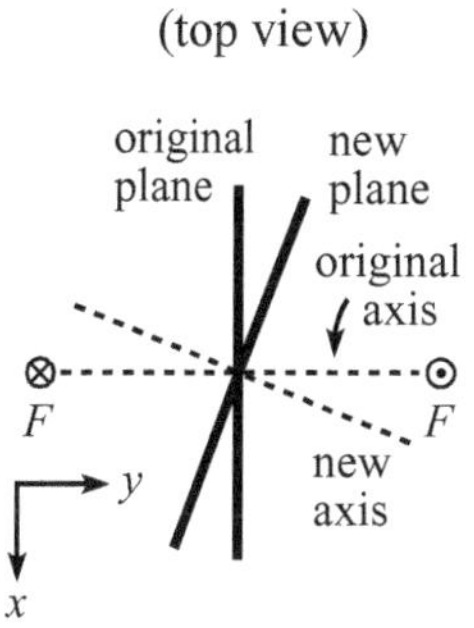

Fig. 9.36

This example illustrates the bizarre fact that if you smack the rotation axis in one direction (vertical here), it will head off in another direction (horizontal).

Impulse applied to a symmetric top

Let's now consider the more complicated situation where instead of the above dumbbell we have a symmetric top, for example, a flat disk with a stick poking through its center. (We'll still ignore gravity for the moment.) Things are more difficult now, because if we apply our impulsive forces to the stick, the plane of the disk can't instantaneously rotate as it did above, because this would involve the "side" points on the disk moving a finite distance in zero time, as they

[20] We're applying equal and opposite forces just so that the CM doesn't move. And we're doing this for no reason other than simplicity. The motion of the CM is irrelevant for the point we want to make in this dumbbell setup.

move from the old plane of rotation to the new plane. So what does the motion look like?

The rough answer is that our symmetric top is vaguely the same type of object as the above dumbbell, so the motion should look roughly the same. In other words, the axis of the top should end up pointing (in one way or another) a little bit in the x direction, as it did with the dumbbell. But the precise answer is that since we have a free top here, we know from the free-top discussion in Section 9.6.2 exactly what happens: the symmetry axis of the disk precesses (along with the angular velocity vector) in a thin cone around the new angular momentum vector, which points slightly out of the page due to the torque in the x direction. So although the symmetry axis of the disk doesn't point in the definite direction along $\mathbf{L}$ as it did with the dumbbell, it points *on average* along $\mathbf{L}$, which is slightly in the x direction.

Let's now bring gravity into the setup. It turns out that we can consider a heavy top's precession to be the result of a succession of many little impulsive blows to a free top that is otherwise in freefall (where gravity provides no torque). The reasoning is as follows.

Imagine that we hold our spinning top (not pivoted anywhere) and move it sideways (perpendicular to the stick) at a given constant speed; this speed will have to be chosen to be a particular value (see Footnote 23 below). And then we let go. Gravity provides a force on the CM, but zero torque around the CM, so the top is simply in freefall with the constant horizontal speed that we gave it. But let's assume that immediately after we let go, we keep the top at a constant height by applying a very quick and very small upward strike to the end of the stick, and then waiting for a very short time while the CM rises and falls in its freefall motion, and then repeating the process indefinitely (let's say we make 100 tiny strikes each second). If we arrange for the time-averaged upward force to equal mg, then the CM stays at (essentially) a constant height.

After each strike, the axis of the top undergoes its free-top precession in a very thin cone, so if the CM of the top weren't moving, the end of the stick would end up pointing slightly to the side, along the direction of the new $\mathbf{L}$.[21] But the sideways motion of the top (due to the properly chosen original speed we gave it) exactly brings the end of the stick back to its original location; see the top view in Fig. 9.37. The dot in the figure represents a fixed point in space. We can imagine this process taking place in two steps. In step 1, the axis changes its direction due to the angular impulse relative to the CM. And in step 2, the top moves sideways (due to the original horizontal speed we gave it) during its rising and falling projectile motion. As the process repeats, the CM moves around in a circle,

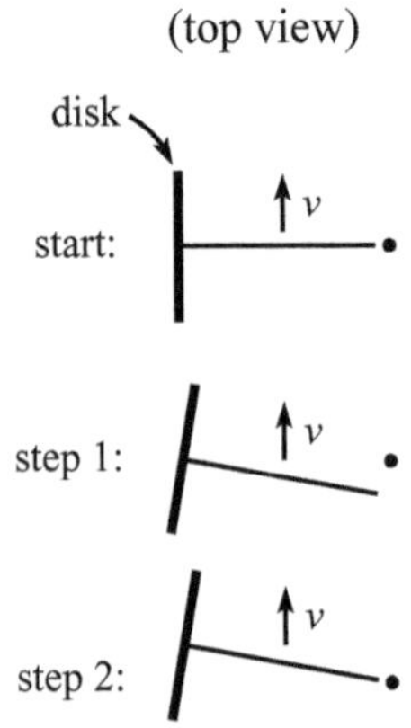

Fig. 9.37

[21] Assume that the stick's tiny precession cone around $\mathbf{L}$ somehow damps out, so that the stick eventually points in the definite direction along $\mathbf{L}$. Since we're eventually going to consider the end of the stick to be located at a pivot, this is a reasonable assumption.

with the end of the stick remaining (essentially) fixed.[22] In other words, we have recreated our precessing top.[23] And since we can consider the continuous upward force that a pivot applies to a heavy top to be a succession of quick infinitesimal strikes, we see that a heavy top precessing around a pivot is essentially the same as the above top precessing around the fixed dot in space in Fig. 9.37.

The above reasoning is only qualitative, but it should make the precession of a heavy top a little more believable. Of course, since we've argued that a heavy top pivoted at a point can be considered to be a free top undergoing a succession of impulsive and free-top motions, we've just shifted the burden of proof to an intuitive understanding of why a free top precesses the way it does. But that makes my head hurt, so I'll stop here. But at least we know that a free top should behave more or less like the dumbbell setup above, for which we showed (by actually looking at the forces) why the axis of rotation shifts to the side when vertical forces are applied.

9.7.7 Nutation

We will now solve Eq. (9.70) in a somewhat more general case, where θ is allowed to vary slightly. That is, we will consider a slight perturbation to the circular motion associated with Eq. (9.75). We will assume that ω_3 is large here, and we will assume that the original circular motion corresponds to the slow precession, so that $\dot{\phi}$ is small. Under these assumptions, we will find that the top will bounce around slightly as it travels (roughly) in a circle. This bouncing is known as *nutation*.

Since $\dot{\phi}$ is small compared with ω_3, we can (to a good approximation) ignore the middle terms on the left-hand sides of Eqs. (9.70) to obtain

$$\begin{aligned} I\ddot{\phi}\sin\theta - \dot{\theta} I_3\omega_3 &= 0, \\ (Mg\ell - I_3\omega_3\dot{\phi})\sin\theta &= I\ddot{\theta}. \end{aligned} \tag{9.84}$$

We must somehow solve these equations for $\theta(t)$ and $\phi(t)$. Taking the derivative of the first equation and dropping the quadratic term (which is negligible for sufficiently small perturbations) gives $\ddot{\theta} = (I\sin\theta/I_3\omega_3)\, d^2\dot{\phi}/dt^2$. Substituting this expression for $\ddot{\theta}$ into the second equation gives

$$\frac{d^2\dot{\phi}}{dt^2} + \omega_{\rm n}^2(\dot{\phi} - \Omega_{\rm s}) = 0, \tag{9.85}$$

[22] We'll need to apply a force directed along the axis, to provide the centripetal acceleration of the precessing CM. But this force produces no torque around the CM, so it doesn't affect any of the other aspects of the problem.

[23] If the initial conditions aren't set up properly, then the top will bounce up and down as it precesses around (this is known as *nutation*, described below in Section 9.7.7). In particular, if you hold the axis at rest and then let go, the top will initially fall straight down. So your intuition works perfectly fine here. But as time goes by, more complicated things start to happen, as explained in Remark 6 in the nutation section below.

where

$$\omega_n \equiv \frac{I_3\omega_3}{I} \quad \text{and} \quad \Omega_s = \frac{Mg\ell}{I_3\omega_3} \tag{9.86}$$

are, respectively, the frequency of nutation (as we shall soon see) and the slow-precession frequency given in Eq. (9.77). Shifting variables to $y \equiv \dot{\phi} - \Omega_s$ in Eq. (9.85) gives us the nice harmonic-oscillator equation, $\ddot{y} + \omega_n^2 y = 0$. Solving this and then shifting back to $\dot{\phi}$ yields

$$\dot{\phi}(t) = \Omega_s + A\cos(\omega_n t + \gamma), \tag{9.87}$$

where A and γ are determined by the initial conditions. Integrating this gives

$$\phi(t) = \Omega_s t + \left(\frac{A}{\omega_n}\right)\sin(\omega_n t + \gamma), \tag{9.88}$$

plus an irrelevant constant.

Now let's solve for $\theta(t)$. Plugging our $\phi(t)$ into the first of Eqs. (9.84) gives

$$\dot{\theta}(t) = -\left(\frac{I\sin\theta}{I_3\omega_3}\right)A\omega_n\sin(\omega_n t + \gamma) = -A\sin\theta\sin(\omega_n t + \gamma), \tag{9.89}$$

where we have used the definition of ω_n in Eq. (9.86). Since $\theta(t)$ doesn't change much, we can set $\sin\theta \approx \sin\theta_0$, where θ_0 is, say, the initial value of $\theta(t)$. Any errors here will be second-order effects in small quantities. Integration then gives

$$\theta(t) = B + \left(\frac{A}{\omega_n}\sin\theta_0\right)\cos(\omega_n t + \gamma), \tag{9.90}$$

where B is a constant of integration. Equations (9.88) and (9.90) show that both ϕ (neglecting the uniform $\Omega_s t$ part) and θ oscillate with frequency ω_n, and with amplitudes inversely proportional to ω_n. Note that Eq. (9.86) says that ω_n grows with ω_3.

Example (Sideways kick): Assume that uniform circular precession is initially taking place with $\theta = \theta_0$ and $\dot{\phi} = \Omega_s$. You then give the top a quick kick along the direction of motion, so that $\dot{\phi}$ suddenly becomes $\Omega_s + \Delta\Omega$ ($\Delta\Omega$ may be positive or negative). Find $\phi(t)$ and $\theta(t)$.

Solution: This is an exercise in initial conditions. We are given the initial values of $\dot{\phi}$, $\dot{\theta}$, and θ (namely $\Omega_s + \Delta\Omega$, 0, and θ_0, respectively), and our goal is to solve for the unknowns A, B, and γ in Eqs. (9.87), (9.89), and (9.90). $\dot{\theta}$ is initially zero, so Eq. (9.89) gives $\gamma = 0$ (or π, but this leads to the same answer). And $\dot{\phi}$ is initially $\Omega_s + \Delta\Omega$, so Eq. (9.87) gives $A = \Delta\Omega$. Finally, θ is initially θ_0, so Eq. (9.90) gives $B = \theta_0 - (\Delta\Omega/\omega_n)\sin\theta_0$. Putting this all together, we have

$$\begin{aligned} \phi(t) &= \Omega_s t + \left(\frac{\Delta\Omega}{\omega_n}\right)\sin\omega_n t, \\ \theta(t) &= \theta_0 - \left(\frac{\Delta\Omega}{\omega_n}\sin\theta_0\right)(1 - \cos\omega_n t). \end{aligned} \tag{9.91}$$

And for future reference (for the problems in this chapter), we'll also list the derivatives,

$$\begin{aligned}\dot{\phi}(t) &= \Omega_s + \Delta\Omega \cos \omega_n t, \\ \dot{\theta}(t) &= -\Delta\Omega \sin\theta_0 \sin \omega_n t.\end{aligned} \tag{9.92}$$

REMARKS:

1. Remember that all of this analysis holds only if $\dot{\theta}$ and $\dot{\phi}$ are small compared with ω_3, and if θ is always near θ_0.
2. For the initial setup we have chosen (that is, for $\dot{\theta} = 0$), Eq. (9.91) shows that θ always stays on one side of θ_0. If $\Delta\Omega > 0$, then $\theta(t) \le \theta_0$ for all t (that is, the top is always at a higher position, since θ is measured from the vertical). If $\Delta\Omega < 0$, then $\theta(t) \ge \theta_0$ for all t (that is, the top is always at a lower position).
3. Consider the point that is a distance ℓ from the origin and that travels in a horizontal circle with angular coordinates given by $(\phi, \theta)_{\text{avg}} = \big(\Omega_s t,\ \theta_0 - (\Delta\Omega/\omega_n)\sin\theta_0\big)$. This is the "average" position of the CM, in the sense that Eq. (9.91) gives the angular coordinates of the CM relative to this point as

$$(\phi, \theta)_{\text{rel}} = \left(\frac{\Delta\Omega}{\omega_n}\right)(\sin\omega_n t,\ \sin\theta_0 \cos\omega_n t). \tag{9.93}$$

The $\sin\theta_0$ in the second coordinate here implies that the amplitude of the θ oscillation is $\sin\theta_0$ times the amplitude of the ϕ oscillation. This is precisely the factor needed to make the CM travel in a circle, as viewed by someone riding along with coordinates $(\phi, \theta)_{\text{avg}}$, because a change in θ causes a displacement of $\ell\, d\theta$, whereas a change in ϕ causes a displacement of $\ell \sin\theta_0\, d\phi$.
4. Figure 9.38 shows plots of $\theta(t)$ vs. $\sin\theta_0\phi(t)$ for various values of $\Delta\Omega$. The dots at the start of each of the nine plots all represent the same starting point at $\theta = \theta_0$. The plots are stacked on top of each other for comparison only; the vertical spacing between them is meaningless. We have chosen the horizontal axis to be $\sin\theta_0\phi(t)$ instead of $\phi(t)$, and we have chosen the vertical axis to have θ increasing downward, so that these plots are exactly the paths you would see the CM trace out in space. We have picked arbitrary values of Ω_s and ω_n to generate the plots (hence no numbers on the axes, since they wouldn't mean much), but no matter what values are picked, the shapes of the plots are still the same (for example, the $\Delta\Omega = \pm\Omega_s$ paths always have cusps). The oscillations have different frequencies and amplitudes, but both axes get scaled by the same amount (as you can check).
5. From the figure, we see that the motions associated with $\Delta\Omega$ and $-\Delta\Omega$ look the same except for being shifted vertically (relative to the starting dots, which represent the same point), and also horizontally by half a cycle. This can be seen from Eq. (9.91); changing $\Delta\Omega$ to $-\Delta\Omega$ has the effect of shifting the constant term in $\theta(t)$, and also shifting the time by half a cycle, because $\Delta\Omega \sin\omega_n t = (-\Delta\Omega)\sin(\omega_n t + \pi)$, and likewise for cosine.
6. In the $\Delta\Omega = -\Omega_s$ case, the CM starts at rest. This corresponds to holding the axis of the top at rest and then dropping it. From Fig. 9.38 we see that initially the CM simply falls straight down, as your intuition suggests. (It might not be clear from the figure, but Problem 9.25 shows that the curve is indeed vertical at the high points in the motion.) But then the strangeness of angular momentum takes over, and the top ends up bouncing and precessing instead of continuing to fall as a point mass would.

   Qualitatively, we can understand this bouncing and precessing as follows. Let the axis and the angular momentum initially point to the right as you hold the axis, as

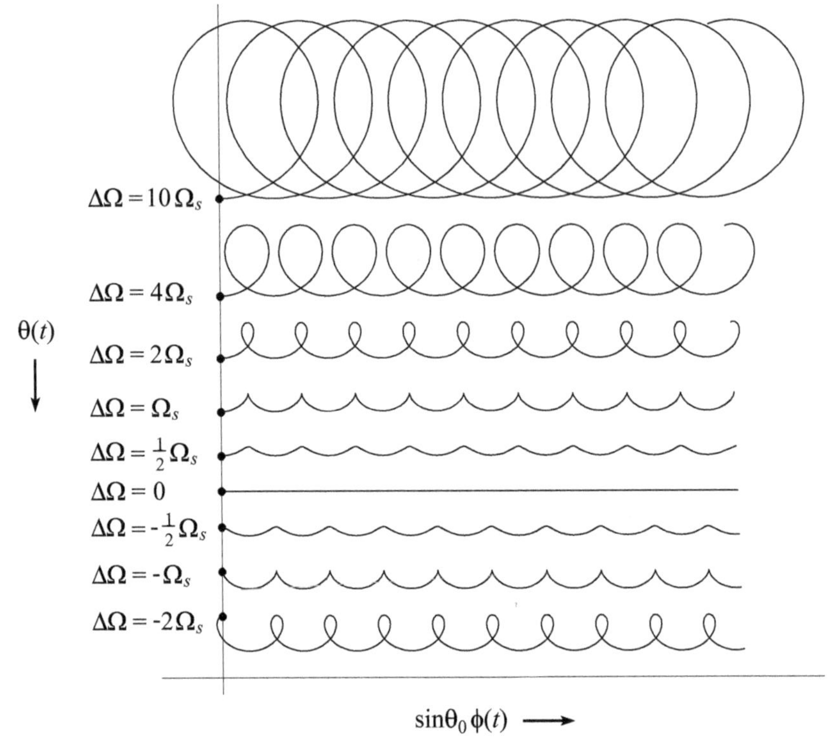

Fig. 9.38

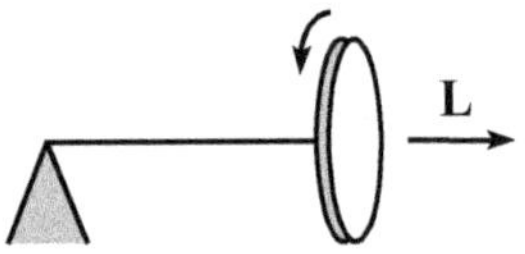

Fig. 9.39

shown in Fig. 9.39. After you let go, a number of things happen: (1) Because of the downward gravitational force, the top starts to fall. The axis therefore now points slightly downward, which means that the angular momentum picks up a downward component. Relative to the CM, the force from the pivot must therefore provide a downward torque. By the right-hand rule, this force must point into the page. (2) From $F = ma$, this inward force causes the top to accelerate into the page. This makes the axis point slightly into the page, so the angular momentum picks up a component in that direction. By the right-hand rule, there must then be an upward force from the pivot to provide the necessary torque for this. (3) This upward force slows down the downward motion and in fact eventually becomes greater than mg and causes the CM to rise back up. This means that the angular momentum increases its vertical component, so by the right-hand rule there must be a force from the pivot that points out of the page to provide the necessary torque for this. (4) This outward force then slows down the motion into the page and eventually stops it, right when the top returns to its original height. The top is momentarily at rest, and then the process repeats itself. Of course, this qualitative reasoning doesn't show that all the details work out correctly, but at least it makes the motion a little more believable.

7. If you start by holding the axis of the top at rest, as in the previous remark, the $\Delta\Omega = 0$ case corresponds to giving the axis the proper initial push into the page (instead of just dropping it, as above) so that the resulting change in angular momentum into the page requires an upward force of mg from the pivot. The CM then stays at the same height and simply moves in a horizontal circle without bouncing. ♣

9.8 Problems

Section 9.1: Preliminaries concerning rotations

9.1. **Many different ω's** *

Consider a particle at the point $(a, 0, 0)$, with velocity $(0, v, 0)$. At this instant, the particle may be considered to be rotating around many different $\boldsymbol{\omega}$ vectors passing through the origin. There isn't just one "correct" $\boldsymbol{\omega}$. Find all the possible $\boldsymbol{\omega}$'s (give their directions and magnitudes).

9.2. **Fixed points on a sphere** **

Consider a transformation of a rigid sphere into itself. Show that two points on the sphere end up where they started.

9.3. **Rolling cone** **

A cone rolls without slipping on a table. The half-angle at the vertex is α, and the axis has length h (see Fig. 9.40). Let the speed of the center of the base, point P in the figure, be v. What is the angular velocity of the cone with respect to the lab frame at the instant shown? There are many ways to do this problem, so you are encouraged to take a look at the three given solutions, even after solving it.

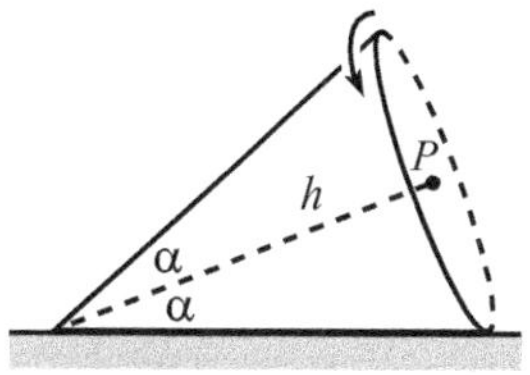

Fig. 9.40

Section 9.2: The inertia tensor

9.4. **Parallel-axis theorem**

Let (X, Y, Z) be the position of an object's CM, and let (x', y', z') be the position relative to the CM. Prove the parallel-axis theorem, Eq. (9.19), by setting $x = X + x'$, $y = Y + y'$, and $z = Z + z'$ in Eq. (9.8).

Section 9.3: Principal axes

9.5. **A nice cylinder** *

What must the ratio of height to radius of a cylinder be so that every axis is a principal axis (with the CM as the origin)?

9.6. **Rotating square** *

Here's an exercise in geometry. Theorem 9.5 says that if the moments of inertia around two principal axes are equal, then any axis in the plane of these axes is a principal axis. This means that the object will happily rotate around any axis in this plane, that is, no torque is needed. Demonstrate this explicitly for four equal masses in the shape of a square, with the center as the origin, which obviously has two moments equal. Assume that the masses are connected by strings to the axis, as shown in Fig. 9.41, and that they all rotate with the same ω around the axis, so that they remain in the shape of a square. Your task is to show that the tensions in the strings

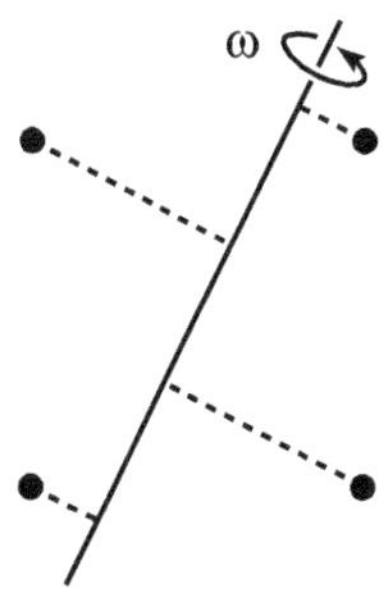

Fig. 9.41

are such that there is no net torque acting on the axis, relative to the center of the square.

9.7. **Existence of principal axes for a pancake** $*$

Given a pancake object in the x-y plane, show that there exist principal axes by considering what happens to the integral $\int xy$ when the coordinate axes are rotated about the origin by an angle of $\pi/2$.

9.8. **Symmetries and principal axes for a pancake** $**$

A rotation of the axes in the x-y plane through an angle θ transforms the coordinates according to (you can accept this)

$$\begin{pmatrix} x' \\ y' \end{pmatrix} = \begin{pmatrix} \cos\theta & \sin\theta \\ -\sin\theta & \cos\theta \end{pmatrix} \begin{pmatrix} x \\ y \end{pmatrix}. \tag{9.94}$$

Use this to show that if a pancake object in the x-y plane has a symmetry under a rotation through $\theta \neq \pi$, then $\int xy = 0$ for any choice of axes, which implies that all axes (through the origin) in the plane are principal axes.

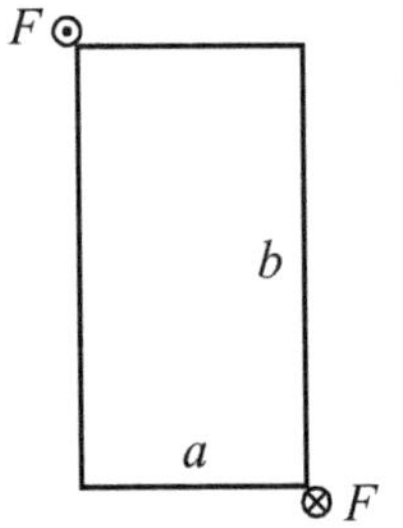

Fig. 9.42

Section 9.4: Two basic types of problems

9.9. **Striking a rectangle** $*$

A flat uniform rectangle with sides of length a and b sits in space, not rotating. You strike the corners at the ends of one diagonal, with equal and opposite forces (see Fig. 9.42). Show that the resulting initial $\boldsymbol{\omega}$ points along the other diagonal.

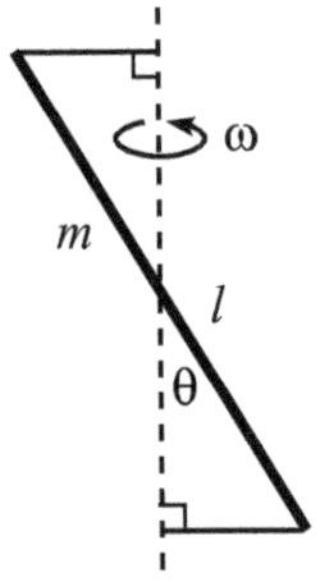

Fig. 9.43

9.10. **Rotating stick** $**$

A stick of mass m and length ℓ spins with frequency ω around an axis, as shown in Fig. 9.43. The stick makes an angle θ with the axis and is kept in its motion by two strings that are perpendicular to the axis. What is the tension in the strings? (Ignore gravity.)

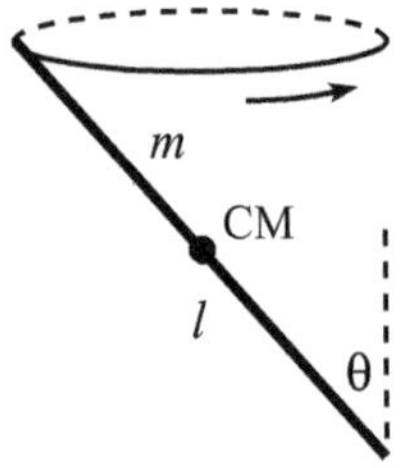

Fig. 9.44

9.11. **Stick under a ring** $**$

A stick of mass m and length ℓ is arranged to have its CM motionless while its top end slides in a circle on a frictionless ring, as shown in Fig. 9.44. The stick makes an angle θ with the vertical. What is the frequency of this motion?

9.12. **Circular pendulum** $**$

Consider a pendulum made of a massless rod of length ℓ with a point mass m on the end. Assume conditions have been set up so that the mass moves in a horizontal circle. Let θ be the constant angle the rod makes with the vertical. Find the frequency, Ω, of this circular motion in three different ways.

(a) Use $\mathbf{F} = m\mathbf{a}$. This method works only if you have a point mass. With an extended object, you have to use one of the following methods involving torque.
(b) Use $\boldsymbol{\tau} = d\mathbf{L}/dt$ with the pendulum pivot as the origin.
(c) Use $\boldsymbol{\tau} = d\mathbf{L}/dt$ with the mass as the origin.

9.13. **Rolling in a cone** **

A fixed cone stands on its tip, with its axis in the vertical direction. The half-angle at the vertex is θ. A small ring of radius r rolls without slipping on the inside surface. Assume that conditions have been set up so that (1) the point of contact between the ring and the cone moves in a circle at height h above the tip, and (2) the plane of the ring is at all times perpendicular to the line joining the point of contact and the tip of the cone (see Fig. 9.45). What is the frequency, Ω, of this circular motion? Work in the approximation where r is much smaller than the radius of the circular motion, $h \tan\theta$.

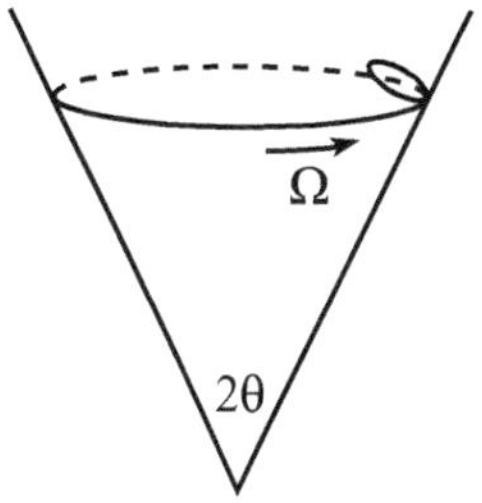

Fig. 9.45

Section 9.5: Euler's equations

9.14. **Tennis racket theorem** ***

If you try to spin a tennis racket (or a book, etc.) around any of its three principal axes, you will find that different things happen with the different axes. Assuming that the principal moments (relative to the CM) are labeled according to $I_1 > I_2 > I_3$ (see Fig. 9.46), you will find that the racket will spin nicely around the $\hat{\mathbf{x}}_1$ and $\hat{\mathbf{x}}_3$ axes, but it will wobble in a rather messy manner around the $\hat{\mathbf{x}}_2$ axis. Verify this claim experimentally with a book (preferably lightweight, and wrapped with a rubber band), or a tennis racket, if you happen to study with one on hand.

Now verify this claim mathematically. The main point here is that you can't start the motion off with $\boldsymbol{\omega}$ pointing *exactly* along a principal axis. Therefore, what you want to show is that the motion around the $\hat{\mathbf{x}}_1$ and $\hat{\mathbf{x}}_3$ axes is *stable* (that is, small errors in the initial conditions remain small), whereas the motion around the $\hat{\mathbf{x}}_2$ axis is *unstable* (that is, small errors in the initial conditions get larger and larger, until the motion eventually doesn't resemble a rotation around the $\hat{\mathbf{x}}_2$ axis).[24] Your task is to use Euler's equations to prove these statements about stability. (Exercise 9.33 gives another derivation of this result.)

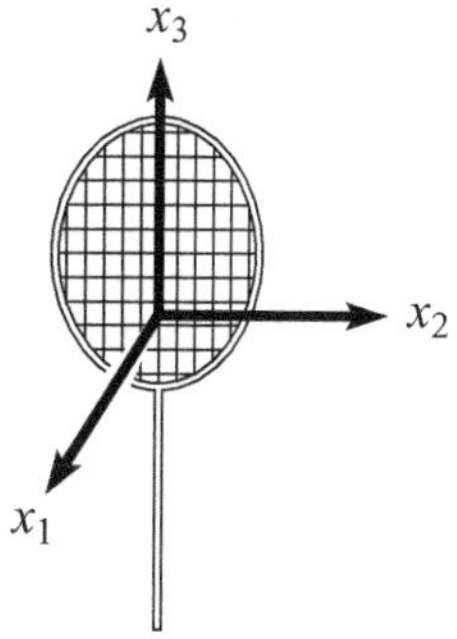

Fig. 9.46

[24] If you try for a long enough time, you'll probably be able to get the initial $\boldsymbol{\omega}$ pointing close enough to $\hat{\mathbf{x}}_2$ so that the book will remain rotating (almost) around $\hat{\mathbf{x}}_2$ for the entire time of its flight. There is, however, undoubtedly a better use for your time, as well as for the book...

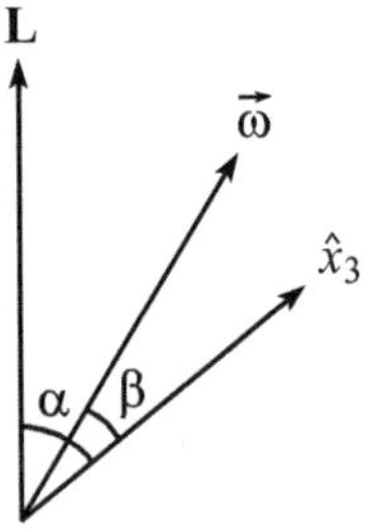

Fig. 9.47

Section 9.6: Free symmetric top

9.15. Free-top angles $*$

In Section 9.6.2, we showed that for a free symmetric top, the angular momentum $\mathbf{L}$, the angular velocity $\boldsymbol{\omega}$, and the symmetry axis $\hat{\mathbf{x}}_3$ all lie in a plane. Let α be the angle between $\hat{\mathbf{x}}_3$ and $\mathbf{L}$, and let β be the angle between $\hat{\mathbf{x}}_3$ and $\boldsymbol{\omega}$ (see Fig. 9.47). Find the relationship between α and β in terms of the principal moments, I and I_3.

9.16. Staying above $**$

A top with $I = nI_3$, where n is a numerical factor, is initially spinning around its x_3 axis with angular speed ω_3. You apply a strike at the bottom point, directed into the page as shown in Fig. 9.48 (imagine that you hit a little peg protruding from the bottom). What is the largest value of n for which the total $\boldsymbol{\omega}$ vector never dips below the horizontal axis in the subsequent motion, no matter how hard your strike is?

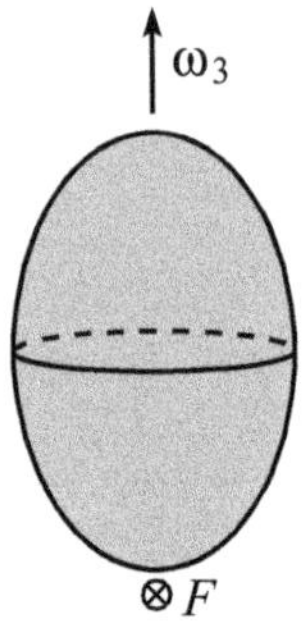

Fig. 9.48

Section 9.7: Heavy symmetric top

9.17. The top $**$

This problem deals with the spinning top example in Section 9.7.5. It uses the result for Ω in Eq. (9.82).

(a) What is the minimum value of ω_3 for which circular precession is possible?

(b) Find approximate expressions for $\Omega_\pm$ when ω_3 is very large. The phrase "very large," however, is rather meaningless. What mathematical statement should replace it?

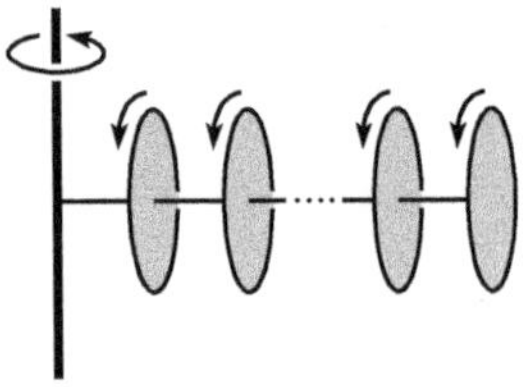

Fig. 9.49

9.18. Many tops $**$

N identical disks and massless sticks are arranged as shown in Fig. 9.49. Each disk is glued to the stick on its left and attached by a pivot to the stick on its right. The leftmost stick is attached by a pivot to a pole. You wish to set up circular precession with the sticks always forming a straight horizontal line. What should the relative angular speeds of the disks be so that this is possible?

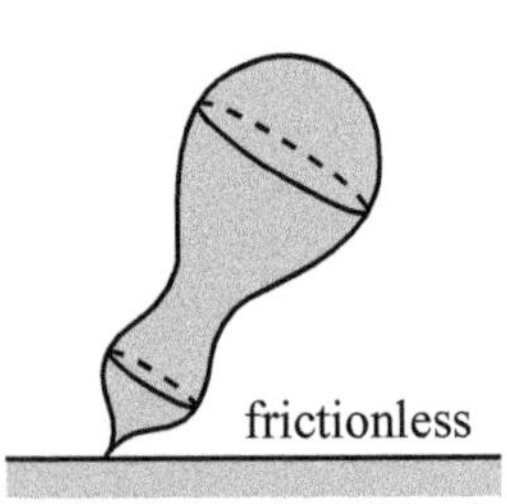

Fig. 9.50

9.19. Heavy top on a slippery table $**$

Solve the problem of a heavy symmetric top spinning on a frictionless table (see Fig. 9.50). You may do this by simply stating what modifications are needed in the derivation in Section 9.7.3 (or Section 9.7.4).

9.20. Fixed highest point **

Consider a top made of a uniform disk of radius R, connected to the origin by a massless stick (which is glued perpendicular to the disk) of length ℓ. Paint a dot on the top at its highest point, and label this as point P (see Fig. 9.51). You wish to set up uniform circular precession, with the stick making a constant angle θ with the vertical (θ can be chosen to be any angle between zero and π), and with P always being the highest point on the top. What is the frequency of precession, Ω? What relation between R and ℓ must be satisfied for this motion to be possible?

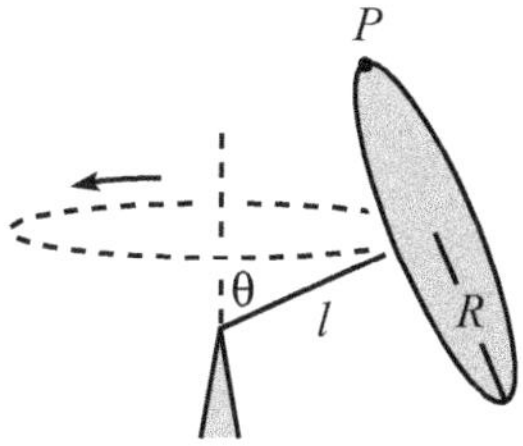

Fig. 9.51

9.21. Basketball on a rim ***

A basketball rolls without slipping around a basketball rim in such a way that the contact points trace out a great circle on the ball, and the CM moves around in a horizontal circle with frequency Ω. The radii of the ball and rim are r and R, respectively, and the ball's radius to the contact point makes an angle θ with the horizontal (see Fig. 9.52). Assume that the ball's moment of inertia around its center is $I = (2/3)mr^2$. Find Ω.

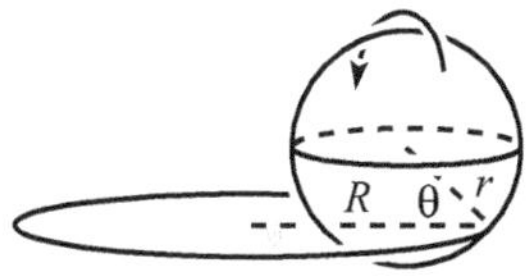

Fig. 9.52

9.22. Rolling lollipop ***

Consider a lollipop made of a solid sphere of mass m and radius r that is radially pierced by a massless stick. The free end of the stick is pivoted on the ground (see Fig. 9.53). The sphere rolls on the ground without slipping, with its center moving in a circle of radius R with frequency Ω. What is the normal force between the ground and the sphere?

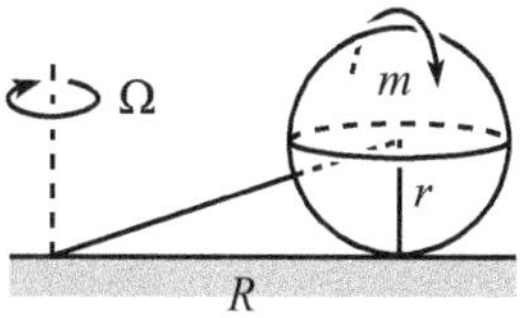

Fig. 9.53

9.23. Rolling coin ***

Initial conditions have been set up so that a coin of radius r rolls around in a circle, as shown in Fig. 9.54. The contact point on the ground traces out a circle of radius R, and the coin makes a constant angle θ with the horizontal. The coin rolls without slipping (assume that the friction with the ground is as large as needed). What is the frequency, Ω, of the circular motion of the contact point on the ground? Show that such motion exists only if $R > (5/6)r\cos\theta$.

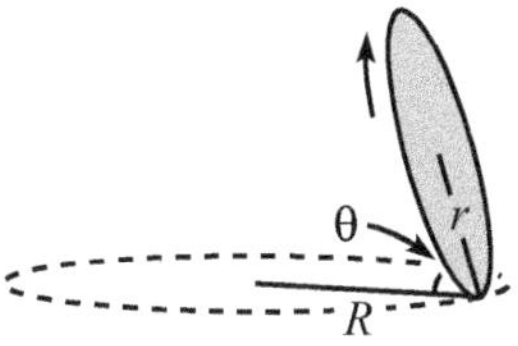

Fig. 9.54

9.24. Wobbling coin ***

If you spin a coin around a vertical diameter on a table, it will slowly lose energy and begin a wobbling motion. The angle between the coin and the table will gradually decrease, and eventually it will come to rest. Assume that this process is slow, and consider the motion when the coin makes an angle θ with the table (see Fig. 9.55). You may assume that the CM is essentially motionless. Let R be the radius of the coin, and let Ω be the frequency at which the contact point on the table traces out its circle. Assume that the coin rolls without slipping.

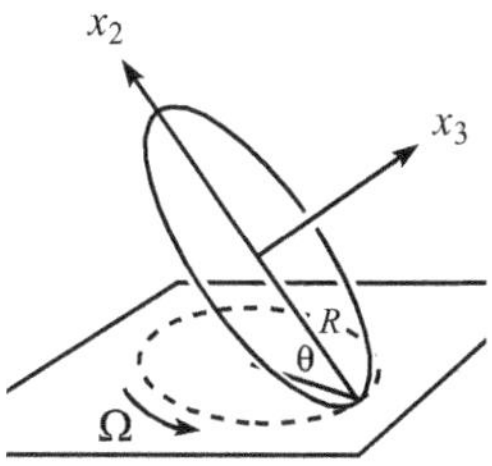

Fig. 9.55

(a) Show that the angular velocity of the coin is $\boldsymbol{\omega} = \Omega \sin\theta\, \hat{\mathbf{x}}_2$, where $\hat{\mathbf{x}}_2$ points upward along the coin, directly away from the contact point.

(b) Show that

$$\Omega = 2\sqrt{\frac{g}{R\sin\theta}}. \tag{9.95}$$

(c) Show that Abe (or Tom, Franklin, George, John, Dwight, Susan, or Sacagawea) appears to rotate, when viewed from above, with frequency

$$2(1-\cos\theta)\sqrt{\frac{g}{R\sin\theta}}. \tag{9.96}$$

9.25. **Nutation cusps** **

(a) Using the notation and initial conditions in the example in Section 9.7.7, prove that kinks occur in nutation if and only if $\Delta\Omega = \pm\Omega_s$. A kink is where the plot of $\theta(t)$ vs. $\phi(t)$ has a discontinuity in its slope.

(b) Prove that these kinks are in fact cusps. A cusp is a kink where the plot reverses direction in the ϕ-θ plane.

9.26. **Nutation circles** **

(a) Using the notation and initial conditions in the example in Section 9.7.7, and assuming that $\omega_3 \gg \Delta\Omega \gg \Omega_s$, find (approximately) the direction of the angular momentum right after the sideways kick takes place.

(b) Use Eq. (9.91) to show that the CM then travels (approximately) in a circle around **L**. And show that this "circular" motion is just what you would expect from the free-top reasoning in Section 9.6.2, in particular, Eq. (9.55).

Additional problems

9.27. **Rolling without slipping** *

The standard way that a ball rolls without slipping on a flat surface is for the contact points on the ball to trace out a vertical great circle on the ball. Are there any other ways that a ball can roll without slipping?

9.28. **Rolling straight?** **

In some situations, such as the rolling-coin setup in Problem 9.23, the velocity of the CM of a rolling object changes direction as time goes by. Consider a uniform sphere that rolls on the ground without slipping (possibly in the nonstandard way described in the solution to Problem 9.27).

Is it possible for the CM's velocity to change direction? Justify your answer rigorously.

9.29. **Ball on paper** ***

A uniform ball rolls without slipping on a table (possibly in the nonstandard way described in the solution to Problem 9.27). It rolls onto a piece of paper, which you then slide around in an arbitrary (horizontal) manner. You may even give the paper abrupt, jerky motions, so that the ball slips with respect to it. After you allow the ball to come off the paper, it will eventually resume rolling without slipping on the table. Show that the final velocity equals the initial velocity.

9.30. **Ball on a turntable** ****

A uniform ball rolls without slipping on a turntable (possibly in the nonstandard way described in the solution to Problem 9.27). As viewed from the inertial lab frame, show that the ball moves in a circle (not necessarily centered at the center of the turntable) with a frequency equal to $2/7$ times the frequency of the turntable.

9.9 Exercises

Section 9.1: Preliminaries concerning rotations

9.31. **Rolling wheel** **

A wheel with spokes rolls without slipping on the ground. A stationary camera takes a picture of it as it rolls by, from the side. Due to the nonzero exposure time of the camera, the spokes generally appear blurred. At what locations in the picture do the spokes *not* appear blurred? *Hint*: A common incorrect answer is that there is only one point.

Section 9.2: The inertia tensor

9.32. **Inertia tensor** *

Calculate the $\mathbf{r} \times (\boldsymbol{\omega} \times \mathbf{r})$ double cross product in Eq. (9.7) by using the vector identity,

$$\mathbf{A} \times (\mathbf{B} \times \mathbf{C}) = \mathbf{B}(\mathbf{A} \cdot \mathbf{C}) - \mathbf{C}(\mathbf{A} \cdot \mathbf{B}). \tag{9.97}$$

Section 9.3: Principal axes

9.33. **Tennis racket theorem** **

Problem 9.14 gives the statement of the "tennis racket theorem," and the solution there involves Euler's equations. Demonstrate the theorem here by writing down the conservation of L^2 and conservation of E statements and then using them in the following way. Produce an equation that says that if ω_2 and ω_3 (or ω_1 and ω_2) start small, then they must remain small. And produce the analogous equation that says that if ω_1 and ω_3 start small, then they need *not* remain small. (It's another matter to show that they actually *won't* remain small. But let's not worry about that here. Anything that *can* happen generally *does* happen in physics.)

9.34. **Moments for a cube** **

In the spirit of Appendix E, calculate the principal moments for a solid cube of mass m and side length ℓ, with the coordinate axes parallel to the edges of the cube, and the origin at a corner.

9.35. **Tilted moments** **

(a) Consider a planar object in the x-y plane. If the x and y axes are principal axes, use the rotation matrix in Eq. (9.94) to show that the moment of inertia around the x' axis, which makes an angle θ with the x axis, is $I_{x'} = I_x \cos^2\theta + I_y \sin^2\theta$.

(b) Consider a general three-dimensional object whose principal axes are the x, y, and z axes. Consider another axis that points along the unit vector (α, β, γ). Show that the moment of inertia around this axis is $\alpha^2 I_x + \beta^2 I_y + \gamma^2 I_z$. *Hint*: The cross product, discussed in Appendix B, provides a nice method of calculating the distance from a point to a line.

9.36. **Quadrupole** **

Consider an arbitrarily shaped body of mass m whose CM is at the origin. Using the law of cosines, the gravitational potential of a mass M at position $\mathbf{R}$ is

$$V(\mathbf{R}) = -\int \frac{GM\,dm}{\sqrt{R^2 + r^2 - 2Rr\cos\beta}}, \tag{9.98}$$

where the integration runs over the volume of the body, and β is the angle that the position vector $\mathbf{r}$ of an arbitrary point in the body makes with the vector $\mathbf{R}$.

(a) Assuming that all points in the body satisfy $r \ll R$, show that an approximate expression for the potential is

$$V(\mathbf{R}) \approx -\frac{GMm}{R} - \frac{GM}{2R^3}\int r^2(3\cos^2\beta - 1)\,dm, \tag{9.99}$$

and then show that this can be written as

$$V(\mathbf{R}) \approx -\frac{GMm}{R} - \frac{GM}{2R^3}(I_1 + I_2 + I_3 - 3I_R), \tag{9.100}$$

where I_1, I_2, and I_3 are the moments around any three orthogonal axes (which we'll take to be principal axes in part (b)), and I_R is the moment around the axis along the $\mathbf{R}$ vector.

(b) Consider now a planet with rotational symmetry around $\hat{\mathbf{x}}_3$, such as the earth which bulges at the equator due to the spinning. Using the result of Exercise 9.35, show that the potential in Eq. (9.100) can be written as

$$V(\mathbf{R}) \approx -\frac{GMm}{R} - \frac{GM}{2R^3}(I_3 - I)(1 - 3\cos^2\theta), \qquad (9.101)$$

where $I \equiv I_1 = I_2$, and θ is the angle that $\mathbf{R}$ makes with $\hat{\mathbf{x}}_3$.

REMARK: The second term here is known as the *quadrupole* term. In electrostatics, a *dipole* consists of equal and opposite charges separated by some distance d. At a given point far away, the forces from these two charges partially cancel. But they don't exactly cancel, because the electrostatic force (which behaves like the gravitational force, with a $1/r^2$ law) depends on the distance and direction to the charges, and the two charges are located at different points. If two dipoles are oriented in opposite directions and then placed side by side, a distance d apart (so that there are charges of q and $-q$ alternating around the corners of a square) then the forces from the dipoles nearly cancel. But again, the cancellation isn't exact, because the dipoles are located in different places. This distribution of charge is called a *quadrupole*, and it is similar to the situation with a spinning (and bulging) planet, because such a planet consists of a spherical ball (which gives rise to the first term in Eq. (9.101)), plus a region of "negative" mass superimposed on the ball at the poles and a region of positive mass superimposed on the ball at the equator. Looking at the above square of charges from far out along the diagonal containing the two negative charges is similar to looking at the earth from far out along its rotation axis. ♣

Section 9.4: Two basic types of problems

9.37. **Sphere and points** *

A uniform sphere of mass m and radius R rotates around the vertical axis with angular speed ω. Two particles of mass $m/2$ are brought close to the sphere at diametrically opposite points, at an angle θ from the vertical, as shown in Fig. 9.56. The masses, which are initially essentially at rest, abruptly stick to the sphere. What angle does the resulting $\boldsymbol{\omega}$ make with the vertical? (If you want, you can check your answer by showing that the θ that makes this angle maximum is $\sin^{-1}\sqrt{7/9} \approx 61.9°$.)

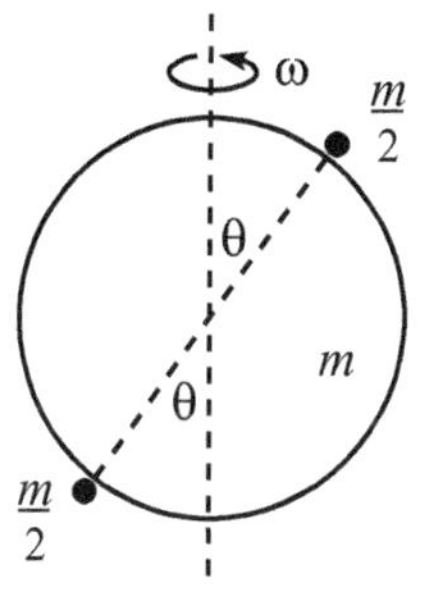

Fig. 9.56

9.38. **Striking a triangle** **

Consider the rigid object in Fig. 9.57. Four masses lie at the points shown on a rigid isosceles right triangle with hypotenuse length $4a$. The mass at the right angle is $3m$, and the other three masses are m. Label them A, B, C, D, as shown. Assume that the object is floating freely in outer space. Mass C is struck with a quick blow, directed into the page. Let the impulse have magnitude $\int F\,dt = P$. What are the velocities of all the masses immediately after the blow?

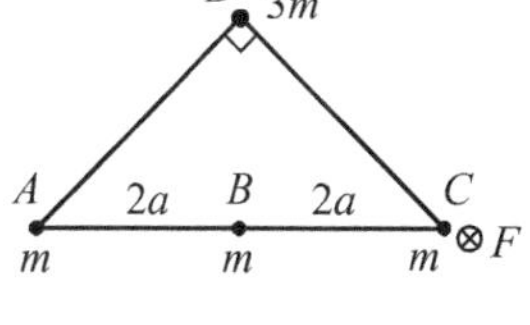

Fig. 9.57

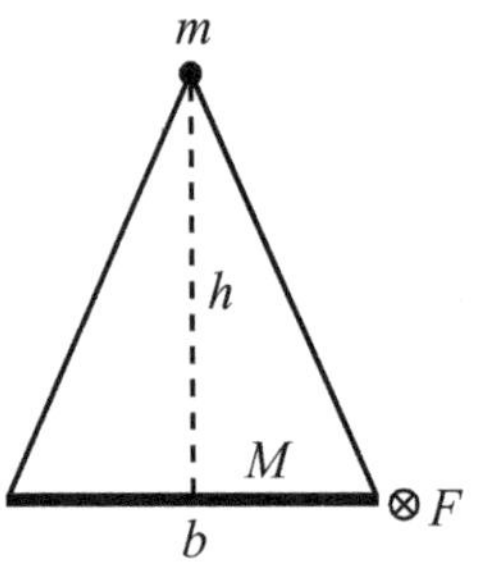

Fig. 9.58

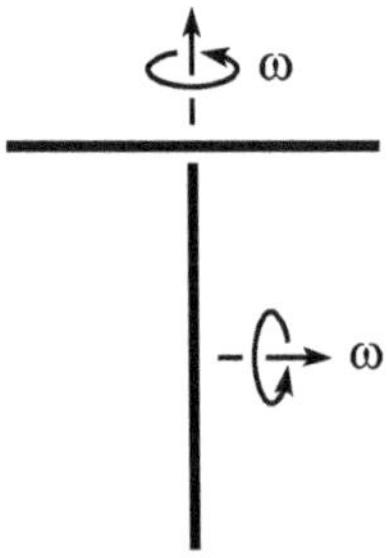

Fig. 9.59

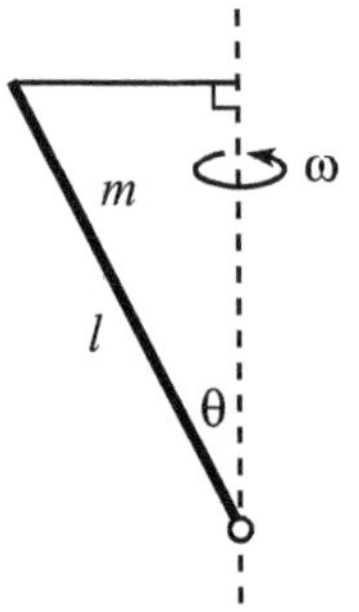

Fig. 9.60

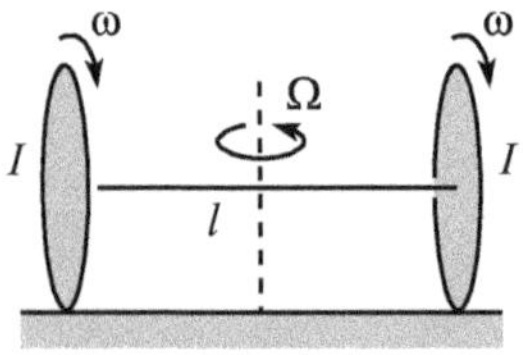

Fig. 9.61

9.39. **Striking another triangle** **

Consider the rigid object in Fig. 9.58. A uniform stick of mass M lies along the base of an isosceles triangle, and a mass m lies at the opposite vertex. The base has length b, and the height is h. Assume that the object is floating freely in outer space. The right end of the stick is struck with a quick blow, directed into the page. Let the impulse have magnitude $\int F\,dt = P$. What is the velocity of mass m immediately after the blow?

9.40. **Sticking sticks** **

Two identical uniform sticks spin around their stationary centers with equal angular speeds, as shown in Fig. 9.59. The bottom stick is slowly raised until its top end collides with the center of the top stick. The sticks stick together to form a rigid "T." Assume that the collision takes place when the top stick lies in the plane of the paper. Immediately after the collision, one point (in addition to the CM) on the T will instantaneously be at rest. Where is this point?

9.41. **Circling stick again** **

Solve the problem in Section 9.4.2 again, but now use the CM as the origin.

9.42. **Pivot and string** **

A stick of mass m and length ℓ spins with frequency ω around an axis, as shown in Fig. 9.60. The stick makes an angle θ with the axis. One end is pivoted on the axis, and the other end is connected to the axis by a string that is perpendicular to the axis. What is the tension in the string, and what is the force that the pivot applies to the stick? (Ignore gravity.)

9.43. **Rotating sheet** **

A uniform flat rectangular sheet of mass m and side lengths a and b rotates with angular speed ω around a diagonal. What torque is required? Given a fixed area A, what should the rectangle look like if you want the required torque to be as large as possible? What is the upper bound on the torque?

9.44. **Rotating axle** **

Two wheels of mass m and moment of inertia I are connected by a massless axle of length ℓ, as shown in Fig. 9.61. The system rests on a frictionless surface, and the wheels rotate with frequency ω around the axle. Additionally, the whole system rotates with frequency Ω around the vertical axis through the center of the axle. What is the largest value of Ω for which both wheels stay on the ground?

9.45. **Stick on a ring** **

(a) A stick of mass m and length $2r$ is arranged to make a constant angle θ with the horizontal, with its bottom end sliding in a circle on a frictionless ring of radius r, as shown in Fig. 9.62. What is the frequency of this motion? It turns out that there is a minimum θ for which this motion is possible; what is it?

(b) If the radius of the ring is now R, what is the largest value of r/R for which this motion is possible for $\theta \to 0$?[25]

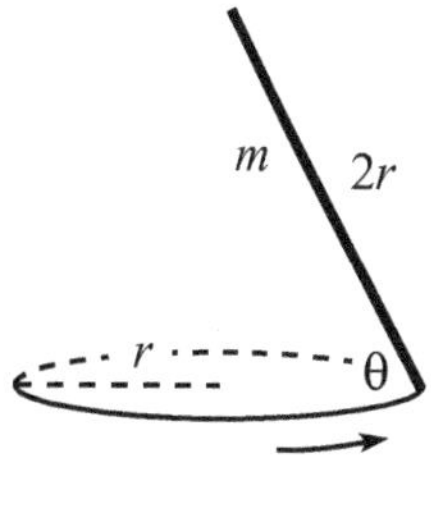

Fig. 9.62

Section 9.6: Free symmetric top

9.46. **Slightly wobbling** *

A coin of mass m and radius R is initially spinning around the axis perpendicular to its plane, with angular speed ω_3. It is supported by a pivot at its center. You apply an infinitesimal downward strike at a point on the rim, as shown in Fig. 9.63, giving the coin an infinitesimal angular velocity component $\omega_\perp$ in the plane of the coin. When the plane of the coin returns to its original plane for the first time, what (approximately) is the orientation of the coin?

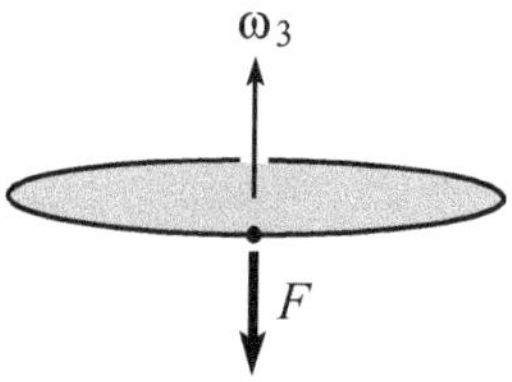

Fig. 9.63

9.47. **Original orientation** **

A coin of mass m and radius R is initially spinning around the axis perpendicular to its plane, with angular speed ω_3. It is supported by a pivot at its center. You apply a (nonzero) downward strike at a point on the rim, as shown in Fig. 9.63, giving the coin an angular velocity component $\omega_\perp$ in the plane of the coin. Consider the nth time the plane of the coin returns to its original plane. What is the minimum value of n for which it is possible for the coin to have *exactly* the same orientation as when it started? What should $\omega_\perp$ be in terms of ω_3 to achieve this?

9.48. **Seeing tails** **

A coin (floating in outer space) of mass m and radius R is initially spinning around the axis perpendicular to its plane, with angular speed ω_3. You view the coin directly from above, and you apply a downward strike at a point on the rim, as shown in Fig. 9.63. What is the minimum impulse, $\int F\,dt$, you must apply in order to be able to barely see the underside of the coin at some later time in its wobbling motion? Assuming that you apply this minimum impulse, how far will the center of the coin have moved by the time you are able to see the underside?

[25] You can also play around with both parts of this problem for the setup where the stick swings around below the ring, with its top end running along the ring.

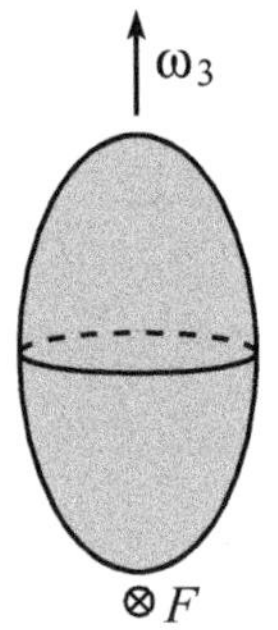

Fig. 9.64

9.49. Flipping a coin **

Imagine flipping an initially horizontal heads-up coin. If the initial angular velocity is horizontal, then the coin will rotate around this horizontal diameter the entire time in the air. The fraction of the flight time that the coin spends heads-up is therefore 1/2. In practice, however, it is impossible to make the initial $\boldsymbol{\omega}$ be *exactly* horizontal, so assume that the initial components are $\omega_\perp$ and ω_3, with $\omega_3 \ll \omega_\perp$. Show that in this limit, the fraction of the flight time that the coin spends heads-up is $1/2 + (4\omega_3^2)/(\pi\omega_\perp^2)$.[26]

9.50. Dipping low **

A top with $I = 3I_3$ floats in outer space and initially spins around its x_3 axis with angular speed ω_3. You apply a strike at the bottom point, directed into the page, as shown in Fig. 9.64, producing an angular velocity component, $\omega_\perp$, directed to the right. What should $\omega_\perp$ be in terms of ω_3 in order to have the total $\boldsymbol{\omega}$ vector dip as far below the horizontal as possible in the subsequent motion?

Section 9.7: Heavy symmetric top

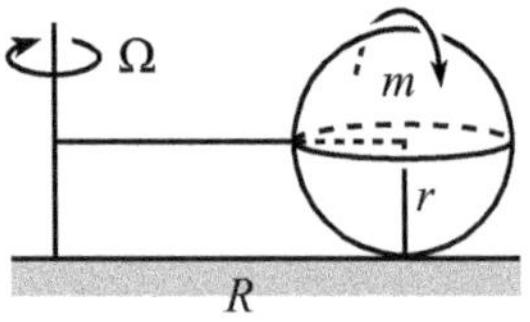

Fig. 9.65

9.51. Rolling lollipop *

Consider a lollipop made of a solid sphere of mass m and radius r that is radially pierced by a massless horizontal stick. The free end of the stick is pivoted on a pole (see Fig. 9.65), and the sphere rolls on the ground without slipping, with its center moving in a circle of radius R with frequency Ω. What is the normal force between the ground and the sphere?

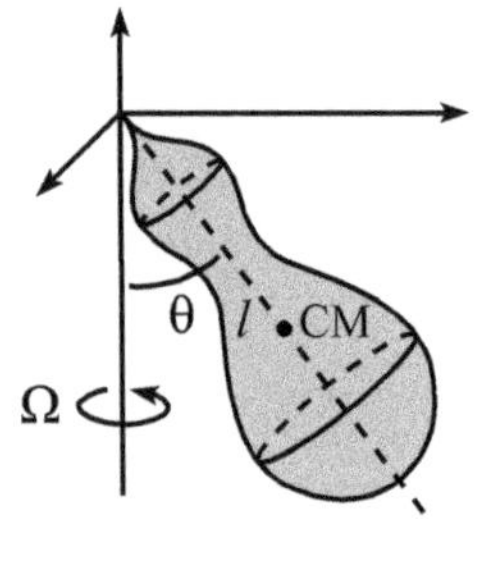

Fig. 9.66

9.52. Horizontal $\boldsymbol{\omega}$ **

A top (with mass m, moments I and I_3, and distance ℓ from the pivot to CM) undergoes uniform precession, with its axis making a constant angle θ with the negative z axis, as shown in Fig. 9.66. If conditions are set up so that the top's $\boldsymbol{\omega}$ is always horizontal, what is the frequency of precession? Can such motion exist if the top is up above the horizontal, making an angle θ with the positive z axis?

[26] If the coin starts truly horizontal (or is at least not biased to tilt in any particular direction on average, which is still a condition that seems difficult to guarantee), and if you catch the coin in your hand to reduce random table-bouncing effects, then the result of this exercise implies that the coin is biased to come up heads. This effect is analyzed in detail in a paper by Persi Diaconis, Susan Holmes, and Richard Montgomery (to be published), who estimate that the probability of obtaining heads for a normally flipped coin is about 0.51.

9.53. **Sliding lollipop** ***

Consider a lollipop made of a solid sphere of mass m and radius r that is radially pierced by a massless stick. The free end of the stick is pivoted on the ground, which is frictionless (see Fig. 9.67). The sphere *slides* along the ground, with the same point on the sphere always touching the ground. The center moves in a circle of radius R with frequency Ω. Show that the normal force between the ground and the sphere is $N = mg + mr\Omega^2$, which is independent of R. Solve this by:

(a) Using an $\mathbf{F} = m\mathbf{a}$ argument.[27]

(b) Using the more complicated $\boldsymbol{\tau} = d\mathbf{L}/dt$ argument.

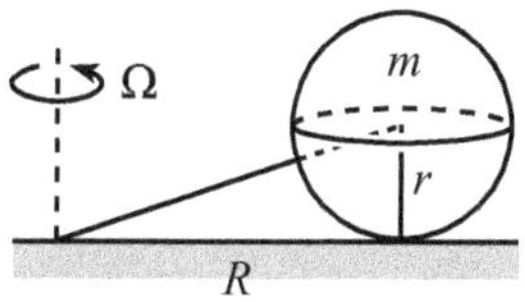

Fig. 9.67

9.54. **Rolling wheel and axle** ***

A massless axle has one end attached to a wheel (a uniform disk of mass m and radius r), with the other end pivoted on the ground (see Fig. 9.68). The wheel rolls on the ground without slipping, with the axle inclined at an angle θ. The point of contact on the ground traces out a circle with frequency Ω.

(a) Show that $\boldsymbol{\omega}$ points horizontally to the right (at the instant shown), with magnitude $\omega = \Omega/\tan\theta$.

(b) Show that the normal force between the ground and the wheel is

$$N = mg\cos^2\theta + mr\Omega^2\left(\frac{1}{4}\cos\theta\sin^2\theta + \frac{3}{2}\cos^3\theta\right). \quad (9.102)$$

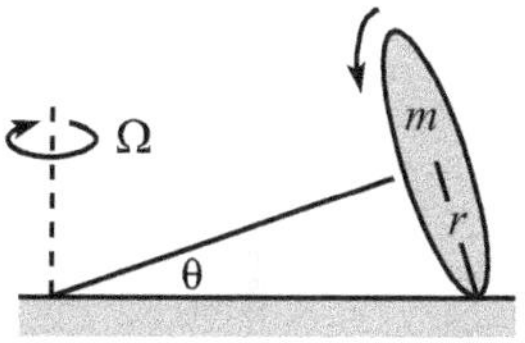

Fig. 9.68

9.55. **Ball under a cone** ***

A hollow ball (with $I = (2/3)mR^2$) rolls without slipping on the inside surface of a fixed cone, whose tip points upward, as shown in Fig. 9.69. The angle at the vertex of the cone is 60°. Initial conditions have been set up so that the contact point on the cone traces out a horizontal circle of radius ℓ at frequency Ω, while the contact point on the ball traces out a circle of radius $R/2$. Assume that the coefficient of friction between the ball and the cone is sufficiently large to prevent slipping. What is the frequency of precession, Ω? What does it reduce to in the limits $\ell \gg R$ and $\ell \to (\sqrt{3}/2)R$ (the ball has to fit inside the cone, of course). What relation between ℓ and R must be satisfied if the setup is to work with a solid ball with $I = (2/5)mR^2$?

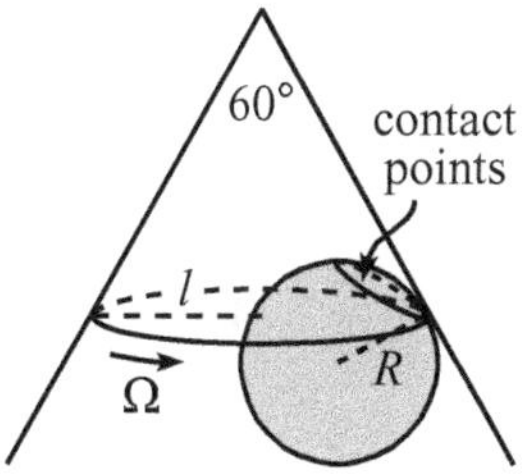

Fig. 9.69

9.56. **Ball in a cone** ****

A ball (with $I = (2/5)MR^2$) rolls without slipping on the inside surface of a fixed cone, whose tip points downward. The half-angle at the vertex

[27] This method happens to work here, due to the unusually nice nature of the sphere's motion. For more general motion (for example, in Problem 9.22, where the sphere is spinning), you must use $\boldsymbol{\tau} = d\mathbf{L}/dt$.

of the cone is θ. Initial conditions have been set up so that the contact point on the cone traces out a horizontal circle of radius $\ell \gg R$, at frequency Ω, while the contact point on the ball traces out a circle of radius r (not necessarily equal to R, as would be the case for a great circle). Assume that the coefficient of friction between the ball and the cone is sufficiently large to prevent slipping. What is the frequency of precession, Ω? It turns out that Ω can be made infinite if r/R takes on a particular value; what is this value? Work in the approximation where $R \ll \ell$.

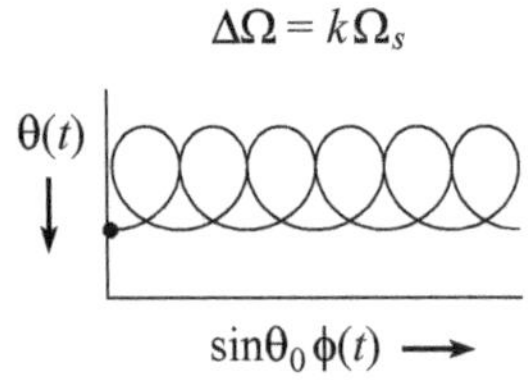

Fig. 9.70

9.57. **Nutation loops** **

In Fig. 9.38, the loops don't quite intersect each other in the $\Delta\Omega = 4\Omega_s$ case, but they very much do in the $\Delta\Omega = 10\Omega_s$ case (a given loop there actually intersects the two on either side). Show that the value of k for which adjacent loops in the $\Delta\Omega = k\Omega_s$ case barely touch each other (as shown in Fig. 9.70) is $k \approx 4.6033$. You will have to solve something numerically. *Hint*: The curve is vertical at the relevant points.

9.10 Solutions

9.1. **Many different ω's**

We want to find all the vectors $\boldsymbol{\omega}$ with the property that $\boldsymbol{\omega} \times a\hat{\mathbf{x}} = v\hat{\mathbf{y}}$. Since $\boldsymbol{\omega}$ is orthogonal to this cross product, $\boldsymbol{\omega}$ must lie in the x-z plane. We claim that if $\boldsymbol{\omega}$ makes an angle θ with the x axis and has magnitude $v/(a\sin\theta)$, then it satisfies $\boldsymbol{\omega} \times a\hat{\mathbf{x}} = v\hat{\mathbf{y}}$. Indeed,

$$\boldsymbol{\omega} \times a\hat{\mathbf{x}} = |\boldsymbol{\omega}||a\hat{\mathbf{x}}|\sin\theta\hat{\mathbf{y}} = v\hat{\mathbf{y}}. \tag{9.103}$$

Alternatively, note that such an $\boldsymbol{\omega}$ can be written as

$$\boldsymbol{\omega} = \frac{v}{a\sin\theta}(\cos\theta, 0, \sin\theta) = \left(\frac{v}{a\tan\theta}, 0, \frac{v}{a}\right). \tag{9.104}$$

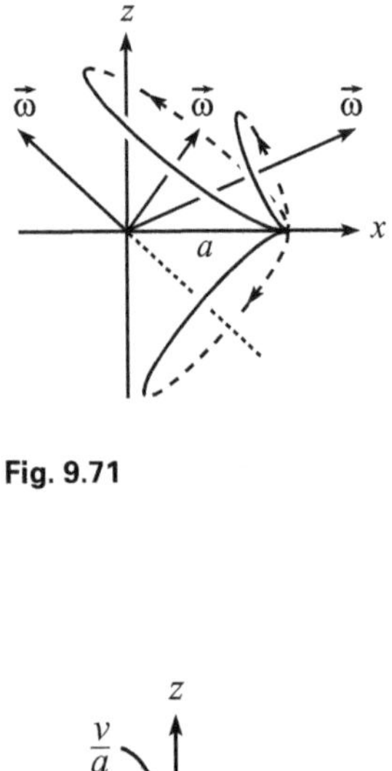

Fig. 9.71

Only the z component here is relevant in the cross product with $a\hat{\mathbf{x}}$, so we have $\boldsymbol{\omega} \times a\hat{\mathbf{x}} = (v/a)\hat{\mathbf{z}} \times a\hat{\mathbf{x}} = v\hat{\mathbf{y}}$. The answer to the problem is therefore that $\boldsymbol{\omega}$ must take the form given in Eq. (9.104).

It makes sense that the magnitude of $\boldsymbol{\omega}$ is $v/(a\sin\theta)$, because if we drop a perpendicular from the particle to the line of $\boldsymbol{\omega}$, we see that the particle may be considered to be instantaneously traveling in a circle of radius $r = a\sin\theta$ around $\boldsymbol{\omega}$, at speed v. And so we have $v = \omega r$, as we should. Whether the particle actually does travel in this circle is irrelevant. The past and future motion doesn't matter in finding the instantaneous $\boldsymbol{\omega}$. All we need to know is the velocity at the given instant.

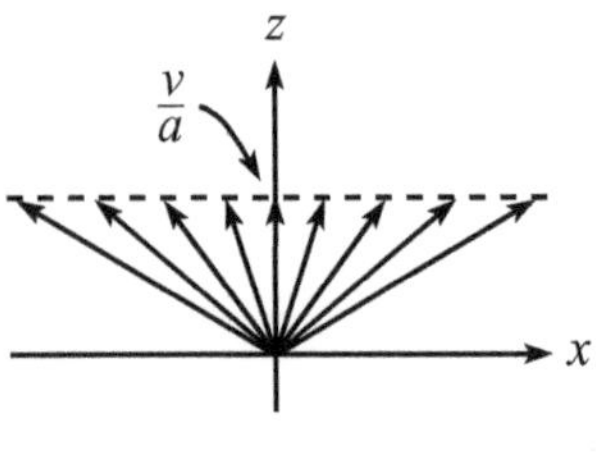

Fig. 9.72

A few possible $\boldsymbol{\omega}$'s are drawn in Fig. 9.71. Technically, it is possible to have $\pi < \theta < 2\pi$, but then the $v/(a\sin\theta)$ coefficient in Eq. (9.104) is negative, which means that $\boldsymbol{\omega}$ really points upward in the x-z plane (physically, $\boldsymbol{\omega}$ must point upward if the particle's velocity is to be in the positive y direction). So we'll assume $0 < \theta < \pi$. And since the v/a value of ω_z in Eq. (9.104) is independent of θ, all of the possible $\boldsymbol{\omega}$'s look like those in Fig. 9.72.

For $\theta = \pi/2$, we have $\omega = v/a$, which makes sense. If θ is very small, then ω is very large, because $\omega \propto 1/\sin\theta$. This also makes sense, because the particle is (instantaneously) traveling around in a very small circle at the given speed v.

REMARK: The point of this problem is that the particle may be in the process of having its position vector trace out a cone around one of many possible axes, or perhaps it may be undergoing a more complicated motion. If we are handed only the given information on position and velocity, then it is impossible to determine which of these motions is happening. And it is likewise impossible to uniquely determine $\boldsymbol{\omega}$. This is true for a collection of points that lie on at most one line through the origin. If the points, along with the origin, span more than a 1-D line, then $\boldsymbol{\omega}$ is in fact uniquely determined (see Footnote 9). ♣

9.2. **Fixed points on a sphere**

FIRST SOLUTION: For the purposes of Theorem 9.1, we need only show that two points end up where they started for an *infinitesimal* transformation. But since it's possible to prove this for a general transformation, we'll consider the general case here.

Consider the point A that ends up farthest away from where it started.[28] Label the ending point B. Draw the great circle, C_{AB}, through A and B. Draw the great circle, C_A, that is perpendicular to C_{AB} at A; and draw the great circle, C_B, that is perpendicular to C_{AB} at B. We claim that the transformation must take C_A to C_B. This is true for the following reason. The image of C_A is certainly a great circle through B. And this great circle must be perpendicular to C_{AB}, because otherwise there would exist another point that ended up farther away from its starting point than A did (see Fig. 9.73). Since there is only one great circle through B that is perpendicular to C_{AB}, the image of C_A must in fact be C_B.

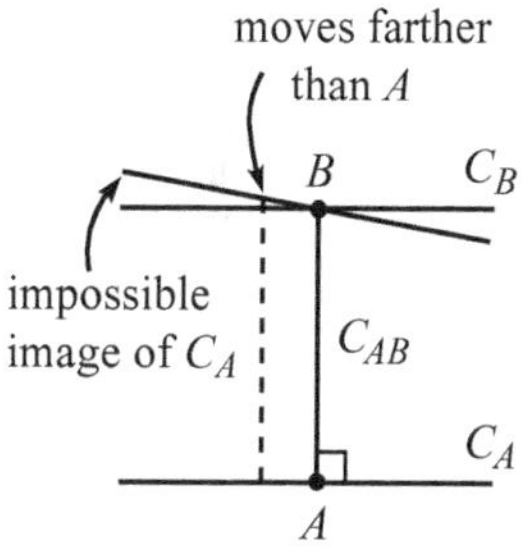

Fig. 9.73

Now consider the two points, P_1 and P_2, where C_A and C_B intersect (any two great circles intersect in two points). Look at P_1. The distances P_1A and P_1B are equal, because C_{AB} makes equal angles (namely 90°) with C_A and C_B. Therefore, the point P_1 is not moved by the transformation. This is true because P_1 ends up on C_B (because C_B is the image of C_A, on which P_1 started), and if it ended up at a point other than P_1, then its final distance from B would be different from its initial distance from A. This would contradict the fact that distances are preserved on the rigid sphere. Likewise for P_2. We have therefore found our two desired points.

Note that for a noninfinitesimal transformation, every point on the sphere may move at some time during the transformation. But what we've just shown is that two points end up back where they started, even if they've moved in the interim.

SECOND SOLUTION: In the spirit of the above solution, we can give a simpler solution, but one that is valid only in the case of infinitesimal transformations.

Pick any point A that moves during the transformation. Draw the great circle that passes through A and is perpendicular to the direction of A's motion. (This direction is well defined, because we are considering an infinitesimal transformation, so A doesn't have time to change direction.) All points on this great circle must move (if they move at all) perpendicular to the great circle, because otherwise their distances from A would change. But they cannot all move in the same direction, because then the center of the great circle, and hence the sphere, would move (but it is assumed to be fixed). Therefore, at least one point on the great circle moves in the direction opposite to the direction in which A moves. Therefore, by continuity, some point (and hence also its diametrically opposite point) on the great circle must remain fixed.

9.3. **Rolling cone**

At the risk of overdoing it, we'll present three solutions. The second and third solutions are the type that might make your head hurt, so you may want to reread them after looking at the discussion of the angular velocity vector in Section 9.7.2.

[28] If there is more than one such point, pick any one of them. Considering that the result of this problem is that the net motion of the sphere is a rotation around an axis, there will in fact be a whole great circle of points that move farthest.

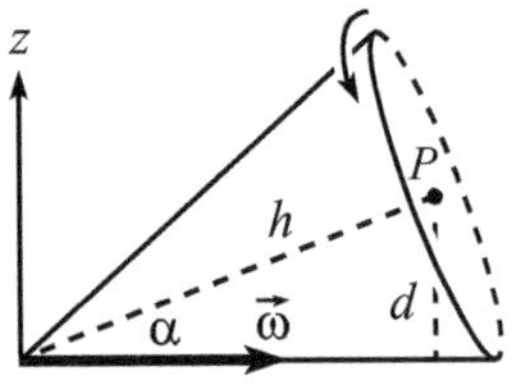

Fig. 9.74

FIRST SOLUTION: Without doing any calculations, we know that $\boldsymbol{\omega}$ points along the line of contact of the cone with the table, because these are the points on the cone that are instantaneously at rest. And we know that as time goes by, $\boldsymbol{\omega}$ rotates around in the horizontal plane with angular speed $v/(h\cos\alpha)$, because point P travels at speed v in a circle of radius $h\cos\alpha$ around the z axis.

The magnitude of $\boldsymbol{\omega}$ can be found as follows. At a given instant, P may be considered to be rotating in a circle of radius $d = h\sin\alpha$ around $\boldsymbol{\omega}$ (see Fig. 9.74). Since P moves with speed v, the angular speed of this rotation is $\omega = v/d$. Therefore,

$$\omega = \frac{v}{h\sin\alpha}. \tag{9.105}$$

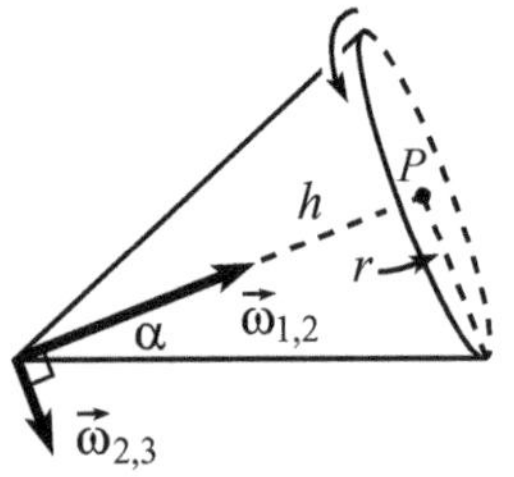

Fig. 9.75

SECOND SOLUTION: We can use Theorem 9.3 with the following frames. S_1 is fixed in the cone; S_3 is the lab frame; and S_2 is the frame that (instantaneously) rotates around the tilted $\boldsymbol{\omega}_{2,3}$ axis shown in Fig. 9.75, at the speed such that the axis of the cone remains fixed in S_2. The tip of $\boldsymbol{\omega}_{2,3}$ traces out a circle as it precesses around the z axis, so after the cone moves a little, we will need to use a new S_2 frame. But at any moment, S_2 instantaneously rotates around the axis perpendicular to the axis of the cone.

In the language of Theorem 9.3, $\boldsymbol{\omega}_{1,2}$ and $\boldsymbol{\omega}_{2,3}$ point in the directions shown. We must find their magnitudes and then add the vectors to determine the angular velocity of S_1 with respect to S_3. First, we have

$$|\boldsymbol{\omega}_{2,3}| = \frac{v}{h}, \tag{9.106}$$

because point P moves (instantaneously) with speed v in a circle of radius h around $\boldsymbol{\omega}_{2,3}$. We now claim that

$$|\boldsymbol{\omega}_{1,2}| = \frac{v}{r} = \frac{v}{h\tan\alpha}, \tag{9.107}$$

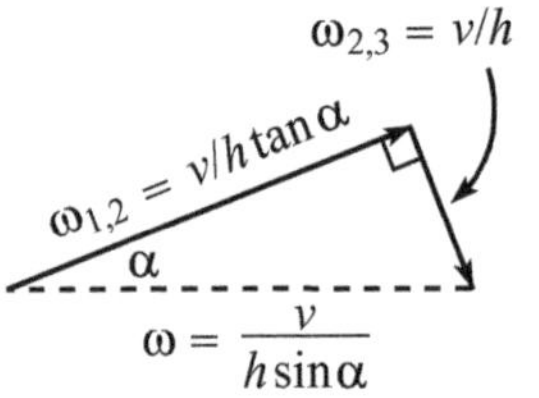

Fig. 9.76

where r is the radius of the base of the cone. This is true because someone fixed in S_2 sees the endpoint of the radius (the one drawn in Fig. 9.75) move "backward" at speed v, because it is stationary with respect to the table. Hence, the cone must be spinning with frequency v/r in S_2.

The addition of $\boldsymbol{\omega}_{1,2}$ and $\boldsymbol{\omega}_{2,3}$ is shown in Fig. 9.76. The result has magnitude $v/(h\sin\alpha)$, and it points horizontally because $|\boldsymbol{\omega}_{2,3}|/|\boldsymbol{\omega}_{1,2}| = \tan\alpha$.

THIRD SOLUTION: We can use Theorem 9.3 with the following frames. S_1 is fixed in the cone; and S_3 is the lab frame (as in the second solution). But now let S_2 be the frame that rotates around the (negative) z axis, at the speed such that the axis of the cone remains fixed in it. Note that we can keep using this same S_2 frame as time goes by, unlike the S_2 frame in the second solution.

$\boldsymbol{\omega}_{1,2}$ and $\boldsymbol{\omega}_{2,3}$ point in the directions shown in Fig. 9.77. As above, we must find their magnitudes and then add the vectors to determine the angular velocity of S_1 with respect to S_3. First, we have

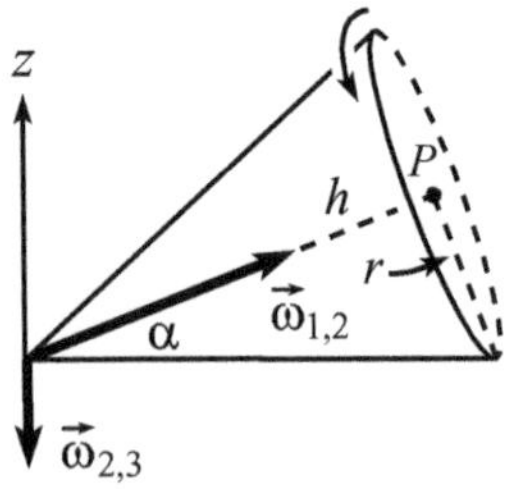

Fig. 9.77

$$|\boldsymbol{\omega}_{2,3}| = \frac{v}{h\cos\alpha}, \tag{9.108}$$

because point P moves with speed v in a circle of radius $h\cos\alpha$ around $\boldsymbol{\omega}_{2,3}$.

It's a little tricker to find $|\boldsymbol{\omega}_{1,2}|$. From the point of view of someone spinning around with S_2, the table rotates backward with frequency $|\boldsymbol{\omega}_{2,3}| = v/(h\cos\alpha)$, from Eq. (9.108). Consider the circle of contact points on the table where the base of the cone touches it. This circle has a radius $h/\cos\alpha$, so someone spinning around with S_2 sees the circle move backward with speed $|\boldsymbol{\omega}_{2,3}|(h/\cos\alpha) = v/\cos^2\alpha$ around the vertical. Since there is no slipping, the contact point on the cone must also move with this speed in S_2 around the axis of the cone (which is fixed in S_2). And since the radius of the base is r, this means that the cone rotates with angular speed $v/(r\cos^2\alpha)$ with respect to S_2. Therefore, using $r = h\tan\alpha$, we have

$$|\omega_{1,2}| = \frac{v}{r\cos^2\alpha} = \frac{v}{h\sin\alpha\cos\alpha}. \qquad (9.109)$$

The addition of $\omega_{1,2}$ and $\omega_{2,3}$ is shown in Fig. 9.78. The result has magnitude $v/(h\sin\alpha)$, and it points horizontally because $|\omega_{2,3}|/|\omega_{1,2}| = \sin\alpha$.

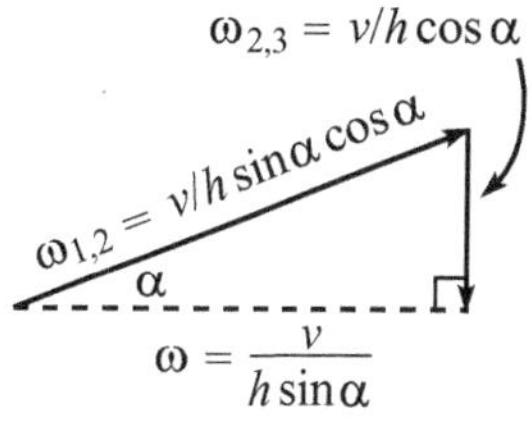

Fig. 9.78

REMARK: The difference between the $\omega_{1,2}$ vectors in the second and third solutions comes down to the following fact. Let Q be the contact point on the base of the cone. Consider the ratio of the distance from Q to $\omega_{2,3}$ to the distance from P to $\omega_{2,3}$. This ratio is 1 in the second solution, but $1/\cos^2\alpha$ in the third solution. This implies that the "backward" speed of Q relative to P, as measured in frame S_2, is a factor $1/\cos^2\alpha$ larger in the third solution. And since Q is the same distance r away from $\omega_{1,2}$ in both cases, we see that $\omega_{1,2}$ is a factor $1/\cos^2\alpha$ larger in the third solution. ♣

9.4. **Parallel-axis theorem**

Consider one of the diagonal entries in **I**, say $I_{xx} \equiv \int(y^2+z^2)$. In terms of the new variables, this equals

$$\begin{aligned} I_{xx} = \int\left((Y+y')^2 + (Z+z')^2\right) &= \int(Y^2+Z^2) + \int(y'^2+z'^2) \\ &= M(Y^2+Z^2) + \int(y'^2+z'^2), \end{aligned} \qquad (9.110)$$

as desired. We have used the fact that the cross terms vanish because, for example, $\int Yy' = Y\int y' = 0$, by definition of the CM. Similarly, consider an off-diagonal entry in **I**, say $I_{xy} \equiv -\int xy$. We have

$$I_{xy} = -\int(X+x')(Y+y') = -\int XY - \int x'y' = -M(XY) - \int x'y', \qquad (9.111)$$

where the cross terms have likewise vanished. We therefore see that all of the terms in **I** take the form of those in Eq. (9.19), as desired.

9.5. **A nice cylinder**

By symmetry, the principal axes are the symmetry axis of the cylinder, along with any diameters in the cross-section circle through the CM. Let the equal moments around the diameters be I. Then by Theorem 9.5, if the moment around the symmetry axis also equals I, then every axis is a principal axis.

Let the mass of the cylinder be M. Let its radius be R and its height be h. Then the moment around the symmetry axis is $MR^2/2$. Let D be a diameter through the CM. The moment around D can be calculated as follows. Slice the cylinder into horizontal disks of thickness dy. Let ρ be the mass per unit height (so $\rho = M/h$). The mass of each disk is then $\rho\,dy$, so the moment around a diameter through the disk is $(\rho\,dy)R^2/4$ (the usual result for a disk). Therefore, by the parallel-axis theorem, the moment of a disk at height y (where $-h/2 \le y \le h/2$) around D is $(\rho\,dy)R^2/4 + (\rho\,dy)y^2$. Hence, the moment of the entire cylinder around D is

$$I = \int_{-h/2}^{h/2}\left(\frac{\rho R^2}{4} + \rho y^2\right)dy = \frac{\rho R^2 h}{4} + \frac{\rho h^3}{12} = \frac{MR^2}{4} + \frac{Mh^2}{12}. \qquad (9.112)$$

We want this to equal $MR^2/2$. Therefore, $h = \sqrt{3}R$.

As an exercise, you can show that if the origin were instead taken to be the center of one of the circular faces, then the answer would be $h = \sqrt{3}R/2$. Note that the I in Eq. (9.112) looks just like the moment of a disk around a diameter plus the moment of a stick around its center. This is no coincidence. The integral that yielded the $Mh^2/12$ term here is the same integral that we would perform for a stick.

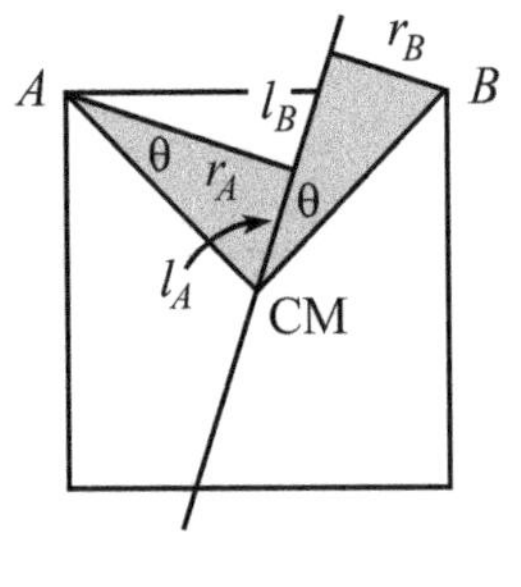

Fig. 9.79

9.6. **Rotating square**

Label two of the masses A and B, as shown in Fig. 9.79. Let ℓ_A be the distance along the axis from the CM to A's string, and let r_A be the length of A's string. Likewise for B. The force, F_A, in A's string must account for the centripetal acceleration of A. Hence, $F_A = mr_A\omega^2$. The torque around the CM due to F_A is therefore

$$\tau_A = mr_A\ell_A\omega^2. \tag{9.113}$$

Likewise, the torque around the CM due to B's string is $\tau_B = mr_B\ell_B\omega^2$, in the opposite direction. But the two shaded triangles in Fig. 9.79 are congruent, because they have the same hypotenuse and the same angle θ. Therefore, $\ell_A = r_B$ and $\ell_B = r_A$. Hence, $\tau_A = \tau_B$, and the torques cancel. The torques from the other two masses likewise cancel. Note that a uniform square is made up of many sets of these squares of point masses, so we've also shown that no torque is needed for a uniform square.

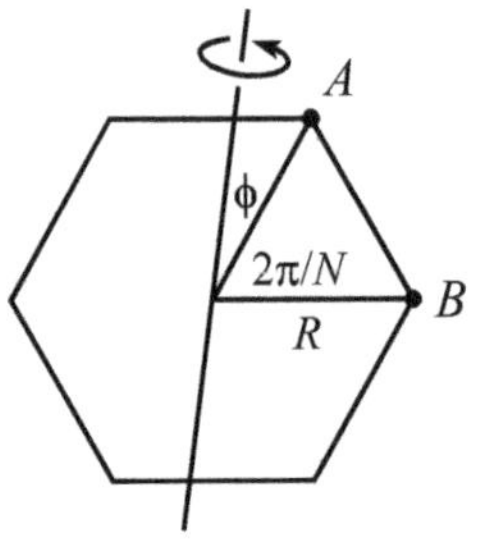

Fig. 9.80

REMARK: For a general N-gon of point masses, Problem 9.8 below shows that any axis in the plane is a principal axis. Let's prove this here by using the above torque argument. We'll use a math trick here that involves writing a trig function (a sine) as the imaginary part of a complex exponential. Using Eq. (9.113), we see that the torque from mass A in Fig. 9.80 is $\tau_A = m\omega^2R^2\sin\phi\cos\phi$. Likewise, the torque from mass B is $\tau_B = m\omega^2R^2\sin(\phi + 2\pi/N)\cos(\phi + 2\pi/N)$, and so on. The total torque around the CM is therefore

$$\begin{aligned}
\tau &= mR^2\omega^2\sum_{k=0}^{N-1}\sin\left(\phi + \frac{2\pi k}{N}\right)\cos\left(\phi + \frac{2\pi k}{N}\right)\\
&= \frac{mR^2\omega^2}{2}\sum_{k=0}^{N-1}\sin\left(2\phi + \frac{4\pi k}{N}\right)\\
&= \frac{mR^2\omega^2}{2}\sum_{k=0}^{N-1}\mathrm{Im}\left(e^{i(2\phi+4\pi k/N)}\right)\\
&= \frac{mR^2\omega^2}{2}\mathrm{Im}\left(e^{2i\phi}\left(1 + e^{4\pi i/N} + e^{8\pi i/N} + \cdots + e^{4(N-1)\pi i/N}\right)\right)\\
&= \frac{mR^2\omega^2}{2}\mathrm{Im}\left(e^{2i\phi}\left(\frac{e^{4N\pi i/N} - 1}{e^{4\pi i/N} - 1}\right)\right)\\
&= 0.
\end{aligned} \tag{9.114}$$

To obtain the fifth line, we summed the geometric series. And to obtain the last line, we used the fact that $e^{4\pi i} = 1$. The one exception to this result is when $N = 2$, because the denominator in the fifth line is zero (but it's hard to have a 2-gon, anyway); this corresponds to the $\theta \neq 180°$ restriction in Theorem 9.6.

Note that what we've done here is show that $\sum r_i\ell_i = 0$, which implies that the total torque is zero (which is one of the definitions of a principal axis). In terms of the chosen axis, this is equivalent to showing that $\sum xy = 0$, that is, showing that the off-diagonal terms in the inertia tensor vanish (which is simply another definition of the principal axes). ♣

9.7. **Existence of principal axes for a pancake**

For a pancake object, the inertia tensor $\mathbf{I}$ takes the form in Eq. (9.8), with $z = 0$. Therefore, if we can find a set of axes for which $\int xy = 0$, then $\mathbf{I}$ will be diagonal, and we will have found our principal axes. We can prove, using a continuity argument, that such a set of axes exists.

Pick a set of axes, and write down the integral $\int xy \equiv I_0$. If $I_0 = 0$, then we are done. If $I_0 \neq 0$, then rotate these axes by an angle of $\pi/2$, so that the new $\hat{\mathbf{x}}$ is the old $\hat{\mathbf{y}}$, and the new $\hat{\mathbf{y}}$ is the old $-\hat{\mathbf{x}}$ (see Fig. 9.81). Write down the new integral $\int xy \equiv I_{\pi/2}$. Since the new and old coordinates are related by $x_{\text{new}} = y_{\text{old}}$ and $y_{\text{new}} = -x_{\text{old}}$, we have $I_{\pi/2} = -I_0$. Therefore, since $\int xy$ switches sign during the (continuous) rotation of the axes, there must exist some intermediate angle for which the integral $\int xy$ is zero.

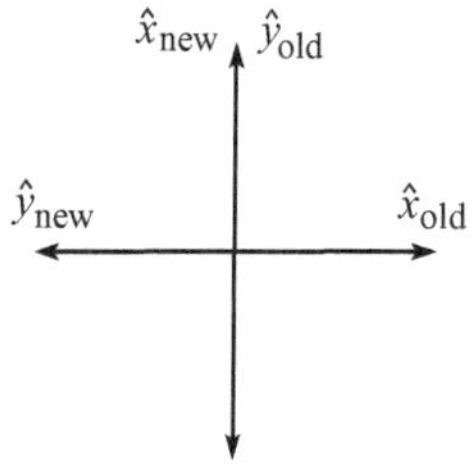

Fig. 9.81

9.8. **Symmetries and principal axes for a pancake**

In view of the form of the inertia tensor in Eq. (9.8), we want to show that if a pancake object has a symmetry under a rotation through $\theta \neq \pi$, then $\int xy = 0$ for any set of axes (through the origin). Take an arbitrary set of axes and rotate them through an angle $\theta \neq \pi$. The new coordinates are $x' = (x\cos\theta + y\sin\theta)$ and $y' = (-x\sin\theta + y\cos\theta)$, so the new matrix entries, in terms of the old ones, are

$$\begin{aligned} I'_{xx} &\equiv \int y'^2 = I_{yy}\sin^2\theta + 2I_{xy}\sin\theta\cos\theta + I_{xx}\cos^2\theta, \\ I'_{yy} &\equiv \int x'^2 = I_{yy}\cos^2\theta - 2I_{xy}\sin\theta\cos\theta + I_{xx}\sin^2\theta, \\ I'_{xy} &\equiv -\int x'y' = I_{yy}\sin\theta\cos\theta + I_{xy}(\cos^2\theta - \sin^2\theta) - I_{xx}\sin\theta\cos\theta. \end{aligned} \tag{9.115}$$

If the object looks exactly like it did before the rotation, then $I'_{xx} = I_{xx}$, $I'_{yy} = I_{yy}$, and $I'_{xy} = I_{xy}$. The first two of these statements are actually equivalent (as you can show), so we'll just use the first and third. Using Eq. (9.116), along with $1 - \cos^2\theta = \sin^2\theta$, these two statements give

$$\begin{aligned} 0 &= -I_{xx}\sin^2\theta + 2I_{xy}\sin\theta\cos\theta + I_{yy}\sin^2\theta, \\ 0 &= -I_{xx}\sin\theta\cos\theta - 2I_{xy}\sin^2\theta + I_{yy}\sin\theta\cos\theta. \end{aligned} \tag{9.116}$$

Multiplying the first of these by $\cos\theta$ and the second by $\sin\theta$, and subtracting, gives

$$2I_{xy}\sin\theta = 0. \tag{9.117}$$

Under the assumption $\theta \neq \pi$ (and $\theta \neq 0$, of course), we must therefore have $I_{xy} = 0$. Our initial axes were arbitrary; hence, any set of axes (through the origin) in the plane is a set of principal axes.

REMARK: If an object is invariant under a rotation through an angle θ, then θ must be of the form $\theta = 2\pi/N$, for some integer N (convince yourself of this). So consider a regular N-gon with "radius" R and with point masses m located at the vertices. Any object that is invariant under a rotation through $\theta = 2\pi/N$ may be considered to be built out of regular point-mass N-gons of various sizes. Theorem 9.5 implies that all axes in the plane of a regular N-gon have the same moment (the two axes in this theorem aren't required to be orthogonal). Let's demonstrate this explicitly for a regular N-gon. This method we'll use here is similar to the one in the remark in Problem 9.6, except that now we'll write a trig function (a cosine) as the real part of a complex exponential. In Fig. 9.82 the distance from mass A to the axis is $r_A = R\sin\phi$. And the distance from B to the axis is $R_B = R\sin(\phi + 2\pi/N)$, and so on for the other masses. The moment of inertia around the axis is therefore

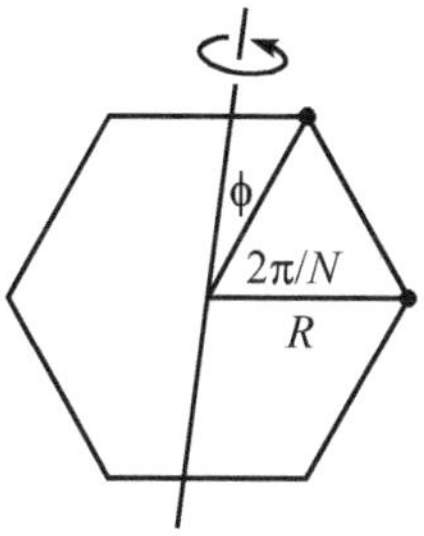

Fig. 9.82

$$\begin{aligned} I_\phi &= mR^2 \sum_{k=0}^{N-1} \sin^2\left(\phi + \frac{2\pi k}{N}\right) \\ &= \frac{mR^2}{2} \sum_{k=0}^{N-1} \left(1 - \cos\left(2\phi + \frac{4\pi k}{N}\right)\right) \end{aligned}$$

$$= \frac{NmR^2}{2} - \frac{mR^2}{2}\sum_{k=0}^{N-1} \text{Re}\Big(e^{i(2\phi+4\pi k/N)}\Big)$$
$$= \frac{NmR^2}{2} - \frac{mR^2}{2} \text{Re}\Big(e^{2i\phi}\Big(1 + e^{4\pi i/N} + e^{8\pi i/N} + \cdots + e^{4(N-1)\pi i/N}\Big)\Big)$$
$$= \frac{NmR^2}{2} - \frac{mR^2}{2} \text{Re}\left(e^{2i\phi}\left(\frac{e^{4N\pi i/N} - 1}{e^{4\pi i/N} - 1}\right)\right), \tag{9.118}$$

where we have summed the geometric series to obtain the last line. The numerator in the parentheses equals $e^{4\pi i} - 1 = 0$. And if $N \neq 2$, the denominator is not zero. Therefore, if $N \neq 2$ (which is equivalent to the $\theta \neq \pi$ restriction), we have

$$I_\phi = \frac{NmR^2}{2}, \tag{9.119}$$

which is independent of ϕ. This result of $NmR^2/2$ for the common value of all the moments makes sense, because the perpendicular-axis theorem says that this common value must be half of the moment around the axis perpendicular to the plane, which is NmR^2. ♣

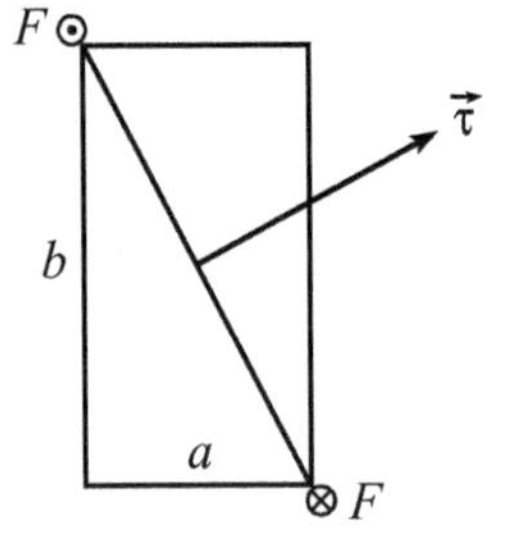

Fig. 9.83

9.9. **Striking a rectangle**

If the force is out of the page at the upper left corner and into the page at the lower right corner, then the torque $\boldsymbol{\tau} = \mathbf{r} \times \mathbf{F}$ points upward to the right, as shown in Fig. 9.83, with $\boldsymbol{\tau} \propto (b, a)$. The angular momentum equals $\int \boldsymbol{\tau}\, dt$, so immediately after the strike, $\mathbf{L}$ is proportional to (b, a). But the angular momentum may also be written as $\mathbf{L} = (I_x\omega_x, I_y\omega_y)$, where $I_x = mb^2/12$ and $I_y = ma^2/12$ are the principal moments. Therefore, we have

$$(I_x\omega_x,\ I_y\omega_y) \propto (b, a) \quad \Longrightarrow \quad (\omega_x, \omega_y) \propto \left(\frac{b}{I_x}, \frac{a}{I_y}\right) \propto \left(\frac{b}{b^2}, \frac{a}{a^2}\right) \propto (a, b), \tag{9.120}$$

which is the direction of the other diagonal. This result checks in the special case $a = b$, and also in the limit where one of a and b is much larger than the other. Basically, if b is much larger than a, then this largeness matters more in the moment of inertia around the x axis (which is quadratic in length) than it does in the torque in the x direction (which is linear in length), so there is very little rotation around the x axis.

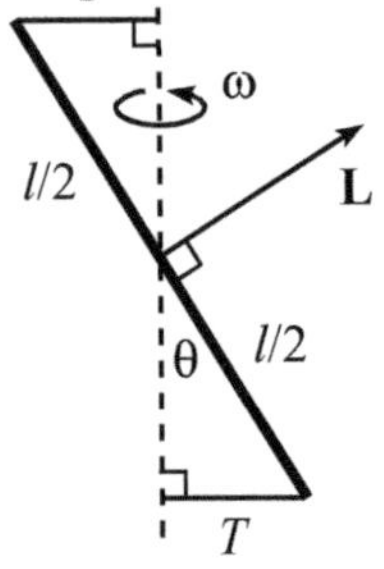

Fig. 9.84

9.10. **Rotating stick**

The angular momentum around the CM can be found as follows. Break $\boldsymbol{\omega}$ up into its components along the principal axes of the stick (which are parallel and perpendicular to the stick). The moment of inertia around the stick is zero. Therefore, only the component of $\boldsymbol{\omega}$ perpendicular to the stick is relevant in calculating $\mathbf{L}$. This component is $\omega \sin\theta$, and the associated moment of inertia is $m\ell^2/12$. Hence, the angular momentum at any time has magnitude

$$L = \frac{1}{12}m\ell^2\omega\sin\theta, \tag{9.121}$$

and it points as shown in Fig. 9.84. The tip of the vector $\mathbf{L}$ traces out a circle in a horizontal plane with frequency ω. The radius of this circle is the horizontal component of $\mathbf{L}$, which is $L_\perp \equiv L\cos\theta$. The rate of change of $\mathbf{L}$ therefore has magnitude

$$\left|\frac{d\mathbf{L}}{dt}\right| = \omega L_\perp = \omega L\cos\theta = \omega\left(\frac{1}{12}m\ell^2\omega\sin\theta\right)\cos\theta, \tag{9.122}$$

and it is directed into the page at the instant shown.

Let the tension in the strings be T. Then the torque due to the strings is $\tau = 2T(\ell/2)\cos\theta$, directed into the page at the instant shown. Therefore, $\boldsymbol{\tau} = d\mathbf{L}/dt$ gives

$$T\ell\cos\theta = \omega\left(\frac{1}{12}m\ell^2\omega\sin\theta\right)\cos\theta \quad \Longrightarrow \quad T = \frac{1}{12}m\ell\omega^2\sin\theta. \qquad (9.123)$$

REMARKS: For $\theta \to 0$, this goes to zero, which makes sense. For $\theta \to \pi/2$, it goes to the finite value $m\ell\omega^2/12$, which isn't so obvious. The lever arm and $|d\mathbf{L}/dt|$ are both small in this limit, and they happen to be proportional to each other as they both head to zero.

Note that if we instead had a massless stick with equal masses $m/2$ on the ends, so that the relevant moment of inertia is now $m\ell^2/4$, then our answer would be $T = (1/4)m\ell\omega^2\sin\theta$. This makes sense if we write it as $T = (m/2)\cdot(\ell/2)\sin\theta\cdot\omega^2$, because each tension is responsible for simply keeping a mass $m/2$ moving in a circle of radius $(\ell/2)\sin\theta$ at frequency ω. This simple force argument doesn't work for the original stick because of internal forces in the stick. ♣

9.11. **Stick under a ring**

As in Problem 9.10, the angular momentum around the CM can be found by breaking $\boldsymbol{\omega}$ up into its components along the principal axes of the stick. The reasoning is exactly the same, so the angular momentum at any time has magnitude

$$L = \frac{1}{12}m\ell^2\omega\sin\theta, \qquad (9.124)$$

and it points as shown in Fig. 9.85. The change in $\mathbf{L}$ comes from its horizontal component. This has length $L\cos\theta$ and travels in a circle at frequency ω. Hence, $|d\mathbf{L}/dt| = \omega L\cos\theta$, and it is directed into the page at the instant shown.

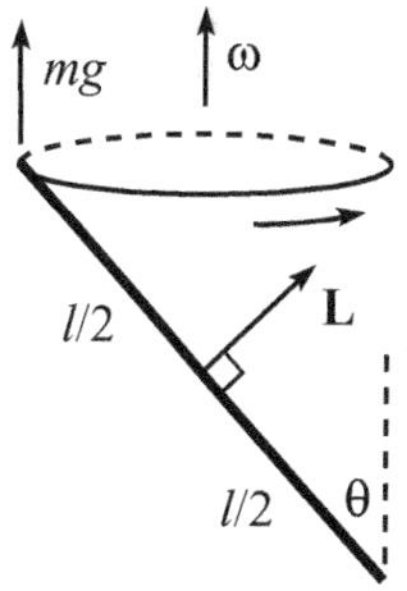

Fig. 9.85

The torque around the CM has magnitude $mg(\ell/2)\sin\theta$, and it points into the page at the instant shown. (This torque arises from the vertical force from the ring. There is no horizontal force from the ring, because the CM does not move.) Therefore, $\boldsymbol{\tau} = d\mathbf{L}/dt$ gives

$$\frac{mg\ell\sin\theta}{2} = \omega\left(\frac{m\ell^2\omega\sin\theta}{12}\right)\cos\theta \quad \Longrightarrow \quad \omega = \sqrt{\frac{6g}{\ell\cos\theta}}. \qquad (9.125)$$

REMARKS:

1. For $\theta \to \pi/2$, this goes to infinity, which makes sense. For $\theta \to 0$, it goes to $\sqrt{6g/\ell}$, which isn't so obvious.
2. The motion in this problem isn't possible if the bottom end of the stick, instead of the top end, slides along the ring. The magnitudes of all quantities are the same as in the original problem, but the torque points in the wrong direction, as you can check. But see Exercise 9.45 for a related setup.
3. Since the middle of the stick is motionless, you might be tempted to treat the bottom half as the stick in the example in Section 9.4.2. That is, you might just want to plug in $\ell/2$ as the length in Eq. (9.38). This, however, does not yield the answer in Eq. (9.125). The error is that there are internal forces in the stick that provide torques. If a pivot were placed at the CM of the stick (connecting the two halves) in this problem, then the stick would not remain straight.
4. If we replace the stick by a massless string with equal masses $m/2$ on the ends, then the relevant moment of inertia is $m\ell^2/4$, and we obtain $\omega = \sqrt{2g/(\ell\cos\theta)}$. This is in fact simply the $\omega = \sqrt{g/[(\ell/2)\cos\theta]}$ answer for a point-mass circular pendulum of length $\ell/2$ (see Problem 9.12), because the middle of the string is motionless, and because the flexible string can't provide the internal torques mentioned in the previous paragraph. ♣

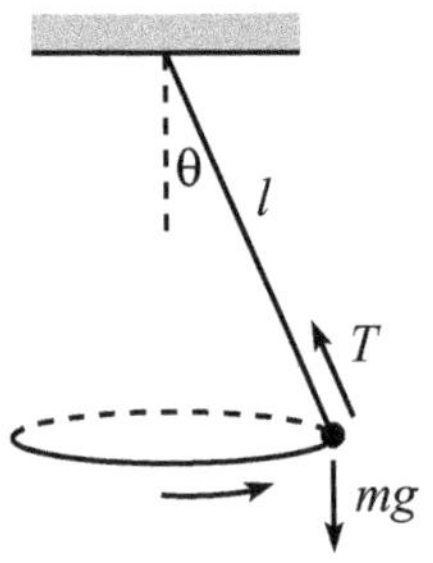

Fig. 9.86

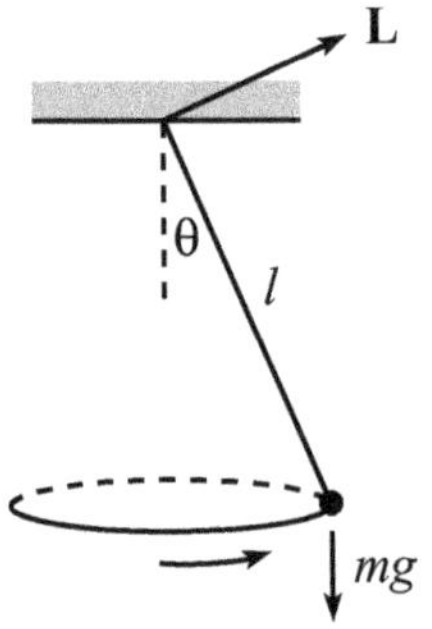

Fig. 9.87

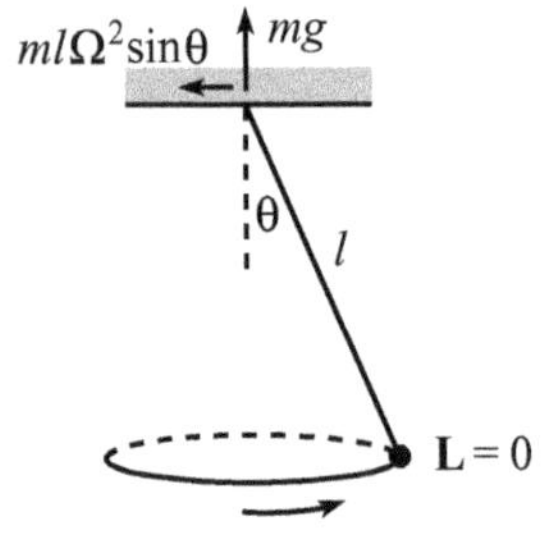

Fig. 9.88

9.12. **Circular pendulum**

(a) The forces on the mass are gravity and the force from the rod (the tension), which points along the rod (see Fig. 9.86).[29] Since there is no vertical acceleration, we have $T\cos\theta = mg$. The unbalanced horizontal force from the tension is therefore $T\sin\theta = mg\tan\theta$. This force accounts for the centripetal acceleration, $m(\ell\sin\theta)\Omega^2$. Hence,

$$\Omega = \sqrt{\frac{g}{\ell\cos\theta}}. \tag{9.126}$$

For $\theta \approx 0$, this is the same as the $\sqrt{g/\ell}$ frequency for a simple pendulum. For $\theta \approx \pi/2$, it goes to infinity, which makes sense. Note that θ must be less than $\pi/2$ for circular motion to be possible. (This restriction doesn't hold for a spinning top with extended mass.)

(b) The only force that applies a torque relative to the pivot is the gravitational force, so the torque is $\tau = mg\ell\sin\theta$, directed into the page at the instant shown in Fig. 9.87.

At this moment, the mass has a speed $(\ell\sin\theta)\Omega$, directed into the page. Therefore, $\mathbf{L} = \mathbf{r}\times\mathbf{p}$ has magnitude $m\ell^2\Omega\sin\theta$ and is directed upward to the right, as shown. The tip of $\mathbf{L}$ traces out a circle of radius $L\cos\theta$, at frequency Ω. Therefore, $d\mathbf{L}/dt$ has magnitude $\Omega L\cos\theta$ and is directed into the page. Hence, $\boldsymbol{\tau} = d\mathbf{L}/dt$ gives $mg\ell\sin\theta = \Omega(m\ell^2\Omega\sin\theta)\cos\theta$, which yields Eq. (9.126).

(c) The only force that applies a torque relative to the mass is the force from the pivot, which has two components (see Fig. 9.88). The vertical piece is mg. Relative to the mass, this provides a torque of $mg(\ell\sin\theta)$, directed into the page. There is also the horizontal piece, which accounts for the centripetal acceleration of the mass, so it equals $m(\ell\sin\theta)\Omega^2$. Relative to the mass, this provides a torque of $m\ell\sin\theta\,\Omega^2(\ell\cos\theta)$, directed out of the page.

Relative to the mass, there is no angular momentum. Therefore, $d\mathbf{L}/dt = 0$, and so there must be no torque. This means that the above two torques must cancel, which gives $mg(\ell\sin\theta) = m\ell\sin\theta\,\Omega^2(\ell\cos\theta)$. This yields Eq. (9.126).

In problems that are more complicated than this one, it is often easier to work with the fixed pivot (if there is one) as the origin, instead of the CM, because then you don't have to worry about the messy pivot forces contributing to the torque.

9.13. **Rolling in a cone**

The forces on the ring are gravity (mg), the normal force (N) from the cone, and the friction force (F) pointing up along the cone (or perhaps down along the cone if F turns out to be negative, but we'll find that it won't). Since there is no net force in the vertical direction, we have

$$N\sin\theta + F\cos\theta = mg. \tag{9.127}$$

The inward horizontal force accounts for the centripetal acceleration, which gives

$$N\cos\theta - F\sin\theta = m(h\tan\theta)\Omega^2. \tag{9.128}$$

Solving the previous two equations for F gives

$$F = mg\cos\theta - m\Omega^2(h\tan\theta)\sin\theta, \tag{9.129}$$

[29] The force from the rod must point along the rod because it is massless. If there were a tangential force on the mass, then Newton's third law would say that there would also be a tangential force on the rod. This would then produce a nonzero torque on the rod (relative to the pivot) and hence an infinite angular acceleration, because there is no gravitational torque on the (massless) rod to counter it.

with upward along the cone taken to be positive (this is just a rearrangement of the $F = ma$ equation along the cone). The torque on the ring (relative to its CM) is due only to this F, because gravity provides no torque, and N points though the center of the ring (by the second assumption in the problem). Therefore, the torque points out of the page with magnitude

$$\tau = rF = r\big(mg\cos\theta - m\Omega^2 h\tan\theta\sin\theta\big). \tag{9.130}$$

We must now find $d\mathbf{L}/dt$. Since we are assuming $r \ll h\tan\theta$, the frequency of the spinning of the ring (call it ω) is much greater than the frequency of precession, Ω. We will therefore neglect the latter in finding $\mathbf{L}$. In this approximation, $\mathbf{L}$ (relative to the CM) has magnitude $mr^2\omega$, and it points downward along the cone (for the direction of motion shown in Fig. 9.45). The horizontal component of $\mathbf{L}$ has magnitude $L_\perp \equiv L\sin\theta$, and it traces out a circle at frequency Ω. Therefore, $d\mathbf{L}/dt$ points out of the page with magnitude

$$\left|\frac{d\mathbf{L}}{dt}\right| = \Omega L_\perp = \Omega L\sin\theta = \Omega(mr^2\omega)\sin\theta. \tag{9.131}$$

The nonslipping condition is $r\omega = (h\tan\theta)\Omega$,[30] which gives $\omega = (h\tan\theta)\Omega/r$. Using this in Eq. (9.131) yields

$$\left|\frac{d\mathbf{L}}{dt}\right| = \Omega^2 mrh\tan\theta\sin\theta. \tag{9.132}$$

Equating this with the torque in Eq. (9.130) gives

$$\Omega = \frac{1}{\tan\theta}\sqrt{\frac{g}{2h}}. \tag{9.133}$$

REMARK: If you consider an object with moment of inertia βmr^2 (our ring has $\beta = 1$), then you can show by the above reasoning that the "2" in Eq. (9.133) is replaced by $(1+\beta)$. This means that if we instead have a particle sliding around a frictionless cone (which is equivalent to a ring with $\beta = 0$), then the frequency is $\sqrt{g/h}/\tan\theta$, as you can check from scratch. ♣

9.14. **Tennis racket theorem**

ROTATION AROUND $\hat{\mathbf{x}}_1$: If the racket is rotated (nearly) around the $\hat{\mathbf{x}}_1$ axis, then the initial ω_2 and ω_3 are much smaller than ω_1. To emphasize this, let's relabel them as $\omega_2 \to \epsilon_2$ and $\omega_3 \to \epsilon_3$. Then Eq. (9.45) becomes (with the torque equal to zero, because gravity provides no torque around the CM)

$$\begin{aligned} 0 &= \dot{\omega}_1 - A\epsilon_2\epsilon_3, \\ 0 &= \dot{\epsilon}_2 + B\omega_1\epsilon_3, \\ 0 &= \dot{\epsilon}_3 - C\omega_1\epsilon_2, \end{aligned} \tag{9.134}$$

where we have defined (for convenience)

$$A \equiv \frac{I_2 - I_3}{I_1}, \quad B \equiv \frac{I_1 - I_3}{I_2}, \quad C \equiv \frac{I_1 - I_2}{I_3}. \tag{9.135}$$

Note that A, B, and C are all positive. This fact will be very important.

Our goal here is to show that if the ϵ's start out small, then they remain small. Assuming that they are small (which is true initially), the first equation says that

[30] This is technically not quite correct, for the same reason that the earth spins around 366 times instead of 365 times in a year. But it's valid enough in the limit of small r.

$\dot{\omega}_1 \approx 0$, to first order in the ϵ's. Therefore, we may assume that ω_1 is essentially constant (when the ϵ's are small). Taking the derivative of the second equation then gives $0 = \ddot{\epsilon}_2 + B\omega_1\dot{\epsilon}_3$. Plugging the value of $\dot{\epsilon}_3$ from the third equation into this yields

$$\ddot{\epsilon}_2 = -\left(BC\omega_1^2\right)\epsilon_2. \tag{9.136}$$

Because of the negative coefficient on the right-hand side, this equation describes simple harmonic motion. Therefore, ϵ_2 oscillates sinusoidally around zero. So if it starts small, it remains small. By the same reasoning, ϵ_3 remains small.

We therefore see that $\boldsymbol{\omega} \approx (\omega_1, 0, 0)$ at all times, which implies that $\mathbf{L} \approx (I_1\omega_1, 0, 0)$ at all times. That is, $\mathbf{L}$ always points (nearly) along the $\hat{\mathbf{x}}_1$ direction (which is fixed in the racket frame). But the direction of $\mathbf{L}$ is fixed in the lab frame, because there is no torque. Therefore, the direction of $\hat{\mathbf{x}}_1$ must also be (nearly) fixed in the lab frame. In other words, the racket doesn't wobble.

ROTATION AROUND $\hat{\mathbf{x}}_3$: This calculation goes through exactly as above, except with "1" and "3" interchanged. We find that if ϵ_1 and ϵ_2 start small, they remain small.

ROTATION AROUND $\hat{\mathbf{x}}_2$: If the racket is rotated (nearly) around the $\hat{\mathbf{x}}_2$ axis, then the initial ω_1 and ω_3 are much smaller than ω_2. As above, let's emphasize this by relabeling them as $\omega_1 \to \epsilon_1$ and $\omega_3 \to \epsilon_3$. Then as above, Eq. (9.45) becomes

$$\begin{aligned} 0 &= \dot{\epsilon}_1 - A\omega_2\epsilon_3, \\ 0 &= \dot{\omega}_2 + B\epsilon_1\epsilon_3, \\ 0 &= \dot{\epsilon}_3 - C\omega_2\epsilon_1. \end{aligned} \tag{9.137}$$

Our goal here is to show that if the ϵ's start out small, then they do *not* remain small. Assuming that they are small (which is true initially), the second equation says that $\dot{\omega}_2 \approx 0$, to first order in the ϵ's. So we may assume that ω_2 is essentially constant (when the ϵ's are small). Taking the derivative of the first equation then gives $0 = \ddot{\epsilon}_1 - A\omega_2\dot{\epsilon}_3$. Plugging the value of $\dot{\epsilon}_3$ from the third equation into this yields

$$\ddot{\epsilon}_1 = \left(AC\omega_2^2\right)\epsilon_1. \tag{9.138}$$

Because of the positive coefficient on the right-hand side, this equation describes an exponentially growing motion, instead of an oscillatory one. Therefore, ϵ_1 grows quickly from its initial small value. So even if it starts small, it becomes large. By the same reasoning, ϵ_3 becomes large. Of course, once the ϵ's become large, then our assumption of $\dot{\omega}_2 \approx 0$ isn't valid anymore. But once the ϵ's become large, we've shown what we wanted to.

We see that $\boldsymbol{\omega}$ does *not* remain (nearly) equal to $(0, \omega_2, 0)$, which implies that $\mathbf{L}$ does *not* remain (nearly) equal to $(0, I_2\omega_2, 0)$. That is, $\mathbf{L}$ does not always point (nearly) along the $\hat{\mathbf{x}}_2$ direction (which is fixed in the racket frame). But the direction of $\mathbf{L}$ is fixed in the lab frame, because there is no torque. Therefore, the direction of $\hat{\mathbf{x}}_2$ must change in the lab frame. In other words, the racket wobbles.

9.15. **Free-top angles**

In terms of the principal axes, $\hat{\mathbf{x}}_1$, $\hat{\mathbf{x}}_2$, $\hat{\mathbf{x}}_3$, we have

$$\begin{aligned} \boldsymbol{\omega} &= (\omega_1\hat{\mathbf{x}}_1 + \omega_2\hat{\mathbf{x}}_2) + \omega_3\hat{\mathbf{x}}_3, \quad \text{and} \\ \mathbf{L} &= I(\omega_1\hat{\mathbf{x}}_1 + \omega_2\hat{\mathbf{x}}_2) + I_3\omega_3\hat{\mathbf{x}}_3. \end{aligned} \tag{9.139}$$

Let $\omega_\perp\hat{\boldsymbol{\omega}}_\perp \equiv (\omega_1\hat{\mathbf{x}}_1 + \omega_2\hat{\mathbf{x}}_2)$ be the component of $\boldsymbol{\omega}$ orthogonal to $\boldsymbol{\omega}_3$. Then we have

$$\tan\beta = \frac{\omega_\perp}{\omega_3}, \quad \text{and} \quad \tan\alpha = \frac{I\omega_\perp}{I_3\omega_3}. \tag{9.140}$$

Therefore,

$$\frac{\tan \alpha}{\tan \beta} = \frac{I}{I_3}. \tag{9.141}$$

If $I > I_3$, then $\alpha > \beta$, and we have the situation shown in Fig. 9.89. A top with this property is called a "prolate top." An example is an American football or a pencil.

If $I < I_3$, then $\alpha < \beta$, and we have the situation shown in Fig. 9.90. A top with this property is called an "oblate top." An example is a coin or a Frisbee.

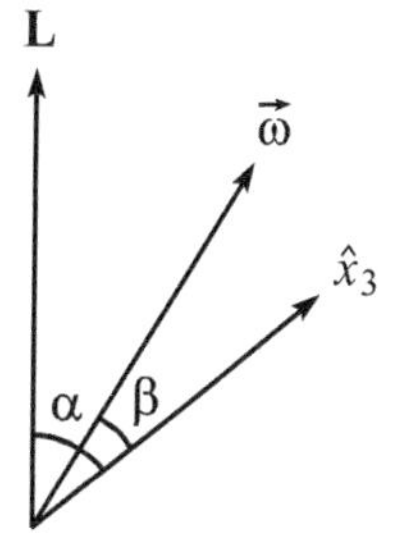

Fig. 9.89

9.16. **Staying above**

If $I < I_3$ (oblate top), then the initial $\hat{\mathbf{x}}_3$, $\boldsymbol{\omega}$, and $\mathbf{L}$ vectors look like those in Fig. 9.91, with $\mathbf{L}$ between $\hat{\mathbf{x}}_3$ and $\boldsymbol{\omega}$. The $\boldsymbol{\omega}$ vector traces out a cone around $\mathbf{L}$ as shown, so it always stays above the horizontal (because it starts above the horizontal).

However, if $I > I_3$ (prolate top), then the initial $\hat{\mathbf{x}}_3$, $\boldsymbol{\omega}$, and $\mathbf{L}$ vectors look like those in Fig. 9.92, with $\boldsymbol{\omega}$ between $\hat{\mathbf{x}}_3$ and $\mathbf{L}$. Depending on the values of I/I_3 and $\omega_\perp/\omega_3$ (where $\omega_\perp$ is the component of $\boldsymbol{\omega}$ orthogonal to $\hat{\mathbf{x}}_3$), the $\boldsymbol{\omega}$ cone might stay above the horizontal axis, or it might dip below it. We are concerned with the cutoff case where it just touches the axis, as shown.

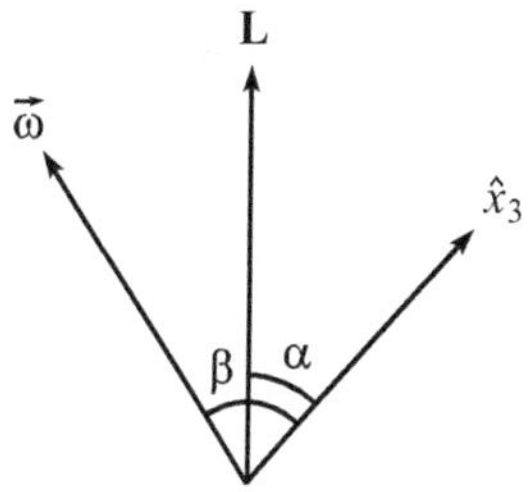

Fig. 9.90

From Fig. 9.92, the half-angle of the cone is $\alpha - \beta$, so if the cone reaches down to the horizontal axis, then we have $\alpha + (\alpha - \beta) = 90°$. But α and β are given by

$$\tan \beta = \frac{\omega_\perp}{\omega_3}, \quad \text{and} \quad \tan \alpha = \frac{L_\perp}{L_3} = \frac{I\omega_\perp}{I_3\omega_3}. \tag{9.142}$$

Therefore, the $2\alpha - \beta = 90°$ condition becomes

$$2\tan^{-1}\left(\frac{I\omega_\perp}{I_3\omega_3}\right) - \tan^{-1}\left(\frac{\omega_\perp}{\omega_3}\right) = 90°. \tag{9.143}$$

With $n \equiv I/I_3$ and $x \equiv \omega_\perp/\omega_3$, this yields

$$\begin{aligned} &2\tan^{-1}(nx) - \tan^{-1}(x) = 90° \\ \Longrightarrow \quad &\tan\Big(2\tan^{-1}(nx)\Big) = \tan\Big(90° + \tan^{-1}(x)\Big) \\ \Longrightarrow \quad &\frac{2nx}{1 - n^2x^2} = -\frac{1}{x} \\ \Longrightarrow \quad &n(n-2)x^2 = 1. \end{aligned} \tag{9.144}$$

We see that if $n \leq 2$, there is no solution for x. But if $n > 2$, there is a solution. Therefore, the answer to the problem is $n = 2$.

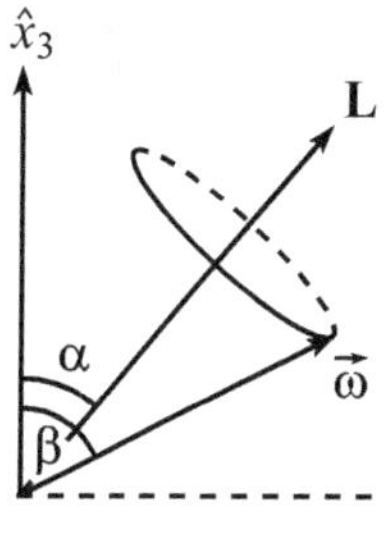

Fig. 9.91

REMARKS: If n is only slightly larger than 2, then the solution for x in Eq. (9.144) (and hence $\omega_\perp$) is large. This means that your strike needs to be large, in order to produce a large $\omega_\perp$.

There are three interesting limits of what can happen in the large-n limit, so let's list them out. For these three cases, picture a thin stick spinning around its vertical x_3 axis, and then imagine striking the bottom end with different impulses.

- If $\omega_\perp \ll \omega_3$ and $L_\perp \ll L_3$, then both $\boldsymbol{\omega}$ and $\mathbf{L}$ point nearly straight upwards, so $\boldsymbol{\omega}$ traces out a very thin cone around $\mathbf{L}$ and is always nearly vertical. $\hat{\mathbf{x}}_3$ is also always nearly vertical.
- If $\omega_\perp \ll \omega_3$ but $L_\perp \gg L_3$ (which is possible if n is sufficiently large), then $\boldsymbol{\omega}$ points nearly straight upwards initially, whereas $\mathbf{L}$ points nearly horizontally. So $\boldsymbol{\omega}$ traces out a very wide cone and swings down almost to the negative vertical direction. $\hat{\mathbf{x}}_3$ also traces out a very wide cone.
- If $\omega_\perp \gg \omega_3$, which implies $L_\perp \gg L_3$, then both $\boldsymbol{\omega}$ and $\mathbf{L}$ point nearly horizontally, so $\boldsymbol{\omega}$ traces out a very thin cone around $\mathbf{L}$ and is always nearly horizontal.

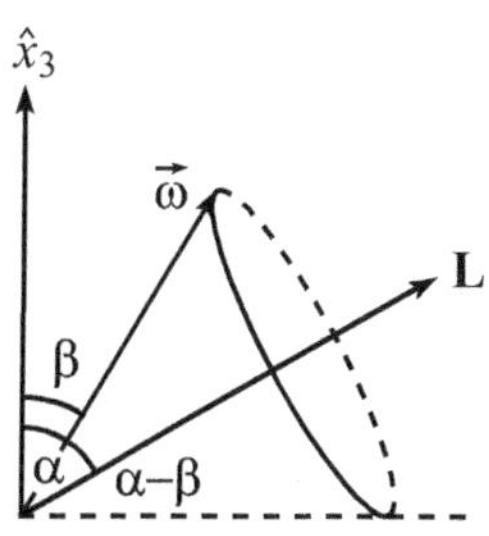

Fig. 9.92

$\hat{\mathbf{x}}_3$ still traces out a very wide cone, because it starts vertical (by assumption), independent of where $\boldsymbol{\omega}$ and $\mathbf{L}$ are. Note that your eye can't distinguish between the second and third cases here, because the $\hat{\mathbf{x}}_3$ axis is doing the same thing in both cases. ♣

9.17. **The top**

(a) In order for there to exist real solutions for Ω in Eq. (9.82), the discriminant must be non-negative. If $\theta > \pi/2$, then $\cos\theta < 0$, so the discriminant is automatically positive, and any value of ω_3 is allowed. But if $\theta < \pi/2$, then the lower limit on ω_3 is

$$\omega_3 \geq \frac{\sqrt{4MIg\ell\cos\theta}}{I_3} \equiv \tilde{\omega}_3. \tag{9.145}$$

The special case of $\theta = \pi/2$ requires a limit to be taken; it turns out that there is only one (noninfinite) solution, as we'll see in part (b). Note that at the critical value of ω_3 in Eq. (9.145), Eq. (9.82) gives

$$\Omega_+ = \Omega_- = \frac{I_3\tilde{\omega}_3}{2I\cos\theta} = \sqrt{\frac{Mg\ell}{I\cos\theta}} \equiv \Omega_0. \tag{9.146}$$

(b) Since ω_3 has units, "large ω_3" is a meaningless description. What we really mean is that the fraction in the square root in Eq. (9.82) is very small compared with 1. That is, $\epsilon \equiv (4MIg\ell\cos\theta)/(I_3^2\omega_3^2) \ll 1$. In this case, we can use $\sqrt{1-\epsilon} \approx 1 - \epsilon/2 + \cdots$ to write

$$\Omega_\pm \approx \frac{I_3\omega_3}{2I\cos\theta}\left(1 \pm \left(1 - \frac{2MIg\ell\cos\theta}{I_3^2\omega_3^2}\right)\right). \tag{9.147}$$

The two solutions for Ω are then, to leading order in ω_3 (or rather, to leading order in ϵ),

$$\Omega_+ \approx \frac{I_3\omega_3}{I\cos\theta}, \quad \text{and} \quad \Omega_- \approx \frac{Mg\ell}{I_3\omega_3}. \tag{9.148}$$

These are known as the "fast" and "slow" frequencies of precession, respectively. Ω_- is the approximate answer we found in Eq. (9.77). It was obtained here under the assumption $\epsilon \ll 1$, which is equivalent to

$$\omega_3 \gg \frac{\sqrt{4MIg\ell\cos\theta}}{I_3} \quad \text{(that is, } \omega_3 \gg \tilde{\omega}_3\text{)}. \tag{9.149}$$

This, therefore, is the condition for the result in Eq. (9.77) to be a good approximation. If I is of the same order as I_3, so that they are both of order $M\ell^2$ (assuming that the top is a reasonably shaped object without any strange tails), and if $\cos\theta$ is of order 1, then this condition can be written as $\omega_3 \gg \sqrt{g/\ell}$, which is the frequency of a pendulum of length ℓ.

REMARKS:

1. The Ω_+ solution is a fairly surprising result. Two strange features of Ω_+ are that it grows with ω_3, and that it is independent of g. To see what is going on with this precession, note that Ω_+ is the value of Ω that makes the $L_\perp$ in Eq. (9.79) essentially equal to zero. So $\mathbf{L}$ points nearly along the vertical axis. The rate of change of $\mathbf{L}$ is the product of a very small horizontal radius (of the tiny circle the tip traces out) and a very large Ω (assuming that ω_3 is large). The product of these equals the "medium sized" torque, $Mg\ell\sin\theta$.
2. In the limit of large ω_3, the fast precession should look basically like the motion of a free top, because the $\mathbf{L}$ here points essentially in a fixed direction, just as it does for a free top. And indeed, Ω_+ is independent of g. More

precisely, we can see that the value of Ω_+ given in Eq. (9.148) agrees with the free-top precession frequency, by the following reasoning. $I_3\omega_3$ is the component of $\mathbf{L}$ along the symmetry axis, which makes an angle θ with the vertical (which is essentially the direction of $\mathbf{L}$). Therefore, $\mathbf{L}$ has magnitude $L \approx I_3\omega_3/\cos\theta$, which means we can write $\Omega_+ \approx L/I$. This is the precession frequency of a free top, given in Eq. (9.55), as viewed from a fixed frame.

3. We can plot the $\Omega_\pm$ in Eq. (9.82) as functions of ω_3. With the definitions of $\tilde{\omega}_3$ and Ω_0 in Eqs. (9.145) and (9.146), we can rewrite Eq. (9.82) as

$$\Omega_\pm = \frac{\omega_3\Omega_0}{\tilde{\omega}_3}\left(1 \pm \sqrt{1 - \frac{\tilde{\omega}_3^2}{\omega_3^2}}\right). \qquad (9.150)$$

Since it is easier to work with dimensionless quantities, let's rewrite this as

$$y_\pm = x \pm \sqrt{x^2 - 1}, \qquad \text{where } y_\pm \equiv \frac{\Omega_\pm}{\Omega_0}, \quad \text{and} \quad x \equiv \frac{\omega_3}{\tilde{\omega}_3}. \qquad (9.151)$$

A rough plot of $y_\pm$ vs. x is shown in Fig. 9.93. The behavior of this graph for large x is found from Eq. (9.151) to be $y_+ \approx 2x$, and $y_- \approx 1/(2x)$. You can show that these are equivalent to the results in Eq. (9.148). ♣

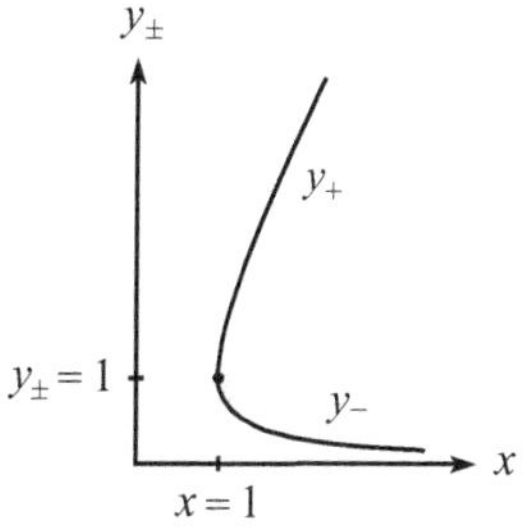

Fig. 9.93

9.18. **Many tops**

The system is made up of N rigid bodies, each consisting of a disk and a massless stick glued to it on its left (see Fig. 9.94). Label these subsystems as S_i, with S_1 being the one closest to the pole. Let each disk have mass m and moment of inertia I, and let each stick have length ℓ. Let the angular speeds be ω_i. The relevant angular momentum of S_i is then $L_i = I\omega_i$, and it points horizontally.[31] Let the desired precession frequency be Ω. Then $d\mathbf{L}_i/dt$ has magnitude $L_i\Omega = (I\omega_i)\Omega$ and points into the page at the instant shown in Fig. 9.49.

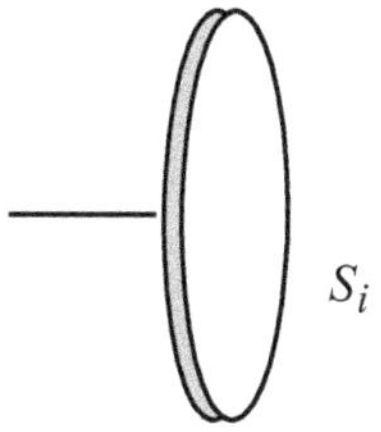

Fig. 9.94

Consider the torque $\boldsymbol{\tau}_i$ on S_i, around its CM. Let's first look at S_1. The pole provides an upward force of Nmg (this force is what keeps all the tops up), so it provides a torque of $Nmg\ell$ (into the page) around the CM of S_1. The downward force from the stick on the right provides no torque around the CM, because it acts at the CM. Likewise, the gravitational force on S_1 provides no torque around the CM. Therefore, $\boldsymbol{\tau}_1 = d\mathbf{L}_1/dt$ gives $Nmg\ell = (I\omega_1)\Omega$, and so

$$\omega_1 = \frac{Nmg\ell}{I\Omega}. \qquad (9.152)$$

Now look at S_2. S_1 provides an upward force of $(N-1)mg$ (this force is what keeps S_2 through S_N up), so it provides a torque of $(N-1)mg\ell$ around the CM of S_2. As with S_1, this is the only torque on S_2. Therefore, $\boldsymbol{\tau}_2 = d\mathbf{L}_2/dt$ gives $(N-1)mg\ell = (I\omega_2)\Omega$, and so

$$\omega_2 = \frac{(N-1)mg\ell}{I\Omega}. \qquad (9.153)$$

Similar reasoning applies to the other S_i, so we arrive at

$$\omega_i = \frac{(N+1-i)mg\ell}{I\Omega}. \qquad (9.154)$$

The ω_i are therefore in the ratio

$$\omega_1 : \omega_2 : \cdots : \omega_{N-1} : \omega_N = N : (N-1) : \cdots : 2 : 1. \qquad (9.155)$$

[31] We're ignoring the angular momentum arising from the precession. This part of $\mathbf{L}$ points vertically (because the tops all point horizontally) and therefore doesn't change. Hence, it doesn't enter into $\boldsymbol{\tau} = d\mathbf{L}/dt$.

Note that we needed to apply $\boldsymbol{\tau} = d\mathbf{L}/dt$ many times, using each CM as the origin. Using only the pivot point on the pole as the origin would have given us only one piece of information, whereas we needed N pieces.

REMARKS: As a double-check, we can verify that the above ω's make $\boldsymbol{\tau} = d\mathbf{L}/dt$ true, where $\boldsymbol{\tau}$ and $\mathbf{L}$ are the total torque and angular momentum relative to the pivot on the pole. The CM of the entire system is $(N+1)\ell/2$ from the pole, so the torque due to gravity is

$$\tau = Nmg\frac{(N+1)\ell}{2}. \tag{9.156}$$

The total angular momentum is, using Eq. (9.154),

$$\begin{aligned} L &= I(\omega_1 + \omega_2 + \cdots + \omega_N) \\ &= \frac{mg\ell}{\Omega}\Big(N + (N-1) + (N-2) + \cdots + 2 + 1\Big) \\ &= \frac{mg\ell}{\Omega}\left(\frac{N(N+1)}{2}\right). \end{aligned} \tag{9.157}$$

Comparing this with Eq. (9.156), we see that $\tau = L\Omega$, that is, $\tau = |d\mathbf{L}/dt|$.

We can also pose this problem for the setup where all the ω_i are equal (call them ω), and the goal is to find the lengths of the sticks that allow circular precession with the sticks always forming a straight horizontal line. We can use the same reasoning as above, and Eq. (9.154) takes the modified form

$$\omega = \frac{(N+1-i)mg\ell_i}{I\Omega}, \tag{9.158}$$

where ℓ_i is the length of the ith stick. Therefore, the ℓ_i are in the ratio

$$\ell_1 : \ell_2 : \cdots : \ell_{N-1} : \ell_N = \frac{1}{N} : \frac{1}{N-1} : \cdots : \frac{1}{2} : 1. \tag{9.159}$$

Again, we can verify that these ℓ's make $\boldsymbol{\tau} = d\mathbf{L}/dt$ true, where $\boldsymbol{\tau}$ and $\mathbf{L}$ are the total torque and angular momentum relative to the pivot on the pole. As an exercise, you can show that the CM happens to be a distance ℓ_N from the pole. So the torque due to gravity is, using Eq. (9.158) to obtain ℓ_N,

$$\tau = Nmg\ell_N = Nmg(\omega I\Omega/mg) = NI\omega\Omega. \tag{9.160}$$

The total angular momentum is simply $L = NI\omega$. So indeed, $\tau = L\Omega = |d\mathbf{L}/dt|$. Note that since the sum $\sum 1/n$ diverges, it is possible to make the setup extend arbitrarily far from the pole. ♣

9.19. **Heavy top on a slippery table**

In Section 9.7.3, we looked at $\boldsymbol{\tau}$ and $\mathbf{L}$ relative to the pivot point. Such quantities are of no use now, because the pivot is accelerating, so it isn't a legal choice of origin for applying $\boldsymbol{\tau} = d\mathbf{L}/dt$. We will therefore look at $\boldsymbol{\tau}$ and $\mathbf{L}$ relative to the CM, which is always a legal origin for applying $\boldsymbol{\tau} = d\mathbf{L}/dt$.

With the CM as our origin, there are two modifications we need to make to the derivation in Section 9.7.3. First, the moments of inertia are now measured with respect to the CM, instead of the pivot point. I_3 is unchanged, but the parallel-axis theorem gives the new I as

$$I' \equiv I - M\ell^2. \tag{9.161}$$

Second, the torque is modified. The only force from the frictionless floor is the normal force N. But N is not necessarily equal to Mg, because the CM may be accelerating in the vertical direction. The vertical $F = ma$ equation is $N - Mg = M\ddot{y}$, where $y = \ell\cos\theta$. Taking the second derivative of y, we obtain $N = Mg - M\ell(\ddot{\theta}\sin\theta + \dot{\theta}^2\cos\theta)$. The torque relative to the CM therefore has magnitude

$$\tau = N\ell\sin\theta = Mg\ell\sin\theta - M\ell^2\sin\theta(\ddot{\theta}\sin\theta + \dot{\theta}^2\cos\theta), \qquad (9.162)$$

and it points in the same direction as in Section 9.7.3. Putting everything together, we see that Eq. (9.69) is unchanged, Eq. (9.70) has I replaced with $I' \equiv I - M\ell^2$, and the second of Eqs. (9.70) also has an extra term and becomes

$$\left(Mg\ell - M\ell^2(\ddot{\theta}\sin\theta + \dot{\theta}^2\cos\theta) + I'\dot{\phi}^2\cos\theta - I_3\omega_3\dot{\phi}\right)\sin\theta = I\ddot{\theta}. \qquad (9.163)$$

Note that if $\dot{\theta} \equiv 0$, the only modification needed is the change in I.

If you prefer to use the Lagrangian method in Section 9.7.4, then I needs to be changed to I' as above, and the other modification comes in the kinetic energy in the Lagrangian. We must add on the kinetic energy of the whole object treated like a point mass at the CM, because so far we have included only the kinetic energy relative to our new origin (the CM). Since the table is frictionless, the CM can move only vertically, so its velocity is $\dot{y} = -\ell\dot{\theta}\sin\theta$. Therefore, we have to add the term $M(\ell\dot{\theta}\sin\theta)^2/2$ to the Lagrangian. You can show that this leads to the extra term in Eq. (9.163).

9.20. **Fixed highest point**

For the desired motion with P always the highest point, the important thing to note is that every point in the top moves in a fixed circle around the $\hat{\mathbf{z}}$ axis. Therefore, $\boldsymbol{\omega}$ points vertically at all times. Hence, if Ω is the frequency of precession, we have $\boldsymbol{\omega} = \Omega\hat{\mathbf{z}}$.

Another way to see that $\boldsymbol{\omega}$ points vertically is to view things in the frame that rotates with angular velocity $\Omega\hat{\mathbf{z}}$. In this frame, the top has no motion at all. It isn't even spinning, because the point P is always the highest point. In the language of Fig. 9.30, we therefore have $\omega' = 0$, so $\boldsymbol{\omega} = \Omega\hat{\mathbf{z}} + \omega'\hat{\mathbf{x}}_3 = \Omega\hat{\mathbf{z}}$.

The principal moments are (with the pivot as the origin; see Fig. 9.95)

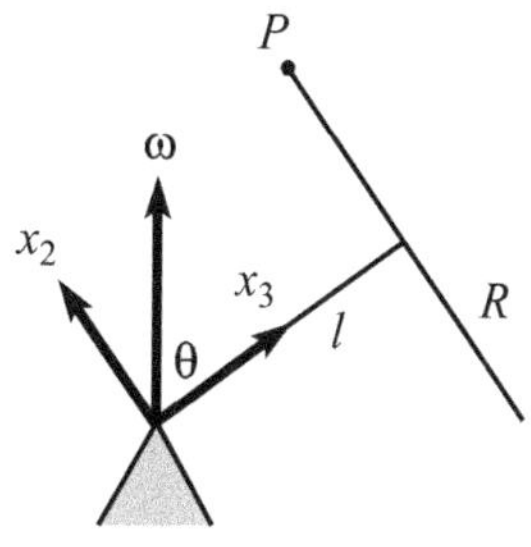

Fig. 9.95

$$I_3 = \frac{MR^2}{2}, \quad \text{and} \quad I \equiv I_1 = I_2 = M\ell^2 + \frac{MR^2}{4}, \qquad (9.164)$$

where we have used the parallel-axis theorem in finding I. The components of $\boldsymbol{\omega}$ along the principal axes are $\omega_3 = \Omega\cos\theta$ and $\omega_2 = \Omega\sin\theta$. Therefore (keeping things in terms of the general moments, I_3 and I, for now),

$$\mathbf{L} = I_3\Omega\cos\theta\,\hat{\mathbf{x}}_3 + I\Omega\sin\theta\,\hat{\mathbf{x}}_2. \qquad (9.165)$$

The horizontal component of $\mathbf{L}$ is then $L_\perp = (I_3\Omega\cos\theta)\sin\theta - (I\Omega\sin\theta)\cos\theta$, so $d\mathbf{L}/dt$ has magnitude

$$\left|\frac{d\mathbf{L}}{dt}\right| = L_\perp\Omega = \Omega^2\sin\theta\cos\theta(I_3 - I), \qquad (9.166)$$

and it is directed into the page (or out of the page, if this quantity is negative). This must equal the torque, which has magnitude $|\boldsymbol{\tau}| = Mg\ell\sin\theta$ and is directed into the page. Therefore,

$$\Omega = \sqrt{\frac{Mg\ell}{(I_3 - I)\cos\theta}}. \qquad (9.167)$$

We see that for a general symmetric top, the desired precessional motion (where the same "side" always points up) is possible only if the product $(I_3 - I)\cos\theta$ is greater than zero. That is,

$$\begin{aligned} \theta < \pi/2 \quad &\Longrightarrow \quad I_3 > I, \\ \theta > \pi/2 \quad &\Longrightarrow \quad I_3 < I. \end{aligned} \tag{9.168}$$

For the problem at hand, I_3 and I are given in Eq. (9.164), so we find

$$\Omega = \sqrt{\frac{4g\ell}{(R^2 - 4\ell^2)\cos\theta}}. \tag{9.169}$$

The necessary condition for such motion to exist is therefore $R > 2\ell$ if $\theta < \pi/2$, or $R < 2\ell$ if $\theta > \pi/2$.

REMARKS:

1. It is intuitively clear that Ω should become very large as $\theta \to \pi/2$, although it is by no means intuitively clear that such motion should exist at all for angles near $\pi/2$.
2. If $\theta > \pi/2$ and $R = 0$, we simply have the circular pendulum discussed in Problem 9.12. And indeed, when $R = 0$, the result in Eq. (9.169) reduces to $\Omega = \sqrt{g/\ell\cos\alpha}$, where $\alpha \equiv \pi - \theta$, which agrees with the result in Eq. (9.126).
3. Ω approaches a nonzero constant as $\theta \to 0$ or $\theta \to \pi$ (depending on the sign of $I_3 - I$), which isn't entirely obvious.
4. If both R and ℓ are scaled up by the same factor, Eq. (9.169) shows that Ω decreases. This also follows from dimensional analysis.
5. Assuming that $\theta < \pi/2$ (the $\theta > \pi/2$ case can be handled in a similar manner), the condition $I_3 > I$ can be understood in the following way. If $I_3 = I$, then $\mathbf{L} \propto \boldsymbol{\omega}$, so $\mathbf{L}$ points vertically along $\boldsymbol{\omega}$. If $I_3 > I$, then $\mathbf{L}$ points somewhere to the right of the z axis (at the instant shown in Fig. 9.95). This means that the tip of $\mathbf{L}$ is moving into the page, along with the top. This is what we need, because $\boldsymbol{\tau}$ points into the page. If, however, $I_3 < I$, then $\mathbf{L}$ points somewhere to the left of the z axis, so $d\mathbf{L}/dt$ points out of the page, and hence cannot be equal to $\boldsymbol{\tau}$. ♣

9.21. **Basketball on a rim**

Assume that the precessional motion is counterclockwise when viewed from above, as indicated in Fig. 9.52. Let's look at things in the frame that has the center of the rim as its origin and that rotates with angular velocity $\Omega\hat{\mathbf{z}}$. In this frame, the center of the ball is at rest, and the rim spins clockwise with angular speed Ω. If the contact points are to form a great circle on the ball, the ball must be spinning around the (negative) $\hat{\mathbf{x}}_3$ axis shown in Fig. 9.96. Let the frequency of this spinning be ω' (in the language of Fig. 9.30). Then the nonslipping condition says that $\omega' r = \Omega R$, and so $\omega' = \Omega R/r$. Therefore, the total angular velocity of the ball with respect to the lab frame is

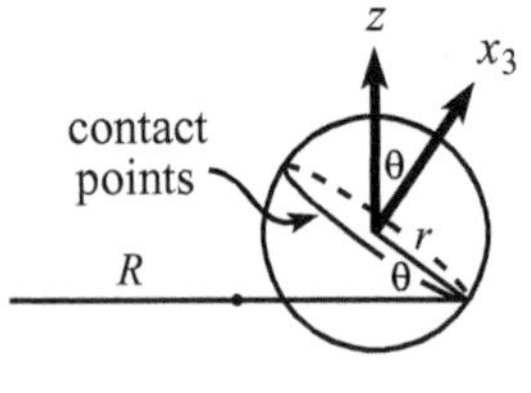

Fig. 9.96

$$\boldsymbol{\omega} = \Omega\hat{\mathbf{z}} - \omega'\hat{\mathbf{x}}_3 = \Omega\hat{\mathbf{z}} - (R/r)\Omega\hat{\mathbf{x}}_3. \tag{9.170}$$

Let us choose the center of the ball as the origin around which we calculate $\boldsymbol{\tau}$ and $\mathbf{L}$. Then every axis in the ball is a principal axis with moment of inertia $I = (2/3)mr^2$. The angular momentum is therefore

$$\mathbf{L} = I\boldsymbol{\omega} = I\Omega\hat{\mathbf{z}} - I(R/r)\Omega\hat{\mathbf{x}}_3. \tag{9.171}$$

Only the $\hat{\mathbf{x}}_3$ piece has a horizontal component which contributes to $d\mathbf{L}/dt$. This component has length $L_\perp = I(R/r)\Omega\sin\theta$. Therefore, $d\mathbf{L}/dt$ has magnitude

$$\left|\frac{d\mathbf{L}}{dt}\right| = \Omega L_\perp = \frac{2}{3}\Omega^2 mrR\sin\theta, \tag{9.172}$$

and it points out of the page.

The torque (relative to the center of the ball) comes from the force at the contact point. There are two components to this force. The vertical component is mg, and the horizontal component is $m(R - r\cos\theta)\Omega^2$ (pointing to the left), because the center of the ball moves in a circle of radius $(R - r\cos\theta)$. The torque therefore has magnitude

$$|\boldsymbol{\tau}| = mg(r\cos\theta) - m(R - r\cos\theta)\Omega^2(r\sin\theta), \tag{9.173}$$

with out of the page taken to be positive. Equating this $|\boldsymbol{\tau}|$ with the $|d\mathbf{L}/dt|$ in Eq. (9.172) gives

$$\Omega^2 = \frac{g}{\frac{5}{3}R\tan\theta - r\sin\theta}. \tag{9.174}$$

REMARKS:

1. $\Omega \to \infty$ as $\theta \to 0$, which makes sense. And $\Omega \to 0$ as $\theta \to \pi/2$, which also makes sense.
2. $\Omega \to \infty$ when $R = (3/5)r\cos\theta$. But this isn't physical, because we must have $R > r\cos\theta$ in order for the other side of the rim to be outside the ball.
3. You can also work out the problem for the case where the contact points trace out a circle other than a great circle (say, one that is inclined at an angle β below the great circle). The expression for the torque in Eq. (9.173) is unchanged, but the value of ω' and the angle of the $\hat{\mathbf{x}}_3$ axis both change, so Eq. (9.172) is modified. If you want to play around with this, one thing you can show (with a bit of work) is that if $R \gg r$ and if you want the ball to travel around the rim infinitely fast, then β must equal $\tan^{-1}((5/2)\tan\theta)$. This is larger than θ, so the contact-point circle actually lies below the horizontal. ♣

9.22. **Rolling lollipop**

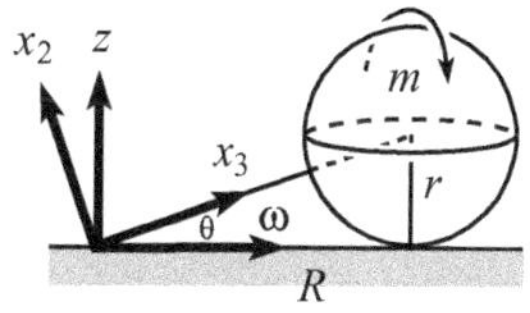

Fig. 9.97

We must first find the angular velocity vector $\boldsymbol{\omega}$. Assume that the precessional motion is clockwise when viewed from above, as indicated in Fig. 9.53. We claim that $\boldsymbol{\omega}$ points horizontally to the right (at the instant shown in Fig. 9.97), with magnitude $(R/r)\Omega$. This can be seen in two ways.

The first way is to recognize that we essentially have the same scenario as in the "Rolling cone" setup of Problem 9.3 (imagine the sphere to be a ball of ice cream in a cone whose tip is located at the left end of the stick). The sphere's contact point with the ground is instantaneously at rest (the nonslipping condition), so $\boldsymbol{\omega}$ must pass through this point. But $\boldsymbol{\omega}$ must also pass through the left end of the stick, because that point is also at rest. Therefore, $\boldsymbol{\omega}$ must be horizontal. To find the magnitude, note that the center of the sphere moves with speed ΩR. But since the center may also be considered to be instantaneously moving with frequency ω in a circle of radius r around the horizontal axis, we have $\omega r = \Omega R$. Therefore, $\omega = (R/r)\Omega$.

The second way is to write $\boldsymbol{\omega}$ as $\boldsymbol{\omega} = -\Omega\hat{\mathbf{z}} + \omega'\hat{\mathbf{x}}_3$ (in the language of Fig. 9.30), where ω' is the frequency of the spinning as viewed by someone rotating around the (negative) $\hat{\mathbf{z}}$ axis with frequency Ω. The contact points form a circle of radius R on the ground. But they also form a circle of radius $r\cos\theta$ on the sphere, where θ is the angle between the stick and the ground (this circle is the circle of points where the above-mentioned ice cream cone touches the sphere). The nonslipping condition then implies that $\Omega R = \omega'(r\cos\theta)$. Therefore, $\omega' = \Omega R/(r\cos\theta)$, and so

$$\boldsymbol{\omega} = -\Omega\hat{\mathbf{z}} + \omega'\hat{\mathbf{x}}_3 = -\Omega\hat{\mathbf{z}} + \left(\frac{\Omega R}{r\cos\theta}\right)(\cos\theta\,\hat{\mathbf{x}} + \sin\theta\,\hat{\mathbf{z}}) = (R/r)\Omega\hat{\mathbf{x}}, \qquad (9.175)$$

where we have used $\tan\theta = r/R$.

Now let's calculate the normal force. Choose the pivot as the origin. The principal axes are then $\hat{\mathbf{x}}_3$ along the stick, along with any two directions orthogonal to the stick. Choose $\hat{\mathbf{x}}_2$ to be in the plane of the paper (see Fig. 9.97). Then the components of $\boldsymbol{\omega}$ along the principal axes are

$$\omega_3 = (R/r)\Omega\cos\theta, \quad \text{and} \quad \omega_2 = -(R/r)\Omega\sin\theta. \qquad (9.176)$$

The principal moments are

$$I_3 = (2/5)mr^2, \quad \text{and} \quad I_2 = (2/5)mr^2 + m(r^2+R^2), \qquad (9.177)$$

where we have used the parallel-axis theorem. The angular momentum is $\mathbf{L} = I_3\omega_3\hat{\mathbf{x}}_3 + I_2\omega_2\hat{\mathbf{x}}_2$, so its horizontal component has length $L_\perp = I_3\omega_3\cos\theta - I_2\omega_2\sin\theta$. Therefore, the magnitude of $d\mathbf{L}/dt$ is

$$\begin{aligned}\left|\frac{d\mathbf{L}}{dt}\right| &= \Omega L_\perp = \Omega(I_3\omega_3\cos\theta - I_2\omega_2\sin\theta)\\ &= \Omega\Big[\Big(\frac{2}{5}mr^2\Big)\Big(\frac{R}{r}\Omega\cos\theta\Big)\cos\theta\\ &\qquad - \Big(\frac{2}{5}mr^2 + m(r^2+R^2)\Big)\Big(-\frac{R}{r}\Omega\sin\theta\Big)\sin\theta\Big]\\ &= \frac{\Omega^2 mR}{r}\left(\frac{2}{5}r^2 + (r^2+R^2)\sin^2\theta\right)\\ &= \frac{7}{5}mrR\Omega^2, \end{aligned}\qquad (9.178)$$

where we have used $\sin\theta = r/\sqrt{r^2+R^2}$. The direction of $d\mathbf{L}/dt$ is out of the page.

The torque (relative to the pivot) is due to the gravitational force acting at the CM, along with the normal force N acting at the contact point. (Any horizontal friction that happens to exist at the contact point produces zero torque relative to the pivot.) Therefore, $\boldsymbol{\tau}$ points out of the page with magnitude $|\boldsymbol{\tau}| = (N - mg)R$. Equating this with the $|d\mathbf{L}/dt|$ in Eq. (9.178) gives

$$N = mg + \frac{7}{5}mr\Omega^2. \qquad (9.179)$$

This has the interesting property of being independent of R, and hence also θ. This independence arises because both $|\boldsymbol{\tau}|$ and $|d\mathbf{L}/dt|$ are proportional to R, a fact which is easier to see via the reasoning in the first remark below.

REMARKS:

1. There is actually a quicker way to calculate the $d\mathbf{L}/dt$ in Eq. (9.178). At a given instant, the sphere is rotating around the horizontal x axis with frequency $\omega = (R/r)\Omega$. The moment of inertia around this axis is $I_x = (7/5)mr^2$, from the parallel-axis theorem. Therefore, the horizontal component of $\mathbf{L}$ has magnitude $L_x = I_x\omega = (7/5)mrR\Omega$. Multiplying this by the frequency (namely Ω) at which $\mathbf{L}$ swings around the z axis gives the result for $|d\mathbf{L}/dt|$ in Eq. (9.178). There is also a vertical component of $\mathbf{L}$ relative to the pivot, but this component doesn't change, so it doesn't come into $d\mathbf{L}/dt$. (The vertical component happens to equal $-mR^2\Omega$. This can be obtained by realizing that the sphere behaves like a point

mass for the present purpose, or by using the inertia tensor relative to the pivot, or by calculating $L_z = I_3\omega_3 \sin\theta + I_2\omega_2 \cos\theta$.)

2. The pivot must provide a downward force of $N - mg = (7/5)mr\Omega^2$, to make the net vertical force on the lollipop equal to zero. This result is slightly larger than the $mr\Omega^2$ result for the "sliding" setup in Exercise 9.53.
3. The sum of the horizontal forces at the pivot and the contact point must equal the required centripetal force of $mR\Omega^2$. But it is impossible to say how this force is divided up, without being given more information. ♣

9.23. **Rolling coin**

Choose the CM as the origin. The principal axes are then $\hat{\mathbf{x}}_2$ and $\hat{\mathbf{x}}_3$ (as shown in Fig. 9.98), along with $\hat{\mathbf{x}}_1$ pointing into the paper. Assume that the precessional motion is counterclockwise when viewed from above, as indicated in Fig. 9.54. Let's look at things in the frame that has the center of the contact-point circle on the ground as its origin, and that rotates with angular velocity $\Omega\hat{\mathbf{z}}$. In this frame, the CM remains fixed, and the coin rotates with frequency ω' (in the language of Fig. 9.30) around the negative $\hat{\mathbf{x}}_3$ axis. The nonslipping condition then says that $\omega' r = \Omega R$, and so $\omega' = \Omega R/r$. Therefore, the total angular velocity of the coin with respect to the lab frame is

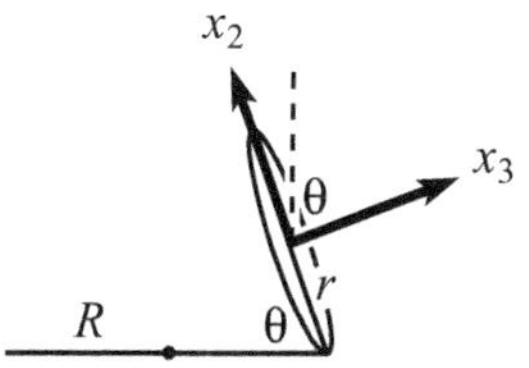

Fig. 9.98

$$\boldsymbol{\omega} = \Omega\hat{\mathbf{z}} - \omega'\hat{\mathbf{x}}_3 = \Omega\hat{\mathbf{z}} - (R/r)\Omega\hat{\mathbf{x}}_3. \quad (9.180)$$

But $\hat{\mathbf{z}} = \sin\theta\,\hat{\mathbf{x}}_2 + \cos\theta\,\hat{\mathbf{x}}_3$, so we can write $\boldsymbol{\omega}$ in terms of the principal axes as

$$\boldsymbol{\omega} = \Omega\sin\theta\hat{\mathbf{x}}_2 - \Omega\left(\frac{R}{r} - \cos\theta\right)\hat{\mathbf{x}}_3. \quad (9.181)$$

The principal moments are

$$I_3 = (1/2)mr^2, \quad \text{and} \quad I_2 = (1/4)mr^2. \quad (9.182)$$

The angular momentum is $\mathbf{L} = I_2\omega_2\hat{\mathbf{x}}_2 + I_3\omega_3\hat{\mathbf{x}}_3$, so its horizontal component has length $L_\perp = I_2\omega_2\cos\theta - I_3\omega_3\sin\theta$, with leftward taken to be positive. Therefore, the magnitude of $d\mathbf{L}/dt$ is

$$\begin{aligned}\left|\frac{d\mathbf{L}}{dt}\right| &= \Omega L_\perp \\ &= \Omega(I_2\omega_2\cos\theta - I_3\omega_3\sin\theta) \\ &= \Omega\left[\left(\frac{1}{4}mr^2\right)\left(\Omega\sin\theta\right)\cos\theta - \left(\frac{1}{2}mr^2\right)\left(-\Omega(R/r - \cos\theta)\right)\sin\theta\right] \\ &= \frac{1}{4}mr\Omega^2\sin\theta(2R - r\cos\theta), \end{aligned} \quad (9.183)$$

with a positive quantity corresponding to $d\mathbf{L}/dt$ pointing out of the page (at the instant shown).

The torque (relative to the CM) comes from the force at the contact point. There are two components to this force. The vertical component is mg, and the horizontal component is $m(R - r\cos\theta)\Omega^2$ leftward, because the CM moves in a circle of radius $(R - r\cos\theta)$. The torque therefore has magnitude

$$|\boldsymbol{\tau}| = mg(r\cos\theta) - m(R - r\cos\theta)\Omega^2(r\sin\theta), \quad (9.184)$$

with out of the page taken to be positive. Equating this $|\boldsymbol{\tau}|$ with the $|d\mathbf{L}/dt|$ in Eq. (9.183) gives

$$\Omega^2 = \frac{g}{\frac{3}{2}R\tan\theta - \frac{5}{4}r\sin\theta}. \quad (9.185)$$

The right-hand side must be positive if a solution for Ω is to exist. Therefore, the condition for the desired motion to be possible is

$$R > \frac{5}{6} r \cos\theta. \tag{9.186}$$

REMARKS:

1. For $\theta \to \pi/2$, Eq. (9.185) gives $\Omega \to 0$, as it should. And for $\theta \to 0$, we obtain $\Omega \to \infty$, which also makes sense.
2. For $r\cos\theta > R > (5/6) r\cos\theta$, the CM of the coin lies to the *left* of the center of the contact-point circle (at the instant shown). The centripetal force, $m(R - r\cos\theta)\Omega^2$, is therefore negative (which means that it is directed radially outward, to the right), but the motion is still possible. As R gets close to $(5/6)r\cos\theta$, the frequency Ω goes to infinity, which means that the radially outward force also goes to infinity. The coefficient of friction between the coin and the ground must therefore be correspondingly large.
3. We can consider a more general coin, whose density depends only on the distance from the center, and whose I_3 equals $\beta m r^2$. For example, a uniform coin has $\beta = 1/2$, and a coin with all its mass on the edge has $\beta = 1$. By the perpendicular-axis theorem, $I_1 = I_2 = (1/2)\beta m r^2$, and you can show that the above methods yield

$$\Omega^2 = \frac{g}{(1+\beta)R\tan\theta - (1+\beta/2)r\sin\theta} \quad \Longrightarrow \quad R > \left(\frac{1+\beta/2}{1+\beta}\right) r\cos\theta. \tag{9.187}$$

The larger β is, the smaller the lower bound on R is. But even if $\beta \to \infty$ (imagine long radial spokes attached to the coin, and eliminate the ground except for the ring of contact points, so it doesn't get in the way), R still can't be any smaller than $(r/2)\cos\theta$. ♣

9.24. **Wobbling coin**

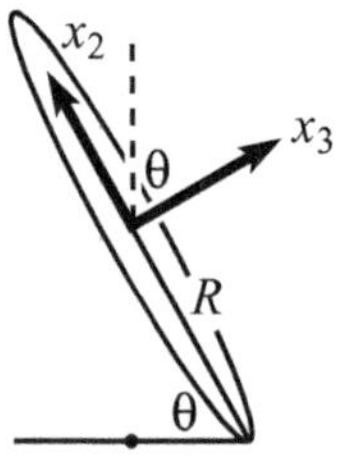

Fig. 9.99

(a) Choose the CM as the origin. The principal axes are then $\hat{\mathbf{x}}_2$ and $\hat{\mathbf{x}}_3$ (as shown in Fig. 9.99), along with $\hat{\mathbf{x}}_1$ pointing into the paper. Assume that the precessional motion is counterclockwise when viewed from above, as indicated in Fig. 9.55. Consider the setup in the frame rotating with angular velocity $\Omega\hat{\mathbf{z}}$. In this frame, the location of the contact point remains fixed, and the coin rotates with frequency ω' (in the language of Fig. 9.30) around the negative $\hat{\mathbf{x}}_3$ axis. The radius of the circle of contact points on the table is $R\cos\theta$. Therefore, the nonslipping condition says that $\omega' R = \Omega(R\cos\theta)$, and so $\omega' = \Omega\cos\theta$. Hence, the total angular velocity of the coin with respect to the lab frame is

$$\boldsymbol{\omega} = \Omega\hat{\mathbf{z}} - \omega'\hat{\mathbf{x}}_3 = \Omega(\sin\theta\,\hat{\mathbf{x}}_2 + \cos\theta\,\hat{\mathbf{x}}_3) - (\Omega\cos\theta)\hat{\mathbf{x}}_3 = \Omega\sin\theta\,\hat{\mathbf{x}}_2. \tag{9.188}$$

In retrospect, it makes sense that $\boldsymbol{\omega}$ must point in the $\hat{\mathbf{x}}_2$ direction. Both the CM and the instantaneous contact point on the coin are at rest, so $\boldsymbol{\omega}$ must lie along the line containing these two points, that is, along the $\hat{\mathbf{x}}_2$ axis.

(b) The principal moment around the $\hat{\mathbf{x}}_2$ axis is $I = mR^2/4$. The angular momentum is $\mathbf{L} = I\omega_2\hat{\mathbf{x}}_2$, so its horizontal component has length $L_\perp = L\cos\theta = (I\omega_2)\cos\theta$. Therefore, $d\mathbf{L}/dt$ has magnitude

$$\left|\frac{d\mathbf{L}}{dt}\right| = \Omega L_\perp = \Omega\left(\frac{mR^2}{4}\right)(\Omega\sin\theta)\cos\theta, \tag{9.189}$$

and it points out of the page.

The torque (relative to the CM) is due to the normal force at the contact point. This normal force is essentially equal to mg, because the CM is assumed to be falling very slowly. Note that there is no sideways friction force at the contact point, because the CM is essentially motionless. The torque therefore has magnitude

$$|\boldsymbol{\tau}| = mgR\cos\theta, \tag{9.190}$$

and it points out of the page. Equating this $|\boldsymbol{\tau}|$ with the $|d\mathbf{L}/dt|$ in Eq. (9.189) gives

$$\Omega = 2\sqrt{\frac{g}{R\sin\theta}}. \tag{9.191}$$

REMARKS:

1. $\Omega \to \infty$ as $\theta \to 0$. This is evident if you do the experiment; the contact point travels very quickly around the circle.
2. $\Omega \to 2\sqrt{g/R}$ as $\theta \to \pi/2$, which isn't so obvious. In this case, $\mathbf{L}$ points nearly vertically, and it traces out a tiny cone due to a tiny torque. In this $\theta \to \pi/2$ limit, Ω is also the frequency at which the plane of the coin spins around the vertical axis. If you spin a coin very fast about a vertical diameter, it initially undergoes a purely spinning motion with only one contact point. But then it gradually loses energy due to friction, until the spinning frequency slows down to $2\sqrt{g/R}$, at which point it begins to wobble (we're assuming that the coin is very thin, so that it can't balance on its edge). In the case where the coin is a US quarter (with $R \approx 0.012\,\text{m}$), this critical frequency of $2\sqrt{g/R}$ turns out to be $\Omega \approx 57\,\text{rad/s}$, which corresponds to about 9 Hz.
3. The result in Eq. (9.191) is a special case of the result in Eq. (9.185) of Problem 9.23. The CM of the coin in Problem 9.23 is motionless if $R = r\cos\theta$. Plugging this into Eq. (9.185) gives $\Omega^2 = 4g/(r\sin\theta)$, which agrees with Eq. (9.191), because r was the coin's radius in Problem 9.23. ♣

(c) Consider one revolution of the contact point around the z axis. Since the radius of the circle on the table is $R\cos\theta$, the contact point moves a distance $2\pi R\cos\theta$ around the coin during this time. Hence, the new contact point on the coin is a distance $2\pi R - 2\pi R\cos\theta$ away from the original contact point. The coin therefore appears to have rotated by a fraction $(1 - \cos\theta)$ of a full turn during this time. The frequency with which you see it turn is therefore

$$(1 - \cos\theta)\Omega = 2(1 - \cos\theta)\sqrt{\frac{g}{R\sin\theta}}. \tag{9.192}$$

REMARKS:

1. If $\theta \approx \pi/2$, then the frequency of Abe's rotation is essentially equal to Ω. This makes sense, because the top of Abe's head will be, say, always near the top of the coin, and this point will trace out a small circle around the z axis, with nearly the same frequency as the contact point.
2. As $\theta \to 0$, Abe appears to rotate with frequency $\theta^{3/2}\sqrt{g/R}$ (using $\sin\theta \approx \theta$ and $\cos\theta \approx 1-\theta^2/2$). Therefore, although the contact point moves infinitely quickly in this limit, we nevertheless see Abe rotating very slowly, as you can experimentally verify.
3. All of the results for frequencies in this problem must be some multiple of $\sqrt{g/R}$, by dimensional analysis. But whether the multiplication factor is zero, infinite, or something in between, is not at all obvious.
4. An incorrect answer for the frequency of Abe's turning (when viewed from above) is that it equals the vertical component of $\boldsymbol{\omega}$, which is

$\omega_z = \omega \sin\theta = (\Omega \sin\theta)\sin\theta = 2(\sin\theta)^{3/2}\sqrt{g/R}$. This does not equal the result in Eq. (9.192). (It agrees at $\theta = \pi/2$ but is off by a factor of 2 for $\theta \to 0$.) This answer is incorrect because there is simply no reason why the vertical component of $\boldsymbol{\omega}$ should equal the frequency of revolution of, say, Abe's nose, around the vertical axis. For example, at moments when $\boldsymbol{\omega}$ passes through the nose, then the nose isn't moving at all, so it certainly cannot be described as moving around the vertical axis with frequency $\omega_z = \omega\sin\theta$. The result in Eq. (9.192) is a sort of average measure of the frequency of rotation. Even though any given point on the coin is not undergoing uniform circular motion, your eye sees the coin essentially rotating uniformly as a whole. ♣

9.25. **Nutation cusps**

(a) Because both $\dot\phi$ and $\dot\theta$ are continuous functions of time, we must have $\dot\phi = \dot\theta = 0$ at a kink. Otherwise, either $d\theta/d\phi = \dot\theta/\dot\phi$ or $d\phi/d\theta = \dot\phi/\dot\theta$ would be well defined at the kink. Let the kink occur at $t = t_0$. Then the second of Eqs. (9.92) gives $\sin(\omega_n t_0) = 0$. Therefore, $\cos(\omega_n t_0) = \pm 1$, and the first of Eqs. (9.92) gives

$$\Delta\Omega = \mp\Omega_s, \tag{9.193}$$

as desired. Note that if $\cos(\omega_n t_0) = 1$, then $\Delta\Omega = -\Omega_s$, so Eq. (9.91) says that the kink occurs at the smallest value of θ, that is, at the highest point of the top's motion. And if $\cos(\omega_n t_0) = -1$, then $\Delta\Omega = \Omega_s$, so Eq. (9.91) again says that the kink occurs at the highest point of the top's motion. These results agree with the $\Delta\Omega = \pm\Omega_s$ plots in Fig. 9.38.

(b) To show that these kinks are cusps, we will show that the slope of the θ vs. ϕ plot is infinite on either side of the kink. That is, we will show that $d\theta/d\phi = \dot\theta/\dot\phi = \pm\infty$. Consider the case where $\cos(\omega_n t_0) = 1$ and $\Delta\Omega = -\Omega_s$ (the $\cos(\omega_n t_0) = -1$ case proceeds similarly). With $\Delta\Omega = -\Omega_s$, Eq. (9.92) gives

$$\frac{\dot\theta}{\dot\phi} = \frac{\sin\theta_0 \sin\omega_n t}{1 - \cos\omega_n t}. \tag{9.194}$$

Let $t = t_0 + \epsilon$. Using $\cos(\omega_n t_0) = 1$ and $\sin(\omega_n t_0) = 0$, and expanding Eq. (9.194) to lowest order in ϵ, we find

$$\frac{\dot\theta}{\dot\phi} = \frac{\sin\theta_0(\omega_n\epsilon)}{\omega_n^2\epsilon^2/2} = \frac{2\sin\theta_0}{\omega_n\epsilon}. \tag{9.195}$$

For infinitesimal ϵ, this switches from $-\infty$ to $+\infty$ as ϵ passes through zero. The point here is that although both $\dot\theta$ and $\dot\phi$ go to zero, $\dot\phi$ goes quadratically, whereas $\dot\theta$ goes only linearly.

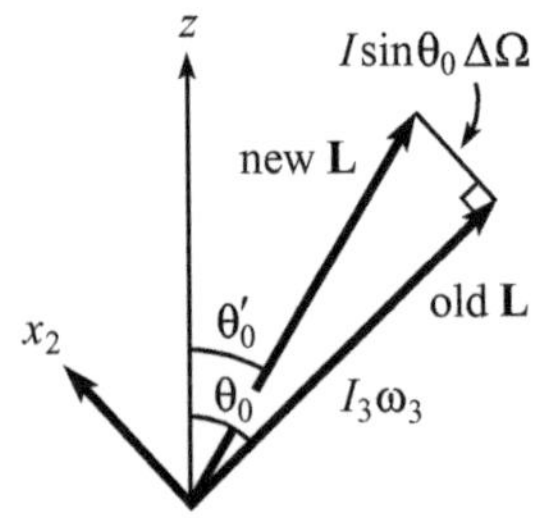

Fig. 9.100

9.26. **Nutation circles**

(a) In the limit $\omega_3 \gg \Omega_s$, the original $\mathbf{L}$ points essentially along the x_3 direction, with magnitude $I_3\omega_3$. So it makes an angle of essentially θ_0 with the vertical z axis. Now consider the quick kick. If $\dot\phi$ (which is the angular speed around the vertical z axis) suddenly increases by $\Delta\Omega$, then this corresponds to a sudden increase in angular speed around the x_2 axis of $\sin\theta_0\,\Delta\Omega$ (see Fig. 9.100). The kick therefore produces an angular momentum component (relative to the pivot) of $I\sin\theta_0\,\Delta\Omega$ in the x_2 direction. So from Fig. 9.100, the angle that the new $\mathbf{L}$ makes with the z axis is (using $\Delta\Omega \ll \omega_3$)

$$\theta_0' \approx \theta_0 - \frac{I\sin\theta_0\,\Delta\Omega}{I_3\omega_3} = \theta_0 - \frac{\sin\theta_0\,\Delta\Omega}{\omega_n}, \tag{9.196}$$

where we have used the definition of ω_n from Eq. (9.86). We see that the effect of the kick is to make $\mathbf{L}$ quickly change its θ value (by only a small amount, because we are assuming $\omega_3 \gg \Delta\Omega$). The ϕ value doesn't immediately change, because immediately after the kick the x_3 axis is still in exactly the same place, because it hasn't had time to move.

(b) After the kick is finished, $\mathbf{L}$ will trace out a cone around the z axis at approximately the same rate as before, namely Ω_s. (None of the relevant quantities in $\boldsymbol{\tau} = d\mathbf{L}/dt$ have changed much from the original circular-precession case, so the precession frequency is basically the same.) So the new $\mathbf{L}$ has its (ϕ, θ) coordinates given by

$$\big(\phi(t), \theta(t)\big)_{\mathbf{L}} = \left(\Omega_s t,\ \theta_0 - \frac{\sin\theta_0\,\Delta\Omega}{\omega_n}\right). \tag{9.197}$$

We're assuming that Ω_s is very small, so if you want, you can ignore the Ω_s term and just consider $\mathbf{L}$ to be fixed, at least on the time scale of the nutations (see the remark below). But this term will cancel anyway in producing the following equation. Looking at Eq. (9.91), we see that the coordinates of the CM relative to $\mathbf{L}$ are

$$\big(\phi(t), \theta(t)\big)_{\text{CM}-\mathbf{L}} = \left(\left(\frac{\Delta\Omega}{\omega_n}\right)\sin\omega_n t,\ \left(\frac{\Delta\Omega}{\omega_n}\sin\theta_0\right)\cos\omega_n t\right). \tag{9.198}$$

The $\sin\theta_0$ factor in $\theta(t)$ is exactly what is needed for the CM to trace out a circle around $\mathbf{L}$, because a change in ϕ corresponds to a CM spatial change of $\ell\Delta\phi\sin\theta_0$, whereas a change in θ corresponds to a CM spatial change of $\ell\Delta\theta$.

Now let's see how this relates to a free top. For $\omega_n \gg \Omega_s$, the time scale on which $\mathbf{L}$ changes due to the gravitational torque (namely $1/\Omega_s$) is very long compared with the time scale of the nutation (namely $1/\omega_n$). Therefore, since $\mathbf{L}$ is essentially motionless on the time scale of $1/\omega_n$, the effects of gravity are negligible on this time scale. Hence, the system should behave like a free top. So let's verify that the results here do indeed agree with the results from Section 9.6.2.

Equation (9.55) in Section 9.6.2 says that the frequency of the precession of $\hat{\mathbf{x}}_3$ around $\mathbf{L}$ for a free top is L/I. But the frequency of the precession of $\hat{\mathbf{x}}_3$ around $\mathbf{L}$ in the present problem is ω_n, so this had better be equal to L/I. And indeed, L is essentially equal to $I_3\omega_3$, so $\omega_n \equiv I_3\omega_3/I = L/I$. Therefore, for short enough time scales (short enough so that $\mathbf{L}$ doesn't move much), a nutating top with $\omega_3 \gg \Delta\Omega \gg \Omega_s$ looks very much like a free top.

REMARK: We need the $\Delta\Omega \gg \Omega_s$ condition so that the nutation motion looks roughly like circles (that is, like the top plot in Fig. 9.38, and not the others). This requirement can be seen by the following reasoning. The time for one period of the nutation motion is $2\pi/\omega_n$. From Eq. (9.91), $\phi(t)$ increases by $\Delta\phi = 2\pi\Omega_s/\omega_n$ during this time. And also from Eq. (9.91), the diameter d of the "circle" along the ϕ axis is roughly $d = 2\Delta\Omega/\omega_n$. The motion looks approximately like a circle if $d \gg \Delta\phi$, that is, if $\Delta\Omega \gg \Omega_s$. ♣

9.27. **Rolling without slipping**

Yes, there are indeed other ways. It isn't necessary for the contact points to form a vertical great circle. They can form a smaller tilted circle, as shown in Fig. 9.101. In this case the angular velocity vector points up to the left. There is indeed no slipping here, and the ball rolls straight. You can convince yourself of this in the following way. Imagine the ball hovering and spinning in place with its center at rest. At all times, the point at the bottom of the ball is moving directly out of the page; the collection of these points is the tilted circle in Fig. 9.101. Let the speed of the instantaneous bottom point be v, and now imagine taking a horizontal sheet of paper and sliding it out of the page with constant speed v, just beneath the ball, barely touching it. Since this is the

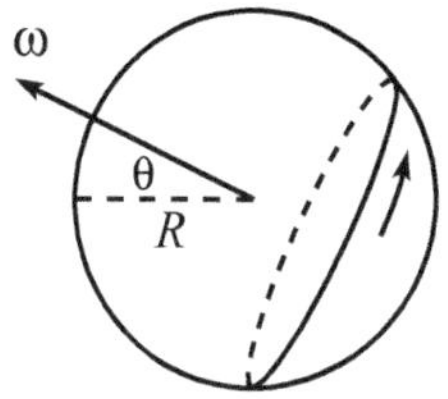

Fig. 9.101

same speed as the bottom point on the ball (which is the contact point, because it's at the bottom), there is no slipping between the ball and the paper. Finally, if we go to the reference frame of the paper, we have the desired motion of a ball rolling without slipping on a flat surface.

If θ is the angle that $\boldsymbol{\omega}$ makes with the horizontal, then the radius of the contact circle is $R\cos\theta$. The linear speed of the ball is therefore $\omega(R\cos\theta)$. If you want, you can think of this as $(\omega\cos\theta)R$. That is, only the horizontal component of $\boldsymbol{\omega}$ leads to translational motion of the ball. The vertical component simply yields a spinning motion around the vertical diameter. If θ approaches $90°$, then the contact circle is very small (see Fig. 9.102), so the speed of the ball is very small (for a given ω).

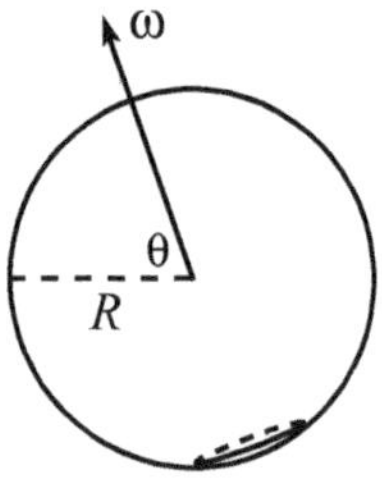

Fig. 9.102

In the real world, the contact area between the ball and the surface isn't an idealized single point, so if the contact circle isn't a vertical great circle, there is inevitably some slipping at the points near the bottom of the ball. The smaller the contact circle, the more the slipping, because the slipping arises from the twisting motion around the vertical diameter, which comes from the vertical component of $\boldsymbol{\omega}$. If you take an actual ball and roll it with a tilted $\boldsymbol{\omega}$, you will find that the rolling motion is noisier the more that $\boldsymbol{\omega}$ is tilted. What you hear is the slipping due to the extended contact region. But in an ideal world (or perhaps in a bowling alley, which is a reasonable approximation thereof), a tilted $\boldsymbol{\omega}$ wouldn't be any noisier.

9.28. **Rolling straight?**

Intuitively, it is fairly clear that the sphere cannot change direction, but it is a little tricky to prove. Qualitatively, we can reason as follows. Assume that there is a nonzero friction force at the contact point. (The normal force is irrelevant here, because it doesn't provide a torque about the center, because the contact point is always directly below the center. This is what is special about a sphere.) Then the sphere accelerates in the direction of this force. However, you can show with the right-hand rule that this force produces a torque that causes the angular momentum to change in a way that corresponds to the sphere accelerating in the direction *opposite* to the friction force (assuming that the sphere is rolling without slipping). There is thus a contradiction, unless the friction force is equal to zero.

Let's now be rigorous. Let the angular velocity of the sphere be $\boldsymbol{\omega}$, which may point diagonally if the contact points don't form a vertical great circle on the sphere. The nonslipping condition says that the sphere's CM velocity equals

$$\mathbf{v} = \boldsymbol{\omega} \times (R\hat{\mathbf{z}}). \tag{9.199}$$

The sphere's angular momentum is

$$\mathbf{L} = I\boldsymbol{\omega}. \tag{9.200}$$

The friction force (if it exists) from the ground at the contact point changes both the momentum and the angular momentum. $\mathbf{F} = d\mathbf{p}/dt$ gives

$$\mathbf{F} = m\frac{d\mathbf{v}}{dt}. \tag{9.201}$$

And $\boldsymbol{\tau} = d\mathbf{L}/dt$ (relative to the center) gives

$$(-R\hat{\mathbf{z}}) \times \mathbf{F} = \frac{d\mathbf{L}}{dt}, \tag{9.202}$$

because the force is applied at position $-R\hat{\mathbf{z}}$ relative to the center. We will now show that the preceding four equations imply that $\dot{\boldsymbol{\omega}} = \mathbf{0}$. From Eq. (9.199), this then implies that $\dot{\mathbf{v}} = \mathbf{0}$, as desired.

Equations (9.199) and (9.201) give $\mathbf{F} = m\dot{\boldsymbol{\omega}} \times (R\hat{\mathbf{z}})$. Plugging this $\mathbf{F}$, along with the $\mathbf{L}$ from Eq. (9.200), into Eq. (9.202) then gives

$$(-R\hat{\mathbf{z}}) \times \big(m\dot{\boldsymbol{\omega}} \times (R\hat{\mathbf{z}})\big) = I\dot{\boldsymbol{\omega}}. \tag{9.203}$$

The vector $\dot{\boldsymbol{\omega}}$ must lie in the horizontal plane, because it equals the cross product of two vectors, one of which is the vertical $\hat{\mathbf{z}}$. This implies that (as you can verify) $\hat{\mathbf{z}} \times (\dot{\boldsymbol{\omega}} \times \hat{\mathbf{z}}) = \dot{\boldsymbol{\omega}}$. Therefore, Eq. (9.203) gives

$$-mR^2\dot{\boldsymbol{\omega}} = \frac{2}{5}mR^2\dot{\boldsymbol{\omega}}, \tag{9.204}$$

and so $\dot{\boldsymbol{\omega}} = \mathbf{0}$, as we wanted to show.

9.29. **Ball on paper**

Our strategy will be to produce, and then equate, two different expressions for the total change in the angular momentum of the ball, relative to its center. The first comes from the effects of the friction force on the ball. The second comes from looking at the general form of the initial and final motion.

To produce our first expression for $\Delta\mathbf{L}$, note that the normal force provides no torque, so we can ignore it. The friction force, $\mathbf{F}$, from the paper changes both $\mathbf{p}$ and $\mathbf{L}$ according to

$$\begin{aligned} \Delta\mathbf{p} &= \int \mathbf{F}\,dt, \\ \Delta\mathbf{L} &= \int \boldsymbol{\tau}\,dt = \int (-R\hat{\mathbf{z}}) \times \mathbf{F}\,dt = (-R\hat{\mathbf{z}}) \times \int \mathbf{F}\,dt. \end{aligned} \tag{9.205}$$

Both of these integrals run over the entire slipping time, which may include time on the table after the ball leaves the paper. In the second line above, we have used the fact that the friction force always acts at the same location, namely $(-R\hat{\mathbf{z}})$, relative to the center of the ball. The two above equations yield

$$\Delta\mathbf{L} = (-R\hat{\mathbf{z}}) \times \Delta\mathbf{p}. \tag{9.206}$$

To produce our second equation for $\Delta\mathbf{L}$, let's examine how $\mathbf{L}_\perp$ (the horizontal component of $\mathbf{L}$)[32] is related to $\mathbf{p}$ when the ball is rolling without slipping, which is the case at both the start and finish. When the ball is not slipping, we have the situation show in Fig. 9.103 (assuming that the ball is rolling to the right). The magnitudes of $\mathbf{p}$ and $\mathbf{L}_\perp$ are given by

$$\begin{aligned} p &= mv, \\ L_\perp &= I\omega_\perp = \frac{2}{5}mR^2\omega_\perp = \frac{2}{5}Rm(R\omega_\perp) = \frac{2}{5}Rmv = \frac{2}{5}Rp, \end{aligned} \tag{9.207}$$

where we have used the nonslipping condition, $v = R\omega_\perp$. (The actual $I = (2/5)mR^2$ value for a solid sphere won't be important for the final result.) We can now combine the directions of $\mathbf{L}_\perp$ and $\mathbf{p}$ in Fig. 9.103 with the above $L_\perp = (2/5)Rp$ scalar relation to write

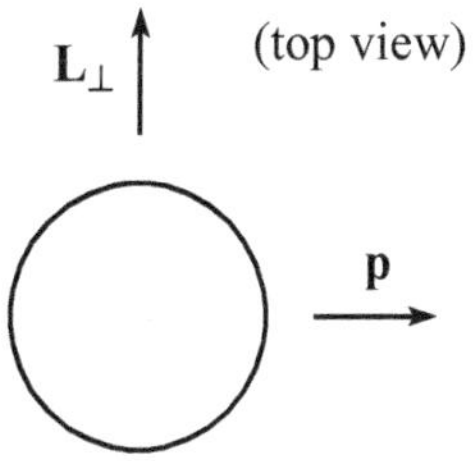

Fig. 9.103

$$\mathbf{L}_\perp = \frac{2}{5}R\hat{\mathbf{z}} \times \mathbf{p}, \tag{9.208}$$

where $\hat{\mathbf{z}}$ points out of the page. Since this relation is true at both the start and the finish, it must also be true for the differences in $\mathbf{L}_\perp$ and $\mathbf{p}$. That is,

$$\Delta\mathbf{L}_\perp = \frac{2}{5}R\hat{\mathbf{z}} \times \Delta\mathbf{p}. \tag{9.209}$$

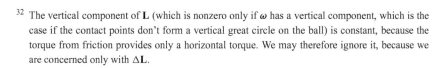

[32] The vertical component of $\mathbf{L}$ (which is nonzero only if $\boldsymbol{\omega}$ has a vertical component, which is the case if the contact points don't form a vertical great circle on the ball) is constant, because the torque from friction provides only a horizontal torque. We may therefore ignore it, because we are concerned only with $\Delta\mathbf{L}$.

But $\Delta\mathbf{L}_{\perp} = \Delta\mathbf{L}$ because the vertical component of $\mathbf{L}$ doesn't change, so Eqs. (9.206) and (9.209) give

$$(-R\hat{\mathbf{z}}) \times \Delta\mathbf{p} = \frac{2}{5}R\hat{\mathbf{z}} \times \Delta\mathbf{p} \quad \Longrightarrow \quad \mathbf{0} = \hat{\mathbf{z}} \times \Delta\mathbf{p}. \tag{9.210}$$

There are three ways this cross product can be zero:

- $\Delta\mathbf{p}$ is parallel to $\hat{\mathbf{z}}$. But it isn't, because $\Delta\mathbf{p}$ lies in the horizontal plane.
- $\hat{\mathbf{z}} = \mathbf{0}$. Not true.
- $\Delta\mathbf{p} = \mathbf{0}$. This is the only possibility, so it must be true. Therefore, $\Delta\mathbf{v} = \mathbf{0}$, as we wanted to show.

Note that the line of the final motion may very well be shifted sideways from the line of the initial motion. But what we've shown is that the final line is parallel to the initial line, and the speed is the same.

REMARKS:

1. As stated in the problem, it's fine if you move the paper in a jerky motion, so that the ball slips around on it. We assumed nothing about the nature of the friction force in the above reasoning. And we used the nonslipping condition only at the initial and final times. The intermediate motion is arbitrary.
2. As a special case, if you start a ball at rest on a piece of paper, then no matter how you choose to (horizontally) slide the paper out from underneath the ball, the ball will be at rest in the end. The final position will most likely be different from the initial position, but the ball will be at rest wherever it ends up.
3. You are encouraged to experimentally verify that these crazy claims are true. Make sure that the paper doesn't wrinkle, because a wrinkle would allow a force to be applied at a point other than the contact point. And balls that don't squish are much better, of course, because the contact region better resembles a single point. ♣

9.30. **Ball on a turntable**

Let the angular velocity of the turntable be $\Omega\hat{\mathbf{z}}$, and let the angular velocity of the ball (with respect to the lab frame) be $\boldsymbol{\omega}$, which may point diagonally if the contact points don't form a vertical great circle on the ball. If the ball is at position $\mathbf{r}$ (with respect to the lab frame), then its CM velocity (with respect to the lab frame) may be broken up into the velocity of the turntable (at position $\mathbf{r}$) plus the ball's velocity with respect to the turntable. The nonslipping condition says that the latter is $\boldsymbol{\omega} \times (a\hat{\mathbf{z}})$, where a is the radius of the ball.[33] The ball's velocity with respect to the lab frame is thus

$$\mathbf{v} = (\Omega\hat{\mathbf{z}}) \times \mathbf{r} + \boldsymbol{\omega} \times (a\hat{\mathbf{z}}). \tag{9.211}$$

The important point to realize in this problem is that the friction force from the turntable is responsible for changing both the ball's linear momentum and its angular momentum. In particular, $\mathbf{F} = d\mathbf{p}/dt$ gives

$$\mathbf{F} = m\frac{d\mathbf{v}}{dt}. \tag{9.212}$$

And the angular momentum of the ball is $\mathbf{L} = I\boldsymbol{\omega}$, so $\boldsymbol{\tau} = d\mathbf{L}/dt$ (relative to the center of the ball) gives

$$(-a\hat{\mathbf{z}}) \times \mathbf{F} = I\frac{d\boldsymbol{\omega}}{dt}, \tag{9.213}$$

because the force is applied at position $-a\hat{\mathbf{z}}$ relative to the center.

[33] The velocity with respect to the turntable is actually $\boldsymbol{\omega}_t \times (a\hat{\mathbf{z}})$, where $\boldsymbol{\omega}_t$ is the angular velocity in the turntable frame. But $\boldsymbol{\omega}_t$ differs from $\boldsymbol{\omega}$ simply by the angular velocity of the turntable, which is $\Omega\hat{\mathbf{z}}$. This implies that $\boldsymbol{\omega}_t \times (a\hat{\mathbf{z}}) = \boldsymbol{\omega} \times (a\hat{\mathbf{z}})$.

We will now use the previous three equations to demonstrate that the ball undergoes circular motion. Our goal will be to produce an equation of the form,

$$\frac{d\mathbf{v}}{dt} = \Omega'\hat{\mathbf{z}} \times \mathbf{v}, \tag{9.214}$$

because this describes circular motion with frequency Ω' (to be determined).[34] We will eliminate $\mathbf{F}$ first, and then $\boldsymbol{\omega}$. Plugging the expression for $\mathbf{F}$ from Eq. (9.212) into Eq. (9.213) gives

$$(-a\hat{\mathbf{z}}) \times \left(m\frac{d\mathbf{v}}{dt}\right) = I\frac{d\boldsymbol{\omega}}{dt} \implies \frac{d\boldsymbol{\omega}}{dt} = -\left(\frac{am}{I}\right)\hat{\mathbf{z}} \times \frac{d\mathbf{v}}{dt}. \tag{9.215}$$

Taking the derivative of Eq. (9.211) gives

$$\begin{aligned}\frac{d\mathbf{v}}{dt} &= \Omega\hat{\mathbf{z}} \times \frac{d\mathbf{r}}{dt} + \frac{d\boldsymbol{\omega}}{dt} \times (a\hat{\mathbf{z}}) \\ &= \Omega\hat{\mathbf{z}} \times \mathbf{v} - \left(\left(\frac{am}{I}\right)\hat{\mathbf{z}} \times \frac{d\mathbf{v}}{dt}\right) \times (a\hat{\mathbf{z}}).\end{aligned} \tag{9.216}$$

Since we know that the vector $d\mathbf{v}/dt$ lies in the horizontal plane, it is easy to work out the cross product in the second term here (or just use the identity $(\mathbf{A} \times \mathbf{B}) \times \mathbf{C} = (\mathbf{A} \cdot \mathbf{C})\mathbf{B} - (\mathbf{B} \cdot \mathbf{C})\mathbf{A}$) to obtain

$$\frac{d\mathbf{v}}{dt} = \Omega\hat{\mathbf{z}} \times \mathbf{v} - \left(\frac{ma^2}{I}\right)\frac{d\mathbf{v}}{dt} \implies \frac{d\mathbf{v}}{dt} = \left(\frac{\Omega}{1 + (ma^2/I)}\right)\hat{\mathbf{z}} \times \mathbf{v}. \tag{9.217}$$

For a uniform sphere, $I = (2/5)ma^2$, so we obtain

$$\frac{d\mathbf{v}}{dt} = \left(\frac{2}{7}\Omega\right)\hat{\mathbf{z}} \times \mathbf{v}. \tag{9.218}$$

Therefore, in view of Eq. (9.214), we see that the ball undergoes circular motion, with a frequency equal to 2/7 times the frequency of the turntable. Note that this result for the frequency doesn't depend on initial conditions. For an extension to this problem, see Weltner (1987) and references therein.

REMARKS:

1. Integrating Eq. (9.218) from the initial time to some later time gives

$$\mathbf{v} - \mathbf{v}_0 = \left(\frac{2}{7}\Omega\right)\hat{\mathbf{z}} \times (\mathbf{r} - \mathbf{r}_0). \tag{9.219}$$

This may be written (as you can verify) in the more suggestive (if not more frightening) form,

$$\mathbf{v} = \left(\frac{2}{7}\Omega\right)\hat{\mathbf{z}} \times \left(\mathbf{r} - \left(\mathbf{r}_0 + \frac{7}{2\Omega}(\hat{\mathbf{z}} \times \mathbf{v}_0)\right)\right). \tag{9.220}$$

This equation describes circular motion, with the center located at the point,

$$\mathbf{r}_c = \mathbf{r}_0 + \frac{7}{2\Omega}(\hat{\mathbf{z}} \times \mathbf{v}_0), \tag{9.221}$$

[34] Equation (9.214) describes circular motion because it implies that $\mathbf{v}$ has constant magnitude (since the change in $\mathbf{v}$ is perpendicular to $\mathbf{v}$), and so the direction of $\mathbf{v}$ changes at the constant angular rate Ω'. The only curve with these properties is a circle. If you want to be more mathematical, you can integrate Eq. (9.214) to obtain $\mathbf{v} = \Omega'\hat{\mathbf{z}} \times \mathbf{r} + \mathbf{C}$, which can be written as $d\mathbf{r}/dt = \Omega'\hat{\mathbf{z}} \times (\mathbf{r} - \mathbf{r}_c)$, which implies that $d(\mathbf{r} - \mathbf{r}_c)/dt = \Omega'\hat{\mathbf{z}} \times (\mathbf{r} - \mathbf{r}_c)$, which means that the length of $\mathbf{r} - \mathbf{r}_c$ doesn't change (because its change is perpendicular to itself). And since the changes are also constrained to be perpendicular to $\hat{\mathbf{z}}$, we must have a horizontal circle.

and with radius,

$$R = |\mathbf{r}_0 - \mathbf{r}_c| = \frac{7}{2\Omega}|\hat{\mathbf{z}} \times \mathbf{v}_0| = \frac{7v_0}{2\Omega}. \tag{9.222}$$

2. There are a few special cases to consider:
   - If $v_0 = 0$ (that is, if the spinning motion of the ball exactly cancels the rotational motion of the turntable, so that the CM of the ball is at rest in the lab frame), then $R = 0$ and the ball remains in the same place, as it should.
   - If the ball is initially not spinning, and just moving along with the turntable, then $v_0 = \Omega r_0$. The radius of the circle is therefore $R = (7/2)r_0$, and its center is located at, from Eq. (9.221),

$$\mathbf{r}_c = \mathbf{r}_0 + \frac{7}{2\Omega}(-\Omega\mathbf{r}_0) = -\frac{5\mathbf{r}_0}{2}. \tag{9.223}$$

     The point on the circle diametrically opposite to the initial point is therefore a distance $r_c + R = (5/2)r_0 + (7/2)r_0 = 6r_0$ from the center of the turntable.
   - If we want the center of the circle to be the center of the turntable, then Eq. (9.221) says that we need $(7/2\Omega)\hat{\mathbf{z}} \times \mathbf{v}_0 = -\mathbf{r}_0$. This implies that $\mathbf{v}_0$ has magnitude $v_0 = (2/7)\Omega r_0$ and points tangentially in the same direction as the turntable moves. That is, the ball moves at $2/7$ times the velocity of the point on the turntable right beneath it.
3. The fact that the frequency $(2/7)\Omega$ is a rational multiple of Ω means that the ball will eventually return to the same point on the turntable. In the lab frame, the ball traces out two circles in the time it takes the turntable to undergo seven revolutions. From the point of view of someone on the turntable, the ball "spirals" around five times before returning to the original position.
4. If we look at a ball with moment of inertia $I = \beta ma^2$ (so a uniform sphere has $\beta = 2/5$), then Eq. (9.217) shows that the "2/7" in Eq. (9.218) gets replaced by "$\beta/(1+\beta)$." If a ball has most of its mass concentrated at its center (so that $\beta \to 0$), then the frequency of the circular motion goes to 0, and the radius goes to ∞ (as long as $v_0 \neq 0$). ♣

Chapter 10
Accelerating frames of reference

Newton's laws hold only in inertial frames of reference. However, there are many noninertial (that is, accelerating) frames that we might reasonably want to consider, such as elevators, merry-go-rounds, and so on. Is there any possible way to modify Newton's laws so that they hold in noninertial frames, or do we have to give up entirely on $\mathbf{F} = m\mathbf{a}$? It turns out that we can in fact hold on to our good friend $\mathbf{F} = m\mathbf{a}$, provided that we introduce some new "fictitious" forces. These are forces that a person in the accelerating frame thinks exist. If she applies $\mathbf{F} = m\mathbf{a}$ while including these new forces, then she will get the correct answer for the acceleration $\mathbf{a}$, as measured with respect to her frame.

To be quantitative about all this, we'll have to spend some time figuring out how the coordinates (and their derivatives) in an accelerating frame relate to those in an inertial frame. But before diving into that, let's look at a simple example that demonstrates the basic idea of fictitious forces.

Example (A train): Imagine that you are standing on a train that is accelerating to the right with acceleration a. If you are to remain at the same spot on the train, then there must be a friction force between the floor and your feet with magnitude $F_{\mathrm{f}} = ma$, pointing to the right. Someone standing in the inertial frame of the ground simply interprets the situation as, "The friction force F_{f}, which equals ma, causes your acceleration a."

How do you interpret the situation in the frame of the train? Assume that there are no windows, so that all you see is the inside of the train. As we will show below in Eq. (10.11), you will feel a fictitious *translation* force, $F_{\text{trans}} = -ma$, pointing to the left. You therefore interpret the situation as, "In my frame (the frame of the train), the friction force $F_{\mathrm{f}} = ma$ pointing to my right exactly cancels the mysterious $F_{\text{trans}} = -ma$ force pointing to my left, resulting in zero acceleration (in my frame)."

Of course, if the floor of the train is frictionless so that there is no force at your feet, then you will say that the net force on you is $F_{\text{trans}} = -ma$, pointing to the left. You will therefore accelerate with acceleration a to the left, with respect to your frame (the train). In other words, you will remain motionless (or move with constant velocity, if you happened to be moving initially) with respect to the inertial frame of the ground, which is all quite obvious to someone standing on the ground.

In the case where the friction force at your feet is nonzero, but not large enough to balance out the whole $F_{\text{trans}} = -ma$ force, you will end up being accelerated toward the back of the train (in the frame of the train), with an acceleration that is less than a. This undesired motion will continue until you make the required adjustments with your feet or hands to balance out all of the F_{trans} force.

Let's now derive the fictitious forces in their full generality. The main task here is to relate the coordinates in an accelerating frame to those in an inertial frame, so this endeavor will require a bit of math.

10.1 Relating the coordinates

Consider an inertial coordinate system with axes $\hat{\mathbf{x}}_{\text{I}}$, $\hat{\mathbf{y}}_{\text{I}}$, and $\hat{\mathbf{z}}_{\text{I}}$, and let there be another (possibly accelerating) coordinate system with axes $\hat{\mathbf{x}}$, $\hat{\mathbf{y}}$, and $\hat{\mathbf{z}}$. These latter axes are allowed to change in an arbitrary manner with respect to the inertial frame. That is, the origin may undergo acceleration, and the axes may rotate (this is the most general possible motion, as we saw in Section 9.1). These axes may be considered to be functions of the inertial axes.

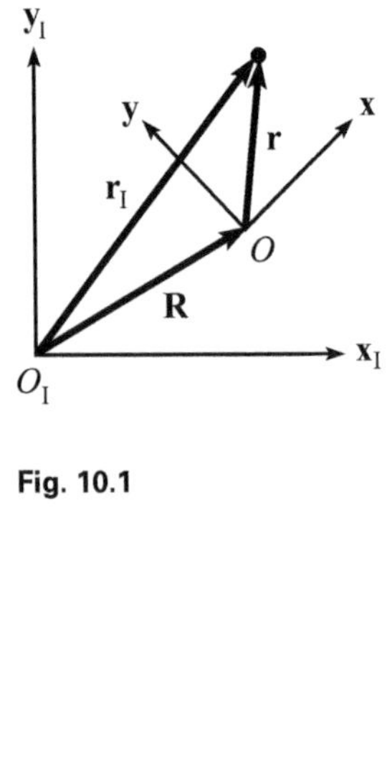

Fig. 10.1

Let O_{I} and O be the origins of the two coordinate systems. Let the vector from O_{I} to O be $\mathbf{R}$, let the vector from O_{I} to a given particle be $\mathbf{r}_{\text{I}}$, and let the vector from O to the particle be $\mathbf{r}$. See Fig. 10.1 for the 2-D case. Then

$$\mathbf{r}_{\text{I}} = \mathbf{R} + \mathbf{r}. \tag{10.1}$$

These vectors have an existence that is independent of any specific coordinate system, but let's write them in terms of some definite coordinates. We may write

$$\begin{aligned} \mathbf{R} &= X\hat{\mathbf{x}}_{\text{I}} + Y\hat{\mathbf{y}}_{\text{I}} + Z\hat{\mathbf{z}}_{\text{I}}, \\ \mathbf{r}_{\text{I}} &= x_{\text{I}}\hat{\mathbf{x}}_{\text{I}} + y_{\text{I}}\hat{\mathbf{y}}_{\text{I}} + z_{\text{I}}\hat{\mathbf{z}}_{\text{I}}, \\ \mathbf{r} &= x\hat{\mathbf{x}} + y\hat{\mathbf{y}} + z\hat{\mathbf{z}}. \end{aligned} \tag{10.2}$$

For reasons that will become clear, we have chosen to write $\mathbf{R}$ and $\mathbf{r}_{\text{I}}$ in terms of the inertial-frame coordinates, and $\mathbf{r}$ in terms of the accelerating-frame coordinates. If desired, Eq. (10.1) may be written as

$$x_{\text{I}}\hat{\mathbf{x}}_{\text{I}} + y_{\text{I}}\hat{\mathbf{y}}_{\text{I}} + z_{\text{I}}\hat{\mathbf{z}}_{\text{I}} = \left(X\hat{\mathbf{x}}_{\text{I}} + Y\hat{\mathbf{y}}_{\text{I}} + Z\hat{\mathbf{z}}_{\text{I}}\right) + \left(x\hat{\mathbf{x}} + y\hat{\mathbf{y}} + z\hat{\mathbf{z}}\right). \tag{10.3}$$

Our goal is to take the second time derivative of Eq. (10.1) and to then interpret the result in an $\mathbf{F} = m\mathbf{a}$ form. The second derivative of $\mathbf{r}_{\text{I}}$ is the acceleration of the particle with respect to the inertial system, so Newton's second law tells us that $\mathbf{F} = m\ddot{\mathbf{r}}_{\text{I}}$. The second derivative of $\mathbf{R}$ is the acceleration of the origin of the moving system. The second derivative of $\mathbf{r}$ is the tricky part. In view of its form in Eq. (10.2), changes in $\mathbf{r}$ can come about in two ways. First, the coordinates (x, y, z) of $\mathbf{r}$, which are measured with respect to the moving axes, may change. And second, the axes $\hat{\mathbf{x}}$, $\hat{\mathbf{y}}$, $\hat{\mathbf{z}}$ themselves may change. The point is that $\mathbf{r}$ is *not* just the ordered triplet (x, y, z). It is the whole expression, $\mathbf{r} = x\hat{\mathbf{x}} + y\hat{\mathbf{y}} + z\hat{\mathbf{z}}$. So even if (x, y, z) are fixed, meaning that $\mathbf{r}$ doesn't change with respect to the moving system, $\mathbf{r}$ can still change with respect to the inertial system if the $\hat{\mathbf{x}}, \hat{\mathbf{y}}, \hat{\mathbf{z}}$ axes are moving. Let's be quantitative about this.

Calculation of $d^2\mathbf{r}/dt^2$

We should clarify our goal here. We would like to obtain $d^2\mathbf{r}/dt^2$ in terms of the coordinates in the *accelerating* frame, because we want to be able to work entirely in terms of these coordinates so that a person in the accelerating frame can write down an $\mathbf{F} = m\mathbf{a}$ equation in terms of her coordinates only, without having to consider the underlying inertial frame at all. In terms of the *inertial* frame, $d^2\mathbf{r}/dt^2$ is simply $d^2(\mathbf{r}_\text{I} - \mathbf{R})/dt^2$, but this isn't very enlightening by itself.

The following exercise in taking derivatives works for a general vector $\mathbf{A} = A_x\hat{\mathbf{x}} + A_y\hat{\mathbf{y}} + A_z\hat{\mathbf{z}}$ in the moving frame; it isn't necessary that it be a position vector. So we'll work with a general $\mathbf{A}$ and then set $\mathbf{A} = \mathbf{r}$ when we're done. To take d/dt of $\mathbf{A} = A_x\hat{\mathbf{x}} + A_y\hat{\mathbf{y}} + A_z\hat{\mathbf{z}}$, we can use the product rule to obtain

$$\frac{d\mathbf{A}}{dt} = \left(\frac{dA_x}{dt}\hat{\mathbf{x}} + \frac{dA_y}{dt}\hat{\mathbf{y}} + \frac{dA_z}{dt}\hat{\mathbf{z}}\right) + \left(A_x\frac{d\hat{\mathbf{x}}}{dt} + A_y\frac{d\hat{\mathbf{y}}}{dt} + A_z\frac{d\hat{\mathbf{z}}}{dt}\right). \quad (10.4)$$

Yes, the product rule works with vectors too. We're doing nothing more than expanding $(A_x + dA_x)(\hat{\mathbf{x}} + d\hat{\mathbf{x}}) - A_x\hat{\mathbf{x}}$, etc., to first order. Note that although we have expressed the vector $\mathbf{A}$ in terms of the coordinate axes of the moving frame, the total derivative $d\mathbf{A}/dt$ is measured with respect to the *inertial* frame. As mentioned above, the total rate of change of $\mathbf{A}$ comes from two effects, namely the two groups of terms in Eq. (10.4). The first group gives the rate of change of $\mathbf{A}$, as measured with respect to the moving frame. We'll denote this quantity by $\delta\mathbf{A}/\delta t$.

The second group arises because the coordinate axes are moving. In what manner are they moving? We have already extracted the motion of the origin of the accelerating system by introducing the vector $\mathbf{R}$, so the only thing left is a rotation about some axis $\boldsymbol{\omega}$ through this origin (see Theorem 9.1). The axis $\boldsymbol{\omega}$ may change with time, but at any instant a unique axis of rotation describes the system. The fact that the axis may change will be relevant in finding the second derivative of $\mathbf{r}$, but not in finding the first derivative.

We saw in Theorem 9.2 that a vector $\mathbf{B}$ that has fixed length (the coordinate axes here do indeed have fixed length) and that rotates with angular velocity $\boldsymbol{\omega} \equiv \omega\hat{\boldsymbol{\omega}}$ changes at a rate $d\mathbf{B}/dt = \boldsymbol{\omega} \times \mathbf{B}$. In particular, $d\hat{\mathbf{x}}/dt = \boldsymbol{\omega} \times \hat{\mathbf{x}}$, etc. So in Eq. (10.4), the $A_x(d\hat{\mathbf{x}}/dt)$ term equals $A_x(\boldsymbol{\omega} \times \hat{\mathbf{x}}) = \boldsymbol{\omega} \times (A_x\hat{\mathbf{x}})$. Likewise for the y and z terms. Combining them gives $\boldsymbol{\omega} \times (A_x\hat{\mathbf{x}} + A_y\hat{\mathbf{y}} + A_z\hat{\mathbf{z}})$, which is just $\boldsymbol{\omega} \times \mathbf{A}$. Therefore, Eq. (10.4) becomes

$$\frac{d\mathbf{A}}{dt} = \frac{\delta\mathbf{A}}{\delta t} + \boldsymbol{\omega} \times \mathbf{A}. \quad (10.5)$$

This agrees with the result obtained in Section 9.5, Eq. (9.41). We've basically given the same proof here, but with a little more mathematical rigor.

We still have to take one more time derivative. Using the product rule, the time derivative of Eq. (10.5) is

$$\frac{d^2\mathbf{A}}{dt^2} = \frac{d}{dt}\left(\frac{\delta \mathbf{A}}{\delta t}\right) + \frac{d\boldsymbol{\omega}}{dt} \times \mathbf{A} + \boldsymbol{\omega} \times \frac{d\mathbf{A}}{dt}. \tag{10.6}$$

Applying Eq. (10.5) to the first term on the right (with $\delta\mathbf{A}/\delta t$ in place of $\mathbf{A}$), and plugging Eq. (10.5) into the third term, gives

$$\begin{aligned}\frac{d^2\mathbf{A}}{dt^2} &= \left(\frac{\delta^2\mathbf{A}}{\delta t^2} + \boldsymbol{\omega} \times \frac{\delta\mathbf{A}}{\delta t}\right) + \left(\frac{d\boldsymbol{\omega}}{dt} \times \mathbf{A}\right) + \boldsymbol{\omega} \times \left(\frac{\delta\mathbf{A}}{\delta t} + (\boldsymbol{\omega} \times \mathbf{A})\right) \\ &= \frac{\delta^2\mathbf{A}}{\delta t^2} + \boldsymbol{\omega} \times (\boldsymbol{\omega} \times \mathbf{A}) + 2\boldsymbol{\omega} \times \frac{\delta\mathbf{A}}{\delta t} + \frac{d\boldsymbol{\omega}}{dt} \times \mathbf{A}. \end{aligned} \tag{10.7}$$

At this point we will set $\mathbf{A} = \mathbf{r}$, which gives

$$\frac{d^2\mathbf{r}}{dt^2} = \mathbf{a} + \boldsymbol{\omega} \times (\boldsymbol{\omega} \times \mathbf{r}) + 2\boldsymbol{\omega} \times \mathbf{v} + \frac{d\boldsymbol{\omega}}{dt} \times \mathbf{r}, \tag{10.8}$$

where $\mathbf{r}$, $\mathbf{v} \equiv \delta\mathbf{r}/\delta t$, and $\mathbf{a} \equiv \delta^2\mathbf{r}/\delta t^2$ are the position, velocity, and acceleration of the particle, as measured *with respect to the accelerating frame*. In other words, if the accelerating frame is enclosed in a windowless box, and if a person inside the box paints a coordinate grid on the floor, then she can (assuming that she has a clock at her disposal) measure $\mathbf{r}$, $\mathbf{v}$, and $\mathbf{a}$ without caring at all about what is going on in the outside world.

You might be concerned that the $\mathbf{r}$, $\mathbf{v}$, and $\mathbf{a}$ on the right-hand side of Eq. (10.8) are measured by someone in the rotating frame, whereas the $d^2\mathbf{r}/dt^2$ on the left-hand side is measured by someone in the inertial frame. But this apparent discrepancy is a nonissue, because Eq. (10.8) is a statement about vectors, and these vectors have an existence that is independent of the coordinate system that is used to describe them. For example, the vector $\mathbf{a}$ points in a certain direction (say, toward the distant star Sirius) and has a certain magnitude, independent of whoever chooses to describe it with whatever coordinates.

10.2 The fictitious forces

From Eq. (10.1) we have

$$\frac{d^2\mathbf{r}}{dt^2} = \frac{d^2\mathbf{r}_\text{I}}{dt^2} - \frac{d^2\mathbf{R}}{dt^2}. \tag{10.9}$$

We can equate this expression for $d^2\mathbf{r}/dt^2$ with the expression in Eq. (10.8), and then multiply through by the mass m of the particle. Recognizing that

$m(d^2\mathbf{r}_\mathrm{I}/dt^2)$ is the force $\mathbf{F}$ acting on the particle ($\mathbf{F}$ may be gravity, a normal force, friction, tension, etc.), we can solve for $m\mathbf{a}$ to obtain

$$\begin{aligned} m\mathbf{a} &= \mathbf{F} - m\frac{d^2\mathbf{R}}{dt^2} - m\boldsymbol{\omega} \times (\boldsymbol{\omega} \times \mathbf{r}) - 2m\boldsymbol{\omega} \times \mathbf{v} - m\frac{d\boldsymbol{\omega}}{dt} \times \mathbf{r} \\ &\equiv \mathbf{F} + \mathbf{F}_{\text{translation}} + \mathbf{F}_{\text{centrifugal}} + \mathbf{F}_{\text{Coriolis}} + \mathbf{F}_{\text{azimuthal}}, \end{aligned} \tag{10.10}$$

where the *fictitious forces* are defined as

$$\begin{aligned} \mathbf{F}_{\text{trans}} &\equiv -m\frac{d^2\mathbf{R}}{dt^2}, \\ \mathbf{F}_{\text{cent}} &\equiv -m\boldsymbol{\omega} \times (\boldsymbol{\omega} \times \mathbf{r}), \\ \mathbf{F}_{\text{cor}} &\equiv -2m\boldsymbol{\omega} \times \mathbf{v}, \\ \mathbf{F}_{\text{az}} &\equiv -m\frac{d\boldsymbol{\omega}}{dt} \times \mathbf{r}. \end{aligned} \tag{10.11}$$

We have taken the liberty of calling these quantities "forces," because the left-hand side of Eq. (10.10) is $m\mathbf{a}$, where $\mathbf{a}$ is measured by someone in the accelerating frame. This person should therefore be able to interpret the right-hand side as some effective force. In other words, if a person in the accelerating frame wishes to calculate $m\mathbf{a}$, she simply needs to take the true force $\mathbf{F}$, and then add on all the other terms on the right-hand side of Eq. (10.10), which she will then quite reasonably interpret as forces (in her frame). She will consider Eq. (10.10) to be an $\mathbf{F} = m\mathbf{a}$ statement in the form,

$$m\mathbf{a} = \sum \mathbf{F}_{\text{acc}}, \tag{10.12}$$

where $\mathbf{F}_{\text{acc}}$ represents all the forces, real or fictitious, in the accelerating frame. All we've really done is transfer some terms to the other side of an equation and reinterpret the result. The last three terms on the right-hand side of Eq. (10.8) are nothing more than various pieces of the acceleration. But if we transfer them to the left-hand side and solve for $\mathbf{a}$, we can then interpret them (after multiplying by m) as forces in Eq. (10.10).

Of course, the extra terms in Eq. (10.10) are not actual forces. The constituents of $\mathbf{F}$ are the only real forces in the problem (and they are the same in any frame). All we are saying is that if our friend in the moving frame assumes that the extra terms are real forces, and if she adds them to $\mathbf{F}$, then she will get the correct answer for $m\mathbf{a}$, as measured in her frame.

For example, consider a box (far away from other objects, in outer space) that accelerates at a rate of $g = 10\,\text{m/s}^2$ in some direction. A person in the box feels a fictitious force of $\mathbf{F}_{\text{trans}} = mg$ down into the floor. For all she knows, she is in a box on the surface of the earth. If she performs various experiments under this assumption, the results will always be what she expects. The surprising fact that no local experiment can distinguish between the fictitious force in the accelerating box and the real gravitational force on the earth is what led Einstein to his Equivalence Principle and his theory of General Relativity (discussed in Chapter 14). These fictitious forces are more meaningful than you might think.

As Einstein explored elevators,
And studied the spinning ice-skaters,
He eyed as suspicious,
The forces, fictitious,
Of gravity's great imitators.

Note that in addition to the sum of all the real forces, $\mathbf{F}$, there are two different types of quantities on the right-hand side of Eq. (10.10). The quantities $\mathbf{r}$ and $\mathbf{v}$ are measured by someone inside the accelerating frame; they depend on what the particle is doing. But $d^2\mathbf{R}/dt^2$ and $\boldsymbol{\omega}$ are properties of the frame. In general, the person inside needs to be given their values, although it is possible for her to figure out what they are in certain special cases (see Problem 10.9).

Let's now look at each of the fictitious forces in detail. The translation and centrifugal forces are fairly easy to understand. The Coriolis force is a little more difficult. And the azimuthal force can be easy or difficult, depending on exactly how $\boldsymbol{\omega}$ is changing.

10.2.1 Translation force: $-md^2\mathbf{R}/dt^2$

This is the most intuitive of the fictitious forces. We've already discussed this force in the train example in the introduction to this chapter. If $\mathbf{R}$ is the position of the train, then $\mathbf{F}_{\text{trans}} \equiv -m\,d^2\mathbf{R}/dt^2$ is the fictitious translation force you feel in the accelerating frame.

10.2.2 Centrifugal force: $-m\boldsymbol{\omega} \times (\boldsymbol{\omega} \times \mathbf{r})$

This force goes hand-in-hand with the $mv^2/r = mr\omega^2$ centripetal acceleration as viewed by someone in an inertial frame.

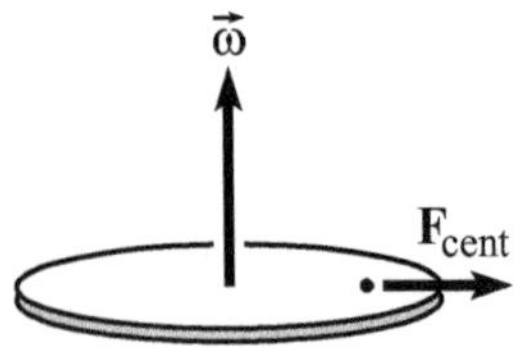

Fig. 10.2

Example (Standing on a carousel): Consider a person standing motionless with respect to a carousel, a distance r from the center. Let the carousel rotate in the x-y plane with angular velocity $\boldsymbol{\omega} = \omega\hat{\mathbf{z}}$ (see Fig. 10.2). What is the centrifugal force felt by the person?

Solution: $\boldsymbol{\omega} \times \mathbf{r}$ has magnitude ωr and points in the tangential direction, in the direction of motion. Therefore, $m\boldsymbol{\omega} \times (\boldsymbol{\omega} \times \mathbf{r})$ has magnitude $mr\omega^2$ and points radially inward. Hence, the centrifugal force $-m\boldsymbol{\omega} \times (\boldsymbol{\omega} \times \mathbf{r})$ has magnitude $mr\omega^2$ and points radially outward.

REMARK: If the person is not moving with respect to the carousel, and if $\boldsymbol{\omega}$ is constant, then the centrifugal force is the only nonzero fictitious force in Eq. (10.10). Since the person is not accelerating in her rotating frame, the net force (as measured in her frame) must be zero. The forces in her frame are (1) gravity pulling downward, (2) the normal force pushing upward (which cancels gravity), (3) the friction force pushing inward at her feet, and (4) the centrifugal force pulling outward. We conclude that the last two of these must cancel. So the friction force must point inward with magnitude $mr\omega^2$.

Someone standing on the ground observes only the first three of these forces, so the net force is not zero. And indeed, there is a centripetal acceleration, $v^2/r = r\omega^2$, due to the friction force. To sum up: in the inertial frame, the friction force exists to provide a centripetal acceleration. In the rotating frame, the friction force exists to balance out the mysterious new centrifugal force, in order to yield zero acceleration. ♣

Example (Effective gravity force, $m\mathbf{g}_{\text{eff}}$): Consider a person standing motionless on the earth, at a polar angle θ; the way we've defined it, θ equals $\pi/2$ minus the latitude angle. In the rotating frame of the earth, the person feels a centrifugal force (directed away from the axis) in addition to the gravitational force, $m\mathbf{g}$; see Fig. 10.3, although this figure is somewhat misleading, as explained in the first remark below. Note that we're using $\mathbf{g}$ to denote the acceleration due solely to the gravitational force. This isn't the "g-value" that the person measures, as we'll shortly see.

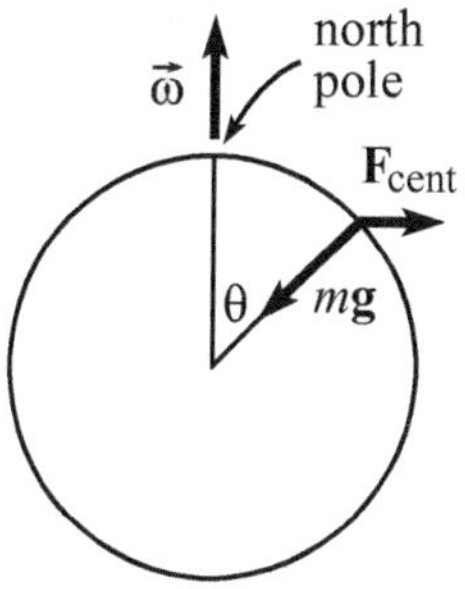

Fig. 10.3

The sum of the gravitational and centrifugal forces (that is, what the person thinks is gravity) doesn't point radially, unless the person is at the equator or at a pole. Let us denote the sum by $m\mathbf{g}_{\text{eff}}$. To calculate $m\mathbf{g}_{\text{eff}}$, we must calculate $\mathbf{F}_{\text{cent}} = -m\boldsymbol{\omega} \times (\boldsymbol{\omega} \times \mathbf{r})$. The $\boldsymbol{\omega} \times \mathbf{r}$ part has magnitude $R\omega \sin\theta$, where R is the radius of the earth, and it is directed tangentially along the latitude circle of radius $R\sin\theta$. So $-m\boldsymbol{\omega} \times (\boldsymbol{\omega} \times \mathbf{r})$ points outward from the axis, with magnitude $mR\omega^2 \sin\theta$, which is just what we expect for something traveling at frequency ω in a circle of radius $R\sin\theta$. Therefore, the effective gravitational force,

$$m\mathbf{g}_{\text{eff}} \equiv m\big(\mathbf{g} - \boldsymbol{\omega} \times (\boldsymbol{\omega} \times \mathbf{r})\big), \tag{10.13}$$

points slightly in the southerly direction (for someone in the northern hemisphere), as shown in Fig. 10.4. The magnitude of the correction term, $mR\omega^2 \sin\theta$, is small compared with g. Using $\omega \approx 7.3 \cdot 10^{-5}\,\text{s}^{-1}$ (that is, one revolution per day, which is 2π radians per 86 400 seconds) and $R \approx 6.4 \cdot 10^6$ m, we find $R\omega^2 \approx 0.03\,\text{m/s}^2$. Therefore, the correction to g is about 0.3% at the equator (but zero at the poles). However, this result does *not* imply that the value of g_{eff} is 0.3% smaller at the equator than at the poles, because the value of g itself (which we've defined to be the acceleration due to the gravitational force only) varies over the surface of the earth, due to its slightly nonspherical shape. (It also varies on a local scale due to density fluctuations and altitude.) When the centrifugal effect is combined with the nonspherical effect, the result is that g_{eff} is about 0.5% smaller at the equator; see Iona (1978).

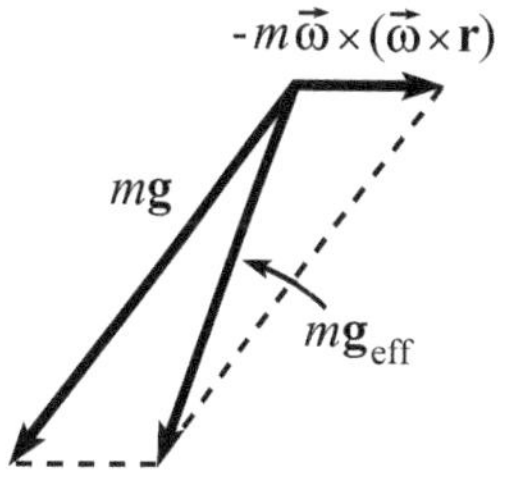

Fig. 10.4

REMARKS:

1. We found above that the sum of the gravitational and centrifugal forces doesn't point radially. Interestingly, there are *two* reasons for this. The first, as we just saw, is that under the (seemingly obvious) assumption that the earth's gravitational force is radial, the centrifugal force causes the sum not to be radial. The second reason is that

the earth's gravitational force actually *isn't* radial; the bulge at the equator causes the gravitational force to point slightly away from the center of the earth (except at the equator or at a pole). This is believable, because the extra mass at the nearby part of the bulge tends to shift the gravitational force in that direction. An exaggerated picture of the situation is shown in Fig. 10.5. The result of the nonradial **g** is that the effective gravitational force points even further from the radial direction than expected. It turns out that the effect from the nonradial **g** on the direction of $\mathbf{g}_{\text{eff}}$ is roughly equal to the effect from the centrifugal force.

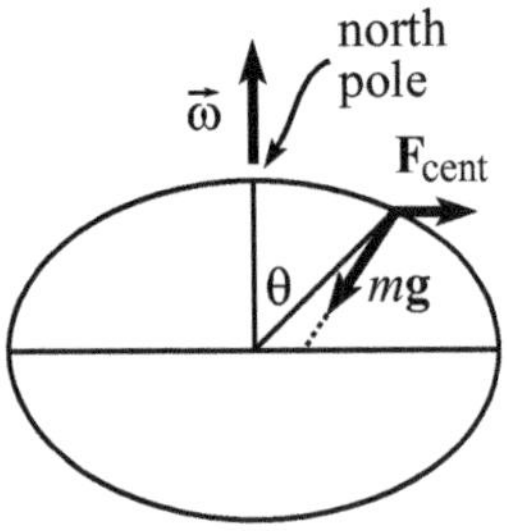

Fig. 10.5

So there is a feedback effect: The spinning of the earth causes $\mathbf{g}_{\text{eff}}$ to point away from the radial, which causes the earth to bulge (because the surface of the earth is, on average, perpendicular to $\mathbf{g}_{\text{eff}}$), which then causes $\mathbf{g}_{\text{eff}}$ to point a little further from the radial, which causes the earth to bulge a little more, and so on. Problem 10.12 works out some of the details of this, but see Mohazzabi and James (2000) for further discussion.

2. In the construction of buildings, and in similar matters, it is $\mathbf{g}_{\text{eff}}$, and not **g**, that determines the "upward" direction in which the building should point. The exact directions of the earth's gravitational force and center of the earth are irrelevant. A plumb bob hanging from the top of a skyscraper touches exactly at the base. Both the bob and the building point in a direction slightly different from both the radial and **g**, but no one cares.
3. If you look in a table and find the value for the acceleration due to gravity in Boston, remember that the number is the g_{eff} value and not the g value (which describes only the gravitational force, in our terminology). The way we've defined it, the g value is the acceleration with which things would fall if the earth kept its same shape but somehow stopped spinning. The exact value of g is therefore generally irrelevant. ♣

10.2.3 Coriolis force: $-2m\boldsymbol{\omega} \times \mathbf{v}$

While the centrifugal force is a very intuitive concept (we've all gone around a corner in a car), the same thing cannot be said about the Coriolis force. This force requires a nonzero velocity **v** relative to the accelerating frame, and people normally don't move with an appreciable **v** with respect to their car while rounding a corner. To get a feel for this force, let's look at two special cases.

Case 1 (Moving radially on a carousel): A carousel rotates counterclockwise with constant angular speed ω. Consider someone walking radially inward on the carousel (imagine a radial line painted on the carousel; the person walks along this line), at speed v with respect to the carousel, at radius r. The angular velocity vector $\boldsymbol{\omega}$ points out of the page, where we've signified the "out" direction in Fig. 10.6 by a little circle with a dot inside.

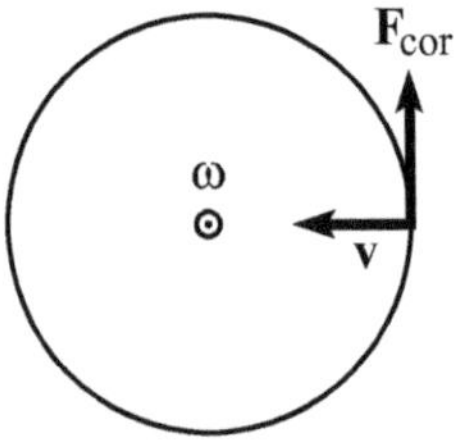

Fig. 10.6

REMARK: This remark might be a little picky, but I'll say it anyway. The direction of rotation is sometimes denoted by a curved arrow pointing tangentially along the circumference of the carousel. But this is technically not correct, because it would imply that the carousel is rotating in the rotating frame, which it isn't; it's just sitting there. And it's understood that Fig. 10.6 is in fact drawn in the rotating frame, and not the lab frame, because it includes a fictitious force, which has nothing to do with the lab frame. (If you wanted to draw things in the lab frame, then you wouldn't draw any fictitious forces, and the velocity **v** would have a tangential component, at least in this setup.) ♣

The Coriolis force, $-2m\boldsymbol{\omega} \times \mathbf{v}$, points tangentially in the direction of the motion of the carousel, that is, to the person's right in our scenario. It has magnitude

$$F_{\text{cor}} = 2m\omega v. \tag{10.14}$$

The person will have to counter this force with a tangential friction force of $2m\omega v$ (pointing to his left) at his feet, so that he continues to walk on the same radial line. Note that there is also the centrifugal force, which is countered by a radial friction force at the person's feet. But this effect won't be important here.

Why does this Coriolis force exist? It exists so that the resultant friction force changes the angular momentum of the person (measured with respect to the lab frame) in the proper way, according to $\tau = dL/dt$. To see this, take d/dt of $L = mr^2\omega$, where ω is the person's angular speed with respect to the lab frame, which is also the carousel's angular speed. Using $dr/dt = -v$, we have

$$\frac{dL}{dt} = -2mr\omega v + mr^2(d\omega/dt). \tag{10.15}$$

But $d\omega/dt = 0$, because the person remains on one radial line, and we are assuming that the carousel is arranged to keep a constant ω. Equation (10.15) then gives $dL/dt = -2mr\omega v$. So the L (with respect to the lab frame) of the person changes at a rate $-(2m\omega v)r$. This is simply the radius times the tangential friction force applied by the carousel. In other words, it is the torque applied to the person.

REMARK: What if the person doesn't apply a tangential friction force at his feet? Then the Coriolis force of $2m\omega v$ produces a tangential acceleration of $2\omega v$ in the rotating frame, and hence also in the lab frame (initially, before the direction of the motion in the rotating frame has a chance to change), because the frames are related by a constant ω. This acceleration exists essentially to keep the person's angular momentum (with respect to the lab frame) constant. (It *is* constant in this scenario, because there are no tangential forces in the lab frame.) To see that this tangential acceleration is consistent with conservation of angular momentum, set $dL/dt = 0$ in Eq. (10.15) to obtain $2\omega v = r(d\omega/dt)$ (this is the person's ω here, which is changing). The right-hand side of this is by definition the tangential acceleration. Therefore, saying that L is conserved is the same as saying that $2\omega v$ is the tangential acceleration (for this situation where the inward radial speed is v). ♣

Case 2 (Moving tangentially on a carousel): Now consider someone walking tangentially on a carousel in the direction of the carousel's motion, with speed v (relative to the carousel) at constant radius r (see Fig. 10.7). The Coriolis force $-2m\boldsymbol{\omega} \times \mathbf{v}$ points radially outward with magnitude $2m\omega v$. Assume that the person applies the friction force necessary to continue moving at radius r.

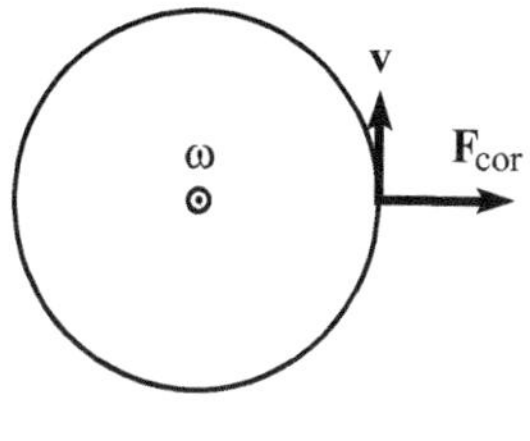

Fig. 10.7

There is a simple way to see why this outward force of $2m\omega v$ exists. Let $V \equiv \omega r$ be the speed of a point on the carousel at radius r, as viewed by an outside observer. If the person moves tangentially (in the same direction as the spinning) with speed v relative to the carousel, then his speed as viewed by the outside observer is $V + v$. The outside observer therefore sees the person walking in a circle of radius r at speed

$V + v$. The acceleration of the person with respect to the ground frame is therefore $(V + v)^2/r$. This acceleration must be caused by an inward-pointing friction force at the person's feet, so

$$F_{\text{friction}} = \frac{m(V+v)^2}{r} = \frac{mV^2}{r} + \frac{2mVv}{r} + \frac{mv^2}{r}. \tag{10.16}$$

This friction force is the same in any frame. How, then, does our person on the carousel interpret the three pieces of the inward-pointing friction force in Eq. (10.16)? The first term balances the outward centrifugal force due to the rotation of the frame, which he always feels. The third term is the inward force his feet must apply if he is to walk in a circle of radius r at speed v, which is exactly what he is doing in the rotating frame. The middle term is the additional inward friction force he must apply to balance the outward Coriolis force of $2m\omega v$ (using $V \equiv \omega r$). Said in an equivalent way, the person on the carousel will write down an $F = ma$ equation of the form (taking radially inward to be positive),

$$\begin{aligned} m\frac{v^2}{r} &= \frac{m(V+v)^2}{r} - \frac{mV^2}{r} - \frac{2mVv}{r} \\ \Longrightarrow \quad m\mathbf{a} &= \mathbf{F}_{\text{friction}} + \mathbf{F}_{\text{cent}} + \mathbf{F}_{\text{cor}}. \end{aligned} \tag{10.17}$$

We see that the net force he feels does indeed equal his ma, where a is measured with respect to the rotating frame. Physically, the difference between the interpretations of Eqs. (10.16) and (10.17) is the existence of fictitious forces in the rotating frame. Mathematically, the difference is simply the rearrangement of terms.

For cases in between the two special cases above, things aren't so clear, but that's the way it goes. Note that no matter what direction you move on a carousel, the Coriolis force always points in the same perpendicular direction relative to your motion. Whether it's to your right or to your left depends on the direction of the rotation. But given $\boldsymbol{\omega}$, you're stuck with the same relative direction of the force.

On a merry-go-round in the night,
Coriolis was shaken with fright.
Despite how he walked,
'Twas like he was stalked
By some fiend always pushing him right.

Let's do some more examples . . .

Example (Dropped ball): A ball is dropped from height h, at a polar angle θ (measured down from the north pole). How far to the east is the ball deflected, by the time it hits the ground?

Solution: The angle between $\boldsymbol{\omega}$ and $\mathbf{v}$ is $\pi - \theta$, so the Coriolis force $-2m\boldsymbol{\omega} \times \mathbf{v}$ is directed eastward with magnitude $2m\omega v \sin\theta$, where $v = gt$ is the speed at time

t (t runs from zero to the usual $\sqrt{2h/g}$).[1] Note that the ball is deflected to the east, independent of which hemisphere it is in. The eastward acceleration at time t is therefore $2\omega gt \sin\theta$. Integrating this to obtain the eastward speed (with an initial eastward speed of zero) gives $v_{\text{east}} = \omega gt^2 \sin\theta$. Integrating again to obtain the eastward deflection (with an initial eastward deflection of zero) gives $d_{\text{east}} = \omega gt^3 \sin\theta/3$. Plugging in $t = \sqrt{2h/g}$ gives

$$d_{\text{east}} = \frac{2\omega h \sin\theta}{3}\sqrt{\frac{2h}{g}}. \tag{10.18}$$

The frequency of the earth's rotation is $\omega \approx 7.3 \cdot 10^{-5}\,\text{s}^{-1}$, so if we pick $\theta = \pi/2$ and $h = 100$ m, for example, then we have $d_{\text{east}} \approx 2$ cm.

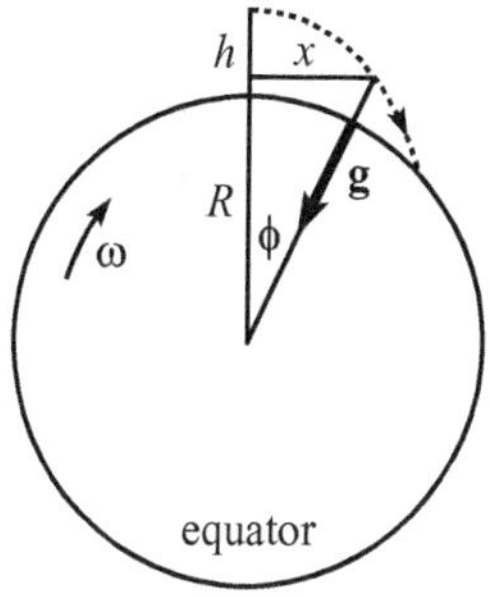

Fig. 10.8

REMARK: We can also solve this problem by working in an inertial frame; see Stirling (1983). Figure 10.8 shows the setup where a ball is dropped from a tower of height h located at the equator (the view is from the south pole). The earth is rotating in the inertial frame, so the initial sideways speed of the ball, $(R+h)\omega$, is larger than the sideways speed of the base of the tower, $R\omega$. This is the basic cause of the eastward deflection.

However, after the ball has moved to the right, the gravitational force on it picks up a component pointing to the left, and this slows down the sideways speed. If the ball has moved a distance x to the right, then the leftward component of gravity equals $g\sin\phi \approx g(x/R)$. Now, to leading order we have $x = R\omega t$, so the sideways acceleration of the ball is $a = -g(R\omega t/R) = -\omega gt$. Integrating this, and using the initial speed of $(R+h)\omega$, gives a rightward speed of $(R+h)\omega - \omega gt^2/2$. Integrating again gives a rightward distance of $(R+h)\omega t - \omega gt^3/6$. Subtracting off the rightward position of the base of the tower (namely $R\omega t$), and using $t \approx \sqrt{2h/g}$ (neglecting higher-order effects such as the curvature of the earth and the variation of g with altitude), we obtain an eastward deflection of $\omega h\sqrt{2h/g}(1 - 1/3) = (2/3)\omega h\sqrt{2h/g}$, relative to the base of the tower. If the ball is dropped at a polar angle θ instead of at the equator, then the only modification is that all speeds are decreased by a factor of $\sin\theta$, so we obtain the result in Eq. (10.18). ♣

Example (Foucault's pendulum): This is the classic example of a consequence of the Coriolis force. It unequivocally shows that the earth rotates. The basic idea is that due to the rotation of the earth, the plane of a swinging pendulum rotates slowly, with a calculable frequency. In the special case where the pendulum is at one of the poles, this rotation is easy to understand. Consider the north pole. An external observer, hovering above the north pole and watching the earth rotate, sees the pendulum's plane stay fixed (with respect to the distant stars) while the earth rotates counterclockwise beneath it.[2] Therefore, to an observer on the earth, the pendulum's plane rotates clockwise (viewed from above). The frequency of this rotation is of course just the frequency of the earth's rotation, so the earth-based observer sees the pendulum's plane make one revolution each day.

[1] Technically, $v = gt$ isn't quite correct. Due to the Coriolis force, the ball will pick up a small eastward velocity component (this is the point of the problem). This component will then produce a second-order Coriolis force that affects the vertical speed (see Exercise 10.21). But we can ignore this small effect in this problem. Also, we really mean g_{eff} instead of g, but any ambiguity in this will have a negligible effect.

[2] Assume that the pivot of the pendulum is a frictionless bearing, so that it can't provide any torque to twist the plane of the pendulum.

What if the pendulum is not at one of the poles? What is the frequency of the precession? Let the pendulum be located at a polar angle θ. We will work in the approximation where the velocity of the pendulum bob is horizontal (with respect to the earth's surface). This is essentially true if the pendulum's string is very long; the correction due to the rising and falling of the bob is negligible. The Coriolis force $-2m\boldsymbol{\omega} \times \mathbf{v}$ points in some complicated direction, but fortunately we are concerned only with the component that lies in the horizontal plane (that is, the plane of the ground). The vertical component serves only to modify the apparent force of gravity and is therefore negligible. Although the frequency of the pendulum does depend on g, the resulting modification is very small. With this in mind, let's break $\boldsymbol{\omega}$ into vertical and horizontal components in a coordinate system located at the pendulum. From Fig. 10.9, we see that

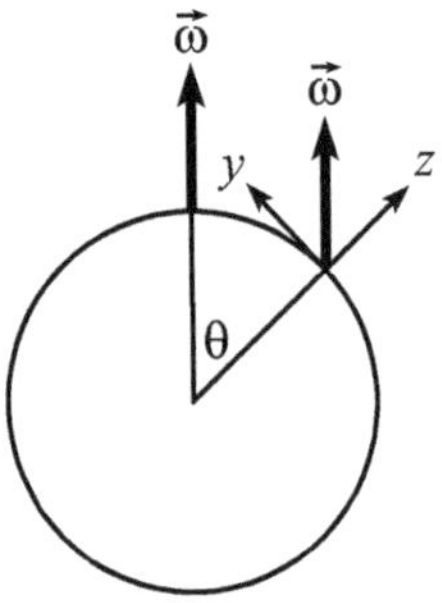

Fig. 10.9

$$\boldsymbol{\omega} = \omega \cos\theta\, \hat{\mathbf{z}} + \omega \sin\theta\, \hat{\mathbf{y}}. \tag{10.19}$$

We'll ignore the y component, because it produces a Coriolis force in the z direction, since $\mathbf{v}$ lies in the horizontal x-y plane. So for our purposes, $\boldsymbol{\omega}$ is essentially equal to $\omega \cos\theta\, \hat{\mathbf{z}}$. From this point on, the problem of finding the frequency of precession can be done in numerous ways. We'll present two solutions.

First solution (The slick way): The horizontal component of the Coriolis force has magnitude

$$F_{\text{cor}}^{\text{horiz}} = |-2m(\omega\cos\theta\hat{\mathbf{z}}) \times \mathbf{v}| = 2m(\omega\cos\theta)v, \tag{10.20}$$

and it is perpendicular to $\mathbf{v}(t)$. Therefore, as far as the pendulum is concerned, it is located at the north pole of a planet called Terra Costhetica which has rotational frequency of $\omega\cos\theta$.[3] But as we saw above, the precessional frequency of a Foucault pendulum located at the north pole of such a planet is simply

$$\omega_{\text{F}} = \omega\cos\theta, \tag{10.21}$$

in the clockwise direction. So that's our answer.

Second solution (In the pendulum's frame): Let's work in the frame of the plane that the Foucault pendulum sweeps through. Our goal is to find the rate of precession of this frame. With respect to a frame fixed on the earth (with axes $\hat{\mathbf{x}}$, $\hat{\mathbf{y}}$, and $\hat{\mathbf{z}}$ as above), we know that this plane rotates with frequency $\boldsymbol{\omega}_{\text{F}} = -\omega\hat{\mathbf{z}}$ if we're at the north pole ($\theta = 0$), and with frequency $\boldsymbol{\omega}_{\text{F}} = \mathbf{0}$ if we're at the equator ($\theta = \pi/2$). So it's a good bet that the general answer is $\boldsymbol{\omega}_{\text{F}} = -\omega\cos\theta\,\hat{\mathbf{z}}$, and that's what we'll now show.

Working in the frame of the plane of the pendulum is useful, because we can take advantage of the fact that the pendulum feels no sideways forces in this frame, because otherwise it would move out of the plane (which it doesn't, by definition). The frame fixed on the earth rotates with frequency $\boldsymbol{\omega} = \omega\cos\theta\hat{\mathbf{z}} + \omega\sin\theta\hat{\mathbf{y}}$, with respect to the inertial frame. Let the pendulum's plane rotate with frequency $\boldsymbol{\omega}_{\text{F}} = \omega_{\text{F}}\hat{\mathbf{z}}$ with

[3] As mentioned above, the setup isn't *exactly* like the one on the new planet. There is also a vertical component of the Coriolis force for the pendulum on the earth, but this effect is negligible.

respect to the earth frame. Then the angular velocity of the pendulum's frame with respect to the inertial frame is

$$\boldsymbol{\omega} + \boldsymbol{\omega}_{\mathrm{F}} = (\omega \cos\theta + \omega_{\mathrm{F}})\hat{\mathbf{z}} + \omega \sin\theta \hat{\mathbf{y}}. \quad (10.22)$$

To find the horizontal component of the Coriolis force in this rotating frame, we care only about the $\hat{\mathbf{z}}$ part of this frequency. The horizontal Coriolis force therefore has magnitude $2m(\omega\cos\theta + \omega_{\mathrm{F}})v$. But in the frame of the pendulum, there must be zero horizontal force, so this must be zero. Therefore,

$$\omega_{\mathrm{F}} = -\omega\cos\theta. \quad (10.23)$$

This agrees with Eq. (10.21), where we just wrote down the magnitude of ω_{F}.

10.2.4 Azimuthal force: $-m(d\boldsymbol{\omega}/dt) \times \mathbf{r}$

In this section, we will restrict ourselves to the simple and intuitive case where $\boldsymbol{\omega}$ changes only in magnitude, that is, not in direction (this more complicated case is the subject of Problem 10.10). The azimuthal force may then be written as

$$\mathbf{F}_{\mathrm{az}} = -m\dot{\omega}\hat{\boldsymbol{\omega}} \times \mathbf{r}. \quad (10.24)$$

This force is easily understood by considering a person standing at rest with respect to a rotating carousel. If the carousel speeds up, then the person must feel a tangential friction force at his feet if he is to remain fixed on the carousel. This friction force equals $ma_{\tan}$, where $a_{\tan} = r\dot{\omega}$ is the tangential acceleration as measured in the ground frame. But from the person's point of view in the rotating frame, he is not moving, so there must be some other mysterious force that balances the friction. This is the azimuthal force. Quantitatively, when $\hat{\boldsymbol{\omega}}$ is orthogonal to $\mathbf{r}$, we have $|\hat{\boldsymbol{\omega}} \times \mathbf{r}| = r$, so the azimuthal force in Eq. (10.24) has magnitude $mr\dot{\omega}$. This is the same as the magnitude of the friction force, as it should be.

What we have here is exactly the same effect that we had with the translation force on the accelerating train. If the floor speeds up beneath you, then you must apply a friction force if you don't want to be thrown backward with respect to the floor. If you shut your eyes and ignore the centrifugal force, then you can't tell if you are on a linearly accelerating train, or on an angularly accelerating carousel. The translation and azimuthal forces both arise from the acceleration of the floor. (Well, for that matter, the centrifugal force does too.)

We can also view things in terms of rotational quantities, instead of the linear $a_{\tan}$ acceleration above. If the carousel speeds up, then a torque must be applied to the person if he is to remain fixed on the carousel, because his angular momentum in the fixed frame increases. Therefore, he must feel a friction force at his feet. Let's show that this friction force, which produces the change in angular momentum of the person in the fixed frame, exactly cancels the azimuthal force in the rotating frame, thereby yielding zero net force in the rotating frame. Since $L = mr^2\omega$, we have $dL/dt = mr^2\dot{\omega}$ (assuming r is fixed). And since $dL/dt = \tau = rF$, we see that the required friction force is $F = mr\dot{\omega}$. And as we saw above, when $\hat{\boldsymbol{\omega}}$ is orthogonal to $\mathbf{r}$, the azimuthal force in Eq. (10.24) also equals $mr\dot{\omega}$, in the direction opposite to the carousel's motion. So the tangential forces in the rotating frame do indeed cancel. This was basically the same calculation as the one above, but with an extra factor of r in the $\tau = dL/dt$ equation.

Example (Spinning ice skater): We have all seen ice skaters increase their angular speed by bringing their arms in close to their body. This can be understood in terms of angular momentum; a smaller moment of inertia requires a larger ω in order to keep L constant. But let's analyze the situation in terms of fictitious forces. We'll idealize things by giving the skater massive hands at the ends of massless arms attached to a massless body.[4] Let the hands have total mass m, and let them be drawn in radially.

Look at things in the skater's frame (which has an increasing ω), defined by the vertical plane containing the hands. The crucial thing to realize here is that the skater always remains in the skater's frame (a fine tautology, indeed). Therefore, the skater must feel zero net tangential force in her frame, because otherwise she would accelerate with respect to it. Her hands are being drawn in by a muscular force that works against the centrifugal force, but there can be no net tangential force on the hands in the skater's frame, by definition.

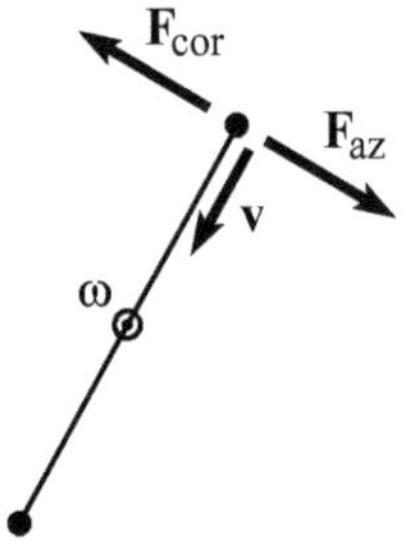

Fig. 10.10

What are the tangential forces in the skater's frame? Let the hands be drawn in at speed v (see Fig. 10.10). Then there is a Coriolis force (in the same direction as the spinning) with magnitude $2m\omega v$. There is also an azimuthal force with magnitude $mr\dot{\omega}$ (in the direction opposite to the spinning, as you can check). Since the net tangential force is zero in the skater's frame, we must have

$$2m\omega v = mr\dot{\omega}. \tag{10.25}$$

Does this relation make sense? Well, let's look at things in the ground frame. The total angular momentum of the hands in the ground frame is constant. Therefore, $d(mr^2\omega)/dt = 0$. Taking this derivative and using $dr/dt \equiv -v$ gives Eq. (10.25).

A word of advice about using fictitious forces: Decide which frame you are going to work in (the lab frame or the accelerating frame), and then stick with it. The common mistake is to work a little in one frame and a little in the other without realizing it. For example, you might introduce a centrifugal force on someone sitting at rest on a carousel, but then also give her a centripetal

[4] This reminds me of a joke about a spherical cow.

acceleration. This is incorrect. In the lab frame, there is a centripetal acceleration (caused by the friction force) and no centrifugal force. In the rotating frame, there is a centrifugal force (which cancels the friction force) and no centripetal acceleration (because the person is sitting at rest on the carousel). In short, if you ever mention the words "centrifugal" or "Coriolis," etc., then you had better be working in an accelerating frame.

10.3 Tides

The tides on the earth exist because the gravitational force from a point mass (or a spherical mass, in particular the moon or the sun) is not uniform; the direction of the force is not constant (the force lines converge to the source), and the magnitude is not constant (it falls off like $1/r^2$). On the earth, these effects cause the oceans to bulge around the earth, producing the observed tides. The study of tides is useful in part because tides are a very real phenomenon in the world, and in part because the following analysis gives us an excuse to use fictitious forces and play around with Taylor series approximations. Before considering the general case of tidal forces, let's look at two special cases.

Longitudinal tidal force

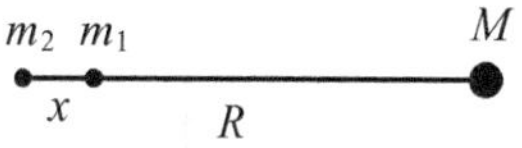

Fig. 10.11

In order to isolate the tidal effect, let's consider a somewhat contrived setup instead of the earth–sun or earth–moon system (we'll eventually get to these). Consider three masses in a line, as shown in Fig. 10.11. The right mass M is very large and exerts a gravitational force on m_1 and m_2. But m_1 and m_2 are sufficiently small so that they don't exert much of a gravitational force on each other. Furthermore, assume $m_2 \gg m_1$. Let R be the distance from M to m_2, and let x be the distance from m_1 to m_2. Assume $x \ll R$.

The masses m_1 and m_2 accelerate radially inward toward M. In the accelerating reference frame of m_2, what is the force on m_1? To answer this, we must add the fictitious translation force on m_1 (which points to the left) to the actual gravitational force on m_1 (which points to the right). So in the accelerating frame of m_2, the net force on m_1 is

$$\begin{aligned}\mathbf{F}_{\text{net}} = \mathbf{F}_{\text{grav}} + \mathbf{F}_{\text{trans}} &= \frac{GMm_1}{(R-x)^2}\hat{\mathbf{x}} - m_1 a_2 \hat{\mathbf{x}} \\ &= \frac{GMm_1}{(R-x)^2}\hat{\mathbf{x}} - \frac{GMm_1}{R^2}\hat{\mathbf{x}}. \end{aligned} \tag{10.26}$$

This is equivalent to the statement that m_2 sees m_1 accelerate away with acceleration

$$\frac{F_{\text{net}}}{m_1} = \frac{GM}{(R-x)^2} - \frac{GM}{R^2} = a_1 - a_2, \tag{10.27}$$

which is just the difference in accelerations you would intuitively expect. Using $x \ll R$ to make suitable approximations in Eq. (10.26), we have

$$\begin{aligned} F_{\text{net}} &\approx \frac{GMm_1}{R^2 - 2Rx} - \frac{GMm_1}{R^2} = \frac{GMm_1}{R^2}\left(\frac{1}{1-2x/R} - 1\right) \\ &\approx \frac{GMm_1}{R^2}\big((1+2x/R) - 1\big) = \frac{2GMm_1x}{R^3}. \end{aligned} \tag{10.28}$$

This is, of course, simply x times the derivative of the gravitational force. It points to the right, so its effect is to increase the separation of the masses.

If you are riding along on m_2, and if a black box encloses m_1 and m_2, then as far as you are concerned, you may as well be in a black box floating freely in outer space (from Einstein's Equivalence Principle, discussed in Chapter 14). But if you are floating in outer space, and if you see m_1 accelerating away from you with acceleration $F_{\text{net}}/m_1 = 2GMx/R^3$, then you will naturally conclude that in your reference frame, there must be a mysterious force of

$$\text{“}F_{\text{tidal}}\text{”} \equiv F_{\text{net}} = \frac{2GMm_1x}{R^3} \tag{10.29}$$

pulling m_1 away from you. This force is called the *tidal force* because it is what causes the tides, as we'll see in detail below. Note that the tidal force is linear in the separation x and inversely proportional to the *cube* of the distance from the source. If x is negative, so that m_1 is on the left side of m_2, then the tidal force is negative, which means that m_1 accelerates leftward away from you. So the effect of the longitudinal tidal force is to increase the separation between the masses, independent of the sign of x.

If you want to keep m_1 at rest with respect to you, then you will have to tie it down with a string, and the tension in the string will be $2GMm_1x/R^3$.[5] Any experiment you do inside the black box will indicate that there is a force pulling m_1 away from you. The only actual force on m_1 is the gravitational force from M equal to $GMm_1/(R-x)^2$, but because M gives m_2 nearly the same acceleration as m_1, you perceive only the small F_{tidal} in Eq. (10.29). It is indeed small, because it is the actual gravitational force times the quantity $2x/R \ll 1$.

Transverse tidal force

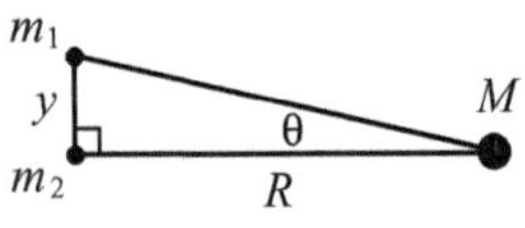

Fig. 10.12

Consider now the setup shown in Fig. 10.12, with m_2 at the corner of a right triangle that has $y \ll R$. The masses m_1 and m_2 accelerate radially inward toward M. In the accelerating frame of m_2, what is the force on m_1? As above, we must add the fictitious translation force on m_1 (which points to the left) to the actual gravitational force on m_1 (which points to the right and slightly downward).

[5] The tension in the string won't affect the acceleration of m_2, because m_2 is assumed to be much larger than m_1. So our accelerating frame is still defined to be the one that accelerates with acceleration $a_2 = GM/R^2$. This is where the $m_2 \gg m_1$ assumption comes in.

In this case, however, the *magnitudes* of the gravitational accelerations of m_1 and m_2 are essentially equal, because they are both a distance R from mass M, up to second-order effects in y/R (by the Pythagorean theorem). The *direction* is the only thing that is different, to first order in y/R. So in the accelerating frame of m_2, the net force (which we'll call the tidal force from now on) on m_1 is

$$\begin{aligned}\mathbf{F}_{\text{tidal}} = \mathbf{F}_{\text{grav}} + \mathbf{F}_{\text{trans}} &\approx \frac{GMm_1}{R^2}(\cos\theta\,\hat{\mathbf{x}} - \sin\theta\,\hat{\mathbf{y}}) - \frac{GMm_1}{R^2}\,\hat{\mathbf{x}} \\ &\approx \frac{GMm_1}{R^2}(-\sin\theta\,\hat{\mathbf{y}}) \approx -\frac{GMm_1 y}{R^3}\hat{\mathbf{y}}, \end{aligned} \qquad (10.30)$$

where we have used $\cos\theta \approx 1$ and $\sin\theta \approx y/R$. This difference is simply the y component of the force on m_1, which is what you would expect. It points along the line joining the masses, and its effect is to pull them together. As in the longitudinal case, the transverse tidal force is linear in the separation and inversely proportional to the cube of the distance from the source.

General tidal force

We will now calculate the tidal force on a mass m located at an arbitrary point on a circle of radius r (for example, this circle could represent a cross section of the earth), due to a mass M located at the vector $-\mathbf{R}$ (so the vector from M to m is $\mathbf{R}+\mathbf{r}$); see Fig. 10.13. We will calculate the tidal force relative to the center of the circle. That is, we will find the net force on m in the accelerating frame whose origin is the center of the circle. As usual, assume $r \ll R$. And for now let's ignore any orbital motion of the circle around M (although it turns out that this isn't relevant to tides anyway; see the third remark below), so the circle just accelerates radially inward toward M.

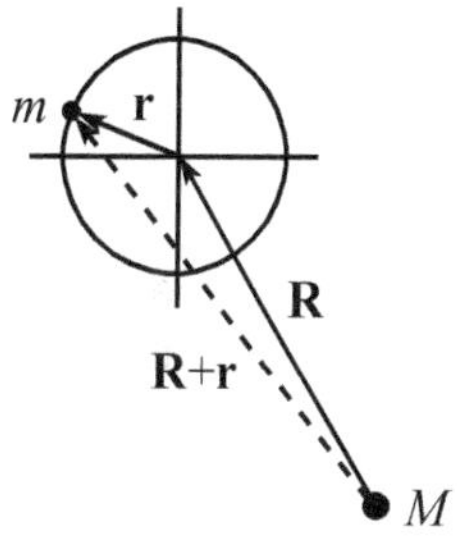

Fig. 10.13

The gravitational force on m may be written as $\mathbf{F}_{\text{grav}} = -GMm(\mathbf{R}+\mathbf{r})/|\mathbf{R}+\mathbf{r}|^3$. The cube is in the denominator because the vector in the numerator contains one power of the distance. As above, adding on the fictitious translation force due to the acceleration of the center of the circle (which is independent of whatever mass we put there) yields a tidal force of

$$\frac{\mathbf{F}_{\text{tidal}}(\mathbf{r})}{GMm} = \frac{-(\mathbf{R}+\mathbf{r})}{|\mathbf{R}+\mathbf{r}|^3} - \frac{-\mathbf{R}}{|\mathbf{R}|^3}. \qquad (10.31)$$

This is the exact expression for the tidal force. However, it is rather useless.[6] Let us therefore make some approximations in Eq. (10.31) and transform it into something technically incorrect (as approximations tend to be), but far more useful. The first thing we need to do is rewrite the $|\mathbf{R}+\mathbf{r}|$ term. We have

[6] This reminds me of a joke about two people lost in a hot-air balloon.

(using $r \ll R$ and ignoring higher-order terms)

$$\begin{aligned}|\mathbf{R}+\mathbf{r}| = \sqrt{(\mathbf{R}+\mathbf{r})\cdot(\mathbf{R}+\mathbf{r})} &= \sqrt{R^2+r^2+2\mathbf{R}\cdot\mathbf{r}} \\ &\approx R\sqrt{1+2\mathbf{R}\cdot\mathbf{r}/R^2} \\ &\approx R\left(1+\frac{\mathbf{R}\cdot\mathbf{r}}{R^2}\right). \end{aligned} \tag{10.32}$$

Therefore (again using $r \ll R$),

$$\begin{aligned}\frac{\mathbf{F}_{\text{tidal}}(\mathbf{r})}{GMm} &\approx -\frac{\mathbf{R}+\mathbf{r}}{R^3(1+\mathbf{R}\cdot\mathbf{r}/R^2)^3}+\frac{\mathbf{R}}{R^3} \\ &\approx -\frac{\mathbf{R}+\mathbf{r}}{R^3(1+3\mathbf{R}\cdot\mathbf{r}/R^2)}+\frac{\mathbf{R}}{R^3} \\ &\approx -\frac{\mathbf{R}+\mathbf{r}}{R^3}\left(1-\frac{3\mathbf{R}\cdot\mathbf{r}}{R^2}\right)+\frac{\mathbf{R}}{R^3}. \end{aligned} \tag{10.33}$$

Letting $\hat{\mathbf{R}} \equiv \mathbf{R}/R$, this finally simplifies to (once again using $r \ll R$)

$$\mathbf{F}_{\text{tidal}}(\mathbf{r}) \approx \frac{GMm}{R^3}(3\hat{\mathbf{R}}(\hat{\mathbf{R}}\cdot\mathbf{r})-\mathbf{r}). \tag{10.34}$$

This is the general expression for the tidal force. We can put it in a simpler form if we let M lie on the positive x axis, which we can arrange for with a rotation of the axes. We then have $\hat{\mathbf{R}} = -\hat{\mathbf{x}}$, and so $\hat{\mathbf{R}}\cdot\mathbf{r} = -x$. Equation (10.34) then tells us that the tidal force on a mass m at position $\mathbf{r} \equiv (x, y)$, due to a mass M at position $(R, 0)$, equals

$$\mathbf{F}_{\text{tidal}}(x,y) \approx \frac{GMm}{R^3}(3x\hat{\mathbf{x}}-(x\hat{\mathbf{x}}+y\hat{\mathbf{y}})) = \frac{GMm}{R^3}(2x,-y). \tag{10.35}$$

Fig. 10.14

This reduces properly in the longitudinal and transverse cases considered above. The tidal forces at various points on the circle are shown in Fig. 10.14. Since our setup with M on the x axis is invariant under rotations around the x axis, this picture of forces is also invariant. And because of the left–right symmetry, the picture would look the same if M were placed on the negative x axis.

The potential energy associated with the tidal force in Eq. (10.35) is proportional to $-x^2+y^2/2$, because the negative gradient of this yields the $(2x,-y)$ vector. Using $x = r\cos\theta$ and $y = r\sin\theta$, the potential can be written as

$$V_{\text{tidal}}(r,\theta) \approx \frac{GMmr^2}{2R^3}(-2\cos^2\theta+\sin^2\theta) = \frac{GMmr^2}{2R^3}(1-3\cos^2\theta). \tag{10.36}$$

If the earth were a rigid body, then the tidal force would have no effect on it. But the water in the oceans is free to slosh around, so it bulges along the line from the earth to the moon, and also along the line from the earth to the sun.

And it also gets drawn in and forms a dip along the directions transverse to the moon and sun.[7] We'll see below that the moon's effect is about twice the sun's. As the earth rotates beneath the bulge and dip, a person on the earth sees them rotate in the other direction relative to the earth. From Fig. 10.14, we see that this produces *two* high tides and two low tides per day.[8] It's actually not exactly two per day, because the moon moves around the earth. But this motion is fairly slow, taking about a month, so it's a reasonable approximation to think of the moon as motionless.

Note that it is *not* the case that the moon pushes the water away on the far side of the earth. It pulls on that water, too; it just does so in a weaker manner than it pulls on the rigid part of the earth. With tides, it's not the force that matters, but rather the *difference* in force. Tides are a *comparative* effect. A child playing with a shovel and a pail on a beach will be able to do so only until the earth rotates to a point where the beach enters the region where the moon pulls on the water sufficiently more (or less) than it pulls on the rigid earth.

Whether sieged by an army more grand,
Or swamped by the tides in the sand,
It's a matter of course
That the difference in force
Is what leads to a castle's last stand.

REMARKS:

1. Consider a mass on the surface of the earth. It turns out that the gravitational force from the sun on it is (much) larger than that from the moon, whereas the tidal force from the sun on it is (slightly) weaker than that from the moon. Quantitatively, the ratio of the gravitational forces is

$$\frac{F_{\rm S}}{F_{\rm M}} = \left(\frac{GM_{\rm S}}{R_{\rm E,S}^2}\right) \Big/ \left(\frac{GM_{\rm M}}{R_{\rm E,M}^2}\right) = \frac{5.9 \cdot 10^{-3}\ {\rm m/s^2}}{3.3 \cdot 10^{-5}\ {\rm m/s^2}} \approx 175. \qquad (10.37)$$

And the ratio of the tidal forces is

$$\frac{F_{\rm t,S}}{F_{\rm t,M}} = \left(\frac{GM_{\rm S}}{R_{\rm E,S}^3}\right) \Big/ \left(\frac{GM_{\rm M}}{R_{\rm E,M}^3}\right) = \frac{3.9 \cdot 10^{-14}\ {\rm s^{-2}}}{8.7 \cdot 10^{-14}\ {\rm s^{-2}}} \approx 0.45. \qquad (10.38)$$

[7] There's actually a lag effect as the earth rotates, due to various complicated things, so the bulge doesn't point directly toward the moon or sun. But let's not worry about that here.

[8] We've made the incorrect but fairly reasonable approximation that the moon lies in the plane of the earth's equator. If you solve the problem exactly by taking into account the position of the moon relative to the equator, you will find that there is a piece of the tides (known as the diurnal tide) that has a period of one day, in addition to the piece we just found (the semidiurnal tide) with a period of half a day. See Horsfield (1976). If you want to derive this result yourself, a hint is that you'll need to use Eqs. (10.36) and (B.7).

Depending on the relative positions of the sun and moon with respect to the earth, their tidal effects may add (when the three bodies are collinear; this is called a "spring" tide), or they may partially cancel (when the sun and moon make a 90° angle in the sky; this is called a "neap" tide).

2. Equation (10.38) shows that the moon's tidal effect is roughly twice the sun's. This has the following interesting implication about the densities of the moon and sun. The tidal force from, say, the moon is proportional to

$$\left(\frac{GM_\text{M}}{R_\text{E,M}^3}\right) = \left(\frac{G(\frac{4}{3}\pi r_\text{M}^3)\rho_\text{M}}{R_\text{E,M}^3}\right) \propto \rho_\text{M}\left(\frac{r_\text{M}}{R_\text{E,M}}\right)^3 \approx \rho_\text{M}\theta_\text{M}^3, \tag{10.39}$$

where θ_M is half of the angular size of the moon in the sky. Likewise for the sun's tidal force. But it just so happens that the angular sizes of the sun and the moon are essentially equal, as you can see by looking at them (preferably quickly in the case of the sun, and through some thick haze), or by noting that total solar eclipses barely exist. Therefore, the combination of Eq. (10.38) and Eq. (10.39) tells us that the moon's density is about twice the sun's. It is quite remarkable that, at least roughly, you can experimentally determine the ratio of the densities of two celestial bodies by spending a few weeks at the beach.

3. There is one effect that we've glossed over, but fortunately it turns out not to be relevant. In the above analysis, we derived the tidal force by subtracting the fictitious translation force from the gravitational force. But in the actual case of the earth, the fictitious force is generally interpreted as a centrifugal force, because the earth is rotating around the earth–moon CM, and this CM is also rotating around the sun. And, of course, the earth is rotating around its axis. Basically, there's a lot of rotation going on.

   Now, because the centrifugal force (from the sum of all the rotations) depends on position, there is a difference between the centrifugal force at the center of the earth and at a point on the surface. Shouldn't this difference come into play when calculating the net force in the accelerating frame whose origin is the center of the earth? Well, yes … and no …

   There are many ways of breaking up the earth's motion, but for the present purposes the most useful way is the following. The motion of any object can be described as the sum of the translation of a nonrotating frame (for example, the seat on a ferris wheel), plus a rotation around a point in that frame (the seat on an old-fashioned ferris wheel wouldn't have any such rotation, but the seat on a more modern scary kind of amusement park ride undoubtedly would).

   The first of these two motions accounts for the motion of the earth as a whole around the earth–moon CM and also for the motion of this CM around the sun.[9] The only relevant fictitious force here is the translation force (at any instant in this motion, every atom in the earth has the same acceleration vector, so there is no need to think in terms of rotation), and this translation force is what we used in the above derivation of the tidal force.[10] The second of the two motions accounts for the rotation of the earth on its axis. This most certainly does create a centrifugal force difference between the center of the earth and a point on the surface, but this difference doesn't contribute to the tides; it contributes only to the uniform bulge of the earth at the equator. Since the force is uniform, it doesn't cause water to slosh around, so it therefore doesn't have anything to do with the tides.

[9] Just imagine grabbing the earth with a very large hand and sliding it around without twisting your hand, so that a given point on the earth always has the same view of stars in the sky.

[10] If you insist on thinking in terms of circular motion as you slide the earth in a circle around the sun without twisting your hand, you can indeed view every point in the earth as moving in a circle. But all of these circles have the same radius; they simply have offset centers. The centrifugal forces are therefore all the same (in both magnitude and direction), and so there are no differences to worry about.

So the answers to the above question are: Yes, because the difference really does contribute to the net force. But no, because it doesn't contribute to the tides, which is what we're concerned with here.

If you work out the actual numbers, you'll see that the centrifugal force completely dominates the tidal force. This is consistent with the fact that the equatorial bulge is on the order of kilometers, whereas the tides are on the order of meters. But the point is that the centrifugal force doesn't affect the tides because it is uniform around the earth. The tidal force, on the other hand, has the nonuniform shape in Fig. 10.14. The football-shaped bulge it creates therefore moves around relative to the earth as the earth rotates beneath it. To sum up: with the tidal force (which we define as the force that produces the tides), we are not concerned with the total difference in force between a point on the earth and the center, but rather only with the part of the difference that isn't invariant under rotations of the earth. ♣

10.4 Problems

Section 10.2: The fictitious forces

10.1. **Which way down?** *

You are floating high up in a balloon, at rest with respect to the earth. Give three quasi-reasonable definitions for which point on the ground is right "below" you.

10.2. **Longjumping in $\mathbf{g}_{\text{eff}}$** *

If a longjumper can jump 8 meters at the north pole, how far can he jump at the equator? Assume that g_{eff} is 0.5% less at the equator than at the north pole (although this is only approximate). Ignore effects of wind resistance, temperature, and runways made of ice.

10.3. **$\mathbf{g}_{\text{eff}}$ vs. g** *

For what angle θ (down from the north pole) is the angle between $\mathbf{g}_{\text{eff}}$ and $\mathbf{g}$ maximum?

10.4. **Lots of circles** *

(a) Two circles in a plane, C_1 and C_2, each rotate with frequency ω relative to an inertia frame. The center of C_1 is fixed in the inertial frame, and the center of C_2 is fixed on C_1, as shown in Fig. 10.15. A mass is fixed on C_2. The position of the mass relative to the center of C_1 is $\mathbf{R}(t)$. Find the fictitious force felt by the mass.

(b) N circles in a plane, C_i, each rotate with frequency ω relative to an inertia frame. The center of C_1 is fixed in the inertial frame, and the center of C_i is fixed on C_{i-1} (for $i = 2, \ldots, N$), as shown in Fig. 10.16. A mass is fixed on C_N. The position of the mass relative to the center of C_1 is $\mathbf{R}(t)$. Find the fictitious force felt by the mass.

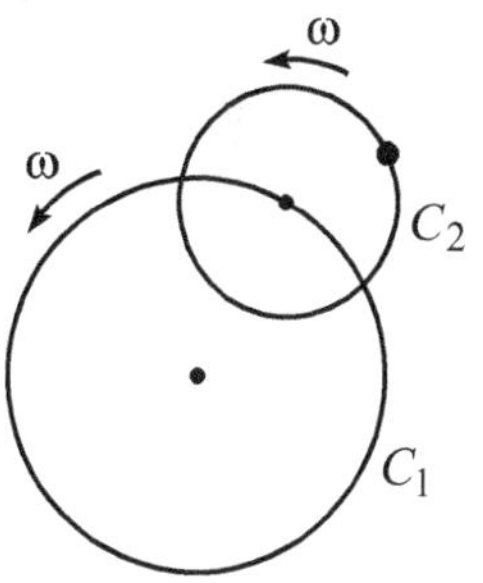

Fig. 10.15

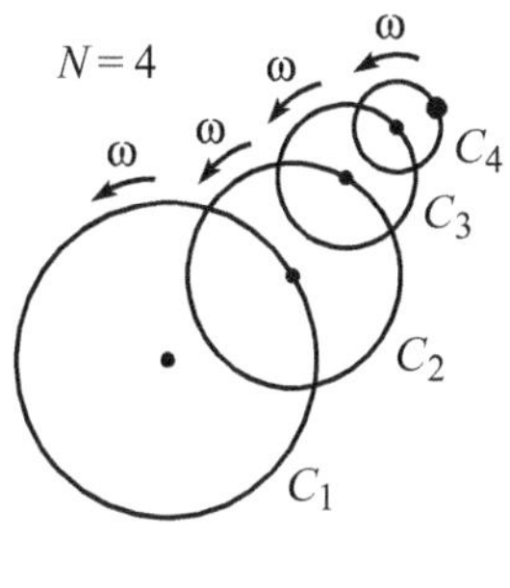

Fig. 10.16

10.5. **Mass on a turntable** *

A mass is at rest with respect to the lab frame, while a frictionless turntable rotates beneath it. The frequency of the turntable is ω, and the mass is located at radius r. In the frame of the turntable, find the forces acting on the mass, and verify that $\mathbf{F} = m\mathbf{a}$.

10.6. **Released mass** *

A mass is bolted down to a frictionless turntable. The frequency of rotation is ω, and the mass is located at radius a. The mass is then released. Viewed from an inertial frame, it travels in a straight line. In the rotating frame, what path does the mass take? Specify $r(t)$ and $\theta(t)$, where θ is the angle with respect to the initial radius, as measured in the rotating frame. Solve this by working in the inertial frame. (Exercise 10.25 deals with the more difficult task of working in the rotating frame.)

10.7. **Coriolis circles** *

A puck slides with speed v on frictionless ice. The surface is "level" in the sense that it is orthogonal to $\mathbf{g}_{\text{eff}}$ at all points. Show that the puck moves in a circle, as seen in the earth's rotating frame. What is the radius of the circle? What is the frequency of the motion? Assume that the radius of the circle is small compared with the radius of the earth.

10.8. $\boldsymbol{\tau} = d\mathbf{L}/dt$ **

In Section 8.4.3, we derived the three conditions under which it is valid to write $\sum \boldsymbol{\tau}_i^{\text{ext}} = d\mathbf{L}/dt$. Rederive these conditions by working entirely in the (possibly) accelerating frame. As in Section 8.4.3, assume that the frame isn't rotating (so at most, the origin is accelerating).

10.9. **Determining your frame** **

Imagine that you are on a large rotating disk, with $\boldsymbol{\omega}$ perpendicular to the disk at its center. Assume that you know that the center of the disk is fixed, and that no real forces act on anything on the disk (except gravity pulling down, which is canceled by the normal force). Assume that $\boldsymbol{\omega}$ changes only in magnitude, and that this rate of change is constant. Is it possible for you to determine ω and $d\omega/dt$, and also the location of the center of the disk, by performing experiments in the rotating frame only?

10.10. **Changing $\boldsymbol{\omega}$'s direction** ***

Consider the special case where a reference frame's $\boldsymbol{\omega}$ changes only in direction (that is, not in magnitude). In particular, consider a cone rolling on a table, which is a natural example of such a situation. Let the origin of the cone frame be the tip of the cone. This point remains fixed in the inertial frame. The instantaneous $\boldsymbol{\omega}$ for a rolling cone points along its line of contact with the table, because these are the points that are instantaneously at rest. This line precesses around the origin. Let the frequency of the precession be Ω.

In order to isolate the azimuthal force, paint a dot on the surface of the cone (call it point P), and consider the moment when the dot lies on the instantaneous $\boldsymbol{\omega}$ (see Fig. 10.17). From Eq. (10.11), we see that there are no contributions from the centrifugal force (because P lies on $\boldsymbol{\omega}$), the Coriolis force (because P is not moving in the cone frame), or the translation force (because the tip of the cone is fixed). The only remaining fictitious force is the azimuthal force, and it exists due to the fact that $\boldsymbol{\omega}$ is changing. Equivalently, it arises from the fact that P is accelerating up away from the table.

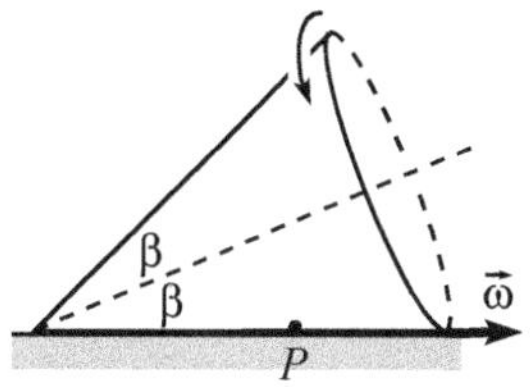

Fig. 10.17

(a) Find the acceleration of P.
(b) Use Eq. (10.11) to calculate the azimuthal force on a mass m located at P, and show that the result is consistent with the acceleration you found in part (a).

10.11. **Unwinding string** ****

A wheel with radius R is placed flat on a table. A massless string with one end attached to the rim of the wheel is wrapped clockwise around the wheel a large number of times. When the string is wrapped completely around the wheel, a point mass m is attached to the free end and glued to the wheel. The wheel is then made to rotate with constant angular speed ω. At some point, the glue on the mass breaks, and the mass and string gradually unwind (with the speed of the wheel kept constant at ω by a motor, if necessary). Show that the length of the unwound string increases at the constant rate $R\omega$, for both the clockwise *and* counterclockwise directions for ω. (The latter is the tricky one.)

10.12. **Shape of the earth** ****

The earth bulges slightly at the equator, due to the centrifugal force in the earth's rotating frame. The goal of this exercise is to find the shape of the earth, first incorrectly, and then correctly.

(a) The common incorrect method is to assume that the gravitation force from the slightly nonspherical earth points toward the center, and to then calculate the equipotential surface (incorporating both the gravitational and centrifugal forces). Show that this method leads to a surface whose height (relative to a spherical earth of the same volume) is given by

$$h(\theta) = R\left(\frac{R\omega^2}{6g}\right)(3\sin^2\theta - 2), \qquad (10.40)$$

where θ is the polar angle (the angle down from the north pole), and R is the radius of the earth.

(b) The above method is incorrect, because the slight distortion of the earth causes the gravitational force to *not* point toward the center of the earth (except at the equator and the poles). This tilt in the force direction then changes the slope of the equipotential surface, and it turns out (although this is by no means obvious) that this effect is of the same order as the slope of the surface found in part (a). Your task: Assuming that the density of the earth is constant,[11] and that the correct height takes the form of some constant factor f times the result found in part (a),[12] show that $f = 5/2$.[13] Do this by demanding that the potential at a pole equals the potential at the equator. Feel free to do things numerically.

10.13. **Southward deflection** ****

A ball is dropped from height h (small compared with the radius of the earth) at a polar angle θ. Assume (incorrectly) that the earth is a perfect sphere. Show that a second-order Coriolis effect leads to a *southward* deflection (in the northern hemisphere) equal to $(2/3)(\omega^2 h^2/g)\sin\theta\cos\theta$.[14]

It turns out that the actual southward deflection is larger than this; it equals $4(\omega^2 h^2/g)\sin\theta\cos\theta$. So apparently there are other effects at work. The main task of this problem is to show how the factor of $2/3$ turns into a factor of 4. In what follows, we will keep terms up to order ω^2 (or technically $\omega^2 R/g$) and order h/R. Also, it will be easiest to calculate southward distances relative to the point on the ground

[11] This isn't true, but it's a very difficult problem to solve with a nonconstant density, assuming that we even know what the density is as a function of radius.

[12] Satellite data show that the general shape found in Eq. (10.40) is essentially correct, up to some factor. This is known as the *quadrupole* shape; the $(3\sin^2\theta - 2)$ term is often written as $(1 - 3\cos^2\theta)$. I can't think of a clean theoretical way to justify the assumption that the shape is of this form, so let's just accept it. It still makes for a very nice problem.

[13] If Δh is the difference between the equatorial and polar radii, then this factor of $5/2$ takes the incorrect Δh of about 11 km (as you can show) in Eq. (10.40) and turns it into about 28 km, which is reasonably close to the observed value of 21.5 km. The discrepancy arises from the fact that the density of the earth isn't constant; it decreases with the radius.

[14] By this we mean that the Coriolis effect causes the ball to land this far south of the point where a plumb bob (hanging from where the ball is dropped) touches the ground. Measuring the deflection relative to the point on the ground along the radius to the dropping point would be impractical, because there is no way of knowing where the center of the earth is.

along the radius to the dropping point; call this point P. But our final goal will be to determine the southward distance relative to a hanging plumb bob.

(a) Show that the distance between the plumb bob and P is $(\omega^2 Rh/g)\sin\theta\cos\theta(1 - h/R)$.

(b) The fact that the gravitational force decreases with height implies that the ball takes more time than the standard $\sqrt{2h/g}$ to hit the ground. Show that the time equals $\sqrt{2h/g}(1 + 5h/6R)$.

(c) Let y be the height above the ground, and let z be the southward distance away from the radial line to the dropping point. Show that the centrifugal force yields a southward acceleration away from the radial line equal to $\ddot{z} = \omega^2(R + y)\sin\theta\cos\theta$.

(d) Show that the gravitational force yields a northward acceleration back toward the radial line equal to $\ddot{z} = -g(z/R)$.

(e) Combine parts (b), (c), and (d) to show that the centrifugal and gravitational forces lead to a southward deflection away from P equal to $(\omega^2 Rh/g)\sin\theta\cos\theta(1 + 7h/3R)$. Adding on the above Coriolis effect and subtracting off the position of the plumb bob then quickly yields the desired factor of 4. (This problem is based on Belorizky and Sivardiere (1987).)

Section 10.3: Tides

10.14. **Bead on a hoop** **

A bead of mass m is constrained to move on a frictionless hoop of radius r that is located a distance R from an object of mass M. Assume that $R \gg r$, and assume that M is much larger than the mass of the hoop, which is much larger than m.

(a) If the hoop is held fixed and the bead is released from a point close to the rightmost point, as shown in Fig. 10.18, what is the frequency of small oscillations?

(b) If the hoop is released and the bead starts at a point close to the rightmost point, what is the frequency of small oscillations? Assume that you grab M and move it to the right to keep it a distance R from the hoop. (The time scale of oscillations turns out to be on the same order as the time scale of the hoop reaching M, if M stays fixed.)

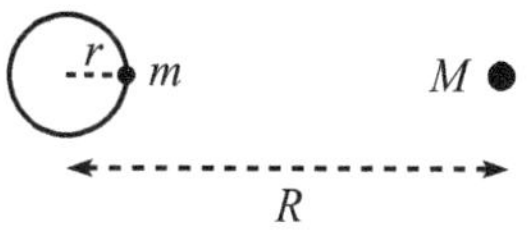

Fig. 10.18

10.15. **Precession of the equinoxes** ***

Because the earth bulges at the equator, and because its axis of rotation is tilted with respect to the plane of the ecliptic (the plane containing the sun and (nearly) the moon), the tidal forces from the sun and the moon produce a torque on the earth, which causes the axis of rotation to precess. The rate of precession is slow; the period is about 26 000 years. Derive (approximately) this result in the following way.[15]

We'll be making some rough approximations here, but our goal is simply to understand intuitively what's going on, and to get an answer that's in the right ballpark. Assume that: (1) the axis of the earth is tilted at 23° with respect to the plane of the ecliptic; (2) the equatorial bulge can be approximated by two point masses located on the equator, at the highest and lowest points relative to the plane of the ecliptic; (3) the mass of each of these effective point masses comes from a patch on the earth with an area of, say, r^2, where r is the radius of the earth (this is just a guess); (4) the patch has an essentially constant height of $h \approx 21$ km, which is the difference between the equatorial and polar radii; (5) the mass density of the patch is roughly the average density of the earth (which isn't true); (6) the earth spends half of its time in the summer/winter orientation, and half in the spring/fall orientation; (7) the moon's tidal effect is twice the sun's. Numerical values of various constants can be found in Appendix J.

10.5 Exercises

Section 10.2: The fictitious forces

10.16. **Swirling down a drain** *

Does the Coriolis force cause the rotation you often see in water going down a drain? That is, does the water rotate in different directions in the northern and southern hemispheres? A rough order-of-magnitude argument is sufficient. See Shapiro (1962) for a discussion of this.

10.17. **Magnitude of $\mathbf{g}_{\text{eff}}$** *

Consider (unrealistically) a perfectly spherical rotating planet whose g value is constant over the surface. What is the magnitude of $\mathbf{g}_{\text{eff}}$ as a function of θ? Give your answer to the leading-order correction in ω.

10.18. **Oscillations across the equator** *

Consider (unrealistically) a perfectly spherical rotating planet whose g value is constant over the surface. A bead lies on a frictionless wire that lies in the north–south direction across the equator. The wire takes the form of an arc of a circle; all points are the same distance from the center of the earth. The bead is released from rest, a short distance from the

[15] It's possible to derive this by taking into account the exact shape of the earth, but this is very complicated. It's also possible to derive it to a good approximation by treating the earth's equatorial bulge as a thin ring of mass at the equator, but this is likewise fairly involved; see Haisch (1981). So we'll use a simplified point-mass model, which will still give a reasonable answer.

equator. Because $\mathbf{g}_{\text{eff}}$ does not point directly toward the earth's center, the bead will head toward the equator and undergo oscillatory motion. What is the frequency of these oscillations?

10.19. **Circular pendulum** *

Consider a circular pendulum of mass m and length ℓ, as shown in Fig. 10.19. The mass swings around in a horizontal circle with the (massless) string always making an angle θ with the vertical. Find the angular speed, ω, of the mass. Solve this in:

(a) The lab frame. Draw the free-body diagram for the mass, and write down the $F = ma$ equations in the vertical and horizontal directions.

(b) The rotating frame of the pendulum. Draw the free-body diagram for the mass, and write down the $F = ma$ equations in the vertical and horizontal directions.

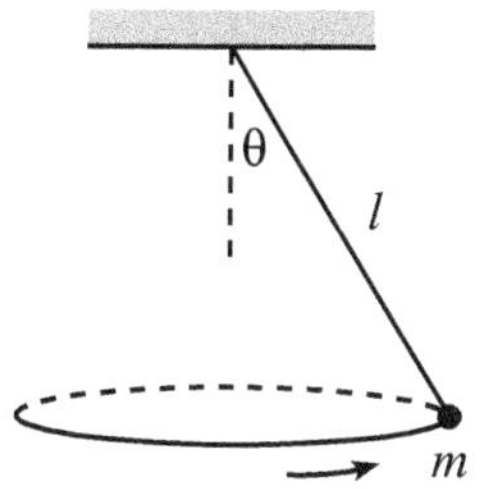

Fig. 10.19

10.20. **Spinning bucket** **

An upright bucket of water is spun at frequency ω around its vertical symmetry axis. If the water is at rest with respect to the bucket, find the shape of the water's surface.

10.21. **Corrections to gravity** **

A mass is dropped from a point directly above the equator. Consider the moment when the object has fallen a distance d. If we consider only the centrifugal force, then you can quickly show that the correction to g_{eff} at this point (relative to the release point) is an increase by $\omega^2 d$. There is, however, also a second-order Coriolis effect. What is the sum of these corrections?[16]

10.22. **Bug on a hoop** **

A hoop of radius R is made to rotate at constant angular speed ω around a diameter, as shown in Fig. 10.20. A small bug of mass m walks at constant angular speed Ω around the hoop. Let $\mathbf{F}$ be the total force that the hoop applies to the bug when the bug is at the angle θ shown, and let $F_\perp$ be the component of $\mathbf{F}$ that is perpendicular to the plane of the hoop. Find $F_\perp$ in two ways (ignore gravity in this problem):

(a) Work in the lab frame: At the angle θ, find the rate of change of the bug's angular momentum around the rotation axis, and then consider the torque on the bug.

(b) Work in the rotating frame of the hoop: At the angle θ, find the relevant fictitious force, and then take it from there.

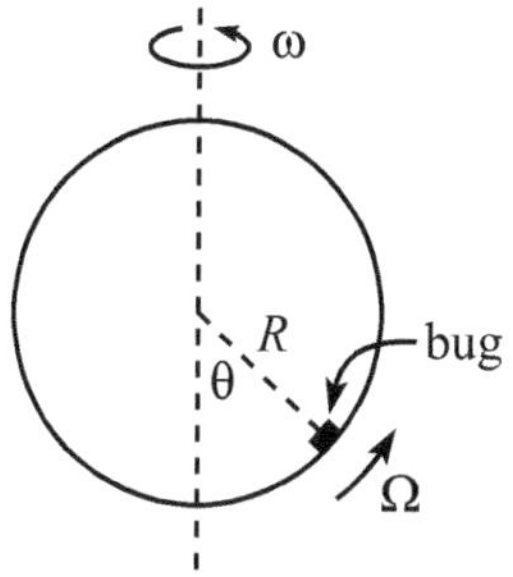

Fig. 10.20

[16] The value of g also varies with height, and this produces an increase in g_{eff} equal to $g(2d/R)$. You can show that this is much larger than the above centrifugal and Coriolis effects.

10.23. **Maximum normal force** **

A frictionless hoop of radius R is made to rotate at constant angular speed ω around a diameter. A bead on the hoop starts on this diameter and is then given a tiny kick. Let $\mathbf{N}$ be the total force that the hoop applies to the bead, and let $N_\perp$ be the component of $\mathbf{N}$ that is perpendicular to the plane of the hoop. Where is $N_\perp$ maximum? What is the magnitude of $\mathbf{N}$ as a function of position? (Ignore gravity in this problem.)

10.24. **Projectile with Coriolis** **

At a polar angle θ, a projectile is fired eastward at an inclination angle α above the ground. Find the westward and southward deflections due to the Coriolis force. In terms of θ, what angle α_{max} yields the maximum total distance of deflection? What is α_{max} when θ equals 60°, 45°, and (approximately) 0°? What about values of θ larger than 60°?

10.25. **Free-particle motion** **

A particle slides on a frictionless turntable that rotates counterclockwise with constant frequency ω. When viewed in an inertial frame, the particle simply travels in a straight line. But in the rotating frame of the turntable, show that the $F = ma$ equations take the form,

$$\begin{aligned} \ddot{x} &= \omega^2 x + 2\omega\dot{y}, \\ \ddot{y} &= \omega^2 y - 2\omega\dot{x}, \end{aligned} \tag{10.41}$$

and verify that the solutions to these differential equations are[17]

$$\begin{aligned} x(t) &= (A + Bt)\cos\omega t + (C + Dt)\sin\omega t, \\ y(t) &= -(A + Bt)\sin\omega t + (C + Dt)\cos\omega t. \end{aligned} \tag{10.42}$$

10.26. **Coin on a turntable** ***

A coin stands upright at an arbitrary point on a rotating turntable, and spins (without slipping) at the required angular speed to make its center remain motionless in the lab frame. In the frame of the turntable, the coin rolls around in a circle with the same frequency as that of the turntable. In the frame of the turntable, show that

(a) $\mathbf{F} = d\mathbf{p}/dt$, and
(b) $\boldsymbol{\tau} = d\mathbf{L}/dt$ (*Hint*: Coriolis).

10.27. **Precession viewed from rotating frame** ***

Consider a top made of a wheel with all its mass on the rim. A massless rod (perpendicular to the plane of the wheel) connects the CM to a pivot.

[17] If you want, you can derive these solutions in the spirit of Chapter 4 by guessing exponential solutions to the $F = ma$ equations. You will find that there are degenerate solutions, which lead to the $(A + Bt)$ type terms that we saw in the critically damped case in Section 4.3.

Initial conditions have been set up so that the top undergoes precession, with the rod always horizontal. In the language of Fig. 9.30, we may write the angular velocity of the top as $\boldsymbol{\omega} = \Omega\hat{\mathbf{z}} + \omega'\hat{\mathbf{x}}_3$ (where $\hat{\mathbf{x}}_3$ is horizontal here).

Consider things in the frame rotating around the $\hat{\mathbf{z}}$ axis with angular speed Ω. In this frame, the top spins with angular speed ω' around its *fixed* symmetry axis. Therefore, in this frame we must have $\boldsymbol{\tau} = 0$, because $\mathbf{L}$ is constant. Verify explicitly that $\boldsymbol{\tau} = 0$ (calculated with respect to the pivot) in this rotating frame (you will need to find the relation between ω' and Ω). In other words, show that the torque due to gravity is exactly canceled by the torque due to the Coriolis force (you can quickly show that the centrifugal force provides no net torque).

Section 10.3: Tides

10.28. **Maximum tangential force** *

At what angle from the horizontal is the tangential component of the tidal force in Eq. (10.35) maximum?

10.29. **Bead on a hoop** **

A bead of mass m is constrained to move on a frictionless hoop of radius r that is located a distance R from an object of mass M. Assume $R \gg r$, and assume that M is much larger than the mass of the hoop, which is much larger than m.

(a) If the hoop is held fixed and the bead is released from the top point, as shown in Fig. 10.21, what is its speed when it gets to the rightmost point on the hoop?
(b) If the hoop is released and the bead starts at a point just slightly to the right of the top point, what is its speed with respect to the hoop when it gets to the rightmost point on the hoop? Assume that you grab M and move it to the right to keep it a distance R from the hoop.

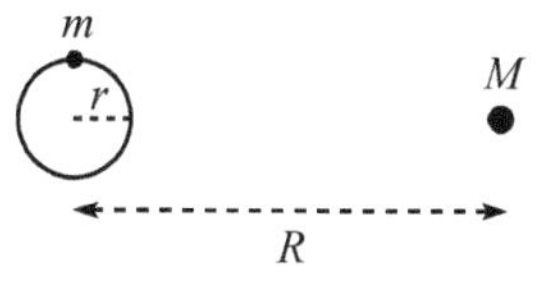

Fig. 10.21

10.30. **Facing the planet** **

A bead of mass m is constrained to move on a frictionless hoop of radius r that orbits a planet of mass M, a distance R from it. Initial conditions have been set up so that the plane of the hoop is always perpendicular to the line from it to the planet. Assume $R \gg r$, and assume that the mass of the hoop is much larger than m. Draw the force lines in the reference frame of the hoop (in the spirit of Fig. 10.14). If the bead is released from a point close to the front point on the hoop, what is the frequency of small oscillations?

10.31. **Roche limit** **

A small spherical rock covered with sand falls in radially toward a planet. Let the planet have radius R and density $\rho_{\rm p}$, and let the rock have density $\rho_{\rm r}$. It turns out that when the rock gets close enough to the planet, the tidal force ripping the sand off the rock will be larger than the gravitational force attracting the sand to the rock. The cutoff distance is called the Roche limit.[18] Show that it is given by (note the lack of dependence on the rock's radius)

$$d = R\left(\frac{2\rho_{\rm p}}{\rho_{\rm r}}\right)^{1/3}. \qquad (10.43)$$

10.32. **Roche limit with rotation** **

If an object orbits a planet in a nonspinning manner (such as the seat of a ferris wheel), then the Roche limit is the same as for the radially falling object in the previous exercise (as you can show). However, show that if an object orbits a planet in such a way that the same side always faces the planet, then the Roche limit is given by

$$d = R\left(\frac{3\rho_{\rm p}}{\rho_{\rm r}}\right)^{1/3}. \qquad (10.44)$$

10.6 Solutions

10.1. **Which way down?**

There are actually (at least) four possible definitions for the point "below" you on the ground: (1) the point that lies along the line between you and the center of the earth, (2) the point that lies along the direction of the earth's gravitational force, (3) the point where a hanging plumb bob rests (that is, the point that lies along the direction of the effective gravitational force), and (4) the point where a dropped object hits the ground.

The third definition is the most reasonable, because it defines the upward direction in which buildings are constructed. At any rate, the third and fourth definitions are the only ones you can make any practical use of. The third differs from the second due to the centrifugal force, which makes $\mathbf{g}_{\rm eff}$ point in a slightly southward direction (in the northern hemisphere) relative to the gravitational force $\mathbf{g}$. It additionally differs from the first due to the fact that $\mathbf{g}$ isn't radial (see the first remark at the end of Section 10.2.2). And it differs from the fourth due to the Coriolis force, which causes a falling object to be deflected slightly eastward. Note that all four definitions are equivalent at the poles. And the first three are equivalent at the equator.

10.2. **Longjumping in $\mathbf{g}_{\rm eff}$**

Let the jumper take off with speed v, at an angle θ. The time to the top of the motion is given by $g_{\rm eff}(t/2) = v\sin\theta$, so the total time is $t = 2v\sin\theta/g_{\rm eff}$. The distance

[18] The Roche limit gives the radial distance below which loose objects won't collect into larger blobs. Our moon (which is a sphere of rock and sand) lies outside the earth's Roche limit. But Saturn's rings (which consist of loose ice particles) lie inside its Roche limit.

traveled is therefore the standard

$$d = v_x t = vt\cos\theta = \frac{2v^2\sin\theta\cos\theta}{g_{\text{eff}}} = \frac{v^2\sin 2\theta}{g_{\text{eff}}}. \tag{10.45}$$

This is maximum when $\theta = \pi/4$, as we well know. So we see that $d \propto 1/g_{\text{eff}}$. Taking $g_{\text{eff}} \approx 10\,\text{m/s}^2$ at the north pole, and $g_{\text{eff}} \approx (10 - 0.05)\,\text{m/s}^2$ at the equator, we find that the jump at the equator is approximately 1.005 times as long as the one at the north pole. So the longjumper gains about four centimeters. This would be completely washed out by even the tiniest wind effect.

REMARK: For a longjumper, the optimal angle of takeoff is undoubtedly not $\pi/4$. The act of changing the direction abruptly from horizontal to such a large angle would entail a significant loss in speed. The optimal angle is some hard-to-determine angle less than $\pi/4$. But this won't change our general $d \propto 1/g_{\text{eff}}$ result (which follows from dimensional analysis). However, we've also made the assumption that the CM of the longjumper starts and ends at the same height, which is definitely not true in longjumping; it is lower at the end. This in fact does change the $d \propto 1/g_{\text{eff}}$ result. But using the result from Problem 3.17, we see that the effect is small (using values of $h \approx 1$ m and $v \approx 10$ m/s). ♣

10.3. **$\mathbf{g}_{\text{eff}}$ vs. g**

The forces $m\mathbf{g}$ and $\mathbf{F}_{\text{cent}}$ are shown in Fig. 10.22. The magnitude of $\mathbf{F}_{\text{cent}}$ is $mR\omega^2\sin\theta$, so the component of $\mathbf{F}_{\text{cent}}$ perpendicular to $m\mathbf{g}$ is $mR\omega^2\sin\theta\cos\theta = mR\omega^2(\sin 2\theta)/2$. For small $\mathbf{F}_{\text{cent}}$, maximizing the angle between $\mathbf{g}_{\text{eff}}$ and $\mathbf{g}$ is equivalent to maximizing this perpendicular component. Therefore, we obtain the maximum angle when $\sin 2\theta = 1 \Longrightarrow \theta = \pi/4$. The maximum angle turns out to be

$$\phi \approx \tan\phi \approx \big(mR\omega^2(\sin\pi/2)/2\big)/mg = R\omega^2/2g \approx 1.7\cdot 10^{-3}, \tag{10.46}$$

which is about 0.1°. Because the earth bulges at the equator, the distance from the axis isn't exactly $R\sin\theta$, the angle of $\mathbf{g}$ isn't exactly θ (see the first remark at the end of Section 10.2.2), and the magnitude of $\mathbf{g}$ isn't exactly constant over the surface of the earth. But these effects are negligible, and the optimal θ is still essentially $\pi/4$.

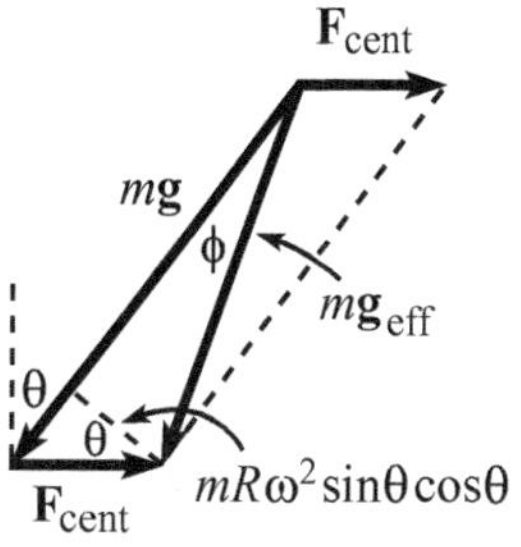

Fig. 10.22

REMARK: The above solution is an approximate one which is valid only when the magnitude of $\mathbf{F}_{\text{cent}}$ is much smaller than mg. We'll now give an exact solution which is valid when the magnitude of $\mathbf{F}_{\text{cent}}$ is comparable to mg, but which is valid only in the unrealistic case where the planet is a perfect sphere, even though it is spinning. No planet would ever take a spherical shape, because planets aren't rigid. But you could imagine a large spherical rock.

To solve the problem exactly, we can break $\mathbf{F}_{\text{cent}}$ into components parallel and perpendicular to $\mathbf{g}$ and make use of the parallel component, in addition to the perpendicular component we used above. If ϕ is the angle between $\mathbf{g}_{\text{eff}}$ and $\mathbf{g}$, then from Fig. 10.22 we have

$$\tan\phi = \frac{mR\omega^2\sin\theta\cos\theta}{mg - mR\omega^2\sin^2\theta}. \tag{10.47}$$

We can then maximize ϕ by taking a derivative. But we need to be careful if $R\omega^2 > g$, in which case maximizing ϕ doesn't mean maximizing $\tan\phi$. You can work this out, and we'll instead give the following slick geometric solution.

In Fig. 10.23, draw the $\mathbf{F}_{\text{cent}}$ vectors for various θ, relative to $m\mathbf{g}$ (so we've chosen $m\mathbf{g}$ always to be vertical in this figure, in contrast with $\mathbf{F}_{\text{cent}}$ always being horizontal in Fig. 10.22). Since the lengths of the $\mathbf{F}_{\text{cent}}$ vectors are proportional to $\sin\theta$, you can show that the tips of the $\mathbf{F}_{\text{cent}}$ vectors form a circle. The maximum ϕ is therefore achieved when $\mathbf{g}_{\text{eff}}$ is tangent to this circle, as shown in Fig. 10.24. In the limit where $g \gg R\omega^2$ (that is, in the limit of a small circle), we want the point of tangency to

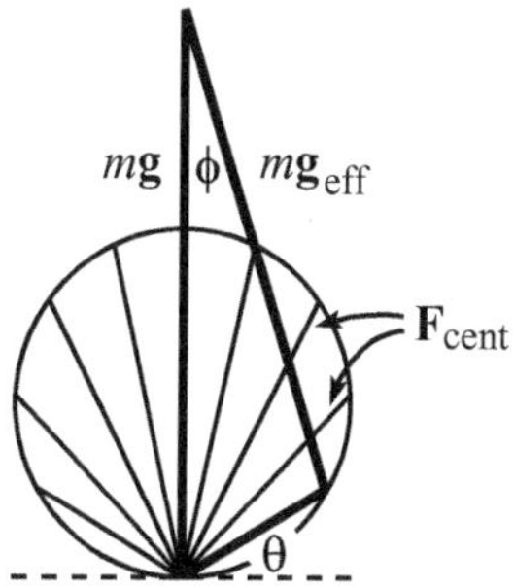

Fig. 10.23

be the rightmost point on the circle, so the maximum ϕ is achieved when $\theta = \pi/4$, in which case $\phi \approx \sin\phi \approx (R\omega^2/2)/g$, as we found above. But in the general case, Fig. 10.24 shows that the maximum ϕ is given by

$$\sin\phi_{\max} = \frac{\frac{1}{2}mR\omega^2}{mg - \frac{1}{2}mR\omega^2}. \tag{10.48}$$

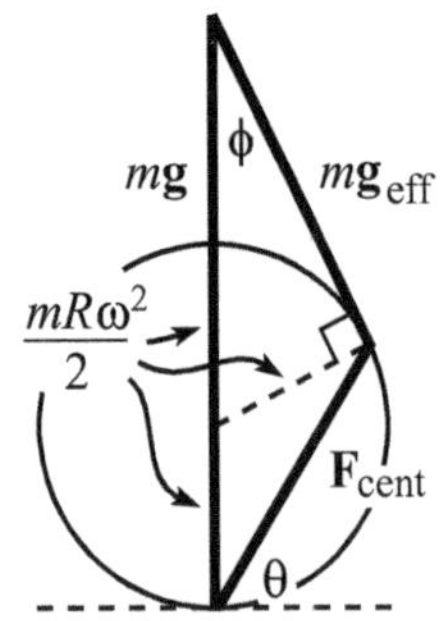

Fig. 10.24

In the limit of small ω, this is approximately $R\omega^2/2g$, as above. Note that this reasoning holds only if $R\omega^2 < g$. In the case where $R\omega^2 > g$ (that is, the circle extends above the top end of the mg segment), the maximum ϕ is simply π, and it is achieved at $\theta = \pi/2$. ♣

10.4. **Lots of circles**

(a) The fictitious force, $\mathbf{F}_f$, on the mass has an $\mathbf{F}_{\text{cent}}$ part and an $\mathbf{F}_{\text{trans}}$ part, because the center of C_2 is moving. So the fictitious force is

$$\mathbf{F}_f = m\omega^2\mathbf{r}_2 + \mathbf{F}_{\text{trans}}, \tag{10.49}$$

where $\mathbf{r}_2$ is the position of the mass in the frame of C_2. But $\mathbf{F}_{\text{trans}}$, which arises from the acceleration of the center of C_2, equals the centrifugal force felt by a point on C_1. Therefore,

$$\mathbf{F}_{\text{trans}} = m\omega^2\mathbf{r}_1, \tag{10.50}$$

where $\mathbf{r}_1$ is the position of the center of C_2 in the frame of C_1. Substituting this into Eq. (10.49) gives

$$\mathbf{F}_f = m\omega^2(\mathbf{r}_2 + \mathbf{r}_1) = m\omega^2\mathbf{R}(t). \tag{10.51}$$

(b) The fictitious force, $\mathbf{F}_f$, on the mass has an $\mathbf{F}_{\text{cent}}$ part and an $\mathbf{F}_{\text{trans}}$ part, because the center of the Nth circle is moving. So the fictitious force is

$$\mathbf{F}_f = m\omega^2\mathbf{r}_N + \mathbf{F}_{\text{trans},N}, \tag{10.52}$$

where $\mathbf{r}_N$ is the position of the mass in the frame of C_N. But $\mathbf{F}_{\text{trans},N}$ equals the centrifugal force felt by a point on the $(N-1)$th circle, plus the translation force coming from the movement of the center of the $(N-1)$th circle. Therefore,

$$\mathbf{F}_{\text{trans},N} = m\omega^2\mathbf{r}_{N-1} + \mathbf{F}_{\text{trans},N-1}. \tag{10.53}$$

Substituting this into Eq. (10.52) and successively rewriting the $\mathbf{F}_{\text{trans},i}$ terms in a similar manner, gives

$$\mathbf{F}_f = m\omega^2(\mathbf{r}_N + \mathbf{r}_{N-1} + \cdots + \mathbf{r}_1) = m\omega^2\mathbf{R}(t). \tag{10.54}$$

This turns out so clean because $\mathbf{F}_{\text{cent}}$ is linear in $\mathbf{r}$, and also because all the ω's are the same.

Remark: A much easier way to see that $\mathbf{F}_f = m\omega^2\mathbf{R}(t)$ is the following. Since all the circles rotate with the same ω, they may as well all be glued together. Such a rigid setup does indeed yield the same ω for all the circles, just as in the case of the moon rotating once on its axis for every revolution it makes around the earth, thereby causing the same side to always face the earth. It is then clear that the mass simply moves in a circle at frequency ω, yielding a fictitious centrifugal force of $m\omega^2\mathbf{R}(t)$. And as a bonus, we see that the magnitude of $\mathbf{R}(t)$ is constant. ♣

10.5. **Mass on turntable**

In the lab frame, the net force on the mass is zero, because it is sitting at rest. (The normal force cancels the gravitational force.) But in the rotating frame, the mass travels in a circle of radius r with frequency ω. So its speed is $v = \omega r$. Therefore, in the rotating frame there must be a net force of $mv^2/r = m\omega^2 r$ inward to account for the centripetal acceleration. And indeed, the mass feels a centrifugal force of $m\omega^2 r$ outward, and a Coriolis force of $2m\omega v = 2m\omega^2 r$ inward, which sum to the desired force (see Fig. 10.25).

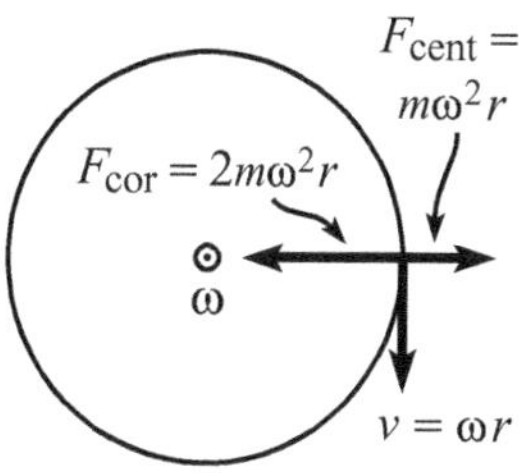

Fig. 10.25

REMARK: The net inward force in this problem is a little different from the force on a person swinging around in a circle of radius r and frequency ω in an inertial frame. If a skater, for example, maintains a circular path by holding a rope whose other end is attached to a pole, then she has to use her muscles to maintain the position of her torso with respect to her arm, and her head with respect to her torso, etc. But if a person takes the place of the mass in the present problem, then she doesn't need to exert any effort at all to keep her body moving in the circle (which is clear, when viewed from the inertial frame), because each atom in her body is moving at (essentially) the same speed and radius, and therefore feels the same centrifugal and Coriolis forces. So she doesn't really *feel* the net force of $m\omega^2 r$, in the same sense that someone doesn't feel gravity when in free-fall with no air resistance, because gravity acts on each bit of mass in the same way. (As mentioned on page 462, this similarity with gravity is what led Einstein to his Equivalence Principle.) ♣

10.6. **Released mass**

Let the x' and y' axes of the rotating frame coincide with the x and y axes of the inertial frame at the moment the mass is released (at $t = 0$). Let the mass initially be located on the x' axis. Then after a time t, the situation looks like that in Fig. 10.26. The speed of the mass is $v = a\omega$, so it has traveled a distance $a\omega t$. The angle that its position vector makes with the inertial x axis is therefore $\tan^{-1} \omega t$, with counterclockwise taken to be positive. Hence, the angle that its position vector makes with the rotating x' axis is

$$\theta(t) = -(wt - \tan^{-1} \omega t). \tag{10.55}$$

And the radius is

$$r(t) = a\sqrt{1 + \omega^2 t^2}. \tag{10.56}$$

For large t, we have $r(t) \approx a\omega t$ and $\theta(t) \approx -wt + \pi/2$, which make sense because the particle approaches an inertial-frame angle of $\pi/2$.

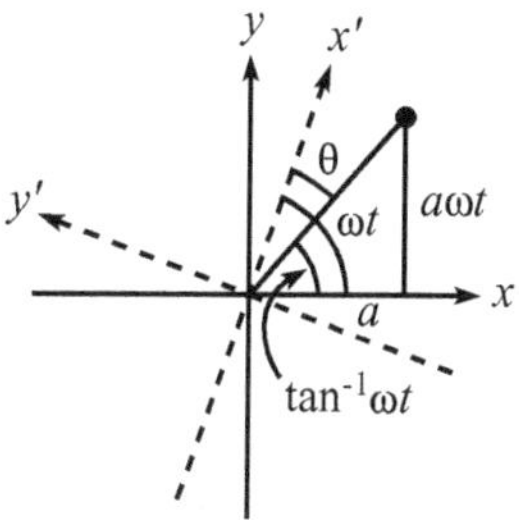

Fig. 10.26

10.7. **Coriolis circles**

By construction (with the surface being orthogonal to $\mathbf{g}_{\text{eff}}$ at all points), the normal force from the ice exactly cancels all effects of the gravitational and centrifugal forces in the rotating frame of the earth. We therefore need only concern ourselves with the Coriolis force, $-2m\boldsymbol{\omega} \times \mathbf{v}$.

Let the angle down from the north pole be θ. We're assuming that the circle is small enough so that θ is essentially constant throughout the motion. The component of the Coriolis force that points horizontally along the surface has magnitude $f = 2mv(\omega \cos\theta)$ and is perpendicular to the direction of motion. (The vertical component of the Coriolis force, which comes from the component of $\boldsymbol{\omega}$ that points along the surface, simply modifies the required normal force.) Because this force is perpendicular to the direction of motion, v does not change. Therefore, $f = 2mv\omega\cos\theta$ is constant. But a constant force perpendicular to the motion of a particle produces a circular path.[19] The radius of the circle is given by

$$2mv\omega\cos\theta = \frac{mv^2}{r} \implies r = \frac{v}{2\omega\cos\theta}. \tag{10.57}$$

The frequency of the circular motion is

$$\omega' = \frac{v}{r} = 2\omega\cos\theta. \tag{10.58}$$

[19] If you want to be mathematical about this, see Footnote 34 in Chapter 9.

To get a rough idea of the size of the circle, you can show (using $\omega \approx 7.3 \cdot 10^{-5}\,\mathrm{s}^{-1}$) that $r \approx 10\,\mathrm{km}$ when $v = 1\,\mathrm{m/s}$ and $\theta = 45°$. Even the tiniest bit of friction will clearly make this effect impossible to see. In the special case of $\theta \approx \pi/2$ (that is, near the equator), the component of the Coriolis force along the surface is negligible, so r becomes large, and ω' goes to 0.

REMARK: In the limit $\theta \approx 0$ (that is, near the north pole), the Coriolis force essentially points along the surface. The above equations give $r \approx v/(2\omega)$, and $\omega' \approx 2\omega$. For the special case where the center of the circle is the north pole, this $\omega' \approx 2\omega$ result might seem incorrect, because you might want to say that the circular motion should be achieved by having the puck remain motionless in the inertial fame, while the earth rotates beneath it (thus making $\omega' = \omega$). The error in this reasoning is that the "level" earth is not spherical, due to the nonradial direction of $\mathbf{g}_{\text{eff}}$. If the puck starts out motionless in the inertial frame, it will be drawn toward the north pole, due to the component of the gravitational force along the nonspherical "level" earth. In order not to fall toward the pole, the puck needs to travel with frequency ω (relative to the inertial frame) in the direction opposite[20] to the earth's rotation. The reason for this is that in the rotating frame of the puck, the puck feels the same centrifugal force that it would feel if it were at rest in the frame of the earth, spinning along with it, because these two frames have the same magnitude of ω; the $\boldsymbol{\omega}$'s point in opposite directions, but this doesn't affect the centrifugal force. The puck therefore happily stays at the same θ value on the "level" surface, just as a puck at rest on the earth does. The angular velocity of the puck with respect to the earth is therefore $(-\omega) - (\omega) = -2\omega$, where the minus sign signifies the backward direction. ♣

10.8. $\boldsymbol{\tau} = d\mathbf{L}/dt$ *

Let $\mathbf{r}'_i$ be the position vector in the accelerating frame. (In terms of the quantities in Section 8.4.3, $\mathbf{r}'_i$ equals $\mathbf{r}_i - \mathbf{r}_0$.) The total angular momentum of an object in the accelerating frame is

$$\mathbf{L} = \sum_i \mathbf{r}'_i \times m_i \dot{\mathbf{r}}'_i. \tag{10.59}$$

Therefore,

$$\begin{aligned}\frac{d\mathbf{L}}{dt} &= \sum_i \dot{\mathbf{r}}'_i \times m_i \dot{\mathbf{r}}'_i + \sum_i \mathbf{r}'_i \times m_i \ddot{\mathbf{r}}'_i \\ &= 0 + \sum_i \mathbf{r}'_i \times \mathbf{F}_i^{\text{total}} \\ &= \sum_i \mathbf{r}'_i \times \left(\mathbf{F}_i^{\text{real,ext}} + \mathbf{F}_i^{\text{real,int}} + \mathbf{F}_i^{\text{fictitious}}\right).\end{aligned} \tag{10.60}$$

The first term, $\sum \mathbf{r}'_i \times \mathbf{F}_i^{\text{real,ext}}$, equals the total external torque measured relative to the origin of the accelerating frame, as desired. The second term, $\sum \mathbf{r}'_i \times \mathbf{F}_i^{\text{real,int}}$, equals the total torque from internal forces, which is zero by the same reasoning as in Section 8.4.3. The third term, $\sum \mathbf{r}'_i \times \mathbf{F}_i^{\text{fictitious}}$, is the tricky one. Since the frame isn't rotating, we have at most the fictitious translation force. So this term equals

$$\sum \mathbf{r}'_i \times (-m_i \ddot{\mathbf{r}}_0) = -\sum m_i \mathbf{r}'_i \times \ddot{\mathbf{r}}_0 = -M\mathbf{r}'_{\text{CM}} \times \ddot{\mathbf{r}}_0, \tag{10.61}$$

where $\mathbf{r}'_{\text{CM}}$ is the position of the object's CM in the accelerating frame, and $\mathbf{r}_0$ is the position of the origin of the frame with respect to the inertial lab frame. This result

[20] Of course, the puck can also move with frequency ω in the *same* direction as the earth's rotation. But in this case, the puck simply sits at one spot on the earth.

agrees with the second term in Eq. (8.45), where $\mathbf{r}'_{\text{CM}}$ is written as $\mathbf{R} - \mathbf{r}_0$. The three conditions under which the third term vanishes are therefore: (1) $\mathbf{r}'_{\text{CM}} = 0$, that is, the CM is located at the origin of the accelerating frame. The fictitious translation force acts just like a gravitational force, so as far as the torque is concerned, the translation force acts at the CM. Therefore, if the CM is located at the origin, the translation force has no lever arm and thus provides no torque. (2) $\ddot{\mathbf{r}}_0 = 0$, that is, the origin is not accelerating, so there is no translation force. (3) $\mathbf{r}'_{\text{CM}}$ is parallel to $\ddot{\mathbf{r}}_0$. This means that if the translation force is considered to be a gravitational force, then the CM lies directly "above" or "below" the origin. So there is no lever arm and thus no torque.

If the frame is rotating in addition to translating, then in general there will be torques from the centrifugal, Coriolis, and azimuthal forces, even if the CM is chosen as the origin. This is true because these fictitious forces involve $\mathbf{r}$ (or $\dot{\mathbf{r}}$), which results in the torque not being linear in $\mathbf{r}$ (because there is already an $\mathbf{r}$ in $\boldsymbol{\tau} = \mathbf{r} \times \mathbf{F}$). This then means that the $\mathbf{r}'_{\text{CM}}$ position vector doesn't arise in the calculation of $d\mathbf{L}/dt$ as it did in Eq. (10.61).

10.9. **Determining your frame**

Yes. You can determine ω and $d\omega/dt$ as follows. Simultaneously measure the forces on particles of mass m at rest at positions $\mathbf{r}_1$ and $\mathbf{r}_2$. The centrifugal and azimuthal are the relevant forces, so the difference in the forces at the two locations is

$$\Delta\mathbf{F} \equiv \mathbf{F}_1 - \mathbf{F}_2 = -m\boldsymbol{\omega} \times (\boldsymbol{\omega} \times (\mathbf{r}_1 - \mathbf{r}_2)) - m\frac{d\boldsymbol{\omega}}{dt} \times (\mathbf{r}_1 - \mathbf{r}_2). \qquad (10.62)$$

Using the fact that the cross product of two vectors is perpendicular to each vector, we see that the magnitudes of the components of $\Delta\mathbf{F}$ parallel and perpendicular to $\mathbf{r}_1 - \mathbf{r}_2$ are $F_\parallel = m\omega^2|\mathbf{r}_1 - \mathbf{r}_2|$ and $F_\perp = m(d\omega/dt)|\mathbf{r}_1 - \mathbf{r}_2|$. So we have

$$\omega = \sqrt{\frac{F_\parallel}{m|\mathbf{r}_1 - \mathbf{r}_2|}}, \quad \text{and} \quad \frac{d\omega}{dt} = \frac{F_\perp}{m|\mathbf{r}_1 - \mathbf{r}_2|}. \qquad (10.63)$$

These expressions give ω and $d\omega/dt$ in terms of measured quantities. Note that we needed to measure the force at two points here. Measuring the force of $-m\boldsymbol{\omega} \times (\boldsymbol{\omega} \times \mathbf{r}) - m(d\boldsymbol{\omega}/dt) \times \mathbf{r}$ at only one point $\mathbf{r}$ doesn't give us anything, because we don't yet know where the origin is, so we don't know the value of $\mathbf{r}$. But with two points, the difference $\mathbf{r}_1 - \mathbf{r}_2$ is independent of the location of the origin.

If you want, you can check the result for ω by finding the difference between the forces on two objects at essentially the same location, one of which is at rest and the other of which is moving with velocity $\mathbf{v}$. Because the $\mathbf{r}$ values are the same, only the Coriolis term survives in the difference. This gives $\omega = |\Delta\mathbf{F}|/(2mv)$.

To find the center of the disk, measure the force on a particle at a given location. Break this force into orthogonal components in the ratio of $d\omega/dt$ to ω^2, by drawing a line at an angle of $\tan^{-1}((d\omega/dt)/\omega^2)$ with respect to the force. Whether this line is to the right or to the left of the force depends on whether the $\mathbf{r}_1 - \mathbf{r}_2$ vector above was to the right or to the left of the $\Delta\mathbf{F}$ vector. This line contains the component proportional to ω^2, which is the radial component. The line therefore passes through the center of the disk. So if we repeat the process with the particle at another location (not along the line) and draw a similar line, then the intersection of the two lines is the center of the disk.

REMARK: If we remove the restriction that there aren't any real forces, then it isn't possible to determine the three desired quantities, because someone could claim that ω, $d\omega/dt$, and the location of the center are different from what you found, and he could then just say that there are a bunch of (contrived) real forces that conspire to make the total force be what you observe. ♣

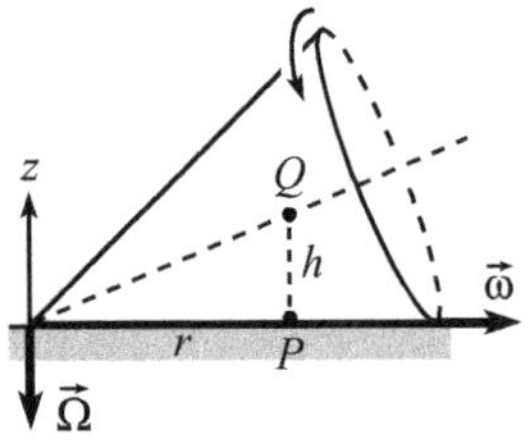

Fig. 10.27

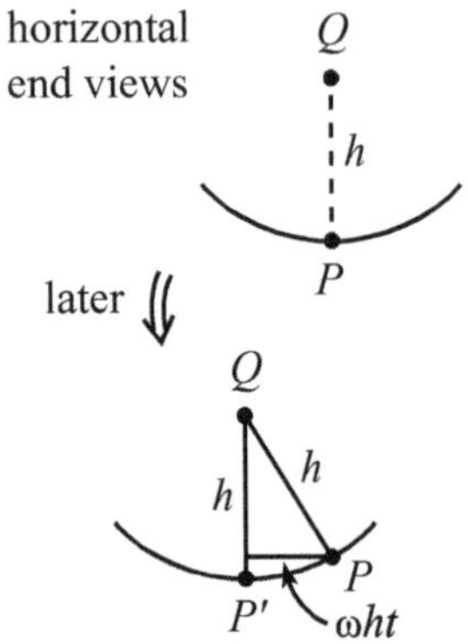

Fig. 10.28

10.10. **Changing ω's direction**

(a) Let Q be the point on the axis of the cone directly above P, and let its height above P be h (see Fig. 10.27). Consider the situation an infinitesimal time t later. Let P' be the point that is now directly below Q (see Fig. 10.28). The angular speed of the cone is ω, so Q moves horizontally with speed $v_Q = \omega h$. Therefore, in the infinitesimal time t, Q moves a distance ωht to the side.

This distance ωht is also (essentially) the horizontal distance between P and P'. Therefore, a little geometry tells us that P is now a height

$$y(t) = h - \sqrt{h^2 - (\omega ht)^2} = h - h\sqrt{1 - (\omega t)^2} \approx \frac{(\omega t)^2 h}{2} = \frac{1}{2}(\omega^2 h)t^2 \tag{10.64}$$

above the table. Since P started on the table with zero speed, this means that P undergoes an acceleration of $\omega^2 h$ in the vertical direction. A mass m located at P must therefore feel a real force (normal force or whatever) of $F_P = m\omega^2 h$ in the upward direction, if it is to remain motionless with respect to the cone.

(b) The precession frequency Ω (which is how fast $\boldsymbol{\omega}$ swings around the origin) is equal to the speed of Q divided by r, where r is the distance from Q to the z axis (that is, the radius of the circle Q travels in). Therefore, $\boldsymbol{\Omega}$ has magnitude $v_Q/r = \omega h/r$, and it points in the $-\hat{\mathbf{z}}$ direction, for the situation shown in Fig. 10.27. Hence, $d\boldsymbol{\omega}/dt = \boldsymbol{\Omega} \times \boldsymbol{\omega}$ has magnitude $\omega^2 h/r$, and it points out of the page in Fig. 10.27 (or to the left in Fig. 10.28). Therefore, $\mathbf{F}_{\text{az}} = -m(d\boldsymbol{\omega}/dt) \times \mathbf{r}$ has magnitude $m\omega^2 h$, and it points in the $-\hat{\mathbf{z}}$ direction.

A person of mass m at point P therefore interprets the situation as, "I am not accelerating with respect to the cone. Therefore, the net force on me in the cone frame is zero. And indeed, the upward normal force F_P from the cone, with magnitude $m\omega^2 h$, is exactly balanced by the mysterious downward force F_{az}, also with magnitude $m\omega^2 h$."

Note that this azimuthal force is still basically the same effect as with the translation force, the centrifugal force, and the simpler case of the azimuthal force discussed in Section 10.2.4. In all of these cases, the "ground" accelerates, so you feel like you are getting flung in the opposite direction with respect to the accelerating frame.

10.11. **Unwinding string**

Consider the clockwise direction. This case is easily solved by working in the lab frame. After the glue breaks, the mass moves in the straight line of the tangent, because the string can't provide a transverse force unless the mass has already diverged from this straight line, which it hasn't. The speed at which the mass starts this straight-line motion (when the glue breaks) is $R\omega$. And it continues to move at this speed because the angular speed of the wheel allows the string to unwind at a rate $R\omega$, which is exactly the rate needed. Note that the string has zero tension in it, so it's just like it's not there.

The counterclockwise direction is tricker, because simple linear motion isn't consistent with the constraint that the mass stays tied to the string. There is now a tension in the string, and the mass undergoes a spiral motion, which makes things more difficult. It's possible to solve this problem with $F = ma$, and also with the Lagrangian method. But we'll solve it here by using a slick argument in a rotating frame. Part of the trickiness of this approach is choosing which rotating frame to use. The most obvious frame is the rotating frame of the wheel, but this frame isn't too helpful, because the mass undergoes a spiral motion which is hard to get a handle on (however, see the fourth remark below). It would be nice to work in a frame where the mass undergoes a simple kind of motion. It turns out that if we consider the frame that rotates counterclockwise at 2ω instead of ω, things turn out to be simple.

The general idea is the following. In this new frame that rotates counterclockwise at 2ω, the wheel rotates *clockwise* at ω. Therefore, if there were no such things as fictitious forces, then we would have exactly the same situation as in the clockwise case we solved above, so we would be done. The bad news, however, is that fictitious forces do exist. And so without any miraculous cancellation, these forces will lead to a transverse force, causing the string's contact point to move one way or another, thereby causing the unwound length to increase at a rate other than $R\omega$. But the good news is that such a miraculous cancellation does indeed occur. Let's see why.

In the frame that rotates counterclockwise at 2ω, consider the moment right after the mass leaves the wheel, as shown in Fig. 10.29. The velocity of the mass at this time is $v = R\omega$, directed to the right. What are the forces on the mass? There may be a tension (it happens to be zero right at the start, but then it grows, as we'll see below), and then there are also the centrifugal and Coriolis forces. The centrifugal force is initially (right when the glue breaks) directed radially upward with magnitude $mR(2\omega)^2 = 4mR\omega^2$. The Coriolis force is (as you can verify) initially directed downward with magnitude $2m(2\omega)v = 2m(2\omega)(R\omega) = 4mR\omega^2$. These two forces cancel, so the mass feels no transverse force, so it continues to move in a straight line to the right.

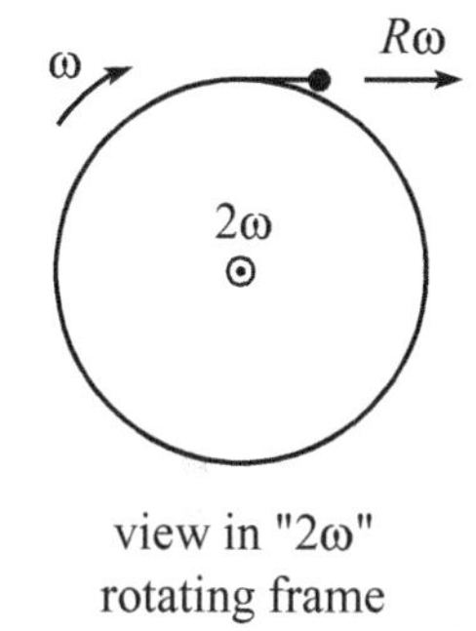

Fig. 10.29

What about a later time? Assume (in an inductive spirit) that the mass is still on the straight line determined by its initial motion and moving with speed $R\omega$. The Coriolis force is still directed downward with magnitude $4mR\omega^2$. And the centrifugal force is directed radially outward with magnitude $mr(2\omega)^2$, where r is the present radius, as shown in Fig. 10.30. The vertical component of this force is obtained by multiplying by R/r, so we again obtain an upward vertical component of $4mR\omega^2$. This therefore cancels the Coriolis force, and we again have the result that there is no transverse force. So we see that if the mass is presently moving in the straight line determined by its initial motion, then it will continue to move in this straight line. And since it is by definition initially moving in this line, we see inductively that it moves in this line for all time, as we wanted to show. Note that the horizontal component of the centrifugal force needs to be canceled by the tension (because the string doesn't stretch), so the tension is nonzero in this scenario, in contrast with the clockwise scenario above. Since the string is always taut, the speed of the mass is determined by the rate at which the wheel rotates, which means that the speed of the mass is always $R\omega$ in this frame. The rate of increase in the unwound string's length is therefore $R\omega$ (in any frame).

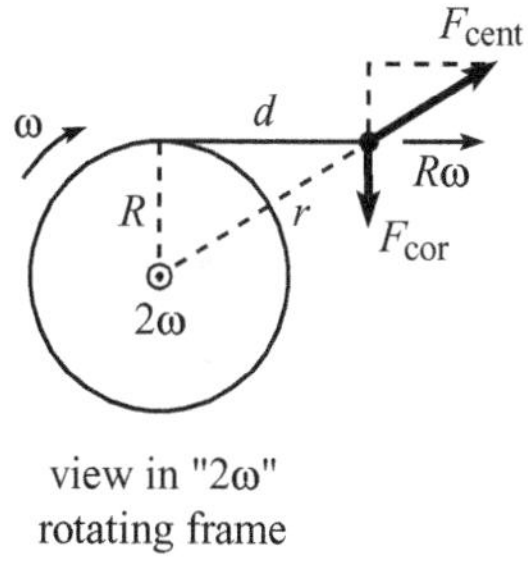

Fig. 10.30

REMARKS:

1. The tension equals the longitudinal component of the centrifugal force. This component is obtained by multiplying the centrifugal force by d/r, where d is the length of unwound string. The tension therefore equals $4md\omega^2$, so it grows linearly with d.
2. Even if the string is massive, and even if the density varies, the unwound length still increases at a rate $R\omega$, because the above reasoning can be used with every atom in the string. This result is a bit surprising, because when viewed in the lab frame, it is by no means obvious that a massive string remains straight.
3. If we hadn't been told in the statement of the problem that the unwound string's length increases at a rate $R\omega$, then the problem would of course be more difficult. For all we'd know, the rate might not even be constant. Aside from making a lucky guess that the rate is $R\omega$, we'd have to solve the problem with $F = ma$ or the Lagrangian method. The former is rather tricky, but the latter isn't so bad.
4. There's actually an even slicker solution for the counterclockwise case than the one we gave above. It goes as follows. First consider the simpler *clockwise* case, and look at the setup in the rotating frame of the wheel. It this case, the centrifugal and Coriolis forces conspire to make the mass move in a backward spiral, the beginning of which is shown in Fig. 10.31. We know from our lab-frame reasoning above that in the lab frame the string provides no force during

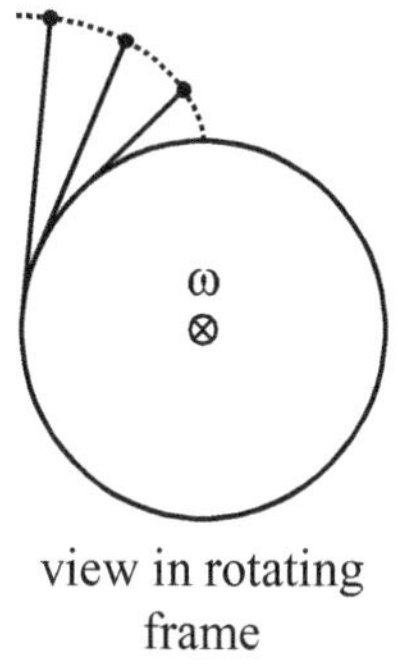

Fig. 10.31

this motion, but that it does in fact remain straight. And the rate of change of the unwound length is $R\omega$.

Now consider the more difficult counterclockwise case in the rotating frame of the wheel. The only difference from the clockwise case is that the $\boldsymbol{\omega}$ vector has switched direction, and now points out of the page. The centrifugal force is therefore still the same function of $\mathbf{r}$, but the Coriolis force has switched signs, as a function of $\mathbf{v}$. However, the centrifugal force is the only force that does work on the mass (because the Coriolis force and the tension are perpendicular to the velocity), so the mass ends up moving at the same speed as a function of $\mathbf{r}$ as in the clockwise case. The string constrains the mass to move along the same path as in the clockwise case (because the string is straight in both cases), and the only effect of the Coriolis force is to increase the tension. The motion is therefore exactly the same (same path, same speed as a function of $\mathbf{r}$), so the rate of change of the unwound length is again $R\omega$. ♣

10.12. **Shape of the earth**

(a) The potential energy function derived from the sum of the gravitational and centrifugal forces must be constant along the surface. Otherwise, a piece of the earth would want to move along the surface, which would mean that we didn't have the correct surface to begin with.

If x is the distance from the earth's axis, then the centrifugal force is $F_c = m\omega^2 x$, directed outward. The potential energy function for this force is $V_c = -m\omega^2 x^2/2$, up to an arbitrary additive constant. Under the assumption that the distortion of the earth doesn't change the gravitational force (which isn't correct, as we'll see below), the potential energy for the gravitation force is mgh, where we've arbitrarily chosen the original spherical surface to correspond to $h = 0$. The equal-potential condition is therefore

$$mgh - \frac{m\omega^2 x^2}{2} = C, \tag{10.65}$$

where C is a constant to be determined. Using $x = r\sin\theta$, we obtain

$$h = \frac{\omega^2 r^2 \sin^2\theta}{2g} + B, \tag{10.66}$$

where $B \equiv C/(mg)$ is another constant. We may replace the r here with the radius of the earth, R, with negligible error.

Depending what the constant B is, this equation describes a whole family of surfaces. We can determine the correct value of B by demanding that the volume of the distorted earth be the same as it would be in its spherical shape if the centrifugal force were turned off. This is equivalent to demanding that the integral of h over the surface of the earth is zero. The integral of $(a\sin^2\theta + b)$ over the surface of the earth is (the integral is easy if we write $\sin^2\theta$ as $1-\cos^2\theta$)

$$\begin{aligned}
&\int_0^\pi \big(a(1-\cos^2\theta) + b\big)\, 2\pi R^2 \sin\theta\, d\theta \\
&= \int_0^\pi \big(-a\cos^2\theta + (a+b)\big)\, 2\pi R^2 \sin\theta\, d\theta \\
&= 2\pi R^2 \left(\frac{a\cos^3\theta}{3} - (a+b)\cos\theta\right)\Bigg|_0^\pi \\
&= 2\pi R^2 \left(-\frac{2a}{3} + 2(a+b)\right).
\end{aligned} \tag{10.67}$$

Therefore, we need $b = -(2/3)a$ for this integral to be zero. Plugging this result into Eq. (10.66) gives

$$h = R\left(\frac{R\omega^2}{6g}\right)(3\sin^2\theta - 2), \tag{10.68}$$

as desired.

(b) For convenience, let the correct height be $h \equiv \beta(3\sin^2\theta - 2)$, with $\beta \equiv fR(R\omega^2/6g)$, where f is the desired fraction.

Consider the earth to be the superposition of the $h = 0$ sphere plus an effective shell of positive or negative mass, depending on the sign of h at a given location. The potential energy of a mass m at a given point on the surface of the distorted earth is the sum of the potentials due to (1) gravity from the sphere, (2) the centrifugal force, and (3) gravity from the shell. From Eq. (10.68), the standard mgh contributions from (1) at a pole and the equator are essentially $mg\beta(-2)$ and $mg\beta(1)$, respectively. The contributions from (2) at a pole and the equator are 0 and $-m\omega^2R^2/2$, respectively. The contributions from (3) are trickier. We need to calculate the integral $-\int Gm\,dM/\ell$, where dM runs over the shell, and ℓ is the distance from m to each dM. The mass of a small element in the shell is

$$dM = \rho\,dV = \rho h\,dA = \rho\beta(3\sin^2\theta - 2)(R\,d\theta)(R\sin\theta\,d\phi). \tag{10.69}$$

The distance from the north pole to a point with polar angle θ equals $\ell = 2R\sin(\theta/2) = R\sqrt{2(1-\cos\theta)}$. Using the Pythagorean theorem, the distance from a point on the equator, say $(R, 0, 0)$, to a point with polar angle θ whose general form is $(R\sin\theta\cos\phi, R\sin\theta\sin\phi, R\cos\theta)$ equals $\ell = R\sqrt{2(1-\sin\theta\cos\phi)}$. Demanding that the total potential at the north pole be equal to the total potential at the equator then gives

$$\begin{aligned} mg\beta(-2) + 0 - \int_0^{\pi}\int_0^{2\pi} \frac{Gm\cdot\rho\beta(3\sin^2\theta - 2)(R\,d\theta)(R\sin\theta\,d\phi)}{R\sqrt{2(1-\cos\theta)}} \\ = mg\beta(1) - \frac{m\omega^2R^2}{2} - \int_0^{\pi}\int_0^{2\pi} \frac{Gm\cdot\rho\beta(3\sin^2\theta - 2)(R\,d\theta)(R\sin\theta\,d\phi)}{R\sqrt{2(1-\sin\theta\cos\phi)}}. \end{aligned} \tag{10.70}$$

Letting $\beta \equiv fR(R\omega^2/6g)$, and using

$$g \equiv \frac{GM_E}{R^2} = \frac{G(4\pi R^3\rho/3)}{R^2} \quad\Longrightarrow\quad G\rho = \frac{3g}{4\pi R}, \tag{10.71}$$

we can rewrite Eq. (10.70) as

$$\begin{aligned} \frac{m\omega^2R^2}{2} &= 3mgfR\left(\frac{R\omega^2}{6g}\right) \\ &- \int_0^{\pi}\int_0^{2\pi}\left(\frac{3g}{4\pi R}\right)mfR\left(\frac{R\omega^2}{6g}\right)(3\sin^2\theta - 2)(R\,d\theta)(R\sin\theta\,d\phi) \\ &\times\left(\frac{1}{R\sqrt{2(1-\sin\theta\cos\phi)}} - \frac{1}{R\sqrt{2(1-\cos\theta)}}\right). \end{aligned} \tag{10.72}$$

After canceling common factors, we end up with

$$\begin{aligned} 1 = f - f\int_0^{\pi}\int_0^{2\pi} \frac{(3\sin^2\theta - 2)\sin\theta}{4\sqrt{2}\pi} \\ \times\left(\frac{1}{\sqrt{1-\sin\theta\cos\phi}} - \frac{1}{\sqrt{1-\cos\theta}}\right)d\phi\,d\theta. \end{aligned} \tag{10.73}$$

Evaluating this integral numerically gives essentially 0.6, which then yields $f = 5/2$, as desired.

REMARK: The $f = 5/2$ result leads to a difference in polar and equatorial radii of $\Delta h = (5/2)(R^2\omega^2/6g)(3) \approx 28\,000$ m $= 28$ km. The fact that this is larger than the correct value of 21.5 km makes sense for the following reason. If all of the earth's mass were concentrated at the center, then the thin distortion shell at the surface would have no effect on the potential, because it would be massless. So the naive calculation in part (a) would in fact be correct, and Δh would be about 11 km. The actual density of the earth decreases with radius, which means that the density lies somewhere between the case of a concentrated center and the case of uniform density. The actual value of Δh should therefore lie somewhere between the corresponding Δh values of 11 km and 28 km. And 21.5 km indeed does. ♣

10.13. **Southward deflection**

The Coriolis force is $2m\omega v \sin\theta$ eastward. But $v \approx gt$, so the eastward acceleration is $2\omega gt \sin\theta$. Integrating this gives an eastward speed of $\omega gt^2 \sin\theta$. This eastward speed produces a Coriolis force in the direction away from the earth's axis, so the acceleration in this direction is $2\omega(\omega gt^2 \sin\theta)$. The component of this acceleration along the surface of the earth (that is, in the southward direction) is $2\omega^2 gt^2 \sin\theta\cos\theta$. Integrating this gives a southward speed of $(2/3)\omega^2 gt^3 \sin\theta\cos\theta$. Integrating again gives a southward deflection of $(1/6)\omega^2 gt^4 \sin\theta\cos\theta$. But $gt^2/2 \approx h \Longrightarrow t^2 \approx 2h/g$. So the southward deflection due to the Coriolis force is $(2/3)(\omega^2 h^2/g)\sin\theta\cos\theta$.

Now for the rest of the problem. To be precise, we'll define θ to be the polar angle at the point P on the radial line to the dropping point. However, we would still obtain the same answer if we defined θ to be the polar angle at the location of the plumb bob, because the difference between these two angles is of order ω^2 (since it is due to the centrifugal contribution to the effective gravity). And this difference yields a negligible effect of order ω^4 in the $4(\omega^2 h^2/g)\sin\theta\cos\theta$ result, because this already contains a factor of ω^2.

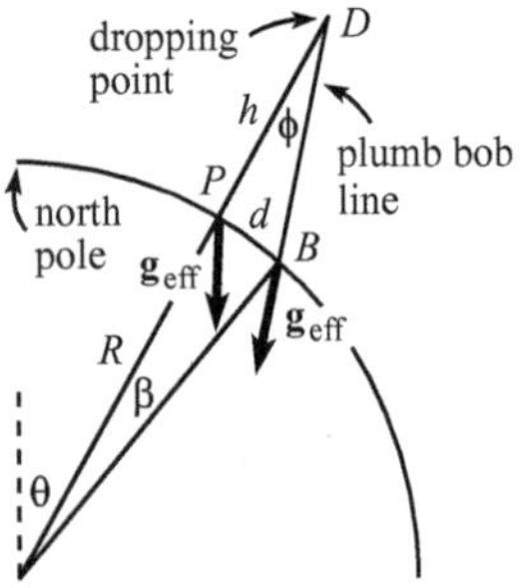

Fig. 10.32

(a) At point P, the magnitude of $\mathbf{F}_{cent}$ is $m\omega^2 R\sin\theta$, so (using Fig. 10.22 in the solution to Problem 10.3) the component of $\mathbf{F}_{cent}$ perpendicular to $m\mathbf{g}$ is $F^{\perp}_{cent} = m\omega^2 R\sin\theta\cos\theta$. The angle that $\mathbf{g}_{eff}$ makes with the radial at P is therefore essentially equal to $F^{\perp}_{cent}/mg = (\omega^2 R/g)\sin\theta\cos\theta$. However, the direction of the plumb bob line is determined by the $\mathbf{g}_{eff}$ vector at the location of the plumb bob (call this point B), and *not* at P; see Fig. 10.32. Now, the angle we just found between $\mathbf{g}_{eff}$ and $\mathbf{g}$ (that is, the radial) at P is the same as the angle between $\mathbf{g}_{eff}$ and $\mathbf{g}$ at B (at least to order ω^2, by the argument in the preceding paragraph). Therefore, since the $\mathbf{g}$ vector at B is tilted at the angle β (shown) relative to the $\mathbf{g}$ vector at P, we see that $\mathbf{g}_{eff}$ at B makes an angle of $\phi = (\omega^2 R/g)\sin\theta\cos\theta - \beta$ relative to the radial at P. The angle β is given by d/R, where d is the distance between P and B. So in the thin triangle DPB, the relation $d \approx h\phi$ gives

$$d \approx h\Big((\omega^2 R/g)\sin\theta\cos\theta - d/R\Big)$$

$$\Longrightarrow \quad d \approx \frac{h(\omega^2 R/g)\sin\theta\cos\theta}{1 + h/R}$$

$$\approx \frac{\omega^2 Rh}{g}\sin\theta\cos\theta\left(1 - \frac{h}{R}\right). \tag{10.74}$$

(b) The gravitational acceleration at height y above the earth is

$$\frac{GM}{(R+y)^2} \approx \frac{GM}{R^2(1+2y/R)} \approx \frac{GM}{R^2}\left(1 - \frac{2y}{R}\right) \equiv g\left(1 - \frac{2y}{R}\right). \tag{10.75}$$

(There are also corrections due to the centrifugal and Coriolis forces, but these are negligible; see Exercise 10.21.) So we have $\ddot{y} = -g(1 - 2y/R)$. Writing $\ddot{y}$ as $v\,dv/dy$, and separating variables and integrating, gives

$$\begin{aligned}\int_0^v v\,dv &= -\int_h^y g\Big(1 - \frac{2y}{R}\Big)\,dy \\ \Longrightarrow \quad v &= -\sqrt{2g(h-y) - (2g/R)(h^2 - y^2)} \\ &\approx -\sqrt{2g(h-y)}\Big(1 - \frac{h+y}{2R}\Big). \end{aligned} \tag{10.76}$$

Writing $v \equiv dy/dt$, and separating variables and integrating, gives

$$\int_0^T dt \approx -\int_h^0 \frac{dy}{\sqrt{2g(h-y)}\big(1 - \frac{h+y}{2R}\big)} \approx -\int_h^0 \frac{1 + \frac{h+y}{2R}}{\sqrt{2g(h-y)}}\,dy. \tag{10.77}$$

You can show that the "1" here gives the leading-order time of $\sqrt{2h/g}$. The additional time comes from the other term, which yields (with $z \equiv y/h$)

$$\Delta t = -\frac{1}{2R\sqrt{2g}} \int_h^0 \frac{h+y}{\sqrt{h-y}}\,dy = -\frac{h\sqrt{h}}{2R\sqrt{2g}} \int_1^0 \frac{1+z}{\sqrt{1-z}}\,dz. \tag{10.78}$$

Looking up this integral gives

$$\Delta t = \frac{h\sqrt{h}}{2R\sqrt{2g}} \cdot \frac{2}{3}(5+z)\sqrt{1-z}\,\Big|_1^0 = \sqrt{\frac{2h}{g}}\Big(\frac{5h}{6R}\Big). \tag{10.79}$$

The total time is therefore $\sqrt{2h/g}(1 + 5h/6R)$, as we wanted to show.

(c) At height y, the distance from the center of the earth is $R + y$. So from the $F_{\text{cent}}^{\perp}$ reasoning in part (a), the acceleration in the southward direction is $\ddot{z} = \omega^2(R+y)\sin\theta\cos\theta$.

(d) If the ball is a distance z away from the radial line through P (call this line L), then the radial line to the ball makes an angle of approximately z/R with respect to L. The component of the gravitational force on the ball perpendicular to L is therefore $\ddot{z} = -g(z/R)$, where the minus sign signifies toward L (that is, northward).

(e) Parts (c) and (d) give

$$\ddot{z} = \omega^2(R+y)\sin\theta\cos\theta - g(z/R). \tag{10.80}$$

The R term here dominates, so to leading order we have $\ddot{z} = \omega^2 R\sin\theta\cos\theta \implies z \approx (\omega^2 R\sin\theta\cos\theta)t^2/2$. Also, to leading order, $y \approx h - gt^2/2$. Plugging these values of z and y into Eq. (10.80) gives

$$\begin{aligned}\ddot{z} &= \omega^2 \sin\theta\cos\theta\left(R + \Big(h - \frac{gt^2}{2}\Big) - \frac{gt^2}{2}\right) \\ \Longrightarrow \quad z &= \omega^2 \sin\theta\cos\theta\left(\frac{Rt^2}{2} + \Big(\frac{ht^2}{2} - \frac{gt^4}{24}\Big) - \frac{gt^4}{24}\right).\end{aligned} \tag{10.81}$$

Plugging in the total time $t = \sqrt{2h/g}(1 + 5h/6R)$ gives, to leading order, a total z value equal to

$$\begin{aligned} z &= \omega^2 \sin\theta\cos\theta\left(\frac{R}{2}\cdot\frac{2h}{g}\left(1+\frac{5h}{3R}\right)\right. \\ &\quad \left.+\left(\frac{h}{2}\cdot\frac{2h}{g}-\frac{g}{24}\cdot\frac{4h^2}{g^2}\right)-\frac{g}{24}\cdot\frac{4h^2}{g^2}\right) \\ &= \frac{\omega^2 Rh}{g}\sin\theta\cos\theta\left(1+\frac{h}{R}\left(\frac{5}{3}+\frac{5}{6}-\frac{1}{6}\right)\right) \\ &= \frac{\omega^2 Rh}{g}\sin\theta\cos\theta\left(1+\frac{7h}{3R}\right). \end{aligned} \tag{10.82}$$

When we subtract off the plumb bob's position in Eq. (10.74), the leading terms proportional to R cancel. Adding on the Coriolis result then gives a total southward deflection (relative to the plumb bob) equal to

$$\frac{\omega^2 h^2}{g}\sin\theta\cos\theta\left(\frac{7}{3}-(-1)+\frac{2}{3}\right)=\frac{4\omega^2 h^2}{g}\sin\theta\cos\theta, \tag{10.83}$$

as we wanted to show. From the third line in Eq. (10.82), we see that the 7/3 result can be broken up into a 5/3 that comes from the extra time it takes the ball to hit the ground due to the decrease in the gravitational force with altitude, a 5/6 that comes from the dependence of the centrifugal force on height, and a −1/6 that comes from the slightly northward component of the gravitational force.

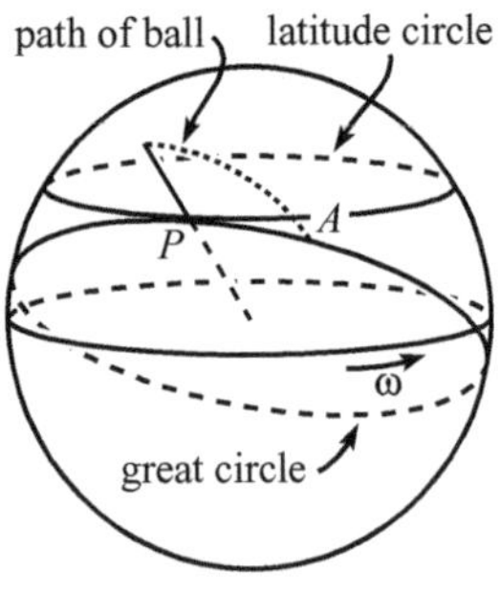

Fig. 10.33

REMARK: We can also solve this problem by working in an inertial frame. In this frame, the ball has an initial sideways motion due to the rotation of the earth, and this motion causes the ball to travel in the path shown in Fig. 10.33. Because only gravity acts on the ball, its path lies in the plane determined by its initial velocity and radius. The intersection of this plane with the surface of the earth is the great circle shown, and the ball hits the earth at a point A on this great circle.

The distance ℓ that the ball travels along the great circle is essentially equal to the distance it travels in the eastward direction (that is, along the latitude circle). This eastward distance equals the distance that point P rotates due to the spinning of the earth, plus the extra Coriolis eastward deflection d_{cor} that we found in Eq. (10.18). (In the spirit of working in an inertial frame, let's assume that we found this deflection by the method given in the remark after Eq. (10.18).) So we have $\ell \approx R\sin\theta\,\omega t + d_{\text{cor}}$. The task now is to determine the distance between A and the latitude circle through P. This distance should equal the result in Eq. (10.82) plus the southward Coriolis deflection. (We'll then subtract off the distance from the pendulum bob to P, given in Eq. (10.74), as we did above.)[21] This task is equivalent to answering the question: If a circle of radius $r = R\sin\theta$ (the latitude circle) sits on a plane (the plane of the great circle) and is inclined at an angle θ with respect to the normal to the plane, how far above the plane is a point on the circle with an "x" coordinate equal to ℓ (with $\ell \ll r$)? Using a Taylor series approximation for the length $\sqrt{r^2-\ell^2}$ in the right triangle in Fig. 10.34, we see that if the circle is perpendicular to the plane, then the desired distance is $\ell^2/2r$. Tilting the circle by an angle θ simply brings in a factor of $\cos\theta$, so the distance is $(\ell^2/2r)\cos\theta = (\ell^2/2R\sin\theta)\cos\theta$.

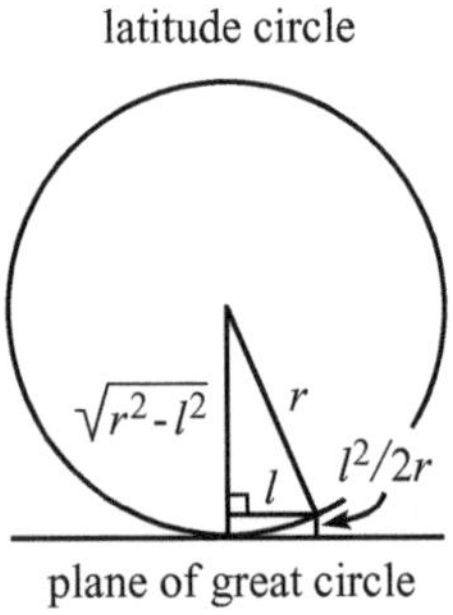

Fig. 10.34

[21] If you want, you can derive the distance from the pendulum bob to P by working in the inertial frame, to maintain the spirit of this solution. The $m\omega^2 R$ type term we found in part (a) above simply shows up through the centripetal acceleration instead of through the centrifugal force.

Using the time of flight obtained above in part (b), and also the d_{cor} from Eq. (10.18), we obtain a southward deflection relative to P equal to (dropping higher order terms)

$$\begin{aligned}
z &= \frac{\ell^2 \cos\theta}{2R\sin\theta} \\
&= \frac{(R\sin\theta\,\omega t + d_{\text{cor}})^2 \cos\theta}{2R\sin\theta} \\
&\approx R\omega^2 \sin\theta\cos\theta\, t^2/2 + \omega d_{\text{cor}} \cos\theta\, t \\
&\approx \frac{R\omega^2 \sin\theta\cos\theta}{2}\cdot\frac{2h}{g}\left(1+\frac{5h}{3R}\right) + \omega\left(\frac{2\omega h\sin\theta}{3}\sqrt{\frac{2h}{g}}\right)\cos\theta\sqrt{\frac{2h}{g}} \\
&\approx \frac{\omega^2 Rh}{g}\sin\theta\cos\theta\left(1+\frac{h}{R}\left(\frac{5}{3}+\frac{4}{3}\right)\right).
\end{aligned} \tag{10.84}$$

As expected, the factor of $9/3$ here equals the $7/3$ from Eq. (10.82) plus the $2/3$ from the southward Coriolis deflection. Subtracting off the position of the plumb bob given in Eq. (10.74) yields the desired southward deflection from the plumb bob, $(4\omega^2 h^2/g)\sin\theta\cos\theta$. ♣

10.14. **Bead on a hoop**

(a) The gravitational force on the bead is essentially $GMm/(R-r)^2$, directed essentially to the right. But to leading order, we can neglect the r here. If the bead is at an angle θ up from the horizontal, then we need to multiply this rightward force by $\sin\theta \approx \theta$ to obtain the force component along the hoop. Therefore, the $F = ma$ equation along the hoop is

$$-\frac{GMm\theta}{R^2} = mr\ddot{\theta} \quad\Longrightarrow\quad \omega = \sqrt{\frac{GM}{rR^2}}. \tag{10.85}$$

This is just the usual $\sqrt{g/r}$ result for a pendulum, in disguised form.

(b) From Eq. (10.35), the tidal force is $(GMm/R^3)(2x, -y)$. The force along the hoop comes not only from the horizontal component of this (when multiplied by negative $\sin\theta$), but also from the vertical component (when multiplied by $\cos\theta$). So the force along the hoop is (using $x = r\cos\theta$ and $y = r\sin\theta$)

$$\frac{GMm}{R^3}\Big(2(r\cos\theta)(-\sin\theta) + (-r\sin\theta)\cos\theta\Big) = -\frac{GMm}{R^3}(3r\sin\theta\cos\theta). \tag{10.86}$$

Using $\sin\theta \approx \theta$ and $\cos\theta \approx 1$, the $F = ma$ equation along the hoop is

$$-\frac{3GMmr\theta}{R^3} = mr\ddot{\theta} \quad\Longrightarrow\quad \omega = \sqrt{\frac{3GM}{R^3}}. \tag{10.87}$$

Note that this is independent of r. It is smaller than the result in part (a) by a factor $\sqrt{3r/R}$.

10.15. **Precession of the equinoxes**

We'll calculate the effect due to the sun, and then multiply by 3 to obtain the total effect, because the moon's effect is twice the sun's. Consider the case where the earth is in the summer/winter orientation. From Eq. (10.35), the tidal force on the effective point mass m is $(GM_S m/R^3)(2x, -y)$. Both components are relevant here, so the tidal forces on the two masses are shown in Fig. 10.35, where r is the radius of the earth, and $k \equiv GM_S m/R^3$. The torque due to these forces has magnitude

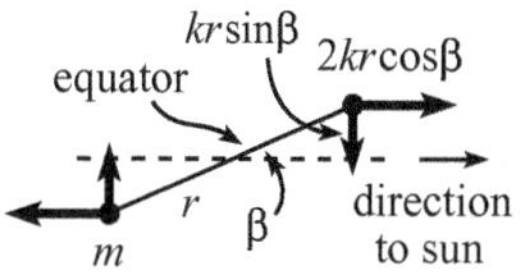

Fig. 10.35

$$2\big(2kr\cos\beta(r\sin\beta) + kr\sin\beta(r\cos\beta)\big) = 6kr^2\sin\beta\cos\beta, \tag{10.88}$$

and it is directed into the page. For the case where the earth is in the spring/fall orientation (that is, where the sun is positioned where your nose is, as you look at Fig. 10.35), there is no torque, because the longitudinal component of the tidal force is zero, and the transverse component is radial. Each of the summer/winter and spring/fall cases is applicable for half the time (by assumption 6), so the time-averaged torque is

$$\overline{\tau}_{\text{sun}} = \frac{1}{2}(6kr^2 \sin\beta\cos\beta + 0) = 3kr^2 \sin\beta\cos\beta. \tag{10.89}$$

Adding on the effect of the moon gives an average total torque of

$$\overline{\tau}_{\text{total}} = 9kr^2 \sin\beta\cos\beta = \frac{9GM_S mr^2 \sin\beta\cos\beta}{R^3}. \tag{10.90}$$

The angular momentum of the earth is $I_3\omega_3$. The "horizontal" component of this is $I_3\omega_3 \sin\beta$, so $|d\mathbf{L}/dt| = \Omega I_3\omega_3 \sin\beta$, where Ω is the frequency of precession. Equating this with the torque gives

$$\Omega = \frac{9GM_S mr^2 \cos\beta}{R^3 I_3\omega_3}. \tag{10.91}$$

But from assumptions 3, 4, and 5, we have $m = \rho r^2 h$, where ρ is the average density of the earth. And $I_3 = (2/5)(4\pi r^3\rho/3)r^2 = (8\pi/15)\rho r^5$, so[22]

$$\Omega = \frac{9GM_S(\rho r^2 h)r^2 \cos\beta}{R^3(8\pi\rho r^5/15)\omega_3} = \frac{135\, GM_S h\cos\beta}{8\pi r R^3\omega_3}. \tag{10.92}$$

Plugging in the numerical values gives

$$\Omega = \frac{135(6.67\cdot 10^{-11}\,\text{m}^3/\text{kg}\,\text{s}^2)(2\cdot 10^{30}\,\text{kg})(21\cdot 10^3\,\text{m})\cos 23^\circ}{8\pi(6.4\cdot 10^6\,\text{m})(1.5\cdot 10^{11}\,\text{m})^3(7.3\cdot 10^{-5}\,\text{s}^{-1})} \approx 8.8\cdot 10^{-12}\,\text{s}^{-1}$$

$$\Longrightarrow T = \frac{2\pi}{\Omega} \approx \frac{2\pi}{8.8\cdot 10^{-12}\,\text{s}^{-1}} \approx 7.1\cdot 10^{11}\,\text{s} \approx 23\,000 \text{ years}. \tag{10.93}$$

This answer actually came out a little too good. We had no right to expect that it would come out so close to 26 000 years, considering the various approximations we made. But we *did* have a right to expect it to come out within a factor of, say, 5 or so, because our approximations could be off by only so much. At any rate, Eq. (10.92) does at least give the correct dependence on the various parameters. Note that Eq. (10.40) says that $h \propto r^2\omega_3^2/g$. Using this along with $g = G(4\pi r^3\rho_E/3)/r^2$ in Eq. (10.92), and ignoring all numerical factors, gives $\Omega \propto \omega_3 M_S/(\rho_E R^3) \propto \omega_3(M_S/M_c)$, where M_c is the mass of a colossal object whose density equals the earth's and whose radius equals the earth–sun distance. (This relation also holds if we consider the moon instead of the sun.)

REMARK: Interestingly, the time of $T = 26\,000$ years is large enough to allow us to treat the axis of rotation as essentially constant, but small enough to have a noticeable effect over the ages. The star we see in the northern direction nowadays is a different northern star from the one people saw, say, 2000 years ago. And the signs of the zodiac that we see now are shifted by approximately one place compared with what they were 2000 years ago. The "precession of the equinoxes" name comes from the fact that if you consider, for example, a distant galaxy that the sun blocks out when the earth is in the spring equinox position now, and if you consider the analogous galaxy 13 000 years ago, then the two galaxies are in opposite directions in the universe, relative to the earth. ♣

[22] The mass m is actually smaller than this, because the earth's crust is less dense than its interior. But I_3 is smaller than this too, because the earth's core is denser than the mantle. These effects somewhat cancel each other, because m appears in the numerator and I_3 appears in the denominator. But we're just doing things roughly anyway.

Chapter 11
Relativity (Kinematics)

We now come to Einstein's theory of relativity. This is where we find out that everything we've done so far in this book has been wrong. Well, perhaps "incomplete" would be a better word. The important point to realize is that Newtonian physics is a limiting case of the more correct relativistic theory. Newtonian physics works perfectly fine when the speeds we're dealing with are much less than the speed of light, which is about $3 \cdot 10^8$ m/s. It would be silly, to put it mildly, to use relativity to solve a problem involving the length of a baseball trajectory. But in problems involving large speeds, or in problems where a high degree of accuracy is required, we must use the relativistic theory.[1] This is the subject of the remainder of this book.

The theory of relativity is certainly one of the most exciting and talked-about topics in physics. It is well known for its "paradoxes," which are quite conducive to discussion. There is, however, nothing at all paradoxical about it. The theory is logically and experimentally sound, and the whole subject is actually quite straightforward, provided that you proceed calmly and keep a firm hold of your wits.

The theory rests upon certain postulates. The one that most people find counterintuitive is that the speed of light has the same value in any inertial (that is, nonaccelerating) reference frame. This speed is much greater than the speed of everyday objects, so most of the consequences of this new theory aren't noticeable. If we instead lived in a world identical to ours except for the fact that the speed of light was 50 mph, then the consequences of relativity would be ubiquitous. We wouldn't think twice about time dilations, length contractions, and so on.

I have included a large number of puzzles and "paradoxes" in the problems and exercises. When attacking these, be sure to follow them through to completion, and don't say, "I could finish this one if I wanted to, but all I'd have to do would be such-and-such, so I won't bother," because the essence of the paradox may

[1] You shouldn't feel too bad about having spent so much time learning about a theory that's just the limiting case of another theory, because you're now going to do it again. Relativity is also the limiting case of another theory (quantum field theory). And likewise, quantum field theory is the limiting case of yet another theory (string theory). And likewise . . . well, you get the idea. Who knows, maybe it really *is* turtles all the way down.

very well be contained in the such-and-such, and you will have missed out on all the fun. Most of the paradoxes arise because different frames of reference *seem* to give different results. Therefore, in explaining a paradox, you not only need to give the correct reasoning, you also need to say what's wrong with incorrect reasoning.

There are two main topics in relativity. One is Special Relativity (which doesn't deal with gravity), and the other is General Relativity (which does). We'll deal mostly with the former, but Chapter 14 contains some of the latter. Special Relativity may be divided into two topics, *kinematics* and *dynamics*. Kinematics deals with lengths, times, speeds, etc. It is concerned only with the space and time coordinates of an abstract particle, and not with masses, forces, energy, momentum, etc. Dynamics, on the other hand, does deal with these quantities. This chapter covers kinematics. Chapter 12 covers dynamics. Most of the fun paradoxes fall into the kinematics part, so the present chapter is the longer of the two. In Chapter 13, we'll introduce the concept of 4-vectors, which ties much of the material in Chapters 11 and 12 together.

11.1 Motivation

Although it was obviously a stroke of genius that led Einstein to his theory of relativity, it didn't just come out of the blue. A number of things going on in nineteenth-century physics suggested that something was amiss. There were many efforts made by many people to explain away the troubles that were arising, and at least a few steps had been taken toward the correct theory. But Einstein was the one who finally put everything together, and he did so in a way that had consequences far beyond the realm of the specific issues that people were trying to understand. Indeed, his theory turned our idea of space and time on its head. But before we get into the heart of the theory, let's look at two of the major problems in late nineteenth-century physics.[2]

11.1.1 Galilean transformations, Maxwell's equations

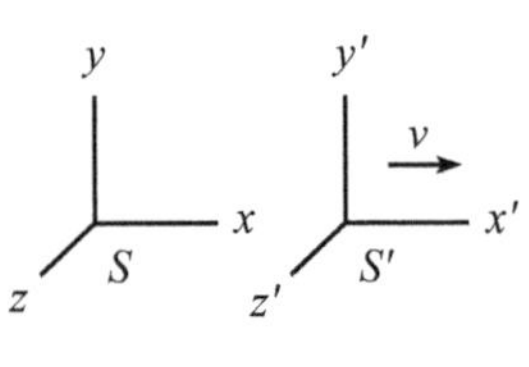

Fig. 11.1

Imagine standing on the ground and watching a train travel by with constant speed v in the x direction. Let the train frame be S' and the ground frame be S, as shown in Fig. 11.1. Consider two events that happen on the train. For example, one person claps her hands, and another person stomps his feet. If the space and time separations between these two events in the frame of the train are $\Delta x'$ and $\Delta t'$, what are the space and time separations, Δx and Δt, in the frame of the ground? Ignoring what we'll be learning about relativity in this chapter, the answers are "obvious" (well, in that incorrectly obvious sort of way, as we'll see

[2] If you can't wait to get to the postulates and results of Special Relativity, you can go straight to Section 11.2. The present section can be skipped on a first reading.

in Section 11.4.1). The time separation, Δt, is the same as on the train, so we have $\Delta t = \Delta t'$. We know from everyday experience that nothing strange happens with time. When you see people exiting a train station, they're not fiddling with their watches, trying to recalibrate them with a ground-based clock.

The spatial separation is a little more exciting, but still nothing too complicated. The train is moving, so everything in it (in particular, the second event) gets carried along at speed v during the time $\Delta t'$ between the two events. So we have $\Delta x = \Delta x' + v\Delta t'$. As a special case, if the two events happen at the same place on the train (so that $\Delta x' = 0$), then we have $\Delta x = v\Delta t'$. This makes sense, because the spot on the train where the events occur simply travels a distance $v\Delta t$ by the time the second event happens. The *Galilean transformations* are therefore

$$\begin{aligned} \Delta x &= \Delta x' + v\Delta t', \\ \Delta t &= \Delta t'. \end{aligned} \tag{11.1}$$

Also, nothing interesting happens in the y and z directions, so we have $\Delta y = \Delta y'$ and $\Delta z = \Delta z'$.

The principle of *Galilean invariance* says that the laws of physics are invariant under the above Galilean transformations. Alternatively, it says that the laws of physics hold in all inertial frames.[3] This is quite believable. For example, Newton's second law holds in all inertial frames, because the constant relative velocity between any two frames implies that the acceleration of a given particle is the same in all frames.

REMARKS: Note that the Galilean transformations aren't symmetric in x and t. This isn't automatically a bad thing, but it turns out that it will in fact be a problem in Special Relativity, where space and time are treated on a more equal footing. We'll find in Section 11.4.1 that the Galilean transformations are replaced by the *Lorentz transformations* (at least in the world we live in), and the latter are indeed symmetric in x and t (up to factors of the speed of light, c).

Note also that Eq. (11.1) deals only with the *differences* in x and t between two events, and not with the values of the coordinates themselves. The values of the coordinates of a single event depend on where you pick your origin, which is an arbitrary choice. The coordinate differences between two events, however, are independent of this choice, and this allows us to make the physically meaningful statement in Eq. (11.1). It makes no sense for a physical result to depend on the arbitrary choice of origin, and so the Lorentz transformations we derive later on will also involve only differences in coordinates. ♣

One of the great triumphs of nineteenth-century physics was the theory of electromagnetism. In 1864, James Clerk Maxwell wrote down a set of equations that collectively described everything that was known about the subject. These equations involve the electric and magnetic fields through their space and time derivatives. We won't worry about the specific form of the equations here,[4] but it

[3] It was assumed prior to Einstein that these two statements say the same thing, but we will soon see that they do not. The second statement is the one that remains valid in relativity.

[4] Maxwell's original formulation involved a large number of equations, but these were later written more compactly, using vectors, as four equations.

turns out that if you transform them from one frame to another via the Galilean transformations, they end up taking a different form. That is, if you've written down Maxwell's equations in one frame (where they take their standard nice-looking form), and if you then replace the coordinates in this frame by those in another frame, using Eq. (11.1), then the equations look different (and not so nice). This presents a major problem. If Maxwell's equations take a nice form in one frame and a not-so-nice form in every other frame, then why is one frame special? Said in another way, Maxwell's equations predict that light moves with a certain speed c. But which frame is this speed measured with respect to? The Galilean transformations imply that if the speed is c with respect to a given frame, then it is *not* c with respect to any other frame. The proposed special frame where Maxwell's equations are nice and the speed of light is c was called the frame of the *ether*. We'll talk in detail about the ether in the next section, but what experiments showed was that light surprisingly moved with speed c in every frame, no matter which way the frame was moving through the supposed ether.

There were therefore two possibilities. Either something was wrong with Maxwell's equations, or something was wrong with the Galilean transformations. Considering how "obvious" the latter are, the natural assumption in the late nineteenth century was that something was wrong with Maxwell's equations, which were quite new, after all. However, after a good deal of effort by many people to make Maxwell's equations fit with the Galilean transformations, Einstein finally showed that the trouble was in fact with the latter. More precisely, in 1905 he showed that the Galilean transformations are a special case of the Lorentz transformations, valid only when the speed involved is much less than the speed of light.[5] As we'll see in Section 11.4.1, the coefficients in the Lorentz transformations depend on both v and the speed of light c, where the c's appear in various denominators. Since c is quite large (about $3 \cdot 10^8$ m/s) compared with everyday speeds v, the parts of the Lorentz transformations involving c are negligible, for any typical v. This is why no one prior to Einstein realized that the transformations had anything to do with the speed of light. Only the terms in Eq. (11.1) were noticeable.

> As he pondered the long futile fight
> To make Galileo's world right,
> In a new variation
> On the old transformation,
> It was Einstein who first saw the light.

[5] It was well known that Maxwell's equations were invariant under the Lorentz transformations (in contrast with their noninvariance under the Galilean transformations), but Einstein was the first to recognize the full meaning of these transformations. Instead of being relevant only to electromagnetism, the Lorentz transformations replaced the Galilean transformations universally.

In short, the reasons why Maxwell's equations were in conflict with the Galilean transformations are: (1) The speed of light is what determines the scale on which the Galilean transformations break down; (2) Maxwell's equations inherently involve the speed of light, because light is an electromagnetic wave.

11.1.2 Michelson–Morley experiment

As mentioned above, it was known in the late nineteenth century, after Maxwell wrote down his equations, that light is an electromagnetic wave and that it moves with a speed of about $3 \cdot 10^8$ m/s. Now, every other wave that people knew about at the time needed a medium to propagate in. Sound waves need air, ocean waves need water, waves on a string of course need the string, and so on. It was therefore natural to assume that light also needed a medium to propagate in. This proposed medium was called the *ether*. However, if light propagates in a given medium, and if the speed in this medium is c, then the speed in a reference frame moving relative to the medium will be different from c. Consider, for example, sound waves in air. If the speed of sound in air is v_{sound}, and if you run toward a sound source with speed v_{you}, then the speed of the sound waves with respect to you (assuming it's a windless day) is $v_{\text{sound}} + v_{\text{you}}$. Equivalently, if you are standing downwind and the speed of the wind is v_{wind}, then the speed of the sound waves with respect to you is $v_{\text{sound}} + v_{\text{wind}}$.

If this ether really exists, then a reasonable thing to do is to try to measure one's speed with respect to it. This can be done in the following way (we'll work in terms of sound waves in air here).[6] Let v_s be the speed of sound in air. Imagine two people standing on the ends of a long platform of length L that moves at speed v_p with respect to the reference frame in which the air is at rest. One person claps, the other person claps immediately when he hears the first clap (assume that the reaction time is negligible), and then the first person records the total time elapsed when she hears the second clap. What is this total time? Well, the answer is that we can't say without knowing in which direction the platform is moving. Is it moving parallel to its length, or transverse to it (or somewhere in between)? Let's look at these two basic cases. For both of these, we'll view the setup and do the calculation in the frame in which the air is at rest.

Consider first the case where the platform moves parallel to its length. In the frame of the air, assume that the person at the rear is the one who claps first. Then it takes a time of $L/(v_s - v_p)$ for the sound to reach the front person. This is true because the sound must close the initial gap of L at a relative speed of $v_s - v_p$,

[6] As we'll soon see, there is no ether, and light travels at the same speed with respect to any frame. This is a rather bizarre fact, and it takes some getting used to. It's hard enough to get away from the old way of thinking, even without any further reminders, so I can't bring myself to work through this method in terms of light waves in an ether. I'll therefore work in terms of sound waves in air.

as viewed in the air frame.[7] By similar reasoning, the time for the sound to return to the rear person is $L/(v_s + v_p)$. The total time is therefore

$$t_1 = \frac{L}{v_s - v_p} + \frac{L}{v_s + v_p} = \frac{2Lv_s}{v_s^2 - v_p^2}. \tag{11.2}$$

This correctly equals $2L/v_s$ when $v_p = 0$, and infinity when $v_p \to v_s$.

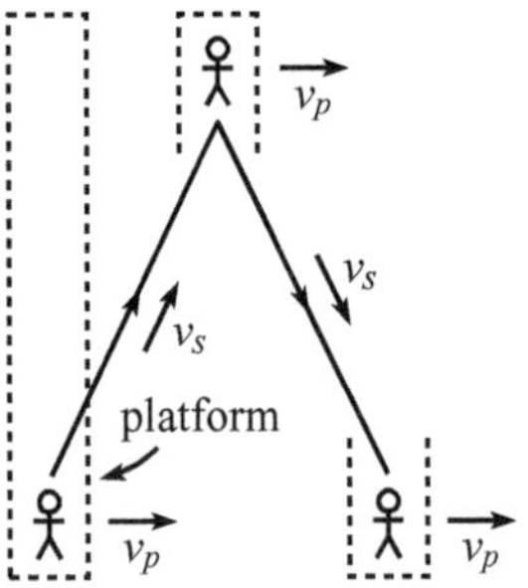

Fig. 11.2

Now consider the case where the platform moves perpendicular to its length. In the frame of the air, we have the situation shown in Fig. 11.2. Since the sound moves diagonally,[8] the "vertical" component is (by the Pythagorean theorem) $\sqrt{v_s^2 - v_p^2}$. This is the relevant component as far as traveling the length of the platform goes, so the total time is

$$t_2 = \frac{2L}{\sqrt{v_s^2 - v_p^2}}. \tag{11.3}$$

Again, this correctly equals $2L/v_s$ when $v_p = 0$, and infinity when $v_p \to v_s$.

The times in Eqs. (11.2) and (11.3) are not equal. As an exercise, you can show that of all the possible orientations of the platform relative to the direction of motion, t_1 is the largest possible time, and t_2 is the smallest. Therefore, if you are on a large surface that is moving with respect to the air, and if you know the values of L and v_s, then if you want to figure out what v_p is (assume that it doesn't occur to you to toss a little piece of paper to at least find the direction of the wind), all you have to do is repeat the above setup with someone standing at various points along the circumference of a given circle around you. If you take the largest total time that occurs and equate it with t_1, then Eq. (11.2) will give you v_p. Alternatively, you can equate the smallest time with t_2, and Eq. (11.3) will yield the same v_p. Note that if $v_p \ll v_s$, we can apply Taylor series approximations to the above two times. For future reference, these approximations give the difference in times as

$$\Delta t = t_1 - t_2 = \frac{2L}{v_s}\left(\frac{1}{1 - v_p^2/v_s^2} - \frac{1}{\sqrt{1 - v_p^2/v_s^2}}\right) \approx \frac{Lv_p^2}{v_s^3}. \tag{11.4}$$

The above setup is the general idea that Michelson and Morley used in 1887 to measure the speed of the earth through the supposed ether.[9] There is, however, a major complication with light that doesn't arise with sound – the speed of light is so large that any time intervals that are individually measured will have inevitable measurement errors that are far larger than the difference

[7] Alternatively, relative to the initial back of the platform, the position of the sound wave is $v_s t$, and the position of the front person is $L + v_p t$. Equating these gives $t = L/(v_s - v_p)$.

[8] The sound actually moves in all directions, of course, but it's only the part of the wave that moves in a particular diagonal direction that ends up hitting the other person.

[9] See Handschy (1982) for the data and analysis of the experiment.

between t_1 and t_2. Therefore, individual time measurements give essentially no information. Fortunately, there is a way out of this impasse.

Consider two of the above "platform" scenarios arranged to be at right angles with respect to each other, with the same starting point. This can be arranged by having a (monochromatic) light beam encounter a beam splitter that sends two beams off at 90° angles. The beams then hit mirrors and bounce back to the beam splitter where they (partially) recombine before hitting a screen, as shown in Fig. 11.3. The fact that light is a wave, which is what got us into this mess in the first place, is now what saves the day. The wave nature of light implies that the recombined light beam produces an interference pattern on the screen. At the center of the pattern, the beams will constructively or destructively interfere (or something in between), depending on whether the two light beams are in phase or out of phase when they recombine. This interference pattern is extremely delicate. The slightest change in travel times of the beams will cause the pattern to noticeably shift.

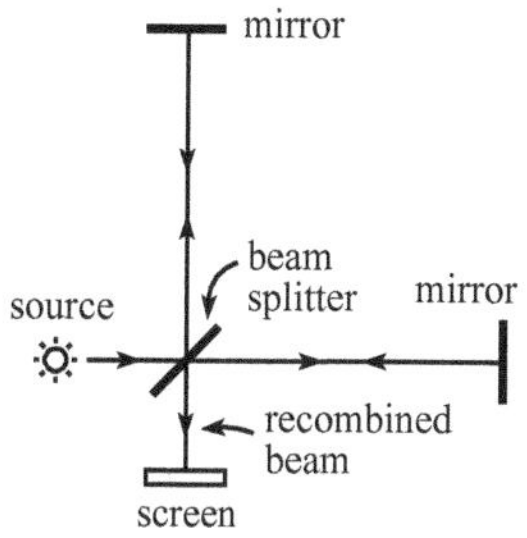

Fig. 11.3

If the whole apparatus is rotated around, so that the experiment is performed at various angles, then the maximum amount that the interference pattern changes can be used to determine the speed of the earth through the ether (v_p in the platform setup above). In one extreme case, the time in one arm is longer than the time in the other by Lv^2/c^3, where we have changed notation in Eq. (11.4) so that $v_p \to v$ is the speed of the earth, and $v_s \to c$ is the speed of light. But in the other extreme case, the time in this arm is shorter by Lv^2/c^3. So the maximal interference shift corresponds to a time difference of $2Lv^2/c^3$.

However, when Michelson and Morley performed their experiment, they observed no interference shift as the apparatus was rotated around. Their setup did in fact allow enough precision to measure a nontrivial earth speed through the ether, if such a speed existed. So if the ether did exist, their results implied that the speed of the earth through it was zero. This result, although improbable, was technically fine; it might simply have been the case that they happened to do their experiment when the relative speed was zero. However, when they performed their experiment half a year later, when the earth's motion around the sun caused it to be moving in the opposite direction, they still measured zero speed. It wasn't possible for both of these results to be zero (assuming that the ether exists), so something must have been wrong with the initial reasoning. Many people over the years tried to explain this null result, but none of the explanations were satisfactory. Some led to incorrect predictions in other setups, and some seemed to work fine but were a bit *ad hoc*.[10] The correct explanation, which followed from

[10] The most successful explanation (and one that was essentially correct, although the reason why it was correct wasn't known until Einstein fully explained things) was the Lorentz–FitzGerald contraction. These two physicists independently proposed that lengths are contracted in the direction of the motion by precisely the right factor, namely $\sqrt{1 - v^2/c^2}$, to make the travel times in the two arms of the Michelson–Morley setup equal, thus yielding the null result.

Einstein's 1905 theory of relativity, was that the ether doesn't exist.[11] In other words, light doesn't need a medium to propagate in; it doesn't move with respect to a certain special reference frame, but rather it moves with respect to whoever is looking at it.

The findings of Michelson–Morley
Allow us to say very surely,
"If this ether is real,
Then it has no appeal,
And shows itself off rather poorly."

REMARKS:

1. We assumed above that the lengths of the two arms in the apparatus were equal. In practice, there is absolutely no hope of constructing the lengths to be equal, up to an error that is sufficiently small compared with the wavelength of the light. But fortunately this doesn't matter. We're concerned not with the difference in the travel times associated with the two arms, but rather with the *difference in these differences* as the apparatus is rotated around. Using Eqs. (11.2) and (11.3) with different lengths L_1 and L_2, you can quickly show that the maximum interference shift corresponds to a time of $(L_1+L_2)v^2/c^3$, assuming $v \ll c$.
2. Assuming that the lengths of the arms are approximately equal, let's plug in some rough numbers to see how much the interference pattern shifts. The Michelson–Morley setup had arms with lengths of about 10 m. And we'll take v to be on the order of the speed of the earth around the sun, which is about $3 \cdot 10^4$ m/s. We then obtain a maximal time change of $t = 2Lv^2/c^3 \approx 7 \cdot 10^{-16}$ s. The large negative exponent here might make us want to throw in the towel, thinking that the effect is hopelessly small. But we should be careful about calling a dimensionful quantity "small." We need to know what other quantity with the dimensions of time we're comparing it with. The distance that light travels in the time t is $ct = (3 \cdot 10^8 \text{ m/s})(7 \cdot 10^{-16} \text{ s}) \approx 2 \cdot 10^{-7}$ m, and this happens to be a perfectly reasonable fraction of a wavelength of visible light, which is around $\lambda = 6 \cdot 10^{-7}$ m, give or take. So we have $ct/\lambda \approx 1/3$. The time we want to compare $2Lv^2/c^3$ with is therefore the time it takes light to travel one wavelength, namely λ/c, and these two times turn out to be roughly the same size. The interference shift of about a third of a cycle was well within the precision of the Michelson–Morley setup. So if the ether had really existed, they definitely would have been able to measure the speed of the earth through it.
3. One proposed explanation of the observed null effect was "frame dragging." Perhaps the earth drags the ether along with it, thereby yielding the observed zero relative speed. This frame dragging is quite plausible, because in the platform example above, the platform drags a thin layer of air along with it. And more mundanely, a car completely drags the air in its interior along with it. But it turns out that frame dragging is inconsistent with *stellar aberration*, which is the following effect.

   Depending on the direction of the earth's instantaneous velocity as it orbits around the sun, a given star might (depending on its location) appear at slightly different places in the sky when viewed at two times, say, six months apart. This is due to the fact that your telescope must be aimed at a slight angle relative to the actual direction to the star, because as the star's light travels down the telescope, the telescope moves slightly in the direction of the earth's motion. The ratio of the earth's speed around the sun to the speed of light is about 10^{-4}, so the effect is small. But it is large enough to be noticeable, and it has

[11] Although we've presented the Michelson–Morley experiment here for pedagogical purposes, the consensus among historians is that Einstein actually wasn't influenced much by the experiment, except indirectly through Lorentz's work on electrodynamics. See Holton (1988).

indeed been measured. Note that it is the velocity of the telescope that matters here, and not its position.[12]

However, if frame dragging were real, then the light from the star would get dragged along with the earth and would therefore travel down a telescope that was pointed directly at the star, in disagreement with the observed fact that the telescope must point at the slight angle mentioned above. Or even worse, the dragging might produce a boundary layer of turbulence which would blur the stars. The existence of stellar aberration therefore implies that frame dragging doesn't occur. ♣

11.2 The postulates

Let's now start from scratch and see what the theory of Special Relativity is all about. We'll take the route that Einstein took and use two postulates as the basis of the theory. We'll start with the speed-of-light postulate:

- *The speed of light has the same value in any inertial frame.*

I don't claim that this statement is obvious, or even believable. But I do claim that it's easy to understand what the statement says (even if you think it's too silly to be true). It says the following. Consider a train moving along the ground at constant velocity. Someone on the train shines a light from one point on the train to another. Let the speed of the light with respect to the train be c ($\approx 3 \cdot 10^8 \text{m/s}$). Then the above postulate says that a person on the ground also sees the light move at speed c.

This is a rather bizarre statement. It doesn't hold for everyday objects. If a baseball is thrown on a train, then the speed of the baseball is different in the different frames. The observer on the ground must add the velocity of the ball (with respect to the train) and the velocity of the train (with respect to the ground) to obtain the velocity of the ball with respect to the ground.[13]

The truth of the speed-of-light postulate cannot be demonstrated from first principles. No statement with any physical content in physics (that is, one that isn't purely mathematical, such as, "two apples plus two apples gives four apples") can be proven. In the end, we must rely on experiment. And indeed, all the consequences of the speed-of-light postulate have been verified countless

[12] This aberration effect is not the same as the *parallax* effect in which the direction of the actual position of an object changes, depending on the location of the observer. For example, people at different locations on the earth see the moon at different angles (that is, they see the moon in line with different distant stars). Although stellar parallax has been measured for nearby stars (as the earth goes around the sun), its angular effect is much smaller than the angular effect from stellar aberration. The former decreases with distance, whereas the latter doesn't. For further discussion of aberration, and of why it is only the earth's velocity (or rather, the change in its velocity) that matters, and not also the star's velocity (since you might think, based on the title of this chapter, that it is the relative velocity that matters), see Eisner (1967).

[13] Actually, this isn't quite true, as the velocity-addition formula in Section 11.5.1 shows. But it's true enough for the point we're making here.

times during the past century. As discussed in the previous section, the most well-known of the early experiments on the speed of light was the one performed by Michelson and Morley. And in more recent years, the consequences of the postulate have been verified continually in high-energy particle accelerators, where elementary particles reach speeds very close to c. The collection of all the data from numerous experiments over the years allows us to conclude with near certainty that our starting assumption of an invariant speed of light is correct (or is at least the limiting case of a more accurate theory).

There is one more postulate in the Special Relativity theory, namely the "relativity" postulate (also called the Principle of Relativity). It is much more believable than the speed-of-light postulate, so you might just take it for granted and forget to consider it. But like any postulate, of course, it is crucial. It can be stated in various ways, but we'll simply word it as:

- *All inertial frames are "equivalent."*

This postulate basically says that a given inertial frame is no better than any other. There is no preferred reference frame. That is, it makes no sense to say that something is moving; it makes sense only to say that one thing is moving with respect to another. This is where the "Relativity" in Special Relativity comes from. There is no absolute frame; the motion of any frame is defined only relative to other frames.

This postulate also says that if the laws of physics hold in one inertial frame (and presumably they do hold in the frame in which I now sit),[14] then they hold in all others. It also says that if we have two frames S and S', then S should see things in S' in exactly the same way as S' sees things in S, because we can just switch the labels of S and S' (we'll get our money's worth out of this statement in the next few sections). It also says that empty space is homogeneous (that is, all points look the same), because we can pick any point to be, say, the origin of a coordinate system. It also says that empty space is isotropic (that is, all directions look the same), because we can pick any axis to be, say, the x axis of a coordinate system.

Unlike the first postulate, this second one is entirely reasonable. We've gotten used to having no special places in the universe. We gave up having the earth as the center, so let's not give any other point a chance, either.

Copernicus gave his reply
To those who had pledged to deny.
"All your addictions
To ancient convictions
Won't bring back your place in the sky."

[14] Technically, the earth is spinning while revolving around the sun, and there are also little vibrations in the floor beneath my chair, etc., so I'm not *really* in an inertial frame. But it's close enough for me.

The second postulate is nothing more than the familiar principle of Galilean invariance, assuming that the latter is written in the "The laws of physics hold in all inertial frames" form, and not in the form that explicitly mentions the Galilean transformations, which are inconsistent with the speed-of-light postulate.

Everything we've said here about the second postulate refers to empty space. If we have a chunk of mass, then there is certainly a difference between the position of the mass and a point a meter away. To incorporate mass into the theory, we would have to delve into General Relativity. But we won't have anything to say about that in this chapter. We will deal only with empty space, containing perhaps a few observant souls sailing along in rockets or floating aimlessly on little spheres. Though it may sound boring at first, it will turn out to be more exciting than you'd think.

REMARK: Given the second postulate, you might wonder if we even need the first. If all inertial frames are equivalent, shouldn't the speed of light be the same in any frame? Well, no. For all we know, light might behave like a baseball. A baseball certainly doesn't have the same speed with respect to different frames, and this doesn't ruin the equivalence of the frames.

It turns out (see Section 11.10) that nearly all of Special Relativity can be derived by invoking *only* the second postulate. The first postulate simply fills in the last bit of necessary information by stating that *something* has the same finite speed in every frame. It's actually not important that this thing happens to be light. It could be mashed potatoes or something else (well, it has to be massless, as we'll see in Chapter 12, so they'd have to be massless potatoes, but whatever), and the theory would come out the same. So to be a little more minimalistic, it's sufficient to state the first postulate as, "There is something that has the same speed in any inertial frame." It just so happens that in our universe this thing is what allows us to see.[15] ♣

11.3 The fundamental effects

The most striking effects of our two postulates are (1) the loss of simultaneity, (2) length contraction, and (3) time dilation. In this section, we'll discuss these three effects using some time-honored concrete examples. In the following section, we'll derive the Lorentz transformations using these three results.

11.3.1 Loss of simultaneity

Consider the following setup. In A's reference frame, a light source is placed midway between two receivers, a distance ℓ' from each (see Fig. 11.4). The light source emits a flash. From A's point of view, the light hits the two receivers at the same time, ℓ'/c seconds after the flash. Now consider another observer, B, who travels to the left at speed v. From her point of view, does the light hit the receivers at the same time? We will show that it does not.

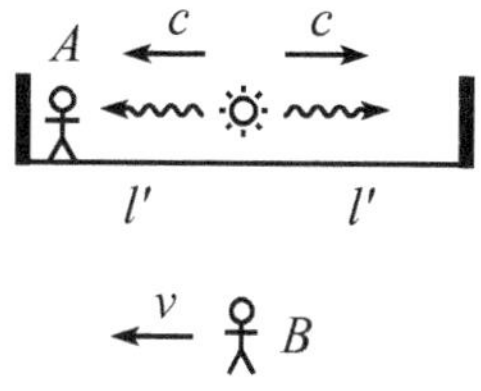

Fig. 11.4

[15] To go a step further, it's actually not even necessary for there to exist something that has the same speed in any frame. The theory will still come out the same if we write the first postulate as, "There is a limiting speed of an object in any frame." (See Section 11.10 for a discussion of this.) There's no need to have something that actually travels at this speed. It's conceivable to have a theory that contains no massless objects, so that everything travels slower than this limiting speed.

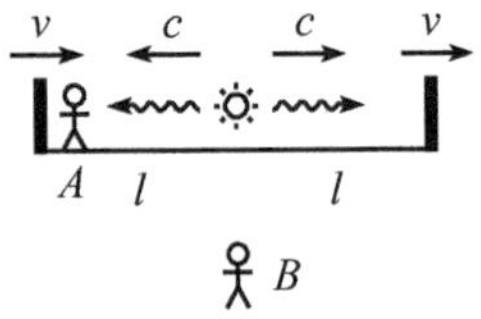

Fig. 11.5

In B's reference frame, the situation looks like that in Fig. 11.5. The receivers (along with everything else in A's frame) move to the right at speed v, and the light travels in both directions at speed c with respect to B (*not* with respect to the light source, as measured in B's frame; this is where the speed-of-light postulate comes into play). Therefore, the relative speed (as viewed by B) of the light and the left receiver is $c + v$, and the relative speed of the light and the right receiver is $c - v$.

REMARK: Yes, it's legal to just add and subtract these speeds to obtain the relative speeds *as viewed by B*. If v equals, say, $2 \cdot 10^8$ m/s, then in one second the left receiver moves $2 \cdot 10^8$ m to the right, while the left ray of light moves $3 \cdot 10^8$ m to the left. This means that they are now $5 \cdot 10^8$ m closer than they were a second ago. In other words, the relative speed (as measured by B) is $5 \cdot 10^8$ m/s, which is simply $c + v$. (Note that this implies that it's perfectly legal for the relative speed of two things, as measured by a third, to take any value up to $2c$.) Both the v and c here are measured with respect to the *same* person, namely B, so our intuition works fine. We don't need to use the "velocity-addition formula," which we'll derive in Section 11.5.1, and which is relevant in a different setup. I include this remark here just in case you've seen the velocity-addition formula and think it's relevant in this setup. But if it didn't occur to you, then never mind. ♣

Let ℓ be the distance from the source to the receivers, as measured by B.[16] Then in B's frame, the light hits the left receiver at t_{l} and the right receiver at t_{r}, where

$$t_{\text{l}} = \frac{\ell}{c+v}, \quad \text{and} \quad t_{\text{r}} = \frac{\ell}{c-v}. \tag{11.5}$$

These are not equal if $v \neq 0$. (The one exception is when $\ell = 0$, in which case the two events happen at the same place and same time in all frames.) The moral of this exercise is that it makes no sense to say that one event happens at the same time as another, unless you state which frame you're talking about. Simultaneity depends on the frame in which the observations are made.

Of the many effects, miscellaneous,
The loss of events, simultaneous,
Allows B to claim
There's no pause in A's frame,

REMARKS:

1. The invariance of the speed of light was used in saying that the two relative speeds above were $c + v$ and $c - v$. If we were talking about baseballs instead of light beams, then the relative speeds wouldn't look like this. If v_{b} is the speed at which the baseballs are thrown

[16] We'll see in Section 11.3.3 that ℓ is not equal to ℓ', due to length contraction. But this won't be important for what we're doing here. The only fact we need for now is that the light source is equidistant from the receivers, as measured by B. This is true because space is homogeneous, which implies that the length-contraction factor must be the same everywhere. More on this in Section 11.3.3.

in A's frame, then B sees the balls move at speeds $v_b - v$ to the left and $v_b + v$ to the right.[17] These are not equal to v_b, as they would be in the case of the light beams. The relative speeds between the balls and the left and right receivers are therefore $(v_b - v) + v = v_b$ and $(v_b + v) - v = v_b$. These are equal, so B sees the balls hit the receivers at the same time, as we know very well from everyday experience.

2. It is indeed legal in Eq. (11.5) to obtain the times by simply dividing ℓ by the relative speeds, $c + v$ and $c - v$. But if you want a more formal method, then you can use this reasoning: In B's frame, the position of the right photon is given by ct, and the position of the right receiver (which had a head start of ℓ) is given by $\ell + vt$. Equating these two positions gives $t_r = \ell/(c - v)$. Likewise for the left photon.
3. There is always a difference between the time an event happens and the time someone *sees* the event happen, because light takes time to travel from the event to the observer. What we calculated above were the times at which the events really happen. If we wanted to, we could calculate the times at which B *sees* the events occur, but such times are rarely important, and in general we won't be concerned with them. They can easily be calculated by adding on a (distance)$/c$ time difference for the path of the photons to B's eye. Of course, if B actually did the above experiment to find t_r and t_l, she would do it by writing down the times at which she saw the events occur, and then subtracting off the relevant (distance)$/c$ time differences to find when the events really happened.

   To sum up, the $t_r \neq t_l$ result in Eq. (11.5) is due to the fact that the events truly occur at different times in B's frame. *It has nothing to do with the time it takes light to travel to your eye.* In this chapter, we will often use sloppy language and say things like, "What time does B see event Q happen?" But we don't really mean, "When do B's eyes register that Q happened?" Instead, we mean, "What time does B *know* that event Q happened in her frame?" If we ever want to use "see" in the former sense, we will explicitly say so (as in Section 11.8 on the Doppler effect). ♣

Where this last line is not so extraneous.

Example (Rear clock ahead): Two clocks are positioned at the ends of a train of length L (as measured in its own frame). They are synchronized in the train frame. The train travels past you at speed v. It turns out that if you observe the clocks at simultaneous times in your frame, you will see the rear clock showing a higher reading than the front clock (see Fig. 11.6). By how much?

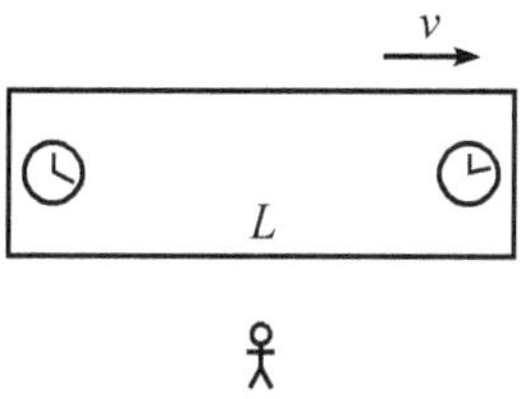

Fig. 11.6

Solution: As above, let's put a light source on the train, but let's now position it so that the light hits the clocks at the ends of the train at the same time in *your frame*. As above, the relative speeds of the photons and the clocks are $c + v$ and $c - v$ (as viewed in your frame). We therefore need to divide the train into lengths in this ratio, in your frame. But since length contraction (discussed in Section 11.3.3) is independent of position, this must also be the ratio in the train frame. So in the train frame, you can quickly show that two numbers that are in this ratio, and that add up to L, are $L(c + v)/2c$ and $L(c - v)/2c$.

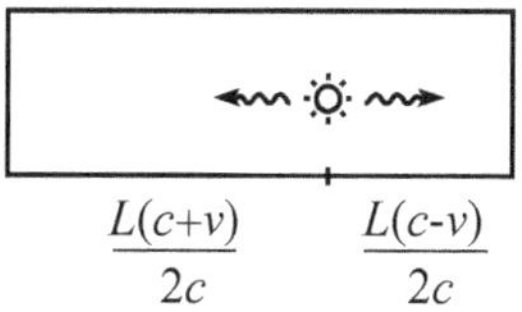

Fig. 11.7

The situation in the train frame therefore looks like that in Fig. 11.7. The light must travel an extra distance of $L(c+v)/2c - L(c-v)/2c = Lv/c$ to reach the rear clock. The light travels at speed c (as always), so the extra time is Lv/c^2. Therefore, the rear

[17] The velocity-addition formula in Section 11.5.1 shows that these formulas aren't actually correct. But they're close enough for our purposes here.

clock reads Lv/c^2 more when it is hit by the backward photon, compared with what the front clock reads when it is hit by the forward photon.

Now, let the instant you look at the clocks be the instant the photons hit them (that's why we chose the hittings to be simultaneous in your frame). Then from the previous paragraph, you observe the rear clock reading more than the front clock by an amount,

$$\text{(difference in readings)} = \frac{Lv}{c^2}\,. \tag{11.6}$$

Note that the L that appears here is the length of the train in its own frame, and not the shortened length that you observe in your frame (see Section 11.3.3). Appendix G gives a number of other derivations of Eq. (11.6), although they rely on material yet to come in this chapter and Chapter 14.

REMARKS:

1. This Lv/c^2 result has nothing to do with the fact that the rear clock takes more time to pass you.
2. The result does *not* say that you see the rear clock ticking at a faster rate than the front clock. They run at the same rate (both have the same time-dilation factor relative to you; see Section 11.3.2). The rear clock is simply a fixed time ahead of the front clock, as seen by you.
3. The fact that the rear clock is *ahead* of the front clock in your frame means that in the train frame the light hits the rear clock *after* it hits the front clock.
4. It's easy to forget which of the clocks is the one that is ahead. But a helpful mnemonic for remembering "rear clock ahead" is that both the first and fourth letters in each word form the same acronym, "rca," which is an anagram for "car," which is sort of like a train. Sure. ♣

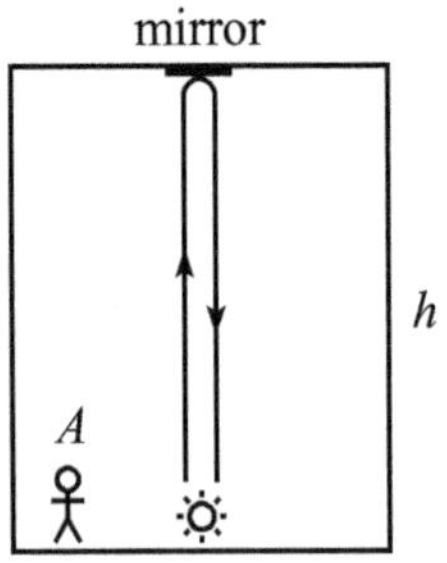

Fig. 11.8

11.3.2 Time dilation

We present here the classic example of a light beam traveling vertically on a train. Let there be a light source on the floor of the train and a mirror on the ceiling, which is a height h above the floor. Let observer A be at rest on the train, and let observer B be at rest on the ground. The speed of the train with respect to the ground is v.[18] A flash of light is emitted. The light travels up to the mirror, bounces off it, and then heads back down. Assume that after the light is emitted, we replace the source with a mirror, so that the light keeps bouncing up and down indefinitely.

In A's frame, the train is at rest. The path of the light is shown in Fig. 11.8. It takes the light a time h/c to reach the ceiling and then a time h/c to return to

[18] Technically, the words, "with respect to ...," should always be included when talking about speeds, because there is no absolute reference frame, and hence no absolute speed. But in the future, when it's clear what we mean (as in the case of a train moving with respect to the ground), we'll occasionally be sloppy and drop the "with respect to ..."

the floor. The roundtrip time is therefore

$$t_A = \frac{2h}{c}. \tag{11.7}$$

In B's frame, the train moves at speed v. The path of the light is shown in Fig. 11.9. The crucial fact to remember is that the speed of light in B's frame is still c. This means that the light travels along its diagonally upward path at speed c. (The vertical component of the speed is *not* c, as would be the case if light behaved like a baseball.) Since the horizontal component of the light's velocity is v,[19] the vertical component must be $\sqrt{c^2 - v^2}$, as shown in Fig. 11.10.[20] The time it takes to reach the mirror is therefore $h/\sqrt{c^2 - v^2}$,[21] so the roundtrip time is

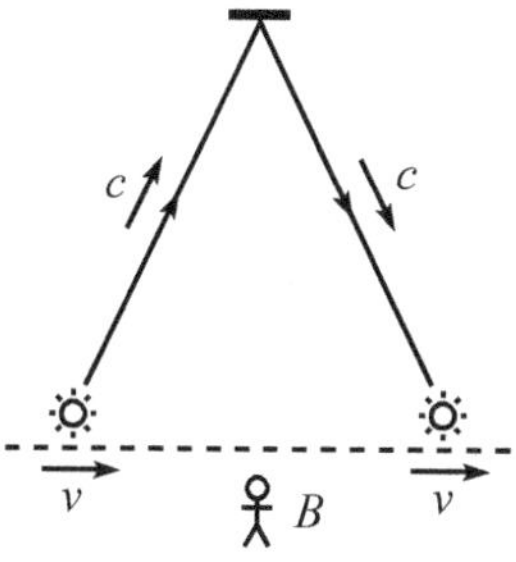

Fig. 11.9

$$t_B = \frac{2h}{\sqrt{c^2 - v^2}}. \tag{11.8}$$

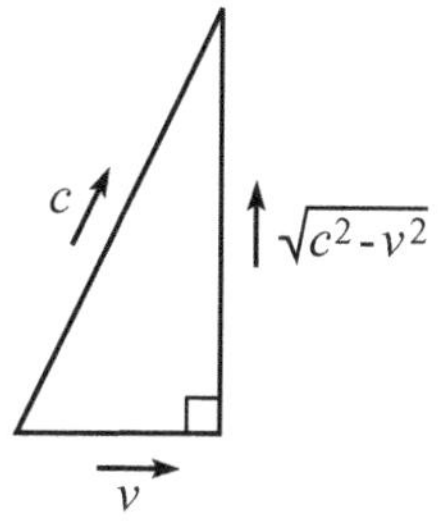

Fig. 11.10

Dividing Eq. (11.8) by Eq. (11.7) gives

$$t_B = \gamma t_A, \tag{11.9}$$

where

$$\gamma \equiv \frac{1}{\sqrt{1 - v^2/c^2}}. \tag{11.10}$$

This γ factor is ubiquitous in Special Relativity. Note that it is always greater than or equal to 1. This means that the roundtrip time is longer in B's frame than in A's frame.

What are the implications of this? For concreteness, let $v/c = 3/5$, so $\gamma = 5/4$. Then we may say the following. If A is standing next to the light source, and if B is standing on the ground, and if A claps his hands at $t_A = 4$ second intervals, then B observes claps at $t_B = 5$ second intervals (after having subtracted off the time for the light to travel to her eye, of course). This is true because both A and B must agree on the number of roundtrips the light beam takes between claps. If we assume, for convenience, that a roundtrip takes one second in A's frame (yes, that would be a tall train), then the four roundtrips between claps take five seconds in B's frame, using Eq. (11.9).

REMARK: We just made the claim that both A and B must agree on the number of roundtrips between claps. However, since A and B disagree on so many things (whether two events are

[19] Yes, it's still v. The light is always located on the vertical line between the source and the mirror. Since both of these objects move horizontally at speed v, the light does also.

[20] The Pythagorean theorem is indeed valid here. It's valid for distances, and since speeds are just distances divided by time, it's valid also for speeds.

[21] We've assumed that the height of the train in B's frame is still h. Although we'll see in Section 11.3.3 that there is length contraction along the direction of motion, there is none in the direction perpendicular to the motion (see Problem 11.1).

simultaneous, the rate at which clocks tick, and the length of things, as we'll see below), you might be wondering if there's *anything* they agree on. Yes, there are still frame-independent statements we can hang on to. If a bucket of paint flies past you and dumps paint on your head, then everyone agrees that you are covered in paint. Likewise, if A is standing next to the light clock and claps when the light reaches the floor, then everyone agrees on this. If the light is actually a strong laser pulse, and if A's clapping motion happens to bring his hands right over the mirror when the pulse gets there, then everyone agrees that his hands get burned by the laser. ♣

What if we have a train that doesn't contain one of our special light clocks? It doesn't matter. We *could* have built one if we wanted to, so the same results concerning the claps must hold. Therefore, light clock or no light clock, B observes A moving strangely slowly. From B's point of view, A's heart beats slowly, his blinks are lethargic, and his sips of coffee are slow enough to suggest that he needs another cup.

The effects of dilation of time
Are magical, strange, and sublime.
In your frame, this verse,
Which you'll see is not terse,
Can be read in the same amount of time it takes someone
else in another frame to read a similar sort of rhyme.

Our assumption that A is at rest on the train was critical in the above derivation. If A is moving with respect to the train, then Eq. (11.9) doesn't hold, because we *cannot* say that both A and B must agree on the number of roundtrips the light beam takes between claps, because of the problem with simultaneity. More precisely, if A is at rest on the train right next to the light source, then there are no issues with simultaneity, because the distance L in Eq. (11.6) is zero. And if A is at rest at a fixed distance from the source, then consider a person A' at rest on the train right next to the source. The distance L between A and A' is nonzero, so from the loss of simultaneity, B sees their two clocks read different times. But this difference is constant, so B sees A's clock tick at the same rate as A''s clock. Equivalently, we can just build a second light clock in a little box and have A hold it, and it will have the same speed v (and thus yield the same γ factor) as the first clock.

However, if A is moving with respect to the train, then we have a problem. If A' is again at rest next to the source, then the distance L between A and A' is changing, so B can't use the reasoning in the previous paragraph to conclude that A's and A''s clocks tick at the same rate. And in fact they do not, because as above, we can build another light clock and have A hold it. In this case, A's speed is what goes into the γ factor in Eq. (11.10), but this is different from A''s speed (which is the speed of the train).

Remarks:

1. The time dilation result derived in Eq. (11.9) is a bit strange, no doubt, but there doesn't seem to be anything downright incorrect about it until we look at the situation from A's

point of view. A sees B flying by at a speed v in the other direction. The ground is no more fundamental than a train, so the same reasoning applies. The time dilation factor, γ, doesn't depend on the sign of v, so A sees the same time dilation factor that B sees. That is, A sees B's clock running slow. But how can this be? Are we claiming that A's clock is slower than B's, and also that B's clock is slower than A's? Well... yes and no.

Remember that the above time-dilation reasoning applies only to a situation where something is motionless in the appropriate frame. In the second situation (where A sees B flying by), the statement $t_A = \gamma t_B$ holds only when the two events (say, two ticks on B's clock) happen at the same place in B's frame. But for two such events, they are certainly not in the same place in A's frame, so the $t_B = \gamma t_A$ result in Eq. (11.9) does *not* hold. The conditions of being motionless in each frame never both hold for a given setup (unless $v = 0$, in which case $\gamma = 1$ and $t_A = t_B$). So, the answer to the question at the end of the previous paragraph is "yes" if you ask the questions in the appropriate frames, and "no" if you think the answer should be frame independent.

2. Concerning the fact that A sees B's clock run slow, *and* B sees A's clock run slow, consider the following statement. "This is a contradiction. It is essentially the same as saying, 'I have two apples on a table. The left one is bigger than the right one, and the right one is bigger than the left one.' " How would you respond to this? Well, it is not a contradiction. Observers A and B are using *different coordinates* to measure time. The times measured in each of their frames are quite different things. The seemingly contradictory time-dilation result is really no stranger than having two people run away from each other into the distance, and having them both say that the other person looks smaller. In short, we are not comparing apples and apples. We are comparing apples and oranges. A more correct analogy would be the following. An apple and an orange sit on a table. The apple says to the orange, "You are a much uglier apple than I am," and the orange says to the apple, "You are a much uglier orange than I am."
3. One might view the statement, "A sees B's clock running slow, and also B sees A's clock running slow," as somewhat unsettling. But in fact it would be a complete disaster for the theory if A and B viewed each other in different ways. A critical fact in the theory of relativity is that A sees B in exactly the same way as B sees A.
4. In everything we've done so far, we've assumed that A and B are in inertial frames, because these are the frames that the postulates of Special Relativity deal with. However, it turns out that the time dilation result in Eq. (11.9) holds even if A is accelerating, as long as B isn't. In other words, if you're looking at a clock that is undergoing a complicated accelerating motion, then to figure out how fast it's ticking in your frame, all you need to know is its speed at any instant; its acceleration is irrelevant (this has plenty of experimental verification). If, however, *you* are accelerating, then all bets are off, and it's not valid for you to use the time dilation result when looking at a clock. It's still possible to get a handle on such situations, but we'll wait until Chapter 14 to do so. ♣

Example (Twin paradox): Twin A stays on the earth, while twin B flies quickly to a distant star and back (see Fig. 11.11). Show that B is younger than A when they meet up again.

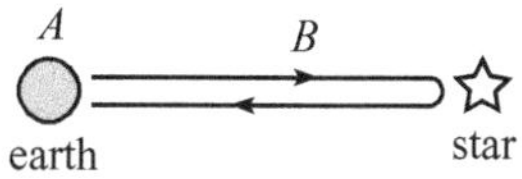

Fig. 11.11

Solution: From A's point of view, B's clock is running slow by a factor γ, on both the outward and return parts of the trip. Therefore, B is younger than A when they meet up again. This is the answer, and that's that. So if getting the right answer is all we care about, then we can pack up and go home. But our reasoning leaves one large point unaddressed. The "paradox" part of this example's title comes from the following alternate reasoning. Someone might say that in B's frame, A's clock is running slow by a factor γ, and so A is younger than B when they meet up again.

It's definitely true that when the two twins are standing next to each other (that is, when they are in the same frame), we can't have both B younger than A, and A younger than B. So what is wrong with the reasoning at the end of the previous paragraph? The error lies in the fact that there is no "one frame" that B is in. The inertial frame for the outward trip is different from the inertial frame for the return trip. The derivation of our time-dilation result applies only to one inertial frame.

Said in a different way, B accelerates when she turns around, and our time-dilation result holds only from the point of view of an *inertial* observer.[22] The symmetry in the problem is broken by the acceleration. If both A and B are blindfolded, they can still tell who is doing the traveling, because B will feel the acceleration at the turnaround. Constant velocity cannot be felt, but acceleration can be. (However, see Chapter 14 on General Relativity. Gravity complicates things.)

The above paragraphs show what is wrong with the "A is younger" reasoning, but they don't show how to modify it quantitatively to obtain the correct answer. There are many different ways of doing this, and you can tackle some of them in the problems (Exercise 11.67, Problems 11.2, 11.19, 11.24, and various problems in Chapter 14). Also, Appendix H gives a list of all the possible resolutions to the twin paradox that I can think of.

Example (Muon decay): Elementary particles called *muons* (which are identical to electrons, except that they are about 200 times as massive) are created in the upper atmosphere when cosmic rays collide with air molecules. The muons have an average lifetime of about $2 \cdot 10^{-6}$ seconds[23] (then they decay into electrons and neutrinos), and move at nearly the speed of light. Assume for simplicity that a certain muon is created at a height of 50 km, moves straight downward, has a speed $v = 0.99998\,c$, decays in exactly $T = 2 \cdot 10^{-6}$ seconds, and doesn't collide with anything on the way down.[24] Will the muon reach the earth before it (the muon!) decays?

Solution: The naive thing to say is that the distance traveled by the muon is $d = vT \approx (3 \cdot 10^8\,\text{m/s})(2 \cdot 10^{-6}\,\text{s}) = 600\,\text{m}$, and that this is less than 50 km, so the muon doesn't reach the earth. This reasoning is incorrect, because of the time-dilation effect. The muon lives longer in the earth frame, by a factor of γ, which is $\gamma = 1/\sqrt{1 - v^2/c^2} \approx 160$ here. The correct distance traveled in the earth frame is therefore $v(\gamma T) \approx 100\,\text{km}$. Hence, the muon travels the 50 km, with room to spare. The real-life fact that we actually do detect muons reaching the surface of the earth in the predicted abundances (while the naive $d = vT$ reasoning would predict that

[22] For the entire outward and return parts of the trip, B *does* observe A's clock running slow, but enough strangeness occurs during the turning-around period to make A end up older. Note, however, that a discussion of acceleration is not required to quantitatively understand the paradox, as Problem 11.2 shows.

[23] This is the "proper" lifetime, that is, the lifetime as measured in the frame of the muon.

[24] In the real world, the muons are created at various heights, move in different directions, have different speeds, decay in lifetimes that vary according to a standard half-life formula, and may very well bump into air molecules. So technically we've got everything wrong here. But that's no matter. This example will work just fine for the present purpose.

we shouldn't see any) is one of the many experimental tests that support the relativity theory.

11.3.3 Length contraction

Consider the following setup. Person A stands on a train which he measures to have length ℓ', and person B stands on the ground. The train moves at speed v with respect to the ground. A light source is located at the back of the train, and a mirror is located at the front. The source emits a flash of light that heads to the mirror, bounces off, then heads back to the source. By looking at how long this process takes in the two reference frames, we can determine the length of the train as measured by B.[25] In A's frame (see Fig. 11.12), the roundtrip time for the light is simply

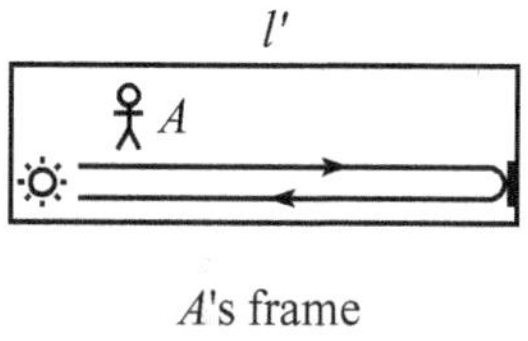

Fig. 11.12

$$t_A = \frac{2\ell'}{c}. \tag{11.11}$$

Things are a little more complicated in B's frame (see Fig. 11.13). Let the length of the train, as measured by B, be ℓ. For all we know at this point, ℓ may equal ℓ', but we'll soon find that it does not. The relative speed (as measured by B) of the light and the mirror during the first part of the trip is $c - v$. The relative speed during the second part is $c + v$. During each part, the light must close a gap with initial length ℓ. Therefore, the total roundtrip time is

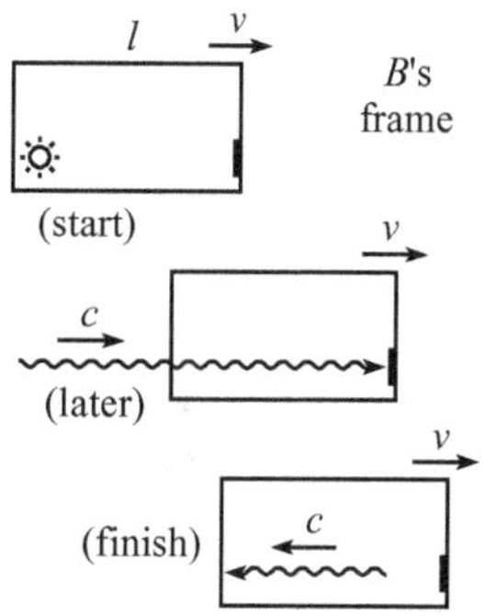

Fig. 11.13

$$t_B = \frac{\ell}{c - v} + \frac{\ell}{c + v} = \frac{2\ell c}{c^2 - v^2} \equiv \frac{2\ell}{c}\gamma^2. \tag{11.12}$$

But we know from Eq. (11.9) that

$$t_B = \gamma t_A. \tag{11.13}$$

This is a valid statement, because the two events we are concerned with (light leaving back, and light returning to back) happen at the same place in the train frame, so it's legal to use the time-dilation result in Eq. (11.9). Substituting the results for t_A and t_B from Eqs. (11.11) and (11.12) into Eq. (11.13), we find

$$\ell = \frac{\ell'}{\gamma}. \tag{11.14}$$

Note that we could not have used this setup to find the length contraction if we had not already found the time dilation in Eq. (11.9).

[25] The third remark below gives another (quicker) derivation of length contraction. But we'll go through the present derivation because the calculation is instructive.

Since $\gamma \geq 1$, we see that B measures the train to be shorter than A measures. The term *proper length* is used to describe the length of an object in its rest frame. So ℓ' is the proper length of the above train, and the length in any other frame is less than or equal to ℓ'. This length contraction is often called the *Lorentz–FitzGerald contraction,* for the reason given in Footnote 10.

Relativistic limericks have the attraction
Of being shrunk by a Lorentz contraction.
But for readers, unwary,
The results may be scary,
When a fraction …

REMARKS:

1. The length-contraction result in Eq. (11.14) is for lengths along the direction of the relative velocity. There is no length contraction in the perpendicular direction, as shown in Problem 11.1.
2. As with time dilation, this length contraction is a bit strange, but there doesn't seem to be anything actually paradoxical about it, until we look at things from A's point of view. To make a nice symmetrical situation, let's say B is standing on an identical train, which is motionless with respect to the ground. A sees B flying by at speed v in the other direction. Neither train is any more fundamental than the other, so the same reasoning applies, and A sees the same length contraction factor that B sees. That is, A measures B's train to be short. But how can this be? Are we claiming that A's train is shorter than B's, and also that B's train is shorter than A's? Is the actual setup the one shown in Fig. 11.14, or is it the one shown in Fig. 11.15? Which does it really look like? Well … it depends.

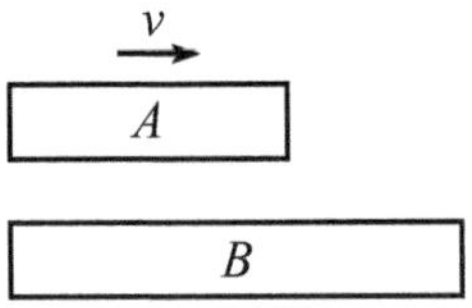

Fig. 11.14

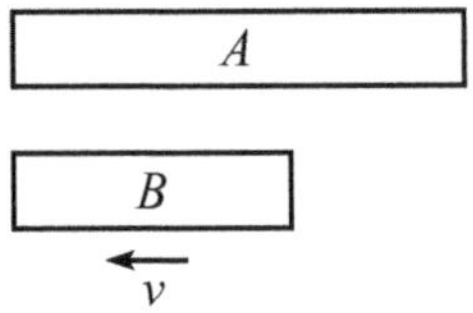

Fig. 11.15

   The word "is" in the above paragraph is a very bad word to use and is generally the cause of all the confusion (but it is ok in this paragraph, thankfully). There is no such thing as "is-ness" when it comes to lengths. It makes no sense to say what the length of the train really *is*. It makes sense only to say what the length is in a given frame. The situation doesn't really *look like* one thing in particular. The look depends on the frame in which the looking is being done.

   Let's be a little more specific. How do you measure a length? You write down the coordinates of the ends of something *measured simultaneously*, and then you take the difference. But the word "simultaneous" here should send up all sorts of red flags. Simultaneous events in one frame are not simultaneous events in another. Stated more precisely, here is what we are claiming: Let B write down simultaneous coordinates of the ends of A's train, and also simultaneous coordinates of the ends of her own train. Then the difference between the former is smaller than the difference between the latter. Likewise, let A write down simultaneous coordinates of the ends of B's train, and also simultaneous coordinates of the ends of his own train. Then the difference between the former is smaller than the difference between the latter. There is no contradiction here, because the times at which A and B are writing down the coordinates don't have much to do with each other, due to the loss of simultaneity. You can be quantitative about this in Problem 11.3. As with time dilation, we are comparing apples and oranges.
3. There is an easy argument to show that time dilation implies length contraction, and vice versa. Let B stand on the ground, next to a stick of length ℓ. Let A fly past the stick at speed v. In B's frame, it takes A a time of ℓ/v to traverse the length of the stick. Therefore (assuming that we have demonstrated the time-dilation result), a watch on A's wrist will advance by a time of only $\ell/\gamma v$ while he traverses the length of the stick.

How does A view the situation? He sees the ground and the stick fly by at speed v. The time between the two ends passing him is $\ell/\gamma v$ (because that is the time elapsed on his watch). To get the length of the stick in his frame, he simply multiplies the speed by the time. That is, he measures the length to be $(\ell/\gamma v)v = \ell/\gamma$, which is the desired contraction. The same argument also shows that length contraction implies time dilation.

4. As mentioned earlier, the length contraction factor γ is independent of position on the object. That is, all parts of the train are contracted by the same amount. This follows from the fact that all points in space are equivalent. Equivalently, we could put a large number of small replicas of the above source–mirror system along the length of the train. They would all produce the same value for γ, independent of the position on the train.
5. If you still want to ask, "Is the contraction really *real*?" then consider the following hypothetical undertaking. Imagine a sheet of paper moving sideways past the Mona Lisa, skimming the surface. A standard sheet of paper is plenty large enough to cover her face, so if the paper is moving slowly, and if you take a photograph at the appropriate time, then in the photo you'll see her entire face covered by the paper. However, if the sheet is flying by sufficiently fast, and if you take a photograph at the appropriate time, then in the photo you'll see a thin vertical strip of paper covering only a small fraction of her face. So you'll still see her smiling at you. ♣

Example (Passing trains): Two trains, A and B, each have proper length L and move in the same direction. A's speed is $4c/5$, and B's speed is $3c/5$. A starts behind B (see Fig. 11.16). How long, as viewed by person C on the ground, does it take for A to overtake B? By this we mean the time between the front of A passing the back of B, and the back of A passing the front of B.

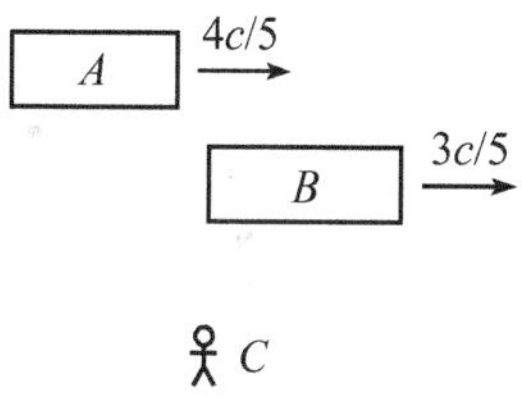

Fig. 11.16

Solution: Relative to C on the ground, the γ factors associated with A and B are $5/3$ and $5/4$, respectively. Therefore, their lengths in the ground frame are $3L/5$ and $4L/5$. While overtaking B, A must travel farther than B, by an excess distance equal to the sum of the lengths of the trains, which is $7L/5$. The relative speed of the two trains (as viewed by C on the ground) is the difference of the speeds, which is $c/5$. The total time is therefore

$$t_C = \frac{7L/5}{c/5} = \frac{7L}{c}. \tag{11.15}$$

Example (Muon decay, again): Consider the "Muon decay" example from Section 11.3.2. From the muon's point of view, it lives for a time of $T = 2 \cdot 10^{-6}$ seconds, and the earth is speeding toward it at $v = 0.99998c$. How, then, does the earth (which travels only $d = vT \approx 600\,\text{m}$ before the muon decays) reach the muon?

Solution: The important point here is that in the muon's frame, the distance to the earth is contracted by a factor $\gamma \approx 160$. Therefore, the earth starts only $50\,\text{km}/160 \approx 300\,\text{m}$ away. Since the earth can travel a distance of 600 m during the muon's lifetime, the earth collides with the muon, with time to spare.

As stated in the third remark above, time dilation and length contraction are intimately related. We can't have one without the other. In the earth's frame, the muon's arrival on the earth is explained by time dilation. In the muon's frame, it is explained by length contraction.

Observe that for muons created,
The dilation of time is related
To Einstein's insistence
Of shrunken-down distance
In the frame where decays aren't belated.

An extremely important strategy in solving relativity problems is to plant yourself in a frame and *stay there*. The only thoughts running through your head should be what *you* observe. That is, don't try to use reasoning along the lines of, "Well, the person I'm looking at in this other frame sees such-and-such." This will almost certainly cause an error somewhere along the way, because you will inevitably end up writing down an equation that combines quantities that are measured in different frames, which is a no-no. Of course, you might want to solve another part of the problem by working in another frame, or you might want to redo the whole problem in another frame. That's fine, but once you decide which frame you're going to use, make sure you put yourself there and stay there.

Another very important strategy is to draw a picture of the setup (in whatever frame you've chosen) at every moment when something significant happens, as we did in Fig. 11.13. Once we drew the pictures there, it was clear what we needed to do. But without the pictures, we almost certainly would have gotten confused.

At this point you might want to look at the "Qualitative relativity questions" in Appendix F, just to make sure we're all on the same page. Some of the questions deal with material we haven't covered yet, but most are relevant to what we've done so far.

This concludes our treatment of the three fundamental effects. In the next section, we'll combine all the information we've gained and use it to derive the Lorentz transformations. But one last comment before we get to those:

Lattice of clocks and meter sticks

In everything we've done so far, we've taken the route of having observers sitting in various frames, making various measurements. But as mentioned earlier, this can cause some ambiguity, because you might think that the time when light reaches the observer is important, whereas what we are generally concerned with is the time when something actually happens.

A way to avoid this ambiguity is to remove the observers and then imagine filling up space with a large rigid lattice of meter sticks and synchronized clocks. Different frames are defined by different lattices; assume that the lattices of different frames can somehow pass freely through each other. All the meter sticks in a given frame are at rest with respect to all the others, so we don't have to worry about issues of length contraction within each frame. But the lattice of a frame moving past you is squashed in the direction of motion, because all the meter sticks in that direction are contracted.

To measure the length of an object in a given frame, we just need to determine where the ends are (at simultaneous times, as measured in that frame) with respect to the lattice. As far as the synchronization of the clocks within each frame goes, this can be accomplished by putting a light source midway between any two clocks and sending out signals, and then setting the clocks to a certain value when the signals hit them. Alternatively, a more straightforward method of synchronization is to start with all the clocks synchronized right next to each other, and then move them very slowly to their final positions. Any time-dilation effects can be made arbitrarily small by moving the clocks sufficiently slowly. This is true because the time-dilation γ factor is second order in v, but the time it takes a clock to reach its final position is only first order in $1/v$.

This lattice way of looking at things emphasizes that observers are not important, and that a frame is defined simply as a lattice of space and time coordinates. Anything that happens (an "event") is automatically assigned a space and time coordinate in every frame, independent of any observer. The concept of an "event" will be very important in the next section.

11.4 The Lorentz transformations

11.4.1 The derivation

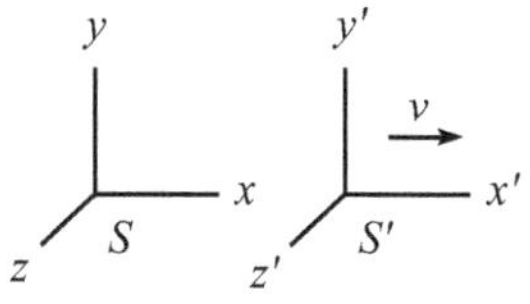

Fig. 11.17

Consider a coordinate system, S', moving relative to another system, S (see Fig. 11.17). Let the constant relative speed of the frames be v. Let the corresponding axes of S and S' point in the same direction, and let the origin of S' move along the x axis of S, in the positive direction. Nothing exciting happens in the y and z directions (see Problem 11.1), so we'll ignore them.

Our goal in this section is to look at two events (an event is anything that has space and time coordinates) in spacetime and relate the Δx and Δt of the coordinates in one frame to the $\Delta x'$ and $\Delta t'$ of the coordinates in another. We therefore want to find the constants A, B, C, and D in the relations,

$$\begin{aligned} \Delta x &= A\,\Delta x' + B\,\Delta t', \\ \Delta t &= C\,\Delta t' + D\,\Delta x'. \end{aligned} \tag{11.16}$$

The four constants here will end up depending on v (which is constant, given the two inertial frames). But we won't explicitly write this dependence, for ease of notation.

REMARKS:

1. We have assumed in Eq. (11.16) that Δx and Δt are linear functions of $\Delta x'$ and $\Delta t'$. And we have also assumed that A, B, C, and D are constants, that is, they depend at most on v, and not on x, t, x', t'.

   The first of these assumptions is justified by the fact that any finite interval can be built up from a series of many infinitesimal ones. But for an infinitesimal interval, any terms such as $(\Delta t')^2$, for example, are negligible compared with the linear terms. Therefore, if we add up all the infinitesimal intervals to obtain a finite one, we will be left with only the linear terms. Equivalently, it shouldn't matter whether we make a measurement with, say, meter sticks or half-meter sticks.

   The second assumption can be justified in various ways. One is that all inertial frames should agree on what "nonaccelerating" motion is. That is, if $\Delta x' = u' \Delta t'$, then we should also have $\Delta x = u \Delta t$, for some constant u. This is true only if the above coefficients are constants, as you can check. Another justification comes from the second of our two relativity postulates, which says that all points in (empty) space are indistinguishable. With this in mind, let's assume that we have a transformation of the form, say, $\Delta x = A\,\Delta x' + B\,\Delta t' + Ex'\,\Delta x'$. The x' in the last term implies that the absolute location in spacetime (and not just the relative position) is important. Therefore, this last term cannot exist.
2. If the relations in Eq. (11.16) turned out to be the usual Galilean transformations (which are the ones that hold for everyday relative speeds v) then we would have $\Delta x = \Delta x' + v\,\Delta t$, and $\Delta t = \Delta t'$ (that is, $A = C = 1$, $B = v$, and $D = 0$). We will find, however, that under the assumptions of Special Relativity, this is *not* the case. The Galilean transformations are not the correct transformations. But we will show below that the correct transformations do indeed reduce to the Galilean transformations in the limit of slow speeds, as they must. ♣

The constants A, B, C, and D in Eq. (11.16) are four unknowns, and we can solve for them by using four facts we found above in Section 11.3. The four facts we will use are:

	Effect	Condition	Result	Eq. in text
1	Time dilation	$x' = 0$	$t = \gamma t'$	(11.9)
2	Length contraction	$t' = 0$	$x' = x/\gamma$	(11.14)
3	Relative v of frames	$x = 0$	$x' = -vt'$	
4	Rear clock ahead	$t = 0$	$t' = -vx'/c^2$	(11.6)

We have taken the liberty of dropping the Δ's in front of the coordinates, lest things get too messy. We will often omit the Δ's, but it should be understood that x really means Δx, etc. We are always concerned with the *difference* between coordinates of two events in spacetime. The actual value of any coordinate is irrelevant, because there is no preferred origin in any frame.

You should pause for a moment and verify that the four "results" in the above table are in fact the proper mathematical expressions for the four effects, given the stated "conditions."[26] My advice is to keep pausing until you're comfortable

[26] We can state the effects in other ways too, by switching the primes and unprimes. For example, time dilation can be written as "$t' = \gamma t$ when $x = 0$." But we've chosen the above ways of writing things because they will allow us to solve for the four unknowns in the most efficient way.

with all the entries in the table. Note that the sign in the "rear clock ahead" effect is indeed correct, because the front clock shows less time than the rear clock. So the clock with the higher x' value is the one with the lower t' value.

We can now use our four facts in the above table to quickly solve for the unknowns A, B, C, and D in Eq. (11.16).

Fact (1) gives $C = \gamma$.
Fact (2) gives $A = \gamma$.
Fact (3) gives $B/A = v \Longrightarrow B = \gamma v$.
Fact (4) gives $D/C = v/c^2 \Longrightarrow D = \gamma v/c^2$.

Equations (11.16), which are known as the *Lorentz transformations*, are therefore given by

$$\begin{aligned} \Delta x &= \gamma(\Delta x' + v\,\Delta t'), \\ \Delta t &= \gamma(\Delta t' + v\,\Delta x'/c^2), \\ \Delta y &= \Delta y', \\ \Delta z &= \Delta z', \end{aligned} \tag{11.17}$$

where

$$\gamma \equiv \frac{1}{\sqrt{1 - v^2/c^2}}. \tag{11.18}$$

We have tacked on the trivial transformations for y and z, but we won't bother writing these in the future. Also, we'll drop the Δ's from now on, but remember that they're always really there.

If we solve for x' and t' in terms of x and t in Eq. (11.17), then we see that the inverse Lorentz transformations are given by

$$\begin{aligned} x' &= \gamma(x - vt), \\ t' &= \gamma(t - vx/c^2). \end{aligned} \tag{11.19}$$

Of course, which ones you label as the "inverse" transformations depends on your point of view. But it's intuitively clear that the only difference between the two sets of equations is the sign of v, because S is simply moving backward with respect to S'.

The reason why the derivation of Eqs. (11.17) was so quick is that we already did most of the work in Section 11.3 when we derived the fundamental effects. If we wanted to derive the Lorentz transformations from scratch, that is, by starting with the two postulates in Section 11.2, then the derivation would be longer. In Appendix I we give such a derivation, where it is clear what information comes from each of the postulates. The procedure there is somewhat cumbersome, but it's worth taking a look at, because we will invoke the result in a very cool way in Section 11.10.

REMARKS:

1. In the limit $v \ll c$ (or more precisely, in the limit $vx'/c^2 \ll t'$, which means that even if v is small, we have to be careful that x' isn't too large), Eqs. (11.17) reduce to $x = x'+vt$ and $t = t'$, which are the good old Galilean transformations. This must be the case, because we know from everyday experience (where $v \ll c$) that the Galilean transformations work just fine.
2. Equations (11.17) exhibit a nice symmetry between x and ct. With $\beta \equiv v/c$, we have

$$x = \gamma\big(x' + \beta(ct')\big),$$
$$ct = \gamma\big((ct') + \beta x'\big). \tag{11.20}$$

Equivalently, in units where $c = 1$ (for example, where one unit of distance equals $3 \cdot 10^8$ meters, or where one unit of time equals $1/(3 \cdot 10^8)$ seconds), Eqs. (11.17) take the symmetric form,

$$x = \gamma(x' + vt'),$$
$$t = \gamma(t' + vx'). \tag{11.21}$$

3. In matrix form, Eq. (11.20) can be written as

$$\begin{pmatrix} x \\ ct \end{pmatrix} = \begin{pmatrix} \gamma & \gamma\beta \\ \gamma\beta & \gamma \end{pmatrix} \begin{pmatrix} x' \\ ct' \end{pmatrix}. \tag{11.22}$$

This looks similar to a rotation matrix. More about this in Section 11.9, and in Problem 11.27.
4. We did the above derivation in terms of a primed and an unprimed system. But when you're doing problems, it's usually best to label your coordinates with subscripts such as A for Alice, or T for train. In addition to being more informative, this notation is less likely to make you think that one frame is more fundamental than the other.
5. It's easy to get confused about the sign on the right-hand side of the Lorentz transformations. To figure out if it should be a plus or a minus, write down $x_A = \gamma(x_B \pm vt_B)$, and then imagine sitting in system A and looking at a fixed point in B. This fixed point satisfies (putting the Δ's back in to avoid any mixup) $\Delta x_B = 0$, which gives $\Delta x_A = \pm\gamma v \Delta t_B$. So if the point moves to the right (that is, if it increases as time increases), then pick the "+." And if it moves to the left, then pick the "−." In other words, the sign is determined by which way A (the person associated with the coordinates on the left-hand side of the equation) sees B (ditto for the right-hand side) moving.
6. One very important thing we must check is that two successive Lorentz transformations (from S_1 to S_2 and then from S_2 to S_3) again yield a Lorentz transformation (from S_1 to S_3). This must be true because we showed that any two frames must be related by Eqs. (11.17). If we composed two L.T.'s (along the same direction) and found that the transformation from S_1 to S_3 was not of the form of Eqs. (11.17), for some new v, then the whole theory would be inconsistent, and we would have to drop one of our postulates.[27] You can show that the combination of an L.T. (with speed v_1) and an L.T. (with speed v_2) does indeed yield an L.T., and it has speed $(v_1 + v_2)/(1 + v_1 v_2/c^2)$. This is the task of Exercise 11.47, and also Problem 11.27 (which is stated in terms of *rapidity*, introduced

[27] This statement is true only for the composition of two L.T.'s in the *same* direction. If we composed an L.T. in the x direction with one in the y direction, the result would interestingly *not* be an L.T. along some new direction, but rather the composition of an L.T. along some direction and a rotation through some angle. This rotation results in what is known as the *Thomas precession*. See the appendix of Muller (1992) for a quick derivation of the Thomas precession. For further discussion, see Costella *et al.* (2001) and Rebilas (2002).

in Section 11.9). This resulting speed is one that we'll see again when we get to the velocity-addition formula in Section 11.5.1. ♣

Example: A train with proper length L moves at speed $5c/13$ with respect to the ground. A ball is thrown from the back of the train to the front. The speed of the ball with respect to the train is $c/3$. As viewed by someone on the ground, how much time does the ball spend in the air, and how far does it travel?

Solution: The γ factor associated with the speed $5c/13$ is $\gamma = 13/12$. The two events we are concerned with are "ball leaving back of train" and "ball arriving at front of train." The spacetime separation between these events is easy to calculate on the train. We have $\Delta x_{\mathrm{T}} = L$, and $\Delta t_{\mathrm{T}} = L/(c/3) = 3L/c$. The Lorentz transformations giving the coordinates on the ground are therefore

$$\begin{aligned} x_{\mathrm{G}} = \gamma(x_{\mathrm{T}} + vt_{\mathrm{T}}) &= \frac{13}{12}\left(L + \left(\frac{5c}{13}\right)\left(\frac{3L}{c}\right)\right) = \frac{7L}{3}, \\ t_{\mathrm{G}} = \gamma(t_{\mathrm{T}} + vx_{\mathrm{T}}/c^2) &= \frac{13}{12}\left(\frac{3L}{c} + \frac{(5c/13)L}{c^2}\right) = \frac{11L}{3c}. \end{aligned} \tag{11.23}$$

In a given problem, such as the above example, one of the frames usually allows for a quick calculation of Δx and Δt, so you simply have to mechanically plug these quantities into the L.T.'s to obtain $\Delta x'$ and $\Delta t'$ in the other frame, where they may not be as obvious.

Relativity is a subject in which there are usually many ways to do a problem. If you're trying to find some Δx's and Δt's, then you can use the L.T.'s, or perhaps the invariant interval (introduced in Section 11.6), or maybe a velocity-addition approach (introduced in Section 11.5.1), or even the sending-of-light-signals strategy used in Section 11.3. Depending on the specific problem and what your personal preferences are, certain approaches will be more enjoyable than others. But no matter which method you choose, you should take advantage of the plethora of possibilities by picking a second method to double-check your answer. Personally, I find the L.T.'s to be the perfect option for this, because the other methods are generally more fun when solving a problem for the first time, while the L.T.'s are usually quick and easy to apply (perfect for a double-check).[28]

The excitement will build in your voice,
As you rise from your seat and rejoice,
"A Lorentz transformation
Provides information,
As an alternate method of choice!"

[28] I would, however, be very wary of solving a problem using only the L.T.'s, with no other check, because it's very easy to mess up a sign in the transformations. And since there's nothing to do except mechanically plug in numbers, there's not much opportunity for an intuitive check, either.

11.4.2 The fundamental effects

Let's now see how the Lorentz transformations imply the three fundamental effects (namely, loss of simultaneity, time dilation, and length contraction) discussed in Section 11.3. Of course, we just used these effects to *derive* the L.T.'s, so we know everything will work out. We'll just be going in circles. But since these fundamental effects are, well, fundamental, let's belabor the point and discuss them one more time, with the starting point being the L.T.'s.

Loss of simultaneity

Let two events occur simultaneously in frame S'. Then the separation between them, as measured by S', is $(x', t') = (x', 0)$. As usual, we are not bothering to write the Δ's in front of the coordinates. Using the second of Eqs. (11.17), we see that the time between the events, as measured by S, is $t = \gamma v x'/c^2$. This is not equal to zero (unless $x' = 0$). Therefore, the events do not occur simultaneously in frame S.

Time dilation

Consider two events that occur in the same place in S'. Then the separation between them is $(x', t') = (0, t')$. Using the second of Eqs. (11.17), we see that the time between the events, as measured by S, is

$$t = \gamma t' \quad (\text{if } x' = 0). \tag{11.24}$$

The factor γ is greater than or equal to 1, so $t \geq t'$. The passing of one second on S''s clock takes more than one second on S's clock. S sees S' drinking his coffee very slowly.

The same strategy works if we interchange S and S'. Consider two events that occur in the same place in S. The separation between them is $(x, t) = (0, t)$. Using the second of Eqs. (11.19), we see that the time between the events, as measured by S', is

$$t' = \gamma t \quad (\text{if } x = 0). \tag{11.25}$$

Therefore, $t' \geq t$. Another way to derive this is to use the first of Eqs. (11.17) to write $x' = -vt'$, and then substitute this into the second equation.

Remark: If we write down the two above equations by themselves, $t = \gamma t'$ and $t' = \gamma t$, they appear to contradict each other. This apparent contradiction arises from the omission of the conditions they are based on. The former equation is based on the assumption that $x' = 0$. The latter equation is based on the assumption that $x = 0$. They have nothing to do with each other. It would perhaps be better to write the equations as

$$(t = \gamma t')_{x'=0}, \quad \text{and} \quad (t' = \gamma t)_{x=0}, \tag{11.26}$$

but this is somewhat cumbersome. ♣

Length contraction

This proceeds just like the time dilation above, except that now we want to set certain time intervals equal to zero, instead of certain space intervals. We want to do this because to measure a length, we calculate the distance between two points whose positions are measured *simultaneously*. That's what a length is.

Consider a stick at rest in S', where it has length ℓ'. We want to find the length ℓ in S. Simultaneous measurements of the coordinates of the ends of the stick in S yield a separation of $(x, t) = (x, 0)$. Using the first of Eqs. (11.19), we have

$$x' = \gamma x \quad (\text{if } t = 0). \tag{11.27}$$

But x is by definition the length in S. And x' is the length in S', because the stick isn't moving in S'.[29] Therefore, $\ell = \ell'/\gamma$. And since $\gamma \geq 1$, we have $\ell \leq \ell'$, so S sees the stick shorter than S' sees it.

Now interchange S and S'. Consider a stick at rest in S, where it has length ℓ. We want to find the length in S'. Measurements of the coordinates of the ends of the stick in S' yield a separation of $(x', t') = (x', 0)$. Using the first of Eqs. (11.17), we have

$$x = \gamma x' \quad (\text{if } t' = 0). \tag{11.28}$$

But x' is by definition the length in S'. And x is the length in S, because the stick is not moving in S. Therefore, $\ell' = \ell/\gamma$, so $\ell' \leq \ell$.

REMARK: As with time dilation, if we write down the two equations by themselves, $\ell = \ell'/\gamma$ and $\ell' = \ell/\gamma$, they appear to contradict each other. But as before, this apparent contradiction arises from the omission of the conditions they are based on. The former equation is based on the assumptions that $t = 0$ and that the stick is at rest in S'. The latter equation is based on the assumptions that $t' = 0$ and that the stick is at rest in S. They have nothing to do with each other. We should really write,

$$(x = x'/\gamma)_{t=0}, \quad \text{and} \quad (x' = x/\gamma)_{t'=0}, \tag{11.29}$$

and then identify x' in the first equation with ℓ' only after invoking the further assumption that the stick is at rest in S'. Likewise for the second equation. But this is a pain. ♣

11.5 Velocity addition

11.5.1 Longitudinal velocity addition

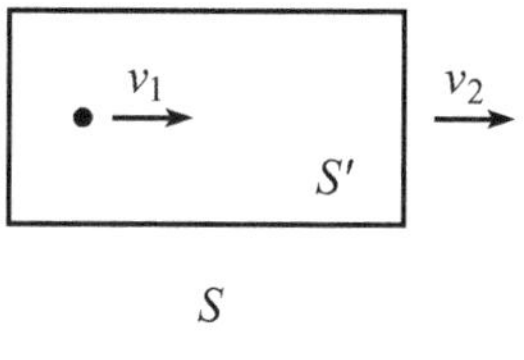

Fig. 11.18

Consider the following setup. An object moves at speed v_1 with respect to frame S'. And frame S' moves at speed v_2 with respect to frame S, in the same direction as the motion of the object (see Fig. 11.18). What is the speed, u, of the object with respect to frame S?

[29] The measurements of the ends made by S are *not* simultaneous in the S' frame. In the S' frame, the separation between the events is (x', t'), where both x' and t' are nonzero. This doesn't satisfy our definition of a length measurement in S' (because $t' \neq 0$), but the stick isn't moving in S', so S' can measure the ends whenever he feels like it, and he will always get the same difference. So x' is indeed the length in the S' frame.

The Lorentz transformations can be used to easily answer this question. The relative speed of the frames is v_2. Consider two events along the object's path (for example, say it makes two beeps). We are given that $\Delta x'/\Delta t' = v_1$. Our goal is to find $u \equiv \Delta x/\Delta t$. The Lorentz transformations from S' to S, Eqs. (11.17), are

$$\Delta x = \gamma_2(\Delta x' + v_2\Delta t'), \quad \text{and} \quad \Delta t = \gamma_2(\Delta t' + v_2\Delta x'/c^2), \tag{11.30}$$

where $\gamma_2 \equiv 1/\sqrt{1 - v_2^2/c^2}$. Therefore,

$$\begin{aligned} u \equiv \frac{\Delta x}{\Delta t} &= \frac{\Delta x' + v_2\Delta t'}{\Delta t' + v_2\Delta x'/c^2} \\ &= \frac{\Delta x'/\Delta t' + v_2}{1 + v_2(\Delta x'/\Delta t')/c^2} \\ &= \frac{v_1 + v_2}{1 + v_1 v_2/c^2}. \end{aligned} \tag{11.31}$$

This is the *velocity-addition formula*, for adding velocities along the same line. Let's look at some of its properties.

- It is symmetric with respect to v_1 and v_2, as it should be, because we could switch the roles of the object and frame S.
- For $v_1 v_2 \ll c^2$, it reduces to $u \approx v_1 + v_2$, which we know holds perfectly well for everyday speeds.
- If $v_1 = c$ or $v_2 = c$, then we find $u = c$, as should be the case, because anything that moves with speed c in one frame moves with speed c in another.
- The maximum (or minimum) of u in the region $-c \le v_1, v_2 \le c$ equals c (or $-c$), which can be seen by noting that $\partial u/\partial v_1$ and $\partial u/\partial v_2$ are never zero in the interior of the region.

If you take any two velocities that are less than c and add them according to Eq. (11.31), then you will obtain a velocity that is again less than c. This shows that no matter how much you keep accelerating an object (that is, no matter how many times you give the object a speed v_1 with respect to the frame moving at speed v_2 that it was just in), you can't bring the speed up to the speed of light. We'll give another argument for this result in Chapter 12 when we discuss energy.

For a bullet, a train, and a gun,
Adding the speeds can be fun.
Take a trip down the path
Paved with Einstein's new math,
Where a half plus a half isn't one.

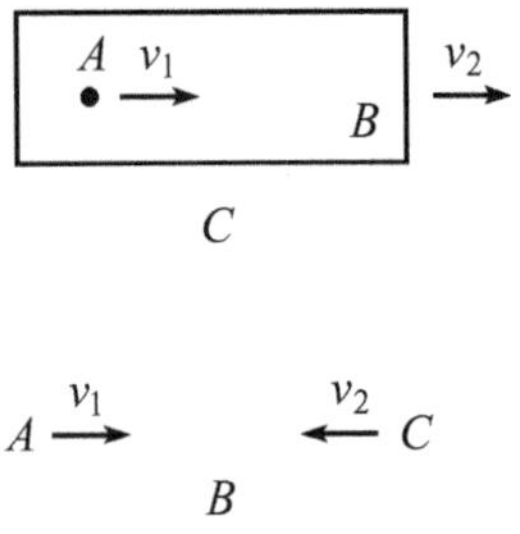

Fig. 11.19

REMARK: Consider the two scenarios shown in Fig. 11.19. If the goal is to find the velocity of A with respect to C, then the velocity-addition formula applies to both scenarios, because the second scenario is the same as the first one, as observed in B's frame.

The velocity-addition formula applies when we ask, "If A moves at v_1 with respect to B, and B moves at v_2 with respect to C (which means, of course, that C moves at speed v_2 with respect to B), then how fast does A move with respect to C?" The formula does *not* apply if we ask the more mundane question, "What is the relative speed of A and C, as viewed by B?" The answer to this is just $v_1 + v_2$.

In short, if the two velocities are given with respect to the *same* observer, say B, and if you are asking for the relative velocity as measured by B, then you simply have to add the velocities.[30] But if you are asking for the relative velocity as measured by A or C, then you have to use the velocity-addition formula. It makes no sense to add velocities that are measured with respect to different observers. Doing so would involve adding things that are measured in different coordinate systems, which is meaningless. In other words, taking the velocity of A with respect to B and adding it to the velocity of B with respect to C, hoping to obtain the velocity of A with respect to C, is invalid. ♣

Example (Passing trains, again): Consider again the scenario in the "Passing trains" example in Section 11.3.3.

(a) How long, as viewed by A and as viewed by B, does it take for A to overtake B?

(b) Let event E_1 be "the front of A passing the back of B", and let event E_2 be "the back of A passing the front of B." Person D walks at constant speed from the back of B to the front (see Fig. 11.20), such that he coincides with both events E_1 and E_2. How long does the "overtaking" process take, as viewed by D?

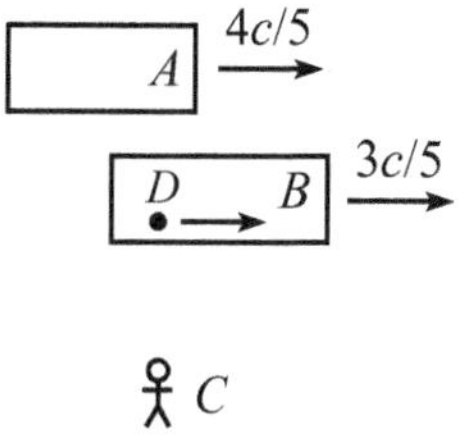

Fig. 11.20

Solution:

(a) First consider B's point of view. From the velocity-addition formula, B sees A move with speed

$$u = \frac{\frac{4c}{5} - \frac{3c}{5}}{1 - \frac{4}{5} \cdot \frac{3}{5}} = \frac{5c}{13}. \tag{11.32}$$

The γ factor associated with this speed is $\gamma = 13/12$. Therefore, B sees A's train contracted to a length $12L/13$. During the overtaking, A must travel a distance equal to the sum of the lengths of the trains in B's frame (see Fig. 11.21), which is $L + 12L/13 = 25L/13$. Since A moves at speed $5c/13$, the total time in B's frame is

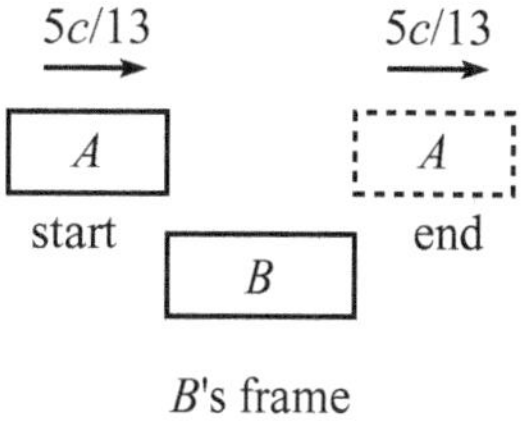

Fig. 11.21

$$t_B = \frac{25L/13}{5c/13} = \frac{5L}{c}. \tag{11.33}$$

The exact same reasoning holds from A's point of view, so we have $t_A = t_B = 5L/c$.

[30] Note that the resulting speed can certainly be greater than c. If I see a ball heading toward me at $0.9c$ from the right, and another one heading toward me at $0.9c$ from the left, then the relative speed of the balls in my frame is $1.8c$. In the frame of one of the balls, however, the relative speed is $(1.8/1.81)c \approx (0.9945)c$, from Eq. (11.31).

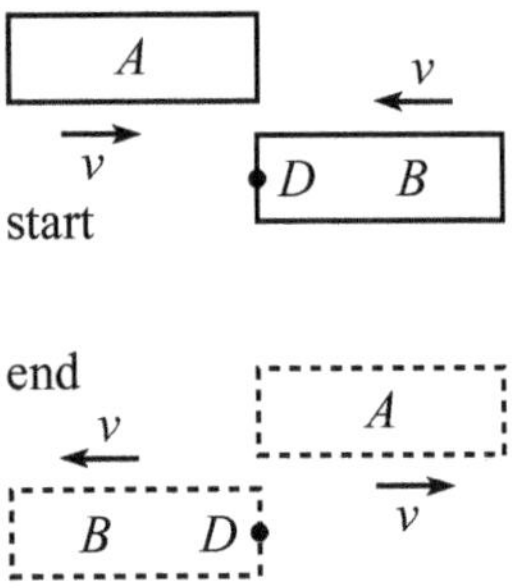

Fig. 11.22

(b) Look at things from D's point of view. D is at rest, and the two trains move with equal and opposite speeds v (see Fig. 11.22), because otherwise the second event E_2 wouldn't be located at D. The relativistic addition of v with itself is the speed of A as viewed by B. But from part (a), we know that this relative speed equals $5c/13$. Therefore,

$$\frac{2v}{1+v^2/c^2} = \frac{5c}{13} \quad \Longrightarrow \quad v = \frac{c}{5}, \tag{11.34}$$

where we have ignored the unphysical solution, $v = 5c$. The γ factor associated with $v = c/5$ is $\gamma = 5/(2\sqrt{6})$. So D sees both trains contracted to a length $2\sqrt{6}L/5$. During the overtaking, each train must travel a distance equal to its length, because both events, E_1 and E_2, take place right at D. The total time in D's frame is therefore

$$t_D = \frac{2\sqrt{6}L/5}{c/5} = \frac{2\sqrt{6}L}{c}. \tag{11.35}$$

REMARKS: There are a few double-checks we can perform. The speed of D with respect to the ground can be obtained either via B's frame by relativistically adding $3c/5$ and $c/5$, or via A's frame by subtracting $c/5$ from $4c/5$. These both give the same answer, namely $5c/7$, as they must. (The $c/5$ speed can in fact be determined by this reasoning, instead of using Eq. (11.34).) The γ factor between the ground and D is therefore $7/2\sqrt{6}$. We can then use time dilation to say that someone on the ground sees the overtaking take a time of $(7/2\sqrt{6})t_D$ (we can say this because both events happen right at D). Using Eq. (11.35), this gives a ground-frame time of $7L/c$, in agreement with Eq. (11.15). Likewise, the γ factor between D and either train is $5/2\sqrt{6}$. So the time of the overtaking as viewed by either A or B is $(5/2\sqrt{6})t_D = 5L/c$, in agreement with Eq. (11.33).

Note that we *cannot* use simple time dilation to relate the ground to A or B, because the two events don't happen at the same place in the train frames. But since both events happen at the same place in D's frame, namely right at D, it's legal to use time dilation to go from D's frame to any other frame. ♣

11.5.2 Transverse velocity addition

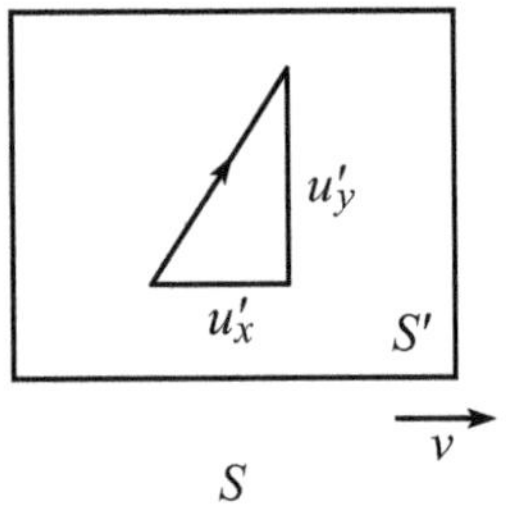

Fig. 11.23

Consider the following general two-dimensional situation. An object moves with velocity (u'_x, u'_y) with respect to frame S'. And frame S' moves with speed v with respect to frame S, in the x direction (see Fig. 11.23). What is the velocity, (u_x, u_y), of the object with respect to frame S?

The existence of motion in the y direction doesn't affect the preceding derivation of the speed in the x direction, so Eq. (11.31) is still valid. In the present notation, it becomes

$$u_x = \frac{u'_x + v}{1 + u'_x v/c^2}. \tag{11.36}$$

To find u_y, we can again make easy use of the Lorentz transformations. Consider two events along the object's path. We are given that $\Delta x'/\Delta t' = u'_x$, and $\Delta y'/\Delta t' = u'_y$. Our goal is to find $u_y \equiv \Delta y/\Delta t$. The relevant Lorentz transformations from S' to S in Eq. (11.17) are

$$\Delta y = \Delta y', \quad \text{and} \quad \Delta t = \gamma(\Delta t' + v\Delta x'/c^2). \tag{11.37}$$

Therefore,

$$\begin{aligned} u_y \equiv \frac{\Delta y}{\Delta t} &= \frac{\Delta y'}{\gamma(\Delta t' + v\Delta x'/c^2)} \\ &= \frac{\Delta y'/\Delta t'}{\gamma(1 + v(\Delta x'/\Delta t')/c^2)} \\ &= \frac{u'_y}{\gamma(1 + u'_x v/c^2)}. \end{aligned} \tag{11.38}$$

REMARK: In the special case where $u'_x = 0$, we have $u_y = u'_y/\gamma$. When u'_y is small and v is large, this result can be seen to be a special case of time dilation, in the following way. Consider a series of equally spaced lines parallel to the x axis (see Fig. 11.24). Imagine that the object's clock ticks once every time it crosses a line. Since u'_y is small, the object's frame is essentially frame S'. So if S flies by to the left, then the object is essentially moving at speed v with respect to S. Therefore, S sees the clock run slow by a factor γ. This means that S sees the object cross the lines at a slower rate, by a factor γ (because the clock still ticks once every time it crosses a line; this is a frame-independent statement). Since distances in the y direction are the same in the two frames, we conclude that $u_y = u'_y/\gamma$. This γ factor will be very important when we deal with momentum in Chapter 12.

To sum up: if you run in the x direction past an object, then its y speed is slower in your frame (or faster, depending on the relative sign of u'_x and v). Strange indeed, but no stranger than other effects we've seen. Problem 11.16 deals with the special case where $u'_x = 0$, but where u'_y is not necessarily small. ♣

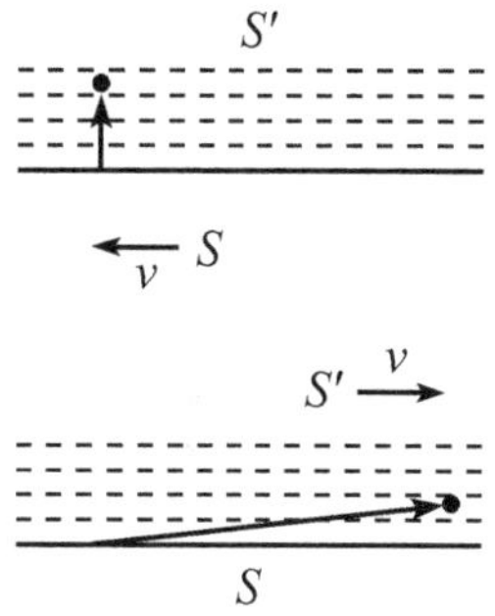

Fig. 11.24

11.6 The invariant interval

Consider the quantity,

$$(\Delta s)^2 \equiv c^2(\Delta t)^2 - (\Delta x)^2. \tag{11.39}$$

Technically, we should also subtract off $(\Delta y)^2$ and $(\Delta z)^2$, but nothing exciting happens in the transverse directions, so we'll ignore them. Using Eq. (11.17), we can write $(\Delta s)^2$ in terms of the S' coordinates, $\Delta x'$ and $\Delta t'$. The result is (dropping the Δ's)

$$\begin{aligned} c^2t^2 - x^2 &= \frac{c^2(t' + vx'/c^2)^2}{1 - v^2/c^2} - \frac{(x' + vt')^2}{1 - v^2/c^2} \\ &= \frac{t'^2(c^2 - v^2) - x'^2(1 - v^2/c^2)}{1 - v^2/c^2} \\ &= c^2t'^2 - x'^2. \end{aligned} \tag{11.40}$$

We see that the Lorentz transformations imply that the quantity $c^2t^2 - x^2$ doesn't depend on the frame. This result is more than we bargained for, for the following reason. The speed-of-light postulate says that if $c^2t'^2 - x'^2 = 0$, then $c^2t^2 - x^2 = 0$. But Eq. (11.40) says that if $c^2t'^2 - x'^2 = b$, then $c^2t^2 - x^2 = b$, for *any* value of b, not just zero. This is, as you might guess, very useful. There are enough things that change when we go from one frame to another, so it's nice to have a frame-independent quantity that we can hang on to. The fact that s^2 is invariant under Lorentz transformations of x and t is exactly analogous to the fact that r^2 is invariant under rotations in the x-y plane. The coordinates themselves change under the transformation, but the special combination of $c^2t^2 - x^2$ for Lorentz transformations, or $x^2 + y^2$ for rotations, remains the same. All inertial observers agree on the value of s^2, independent of what they measure for the actual coordinates.

> "Potato?! Pot*ah*to!" said she,
> "And of *course* it's tom*ah*to, you see.
> But the square of ct
> Minus x^2 will be
> Always something on which we agree."

A note on terminology: The separation in the coordinates, $(c\Delta t, \Delta x)$, is usually referred to as the *spacetime interval*, while the quantity $(\Delta s)^2 \equiv c^2(\Delta t)^2 - (\Delta x)^2$ is referred to as the *invariant interval* (or technically the square of the invariant interval). At any rate, just call it s^2, and people will know what you mean. The invariance of s^2 is actually just a special case of more general results involving inner products and 4-vectors, which we'll discuss in Chapter 13. Let's now look at the physical significance of $s^2 \equiv c^2t^2 - x^2$; there are three cases to consider.

Case 1: $s^2 > 0$ (timelike separation)

In this case, we say that the two events are *timelike* separated. We have $c^2t^2 > x^2$, and so $|x/t| < c$. Consider a frame S' moving at speed v with respect to S. The Lorentz transformation for x is

$$x' = \gamma(x - vt). \tag{11.41}$$

Since $|x/t| < c$, there exists a v that is less than c (namely $v = x/t$) that makes $x' = 0$. In other words, if two events are timelike separated, it is possible to find a frame S' in which the two events happen at the same place. (In short, the condition $|x/t| < c$ means that it is possible for a particle to travel from one event to the other.) The invariance of s^2 then gives $s^2 = c^2t'^2 - x'^2 = c^2t'^2$. So we see that s/c is the time between the events in the frame in which the events occur at the same place. This time is called the *proper time*.

Case 2: $s^2 < 0$ (spacelike separation)

In this case, we say that the two events are *spacelike* separated.[31] We have $c^2t^2 < x^2$, and so $|t/x| < 1/c$. Consider a frame S' moving at speed v with respect to S. The Lorentz transformation for t' is

$$t' = \gamma(t - vx/c^2). \tag{11.42}$$

Since $|t/x| < 1/c$, there exists a v that is less than c (namely $v = c^2t/x$) that makes $t' = 0$. In other words, if two events are spacelike separated, it is possible to find a frame S' in which the two events happen at the same time. (This statement is not as easy to see as the corresponding one in the timelike case above. But if you draw a Minkowski diagram, described in the next section, it becomes clear.) The invariance of s^2 then gives $s^2 = c^2t'^2 - x'^2 = -x'^2$. So we see that $|s|$ is the distance between the events in the frame in which the events occur at the same time. This distance is called the *proper distance*, or *proper length*.

Case 3: $s^2 = 0$ (lightlike separation)

In this case, we say that the two events are *lightlike* separated. We have $c^2t^2 = x^2$, and so $|x/t| = c$. This holds in every frame, so in every frame a photon emitted at one of the events will arrive at the other. It is not possible to find a frame S' in which the two events happen at the same place or the same time, because the frame would have to travel at the speed of light.

Example (Time dilation): An illustration of the usefulness of the invariance of s^2 is a derivation of time dilation. Let frame S' move at speed v with respect to frame S. Consider two events at the origin of S', separated by time t'. The separation between the events is

$$\begin{aligned} \text{in } S'&: \ (x', t') = (0, t'), \\ \text{in } S&: \ (x, t) = (vt, t). \end{aligned} \tag{11.43}$$

The invariance of s^2 implies $c^2t'^2 - 0 = c^2t^2 - v^2t^2$. Therefore,

$$t = \frac{t'}{\sqrt{1 - v^2/c^2}}. \tag{11.44}$$

This method makes it clear that the time-dilation result rests on the assumption that $x' = 0$.

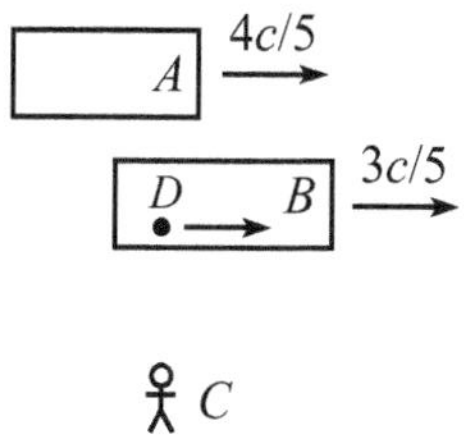

Fig. 11.25

Example (Passing trains, yet again): Consider again the scenario in the "Passing trains" examples in Sections 11.3.3 and 11.5.1. Verify that the s^2 between the events E_1 and E_2 is the same in all of the frames, A, B, C, and D (see Fig. 11.25).

[31] It's fine that s^2 is negative in this case, which means that s is imaginary. We can take the absolute value of s if we want to obtain a real number.

Solution: The only quantity that we'll need that we haven't already found in the two examples above is the distance between E_1 and E_2 in C's frame (the ground frame). In this frame, train A travels at a rate $4c/5$ for a time $t_C = 7L/c$, covering a distance of $28L/5$. But event E_2 occurs at the back of the train, which is a distance $3L/5$ behind the front end (this is the contracted length in the ground frame). Therefore, the distance between events E_1 and E_2 in the ground frame is $28L/5 - 3L/5 = 5L$. You can also apply the same line of reasoning using train B, in which the $5L$ result takes the form, $(3c/5)(7L/c) + 4L/5$.

Putting the previous results together, we have the following separations between the events in the various frames:

	A	B	C	D
Δt	$5L/c$	$5L/c$	$7L/c$	$2\sqrt{6}L/c$
Δx	$-L$	L	$5L$	0

From the table, we see that $\Delta s^2 \equiv c^2\Delta t^2 - \Delta x^2 = 24L^2$ for all four frames, as desired. We could have worked backwards, of course, and used the $s^2 = 24L^2$ result from frames A, B, or D, to deduce that $\Delta x = 5L$ in frame C. In Problem 11.10, you are asked to perform the tedious task of checking that the values in the above table satisfy the Lorentz transformations between the six different pairs of frames.

11.7 Minkowski diagrams

Minkowski diagrams (sometimes called "spacetime" diagrams) are extremely useful in seeing how coordinates transform between different reference frames. If you want to produce exact numbers in a problem, you'll probably have to use one of the strategies we've encountered so far. But as far as getting an overall intuitive picture of a setup goes (if there is in fact any such thing as intuition in relativity), there is no better tool than a Minkowski diagram. Here's how you make one.

Let frame S' move at speed v with respect to frame S (along the x axis, as usual, and ignore the y and z components). Draw the x and ct axes of frame S.[32] What do the x' and ct' axes of S' look like, superimposed on this diagram? That is, at what angles are the axes inclined, and what is the size of one unit on these axes? (There is no reason why one unit on the x' and ct' axes should have the same length on the paper as one unit on the x and ct axes.) We can answer these questions by using the Lorentz transformations, Eqs. (11.17). We'll first look at the ct' axis, and then the x' axis.

[32] We choose to plot ct instead of t on the vertical axis, so that the trajectory of a light beam lies at a nice 45° angle. Alternatively, we could choose units where $c = 1$.

ct'-axis angle and unit size

Look at the point $(x', ct') = (0, 1)$, which lies on the ct' axis, one ct' unit from the origin (see Fig. 11.26). Equations (11.17) tell us that this point is the point $(x, ct) = (\gamma v/c, \gamma)$. The angle between the ct' and ct axes is therefore given by $\tan\theta_1 = x/ct = v/c$. With $\beta \equiv v/c$, we have

$$\tan\theta_1 = \beta. \qquad (11.45)$$

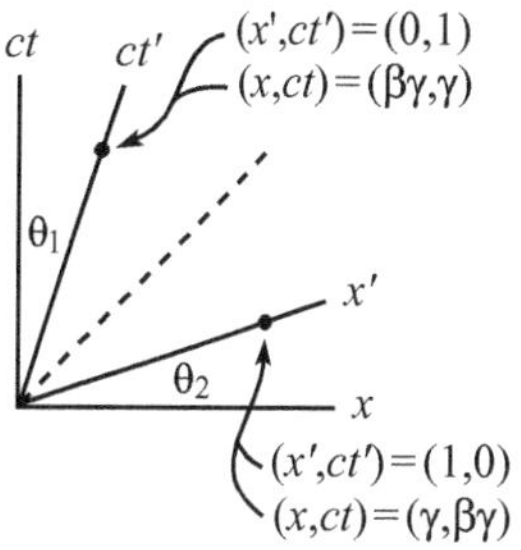

Fig. 11.26

Alternatively, the ct' axis is simply the "worldline" of the origin of S'. (A worldline is the path an object takes as it travels through spacetime.) The origin moves at speed v with respect to S. Therefore, points on the ct' axis satisfy $x/t = v$, or $x/ct = v/c$.

On the paper, the point $(x', ct') = (0, 1)$, which we just found to be the point $(x, ct) = (\gamma v/c, \gamma)$, is a distance $\gamma\sqrt{1 + v^2/c^2}$ from the origin. Therefore, using the definitions of β and γ, we see that

$$\frac{\text{one } ct' \text{ unit}}{\text{one } ct \text{ unit}} = \sqrt{\frac{1+\beta^2}{1-\beta^2}}, \qquad (11.46)$$

as measured on a grid where the x and ct axes are orthogonal. This ratio approaches infinity as $\beta \to 1$. And it of course equals 1 if $\beta = 0$.

x'-axis angle and unit size

The same basic argument holds here. Look at the point $(x', ct') = (1, 0)$, which lies on the x' axis, one x' unit from the origin (see Fig. 11.26). Equations (11.17) tell us that this point is the point $(x, ct) = (\gamma, \gamma v/c)$. The angle between the x' and x axes is therefore given by $\tan\theta_2 = ct/x = v/c$. So, as in the ct'-axis case,

$$\tan\theta_2 = \beta. \qquad (11.47)$$

On the paper, the point $(x', ct') = (1, 0)$, which we just found to be the point $(x, ct) = (\gamma, \gamma v/c)$, is a distance $\gamma\sqrt{1 + v^2/c^2}$ from the origin. So, as in the ct'-axis case,

$$\frac{\text{one } x' \text{ unit}}{\text{one } x \text{ unit}} = \sqrt{\frac{1+\beta^2}{1-\beta^2}}, \qquad (11.48)$$

as measured on a grid where the x and ct axes are orthogonal. Both the x' and ct' axes are therefore stretched by the same factor, and tilted in by the same angle, relative to the x and ct axes. This "squeezing in" of the axes in a Lorentz transformation is different from what happens in a rotation, where both axes rotate in the same direction.

REMARKS: If $v/c \equiv \beta = 0$, then $\theta_1 = \theta_2 = 0$, so the ct' and x' axes coincide with the ct and x axes, as they should. If β is very close to 1, then the x' and ct' axes are both very close to the 45° light-ray line. Note that since $\theta_1 = \theta_2$, the light-ray line bisects the x' and ct' axes. Therefore (as we verified above), the scales on these axes must be the same, because a light ray must satisfy $x' = ct'$. ♣

We now know what the x' and ct' axes look like. Given any two points in a Minkowski diagram (that is, given any two events in spacetime), we can just read off the Δx, Δct, $\Delta x'$, and $\Delta ct'$ quantities that our two observers measure, assuming that our graph is accurate enough. Although these quantities must of course be related by the Lorentz transformations, the advantage of a Minkowski diagram is that you can actually see geometrically what's going on.

There are very useful physical interpretations of the ct' and x' axes. If you stand at the origin of S', then the ct' axis is the "here" axis, and the x' axis is the "now" axis (the line of simultaneity). That is, all events on the ct' axis take place at your position (the ct' axis is your worldline, after all), and all events on the x' axis take place simultaneously (they all have $t' = 0$).

Example (Length contraction): For both parts of this problem, use a Minkowski diagram where the axes in frame S are orthogonal.

(a) The relative speed of S' and S is v (along the x direction). A meter stick lies along the x' axis and is at rest in S'. If S measures its length, what is the result?

(b) Now let the meter stick lie along the x axis and be at rest in S. If S' measures its length, what is the result?

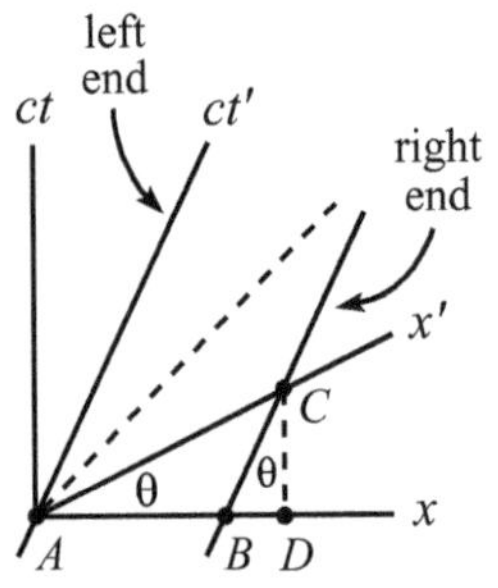

Fig. 11.27

Solution:

(a) Without loss of generality, pick the left end of the stick to be at the origin in S'. Then the worldlines of the two ends are shown in Fig. 11.27. The distance AC is 1 meter in the S' frame, because A and C are the endpoints of the stick at simultaneous times in the S' frame; this is how a length is measured. And since one unit on the x' axis has length $\sqrt{1+\beta^2}/\sqrt{1-\beta^2}$, this is the length on the paper of the segment AC.

How does S measure the length of the stick? He writes down the x coordinates of the ends at simultaneous times (as measured by him, of course), and takes the difference. Let the time he makes the measurements be $t = 0$. Then he measures the ends to be at the points A and B.[33] Now it's time to do some geometry. We have to find the length of segment AB in Fig. 11.27, given that segment AC has length $\sqrt{1+\beta^2}/\sqrt{1-\beta^2}$. We know that the primed axes are tilted at an angle θ, where $\tan\theta = \beta$. Therefore, $CD = (AC)\sin\theta$. And since

[33] If S measures the ends in a dramatic fashion by, say, blowing them up, then S' will see the right end blow up first (the event at B has a negative t' coordinate, because it lies below the x' axis), and then a little while later S' will see the left end blow up (the event at A has $t' = 0$). So S measures the ends at different times in the S' frame. This is part of the reason why S' should not be at all surprised that S's measurement is smaller than one meter.

$\angle BCD = \theta$, we have $BD = (CD)\tan\theta = (AC)\sin\theta\tan\theta$. Therefore (using $\tan\theta = \beta$),

$$\begin{aligned} AB &= AD - BD = (AC)\cos\theta - (AC)\sin\theta\tan\theta \\ &= (AC)\cos\theta(1-\tan^2\theta) \\ &= \sqrt{\frac{1+\beta^2}{1-\beta^2}}\frac{1}{\sqrt{1+\beta^2}}(1-\beta^2) \\ &= \sqrt{1-\beta^2}. \end{aligned} \tag{11.49}$$

Therefore, S measures the meter stick to have length $\sqrt{1-\beta^2}$, which is the standard length-contraction result.

(b) The stick is now at rest in S, and we want to find the length that S' measures. Pick the left end of the stick to be at the origin in S. Then the worldlines of the two ends are shown in Fig. 11.28. The distance AB is 1 meter in the S frame.

In measuring the length of the stick, S' writes down the x' coordinates of the ends at simultaneous times (as measured by him), and takes the difference. Let the time he makes the measurements be $t' = 0$. Then he measures the ends to be at the points A and E. Now we do the geometry, which is easy in this case. The length of AE is simply $1/\cos\theta = \sqrt{1+\beta^2}$. But since one unit along the x' axis has length $\sqrt{1+\beta^2}/\sqrt{1-\beta^2}$ on the paper, we see that AE is $\sqrt{1-\beta^2}$ of one unit in the S' frame. Therefore, S' measures the meter stick to have length $\sqrt{1-\beta^2}$, which again is the standard length-contraction result.

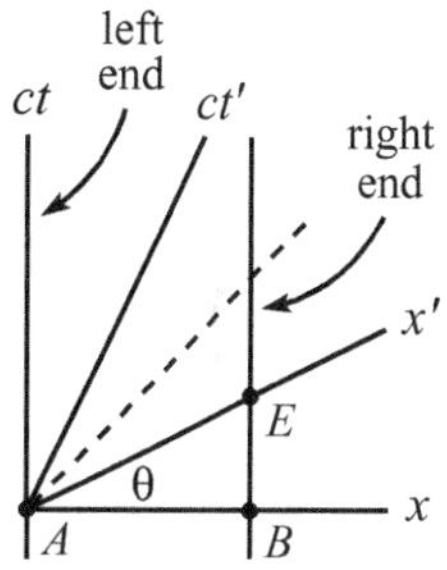

Fig. 11.28

The analysis used in the above example also works for time intervals. The derivation of time dilation, using a Minkowski diagram, is the task of Exercise 11.62. And the derivation of the Lv/c^2 rear-clock-ahead result is the task of Exercise 11.63.

11.8 The Doppler effect

11.8.1 Longitudinal Doppler effect

Consider a source that emits flashes at frequency f' (in its own frame) while moving directly toward you at speed v, as shown in Fig. 11.29. With what frequency do the flashes hit your eye? In these Doppler-effect problems, you must be careful to distinguish between the time at which an event *occurs* in your frame, and the time at which you *see* the event occur. This is one of the few situations where we are concerned with the latter.

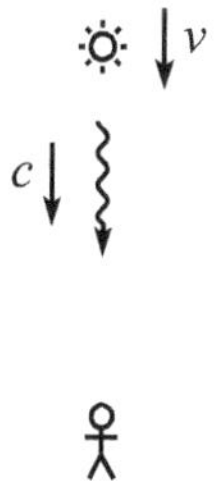

Fig. 11.29

There are two effects contributing to the longitudinal Doppler effect. The first is relativistic time dilation. There is more time between the flashes in your frame, which means that they occur at a smaller frequency. The second is the everyday

Doppler effect (as with sound), arising from the motion of the source. Successive flashes have a smaller (or larger, if v is negative) distance to travel to reach your eye. This effect increases (or decreases, if v is negative) the frequency at which the flashes hit your eye.

Let's now be quantitative and find the observed frequency. The time between emissions in the source's frame is $\Delta t' = 1/f'$. The time between emissions in your frame is then $\Delta t = \gamma \Delta t'$, by the usual time dilation. So the photons of one flash have traveled a distance (in your frame) of $c\Delta t = c\gamma \Delta t'$ by the time the next flash occurs. During this time between emissions, the source has traveled a distance $v\Delta t = v\gamma \Delta t'$ toward you in your frame. Therefore, at the instant the next flash occurs, the photons of this next flash are a distance (in your frame) of $c\Delta t - v\Delta t = (c - v)\gamma \Delta t'$ behind the photons of the previous flash. This result holds for all adjacent flashes. The time, ΔT, between the arrivals of the flashes at your eye is $1/c$ times this distance, so we have

$$\Delta T = \frac{1}{c}(c - v)\gamma \Delta t' = \frac{1-\beta}{\sqrt{1-\beta^2}}\,\Delta t' = \sqrt{\frac{1-\beta}{1+\beta}}\left(\frac{1}{f'}\right), \tag{11.50}$$

where $\beta = v/c$. Therefore, the frequency you see is

$$f = \frac{1}{\Delta T} = \sqrt{\frac{1+\beta}{1-\beta}}\,f'. \tag{11.51}$$

If $\beta > 0$ (that is, the source is moving toward you), then $f > f'$. The everyday Doppler effect wins out over the time-dilation effect. In this case we say that the light is "blueshifted," because blue light is at the high-frequency end of the visible spectrum. The light need not have anything to do with the color blue, of course; by "blue" we just mean that the frequency is increased. If $\beta < 0$ (that is, the source is moving away from you), then $f < f'$. Both effects serve to decrease the frequency. In this case we say that the light is "redshifted," because red light is at the low-frequency end of the visible spectrum.

REMARK: We can also derive Eq. (11.51) by working in the frame of the source. In this frame, the distance between successive flashes is $c\Delta t'$. And since you are moving toward the source at speed v, the relative speed of you and a given flash is $c + v$. So the time between your running into successive flashes is $c\Delta t'/(c + v) = \Delta t'/(1 + \beta)$, as measured in the frame of the source. But your clock runs slow in this frame, so a time of only $\Delta T = (1/\gamma)\Delta t'/(1 + \beta)$ elapses on your watch, which you can show agrees with the time in Eq. (11.50).

We certainly needed to obtain the same result by working in the frame of the source, because the principle of relativity states that the result can't depend on which object we consider to be the one at rest; there is no preferred frame. This is different from the situation with the standard nonrelativistic Doppler effect (relevant to a siren moving toward you), because the frequency there *does* depend on whether you or the source is the one that is moving. The reason for this is that when we say "moving" here, we mean moving with respect to the rest frame of the air, which is the medium that sound travels in. We therefore do in fact have a preferred frame of reference, unlike in relativity (where there is no "ether"). Using the arguments given above for the two different frames, but without the γ factors, you can show that the two nonrelativistic Doppler results are $f = f'/(1-\beta)$ if the source is moving toward a stationary you, and $f = (1+\beta)f'$ if you are moving toward the stationary source. Here β is the ratio of the speed of the moving object to the speed of sound. In view of these two different results, the relativistic Doppler effect can be considered to be a simpler effect, in the sense that there is only one frequency to remember. ♣

11.8.2 Transverse Doppler effect

Let's now consider a two-dimensional situation. Consider a source that emits flashes at frequency f' (in its own frame), while moving across your field of vision at speed v. There are two reasonable questions we can ask about the frequency you observe:

- **Case 1:** At the instant the source is at its closest approach to you, with what frequency do the flashes hit your eye?
- **Case 2:** When you *see* the source at its closest approach to you, with what frequency do the flashes hit your eye?

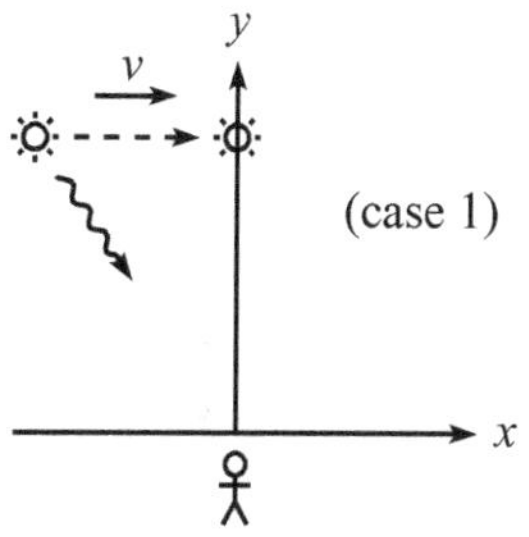

Fig. 11.30

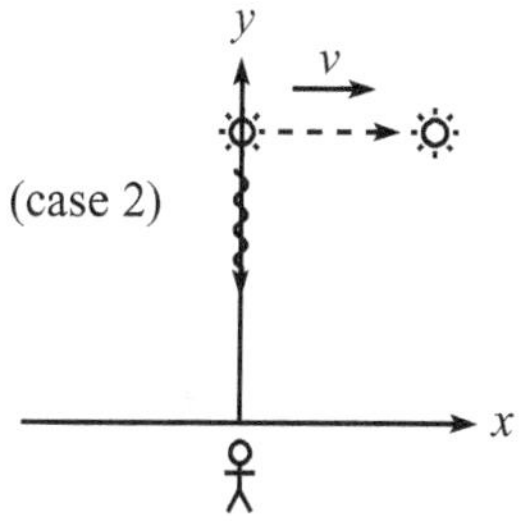

Fig. 11.31

The difference between these two scenarios is shown in Fig. 11.30 and Fig. 11.31, where the source's motion is taken to be parallel to the x axis. In the first case, the photons you see must have been emitted at an earlier time, because the source moves during the nonzero time it takes the light to reach you. In this scenario, we are dealing with photons that hit your eye *when* (as measured in your frame) the source crosses the y axis. You therefore see the photons come in at an angle with respect to the y axis.

In the second case, you see the photons come in along the y axis (by the definition of this scenario). At the instant you observe one of these photons, the source is at a position past the y axis. Let's find the observed frequencies in these two cases.

Case 1: Let your frame be S, and let the source's frame be S'. Consider the situation from S''s point of view. S' sees you moving across his field of vision at speed v. The relevant photons hit your eye when you cross the y' axis (defined to be the axis that passes through the source) of the S' frame. S' sees you get hit by a flash every $\Delta t' = 1/f'$ seconds in his frame. (This is true because when you are very close to the y' axis, all points on your path are essentially equidistant from the source. So we don't have to worry about any longitudinal effects, in the S' frame.) This means that you get hit by a flash every $\Delta T = \Delta t'/\gamma = 1/(f'\gamma)$ seconds in your frame, because S' sees your clock running slow. Therefore, the frequency in your frame is

$$f = \frac{1}{\Delta T} = \gamma f' = \frac{f'}{\sqrt{1-\beta^2}}. \qquad (11.52)$$

Hence, f is greater than f'. You see the flashes at a higher frequency than S' emits them.

Case 2: Again, let your frame be S, and let the source's frame be S'. Consider the situation from your point of view. Because of time dilation, a clock on the source runs slow (in your frame) by a factor of γ. So you get hit by a flash every $\Delta T = \gamma \Delta t' = \gamma / f'$ seconds in your frame. (We have used the fact that the relevant photons are emitted from points that are essentially equidistant from you. So they all travel the same distance, and we don't have to worry about any longitudinal effects, in your frame.) When you see the source cross the y axis, you therefore observe a frequency of

$$f = \frac{1}{\Delta T} = \frac{1}{\gamma \Delta t'} = \frac{f'}{\gamma} = f'\sqrt{1-\beta^2}. \tag{11.53}$$

Hence, f is smaller than f'. You see the flashes at a lower frequency than S' emits them.

When people talk about the "transverse Doppler effect," they sometimes mean Case 1, and they sometimes mean Case 2. The title "transverse Doppler" is therefore ambiguous, so you should remember to state exactly which scenario you are talking about. Other cases that are "in between" these two can also be considered. But they get a bit messy.

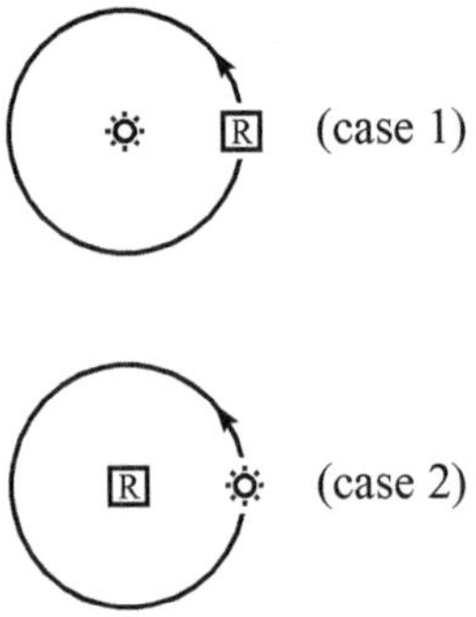

Fig. 11.32

REMARKS:

1. The two scenarios may alternatively be described, respectively (as you can convince yourself), in the following ways (see Fig. 11.32).
   - **Case 1:** A receiver moves with speed v in a circle around a source. What frequency does the receiver register?
   - **Case 2:** A source moves with speed v in a circle around a receiver. What frequency does the receiver register?

   These two setups make it clear that the results in Eqs. (11.52) and (11.53) arise from a simple time-dilation argument used by the inertial object at the center of each circle. These setups involve accelerating objects. We must therefore invoke the fact (which has plenty of experimental verification) that if an inertial observer looks at a moving clock, then only the instantaneous speed of the clock is important in computing the time-dilation factor. The acceleration is irrelevant.[34]
2. Beware of the following incorrect reasoning for Case 1, leading to an incorrect version of Eq. (11.52). "S sees things in S' slowed down by a factor γ (that is, $\Delta t = \gamma \Delta t'$), by the usual time-dilation effect. Hence, S sees the light flashing at a slower pace. Therefore, $f = f'/\gamma$." This reasoning puts the γ in the wrong place. Where is the error? The error lies in confusing the time at which an event *occurs* in S's frame, with the time at which S *sees* (with his eyes) the event occur. The flashes certainly *occur* at a lower frequency in S, but due to the motion of S' relative to S, it turns out that the flashes meet S's eye at a faster rate, because the source is moving slightly toward S while it is emitting the relevant photons. You can work out the details from S's point of view.[35]

[34] The acceleration is, however, very important if things are considered from an accelerating object's point of view. But we'll wait until Chapter 14 on General Relativity to talk about this.

[35] This is a fun exercise (Exercise 11.66), but it should convince you that it's much easier to look at things in the frame in which there are no longitudinal effects, as we did in our solutions above.

Alternatively, the error can be stated as follows. The time dilation result, $\Delta t = \gamma \Delta t'$, rests on the assumption that the $\Delta x'$ between the two events is zero. This applies fine to two emissions of light from the source. However, the two events in question are the absorption of two light flashes by your eye (which is moving in S'), so $\Delta t = \gamma \Delta t'$ is not applicable. Instead, $\Delta t' = \gamma \Delta t$ is the relevant result, valid when $\Delta x = 0$. (But we still need to invoke the fact that all the relevant photons travel the same distance, which means that we don't have to worry about any longitudinal effects.) ♣

11.9 Rapidity

11.9.1 Definition

Let us define the *rapidity*, ϕ, by

$$\tanh \phi \equiv \beta \equiv \frac{v}{c}. \tag{11.54}$$

A few properties of the hyperbolic trig functions are given in Appendix A. In particular, $\tanh \phi \equiv (e^{\phi} - e^{-\phi})/(e^{\phi} + e^{-\phi})$. The rapidity defined in Eq. (11.54) is very useful in relativity because many of our expressions take on a particularly nice form when written in terms of it. Consider, for example, the velocity-addition formula. Let $\beta_1 = \tanh \phi_1$ and $\beta_2 = \tanh \phi_2$. Then if we add β_1 and β_2 using the velocity-addition formula, Eq. (11.31), we obtain

$$\frac{\beta_1 + \beta_2}{1 + \beta_1 \beta_2} = \frac{\tanh \phi_1 + \tanh \phi_2}{1 + \tanh \phi_1 \tanh \phi_2} = \tanh(\phi_1 + \phi_2), \tag{11.55}$$

where we have used the addition formula for $\tanh \phi$, which you can prove by writing things in terms of the exponentials, $e^{\pm\phi}$. Therefore, while the velocities add in the strange manner of Eq. (11.31), the rapidities add by standard addition.

The Lorentz transformations also take a nice form when written in terms of the rapidity. Our friendly γ factor can be written as

$$\gamma \equiv \frac{1}{\sqrt{1 - \beta^2}} = \frac{1}{\sqrt{1 - \tanh^2 \phi}} = \cosh \phi. \tag{11.56}$$

Also,

$$\gamma \beta \equiv \frac{\beta}{\sqrt{1 - \beta^2}} = \frac{\tanh \phi}{\sqrt{1 - \tanh^2 \phi}} = \sinh \phi. \tag{11.57}$$

Therefore, the Lorentz transformations in matrix form, Eqs. (11.22), become

$$\begin{pmatrix} x \\ ct \end{pmatrix} = \begin{pmatrix} \cosh \phi & \sinh \phi \\ \sinh \phi & \cosh \phi \end{pmatrix} \begin{pmatrix} x' \\ ct' \end{pmatrix}. \tag{11.58}$$

This transformation looks similar to a rotation in a plane, which is given by

$$\begin{pmatrix} x \\ y \end{pmatrix} = \begin{pmatrix} \cos \theta & \sin \theta \\ -\sin \theta & \cos \theta \end{pmatrix} \begin{pmatrix} x' \\ y' \end{pmatrix}, \tag{11.59}$$

except that we now have hyperbolic trig functions instead of trig functions. The fact that the interval $s^2 \equiv c^2t^2 - x^2$ does not depend on the frame is clear from Eq. (11.58), because the cross terms in the squares cancel, and $\cosh^2\phi - \sinh^2\phi = 1$. (Compare with the invariance of $r^2 \equiv x^2 + y^2$ for rotations in a plane, where the cross terms from Eq. (11.59) likewise cancel, and $\cos^2\theta + \sin^2\theta = 1$.)

The quantities associated with a Minkowski diagram also take a nice form when written in terms of the rapidity. The angle between the S and S' axes satisfies

$$\tan\theta = \beta = \tanh\phi. \tag{11.60}$$

And the size of one unit on the x' or ct' axes is, from Eq. (11.46),

$$\sqrt{\frac{1+\beta^2}{1-\beta^2}} = \sqrt{\frac{1+\tanh^2\phi}{1-\tanh^2\phi}} = \sqrt{\cosh^2\phi + \sinh^2\phi} = \sqrt{\cosh 2\phi}. \tag{11.61}$$

For large ϕ, this is approximately equal to $e^{\phi}/\sqrt{2}$.

11.9.2 Physical meaning

The fact that the rapidity makes many of our formulas look nice and pretty is reason enough to consider it. But in addition, it turns out to have a very meaningful physical interpretation. Consider the following setup. A spaceship is initially at rest in the lab frame. At a given instant, it starts to accelerate. Let a be the *proper acceleration*, which is defined as follows. Let t be the time coordinate in the spaceship's frame.[36] If the proper acceleration is a, then at time $t + dt$, the spaceship is moving at speed $a\,dt$ relative to the frame it was in at time t. An equivalent definition is that the astronaut feels a force of ma applied to his body by the spaceship. If he is standing on a scale, then the scale shows a reading of $F = ma$.

What is the relative speed of the spaceship and the lab frame at (the spaceship's) time t? We can answer this question by considering two nearby times and using the velocity-addition formula, Eq. (11.31). From the definition of a, Eq. (11.31) gives, with $v_1 \equiv a\,dt$ and $v_2 \equiv v(t)$,

$$v(t+dt) = \frac{v(t) + a\,dt}{1 + v(t)a\,dt/c^2}\,. \tag{11.62}$$

[36] This frame is of course changing as time goes by, because the spaceship is accelerating. The time t is simply the spaceship's proper time. Normally, we would denote this by t', but we don't want to have to keep writing the primes over and over in the following calculation.

Expanding this to first order in dt yields[37]

$$\frac{dv}{dt} = a\left(1 - \frac{v^2}{c^2}\right) \implies \int_0^v \frac{dv}{1 - v^2/c^2} = \int_0^t a\,dt. \tag{11.63}$$

Separating variables and integrating gives, using $\int dz/(1-z^2) = \tanh^{-1} z$,[38] and assuming that a is constant,

$$v(t) = c\tanh(at/c). \tag{11.64}$$

For small a or small t (more precisely, for $at/c \ll 1$), we obtain $v(t) \approx at$, as we should (because $\tanh z \approx z$ for small z, which follows from the exponential form of tanh). And for $at/c \gg 1$, we obtain $v(t) \approx c$, as we should. If a happens to be a function of time, $a(t)$, then we can't take the a outside the integral in Eq. (11.63), so we instead end up with the general formula,

$$v(t) = c\tanh\left(\frac{1}{c}\int_0^t a(t)\,dt\right). \tag{11.65}$$

The rapidity ϕ, as defined in Eq. (11.54), is therefore given by

$$\phi(t) \equiv \frac{1}{c}\int_0^t a(t)\,dt. \tag{11.66}$$

Note that whereas v has c as a limiting value, ϕ can become arbitrarily large. Looking at Eq. (11.65) we see that the ϕ associated with a given v is $1/mc$ times the time integral of the force (felt by the astronaut) that is needed to bring the astronaut up to speed v. By applying a force for an arbitrarily long time, we can make ϕ arbitrarily large.

The integral $\int a(t)\,dt$ may be described as the naive, incorrect speed. That is, it is the speed that the astronaut might think he has, if he has his eyes closed and knows nothing about the theory of relativity. And indeed, his thinking would be essentially correct for small speeds. This quantity $\int a(t)\,dt = \int F(t)\,dt/m$ looks like a reasonably physical thing, so it seems like it should have *some* meaning. And indeed, although it doesn't equal v, all you have to do to get v is take a tanh and throw in some factors of c.

The fact that rapidities add via simple addition when using the velocity-addition formula, as we saw in Eq. (11.55), is evident from Eq. (11.65). There is really nothing more going on here than the fact that

$$\int_{t_0}^{t_2} a(t)\,dt = \int_{t_0}^{t_1} a(t)\,dt + \int_{t_1}^{t_2} a(t)\,dt. \tag{11.67}$$

[37] Equivalently, just take the derivative of $(v+w)/(1+vw/c^2)$ with respect to w, and then set $w = 0$.

[38] Alternatively, you can use $1/(1-z^2) = 1/(2(1-z)) + 1/(2(1+z))$, and then integrate to obtain some logs, which then yield the tanh. You can also use the result of Problem 11.17 to find $v(t)$. See the remark in the solution to that problem (after trying to solve it, of course!).

To be explicit, let a force be applied from t_0 to t_1 that brings a mass up to a speed of (dropping the c's) $\beta_1 = \tanh\phi_1 = \tanh\big(\int_{t_0}^{t_1} a\,dt\big)$, and then let an additional force be applied from t_1 to t_2 that adds on an additional speed of $\beta_2 = \tanh\phi_2 = \tanh\big(\int_{t_1}^{t_2} a\,dt\big)$, relative to the frame at t_1. Then the resulting speed may be looked at in two ways: (1) it is the result of relativistically adding the speeds $\beta_1 = \tanh\phi_1$ and $\beta_2 = \tanh\phi_2$, and (2) it is the result of applying the force from t_0 to t_2 (you get the same final speed, of course, whether or not you bother to record the speed along the way at t_1), which is $\beta = \tanh\big(\int_{t_0}^{t_2} a\,dt\big) = \tanh(\phi_1 + \phi_2)$, where the second equality comes from the statement, Eq. (11.67), that integrals simply add. Therefore, the relativistic addition of $\tanh\phi_1$ and $\tanh\phi_2$ gives $\tanh(\phi_1 + \phi_2)$, as we wanted to show. Note that this reasoning doesn't rely on the fact that the function here is a tanh. It could be anything. If we lived in a world where the speed were given by, for example, $\beta = \tan\big(\int a\,dt\big)$, then the rapidities would still add via simple addition. It's just that in our world we have a tanh. (We'll see in the next section that a "tan" world would have issues, anyway.)

11.10 Relativity without *c*

In Section 11.2, we introduced the two postulates of Special Relativity, namely the speed-of-light postulate and the relativity postulate. Appendix I gives a derivation of the Lorentz transformations that works directly from these two postulates and doesn't use the three fundamental effects, which were the basis of the derivation in Section 11.4.1. It's interesting to see what happens if we relax these postulates. It's hard to imagine a reasonable (empty) universe where the relativity postulate doesn't hold, but it's easy to imagine a universe where the speed of light depends on the reference frame. Light could behave like a baseball, for example. So let's drop the speed-of-light postulate now and see what we can say about the coordinate transformations between frames, using only the relativity postulate. For further discussion of this topic, see Lee and Kalotas (1975) and references therein.

In Appendix I, the form of the transformations, just prior to invoking the speed-of-light postulate, is given in Eq. (I.8) as

$$\begin{aligned} x &= A_v(x' + vt'), \\ t &= A_v\left(t' + \frac{1}{v}\Big(1 - \frac{1}{A_v^2}\Big)x'\right). \end{aligned} \tag{11.68}$$

We'll put a subscript on A in this section, to remind us of the v dependence. Can we say anything about A_v without invoking the speed-of-light postulate? Indeed we can. Define V_v by

$$\frac{1}{V_v^2} \equiv \frac{1}{v^2}\left(1 - \frac{1}{A_v^2}\right) \implies A_v = \frac{1}{\sqrt{1 - v^2/V_v^2}}. \tag{11.69}$$

We have picked the positive square root because when $v = 0$ we should have $x = x'$ and $t = t'$. Equations (11.68) now become

$$\begin{aligned} x &= \frac{1}{\sqrt{1 - v^2/V_v^2}}(x' + vt'), \\ t &= \frac{1}{\sqrt{1 - v^2/V_v^2}}\left(\frac{v}{V_v^2}x' + t'\right). \end{aligned} \tag{11.70}$$

All we've done so far is make a change of variables. But we now make the following claim.

Claim 11.1 *V_v^2 is independent of v.*

Proof: As stated in the last remark in Section 11.4.1, we know that two successive applications of the transformations in Eqs. (11.70) must again yield a transformation of the same form. Consider a transformation characterized by velocity v_1, and another one characterized by velocity v_2. For simplicity, define

$$\begin{aligned} &V_1 \equiv V_{v_1}, \quad V_2 \equiv V_{v_2}, \\ &\gamma_1 \equiv \frac{1}{\sqrt{1 - v_1^2/V_1^2}}, \quad \gamma_2 \equiv \frac{1}{\sqrt{1 - v_2^2/V_2^2}}. \end{aligned} \tag{11.71}$$

To calculate the composite transformation, it is easiest to use matrix notation. Looking at Eqs. (11.70), we see that the composite transformation is given by the matrix

$$\begin{pmatrix} \gamma_2 & \gamma_2 v_2 \\ \gamma_2 \frac{v_2}{V_2^2} & \gamma_2 \end{pmatrix} \begin{pmatrix} \gamma_1 & \gamma_1 v_1 \\ \gamma_1 \frac{v_1}{V_1^2} & \gamma_1 \end{pmatrix} = \gamma_1 \gamma_2 \begin{pmatrix} 1 + \frac{v_1 v_2}{V_1^2} & v_2 + v_2 \\ \frac{v_1}{V_1^2} + \frac{v_2}{V_2^2} & 1 + \frac{v_1 v_2}{V_2^2} \end{pmatrix}. \tag{11.72}$$

The composite transformation must still be of the form of Eqs. (11.70). But this implies that the upper-left and lower-right entries of the composite matrix must be equal. Therefore, $V_1^2 = V_2^2$. Since this holds for arbitrary v_1 and v_2, we see that V_v^2 must be a constant. So it is independent of v. ■

Denote the constant value of V_v^2 by V^2. Then the coordinate transformations in Eq. (11.70) become

$$\begin{aligned} x &= \frac{1}{\sqrt{1 - v^2/V^2}}(x' + vt'), \\ t &= \frac{1}{\sqrt{1 - v^2/V^2}}\left(t' + \frac{v}{V^2}x'\right). \end{aligned} \tag{11.73}$$

We have obtained this result using only the relativity postulate. These transformations have the same form as the Lorentz transformations in Eqs. (11.17).

The only extra information in Eqs. (11.17) is that V equals the speed of light, c. It is remarkable that we were able to prove so much by using only the relativity postulate.

We can say a few more things. There are four basic possibilities for the value of V^2. However, two of these are not physical.

- $V^2 = \infty$: This gives the Galilean transformations, $x = x' + vt'$ and $t = t'$.
- $0 < V^2 < \infty$: This gives transformations of the Lorentz type. V is the limiting speed of an object. Experiments show that this case is the one that corresponds to the world we live in.
- $V^2 = 0$: This case isn't physical, because any nonzero value of v makes the γ factor imaginary (and infinite). Nothing could ever move.
- $V^2 < 0$: It turns out that this case is also not physical. You might be concerned that the square of V is less than zero, but this is fine because V appears in the transformations (11.73) only through its square. There is no need for V to actually be the speed of anything. The trouble is that the nature of Eqs. (11.73) implies the possibility of time reversal. This opens the door for causality violation and all the other problems associated with time reversal. We therefore reject this case. To be more explicit, define $b^2 \equiv -V^2$, where b is a positive number. Then Eqs. (11.73) may be written in the form,

$$\begin{aligned} x &= x' \cos\theta + (bt')\sin\theta, \\ bt &= -x' \sin\theta + (bt')\cos\theta, \end{aligned} \tag{11.74}$$

where $\tan\theta \equiv v/b$. Also, $-\pi/2 \le \theta \le \pi/2$, because the positive sign of the x' and t' coefficients in Eqs. (11.73) implies that the $\cos\theta$ in Eqs. (11.74) satisfies $\cos\theta \ge 0$. We have the normal trig functions in Eqs. (11.74) instead of the hyperbolic trig functions in the Lorentz transformation in Eqs. (11.58). This present transformation is simply a rotation of the axes through an angle θ in the plane. The S axes are rotated counterclockwise (as you can check) by an angle θ relative to the S' axes. Equivalently, the S' axes are rotated clockwise by an angle θ relative to the S axes.

Equations (11.74) satisfy the requirement that the composition of two transformations is again a transformation of the same form. Rotation by θ_1, and then by θ_2, yields a rotation by $\theta_1 + \theta_2$. However, if θ_1 and θ_2 are positive, and if the resulting rotation is through an angle θ that is greater than 90°, then we have a problem. The tangent of such an angle is negative. Therefore, $\tan\theta = v/b$ implies that v is negative.

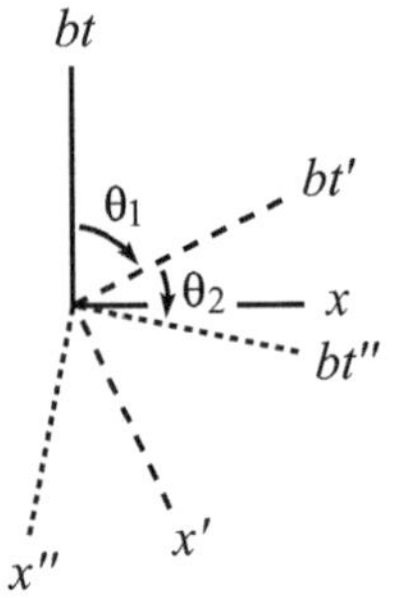

Fig. 11.33

This situation is shown in Fig. 11.33. Frame S'' moves at velocity $v_2 > 0$ with respect to frame S', which moves at velocity $v_1 > 0$ with respect to frame S. From the figure, we see that someone standing at rest in frame S'' (that is, someone whose worldline is the bt'' axis) is going to have some serious issues in frame S. For one, the bt'' axis has a negative slope in frame S, which means that as t increases, x decreases. The person is therefore moving at a *negative* velocity with respect to S. Adding two positive velocities and obtaining a negative one is clearly absurd. But even worse, someone standing at rest in S'' is moving in the positive direction along the bt'' axis, which means that he is traveling *backwards* in time in S. That is, he will die before he is born. This is not good.

An equivalent method of dismissing this case, given in Lee and Kalotas (1977), but one that doesn't specifically refer to causality violation, is to note that the transformations in Eqs. (11.74) don't form a closed group. In other words, successive applications of the transformations can eventually yield a transformation that isn't of the form in Eqs. (11.74), due to the $-\pi/2 \leq \theta \leq \pi/2$ restriction. (In contrast, the rotations in a plane form a closed group, because there are no restrictions on what θ can be.) This argument is equivalent to the time-reversal argument above, because $-\pi/2 \leq \theta \leq \pi/2$ is equivalent to $\cos\theta \geq 0$, which is equivalent to the statement that the coefficients of t and t' in Eqs. (11.74) have the same sign.

Note that all of the finite $0 < V^2 < \infty$ possibilities are essentially the same. Any difference in the numerical value of V can be absorbed into the definitions of the unit sizes for x and t. Given that V is finite, it has to be *something*, so it doesn't make sense to put much importance on its numerical value. There is therefore only one decision to be made when constructing the spacetime structure of an (empty) universe. You just have to say whether V is finite or infinite, that is, whether the universe is Lorentzian or Galilean. Equivalently, all you have to say is whether or not there is an upper limit on the speed of any object. If there is, then you can simply postulate the existence of something that moves with this limiting speed. In other words, to create your universe, you simply have to say, "Let there be light."

11.11 Problems

Section 11.3: The fundamental effects

11.1. **No transverse length contraction** *

Two meter sticks, A and B, move past each other as shown in Fig. 11.34. Stick A has paint brushes on its ends. Use this setup to show that in the frame of one stick, the other stick still has a length of one meter.

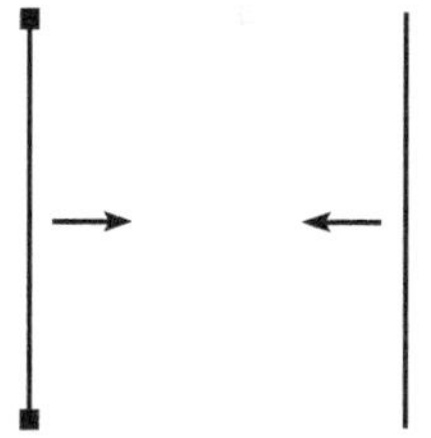

Fig. 11.34

11.2. **Explaining time dilation** **

Two planets, A and B, are at rest with respect to each other, a distance L apart, with synchronized clocks. A spaceship flies at speed v past planet A toward planet B and synchronizes its clock with A's right when it passes A (they both set their clocks to zero). The spaceship eventually flies past planet B and compares its clock with B's. We know, from working in the planets' frame, that when the spaceship reaches B, B's clock reads L/v. And the spaceship's clock reads $L/\gamma v$, because it runs slow by a factor of γ when viewed in the planets' frame.

How would someone on the spaceship quantitatively explain to you why B's clock reads L/v (which is *more* than its own $L/\gamma v$), considering that the spaceship sees B's clock running *slow*?

11.3. **Explaining length contraction** **

Two bombs lie on a train platform, a distance L apart. As a train passes by at speed v, the bombs explode simultaneously (in the platform frame) and leave marks on the train. Due to the length contraction of the train, we know that the marks on the train are a distance γL apart when viewed in the train's frame, because this distance is what is length-contracted down to the given distance L in the platform frame.

How would someone on the train quantitatively explain why the marks are a distance γL apart, considering that the bombs are a distance of only L/γ apart in the train frame?

11.4. **A passing stick** **

A stick of length L moves past you at speed v. There is a time interval between the front end coinciding with you and the back end coinciding with you. What is this time interval in

(a) your frame? (Calculate this by working in your frame.)
(b) your frame? (Work in the stick's frame.)
(c) the stick's frame? (Work in your frame. This is the tricky one.)
(d) the stick's frame? (Work in the stick's frame.)

11.5. **Rotated square** **

A square with side L flies past you at speed v, in a direction parallel to two of its sides. You stand in the plane of the square. When you see the square at its nearest point to you, show that it *looks* to you like it is rotated, instead of contracted. (Assume that L is small compared with the distance between you and the square.)

11.6. **Train in a tunnel** **

A train and a tunnel both have proper lengths L. The train moves toward the tunnel at speed v. A bomb is located at the front of the train. The bomb is designed to explode when the front of the train passes the far end of the tunnel. A deactivation sensor is located at the back of the train. When the back of the train passes the near end of the tunnel, the sensor tells the bomb to disarm itself. Does the bomb explode?

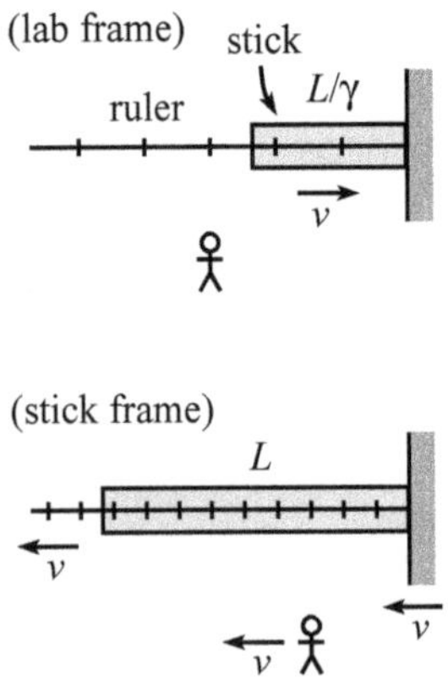

Fig. 11.35

11.7. **Seeing behind the stick** **

A ruler is positioned perpendicular to a wall, and you stand at rest with respect to the ruler and the wall. A stick of length L flies by at speed v. It travels in front of the ruler, so that it obscures part of the ruler from your view. When the stick hits the wall it stops. Which of the following two reasonings is correct (and what is wrong with the incorrect one)?

In your reference frame, the stick is shorter than L. Therefore, right before it hits the wall, you are able to see a mark on the ruler that is closer than L units to the wall (see Fig. 11.35).

But in the stick's frame, the marks on the ruler are closer together. Therefore, when the wall hits the stick, the closest mark to the wall that you can see on the ruler is greater than L units (see Fig. 11.35).

11.8. **Cookie cutter** **

Cookie dough (chocolate chip, of course) lies on a conveyor belt that moves at speed v. A circular stamp stamps out cookies as the dough rushes by beneath it. When you buy these cookies in a store, what shape are they? That is, are they squashed in the direction of the belt, stretched in that direction, or circular?

11.9. **Getting shorter** **

Two balls move with speed v along a line toward two people standing along the same line. The proper distance between the balls is γL, and the proper distance between the people is L. Due to length contraction, the people measure the distance between the balls to be L, so the balls pass the people simultaneously (as measured by the people), as shown in Fig. 11.36. Assume that the people's clocks both read zero at this time. If the people catch the balls, then the resulting proper distance between the balls becomes L, which is shorter than the initial proper distance of γL. Your task: By working in the frame in which the balls are initially at rest, explain how the proper distance between the balls decreases from γL to L. Do this in the following way.

Fig. 11.36

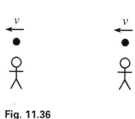

(a) Draw the beginning and ending pictures for the process. Indicate the readings on both clocks in the two pictures, and label all relevant lengths.
(b) Using the distances labeled in your pictures, how far do the people travel? Using the times labeled in your pictures, how far do the people travel? Show that these two methods give the same result.
(c) Explain in words how the proper distance between the balls decreases from γL to L.

Section 11.4: The Lorentz transformations

11.10. **A bunch of L.T.'s** *

Verify that the values of Δx and Δt in the table in the "Passing trains" example in Section 11.6 satisfy the L.T.'s between the six pairs of frames, namely AB, AC, AD, BC, BD, and CD (see Fig. 11.37).

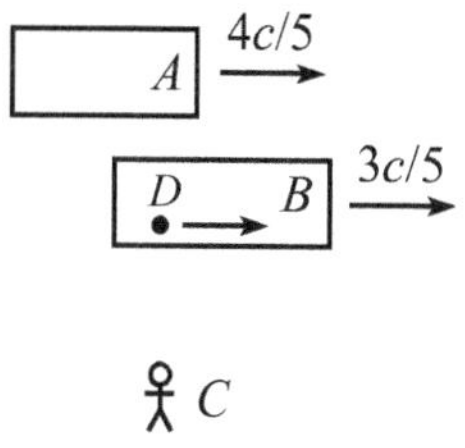

Fig. 11.37

Section 11.5: Velocity addition

11.11. **Equal speeds** *

A and B travel at $4c/5$ and $3c/5$ with respect to the ground, as shown in Fig. 11.38. How fast should C travel so that she sees A and B approaching her at the same speed? What is this speed?

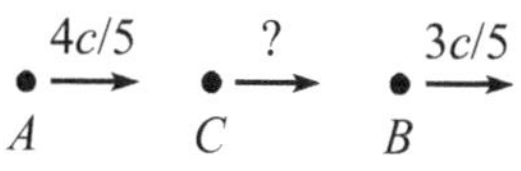

Fig. 11.38

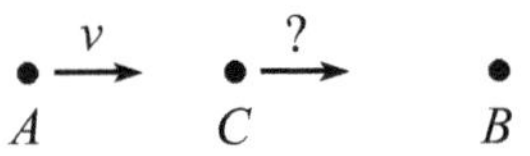

Fig. 11.39

11.12. **More equal speeds** $**$

A travels at speed v with respect to the ground, and B is at rest, as shown in Fig. 11.39. How fast should C travel so that she sees A and B approaching her at the same speed? In the ground frame (B's frame), what is the ratio of the distances CB and AC (assume that A and C arrive at B at the same time)? The answer to this is very nice and clean. Can you think of a simple intuitive explanation for the result?

11.13. **Equal transverse speeds** $*$

In the lab frame, an object moves with velocity (u_x, u_y), and you move with velocity v in the x direction. What should v be so that you also see the object move with velocity u_y in your y direction? One solution is of course $v = 0$. Find the other one.

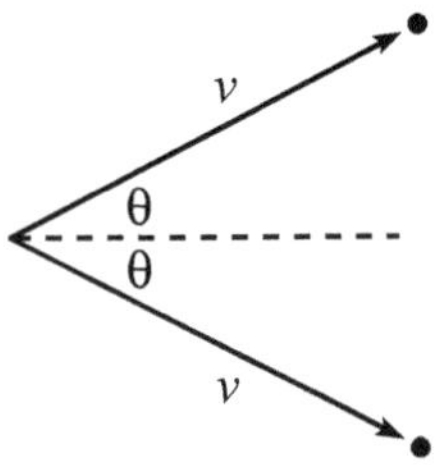

Fig. 11.40

11.14. **Relative speed** $*$

In the lab frame, two particles move with speed v along the paths shown in Fig. 11.40. The angle between the trajectories is 2θ. What is the speed of one particle, as viewed by the other? (*Note*: This problem is posed again in Chapter 13, where it can be solved in a much simpler way, using 4-vectors.)

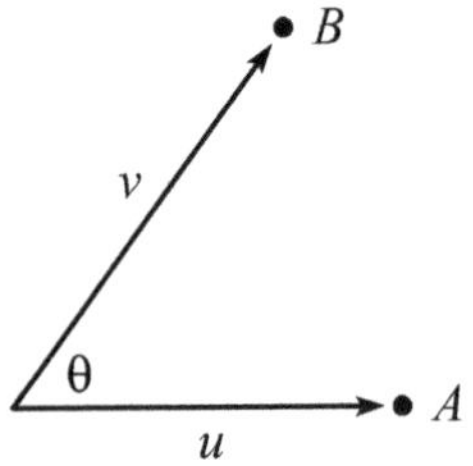

Fig. 11.41

11.15. **Another relative speed** $**$

In the lab frame, particles A and B move with speeds u and v along the paths shown in Fig. 11.41. The angle between the trajectories is θ. What is the speed of one particle, as viewed by the other? (*Note*: This problem is posed again in Chapter 13, where it can be solved in a much simpler way, using 4-vectors.)

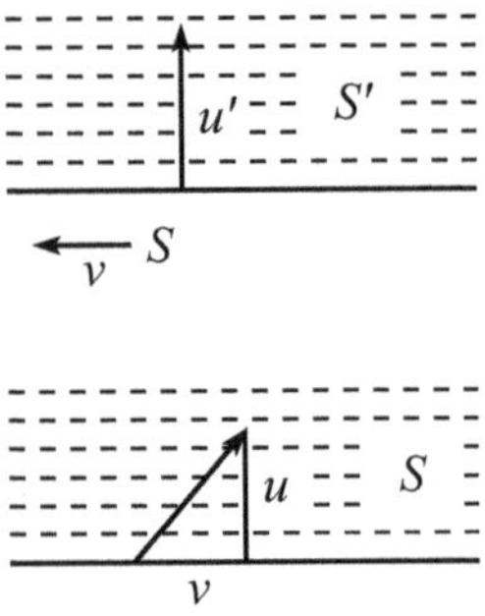

Fig. 11.42

11.16. **Transverse velocity addition** $**$

For the special case of $u'_x = 0$, the transverse velocity-addition formula, Eq. (11.38), yields $u_y = u'_y/\gamma$. Derive this in the following way: In frame S', a particle moves with velocity $(0, u')$, as shown in the first picture in Fig. 11.42. Frame S moves to the left with speed v, so the situation in S looks like what is shown in the second picture in Fig. 11.42, with the y speed now u. Consider a series of equally spaced dotted lines, as shown. The ratio of the times between passes of the dotted lines in frames S and S' is $T_S/T_{S'} = u'/u$. Derive another expression for this ratio by using time-dilation arguments, and then equate the two expressions to solve for u in terms of u' and v.

11.17. **Many velocity additions** $**$

An object moves at speed $v_1/c \equiv \beta_1$ with respect to S_1, which moves at speed β_2 with respect to S_2, which moves at speed β_3 with respect to S_3, and so on, until finally S_{N-1} moves at speed β_N with respect to S_N

(see Fig. 11.43). Show by induction that the speed, call it $\beta_{(N)}$, of the object with respect to S_N can be written as

$$\beta_{(N)} = \frac{P_N^+ - P_N^-}{P_N^+ + P_N^-},$$

$$\text{where } P_N^+ \equiv \prod_{i=1}^{N}(1+\beta_i), \quad \text{and } P_N^- \equiv \prod_{i=1}^{N}(1-\beta_i). \tag{11.75}$$

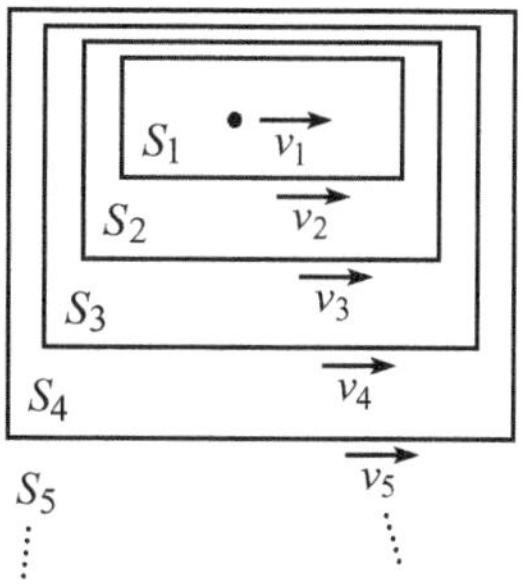

Fig. 11.43

11.18. **Velocity addition from scratch** **

A ball moves at speed v_1 with respect to a train. The train moves at speed v_2 with respect to the ground. What is the speed of the ball with respect to the ground? Solve this problem (that is, derive the velocity-addition formula, Eq. (11.31)) in the following way (don't use time dilation, length contraction, etc; use only the relativity postulate and the fact that the speed of light is the same in any inertial frame):

Let the ball be thrown from the back of the train. At the same instant, a photon is released next to it (see Fig. 11.44). The photon heads to the front of the train, bounces off a mirror, heads back, and eventually runs into the ball. In both the frame of the train and the frame of the ground, calculate the fraction of the way along the train where the meeting occurs, and then equate these fractions.

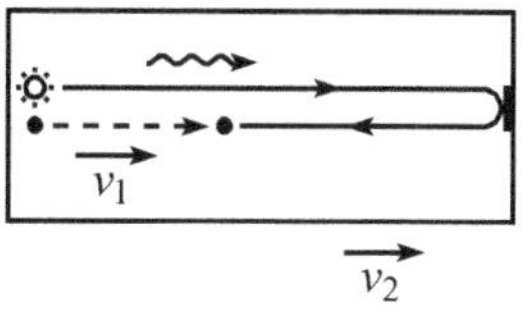

Fig. 11.44

11.19. **Modified twin paradox** ***

Consider the following variation of the twin paradox. A, B, and C each have a clock. In A's reference frame, B flies past A with speed v to the right. When B passes A, they both set their clocks to zero. Also, in A's reference frame, C starts far to the right and moves to the left with speed v. When B and C pass each other, C sets his clock to read the same as B's. Finally, when C passes A, they compare the readings on their clocks. At this moment, let A's clock read T_A, and let C's clock read T_C.

(a) Working in A's frame, show that $T_C = T_A/\gamma$, where $\gamma = 1/\sqrt{1-v^2/c^2}$.
(b) Working in B's frame, show again that $T_C = T_A/\gamma$.
(c) Working in C's frame, show again that $T_C = T_A/\gamma$.

Section 11.6: The invariant interval

11.20. **Throwing on a train** **

A train with proper length L moves at speed $c/2$ with respect to the ground. A ball is thrown from the back to the front, at speed $c/3$ with respect to the train. How much time does this take, and what distance does the ball cover, in:

(a) The train frame?
(b) The ground frame? Solve this by
  i. Using a velocity-addition argument.
  ii. Using the Lorentz transformations to go from the train frame to the ground frame.
(c) The ball frame?
(d) Verify that the invariant interval is indeed the same in all three frames.
(e) Show that the times in the ball frame and ground frame are related by the relevant γ factor.
(f) Ditto for the ball frame and train frame.
(g) Show that the times in the train frame and ground frame are *not* related by the relevant γ factor. Why not?

Section 11.7: Minkowski diagrams

11.21. **A new frame** *

In a given reference frame, Event 1 happens at $x = 0$, $ct = 0$, and Event 2 happens at $x = 2$, $ct = 1$ (in units of some specified length). Find a frame in which the two events are simultaneous.

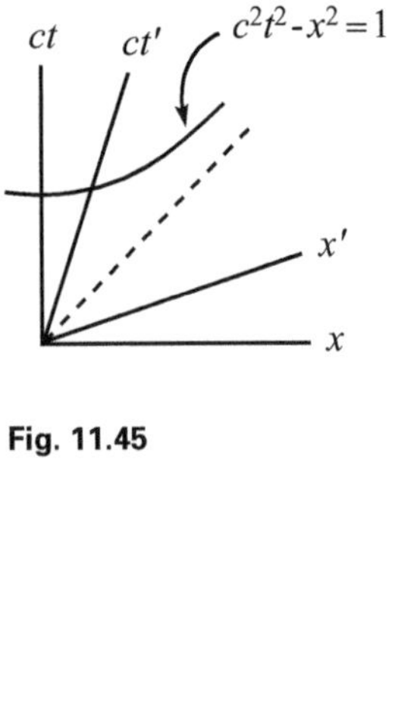

Fig. 11.45

11.22. **Minkowski diagram units** *

Consider the Minkowski diagram in Fig. 11.45. In frame S, the hyperbola $c^2t^2 - x^2 = 1$ is drawn. Also drawn are the axes of frame S', which moves past S with speed v. Use the invariance of the interval $s^2 = c^2t^2 - x^2$ to derive the ratio of the unit sizes on the ct' and ct axes, and check your result with Eq. (11.46).

11.23. **Velocity addition via Minkowski** **

An object moves at speed v_1 with respect to frame S'. Frame S' moves at speed v_2 with respect to frame S (in the same direction as the motion of the object). What is the speed, u, of the object with respect to frame S? Solve this problem (that is, derive the velocity-addition formula) by drawing a Minkowski diagram with frames S and S', drawing the worldline of the object, and doing some geometry.

11.24. **Clapping both ways** **

Twin A stays on the earth, and twin B flies to a distant star and back. For both of the following setups, draw a Minkowski diagram that explains what is happening.

(a) Throughout the trip, B claps in such a way that his claps occur at equal time intervals Δt in A's frame. At what time intervals do the claps occur in B's frame?

(b) Now let A clap in such a way that his claps occur at equal time intervals Δt in B's frame. At what time intervals do the claps occur in A's frame? (Be careful on this one. The sum of all the time intervals must equal the increase in A's age, which is greater than the increase in B's age, as we know from the usual twin paradox.)

11.25. **Acceleration and redshift** ***

Use a Minkowski diagram to solve the following problem: Two people stand a distance d apart. They simultaneously start accelerating in the same direction (along the line between them) with the same proper acceleration a. At the instant they start to move, how fast does each person's clock tick in the (changing) frame of the other person?

11.26. **Break or not break?** ***

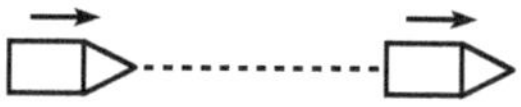

Fig. 11.46

Two spaceships float in space and are at rest relative to each other. They are connected by a string (see Fig. 11.46). The string is strong, but it cannot withstand an arbitrary amount of stretching. At a given instant, the spaceships simultaneously (with respect to their initial inertial frame) start accelerating in the same direction (along the line between them) with the same constant proper acceleration. In other words, assume they bought identical engines from the same store, and they put them on the same setting. Will the string eventually break?

Section 11.9: Rapidity

11.27. **Successive Lorentz transformations**

The Lorentz transformation in Eq. (11.58) may be written in matrix form as

$$\begin{pmatrix} x \\ ct \end{pmatrix} = \begin{pmatrix} \cosh\phi & \sinh\phi \\ \sinh\phi & \cosh\phi \end{pmatrix} \begin{pmatrix} x' \\ ct' \end{pmatrix}. \qquad (11.76)$$

Show that if you apply an L.T. with $v_1 = \tanh\phi_1$, and then another one with $v_2 = \tanh\phi_2$, the result is an L.T. with $v = \tanh(\phi_1 + \phi_2)$.

11.28. **Accelerator's time** **

A spaceship is initially at rest in the lab frame. At a given instant, it starts to accelerate. Let this happen when the lab clock reads $t = 0$ and the spaceship clock reads $t' = 0$. The proper acceleration is a. (That is, at time $t' + dt'$, the spaceship is moving at speed $a\,dt'$ relative to the frame it was in at time t'.) Later on, a person in the lab measures t and t'. What is the relation between them?

11.12 Exercises

Section 11.3: The fundamental effects

11.29. **Effectively speed c** $*$

A rocket flies between two planets that are one light-year apart. What should the rocket's speed be so that the time elapsed on the captain's watch is one year?

11.30. **A passing train** $*$

A train of length 15 cs moves at speed $3c/5$.[39] How much time does it take to pass a person standing on the ground (as measured by that person)? Solve this by working in the frame of the person, and then again by working in the frame of the train.

11.31. **Overtaking a train** $*$

Train A has length L. Train B moves past A (on a parallel track, facing the same direction) with relative speed $4c/5$. The length of B is such that A says that the fronts of the trains coincide at exactly the same time as the backs coincide. What is the time difference between the fronts coinciding and the backs coinciding, as measured by B?

11.32. **Walking on a train** $*$

A train of proper length L and speed $3c/5$ approaches a tunnel of length L. At the moment the front of the train enters the tunnel, a person leaves the front of the train and walks (briskly) toward the back. She arrives at the back of the train right when it (the back) leaves the tunnel.

(a) How much time does this take in the ground frame?
(b) What is the person's speed with respect to the ground?
(c) How much time elapses on the person's watch?

11.33. **Simultaneous waves** $*$

Alice flies past Bob at speed v. Right when she passes, they both set their watches to zero. When Alice's watch shows a time T, she waves to Bob. Bob then waves to Alice simultaneously (as measured by him) with Alice's wave (so this is before he actually *sees* her wave). Alice then waves to Bob simultaneously (as measured by her) with Bob's wave. Bob then waves to Alice simultaneously (as measured by him)

[39] 1 cs is one "light-second." It equals $(1)(3\cdot 10^8\text{ m/s})(1\text{ s}) = 3\cdot 10^8$ m.

with Alice's second wave. And so on. What are the readings on Alice's watch for all the times she waves? And likewise for Bob?

11.34. **Here and there** *

A train of proper length L travels past you at speed v. A person on the train stands at the front, next to a clock that reads zero. At this moment in time (as measured by you), a clock at the back of the train reads Lv/c^2. How would you respond to the following statement:

"The person at the front of the train can leave the front right after the clock there reads zero, and then run to the back and get there right before the clock there reads Lv/c^2. You (on the ground) will therefore see the person simultaneously at *both* the front and the back of the train when the clocks there read zero and Lv/c^2, respectively."

11.35. **Photon on a train** *

A train of proper length L has clocks at the front and back. A photon is fired from the back of the train to the front. Working in the train frame, we can easily say that if the photon leaves the back of the train when the clock there reads zero, then it arrives at the front when the clock there reads L/c.

Now consider this setup in the ground frame, where the train travels by at speed v. Rederive the above result (that the difference in the readings of the two clocks is L/c) by working *only* in the ground frame.

11.36. **Triplets** *

Triplet A stays on the earth. Triplet B travels at speed $4c/5$ to a planet (a distance L away) and back. Triplet C travels out to the planet at speed $3c/4$, and then returns at the necessary speed to arrive back exactly when B does. How much does each triplet age during this process? Who is youngest?

11.37. **Seeing the light** **

A and B leave from a common point (with their clocks both reading zero) and travel in opposite directions with relative speed v. When B's clock reads T, he (B) sends out a light signal. When A receives the signal, what time does his (A's) clock read? Answer this question by doing the calculation entirely in (a) A's frame, and then (b) B's frame. (This problem is basically a derivation of the longitudinal Doppler effect, discussed in Section 11.8.1.)

11.38. **Two trains and a tree** **

Two trains of proper length L move toward each other in opposite directions on parallel tracks. They both move at speed v with respect to the ground. Both trains have clocks at the front and back, and these clocks are synchronized as usual in the frame of the train they are in. A tree is located on the ground at the place where the fronts of the trains pass each other. The clocks at the fronts of the trains both read zero when they pass. Find the reading on the clocks at the backs of the trains when they (the backs) pass each other at the tree. Do this in three different ways:

(a) Imagine that you stand next to the tree on the ground, and you observe what one of the rear clocks is doing.
(b) Imagine that you are on one of the trains, and you observe what your own rear clock is doing during the time the tree travels the relevant distance.
(c) Imagine that you are on one of the trains, and you observe what the other train's rear clock is doing during the time the tree travels the relevant distance. (You'll need to use velocity addition.)

11.39. **Twice simultaneous** **

A train of proper length L moves at speed v with respect to the ground. When the front of the train passes a tree on the ground, a ball is simultaneously (as measured in the *ground frame*) thrown from the back of the train toward the front, with speed u with respect to the train. What should u be so that the ball hits the front simultaneously (as measured in the *train frame*) with the tree passing the back of the train? Show that in order for a solution for u to exist, we must have $v/c < (\sqrt{5} - 1)/2$, which happens to be the inverse of the golden ratio.

11.40. **People clapping** **

Two people stand a distance L apart along an east–west road. They both clap their hands at precisely noon in the ground frame. You are driving eastward down this road at speed $4c/5$. You notice that you are next to the western person at the same instant (as measured in your frame) that the eastern person claps. Later on, you notice that you are next to a tree at the same instant (as measured in your frame) that the western person claps. Where is the tree along the road? (Describe its location in the ground frame.)

11.41. **Photon, tree, and house** **

(a) A train of proper length L moves at speed v with respect to the ground. At the instant the back of the train passes a certain tree, someone at the back of the train shines a photon toward the front. The photon happens to hit the front of the train at the instant the front passes a certain house. As measured in the ground frame, how far apart are the tree and the house? Solve this by working in the ground frame.

(b) Now look at the setup from the point of view of the train frame. Using your result for the tree–house distance from part (a), verify that the house meets the front of the train at the same instant the photon meets it.

11.42. **Tunnel fraction** **

A person runs with speed v toward a tunnel of length L. A light source is located at the far end of the tunnel. At the instant the person enters the tunnel, the light source simultaneously (as measured in the tunnel frame) emits a photon that travels down the tunnel toward the person. When the person and the photon eventually meet, the person's location is a fraction f along the tunnel. What is f? Solve this by working in the tunnel frame, and then again by working in the person's frame.

11.43. **Overlapping trains** **

An observer on the ground sees two trains, A and B, both of proper length L, move in opposite directions at speed v with respect to the ground. She notices that when the trains "overlap," clocks at the front of A and rear of B both read zero. From the "rear clock ahead" effect, she therefore also notices that clocks at the rear of A and front of B read Lv/c^2 and $-Lv/c^2$, respectively (as shown in Fig. 11.47). Now imagine riding along on A. When the rear of B passes the front of your train (A), clocks at both of these places read zero (as stated above). Explain, by working *only* in the frame of A, why clocks at the back of A and the front of B read Lv/c^2 and $-Lv/c^2$, respectively, when these points coincide. (You'll need to use velocity addition.)

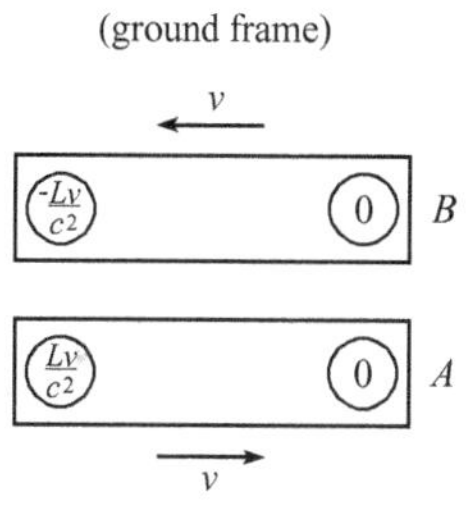

Fig. 11.47

11.44. **Bouncing stick** **

A stick, oriented horizontally, falls and bounces off the ground. Qualitatively, what does this look like in the frame of someone running by at speed v?

11.45. **Through the hole?** **

A stick of proper length L moves at speed v in the direction of its length. It passes over an infinitesimally thin sheet that has a hole of diameter L cut in it. As the stick passes over the hole, the sheet is raised so that the stick passes through the hole and ends up underneath the sheet. Well, maybe . . .

In the lab frame, the stick's length is contracted to L/γ, so it appears to easily make it through the hole. But in the stick frame, the hole is contracted to L/γ, so it appears that the stick does *not* make it through the hole (or rather, the hole doesn't make it around the stick, since the hole is what is moving in the stick frame). So the question is: Does the stick end up on the other side of the sheet or not?

11.46. **Short train in a tunnel** **

Consider the scenario in Problem 11.6, with the only change being that the train now has length rL, where r is some numerical factor. What is the largest value of r, in terms of v, for which it is possible for the bomb not to explode? Verify that you obtain the same answer working in the train frame and working in the tunnel frame.

Section 11.4: The Lorentz transformations

11.47. **Successive L.T.'s** **

Show that the combination of an L.T. (with speed v_1) and an L.T. (with speed v_2) yields an L.T. with speed $u = (v_1 + v_2)/(1 + v_1 v_2/c^2)$.

11.48. **Loss of simultaneity** **

A train moves at speed v with respect to the ground. Two events occur simultaneously, a distance L apart, in the train frame. What are the time and space separations in the ground frame? Solve this by:

(a) Using the Lorentz transformations.
(b) Using only the results in Section 11.3. Do this by working in the ground frame, and then again by working in the train frame.

Section 11.5: Velocity addition

11.49. **Some γ's** *

Show that the relativistic addition (or subtraction) of the velocities u and v has a γ factor given by $\gamma = \gamma_u \gamma_v (1 \pm uv)$.

11.50. **Slanted time dilation** *

A clock moves vertically with speed u in a given frame, and you run horizontally with speed v with respect to this frame. Show that you see the clock run slow by the nice simple factor, $\gamma_u \gamma_v$.

11.51. **Pythagorean triples** *

Let (a, b, h) be a Pythagorean triple. (We'll use h to denote the hypotenuse, instead of c, for obvious reasons.) Consider the relativistic addition or subtraction of the two speeds, $\beta_1 = a/h$ and $\beta_2 = b/h$. Show that the numerator and denominator of the result are the leg and hypotenuse of another Pythagorean triple, and find the other leg. What is the associated γ factor?

11.52. **Running away** *

A and B both start at the origin and simultaneously head off in opposite directions at speed $3c/5$ with respect to the ground. A moves to the right, and B moves to the left. Consider a mark on the ground at $x = L$.

As viewed in the ground frame, A and B are a distance $2L$ apart when A passes this mark. As viewed by A, how far away is B when A coincides with the mark?

11.53. **Angled photon** *

A photon moves at an angle θ with respect to the x' axis in frame S'. Frame S' moves at speed v with respect to frame S (along the x' axis). Calculate the components of the photon's velocity in S, and verify that the speed is c.

11.54. **Running on a train** **

A train of proper length L moves at speed v_1 with respect to the ground. A passenger runs from the back of the train to the front at speed v_2 with respect to the train. How much time does this take, as viewed by someone on the ground? Solve this in two different ways:

(a) Find the relative speed of the passenger and the train (as viewed by someone on the ground), and then find the time it takes for the passenger to erase the initial "head start" that the front of the train had.

(b) Find the time elapsed on the passenger's clock (by working in whatever frame you want), and then use time dilation to get the time elapsed on a ground clock.

11.55. **Velocity addition** **

The fact that the previous exercise can be solved in two different ways suggests a method of deriving the velocity-addition formula: A train of proper length L moves at speed v_1 with respect to the ground. A ball is thrown from the back of the train to the front at speed v_2 with respect to the train. Let the speed of the ball with respect to the ground be V. Calculate the time of the ball's journey, as measured by an observer on the ground, in the following two different ways, and then set them equal to solve for V in terms of v_1 and v_2. (This gets a bit messy. And yes, you have to solve a quadratic.)

(a) First way: Find the relative speed of the ball and the train (as viewed by someone on the ground), and then find the time it takes for the ball to erase the initial "head start" that the front of the train had.

(b) Second way: Find the time elapsed on the ball's clock (by working in whatever frame you want), and then use time dilation to get the time elapsed on a ground clock.

11.56. **Velocity addition again** **

A train of proper length L moves at speed v with respect to the ground. A ball is thrown from the back of the train to the front at speed u with respect to the train. When finding the time of this process in the ground frame, a common error is to use time dilation to go from the train frame to the ground frame, which gives an incorrect answer of $\gamma_v(L/u)$. This is incorrect because time dilation is valid only if the two relevant events occur at the same place in one of the frames; otherwise simultaneity becomes an issue.

(a) Find the total time in the ground frame correctly by looking at how much a clock at rest in the train frame advances (for example, a clock at the front of the train), and by applying time dilation to this clock.

(b) Find the total time in the ground frame by applying time dilation to the ball's clock. Your answer will contain the unknown speed V of the ball with respect to the ground.

(c) Equate your results from parts (a) and (b) to show that $\gamma_V = \gamma_u \gamma_v(1 + uv/c^2)$. Then solve for V to produce the velocity-addition formula.

11.57. **Bullets on a train** **

A train moves at speed v. Bullets are successively fired at speed u (relative to the train) from the back of the train to the front. A new bullet is fired at the instant (as measured in the train frame) the previous bullet hits the front. In the frame of the ground, what fraction of the way along the train is a given bullet, at the instant (as measured in the ground frame) the next bullet is fired? What is the maximum number of bullets that are in flight at a given instant, in the ground frame?

11.58. **Time dilation and Lv/c^2** **

A person walks very slowly at speed u from the back of a train of proper length L to the front. The time-dilation effect in the train frame can be made arbitrarily small by picking u to be sufficiently small (because the effect is second order in u). Therefore, if the person's watch agrees with a clock at the back of the train when he starts, then it also (essentially) agrees with a clock at the front when he finishes.

Now consider this setup in the ground frame, where the train moves at speed v. The rear clock reads Lv/c^2 more than the front, so in view of the preceding paragraph, the time gained by the person's watch during the process must be Lv/c^2 less than the time gained by the front clock. By working in the ground frame, explain why this is the case. Assume $u \ll v$.[40]

[40] If you line up a collection of these train systems around the circumference of a rotating platform, then the above result implies the following fact. Let person A be at rest on the platform, and let person B walk arbitrarily slowly around the circumference. Then when B returns to A, B's clock will read less than A's. This is true because the above reasoning shows (as you will figure out)

Section 11.6: The invariant interval

11.59. **Passing a train** **

Person A stands on the ground, train B with proper length L moves to the right at speed $3c/5$, and person C runs to the right at speed $4c/5$. C starts behind the train and eventually passes it. Let event E_1 be "C coincides with the back of the train," and let event E_2 be "C coincides with the front of the train." Find the Δt and Δx between the events E_1 and E_2 in the frames of A, B, and C, and show that $c^2\Delta t^2 - \Delta x^2$ is the same in all three frames.

11.60. **Passing trains** **

Train A with proper length L moves eastward at speed v, while train B with proper length $2L$ moves westward also at speed v. How much time does it take for the trains to pass each other (defined as the time between the fronts coinciding and the backs coinciding):

(a) In A's frame?
(b) In B's frame?
(c) In the ground frame?
(d) Verify that the invariant interval is the same in all three frames.

11.61. **Throwing on a train** **

A train with proper length L moves at speed $3c/5$ with respect to the ground. A ball is thrown from the back to the front, at speed $c/2$ with respect to the train. How much time does this take, and what distance does the ball cover, in:

(a) The train frame?
(b) The ground frame? Solve this by
   i. Using a velocity-addition argument.
   ii. Using the Lorentz transformations to go from the train frame to the ground frame.
(c) The ball frame?
(d) Verify that the invariant interval is indeed the same in all three frames.
(e) Show that the times in the ball frame and ground frame are related by the relevant γ factor.

that an inertial observer sees B's clock running slower than A's. This result, that you can walk arbitrarily slowly in a particular reference frame and have your clock lose synchronization with other clocks, is a consequence of the fact that in some accelerating reference frames it is impossible to produce a consistent method (that is, one without a discontinuity) of clock synchronization. See Cranor *et al.* (2000) for more details.

(f) Ditto for the ball frame and train frame.

(g) Show that the times in the train frame and ground frame are *not* related by the relevant γ factor. Why not?

Section 11.7: Minkowski diagrams

11.62. **Time dilation via Minkowski** *

In the spirit of the example in Section 11.7, use a Minkowski diagram to derive the time-dilation result between frames S and S' (in both directions, as in the example).

11.63. **Lv/c^2 via Minkowski** *

In the spirit of the example in Section 11.7, use a Minkowski diagram to derive the Lv/c^2 rear-clock-ahead result for frames S and S' (in both directions, as in the example).

11.64. **Simultaneous waves again** **

Solve Exercise 11.33 by using a Minkowski diagram from the point of view of someone who sees Alice and Bob moving with equal and opposite speeds.

11.65. **Short train in a tunnel again** ***

Solve Exercise 11.46 by using a Minkowski diagram from the point of view of the train, and also of the tunnel.

Section 11.8: The Doppler effect

11.66. **Transverse Doppler** **

As mentioned in Remark 2 of Section 11.8.2, it is possible to solve the transverse Doppler effect for Case 1 by working in the frame of the observer, provided that you account for the longitudinal component of the source's motion. Solve the problem this way and reproduce Eq. (11.52).

11.67. **Twin paradox via Doppler** **

Twin A stays on the earth, and twin B flies at speed v to a distant star and back. The star is a distance L from the earth in the earth–star frame. Use the Doppler effect to show that B is younger by a factor γ when she returns (don't use any time dilation or length contraction results). Do this in the following two ways; both are doable by working in either A's frame or B's frame(s), so take your pick.

(a) A sends out flashes at intervals of t seconds (as measured in his frame). By considering the numbers of redshifted and blueshifted flashes that B receives, show that $T_B = T_A/\gamma$.

(b) B sends out flashes at intervals of t seconds (as measured in her frame). By considering the numbers of redshifted and blueshifted flashes that A receives, show that $T_B = T_A/\gamma$.

Section 11.9: Rapidity

11.68. **Time of travel** **

Consider the setup in Problem 11.28 (and feel free to use the results from that problem in this exercise). Let the spaceship travel to a planet a distance L from the earth.

(a) By working in the frame of the earth, find the time of the journey, as measured by the earth. Check the large and small L (compared with c^2/a) limits.

(b) By working in the (changing) frame of the spaceship, find the time of the journey, as measured by the spaceship (an implicit equation is fine). Check the small L limit. How does the time behave for large L?

11.13 Solutions

11.1. **No transverse length contraction**

Assume that the paint brushes are capable of leaving marks on stick B if B is long enough, or if A is short enough. The key fact we need here is the second postulate of relativity, which says that the frames of the sticks are equivalent. That is, if A sees B shorter than (or longer than, or equal to) itself, then B also sees A shorter than (or longer than, or equal to) itself. The contraction factor must be the same when going each way between the frames.

Assume (in search of a contradiction) that A sees B shortened. Then B won't extend out to the ends of A, so there will be no marks on B. But in this case, B must *also* see A shortened, so there *will* be marks on B (see Fig. 11.48). This is a contradiction. Likewise, if we assume that A sees B lengthened, we also reach a contradiction. We are therefore left with only the third possibility, namely that each stick sees the other stick as one meter long.

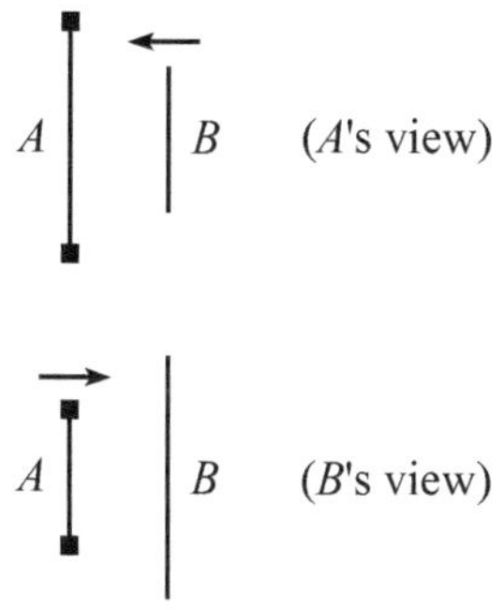

Fig. 11.48

11.2. **Explaining time dilation**

The resolution to the apparent paradox is the "head start" that B's clock has over A's clock, as seen in the spaceship frame. From Eq. (11.6), we know that in the spaceship frame, B's clock reads Lv/c^2 more than A's. (The two planets may be considered to be at the ends of the train in the example in Section 11.3.1.)

Therefore, what a person on the spaceship says is: "My clock advances by $L/\gamma v$ during the whole trip. I see B's clock running slow by a factor γ, so I see B's clock advance by only $(L/\gamma v)/\gamma = L/\gamma^2 v$. However, B's clock started not at zero but at Lv/c^2. Therefore, the final reading on B's clock when I get there is

$$\frac{Lv}{c^2} + \frac{L}{\gamma^2 v} = \frac{L}{v}\left(\frac{v^2}{c^2} + \frac{1}{\gamma^2}\right) = \frac{L}{v}\left(\frac{v^2}{c^2} + \left(1 - \frac{v^2}{c^2}\right)\right) = \frac{L}{v}, \tag{11.77}$$

as we wanted to show."

11.3. **Explaining length contraction**

The resolution to the apparent paradox is that the explosions do not occur simultaneously in the train frame. As the platform rushes past the train, the "rear" bomb explodes before the "front" bomb explodes.[41] The front bomb then gets to travel farther by the time it explodes and leaves its mark. The distance between the marks is therefore larger than you might naively expect. Let's be quantitative about this.

Let the two bombs contain clocks that read zero when they explode (they are synchronized in the platform frame). Then in the train frame, the front bomb's clock reads only $-Lv/c^2$ when the rear bomb explodes when showing a time of zero. (This is the "rear clock ahead" result from Eq. (11.6).) The front bomb's clock must therefore advance by a time of Lv/c^2 before it explodes. But the train sees the bombs' clocks running slow by a factor γ, so in the train frame the front bomb explodes a time $\gamma Lv/c^2$ after the rear bomb explodes. During this time of $\gamma Lv/c^2$, the platform moves a distance $(\gamma Lv/c^2)v$ relative to the train.

Therefore, what a person on the train says is: "Due to length contraction, the distance between the bombs is L/γ. The front bomb is therefore a distance L/γ ahead of the rear bomb when the latter explodes. The front bomb then travels an additional distance of $\gamma Lv^2/c^2$ by the time it explodes, at which point it is a distance of

$$\frac{L}{\gamma} + \frac{\gamma Lv^2}{c^2} = \gamma L\left(\frac{1}{\gamma^2} + \frac{v^2}{c^2}\right) = \gamma L\left(\left(1 - \frac{v^2}{c^2}\right) + \frac{v^2}{c^2}\right) = \gamma L \qquad (11.78)$$

ahead of the rear bomb's mark, as we wanted to show."

11.4. **A passing stick**

(a) The stick has length L/γ in your frame, and it moves with speed v. Therefore, the time taken in your frame to cover the distance L/γ is $L/\gamma v$.

(b) The stick sees you fly by at speed v. The stick has length L in its own frame, so the time elapsed in the stick frame is L/v. During this time, the stick sees the watch on your wrist run slow by a factor γ. Therefore, a time of $L/\gamma v$ elapses on your watch, in agreement with part (a).

Logically, the two solutions in (a) and (b) differ in that one uses length contraction and the other uses time dilation. Mathematically, they differ simply in the order in which the divisions by γ and v occur.

(c) You see the rear clock on the stick showing a time of Lv/c^2 more than the front clock. In addition to this head start, more time will of course elapse on the rear clock by the time it reaches you. The time in your frame is $L/\gamma v$ (because the stick has length L/γ in your frame). But the stick's clocks run slow, so a time of only $L/\gamma^2 v$ will elapse on the rear clock by the time it reaches you. The total additional time (compared with the front clock's reading when it passed you) that the rear clock shows is therefore

$$\frac{Lv}{c^2} + \frac{L}{\gamma^2 v} = \frac{L}{v}\left(\frac{v^2}{c^2} + \frac{1}{\gamma^2}\right) = \frac{L}{v}\left(\frac{v^2}{c^2} + \left(1 - \frac{v^2}{c^2}\right)\right) = \frac{L}{v}, \qquad (11.79)$$

in agreement with the quick calculation that follows in part (d).

(d) The stick sees you fly by at speed v. The stick has length L in its own frame, so the time elapsed in the stick frame is L/v.

[41] Since we'll be working in the train frame here, we'll use the words "rear" and "front" in the way that someone on the train uses them as she watches the platform rush by. That is, if the train is heading east with respect to the platform, then from the point of view of the train, the platform is heading west, so the eastern bomb on the platform is the rear one, and the western bomb is the front one. So they have the opposite orientation compared with the way that someone on the platform labels the rear and front of the train. Using the same orientation would entail writing the phrase "front clock ahead" below, which would make me cringe.

11.5. **Rotated square**

Figure 11.49 shows a top view of the square at the instant (in your frame) when it is closest to you. Its length is contracted along the direction of motion, so it takes the shape of a rectangle with sides L and L/γ. This is what the shape *is* in your frame (where *is*-ness is defined by where all the points of an object are at simultaneous times). But what does the square *look* like to you? That is, what is the nature of the photons hitting your eye at a given instant?[42]

Photons from the far side of the square have to travel an extra distance L to get to your eye, compared with photons from the near side. So they need an extra time L/c of flight. During this time L/c, the square moves a distance $Lv/c \equiv L\beta$ sideways. Therefore, referring to Fig. 11.50, a photon emitted at point A reaches your eye at the same time as a photon emitted from point B. This means that the trailing side of the square spans a distance $L\beta$ across your field of vision, while the near side spans a distance $L/\gamma = L\sqrt{1-\beta^2}$ across your field of vision. But this is exactly what a rotated square of side L looks like, as shown in Fig. 11.51, where the angle of rotation satisfies $\sin\theta = \beta$. For the case of a circle instead of a square, see Hollenbach (1976).

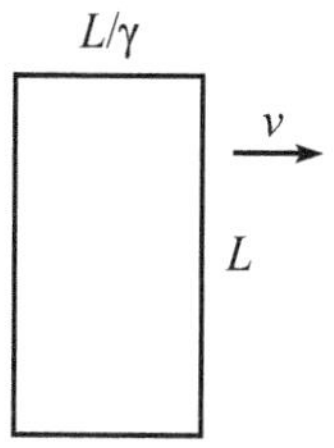

Fig. 11.49

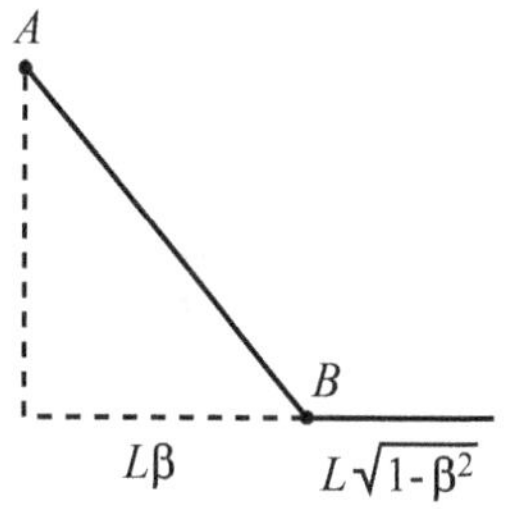

Fig. 11.50

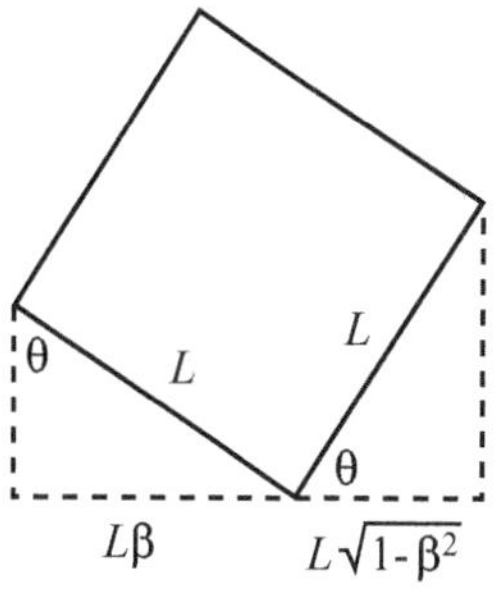

Fig. 11.51

11.6. **Train in a tunnel**

Yes, the bomb explodes. This is clear in the frame of the train (see Fig. 11.52). In this frame, the train has length L, and the tunnel speeds past it. The tunnel is length-contracted down to L/γ. Therefore, the far end of the tunnel passes the front of the train before the near end passes the back, so the bomb explodes.

We can, however, also look at things in the frame of the tunnel (see Fig. 11.53). Here the tunnel has length L, and the train is length-contracted down to L/γ. Therefore, the deactivation device gets triggered *before* the front of the train passes the far end of the tunnel, so you might think that the bomb does *not* explode. We appear to have a paradox.

The resolution to this paradox is that the deactivation device cannot instantaneously tell the bomb to deactivate itself. It takes a finite time for the signal to travel the length of the train from the sensor to the bomb. And it turns out that this transmission time makes it impossible for the deactivation signal to get to the bomb before the bomb gets to the far end of the tunnel, no matter how fast the train is moving. Let's show this.

The signal has the best chance of winning this "race" if it has speed c, so let's assume this is the case. Now, the signal gets to the bomb before the bomb gets to the far end of the tunnel if and only if a light pulse emitted from the near end of the tunnel (at the instant the back of the train goes by) reaches the far end of the tunnel before the front of the train does. The former takes a time L/c. The latter takes a time $L(1-1/\gamma)/v$, because the front of the train is already a distance L/γ through the tunnel. So if the bomb is *not* to explode, we must have

$$\begin{aligned} L/c &< L(1-1/\gamma)/v \\ \iff \quad \beta &< 1-\sqrt{1-\beta^2} \\ \iff \quad \sqrt{1-\beta^2} &< 1-\beta \\ \iff \quad \sqrt{1+\beta} &< \sqrt{1-\beta}. \end{aligned} \tag{11.80}$$

This is never true. Therefore, the signal always arrives too late, and the bomb always explodes.

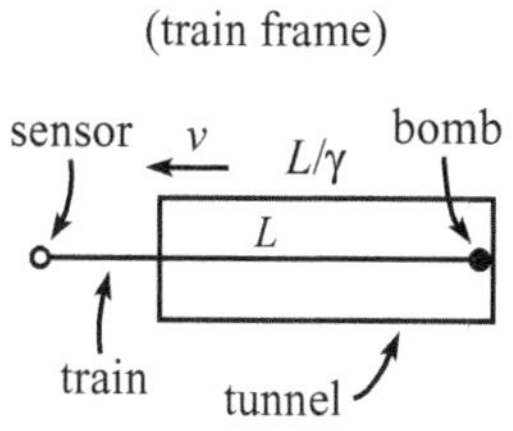

Fig. 11.52

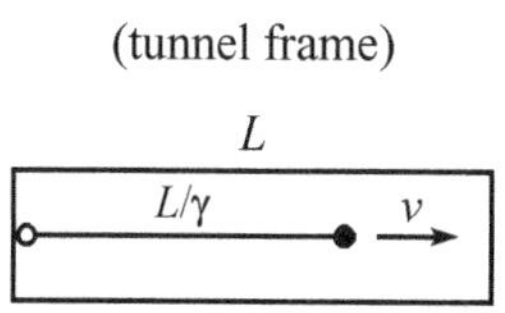

Fig. 11.53

[42] In relativity problems, we virtually always subtract off the time it takes light to travel from the object to your eye (that is, we find out what really *is*). But along with the Doppler effect discussed in Section 11.8, this problem is one of the few exceptions where we actually want to determine what your eye registers.

11.7. **Seeing behind the stick**

The first reasoning is correct. You will be able to see a mark on the ruler that is less than L units from the wall. You will actually be able to see a mark even closer to the wall than L/γ, as we'll show below.

The error in the second reasoning (in the stick's frame) is that the second picture in Fig. 11.35 is *not* what you see. This second picture shows where things are at simultaneous times in the stick's frame, which are not simultaneous times in your frame. Alternatively, the error is the implicit assumption that signals travel instantaneously. But in fact the back of the stick cannot know that the front of the stick has been hit by the wall until a finite time has passed. During this time, the ruler (and the wall and you) travels farther to the left, allowing you to see more of the ruler. Let's be quantitative about this and calculate (in both frames) the closest mark to the wall that you can see.

Consider your reference frame. The stick has length L/γ. Therefore, when the stick hits the wall, you can see a mark a distance L/γ from the wall. You will, however, be able to see a mark even closer to the wall, because the back end of the stick will keep moving forward, since it doesn't know yet that the front end has hit the wall. The stopping signal (shock wave, etc.) takes time to travel.

Let's assume that the stopping signal travels along the stick at speed c. (We could instead work with a general speed u, but the speed c is simpler, and it yields an upper bound on the closest mark you can see.) Where will the signal reach the back end? Starting from the time the stick hits the wall, the signal travels backward from the wall at speed c, while the back end of the stick travels forward at speed v (from a point L/γ away from the wall). So the relative speed (as viewed by you) of the signal and the back end is $c + v$. Therefore, the signal hits the back end after a time $(L/\gamma)/(c + v)$. During this time, the signal has traveled a distance $c(L/\gamma)/(c + v)$ from the wall. The closest point to the wall that you can see on the ruler is therefore the mark with the value

$$\frac{L}{\gamma(1+\beta)} = L\sqrt{\frac{1-\beta}{1+\beta}}. \tag{11.81}$$

Now consider the stick's reference frame. The wall is moving to the left toward the stick at speed v. After the wall hits the right end of the stick, the signal moves to the left with speed c, while the wall keeps moving to the left with speed v. Where is the wall when the signal reaches the left end? The wall travels v/c as fast as the signal, so it travels a distance Lv/c in the time that the signal travels the distance L. This means that the wall is $L(1 - v/c)$ away from the left end of the stick. In the stick's frame, this corresponds to a distance $\gamma L(1 - v/c)$ on the ruler, because the ruler is length contracted. So the left end of the stick is at the mark with the value

$$L\gamma(1-\beta) = L\sqrt{\frac{1-\beta}{1+\beta}}, \tag{11.82}$$

in agreement with Eq. (11.81).

11.8. **Cookie cutter**

Let the diameter of the cookie cutter be L, and consider the two following reasonings.

- In the lab frame, the dough is length contracted, so the diameter L corresponds to a distance larger than L (namely γL) in the dough's frame. Therefore, when you buy a cookie, it is stretched out by a factor γ in the direction of the belt.[43]

[43] The shape is an ellipse, since that's what a stretched-out circle is. The eccentricity of an ellipse is the focal distance divided by the semi-major axis length. As an exercise, you can show that this equals $\beta \equiv v/c$ here.

- In the frame of the dough, the cookie cutter is length contracted down to L/γ in the direction of motion. So in the frame of the dough, the cookies have a length of only L/γ. Therefore, when you buy a cookie, it is squashed by a factor γ in the direction of the belt.

Which reasoning is correct? The first one is. The cookies are stretched out. The fallacy in the second reasoning is that the various parts of the cookie cutter do *not* strike the dough simultaneously in the dough frame. What the dough sees is this: Assuming that the cutter moves to the left, the right side of the cutter stamps the dough, then nearby parts of the cutter stamp it, and so on, until finally the left side of the cutter stamps the dough. But by this time the front (that is, the left) of the cutter has moved farther to the left. So the cookie turns out to be longer than L. It takes a little work to demonstrate (by working in the dough frame) that the length is actually γL, but let's do that now.

Consider the moment when the rightmost point of the cutter strikes the dough. In the dough frame, a clock at the rear (the right side) of the cutter reads Lv/c^2 more than a clock at the front (the left side). The front clock must therefore advance by Lv/c^2 by the time it strikes the dough. (This is true because all points on the cutter strike the dough simultaneously in the cutter frame. Hence, all cutter clocks read the same when they strike.) But due to time dilation, this takes a time $\gamma(Lv/c^2)$ in the dough frame. During this time, the cutter travels a distance $v(\gamma Lv/c^2)$. Since the front of the cutter was initially a distance L/γ (due to length contraction) ahead of the back, the total length of the cookie in the dough frame is

$$\ell = \frac{L}{\gamma} + v\left(\frac{\gamma Lv}{c^2}\right) = \gamma L\left(\frac{1}{\gamma^2} + \frac{v^2}{c^2}\right) = \gamma L\left(\left(1 - \frac{v^2}{c^2}\right) + \frac{v^2}{c^2}\right) = \gamma L,$$

as we wanted to show. If the dough is then slowly decelerated, the shape of the cookies won't change. So this is the shape you see in the store.

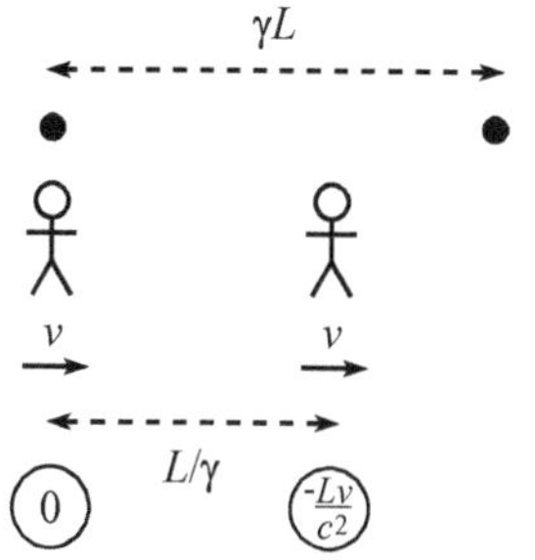

Fig. 11.54

11.9. **Getting shorter**

(a) In the frame in which the balls are initially at rest, the beginning picture (when the left person catches the left ball when his clock reads zero) is shown in Fig. 11.54. The balls are a distance γL apart, and the people's separation is length contracted down to L/γ. The front person's clock is Lv/c^2 behind the back person's clock, so it reads $-Lv/c^2$.

The ending picture (when the right person catches the right ball when his clock reads zero) is shown in Fig. 11.55. By the time the right person catches the ball, the left person has moved to the right while holding the left ball. The left person's clock is ahead, so it reads Lv/c^2.

(b) By looking at the distances in the figures, we see that the people travel a distance $\gamma L - L/\gamma$.

Let's now use the clock readings to get the distance. The total time for the process is $\gamma(Lv/c^2)$ because each person's clock advances by Lv/c^2, but these clocks run slow in the frame we're working in. Since the speed of the people is v, the distance they travel is $v(\gamma Lv/c^2)$. This had better be equal to $\gamma L - L/\gamma$. And it is, because

$$\gamma L - \frac{L}{\gamma} = \gamma L\left(1 - \frac{1}{\gamma^2}\right) = \gamma L\left(1 - \left(1 - \frac{v^2}{c^2}\right)\right) = \frac{\gamma L v^2}{c^2}. \quad (11.83)$$

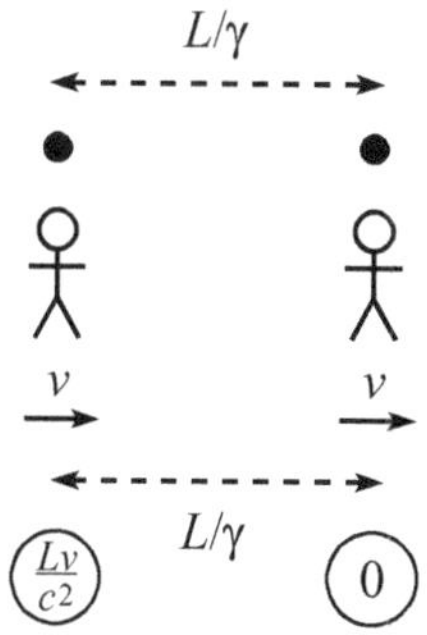

Fig. 11.55

If we then shift to the frame in which everything is at rest, we see that the proper distance between the balls is L, as we wanted to show.

(c) To sum up, the proper distance between the balls decreases because in the frame in which the balls are initially at rest, the left person catches the left ball first and then drags it closer to the right ball by the time the right person catches that one. So it all comes down to the loss of simultaneity.

11.10. A bunch of L.T.'s

Using the results from the "Passing trains" examples in Sections 11.5.1 and 11.6, the relative speeds and the associated γ factors for the six pairs of frames are

	AB	AC	AD	BC	BD	CD
v	$5c/13$	$4c/5$	$c/5$	$3c/5$	$c/5$	$5c/7$
γ	$13/12$	$5/3$	$5/2\sqrt{6}$	$5/4$	$5/2\sqrt{6}$	$7/2\sqrt{6}$

From the example in Section 11.6, the separations between the two events in the four frames are

	A	B	C	D
Δx	$-L$	L	$5L$	0
Δt	$5L/c$	$5L/c$	$7L/c$	$2\sqrt{6}L/c$

The Lorentz transformations are

$$\begin{aligned} \Delta x &= \gamma(\Delta x' + v\,\Delta t'), \\ \Delta t &= \gamma(\Delta t' + v\,\Delta x'/c^2). \end{aligned} \tag{11.84}$$

For each of the six pairs, we'll transform from the faster frame to the slower frame. This means that the coordinates of the faster frame will be on the right-hand side of the L.T.'s. The sign on the right-hand side of the L.T.'s will therefore always be a "+." In the AB case, for example, we'll write "Frames B and A," in that order, to signify that the B coordinates are on the left-hand side and the A coordinates are on the right-hand side. We'll simply list the L.T.'s for the six cases, and you can check that they do indeed all work out.

$$\begin{aligned}
\text{Frames } B \text{ and } A: \quad & L = \frac{13}{12}\left(-L + \left(\frac{5c}{13}\right)\left(\frac{5L}{c}\right)\right), \\
& \frac{5L}{c} = \frac{13}{12}\left(\frac{5L}{c} + \frac{\frac{5c}{13}(-L)}{c^2}\right). \\
\text{Frames } C \text{ and } A: \quad & 5L = \frac{5}{3}\left(-L + \left(\frac{4c}{5}\right)\left(\frac{5L}{c}\right)\right), \\
& \frac{7L}{c} = \frac{5}{3}\left(\frac{5L}{c} + \frac{\frac{4c}{5}(-L)}{c^2}\right). \\
\text{Frames } D \text{ and } A: \quad & 0 = \frac{5}{2\sqrt{6}}\left(-L + \left(\frac{c}{5}\right)\left(\frac{5L}{c}\right)\right), \\
& \frac{2\sqrt{6}L}{c} = \frac{5}{2\sqrt{6}}\left(\frac{5L}{c} + \frac{\frac{c}{5}(-L)}{c^2}\right). \\
\text{Frames } C \text{ and } B: \quad & 5L = \frac{5}{4}\left(L + \left(\frac{3c}{5}\right)\left(\frac{5L}{c}\right)\right), \\
& \frac{7L}{c} = \frac{5}{4}\left(\frac{5L}{c} + \frac{\frac{3c}{5}L}{c^2}\right).
\end{aligned} \tag{11.85}$$

$$\text{Frames } B \text{ and } D: \qquad L = \frac{5}{2\sqrt{6}}\left(0 + \left(\frac{c}{5}\right)\left(\frac{2\sqrt{6}L}{c}\right)\right),$$

$$\frac{5L}{c} = \frac{5}{2\sqrt{6}}\left(\frac{2\sqrt{6}L}{c} + \frac{\frac{c}{5}(0)}{c^2}\right).$$

$$\text{Frames } C \text{ and } D: \qquad 5L = \frac{7}{2\sqrt{6}}\left(0 + \left(\frac{5c}{7}\right)\left(\frac{2\sqrt{6}L}{c}\right)\right),$$

$$\frac{7L}{c} = \frac{7}{2\sqrt{6}}\left(\frac{2\sqrt{6}L}{c} + \frac{\frac{5c}{7}(0)}{c^2}\right).$$

11.11. **Equal speeds**

FIRST SOLUTION: Let C move at speed v with respect to the ground, and let the relative speed of C and both A and B be u (as viewed by C). Then two different expressions for u are the relativistic subtraction of v from $4c/5$, and the relativistic subtraction of $3c/5$ from v. Therefore (dropping the c's),

$$\frac{\frac{4}{5} - v}{1 - \frac{4}{5}v} = u = \frac{v - \frac{3}{5}}{1 - \frac{3}{5}v}. \tag{11.86}$$

This gives $0 = 35v^2 - 74v + 35 = (5v-7)(7v-5)$. Since the $v = 7/5$ root represents a speed larger than c, we must have

$$v = \frac{5}{7}c. \tag{11.87}$$

Plugging this back into Eq. (11.86) gives $u = c/5$.

SECOND SOLUTION: With v and u defined as above, two different expressions for v are the relativistic subtraction of u from $4c/5$, and the relativistic addition of u to $3c/5$. Therefore,

$$\frac{\frac{4}{5} - u}{1 - \frac{4}{5}u} = v = \frac{\frac{3}{5} + u}{1 + \frac{3}{5}u}. \tag{11.88}$$

This gives $0 = 5u^2 - 26u + 5 = (5u - 1)(u - 5)$. Since the $u = 5$ root represents a speed larger than c, we must have

$$u = \frac{c}{5}. \tag{11.89}$$

Plugging this back into Eq. (11.88) gives $v = 5c/7$.

THIRD SOLUTION: The relative speed of A and B is

$$\frac{\frac{4}{5} - \frac{3}{5}}{1 - \frac{4}{5}\cdot\frac{3}{5}} = \frac{5}{13}. \tag{11.90}$$

From C's point of view, this $5/13$ is the result of relativistically adding u with another u. Therefore,

$$\frac{5}{13} = \frac{2u}{1 + u^2} \quad \Longrightarrow \quad 5u^2 - 26u + 5 = 0, \tag{11.91}$$

as in the second solution.

11.12. **More equal speeds**

Let u be the speed at which C sees A and B approaching her. So u is the desired speed of C with respect to B, that is, the ground. From C's point of view, the given speed v is the result of relativistically adding u with another u. Therefore (dropping the c's),

$$v = \frac{2u}{1 + u^2} \quad \Longrightarrow \quad u = \frac{1 - \sqrt{1 - v^2}}{v}. \tag{11.92}$$

The quadratic equation for u also has a solution with a plus sign in front of the square root, but this solution cannot be correct, because it is greater than 1 (and in fact goes to infinity as v goes to zero). The above solution for u has the proper limit as v goes to zero, namely $u \to v/2$, which can be obtained by using the Taylor expansion for the square root.

The ratio of the distances CB and AC in the lab frame is the same as the ratio of the differences in the velocities in the lab frame (because both A and C run into B at the same time, so you could imagine running the scenario backwards in time). Therefore,

$$\begin{aligned}\frac{CB}{AC} = \frac{V_C - V_B}{V_A - V_C} &= \frac{\frac{1-\sqrt{1-v^2}}{v} - 0}{v - \frac{1-\sqrt{1-v^2}}{v}} \\ &= \frac{1-\sqrt{1-v^2}}{\sqrt{1-v^2} - (1-v^2)} \\ &= \frac{1}{\sqrt{1-v^2}} \equiv \gamma. \end{aligned} \tag{11.93}$$

We see that C is γ times as far from B as she is from A, as measured in the lab frame. Note that for nonrelativistic speeds, we have $\gamma \approx 1$, so C is midway between A and B, as expected. An intuitive reason for the simple factor of γ is the following. Imagine that A and B are carrying identical jousting sticks as they run toward C. Consider what the situation looks like when the tips of the sticks reach C. In the lab frame (in which B is at rest), B's stick is uncontracted, but A's stick is length contracted by a factor γ. Therefore, in the lab frame, A is closer to C than B is, by a factor γ.

11.13. **Equal transverse speeds**

From your point of view, the lab frame is moving with speed v in the negative x direction. The transverse velocity-addition formula, Eq. (11.38), therefore gives the y speed in your frame as $u_y/\gamma(1 - u_x v)$. Demanding that this equals u_y gives

$$\gamma(1 - u_x v) = 1 \quad \Longrightarrow \quad \sqrt{1-v^2} = (1 - u_x v) \quad \Longrightarrow \quad v = \frac{2u_x}{1+u_x^2}, \tag{11.94}$$

or $v = 0$, of course. This v is simply the relativistic addition of u_x with itself. This makes sense, because it means that both your frame and the original lab frame move with speed u_x (but in opposite directions) relative to the frame in which the object has no speed in the x direction. By symmetry, therefore, the y speed of the object must be the same in your frame and in the lab frame.

11.14. **Relative speed**

Consider the frame S' that travels along with the point P midway between the particles. S' moves at speed $v\cos\theta$, so the γ factor relating it to the lab frame is

$$\gamma = \frac{1}{\sqrt{1 - v^2\cos^2\theta}}. \tag{11.95}$$

Let's find the vertical speeds of the particles in S'. Since the particles have $u'_x = 0$, the transverse velocity-addition formula, Eq. (11.38), gives $v\sin\theta = u'_y/\gamma$. Therefore, in S' each particle moves along the vertical axis away from P with speed

$$u'_y = \gamma v \sin\theta. \tag{11.96}$$

The speed of one particle as viewed by the other can now be found via the longitudinal velocity-addition formula,

$$V = \frac{2u'_y}{1 + {u'_y}^2} = \frac{\frac{2v\sin\theta}{\sqrt{1-v^2\cos^2\theta}}}{1 + \frac{v^2\sin^2\theta}{1-v^2\cos^2\theta}} = \frac{2v\sin\theta\sqrt{1-v^2\cos^2\theta}}{1 - v^2\cos 2\theta}. \tag{11.97}$$

If desired, this can be written as (for future reference in Chapter 13)

$$V=\sqrt{1-\frac{(1-v^2)^2}{(1-v^2\cos 2\theta)^2}}\,. \tag{11.98}$$

REMARK: If $2\theta=180°$, then $V=2v/(1+v^2)$, as it should. And if $\theta=0°$, then $V=0$, as it should. If θ is very small, then the result reduces to $V\approx 2v\sin\theta/\sqrt{1-v^2}$, which is simply the nonrelativistic addition of (essentially) the speed in Eq. (11.96) with itself, as it should be. ♣

11.15. **Another relative speed**

Let the velocity of A point in the x direction, as shown in Fig. 11.56. Let S' be the lab frame, and let S be A's frame (so frame S' moves at velocity $-u$ with respect to S). The x and y speeds of B in frame S' are $v\cos\theta$ and $v\sin\theta$. Therefore, the longitudinal and transverse velocity-addition formulas, Eqs. (11.31) and (11.38), give the components of B's velocity in S as

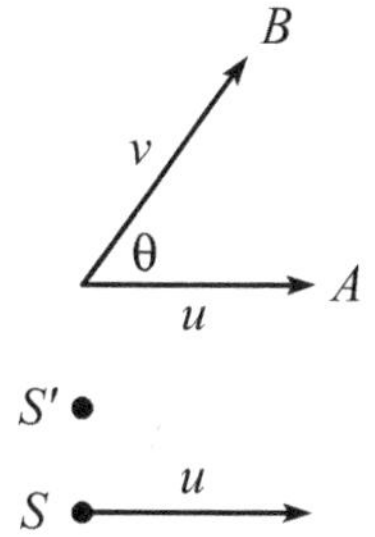

Fig. 11.56

$$\begin{aligned} V_x &= \frac{v\cos\theta-u}{1-uv\cos\theta}\,,\\ V_y &= \frac{v\sin\theta}{\gamma_u(1-uv\cos\theta)}=\frac{\sqrt{1-u^2}\;v\sin\theta}{1-uv\cos\theta}\,. \end{aligned} \tag{11.99}$$

The total speed of B in frame S (that is, from A's point of view) is therefore

$$\begin{aligned} V &= \sqrt{V_x^2+V_y^2}=\sqrt{\left(\frac{v\cos\theta-u}{1-uv\cos\theta}\right)^2+\left(\frac{\sqrt{1-u^2}\;v\sin\theta}{1-uv\cos\theta}\right)^2}\\ &= \frac{\sqrt{u^2+v^2-2uv\cos\theta-u^2v^2\sin^2\theta}}{1-uv\cos\theta}\,. \end{aligned} \tag{11.100}$$

If desired, this can be written as

$$V=\sqrt{1-\frac{(1-u^2)(1-v^2)}{(1-uv\cos\theta)^2}}\,. \tag{11.101}$$

The reason why this can be written in such an organized form will become clear in Chapter 13.

REMARK: If $u=v$, this reduces to the result of the previous problem (if we replace θ by 2θ). If $\theta=180°$, then $V=(u+v)/(1+uv)$, as it should. And if $\theta=0°$, then $V=|v-u|/(1-uv)$, as it should. ♣

11.16. **Transverse velocity addition**

Assume that a clock on the particle shows a time T between successive passes of the dotted lines. In frame S', the speed of the particle is u', so the time-dilation factor is $\gamma'=1/\sqrt{1-u'^2}$. The time between successive passes of the dotted lines is therefore $T_{S'}=\gamma' T$.

In frame S, the speed of the particle is $\sqrt{v^2+u^2}$. (Yes, the Pythagorean theorem holds for these speeds, because both speeds are measured with respect to the same frame.) Hence, the time-dilation factor is $\gamma=1/\sqrt{1-v^2-u^2}$. The time between successive passes of the dotted lines is therefore $T_S=\gamma T$. Equating our two expressions for $T_S/T_{S'}$ gives

$$\frac{u'}{u}=\frac{T_S}{T_{S'}}=\frac{\sqrt{1-u'^2}}{\sqrt{1-v^2-u^2}}\,. \tag{11.102}$$

Solving for u gives the desired result,

$$u = u'\sqrt{1-v^2} \equiv \frac{u'}{\gamma_v}. \tag{11.103}$$

REMARK: A slightly quicker method is the following. Imagine a clock at rest in S', with the same x' value as the particle. Let this clock tick simultaneously (as seen in S') with the particle's crossings of the dotted lines. Then the clock also ticks simultaneously with the particle's crossings of the dotted lines in S, because the clock and the particle have the same x' values. But the clock ticks slower in S by a factor $\sqrt{1-v^2}$. Therefore, the y speed is smaller in S by this factor, as we wanted to show. ♣

11.17. **Many velocity additions**

Let's first check the formula for $N = 1$ and $N = 2$. When $N = 1$, the formula gives

$$\beta_{(1)} = \frac{P_1^+ - P_1^-}{P_1^+ + P_1^-} = \frac{(1+\beta_1) - (1-\beta_1)}{(1+\beta_1) + (1-\beta_1)} = \beta_1, \tag{11.104}$$

as it should. And when $N = 2$, the formula gives

$$\beta_{(2)} = \frac{P_2^+ - P_2^-}{P_2^+ + P_2^-} = \frac{(1+\beta_1)(1+\beta_2) - (1-\beta_1)(1-\beta_2)}{(1+\beta_1)(1+\beta_2) + (1-\beta_1)(1-\beta_2)} = \frac{\beta_1+\beta_2}{1+\beta_1\beta_2}, \tag{11.105}$$

in agreement with the velocity-addition formula.

Let's now prove the formula for general N. We will use induction. That is, we will assume that the result holds for N and then show that it holds for $N+1$. To find the speed, $\beta_{(N+1)}$, of the object with respect to S_{N+1}, we can relativistically add the speed of the object with respect to S_N (which is $\beta_{(N)}$) with the speed of S_N with respect to S_{N+1} (which is β_{N+1}). This gives

$$\beta_{(N+1)} = \frac{\beta_{N+1} + \beta_{(N)}}{1 + \beta_{N+1}\beta_{(N)}}. \tag{11.106}$$

Under the assumption that our formula holds for N, this becomes

$$\begin{aligned}\beta_{(N+1)} = \frac{\beta_{N+1} + \frac{P_N^+ - P_N^-}{P_N^+ + P_N^-}}{1 + \beta_{N+1}\frac{P_N^+ - P_N^-}{P_N^+ + P_N^-}} &= \frac{\beta_{N+1}(P_N^+ + P_N^-) + (P_N^+ - P_N^-)}{(P_N^+ + P_N^-) + \beta_{N+1}(P_N^+ - P_N^-)}\\ &= \frac{P_N^+(1+\beta_{N+1}) - P_N^-(1-\beta_{N+1})}{P_N^+(1+\beta_{N+1}) + P_N^-(1-\beta_{N+1})}\\ &\equiv \frac{P_{N+1}^+ - P_{N+1}^-}{P_{N+1}^+ + P_{N+1}^-},\end{aligned} \tag{11.107}$$

as we wanted to show. We have therefore shown that if the result holds for N, then it also holds for $N+1$. Since we know that the result does indeed hold for $N = 1$, it therefore holds for all N.

The expression for $\beta_{(N)}$ has some expected properties. It is symmetric in the β_i. And if the given object is a photon with $\beta_1 = 1$, then $P_N^- = 0$, which gives $\beta_{(N)} = 1$ as it should. And if the given object is a photon with $\beta_1 = -1$, then $P_N^+ = 0$, which gives $\beta_{(N)} = -1$ as it should.

REMARK: We can use the result of this problem to derive the $v(t)$ given in Eq. (11.64). First, note that if all the β_i here are equal, and if their common value is sufficiently small, then

$$\beta_{(N)} = \frac{(1+\beta)^N - (1-\beta)^N}{(1+\beta)^N + (1-\beta)^N} \approx \frac{e^{\beta N} - e^{-\beta N}}{e^{\beta N} + e^{-\beta N}} = \tanh(\beta N). \quad (11.108)$$

Let β equal $a\,dt/c$, which is the relative speed of two frames at nearby times in the spaceship scenario leading up to Eq. (11.64). If we let $N = t/dt$ be the number of frames (and if we take the limit $dt \to 0$), then we have reproduced the spaceship scenario. Therefore, the $\beta_{(N)}$ in Eq. (11.108) should equal the $v(t)$ in Eq. (11.64). And indeed, with $\beta = a\,dt/c$ and $N = t/dt$, Eq. (11.108) gives $\beta_{(N)} = \tanh(at/c)$, as desired. ♣

11.18. **Velocity addition from scratch**

As stated in the problem, we will use the fact that the meeting of the photon and the ball occurs at the same fraction of the way along the train, independent of the frame. This is true because although distances may change depending on the frame, fractions remain the same, because length contraction doesn't depend on position. We'll compute the desired fraction in the train frame S', and then in the ground frame S.

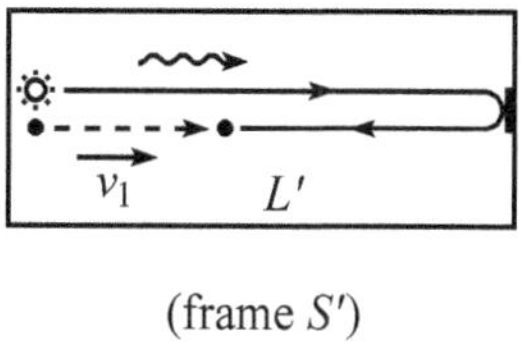

Fig. 11.57

TRAIN FRAME: Let the train have length L' in the train frame. Let's first find the time at which the photon meets the ball (see Fig. 11.57). From the figure, we see that the sum of the distances traveled by the ball and the photon, which is $v_1 t' + ct'$, must equal twice the length of the train, which is $2L'$. The time of the meeting is therefore

$$t' = \frac{2L'}{c + v_1}. \quad (11.109)$$

The distance the ball has traveled is then $v_1 t' = 2v_1 L'/(c + v_1)$, and the desired fraction F' is

$$F' = \frac{2v_1}{c + v_1}. \quad (11.110)$$

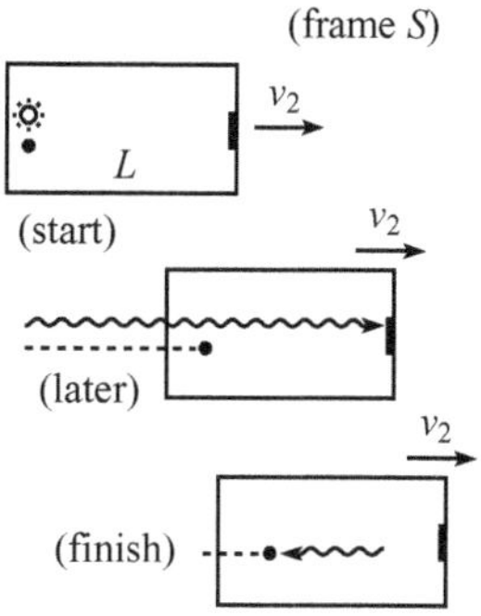

Fig. 11.58

GROUND FRAME: Let the speed of the ball with respect to the ground be v, and let the train have length L in the ground frame (L equals L'/γ, but we're not going to use this). Again, let's first find the time at which the photon meets the ball (see Fig. 11.58). Light takes a time $L/(c - v_2)$ to reach the mirror, because the mirror is receding at speed v_2. At this time, the light has traveled a distance $cL/(c - v_2)$. From the figure, we see that we can use the same reasoning as in the train-frame case, but now with the sum of the distances traveled by the ball and the photon, which is $vt + ct$, equaling $2cL/(c - v_2)$. The time of the meeting is therefore

$$t = \frac{2cL}{(c - v_2)(c + v)}. \quad (11.111)$$

The relative speed of the ball and the back of the train (as viewed in the ground frame) is $v - v_2$, so the distance between the ball and the back of the train at this time is $2(v - v_2)cL/[(c - v_2)(c + v)]$. The desired fraction F is therefore

$$F = \frac{2(v - v_2)c}{(c - v_2)(c + v)}. \quad (11.112)$$

We can now equate the above expressions for F' and F. For convenience, define $\beta \equiv v/c$, $\beta_1 \equiv v_1/c$, and $\beta_2 \equiv v_2/c$. Then $F' = F$ yields

$$\frac{\beta_1}{1 + \beta_1} = \frac{\beta - \beta_2}{(1 - \beta_2)(1 + \beta)}. \quad (11.113)$$

Solving for β in terms of β_1 and β_2 gives

$$\beta = \frac{\beta_1 + \beta_2}{1 + \beta_1 \beta_2}, \quad (11.114)$$

as desired. This problem is solved in Mermin (1983).

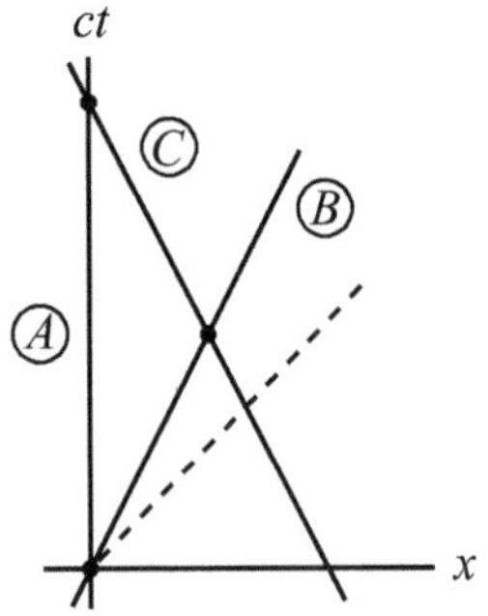

Fig. 11.59

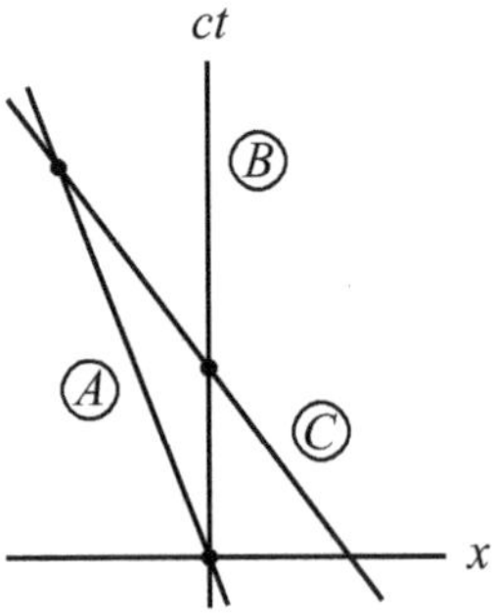

Fig. 11.60

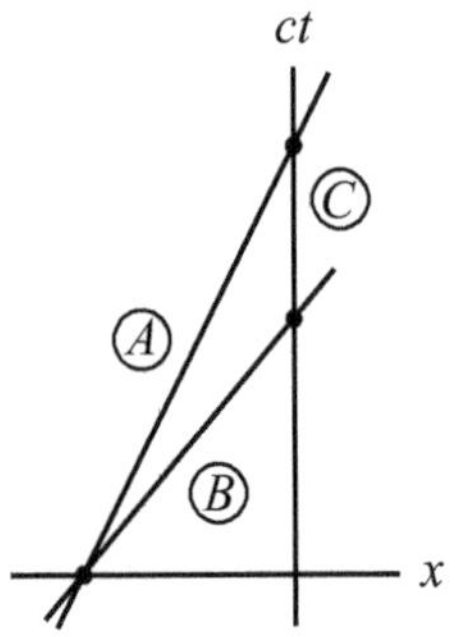

Fig. 11.61

11.19. **Modified twin paradox**

(a) In A's reference frame, the worldlines of A, B, and C are shown in Fig. 11.59. B's clock runs slow by a factor $1/\gamma$. Therefore, if A's clock reads t when B meets C, then B's clock reads t/γ when he meets C. So the time he gives to C is t/γ.

In A's reference frame, the time between this event and the event where C meets A is again t, because B and C travel at the same speed. But A sees C's clock run slow by a factor $1/\gamma$, so A sees C's clock increase by t/γ. Therefore, when A and C meet, A's clock reads $2t$, and C's clock reads $2t/\gamma$. In other words, $T_C = T_A/\gamma$.

(b) Let's now look at things in B's frame. The worldlines of A, B, and C are shown in Fig. 11.60. From B's point of view, there are two competing effects that lead to the relation $T_C = T_A/\gamma$. The first is that B sees A's clock run slow, so the time he hands off to C is *larger* than the time A's clock reads at this moment. The second effect is that from this point on, B sees C's clock run *slower* than A's (because the relative speed of C and B is greater than the relative speed of A and B). It turns out that this slowness wins out over the head start that C's clock had over A's. So in the end, C's clock reads a smaller time than A's. Let's be quantitative about this.

Let B's clock read t_B when he meets C. Then when B hands off this time to C, A's clock reads only t_B/γ. We'll find all relevant times below in terms of t_B. We must determine how much additional time elapses on A's clock and C's clock, by the time they meet. From the velocity-addition formula, B sees C move to the left at speed $2v/(1+v^2)$. He also sees A move to the left at speed v. But A had a head start of vt_B in front of C, so if t is the time (as viewed from B) between the meeting of B and C and the meeting of A and C, then

$$\frac{2vt}{1+v^2} = vt + vt_B \quad \Longrightarrow \quad t = t_B\left(\frac{1+v^2}{1-v^2}\right). \tag{11.115}$$

During this time, B sees A's and C's clocks increase by t divided by the relevant time-dilation factor. For A, this factor is $\gamma = 1/\sqrt{1-v^2/c^2}$. For C, it is

$$\frac{1}{\sqrt{1-\left(\frac{2v}{1+v^2}\right)^2}} = \frac{1+v^2}{1-v^2}. \tag{11.116}$$

Therefore, the total time shown on A's clock when A and C meet is

$$\begin{aligned} T_A &= \frac{t_B}{\gamma} + t\sqrt{1-v^2} = t_B\sqrt{1-v^2} + t_B\left(\frac{1+v^2}{1-v^2}\right)\sqrt{1-v^2} \\ &= \frac{2t_B}{\sqrt{1-v^2}}. \end{aligned} \tag{11.117}$$

And the total time shown on C's clock when A and C meet is

$$T_C = t_B + t\left(\frac{1-v^2}{1+v^2}\right) = t_B + t_B\left(\frac{1+v^2}{1-v^2}\right)\left(\frac{1-v^2}{1+v^2}\right) = 2t_B. \tag{11.118}$$

Therefore, $T_C = T_A\sqrt{1-v^2} \equiv T_A/\gamma$.

(c) Let's now work in C's frame. The worldlines of A, B, and C are shown in Fig. 11.61. As in part (b), the relative speed of B and C is $2v/(1+v^2)$, and the time-dilation factor between B and C is $(1+v^2)/(1-v^2)$. Also, as in part (b), let B and C meet when B's clock reads t_B. So this is the time that B hands off to C. We'll find all relevant times below in terms of t_B.

C sees B's clock running slow, so from C's point of view, B travels for a time $t_B(1+v^2)/(1-v^2)$ after his meeting with A. In this time, B covers a distance in C's frame equal to

$$d = t_B\left(\frac{1+v^2}{1-v^2}\right)\frac{2v}{1+v^2} = \frac{2vt_B}{1-v^2}\,. \tag{11.119}$$

A must travel this same distance (from where he met B) to meet up with C. We can now find T_A. The time (as viewed by C) that it takes A to travel the distance d to reach C is $d/v = 2t_B/(1-v^2)$. But since C sees A's clock running slow by a factor $\sqrt{1-v^2}$, A's clock will read only

$$T_A = \frac{2t_B}{\sqrt{1-v^2}}\,. \tag{11.120}$$

Now let's find T_C. To find T_C, we must take t_B and add to it the extra time it takes A to reach C, compared with the time it takes B to reach C. From above, this extra time is $2t_B/(1-v^2) - t_B(1+v^2)/(1-v^2) = t_B$. Therefore, C's clock reads

$$T_C = 2t_B. \tag{11.121}$$

Hence, $T_C = T_A\sqrt{1-v^2} \equiv T_A/\gamma$.

11.20. **Throwing on a train**

(a) In the train frame, the distance is simply $d = L$. And the time is $t = L/(c/3) = 3L/c$.

(b) i. The velocity of the ball with respect to the ground is (with $u = c/3$ and $v = c/2$)

$$V_g = \frac{u+v}{1+\frac{uv}{c^2}} = \frac{\frac{c}{3}+\frac{c}{2}}{1+\frac{1}{3}\cdot\frac{1}{2}} = \frac{5c}{7}\,. \tag{11.122}$$

The length of the train in the ground frame is $L/\gamma_{1/2} = \sqrt{3}L/2$. Therefore, at time t the position of the front of the train is $\sqrt{3}L/2 + vt$. And the position of the ball is $V_g t$. These two positions are equal when

$$(V_g - v)t = \frac{\sqrt{3}L}{2} \implies t = \frac{\frac{\sqrt{3}L}{2}}{\frac{5c}{7}-\frac{c}{2}} = \frac{7L}{\sqrt{3}c}\,. \tag{11.123}$$

Equivalently, this time is obtained by noting that the ball closes the initial head start of $\sqrt{3}L/2$ that the front of the train had, at a relative speed of $V_g - v$. The distance the ball travels is $d = V_g t = (5c/7)(7L/\sqrt{3}c) = 5L/\sqrt{3}$.

ii. In the train frame, the space and time intervals are $x' = L$ and $t' = 3L/c$, from part (a). The γ factor between the frames is $\gamma_{1/2} = 2/\sqrt{3}$, so the Lorentz transformations give the coordinates in the ground frame as

$$\begin{aligned} x &= \gamma(x' + vt') = \frac{2}{\sqrt{3}}\left(L + \frac{c}{2}\left(\frac{3L}{c}\right)\right) = \frac{5L}{\sqrt{3}}\,, \\ t &= \gamma(t' + vx'/c^2) = \frac{2}{\sqrt{3}}\left(\frac{3L}{c} + \frac{\frac{c}{2}(L)}{c^2}\right) = \frac{7L}{\sqrt{3}c}\,, \end{aligned} \tag{11.124}$$

in agreement with the above results.

(c) In the ball frame, the train has length, $L/\gamma_{1/3} = \sqrt{8}L/3$. Therefore, the time it takes the train to fly past the ball at speed $c/3$ is $t = (\sqrt{8}L/3)/(c/3) = 2\sqrt{2}L/c$. And the distance is $d = 0$, of course, because the ball doesn't move in the ball frame.

(d) The values of $c^2t^2 - x^2$ in the three frames are:

$$\text{Train frame:} \quad c^2t^2 - x^2 = c^2(3L/c)^2 - L^2 = 8L^2.$$

$$\text{Ground frame:} \quad c^2t^2 - x^2 = c^2(7L/\sqrt{3}c)^2 - (5L/\sqrt{3})^2 = 8L^2.$$

$$\text{Ball frame:} \quad c^2t^2 - x^2 = c^2(2\sqrt{2}L/c)^2 - (0)^2 = 8L^2.$$

These are all equal, as they should be.

(e) The relative speed of the ball frame and the ground frame is $5c/7$. Therefore, $\gamma_{5/7} = 7/2\sqrt{6}$, and the times are indeed related by

$$t_g = \gamma t_b \quad \Longleftrightarrow \quad \frac{7L}{\sqrt{3}c} = \frac{7}{2\sqrt{6}}\left(\frac{2\sqrt{2}L}{c}\right), \quad \text{which is true.} \tag{11.125}$$

(f) The relative speed of the ball frame and the train frame is $c/3$. Therefore, $\gamma_{1/3} = 3/2\sqrt{2}$, and the times are indeed related by

$$t_t = \gamma t_b \quad \Longleftrightarrow \quad \frac{3L}{c} = \frac{3}{2\sqrt{2}}\left(\frac{2\sqrt{2}L}{c}\right), \quad \text{which is true.} \tag{11.126}$$

(g) The relative speed of the train frame and the ground frame is $c/2$. Therefore, $\gamma_{1/2} = 2/\sqrt{3}$, and the times are *not* related by a simple time-dilation factor, because

$$t_g \neq \gamma t_t \quad \Longleftrightarrow \quad \frac{7L}{\sqrt{3}c} \neq \frac{2}{\sqrt{3}}\left(\frac{3L}{c}\right). \tag{11.127}$$

We don't obtain an equality because time dilation is legal to use only if the two events happen at the *same place* in one of the frames. Mathematically, the Lorentz transformation $\Delta t = \gamma\big(\Delta t' + (v/c^2)\Delta x'\big)$ leads to $\Delta t = \gamma \Delta t'$ only if $\Delta x' = 0$. In this problem, the "ball leaving the back" and "ball hitting the front" events happen at the same place in the ball frame, but in neither the train frame nor the ground frame. Equivalently, neither the train frame nor the ground frame is any more special than the other, as far as these two events go. So if someone insisted on trying to use time dilation, he would have a hard time deciding which side of the equation the γ should go on.

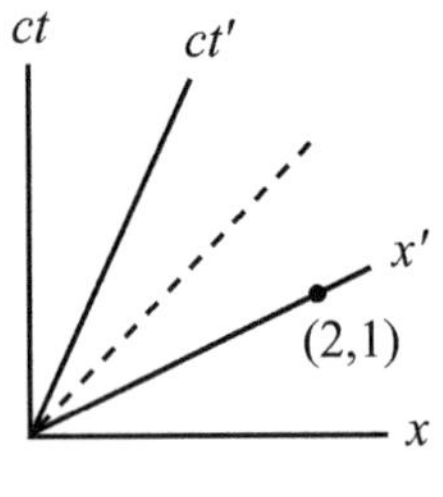

Fig. 11.62

11.21. A new frame

FIRST SOLUTION: Consider the Minkowski diagram in Fig. 11.62. In frame S, Event 1 is at the origin, and Event 2 is at the point $(2, 1)$. Consider now the frame S' whose x' axis passes through the point $(2, 1)$. Since all points on the x' axis are simultaneous in the frame S' (they all have $t' = 0$), we see that S' is the desired frame. From Eq. (11.47), the slope of the x' axis equals $\beta \equiv v/c$. Since the slope is $1/2$, we have $v = c/2$. Note that by looking at our Minkowski diagram, it is clear that if the relative speed of S and S' is greater than $c/2$, then Event 2 occurs before Event 1 in S'. And if it is less than $c/2$, then Event 2 occurs after Event 1 in S'.

SECOND SOLUTION: Let the original frame be S, and let the desired frame be S'. Let S' move at speed v (in the positive direction) with respect to S. Our goal is to find v. The Lorentz transformations from S to S' are

$$\Delta x' = \gamma(\Delta x - v\Delta t), \quad \Delta t' = \gamma(\Delta t - v\,\Delta x/c^2). \tag{11.128}$$

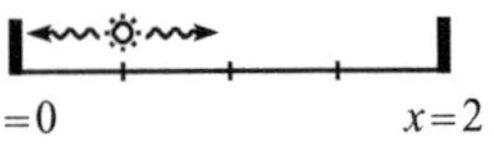

Fig. 11.63

We want to make $\Delta t'$ equal to zero, so the second of these equations yields $\Delta t - v\,\Delta x/c^2 = 0$, or $v = c^2\Delta t/\Delta x$. We are given $\Delta x = 2$ and $\Delta t = 1/c$, so the desired v is $c/2$.

THIRD SOLUTION: Consider the setup in Fig. 11.63, which explicitly constructs the two given events. Receivers are located at $x = 0$ and $x = 2$, and a light source is located at $x = 1/2$. The source emits a flash, and when the light hits a receiver we will say an event has occurred. So the left event happens at $x = 0$, $ct = 1/2$. And the right event happens at $x = 2$, $ct = 3/2$. If we want, we can shift our clocks by $-1/(2c)$ in order to make the events happen at $ct = 0$ and $ct = 1$, but this shift is irrelevant because all we are concerned with is differences in time.

Now consider an observer moving to the right at speed v. She sees the apparatus moving to the left at speed v (see Fig. 11.64). Our goal is to find the v for which the photons hit the receivers at the same time in her frame. Consider the photons moving to the left. She sees them moving at speed c, but the left-hand receiver is receding at speed v. So the relative speed (as measured by her) of the photons and the left-hand receiver is $c - v$. By similar reasoning, the relative speed of the photons and the right-hand receiver is $c + v$. The light source is three times as far from the right-hand receiver as it is from the left-hand receiver. Therefore, if the light is to reach the two receivers at the same time, we must have $c + v = 3(c - v)$. This gives $v = c/2$.

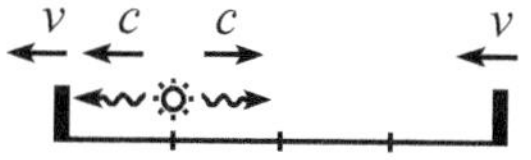

Fig. 11.64

11.22. **Minkowski diagram units**

All points on the ct' axis have the property that $x' = 0$. All points on the hyperbola have the property that $c^2t'^2 - x'^2 = 1$, due to the invariance of s^2. So the ct' value at the intersection point A equals 1. Therefore, we simply have to determine the distance on the paper from A to the origin (see Fig. 11.65). We'll do this by finding the (x, ct) coordinates of A. We know that $\tan\theta = \beta$. But $\tan\theta = x/ct$. Therefore, $x = \beta(ct)$ (which is just the statement that $x = vt$). Plugging this into the given information, $c^2t^2 - x^2 = 1$, we find $ct = 1/\sqrt{1-\beta^2}$. The distance from A to the origin is then

$$\sqrt{c^2t^2 + x^2} = ct\sqrt{1+\beta^2} = \sqrt{\frac{1+\beta^2}{1-\beta^2}}. \qquad (11.129)$$

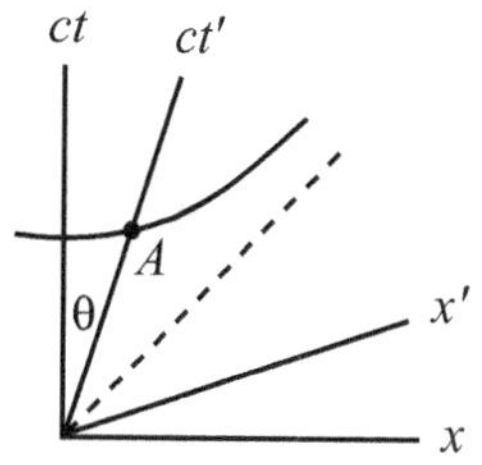

Fig. 11.65

This quantity is therefore the ratio of the unit sizes on the ct' and ct axes, in agreement with Eq. (11.46). Exactly the same analysis holds for the x-axis unit size ratio.

11.23. **Velocity addition via Minkowski**

Pick a point P on the object's worldline. Let the coordinates of P in frame S be (x, ct). Our goal is to find the speed $u = x/t$. Throughout this problem, it will be easier to work with the quantities $\beta \equiv v/c$, so our goal is then to find $\beta_u \equiv x/(ct)$.

The coordinates of P in S', namely (x', ct'), are indicated by the parallelogram in Fig. 11.66. For convenience, let ct' have length a on the paper. Then from the given information, we have $x' = v_1t' \equiv \beta_1(ct') = \beta_1 a$. This is the distance from A to P on the paper. In terms of a, we can now determine the coordinates (x, ct) of P. The coordinates of point A are

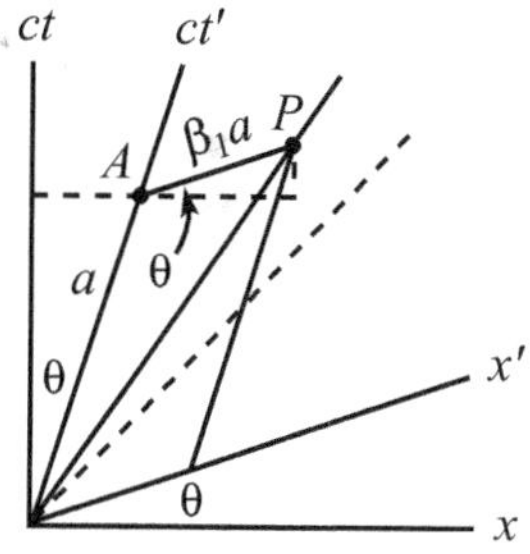

Fig. 11.66

$$(x, ct)_A = (a\sin\theta, a\cos\theta). \qquad (11.130)$$

The coordinates of P, relative to A, are

$$(x, ct)_{P-A} = (\beta_1 a\cos\theta, \beta_1 a\sin\theta). \qquad (11.131)$$

Adding these two sets of coordinates gives the coordinates of point P as

$$(x, ct)_P = (a\sin\theta + \beta_1 a\cos\theta, a\cos\theta + \beta_1 a\sin\theta). \qquad (11.132)$$

The ratio of x to ct at the point P is therefore

$$\beta_u \equiv \frac{x}{ct} = \frac{\sin\theta + \beta_1\cos\theta}{\cos\theta + \beta_1\sin\theta} = \frac{\tan\theta + \beta_1}{1 + \beta_1\tan\theta} = \frac{\beta_2 + \beta_1}{1 + \beta_1\beta_2}, \qquad (11.133)$$

where we have used $\tan\theta = v_2/c \equiv \beta_2$, because S' moves at speed v_2 with respect to S. If we change from the β's back to the v's, the result is $u = (v_2+v_1)/(1+v_1v_2/c^2)$, as expected.

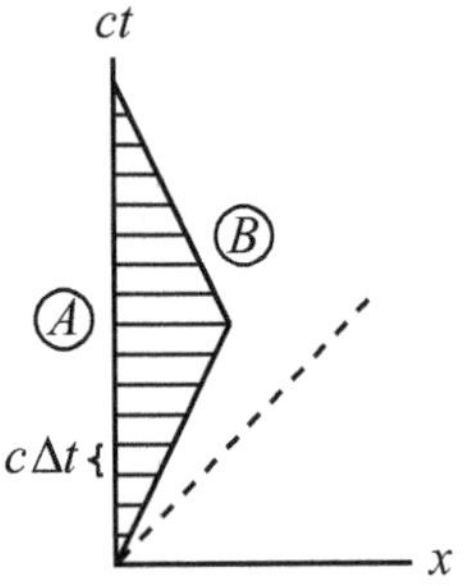

Fig. 11.67

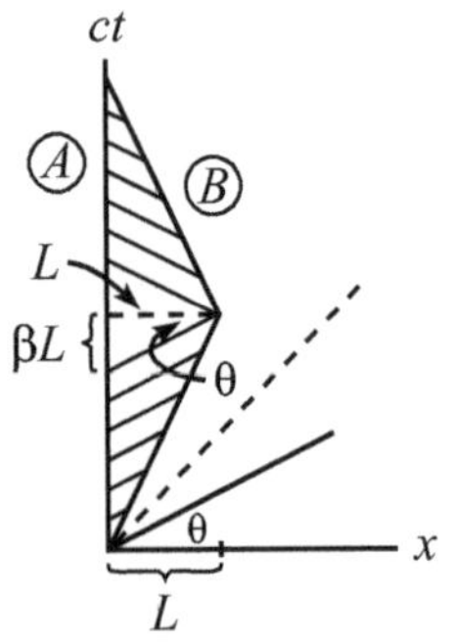

Fig. 11.68

11.24. **Clapping both ways**

(a) From the usual time-dilation result, A sees B's clock run slow, so B must clap his hands at time intervals of $\Delta t/\gamma$ in order for the intervals to be Δt in A's frame. The relevant Minkowski diagram is shown in Fig. 11.67. Let B clap at the spacetime locations where the horizontal lines (lines of simultaneity in A's frame) intersect B's worldline. From the given information, the vertical spacing between the lines is $c\,\Delta t$. Therefore, the spacing along B's tilted worldline (which has a slope of $\pm 1/\beta$) is $\sqrt{1+\beta^2}\,c\,\Delta t$. But we know from Eq. (11.46) that the unit size of B's ct axis on the paper is $\sqrt{(1+\beta^2)/(1-\beta^2)}$ times the unit size of A's ct axis. Therefore, the time interval between claps in B's frame is

$$\frac{\sqrt{1+\beta^2}\,\Delta t}{\sqrt{(1+\beta^2)/(1-\beta^2)}} = \sqrt{1-\beta^2}\,\Delta t \equiv \frac{\Delta t}{\gamma}, \tag{11.134}$$

as above.

(b) From the usual time-dilation result, B sees A's clock run slow, so A must clap his hands at time intervals of $\Delta t/\gamma$ in order for the intervals to be Δt in B's frame. However, B can invoke this standard time-dilation reasoning only during the parts of the trip where he is in an inertial frame. That is, he cannot invoke it during the turnaround period. We therefore have the situation shown in Fig. 11.68. Let A clap at the spacetime locations where the tilted lines (lines of simultaneity in B's frame) intersect A's worldline. From the given information, the tilted spacing between the lines along B's worldline is $c\,\Delta t$ time units, which has a length $\sqrt{(1+\beta^2)/(1-\beta^2)}\,c\,\Delta t$ on the paper. But then by the same type of geometry reasoning that led to Eq. (11.49), the vertical spacing between the lines along A's worldline is (as you should verify) $\sqrt{1-\beta^2}\,c\,\Delta t$, which gives a time interval of $\Delta t/\gamma$, as above.

But the critical point here is that since the slope of the tilted lines abruptly changes when B turns around, there is a large interval of time in the middle of A's worldline where the tilted lines don't hit it. The result is that A claps frequently for a while, then doesn't clap at all for a while, then claps frequently again. The overall result, as we will now show, is that more time elapses in A's frame (it will turn out to be $2L/v$, of course, where L is the distance in A's frame to the star) than in B's frame ($2L/\gamma v$, from the usual time-dilation result).

Since the time between claps is shorter in A's frame than in B's frame (except in the middle region), the time elapsed on A's clock while he is clapping is $1/\gamma$ times the total time elapsed on B's clock, which gives $(2L/\gamma v)/\gamma = 2L/\gamma^2 v$. But as shown in Fig. 11.68, the length of the region on A's ct axis where there is no clapping is $2\beta L$, which corresponds to a time of $2vL/c^2$. The total time elapsed on A's clock during the entire process is therefore

$$\frac{2L}{\gamma^2 v} + \frac{2vL}{c^2} = \frac{2L}{v}\left(\frac{1}{\gamma^2} + \frac{v^2}{c^2}\right) = \frac{2L}{v}\left(\left(1-\frac{v^2}{c^2}\right) + \frac{v^2}{c^2}\right) = \frac{2L}{v}, \tag{11.135}$$

as expected.

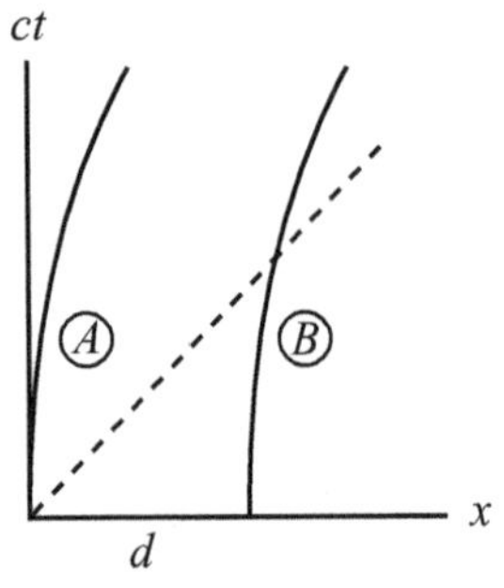

Fig. 11.69

11.25. **Acceleration and redshift**

There are various ways to solve this problem, for example, by sending photons between the people, or by invoking the Equivalence Principle in General Relativity. We'll do it here by using a Minkowski diagram, to demonstrate that this redshift result can be derived perfectly fine using only basic Special Relativity.

Draw the world lines of the two people, A and B, as seen by an observer, C, in the frame where they were both initially at rest. We have the situation shown in Fig. 11.69. Consider an infinitesimal time Δt, as measured by C. At this time (in C's frame),

A and B are both moving at speed $a\Delta t$. The axes of the A frame are shown in Fig. 11.70. Both A and B have moved a distance $a(\Delta t)^2/2$, which can be neglected because Δt is small (it will turn out that the leading-order terms in the result are of order Δt, so any $(\Delta t)^2$ terms can be ignored.) Also, the special-relativistic time-dilation factor between any of the A, B, C frames can be neglected, because the relative speeds are at most $v = a\Delta t$, so the time-dilation factors differ from 1 by order $(\Delta t)^2$. Let A make a little explosion (call this event E_1) at time Δt in C's frame. Then Δt is also the time of the explosion as measured by A, up to an error of order $(\Delta t)^2$.

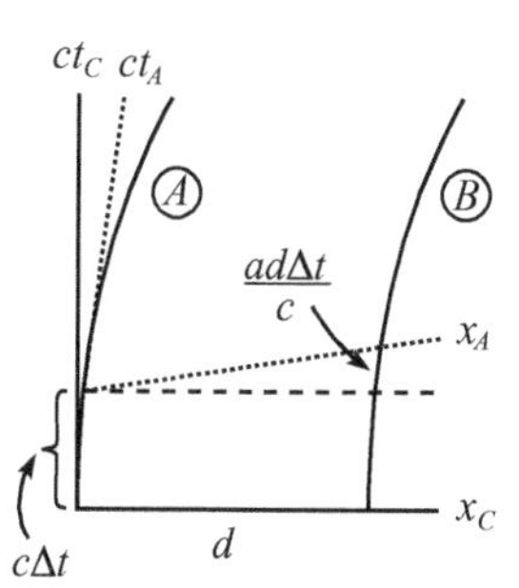

Fig. 11.70

Let's figure out where A's x axis (that is, the "now" axis in A's frame) meets B's worldline. The slope of A's x axis in the figure is $v/c = a\,\Delta t/c$. So the axis starts at a height $c\,\Delta t$, and then climbs up by the amount $ad\,\Delta t/c$, over the distance d (the distance is indeed d, up to corrections of order $(\Delta t)^2$). Therefore, the axis meets B's worldline at a height $c\,\Delta t + ad\,\Delta t/c$ as viewed by C, that is, at a time $\Delta t + ad\,\Delta t/c^2$, as viewed by C. But C's time is the same as B's time (up to order $(\Delta t)^2$), so B's clock reads $\Delta t(1 + ad/c^2)$. Let's say that B makes a little explosion (event E_2) at this time.

Events E_1 and E_2 both occur at the same time in A's frame, because they both lie along a line of constant time in A's frame. This means that in A's frame, B's clock reads $\Delta t(1 + ad/c^2)$ when A's clock reads Δt. Therefore, in A's (changing) frame, B's clock is sped up by a factor,

$$\frac{\Delta t_B}{\Delta t_A} = 1 + \frac{ad}{c^2}\,. \tag{11.136}$$

We can perform the same procedure to see how A's clock behaves in B's frame. Drawing B's x axis at time Δt, we quickly find that in B's (changing) frame, A's clock is slowed down by a factor,

$$\frac{\Delta t_A}{\Delta t_B} = 1 - \frac{ad}{c^2}\,. \tag{11.137}$$

We'll see much more of these results in Chapter 14.

REMARKS:

1. In the usual Special Relativity situation where two observers fly past each other with relative speed v, they *both* see the other person's time slowed down by the same factor. This had better be the case, because the situation is symmetric between the observers. But in this problem, A sees B's clock sped up, and B sees A's clock slowed down. This difference is possible because the situation is *not* symmetric between A and B. The acceleration vector determines a direction in space, and one person (namely B) is farther along this direction than the other person.
2. Another derivation of this ad/c^2 result is the following. Consider the setup a short time after the start. An outside observer sees A's and B's clocks showing the same time. Therefore, by the usual vd/c^2 rear-clock-ahead result in Special Relativity, B's clock must read vd/c^2 more than A's, in the moving frame. The increase per unit time, as viewed by A, is therefore $(vd/c^2)/t = ad/c^2$. Note that any special-relativistic time-dilation or length-contraction effects are of second order in v/c, and hence negligible since v is small here. At any later time, we can repeat (roughly) this derivation in the instantaneous rest frame of A. ♣

11.26. **Break or not break?**

There are two possible reasonings, so we seem to have a paradox:

- To an observer in the original rest frame, the spaceships stay the same distance d apart. Therefore, in the frame of the spaceships, the distance between them, d', must equal γd. This is true because d' is the distance that gets length-contracted down to d. After a long enough time, γ will differ appreciably from 1, so the string will be stretched by a large factor. Therefore, it will break.

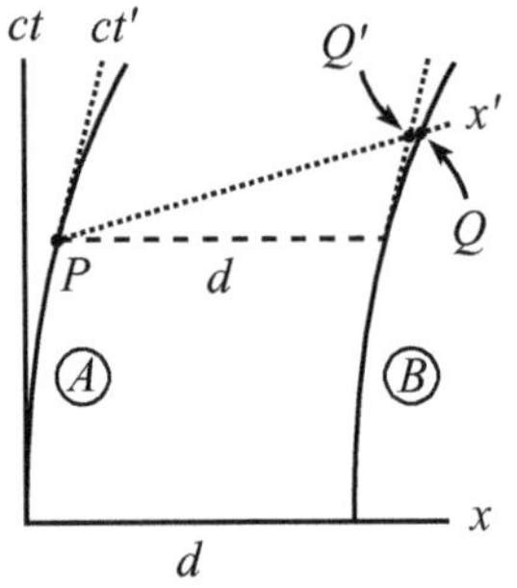

Fig. 11.71

- Let A be the rear spaceship, and let B be the front spaceship. From A's point of view, it looks like B is doing exactly what he is doing (and vice versa). A says that B has the same acceleration that he has. So B should stay the same distance ahead of him. Therefore, the string shouldn't break.

The first reasoning is correct (or mostly correct; see the first remark below). The string *will* break. So that's the answer to our problem. But as with any good relativity paradox, we shouldn't feel at ease until we've explained what's wrong with the incorrect reasoning.

The problem with the second reasoning is that A does *not* see B doing exactly what he is doing. Rather, we know from Problem 11.25 that A sees B's clock running fast (and B sees A's clock running slow). A therefore sees B's engine running faster, and so B pulls away from A. Therefore, the string eventually breaks.[44]

Things become more clear if we draw a Minkowski diagram. Figure 11.71 shows the x' and ct' axes of A's frame. The x' axis is tilted up, so it meets B's worldline farther to the right than you might think. The distance PQ along the x' axis is the distance that A measures the string to be. Although it isn't obvious that this distance in A's frame is larger than d (because the unit size on the x' axis is larger than that in the original frame), we can demonstrate this as follows. In A's frame, the distance PQ is greater than the distance PQ'. But PQ' is simply the length of something in A's frame that has length d in the original frame. So PQ' is γd in A's frame. And since $PQ > \gamma d > d$ in A's frame, the string breaks.

Remarks:

1. There is one slight (inconsequential) flaw in the first reasoning above. There isn't one "frame of the spaceships." Their frames differ, because they measure a relative speed between themselves. Therefore, it isn't clear exactly what is meant by the "length" of the string, because it isn't clear what frame the measurement should take place in. This ambiguity, however, does not change the fact that A and B observe their separation to be (roughly) γd.

   If we want there to eventually be a well-defined "frame of the spaceships," we can modify the problem by stating that after a while the spaceships stop accelerating simultaneously, as measured by someone in the original inertial frame. Equivalently, A and B turn off their engines after equal proper times. What A sees is the following. B pulls away from A. B then turns off his engine. The gap continues to widen. But A continues to fire his engine until be reaches B's speed. They then sail onward, in a common frame, keeping a constant separation (which is greater than the original separation, by a factor γ).
2. The main issue in this problem is that it depends on exactly how we choose to accelerate an extended object. If we accelerate a stick by pushing on the back end (or by pulling on the front end), its length will remain essentially the same in its own frame, and it will become shorter in the original frame. But if we arrange for each end (or perhaps a number of points on the stick) to speed up in such a way that they always move at the same speed with respect to the original frame, then the stick will be torn apart.
3. This problem gives the key to the classic problem of the relativistic wagon wheel, which can be stated as follows (you may want to cover up the following paragraph, so you can solve the problem on your own). A wheel is spun faster

[44] This also follows from the Equivalence Principle and the general-relativistic time-dilation effect which we'll discuss in Chapter 14. Since A and B are accelerating, they may be considered (by the Equivalence Principle) to be in a gravitational field, with B "higher" in the field. But high clocks run fast in a gravitational field. Therefore, A sees B's clock running fast (and B sees A's clock running slow).

and faster, until the points on the rim move at a relativistic speed. In the lab frame, the circumference is length contracted, but the spokes aren't (because they always lie perpendicular to the direction of motion). So if the rim has length $2\pi r$ in the lab frame, then it has length $2\pi\gamma r$ in the wheel frame. Therefore, in the wheel frame, the ratio of the circumference to the diameter is larger than π. So the question is: Is this really true? And if so, how does the circumference become longer in the wheel frame?

I'm putting this sentence here just in case you happened to see the first sentence of this paragraph as you were trying to cover it up and solve the problem on your own; it should probably even be a little longer, maybe like ... this. The answer is that it is indeed true. If we imagine little rocket engines placed around the rim, and if we have them all accelerate with the same proper acceleration, then from the above results, the separation between the engines will gradually increase, thereby increasing the length of the circumference. Assuming that the material in the rim can't stretch indefinitely, the rim will eventually break between each engine. In the rotating (and hence accelerating) wheel frame, the ratio of the circumference to the diameter is indeed larger than π. In other words, space is curved in the wheel frame. This is consistent with the facts that (1) the Equivalence Principle (discussed in Chapter 14) states that acceleration is equivalent to gravity, and (2) gravitational fields in General Relativity are associated with curved space. ♣

11.27. **Successive Lorentz transformations**

It isn't necessary, of course, to use matrices in this problem, but things look nicer if we do. The desired composite L.T. is obtained by multiplying the matrices for the individual L.T.'s. So we have

$$\begin{aligned}
L &= \begin{pmatrix} \cosh\phi_2 & \sinh\phi_2 \\ \sinh\phi_2 & \cosh\phi_2 \end{pmatrix} \begin{pmatrix} \cosh\phi_1 & \sinh\phi_1 \\ \sinh\phi_1 & \cosh\phi_1 \end{pmatrix} \\
&= \begin{pmatrix} \cosh\phi_1\cosh\phi_2 + \sinh\phi_1\sinh\phi_2 & \sinh\phi_1\cosh\phi_2 + \cosh\phi_1\sinh\phi_2 \\ \cosh\phi_1\sinh\phi_2 + \sinh\phi_1\cosh\phi_2 & \sinh\phi_1\sinh\phi_2 + \cosh\phi_1\cosh\phi_2 \end{pmatrix} \\
&= \begin{pmatrix} \cosh(\phi_1+\phi_2) & \sinh(\phi_1+\phi_2) \\ \sinh(\phi_1+\phi_2) & \cosh(\phi_1+\phi_2) \end{pmatrix}. \qquad (11.138)
\end{aligned}$$

This is an L.T. with $v = \tanh(\phi_1 + \phi_2)$, as desired. Except for a few minus signs, this proof is just like the one for successive rotations in a plane.

11.28. **Accelerator's time**

Equation (11.64) gives the speed as a function of the spaceship's time (which we are denoting by t' here) as

$$\beta(t') \equiv \frac{v(t')}{c} = \tanh(at'/c). \qquad (11.139)$$

The person in the lab frame sees the spaceship's clock slowed down by a factor $1/\gamma = \sqrt{1-\beta^2}$, which means that $dt = dt'/\sqrt{1-\beta^2}$. So we have

$$t = \int_0^t dt = \int_0^{t'} \frac{dt'}{\sqrt{1-\beta(t')^2}} = \int_0^{t'} \cosh(at'/c)\,dt' = \frac{c}{a}\sinh(at'/c). \qquad (11.140)$$

For small a or t' (more precisely, for $at'/c \ll 1$), we obtain $t \approx t'$, as we should. For very large times, we essentially have

$$t \approx \frac{c}{2a}e^{at'/c}, \quad \text{or} \quad t' = \frac{c}{a}\ln(2at/c). \qquad (11.141)$$

The lab frame will see the astronaut read all of *Moby Dick*, but it will take an exponentially long time (not that it doesn't already).

Chapter 12
Relativity (Dynamics)

In the previous chapter, we dealt only with abstract particles flying through space and time. We didn't concern ourselves with the nature of the particles, how they got to be moving, or what would happen if various particles interacted. In this chapter we will deal with these issues. That is, we will discuss masses, forces, energy, momentum, etc. The two main results of this chapter are that the momentum and energy of a particle are given by

$$\mathbf{p} = \gamma m\mathbf{v}, \quad \text{and} \quad E = \gamma mc^2, \tag{12.1}$$

where $\gamma \equiv 1/\sqrt{1 - v^2/c^2}$, and m is the mass of the particle.[1] When $v \ll c$, the expression for $\mathbf{p}$ reduces to $\mathbf{p} = m\mathbf{v}$, as it should for a nonrelativistic particle. When $v = 0$, the expression for E reduces to the well-known $E = mc^2$.

12.1 Energy and momentum

In this section, we'll give some justification for Eqs. (12.1). The reasoning here should convince you of their truth. An alternative, and perhaps more convincing, motivation comes from the 4-vector formalism in Chapter 13. In the end, however, the justification for Eqs. (12.1) is obtained through experiments. And indeed, experiments in high-energy accelerators are continually verifying the truth of these expressions. (More precisely, they are verifying that these energy and momenta are *conserved* in any type of collision.) We therefore conclude, with reasonable certainty, that Eqs. (12.1) give the correct expressions for energy and momentum. But actual experiments aside, let's consider a few thought-experiments that motivate the above expressions.

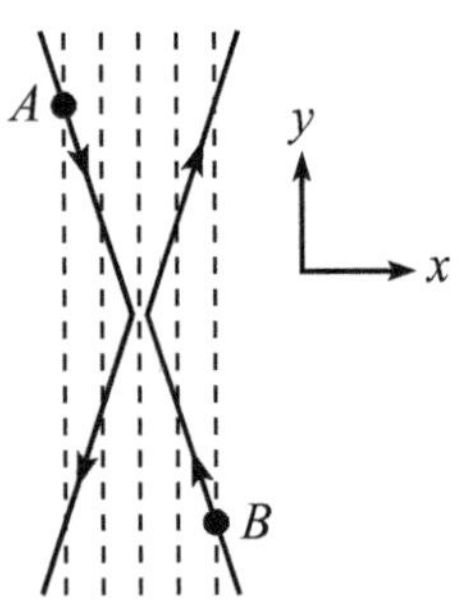

Fig. 12.1

12.1.1 Momentum

Consider the following system. In the lab frame, identical particles A and B move as shown in Fig. 12.1. They move with equal and opposite small speeds in the

[1] There are various ways that people use the word "mass" in relativity. In particular, some people talk about "rest mass" and "relativistic mass." But we won't use these terms. We'll use "mass" to refer only to what other people would call "rest mass." See the discussion on page 590 for more on this.

x direction, and with equal and opposite large speeds in the y direction. Their paths are arranged so that they glance off each other and reverse their motion in the x direction. Imagine a series of equally spaced vertical lines for reference. Assume that A and B have identical clocks that tick every time they cross one of these lines.

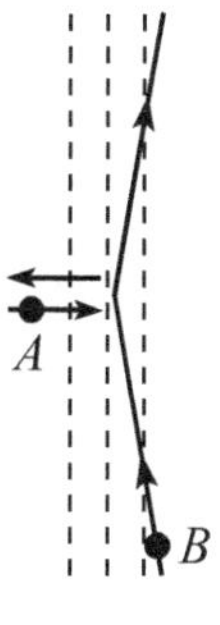

Fig. 12.2

Consider now the reference frame that moves in the y direction with the same v_y as A. In this frame, the situation is shown in Fig. 12.2. The collision simply changes the sign of the x velocities of the particles. Therefore, the x momenta of the two particles must be the same. This is true because if, say, A's p_x were larger than B's p_x, then the total p_x would point to the right before the collision, and to the left after the collision. Since momentum is something we want to be conserved, this cannot be the case.

However, the x speeds of the two particles are *not* the same in this frame. A is essentially at rest in this frame, and B is moving with a very large speed, v. Therefore, B's clock is running slower than A's, by a factor essentially equal to $1/\gamma \equiv \sqrt{1 - v^2/c^2}$. And since B's clock ticks once for every vertical line it crosses (this fact is independent of the frame), B must therefore be moving slower in the x direction, by a factor of $1/\gamma$. Therefore, the Newtonian expression, $p_x = mv_x$, cannot be the correct one for momentum, because B's momentum would be smaller than A's (by a factor of $1/\gamma$), due to their different v_x's. But the γ factor in

$$p_x = \gamma m v_x \equiv \frac{m v_x}{\sqrt{1 - v^2/c^2}} \tag{12.2}$$

precisely takes care of this problem, because $\gamma \approx 1$ for A, and $\gamma = 1/\sqrt{1 - v^2/c^2}$ for B, which cancels the effect of B's smaller v_x.

To obtain the three-dimensional form of **p**, we can use the fact that the vector **p** must point in the same direction as the vector **v**, because any other direction for **p** would violate rotation invariance. If someone claims that **p** points in the direction shown in Fig. 12.3, then he would be hard-pressed to explain why it doesn't instead point along the direction **p**$'$ shown. In short, the direction of **v** is the only preferred direction in space. Therefore, Eq. (12.2) implies that the momentum vector must be

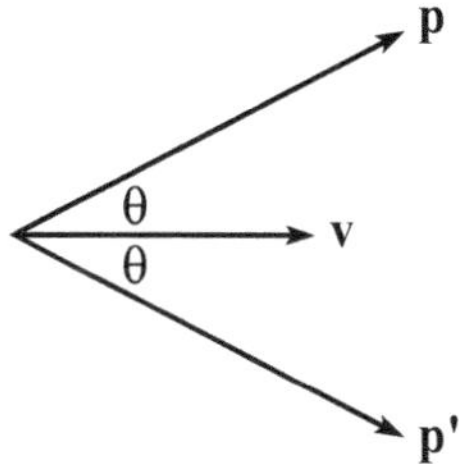

Fig. 12.3

$$\mathbf{p} = \gamma m \mathbf{v} \equiv \frac{m\mathbf{v}}{\sqrt{1 - v^2/c^2}}, \tag{12.3}$$

in agreement with Eqs. (12.1). Note that all the components of **p** have the same denominator, which involves the whole speed, $v^2 = v_x^2 + v_y^2 + v_z^2$. The denominator of, say, p_x is *not* $\sqrt{1 - v_x^2/c^2}$. This would yield a **p** that doesn't point along **v**.

The above setup is only one specific type of collision among an infinite number of possible types of collisions. What we've shown with this setup is that the only possible vector of the form $f(v)m\mathbf{v}$ (where f is some function) that has any chance at being conserved in all collisions is $\gamma m\mathbf{v}$ (or some constant multiple of this). We haven't proved that it actually *is* conserved in all collisions. This is where the gathering of data from experiments comes in. But we've shown above that it would be a waste of time to consider, for example, the vector $\gamma^5 m\mathbf{v}$.

12.1.2 Energy

Having given some justification for the momentum expression, $\mathbf{p} = \gamma m\mathbf{v}$, let's now try to justify the energy expression,

$$E = \gamma mc^2. \tag{12.4}$$

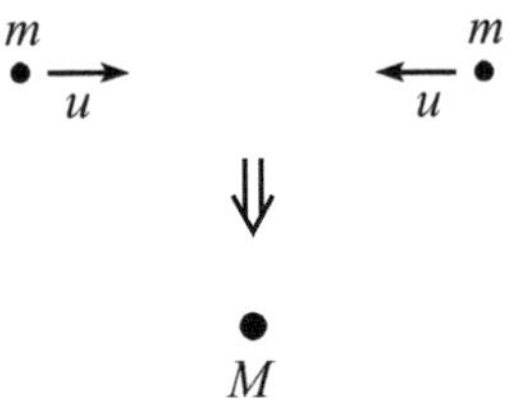

Fig. 12.4

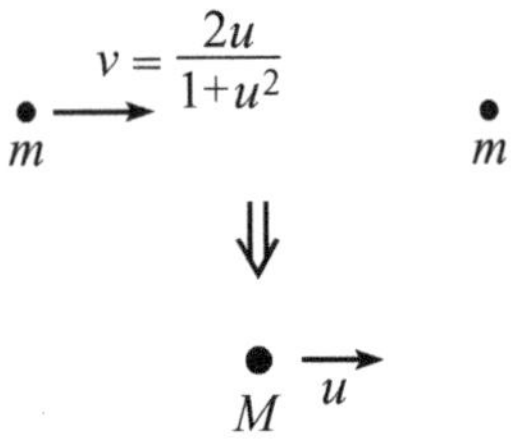

Fig. 12.5

More precisely, we'll show that the above form of the momentum implies that γmc^2 is conserved in interactions (or at least in the specific interaction below). There are various ways to do this. The best way, perhaps, is to use the 4-vector formalism in Chapter 13. But we'll study one simple setup here that should do the job.

Consider the following system. Two identical particles of mass m head toward each other, both with speed u, as shown in Fig. 12.4. They stick together and form a particle of mass M. M is at rest, due to the symmetry of the situation. At the moment we can't assume anything about the size of M, but we'll find below that it does *not* equal the naive value of $2m$. This setup is a fairly uninteresting one (conservation of momentum gives $0 = 0$), so let's instead consider the less trivial view of it in the frame moving to the left at speed u. This situation is shown in Fig. 12.5. The right mass is at rest, the left mass moves to the right at speed $v = 2u/(1+u^2)$ from the velocity-addition formula,[2] and the final mass M moves to the right at speed u. Note that the γ factor associated with the speed v is

$$\gamma_v \equiv \frac{1}{\sqrt{1-v^2}} = \frac{1}{\sqrt{1-\left(\frac{2u}{1+u^2}\right)^2}} = \frac{1+u^2}{1-u^2}. \tag{12.5}$$

Conservation of momentum in this collision then gives

$$\begin{aligned} \gamma_v mv + 0 = \gamma_u Mu \quad &\Longrightarrow \quad m\left(\frac{1+u^2}{1-u^2}\right)\left(\frac{2u}{1+u^2}\right) = \frac{Mu}{\sqrt{1-u^2}} \\ &\Longrightarrow \quad M = \frac{2m}{\sqrt{1-u^2}}. \end{aligned} \tag{12.6}$$

Conservation of momentum therefore tells us that M does *not* equal $2m$. But if u is very small, then M is approximately equal to $2m$, as we know from everyday experience.

[2] We're going to set $c = 1$ for a little while here, because this calculation would get a bit messy if we kept in the c's. We'll discuss the issue of setting $c = 1$ in more detail at the end of this section.

Using the value of M from Eq. (12.6), let's now check that our candidate for energy, $E = \gamma mc^2$, is conserved in this collision. There is no freedom left in any of the parameters, so γmc^2 is either conserved or it isn't. In the original frame where M is at rest, E is conserved if

$$\gamma_0 Mc^2 = 2(\gamma_u mc^2) \quad \Longleftrightarrow \quad \frac{2m}{\sqrt{1-u^2}} = 2\left(\frac{1}{\sqrt{1-u^2}}\right) m , \qquad (12.7)$$

which is indeed true. Let's also check that E is conserved in the frame where the right mass is at rest. E is conserved if

$$\begin{aligned} &\gamma_v mc^2 + \gamma_0 mc^2 = \gamma_u Mc^2 \\ \Longleftrightarrow \quad &\left(\frac{1+u^2}{1-u^2}\right) m + m = \frac{M}{\sqrt{1-u^2}} \\ \Longleftrightarrow \quad &\frac{2m}{1-u^2} = \left(\frac{2m}{\sqrt{1-u^2}}\right) \frac{1}{\sqrt{1-u^2}} , \end{aligned} \qquad (12.8)$$

which is indeed true. So E is also conserved in this frame. This example should convince you that γmc^2 is at least a believable expression for the energy of a particle. But just as in the case of momentum, we haven't proved that γmc^2 actually *is* conserved in all collisions. This is the duty of experiments. But we've shown that it would be a waste of time to consider, for example, the quantity $\gamma^4 mc^2$.

One thing that we certainly need to check is that if E and p are conserved in one reference frame, then they are conserved in any other. We'll demonstrate this in Section 12.2. A conservation law shouldn't depend on what frame you're in, after all.

REMARKS:

1. As mentioned above, we're technically not trying to justify Eqs. (12.1) here. These two equations by themselves are devoid of any meaning. All they do is define the letters $\mathbf{p}$ and E. Our goal is to make a meaningful physical statement, not just a definition.
   The meaningful physical statement that we want to make is that the quantities $\gamma m\mathbf{v}$ and γmc^2 are *conserved* in an interaction among particles (and this is what we tried to justify above). This fact then makes these quantities worthy of special attention, because conserved quantities are very helpful in understanding what is happening in a given physical situation. And anything worthy of special attention certainly deserves a label, so we may then attach the names "momentum" and "energy" to $\gamma m\mathbf{v}$ and γmc^2. Any other names would work just as well, of course, but we choose these because in the limit of small speeds, $\gamma m\mathbf{v}$ and γmc^2 reduce (as we will soon show) to some other nicely conserved quantities, which someone already tagged with the labels "momentum" and "energy" long ago.
2. As we've noted, the fact of the matter is that we can't *prove* that $\gamma m\mathbf{v}$ and γmc^2 are conserved. In Newtonian physics, conservation of $\mathbf{p} \equiv m\mathbf{v}$ is basically postulated by Newton's third law, and we're not going to be able to do any better than that here. All we can hope to do as physicists is provide some motivation for considering $\gamma m\mathbf{v}$ and γmc^2, then show that it is consistent for $\gamma m\mathbf{v}$ and γmc^2 to be conserved during an interaction, and then gather a large amount of experimental evidence, all of which is consistent with

$\gamma m\mathbf{v}$ and γmc^2 being conserved. As far as the experimental evidence goes, suffice it to say that high-energy accelerators, cosmological observations, and many other forums are continually verifying everything that we think is true about relativistic dynamics. If the theory isn't correct, then we know that it must be the limiting case of a more correct theory.[3] But all this experimental induction has to count for something ...

> "To three, five, and seven, assign
> A name," the prof said, "We'll define."
> But he botched the instruction
> With woeful induction
> And told us the next prime was nine.

3. Conservation of energy in relativistic mechanics is actually a much simpler concept than in nonrelativistic mechanics, because $E = \gamma m$ is conserved, period. We don't have to worry about the generation of internal energy (heat or internal potential energy), which ruins conservation of the nonrelativistic $E = mv^2/2$. The internal energy is simply built into the total energy. In the above example, the two m's collide and generate internal energy (heat) in the resulting mass M. This internal energy shows up as an increase in mass, which makes M larger than $2m$. The energy that corresponds to the increase in mass comes from the initial kinetic energy of the two m's.
4. Problem 12.1 gives an alternate derivation of the energy and momentum expressions in Eq. (12.1). This derivation uses additional facts, namely that the energy and momentum of a photon are given by $E = h\nu$ and $p = h\nu/c$, where ν is the frequency of the light, and h is Planck's constant. ♣

Any multiple of γmc^2 is also conserved, of course. Why did we pick γmc^2 to label as "E" instead of, say, $5\gamma mc^3$? Consider the approximate form that γmc^2 takes in the Newtonian limit, that is, in the limit $v \ll c$. We have, using the Taylor series expansion for $(1-x)^{-1/2}$,

$$
\begin{aligned}
E \equiv \gamma mc^2 &= \frac{mc^2}{\sqrt{1-v^2/c^2}} \\
&= mc^2\left(1 + \frac{v^2}{2c^2} + \frac{3v^4}{8c^4} + \cdots\right) \\
&= mc^2 + \frac{1}{2}mv^2 + \cdots . \qquad (12.9)
\end{aligned}
$$

The dots represent higher-order terms in v^2/c^2, which may be neglected if $v \ll c$. In an elastic collision in Newtonian physics, no heat is generated, so mass is conserved. That is, the quantity mc^2 has a fixed value. We therefore see that conservation of $E \equiv \gamma mc^2$ reduces to the familiar conservation of Newtonian kinetic energy, $mv^2/2$, for elastic collisions in the limit of slow speeds. This is an example of the *correspondence principle*, which says that relativistic formulas must reduce to the familiar nonrelativistic ones in the nonrelativistic limit. Likewise, for the momentum we picked $\mathbf{p} \equiv \gamma m\mathbf{v}$ instead of, say, $6\gamma mc^4\mathbf{v}$, because

[3] And in fact it isn't correct, because it doesn't incorporate quantum mechanics. The more complete theory that includes both relativity and quantum mechanics is quantum field theory.

the former reduces to the familiar Newtonian momentum, $m\mathbf{v}$, in the limit of slow speeds.

Whether abstract, profound, or just mystic,
Or boring, or somewhat simplistic,
A theory must lead
To results that we need
In limits, nonrelativistic.

Whenever we use the term "energy," we will mean the total energy, γmc^2. If we use the term "kinetic energy," we will mean a particle's excess energy over the energy it has when it is motionless. In other words, the kinetic energy is $\gamma mc^2 - mc^2$. Kinetic energy is *not* necessarily conserved in a collision, because mass is not necessarily conserved, as we saw in Eq. (12.6) in the above example. In the CM frame, there was kinetic energy before the collision, but none after. Kinetic energy is a rather artificial concept in relativity. You virtually always want to use the total energy, γmc^2, when solving a problem.

Note the following important relation,

$$\begin{aligned} E^2 - |\mathbf{p}|^2c^2 &= \gamma^2m^2c^4 - \gamma^2m^2|\mathbf{v}|^2c^2 \\ &= \gamma^2m^2c^4\left(1 - \frac{v^2}{c^2}\right) \\ &= m^2c^4. \end{aligned} \tag{12.10}$$

This is a primary ingredient in solving relativistic collision problems, as we'll soon see. It replaces the $K = p^2/2m$ relation between kinetic energy and momentum in Newtonian physics. It can be derived in more profound ways, as we'll see in Chapter 13. It's so important that I like to call it the Very Important Relation. Let's put it in a box:

$$\boxed{E^2 = p^2c^2 + m^2c^4}. \tag{12.11}$$

Whenever you know two of the three quantities E, p, and m, this equation gives you the third. In the case where $m = 0$ (as with photons), Eq. (12.11) says that

$$E = pc \quad \text{(for photons)}. \tag{12.12}$$

This is the key equation for massless objects. For photons, the two equations, $\mathbf{p} = \gamma m\mathbf{v}$ and $E = \gamma mc^2$, don't tell us much, because $m = 0$ and $\gamma = \infty$, so their product is undetermined. But $E^2 - |\mathbf{p}|^2c^2 = m^2c^4$ still holds, and we conclude that $E = pc$. Note that any massless particle must have $\gamma = \infty$. That is, it must travel at speed c. If this weren't the case, then $E = \gamma mc^2$ would equal zero, in which case the particle wouldn't be much of a particle. We'd have a hard time observing something with no energy.

Another nice relation, which follows from Eqs. (12.1) and holds for particles of any mass, is

$$\frac{\mathbf{p}}{E} = \frac{\mathbf{v}}{c^2}. \tag{12.13}$$

Given p and E, this is definitely the quickest way to get v.

The general size of mc²

What is the general size of mc^2?[4] If we let $m = 1$ kg, then we have $mc^2 = (1\,\text{kg})(3 \cdot 10^8\,\text{m/s})^2 \approx 10^{17}$ J. How big is this? A typical household electric bill might be around \$50 per month, or \$600 per year. At about 10 cents per kilowatt-hour, this translates to 6000 kilowatt-hours per year. Since there are 3600 seconds in an hour, this converts to $(6000)(10^3)(3600) \approx 2 \cdot 10^{10}$ watt-seconds. That is, $2 \cdot 10^{10}$ joules per year. We therefore see that if one kilogram were converted completely into usable energy (that is, kinetic energy, which can then be used to drive a turbine), it would be enough to provide electricity to about $10^{17}/(2 \cdot 10^{10})$, or 5 million, homes for a year. That's a lot.

In a nuclear reactor, only a small fraction of the mass is converted into usable energy (heat). Most of the mass remains in the final products, which doesn't help in lighting your home. If a particle were to combine with its antiparticle, then it would be possible for all of the mass energy to be converted into usable energy. But we're a while away from being able to do this productively. However, even a small fraction of the very large quantity, $E = mc^2$, can be large, as evidenced by the use of nuclear power and nuclear weapons. Any quantity with a few factors of c is bound to change the face of the world.

Mass

Some treatments of relativity use the term "rest mass," m_0, to refer to the mass of a motionless particle, and "relativistic mass," m_{rel}, to refer to the quantity γm_0 of a moving particle. We won't use this terminology here. The only thing we'll call "mass" is what the above treatments call "rest mass." (For example, the mass of an electron is $9.11 \cdot 10^{-31}$ kg, and the mass of a liter of water is 1 kg, independent of their speed.) And since we'll refer to only one type of mass, there is no need to use the qualifier "rest" or the subscript "0." We'll therefore simply use the notation "m."

Of course, you can *define* the quantity γm to be anything you want. There's nothing wrong with calling it "relativistic mass." But the point is that γm already goes by another name. It's just the energy, up to factors of c. The use of the word "mass" for this quantity, although quite permissible, is certainly not needed.

Furthermore, the word "mass" is used to describe what is on the right-hand side of the equation, $E^2 - |\mathbf{p}|^2c^2 = m^2$. The m^2 here is an *invariant*, that is,

[4] See Fadner (1988) for a history of the mass–energy relation.

it is something that is independent of the frame of reference. Although E and $\mathbf{p}$ depend on the frame because they involve v, this v dependence cancels out when taking the difference of their squares, yielding a frame-independent m^2. If "mass" is to be used in this definite way to describe an invariant, then it doesn't make sense to also use it to describe the quantity γm, which is frame-dependent. The prefixes "rest" and "relativistic" are introduced to avoid this problem, but the result of this is just a watering down of the very important invariance quality of "mass."

However, since there is in fact nothing actually wrong with using "relativistic mass," it mainly comes down to personal preference whether or not you like the term. Certainly one motivation for labeling γm as some kind of mass is that the expression for momentum looks like $p = \gamma m_0 v \equiv m_{\text{rel}} v$, which mimics the Newtonian expression. And the expression for energy takes the nice form, $E = \gamma m_0 c^2 \equiv m_{\text{rel}} c^2$. But considering that Newtonian physics is only a limiting case of relativistic physics, it is questionable what is gained by molding a theory that is more correct into one that is less correct. At any rate, my view is that these nice formulas don't outweigh the usefulness of having the word "mass" mean a very specific invariant quantity, with the word "energy" referring to the frame-dependent quantity γm. Invariant quantities have a certain sacred place in physics, so they should be given a name that doesn't have any noninvariant connotations.

Lagrangian method

Another way to motivate the above expressions for E and p is to use the Lagrangian method. To take this route, we must of course figure out what the relativistic Lagrangian is. This might seem like a daunting task, considering that the Lagrangians we dealt with in Chapter 6 involved the kinetic energy, and we have no idea yet what the energy is here (that's one of our goals, after all).

It turns out, however, that it's fairly easy to write down the relativistic Lagrangian, provided that we think outside the "$T - V$" box and concentrate instead on the all-important fact that the action is something that is stationary for the classical path. Combining this fact with the relativity postulate, which states that there is no preferred reference frame, we see that the action must be stationary in all frames. A sensible property to require of the action is therefore that it be *independent* of the frame. This way, if it is stationary in one frame, then it is stationary in any other.[5]

What invariant quantities do we know of? Well, there are certainly m and c. But the more interesting one that we're familiar with is the invariant interval (the proper time), $c\,d\tau \equiv \sqrt{c^2\,dt^2 - dx^2}$. This is pretty much all we have to work

[5] I suppose you could imagine a quantity that depends on the frame, but that is still stationary in any frame. But clearly the simplest thing to try (which will in fact work) is something that is independent of the frame.

with, so let's try an action of the form (we'll just deal with a free particle),

$$S = -mc \int c\, d\tau = -mc \int \sqrt{c^2\, dt^2 - dx^2} = -mc \int \sqrt{c^2 - \dot{x}^2}\, dt. \quad (12.14)$$

We have put the mc out front to give S the usual units of energy times time. And the minus sign (which doesn't affect the stationary property) is included so that we will obtain the correct signs for the energy and momentum. Our Lagrangian is therefore

$$L = -mc\sqrt{c^2 - \dot{x}^2}. \quad (12.15)$$

The Euler–Lagrange equation is thus

$$\frac{d}{dt}\left(\frac{\partial L}{\partial \dot{x}}\right) = \frac{\partial L}{\partial x} \quad \Longrightarrow \quad \frac{d}{dt}\left(\frac{mc\dot{x}}{\sqrt{c^2 - \dot{x}^2}}\right) = 0. \quad (12.16)$$

The quantity in the parentheses can be written as

$$p \equiv \frac{mv}{\sqrt{1 - v^2/c^2}} \equiv \gamma m v. \quad (12.17)$$

This is simply the momentum we defined earlier, which, in view of Eq. (12.16), is conserved, as it should be. What about the energy? The energy associated with our Lagrangian is given by (using Eq. (6.52))

$$\begin{aligned} E = \dot{x}\frac{\partial L}{\partial \dot{x}} - L &= \dot{x}\left(\frac{mc\dot{x}}{\sqrt{c^2 - \dot{x}^2}}\right) - \left(- mc\sqrt{c^2 - \dot{x}^2}\right) \\ &= \frac{mc^3}{\sqrt{c^2 - \dot{x}^2}} = \frac{mc^2}{\sqrt{1 - v^2/c^2}} \equiv \gamma mc^2, \end{aligned} \quad (12.18)$$

which agrees with our earlier expression for E. We can also calculate the angular momentum. In polar coordinates, the Lagrangian can be written as

$$L = -mc\sqrt{c^2 - (\dot{r}^2 + r^2\dot{\theta}^2)}. \quad (12.19)$$

The Euler–Lagrange equation obtained from varying θ is therefore

$$\frac{d}{dt}\left(\frac{\partial L}{\partial \dot{\theta}}\right) = \frac{\partial L}{\partial \theta} \quad \Longrightarrow \quad \frac{d}{dt}\left(\frac{mcr^2\dot{\theta}}{\sqrt{c^2 - (\dot{r}^2 + r^2\dot{\theta}^2)}}\right) = 0. \quad (12.20)$$

The quantity in the parentheses, which we see is conserved, can be written as $\gamma m r^2\dot{\theta}$. Because it is associated with the angle θ, we call it the angular momentum. And as expected, it reduces properly to the nonrelativistic result when $\gamma = 1$.

Of course, this Lagrangian justification for E and p might seem a little silly, because we're just dealing with a free particle, and *any* quantity involving v is conserved for a free particle. But the fact that the above E and p can be derived from a Lagrangian involving an invariant quantity (which is the only nontrivial invariant quantity we know of) gives them a reasonable amount of credence.

If we want to go beyond a free particle and incorporate a potential energy into the system, things get more complicated; see Brehme (1971). The first thought that comes to mind is to simply subtract off a potential $V(x)$ in the Lagrangian, as we did in the nonrelativistic case. We then have $L = -mc\sqrt{c^2 - \dot{x}^2} - V(x)$, and the Euler–Lagrange equation becomes

$$\frac{d}{dt}\left(\frac{\partial L}{\partial \dot{x}}\right) = \frac{\partial L}{\partial x} \quad \Longrightarrow \quad \frac{d}{dt}\left(\frac{mc\dot{x}}{\sqrt{c^2 - \dot{x}^2}}\right) = -\frac{\partial V}{\partial x} \quad \Longrightarrow \quad \frac{dp}{dt} = F(x). \tag{12.21}$$

This is the correct $F = dp/dt$ statement, so it seems that we picked the right Lagrangian. However, there is something amiss here; the action now involves the quantity $\int V(x)\,dt$ which isn't invariant, because the time in one frame doesn't equal the time in another. Although the above Lagrangian does indeed yield the correct equation of motion in the particular reference frame in which the function $V(x)$ is defined, if we switch to another frame there is no guarantee that it will do so again. In order to maintain an invariant action, we must construct it from 4-vectors (discussed in Chapter 13), or other tensors, multiplied in an appropriate way. The classic system involving 4-vectors is a charged particle in an electromagnetic field. Other systems involving more complicated tensors occur in General Relativity. But we're not going to get into any of that here, because our goal was only to motivate the expressions for E and p.[6]

Setting $c = 1$

For the remainder of our treatment of relativity, we will invariably work in units where $c = 1$. For example, instead of one meter being the unit of distance, we can make $3 \cdot 10^8$ meters equal to one unit. Or, we can keep the meter as is, and make $1/(3 \cdot 10^8)$ seconds the unit of time. In such units, our various expressions become

$$\mathbf{p} = \gamma m \mathbf{v}, \quad E = \gamma m, \quad E^2 = p^2 + m^2, \quad \frac{\mathbf{p}}{E} = \mathbf{v}. \tag{12.22}$$

Said in another way, you can simply ignore all the c's in your calculations (which will generally save you a lot of strife), and then put them back into your final answer to make the units correct. For example, let's say the goal of a certain

[6] In multiparticle systems, it turns out that in general it is impossible to create a consistent relativistic Lagrangian formalism. The only way to proceed is to work in terms of fields instead of particles. But this is well beyond the scope of what we're doing here. See Goldstein *et al.* (2002).

problem is to find the time of some event. If your answer comes out to be ℓ, where ℓ is a given length, then you know that the correct answer (in terms of the usual mks units) has to be ℓ/c, because this has units of time. In order for this procedure to work, there must be only one way to put the c's back in at the end. This is always the case, because if there were two ways, then we would have $c^a = c^b$, for some numbers $a \neq b$. But this is impossible, because c has units.

12.2 Transformations of *E* and *p*

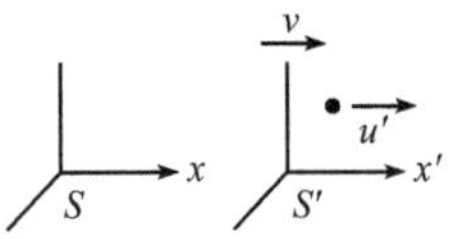

Fig. 12.6

Consider the following one-dimensional situation, where all the motion is along the x axis. A particle has energy E' and momentum p' in frame S'. Frame S' moves at speed v with respect to frame S, in the positive x direction (see Fig. 12.6). What are E and p in S?

Let u' be the particle's speed in S'. From the velocity-addition formula, the particle's speed in S is (dropping the factors of c)

$$u = \frac{u' + v}{1 + u'v}. \tag{12.23}$$

This is all we need to know, because a particle's velocity determines its energy and momentum. But we'll need to go through a little algebra to make things look nice and pretty. The γ factor associated with the speed u is

$$\gamma_u = \frac{1}{\sqrt{1 - \left(\frac{u'+v}{1+u'v}\right)^2}} = \frac{1 + u'v}{\sqrt{(1 - u'^2)(1 - v^2)}} \equiv \gamma_{u'}\gamma_v(1 + u'v). \tag{12.24}$$

The energy and momentum in S' are

$$E' = \gamma_{u'}m, \quad \text{and} \quad p' = \gamma_{u'}mu', \tag{12.25}$$

while the energy and momentum in S are, using Eq. (12.24),

$$\begin{aligned} E &= \gamma_u m = \gamma_{u'}\gamma_v(1 + u'v)m, \\ p &= \gamma_u mu = \gamma_{u'}\gamma_v(1 + u'v)m\left(\frac{u' + v}{1 + u'v}\right) = \gamma_{u'}\gamma_v(u' + v)m. \end{aligned} \tag{12.26}$$

Using the E' and p' from Eqs. (12.25), we can rewrite E and p as (with $\gamma \equiv \gamma_v$)

$$\begin{aligned} E &= \gamma(E' + vp'), \\ p &= \gamma(p' + vE'). \end{aligned} \tag{12.27}$$

These are transformations for E and p between frames. If you want to put the factors of c back in, then the vE' term becomes vE'/c^2, to make the units correct. These transformations are easy to remember, because they look *exactly* like the

Lorentz transformations for the coordinates t and x in Eq. (11.21). More precisely with the c's included, E and pc transform like ct and x, respectively.[7] This is no coincidence, as we will see in Chapter 13. As a check on Eqs. (12.27), if $u' = 0$ (so that $p' = 0$ and $E' = m$), then $E = \gamma m$ and $p = \gamma mv$, as expected. Also, if $u' = -v$ (so that $p' = -\gamma mv$ and $E' = \gamma m$), then $E = m$ and $p = 0$, as expected.

Because the transformations in Eqs. (12.27) are linear, they also hold if E and p represent the total energy and momentum of a collection of particles. That is,

$$\begin{aligned} \sum E &= \gamma\left(\sum E' + v\sum p'\right), \\ \sum p &= \gamma\left(\sum p' + v\sum E'\right). \end{aligned} \tag{12.28}$$

In fact, any (corresponding) linear combinations of the energies and momenta are valid here, in place of the sums. For example, we can use the combinations $(E_1^b + 3E_2^a - 7E_5^b)$ and $(p_1^b + 3p_2^a - 7p_5^b)$ in Eqs. (12.27), where the subscripts indicate which particle, and the superscripts indicate before or after a collision. You can verify this by simply taking the appropriate linear combination of Eqs. (12.27) for the various particles. This consequence of linearity is a very important and useful result, as will become clear in the remarks below.

You can use Eqs. (12.27) to show that

$$E^2 - p^2 = E'^2 - p'^2, \tag{12.29}$$

just as we showed that $t^2 - x^2 = t'^2 - x'^2$ in Eq. (11.40). The proof there was based on the fact that t and x transform under the Lorentz transformations, so exactly the same proof works here with E and p, in view of the Lorentz transformations in Eqs. (12.27). The E's and p's in Eq. (12.29) can represent any (corresponding) linear combinations of the E's and p's of the various particles (for example, the total E and p of the particles), due to the linearity of Eqs. (12.27). For one particle, we already know that Eq. (12.29) is true, because both sides are equal to m^2, from Eq. (12.10). For many particles, the invariant quantity $E_{\text{total}}^2 - p_{\text{total}}^2$ equals the square of the total energy in the CM frame (which reduces to m^2 for one particle), because $p_{\text{total}} = 0$ in the CM frame, by definition.

REMARKS:

1. In the previous section, we said that we needed to show that if E and p are conserved in one reference frame during a collision, then they are conserved in any other frame (because a conservation law shouldn't depend on what frame you're in). This can be shown as follows. The total ΔE is a linear combination of the initial and final E's, and likewise for

[7] Since the Lorentz transformations are symmetric in ct and x, and likewise in E and pc, it isn't clear which coordinate in one transformation corresponds to which coordinate in the other. But since x and p are the components of vectors, while t and E aren't, the correct correspondence must be $ct \longleftrightarrow E$ and $x \longleftrightarrow pc$.

Δp. Therefore, since Eq. (12.27) is a linear equation in the E's and p's, it also holds for the ΔE's and Δp's. That is,

$$\Delta E = \gamma(\Delta E' + v\Delta p'), \quad \text{and} \quad \Delta p = \gamma(\Delta p' + v\Delta E'). \tag{12.30}$$

So if the total $\Delta E'$ and $\Delta p'$ in S' are zero, then the total ΔE and Δp in S must also be zero.

2. Equation (12.30) makes it clear that if you accept the fact that $p = \gamma mv$ is conserved in all frames, then you must also accept the fact that $E = \gamma m$ is conserved in all frames (and vice versa). This is true because the second of Eqs. (12.30) says that if Δp and $\Delta p'$ are both zero, then $\Delta E'$ must also be zero. E and p have no choice but to go hand in hand. ♣

Equation (12.27) applies to the x component of the momentum. How do the transverse components, p_y and p_z, transform? Just as with the y and z coordinates in the Lorentz transformations, p_y and p_z don't change between frames. The analysis in Chapter 13 makes this obvious, so for now we'll simply state that if the relative velocity between the frames is in the x direction, then

$$p_y = p'_y, \quad \text{and} \quad p_z = p'_z. \tag{12.31}$$

If you really want to show explicitly that the transverse components don't change between frames, or if you're worried that a nonzero speed in the y direction will mess up the relationship between p_x and E that we calculated in Eq. (12.27), then Exercise 12.23 is for you. But it's a bit tedious, so feel free to settle for the much cleaner reasoning in Chapter 13.

12.3 Collisions and decays

The strategy for studying relativistic collisions is the same as that for studying nonrelativistic ones. You just have to write down all the conservation of energy and momentum equations, and then solve for whatever variables you want to solve for. The conservation principles are the same as they've always been. The only difference is that now the energy and momentum take the new forms in Eqs. (12.1). In writing down the conservation of energy and momentum equations, it proves extremely useful to put E and $\mathbf{p}$ together into one four-component vector,

$$P \equiv (E, \mathbf{p}) \equiv (E, p_x, p_y, p_z). \tag{12.32}$$

This is called the *energy–momentum 4-vector*, or the *4-momentum*, for short. If we were keeping in the factors of c, then the first term would be E/c, although some people instead multiply the $\mathbf{p}$ by c; either convention is fine. Our notation in this chapter will be to use an uppercase P to denote a 4-momentum and a lowercase $\mathbf{p}$ or p to denote a spatial momentum. The components of a 4-momentum are usually indexed from 0 to 3, so that $P_0 \equiv E$, and $(P_1, P_2, P_3) \equiv \mathbf{p}$. For one particle, we have

$$P = (\gamma m, \gamma m v_x, \gamma m v_y, \gamma m v_z). \tag{12.33}$$

The 4-momentum for a collection of particles consists of the total E and total $\mathbf{p}$ of all the particles. There are deep reasons for considering the 4-momentum, as we'll see in Chapter 13, but for now we'll just view it as a matter of convenience. If nothing else, it helps with the bookkeeping. Conservation of energy and momentum in a collision reduce to the concise statement,

$$P_{\text{before}} = P_{\text{after}}, \tag{12.34}$$

where these are the total 4-momenta of all the particles.

If we define the *inner product* between two 4-momenta, $A \equiv (A_0, A_1, A_2, A_3)$ and $B \equiv (B_0, B_1, B_2, B_3)$, to be

$$A \cdot B \equiv A_0B_0 - A_1B_1 - A_2B_2 - A_3B_3, \tag{12.35}$$

then the Very Important Relation in Eq. (12.11), $E^2 - p^2 = m^2$, which is true for one particle, may be concisely written as

$$P \cdot P = m^2, \quad \text{or} \quad P^2 = m^2, \tag{12.36}$$

where $P^2 \equiv P \cdot P$. In other words, the square of a particle's 4-momentum equals the square of its mass. This relation will prove to be very useful in collision problems. Note that it is frame-independent, as we saw in Eq. (12.29).

This inner product is different from the one we're used to in three-dimensional space. It has one positive sign and three negative signs, in contrast with the usual three positive signs. But we are free to define it however we wish, and we did indeed pick a good definition, because our inner product is invariant under Lorentz transformations, just as the usual 3-D inner product is invariant under rotations. For the inner product of a 4-momentum with itself (which could be any linear combination of 4-momenta of various particles), this invariance is the statement in Eq. (12.29). For the inner product of two different 4-momenta, we'll prove the invariance in Section 13.3.

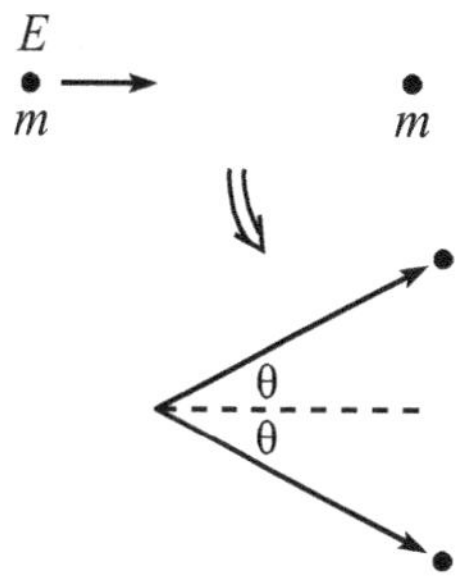

Fig. 12.7

Example (Relativistic billiards): A particle with mass m and energy E approaches an identical particle at rest. They collide elastically[8] in such a way that they both scatter at an angle θ relative to the incident direction (see Fig. 12.7). What is θ in terms of E and m? What is θ in the relativistic and nonrelativistic limits?

Solution: The first thing we should do is write down the 4-momenta. The 4-momenta before the collision are

$$P_1 = (E, p, 0, 0), \quad P_2 = (m, 0, 0, 0), \tag{12.37}$$

[8] An elastic collision in nonrelativistic physics is defined as one in which no heat is generated. In relativity, heat shows up as mass. So an elastic collision in relativistic physics is defined as one in which none of the masses change.

where $p = \sqrt{E^2 - m^2}$. The 4-momenta after the collision are (primes now denote "after")

$$P'_1 = (E', p' \cos\theta, p' \sin\theta, 0), \quad P'_2 = (E', p' \cos\theta, -p' \sin\theta, 0), \qquad (12.38)$$

where $p' = \sqrt{E'^2 - m^2}$. Conservation of energy gives $E' = (E + m)/2$, and conservation of p_x gives $p' \cos\theta = p/2$. Therefore, the 4-momenta after the collision are

$$P'_{1,2} = \left(\frac{E+m}{2}, \frac{p}{2}, \pm\frac{p}{2} \tan\theta, 0 \right). \qquad (12.39)$$

From Eq. (12.36), the squares of these 4-momenta must be m^2. Therefore,

$$\begin{aligned} m^2 &= \left(\frac{E+m}{2} \right)^2 - \left(\frac{p}{2} \right)^2 (1 + \tan^2\theta) \\ \Longrightarrow \quad 4m^2 &= (E+m)^2 - \frac{(E^2 - m^2)}{\cos^2\theta} \\ \Longrightarrow \quad \cos^2\theta &= \frac{E^2 - m^2}{E^2 + 2Em - 3m^2} = \frac{E+m}{E+3m}. \end{aligned} \qquad (12.40)$$

The relativistic limit is $E \gg m$, which yields $\cos\theta \approx 1$. This means that both particles scatter almost directly forward. You can convince yourself that θ should be small by looking at the collision in the CM frame and then shifting back to the lab frame. The transverse speeds decrease during this shift of frames.

The nonrelativistic limit is $E \approx m$ (it's *not* $E \approx 0$), which yields $\cos\theta \approx 1/\sqrt{2}$. So $\theta \approx 45°$, and the particles scatter with a 90° angle between them. This agrees with the result from the "Billiards" example in Section 5.7.2, a result which pool players are very familiar with.

Note that we never wrote down any v's in the above solution. That is, we never used the relations $E = \gamma mc^2$ and $p = \gamma mv$. There was no need to find the velocities; doing so would have essentially involved going in circles. Using $E = \gamma mc^2$ and $p = \gamma mv$ certainly provides a valid way of solving the problem, but in many situations like this one, where velocities aren't given or explicitly asked for, it leads to a very messy solution. A far cleaner method is to use the $E^2 - p^2 = m^2$ relation.

If γmv yields frustration,
And the similar E, irritation,
Just ditch all the v's,
And use (won't you *please*)
The Very Important Relation!

Let's now look at a decay. Decays are basically the same as collisions. All you have to do is conserve energy and momentum, as the following example shows.

Example (Decay at an angle): A particle with mass M and energy E decays into two identical particles. In the lab frame, one of them is emitted at a 90° angle, as shown in Fig. 12.8. What are the energies of the created particles? We'll give two solutions. The second one shows how 4-momenta can be used in a very clever and time-saving way.

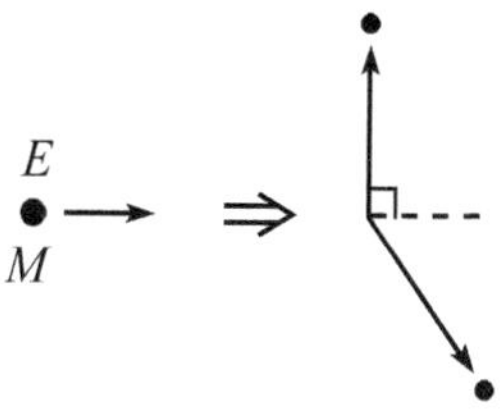

Fig. 12.8

First solution: The 4-momentum before the decay is

$$P = (E, p, 0, 0), \tag{12.41}$$

where $p = \sqrt{E^2 - M^2}$. Let the created particles have mass m, and let the second particle make an angle θ with the x axis. The 4-momenta after the decay are

$$P_1 = (E_1, 0, p_1, 0), \quad P_2 = (E_2, p_2 \cos\theta, -p_2 \sin\theta, 0). \tag{12.42}$$

Conservation of p_x immediately gives $p_2 \cos\theta = p$, which then implies that $p_2 \sin\theta = p \tan\theta$. Conservation of p_y says that the final p_y's are opposites. Therefore, the 4-momenta after the decay are

$$P_1 = (E_1, 0, p \tan\theta, 0), \quad P_2 = (E_2, p, -p \tan\theta, 0). \tag{12.43}$$

Conservation of energy gives $E = E_1 + E_2$. Writing E_1 and E_2 in terms of the momenta and masses, this becomes

$$E = \sqrt{p^2 \tan^2\theta + m^2} + \sqrt{p^2(1 + \tan^2\theta) + m^2}. \tag{12.44}$$

Putting the first radical on the left side, squaring, and solving for that radical (which is E_1) gives

$$E_1 = \frac{E^2 - p^2}{2E} = \frac{M^2}{2E}. \tag{12.45}$$

In a similar manner, we find that E_2 equals

$$E_2 = \frac{E^2 + p^2}{2E} = \frac{2E^2 - M^2}{2E}. \tag{12.46}$$

These add up to E, as they should.

Second solution: With the 4-momenta defined as in Eqs. (12.41) and (12.42), conservation of energy and momentum can be combined into the statement, $P = P_1 + P_2$. Therefore,

$$\begin{aligned} P - P_1 &= P_2, \\ \Longrightarrow \quad (P - P_1) \cdot (P - P_1) &= P_2 \cdot P_2, \\ \Longrightarrow \quad P^2 - 2P \cdot P_1 + P_1^2 &= P_2^2, \\ \Longrightarrow \quad M^2 - 2EE_1 + m^2 &= m^2, \\ \Longrightarrow \quad E_1 &= \frac{M^2}{2E}. \end{aligned} \tag{12.47}$$

And then $E_2 = E - E_1 = (2E^2 - M^2)/2E$. This solution should convince you that 4-momenta can save you a lot of work. What happened here was that the expression for P_2 was fairly messy, but we arranged things so that it appeared only in the form of P_2^2, which is simply m^2. 4-momenta provide a remarkably organized method for sweeping unwanted garbage under the rug.

12.4 Particle-physics units

A branch of physics that uses relativity as one of its main ingredients is elementary-particle physics, which is the study of the building blocks of matter (electrons, quarks, neutrinos, etc.). It's unfortunately the case that most of the elementary particles we want to study don't exist naturally in the world. We therefore have to create them in particle accelerators by colliding other particles together at very high energies. The high speeds involved require the use of relativistic dynamics. Newtonian physics is essentially useless.

What is a typical size of a rest energy, mc^2, of an elementary particle? The rest energy of a proton (which isn't really elementary; it's made up of quarks, but never mind) is

$$E_{\rm p} = m_{\rm p}c^2 = (1.67 \cdot 10^{-27}\,\text{kg})(3 \cdot 10^8\,\text{m/s})^2 = 1.5 \cdot 10^{-10}\,\text{joules}. \quad (12.48)$$

This is very small, of course. So a joule is probably not the best unit to work with. We would get very tired of writing the negative exponents over and over. We could perhaps work with "nanojoules," but particle physicists like to work instead with the "eV," the *electron-volt*. This is the change in energy of an electron when it passes through a potential of one volt. The electron charge is (negative) $e = 1.602 \cdot 10^{-19}\,\text{C}$, and a volt is defined as $1\,\text{V} = 1\,\text{J/C}$. So the conversion from eV to joules is[9]

$$1\,\text{eV} = (1.602 \cdot 10^{-19}\,\text{C})(1\,\text{J/C}) = 1.602 \cdot 10^{-19}\,\text{J}. \quad (12.49)$$

Therefore, in terms of eV, the rest energy of a proton is $938 \cdot 10^6$ eV. We now have the opposite problem of having a large exponent hanging around. But this is easily remedied by the prefix "M," which stands for "mega" or "million." So we finally have a proton rest energy of

$$E_{\rm p} = 938\,\text{MeV}. \quad (12.50)$$

[9] This is getting a little picky, but "eV" should technically be written as "*e*V," because when people write "eV," they actually mean that two things are being multiplied together (in contrast with, for example, the "kg" symbol for "kilogram"). One of these things is the electron charge, which is usually denoted by *e*.

You can work out for yourself that an electron has a rest energy of $E_e = 0.511$ MeV. The rest energies of various particles are listed in the table below. The ones preceded by a "$\approx$" are the averages of differently charged particles, whose energies differ by a few MeV. These (and the many other) elementary particles have specific properties (spin, charge, etc.), but for the present purposes they need only be thought of as point objects having a definite mass.

Particle	Rest energy (MeV)
electron (e)	0.511
muon (μ)	105.7
tau (τ)	1784
proton (p)	938.3
neutron (n)	939.6
lambda (Λ)	1115.6
sigma (Σ)	$\approx$1193
delta (Δ)	$\approx$1232
pion (π)	$\approx$137
kaon (K)	$\approx$496

For higher energies, the prefixes "G" (for "giga," meaning 10^9) and "T" (for "tera," meaning 10^{12}) are used.

We now come to a slight abuse of language. When particle physicists talk about masses, they say things like, "The mass of a proton is 938 MeV." This, of course, makes no sense, because the units are wrong; a mass can't equal an energy. But what they mean is that if you take this energy and divide it by c^2, then you get the mass. It would truly be a pain to keep saying, "The mass is such-and-such an energy, divided by c^2." For a quick conversion back to kilograms, you can show that

$$1\ \text{MeV}/c^2 = 1.783 \cdot 10^{-30}\ \text{kg}. \tag{12.51}$$

12.5 Force

12.5.1 Force in one dimension

In nonrelativistic physics, Newton's second law is $F = dp/dt$, which reduces to $F = ma$ if the mass is constant. We'll carry this law over to relativity and continue to write (we'll just deal with one-dimensional motion for now)

$$F = \frac{dp}{dt}. \tag{12.52}$$

However, in relativity we have $p = \gamma mv$, and γ can change with time. This complicates things, and it turns out that F does *not* equal ma. But if dp/dt and ma are different, why does F equal dp/dt instead of ma? Perhaps the best reason arises from the 4-vector formalism in Chapter 13. But another reason is that the F in Eq. (12.52) leads to a familiar work–energy theorem, as we'll see in Eq. (12.57).

To see what form the F in Eq. (12.52) takes in terms of the acceleration $a \equiv \dot{v}$, let's first calculate $d\gamma/dt$:

$$\frac{d\gamma}{dt} \equiv \frac{d}{dt}\left(\frac{1}{\sqrt{1-v^2}}\right) = \frac{v\dot{v}}{(1-v^2)^{3/2}} \equiv \gamma^3 va. \tag{12.53}$$

Assuming that m is constant, we therefore have

$$F = \frac{d(\gamma mv)}{dt} = m(\dot{\gamma}v + \gamma\dot{v}) = ma\gamma(\gamma^2v^2 + 1) = \gamma^3 ma. \tag{12.54}$$

This doesn't look as nice as $F = ma$, but that's the way it goes. However, F correctly reduces to ma in the limit of small speeds (where $\gamma \approx 1$), as it must.

They *said*, "F is ma, bar none."
What they *meant* wasn't quite as much fun.
It's dp by dt,
Which just happens to be
Good ol' "ma" when γ is 1.

Consider now the quantity dE/dx, where E is the energy, $E = \gamma m$. We have

$$\frac{dE}{dx} = \frac{d(\gamma m)}{dx} = m\frac{d\left(1/\sqrt{1-v^2}\right)}{dx} = \gamma^3 mv\frac{dv}{dx}. \tag{12.55}$$

But $v(dv/dx) = (dx/dt)(dv/dx) = dv/dt = a$. Therefore, $dE/dx = \gamma^3 ma$. Combining this with Eq. (12.54) gives

$$F = \frac{dE}{dx}. \tag{12.56}$$

Note that Eqs. (12.52) and (12.56) take exactly the same form as in nonrelativistic physics. The only new thing in relativity is that the expressions for p and E are modified.

Remark: The result in Eq. (12.56) suggests another way to motivate the $E = \gamma m$ expression. The reasoning is exactly the same as in the derivation of nonrelativistic energy conservation in Section 5.1. Define F, as we have done, through Eq. (12.52). Then integrate Eq. (12.54) from x_1 to x_2 to obtain

$$\int_{x_1}^{x_2} F\,dx = \int_{x_1}^{x_2} (\gamma^3 ma)\,dx = \int_{x_1}^{x_2} \left(\gamma^3 mv\frac{dv}{dx}\right) dx = \int_{v_1}^{v_2} \gamma^3 mv\,dv = \gamma m\Big|_{v_1}^{v_2}, \tag{12.57}$$

where we have used the $d\gamma = \gamma^3 v\,dv$ relation from Eq. (12.55). We see that if we define the energy as $E = \gamma m$, then the work–energy theorem, $\int F\,dx = \Delta E$, holds in relativity just as it does in Newtonian physics. The only difference is that E is γm instead of $mv^2/2$.[10] ♣

12.5.2 Force in two dimensions

In two dimensions, the concept of force becomes a little strange. In particular, as we'll see, the acceleration of an object need not point in the same direction as the force. We start with

$$\mathbf{F} = \frac{d\mathbf{p}}{dt}. \tag{12.58}$$

This is a vector equation. Without loss of generality, we'll deal with only two spatial dimensions. Consider a particle moving in the x direction, and let us apply a force, $\mathbf{F} = (F_x, F_y)$. The particle's momentum is

$$\mathbf{p} = \frac{m(v_x, v_y)}{\sqrt{1 - v_x^2 - v_y^2}}. \tag{12.59}$$

Taking the derivative of this, and using the fact that v_y is initially zero, we obtain

$$\begin{aligned}
\mathbf{F} &= \left.\frac{d\mathbf{p}}{dt}\right|_{v_y=0} \\
&= m\left.\left(\frac{\dot{v}_x}{\sqrt{1-v^2}} + \frac{v_x(v_x\dot{v}_x + v_y\dot{v}_y)}{(\sqrt{1-v^2})^3}, \frac{\dot{v}_y}{\sqrt{1-v^2}} + \frac{v_y(v_x\dot{v}_x + v_y\dot{v}_y)}{(\sqrt{1-v^2})^3}\right)\right|_{v_y=0} \\
&= m\left(\frac{\dot{v}_x}{\sqrt{1-v^2}}\left(1 + \frac{v^2}{1-v^2}\right), \frac{\dot{v}_y}{\sqrt{1-v^2}}\right) \\
&= m\left(\frac{\dot{v}_x}{(\sqrt{1-v^2})^3}, \frac{\dot{v}_y}{\sqrt{1-v^2}}\right) \\
&\equiv m(\gamma^3 a_x, \gamma a_y).
\end{aligned} \tag{12.60}$$

This is *not* proportional to (a_x, a_y). The first component agrees with Eq. (12.54), but the second component has only one factor of γ. The difference comes from the fact that γ has a first-order change if v_x changes, but not if v_y changes, assuming that v_y is initially zero. The particle therefore responds differently to forces in the x and y directions. It is easier to accelerate something in the transverse direction.

[10] Actually, this reasoning suggests only that E is given by γm up to an additive constant. For all we know, E might take the form, $E = \gamma m - m$, which would make the energy of a motionless particle equal to zero. An argument along the lines of Section 12.1.2 is required to show that the additive constant is zero.

12.5.3 Transformation of forces

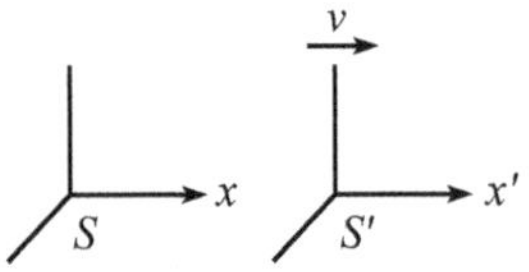

Fig. 12.9

Let a force act on a particle. How are the components of the force in the particle's frame, S', related to the components of the force in another frame, S?[11] Let the relative motion be along the x and x' axes, as shown in Fig. 12.9. In frame S, Eq. (12.60) says

$$(F_x, F_y) = m(\gamma^3 a_x, \gamma a_y). \tag{12.61}$$

And in frame S', the γ factor for the particle equals 1, so Eq. (12.60) reduces to the usual expression,

$$(F'_x, F'_y) = m(a'_x, a'_y). \tag{12.62}$$

Let's now relate these two forces, by writing the primed accelerations on the right-hand side of Eq. (12.62) in terms of the unprimed accelerations.

First, we have $a'_y = \gamma^2 a_y$. This is true because transverse distances are the same in the two frames, but times are shorter in S' by a factor γ. That is, $dt' = dt/\gamma$. We have indeed put the γ in the right place here, because the particle is essentially at rest in S', so the usual time dilation holds. Therefore, $a'_y \equiv d^2y'/dt'^2 = d^2y/(dt/\gamma)^2 \equiv \gamma^2 a_y$.

Second, we have $a'_x = \gamma^3 a_x$. In short, this is true because time dilation brings in two factors of γ (as in the a_y case), and length contraction brings in one. In more detail: Let the particle move from one point to another in frame S', as it accelerates from rest in S'. Mark these two points, which are essentially a distance $a'_x(dt')^2/2$ apart, in S'. As S' flies past S, the distance between the two marks is length contracted by a factor γ, as viewed by S. This distance (which is the excess distance the particle travels over what it would have traveled if there were no acceleration) is what S calls $a_x(dt)^2/2$. Therefore,

$$\frac{1}{2}a_x\,dt^2 = \frac{1}{\gamma}\left(\frac{1}{2}a'_x\,dt'^2\right) \implies a'_x = \gamma a_x\left(\frac{dt}{dt'}\right)^2 = \gamma^3 a_x. \tag{12.63}$$

Equation (12.62) may now be written as

$$(F'_x, F'_y) = m(\gamma^3 a_x, \gamma^2 a_y). \tag{12.64}$$

Comparing Eqs. (12.61) and (12.64) then gives

$$F_x = F'_x, \quad \text{and} \quad F_y = \frac{F'_y}{\gamma}. \tag{12.65}$$

We see that the longitudinal force is the same in the two frames, but the transverse force is larger by a factor of γ in the particle's frame.

[11] To be more precise, S' is the instantaneous inertial frame of the particle. Once the force is applied, the particle will accelerate and will therefore no longer be at rest in S'. But for a very small elapsed time, the particle will still essentially be in S'.

REMARKS:

1. What if someone comes along and switches the labels of the primed and unprimed frames in Eq. (12.65), and concludes that the transverse force is *smaller* in the particle's frame? He certainly can't be correct, given that Eq. (12.65) is true, but where is the error? The error lies in the fact that we (correctly) used $dt' = dt/\gamma$ above, because this is the relevant expression concerning two events along the particle's worldline. We are interested in two such events, because we want to see how the particle moves. The inverted expression, $dt = dt'/\gamma$, deals with two events located at the same position in S, and therefore has nothing to do with the situation at hand. Similar reasoning holds for the relation between dx and dx'. Because we are dealing with a given particle, there is indeed one frame that is special among all possible frames, namely the particle's instantaneous inertial frame.
2. If you want to compare forces in two frames, neither of which is the particle's rest frame, then just use Eq. (12.65) twice and relate each of the forces to the rest-frame forces. It quickly follows that for another frame S'', we have $F''_x = F_x$, and $\gamma'' F''_y = \gamma F_y$, where the γ's are measured relative to the rest frame S'. ♣

Example (Bead on a rod): A spring with a tension has one end attached to the end of a rod, and the other end attached to a bead that is constrained to move along the rod. The rod makes an angle θ' with respect to the x' axis, and is fixed at rest in the S' frame (see Fig. 12.10). The bead is released and is pulled along the rod by the spring. Right after the bead is released, what does the situation look like in the frame, S, of someone moving to the left at speed v? In answering this, draw the directions of

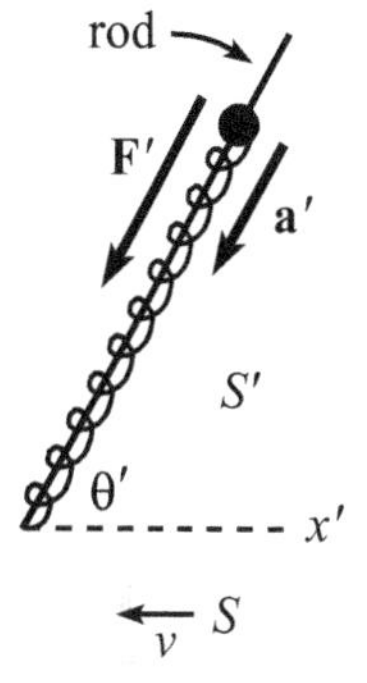

Fig. 12.10

(a) the rod,
(b) the acceleration of the bead,
(c) the force on the bead.

In frame S, does the wire exert a force of constraint?

Solution: In frame S:

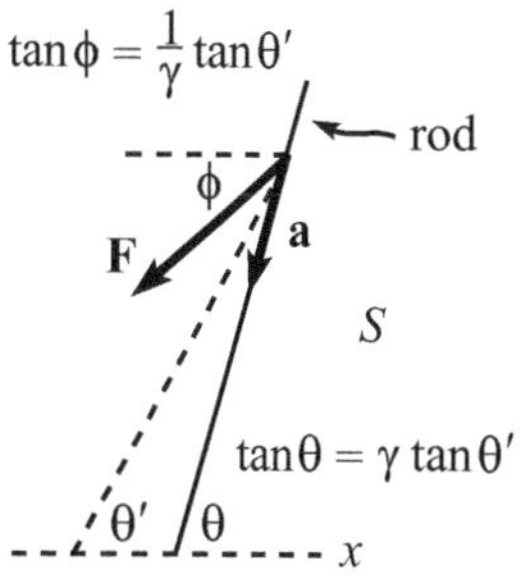

Fig. 12.11

(a) The horizontal span of the rod is decreased by a factor γ, due to length contraction, and the vertical span is unchanged, so we have $\tan\theta = \gamma\tan\theta'$, as shown in Fig. 12.11.
(b) The acceleration must point along the rod, because the bead always lies on the rod, and because the rod moves at constant speed in S. Quantitatively, the position of the bead in S takes the form of $(x, y) = (vt - a_x t^2/2, -a_y t^2/2)$, by the definition of acceleration. The position relative to the starting point on the rod, which has coordinates $(vt, 0)$, is then $(\Delta x, \Delta y) = (-a_x t^2/2, -a_y t^2/2)$. The condition for the bead to stay on the rod is that the ratio of these coordinates equals the slope of the rod in S. Therefore, $a_y/a_x = \tan\theta$, and the acceleration points along the rod.
(c) The y component of the force on the bead is decreased by a factor γ, by Eq. (12.65). Therefore, since the force points at an angle θ' in S' (because it points along the rod in S'), the angle in S is given by $\tan\phi = (1/\gamma)\tan\theta'$, as shown in the figure.

As a double-check that **a** does indeed point along the rod, we can use Eq. (12.60) to write $a_y/a_x = \gamma^2 F_y/F_x$. Then Eq. (12.65) gives $a_y/a_x = \gamma F'_y/F'_x = \gamma \tan\theta' = \tan\theta$, which is the direction of the wire.

The wire does *not* exert a force of constraint. The bead doesn't need to touch the wire in S', so it doesn't need to touch it in S. There is no need to have an extra force to combine with **F** to make the result point along **a**, because **F** simply doesn't have to be collinear with **a**.

12.6 Rocket motion

Up to this point, we have dealt with situations where the masses of our particles are constant, or where they change abruptly (as in a decay, where the sum of the masses of the products is less than the mass of the initial particle). But in many setups, the mass of an object changes continuously. A rocket is the classic example of this, so we'll use the term "rocket motion" to describe the general class of problems where the mass changes continuously.

The relativistic rocket itself encompasses all of the important ideas, so we'll study that example here. Many more examples are left for the problems. We'll present three solutions to the rocket problem, the last of which is rather slick. In the end, the solutions are all basically the same, but it should be helpful to see the various ways of looking at things.

Example (Relativistic rocket): Assume that a rocket propels itself by continually converting mass into photons and firing them out the back. Let m be the instantaneous mass of the rocket, and let v be the instantaneous speed with respect to the ground. Show that

$$\frac{dm}{m} + \frac{dv}{1 - v^2} = 0. \tag{12.66}$$

If the initial mass is M, and the initial v is zero, integrate Eq. (12.66) to obtain

$$m = M\sqrt{\frac{1-v}{1+v}}. \tag{12.67}$$

First solution: The strategy of this solution will be to use conservation of momentum in the ground frame. Consider the effect of a small mass being converted into photons. The mass of the rocket goes from m to $m + dm$ (where dm is negative). So in the frame of the rocket, photons with total energy $E_r = -dm$ (which is positive) are fired out the back. In the frame of the rocket, these photons have momentum $p_r = dm$ (which is negative). We'll drop the c's here.

Let the rocket move at speed v with respect to the ground. Then the momentum of the photons in the ground frame, p_g, can be found via the Lorentz transformation,

$$p_g = \gamma(p_r + vE_r) = \gamma\big(dm + v(-dm)\big) = \gamma(1-v)\,dm. \tag{12.68}$$

This is still negative, of course.

Remark: A common error is to say that the converted mass $(-dm)$ takes the form of photons of energy $(-dm)$ in the ground frame. This is incorrect, because although the photons have energy $(-dm)$ in the rocket frame, they are redshifted (due to the Doppler effect) in the ground frame. From Eq. (11.51), we see that the frequency (and hence the energy) of the photons decreases by a factor of $\sqrt{(1-v)/(1+v)}$ when going from the rocket frame to the ground frame. This factor equals the $\gamma(1-v)$ factor in Eq. (12.68). ♣

We can now use conservation of momentum in the ground frame to say that

$$(m\gamma v)_{\text{old}} = \gamma(1-v)\,dm + (m\gamma v)_{\text{new}} \quad\Longrightarrow\quad \gamma(1-v)\,dm + d(m\gamma v) = 0. \tag{12.69}$$

The $d(m\gamma v)$ term may be expanded to give

$$\begin{aligned} d(m\gamma v) &= (dm)\gamma v + m(d\gamma)v + m\gamma(dv) \\ &= \gamma v\,dm + m(\gamma^3 v\,dv)v + m\gamma\,dv \\ &= \gamma v\,dm + m\gamma(\gamma^2 v^2 + 1)\,dv \\ &= \gamma v\,dm + m\gamma^3\,dv. \end{aligned} \tag{12.70}$$

Therefore, Eq. (12.69) gives

$$\begin{aligned} 0 &= \gamma(1-v)\,dm + \gamma v\,dm + m\gamma^3 dv \\ &= \gamma\,dm + m\gamma^3\,dv. \end{aligned} \tag{12.71}$$

Hence,

$$\frac{dm}{m} + \frac{dv}{1-v^2} = 0, \tag{12.72}$$

in agreement with Eq. (12.66). We must now integrate this. With the given initial values, we have

$$\int_M^m \frac{dm}{m} + \int_0^v \frac{dv}{1-v^2} = 0. \tag{12.73}$$

We can look up the dv integral in a table, but let's instead do it from scratch.[12] Writing $1/(1-v^2)$ as the sum of two fractions gives

$$\begin{aligned} \int_0^v \frac{dv}{1-v^2} &= \frac{1}{2}\int_0^v \left(\frac{1}{1+v} + \frac{1}{1-v}\right) dv \\ &= \frac{1}{2}\Big(\ln(1+v) - \ln(1-v)\Big)\Big|_0^v \\ &= \frac{1}{2}\ln\left(\frac{1+v}{1-v}\right). \end{aligned} \tag{12.74}$$

[12] Tables often list the integral of $1/(1-v^2)$ as $\tanh^{-1}(v)$. You can show that this is equivalent to the result in Eq. (12.74).

Equation (12.73) therefore gives

$$\ln\left(\frac{m}{M}\right) = -\frac{1}{2}\ln\left(\frac{1+v}{1-v}\right) \quad \Longrightarrow \quad m = M\sqrt{\frac{1-v}{1+v}}, \tag{12.75}$$

in agreement with Eq. (12.67). This result is independent of the rate at which the mass is converted into photons. It is also independent of the frequency of the emitted photons. Only the total mass expelled matters. Note that Eq. (12.75) quickly tells us that the energy of the rocket, as a function of velocity, is

$$E = \gamma m = \gamma M\sqrt{\frac{1-v}{1+v}} = \frac{M}{1+v}. \tag{12.76}$$

This has the interesting property of approaching $M/2$ as $v \to c$. In other words, half of the initial energy remains with the rocket, and half ends up as photons (see Exercise 12.38).

Remark: From Eq. (12.68), or from the previous remark, we see that the ratio of the energy of the photons in the ground frame to that in the rocket frame is $\sqrt{(1-v)/(1+v)}$. This factor is the same as the factor in Eq. (12.75). In other words, the photons' energy in the ground frame decreases in exactly the same manner as the mass of the rocket (assuming that the photons are ejected with the same frequency in the rocket frame throughout the process). Therefore, in the ground frame, the ratio of the photons' energy to the mass of the rocket doesn't change with time. There must be a nice intuitive explanation for this, but it eludes me. ♣

Second solution: The strategy of this solution will be to use $F = dp/dt$ in the ground frame. Let τ denote the time in the rocket frame. Then in the rocket frame, $dm/d\tau$ is the rate at which the mass of the rocket decreases and is converted into photons (dm is negative). The photons therefore acquire momentum at the rate $dp/d\tau = dm/d\tau$ in the rocket frame. Since force is the rate of change in momentum, we see that a force of $dm/d\tau$ pushes the photons backward, and so an equal and opposite force of $F = -dm/d\tau$ pushes the rocket forward in the rocket frame.

Now go to the ground frame. We know from Eq. (12.65) that the longitudinal force is the same in both frames, so $F = -dm/d\tau$ is also the force on the rocket in the ground frame. And since $dt = \gamma d\tau$, where t is the time on the ground (the photon emissions occur at the same place in the rocket frame, so we have indeed put the time-dilation factor of γ in the right place), we have

$$F = -\gamma\frac{dm}{dt}. \tag{12.77}$$

Remark: We can also calculate the force on the rocket by working entirely in the ground frame. Consider a mass $(-dm)$ that is converted into photons. Initially, this mass is traveling along with the rocket, so it has momentum $(-dm)\gamma v$. After it is converted into photons, it has momentum $\gamma(1-v)\,dm$ (from the first solution above). The change in momentum is therefore $\gamma(1-v)\,dm - (-dm)\gamma v = \gamma\,dm$. Since force is the rate of change in momentum, a force of $\gamma\,dm/dt$ pushes the photons backward, and so an equal and opposite force of $F = -\gamma\,dm/dt$ pushes the rocket forward. ♣

Now things get a little tricky. It is tempting to write down $F = dp/dt = d(m\gamma v)/dt$, which gives $F = (dm/dt)\gamma v + m\,d(\gamma v)/dt$. This, however, is incorrect, because the

dm/dt term isn't relevant here. When the force is applied to the rocket at an instant when the rocket has mass m, the only thing the force cares about is that the mass of the rocket at the given instant is m. It doesn't care that m is changing.[13] Therefore, the correct expression we want is

$$F = m\frac{d(\gamma v)}{dt}. \qquad (12.78)$$

As in the first solution above, or in Eq. (12.54), we have $d(\gamma v)/dt = \gamma^3\, dv/dt$. Using the F from Eq. (12.77), we arrive at

$$-\gamma\frac{dm}{dt} = m\gamma^3\frac{dv}{dt}, \qquad (12.79)$$

which is equivalent to Eq. (12.71). The solution proceeds as above.

Third solution: The strategy of this solution will be to use conservation of energy and momentum in the ground frame, in a slick way. Consider a clump of photons fired out the back. The energy and momentum of these photons are equal in magnitude and opposite in sign (with the convention that the photons are fired in the negative direction). By conservation of energy and momentum, the same statement must be true about the changes in energy and momentum of the rocket. That is,

$$d(\gamma m) = -d(\gamma m v) \implies d(\gamma m + \gamma m v) = 0. \qquad (12.80)$$

Therefore, $\gamma m(1 + v)$ is a constant. We are given that $m = M$ when $v = 0$. Hence, the constant equals M. Therefore,

$$\gamma m(1+v) = M \implies m = M\sqrt{\frac{1-v}{1+v}}. \qquad (12.81)$$

Now, *that's* a quick solution, if there ever was one!

12.7 Relativistic strings

Consider a "massless" string with a tension that is constant, that is, independent of length.[14] We call such objects *relativistic strings*, and we will study them for two reasons. First, these strings, or reasonable approximations thereof, actually do occur in nature. For example, the gluon force which holds quarks together is approximately constant over distance. And second, they open the door to a whole new class of setups we can study, such as the two below. Relativistic strings might

[13] Said in a different way, the momentum associated with the ejected mass still exists. It's just that it's not part of the rocket anymore; it's in the photons. This issue is expanded on in Appendix C.

[14] By "massless," we mean that the string has no mass in its unstretched (that is, zero-length) state. Once it is stretched, it will have energy in its rest frame, and hence mass.

seem a bit strange, but any one-dimensional problem involving them basically comes down to the two equations,

$$F = \frac{dp}{dt}, \quad \text{and} \quad F = \frac{dE}{dx}. \tag{12.82}$$

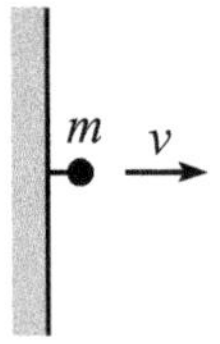

Fig. 12.12

Example (Mass connected to a wall): A mass m is connected to a wall by a relativistic string with tension T. The mass starts next to the wall and has an initial speed v away from it (see Fig. 12.12). How far from the wall does the mass get? How much time does it take to reach this point?

Solution: Let ℓ be the maximum distance from the wall. The initial energy of the mass is $E = \gamma m$. The final energy at $x = \ell$ is simply m, because the mass is instantaneously at rest there. Integrating $F = dE/dx$, and using the fact that the force always equals $-T$, gives

$$F\Delta x = \Delta E \implies (-T)\ell = m - \gamma m \implies \ell = \frac{m(\gamma - 1)}{T}. \tag{12.83}$$

Let t be the time it takes to reach this point. The initial momentum of the mass is $p = \gamma m v$. Integrating $F = dp/dt$, and using the fact that the force always equals $-T$, gives

$$F\Delta t = \Delta p \implies (-T)t = 0 - \gamma m v \implies t = \frac{\gamma m v}{T}. \tag{12.84}$$

Note that we *cannot* use $F = ma$ to do this problem. F does not equal ma. It equals dp/dt (and also dE/dx).

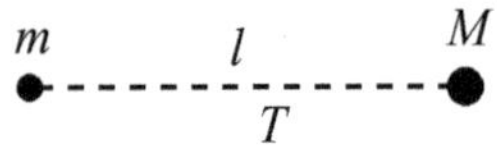

Fig. 12.13

Example (Where the masses meet): A relativistic string of length ℓ and tension T connects a mass m and a mass M (see Fig. 12.13). The masses are released from rest. Where do they meet?

Solution: Let the masses meet at a distance x from the initial position of m. At this meeting point, $F = dE/dx$ tells us that the energy of m is $m + Tx$, and the energy of M is $M + T(\ell - x)$. Using $p = \sqrt{E^2 - m^2}$ we see that the magnitudes of the momenta at the meeting point are

$$p_m = \sqrt{(m + Tx)^2 - m^2} \quad \text{and} \quad p_M = \sqrt{(M + T(\ell - x))^2 - M^2}. \tag{12.85}$$

But $F = dp/dt$ tells us that these must be equal, because the same force (in magnitude, but opposite in direction) acts on the two masses for the same time. Equating the above p's yields

$$x = \frac{\ell\big(M + T(\ell/2)\big)}{M + m + T\ell}. \tag{12.86}$$

This result is reassuring, because it is the location of the initial center of mass, with the string being treated (quite correctly) like a stick of length ℓ and mass $T\ell$ (divided by c^2).

REMARK: Let's check a few limits. In the limit of large T or ℓ (more precisely, in the limit $T\ell \gg Mc^2$ and $T\ell \gg mc^2$), we have $x = \ell/2$. This makes sense, because in this case the masses are negligible and therefore both move at essentially speed c, and hence meet in the middle. In the limit of small T or ℓ (more precisely, in the limit $T\ell \ll Mc^2$ and $T\ell \ll mc^2$), we have $x = M\ell/(M+m)$, which is simply the Newtonian result for an everyday-strength massless spring. ♣

12.8 Problems

Section 12.1: Energy and momentum

12.1. **Deriving *E* and *p*** **

Accepting the facts that the energy and momentum of a photon are $E = h\nu$ and $p = h\nu/c$ (where ν is the frequency of the light wave, and h is Planck's constant), derive the relativistic formulas for the energy and momentum of a massive particle, $E = \gamma mc^2$ and $p = \gamma mv$. *Hint*: Consider a mass m that decays into two photons. Look at this decay in the rest frame of the mass, and then in a frame where the mass has speed v. You'll need to use the Doppler effect.

Section 12.3: Collisions and decays

12.2. **Colliding photons** *

Two photons each have energy E. They collide at an angle θ and create a particle of mass M. What is M?

12.3. **Increase in mass** *

A large mass M, moving at speed V, collides and sticks to a small mass m, initially at rest. What is the mass of the resulting object? Work in the approximation where $M \gg m$.

12.4. **Two-body decay** *

A stationary mass M_A decays into masses M_B and M_C. What are the energies of M_B and M_C? What are their momenta?

12.5. **Threshold energy** **

A particle of mass m and energy E collides with an identical stationary particle. What is the threshold energy for a final state containing N particles of mass m? ("Threshold energy" is the minimum energy for which the process can occur.)

12.6. **Head-on collision** **

A ball of mass M and energy E collides head-on elastically with a stationary ball of mass m. Show that the final energy of mass M is

$$E' = \frac{2mM^2 + E(m^2 + M^2)}{2Em + m^2 + M^2}. \tag{12.87}$$

Hint: This problem is a little messy, but you can save yourself a lot of trouble by noting that $E' = E$ must be a root of the equation you get for E'. (Why?) As a reward for trudging through the mess, there are lots of interesting limits you can take.

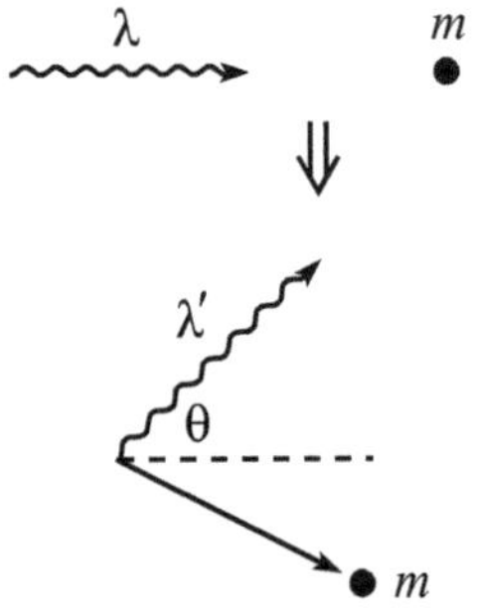

Fig. 12.14

12.7. **Compton scattering** **

A photon collides with a stationary electron. If the photon scatters at an angle θ (see Fig. 12.14), show that the resulting wavelength, λ', is given in terms of the original wavelength, λ, by

$$\lambda' = \lambda + \frac{h}{mc}(1 - \cos\theta), \tag{12.88}$$

where m is the mass of the electron. *Note*: The energy of a photon is $E = h\nu = hc/\lambda$.

Section 12.5: Force

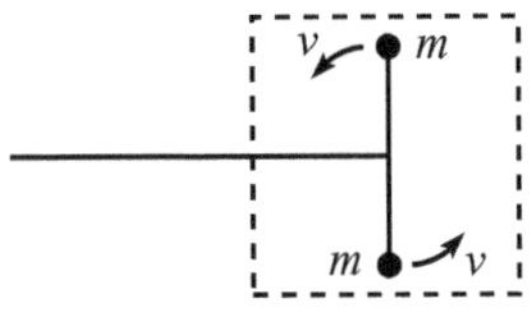

Fig. 12.15

12.8. **System of masses** **

Consider a dumbbell made of two equal masses, m. The dumbbell spins around, with its center pivoted at the end of a stick (see Fig. 12.15). If the speed of the masses is v, then the energy of the system is $2\gamma m$. Treated as a whole, the system is at rest. Therefore, the mass of the system must be $2\gamma m$. (Imagine enclosing it in a box, so that you can't see what's going on inside.) Convince yourself that the system does indeed behave like a mass of $M = 2\gamma m$, by pushing on the stick (when the dumbbell is in the "transverse" position shown in the figure) and showing that $F \equiv dp/dt = Ma$.

12.9. **Relativistic harmonic oscillator** **

A particle of mass m moves along the x axis under a force $F = -m\omega^2 x$. The amplitude is b. Show that the period is given by

$$T = \frac{4}{c}\int_0^b \frac{\gamma}{\sqrt{\gamma^2 - 1}}\,dx, \quad \text{where} \quad \gamma = 1 + \frac{\omega^2}{2c^2}(b^2 - x^2). \tag{12.89}$$

Section 12.6: Rocket motion

12.10. **Relativistic rocket** **

Consider the relativistic rocket in Section 12.6. Let mass be converted into photons at a rate σ in the rest frame of the rocket. Find the time t in the ground frame as a function of v. (Alas, it isn't possible to invert this, to obtain v as a function of t.) You'll need to evaluate a slightly tricky integral. Pick your favorite method – pencil, book, or computer.

12.11. **Relativistic dustpan I** *

A dustpan of mass M is given an initial relativistic speed. It gathers up dust with mass density λ per unit length on the floor (as measured in the lab frame). At the instant the speed is v, find the rate (as measured in the lab frame) at which the mass of the dustpan-plus-dust-inside system is increasing.

12.12. **Relativistic dustpan II** **

Consider the setup in Problem 12.11. If the initial speed of the dustpan is V, find $v(x)$, $v(t)$, and $x(t)$. All quantities here are measured with respect to the lab frame.

12.13. **Relativistic dustpan III** **

Consider the setup in Problem 12.11. Calculate, in both the dustpan frame and the lab frame, the force on the dustpan-plus-dust-inside system (due to the newly acquired dust particles smashing into it) as a function of v, and show that the results are equal.

12.14. **Relativistic cart I** ****

A long cart moves at relativistic speed v. Sand is dropped into the cart at a rate $dm/dt = \sigma$ in the ground frame. Assume that you stand on the ground next to where the sand falls in, and you push on the cart to keep it moving at constant speed v. What is the force between your feet and the ground? Calculate this force in both the ground frame (your frame) and the cart frame, and show that the results are equal.

12.15. **Relativistic cart II** ****

A long cart moves at relativistic speed v. Sand is dropped into the cart at a rate $dm/dt = \sigma$ in the ground frame. Assume that you grab the front of the cart and pull on it to keep it moving at constant speed v (while running with it). What force does your hand apply to the cart? (Assume that the cart is made of the most rigid material possible.) Calculate this force in both the ground frame and the cart frame (your frame), and show that the results are equal.

Section 12.7: Relativistic strings

12.16. **Different frames** $**$

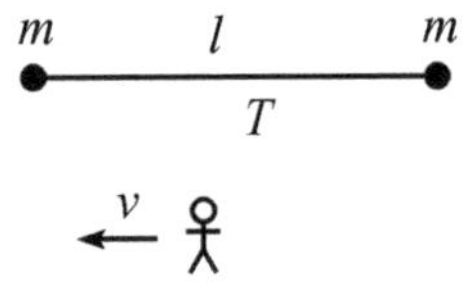

Fig. 12.16

(a) Two masses m are connected by a string of length ℓ and constant tension T. The masses are released simultaneously, and they collide and stick together. What is the mass, M, of the resulting blob?

(b) Consider this scenario from the point of view of a frame moving to the left at speed v (see Fig. 12.16). The energy of the resulting blob must be γMc^2, from part (a). Show that you obtain this same result by computing the work done on the two masses.

12.17. **Splitting mass** $**$

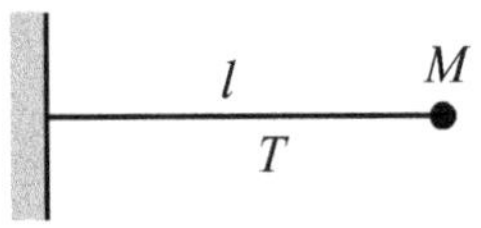

Fig. 12.17

A massless string with constant tension T has one end attached to a wall and the other end attached to a mass M. The initial length of the string is ℓ (see Fig. 12.17). The mass is released. Halfway to the wall, the back half of the mass breaks away from the front half (with zero initial relative speed). What is the total time it takes the front half to reach the wall?

12.18. **Relativistic leaky bucket** $***$

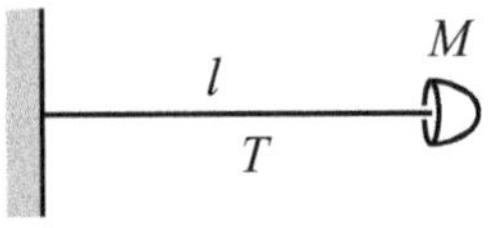

Fig. 12.18

Let the mass M in Problem 12.17 be replaced by a massless bucket containing an initial mass M of sand (see Fig. 12.18). On the way to the wall, the bucket leaks sand at a rate $dm/dx = M/\ell$, where m denotes the mass at later positions. Note that dm and dx are both negative here.

(a) What is the energy of the bucket, as a function of distance from the wall? What is its maximum value? What is the maximum value of the kinetic energy?

(b) What is the momentum of the bucket, as a function of distance from the wall? Where is it maximum?

12.19. **Relativistic bucket** $***$

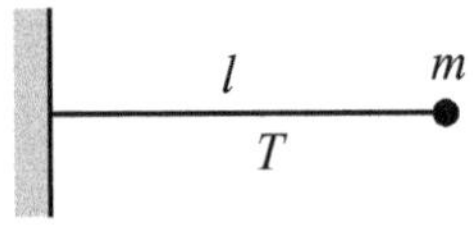

Fig. 12.19

(a) A massless string with constant tension T has one end attached to a wall and the other end attached to a mass m. The initial length of the string is ℓ (see Fig. 12.19). The mass is released. How long does it take to reach the wall?

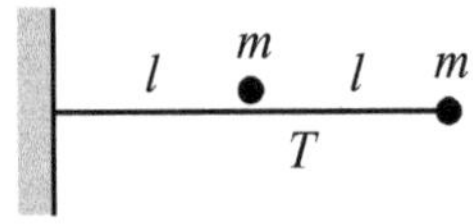

Fig. 12.20

(b) Let the string now have length 2ℓ, with a mass m on the end. Let another mass m be positioned next to the ℓ mark on the string, but not touching it (see Fig. 12.20). The right mass is released. It heads toward the wall (while the left mass remains motionless) and then sticks to the left mass to make one large blob, which then heads toward the wall.[15] How much time does this whole process take? *Hint*: You can solve this in various ways, but one method that generalizes nicely for part (c) is to show that the change in p^2 from the start to a point right before the wall is $\Delta(p^2) = (E_2^2 - E_1^2) + (E_4^2 - E_3^2)$, where the energies of the moving object (that is, the initial m or the resulting blob) are: E_1 right at the start, E_2 just before the collision, E_3 just after the collision, and E_4 right before the wall. Note that this method doesn't require knowledge of the mass of the blob (which is *not* $2m$).

(c) Let there now be N masses and a string of length $N\ell$ (see Fig. 12.21). How much time does this whole process take?

(d) Consider now a massless bucket at the end of the string (of length L) that gathers up a continuous stream of sand (of total mass M) as it gets pulled to the wall (see Fig. 12.22). How much time does this whole process take? What are the mass and speed of the bucket right before it hits the wall?

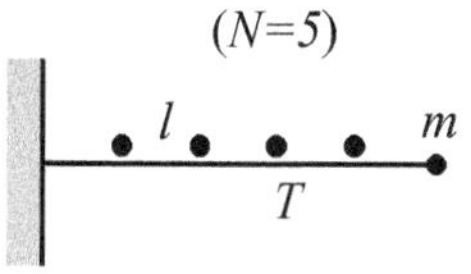

Fig. 12.21

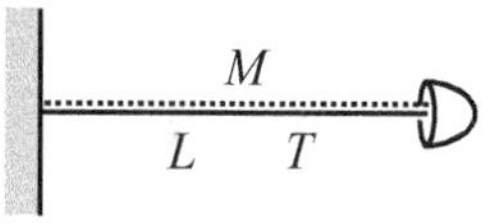

Fig. 12.22

12.9 Exercises

Section 12.2: Transformations of E and p

12.20. **Energy of two masses** $*$

Two masses M move at speed V, one to the east and one to the west. What is the total energy of the system? Now consider the setup as viewed in a frame moving to the west at speed u. Find the energy of each mass in this frame. Is the total energy larger or smaller than the total energy in the lab frame?

12.21. **System of particles** $*$

Given p_{total} and E_{total} for a system of particles, use a Lorentz transformation to find the velocity of the CM. More precisely, find the speed of the frame in which the total momentum is zero.

12.22. **CM frame** $**$

A mass m travels at speed $3c/5$, and another mass m sits at rest.

(a) Find the energy and momentum of the two particles in the lab frame.

(b) Find the speed of the CM of the system, by using a velocity-addition argument.

[15] The left mass could actually be attached to the string, and we would still have the same situation. The mass wouldn't move during the first part of the process, because there would be equal tensions T on both sides of it.

(c) Find the energy and momentum of the two particles in the CM frame, without using the Lorentz transformations.
(d) Verify that the E's and p's are related by the relevant Lorentz transformations.
(e) Verify that $E^2 - p^2c^2$ for each mass is the same in both frames. Likewise for $E^2_{\text{total}} - p^2_{\text{total}}c^2$.

12.23. **Transformations for 2-D motion** **

A particle has velocity (u'_x, u'_y) in frame S', which travels at speed v in the x direction relative to frame S. Use the velocity-addition formulas in Section 11.5.2 (Eqs. (11.36) and (11.38)) to show that

$$\gamma_u = \gamma_{u'}\gamma_v(1 + u'_x v), \quad \text{where } u = \sqrt{u_x^2 + u_y^2} \text{ and } u' = \sqrt{u_x'^2 + u_y'^2} \qquad (12.90)$$

are the speeds in the two frames. Then verify that E and p_x transform according to Eqs. (12.27), and also that $p_y = p'_y$.

Section 12.3: Collisions and decays

12.24. **Photon and mass collision** *

A photon with energy E collides with a stationary mass m. They combine to form one particle. What is the mass of this particle? What is its speed?

12.25. **A decay** *

A stationary mass M decays into a particle and a photon. If the speed of the particle is v, what is its mass? What is the energy of the photon?

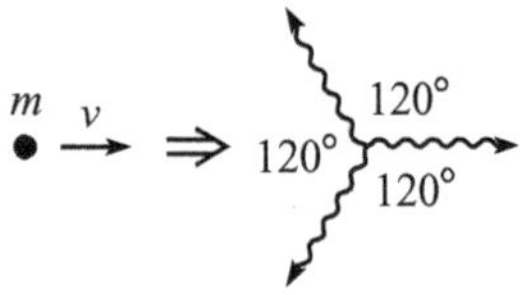

Fig. 12.23

12.26. **Three photons** *

A mass m moves at speed v. It decays into three photons, one of which travels in the forward direction, and the other two of which move at angles of 120° (in the lab frame) as shown in Fig. 12.23. What are the energies of these three photons?

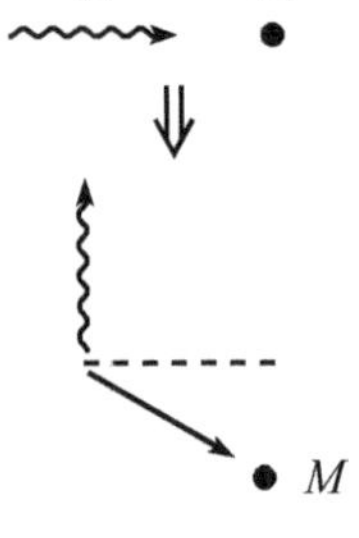

Fig. 12.24

12.27. **Perpendicular photon** *

A photon with energy E collides with a mass M. The mass M scatters at an angle. If the resulting photon moves perpendicular to the incident photon's direction, as shown in Fig. 12.24, what is its energy?

12.28. **Another perpendicular photon** *

A mass m moving at speed $4c/5$ collides with another mass m at rest. The collision produces a photon with energy E traveling perpendicular to the original direction, and a mass M traveling in another direction, as shown in Fig. 12.25. In terms of E and m, what is M? What is the largest value of E (in terms of m) for which this setup is possible?

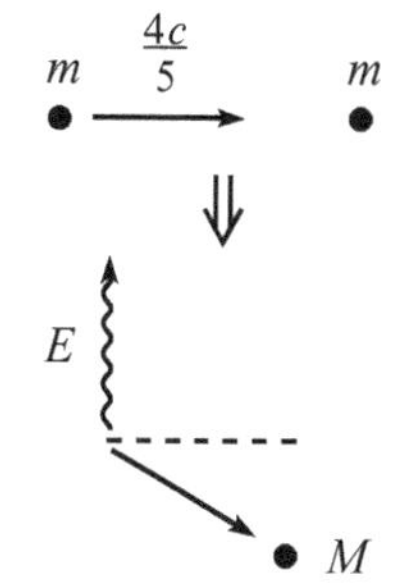

Fig. 12.25

12.29. **Decay into photons** $*$

A mass m moving at speed v decays into two photons. One photon moves perpendicular to the original direction, and the other photon moves off at an angle θ, as shown in Fig. 12.26. Show that if $\tan\theta = 1/2$, then $v/c = (\sqrt{5} - 1)/2$, which just happens to be the inverse of the golden ratio.

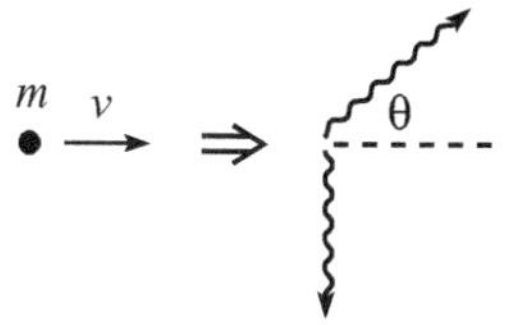

Fig. 12.26

12.30. **Maximum mass** $*$

A photon and a mass m move directly toward each other. They collide head-on and create a new particle. If the total energy of the system is E, how should it be divided between the photon and the mass m so that the mass of the resulting particle is as large as possible?

12.31. **Equal angles** $*$

A photon with energy E collides with a stationary mass m. If the mass m and the resulting photon (with unknown energy) scatter at equal angles θ with respect to the initial photon direction, as shown in Fig. 12.27, what is θ in terms of E and m? What is θ in the limit $E \ll mc^2$?

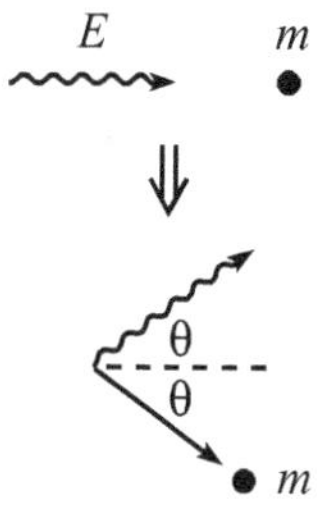

Fig. 12.27

Section 12.4: Particle-physics units

12.32. **Pion–muon race** $*$

A pion and a muon have a 100 m race. If they both have an energy of 10 GeV, by how much distance does the muon win?

12.33. **Higgs production** $*$

The *Higgs boson* is a proposed elementary particle that should be experimentally detected in a few years, assuming it exists. One strategy for producing it in a high-energy particle accelerator is to collide protons and antiprotons together. Taking the rest energy of a proton (and antiproton) to be about 1 GeV, and assuming that the rest energy of the Higgs is about 100 GeV, how much energy is required to produce the Higgs if:

(a) A moving proton collides with a stationary antiproton?
(b) A proton and antiproton have equal and opposite momenta?

12.34. **Maximum energy** $**$

(a) A particle of mass M decays into a number of particles, some of which may be photons. If one of the particles has mass m, and if the sum of the masses of all the other products is μ, what is the maximum possible energy that m can have? *Hint*: Write the conservation of energy and momentum statements as $P_M - P_m = P_\mu$, where P_μ is the total 4-momentum of the other products, and then square. The technique of Problem 12.5 may be useful.

(b) In beta decay, a neutron decays into a proton, an electron, and a neutrino (which is essentially a photon, for the present purposes). The rest energies are $E_{\text{n}} = 939.6$ MeV, $E_{\text{p}} = 938.3$ MeV, $E_{\text{e}} = 0.5$ MeV, and $E_\nu \approx 0$. What is the maximum energy that the electron can have? The neutrino? Interpret your results.

Section 12.5: Force

12.35. **Force and a collision** $*$

Two identical masses m are initially at rest, a distance x apart. A constant force F accelerates one of them toward the other until they collide and stick together. What is the mass of the resulting particle?

12.36. **Pushing on a mass** $**$

(a) A mass m starts at rest. You push on it with a constant force F. How much time t does it take for the mass to move a distance x? (Both t and x here are measured in the lab frame.)

(b) After a very long time, m's speed will approach c. It turns out that it approaches c sufficiently fast so that after a very long time, m will remain (approximately) a constant distance (as measured in the lab frame) behind a photon that was emitted at $t = 0$ from the starting position of m. What is this distance?

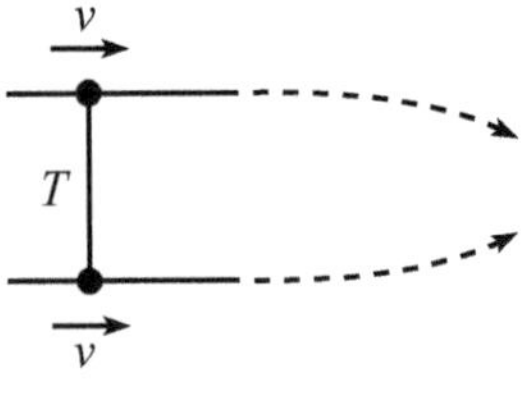

Fig. 12.28

12.37. **Momentum paradox** $****$

Two equal masses are connected by a massless string with tension T. The masses are constrained to move with speed v along parallel lines, as shown in Fig. 12.28. The constraints are then removed, and the masses are drawn together. They collide and make one blob which continues to move to the right. Is the following reasoning correct? If your answer is "no," state what is invalid about whichever of the four sentences is/are invalid.

"The forces on the masses point in the y direction. Therefore, there is no change in the momentum of the masses in the x direction. But the mass of the resulting blob is greater than the sum of the initial masses (because they collide with some relative speed). Therefore, the speed of the resulting blob must be less than v (to keep p_x constant), so the whole apparatus slows down in the x direction."

Section 12.6: Rocket motion

12.38. **Rocket energy** $**$

As mentioned near the end of the first solution to the rocket problem in Section 12.6, the energy of the rocket in the ground frame equals $M/(1+v)$. Derive this result again, by integrating up the amount of energy that the photons have in the ground frame.

Section 12.7: Relativistic strings

12.39. **Two masses** $*$

A mass m is placed right in front of an identical one. They are connected by a relativistic string with tension T. The front mass suddenly acquires a speed $3c/5$. How far from the starting point will the masses collide with each other?

12.40. **Relativistic bucket** $**$

One of the results in part (d) of Problem 12.19 is that the bucket moves toward the wall at constant speed $\sqrt{T/(T+\rho)}$. Derive this again, without using the technique of taking the $N \to \infty$ limit of many masses.

12.10 Solutions

12.1. **Deriving E and p**

Let's derive the energy formula, $E = \gamma mc^2$, first. Let the given mass decay into two photons, and let E_0 be the energy of the mass in its rest frame. Then each of the resulting photons has energy $E_0/2$ in this frame.

Now look at the decay in a frame where the mass moves at speed v. From Eq. (11.51), the frequencies of the photons are Doppler-shifted by the factors $\sqrt{(1+v)/(1-v)}$ and $\sqrt{(1-v)/(1+v)}$. Since the energies of the photons are given by $E = h\nu$, they are shifted by the same Doppler factors, relative to the $E_0/2$ value in the original frame. The total energy of the photons in the frame where the mass moves at speed v is therefore

$$E = \frac{E_0}{2}\sqrt{\frac{1+v}{1-v}} + \frac{E_0}{2}\sqrt{\frac{1-v}{1+v}} = \frac{E_0}{\sqrt{1-v^2}} \equiv \gamma E_0. \tag{12.91}$$

By conservation of energy, this is the energy of the mass m moving at speed v. So we see that a moving mass has an energy that is γ times its rest energy.

We can now use the correspondence principle (which says that relativistic formulas must reduce to the familiar nonrelativistic ones in the nonrelativistic limit) to find E_0 in terms of m and c. We just found that the difference between the energies of a moving mass and a stationary mass is $\gamma E_0 - E_0$. This must reduce to the familiar kinetic energy, $mv^2/2$, in the limit $v \ll c$. In other words,

$$\frac{mv^2}{2} \approx \frac{E_0}{\sqrt{1-v^2/c^2}} - E_0 \approx E_0\left(1+\frac{v^2}{2c^2}\right) - E_0 = \left(\frac{E_0}{c^2}\right)\frac{v^2}{2}, \tag{12.92}$$

where we have used the Taylor series, $1/\sqrt{1-\epsilon} \approx 1+\epsilon/2$. Therefore $E_0 = mc^2$, and so $E = \gamma mc^2$.

We can derive the momentum formula, $p = \gamma m v$, in a similar way. Let the magnitude of the photons' (equal and opposite) momenta in the particle's rest frame be $p_0/2$.[16] Since the photons' momenta are given by $E = h\nu/c$, we can use the Doppler-shifted frequencies as we did above to say that the total momentum of the photons in the frame where the mass moves at speed v is

$$p = \frac{p_0}{2}\sqrt{\frac{1+v}{1-v}} - \frac{p_0}{2}\sqrt{\frac{1-v}{1+v}} = \gamma p_0 v. \tag{12.93}$$

Putting the c's back in, we have $p = \gamma p_0 v/c$. By conservation of momentum, this is the momentum of the mass m moving at speed v.

We can now use the correspondence principle to find p_0 in terms of m and c. If $p = \gamma(p_0/c)v$ is to reduce to the familiar $p = mv$ result in the limit $v \ll c$, then we must have $p_0 = mc$. Therefore, $p = \gamma m v$.

12.2. **Colliding photons**

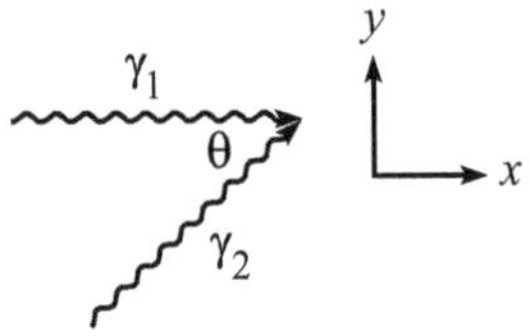

Fig. 12.29

The 4-momenta of the photons are (see Fig. 12.29)

$$P_{\gamma_1} = (E, E, 0, 0), \quad \text{and} \quad P_{\gamma_2} = (E, E\cos\theta, E\sin\theta, 0). \tag{12.94}$$

Energy and momentum are conserved, so the 4-momentum of the final particle is $P_M = (2E, E + E\cos\theta, E\sin\theta, 0)$. Therefore,

$$M^2 = P_M \cdot P_M = (2E)^2 - (E + E\cos\theta)^2 - (E\sin\theta)^2, \tag{12.95}$$

which gives

$$M = E\sqrt{2(1-\cos\theta)}\,. \tag{12.96}$$

If $\theta = 180°$ then $M = 2E$, as it should (none of the final energy is kinetic). And if $\theta = 0°$ then $M = 0$, as it should (all of the final energy is kinetic; we simply have a photon with twice the energy).

12.3. **Increase in mass**

In the lab frame, the energy of the resulting object is $\gamma M + m$, and the momentum is $\gamma M V$. The mass of the object is therefore

$$M' = \sqrt{(\gamma M + m)^2 - (\gamma M V)^2} = \sqrt{M^2 + 2\gamma M m + m^2}\,. \tag{12.97}$$

The m^2 term is negligible compared with the other two terms, so we may approximate M' as

$$M' \approx M\sqrt{1 + \frac{2\gamma m}{M}} \approx M\left(1 + \frac{\gamma m}{M}\right) = M + \gamma m, \tag{12.98}$$

where we have used the Taylor series, $\sqrt{1+\epsilon} \approx 1 + \epsilon/2$. Therefore, the increase in mass is γ times the mass of the stationary object. This increase is greater than the nonrelativistic answer of "m," because heat is generated during the collision, and this heat shows up as mass in the final object.

REMARK: The γm result is clear if we work in the frame where M is initially at rest. In this frame, the mass m comes flying in with energy γm, and then essentially all of this energy shows up as mass in the final object. That is, essentially none of it shows up as overall kinetic energy of the object. This negligible-kinetic-energy result is a general one whenever a small object hits a stationary large one. It follows from the fact that the speed of the large object is proportional to m/M, by momentum

[16] With the given information that a photon has $E = h\nu$ and $p = h\nu/c$, we can use the preceding $E_0 = mc^2$ result to quickly conclude that $p_0 = mc$. But let's pretend that we haven't found E_0 yet. This will give us an excuse to use the correspondence principle again.

conservation (there's a factor of γ if things are relativistic), so the kinetic energy goes like $Mv^2 \propto M(m/M)^2 \approx 0$, if $M \gg m$. In other words, the smallness of v wins out over the largeness of M. When a snowball hits a tree, all (essentially) of the initial energy goes into heat. None of it goes into changing the kinetic energy of the earth. ♣

12.4. **Two-body decay**

B and C have equal and opposite momenta. Therefore,

$$E_B^2 - M_B^2 = p^2 = E_C^2 - M_C^2. \tag{12.99}$$

Also, conservation of energy gives

$$E_B + E_C = M_A. \tag{12.100}$$

Solving the two previous equations for E_B and E_C gives (using the shorthand $a \equiv M_A$, etc.)

$$E_B = \frac{a^2 + b^2 - c^2}{2a}, \quad \text{and} \quad E_C = \frac{a^2 + c^2 - b^2}{2a}. \tag{12.101}$$

Equation (12.99) then gives the momentum of the particles as

$$p = \frac{1}{2a}\sqrt{a^4 + b^4 + c^4 - 2a^2b^2 - 2a^2c^2 - 2b^2c^2}. \tag{12.102}$$

REMARK: It turns out that the quantity under the radical can be factored into

$$(a + b + c)(a + b - c)(a - b + c)(a - b - c). \tag{12.103}$$

This makes it clear that if $a = b + c$, then $p = 0$, because there is no leftover energy for the particles to be able to move. Interestingly, the form of p in Eq. (12.102) looks very much like the area of a triangle with sides a, b, c, which is given by Heron's formula as

$$A = \frac{1}{4}\sqrt{2a^2b^2 + 2a^2c^2 + 2b^2c^2 - a^4 - b^4 - c^4}. \quad ♣ \tag{12.104}$$

12.5. **Threshold energy**

The initial 4-momenta in the lab frame are

$$(E, p, 0, 0), \quad \text{and} \quad (m, 0, 0, 0), \tag{12.105}$$

where $p = \sqrt{E^2 - m^2}$. Therefore, the total 4-momentum of the final particles in the lab frame is $(E + m, p, 0, 0)$. The quantity $E_{\text{total}}^2 - p_{\text{total}}^2$ is frame independent, and it equals the square of the energy in the CM frame (where $p = 0$). So we have, with the subscript "f" for "final,"

$$(E + m)^2 - \left(\sqrt{E^2 - m^2}\right)^2 = \left(E_{\text{f}}^{\text{CM}}\right)^2$$
$$\Longrightarrow \quad 2Em + 2m^2 = \left(E_{\text{f}}^{\text{CM}}\right)^2. \tag{12.106}$$

We see that minimizing E is equivalent to minimizing E_{f}^{CM}. But E_{f}^{CM} is clearly minimized when all the final particles are at rest in the CM frame, so that there is no kinetic energy added to the rest energy. So the energy in the CM frame at threshold is simply the sum of the rest energies. In other words, $E_{\text{f}}^{\text{CM}} = Nm$. We therefore have

$$2Em + 2m^2 = (Nm)^2 \quad \Longrightarrow \quad E = \left(\frac{N^2}{2} - 1\right)m. \tag{12.107}$$

Since there is no relative motion among the final particles in the CM frame at threshold there is also no relative motion in any other frame. This means that at threshold the N masses travel together as a blob in the lab frame. The threshold E is larger than the naive answer of $(N-1)m$, because the final state has inevitable "wasted" energy in the form of kinetic energy (which is necessitated by conservation of momentum) in the lab frame. Note that $E \propto N^2$ for large N.

12.6. **Head-on collision**

The 4-momenta before the collision are

$$P_M = (E, p, 0, 0), \quad P_m = (m, 0, 0, 0), \tag{12.108}$$

where $p = \sqrt{E^2 - M^2}$. The 4-momenta after the collision are

$$P'_M = (E', p', 0, 0), \quad P'_m = \text{(we won't need this)}, \tag{12.109}$$

where $p' = \sqrt{E'^2 - M^2}$. Conservation of energy and momentum give $P_M + P_m = P'_M + P'_m$. Therefore,

$$\begin{aligned}
P'^2_m &= (P_M + P_m - P'_M)^2 \qquad (12.110)\\
\Longrightarrow \quad P'^2_m &= P^2_M + P^2_m + P'^2_M + 2P_m \cdot (P_M - P'_M) - 2P_M \cdot P'_M\\
\Longrightarrow \quad m^2 &= M^2 + m^2 + M^2 + 2m(E - E') - 2(EE' - pp')\\
\Longrightarrow \quad -pp' &= M^2 - EE' + m(E - E')\\
\Longrightarrow \quad 0 &= \Big((M^2 - EE') + m(E - E')\Big)^2 - \Big(\sqrt{E^2 - M^2}\sqrt{E'^2 - M^2}\Big)^2\\
\Longrightarrow \quad 0 &= M^2(E^2 - 2EE' + E'^2) + 2(M^2 - EE')m(E - E') + m^2(E - E')^2.
\end{aligned}$$

As claimed, $E' = E$ is a root of this equation, as it must be, because $E' = E$ and $p' = p$ certainly satisfy conservation of energy and momentum with the initial conditions, by definition. Dividing through by $(E - E')$ gives

$$M^2(E - E') + 2m(M^2 - EE') + m^2(E - E') = 0. \tag{12.111}$$

Solving for E' gives the desired result,

$$E' = \frac{2mM^2 + E(m^2 + M^2)}{2Em + m^2 + M^2}. \tag{12.112}$$

REMARK: Let's look at some limits:

1. $E \approx M$ (barely moving): then $E' \approx M$, because M is still barely moving.
2. $M = m$: then $E' = M$, because M stops, and m picks up all the energy that M had.
3. $m \gg E\ (> M)$ (brick wall): then $E' \approx E$, because the heavy mass m picks up essentially no energy.
4. $(E >)\ M \gg m$ and $M^2 \gg Em$: then $E' \approx E$, because it's essentially like m isn't there.
5. $(E >)\ M \gg m$ but $Em \gg M^2$: then $E' \approx M^2/2m$. This isn't obvious, but it's interesting that it doesn't depend on E. This means that no matter how fast you throw a big object at a small one (head-on), the big one will always (as long as you throw it hard enough) end up with the same energy, $M^2/2m$. And since

we're assuming $M^2 \ll Em$, this resulting energy is much less than E. So most of the (very large) initial energy ends up in m.

6. $E \gg m \gg M$: then $E' \approx m/2$. This isn't obvious, but it's similar to an analogous limit in the Compton scattering in Problem 12.7. As above, no matter how fast you throw a small object at a big one (head-on), the small one will always (as long as you throw it hard enough) end up with the same energy, $m/2$. ♣

12.7. **Compton scattering**

The 4-momenta before the collision are (see Fig. 12.30)

$$P_\gamma = \left(\frac{hc}{\lambda}, \frac{hc}{\lambda}, 0, 0\right), \quad P_m = (mc^2, 0, 0, 0). \tag{12.113}$$

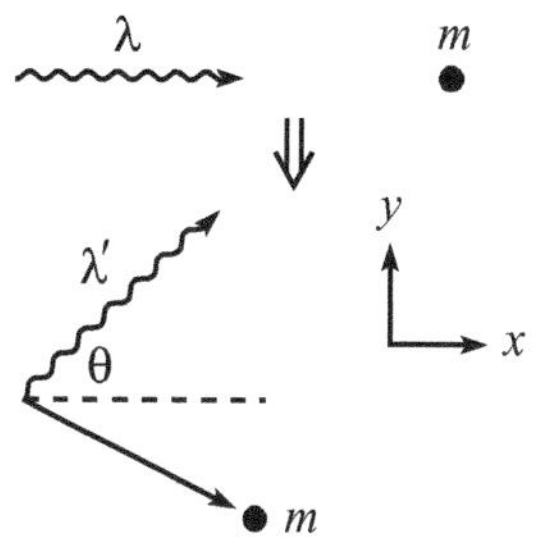

Fig. 12.30

The 4-momenta after the collision are

$$P'_\gamma = \left(\frac{hc}{\lambda'}, \frac{hc}{\lambda'}\cos\theta, \frac{hc}{\lambda'}\sin\theta, 0\right), \quad P'_m = (\text{we won't need this}). \tag{12.114}$$

If we wanted to, we could write P'_m in terms of its momentum and scattering angle. But we're not concerned with these quantities, and the nice thing about the following method is that we don't need to introduce them. Conservation of energy and momentum give $P_\gamma + P_m = P'_\gamma + P'_m$. Therefore,

$$\begin{aligned} P'^2_m &= (P_\gamma + P_m - P'_\gamma)^2 \\ \Longrightarrow \quad P'^2_m &= P^2_\gamma + P^2_m + P'^2_\gamma + 2P_m \cdot (P_\gamma - P'_\gamma) - 2P_\gamma \cdot P'_\gamma \\ \Longrightarrow \quad m^2c^4 &= 0 + m^2c^4 + 0 + 2mc^2\left(\frac{hc}{\lambda} - \frac{hc}{\lambda'}\right) - 2\frac{hc}{\lambda}\frac{hc}{\lambda'}(1-\cos\theta). \end{aligned} \tag{12.115}$$

Canceling the m^2c^4 terms and multiplying through by $\lambda\lambda'/(2hmc^3)$ gives the desired result,

$$\lambda' = \lambda + \frac{h}{mc}(1-\cos\theta). \tag{12.116}$$

The nice thing about this solution is that all the unknown garbage in P'_m disappeared when we squared it.

REMARK: Let's look at some limits:

1. If $\theta \approx 0$ (that is, not much scattering), then $\lambda' \approx \lambda$, as expected.
2. If $\theta = \pi$ (that is, backward scattering), then $\lambda' = \lambda + 2h/mc$.
3. If $\theta = \pi$ and additionally $\lambda \ll h/mc$ (that is, $mc^2 \ll hc/\lambda = E_\gamma$, so the photon's energy is much larger than the electron's rest energy), then $\lambda' \approx 2h/mc$, so

$$E'_\gamma = \frac{hc}{\lambda'} \approx \frac{hc}{2h/mc} = \frac{1}{2}mc^2. \tag{12.117}$$

Therefore, the photon bounces back with an essentially fixed E'_γ, independent of the initial E_γ (as long as E_γ is large enough). This isn't obvious. In fact, it's not even obvious that the photon *can* bounce straight back. You might think that if it has enough energy, it will end up moving forward along with the electron. However, in the CM frame, the photon bounces backward in a head-on collision. So it must bounce backward in every frame, because a photon's direction can't switch when going from one frame to another. ♣

12.8. **System of masses**

Let the speed of the stick go from zero to ϵ, where $\epsilon \ll v$. Then the final speeds of the two masses are obtained by relativistically adding or subtracting ϵ from v. (Assume that the time involved is small, so that the masses are still essentially moving horizontally.) Repeating the derivation leading to Eq. (12.26), we see that the final momenta of the two masses have magnitudes $\gamma_v \gamma_\epsilon (v \pm \epsilon)m$. But since ϵ is small, we can set $\gamma_\epsilon \approx 1$, to first order. Therefore, the forward-moving mass has momentum $\gamma_v(v + \epsilon)m$, and the backward-moving mass has momentum $-\gamma_v(v - \epsilon)m$. So the net increase in momentum is $\Delta p = 2\gamma m\epsilon$, where $\gamma \equiv \gamma_v$. Hence,

$$F \equiv \frac{\Delta p}{\Delta t} = 2\gamma m \frac{\epsilon}{\Delta t} \equiv 2\gamma m a = Ma. \qquad (12.118)$$

12.9. **Relativistic harmonic oscillator**

$F = dp/dt$ gives $-m\omega^2 x = d(m\gamma v)/dt$. Using Eq. (12.54), we have

$$-\omega^2 x = \gamma^3 \frac{dv}{dt}. \qquad (12.119)$$

We must somehow solve this differential equation. A helpful thing to do is to multiply both sides by v (which is equivalent to rewriting dv/dt as $v\,dv/dx$) to obtain $-\omega^2 x\dot{x} = \gamma^3 v\dot{v}$. But from Eq. (12.53), the right-hand side of this is $d\gamma/dt$. Integration then gives $-\omega^2 x^2/2 + C = \gamma$, where C is a constant of integration. We know that $\gamma = 1$ when $x = b$, so we find

$$\gamma = 1 + \frac{\omega^2}{2c^2}(b^2 - x^2), \qquad (12.120)$$

where we have put the c's back in to make the units right. The period is given by

$$T = 4\int_0^b \frac{dx}{v}. \qquad (12.121)$$

But $\gamma \equiv 1/\sqrt{1 - v^2/c^2}$, which gives $v = c\sqrt{\gamma^2 - 1}/\gamma$. Therefore,

$$T = \frac{4}{c}\int_0^b \frac{\gamma}{\sqrt{\gamma^2 - 1}}\,dx. \qquad (12.122)$$

REMARK: In the limit $\omega b \ll c$ (so that $\gamma \approx 1$, from Eq. (12.120), which means that the speed is always small), we must recover the Newtonian limit. And indeed, to lowest order, $\gamma^2 \approx 1 + (\omega^2/c^2)(b^2 - x^2)$, so Eq. (12.122) gives

$$T \approx \frac{4}{c}\int_0^b \frac{dx}{(\omega/c)\sqrt{b^2 - x^2}}. \qquad (12.123)$$

This is the correct result, because conservation of energy for a nonrelativistic spring gives

$$\frac{1}{2}k(b^2 - x^2) = \frac{1}{2}mv^2 \implies \omega^2(b^2 - x^2) = v^2. \qquad (12.124)$$

Using this v in the general expression for T given in Eq. (12.121) yields Eq. (12.123). ♣

12.10. **Relativistic rocket**

The relation between m and v obtained in Eq. (12.67) is independent of the rate at which mass is converted into photons. The point of this problem is to assume a certain rate, in order to obtain a relation between v and t.

In the frame of the rocket, we are given $dm = -\sigma\, d\tau$. The usual time-dilation effect gives $dt = \gamma\, d\tau$, so we have $dm = -(\sigma/\gamma)\, dt$ in the ground frame. Differentiating Eq. (12.67) to obtain another expression for dm, we have

$$dm = \frac{-M\, dv}{(1+v)\sqrt{1-v^2}}\,. \tag{12.125}$$

Equating the two expressions for dm gives

$$\int_0^t \frac{\sigma\, dt}{M} = \int_0^v \frac{dv}{(1+v)(1-v^2)}\,. \tag{12.126}$$

We could use a computer to do this dv integral, but let's do it from scratch. Using a few partial-fraction tricks, we have

$$\begin{aligned}
\int \frac{dv}{(1+v)(1-v^2)} &= \int \frac{dv}{(1+v)(1-v)(1+v)} \\
&= \frac{1}{2}\int \left(\frac{1}{1+v} + \frac{1}{1-v}\right)\frac{dv}{1+v} \\
&= \frac{1}{2}\int \frac{dv}{(1+v)^2} + \frac{1}{4}\int \left(\frac{1}{1+v} + \frac{1}{1-v}\right) dv \\
&= -\frac{1}{2(1+v)} + \frac{1}{4}\ln\left(\frac{1+v}{1-v}\right).
\end{aligned} \tag{12.127}$$

Equation (12.126) therefore gives

$$\frac{\sigma t}{M} = \frac{1}{2} - \frac{1}{2(1+v)} + \frac{1}{4}\ln\left(\frac{1+v}{1-v}\right) = \frac{v}{2(1+v)} + \frac{1}{4}\ln\left(\frac{1+v}{1-v}\right). \tag{12.128}$$

REMARKS: If $v \ll 1$ (or rather, if $v \ll c$), we can Taylor-expand the two terms in Eq. (12.128) to obtain $\sigma t/M \approx v$, which can be written as $\sigma \approx M(v/t) \equiv Ma$. But σ equals the force acting on the rocket (or rather σc, to make the units correct), because $-\sigma$ is the rate of change in momentum of the photons (since their momentum is $p = -E/c = -(dm\, c^2)/c$). We therefore obtain the expected nonrelativistic $F = ma$ equation.

If $v = 1 - \epsilon$, where ϵ is very small (that is, if v is very close to c), then we can make approximations in Eq. (12.128) to obtain $\epsilon \approx (2e)e^{-4\sigma t/M}$. We see that the difference between v and 1 decreases exponentially with t. ♣

12.11. **Relativistic dustpan I**

This problem is essentially the same as Problem 12.3. Let M' be the mass of the dustpan-plus-dust-inside system (which we'll label "S") when its speed is v. After a small time dt in the lab frame, S has moved a distance $v\, dt$, so it has basically collided with an infinitesimal mass $\lambda v\, dt$. Its energy therefore increases to $\gamma M + \lambda v\, dt$. Its momentum is still $\gamma M v$, so its mass is now

$$M' = \sqrt{(\gamma M + \lambda v\, dt)^2 - (\gamma M v)^2} \approx \sqrt{M^2 + 2\gamma M \lambda v\, dt}\,, \tag{12.129}$$

where we have dropped the second-order dt^2 terms. Using the Taylor series $\sqrt{1+\epsilon} \approx 1 + \epsilon/2$, we can approximate M' as

$$M' \approx M\sqrt{1 + \frac{2\gamma\lambda v\,dt}{M}} \approx M\left(1 + \frac{\gamma\lambda v\,dt}{M}\right) = M + \gamma\lambda v\,dt. \qquad (12.130)$$

The rate of increase in S's mass is therefore $\gamma\lambda v$. As in Problem 12.3, this increase is greater than the nonrelativistic answer of "λv," because heat is generated during the collision, and this heat shows up as mass in the final object.

REMARKS: As explained in the remark in the solution to Problem 12.3, this result is clear if we work in the dustpan frame. In this frame, the above-mentioned infinitesimal mass $\lambda v\,dt$ comes flying in with energy $\gamma(\lambda v\,dt)$, and essentially all of this energy shows up as mass in the final object.

Note that the rate at which the mass increases, as measured in the dustpan frame, is $\gamma^2\lambda v$, due to time dilation. The dust-entering-dustpan events happen at the same location in the dustpan frame, so we have indeed put the extra γ factor in the correct place. Alternatively, you can think in terms of length contraction. S sees the dust contracted, so its density is increased to $\gamma\lambda$. And the other γ factor comes from the fact that the dust is moving in the dustpan frame, so there is a γ factor in the energy. ♣

12.12. **Relativistic dustpan II**

The initial momentum is $\gamma_V MV \equiv P$. There are no external forces, so the momentum of the dustpan-plus-dust-inside system (denoted by "S") always equals P. That is, $\gamma mv = P$, where m and v are the mass and speed of S at any later time.

Let's find $v(x)$ first. The energy of S, namely γm, increases due to the acquisition of new dust. Therefore, $d(\gamma m) = \lambda\,dx$, which we can write as

$$d\left(\frac{P}{v}\right) = \lambda\,dx. \qquad (12.131)$$

Integrating this, and using the fact that the initial speed is V, gives $P/v - P/V = \lambda x$. Therefore,

$$v(x) = \frac{V}{1 + (V\lambda x/P)}. \qquad (12.132)$$

For large x, this approaches $P/(\lambda x)$. This makes sense, because the mass of S is essentially equal to λx, and it is moving at a slow, nonrelativistic speed.

To find $v(t)$, write the dx in Eq. (12.131) as $v\,dt$ to obtain $(-P/v^2)\,dv = \lambda v\,dt$. This gives

$$-\int_V^v \frac{P\,dv}{v^3} = \int_0^t \lambda\,dt \Longrightarrow \frac{P}{v^2} - \frac{P}{V^2} = 2\lambda t \Longrightarrow v(t) = \frac{V}{\sqrt{1 + (2V^2\lambda t/P)}}. \qquad (12.133)$$

Integrating this to obtain $x(t)$ gives

$$x(t) = \frac{P}{V\lambda}\left(\sqrt{1 + \frac{2V^2\lambda t}{P}} - 1\right). \qquad (12.134)$$

If you want, you can rewrite all of these answers in terms of M via $P \equiv \gamma_V MV$.

REMARKS:

1. You can also obtain the result in Eq. (12.134) by equating the expressions for v in Eqs. (12.132) and (12.133). Or you can write the v in Eq. (12.132) as dx/dt, and then separate variables and integrate.
2. For small t, you can show that Eq. (12.134) reduces to $x = Vt$, as it should. For large t, x has the interesting property of being proportional to $\sqrt{t}$.
3. Given P, all the results in this problem (when expressed in terms of P) are actually the same as the ones obtained in the nonrelativistic case, because Eq. (12.131) is still true there; it's simply the expression for how the mass changes in the nonrelativistic case. From that point on, relativity never entered into the reasoning. ♣

12.13. **Relativistic dustpan III**

DUSTPAN FRAME: Let S denote the dustpan-plus-dust-inside system at a given time, and consider a small bit of dust (call this system s) that enters the dustpan. In S's frame, the density of the dust is $\gamma\lambda$, due to length contraction. Therefore, in a time $d\tau$ (where τ is the time in the dustpan frame), a little s system of dust with mass $\gamma\lambda v\,d\tau$ crashes into S and loses its negative momentum of $-\gamma(\gamma\lambda v\,d\tau)v = -\gamma^2 v^2 \lambda\,d\tau$. The force on s is therefore $F = dp/d\tau = \gamma^2 v^2 \lambda$. The desired force on S is equal and opposite to this, so

$$F = -\gamma^2 v^2 \lambda. \tag{12.135}$$

LAB FRAME: In a time dt, where t is the time in the lab frame, a little s system of dust with mass $\lambda v\,dt$ gets picked up by the dustpan. What is the change in momentum of s? It is tempting to say that it is $\gamma(\lambda v\,dt)v$, but this would lead to a force of $-\gamma v^2 \lambda$ on the dustpan, which doesn't agree with the result we found above in the dustpan frame. This would be a problem, because longitudinal forces should be the same in different frames.

The key point to realize is that the mass of whatever is moving increases at a rate $\gamma\lambda v$, and not λv (see Problem 12.11). We therefore see that the change in momentum of the additional moving mass is $\gamma(\gamma\lambda v\,dt)v = \gamma^2 v^2 \lambda\,dt$. The original moving system S therefore loses this much momentum, and so the force on it is $F = dp/dt = -\gamma^2 v^2 \lambda$, in agreement with the result in the dustpan frame.

12.14. **Relativistic cart I**

GROUND FRAME (YOUR FRAME): Using reasoning similar to that in Problem 12.3 and Problem 12.11, we see that the mass of the cart-plus-sand-inside system increases at a rate $\gamma\sigma$. Therefore, its momentum increases at a rate (using the fact that v is constant)

$$\frac{dp}{dt} = \gamma\left(\frac{dm}{dt}\right)v = \gamma(\gamma\sigma)v = \gamma^2 \sigma v. \tag{12.136}$$

Since $F = dp/dt$, this is the force that you exert on the cart. Therefore, it is also the force that the ground exerts on your feet, because the net force on you is zero (since your momentum is constant, and in fact zero).

CART FRAME: The sand-entering-cart events happen at the same location in the ground frame, so time dilation says that the sand enters the cart at a slower rate in the cart frame, that is, at a rate σ/γ. The sand flies in at speed v, and then eventually comes to rest on the cart, so its momentum decreases at a rate $\gamma(\sigma/\gamma)v = \sigma v$. This must therefore be the force that your hand applies to the cart.

If this were the only change in momentum in the problem, then we would have a problem, because the force on your feet would be σv in the cart frame, whereas we found above that it is $\gamma^2\sigma v$ in the ground frame. This would contradict the fact that longitudinal forces are the same in different frames. What is the resolution to this apparent paradox? The resolution is that while you are pushing on the cart, *your mass is decreasing*. You are moving at speed v in the cart frame, and mass is continually being transferred from you (who are moving) to the cart (which is at rest), as we will show. This is the missing change in momentum that we need. The quantitative reasoning is as follows.

Go back to the ground frame for a moment. We saw above that the mass of the cart-plus-sand-inside system (call this system "C") increases at a rate $\gamma\sigma$ in the ground frame. Therefore, the energy of C increases at a rate $\gamma(\gamma\sigma)$ in the ground frame. The sand provides σ of this energy, so you must provide the remaining $(\gamma^2 - 1)\sigma$ part. Therefore, since you are losing energy at this rate, you must also be losing mass at this rate in the ground frame (because you are at rest there).

Now go back to the cart frame. Due to time dilation, you lose mass at a rate of only $(\gamma^2 - 1)\sigma/\gamma$. This mass goes from moving at speed v (that is, along with you), to speed zero (that is, at rest on the cart). Therefore, the rate of decrease in momentum of this mass is $\gamma\big((\gamma^2 - 1)\sigma/\gamma\big)v = (\gamma^2 - 1)\sigma v$. Adding this result to the σv result we found for the sand, we see that the total rate of decrease in momentum is $\gamma^2\sigma v$. This is therefore the force that the ground applies to your feet, in agreement with the above calculation in the ground frame.

Note that the reason why we didn't have to worry about your changing mass when doing the calculation in the ground frame was that your speed there was zero. Your momentum was therefore always zero, independent of what was happening to your mass.

12.15. **Relativistic cart II**

GROUND FRAME: Using reasoning similar to that in Problem 12.3 and Problem 12.11, we see that the mass of the cart-plus-sand-inside system increases at a rate $\gamma\sigma$. Therefore, its momentum increases at a rate $\gamma(\gamma\sigma)v = \gamma^2\sigma v$. However, this is *not* the force that your hand exerts on the cart. The reason is that your hand is receding from the location where the sand enters the cart, so your hand cannot immediately be aware of the need for additional momentum. No matter how rigid the cart is, it can't transmit information faster than c. In a sense, there is a sort of Doppler effect going on, and your hand needs to be responsible for only a certain fraction of the momentum increase. Let's be quantitative about this.

Consider two grains of sand that enter the cart a time t apart. What is the difference between the two times that your hand becomes aware that the grains have entered the cart? Assuming maximal rigidity (that is, assuming that signals propagate along the cart at speed c), the relative speed (as measured by someone on the ground) of the signals and your hand is $c - v$. The distance between the two signals is ct. Therefore, they arrive at your hand separated by a time of $ct/(c - v)$. In other words, the rate at which you feel sand entering the cart is $(c - v)/c$ times the given σ rate. This is the factor by which we must multiply the naive $\gamma^2\sigma v$ result for the force we found above. The force you apply is therefore (dropping the c's)

$$F = \left(1 - \frac{v}{c}\right)\gamma^2\sigma v = \frac{\sigma v}{1 + v}. \qquad (12.137)$$

CART FRAME (YOUR FRAME): The sand-entering-cart events happen at the same location in the ground frame, so time dilation says that the sand enters the cart at a slower rate in the cart frame, that is, at a rate σ/γ. The sand flies in at speed v, and then eventually comes to rest on the cart, so its momentum decreases at a rate $\gamma(\sigma/\gamma)v = \sigma v$. But again, this is *not* the force that your hand exerts on the cart. As above, the sand enters the cart at a location that is receding from your hand, so your hand cannot immediately be aware of the need for additional momentum. Let's be quantitative about this.

Consider two grains of sand that enter the cart a time t apart. What is the difference between the two times that your hand becomes aware that the grains have entered the cart? Assuming maximal rigidity (that is, assuming that signals propagate along the cart at speed c), the relative speed (as measured by someone on the cart) of the signals and your hand is c, because you are at rest. The distance between the two signals is $ct + vt$, because the sand source is moving away from you at speed v. Therefore, the signals arrive at your hand separated by a time of $(c + v)t/c$. In other words, the rate at which you feel sand entering the cart is $c/(c + v)$ times the time-dilated σ/γ rate. This is the factor by which we must multiply the naive σv result for the force we found above. The force you apply is therefore (dropping the c's)

$$F = \left(\frac{1}{1 + v/c}\right)\sigma v = \frac{\sigma v}{1 + v}, \qquad (12.138)$$

in agreement with Eq. (12.137).

In a nutshell, the two naive results in the two frames, $\gamma^2\sigma v$ and σv, differ by two factors of γ. But the ratio of the two "Doppler-effect" factors (which arose from the impossibility of absolute rigidity) precisely remedies this discrepancy. The reason why we didn't need to consider this Doppler effect in Problem 12.14 is that there your hand is always right next to the point where the sand enters the cart.

12.16. **Different frames**

(a) The energy of the resulting blob is $2m + T\ell$. Since the blob is at rest, we have

$$M = 2m + T\ell. \qquad (12.139)$$

(b) Let the new frame be S. Let the original frame be S'. The critical point to realize is that in frame S, the left mass starts to accelerate before the right mass does. This is due to the loss of simultaneity between the frames.

Consider the two events at which the two masses start to move. Let the left mass and right mass start moving at positions x_l and x_r in S. The Lorentz transformation $\Delta x = \gamma(\Delta x' + v\Delta t')$ tells us that $x_r - x_l = \gamma\ell$, because $\Delta x' = \ell$ and $\Delta t' = 0$ for these events. Alternatively, this follows from length contraction, because if we picture things in the original S' frame, then the length $\gamma\ell$ in S is what is length contracted down to ℓ in S'.

Let the masses collide at position x_c in S. Then the gain in energy of the left mass is $T(x_c - x_l)$, and the gain in energy of the right mass is $(-T)(x_c - x_r)$ which is negative if $x_c > x_r$. We have used the fact that the longitudinal force is the same in the two frames, so the masses still feel a tension T in frame S. The gain in the sum of the energies of the two masses is therefore

$$\Delta E = T(x_c - x_l) + (-T)(x_c - x_r) = T(x_r - x_l) = T\gamma\ell. \qquad (12.140)$$

The initial sum of the energies was $2\gamma m$, so the final energy is

$$E = 2\gamma m + T\gamma\ell = \gamma M, \qquad (12.141)$$

as desired.

12.17. Splitting mass

We'll calculate the times for the two parts of the process to occur. The energy of the mass right before it splits is (with the subscript b for "before") $E_b = M + T(\ell/2)$, so the momentum is $p_b = \sqrt{E_b^2 - M^2} = \sqrt{MT\ell + T^2\ell^2/4}$. Using $F = dp/dt \Longrightarrow t = \Delta p/T$, the time for the first part of the process is

$$t_1 = \frac{\sqrt{MT\ell + T^2\ell^2/4}}{T} = \sqrt{\frac{M\ell}{T} + \frac{\ell^2}{4}}. \qquad (12.142)$$

The momentum of the front half of the mass immediately after it splits is $p_a = p_b/2 = (1/2)\sqrt{MT\ell + T^2\ell^2/4}$. The energy at the wall is $E_w = E_b/2 + T(\ell/2) = M/2 + 3T\ell/4$, so the momentum at the wall is $p_w = \sqrt{E_w^2 - (M/2)^2} = (1/2)\sqrt{3MT\ell + 9T^2\ell^2/4}$. The change in momentum during the second part of the process is therefore

$$\Delta p = p_w - p_a = (1/2)\sqrt{3MT\ell + 9T^2\ell^2/4} - (1/2)\sqrt{MT\ell + T^2\ell^2/4}. \qquad (12.143)$$

Since we have $t = \Delta p/T$, the time for the second part is

$$t_2 = \frac{1}{2}\sqrt{\frac{3M\ell}{T} + \frac{9\ell^2}{4}} - \frac{1}{2}\sqrt{\frac{M\ell}{T} + \frac{\ell^2}{4}}. \qquad (12.144)$$

The total time is $t_1 + t_2$, which simply changes the minus sign in this expression to a plus sign.

12.18. Relativistic leaky bucket

(a) Let the wall be at $x = 0$, and let the initial position be $x = \ell$. Consider a small interval during which the bucket moves from x to $x + dx$ (where dx is negative). The bucket's energy changes by $(-T)\,dx$ due to the string (this is positive), and also changes by a fraction dx/x, due to the leaking (this is negative). Therefore, $dE = (-T)\,dx + E\,dx/x$, or

$$\frac{dE}{dx} = -T + \frac{E}{x}. \qquad (12.145)$$

In solving this differential equation, it is convenient to introduce the variable $y \equiv E/x$. With this definition, we have $E' = (xy)' = xy' + y$, where a prime denotes differentiation with respect to x. Equation (12.145) then becomes $xy' = -T$, or $dy = -T\,dx/x$. Integration gives $y = -T\ln x + C$, which we can write as $y = -T\ln(x/\ell) + B$, in order to have a dimensionless argument in the log. Since $E = xy$, we therefore have

$$E(x) = Bx - Tx\ln(x/\ell), \qquad (12.146)$$

where B is a constant of integration. The reasoning up to this point is valid for *both* the total energy and the kinetic energy, because they both change in the two ways described above. Let's look at each case.

TOTAL ENERGY: Equation (12.146) gives

$$E = M(x/\ell) - Tx\ln(x/\ell), \qquad (12.147)$$

where the constant of integration, B, has been chosen to be M/ℓ so that $E = M$ when $x = \ell$. In terms of the fraction $z \equiv x/\ell$, we have $E = Mz - T\ell z\ln z$. Setting $dE/dz = 0$ to find the maximum gives

$$\ln z_{\max} = \frac{M}{T\ell} - 1 \quad \Longrightarrow \quad E_{\max} = \frac{T\ell}{e}e^{M/T\ell}. \qquad (12.148)$$

The fraction z must satisfy $z \le 1$, so we must have $\ln z \le 0$. Therefore, a solution for z exists only if $M \le T\ell$. If $M \ge T\ell$, then the total energy decreases all the way to the wall.

REMARKS: If M is slightly less then $T\ell$, then z_{max} is slightly less than 1, so E quickly achieves a maximum of slightly more than M, then decreases for the rest of the way to the wall.

If $M \ll T\ell$, then E achieves its maximum at $z_{\text{max}} \approx 1/e$, where it has the value $T\ell/e$. Essentially all of the energy is kinetic in this case, so this result must agree with the result for the kinetic energy below, which it indeed will. ♣

KINETIC ENERGY: Equation (12.146) gives

$$K = -Tx\ln(x/\ell), \tag{12.149}$$

where the constant of integration, B, has been chosen to be zero so that $K = 0$ when $x = \ell$. Equivalently, $E - K$ must equal the mass $M(x/\ell)$. In terms of the fraction $z \equiv x/\ell$, we have $K = -T\ell z \ln z$. Setting $dK/dz = 0$ to find the maximum gives

$$z_{\text{max}} = \frac{1}{e} \implies K_{\text{max}} = \frac{T\ell}{e}, \tag{12.150}$$

which is independent of M. This result must reduce properly in the nonrelativistic limit. But since there's nothing that needs reducing (there aren't any terms that are small compared with others when $v \ll c$), this result must exactly equal the nonrelativistic result. And indeed, the analogous "Leaky bucket" problem in Chapter 5 (Problem 5.17) gives the same answer.

(b) With $z \equiv x/\ell$, the momentum of the bucket is $p = \sqrt{E^2 - m^2} = \sqrt{E^2 - (Mz)^2}$, so Eq. (12.147) gives

$$p = \sqrt{(Mz - T\ell z\ln z)^2 - (Mz)^2} = \sqrt{-2MT\ell z^2 \ln z + T^2\ell^2 z^2 \ln^2 z}\,. \tag{12.151}$$

Setting the derivative equal to zero gives $T\ell \ln^2 z + (T\ell - 2M)\ln z - M = 0$. The maximum momentum therefore occurs at

$$\ln z_{\text{max}} = \frac{2M - T\ell - \sqrt{T^2\ell^2 + 4M^2}}{2T\ell}. \tag{12.152}$$

We have ignored the other root, because it gives $\ln z > 0 \implies z > 1$.

REMARKS:

1. If $M \ll T\ell$, then $\ln z_{\text{max}} \approx -1 \implies z_{\text{max}} \approx 1/e$. In this case, the bucket immediately moves with $v \approx c$, so we have $E \approx pc$. Therefore, E and p should achieve their maxima at the same location. And indeed, we saw above that E_{max} occurs at $z_{\text{max}} \approx 1/e$.
2. If $M \gg T\ell$, then $\ln z_{\text{max}} \approx -1/2 \implies z_{\text{max}} \approx 1/\sqrt{e}$. In this case, the bucket is nonrelativistic, so this result should agree with Problem 5.17, which it does.
3. If $M = T\ell$, then $\ln z_{\text{max}} = (1 - \sqrt{5})/2$, which is the negative of the inverse of the golden ratio. ♣

12.19. **Relativistic bucket**

(a) The energy of the mass right before it hits the wall is $E = m + T\ell$. Therefore, the momentum right before it hits the wall is $p = \sqrt{E^2 - m^2} = \sqrt{2mT\ell + T^2\ell^2}$. So $F = dp/dt$ gives (using the fact that the tension is constant)

$$\Delta t = \frac{\Delta p}{F} = \frac{\sqrt{2mT\ell + T^2\ell^2}}{T}. \tag{12.153}$$

If $m \ll T\ell$, then $\Delta t \approx \ell$ (or ℓ/c in normal units), which makes sense, because the mass travels at essentially speed c. And if $m \gg T\ell$, then $\Delta t \approx \sqrt{2m\ell/T}$. This is the nonrelativistic limit, and it agrees with the result obtained from the familiar expression, $\ell = at^2/2$, where $a = T/m$ is the acceleration.

(b) STRAIGHTFORWARD METHOD: The energy of the blob right before it hits the wall is $E_{\rm w} = 2m + 2T\ell$ (with the subscript w for "wall"). If we can find the mass, M, of the blob, then we can use $p = \sqrt{E^2 - M^2}$ to get the momentum, and then use $\Delta t = \Delta p/F$ to get the time.[17]

From part (a), the momentum right before the collision is $p_{\rm b} = \sqrt{2mT\ell + T^2\ell^2}$, and this is also the momentum of the blob right after the collision, $p_{\rm a}$. The energy of the blob right after the collision is $E_{\rm a} = 2m + T\ell$. So the mass of the blob after the collision is $M = \sqrt{E_{\rm a}^2 - p_{\rm a}^2} = \sqrt{4m^2 + 2mT\ell}$. Therefore, the momentum at the wall is $p_{\rm w} = \sqrt{E_{\rm w}^2 - M^2} = \sqrt{6mT\ell + 4T^2\ell^2}$, and so

$$\Delta t = \frac{\Delta p}{F} = \frac{\sqrt{6mT\ell + 4T^2\ell^2}}{T}. \tag{12.154}$$

If $m = 0$ then $\Delta t = 2\ell$, as expected.

BETTER METHOD: In the notation in the hint in the statement of the problem, the change in p^2 from the start to just before the collision is $\Delta(p^2) = E_2^2 - E_1^2$. This is true because

$$E_1^2 - m^2 = p_1^2, \quad \text{and} \quad E_2^2 - m^2 = p_2^2, \tag{12.155}$$

and since m is the same throughout the first half of the process, we have $\Delta(E^2) = \Delta(p^2)$. Likewise, the change in p^2 during the second half of the process is $\Delta(p^2) = E_4^2 - E_3^2$, because

$$E_3^2 - M^2 = p_3^2, \quad \text{and} \quad E_4^2 - M^2 = p_4^2, \tag{12.156}$$

and since M is the same throughout the second half of the process,[18] we have $\Delta(E^2) = \Delta(p^2)$. The total change in p^2 is the sum of the above two changes, so the final p^2 is

$$\begin{aligned} p^2 &= (E_2^2 - E_1^2) + (E_4^2 - E_3^2) \\ &= \big((m + T\ell)^2 - m^2\big) + \big((2m + 2T\ell)^2 - (2m + T\ell)^2\big) \\ &= 6mT\ell + 4T^2\ell^2, \end{aligned} \tag{12.157}$$

as in Eq. (12.154). The first solution in effect performed the same calculation, but in a more obscure manner.

(c) The reasoning in part (b) tells us that the final p^2 equals the sum of the $\Delta(E^2)$ terms over the N parts of the process. So we have, using an indexing notation

[17] Although the tension T acts on two different things (the mass m initially, and then the blob), it is valid to use the total Δp to obtain the total time via $\Delta t = \Delta p/F$, because if we wanted to, we could break up the Δp into its two parts, and then find the two partial times, and then add them back together to get the total Δt.

[18] From the first solution, M happens to be $\sqrt{4m^2 + 2mT\ell}$, but the nice thing about this solution is that we don't need to know this. All we need to know is that it is constant.

analogous to that in part (b),

$$\begin{aligned} p^2 &= \sum_{k=1}^{N} (E_{2k}^2 - E_{2k-1}^2) = \sum_{k=1}^{N} \Big((km + kT\ell)^2 - \big(km + (k-1)T\ell\big)^2 \Big) \\ &= \sum_{k=1}^{N} \Big(2kmT\ell + \big(k^2 - (k-1)^2\big)T^2\ell^2 \Big) \\ &= N(N+1)mT\ell + N^2T^2\ell^2. \end{aligned} \tag{12.158}$$

Therefore,

$$\Delta t = \frac{\Delta p}{F} = \frac{\sqrt{N(N+1)mT\ell + N^2T^2\ell^2}}{T}. \tag{12.159}$$

This agrees with the results from parts (a) and (b), for $N = 1$ and 2.

(d) We want to take the limit $N \to \infty$, $\ell \to 0$, $m \to 0$, with the restrictions that $N\ell = L$ and $Nm = M$. Written in terms of M and L, Eq. (12.159) becomes

$$\Delta t = \frac{\sqrt{(1 + 1/N)MTL + T^2L^2}}{T} \quad \longrightarrow \quad \frac{\sqrt{MTL + T^2L^2}}{T}, \tag{12.160}$$

as $N \to \infty$. This Δt is the same as the time it takes for one particle of mass $m = M/2$ to reach the wall, from part (a). The mass of the bucket at the wall is

$$\begin{aligned} M_{\mathrm{w}} &= \sqrt{E_{\mathrm{w}}^2 - p_{\mathrm{w}}^2} = \sqrt{(M + TL)^2 - (MTL + T^2L^2)} \\ &= \sqrt{M^2 + MTL}. \end{aligned} \tag{12.161}$$

If $TL \ll M$, then $M_{\mathrm{w}} \approx M$, which makes sense. If $M \ll TL$, then $M_{\mathrm{w}} \approx \sqrt{MTL}$, which means that M_{w} is the geometric mean between the given mass and the energy stored in the string. This isn't entirely obvious. The speed of the bucket right before it hits the wall is

$$\begin{aligned} v_{\mathrm{w}} &= \frac{p_{\mathrm{w}}}{E_{\mathrm{w}}} = \frac{\sqrt{MTL + T^2L^2}}{M + TL} \\ &= \sqrt{\frac{TL}{M + TL}} = \sqrt{\frac{T}{T + \rho}} \longrightarrow c\sqrt{\frac{T}{T + \rho c^2}}, \end{aligned} \tag{12.162}$$

where $\rho \equiv M/L$ is the mass density.

REMARK: Note that v_{w} depends only on T and ρ. This means that if we change L by moving the wall to any other position, the speed at the wall will still be v_{w}. In other words, the bucket moves toward the wall at the *constant* speed v_{w}. (The task of Exercise 12.40 is to derive this result without taking the $N \to \infty$ limit of many masses.) This constant-speed result must also hold in the nonrelativistic limit (that is, $T \ll \rho c^2$), for which we have $v_{\mathrm{w}} \approx \sqrt{T/\rho}$. And indeed, this agrees with the result for Problem 5.27 ("Pulling a chain again"), which is essentially the same problem, in different language. ♣

Chapter 13
4-vectors

We now come to a very powerful concept in relativity, that of *4-vectors*. Although it's possible to derive everything in Special Relativity without the use of 4-vectors (and indeed, this is the route, give or take, that we've taken in the previous two chapters), they are extremely helpful in making calculations simpler and concepts more transparent.

I have chosen to postpone the full introduction to 4-vectors until now, in order to make it clear that everything in Special Relativity can be derived without them. In encountering relativity for the first time, it's nice to know that no "advanced" techniques are required. But now that you've seen everything once, let's go back and derive various things in an easier way.

Although Special Relativity doesn't require knowledge of 4-vectors, the subject of General Relativity definitely requires a firm understanding of *tensors*, which are the generalization of 4-vectors. We won't have time to go very deeply into GR in Chapter 14, so you'll just have to accept this fact. But suffice it to say that an eventual understanding of GR requires a solid foundation in the 4-vectors of Special Relativity. So let's see what they're all about.

13.1 Definition of 4-vectors

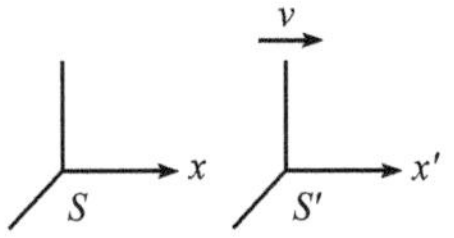

Fig. 13.1

Definition 13.1 *The 4-tuplet, $A = (A_0, A_1, A_2, A_3)$, is a "4-vector" if the A_i transform under a Lorentz transformation in the same way as $(c\,dt, dx, dy, dz)$ do. In other words, A is a 4-vector if it transforms like (assuming the Lorentz transformation is along the x direction; see Fig. 13.1):*

$$\begin{aligned} A_0 &= \gamma(A_0' + (v/c)A_1'), \\ A_1 &= \gamma(A_1' + (v/c)A_0'), \\ A_2 &= A_2', \\ A_3 &= A_3'. \end{aligned} \tag{13.1}$$

REMARKS:

1. Similar equations must hold, of course, for Lorentz transformations in the y and z directions.
2. Additionally, the last three components must be a vector in 3-space. That is, they must transform like a usual vector under rotations in 3-space. So the full definition of a 4-

vector is that it must transform like $(c\,dt, dx, dy, dz)$ under Lorentz transformations *and* rotations.

3. We'll use a capital italic letter to denote a 4-vector. A bold-face letter will denote, as usual, a vector in 3-space.
4. Lest we get tired of writing the c's over and over, we'll work in units where $c = 1$ from now on.
5. The first component of a 4-vector is called the "time" component. The other three are the "space" components.
6. The components in (dt, dx, dy, dz) are sometimes referred to as (dx_0, dx_1, dx_2, dx_3). Also, some treatments use the indices "1" through "4," with "4" being the "time" component. But we'll use "0" through "3."
7. The A_i may be functions of v, the dx_i, the x_i and their derivatives, and any invariants (that is, frame-independent quantities) such as the mass m.
8. 4-vectors are the obvious generalization of vectors in regular space. A vector in three dimensions, after all, is something that transforms under a rotation just like (dx, dy, dz) does. We have simply generalized a 3-D rotation to a 4-D Lorentz transformation. ♣

13.2 Examples of 4-vectors

So far, we have only one 4-vector at our disposal, namely (dt, dx, dy, dz). What are some others? Well, $(7dt, 7dx, 7dy, 7dz)$ certainly works, as does any other constant multiple of (dt, dx, dy, dz). Indeed, $m(dt, dx, dy, dz)$ is a 4-vector, because m is an invariant. But how about $A = (dt, 2dx, dy, dz)$? No, this isn't a 4-vector, because on one hand it must transform (assuming it's a 4-vector) like

$$\begin{aligned}
dt &\equiv A_0 = \gamma(A'_0 + vA'_1) \equiv \gamma\left(dt' + v(2\,dx')\right), \\
2\,dx &\equiv A_1 = \gamma(A'_1 + vA'_0) \equiv \gamma\left((2\,dx') + v\,dt'\right), \\
dy &\equiv A_2 = A'_2 \equiv dy', \\
dz &\equiv A_3 = A'_3 \equiv dz',
\end{aligned} \tag{13.2}$$

from the definition of a 4-vector. But on the other hand, it must transform like

$$\begin{aligned}
dt &= \gamma(dt' + v\,dx'), \\
2\,dx &= 2\gamma(dx' + v\,dt'), \\
dy &= dy', \\
dz &= dz',
\end{aligned} \tag{13.3}$$

because this is how the dx_i transform. The two preceding sets of equations are inconsistent, so $A = (dt, 2dx, dy, dz)$ is not a 4-vector. Note that if we had instead considered the 4-tuplet, $A = (dt, dx, 2dy, dz)$, then the two preceding equations would have been consistent. But if we had then looked at how A transforms under a Lorentz transformation in the y direction, we would have found that it is not a 4-vector.

The moral of this story is that the above definition of a 4-vector is a nontrivial one because there are two possible ways that a 4-tuplet can transform. It can transform according to the 4-vector definition, as in Eq. (13.2). Or it can transform simply by having each of the A_i separately transform (using how the dx_i, or whatever else they may be made of, transform), as in Eq. (13.3). Only for certain special 4-tuplets do these two methods give the same result. By definition, we label these special 4-tuplets as 4-vectors.

Let's now construct some less trivial examples of 4-vectors. In constructing these, we'll make abundant use of the fact that the proper-time interval, $d\tau \equiv \sqrt{dt^2 - d\mathbf{r}^2}$, is an invariant.

- **Velocity 4-vector:** We can divide (dt, dx, dy, dz) by $d\tau$, where $d\tau$ is the proper time between two events (the same two events that yielded the dt, etc.). The result is indeed a 4-vector, because $d\tau$ is independent of the frame in which it is measured. Using $d\tau = dt/\gamma$, we obtain

$$V \equiv \frac{1}{d\tau}(dt, dx, dy, dz) = \gamma\left(1, \frac{dx}{dt}, \frac{dy}{dt}, \frac{dz}{dt}\right) = (\gamma, \gamma\mathbf{v}). \tag{13.4}$$

This is known as the *velocity 4-vector*. In the rest frame of the object we have $\mathbf{v} = \mathbf{0}$, so V reduces to $V = (1, 0, 0, 0)$. With the c's, we have $V = (\gamma c, \gamma\mathbf{v})$.

- **Energy–momentum 4-vector:** If we multiply the velocity 4-vector by the invariant m, we obtain another 4-vector,

$$P \equiv mV = (\gamma m, \gamma m\mathbf{v}) = (E, \mathbf{p}), \tag{13.5}$$

which is known as the *energy–momentum 4-vector* (or the *4-momentum* for short), for obvious reasons. In the rest frame of the object, P reduces to $P = (m, 0, 0, 0)$. With the c's, we have $P = (\gamma mc, \gamma m\mathbf{v}) = (E/c, \mathbf{p})$. Some treatments multiply through by c, so that the 4-momentum is $(E, \mathbf{p}c)$.

- **Acceleration 4-vector:** We can also take the derivative of the velocity 4-vector with respect to τ. The result is indeed a 4-vector, because taking the derivative entails taking the (infinitesimal) difference between two 4-vectors (which results in a 4-vector because Eq. (13.1) is linear), and then dividing by the invariant $d\tau$ (which again results in a 4-vector). Using $d\tau = dt/\gamma$, we obtain

$$A \equiv \frac{dV}{d\tau} = \frac{d}{d\tau}(\gamma, \gamma\mathbf{v}) = \gamma\left(\frac{d\gamma}{dt}, \frac{d(\gamma\mathbf{v})}{dt}\right). \tag{13.6}$$

Using $d\gamma/dt = v\dot{v}/(1 - v^2)^{3/2} = \gamma^3 v\dot{v}$, we have

$$A = (\gamma^4 v\dot{v}, \gamma^4 v\dot{v}\mathbf{v} + \gamma^2\mathbf{a}), \tag{13.7}$$

where $\mathbf{a} \equiv d\mathbf{v}/dt$. A is known as the *acceleration 4-vector*. In the rest frame of the object (or, rather, in the instantaneous inertial frame), A reduces to $A = (0, \mathbf{a})$. As we usually do, we'll pick the velocity $\mathbf{v}$ to point in the x direction. That is, $\mathbf{v} = (v_x, 0, 0)$. This means that $v = v_x$, and also that $\dot{v} = \dot{v}_x \equiv a_x$.[1] Equation (13.7) then becomes

$$\begin{aligned} A &= (\gamma^4 v_x a_x,\ \gamma^4 v_x^2 a_x + \gamma^2 a_x,\ \gamma^2 a_y,\ \gamma^2 a_z) \\ &= (\gamma^4 v_x a_x,\ \gamma^4 a_x,\ \gamma^2 a_y,\ \gamma^2 a_z). \end{aligned} \tag{13.8}$$

We can keep taking derivatives with respect to τ to create other 4-vectors, but these have little relevance in the real world.

- **Force 4-vector:** We define the *force 4-vector* as

$$F \equiv \frac{dP}{d\tau} = \gamma\left(\frac{dE}{dt}, \frac{d\mathbf{p}}{dt}\right) = \gamma\left(\frac{dE}{dt}, \mathbf{f}\right), \tag{13.9}$$

where $\mathbf{f} \equiv d(\gamma m\mathbf{v})/dt$ is the usual 3-force. We'll use $\mathbf{f}$ instead of $\mathbf{F}$ in this chapter, to avoid confusion with the 4-force, F. In the case where m is constant,[2] F can be written as $F = d(mV)/d\tau = m\,dV/d\tau = mA$. We therefore still have a nice "F equals mA" law of physics, but it's now a 4-vector equation instead of the old 3-vector one. In terms of the acceleration 4-vector, we can use Eqs. (13.7) and (13.8) to write (if m is constant)

$$\begin{aligned} F &= mA = m(\gamma^4 v\dot{v},\ \gamma^4 v\dot{v}\mathbf{v} + \gamma^2\mathbf{a}) \\ &= m(\gamma^4 v_x a_x,\ \gamma^4 a_x,\ \gamma^2 a_y,\ \gamma^2 a_z). \end{aligned} \tag{13.10}$$

Combining this with Eq. (13.9), we see that the 3-force is

$$\mathbf{f} = m(\gamma^3 a_x\,,\ \gamma a_y\,,\ \gamma a_z), \tag{13.11}$$

in agreement with Eq. (12.60). In the rest frame of the object (or, rather, the instantaneous inertial frame), the F in Eq. (13.9) reduces to $F = (0, \mathbf{f})$, because $dE/dt = 0$ when $v = 0$, as you can verify. Also, in the rest frame of the object, the mA in Eq. (13.10) reduces to $mA = (0, m\mathbf{a})$. So $F = mA$ reduces to the familiar $\mathbf{f} = m\mathbf{a}$.

13.3 Properties of 4-vectors

The appealing thing about 4-vectors is that they have many useful properties. Let's look at some of these.

- **Linear combinations:** If A and B are 4-vectors, then $C \equiv aA + bB$ is also a 4-vector. This is true because the transformations in Eq. (13.1) are linear (as we noted above

[1] The acceleration vector $\mathbf{a}$ is free to point in any direction, but you can check that the 0's in $\mathbf{v}$ lead to $\dot{v} = a_x$. See Exercise 13.5.

[2] The mass m wouldn't be constant if the object were being heated, or if extra mass were being added to it. We won't concern ourselves with such cases here.

when deriving the acceleration 4-vector). This linearity implies that the transformation of, say, the time component is

$$\begin{aligned} C_0 \equiv (aA + bB)_0 = aA_0 + bB_0 &= a(A'_0 + vA'_1) + b(B'_0 + vB'_1) \\ &= (aA'_0 + bB'_0) + v(aA'_1 + bB'_1) \\ &\equiv C'_0 + vC'_1, \end{aligned} \tag{13.12}$$

which is the proper transformation for the time component of a 4-vector. Likewise for the other components. This property holds, of course, just as it does for linear combinations of vectors in 3-space.

- **Inner-product invariance:** Consider two arbitrary 4-vectors, A and B. Define their inner product to be

$$A \cdot B \equiv A_0B_0 - A_1B_1 - A_2B_2 - A_3B_3 \equiv A_0B_0 - \mathbf{A} \cdot \mathbf{B}. \tag{13.13}$$

Then $A \cdot B$ is invariant. That is, it is independent of the frame in which it is calculated. This can be shown by direct calculation, using the transformations in Eq. (13.1):

$$\begin{aligned} A \cdot B &\equiv A_0B_0 - A_1B_1 - A_2B_2 - A_3B_3 \\ &= \big(\gamma(A'_0 + vA'_1)\big)\big(\gamma(B'_0 + vB'_1)\big) - \big(\gamma(A'_1 + vA'_0)\big)\big(\gamma(B'_1 + vB'_0)\big) \\ &\quad - A'_2B'_2 - A'_3B'_3 \\ &= \gamma^2\big(A'_0B'_0 + v(A'_0B'_1 + A'_1B'_0) + v^2A'_1B'_1\big) \\ &\quad - \gamma^2\big(A'_1B'_1 + v(A'_1B'_0 + A'_0B'_1) + v^2A'_0B'_0\big) - A'_2B'_2 - A'_3B'_3 \\ &= A'_0B'_0(\gamma^2 - \gamma^2v^2) - A'_1B'_1(\gamma^2 - \gamma^2v^2) - A'_2B'_2 - A'_3B'_3 \\ &= A'_0B'_0 - A'_1B'_1 - A'_2B'_2 - A'_3B'_3 \\ &\equiv A' \cdot B'. \end{aligned} \tag{13.14}$$

The importance of this result cannot be overstated. This invariance is analogous to the invariance of the inner product, $\mathbf{A} \cdot \mathbf{B}$, for rotations in 3-space. The above inner product is also invariant under rotations in 3-space, because it involves the combination $\mathbf{A} \cdot \mathbf{B}$. The minus signs in the inner product may seem a little strange. But what we want is a combination of two arbitrary vectors that is invariant under a Lorentz transformation, because such combinations are very useful in seeing what's going on in a system. The nature of the Lorentz transformations demands that there be opposite signs in the inner product, so that's the way it is.

- **Norm:** As a corollary to the invariance of the inner product, we can look at the inner product of a 4-vector with itself, which is by definition the square of the norm. We see that

$$A^2 \equiv A \cdot A \equiv A_0A_0 - A_1A_1 - A_2A_2 - A_3A_3 = A_0^2 - |\mathbf{A}|^2 \tag{13.15}$$

is invariant. This is analogous to the invariance of the norm $\sqrt{\mathbf{A}\cdot\mathbf{A}}$ for rotations in 3-space. Special cases of the invariance of the 4-vector norm are the invariance of $c^2dt^2 - dx^2$ and of $E^2 - p^2c^2$.

- **A theorem:** Here's a nice little theorem:

  If a certain one of the components of a 4-vector is 0 in every frame, then all four components are 0 in every frame.

  Proof: If one of the space components (say, A_1) is 0 in every frame, then the other space components must also be 0 in every frame, because otherwise a rotation would make $A_1 \neq 0$. Also, the time component A_0 must be 0 in every frame, because otherwise a Lorentz transformation in the x direction would make $A_1 \neq 0$.

  If the time component, A_0, is 0 in every frame, then the space components must also be 0 in every frame, because otherwise a Lorentz transformation in the appropriate direction would make $A_0 \neq 0$. ■

  If someone comes along and says that she has a vector in 3-space that has no x component, no matter how you rotate the axes, then you would certainly say that the vector must obviously be the zero vector. The situation in Lorentzian 4-space is the same, because all the coordinates get intertwined with each other in the Lorentz (and rotation) transformations.

13.4 Energy, momentum

13.4.1 Norm

Many useful things arise from the fact that the P in Eq. (13.5) is a 4-vector. The invariance of the norm implies that $P \cdot P = E^2 - |\mathbf{p}|^2$ is invariant. If we are dealing with only one particle, we can determine the value of P^2 by conveniently working in the rest frame of the particle (so that $\mathbf{v} = \mathbf{0}$), which gives

$$E^2 - p^2 = m^2, \tag{13.16}$$

or $E^2 - p^2c^2 = m^2c^4$, with the c's. We already knew this, of course, from just writing out $E^2 - p^2 = \gamma^2 m^2 - \gamma^2 m^2 v^2 = m^2$.

For a collection of particles, knowledge of the norm is very useful. If a process involves many particles, then we can say that for *any* subset of the particles,

$$\left(\sum E\right)^2 - \left(\sum \mathbf{p}\right)^2 \quad \text{is invariant,} \tag{13.17}$$

because this is the norm of the sum of the energy–momentum 4-vectors of the chosen particles. The sum is again a 4-vector, due to the linearity of Eq. (13.1). What is the value of the invariant in Eq. (13.17)? The most concise description (which is basically a tautology) is that it is the square of the energy in the CM frame, that is, in the frame in which $\sum \mathbf{p} = \mathbf{0}$. For one particle, this reduces to m^2. Note that the sums are taken before squaring in Eq. (13.17). Squaring before adding would simply give the sum of the squares of the masses.

13.4.2 Transformations of E and p

We already know how the energy and momentum transform (see Section 12.2), but let's derive the transformation again here in a very quick and easy manner. We know that (E, p_x, p_y, p_z) is a 4-vector. So it must transform according to Eq. (13.1). Therefore, for a Lorentz transformation in the x direction, we have

$$\begin{aligned} E &= \gamma(E' + vp'_x), \\ p_x &= \gamma(p'_x + vE'), \\ p_y &= p'_y, \\ p_z &= p'_z, \end{aligned} \tag{13.18}$$

in agreement with Eq. (12.27). That's all there is to it. The fact that E and $\mathbf{p}$ are part of the same 4-vector provides an easy way to see that if one of them is conserved (in every frame) in a collision, then the other is also. Consider an interaction among a set of particles, and look at the 4-vector, $\Delta P \equiv P_{\text{after}} - P_{\text{before}}$. If E is conserved in every frame, then the time component of ΔP is 0 in every frame. But then the theorem in Section 13.3 says that all four components of ΔP are 0 in every frame. Therefore, $\mathbf{p}$ is conserved. Likewise for the case where one of the p_i is known to be conserved.

13.5 Force and acceleration

Throughout this section, we'll deal with objects with constant mass, which we'll call "particles." The treatment here can be generalized to cases where the mass changes (for example, the object is being heated, or extra mass is being dumped on it), but we won't concern ourselves with these.

13.5.1 Transformation of forces

Let's first look at the force 4-vector in the instantaneous inertial frame of a given particle (frame S'). Equation (13.9) gives

$$F' = \gamma\left(\frac{dE'}{dt}, \mathbf{f}'\right) = (0, \mathbf{f}'). \tag{13.19}$$

The first component is zero because $dE'/dt = d\big(m/\sqrt{1-v'^2}\big)/dt$, and this contains a factor of v', which is zero in this frame. Equivalently, you can just use Eq. (13.10), with a speed of zero.

We can now write down two expressions for the 4-force, F, in another frame, S, in which the particle moves with speed v in the x direction. First, since F

is a 4-vector, it transforms according to Eq. (13.1). We therefore have, using Eq. (13.19),

$$\begin{aligned} F_0 &= \gamma(F'_0 + vF'_1) = \gamma v f'_x, \\ F_1 &= \gamma(F'_1 + vF'_0) = \gamma f'_x, \\ F_2 &= F'_2 = f'_y, \\ F_3 &= F'_3 = f'_z. \end{aligned} \tag{13.20}$$

But second, from the definition in Eq. (13.9), we also have

$$\begin{aligned} F_0 &= \gamma\, dE/dt, \\ F_1 &= \gamma f_x, \\ F_2 &= \gamma f_y, \\ F_3 &= \gamma f_z. \end{aligned} \tag{13.21}$$

Combining Eqs. (13.20) and (13.21), we obtain

$$\begin{aligned} dE/dt &= v f'_x, \\ f_x &= f'_x, \\ f_y &= f'_y/\gamma, \\ f_z &= f'_z/\gamma. \end{aligned} \tag{13.22}$$

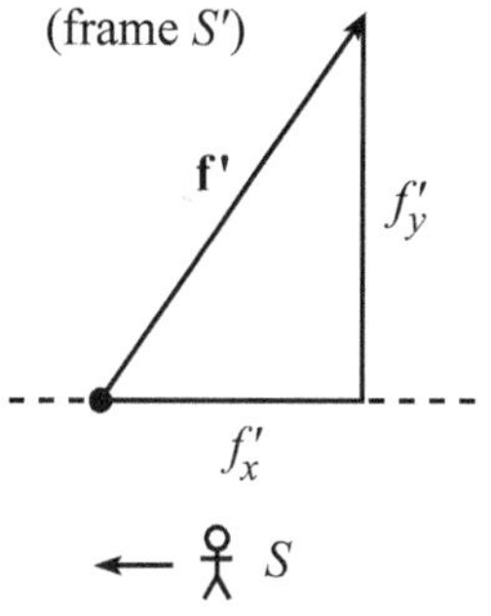

Fig. 13.2

We therefore recover the results of Section 12.5.3. The longitudinal force is the same in both frames, but the transverse forces are larger by a factor of γ in the particle's frame. Hence, f_y/f_x decreases by a factor of γ when going from the particle's frame to the lab frame (see Fig. 13.2 and Fig. 13.3). And as a bonus, the F_0 component in Eq. (13.22) tells us (after multiplying through by dt) that $dE = f_x\, dx$, which is the work–energy result. In other words, using $f_x \equiv dp_x/dt$, we have just derived again the result, $dE/dx = f_x = dp/dt$, that we derived in Section 12.5.1.

As noted in the first remark in Section 12.5.3, we can't switch the S and S' frames and write $f'_y = f_y/\gamma$. When talking about the forces on a particle, there is indeed one preferred reference frame, namely that of the particle. All frames are not equivalent here. When forming all of our 4-vectors in Section 13.2, we explicitly used the $d\tau$, dt, dx, etc., from two events, and it was understood that these two events were located at the particle.

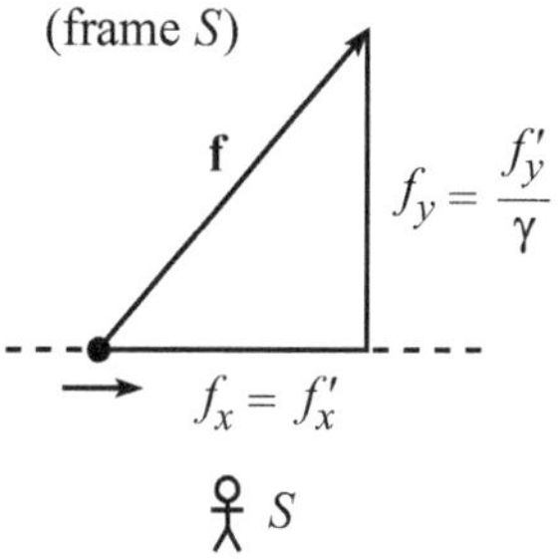

Fig. 13.3

13.5.2 Transformation of accelerations

The procedure here is similar to the above treatment of the force. Let's first look at the acceleration 4-vector in the instantaneous inertial frame of a given particle

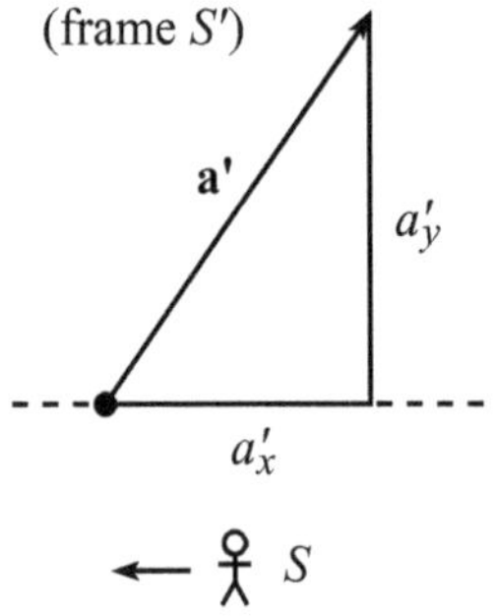

Fig. 13.4

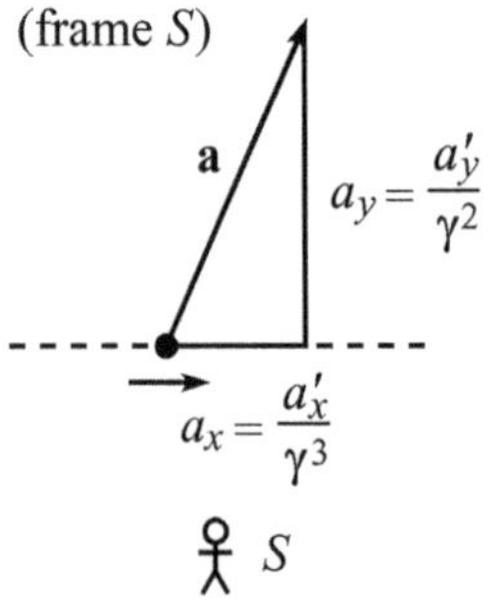

Fig. 13.5

(frame S'). Equation (13.7) or Eq. (13.8) gives

$$A' = (0, \mathbf{a}'), \tag{13.23}$$

because $v' = 0$ in S'.

We can now write down two expressions for the 4-acceleration, A, in another frame, S. First, since A is a 4-vector, it transforms according to Eq. (13.1). So we have, using Eq. (13.23),

$$\begin{aligned} A_0 &= \gamma(A_0' + vA_1') = \gamma v a_x', \\ A_1 &= \gamma(A_1' + vA_0') = \gamma a_x', \\ A_2 &= A_2' = a_y', \\ A_3 &= A_3' = a_z'. \end{aligned} \tag{13.24}$$

But second, from the expression in Eq. (13.8), we also have

$$\begin{aligned} A_0 &= \gamma^4 v a_x, \\ A_1 &= \gamma^4 a_x, \\ A_2 &= \gamma^2 a_y, \\ A_3 &= \gamma^2 a_z. \end{aligned} \tag{13.25}$$

Combining Eqs. (13.24) and (13.25), we obtain

$$\begin{aligned} a_x &= a_x'/\gamma^3, \\ a_x &= a_x'/\gamma^3, \\ a_y &= a_y'/\gamma^2, \\ a_z &= a_z'/\gamma^2. \end{aligned} \tag{13.26}$$

(The first two equations here are redundant.) We therefore again recover the results of Section 12.5.3. We see that a_y/a_x increases by a factor of $\gamma^3/\gamma^2 = \gamma$ when going from the particle's frame to the lab frame (see Fig. 13.4 and Fig. 13.5). This is the opposite of the effect on f_y/f_x.[3] This difference makes it clear that an $\mathbf{f} = m\mathbf{a}$ law wouldn't make any sense. If it were true in one frame, then it wouldn't be true in another. Note that the increase in a_y/a_x in going to the lab frame is consistent with length contraction, as the "Bead on a rod" example in Section 12.5.3 showed.

[3] In a nutshell, this difference is due to the fact that γ changes with time. When talking about the acceleration 4-vector, there are γ's that we have to differentiate; see Eq. (13.6). This isn't the case with the force 4-vector, because the γ is absorbed into the definition of $\mathbf{p} \equiv \gamma m\mathbf{v}$; see Eq. (13.9). This is what leads to the different powers of γ in Eq. (13.25), in contrast with the identical powers in Eq. (13.21).

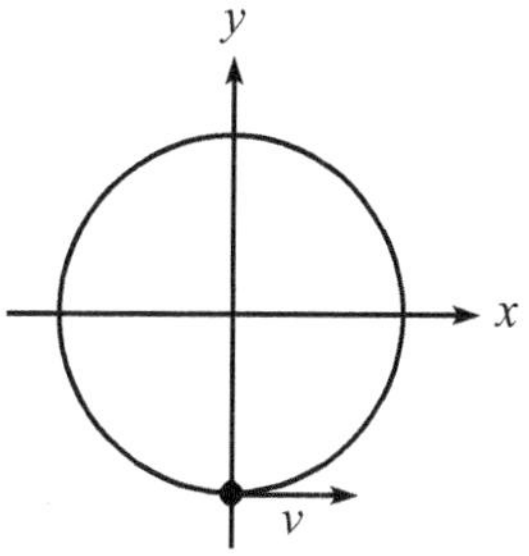

Fig. 13.6

Example (Acceleration for circular motion): A particle moves with constant speed v around the circle $x^2 + y^2 = r^2$, $z = 0$, in the lab frame. At the instant the particle crosses the negative y axis (see Fig. 13.6), find the 3-acceleration and 4-acceleration in both the lab frame and the instantaneous inertial frame of the particle (with axes chosen parallel to the lab's axes).

Solution: Let the lab frame be S, and let the particle's instantaneous inertial frame be S' when it crosses the negative y axis. Then S and S' are related by a Lorentz transformation in the x direction. The 3-acceleration in S is simply

$$\mathbf{a} = (0, v^2/r, 0). \tag{13.27}$$

There's nothing fancy going on here; the nonrelativistic proof of $a = v^2/r$ works just fine again in the relativistic case. Equation (13.7) or (13.8) then gives the 4-acceleration in S as

$$A = (0, 0, \gamma^2 v^2/r, 0). \tag{13.28}$$

To find the acceleration vectors in S', we can use the fact that S' and S are related by a Lorentz transformation in the x direction. This means that the A_2 component of the 4-acceleration is unchanged. So the 4-acceleration in S' is also

$$A' = A = (0, 0, \gamma^2 v^2/r, 0). \tag{13.29}$$

In the particle's frame, $\mathbf{a}'$ is the space part of A (using Eq. (13.7) or (13.8), with $v = 0$ and $\gamma = 1$). Therefore, the 3-acceleration in S' is

$$\mathbf{a}' = (0, \gamma^2 v^2/r, 0). \tag{13.30}$$

Note that our results for $\mathbf{a}$ and $\mathbf{a}'$ are consistent with Eq. (13.26).

REMARK: We can also arrive at the two factors of γ in $\mathbf{a}'$ by using a simple time-dilation argument. We have

$$a'_y = \frac{d^2y'}{d\tau^2} = \frac{d^2y'}{d(t/\gamma)^2} = \gamma^2\frac{d^2y}{dt^2} = \gamma^2\frac{v^2}{r}, \tag{13.31}$$

where we have used the fact that transverse lengths are the same in the two frames. ♣

13.6 The form of physical laws

One of the postulates of Special Relativity is that all inertial frames are equivalent. Therefore, if a physical law holds in one frame, then it must hold in all frames. Otherwise, it would be possible to differentiate between frames. As noted in the previous section, the statement "$\mathbf{f} = m\mathbf{a}$" cannot be a physical law. The two sides of the equation transform differently when going from one frame to another, so the statement cannot be true in all frames. If a statement has any chance of being true in all frames, it must involve only 4-vectors. Consider a 4-vector equation (say, "$A = B$") that is true in frame S. Then if we apply to this equation a Lorentz transformation (call it $\mathcal{M}$) from S to another frame S', we have

$$\begin{aligned} & A = B \\ \Longrightarrow \quad & \mathcal{M}A = \mathcal{M}B \\ \Longrightarrow \quad & A' = B'. \end{aligned} \tag{13.32}$$

The law is therefore also true in frame S'. Of course, there are many 4-vector equations that are simply not true in any frame (for example, $F = P$, or $2P = 3P$). Only a small set of such equations (for example, $F = mA$) are true in at least one frame, and hence in all frames.

Physical laws may also take the form of scalar equations, such as $P \cdot P = m^2$. A scalar is by definition a quantity that is frame-independent (as we have shown the inner product to be). So if a scalar statement is true in one inertial frame, then it is true in all inertial frames. Physical laws may also be higher-rank "tensor" equations, such as the ones that arise in electromagnetism and General Relativity. We won't discuss tensors here, but suffice it to say that they may be thought of as things built up from 4-vectors. Scalars and 4-vectors are special cases of tensors.

All of this is exactly analogous to the situation in 3-D space. In Newtonian mechanics, $\mathbf{f} = m\mathbf{a}$ is a possible law, because both sides are 3-vectors. But $\mathbf{f} = m(2a_x, a_y, a_z)$ is not a possible law, because the right-hand side is not a 3-vector; it depends on which axis you label as the x axis. If a particle has acceleration a in the eastward direction, and if you happen to pick this as your x direction, then the force is $2ma$ eastward. But if you happen to pick eastward as your y direction, then the force is ma eastward. It makes no sense for a law to give two different results based on your arbitrary choice of axis labels. An example of a frame-independent statement (under rotations) is the claim that a given stick has a length of 2 meters. This is fine, because it involves the norm, which is a scalar. But if you say that the stick has an x component of 1.7 meters, then this cannot be true in all frames.

God said to his cosmos directors,
"I've added some stringent selectors.
One is the clause
That your physical laws
Shall be written in terms of 4-vectors."

13.7 Problems

13.1. **Velocity addition** *

In A's frame, B moves to the right with speed u, and C moves to the left with speed v. What is the speed of B with respect to C? In other words, use 4-vectors to derive the velocity-addition formula.

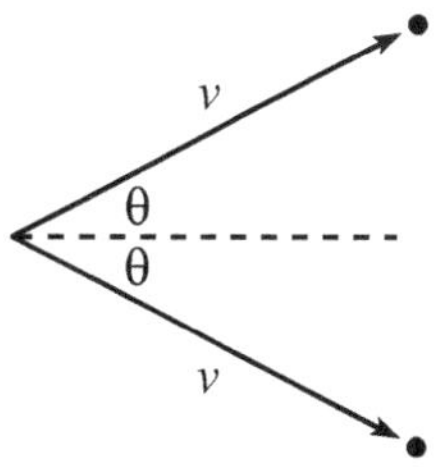

Fig. 13.7

13.2. **Relative speed** *

In the lab frame, two particles move with speed v along the paths shown in Fig. 13.7. The angle between the trajectories is 2θ. What is the speed of one particle, as viewed by the other?

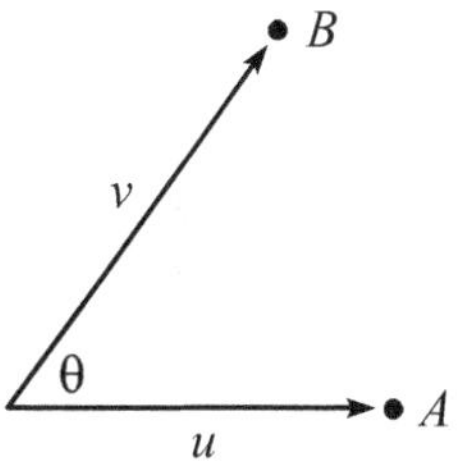

Fig. 13.8

13.3. **Another relative speed** *

In the lab frame, particles A and B move with speeds u and v along the paths shown in Fig. 13.8. The angle between the trajectories is θ. What is the speed of one particle, as viewed by the other?

13.4. **Acceleration for linear motion** *

A spaceship starts at rest with respect to frame S and accelerates with constant proper acceleration a. In Section 11.9, we showed that the speed of the spaceship with respect to S is given by $v(\tau) = \tanh(a\tau)$, where τ is the spaceship's proper time (we have dropped the c's). Let V be the spaceship's 4-velocity, and let A be its 4-acceleration. In terms of the proper time τ:

(a) Find V and A in frame S, by explicitly using $v(\tau) = \tanh(a\tau)$.
(b) Write down V' and A' in the spaceship's frame, S'.
(c) Verify that V and V' transform like 4-vectors between the two frames. Likewise for A and A'.

13.8 Exercises

13.5. **Acceleration at rest**

Show that the derivative of $v \equiv \sqrt{v_x^2 + v_y^2 + v_z^2}$ equals a_x, independent of how the various v_i's are changing, provided that $v_y = v_z = 0$ at the moment in question.

13.6. **Linear acceleration** *

A particle's velocity and acceleration both point in the x direction, with magnitudes v and $\dot{v}$, respectively (as measured in the lab frame). In the spirit of the example in Section 13.5.2, find the 3-acceleration and 4-acceleration in both the lab frame and the instantaneous inertial frame of the particle. Verify that the 3-accelerations are related according to Eq. (13.26).

13.7. **Linear force** *

For the setup in the previous exercise, find the 3-force and 4-force in both the lab frame and the instantaneous inertial frame of the particle. Verify that the 3-forces are related according to Eq. (13.22).

13.8. **Circular motion force** *

For the setup in the example in Section 13.5.2, find the 3-force and 4-force in both the lab frame and the instantaneous inertial frame of the particle. Verify that the 3-forces are related according to Eq. (13.22). (Solve this from scratch; don't just use the results from the example.)

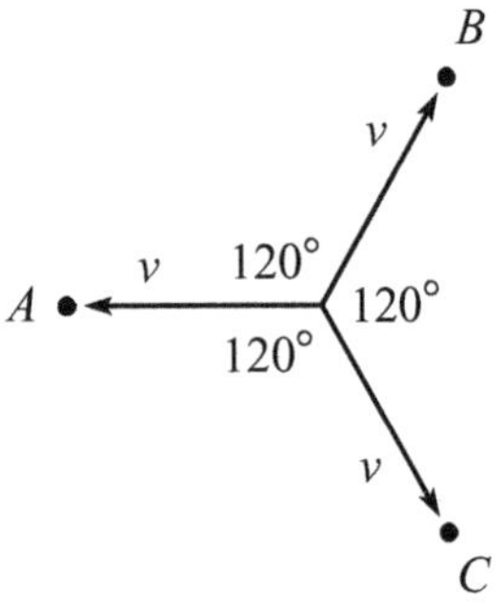

Fig. 13.9

13.9. **Same speed** *

Consider the setup in Problem 13.2. Given v, what should θ be so that the speed of one particle, as viewed by the other, is also v? (The first form of the speed in Eq. (13.38) will make your calculations the simplest.) Do your answers make sense for $v \approx 0$ and $v \approx c$?

13.10. **Doppler effect** *

Consider a photon traveling in the x direction. Ignoring the y and z components, and setting $c = 1$, the 4-momentum is (p, p). In matrix notation, what are the Lorentz transformations for the frames traveling to the left and to the right at speed v? What is the new 4-momentum of the photon in these new frames? Accepting the fact that a photon's energy is proportional to its frequency, verify that your results are consistent with the Doppler results in Section 11.8.1.

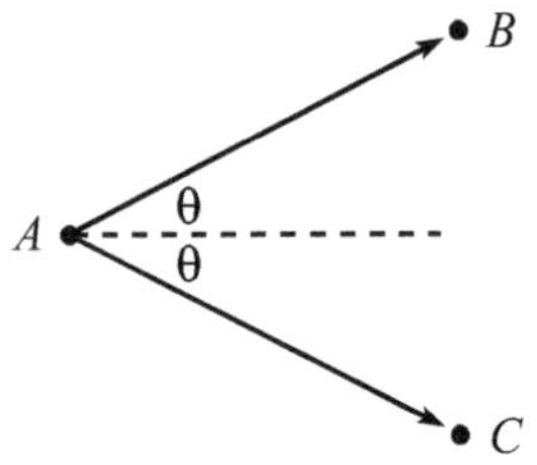

Fig. 13.10

13.11. **Three particles** **

Three particles head off with equal speeds v, at 120° with respect to each other, as shown in Fig. 13.9. What is the inner product of any two of the 4-velocities in any frame? Use your result to find the angle θ (see Fig. 13.10) at which two particles travel in the frame of the third.

13.9 Solutions

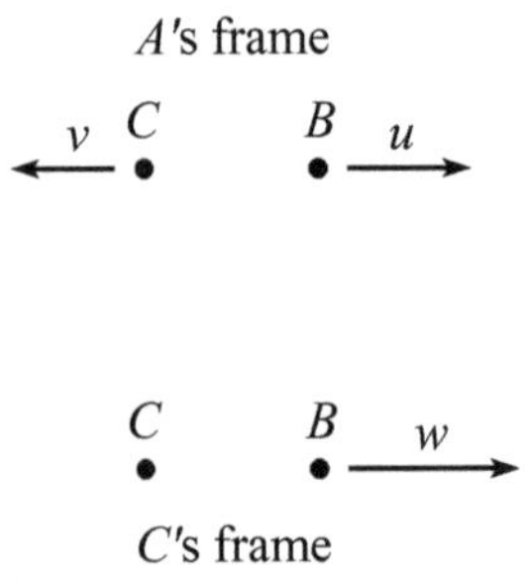

Fig. 13.11

13.1. **Velocity addition**

Let the desired speed of B with respect to C be w (see Fig. 13.11). In A's frame, the 4-velocity of B is $(\gamma_u, \gamma_u u)$, and the 4-velocity of C is $(\gamma_v, -\gamma_v v)$, where we have suppressed the y and z components. In C's frame, the 4-velocity of B is $(\gamma_w, \gamma_w w)$, and the 4-velocity of C is $(1, 0)$. The invariance of the inner product implies that

$$\begin{aligned}(\gamma_u, \gamma_u u) \cdot (\gamma_v, -\gamma_v v) &= (\gamma_w, \gamma_w w) \cdot (1, 0) \\ \Longrightarrow \quad \gamma_u \gamma_v (1 + uv) &= \gamma_w \\ \Longrightarrow \quad \frac{1 + uv}{\sqrt{1-u^2}\sqrt{1-v^2}} &= \frac{1}{\sqrt{1-w^2}}.\end{aligned} \tag{13.33}$$

Squaring and then solving for w gives

$$w = \frac{u+v}{1+uv}. \tag{13.34}$$

13.2. **Relative speed**

In the lab frame, the 4-velocities of the particles are (suppressing the z component)

$$(\gamma_v, \gamma_v v\cos\theta, \gamma_v v\sin\theta) \quad \text{and} \quad (\gamma_v, \gamma_v v\cos\theta, -\gamma_v v\sin\theta). \tag{13.35}$$

Let w be the desired speed of one particle as viewed by the other. Then in the frame of one particle, the 4-velocities are (suppressing two spatial components)

$$(\gamma_w, \gamma_w w) \quad \text{and} \quad (1, 0), \tag{13.36}$$

where we have rotated the axes so that the relative motion is along the x axis in this frame. Since the 4-vector inner product is invariant under Lorentz transformations and rotations, we have (using $\cos 2\theta = \cos^2\theta - \sin^2\theta$)

$$\begin{aligned}(\gamma_v, \gamma_v v\cos\theta, \gamma_v v\sin\theta) \cdot (\gamma_v, \gamma_v v\cos\theta, -\gamma_v v\sin\theta) &= (\gamma_w, \gamma_w w) \cdot (1, 0)\\ \Longrightarrow \quad \gamma_v^2(1 - v^2\cos 2\theta) &= \gamma_w. \end{aligned} \tag{13.37}$$

Using the definitions of the γ's, squaring, and solving for w gives

$$w = \sqrt{1 - \frac{(1-v^2)^2}{(1-v^2\cos 2\theta)^2}} = \frac{\sqrt{2v^2(1-\cos 2\theta) - v^4\sin^2 2\theta}}{1 - v^2\cos 2\theta}. \tag{13.38}$$

If desired, this can be rewritten (using some double-angle formulas) in the form,

$$w = \frac{2v\sin\theta\sqrt{1 - v^2\cos^2\theta}}{1 - v^2\cos 2\theta}. \tag{13.39}$$

See the solution to Problem 11.14 for some limiting cases.

13.3. **Another relative speed**

In the lab frame, the 4-velocities of the particles are (suppressing the z component)

$$V_A = (\gamma_u, \gamma_u u, 0) \quad \text{and} \quad V_B = (\gamma_v, \gamma_v v\cos\theta, \gamma_v v\sin\theta). \tag{13.40}$$

Let w be the desired speed of one particle as viewed by the other. Then in the frame of one particle, the 4-velocities are (suppressing two spatial components)

$$(\gamma_w, \gamma_w w) \quad \text{and} \quad (1, 0), \tag{13.41}$$

where we have rotated the axes so that the relative motion is along the x axis in this frame. Since the 4-vector inner product is invariant under Lorentz transformations and rotations, we have

$$\begin{aligned}(\gamma_u, \gamma_u u, 0) \cdot (\gamma_v, \gamma_v v\cos\theta, \gamma_v v\sin\theta) &= (\gamma_w, \gamma_w w) \cdot (1, 0)\\ \Longrightarrow \quad \gamma_u\gamma_v(1 - uv\cos\theta) &= \gamma_w. \end{aligned} \tag{13.42}$$

Using the definitions of the γ's, squaring, and solving for w gives

$$w = \sqrt{1 - \frac{(1-u^2)(1-v^2)}{(1-uv\cos\theta)^2}} = \frac{\sqrt{u^2 + v^2 - 2uv\cos\theta - u^2v^2\sin^2\theta}}{1 - uv\cos\theta}. \tag{13.43}$$

See the solution to Problem 11.15 for some limiting cases.

13.4. **Acceleration for linear motion**

(a) Using $v(\tau) = \tanh(a\tau)$, we have $\gamma = 1/\sqrt{1-v^2} = \cosh(a\tau)$. Therefore,

$$V = (\gamma, \gamma v) = \big(\cosh(a\tau), \sinh(a\tau)\big), \tag{13.44}$$

where we have suppressed the two transverse components. We then have

$$A = \frac{dV}{d\tau} = a\big(\sinh(a\tau), \cosh(a\tau)\big). \tag{13.45}$$

(b) The spaceship is at rest in its instantaneous inertial frame, so

$$V' = (1, 0) \quad \text{and} \quad A' = (0, a). \tag{13.46}$$

Equivalently, these are obtained by setting $\tau = 0$ in the results from part (a), because the spaceship hasn't started moving at $\tau = 0$, as is always the case in the instantaneous rest frame.

(c) The Lorentz transformation matrix from S' to S is

$$\mathcal{M} = \begin{pmatrix} \gamma & \gamma v \\ \gamma v & \gamma \end{pmatrix} = \begin{pmatrix} \cosh(a\tau) & \sinh(a\tau) \\ \sinh(a\tau) & \cosh(a\tau) \end{pmatrix}. \tag{13.47}$$

We must check that

$$\begin{pmatrix} V_0 \\ V_1 \end{pmatrix} = \mathcal{M} \begin{pmatrix} V'_0 \\ V'_1 \end{pmatrix} \quad \text{and} \quad \begin{pmatrix} A_0 \\ A_1 \end{pmatrix} = \mathcal{M} \begin{pmatrix} A'_0 \\ A'_1 \end{pmatrix}. \tag{13.48}$$

These are easily seen to be true.

Chapter 14

General Relativity

This will be somewhat of a strange chapter, because we won't have enough time to get to the heart of General Relativity (GR).[1] But we'll still be able to get a flavor of the subject and derive a few interesting GR results. One crucial idea in GR is the Equivalence Principle. This says that gravity is equivalent to acceleration. Or in more practical terms, it says that you can't tell the difference. We'll have much to say about this in the sections below. Another crucial concept in GR is that of coordinate independence: the laws of physics can't depend on what coordinate system you choose. This seemingly innocuous statement has surprisingly far-reaching consequences. However, a discussion of this topic is one of the many things we won't have time for. We would need a whole book on GR to do it justice. But fortunately it's possible to get a sense of the nature of GR without having to master such things. This is the route we'll take in this chapter.

14.1 The Equivalence Principle

Einstein's Equivalence Principle says that it is impossible to locally distinguish between gravity and acceleration. This may be stated more precisely in (at least) three ways.

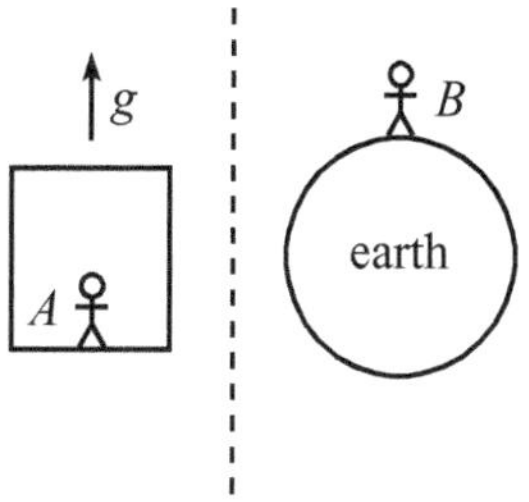

Fig. 14.1

- Let person A be enclosed in a small box, far from any massive objects, that undergoes uniform acceleration (say, g). Let person B stand at rest on the earth (see Fig. 14.1). The Equivalence Principle says that there are no local experiments these two people can perform that will tell them which of the two settings they are in. The physics of each setting is the same.
- Let person A be enclosed in a small box that is in free-fall near a planet. Let person B float freely in space, far away from any massive objects (see Fig. 14.2). The Equivalence Principle says that there are no local experiments these two people can perform that will tell them which of the two settings they are in. The physics of each setting is the same.

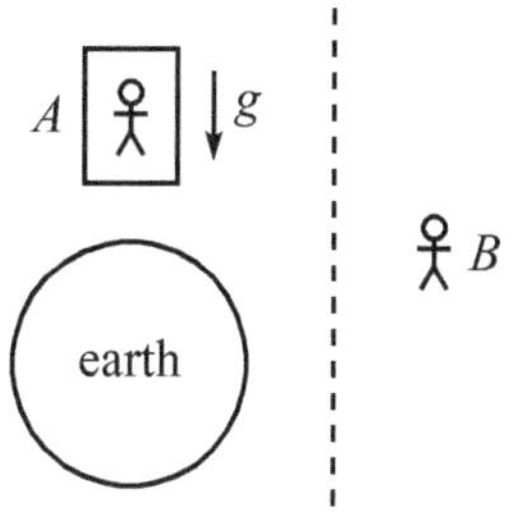

Fig. 14.2

[1] Ten years after his 1905 paper on the Special Theory of Relativity, Einstein completed his General Theory of Relativity in 1915, collaborating with Marcel Grossmann during the latter part of this period. David Hilbert also developed many of the final pieces of the theory in parallel with Einstein; see Medicus (1984). For a wonderful account of other historical developments of the theory, see Chandrasekhar (1979).

- "Gravitational" mass is equal to (or proportional to) "inertial" mass. Gravitational mass is the m_g that appears in the formula, $F = GMm_g/r^2 \equiv m_g g$. Inertial mass is the m_i that appears in the formula, $F = m_i a$. There is no a-priori reason why these two m's should be the same (or proportional). An object that is dropped on the earth has acceleration $a = (m_g/m_i)g$. For all we know, the ratio m_g/m_i for plutonium might be different from that for copper. But experiments with various materials have detected no difference in the ratios. The Equivalence Principle states that the ratios are equal for any type of mass.

  This definition of the Equivalence Principle is equivalent to, say, the second one above for the following reason. Two different masses that start at rest near B will stay right where they are as they float freely in space. But two different masses that start at rest near A will stay next to each other if and only if their accelerations are equal, that is, if and only if the ratio m_g/m_i is the same for both. If this ratio is different for the two masses, then they will diverge from each other, which means that it is possible to distinguish between the two settings.

These statements are all quite believable. Consider the first one, for example. When standing on the earth, you have to keep your legs firm to avoid falling down. When standing in the accelerating box, you have to keep your legs firm to maintain the same position relative to the floor (that is, to avoid "falling down"). You certainly can't naively tell the difference between the two scenarios. The Equivalence Principle says that it's not just that you're too inept to figure out a way to differentiate between them, but instead that there is no possible local experiment you can perform to tell the difference, no matter how clever you are.

REMARK: Note the inclusion of the words "small box" and "local" above. On the surface of the earth, the lines of the gravitational force are not parallel; they converge to the center. The gravitational force also varies with height. Therefore, an experiment performed over a non-negligible distance (for example, dropping two balls next to each other, and watching them converge; or dropping two balls on top of each other and watching them diverge) will have different results from the same experiment in the accelerating box. The Equivalence Principle says that if your laboratory is small enough, or if the gravitational field is sufficiently uniform, then the two scenarios look essentially the same. ♣

14.2 Time dilation

The Equivalence Principle has a striking consequence concerning the behavior of clocks in a gravitational field. It implies that higher clocks run faster than lower clocks. If someone puts a watch on top of a tower, and if you stand on the ground, then you will see the watch on the tower tick faster than an identical watch on your wrist. If the watch on the tower is taken down and you compare it with the one on your wrist, it will show more time elapsed.[2] Likewise, someone standing

[2] This is true only if the watch on the tower is kept there for a long enough time (assuming that the watches show the same time when you start observing), because the movement of the watch when

on top of the tower will see a clock on the ground run slow. To be quantitative about this, let's consider the following two scenarios.

- A light source on top of a tower of height h emits flashes at time intervals t_s. A receiver on the ground receives the flashes at time intervals t_r (see Fig. 14.3). What is t_r in terms of t_s?
- A rocket of length h accelerates with acceleration g. A light source at the front end emits flashes at time intervals t_s. A receiver at the back end receives the flashes at time intervals t_r (see Fig. 14.4). What is t_r in terms of t_s?

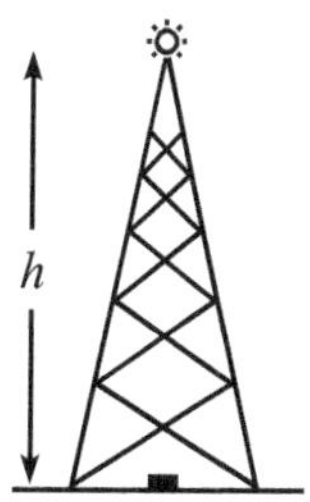

Fig. 14.3

The Equivalence Principle tells us that these two scenarios are exactly the same, as far as the sources and receivers are concerned. Hence, the relation between t_r and t_s must be the same in each. Therefore, to find out what is going on in the first scenario, we will study the second scenario (because we can figure out how this one behaves).

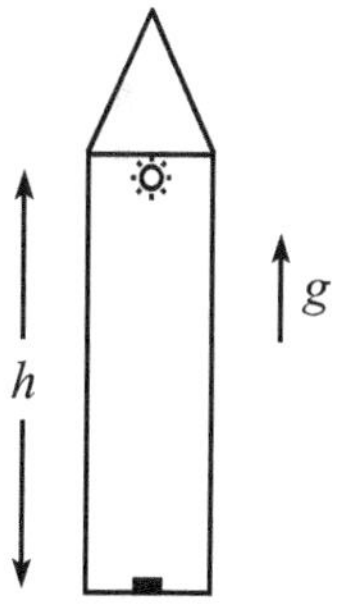

Fig. 14.4

Consider an instantaneous inertial frame, S, of the rocket. In this frame, the rocket is momentarily at rest (at, say, $t = 0$), and then it accelerates out of the frame with acceleration g. The following discussion will be made with respect to the frame S. Consider a series of quick light pulses emitted from the source, starting at $t = 0$. The distance the rocket has traveled out of S at time t is $gt^2/2$, so if we assume that t_s is very small, then we may say that many light pulses are emitted before the rocket moves appreciably. Likewise, the speed of the source, namely gt, is also very small. We may therefore ignore the motion of the rocket, as far as the light source is concerned.

However, the light takes a finite time to reach the receiver, and by then the receiver will be moving. We therefore *cannot* ignore the motion of the rocket when dealing with the receiver. The time it takes the light to reach the receiver is h/c, at which point the receiver has a speed of $v = g(h/c)$.[3] Therefore, by the usual classical Doppler effect, the time between the received pulses is[4]

$$t_r = \frac{t_s}{1 + (v/c)}. \tag{14.1}$$

it is taken down causes it to run slow, due to the usual special-relativistic time dilation. But the speeding-up effect due to the height can be made arbitrarily large compared with the slowing-down effect due to the motion, by simply keeping the watch on the tower for an arbitrarily long time.

[3] The receiver moves a tiny bit during this time, so the "h" here should really be replaced by a slightly smaller distance. But this yields a negligible second-order effect in the small quantity gh/c^2, as you can show. To sum up, the displacement of the source, the speed of the source, and the displacement of the receiver are all negligible. But the speed of the receiver is quite relevant.

[4] Quick proof of the classical Doppler effect (for a moving receiver): As seen in frame S, when the receiver and a particular pulse meet, the next pulse is a distance ct_s behind. The receiver and this next pulse then travel toward each other at relative speed $c + v$ (as measured by someone in S). The time difference between receptions is therefore $t_r = ct_s/(c + v)$.

So the frequencies, $f_r = 1/t_r$ and $f_s = 1/t_s$, are related by

$$f_r = \left(1 + \frac{v}{c}\right) f_s = \left(1 + \frac{gh}{c^2}\right) f_s. \tag{14.2}$$

Returning to the clock-on-tower scenario, the Equivalence Principle tells us that an observer on the ground must see the clock on the tower running fast, by a factor $1 + gh/c^2$. This means that the upper clock really *is* running fast, compared with the lower clock.[5] That is,

$$\Delta t_h = \left(1 + \frac{gh}{c^2}\right) \Delta t_0. \tag{14.3}$$

A twin from Denver will be older than his twin from Boston when they meet up at a family reunion (all other things being equal).

Greetings! Dear brother from Boulder,
I hear that you've gotten much older.
And please tell me why
My lower left thigh
Hasn't aged quite as much as my shoulder!

Note that the gh in Eq. (14.3) is the gravitational potential energy divided by m.

REMARK: You might object to the above derivation, because t_r is the time measured by someone in the inertial frame S. And since the receiver is eventually moving with respect to S, we should multiply the f_r in Eq. (14.2) by the usual special-relativistic time-dilation factor, $1/\sqrt{1-(v/c)^2}$ (because the receiver's clocks are running slow relative to S, so the frequency measured by the receiver is greater than that measured in S). However, this is a second-order effect in the small quantity $v/c = gh/c^2$. We already dropped other effects of the same order, so we have no right to keep this one. Of course, if the leading effect in our final answer turned out to be second order in v/c, then we would know that our answer was garbage. But the leading effect happens to be first order, so we can afford to be careless with the second-order effects. ♣

After a finite time has passed, the frame S will no longer be of any use to us. But we can always pick a new instantaneous rest frame of the rocket, so we can repeat the above analysis at any later time. Therefore, the result in Eq. (14.2) holds at all times.

This gravitational time-dilation effect was first measured by R. Pound and G. Rebka in 1960. By sending gamma rays both up and down a 22 m tower, they were able to measure the redshift (that is, the decrease in frequency) at the top. This was a notable feat indeed, considering that they were able to measure a frequency shift of gh/c^2 (which is only a few parts in 10^{15}) to within 10% accuracy. In 1964, R. Pound and J. Snider improved the accuracy to 1%.

[5] Unlike the situation where two people fly past each other (as with the usual twin paradox), we can say here that what an observer *sees* is also what actually *is*. We can say this because everyone here is in the same frame. The "turnaround" effect that was present in the twin paradox isn't present now. The two clocks can be slowly moved together without anything exciting or drastic happening to their readings.

14.3 Uniformly accelerating frame

Before reading this section, you should think carefully about the "Break or not break" problem in Chapter 11 (Problem 11.26). Don't look at the solution too soon, because chances are you will change your answer after a few more minutes of thought. This is a classic problem, so don't waste it by peeking!

Technically, the uniformly accelerating frame we'll construct here has nothing to do with GR. We won't need to leave the realm of Special Relativity for the analysis in this section. But the reason we choose to study this special-relativistic setup in detail is that it shows many similarities to genuine GR situations, such as black holes.

14.3.1 Uniformly accelerating point particle

In order to understand a uniformly accelerating frame, we first need to understand a uniformly accelerating point particle. In Section 11.9, we briefly discussed the motion of a uniformly accelerating particle, that is, one that feels a constant force in its instantaneous rest frame. We'll now take a closer look at such a particle. Let the particle's instantaneous rest frame be S', and let it start at rest in the inertial frame S. Let its mass be m. We know from Section 12.5.3 that the longitudinal force is the same in the two frames. Therefore, since it is constant in S', it is also constant in S. Call it f. If we let $g \equiv f/m$ (so g is the proper acceleration felt by the particle), then in S we have, using the fact that f is constant,

$$f = \frac{dp}{dt} = \frac{d(m\gamma v)}{dt} \quad \Longrightarrow \quad gt = \gamma v \quad \Longrightarrow \quad v = \frac{gt}{\sqrt{1+(gt)^2}}, \tag{14.4}$$

where we have set $c = 1$. As a double-check, this has the correct behavior for $t \to 0$ and $t \to \infty$. If you want to keep the c's in, then $(gt)^2$ becomes $(gt/c)^2$ to make the units right. Having found the speed in S at time t, the position in S at time t is given by

$$x = \int_0^t v\,dt = \int_0^t \frac{gt\,dt}{\sqrt{1+(gt)^2}} = \frac{1}{g}\left(\sqrt{1+(gt)^2} - 1\right). \tag{14.5}$$

For convenience, let P be the point (see Fig. 14.5)

$$(x_P, t_P) = (-1/g, 0). \tag{14.6}$$

Then Eq. (14.5) yields

$$(x - x_P)^2 - t^2 = \frac{1}{g^2}. \tag{14.7}$$

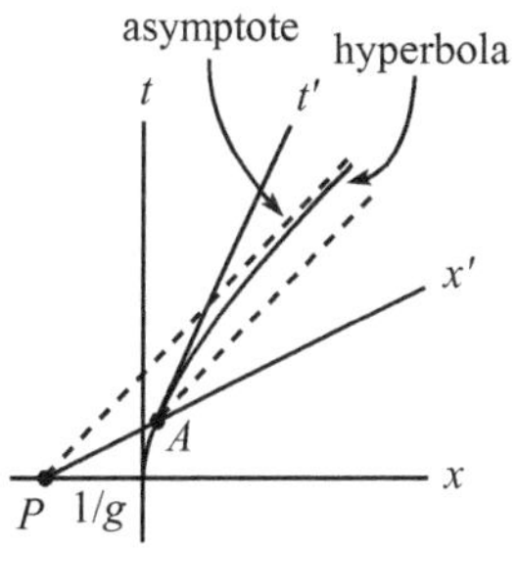

Fig. 14.5

This is the equation for a hyperbola with its center (defined as the intersection of the asymptotes) at point P. For a large acceleration g, the point P is very close to the particle's starting point. For a small acceleration, it is far away.

Everything has been fairly normal up to this point, but now the fun begins. Consider a point A on the particle's hyperbolic worldline at time t. From Eq. (14.5), A has coordinates

$$(x_A, t_A) = \left(\frac{1}{g}\left(\sqrt{1+(gt)^2}-1\right), t\right). \tag{14.8}$$

The slope of the line PA is therefore

$$\frac{t_A - t_P}{x_A - x_P} = \frac{gt}{\sqrt{1+(gt)^2}}. \tag{14.9}$$

Looking at Eq. (14.4), we see that this slope equals the speed v of the particle at point A. But we know very well that the speed v is the slope of the particle's instantaneous x' axis; see Eq. (11.47). Therefore, the line PA and the particle's x' axis are the same line. This holds for any arbitrary time t. So we may say that at any point along the particle's worldline, the line PA is the instantaneous x' axis of the particle. Or, said another way, no matter where the particle is, the event at P is simultaneous with an event located at the particle, as measured in the instantaneous frame of the particle. In other words, the particle always says that P happens "now."[6]

Here is another strange fact. What is the distance from P to A, as measured in an instantaneous rest frame, S', of the particle? The γ factor between frames S and S' is, using Eq. (14.4), $\gamma = \sqrt{1+(gt)^2}$. The distance between P and A in frame S is $x_A - x_P = \sqrt{1+(gt)^2}\,/g$. So the distance between P and A in frame S' is (using the Lorentz transformation $\Delta x = \gamma(\Delta x' + v\Delta t')$, with $\Delta t' = 0$)

$$x'_A - x'_P = \frac{1}{\gamma}(x_A - x_P) = \frac{1}{g}. \tag{14.10}$$

This has the unexpected property of being independent of t. Therefore, not only do we find that P is always simultaneous with the particle, as measured in the particle's instantaneous rest frame; we also find that P is always the same distance (namely $1/g$) away from the particle, in the particle's frame. This is rather strange. The particle accelerates away from point P, but it doesn't get farther away from it, as measured in its own frame.

REMARK: We can give a continuity argument that shows that such a point P must exist. If a point is close to you, and if you accelerate away from it, then of course you get farther away from it. Everyday experience is quite valid here. But if a point is sufficiently far away from you, and if you accelerate away from it, then the $at^2/2$ distance you travel away from it can easily be compensated by the decrease in distance due to length contraction (brought about by your newly acquired velocity). This effect grows with distance, so we simply need to pick the

[6] The point P is very much like the event horizon of a black hole. Time seems to stand still at P in this treatment. And if we went more deeply into GR, we would find that time seems to stand still at the edge of a black hole, too (as viewed by someone far away).

point to be sufficiently far away. What this means is that every time you get out of your chair and walk to the door, there are stars very far away behind you that get closer to you as you walk away from them (as measured in your instantaneous rest frame). By continuity, then, there must exist a point P that remains the same distance from you (in your frame) as you accelerate away from it. ♣

14.3.2 Uniformly accelerating frame

Let's now put a collection of uniformly accelerating particles together to make a uniformly accelerating frame. Our goal will be to create a frame in which the distances between particles (as measured in any particle's instantaneous rest frame) remain constant. Why is this our goal? We know from the "Break or not break" problem in Chapter 11 that if all the particles accelerate with the same proper acceleration, g, then the distances (as measured in a particle's instantaneous rest frame) grow larger. While this is a perfectly possible frame to construct, it is not desirable, for the following reason. Einstein's Equivalence Principle states that an accelerating frame is equivalent to a frame sitting on, say, the earth. We can therefore study the effects of gravity by studying an accelerating frame. But if we want this frame to look anything like the surface of the earth, we certainly can't have distances that change over time. We therefore want to construct a *static* frame, that is, one in which distances do not change (as measured in the frame). This will allow us to say that if we enclose the frame in a windowless box, then for all a person inside knows, he is standing motionless in a static gravitational field (which has a certain definite form, as we shall see).

Let's figure out how to construct the frame. We'll discuss the acceleration of just two particles here. Others can be added in a similar manner. In the end, the desired frame as a whole is constructed by accelerating each atom in the floor of the frame with a specific proper acceleration. From Section 14.3.1, we already have a particle A that is "centered" around the point P. (This will be our shorthand notation for "traveling along a branch of a hyperbola whose center is the point P.") We claim, for reasons that will become clear, that every other particle in the frame should also be "centered" around the same point P.

Consider another particle, B. Let a and b be the initial distances from P to A and B. If both particles are to be centered around P, then their proper accelerations must be, from Eq. (14.6),

$$g_A = \frac{1}{a}, \quad \text{and} \quad g_B = \frac{1}{b}. \tag{14.11}$$

Therefore, in order to have all points in the frame be centered around P, we simply have to make their proper accelerations inversely proportional to their initial distances from P.

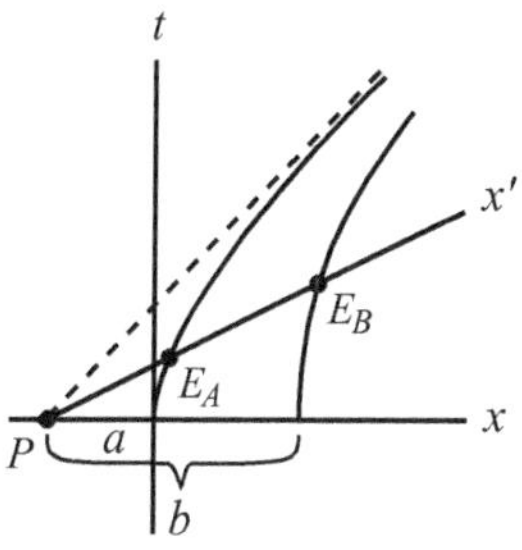

Fig. 14.6

Why do we want every particle to be centered around P? Consider two events, E_A and E_B, such that P, E_A, and E_B are collinear in Fig. 14.6. From Section 14.3.1, we know that the line PE_AE_B is the x' axis for both particle

A and particle B, at the positions shown. We also know that A is always a distance a from P, and B is always a distance b from P (in their frame). Combining these facts with the fact that A and B measure their distances along the x' axis of the same frame (at the events shown in the figure), we see that both A and B measure the distance between them to be $b - a$. This is independent of t, so A and B measure a constant distance between them. We have therefore constructed our desired static frame. This frame is often called a "Rindler space." If a person walks around in the frame, he will think that he lives in a static world where the acceleration due to gravity takes the form $g(z) \propto 1/z$, where z is the distance to a certain magical point which is located at the end of the known "universe."

What if a person releases himself from the accelerating frame, so that he forever sails through space at constant speed? His view is that he is falling. He falls past the "magical point" P in a finite proper time, because the hyperbolic worldline of a point infinitesimally close to P is essentially the asymptote of all the hyperbolas, and the person's straight-line worldline intersects this line. But his friends who are still in the frame will see him take an infinitely long time to get to P, because the x' axes of points in the frame never quite swing up to the asymptote, even after an infinite amount of time. So the "now" line of any point in the frame never quite passes through the event where the person crosses the asymptote. This is similar to the situation with a black hole. An outside observer will see it take an infinitely long time for a falling person to reach the "boundary" of a black hole, even though it takes a finite proper time for the person.

Our analysis shows that A and B feel different proper accelerations, because $a \neq b$. There is no way to construct a static frame where all points feel the same proper acceleration, so it is impossible to mimic a constant gravitational field (over a finite distance) by using an accelerating frame. The problems and exercises for this chapter offer plenty of opportunity for you to play around with the properties of our uniformly accelerating frame.

14.4 Maximal-proper-time principle

The maximal-proper-time principle in General Relativity says: Given two events in spacetime, a particle under the influence of only gravity takes the path in spacetime that maximizes the proper time. For example, if you throw a ball from given coordinates $(\mathbf{r}_1, t_1)$, and it lands at given coordinates $(\mathbf{r}_2, t_2)$, then the claim is that the ball takes the path that maximizes its proper time.[7]

This is clear for a freely moving ball in outer space, far from any massive objects. The ball travels at constant speed from one point to another, and we

[7] The principle is actually the "*stationary*-proper-time principle," because as with the Lagrangian formalism in Chapter 6, any type of stationary point (a maximum, minimum, or saddle point) is allowed. But although we were very careful about stating things properly in Chapter 6, we'll be a little sloppy here and just use the word "maximum," because that's what it will generally turn out to be in the situations we'll look at. However, see Problem 14.8.

know that this constant-speed motion is the motion with the maximal proper time. This is true because a ball, A, moving at constant speed sees the clock on any other ball, B, running slow due to the special-relativistic time dilation, if there is a relative speed between them. (We're assuming that B's nonuniform velocity is caused by a nongravitational force acting on it.) B therefore shows a shorter elapsed time. This argument doesn't work the other way around, because B isn't in an inertial frame and therefore can't use the special-relativistic time-dilation result.

Consistency with Newtonian physics

The maximal-proper-time principle sounds like a plausible idea, but we already know from Chapter 6 that the path an object takes is the one that yields a stationary value of the classical action, $\int (T - V)$. We must therefore demonstrate that the maximal-proper-time principle reduces to the stationary-action principle, in the limit of small velocities. If this weren't the case, then we'd have to throw out the maximal-proper-time principle.

Consider a ball thrown vertically on the earth. Assume that the initial and final coordinates are fixed to be (y_1, t_1) and (y_2, t_2). Our plan will be to assume that the maximal-proper-time principle holds, and to then show that this leads to the stationary-action principle. Before being quantitative, let's get a qualitative idea about what's going on with the ball. There are two competing effects, as far as maximizing the proper time goes. On one hand, the ball wants to climb very high, because its clock will run faster there, due to the GR time dilation. But on the other hand, if it climbs very high, then it must move very fast to get there (because the total time, $t_2 - t_1$, is fixed), and this will make its clock run slow, due to the SR time dilation. So there's a tradeoff. Let's now look quantitatively at the implications of this tradeoff. The goal is to maximize

$$\tau = \int_{t_1}^{t_2} d\tau. \tag{14.12}$$

Due to the motion of the ball, we have the usual time dilation, $d\tau = \sqrt{1 - v^2/c^2}\, dt$. But due to the height of the ball, we also have the gravitational time dilation, $d\tau = (1 + gy/c^2)\, dt$. Combining these effects gives[8]

$$d\tau = \sqrt{1 - \frac{v^2}{c^2}} \left(1 + \frac{gy}{c^2}\right) dt. \tag{14.13}$$

[8] This result is technically incorrect, because the two effects are intertwined in a somewhat more complicated manner (see Exercise 14.20). But it's valid up to first order in v^2/c^2 and gy/c^2, which is all we're concerned with here.

Using the Taylor expansion for $\sqrt{1-\epsilon}$, and dropping terms of order $1/c^4$ and smaller, we see that we want to maximize

$$\int_{t_1}^{t_2} d\tau \approx \int_{t_1}^{t_2} \left(1 - \frac{v^2}{2c^2}\right)\left(1 + \frac{gy}{c^2}\right) dt \approx \int_{t_1}^{t_2} \left(1 - \frac{v^2}{2c^2} + \frac{gy}{c^2}\right) dt. \quad (14.14)$$

The "1" term gives a constant, so maximizing this integral is equivalent to minimizing

$$mc^2 \int_{t_1}^{t_2} \left(\frac{v^2}{2c^2} - \frac{gy}{c^2}\right) dt = \int_{t_1}^{t_2} \left(\frac{mv^2}{2} - mgy\right) dt, \quad (14.15)$$

which is the classical action, as desired. For a one-dimensional gravitational problem such as this one, the action is always a (global) minimum, and the proper time is always a (global) maximum, as you can show by considering the second-order change in the action (see Exercise 14.23, which is the same as Exercise 6.32).

In retrospect, it isn't surprising that the kinetic energy here works out. The factor of $1/2$ comes about in exactly the same way as in the derivation in Eq. (12.9), where we showed that the relativistic form of energy reduces to the familiar Newtonian expression. As far as the potential energy goes, the gy here can be traced to the acceleration times a certain time, where this time is proportional to the distance; see the paragraph before Eq. (14.1). This yields (when multiplied by mc^2) the usual force times distance expression for the potential.

The maximal-proper-time principle here is equivalent to the statement that the action for the Lagrangian we introduced near the end of Section 12.1 is $S = -mc\int d\tau$. In both cases we want to extremize the integral of the proper time. In Section 12.1 we dealt only with a free particle, so the treatment here is a little more general because it includes the effect of gravity. But note that although Eq. (14.15) has the obvious interpretation of including the gravitational potential, the starting point in Eq. (14.12) makes no mention of a gravitational force. In the present treatment, gravity shouldn't be thought of as a force, but instead as something that affects the proper time.[9]

14.5 Twin paradox revisited

Let's take another look at the standard twin paradox, this time from the perspective of General Relativity. We should emphasize that GR is by no means necessary for an understanding of the original formulation of the paradox (the first scenario below). We were able to solve it in Section 11.3.2, after all. The present discussion is given simply to show that the answer to an alternative formulation (the second

[9] For an interesting article on how the classical action, along with a few ingredients from differential geometry, can lead to General Relativity, see Rindler (1994).

scenario below) is consistent with what we've learned about GR. Consider the two following twin-paradox setups.

- Twin A floats freely in outer space. Twin B flies past A in a spaceship, with speed v_0 (see Fig. 14.7). At the instant they are next to each other, they both set their clocks to zero. At this same instant, B turns on the reverse thrusters of his spaceship and decelerates with proper deceleration g. B eventually reaches a farthest point from A and then accelerates back toward A, finally passing him with speed v_0 again. When they are next to each other, they compare the readings on their clocks. Which twin is younger?
- Twin B stands on the earth. Twin A is thrown upward with speed v_0 (let's say he is fired from a cannon in a hole in the ground; see Fig. 14.8). At the instant they are next to each other, they both set their clocks to zero. A rises up and then falls back down, finally passing B with speed v_0 again. When they are next to each other, they compare the readings on their clocks. Which twin is younger?

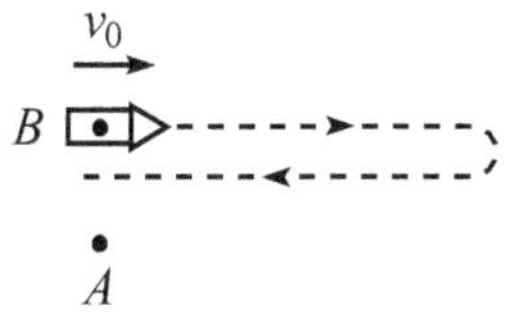

Fig. 14.7

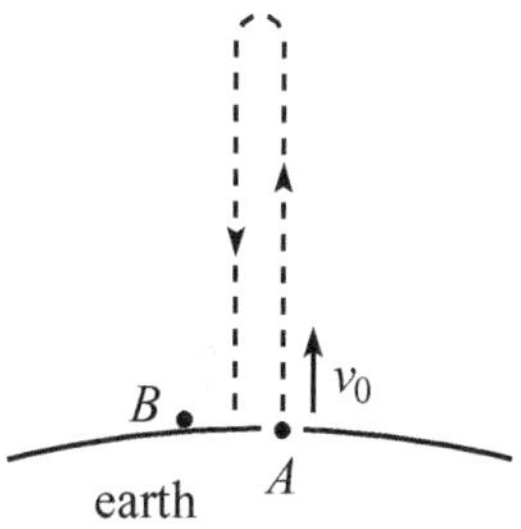

Fig. 14.8

The first scenario is easily solved using Special Relativity. Since A is in an inertial frame, he can apply the results of Special Relativity. In particular, he sees B's clock run slow, due to the usual special-relativistic time dilation. Therefore, B ends up younger at the end. B cannot use the reverse reasoning, because she is not in an inertial frame.

What about the second scenario? The key point to realize is that the Equivalence Principle says that these two scenarios are exactly the same (ignoring the nonuniformity of the earth's gravitational force), as far as the twins are concerned. Twin B has no way of knowing whether she is in a spaceship accelerating at g or on the surface of the earth. And A has no way of knowing whether he is floating freely in outer space or in free-fall in a gravitational field.[10] We therefore conclude that B must be younger in the second scenario, too.

At first glance, this seems incorrect, because in the second scenario, B is sitting motionless while A is the one who is moving. It seems that B should see A's clock running slow, due to the usual special-relativistic time dilation, and hence A should be younger. This reasoning is incorrect because it fails to take into account the gravitational time dilation. The fact of the matter is that A is higher in the gravitational field, and therefore his clock runs faster. This effect does indeed win out over the special-relativistic time dilation, and A ends up older. You can explicitly show this in Problem 14.11.

The reasoning in this section yields another way to conclude that the Equivalence Principle implies that higher clocks must run faster (in one way or another). The Equivalence Principle implies that A must be older in the second scenario, which means that there must be some height effect that makes A's clock run

[10] As mentioned in Section 14.1, this fact is made possible by the equivalence of inertial and gravitational mass. Were it not for this, different parts of A's body would want to accelerate at different rates in the gravitational field in the second scenario. This would certainly clue him in to the fact that he wasn't floating freely in space.

fast (fast enough to win out over the special-relativistic time dilation). But it takes some more work to show that the factor is actually $1 + gh/c^2$. Note that the fact that A is older is consistent with the maximal-proper-time principle. In both scenarios, A is under the influence of only gravity (zero gravity in the first scenario), whereas B feels a normal force from either the spaceship's floor or the ground.

14.6 Problems

Section 14.2: Time dilation

14.1. **Airplane's speed** *

A plane flies at constant height h. What should its speed be so that an observer on the ground sees the plane's clock tick at the same rate as a ground clock? (Assume $v \ll c$.)

14.2. **Clock on a tower** **

A clock starts on the ground and then moves up a tower at constant speed v. It sits on top of the tower for a time T and then descends at constant speed v. If the tower has height h, how long should the clock sit at the top so that it comes back showing the same time as a clock that remained on the ground? (Assume $v \ll c$.)

14.3. **Circular motion** **

B moves at speed v (with $v \ll c$) in a circle of radius r around A, far from any masses. By what fraction does B's clock run slower than A's? Calculate this in three ways. Work in:

(a) A's frame.
(b) The frame whose origin is B and whose axes remain parallel to an inertial set of axes.
(c) The rotating frame that is centered at A and rotates with the same frequency as B.

14.4. **More circular motion** **

A and B move at speed v ($v \ll c$) in a circle of radius r, at diametrically opposite points, far from any masses. They both see their clocks ticking at the same rate. Show this in three ways. Work in:

(a) The lab frame (the inertial frame whose origin is the center of the circle).
(b) The frame whose origin is B and whose axes remain parallel to an inertial set of axes.
(c) The rotating frame that is centered at the origin and rotates with the same frequency as A and B.

Section 14.3: Uniformly accelerating frame

14.5. **Accelerator's point of view** ***

A rocket starts at rest relative to a planet, a distance ℓ away. It accelerates toward the planet with proper acceleration g. Let τ and t be the readings on the rocket's and planet's clocks, respectively.

(a) Show that when the astronaut's clock reads τ, he observes the rocket–planet distance x (as measured in his instantaneous inertial frame) to be given by

$$1 + gx = \frac{1 + g\ell}{\cosh(g\tau)}. \tag{14.16}$$

(b) Show that when the astronaut's clock reads τ, he observes the time t on the planet's clock to be given by

$$gt = (1 + g\ell)\tanh(g\tau). \tag{14.17}$$

The results from Exercises 14.16 and 14.20 will be useful here.

14.6. **Getting way ahead** ****

A rocket with proper length L accelerates from rest, with proper acceleration g (where $gL \ll c^2$). Clocks are located at the front and back of the rocket. If we look at this setup in the rocket frame, then the GR time-dilation effect tells us that the times on the two clocks are related by $t_{\text{f}} = (1 + gL/c^2)t_{\text{b}}$. Therefore, if we look at things in the ground frame, then the times on the two clocks are related by

$$t_{\text{f}} = t_{\text{b}}\left(1 + \frac{gL}{c^2}\right) - \frac{Lv}{c^2}, \tag{14.18}$$

where the last term comes from the SR "rear clock ahead" result. Derive the above relation by working entirely in the ground frame.[11]

14.7. ***Lv/c^2* revisited** **

You stand at rest relative to a rocket that has synchronized clocks at its ends. It is then arranged for you and the rocket to move with relative speed v. A reasonable question to ask is: As viewed by you, what is the difference in readings on the clocks located at the ends of the rocket?

[11] You might find this relation surprising, because it implies that the front clock will eventually be an arbitrarily large time ahead of the back clock, in the ground frame. (The subtractive Lv/c^2 term is bounded by L/c and will therefore eventually become negligible compared with the additive, and unbounded, $(gL/c^2)t_{\text{b}}$ term.) But both clocks seem to be doing basically the same thing relative to the ground frame, so how can they end up differing by so much? Your job is to find out.

It turns out that this question cannot be answered without further information on how you and the rocket got to be moving with relative speed v. There are two basic ways this relative speed can come about. The rocket can accelerate while you sit there, or you can accelerate while the rocket sits there. Using the results from Problems 14.5 and 14.6, explain what the answers to the above question are in these two cases.

Section 14.4: Maximal-proper-time principle

14.8. **Circling the earth** **

Clock A sits at rest on the earth, and clock B circles the earth in an orbit that skims along the ground. Both A and B are essentially at the same radius, so the GR time-dilation effect yields no difference in their times. But B is moving relative to A, so A sees B running slow, due to the usual SR time-dilation effect. The orbiting clock, B, therefore shows a *smaller* elapsed proper time each time it passes A. In other words, the clock under the influence of only gravity (B) does *not* show the maximal proper time, in conflict with what we have been calling the maximal-proper-time principle. Explain.

Section 14.5: Twin paradox revisited

14.9. **Twin paradox** *

A spaceship travels at speed v ($v \ll c$) to a distant star. Upon reaching the star, it decelerates and then accelerates back up to speed v in the opposite direction (uniformly, and in a short time compared with the total journey time). By what fraction does the traveler age less than her twin on the earth? (Ignore the gravity from the earth.) Work in:

(a) The earth frame.
(b) The spaceship frame.

14.10. **Twin paradox again** **

(a) Answer part (b) of the previous problem, except now let the spaceship turn around by moving in a small semicircle while maintaining speed v.
(b) Answer part (b) of the previous problem, except now let the spaceship turn around by moving in an arbitrary manner. The only constraints are that the turnaround is done quickly (compared with the total journey time) and that it is contained in a small region of space (compared with the earth–star distance).

14.11. **Twin paradox times** ***

(a) In the first scenario in Section 14.5, calculate the ratio of B's elapsed time to A's, in terms of v_0 and g. Work in A's frame. Assume $v_0 \ll c$, and drop high-order terms.

(b) Do the same for the second scenario in Section 14.5. Do this from scratch using the time dilations, and then check that your answer agrees (within the accuracy of the calculations) with part (a), as the Equivalence Principle demands. Work in B's frame.

14.7 Exercises

Section 14.2: Time dilation

14.12. **Driving on a hill** *

You drive up and down a hill of height h at constant speed. The hill is in the shape of an isosceles triangle with altitude h. What should your speed be so that you age the same amount as someone standing at the base of the hill? (Assume $v \ll c$.)

14.13. **Lv/c^2 and gh/c^2** *

The familiar special-relativistic "rear clock ahead" result, Lv/c^2, looks rather similar to the gh/c^2 term in the GR time-dilation result, Eq. (14.3). Imagine standing on the ground near the front of a train of length L. For small v, devise a thought experiment that explains how the Lv/c^2 result follows from the gh/c^2 result.

14.14. **Both points of view** **

A and B are initially a distance L apart, at rest with respect to each other. At a given time, B accelerates toward A with constant proper acceleration a. Assume $aL \ll c^2$.

(a) Working in A's frame, calculate the difference in readings on A's and B's clocks when B reaches A.

(b) Do the same by working in B's frame, and show that the result agrees (neglecting higher-order effects) with the result from part (a).

14.15. **Opposite circular motion** ****

A and B move at speed v ($v \ll c$) in opposite directions around a circle of radius r (so they pass each other after each half-revolution), far from any masses. They both see their two clocks ticking, on average, at the same rate. That is, if they compare their clocks each time they pass each other, both clocks show the same time elapsed. Demonstrate this in three ways. Work in:

(a) The lab frame (the inertial frame whose origin is the center of the circle).
(b) The frame whose origin is B and whose axes remain parallel to an inertial set of axes.
(c) The rotating frame that is centered at the origin and rotates along with B. This part is very tricky; the solution is given in Cranor *et al.* (2000), but don't peek too soon.

Section 14.3: Uniformly accelerating frame

14.16. **Various quantities** $*$

A particle starts at rest and accelerates with proper acceleration g. Let τ be the time on the particle's clock. Starting with the v from Eq. (14.4), use time dilation to show that the time t in the original inertial frame, the speed of the particle, and the associated γ factor are given by (with $c = 1$)

$$gt = \sinh(g\tau), \quad v = \tanh(g\tau), \quad \gamma = \cosh(g\tau). \tag{14.19}$$

14.17. **Using rapidity** $*$

Another way to derive the v in Eq. (14.4) is to use the $v = \tanh(g\tau)$ rapidity result (where τ is the particle's proper time) from Section 11.9. Use time dilation to show that this implies $gt = \sinh(g\tau)$, which then implies Eq. (14.4).

14.18. **Speed in an accelerating frame** $*$

In the setup in Problem 14.5, use Eq. (14.16) to find the speed of the planet in the rocket's accelerating frame, $|dx/d\tau|$, as a function of τ. What is the maximum value of this speed, in terms of g and the initial distance ℓ?

14.19. **Redshift, blueshift** $**$

We found in Section 14.2 that a clock at the rear of a rocket sees a clock at the front run fast by a factor $1 + gh/c^2$. However, we ignored higher-order effects in $1/c^2$, so for all we know, we found only the first term in the Taylor series, and the factor is actually something like e^{gh/c^2} or perhaps $1 + \ln(1 + gh/c^2)$.

(a) For the uniformly accelerating frame in Section 14.3.2, show that the factor is in fact exactly $1 + g_r h/c^2$, where g_r is the acceleration of the rear of the rocket. Show this by lining up a series of clocks and looking at the successive factors between them. (*Hint*: Take the log of the product of the factors.) The overall factor takes a very nice form when written in terms of the a and b in Fig. 14.6; what is it?

(b) By the same reasoning as in part (a), it follows that the front clock sees the rear clock run slow by a factor $1 - g_f h/c^2$, where g_f is the acceleration of the front. Show explicitly that $(1 + g_r h/c^2)(1 - g_f h/c^2) = 1$, as must be the case, because a clock can't gain time with respect to itself (and because the two clocks are at rest in the frame of the rocket; this reasoning wouldn't apply to two clocks flying past each other).

14.20. **Gravity and speed combined** **

In the spirit of Problem 11.25 ("Acceleration and redshift"), use a Minkowski diagram to solve this problem. A rocket accelerates with proper acceleration g toward a planet. As measured in the instantaneous inertial frame of the rocket, the planet is a distance x away and moves at speed v. Everything is in one dimension here. As measured in the *accelerating* frame of the rocket, show that the planet's clock runs at a rate (with $c = 1$),

$$dt_p = dt_r(1 + gx)\sqrt{1 - v^2}\,, \tag{14.20}$$

and show that the planet's speed is

$$V = (1 + gx)v. \tag{14.21}$$

Note that if we combine these two results and eliminate v, and if we then invoke the Equivalence Principle, we arrive at the result that a clock moving at speed V at height h in a gravitational field is seen by someone on the ground to run at a rate (putting the c's back in),

$$\sqrt{\left(1 + \frac{gh}{c^2}\right)^2 - \frac{V^2}{c^2}}\,. \tag{14.22}$$

14.21. **Length contraction** **

A pencil points directly at you. From some distance away from it, you start at rest and accelerate toward it with acceleration a. After a time t, the pencil is length contracted in your frame by a factor $\sqrt{1 - v^2/c^2} \approx 1 - (at)^2/2c^2$, for small t (more precisely, for $at \ll c$). But the only way the pencil can shrink is for the back to move faster than the front, as measured by you in your accelerating frame. Using Eq. (14.21), show that the contraction factor works out as it should, for small t.

14.22. **Accelerating stick's length** **

Consider a uniformly accelerating frame consisting of a stick, the ends of which have worldlines given by the curves in Fig. 14.6 (so the stick has proper length $b - a$). At time t in the lab frame, we know from Eqs. (14.5) and (14.6) that a point that undergoes acceleration g has position $\sqrt{1 + (gt)^2}/g$ relative to the point P in Fig. 14.6. An observer in the original inertial frame sees the stick being length contracted by different factors along its length, because different points move with different speeds (at a given time in the original frame). Show, by doing the appropriate integral, that the inertial observer concludes that the stick always has proper length $b - a$.

Section 14.4: Maximal-proper-time principle

14.23. **Maximum proper time** *

Show that the stationary value of the gravitational action in Eq. (14.15) is always a minimum (which means that the proper time is always a maximum). Do this by considering a function, $y(t) = y_0(t) + \xi(t)$, where y_0 is the path that yields the stationary value, and ξ is a small variation.

Section 14.5: Twin paradox revisited

14.24. **Symmetric twin nonparadox** **

Two twins travel toward each other, both at speed v ($v \ll c$) with respect to an inertial observer. They synchronize their clocks when they pass each other. They travel to stars located at positions $\pm\ell$, and then decelerate and accelerate back up to speed v in the opposite direction (uniformly, and in a short time compared with the total journey time). In the frame of the inertial observer, it is clear (by symmetry) that both twins age the same amount by the time they pass each other again. Reproduce this result by working in the frame of one of the twins.

14.8 Solutions

14.1. **Airplane's speed**

An observer on the ground sees the plane's clock run slow by a factor $\sqrt{1 - v^2/c^2}$ due to SR time dilation. But he also sees it run fast by a factor $(1 + gh/c^2)$ due to GR time dilation. We therefore want the product of these two factors to equal 1. Using the standard Taylor-series approximation for slow speeds in the first factor, we find

$$\left(1 - \frac{v^2}{2c^2}\right)\left(1 + \frac{gh}{c^2}\right) = 1 \quad \Longrightarrow \quad 1 - \frac{v^2}{2c^2} + \frac{gh}{c^2} - \mathcal{O}\left(\frac{1}{c^4}\right) = 1. \qquad (14.23)$$

Neglecting the small $1/c^4$ term, and canceling the 1's, we obtain $v = \sqrt{2gh}$. Interestingly, $\sqrt{2gh}$ is also the answer to a standard question from Newtonian physics, namely, how fast do you need to throw a ball straight up so that it reaches a height h?

14.2. **Clock on a tower**

The SR time-dilation factor is $\sqrt{1 - v^2/c^2} \approx 1 - v^2/2c^2$. The clock therefore loses a fraction $v^2/2c^2$ of the time elapsed during its motion up and down the tower. The upward journey takes a time h/v, and likewise for the downward trip, so the time loss due to the SR effect is

$$\left(\frac{v^2}{2c^2}\right)\left(\frac{2h}{v}\right) = \frac{vh}{c^2}. \tag{14.24}$$

Our goal is to balance this time loss with the time gain due to the GR time-dilation effect. If the clock sits on top of the tower for a time T, then the time gain is $(gh/c^2)T$.

But we must not forget the increase in time due to the height gained while the clock is in motion. During its motion, the clock's average height is $h/2$. The total time in motion is $2h/v$, so the GR time gain while the clock is moving is (we can use the average height here, because the GR effect is linear in h)

$$\left(\frac{g(h/2)}{c^2}\right)\left(\frac{2h}{v}\right) = \frac{gh^2}{c^2 v}. \tag{14.25}$$

Setting the total change in the clock's time equal to zero gives

$$-\frac{vh}{c^2} + \frac{gh}{c^2}T + \frac{gh^2}{c^2 v} = 0 \quad \Longrightarrow \quad -v + gT + \frac{gh}{v} = 0 \quad \Longrightarrow \quad T = \frac{v}{g} - \frac{h}{v}. \tag{14.26}$$

REMARKS: Note that we must have $v > \sqrt{gh}$ in order for a positive solution for T to exist. If $v < \sqrt{gh}$, then the SR effect is too small to cancel out the GR effect, even if the clock spends no time sitting at the top. If $v = \sqrt{gh}$, then $T = 0$, and we essentially have the same situation as in Exercise 14.12. Note also that if v is very large compared with $\sqrt{gh}$ (but still small compared with c, so that our $\sqrt{1 - v^2/c^2} \approx 1 - v^2/2c^2$ approximation is valid), then $T \approx v/g$, which is independent of h. ♣

14.3. **Circular motion**

(a) In A's frame, there is only the SR time-dilation effect. A sees B move at speed v, so B's clock runs slow by a factor of $\sqrt{1 - v^2/c^2}$. And since $v \ll c$, we may use the Taylor series to approximate this as $1 - v^2/2c^2$.

(b) In this frame, there are both SR and GR time-dilation effects. A moves at speed v with respect to B in this frame, so the SR effect is that A's clock runs slow by a factor $\sqrt{1 - v^2/c^2} \approx 1 - v^2/2c^2$. But B undergoes an acceleration of $a = v^2/r$ toward A, so for all B knows, he lives in a world where the acceleration due to gravity is v^2/r. A is "higher" in the gravitational field, so the GR effect is that A's clock runs fast by a factor $1 + ar/c^2 = 1 + v^2/c^2$. Multiplying the SR and GR effects together, we find (to lowest order) that A's clock runs fast by a factor $1 + v^2/2c^2$. This means (to lowest order) that B's clock runs slow by a factor $1 - v^2/2c^2$, in agreement with the answer to part (a).

(c) In this frame, there is no relative motion between A and B, so there is only the GR time-dilation effect. The gravitational field (that is, the centripetal acceleration) at a distance x from the center is $g_x = x\omega^2$. Imagine lining up a series of clocks along a radius, with separation dx. Then the GR time-dilation result tells us that each clock loses a fraction $g_x\,dx/c^2 = x\omega^2\,dx/c^2$ of time relative to the next clock inward. Integrating these fractions from $x = 0$ to $x = r$ shows that B's clock loses a fraction $r^2\omega^2/2c^2 = v^2/2c^2$, compared with A's clock. This agrees with the results in parts (a) and (b).

REMARK: If you want to imagine an analogous line of clocks in part (b), you can imagine lining a stick with them, with B at one end of the stick. The sensible thing to do with the stick is to have it be motionless with respect to B's frame (as the line of clocks was in part (c)). But this means that the stick doesn't rotate with respect to the inertial axes, because B's axes don't rotate. So all of the clocks feel the same acceleration $r\omega^2 = v^2/r$, in contrast with the decreasing accelerations in part (c). The integral of all the fractions therefore doesn't pick up the factor of 2 as it did in part (c), and so we simply end up with v^2/c^2. The SR effect then arises in part (b) because A is flying past the clock on the other end of the stick, whereas in part (c) A was at rest. ♣

14.4. **More circular motion**

(a) In the lab frame, the situation is symmetric with respect to A and B. Therefore, if A and B are decelerated in a symmetric manner and brought together, their clocks must read the same time.

Assume (in the interest of obtaining a contradiction) that A sees B's clock run slow. Then after an arbitrarily long time, A will see B's clock an arbitrarily large time behind his. Now bring A and B to a stop. There is no possible way that the stopping motion can make B's clock gain an arbitrarily large amount of time, as seen by A. This is true because everything takes place in a finite region of space, so there is an upper bound on the GR time-dilation effect (because it behaves like gh/c^2, and h is bounded). Therefore, A will end up seeing B's clock reading less. This contradicts the result of the previous paragraph. Likewise for the case where A sees B's clock run fast.

REMARK: Note how this problem differs from the problem where A and B move with equal speeds directly away from each other, and then reverse directions and head back to meet up again. For this new "linear" problem, the symmetry reasoning in the first paragraph above still holds, so A and B will indeed have the same clock readings when they meet up again. But the reasoning in the second paragraph does not hold (it had better not, because each person does *not* see the other person's clock running at the same rate). The error is that in this linear scenario, the experiment is not contained in a small region of space, so the turning-around effects of order gh/c^2 become arbitrarily large as the time of travel becomes arbitrarily large, because h grows with time (see Problem 14.9). ♣

(b) In this frame, there are both SR and GR time-dilation effects. A moves at speed $2v$ with respect to B in this frame (we don't need to use the relativistic velocity-addition formula, because $v \ll c$), so the SR effect is that A's clock runs slow by a factor $\sqrt{1-(2v)^2/c^2} \approx 1 - 2v^2/c^2$. But B undergoes an acceleration of $a = v^2/r$ toward A, so the GR effect is that A's clock runs fast by a factor $1 + a(2r)/c^2 = 1 + 2v^2/c^2$ (because they are separated by a distance $2r$). Multiplying the SR and GR effects together, we find (to lowest order) that the two clocks run at the same rate.

(c) In this frame, there is no relative motion between A and B, so there is only (at most) the GR effect. But A and B are both at the same gravitational potential, because they are at the same radius. Therefore, they both see the clocks running at the same rate. If you want, you can line up a series of clocks along the diameter between A and B, as we did along a radius in part (c) of Problem 14.3. The clocks will gain time as you march in toward the center, and then lose the same amount of time as you march back out to the diametrically opposite point.

14.5. **Accelerator's point of view**

(a) FIRST SOLUTION: Equation (14.5) says that the distance traveled by the rocket (as measured in the original inertial frame), as a function of the time in the inertial frame, is

$$d = \frac{1}{g}\left(\sqrt{1+(gt)^2} - 1\right). \tag{14.27}$$

An inertial observer on the planet therefore measures the rocket–planet distance to be

$$x = \ell - \frac{1}{g}\left(\sqrt{1+(gt)^2} - 1\right). \tag{14.28}$$

The rocket observer sees this length contracted by a factor γ. Using the result of Exercise 14.16, we have $\gamma = \sqrt{1+(gt)^2} = \cosh(g\tau)$. So the rocket–planet distance, as measured in the instantaneous inertial frame of the rocket, is

$$x = \frac{\ell - \frac{1}{g}\big(\cosh(g\tau) - 1\big)}{\cosh(g\tau)} \quad\Longrightarrow\quad 1 + gx = \frac{1+g\ell}{\cosh(g\tau)}. \tag{14.29}$$

SECOND SOLUTION: Equation (14.21) gives the speed of the planet in the accelerating frame of the rocket. Using the results of Exercise 14.16 to write v in terms of τ, we have (with $c = 1$)

$$\frac{dx}{d\tau} = -(1+gx)\tanh(g\tau). \tag{14.30}$$

Separating variables and integrating gives

$$\int \frac{dx}{1+gx} = -\int \tanh(g\tau)\,d\tau \quad\Longrightarrow\quad \ln(1+gx) = -\ln\big(\cosh(g\tau)\big) + C$$

$$\Longrightarrow\quad 1 + gx = \frac{A}{\cosh(g\tau)}. \tag{14.31}$$

Since the initial condition is $x = \ell$ when $\tau = 0$, we must have $A = 1 + g\ell$, which gives Eq. (14.16), as desired.

(b) Equation (14.20) says that the planet's clock runs fast (or slow) according to

$$dt = d\tau\,(1+gx)\sqrt{1-v^2}. \tag{14.32}$$

The results of Exercise 14.16 yield $\sqrt{1-v^2} = 1/\cosh(g\tau)$. Combining this with the result for $1 + gx$ above, and integrating, gives

$$\int dt = \int \frac{(1+g\ell)\,d\tau}{\cosh^2(g\tau)} \quad\Longrightarrow\quad gt = (1+g\ell)\tanh(g\tau). \tag{14.33}$$

14.6. **Getting way ahead**

The explanation of why the two clocks show different times in the ground frame is the following. The rocket becomes increasingly length contracted in the ground frame, which means that the front end isn't traveling quite as fast as the back end. Therefore, the time-dilation factor for the front clock isn't as large as that for the back clock. So the front clock loses less time relative to the ground, and hence ends up ahead of the back clock. Of course, it's not at all obvious that everything works out quantitatively and that the front clock eventually ends up an arbitrarily large time ahead of the back clock. In fact, it's quite surprising that this is the case, because the above difference in speeds is very small. But let's now show that the above explanation does indeed account for the difference in the clock readings.

Let the back of the rocket be located at position x. Then the front is located at position $x + L\sqrt{1-v^2}$, due to the length contraction. Taking the time derivatives of the two positions, we see that the speeds of the back and front are (with $v \equiv dx/dt$)[12]

$$v_b = v, \quad \text{and} \quad v_f = v(1 - L\gamma\dot{v}). \tag{14.34}$$

If we assume that the back is the part that accelerates at g (it doesn't matter which point we pick, to leading order), then we can just invoke the result in Eq. (14.4),

$$v_b = v = \frac{gt}{\sqrt{1+(gt)^2}}, \tag{14.35}$$

where t is the time in the ground frame. Having written down v, we must now find the γ factors associated with the speeds of the front and back. The γ factor associated with the speed of the back, namely v, is

$$\gamma_b = \frac{1}{\sqrt{1-v^2}} = \sqrt{1+(gt)^2}. \tag{14.36}$$

The γ factor associated with the speed of the front, $v_f = v(1 - L\gamma\dot{v})$, is a bit more complicated. We must first calculate $\dot{v}$. From Eq. (14.35), we obtain $\dot{v} = g/(1+g^2t^2)^{3/2}$, which gives

$$v_f = v(1 - L\gamma\dot{v}) = \frac{gt}{\sqrt{1+(gt)^2}}\left(1 - \frac{gL}{1+g^2t^2}\right). \tag{14.37}$$

The γ factor (or rather $1/\gamma$, which is what we'll be concerned with) associated with this speed is given as follows. In the first line below, we ignore the higher-order $(gL)^2$ term, because it is really $(gL/c^2)^2$, and we are assuming that gL/c^2 is small. And in obtaining the third line, we use the Taylor-series approximation, $\sqrt{1-\epsilon} \approx 1 - \epsilon/2$.

$$\begin{aligned}\frac{1}{\gamma_f} = \sqrt{1-v_f^2} &\approx \sqrt{1 - \frac{g^2t^2}{1+g^2t^2}\left(1 - \frac{2gL}{1+g^2t^2}\right)} \\ &= \frac{1}{\sqrt{1+g^2t^2}}\sqrt{1 + \frac{2g^3t^2L}{1+g^2t^2}} \\ &\approx \frac{1}{\sqrt{1+g^2t^2}}\left(1 + \frac{g^3t^2L}{1+g^2t^2}\right).\end{aligned} \tag{14.38}$$

We can now calculate the time that each clock shows, at time t in the ground frame. The time on the back clock changes according to $dt_b = dt/\gamma_b$, so Eq. (14.36) gives

$$t_b = \int_0^t \frac{dt}{\sqrt{1+g^2t^2}}. \tag{14.39}$$

The integral of $dx/\sqrt{1+x^2}$ is $\sinh^{-1} x$ (to derive this, make the substitution $x \equiv \sinh\theta$). Letting $x \equiv gt$, this gives

$$gt_b = \sinh^{-1}(gt). \tag{14.40}$$

The time on the front clock changes according to $dt_f = dt/\gamma_f$, so Eq. (14.38) gives

$$t_f = \int_0^t \frac{dt}{\sqrt{1+g^2t^2}} + \int_0^t \frac{g^3t^2L\,dt}{(1+g^2t^2)^{3/2}}. \tag{14.41}$$

[12] Since these two speeds aren't equal, there is an ambiguity concerning which speed we should use in the length-contraction factor, $\sqrt{1-v^2}$. Equivalently, the rocket doesn't have a single inertial frame that describes all of it. But you can show that any differences arising from this ambiguity are of higher order in gL/c^2 than we need to be concerned with.

The integral of $x^2\,dx/(1+x^2)^{3/2}$ is $\sinh^{-1} x - x/\sqrt{1+x^2}$ (to derive this, make the substitution $x \equiv \sinh\theta$, and use $\int d\theta/\cosh^2\theta = \tanh\theta$). Letting $x \equiv gt$, this gives

$$gt_{\rm f} = \sinh^{-1}(gt) + (gL)\left(\sinh^{-1}(gt) - \frac{gt}{\sqrt{1+g^2t^2}}\right). \qquad (14.42)$$

Using Eqs. (14.35) and (14.40), we may rewrite this as

$$gt_{\rm f} = gt_{\rm b}(1+gL) - gLv. \qquad (14.43)$$

Dividing by g, and putting the c's back in to make the units correct, we finally have

$$t_{\rm f} = t_{\rm b}\left(1+\frac{gL}{c^2}\right) - \frac{Lv}{c^2}, \qquad (14.44)$$

as we wanted to show. If we look at this calculation from the reverse point of view, we see that by using only Special Relativity concepts, we have demonstrated that someone at the back of a rocket sees a clock at the front running fast by a factor $(1+gL/c^2)$. There are, however, much easier ways of deriving this, as we saw in Section 14.2 and in Problem 11.25 ("Acceleration and redshift").

14.7. ***Lv/c*² revisited**

Consider first the case where the rocket accelerates while you sit there. Problem 14.6 is exactly relevant here, and it tells us that in your frame the clock readings are related by

$$t_{\rm f} = t_{\rm b}\left(1+\frac{gL}{c^2}\right) - \frac{Lv}{c^2}. \qquad (14.45)$$

You will eventually see the front clock an arbitrarily large time ahead of the back clock. But note that for small times (before things become relativistic), the standard Newtonian result, $v \approx gt_{\rm b}$, is valid, so we have

$$t_{\rm f} \approx \left(t_{\rm b} + \frac{Lv}{c^2}\right) - \frac{Lv}{c^2} = t_{\rm b}. \qquad (14.46)$$

So in this setup where the rocket is the one that accelerates, both clocks show essentially the same time near the start. This makes sense; both clocks have essentially the same speed at the beginning, so to lowest order their γ factors are the same, so the clocks run at the same rate. But eventually the front clock will get ahead of the back clock.

Now consider the case where you accelerate while the rocket sits there. Problem 14.5 is relevant here, if we let the rocket in that problem now become you, and if we let two planets a distance L apart become the two ends of the rocket. The times you observe on the front and back clocks on the rocket are then, using Eq. (14.33) and assuming that you are accelerating toward the rocket,

$$gt_{\rm f} = (1+g\ell)\tanh(g\tau), \quad \text{and} \quad gt_{\rm b} = \big(1+g(\ell+L)\big)\tanh(g\tau). \qquad (14.47)$$

But from Exercise 14.16, we know that your speed relative to the rocket is $v = \tanh(g\tau)$. Equation (14.47) therefore gives $t_{\rm b} = t_{\rm f} + Lv$, or $t_{\rm b} = t_{\rm f} + Lv/c^2$ with the c^2. So in this case we arrive at the standard Lv/c^2 "rear clock ahead" result.

The point here is that in this second case, the clocks are synchronized in the rocket frame, and this is the assumption that went into our derivation of the Lv/c^2 result in Chapter 11. In the first case above where the rocket accelerates, the clocks are *not* synchronized in the rocket frame (except right at the start), so it isn't surprising that we don't obtain the Lv/c^2 result.

14.8. **Circling the earth**

This is one setup where we really need to use the correct term, *stationary*-proper-time principle. It turns out that B's path yields a saddle point for the proper time. The value at this saddle point is less than A's proper time, but this is irrelevant, because we care only about local stationary points, not about global extrema.

To show that we have a saddle point, we must show that (1) the differences in the proper time are second order, and (2) there exist nearby paths that give both a larger and a smaller proper time. The first of these is true because the first-order differences vanish due to the fact that the path satisfies the Euler–Lagrange equations for the Lagrangian in Eq. (14.15), because the path is indeed a physical one.

The second is true because the proper time can be made smaller by having B speed up and slow down. This will cause a net increase in the time-dilation effect as viewed by A, thereby yielding a smaller proper time.[13] And the proper time can be made larger by having B take a nearby path that doesn't quite form a great circle on the earth. (Imagine the curve traced out by a rubber band that has just begun to slip away from a great-circle position.) This path is shorter, so B won't have to travel as fast to get back in a given time, so the time-dilation effect will be smaller as viewed by A, thereby yielding a larger proper time.

14.9. **Twin paradox**

(a) In the earth frame, the spaceship travels at speed v for essentially the whole time. Therefore, the traveler ages less by a fraction $\sqrt{1-v^2/c^2} \approx 1-v^2/2c^2$. The fractional loss of time is thus $v^2/2c^2$. The time-dilation effect is different during the short turning-around period, but this is negligible.

(b) Let the distance to the star be ℓ, as measured in the frame of the earth (but the difference in lengths in the two frames is negligible in this problem), and let the turnaround take a time T. Then the given information says that $T \ll (2\ell)/v$.

During the constant-speed part of the trip, the traveler sees the earth clock running slow by a fraction $\sqrt{1-v^2/c^2} \approx 1-v^2/2c^2$. The time for this constant-speed part is (essentially) $2\ell/v$, so the earth clock loses a time of $(v^2/2c^2)(2\ell/v) = v\ell/c^2$.

However, during the turnaround time, the spaceship is accelerating toward the earth, so the traveler sees the earth clock running fast, due to the GR time dilation. The magnitude of the acceleration is $a = 2v/T$, because the spaceship goes from velocity v to $-v$ in time T. The earth clock therefore runs fast by a factor $1 + a\ell/c^2 = 1 + 2v\ell/Tc^2$. This happens for a time T, so the earth clock gains a time of $(2v\ell/Tc^2)T = 2v\ell/c^2$.

Combining the results of the previous two paragraphs, we see that the earth clock gains a time of $2v\ell/c^2 - v\ell/c^2 = v\ell/c^2$. This is a fraction $(v\ell/c^2)/(2\ell/v) = v^2/2c^2$ of the total time, in agreement with part (a).

14.10. **Twin paradox again**

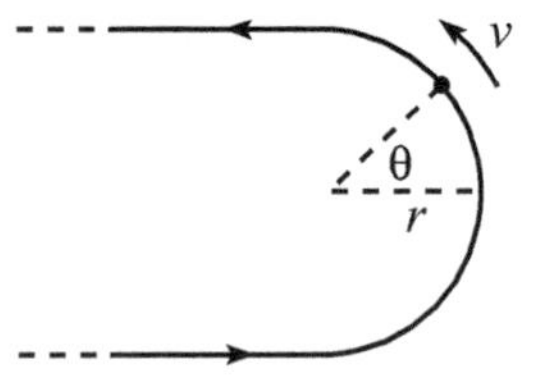

Fig. 14.9

(a) The only difference between this problem and the previous one is the nature of the turnaround, so all we need to show here is that the traveler still sees the earth clock gain a time of $2v\ell/c^2$ during the turnaround.

Let the radius of the semicircle be r. Then the magnitude of the acceleration is $a = v^2/r$. Let θ be the angle shown in Fig. 14.9. For a given θ, the earth is at

[13] This is true for the same reason that a person who travels at constant speed in a straight line between two points shows a larger proper time than a second person who speeds up and slows down. This follows directly from SR time dilation, as viewed by the first person. If you want, you can imagine unrolling B's circular orbit into a straight line, and then invoke the result just mentioned. As far as SR time-dilation effects from clock A's point of view go, it doesn't matter if the circle is unrolled into a straight line.

a height of essentially $\ell \cos\theta$ in the gravitational field felt by the spaceship. The fractional time that the earth gains while the traveler is at an angle θ is therefore $ah/c^2 = (v^2/r)(\ell\cos\theta)/c^2$. Integrating this over the time of the turnaround, and using $dt = r\,d\theta/v$, we see that the earth gains the desired time of

$$\Delta t = \int_{-\pi/2}^{\pi/2} \left(\frac{v^2\ell\cos\theta}{rc^2}\right)\left(\frac{r\,d\theta}{v}\right) = \frac{2v\ell}{c^2}. \qquad (14.48)$$

(b) Let the acceleration vector at a given instant be $\mathbf{a}$, and let $\boldsymbol{\ell}$ be the vector from the spaceship to the earth. Note that since the turnaround is done in a small region of space, $\boldsymbol{\ell}$ is essentially constant here. The earth is at a height of $\hat{\mathbf{a}}\cdot\boldsymbol{\ell}$ in the gravitational field felt by the spaceship; the dot product just gives the cosine term in the above solution in part (a). The fractional time gain, ah/c^2, is therefore equal to $|\mathbf{a}|(\hat{\mathbf{a}}\cdot\boldsymbol{\ell})/c^2 = \mathbf{a}\cdot\boldsymbol{\ell}/c^2$. Integrating this over the time of the turnaround, we see that the earth gains a time of

$$\begin{aligned}\Delta t = \int_{t_i}^{t_f} \frac{\mathbf{a}\cdot\boldsymbol{\ell}}{c^2}\,dt &= \frac{\boldsymbol{\ell}}{c^2}\cdot\int_{t_i}^{t_f}\mathbf{a}\,dt \\ &= \frac{\boldsymbol{\ell}}{c^2}\cdot(\mathbf{v}_f - \mathbf{v}_i) \\ &= \frac{\boldsymbol{\ell}\cdot(2\mathbf{v}_f)}{c^2} \\ &= \frac{2v\ell}{c^2},\end{aligned} \qquad (14.49)$$

as we wanted to show. The point here is that no matter how complicated the motion is during the turnaround, the total effect is simply to change the velocity from $\mathbf{v}$ outward to $\mathbf{v}$ inward.

14.11. **Twin paradox times**

(a) As viewed by A, the relation between the twins' times is

$$dt_B = \sqrt{1-v^2}\,dt_A. \qquad (14.50)$$

Assuming $v_0 \ll c$, we may say that $v(t_A)$ is essentially equal to $v_0 - gt_A$, so the out and back parts of the trip each take a time of essentially v_0/g in A's frame. The total elapsed time on B's clock is therefore

$$\begin{aligned}T_B = \int dt_B &\approx 2\int_0^{v_0/g} \sqrt{1-v^2}\,dt_A \\ &\approx 2\int_0^{v_0/g}\left(1-\frac{v^2}{2}\right)dt_A \\ &\approx 2\int_0^{v_0/g}\left(1-\frac{1}{2}(v_0-gt)^2\right)dt \\ &= 2\left(t+\frac{1}{6g}(v_0-gt)^3\right)\Bigg|_0^{v_0/g} \\ &= \frac{2v_0}{g} - \frac{v_0^3}{3gc^2},\end{aligned} \qquad (14.51)$$

where we have put the c's back in to make the units right. The ratio of B's elapsed time to A's is therefore

$$\frac{T_B}{T_A} \approx \frac{T_B}{2v_0/g} \approx 1 - \frac{v_0^2}{6c^2}. \qquad (14.52)$$

(b) As viewed by B, the relation between the twins' times is given by Eq. (14.13),

$$dt_A = \sqrt{1 - \frac{v^2}{c^2}}\left(1 + \frac{gy}{c^2}\right) dt_B. \qquad (14.53)$$

Assuming $v_0 \ll c$, we may say that $v(t_B)$ is essentially equal to $v_0 - gt_B$, and A's height is essentially equal to $v_0 t_B - gt_B^2/2$. The up and down parts of the trip each take a time of essentially v_0/g in B's frame. Therefore, the total elapsed time on A's clock is (using the approximation in Eq. (14.14), and dropping the c's)

$$\begin{aligned}
T_A = \int dt_A &\approx 2\int_0^{v_0/g}\left(1 - \frac{v^2}{2} + gy\right) dt_B. \\
&\approx 2\int_0^{v_0/g}\left(1 - \frac{1}{2}(v_0 - gt)^2 + g(v_0 t - gt^2/2)\right) dt. \\
&= 2\left(t + \frac{1}{6g}(v_0 - gt)^3 + g\left(\frac{v_0 t^2}{2} - \frac{gt^3}{6}\right)\right)\Bigg|_0^{v_0/g} \\
&= \frac{2v_0}{g} - \frac{v_0^3}{3g} + g\left(\frac{v_0^3}{g^2} - \frac{v_0^3}{3g^2}\right) \\
&= \frac{2v_0}{g} + \frac{v_0^3}{3gc^2}, \qquad (14.54)
\end{aligned}$$

where we have put the c's back in to make the units right. We therefore have

$$\frac{T_A}{T_B} \approx \frac{T_A}{2v_0/g} \approx 1 + \frac{v_0^2}{6c^2} \quad \Longrightarrow \quad \frac{T_B}{T_A} \approx 1 - \frac{v_0^2}{6c^2}, \qquad (14.55)$$

up to higher-order corrections. This agrees with the result found in part (a), as the Equivalence Principle requires.

Appendix A Useful formulas

A.1 Taylor series

The general form of a Taylor series is

$$f(x_0 + x) = f(x_0) + f'(x_0)x + \frac{f''(x_0)}{2!}x^2 + \frac{f'''(x_0)}{3!}x^3 + \cdots, \tag{A.1}$$

which can be verified by taking derivatives and then setting $x = 0$. For example, taking the first derivative and then setting $x = 0$ yields $f'(x_0)$ on the left, and also $f'(x_0)$ on the right, because the first term is a constant and gives zero, the second term gives $f'(x_0)$, and all the rest of the terms give zero once we set $x = 0$ because they all have at least one power of x left in them. Likewise, if we take the second derivative of each side and then set $x = 0$, we obtain $f''(x_0)$ on both sides. And so on for all derivatives. Therefore, since the two functions on each side of the above equation are equal at $x = 0$ and also have their nth derivatives equal at $x = 0$ for all n, they must in fact be the same function (assuming that they're nicely behaved functions, which we generally assume in physics).

Some specific Taylor series that come up often are listed below. They are all derivable via Eq. (A.1), but sometimes there are quicker ways of obtaining them. For example, Eq. (A.3) is most easily obtained by taking the derivative of Eq. (A.2), which itself is simply the sum of a geometric series.

$$\frac{1}{1-x} = 1 + x + x^2 + x^3 + \cdots \tag{A.2}$$

$$\frac{1}{(1-x)^2} = 1 + 2x + 3x^2 + 4x^3 + \cdots \tag{A.3}$$

$$\ln(1-x) = -x - \frac{x^2}{2} - \frac{x^3}{3} - \cdots \tag{A.4}$$

$$e^x = 1 + x + \frac{x^2}{2!} + \frac{x^3}{3!} + \cdots \tag{A.5}$$

$$\sin x = x - \frac{x^3}{3!} + \frac{x^5}{5!} - \cdots \tag{A.6}$$

$$\cos x = 1 - \frac{x^2}{2!} + \frac{x^4}{4!} - \cdots \tag{A.7}$$

$$\sqrt{1+x} = 1 + \frac{x}{2} - \frac{x^2}{8} + \cdots \tag{A.8}$$

$$\frac{1}{\sqrt{1+x}} = 1 - \frac{x}{2} + \frac{3x^2}{8} + \cdots \tag{A.9}$$

$$(1+x)^n = 1 + nx + \binom{n}{2}x^2 + \binom{n}{3}x^3 + \cdots \tag{A.10}$$

A.2 Nice formulas

The first formula here can be quickly proved by showing that the Taylor series for both sides are equal.

$$e^{i\theta} = \cos\theta + i\sin\theta \tag{A.11}$$

$$\cos\theta = \frac{1}{2}(e^{i\theta} + e^{-i\theta}), \quad \sin\theta = \frac{1}{2i}(e^{i\theta} - e^{-i\theta}) \tag{A.12}$$

$$\cos\frac{\theta}{2} = \pm\sqrt{\frac{1+\cos\theta}{2}}, \quad \sin\frac{\theta}{2} = \pm\sqrt{\frac{1-\cos\theta}{2}} \tag{A.13}$$

$$\tan\frac{\theta}{2} = \pm\sqrt{\frac{1-\cos\theta}{1+\cos\theta}} = \frac{1-\cos\theta}{\sin\theta} = \frac{\sin\theta}{1+\cos\theta} \tag{A.14}$$

$$\sin 2\theta = 2\sin\theta\cos\theta, \quad \cos 2\theta = \cos^2\theta - \sin^2\theta \tag{A.15}$$

$$\sin(\alpha+\beta) = \sin\alpha\cos\beta + \cos\alpha\sin\beta \tag{A.16}$$

$$\cos(\alpha+\beta) = \cos\alpha\cos\beta - \sin\alpha\sin\beta \tag{A.17}$$

$$\tan(\alpha+\beta) = \frac{\tan\alpha + \tan\beta}{1 - \tan\alpha\tan\beta} \tag{A.18}$$

$$\cosh x = \frac{1}{2}(e^x + e^{-x}), \quad \sinh x = \frac{1}{2}(e^x - e^{-x}) \tag{A.19}$$

$$\cosh^2 x - \sinh^2 x = 1 \tag{A.20}$$

$$\frac{d}{dx}\cosh x = \sinh x, \quad \frac{d}{dx}\sinh x = \cosh x \tag{A.21}$$

A.3 Integrals

$$\int \ln x \, dx = x \ln x - x \tag{A.22}$$

$$\int x \ln x \, dx = \frac{x^2}{2}\ln x - \frac{x^2}{4} \tag{A.23}$$

$$\int x e^x \, dx = e^x(x-1) \tag{A.24}$$

$$\int \frac{dx}{1+x^2} = \tan^{-1}x \quad \text{or} \quad -\cot^{-1}x \tag{A.25}$$

$$\int \frac{dx}{x(1+x^2)} = \frac{1}{2}\ln\left(\frac{x^2}{1+x^2}\right) \tag{A.26}$$

$$\int \frac{dx}{1-x^2} = \frac{1}{2}\ln\left(\frac{1+x}{1-x}\right) \quad \text{or} \quad \tanh^{-1}x \qquad (x^2 < 1) \tag{A.27}$$

$$\int \frac{dx}{1-x^2} = \frac{1}{2}\ln\left(\frac{x+1}{x-1}\right) \quad \text{or} \quad \coth^{-1}x \qquad (x^2 > 1) \tag{A.28}$$

$$\int \sqrt{1+x^2}\, dx = \frac{1}{2}\left(x\sqrt{1+x^2} + \ln\left(x + \sqrt{1+x^2}\right)\right) \tag{A.29}$$

$$\int \frac{1+x}{\sqrt{1-x}}\, dx = -\frac{2}{3}(5+x)\sqrt{1-x} \tag{A.30}$$

$$\int \frac{dx}{\sqrt{1-x^2}} = \sin^{-1}x \quad \text{or} \quad -\cos^{-1}x \tag{A.31}$$

$$\int \frac{dx}{\sqrt{x^2+1}} = \ln\left(x+\sqrt{x^2+1}\right) \quad \text{or} \quad \sinh^{-1} x \tag{A.32}$$

$$\int \frac{dx}{\sqrt{x^2-1}} = \ln\left(x+\sqrt{x^2-1}\right) \quad \text{or} \quad \cosh^{-1} x \tag{A.33}$$

$$\int \frac{dx}{x\sqrt{x^2-1}} = \sec^{-1} x \quad \text{or} \quad -\csc^{-1} x \tag{A.34}$$

$$\int \frac{dx}{x\sqrt{1+x^2}} = -\ln\left(\frac{1+\sqrt{1+x^2}}{x}\right) \quad \text{or} \quad -\operatorname{csch}^{-1} x \tag{A.35}$$

$$\int \frac{dx}{x\sqrt{1-x^2}} = -\ln\left(\frac{1+\sqrt{1-x^2}}{x}\right) \quad \text{or} \quad -\operatorname{sech}^{-1} x \tag{A.36}$$

$$\int \frac{dx}{\cos x} = \ln\left(\frac{1+\sin x}{\cos x}\right) \tag{A.37}$$

$$\int \frac{dx}{\sin x} = \ln\left(\frac{1-\cos x}{\sin x}\right) \tag{A.38}$$

$$\int \frac{dx}{\sin x \cos x} = -\ln\left(\frac{\cos x}{\sin x}\right) \tag{A.39}$$

Appendix B Multivariable, vector calculus

This appendix gives a brief review of multivariable calculus, also known as vector calculus. The first three topics below (dot product, cross product, partial derivatives) are used often in this book, so if you haven't seen them before, you should read these parts carefully. But the last three topics (gradient, divergence, curl) are used only occasionally, so it isn't crucial that you master these (for this book, at least). For all of the topics, it's possible to go much deeper into them, but I'll present just the basics here. If you want further material, any book on multivariable calculus should do the trick.

B.1 Dot product

The *dot product*, or *scalar product*, between two vectors is defined to be

$$\mathbf{a} \cdot \mathbf{b} \equiv a_x b_x + a_y b_y + a_z b_z. \tag{B.1}$$

The dot product takes two vectors and produces a scalar, which is just a number. You can quickly use Eq. (B.1) to show that the dot product is commutative and distributive. That is, $\mathbf{a} \cdot \mathbf{b} = \mathbf{b} \cdot \mathbf{a}$, and $(\mathbf{a} + \mathbf{b}) \cdot \mathbf{c} = \mathbf{a} \cdot \mathbf{c} + \mathbf{b} \cdot \mathbf{c}$. Note that the dot product of a vector with itself is $\mathbf{a} \cdot \mathbf{a} = a_x^2 + a_y^2 + a_z^2$, which is just its length squared, $|\mathbf{a}|^2 \equiv a^2$.

Taking the sum of the products of the corresponding components of two vectors, as we did in Eq. (B.1), might seem like a silly and arbitrary thing to do. Why don't we instead look at the sum of the cubes of the products of the corresponding components? The reason is that the dot product as we've defined it has many nice properties, the most useful of which is that it can be written as

$$\mathbf{a} \cdot \mathbf{b} = |\mathbf{a}||\mathbf{b}| \cos\theta \equiv ab \cos\theta, \tag{B.2}$$

where θ is the angle between the two vectors. We can demonstrate this as follows. Consider the dot product of the vector $\mathbf{c} \equiv \mathbf{a} + \mathbf{b}$ with itself, which is simply the

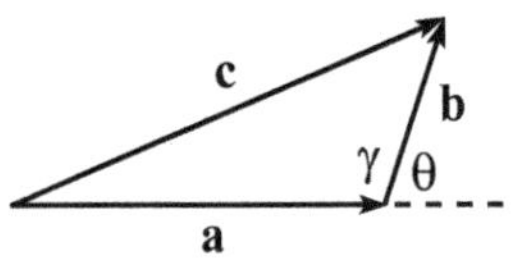

Fig. B.1

square of the length of **c**. Using the commutative and distributive properties, we have

$$c^2 = (\mathbf{a}+\mathbf{b})\cdot(\mathbf{a}+\mathbf{b}) = \mathbf{a}\cdot\mathbf{a} + 2\mathbf{a}\cdot\mathbf{b} + \mathbf{b}\cdot\mathbf{b}$$
$$= a^2 + 2\mathbf{a}\cdot\mathbf{b} + b^2. \qquad \text{(B.3)}$$

But from the law of cosines applied to the triangle in Fig. B.1, we have

$$c^2 = a^2 + b^2 - 2ab\cos\gamma = a^2 + b^2 + 2ab\cos\theta, \qquad \text{(B.4)}$$

because $\gamma = \pi - \theta$. Comparing this with Eq. (B.3) yields $\mathbf{a}\cdot\mathbf{b} = ab\cos\theta$, as desired. The angle between two vectors is therefore given by

$$\cos\theta = \frac{\mathbf{a}\cdot\mathbf{b}}{|\mathbf{a}||\mathbf{b}|}. \qquad \text{(B.5)}$$

A nice corollary of this result is that if the dot product of two vectors is zero, then $\cos\theta = 0$, which means that the vectors are perpendicular. If someone gives you the vectors $(1,-2,3)$ and $(4,5,2)$, it's by no means obvious that they're perpendicular. But you know from Eq. (B.5) that they indeed are.

Geometrically, the dot product $\mathbf{a}\cdot\mathbf{b} = ab\cos\theta$ equals the length of **a** times the component of **b** along **a**. Or vice versa, depending on which length you want to group the $\cos\theta$ factor with. If we rotate our coordinate system, the dot product of two vectors remains the same, because it depends only on their lengths and the angle between them, and these are unaffected by the rotation. In other words, the dot product is a scalar. This certainly isn't obvious from looking at the original definition in Eq. (B.1), because the coordinates get all messed up during the rotation.

Example (Distance on the earth): Given the longitude angle ϕ and the polar angle θ (measured down from the north pole, so θ is 90° minus the latitude angle) for two points on the earth, what is the distance between them, as measured along the earth?

Solution: Our goal is to find the angle, β, between the radii vectors to the two points, because the desired distance is then $R\beta$. This would be a tricky problem if we didn't have the dot product at our disposal, but things are easy if we make use of Eq. (B.5) to say that $\cos\beta = \mathbf{r}_1\cdot\mathbf{r}_2/R^2$. The problem then reduces to finding $\mathbf{r}_1\cdot\mathbf{r}_2$. The Cartesian components of these vectors are

$$\begin{aligned}\mathbf{r}_1 &= R(\sin\theta_1\cos\phi_1,\ \sin\theta_1\sin\phi_1,\ \cos\theta_1),\\ \mathbf{r}_2 &= R(\sin\theta_2\cos\phi_2,\ \sin\theta_2\sin\phi_2,\ \cos\theta_2).\end{aligned} \qquad \text{(B.6)}$$

The desired distance is then $R\beta = R\cos^{-1}(\mathbf{r}_1\cdot\mathbf{r}_2/R^2)$, where

$$\mathbf{r}_1\cdot\mathbf{r}_2/R^2 = \sin\theta_1\sin\theta_2(\cos\phi_1\cos\phi_2 + \sin\phi_1\sin\phi_2) + \cos\theta_1\cos\theta_2$$
$$= \sin\theta_1\sin\theta_2\cos(\phi_2-\phi_1) + \cos\theta_1\cos\theta_2. \qquad \text{(B.7)}$$

We can check some limits: If $\phi_1 = \phi_2$, then this gives $\beta = \theta_2 - \theta_1$ (or $\theta_1 - \theta_2$, depending on which is larger), as expected. And if $\theta_1 = \theta_2 = 90°$, then it gives $\beta = \phi_2 - \phi_1$ (or $\phi_1 - \phi_2$), as expected.

B.2 Cross product

The *cross product*, or *vector product*, between two vectors is defined via a determinant to be

$$\mathbf{a} \times \mathbf{b} \equiv \begin{vmatrix} \hat{\mathbf{x}} & \hat{\mathbf{y}} & \hat{\mathbf{z}} \\ a_x & a_y & a_z \\ b_x & b_y & b_z \end{vmatrix}$$
$$= \hat{\mathbf{x}}(a_y b_z - a_z b_y) + \hat{\mathbf{y}}(a_z b_x - a_x b_z) + \hat{\mathbf{z}}(a_x b_y - a_y b_x). \qquad \text{(B.8)}$$

The cross product takes two vectors and produces another vector. As with the dot product, you can show that the cross product is distributive. However, it is *anti*-commutative (that is, $\mathbf{a} \times \mathbf{b} = -\mathbf{b} \times \mathbf{a}$), which is evident from Eq. (B.8). So the cross product of any vector with itself is zero.

As with the dot product, the reason why we study this particular combination of components is that it has many nice properties, the most useful of which are that its direction is perpendicular to both **a** and **b** (in the orientation determined by the right-hand rule; see below), and its magnitude is

$$|\mathbf{a} \times \mathbf{b}| = |\mathbf{a}||\mathbf{b}| \sin\theta \equiv ab \sin\theta. \qquad \text{(B.9)}$$

Let's first show that $\mathbf{a} \times \mathbf{b}$ is indeed perpendicular to both **a** and **b**. We'll do this by making use of the above handy fact that if the dot product of two vectors is zero, then the vectors are perpendicular. We have

$$\mathbf{a} \cdot (\mathbf{a} \times \mathbf{b}) = a_x(a_y b_z - a_z b_y) + a_y(a_z b_x - a_x b_z) + a_z(a_x b_y - a_y b_x) = 0, \qquad \text{(B.10)}$$

as desired. Likewise for **b**. There is still an ambiguity, however, because although we know that $\mathbf{a} \times \mathbf{b}$ points along the direction perpendicular to the plane spanned by **a** and **b**, there are two possible directions along this line. Assuming that our coordinate system has been chosen to be "right-handed" (that is, if you point the fingers of your right hand in the direction of $\hat{\mathbf{x}}$ and then swing them to $\hat{\mathbf{y}}$, your thumb points along $\hat{\mathbf{z}}$), then the direction of $\mathbf{a} \times \mathbf{b}$ is determined by the right-hand rule. That is, if you point the fingers of your right hand in the direction of **a** and then swing them to **b** (through the angle that is less than 180°), your thumb points along $\mathbf{a} \times \mathbf{b}$. This is consistent with the fact that Eq. (B.8) gives $(1, 0, 0) \times (0, 1, 0) = (0, 0, 1)$, or $\hat{\mathbf{x}} \times \hat{\mathbf{y}} = \hat{\mathbf{z}}$.

Let's now demonstrate the $|\mathbf{a} \times \mathbf{b}| = ab \sin\theta$ result, which is equivalent to $|\mathbf{a} \times \mathbf{b}|^2 = a^2b^2(1 - \cos^2\theta)$, which is equivalent to $|\mathbf{a} \times \mathbf{b}|^2 = a^2b^2 - (\mathbf{a} \cdot \mathbf{b})^2$. Written in terms of the components, this last equation is

$$\begin{aligned}(a_yb_z - a_zb_y)^2 &+ (a_zb_x - a_xb_z)^2 + (a_xb_y - a_yb_x)^2 \\ &= (a_x^2 + a_y^2 + a_z^2)(b_x^2 + b_y^2 + b_z^2) - (a_xb_x + a_yb_y + a_zb_z)^2. \qquad \text{(B.11)}\end{aligned}$$

If you stare at this long enough, you'll see that it's true. The three different types of terms agree on both sides. For example, both sides have an $a_y^2b_z^2$ term, a $-2a_yb_ya_zb_z$ term, and no $a_x^2b_x^2$ term.

B.3 Partial derivatives

When dealing with a function of only one variable, there is no ambiguity when taking a derivative. However, with a function of many variables, we have to specify which one of the variables we're differentiating with respect to. If we have a function of, say, two variables, $f(x, y)$, and if we want to take the derivative with respect to x, then we use the terminology "*partial derivative* with respect to x," with the notation $\partial f/\partial x$. To evaluate this partial derivative, we don't have to do anything fancy. We just take a regular derivative with respect to x, while assuming that y is constant. For example, if $f(x,y) = x - 2y + x^2y^3$, then $\partial f/\partial x = 1 + 2xy^3$, and $\partial f/\partial y = -2 + 3x^2y^2$. If we plot the value of f as the height above the x-y plane, then when we take the partial derivative with respect to x, we're simply finding the slope of the curve formed by the intersection of the function's surface with the vertical plane parallel to the x axis and passing through the point in question. Similarly for y.

If we want to maximize or minimize a function of more than one variable, we need to set all the partial derivatives equal to zero. This is true because if the partial derivative with respect to a certain variable isn't zero, then the slope of the function in that direction is nonzero, which means that the point can't be a local maximum or minimum. This argument is the same as in the single-variable case. It's just that now we can make the argument for each of the variables independently.

Demanding that all the partial derivatives equal zero doesn't actually guarantee having a local maximum or minimum. The point in question might be a *saddle point*, which means that the function is a local maximum in some directions and a local minimum in others (so in two dimensions the function looks like a saddle; hence the name). For example, consider the function of two variables, $f(x,y) = 3x^2 - y^2$. Then the point $(0, 0)$ is a local minimum in the x direction and a local maximum in the y direction.

For two variables, if the second partial derivatives have opposite signs at a point where the first partial derivatives are zero, then we have a saddle point, because there is an upward parabola in one direction and a downward parabola in the other. However, we might have a saddle point even if the second partial derivatives have the same sign. For example, if we make the change of variables $x \equiv w - z$ and $y \equiv w + z$ in $f(x,y) = 3x^2 - y^2$, then it becomes $f(w,z) = 2w^2 + 2z^2 - 8wz$. If we didn't already know from the $f(x,y)$ form that $(0,0)$ is a saddle point, we could deduce this in the following way. Imagine that z is given, and then solve for the w that makes $f(w,z) = 0$. The result of solving this quadratic equation is that w takes the form of some multiple of z. That is, $w = Az$, where A happens to be $2 \pm \sqrt{3}$ in the present case. Since there are two (real) solutions for A here, there are two lines, namely $w = (2 \pm \sqrt{3})z$, emanating from $(0,0)$ for which $f(w,z) = 0$. So $(0,0)$ can't be a local maximum or minimum. It must therefore be a saddle point.[1]

In general, there are two real solutions for A if and only if the discriminant of the above quadratic equation is positive. For an arbitrary function of two variables, its shape in the vicinity of a point (which we will take to be $(0,0)$ after a shift in the coordinates) where both first partial derivatives are zero can be approximated by the Taylor series to second order (which you can verify by taking various derivatives, just as you would do with a function of one variable),

$$f(x,y) = C + \frac{1}{2}\left(\frac{\partial^2 f}{\partial x^2}\right)x^2 + \frac{1}{2}\left(\frac{\partial^2 f}{\partial y^2}\right)y^2 + \left(\frac{\partial^2 f}{\partial x\,\partial y}\right)xy + \cdots, \tag{B.12}$$

where it is understood that the partial derivatives here are evaluated at $(0,0)$. The condition that the discriminant is positive is therefore

$$\left(\frac{\partial^2 f}{\partial x\,\partial y}\right)^2 - \left(\frac{\partial^2 f}{\partial x^2}\right)\left(\frac{\partial^2 f}{\partial y^2}\right) > 0. \tag{B.13}$$

If this is true, then the point is a saddle point. If the left side is less than zero, then the point is a local maximum or minimum, because there are no nearby points for which $f(x,y) = C$; they are all either greater than C or less than C. If the left side equals zero, then the function looks like a trough, at least in the vicinity of the point in question (assuming that there is at least some quadratic dependence in the function).

[1] This is true because along these two lines, the first partial derivatives aren't zero, which means that the plane of the function's surface is tilted there. So the function is positive on one side of each line and negative on the other, which is exactly what happens with a saddle. There is, however, the special case where the discriminant of the quadratic equation is zero (as with, for example, $f(x, y) = (x - y)^2$), in which case there is only one solution for A and thus only one line for which the function is zero. In this case, the function looks like a (possibly upside down) trough. It is zero (or some given constant) along the line, at least to second order. And it curves up (or down) quadratically as you move away from the line (assuming that there is at least some quadratic dependence in the function).

B.4 Gradient

Given a function $f(x, y, z)$ (we'll work mainly with three variables from now on), we can form the vector whose components are the partial derivatives of f, namely $(\partial f/\partial x,\ \partial f/\partial y,\ \partial f/\partial z)$. This vector is called the *gradient.* If we define the differential vector operator ∇ (usually called "del") to be $\nabla \equiv (\partial/\partial x,\ \partial/\partial y,\ \partial/\partial z)$, then the gradient is simply

$$\nabla f = \left(\frac{\partial f}{\partial x}, \frac{\partial f}{\partial y}, \frac{\partial f}{\partial z}\right). \tag{B.14}$$

The gradient takes a function and produces a vector. For example, if $f(x, y, z) = xy^2 - yz^3$, then $\nabla f = (y^2,\ 2xy - z^3,\ -3yz^2)$. We call ∇ an "operator" because it needs to operate on a function to produce the gradient vector.

What is the physical meaning of the gradient? The gradient gives the direction you should march in if you want f to increase at the greatest rate. The reason for this is the following. Consider the value of a function $f(x,\ y,\ z)$ at a certain point, and then look at the value at a nearby point, displaced by the vector (dx, dy, dz). What (approximately, to first order) is the change in f between the two points? Well, as you march a distance dx in the x direction, the (first-order) change in f is $(\partial f/\partial x)\,dx$, by the definition of the partial derivative (just as in the one-variable case). If you then march a distance dy in the y direction, the function changes by an additional amount of $(\partial f/\partial y)\,dy$. And likewise the z direction gives a change of $(\partial f/\partial z)\,dz$. Adding up these three changes in f, we see that the total first-order change in f is[2]

$$df = \frac{\partial f}{\partial x}\,dx + \frac{\partial f}{\partial y}\,dy + \frac{\partial f}{\partial z}\,dz. \tag{B.15}$$

Using the dot product, this can be written concisely as

$$df = \left(\frac{\partial f}{\partial x}, \frac{\partial f}{\partial y}, \frac{\partial f}{\partial z}\right) \cdot (dx, dy, dz) \equiv \nabla f \cdot d\mathbf{r}. \tag{B.16}$$

We can now make use of Eq. (B.2) to say that the change in f is $df = |\nabla f||d\mathbf{r}|\cos\theta$, where θ is the angle between ∇f and $d\mathbf{r}$. The meaning of this is the following. Consider a given point (x, y, z). The ∇f gradient vector at this point is a particular vector. Imagine marching along little $d\mathbf{r}$ vectors in various directions and seeing how much f changes (assume that all the $d\mathbf{r}$ vectors have the same length, to be consistent). What direction should you march in, in order to have f change the most? Or not change at all? In the $df = |\nabla f||d\mathbf{r}|\cos\theta$

[2] There is technically an ambiguity in where each of these partial derivative is evaluated, because the three little steps you took started at three different points. But these ambiguities involve first-order corrections to the partial derivatives, and since these partial derivatives are already being multiplied by the first-order dx, dy, and dz terms in Eq. (B.15), any ambiguities will result in second-order effects and can therefore be ignored.

expression, $|\nabla f|$ has a definite value at the point in question, and we're assuming that we pick $|d\mathbf{r}|$ to always be the same, so it comes down to the $\cos\theta$ factor. Therefore, if you march directly along the ∇f gradient vector at the point, then f increases the most. And if you march in any direction in the plane perpendicular to ∇f, then f doesn't change at all (to first order). And if you march in the direction anti-parallel to ∇f, then f decreases the most.

This is more easily visualized in the case of a function of only two variables, $f(x,y)$, because then we can picture the value of f as being the height in the z direction. The graph of f is just a surface of mountains and valleys above (or below) the x-y plane. The gradient $\nabla f = (\partial f/\partial x,\ \partial f/\partial y)$ then gives the direction of steepest ascent. That is, f changes at the greatest rate if you march in the direction of ∇f in the x-y plane; the slope of the surface in this direction in the x-y plane is larger than in any other direction. And if you march in either direction along the line in the x-y plane that is perpendicular to ∇f, then f has zero change. By continuing to march along the direction perpendicular to ∇f wherever you are, you will form a curve in the x-y plane for which all the points give the same value of f. In other words, if you slice the surface of f with a horizontal plane whose height equals this particular value of f, and if you look at the intersection of this plane with the surface, then the projection of this intersection onto the x-y plane is the above curve you formed in the x-y plane.

B.5 Divergence

Consider a vector whose components are functions of the coordinates. For example, let $\mathbf{F} = (F_x, F_y, F_z) = (3xz,\ 2y^2 + xyz,\ x^2 + z^3)$. Then the *divergence* of $\mathbf{F}$ is defined to be the dot product of the ∇ operator with $\mathbf{F}$, that is,

$$\nabla \cdot \mathbf{F} = \frac{\partial F_x}{\partial x} + \frac{\partial F_y}{\partial y} + \frac{\partial F_z}{\partial z}\,. \tag{B.17}$$

The divergence takes a vector and produces a number. The above $\mathbf{F}$ has a divergence of $(3z) + (4y + xz) + (3z^2)$.

What is the physical meaning of the divergence? Consider an infinitesimal box with sides of length dx, dy, and dz. Then the divergence measures the net flux of the vector field out of the box, divided by the volume of the box. (The flux through a surface is defined to be the integral of the area times the component of the vector perpendicular to the surface.) For example, if a certain vector field gives the velocity at each point in a fluid flow, and if the divergence is nonzero, then there must be a source (or sink) that creates (or destroys) fluid, because otherwise whatever fluid goes into the little box would have to come out of it somewhere, yielding zero net flux.

Let's see why the divergence equals the flux per volume. Consider the "left" $dy \times dz$ face of the little box. The amount of flux from the vector field into the box through this face equals the area $dy\,dz$ times the F_x component. (The F_y and

F_z components are parallel to this face and therefore contribute nothing to the flux through it.) The amount of flux *out* of the "right" $dy \times dz$ face equals the area $dy\,dz$ times the value of the F_x component there. But this value equals (to first order) the original F_x plus $(\partial F_x/\partial x)\,dx$, by the definition of the partial derivative. The F_x part of this cancels the inward flux from the left face, so the net flux out of the little box through these two faces is $\big((\partial F_x/\partial x)\,dx\big)\,dy\,dz$. Similar calculations work for the other two pairs of parallel faces, so the total flux out of the box is

$$\text{Net flux} = \left(\frac{\partial F_x}{\partial x} + \frac{\partial F_y}{\partial y} + \frac{\partial F_z}{\partial z}\right) dx\,dy\,dz. \tag{B.18}$$

Therefore, as promised, the net flux per volume equals the divergence. The integrated form of this result is the *divergence theorem,* or *Gauss' theorem,* which takes the form,

$$\int_V \nabla \cdot \mathbf{F}\,dV = \int_S \mathbf{F} \cdot d\mathbf{A}. \tag{B.19}$$

The integral on the left runs over a given volume, and the integral on the right runs over the surface that encloses this volume. The vector $d\mathbf{A}$ has a magnitude equal to an infinitesimal piece of area of S and a direction defined to be perpendicular to the plane containing this piece (with the positive direction being outward from the volume). Dotting $d\mathbf{A}$ with $\mathbf{F}$ has the effect of picking out only the component of $\mathbf{F}$ that is perpendicular to the piece (which is what is relevant in calculating the flux).

We'll skip the fine details here, but the basic idea of the proof of Eq. (B.19) is to divide the volume up into many infinitesimal cubes and to look at the total flux through all the cubes. From Eq. (B.18), the integral of the divergence over one little cube (which is essentially the divergence times the volume, because the divergence is essentially constant over the tiny volume) equals the flux through that cube. The integral of the divergence over the whole volume therefore equals the sum of the fluxes through all the cubes. But all the faces of the cubes that lie in the interior of the volume are shared by two cubes, so the flux through these faces cancels when taking the sum (because the flux through a given face is counted positive for one cube and negative for the other). So we are left with only the flux through the faces on the boundary of the volume (because these faces appear only once in the total integral). We are therefore left with the flux through the surface S, which is what appears on the right-hand side of Eq. (B.19).

Example (Flux through a sphere): Verify the divergence theorem in the case where the surface is a sphere of radius R centered at the origin, and $\mathbf{F} = (x, y, z)$.

Solution: On the left-hand side of Eq. (B.19), the divergence of (x, y, z) is 1+1+1=3, so the integral of this over the volume of the sphere is simply $3(4\pi R^3/3) = 4\pi R^3$. On the right-hand side, the unit vector perpendicular to the surface is $(x, y, z)/R$, so

$d\mathbf{A} = (dA)(x,y,z)/R$. The dot product of this with (x,y,z) is $(dA)(x^2+y^2+z^2)/R = (dA)R$. The integral of this over the surface of the sphere is just $(4\pi R^2)R = 4\pi R^3$. The two sides are therefore equal, as we wanted to show.

B.6 Curl

Consider a vector whose components are functions of the coordinates. For example, let $\mathbf{F} = (F_x, F_y, F_z) = (3xz,\ x^2yz,\ x+z)$. Then the *curl* of $\mathbf{F}$ is defined to be the cross product of the ∇ operator with $\mathbf{F}$, that is,

$$\nabla \times \mathbf{F} \equiv \begin{vmatrix} \hat{\mathbf{x}} & \hat{\mathbf{y}} & \hat{\mathbf{z}} \\ \partial/\partial x & \partial/\partial y & \partial/\partial z \\ F_x & F_y & F_z \end{vmatrix} = \left(\frac{\partial F_z}{\partial y} - \frac{\partial F_y}{\partial z},\ \frac{\partial F_x}{\partial z} - \frac{\partial F_z}{\partial x},\ \frac{\partial F_y}{\partial x} - \frac{\partial F_x}{\partial y} \right). \quad \text{(B.20)}$$

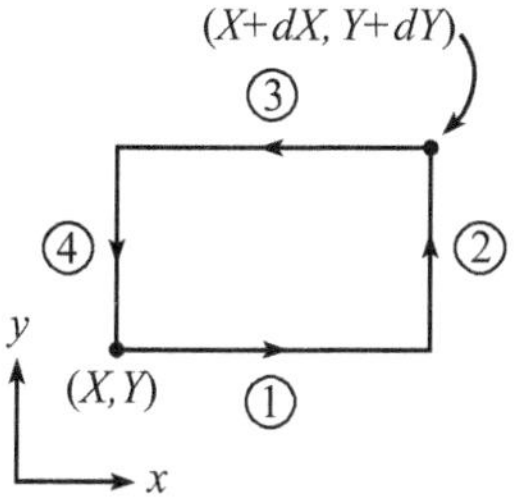

Fig. B.2

The curl takes a vector and produces another vector. The above $\mathbf{F}$ has a curl of $(-x^2y, 3x-1, 2xyz)$.

What is the physical meaning of the curl? Consider the infinitesimal rectangle shown in Fig. B.2. This rectangle lies in the x-y plane, so for the moment we will suppress the z component of all coordinates, for convenience. It turns out that the z component of the curl equals the counterclockwise integral $\int \mathbf{F} \cdot d\mathbf{r}$ around the closed loop, divided by the area of the loop (similar statements hold for the y and x components and the associated little rectangles in the x-z and y-z planes). Let's see why this is true.

The total counterclockwise integral of $\mathbf{F} \cdot d\mathbf{r}$ around the loop entails moving to the right on segment 1 and to the left on 3, and up on segment 2 and down on 4. On segments 1 and 3, both dy and dz are zero, so only the $F_x\,dx$ term survives in the dot product $\mathbf{F} \cdot d\mathbf{r}$. Likewise, $F_y\,dy$ is the only nonzero term on segments 2 and 4. If we pair up the two pairs of parallel sides, the total counterclockwise integral is

$$\int \mathbf{F} \cdot d\mathbf{r} = \int_X^{X+dX} \big(F_x(x,Y) - F_x(x,Y+dY)\big)\,dx + \int_Y^{Y+dY} \big(F_y(X+dX,y) - F_y(X,y)\big)\,dy. \quad \text{(B.21)}$$

Let's approximate the differences in these parentheses. To first order, we have

$$F_x(x,Y+dY) - F_x(x,Y) \approx dY \left.\frac{\partial F_x(x,y)}{\partial y}\right|_{(x,Y)} \approx dY \left.\frac{\partial F_x(x,y)}{\partial y}\right|_{(X,Y)}. \quad \text{(B.22)}$$

The first approximation here is valid due to the definition of the partial derivative. The second approximation (replacing x with X) is valid because our rectangle is small enough so that x is essentially equal to X. Any error in this approximation is second-order small, because we already have a factor of dY in our term. A similar treatment works for the F_y terms, so Eq. (B.21) becomes

$$\int \mathbf{F} \cdot d\mathbf{r} = \int_Y^{Y+dY} dX \frac{\partial F_y(x,y)}{\partial x}\bigg|_{(X,Y)} dy - \int_X^{X+dX} dY \frac{\partial F_x(x,y)}{\partial y}\bigg|_{(X,Y)} dx. \tag{B.23}$$

The integrands are constants, so we can quickly perform the integrals to obtain

$$\int \mathbf{F} \cdot d\mathbf{r} = dX\, dY \left(\frac{\partial F_y(x,y)}{\partial x} - \frac{\partial F_x(x,y)}{\partial y} \right)\bigg|_{(X,Y)}. \tag{B.24}$$

As promised, the z component of the curl equals the counterclockwise integral $\int \mathbf{F} \cdot d\mathbf{r}$ around the closed loop, divided by the area of the loop. The preceding analysis also works, of course, for little rectangles in the x-z and y-z planes. We therefore obtain the two other components of the curl.

The generalization of the above result to tilted and wavy surfaces is *Stokes' theorem*, which states that

$$\int_S (\nabla \times \mathbf{F}) \cdot d\mathbf{A} = \int_C \mathbf{F} \cdot d\mathbf{r}. \tag{B.25}$$

The integral on the left runs over a given surface, and the integral on the right runs over the curve that is the boundary of this surface. The vector $d\mathbf{A}$ has a magnitude equal to an infinitesimal piece of area of S and a direction defined to be perpendicular to the plane containing this piece (with its orientation defined via the orientation along C and the right-hand rule). We'll skip the details here, but the basic idea of the proof is similar to the idea behind the divergence theorem above, except with certain words replaced by other words ("volume" becomes "surface," and "surface" becomes "curve," etc.). We'll divide the surface up into many infinitesimal rectangles and look at the total integral around all the rectangles. For simplicity, let's just deal with a flat surface in the x-y plane.

From above, the integral of the z component of the curl over one little rectangle equals the counterclockwise integral of $\mathbf{F} \cdot d\mathbf{r}$ around the edges of the rectangle. The integral of the z component of the curl over the whole surface therefore equals the sum of the integrals around all the rectangles. But all the edges of the rectangles that lie in the interior of the surface are shared by two rectangles, so the integral along these edges cancels when taking the sum (because the integral along a given edge is counted positive for one rectangle and negative for the other). So we are left with only the integral along the edges on the boundary of the surface (because these edges appear only once in the total integral). We are therefore left with the integral along the curve C, which is what appears on the right-hand side of Eq. (B.25).

Note that if the surface is closed, so that it has no boundary (in other words, there is no curve C), then the right-hand side of Eq. (B.25) is zero, and hence the left-hand side is also. Such is the case with, for example, a sphere. Having no boundary means that if you're a little bug walking on the surface, you can't walk off it.

Example (Integral around a circle): Verify Stokes' theorem in the case where the curve is a circle of radius R in the x-y plane, centered at the origin, and $\mathbf{F} = (-y, x, 0)$.

Solution: On the left-hand side of Eq. (B.25), the curl of $(-y, x, 0)$ is $(0, 0, 2)$. The $d\mathbf{A}$ vector also points in the z direction, so the dot product is just $2(dA)$. The integral of this over the interior of the circle is simply $2(\pi R^2)$. On the right-hand side, the dot product equals $-y\,dx + x\,dy$. Integrating this along the circumference of the circle is most easily done in polar coordinates. With $x = R\cos\theta$ and $y = R\sin\theta$, we have $dx = -R\sin\theta\,d\theta$ and $dy = R\cos\theta\,d\theta$. So $-y\,dx + x\,dy = R^2\,d\theta$. The integral of this as θ ranges from 0 to 2π is $R^2(2\pi)$. The two sides are therefore equal, as desired.

There are some handy facts that deal with combinations of the gradient, divergence, and curl. One is that the curl of a gradient is identically zero. That is, $\nabla \times \nabla f = 0$. You can verify this explicitly by using the definitions of the curl and the gradient, and also the fact that partial differentiation is commutative (that is, $\partial^2 f/\partial x\,\partial y = \partial^2 f/\partial y\,\partial x$). Alternatively, you can let $\mathbf{F} \equiv \nabla f$ in Stokes' theorem, which gives $\int_S (\nabla \times \nabla f) \cdot d\mathbf{A} = \int_C \nabla f \cdot d\mathbf{r}$. The right-hand side of this is simply the net change in the function f around the closed curve C, which is always zero. The integrand on the left-hand side must therefore identically be zero.

Also, the divergence of a curl is identically zero. That is, $\nabla \cdot (\nabla \times \mathbf{F}) = 0$. Again, you can verify this explicitly by using the definitions of the divergence and the curl, and also the fact that partial differentiation is commutative. Alternatively, you can combine Gauss' theorem and Stokes' theorem to write $\int_V \nabla \cdot (\nabla \times \mathbf{F})\,dV = \int_C \mathbf{F} \cdot d\mathbf{r}$. The right-hand side is always zero, because the boundary surface S of any given volume V is closed, so there is no curve C. The integrand on the left-hand side must therefore identically be zero.

Appendix C $F = ma$ vs. $F = dp/dt$

In nonrelativistic mechanics,[1] the equations $F = ma$ and $F = dp/dt$ say exactly the same thing if m is constant. But if m is not constant, then $dp/dt = d(mv)/dt = ma + (dm/dt)v$, which doesn't equal ma. So if a system has a changing mass, should we use $F = ma$ or $F = dp/dt$? Which equation correctly describes the physics? The answer to this depends on what you label as the system to which you associate the quantities m, p, and a. You can generally do a problem using either $F = ma$ or $F = dp/dt$, but you must be very careful about how you label things and how you treat them. The subtleties are best understood through two examples.

Example 1 (Sand dropping into a cart): Consider a cart into which sand is dropped vertically at a rate $dm/dt = \sigma$. With what force must you push on the cart to keep it moving horizontally at a constant speed v? (This was the setup in the first example in Section 5.8.)

First solution: Let $m(t)$ be the mass of the cart-plus-sand-inside system (which we'll just call the "cart"). If we use $F = ma$ (where a is the acceleration of the cart, which is zero), then we obtain $F = 0$, which is incorrect. The correct expression to use is $F = dp/dt$. This gives

$$F = \frac{dp}{dt} = ma + \frac{dm}{dt}v = 0 + \sigma v. \tag{C.1}$$

This makes sense, because your force is what increases the momentum of the cart, and this momentum increases simply because the mass of the cart increases.

Second solution: It is possible to solve this problem by using $F = ma$ if we let our system be a small piece of mass that is being added to the cart. Your force is what accelerates this mass from rest to speed v. Consider a mass Δm that falls into the cart during a time Δt. Imagine that it falls into the cart in one lump at the start of the Δt, and then accelerates up to speed v during the time Δt. (It accelerates due to friction. But if you want, you can eliminate the cart as the intermediate object and just push directly on the mass.) This process repeats during each successive Δt interval. We

[1] We won't bother with relativity in this appendix, because nonrelativistic mechanics contains all the critical aspects we want to address.

can use $F = ma$ here because the mass of the small piece is constant. So we have $F = ma = \Delta m(v/\Delta t)$. Writing this as $(\Delta m/\Delta t)v$ gives the σv result we found above.

Third solution: As in the second solution, let's imagine the process occurring in discrete steps, but now with the cart as the system. Assume that a mass Δm falls into the cart and instantaneously decreases its speed (by conservation of momentum) to $v' = mv/(m+\Delta m)$, which is $\Delta v = v - v' = v\Delta m/(m+\Delta m)$ less than the original v. Assume that you then push on the cart for a time Δt (during which the mass remains constant at $m + \Delta m$, so $F = ma$ is the relevant expression) and bring it back up to speed v. The acceleration is $a = \Delta v/\Delta t = v(\Delta m/\Delta t)/(m + \Delta m) = \sigma v/(m + \Delta m)$, so your force is

$$F = (m + \Delta m)a = (m + \Delta m)\left(\frac{\sigma v}{m + \Delta m}\right) = \sigma v. \tag{C.2}$$

Example 2 (Sand leaking from a cart): Consider a cart that leaks sand out of the bottom at a rate $dm/dt = \sigma$. If you apply a force F to the cart, what is its acceleration?

Solution: Let $m(t)$ be the mass of the cart-plus-sand-inside system (the "cart"). In this example, the correct expression to use is $F = ma$, so the acceleration is

$$a = \frac{F}{m}. \tag{C.3}$$

Note that since m decreases with time, a increases with time. We used $F = ma$ here because at any instant, the mass m is what is being accelerated by the force F. As above, you can imagine the process occurring in discrete steps: You push on the mass for a short period of time, then a little piece instantaneously leaks out; then you push again on the new (smaller) mass, then another little piece leaks out; and so on. In this discretized scenario, it is clear that $F = ma$ is the appropriate formula, because it holds for each step in the process. The only ambiguity is whether to use m or $m + dm$ at a certain time, but this yields a negligible error.

REMARKS: It is still true that $F = dp/dt$ in this second example, provided that you let F be *total* force, and let p be the *total* momentum. In this example, F is the only force. However, the total momentum consists of both the sand in the cart and the sand that has leaked out and is falling through the air.[2] A common mistake is to use $F = dp/dt$, with p being only the cart's momentum. The leaked sand still has momentum.

There is a simple example that demonstrates why $F = dp/dt$ doesn't work when p refers only to the cart. Choose $F = 0$, so that the cart moves with constant speed v. Cut the cart in half, and label the back part as the "leaked sand" and the front part as the "cart." If you want the cart's p to have $dp/dt = F = 0$, then the cart's speed must double if its mass gets cut in half. But this is nonsense. Both halves simply continue to move at the same rate. ♣

[2] If there were air resistance, we would have to worry about its effect on the falling sand if we wanted to use $F = dp/dt$ to solve the problem, where p is the total momentum. This is clearly not the best way to do the problem. If complicated things happen with the sand in the air, it would be foolish to consider this part of the sand when we don't have to.

To sum up, $F = dp/dt$ is always valid, provided that you use the *total* force and *total* momentum of a given system of particles. This approach, however, can get messy in certain situations. So in some cases it is easier to use an $F = ma$ argument, but you must be careful to correctly identify the system that is being accelerated by the force. The asymmetry in the above two examples is that in the first example, the force does indeed accelerate the incoming sand. But in the second example, the force does *not* accelerate (or decelerate) the outgoing sand. F has nothing to do with the leaked sand.

Appendix D Existence of principal axes

In this appendix, we will prove Theorem 9.4. That is, we will show that an orthonormal set of principal axes does indeed exist for any object, and for any choice of origin. It isn't crucial that you study this proof. If you want to just accept the fact that principal axes exist, that's perfectly fine. But the method we will use in this proof is one you will see again and again in your physics studies, in particular when you study quantum mechanics (see the remark following the proof).

Theorem D.1 *Given a real symmetric* 3×3 *matrix,* $\mathbf{I}$*, there exist three orthonormal real vectors,* $\hat{\boldsymbol{\omega}}_k$*, and three real numbers,* I_k*, with the property that*

$$\mathbf{I}\hat{\boldsymbol{\omega}}_k = I_k \hat{\boldsymbol{\omega}}_k. \tag{D.1}$$

Proof: This theorem holds more generally with 3 replaced by N (all the steps below easily generalize), but we'll work with $N = 3$, to be concrete. Consider a general 3×3 matrix, $\mathbf{I}$ (we don't need to assume yet that it's real or symmetric). Assume that $\mathbf{Iu} = I\mathbf{u}$ for some vector $\mathbf{u}$ and some number I.[1] This may be rewritten as

$$\begin{pmatrix} (I_{xx} - I) & I_{xy} & I_{xz} \\ I_{yx} & (I_{yy} - I) & I_{yz} \\ I_{zx} & I_{zy} & (I_{zz} - I) \end{pmatrix} \begin{pmatrix} u_x \\ u_y \\ u_z \end{pmatrix} = \begin{pmatrix} 0 \\ 0 \\ 0 \end{pmatrix}. \tag{D.2}$$

In order for there to be a nontrivial solution for the vector $\mathbf{u}$ (that is, one where $\mathbf{u} \neq (0, 0, 0)$), the determinant of this matrix must be zero.[2] Taking the determinant, we see that we get an equation for I of the form

$$aI^3 + bI^2 + cI + d = 0. \tag{D.3}$$

[1] Such a vector $\mathbf{u}$ is called an *eigenvector* of $\mathbf{I}$, and the associated I is called an *eigenvalue*. But don't let these names scare you. They're just definitions.

[2] If the determinant were not zero, then we could explicitly construct the inverse of the matrix, which involves cofactors divided by the determinant. Multiplying both sides by this inverse would yield $\mathbf{u} = \mathbf{0}$.

The constants a, b, c, and d are functions of the matrix entries I_{ij}, but we won't need their precise form to prove this existence theorem. The only thing we need this equation for is to say that there do exist three (generally complex) solutions for I, because the equation is a cubic.

We will now show that the solutions for I are real. This will imply that there exist three real vectors $\mathbf{u}$ satisfying $\mathbf{I}\mathbf{u} = I\mathbf{u}$, because we can plug the real I's back into Eq. (D.2) and solve for the real components u_x, u_y, and u_z, up to an overall constant. We will then show that these vectors are orthogonal.

- *Proof that the I's are real:* This follows from the real and symmetric conditions on $\mathbf{I}$. Start with the equation $\mathbf{I}\mathbf{u} = I\mathbf{u}$, and take the dot product with $\mathbf{u}^*$ to obtain

$$\begin{aligned}\mathbf{u}^* \cdot \mathbf{I}\mathbf{u} &= \mathbf{u}^* \cdot I\mathbf{u} \\ &= I\mathbf{u}^* \cdot \mathbf{u}.\end{aligned} \tag{D.4}$$

The vector $\mathbf{u}^*$ is the vector obtained by complex conjugating each component of $\mathbf{u}$ (we don't know yet that $\mathbf{u}$ can be chosen to be real). On the right side, I is a scalar, so we can take it out from between the $\mathbf{u}^*$ and $\mathbf{u}$. The fact that $\mathbf{I}$ is real implies that if we complex conjugate the equation $\mathbf{I}\mathbf{u} = I\mathbf{u}$, we obtain $\mathbf{I}\mathbf{u}^* = I^*\mathbf{u}^*$ (we know that $\mathbf{I}$ is real, but we don't know yet that I is real). If we then take the dot product of this equation with $\mathbf{u}$, we obtain

$$\mathbf{u} \cdot \mathbf{I}\mathbf{u}^* = I^*\mathbf{u} \cdot \mathbf{u}^*. \tag{D.5}$$

We now claim that if $\mathbf{I}$ is symmetric, then $\mathbf{a} \cdot \mathbf{I}\mathbf{b} = \mathbf{b} \cdot \mathbf{I}\mathbf{a}$, for any vectors $\mathbf{a}$ and $\mathbf{b}$. (We'll leave this for you to show by simply multiplying each side out.) In particular, $\mathbf{u}^* \cdot \mathbf{I}\mathbf{u} = \mathbf{u} \cdot \mathbf{I}\mathbf{u}^*$, so Eqs. (D.4) and (D.5) give

$$(I - I^*)\mathbf{u} \cdot \mathbf{u}^* = 0. \tag{D.6}$$

And since $\mathbf{u} \cdot \mathbf{u}^* = |u_1|^2 + |u_2|^2 + |u_3|^2 \neq 0$, we must have $I = I^*$. Therefore, I is real.

- *Proof that the* $\mathbf{u}$ *are orthogonal:* This follows from the symmetric condition on $\mathbf{I}$. Let $\mathbf{I}\mathbf{u}_1 = I_1\mathbf{u}_1$, and $\mathbf{I}\mathbf{u}_2 = I_2\mathbf{u}_2$. Take the dot product of the former equation with $\mathbf{u}_2$ to obtain

$$\mathbf{u}_2 \cdot \mathbf{I}\mathbf{u}_1 = I_1\mathbf{u}_2 \cdot \mathbf{u}_1, \tag{D.7}$$

and take the dot product of the latter equation with $\mathbf{u}_1$ to obtain

$$\mathbf{u}_1 \cdot \mathbf{I}\mathbf{u}_2 = I_2\mathbf{u}_1 \cdot \mathbf{u}_2. \tag{D.8}$$

As above, the symmetric condition on $\mathbf{I}$ implies that the left-hand sides of Eqs. (D.7) and (D.8) are equal. Therefore,

$$(I_1 - I_2)\mathbf{u}_1 \cdot \mathbf{u}_2 = 0. \tag{D.9}$$

There are two possibilities here: (1) If $I_1 \neq I_2$, then we are done, because $\mathbf{u}_1 \cdot \mathbf{u}_2 = 0$, which says that $\mathbf{u}_1$ and $\mathbf{u}_2$ are orthogonal. (2) If $I_1 = I_2 \equiv I$, then we have $\mathbf{I}(a\mathbf{u}_1 + b\mathbf{u}_2) = I(a\mathbf{u}_1 + b\mathbf{u}_2)$, for any a and b. So any linear combination of $\mathbf{u}_1$ and $\mathbf{u}_2$ has the same property that $\mathbf{u}_1$ and $\mathbf{u}_2$ have (namely, that applying $\mathbf{I}$ is the same as just multiplying by I). We therefore have a whole plane of such vectors, so we can pick any two orthogonal vectors in this plane to be called $\mathbf{u}_1$ and $\mathbf{u}_2$. ■

This theorem proves the existence of principal axes, because the inertia tensor in Eq. (9.8) is indeed a real and symmetric matrix.

REMARK: (Warning: This remark has nothing to do with classical mechanics. It's simply an ill-disguised excuse to get a limerick on quantum mechanics into the book.) In quantum mechanics, it turns out that any observable quantity, such as position, energy, momentum, angular momentum, etc., can be represented by a *Hermitian* matrix, with the observed value being an eigenvalue of the matrix. A Hermitian matrix is a (generally complex) matrix with the property that the transpose of the matrix equals the complex conjugate of itself. For example, a 2×2 Hermitian matrix must be of the form,

$$\begin{pmatrix} a & b+ic \\ b-ic & d \end{pmatrix}, \tag{D.10}$$

for real numbers a, b, c, and d. Now, if observed values are to be given by the eigenvalues of such a matrix, then the eigenvalues had better be real, because no one (in this world, at least) is about to go for a jog of $4 + 3i$ miles, or pay an electric bill for $17 - 43i$ kilowatt-hours. And indeed, you can show, via a slightly modified version of the above "Proof that the I's are real" procedure, that the eigenvalues of any Hermitian matrix are in fact real. (And likewise, the eigenvectors are orthogonal.) This is, to say the least, very fortunate.

God's first tries were hardly ideal,
For complex worlds have no appeal.
So in the present edition,
He made things Hermitian,
And *this* world, it seems, is quite real. ♣

Appendix E Diagonalizing matrices

This appendix is relevant to Section 9.3, which covers principal axes. The process of diagonalizing matrices (that is, finding the *eigenvectors* and *eigenvalues*, defined below) has countless applications in a wide variety of subjects. We'll describe the process here as it applies to principal axes and moments of inertia.

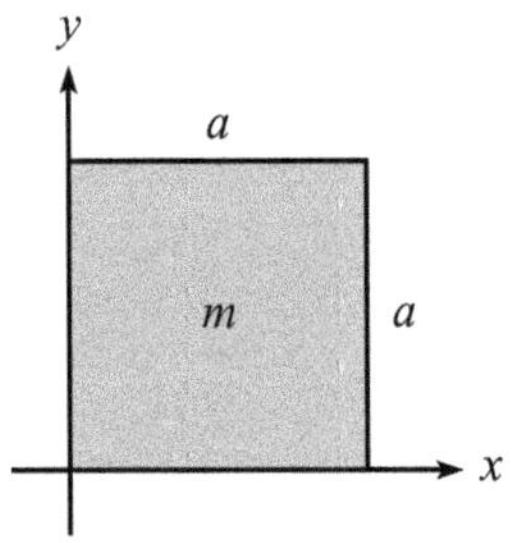

Fig. E.1

Let's find the three principal axes and moments of inertia for a square with side length a, mass m, and one corner at the origin. The square lies in the x-y plane, with sides along the x and y axes (see Fig. E.1). We'll choose the given x, y, and z axes as our initial basis axes. Using Eq. (9.8), you can show that the matrix $\mathbf{I}$ (with respect to this initial basis) is

$$\mathbf{I} = \rho \begin{pmatrix} \int y^2 & -\int xy & 0 \\ -\int xy & \int x^2 & 0 \\ 0 & 0 & \int (x^2 + y^2) \end{pmatrix} = ma^2 \begin{pmatrix} 1/3 & -1/4 & 0 \\ -1/4 & 1/3 & 0 \\ 0 & 0 & 2/3 \end{pmatrix}, \tag{E.1}$$

where ρ is the mass per unit area, so that $a^2\rho = m$. We have used the fact that $z = 0$, and we have not bothered to write the $dx\,dy$ in the integrals.

Our goal is to find the basis in which $\mathbf{I}$ is diagonal. That is, we want to find three solutions[1] for $\mathbf{u}$ (and I) for the equation $\mathbf{Iu} = I\mathbf{u}$. Letting $I \equiv \lambda ma^2$ to make things look a little cleaner, and using the above explicit form of $\mathbf{I}$, the equation $(\mathbf{I} - I)\mathbf{u} = 0$ becomes

$$ma^2 \begin{pmatrix} 1/3 - \lambda & -1/4 & 0 \\ -1/4 & 1/3 - \lambda & 0 \\ 0 & 0 & 2/3 - \lambda \end{pmatrix} \begin{pmatrix} u_x \\ u_y \\ u_z \end{pmatrix} = \begin{pmatrix} 0 \\ 0 \\ 0 \end{pmatrix}. \tag{E.2}$$

In order for there to be a nonzero solution for the components u_x, u_y, u_z, the determinant of this matrix must be zero (see Footnote D.2). The resulting cubic

[1] One obvious solution is $\mathbf{u} = \hat{\mathbf{z}}$, because $\mathbf{I}\hat{\mathbf{z}} = (2/3)ma^2\hat{\mathbf{z}}$. From the orthogonality result of Theorem 9.4, we know that the other two vectors must lie in the x-y plane. So we could quickly reduce this problem to a two-dimensional one, but let's forge ahead with the general method.

equation for λ is easy to solve, because the determinant is $[(1/3-\lambda)^2-(1/4)^2](2/3-\lambda)=0$. The solutions are $\lambda = 1/3 \pm 1/4$, and $\lambda = 2/3$. So our three principal moments, $I \equiv \lambda ma^2$, are

$$I_1 = \frac{7}{12}ma^2, \quad I_2 = \frac{1}{12}ma^2, \quad I_3 = \frac{2}{3}ma^2. \tag{E.3}$$

These are the *eigenvalues* of $\mathbf{I}$.

What are the vectors, $\mathbf{u}_1$, $\mathbf{u}_2$, and $\mathbf{u}_3$, associated with each of these I's? Plugging $\lambda = 7/12$ into Eq. (E.2) gives the three equations (one for each component), $-u_x - u_y = 0$, $-u_x - u_y = 0$, and $u_z = 0$. These are redundant equations (that was the point of setting the determinant equal to zero). So $u_x = -u_y$, and $u_z = 0$. The vector may therefore be written as $\mathbf{u}_1 = (c, -c, 0)$, where c is any constant.[2] If we want a normalized vector, then $c = 1/\sqrt{2}$. In a similar manner, plugging $\lambda = 1/12$ into Eq. (E.2) gives $\mathbf{u}_2 = (c, c, 0)$. And finally, plugging $\lambda = 2/3$ into Eq. (E.2) gives $\mathbf{u}_3 = (0, 0, c)$, as claimed in the above footnote. Our three orthonormal principal axes corresponding to the moments in Eq. (E.3) are therefore

$$\hat{\omega}_1 = \left(\frac{1}{\sqrt{2}}, -\frac{1}{\sqrt{2}}, 0\right), \quad \hat{\omega}_2 = \left(\frac{1}{\sqrt{2}}, \frac{1}{\sqrt{2}}, 0\right), \quad \hat{\omega}_3 = (0, 0, 1). \tag{E.4}$$

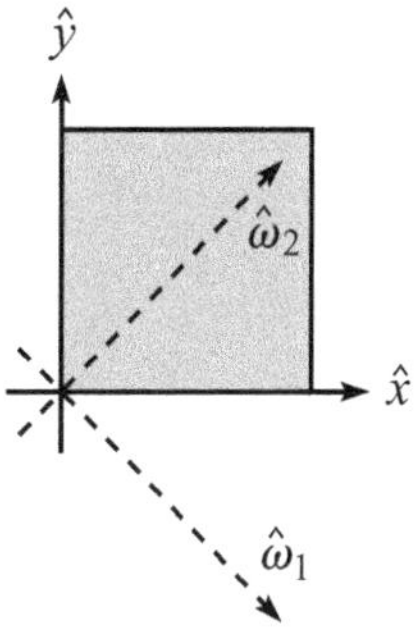

Fig. E.2

These are the *eigenvectors* of $\mathbf{I}$. They are shown in Fig. E.2. In the new basis of the principal axes (where $\hat{\omega}_1 = (1, 0, 0)$, etc.), the matrix $\mathbf{I}$ takes the form,

$$\mathbf{I} = ma^2 \begin{pmatrix} 7/12 & 0 & 0 \\ 0 & 1/12 & 0 \\ 0 & 0 & 2/3 \end{pmatrix}. \tag{E.5}$$

In other words, we have "diagonalized" the matrix. The basic idea is that from now on we should use the principal axes as our basis vectors. We can forget that we ever had anything to do with the original x, y, and z axes.

REMARKS:

1. $I_1 + I_2 = I_3$, as the perpendicular-axis theorem demands.
2. I_2 is the moment around one diagonal through the center of the square, which of course equals the moment around the other diagonal through the center. But the latter is related to I_1 by the parallel-axis theorem. And indeed, $I_1 = I_2 + m(a/\sqrt{2})^2$.
3. Any axis through the center of the square, in the plane of the square, has the same moment (by Theorem 9.5 or 9.6). So I_2 equals the moment around an axis through the center, parallel to a side. But this is the same as the moment of a stick of length a around its center (the extent of the square in the direction of the axis is irrelevant). Hence the factor of $1/12$ in I_2. ♣

[2] We can solve for $\mathbf{u}$ only up to an overall constant, because if $\mathbf{Iu} = I\mathbf{u}$ is true for a certain $\mathbf{u}$, then it is also true that $\mathbf{I}(c\mathbf{u}) = I(c\mathbf{u})$, where c is any constant.

Appendix F Qualitative relativity questions

1. Is there such a thing as a perfectly rigid object?

   Answer: No. Since information can move no faster than the speed of light, it takes time for the atoms in the object to communicate with each other. If you push on one end of a rod, then the other end won't move right away. If it did move right away, then these "pushing" and "moving" events would be spacelike separated (see Section 11.6), which would mean that there would exist a frame in which the "moving" event occurred before the "pushing" event. This is a violation of causality, so we conclude that the other end doesn't move right away.

2. How do you synchronize two clocks that are at rest with respect to each other?

   Answer: One way is to put a light source midway between the two clocks and send out signals, and then set the clocks to a certain value when the signals hit them. Another way is to put a watch right next to one of the clocks and synchronize it with this clock, and then move the watch very slowly over to the other clock and synchronize that clock with it. Any time-dilation effects can be made arbitrarily small by moving the watch sufficiently slowly, because the time-dilation effect is second order in v.

3. Moving clocks run slow. Does this result have anything to do with the time it takes light to travel from the clock to your eye?

   Answer: No. When we talk about how fast a clock is running in a given frame, we are referring to what the clock actually reads in that frame. It will certainly take time for the light from the clock to reach an observer's eye, but it is understood that the observer subtracts off this transit time in order to calculate the time at which the clock actually shows a particular reading. Likewise, other relativistic effects, such as length contraction and the loss of simultaneity, have nothing to do with the time it takes light to reach your eye. They deal only with what really *is*, in your frame. One way to avoid the complication of the travel time of light is to use the lattice of clocks and meter sticks described at the end of Section 11.3.3.

4. Does time dilation depend on whether a clock is moving across your vision or directly away from you?

Answer: No. A moving clock runs slow, no matter which way it is moving. This is clearer if you think in terms of the lattice of clocks and meter sticks from Section 11.3.3. If you imagine a million people standing at the points of the lattice, then they all observe the clock running slow. Time dilation is an effect that depends on the *frame* and the speed of a clock with respect to it. It doesn't matter where you are in the frame (as long as you're at rest in it).

5. Does special-relativistic time dilation depend on the acceleration of the moving clock?

Answer: No. The time-dilation factor is $\gamma = 1/\sqrt{1 - v^2/c^2}$, which doesn't depend on a. The only relevant quantity is the v at a given instant. It doesn't matter if v is changing. But if *you* are accelerating, then you can't naively apply the results of Special Relativity. (To do things correctly, it is perhaps easiest to think in terms of General Relativity. But GR is actually not required; see Chapter 14 for a discussion of this.) But as long as you represent an inertial frame, then the clock you are viewing can undergo whatever motion it wants, and you will observe it running slow by the simple factor of γ.

6. Someone says, "A stick that is length-contracted isn't *really* shorter, it just *looks* shorter." How do you respond?

Answer: The stick really *is* shorter in your frame. Length contraction has nothing to do with how things look. It has to do with where the ends of the stick are at simultaneous times in your frame. (This is, after all, how you measure the length of something.) At a given instant in time in your frame, the distance between the ends of the stick is indeed less than the proper length of the stick.

7. Consider a stick that moves in the direction in which it points. Does its length contraction depend on whether this direction is across your vision or directly away from you?

Answer: No. The stick is length-contracted in both cases. Of course, if you look at the stick in the latter case, then all you see is the end, which is just a dot. But the stick is indeed shorter in your reference frame. As in Question 4 above, length contraction depends on the frame, not where you are in it.

8. A mirror moves toward you at speed v. You shine a light toward it and the light beam bounces back at you. What is the speed of the reflected beam?

Answer: The speed is c, as always. You will observe the light having a higher frequency, due to the Doppler effect. But the speed is still c.

9. In relativity, the order of two events in one frame may be reversed in another frame. Does this imply that there exists a frame in which I get off a bus before I get on it?

   Answer: No. The order of two events can be reversed in another frame only if the events are spacelike separated. That is, if $\Delta x > c\Delta t$ (in other words, the events are too far apart for even light to get from one to the other). The two relevant events here (getting on the bus, and getting off the bus) are not spacelike separated, because the bus travels at a speed less than c, of course. They are timelike separated. Therefore, in all frames it is the case that I get off the bus after I get on it.

   There would be causality problems if there existed a frame in which I got off the bus before I got on it. If I break my ankle getting off a bus, then I wouldn't be able to make the mad dash that I made to catch the bus in the first place, in which case I wouldn't have the opportunity to break my ankle getting off the bus, in which case I could have made the mad dash to catch the bus and get on, and, well, you get the idea.

10. You are in a spaceship sailing along in outer space. Is there any way you can measure your speed without looking outside?

    Answer: There are two points to be made here. First, the question is meaningless, because absolute speed doesn't exist. The spaceship doesn't have a speed; it only has a speed relative to something else. Second, even if the question asked for the speed with respect to, say, a given piece of stellar dust, the answer would be "no." Uniform speed is not measurable from within the spaceship. Acceleration, on the other hand, is measurable (assuming there is no gravity around to confuse it with).

11. If you move at the speed of light, what shape does the universe take in your frame?

    Answer: The question is meaningless, because it's impossible for you to move at the speed of light. A meaningful question to ask is: What shape does the universe take if you move at a speed very close to c? The answer is that in your frame everything would be squashed along the direction of your motion, due to length contraction. Any given region of the universe would be squashed down to a pancake.

12. Two objects fly toward you, one from the east with speed u, and the other from the west with speed v. Is it correct that their relative speed, as measured by you, is $u + v$? Or should you use the velocity-addition formula, $V = (u + v)/(1 + uv/c^2)$? Is it possible for their relative speed, as measured by you, to exceed c?

    Answer: Yes, no, yes, to the three questions. It is legal to simply add the two speeds to obtain $u + v$. There is no need to use the velocity-addition formula, because both speeds here are measured with respect to the *same* *thing*, namely you. It's perfectly legal for the result to be greater than c, but it must be less than (or equal to, for photons) $2c$.

You need to use the velocity-addition formula when, for example, you are given the speed of a ball with respect to a train, and also the speed of the train with respect to the ground, and your goal is to find the speed of the ball with respect to the ground. The point is that now the two given speeds are measured with respect to *different* things, namely the train and the ground.

13. Two clocks at the ends of a train are synchronized with respect to the train. If the train moves past you, which clock shows the higher time?

    Answer: The rear clock shows the higher time. It shows Lv/c^2 more than the front clock, where L is the proper length of the train.

14. A train moves at speed $4c/5$. A clock is thrown from the back of the train to the front. As measured in the ground frame, the time of flight is 1 second. Is the following reasoning correct? "The γ factor between the train and the ground is $\gamma = 1/\sqrt{1-(4/5)^2} = 5/3$. And since moving clocks run slow, the time elapsed on the clock during the flight is $3/5$ of a second."

    Answer: No. It is incorrect, because the time-dilation result holds only for two events that happen at the *same place* in the relevant reference frame (the train, here). The clock moves with respect to the train, so the above reasoning is invalid.

    Another way of seeing why it must be incorrect is the following. A certainly valid way to calculate the clock's elapsed time is to find the speed of the clock with respect to the ground (more information would have to be given to determine this), and to then apply time dilation with the associated γ factor to arrive at the answer of $(1\text{ s})/\gamma$. Since the clock's v is definitely not $4c/5$, the correct answer is definitely not $3/5$ s.

15. Person A chases person B. As measured in the ground frame, they have speeds $4c/5$ and $3c/5$, respectively. If they start a distance L apart (as measured in the ground frame), how much time will it take (as measured in the ground frame) for A to catch B?

    Answer: As measured in the ground frame, the relative speed is $4c/5 - 3c/5 = c/5$. Person A must close the initial gap of L, so the time it takes is $L/(c/5) = 5L/c$. There is no need to use any fancy velocity-addition or length-contraction formulas, because all quantities in this problem are measured with respect to the *same* frame. So it quickly reduces to a simple "(rate)(time) = (distance)" problem.

16. Is the "the speed of light is the same in all inertial frames" postulate really necessary? That is, is it not already implied by the "the laws of physics are the same in all inertial frames" postulate?

Answer: Yes, it is necessary. The speed-of-light postulate is definitely not implied by the laws-of-physics postulate. The latter doesn't imply that baseballs have the same speed in all inertial frames, so it likewise doesn't imply it for light.

It turns out that nearly all the results in Special Relativity can be deduced by using only the laws-of-physics postulate. What you can find (with some work) is that there is some limiting speed, which may or may not be infinite (see Section 11.10). But you still have to say whether this speed is finite or infinite. The speed-of-light postulate does the trick.

17. Imagine closing a very large pair of scissors. It is quite possible for the point of intersection of the blades to move faster than the speed of light. Does this violate anything in relativity?

Answer: No. If the angle between the blades is small enough, then the tips of the blades (and all the other atoms in the scissors) can move at a speed well below c, while the intersection point moves faster than c. But this doesn't violate anything in relativity. The intersection point is not an actual object, so there is nothing wrong with it moving faster than c.

You might be worried that this result allows you to send a signal down the scissors at a speed faster than c. However, since there is no such thing as a rigid body, it is impossible to get the far end of the scissors to move right away, when you apply a force at the handle. The scissors would have to already be moving, in which case the motion is independent of any decision you make at the handle to change the motion of the blades.

18. Two twins travel away from each other at relativistic speed. The time-dilation result says that each twin sees the other twin's clock running slow, so each says the other has aged less. How would you reply to someone who asks, "But which twin really *is* younger?"

Answer: It makes no sense to ask which twin really is younger, because the two twins aren't in the same reference frame; they are using different coordinates to measure time. It's as silly as having two people run away from each other into the distance (so that each person sees the other become small), and then asking: Who is really smaller?

19. A particular event has coordinates (x, t) in one frame. How do you use a Lorentz transformation to find the coordinates of this event in another frame?

Answer: You don't. Lorentz transformations have nothing to do with single events. They deal only with *pairs* of events and the *separation* between them. As far as a single event goes, its coordinates in another frame can be anything you want, simply by defining your origin to be wherever and whenever you please. But for pairs of events, their separation is a well-defined quantity, independent of any definition of origin. It is therefore a meaningful question to ask how the separations in two different frames are related, and the Lorentz transformations answer this question.

20. When using the Lorentz transformations, how do you tell which frame is the moving "primed" frame?

Answer: You don't. There is no preferred frame, so it doesn't make sense to ask which frame is moving. We used the "primed" notation in the derivation in Section 11.4.1 for ease of notation, but don't take it to imply that there is a preferred frame S and a less fundamental frame S'. In general, a better notation is to use subscripts that describe the two frames, such as "g" for ground and "t" for train. For example, if you know the values of Δt_t and Δx_t on a train (which we'll assume is moving in the positive x direction with respect to the ground), and if you want to find the values of Δt_g and Δx_g on the ground, then you can write down:

$$\begin{aligned} \Delta x_g &= \gamma(\Delta x_t + v\,\Delta t_t), \\ \Delta t_g &= \gamma(\Delta t_t + v\,\Delta x_t/c^2). \end{aligned} \tag{F.1}$$

The sign is a "+" because the frame associated with the left side of the equation (the ground) sees the frame associated with the right side (the train) moving to the right. If instead you know the intervals on the ground and you want to find them on the train, then you just need to switch the subscripts "g" and "t" and change the sign to "−", by the reasoning in the previous sentence.

21. The momentum of an object with mass m and speed v is $p = \gamma m v$. "A photon has zero mass, so it should have zero momentum." Correct or incorrect?

Answer: Incorrect. True, m is zero, but the γ factor is infinite because $v = c$. Infinity times zero is undefined. A photon does indeed have momentum, and it equals E/c (which happens to equal $h\nu/c$, where ν is the frequency of the light).

22. It is not necessary to postulate the impossibility of accelerating an object to speed c. It follows as a consequence of the relativistic form of energy. Explain.

Answer: $E = \gamma mc^2$, so if $v = c$ then $\gamma = \infty$, and the object must have an infinite amount of energy (unless $m = 0$, as for a photon). All the energy in the universe, let alone all the king's horses and all the king's men, can't accelerate something to speed c.

Appendix G Derivations of the Lv/c^2 result

In Section 11.3.1, we showed that if a train with proper length L moves at speed v with respect to the ground, then in the ground frame the rear clock reads Lv/c^2 more than the front clock (assuming that the clocks are synchronized in the train frame). There are various other ways to derive this result, so for the fun of it I've listed here all the derivations I can think of. The explanations are terse, but I refer you to the specific problem or section in the text where things are discussed in more detail. Many of these derivations are slight variations on each other, so perhaps they shouldn't all count as separate ones, but here's my list:

1. **Light source on train**: Put a light source on a train, at distances $d_\text{f} = L(c - v)/2c$ from the front and $d_\text{b} = L(c + v)/2c$ from the back. You can show that the photons hit the ends of the train simultaneously in the ground frame. But they hit the ends at different times in the train frame; the difference in the readings on clocks at the ends when the photons run into them is $(d_\text{b} - d_\text{f})/c = Lv/c^2$. Therefore, at a given instant in the ground frame (for example, the moment when the clocks at the ends are simultaneously illuminated by the photons), a person on the ground sees the rear clock read Lv/c^2 more than the front clock. (See the example in Section 11.3.1.)
2. **Lorentz transformation**: The second of Eqs. (11.17) is $\Delta t_\text{g} = \gamma(\Delta t_\text{t} + v\,\Delta x_\text{t}/c^2)$, where the subscripts refer to the ground and train frames. If two events (for example, two clocks flashing their times) located at the ends of the train are simultaneous in the ground frame, then we have $\Delta t_\text{g} = 0$. And $\Delta x_\text{t} = L$, of course. The above Lorentz transformation therefore gives $\Delta t_\text{t} = -Lv/c^2$. The minus sign here means that the event with the larger x_t value has the smaller t_t value. In other words, the front clock reads Lv/c^2 less time than the rear clock, at a given instant in the ground frame.
3. **Invariant interval**: This is actually just a partial derivation, because it determines only the magnitude of the Lv/c^2 result, and not the sign. The invariant interval says that $c^2\Delta t_\text{g}^2 - \Delta x_\text{g}^2 = c^2\Delta t_\text{t}^2 - \Delta x_\text{t}^2$, where the subscripts refer to the ground and train frames. If two events (for example, two clocks flashing their times) located at the ends of the train are simultaneous in the ground frame, then we have $\Delta t_\text{g} = 0$. And $\Delta x_\text{t} = L$, of course. And we also know from length contraction that $\Delta x_\text{g} = L/\gamma$. The invariant interval then gives $c^2(0)^2 - (L/\gamma)^2 = c^2\Delta t_\text{t}^2 - L^2$, which yields

$c^2\Delta t_t^2 = L^2(1 - 1/\gamma^2) \Longrightarrow c^2\Delta t_t^2 = L^2v^2/c^2 \Longrightarrow \Delta t_t = \pm Lv/c^2$. As mentioned above, the sign isn't determined by this method.

4. **Minkowski diagram**: The task of Exercise 11.63 is to use a Minkowski diagram to derive the Lv/c^2 result. The basic goal is to determine how many ct' units fit in the segment BC in Figure 11.27, and also how many ct units fit in the segment BE in Figure 11.28.
5. **Walking slowly on a train**: In Exercise 11.58, a person walks very slowly at speed u from the back of a train of proper length L to the front. In the frame of the train, the time-dilation effect is second order in u/c and therefore negligible (because the total time is only first order in $1/u$). But in the frame of the ground, the time-dilation effect is (as you can show) *first* order in u/c and therefore has a nonzero effect; an observer on the ground sees the person's clock advance by less than a clock that is fixed on the train. Now, the person's clock agrees with clocks at the rear and front at the start and finish, because of the negligible time dilation in the train frame. Therefore, since less time elapses on the person's clock than on the front clock (in the ground frame), the person's clock must have started out reading more time than the front clock (in the ground frame). This then implies that the rear clock must show more time than the front clock. A quantitative analysis shows that this excess time is in fact Lv/c^2.
6. **Consistency arguments**: There are many setups (a few examples are Problems 11.2, 11.3, 11.8, and Exercise 11.35) where the Lv/c^2 result is an ingredient in explaining a result. Without it, you would encounter a contradiction, such as two different frames giving two different answers to a frame-independent question. So if you wanted to, you could work backwards (under the assumption that everything is consistent in relativity) and let the rear-clock-ahead effect be some unknown time T (which might be zero, for all you know), and then solve for the T that makes everything consistent. You would arrive at $T = Lv/c^2$.
7. **Gravitational time dilation**: The task of Exercise 14.13 is to derive (for small v) the Lv/c^2 result by making use of the fact that Lv/c^2 looks a lot like the gh/c^2 term in the GR time-dilation result. If you stand on the ground near the front of a train of length L and then accelerate toward the back with acceleration g, you will see a clock at the back running faster by a factor $(1 + gL/c^2)$, which will cause it to read $(gL/c^2)t = Lv/c^2$ more than a clock at the front. (Assume that you accelerate for a short period of time, so that $v \approx gt$ and that the distance in the gL/c^2 term remains essentially L.)
8. **Accelerating rocket**: The task of Problem 14.6 is to show that if a rocket with proper length L accelerates at g and reaches a speed v, then in the ground frame the readings on the front and rear clocks are related by $t_f = t_b(1+gL/c^2) - Lv/c^2$. In other words, the front clock reads $t_b(1+gL/c^2) - Lv/c^2$ simultaneously with the rear clock reading t_b, in the ground frame. But in the rocket frame, gravitational time dilation tells us that the front clock reads $t_b(1 + gL/c^2)$ simultaneously with the rear clock reading t_b. The difference in clock readings (front minus rear) is therefore smaller in the ground frame than in the rocket frame, by an amount equal to Lv/c^2. This is the desired result.

Appendix H Resolutions to the twin paradox

The twin paradox appeared in Chapters 11 and 14, both in the text and in various problems. To summarize, the twin paradox deals with twin A who stays on the earth,[1] and twin B who travels quickly to a distant star and back. When they meet up again, they discover that B is younger. This is true because A can use the standard special-relativistic time-dilation result to say that B's clock runs slow by a factor γ.

The "paradox" arises from the fact that the situation seems symmetrical. That is, it seems as though each twin should be able to consider herself to be at rest, so that she sees the other twin's clock running slow. So why does B turn out to be younger? The resolution to the paradox is that the setup is in fact *not* symmetrical, because B must turn around and thus undergo acceleration. She is therefore not always in an inertial frame, so she cannot always apply the simple special-relativistic time-dilation result.

While the above reasoning is sufficient to get rid of the paradox, it isn't quite complete, because (a) it doesn't explain how the result from B's point of view quantitatively agrees with the result from A's point of view, and (b) the paradox can actually be formulated without any mention of acceleration, in which case slightly different reasoning applies.

Below is a list of all the complete resolutions I can think of. The descriptions are terse, but I refer you to the specific problem or section in the text where things are discussed in more detail. As with the Lv/c^2 derivations in Appendix G, many of these resolutions are slight variations on each other, so perhaps they shouldn't all count as separate ones, but here's my list:

1. **Rear-clock-ahead effect:** Let the distant star be labeled as C. Then on the outward part of the journey, B sees C's clock ahead of A's by Lv/c^2, because C is the rear clock in the universe as the universe flies by. But after B turns around, A becomes the rear clock and is therefore now ahead of C. This means that A's clock must jump forward very quickly, from B's point of view. (See Section 11.3.1 and Problem 11.2.)

[1] We should actually have A floating in space, to avoid any GR time-dilation effects from the earth's gravity. But if B travels quickly enough, the SR effects will dominate the gravitational ones.

2. **Looking out of the portholes:** Imagine many clocks lined up between the earth and the star, all synchronized in the earth–star frame. And imagine looking out of the portholes of the spaceship and making a movie of the clocks as you fly past them. Although you see each individual clock running slow, you see the "effective" clock in the movie (which is really many successive clocks) running fast. This effect is just a series of small applications (see Problem 11.2) of the rear-clock-ahead effect mentioned above.
3. **Minkowski diagram:** Draw a Minkowski diagram with the axes in A's frame perpendicular. Then the lines of simultaneity (that is, the successive x axes) in B's frame are titled in different directions for the outward and inward parts of the journey. The change in the tilt at the turnaround causes a large amount of time to advance on A's clock, as measured in B's frame. (See Section 11.7 and Figure 11.68.)
4. **General-relativistic turnaround effect:** The acceleration that B feels when she turns around may equivalently be thought of as a gravitational field. Twin A on the earth is high up in the gravitational field, so B sees A's clock run very fast during the turnaround. This causes A's clock to show more time in the end. (See Problem 14.9.)
5. **Doppler effect:** By equating the total number of signals one twin sends out with the total number of signals the other twin receives, we can relate the total times on their clocks. (See Exercise 11.67.)

Appendix I Lorentz transformations

In this appendix, we will give an alternate derivation of the Lorentz transformations in Eq. (11.17). The goal here is to derive them from scratch, using only the two postulates of relativity. We will *not* use any of the results derived in Section 11.3. Our strategy will be to use the relativity postulate ("all inertial frames are equivalent") to figure out as much as we can, and to then invoke the speed-of-light postulate at the end. The main reason for doing things in this order is that it will allow us to derive a very interesting result in Section 11.10.

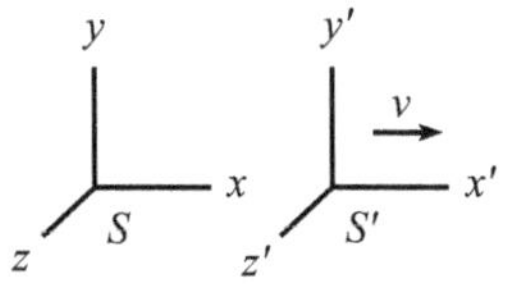

Fig. I.1

As in Section 11.4, consider a coordinate system S' moving relative to another system S (see Fig. I.1). Let the constant relative speed between the frames be v. Let the corresponding axes of S and S' point in the same direction, and let the origin of S' move along the x axis of S, in the positive direction. As in Section 11.4, we want to find the constants, A, B, C, and D, in the relations,

$$\begin{aligned} \Delta x &= A\,\Delta x' + B\,\Delta t', \\ \Delta t &= C\,\Delta t' + D\,\Delta x'. \end{aligned} \tag{I.1}$$

The four constants will end up depending on v (which is constant, given the two inertial frames). Since we have four unknowns, we need four facts. The facts we have at our disposal (using only the two postulates of relativity) are the following.

1. The physical setup: S' travels with velocity v with respect to S.
2. The principle of relativity: S should see things in S' in exactly the same way as S' sees things in S (except perhaps for a minus sign in some relative positions, but this just depends on our arbitrary choice of directional signs for the axes).
3. The speed-of-light postulate: A light pulse with speed c in S' also has speed c in S.

The second statement here contains two independent bits of information. (It contains at least two, because we will indeed be able to solve for our four unknowns. And it contains no more than two, because then our four unknowns would be over-constrained.) The two bits that are used depend on personal preference. Three that are commonly used are: (a) the relative speed looks the same from either frame, (b) time dilation (if any) looks the same from either frame, and

(c) length contraction (if any) looks the same from either frame. It is also common to recast the second statement in the form: The Lorentz transformations are the same as their inverse transformations (up to a possible minus sign). We'll choose to work with (a) and (b). Our four independent facts are then:

1. S' travels with velocity v with respect to S.
2. S travels with velocity $-v$ with respect to S'. The minus sign here is due to the convention that we picked the positive x axes of the two frames to point in the same direction.
3. Time dilation (if any) looks the same from either frame.
4. A light pulse with speed c in S' also has speed c in S.

Let's see what these imply, in the above order.[1]

- (1) says that a given point in S' moves with velocity v with respect to S. Letting $x' = 0$ (which is understood to be $\Delta x' = 0$, but we'll drop the Δ's from here on) in Eqs. (I.1) and dividing them gives $x/t = B/C$. This must equal v. Therefore, $B = vC$, and the transformations become

$$\begin{aligned} x &= Ax' + vCt', \\ t &= Ct' + Dx'. \end{aligned} \tag{I.2}$$

- (2) says that a given point in S moves with velocity $-v$ with respect to S'. Letting $x = 0$ in the first of Eqs. (I.2) gives $x'/t' = -vC/A$. This must equal $-v$. Therefore, $C = A$, and the transformations become

$$\begin{aligned} x &= Ax' + vAt', \\ t &= At' + Dx'. \end{aligned} \tag{I.3}$$

  Note that these are consistent with the Galilean transformations, which have $A = 1$ and $D = 0$.
- (3) can be used in the following way. How fast does a person in S see a clock in S' tick? (The clock is assumed to be at rest with respect to S'.) Let our two events be two successive ticks of the clock. Then $x' = 0$, and the second of Eqs. (I.3) gives

$$t = At'. \tag{I.4}$$

  In other words, one second on S''s clock takes a time of A seconds in S's frame.

  Consider the analogous situation from S''s point of view. How fast does a person in S' see a clock in S tick? (The clock is now assumed to be at rest with respect to S, in order to create the analogous setup. This is important.) If we invert Eqs. (I.3) to solve

[1] In what follows, we could obtain the final result a little quicker if we invoked the speed-of-light fact prior to the time-dilation one. But we'll do things in the above order so that we can easily carry over the results of this appendix to the discussion in Section 11.10.

for x' and t' in terms of x and t, we find

$$\begin{aligned} x' &= \frac{x - vt}{A - Dv}, \\ t' &= \frac{At - Dx}{A(A - Dv)}. \end{aligned} \tag{I.5}$$

Two successive ticks of the clock in S satisfy $x = 0$, so the second of Eqs. (I.5) gives

$$t' = \frac{t}{A - Dv}. \tag{I.6}$$

In other words, one second on S's clock takes a time of $1/(A - Dv)$ seconds in S''s frame.

Both Eqs. (I.4) and (I.6) apply to the same situation (someone looking at a clock flying by). Therefore, the factors on the right-hand sides must be equal, that is,

$$A = \frac{1}{A - Dv} \quad \Longrightarrow \quad D = \frac{1}{v}\left(A - \frac{1}{A}\right). \tag{I.7}$$

Our transformations in Eq. (I.3) therefore take the form

$$\begin{aligned} x &= A(x' + vt'), \\ t &= A\left(t' + \frac{1}{v}\left(1 - \frac{1}{A^2}\right)x'\right). \end{aligned} \tag{I.8}$$

These are consistent with the Galilean transformations, which have $A = 1$.

- (4) may now be used to say that if $x' = ct'$, then $x = ct$. In other words, if $x' = ct'$, then

$$c = \frac{x}{t} = \frac{A\big((ct') + vt'\big)}{A\left(t' + \frac{1}{v}\left(1 - \frac{1}{A^2}\right)(ct')\right)} = \frac{c + v}{1 + \frac{c}{v}\left(1 - \frac{1}{A^2}\right)}. \tag{I.9}$$

Solving for A gives

$$A = \frac{1}{\sqrt{1 - v^2/c^2}}. \tag{I.10}$$

We have chosen the positive square root so that the positive x and x' axes point in the same direction. The transformations are now no longer consistent with the Galilean transformations, because c is not infinite, which means that A is not 1.

The constant A is commonly denoted by γ, so we may finally write our Lorentz transformations, Eqs. (I.8), in the form,

$$\begin{aligned} x &= \gamma(x' + vt'), \\ t &= \gamma(t' + vx'/c^2), \end{aligned} \tag{I.11}$$

where

$$\gamma \equiv \frac{1}{\sqrt{1 - v^2/c^2}}, \tag{I.12}$$

in agreement with Eq. (11.17).

Appendix J Physical constants and data

Earth

Mass	$M_E = 5.97 \cdot 10^{24}$ kg
Mean radius	$R_E = 6.37 \cdot 10^6$ m
Mean density	5.52 g/cm^3
Surface acceleration	$g = 9.81\ m/s^2$
Mean distance from sun	$1.5 \cdot 10^{11}$ m
Orbital speed	29.8 km/s
Period of rotation	23 h 56 min 4 s $= 8.6164 \cdot 10^4$ s
Period of orbit	365 days 6 h $= 3.16 \cdot 10^7\ s \approx \pi \cdot 10^7$ s

Moon

Mass	$M_M = 7.35 \cdot 10^{22}$ kg
Radius	$R_M = 1.74 \cdot 10^6$ m
Mean density	3.34 g/cm^3
Surface acceleration	$1.62\ m/s^2 \approx g/6$
Mean distance from earth	$3.84 \cdot 10^8$ m
Orbital speed	1.0 km/s
Period of rotation	27.3 days $= 2.36 \cdot 10^6$ s
Period of orbit	27.3 days $= 2.36 \cdot 10^6$ s

Sun

Mass	$M_S \equiv M_\odot = 1.99 \cdot 10^{30}$ kg
Radius	$R_S = 6.96 \cdot 10^8$ m
Mean density	1.41 g/cm^3
Surface acceleration	$274\ m/s^2 \approx 28g$

Fundamental constants

Speed of light	$c = 2.998 \cdot 10^8$ m/s
Gravitational constant	$G = 6.674 \cdot 10^{-11}\ m^3/kg\,s^2$

Planck's constant	$h = 6.63 \cdot 10^{-34}$ J s
	$\hbar \equiv h/2\pi = 1.05 \cdot 10^{-34}$ J s
Electron charge	$-e = -1.602 \cdot 10^{-19}$ C
Electron mass	$m_{\mathrm{e}} = 9.11 \cdot 10^{-31}$ kg $= 0.511$ MeV$/c^2$
Proton mass	$m_{\mathrm{p}} = 1.673 \cdot 10^{-27}$ kg $= 938.3$ MeV$/c^2$
Neutron mass	$m_{\mathrm{n}} = 1.675 \cdot 10^{-27}$ kg $= 939.6$ MeV$/c^2$

References

Adler, C. G. and Coulter, B. L. (1978). Galileo and the Tower of Pisa experiment. *American Journal of Physics*, **46**, 199–201.

Anderson, J. L. (1990). Newton's first two laws of motion are not definitions. *American Journal of Physics*, **58**, 1192–1195.

Aravind, P. K. (2007). The physics of the space elevator. *American Journal of Physics*, **75**, 125–130.

Atwood, G. (1784). *A Treatise on the Rectilinear Motion and Rotation of Bodies*, Cambridge: Cambridge University Press.

Belorizky, E. and Sivardiere, J. (1987). Comments on the horizontal deflection of a falling object. *American Journal of Physics*, **55**, 1103–1104.

Billah, K. Y. and Scanlan, R. H. (1991). Resonance, Tacoma Narrows bridge failure, and undergraduate physics textbooks. *American Journal of Physics*, **59**, 118–124.

Brehme, R. W. (1971). The relativistic Lagrangian. *American Journal of Physics*, **39**, 275–280.

Brown, L. S. (1978). Forces giving no orbit precession. *American Journal of Physics*, **46**, 930–931.

Brush, S. G. (1980). Discovery of the Earth's core. *American Journal of Physics*, **48**, 705–724.

Buckmaster, H. A. (1985). Ideal ballistic trajectories revisited. *American Journal of Physics*, **53**, 638–641.

Butikov, E. I. (2001). On the dynamic stabilization of an inverted pendulum. *American Journal of Physics*, **69**, 755–768.

Calkin, M. G. (1989). The dynamics of a falling chain: II. *American Journal of Physics*, **57**, 157–159.

Calkin, M. G. and March, R. H. (1989). The dynamics of a falling chain: I. *American Journal of Physics*, **57**, 154–157.

Castro, A. S. de (1986). Damped harmonic oscillator: A correction in some standard textbooks. *American Journal of Physics*, **54**, 741–742.

Celnikier, L. M. (1983). Weighing the Earth with a sextant. *American Journal of Physics*, **51**, 1018–1020.

Chandrasekhar, S. (1979). Einstein and general relativity: Historical perspectives. *American Journal of Physics*, **47**, 212–217.

Clotfelter, B. E. (1987). The Cavendish experiment as Cavendish knew it. *American Journal of Physics*, **55**, 210–213.

Cohen, J. E. and Horowitz, P. (1991). Paradoxical behaviour of mechanical and electrical networks. *Nature*, **352**, 699–701.

Costella, J. P., McKellar, B. H. J., Rawlinson, A. A., and Stephenson, G. J. (2001). The Thomas rotation. *American Journal of Physics*, **69**, 837–847.

Cranor, M. B., Heider, E. M., and Price R. H. (2000). A circular twin paradox. *American Journal of Physics*, **68**, 1016–1020.

Ehrlich, R. (1994). "Ruler physics:" Thirty-four demonstrations using a plastic ruler. *American Journal of Physics*, **62**, 111–120.

Eisenbud, L. (1958). On the classical laws of motion. *American Journal of Physics*, **26**, 144–159.

Eisner, E. (1967). Aberration of light from binary stars – a paradox? *American Journal of Physics*, **35**, 817–819.

Fadner, W. L. (1988). Did Einstein really discover "$E = mc^2$"? *American Journal of Physics*, **56**, 114–122.

Feng, S. (1969). Discussion on the criterion for conservative fields. *American Journal of Physics*, **37**, 616–618.

Feynman, R. P. (2006). *QED: The Strange Theory of Light and Matter*, Princeton: Princeton University Press.

Goldstein, H., Poole, C., and Safko, J. (2002). *Classical Mechanics*, 3rd edn., New York: Addison Wesley, Sections 7.9 and 13.5.

Goodstein, D. and Goodstein, J. (1996). *Feynman's Lost Lecture*, New York: W. W. Norton.

Green, D. and Unruh, W. G. (2006). The failure of the Tacoma Bridge: A physical model. *American Journal of Physics*, **74**, 706–716.

Greenslade, T. B. (1985). Atwood's Machine. *The Physics Teacher*, **23**, 24–28.

Gross, R. S. (2000). The excitation of the Chandler wobble. *Geophysical Research Letters*, **27**, 2329–2332.

Haisch, B. M. (1981). Astronomical precession: A good and a bad first-order approximation. *American Journal of Physics*, **49**, 636–640.

Hall, D. E. (1981). The difference between difference tones and rapid beats. *American Journal of Physics*, **49**, 632–636.

Hall, J. F. (2005). Fun with stacking blocks. *American Journal of Physics*, **73**, 1107–1116.

Handschy, M. A. (1982). Re-examination of the 1887 Michelson–Morley experiment. *American Journal of Physics*, **50**, 987–990.

Hendel, A. Z. and Longo, M. J. (1988). Comparing solutions for the solar escape problem. *American Journal of Physics*, **56**, 82–85.

Hollenbach, D. (1976). Appearance of a rapidly moving sphere: A problem for undergraduates. *American Journal of Physics*, **44**, 91–93.

Holton, G. (1988). Einstein, Michelson, and the "Crucial" Experiment. In *Thematic Origins of Scientific Thought, Kepler to Einstein*, Cambridge: Harvard University Press.

Horsfield, E. (1976). Cause of the earth tides. *American Journal of Physics*, **44**, 793–794.

Iona, M. (1978). Why is g larger at the poles? *American Journal of Physics*, **46**, 790–791.

Keller, J. B. (1987). Newton's second law. *American Journal of Physics*, **55**, 1145–1146.

Krane, K. S. (1981). The falling raindrop: Variations on a theme of Newton. *American Journal of Physics*, **49**, 113–117.

Lee, A. R. and Kalotas, T. M. (1975). Lorentz transformations from the first postulate. *American Journal of Physics*, **43**, 434–437.

Lee, A. R. and Kalotas, T. M. (1977). Causality and the Lorentz transformation. *American Journal of Physics*, **45**, 870.

Madsen, E. L. (1977). Theory of the chimney breaking while falling. *American Journal of Physics*, **45**, 182–184.

Mallinckrodt, A. J. and Leff, H. S. (1992). All about work. *American Journal of Physics*, **60**, 356–365.

Medicus, H. A. (1984). A comment on the relations between Einstein and Hilbert. *American Journal of Physics*, **52**, 206–208.

Mermin, N. D. (1983). Relativistic addition of velocities directly from the constancy of the velocity of light. *American Journal of Physics*, **51**, 1130–1131.

Mohazzabi, P. and James, M. C. (2000). Plumb line and the shape of the earth. *American Journal of Physics*, **68**, 1038–1041.

Muller, R. A. (1992). Thomas precession: Where is the torque? *American Journal of Physics*, **60**, 313–317.

O'Sullivan, C. T. (1980). Newton's laws of motion: Some interpretations of the formalism. *American Journal of Physics*, **48**, 131–133.

Peterson, M. A. (2002). Galileo's discovery of scaling laws. *American Journal of Physics*, **70**, 575–580.

Pound, R. V. and Rebka, G. A. (1960). Apparent weight of photons. *Physical Review Letters*, **4**, 337–341.

Prior, T. and Mele, E. J. (2007). A block slipping on a sphere with friction: Exact and perturbative solutions. *American Journal of Physics*, **75**, 423–426.

Rawlins, D. (1979). Doubling your sunsets or how anyone can measure the earth's size with wristwatch and meterstick. *American Journal of Physics*, **47**, 126–128.

Rebilas, K. (2002). Comment on "The Thomas rotation," by John P. Costella *et al.* *American Journal of Physics*, **70**, 1163–1165.

Rindler, W. (1994). General relativity before special relativity: An unconventional overview of relativity theory. *American Journal of Physics*, **62**, 887–893.

Shapiro, A. H. (1962). Bath-tub vortex. *Nature*, **196**, 1080–1081.

Sherwood, B. A. (1984). Work and heat transfer in the presence of sliding friction. *American Journal of Physics*, **52**, 1001–1007.

Stirling, D. R. (1983). The eastward deflection of a falling object. *American Journal of Physics*, **51**, 236.

Varieschi, G. and Kamiya, K. (2003). Toy models for the falling chimney. *American Journal of Physics*, **71**, 1025–1031.

Weltner, K. (1987). Central drift of freely moving balls on rotating disks: A new method to measure coefficients of rolling friction. *American Journal of Physics*, **55**, 937–942.

Zaidins, C. S. (1972). The radial variation of g in a spherically symmetric mass with nonuniform density. *American Journal of Physics*, **40**, 204–205.

Index